THE CAMBRIDGE HISTORY OF SCIENCE

VOLUME 3

Early Modern Science

Volume 3 offers a broad and detailed account of how the study of nature was transformed in Europe between ca. 1500 and ca. 1700. Chapters on how nature was studied, where, and by whom cover disciplines from astronomy and astrology to magic and natural history, sites of knowledge from the laboratory and the battlefield to the library and the marketplace, and types of knowers, from university professors and apothecaries to physicians and instrument makers. Separate sections on "The New Nature" and "Cultural Meanings of Natural Knowledge" address the impact of the new natural knowledge on conceptions of nature, experience, explanation, and evidence and on religion, art, literature, gender, and European self-definition, respectively. Contributions are written in clear, accessible prose, with extensive bibliographical notes, by noted specialists. The volume offers to scholars and general readers a synoptic overview of the research on early modern science that has challenged the traditional view of the "Scientific Revolution" while emphasizing profound but diverse changes in natural knowledge during this key epoch in European history.

Katharine Park is Samuel Zemurray, Jr., and Doris Zemurray Stone Radcliffe Professor of the History of Science and of the Studies of Women, Gender, and Sexuality at Harvard University. In addition to *Wonders and the Order of Nature* (1998), she is the author of *Doctors and Medicine in Early Renaissance Florence* (1985) and *The Secrets of Women: Gender, Generation, and the Origins of Human Dissection* (2006).

Lorraine Daston is Director at the Max Planck Institute for the History of Science and Honorary Professor at the Humboldt-Universität zu Berlin. She is the author of *Classical Probability in the Enlightenment* (1988), *Wonders and the Order of Nature, 1150–1750* (1998, with Katharine Park), *Wunder, Beweise und Tatsachen: Zur Geschichte der Rationalität* (2001), and *Images of Objectivity* (2006, with Peter Galison).

THE CAMBRIDGE HISTORY OF SCIENCE

General editors
David C. Lindberg and Ronald L. Numbers

VOLUME 1: *Ancient Science*
Edited by Alexander Jones

VOLUME 2: *Medieval Science*
Edited by David C. Lindberg and Michael H. Shank

VOLUME 3: *Early Modern Science*
Edited by Katharine Park and Lorraine Daston

VOLUME 4: *Eighteenth-Century Science*
Edited by Roy Porter

VOLUME 5: *The Modern Physical and Mathematical Sciences*
Edited by Mary Jo Nye

VOLUME 6: *The Modern Biological and Earth Sciences*
Edited by Peter Bowler and John Pickstone

VOLUME 7: *The Modern Social Sciences*
Edited by Theodore M. Porter and Dorothy Ross

VOLUME 8: *Modern Science in National and International Context*
Edited by David N. Livingstone and Ronald L. Numbers

David C. Lindberg is Hilldale Professor Emeritus of the History of Science at the University of Wisconsin–Madison. He has written or edited a dozen books on topics in the history of medieval and early modern science, including *The Beginnings of Western Science* (1992). He and Ronald L. Numbers have previously coedited *God and Nature: Historical Essays on the Encounter between Christianity and Science* (1986) and *Science and the Christian Tradition: Twelve Case Histories* (2003). A Fellow of the American Academy of Arts and Sciences, he has been a recipient of the Sarton Medal of the History of Science Society, of which he is also past president (1994–5).

Ronald L. Numbers is Hilldale and William Coleman Professor of the History of Science and Medicine at the University of Wisconsin–Madison, where he has taught since 1974. A specialist in the history of science and medicine in America, he has written or edited more than two dozen books, including *The Creationists* (1992) and *Darwinism Comes to America* (1998). A Fellow of the American Academy of Arts and Sciences and a former editor of *Isis*, the flagship journal of the history of science, he has served as the president of both the American Society of Church History (1999–2000) and the History of Science Society (2000–1).

THE CAMBRIDGE HISTORY OF SCIENCE

VOLUME 3

Early Modern Science

Edited by
KATHARINE PARK
LORRAINE DASTON

CAMBRIDGE UNIVERSITY PRESS
Cambridge, New York, Melbourne, Madrid, Cape Town, Singapore, São Paulo

Cambridge University Press
40 West 20th Street, New York, NY 10011-4211, USA

www.cambridge.org
Information on this title: www.cambridge.org/9780521572446

© Cambridge University Press 2006

This publication is in copyright. Subject to statutory exception
and to the provisions of relevant collective licensing agreements,
no reproduction of any part may take place without
the written permission of Cambridge University Press.

First published 2006

Printed in the United States of America

A catalog record for this publication is available from the British Library.

Library of Congress Cataloging in Publication Data

(Revised for volume 3)
The Cambridge history of science
p. cm.
Includes bibliographical references and indexes.
Contents: – v. 3. Early modern science / edited by Katharine Park and Lorraine Daston
v. 4. Eighteenth-century science / edited by Roy Porter
v. 5. The modern physical and mathematical sciences / edited by Mary Jo Nye
v. 7. The modern social sciences / edited by Theodore M. Porter and Dorothy Ross
ISBN 0-521-57244-4 (v. 3)
ISBN 0-521-57243-6 (v. 4)
ISBN 0-521-57199-5 (v. 5)
ISBN 0-521-59442-1 (v. 7)
1. Science – History. I. Lindberg, David C. II. Numbers, Ronald L.
Q125C32 2001
509 – dc21
2001025311

ISBN-13 978-0-521-57244-6 hardback
ISBN-10 0-521-57244-4 hardback

Cambridge University Press has no responsibility for
the persistence or accuracy of URLs for external or
third-party Internet Web sites referred to in this publication
and does not guarantee that any content on such
Web sites is, or will remain, accurate or appropriate.

CONTENTS

List of Illustrations	*page* xv
Notes on Contributors	xvii
General Editors' Preface	xxiii
Acknowledgments	xxvii

1 **Introduction: The Age of the New**   1
    KATHARINE PARK AND LORRAINE DASTON   1

PART I. THE NEW NATURE

2 **Physics and Foundations**   21
    DANIEL GARBER

Foundations	22
The Aristotelian Framework	25
Renaissance Anti-Aristotelianisms: Chymical Philosophies	29
Renaissance Anti-Aristotelianisms: The Italian Naturalists	33
Renaissance Anti-Aristotelianisms: Mathematical Order and Harmony	36
The Rise of the Mechanical and Corpuscular Philosophy	43
The Mechanical Philosophy: Theories of Matter	47
The Mechanical Philosophy: Space, Void, and Motion	52
The Mechanical Philosophy: Spirit, Force, and Activity	59
The Mechanical Philosophy: God and Final Causes	63
Beyond the Mechanical Philosophy: Newton	66
Conclusion: Beyond Foundations	68

3 **Scientific Explanation from Formal Causes to Laws of Nature**   70
    LYNN S. JOY
    Three Notable Changes in Early Modern Scientific Explanations   70

	Causality in the Aristotelian Tradition	73
	God as a Final Cause and the Emergence of Laws of Nature	77
	Intrinsic versus Extrinsic Efficient Causes among the Aristotelian Reformers	82
	Intrinsic versus Extrinsic Efficient Causes among the Corpuscular Physicists	87
	Active and Passive Principles as a Model for Cause and Effect	93
4	**The Meanings of Experience**	106
	PETER DEAR	
	Experience and the Natural Philosophy of Aristotle in Early Modern Europe	108
	Experiences of Life and Health	111
	Experience and Natural History: Individuals, Species, and Taxonomy	115
	Experience and the Mathematical Sciences	119
	Event Experiments and "Physico-mathematics"	124
	Newtonian Experience	126
	Conclusion	130
5	**Proof and Persuasion**	132
	R. W. SERJEANTSON	
	Disciplinary Decorum	134
	Theories of Proof and Persuasion	138
	Disciplinary Reconfigurations	150
	Mathematical Traditions	154
	Experiment	157
	Probability and Certainty	162
	Proof and Persuasion in the Printed Book	164
	Proof, Persuasion, and Social Institutions	168
	Conclusion	174

PART II. PERSONAE AND SITES OF NATURAL KNOWLEDGE

6	**The Man of Science**	179
	STEVEN SHAPIN	
	The University Scholar	182
	The Medical Man	186
	The Gentleman	188
7	**Women of Natural Knowledge**	192
	LONDA SCHIEBINGER	
	Learned Elites	193
	Artisans	199
	Colonial Connections	201

8 Markets, Piazzas, and Villages ... 206
WILLIAM EAMON
Markets and Shops ... 207
Natural Knowledge in the Piazza ... 213
Natural Knowledge in the Countryside and Villages ... 217
Conclusion: Popular Culture and the New Philosophy ... 221

9 Homes and Households ... 224
ALIX COOPER
Domestic Spaces ... 226
Natural Inquiry as a Family Project ... 229
Dividing Labor in the Scientific Household ... 233

10 Libraries and Lecture Halls ... 238
ANTHONY GRAFTON
The Classroom ... 240
The Library ... 244

11 Courts and Academies ... 251
BRUCE T. MORAN
Science at Court ... 253
Cabinets and Workshops ... 263
From Court to Academy ... 267

12 Anatomy Theaters, Botanical Gardens, and Natural History Collections ... 272
PAULA FINDLEN
Anatomizing ... 274
Botanizing ... 280
Collecting ... 283

13 Laboratories ... 290
PAMELA H. SMITH
Theory and Practice ... 293
Toward a New Epistemology ... 295
Evolution of Laboratory Spaces ... 300
Experiment in the Laboratory ... 302
Academic Institutionalization of the Laboratory ... 304

14 Sites of Military Science and Technology ... 306
KELLY DEVRIES
Offensive Technologies: Gunpowder and Guns ... 307
Defensive Technologies: Armor and Fortification ... 313
Courtly Engineers and Gentleman Practitioners ... 317

15 Coffeehouses and Print Shops ... 320
ADRIAN JOHNS
Print ... 322

		Coffee	332
		Audiences and Arguments	339
16		**Networks of Travel, Correspondence, and Exchange**	341
		STEVEN J. HARRIS	
		The Expanding Horizon of Scientific Engagement	341
		The Metrics of Scientific Practice	344
		Correspondence Networks, Long-Distance Travel, and Printing	347
		Virtual Spaces and Their Extension	355
		Conclusion	360

PART III. DIVIDING THE STUDY OF NATURE

17	**Natural Philosophy**	365
	ANN BLAIR	
	The University Context of Natural Philosophy	366
	Aristotelianism and the Innovations of the Renaissance	372
	The Impact of the Reformations and Religious Concerns	379
	New Observations and Practices	384
	Resistance to Radical Innovation	390
	Forces for Change in the Seventeenth Century	393
	The Origins of the Mechanical Philosophy	395
	The Transformation of Natural Philosophy by Empirical and Mathematical Methods	399
	The Social Conventions of the New Natural Philosophy	403
	Conclusion	405
18	**Medicine**	407
	HAROLD J. COOK	
	The Science of Physic	408
	New Worlds, New Diseases, New Remedies	416
	Toward Materialism	424
	Conclusion	432
19	**Natural History**	435
	PAULA FINDLEN	
	The Revival of an Ancient Tradition	437
	Words and Things	442
	Things Without Names	448
	Sharing Information	454
	The Emergence of the Naturalist	459
20	**Cosmography**	469
	KLAUS A. VOGEL	
	Translated by ALISHA RANKIN	
	Cosmography before 1490	472

	Globus mundi: Discoveries at Sea and the Cosmographic Revolution (1490–1510)	476
	Cosmographia universalis: Cosmography as a Leading Science (1510–1600)	480
	Geographia generalis: Toward a Science of Description and Measurement (1600–1700)	491
	Experience and Progress: Contemporary Views of the Emergence of Geography	494
21	**From Alchemy to "Chymistry"**	497
	WILLIAM R. NEWMAN	
	The Early Sixteenth Century	499
	Paracelsus	502
	Reaction to and Influence of Paracelsus	506
	Transmutation and Matter Theory	510
	Schools of Thought in Early Modern Chymistry	513
22	**Magic**	518
	BRIAN P. COPENHAVER	
	Agrippa's Magic Manual	519
	The Credibility of Magic: Text, Image, and Experience	526
	Magic on Trial	529
	Virtues Dormitive and Visual	532
	Magic Out of Sight	538
23	**Astrology**	541
	H. DARREL RUTKIN	
	Astrology circa 1500: Intellectual and Institutional Structures	542
	Astrological Reforms	547
	The Fate of Astrology	552
	The Eighteenth Century and Beyond	558
24	**Astronomy**	562
	WILLIAM DONAHUE	
	Astronomical Education in the Early Sixteenth Century	564
	Renaissance Humanism and *renovatio*	565
	Cracks in the Structure of Learning	569
	The Reformation and the Status of Astronomy	573
	Astrology	577
	Kepler's Revolution	581
	Galileo	584
	Descartes' Cosmology	586
	The Situation circa 1650: The Reception of Kepler, Galileo, and Descartes	587
	Novae, Variable Stars, and the Development of Stellar Astronomy	590
	Newton	592
	Conclusion	594

25 Acoustics and Optics — 596
PAOLO MANCOSU

Music Theory and Acoustics in the Early Modern Period — 597
The Sixteenth Century: Pythagorean and Aristoxenian Traditions — 598
The Birth of Acoustics in the Early Seventeenth Century — 604
Developments in Acoustics in the Second Half of the Seventeenth Century — 608
Optics in the Early Modern Period: An Overview — 611
Optics in the Sixteenth Century — 612
Kepler's Contributions to Optics — 613
Refraction and Diffraction — 618
Geometrical Optics and Image Location — 623
The Nature of Light and Its Speed — 624
Newton's Theory of Light and Colors — 626
Conclusion — 630

26 Mechanics — 632
DOMENICO BERTOLONI MELI

Mechanical Traditions — 634
Studies on Motion — 636
Motion and Mechanics in the Sixteenth Century — 638
Galileo — 640
Reading Galileo: From Torricelli to Mersenne — 649
Descartes' Mechanical Philosophy and Mechanics — 653
Reading Descartes and Galileo: Huygens and the Age of Academies — 659
Newton and a New World System — 664
Reading Newton and Descartes: Leibniz and His School — 668

27 The Mechanical Arts — 673
JIM BENNETT

The Mechanical Arts in 1500 — 677
Clocks and Other Celestial Instruments — 679
Mathematical and Optical Instruments — 683
Navigation, Surveying, Warfare, and Cartography — 686
Art and Nature — 693

28 Pure Mathematics — 696
KIRSTI ANDERSEN AND HENK J. M. BOS

The Social Context — 697
Stimuli: Methods and Problems — 702
The Inherited Algebra and an Inherited Challenge — 708
The Reception of Euclid's *Elements* — 710
The Response to Advanced Greek Mathematics: The Apollonian, Archimedean, and Diophantine Traditions — 712
The Merging of Algebra and Geometry — 714

The Calculus   718
Conclusion: Modernity and Context   722

PART IV. CULTURAL MEANINGS OF NATURAL KNOWLEDGE

29 Religion   727
RIVKA FELDHAY

Theological and Intellectual Contexts: Sacred Message and
  Bodies of Knowledge   730
Religious Identities and Educational Reforms   735
From Copernicus to Galileo: Scientific Objects, Boundaries, and
  Authority   740
Authorization and Legitimation: Science, Religion, and Politics
  in the Seventeenth Century   748
Conclusion   753

30 Literature   756
MARY BAINE CAMPBELL

Language   759
Telescope, Microscope, and Realism   762
Plurality of Worlds: From Astronomy to Sociology   764
Geography, Ethnography, Fiction, and the World of Others   766
Antagonisms   770
Conclusion   771

31 Art   773
CARMEN NIEKRASZ AND CLAUDIA SWAN

Naturalism   775
Scientific Illustration   779
Anatomy Lessons   782
The Artist as Scientist   786
Scientific Naturalism   791

32 Gender   797
DORINDA OUTRAM

Sex and Gender Difference in the Early Modern Period   801
The Problem of Nature   810
Conclusion   815

33 European Expansion and Self-Definition   818
KLAUS A. VOGEL
Translated by ALISHA RANKIN

Natural Knowledge and Colonial Science: Colleges of Higher
  Education and the Real y Pontificia Universidad de México
  (1553)   821

Natural Knowledge and the Christian Mission: The Jesuits in
  Japan and China 827
Natural Knowledge in European Self-Definition and Hegemony 836

Index 841

ILLUSTRATIONS

1.1	Title page from *Nova reperta*	*page* 2
1.2	*Iron Clocks* from *Nova reperta*	5
1.3	*America* from *Nova reperta*	17
2.1	Robert Fludd's representation of the cosmos in terms of a monochord	38
2.2	Robert Fludd's alternative representation of the cosmos in terms of interpenetrating pyramids	39
2.3	Athanasius Kircher's representation of the cosmos in terms of an organ	40
9.1	Johannes Hevelius's house in Danzig	228
10.1	*Bibliotheca publica* in Leiden	239
11.1	Vladislav Hall, Hradschin Castle, Prague	256
12.1	The Padua anatomy theater designed by Hieronymus Fabricius	279
12.2	The botanical garden at the University of Leiden	284
12.3	Ferrante Imperato's natural history museum in Naples	288
13.1	Idealized version of an alchemical laboratory	291
15.1	A printing house in Holland	324
15.2	A coffeehouse in London	334
15.3	Coffee plant	338
20.1	The spheres of earth and water	475
20.2	Globe of the Old World in the Ptolemaic style	481
20.3	*The Ambassadors* by Hans Holbein the Younger	482
20.4	Europa as Queen of Cosmography	490
21.1	Cosmology of the *Emerald Tablet* of Hermes	503
22.1	Heinrich Cornelius Agrippa von Nettesheim's lunar dragon	525
22.2	Dragons, horses, and cats by Leonardo da Vinci	530
22.3	René Descartes' illustration of magnetic action	535
24.1	Ptolemaic planetary model	565
24.2	Diurnal parallax of the moon	571
24.3	Tychonic system	575
24.4	Annual parallax	590

25.1	Pythagoras with musical devices	600
25.2	Monochord with two movable bridges	601
25.3	Johannes Kepler's model of radiation through small apertures	615
25.4	René Descartes' illustration of Kepler's theory of vision	617
25.5	Deflection of ball's trajectory passing from air to water	620
25.6	Refraction of light ray passing from air to water	621
25.7	Francesco Maria Grimaldi's illustration of diffraction through one small aperture with needle point	622
25.8	Francesco Maria Grimaldi's illustration of diffraction through two small apertures	622
25.9	Christiaan Huygens's illustration of his principle of secondary wavelets	626
25.10	Isaac Newton's prism experiment on spectral colors	628
28.1	Analytical solution of a geometrical construction problem	703
28.2	René Descartes' illustration of an ellipse and its equation	705
28.3	Formula as written by Rafael Bombelli	709
28.4	François Viète's algebraic notation	715
28.5	Apollonian construction of an ellipse	716
28.6	An area whose quadrature is wanted	719
28.7	The area of Figure 28.6 with circumscribed rectangles	719
31.1	*Iris bulbosa*	774
31.2	*The Young Hare* by Albrecht Dürer	778
31.3	Artists' portraits from Leonhart Fuchs's illustrated herbal	781
31.4	*The Anatomy Lesson of Dr. Nicolaes Tulp* by Rembrandt van Rijn	785
31.5	*The Invention of Oil Paint* from *Nova reperta*	789

NOTES ON CONTRIBUTORS

KIRSTI ANDERSEN teaches the history of mathematics in the History of Science Department at the University of Aarhus, Denmark. She has published on the developments leading to Newton's and Leibniz's creation of the calculus and is currently finishing *The Geometry of an Art: The History of the Mathematical Theory of Perspective from Alberti to Monge*.

JIM BENNETT is Director of the Museum of the History of Science, University of Oxford. He has published on a wide range of topics in the history of practical mathematics, astronomy, and scientific instruments from the sixteenth to the nineteenth century.

DOMENICO BERTOLONI MELI teaches the history of science at Indiana University, Bloomington. He is the author of *Equivalence and Priority* (1993, paperback, 1997) and the editor of *Marcello Malpigh: Anatomist and Physician* (1997). His forthcoming book *Thinking with Objects: The Transformations of Mechanics in the Seventeenth Century* is to be published by Johns Hopkins University Press. His current research concerns mechanistic anatomy in the seventeenth century.

ANN BLAIR teaches in the History Department at Harvard University. She is the author of *The Theater of Nature: Jean Bodin and Renaissance Science* (1997) and is currently working on a project entitled "Coping with Information Overload in Early Modern Europe."

HENK J. M. BOS is Professor of the History of Mathematics in the Department of Mathematics at Utrecht University. He has published on Huygens's mathematical and scientific work, the fundamental concepts of the Leibnizian calculus, and Descartes' geometry, including the monograph *Redefining Geometrical Exactness: Descartes' Transformation of the Early Modern Concept of*

Construction (2001). He is editor, together with Jed Buchwald, of the *Archive for History of Exact Sciences*.

MARY BAINE CAMPBELL is Professor of English and American Literature at Brandeis University. She is the author of *The Witness and the Other World: Exotic European Travel Writing, 400–1600* (1988), and *Wonder and Science: Imagining Worlds in Early Modern Europe* (1999), as well as two collections of poetry. She is currently studying early modern dreams and dream theories in relation to the fate of metaphor and the reorganization of knowledge in that period.

HAROLD J. COOK is Professor of the History of Medicine and Director of the Wellcome Trust Centre for the History of Medicine at University College London. He has published many essays and articles on early modern medicine and is the author of *The Decline of the Old Medical Regime* (1986) and *Trials of an Ordinary Doctor: Joannes Groenevelt in Seventeenth-Century London* (1994), which won the Welch Medal (of the American Association of the History of Medicine). He is currently working on a book about medicine and natural history in the Dutch Golden Age.

ALIX COOPER teaches early modern European history, history of science, and environmental history in the History Department at the State University of New York, Stony Brook. She is currently preparing for publication *Inventing the Indigenous: Local Knowledge and the Inventory of Nature in Early Modern Europe*.

BRIAN P. COPENHAVER is Professor of Philosophy and History at the University of California, Los Angeles. His books include *Renaissance Philosophy* (1992), *Hermetica* (1992), and *Polydore Vergil, On Discovery* (2002), in addition to chapters in *The Cambridge History of Philosophy* on magic and science and many related articles. His current research focuses on Giovanni Pico della Mirandola.

LORRAINE DASTON is Director at the Max Planck Institute for the History of Science and Honorary Professor at the Humboldt-Universität zu Berlin. She is the author of *Classical Probability in the Enlightenment* (1988), *Wonders and the Order of Nature, 1150–1750* (1998, with Katharine Park), and *Wunder, Beweise und Tatsachen: Zur Geschichte der Rationalität* (2001) and editor of *Biographies of Scientific Objects* (2000), *The Moral Authority of Nature* (2003, with Fernando Vidal), and *Things that Talk: Object Lessons from Art and Science* (2004). With Peter Galison, she is completing *Images of Objectivity*.

PETER DEAR teaches in the departments of History and Science and Technology Studies at Cornell University. He is the author of *Mersenne and the Learning of the Schools* (1988), *Discipline and Experience: The Mathematical Way in the Scientific Revolution* (1995), and *Revolutionizing the Sciences:*

European Knowledge and Its Ambitions, 1500–1700 (2001), as well as a forthcoming book on intelligibility in science.

KELLY DEVRIES is Professor of History at Loyola College in Maryland. His books include *Medieval Military Technology* (1992), *Infantry Warfare in the Early Fourteenth Century: Discipline, Tactics, and Technology* (1996), *The Norwegian Invasion of England in 1066* (1999), *Joan of Arc: A Military History* (1999), *A Cumulative Bibliography of Medieval Military History and Technology* (2002), and *Guns and Men in Medieval Europe, 1200–1500: Studies in Military History and Technology* (2002). *The Artillery of the Dukes of Burgundy, 1363–1477*, coauthored with Robert D. Smith, is to be published shortly. He edits the *Journal of Medieval Military History* and is the series editor for the History of Warfare series of Brill Publishing.

WILLIAM DONAHUE is Co-Director of Green Lion Press in Santa Fe, New Mexico, and translator of Kepler's *Astronomia nova* and Kepler's *Optics*. He is completing a guidebook to Kepler's planetary theory as developed in the *Astronomia nova*.

WILLIAM EAMON is Regents Professor of History at New Mexico State University, where he teaches the history of science and medicine and early modern history. His research concerns science and popular culture in early modern Europe as well as the history of science in early modern Italy and Spain. He is the author of *Science and the Secrets of Nature: Books of Secrets in Medieval and Early Modern Culture* (1994) and *The Charlatan's Tale: A Renaissance Surgeon's World* (forthcoming). He is at work on a book titled *Science and Everyday Life in Early Modern Europe, 1500–1750*.

RIVKA FELDHAY is Professor of History of Science and Ideas at Tel Aviv University. Her publications include *Galileo and the Church: Political Inquisition or Critical Dialogue?* (1995, reprint 1999); "The Use and Abuse of Mathematical Entities: Galileo and the Jesuits Revisited," in P. Machamer (ed.), *A Companion to Galileo* (1998); "The Cultural Field of Jesuit Sciences," in J. O'Malley, S.J. et al. (eds.), *The Jesuits: Cultures, Sciences, and the Arts, 1540–1773* (1999); "Giordano Bruno Nolanus: Authoritarian Sage and Martyr for Free Speech," in Lord Dahrendorf et al. (eds.), *The Paradoxes of Unintended Consequences* (2000); "Strangers to Ourselves: Identity Construction and Historical Research," in M. Zuckermann (ed.), *Psychoanalyse und Geschichte in Tel Aviver Jahrbuch fuer deutsche Geshchichte* (2004).

PAULA FINDLEN is Ubaldo Pierotti Professor of Italian History at Stanford University. She is the author of *Possessing Nature: Museums, Collecting, and Scientific Culture in Early Modern Italy* (1994) and other studies of science and culture in the early modern period. She has coedited *Merchants and Marvels: Commerce, Science, and Art in Early Modern Europe* (2002, with Pamela H.

Smith) and edited *Athanasius Kircher: The Last Man Who Knew Everything* (2004).

DANIEL GARBER is Professor of Philosophy at Princeton University and Associated Faculty in the Program in the History of Science. He is the author of *Descartes' Metaphysical Physics* (1992) and *Descartes Embodied* (2001), and he is the coeditor of *The Cambridge History of Seventeenth-Century Philosophy* (1998, with Michael Ayers). He is working on early seventeenth-century Aristotelianisms and anti-Aristotelianisms and on a monograph on Leibniz's conception of the physical world.

ANTHONY GRAFTON teaches history and history of science at Princeton University. He has written widely on the cultural history of Renaissance Europe, the history of books and readers, the history of scholarship and education in the West, and the history of science. Among other books, he is the author of *Joseph Scaliger* (1983–93), *Leon Battista Alberti* (2001), and *Bring Out Your Dead* (2002).

STEVEN J. HARRIS has taught at Harvard University, Brandeis University, and Wellesley College. His main research interest has concerned the scientific activities of members of the Society of Jesus. He is coeditor of two volumes on Jesuit cultural history, *The Jesuits: Cultures, Sciences, and the Arts, 1540–1773* (1999, second volume to appear in 2006). His current work is on the history of early modern cosmography.

ADRIAN JOHNS teaches in the Department of History and the Committee on Conceptual and Historical Studies of Science at the University of Chicago. He is the author of *The Nature of the Book: Print and Knowledge in the Making* (1998). He is currently working on a history of intellectual piracy from the invention of print to the present.

LYNN S. JOY, Professor of Philosophy at the University of Notre Dame, teaches modern philosophy, ethics, and philosophy of science. She is the author of *Gassendi the Atomist: Advocate of History in an Age of Science* (1987/2002). She currently writes on contemporary meta-ethics as well as the history of ethics. Her work-in-progress includes *Making Sense of Normativity*, a book on the role of natural dispositions in explaining moral norms and values, and articles such as "Hume on Natural and Moral Dispositions" and "Newtonianism without God: Hume as a Philosophical Critic."

PAOLO MANCOSU is Associate Professor of Philosophy at the University of California, Berkeley. His main interests are in mathematical logic and the history and philosophy of mathematics. He is the author of *Philosophy of Mathematics and Mathematical Practice in the Seventeenth Century* (1996) and

From Brouwer to Hilbert (1998). He has coedited the volume *Explanation, Visualizations and Reasoning Styles in Mathematics* (2005).

BRUCE T. MORAN is Professor of History at the University of Nevada at Reno, where he teaches the history of science and early medicine. In addition to other books and articles, he is the author of *Distilling Knowledge: Alchemy, Chemistry and Scientific Revolution* (2005) and is completing another study, "Chemists and Cultures in Early Modern Germany: The Torments and Tempests of Andrea Libavius."

WILLIAM R. NEWMAN is Ruth N. Halls Professor in the Department of History and Philosophy of Science at Indiana University. He works on the history of medieval and early modern alchemy, natural philosophy, and matter theory. His most recent books are *Alchemy Tried in the Fire* (2002, with Lawrence M. Principe), *Promethean Ambitions: Alchemy and the Quest to Perfect Nature* (2004), and *Atoms and Alchemy: Geber, Sennert, Boyle, and the Experimental Origins of the Scientific Revolution* (forthcoming, 2006). He is also researching the "chymistry" of Isaac Newton.

CARMEN NIEKRASZ is a doctoral student in the Art History Department at Northwestern University. Her dissertation title is "Flemish Tapestry and Natural History, 1550–1600."

DORINDA OUTRAM is Franklin I. Clark Professor of History at the University of Rochester. She has published widely on the history of science, the Enlightenment, and the history of exploration and culture contact in the same period. She is the author of *The Body and the French Revolution: Sex, Class and Political Culture* (1989) and *The Enlightenment* (1995) and is currently working on a project on the history of foolishness.

KATHARINE PARK is Samuel Zemurray, Jr., and Doris Zemurray Stone Radcliffe Professor of the History of Science and of the Studies of Women, Gender, and Sexuality at Harvard University. She studies the history of science and medicine in late medieval and Renaissance Europe and the history of women, gender, and the body. Her books include *Doctors and Medicine in Early Renaissance Florence* (1985), *Wonders and the Order of Nature, 1150–1750* (1998, with Lorraine Daston), and *The Secrets of Women: Gender, Generation, and the Origins of Human Dissection* (2006).

H. DARREL RUTKIN is currently a Hanna Kiel Fellow at the Harvard University Center of Italian Renaissance Studies at Villa I Tatti, Florence. He researches the complex roles of astrology in premodern Western science and culture, circa 1250–1750.

LONDA SCHIEBINGER is Barbara D. Finberg Director of the Institute for Research on Women and Gender and Professor of History of Science at Stanford University. She is author of *The Mind Has No Sex? Women in the*

Origins of Modern Science (1989), *Nature's Body: Gender in the Making of Modern Science* (1993, 2nd ed. 2004), *Has Feminism Changed Science?* (1999), and *Plants and Empire: Colonial Bioprospecting in the Atlantic World* (2004). She is the editor of *Feminism and the Body* (2000), section editor of the *Oxford Companion to the Body* (2001), coeditor of *Feminism in Twentieth-Century Science, Technology, and Medicine* (2001, with Angela Creager and Elizabeth Lunbeck), and coeditor of *Colonial Botany: Science, Commerce, and Politics* (2004, with Claudia Swan). She is currently working on race and health in eighteenth-century colonial science.

R. W. SERJEANTSON is a Fellow of Trinity College, Cambridge, where he teaches history and history of science. He is the editor of *Generall Learning* by Meric Casaubon (1999).

STEVEN SHAPIN is Franklin L. Ford Professor of History of Science at Harvard University. His books include *Leviathan and the Air-Pump: Hobbes, Boyle, and the Experimental Life* (1985, with Simon Schaffer), *A Social History of Truth: Civility and Science in Seventeenth-Century England* (1994), and *The Scientific Revolution* (1996).

PAMELA H. SMITH is Professor of History at Columbia University. She is author of *The Business of Alchemy: Science and Culture in the Holy Roman Empire* (1994) and *The Body of the Artisan: Art and Experience in the Scientific Revolution* (2004).

CLAUDIA SWAN is Associate Professor in the Art History Department at Northwestern University, where she is also a founding director of the Program in the Study of Imagination. She is the author of *The Clutius Botanical Watercolors: Plants and Flowers of the Renaissance* (1998) and *Art, Science, and Witchcraft in Early Modern Holland: Jacques de Gheyn II (1565–1629)* (2005). She has coedited *Colonial Botany: Science, Commerce, and Politics in the Early Modern World* (2004, with Londa Schiebinger).

KLAUS A. VOGEL is a historian and merchant marine captain. He has been a researcher at the Max Planck Institute for History, Göttingen, and a lecturer at the University of Göttingen. He is the author of *Sphaera terrae: Das mittelalterliche Bild der Erde und die kosmographische Revolution* (1995) and editor of the *Pirckheimer Jahrbuch für Renaissance- und Humanismusforschung* (1995–2000). Since 2000, he has been working on ocean-going container vessels for the Claus Peter Offen Shipping Company, Hamburg.

GENERAL EDITORS' PREFACE

In 1993, Alex Holzman, former editor for the history of science at Cambridge University Press, invited us to submit a proposal for a history of science that would join the distinguished series of Cambridge histories launched nearly a century ago with the publication of Lord Acton's fourteen-volume *Cambridge Modern History* (1902–12). Convinced of the need for a comprehensive history of science and believing that the time was auspicious, we accepted the invitation.

Although reflections on the development of what we call "science" date back to antiquity, the history of science did not emerge as a distinctive field of scholarship until well into the twentieth century. In 1912, the Belgian scientist-historian George Sarton (1884–1956), who contributed more than any other single person to the institutionalization of the history of science, began publishing *Isis*, an international review devoted to the history of science and its cultural influences. Twelve years later, he helped to create the History of Science Society, which by the end of the century had attracted some 4,000 individual and institutional members. In 1941, the University of Wisconsin established a department of the history of science, the first of dozens of such programs to appear worldwide.

Since the days of Sarton, historians of science have produced a small library of monographs and essays, but they have generally shied away from writing and editing broad surveys. Sarton himself, inspired in part by the Cambridge histories, planned to produce an eight-volume *History of Science*, but he completed only the first two installments (1952, 1959), which ended with the birth of Christianity. His mammoth three-volume *Introduction to the History of Science* (1927–48), a reference work more than a narrative history, never got beyond the Middle Ages. The closest predecessor to *The Cambridge History of Science* is the three-volume (four-book) *Histoire générale des sciences* (1957–64), edited by René Taton, which appeared in an English translation under the title *General History of the Sciences* (1963–4). Edited just before the late twentieth-century boom in the history of science, the Taton set quickly became dated. During the 1990s, Roy Porter began editing the very useful Fontana History of Science (published in the United States as the Norton History of Science), with volumes devoted to a single discipline and written by a single author.

The Cambridge History of Science comprises eight volumes, the first four arranged chronologically from antiquity through the eighteenth century, the latter four organized thematically and covering the nineteenth and twentieth centuries. Eminent scholars from Europe and North America, who together form the editorial board for the series, edit the respective volumes:

Volume 1: *Ancient Science*, edited by Alexander Jones, University of Toronto
Volume 2: *Medieval Science*, edited by David C. Lindberg and Michael H. Shank, University of Wisconsin–Madison
Volume 3: *Early Modern Science*, edited by Katharine Park, Harvard University, and Lorraine Daston, Max Planck Institute for the History of Science, Berlin
Volume 4: *Eighteenth-Century Science*, edited by Roy Porter, late of Wellcome Trust Centre for the History of Medicine at University College London
Volume 5: *The Modern Physical and Mathematical Sciences*, edited by Mary Jo Nye, Oregon State University
Volume 6: *The Modern Biological and Earth Sciences*, edited by Peter Bowler, Queen's University of Belfast, and John Pickstone, University of Manchester
Volume 7: *The Modern Social Sciences*, edited by Theodore M. Porter, University of California, Los Angeles, and Dorothy Ross, Johns Hopkins University
Volume 8: *Modern Science in National and International Context*, edited by David N. Livingstone, Queen's University of Belfast, and Ronald L. Numbers, University of Wisconsin–Madison

Our collective goal is to provide an authoritative, up-to-date account of science – from the earliest literate societies in Mesopotamia and Egypt to the beginning of the twenty-first century – that even nonspecialist readers will find engaging. Written by leading experts from every inhabited continent, the essays in *The Cambridge History of Science* explore the systematic investigation of nature and society, whatever it was called. (The term "science" did not acquire its present meaning until early in the nineteenth century.) Reflecting the ever-expanding range of approaches and topics in the history of science, the contributing authors explore non-Western as well as Western science, applied as well as pure science, popular as well as elite science, scientific practice as well as scientific theory, cultural context as well as intellectual content, and the dissemination and reception as well as the production of scientific knowledge. George Sarton would scarcely recognize this collaborative effort as the history of science, but we hope we have realized his vision.

David C. Lindberg
Ronald L. Numbers

ACKNOWLEDGMENTS

It is a pleasure to thank Josephine Fenger, Nathalie Huet, John Kuczwara, Carola Kuntze, and Alisha Rankin for their help in preparing this volume. The project has extended over a decade and two continents, and without their patient assistance in keeping track of drafts, correspondence, figures, and a swarm of editorial details, this volume would have taken even longer to appear. We are also grateful to Harvard University, especially the Radcliffe Institute for Advanced Study, and the Max Planck Institute for the History of Science, Berlin, for substantial institutional support. At Cambridge University Press, we were fortunate to be in the capable editorial hands of Alex Holzman and Helen Wheeler. As the General Editor responsible for our volume, David Lindberg read though the entire manuscript; we profited greatly from his characteristically sharp eye for argument and style. Our authors were models of learning and forbearance, and occasionally even of punctuality. Martin Brody, Gerd Gigerenzer, and Thalia Gigerenzer cheered us on and up throughout; we thank them from the heart.

Katharine Park
Lorraine Daston

I

INTRODUCTION
The Age of the New

Katharine Park and Lorraine Daston

This volume of the *Cambridge History of Science* covers the period from roughly 1490 to 1730, which is known to anglophone historians of Europe as the "early modern" era,[1] a term pregnant with expectations of things to come. These things were of course mostly unknown and unanticipated by the Europeans who lived during those years, and had they been asked to give their own epoch a name, they would perhaps have called it "the new age" (*aetas nova*). New worlds, East and West, had been discovered, new devices such as the printing press had been invented, new faiths propagated, new stars observed in the heavens with new instruments, new forms of government established and old ones overthrown, new artistic techniques exploited, new markets and trade routes opened, new philosophies advanced with new arguments, and new literary genres created whose very names, such as "news" and "novel," advertised their novelty.

Some of the excitement generated by this ferment is captured in *Nova reperta* (New Discoveries), a series of engravings issued in Antwerp in the early seventeenth century, after the late sixteenth-century designs of the Flemish painter and draftsman Jan van der Straet (1523–1605).[2] The title page shows numbered icons of the first nine discoveries celebrated in the series: of the Americas, the compass, gunpowder, printing, the mechanical clock, guaiacum (an American wood used in the treatment of the French

---

[1] Among anglophone historians, this term is used to cover the period between roughly 1500 and 1750; historians writing in Italian, French, and German define the period differently, beginning as early as 1350 (the Italians) and ending as late as 1815 (the Germans). Moreover, depending on national historiographic traditions, period designations such as the Renaissance, the Baroque, or *l'âge classique* are preferred over "early modern": see Ilja Micek, "Die Frühe Neuzeit: Definitionsprobleme, Methodendiskussion, Forschungstendenzen," in *Die Frühe Neuzeit in der Geschichtswissenschaft: Forschungstendenzen und Forschungserträge*, ed. Nada Boskovska Leimgruber (Paderborn: Ferdinand Schöningh, 1997), pp. 17–38.

[2] See Alessandra Baroni Vannucci, *Jan van der Straet detto Giovanni Stradano: Flandrus pictor et inventor* (Milan: Jandi Sapi, 1997), pp. 397–400. Reproductions are on the Web site of the University of Liège, http://www.ulg.ac.be/wittert/fr/flori/opera/vanderstraet/vanderstraet_reperta.html. The original designs date from the 1580s.

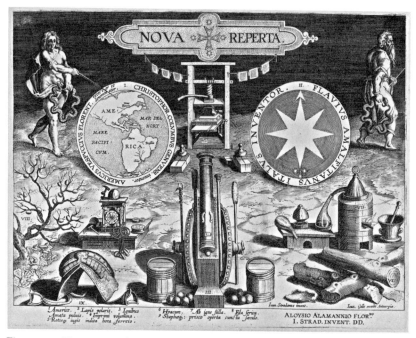

Figure 1.1. *Nova reperta* (New Discoveries). Jan Galle after Joannes Stradanus (Jan van der Straet), ca. 1580, title page of *Nova reperta*. In *Speculum diuersarum imaginum speculatiuarum a varijs viris doctis adinuentarum, atq[ue] insignibus pictoribus ac sculptoribus delineatarum* ... (Antwerp: Jan Galle, 1638). Reproduced by permission of the Print Collection, Miriam and Ira D. Wallach Division of Art, Prints and Photographs, The New York Public Library, Astor, Lenox and Tilden Foundations.

disease, or syphilis, distillation, the cultivation of silkworms, and the harnessing of horses (Figure 1.1). Later editions of the series include depictions of the manufacture of cane sugar, the discovery of a method for finding longitude by the declination of the compass, and the invention of the techniques of painting using oil glazes and of copper engraving itself. Although a number of these innovations predated the early modern period, most were closely identified with it, if not because they were the work of early modern Europeans, then because their effects were perceived as having transformed early modern European culture. Certainly, the aggregate effect of the *Nova reperta* engravings, which depict sixteenth-century landscapes, workshops, ships, and domestic spaces, is to portray the period as one of extraordinary fertility, creative ambition, and innovation.

This book concerns one particularly dynamic field of innovation in early modern Europe; for the sake of convenience, this field is usually (albeit anachronistically) subsumed under the portmanteau term "science," taken in its sense (since the nineteenth century) of disciplined inquiry into the

Introduction: The Age of the New

phenomena and order of the natural world.[3] This modern category had no single, coherent counterpart in the sixteenth and seventeenth centuries. Indeed, one of the most striking innovations tracked by the chapters in this volume is the gradual emergence of a new domain of inquiry, which had some – but by no means all – of the features of natural science since about 1850. This domain embraced both intellectual and technical approaches and was composed of what had previously been disparate disciplines and pursuits, practiced by people in different professions in different institutions at different sites.

A glance at library classification systems of the period makes this shift vivid. In 1584, a classification system was proposed for the some 10,000 books in the library of French king Henry III, which envisaged separate sections for books on medicine, philosophy (including natural philosophy), mathematics (including optics and astronomy as well as geometry and arithmetic), alchemy, music, and the "vile and mechanical arts," as well as other "arts and sciences," which included theology, jurisprudence, grammar, poetry, and the art of oratory.[4] About a century later, the much-imitated classification of the library of Charles Maurice le Tellier, Archbishop of Reims, lumped together under the rubric of philosophy the following previously disparate fields: natural history, medicine (including anatomy, surgery, pharmacy, and chemistry), the mathematical disciplines (including astronomy and astrology, architecture, and military science and navigation), and the mechanical arts.[5] A new constellation had become visible in the firmament of knowledge, composed of stars that had earlier belonged to quite distinct constellations.

What were these older constellations? To map them accurately, attention must be paid to the sites where the various types of knowledge were cultivated, and by whom, as well as to more formal classifications of knowledge. Names alone (especially when mechanically matched to cognates in modern vernacular languages) are often unreliable guides. The medieval Latin *scientia*, although cognate with the modern English "science," referred to any rigorous and certain body of knowledge that could be organized (in precept though not always in practice) in the form of syllogistic demonstrations from self-evident premises. Under this description, rational theology belonged to *scientia* – indeed, it was the "queen of sciences" – because its premises were the highest and most certain. Excluded, however, were disciplines that studied empirical particulars, such as medical therapeutics, natural history, and

---

[3] See Andrew Cunningham and Perry Williams, "De-Centring the 'Big Picture': *The Origins of Modern Science* and the Modern Origins of Science," *British Journal for the History of Science*, 26 (1993), 407–32.

[4] Henri-Jean Martin, "Classements et conjonctures," in *Histoire de l'édition française*, ed. Henri-Jean Martin and Roger Chartier, 4 vols. (Paris: Promodis, 1982–6), 1: 429–57, at p. 435.

[5] [Philippe Dubois], *Bibliotheca Telleriana, sive catalogus librorum bibliothecae illustrissimi ac reverendissimi D. D. Caroli Mauritii Le Tellier* (Paris: Typographia Regia, 1693), [Introduction], n.p. On the influence of this classification scheme, see Archer Taylor, *Book Catalogues: Their Varieties and Uses* (Chicago: The Newberry Library, 1957), pp. 157–8.

alchemy, because there can be no absolute certainty about particular phenomena.[6]

The kind of *scientia* that covered topics closer but by no means identical to those treated by modern science was natural philosophy – *philosophia naturalis*, sometimes known as *scientia naturalis* – which studied the material world as it was visible to the senses. Natural philosophy examined change of all kinds, organic and physical, including motion, as well as the principles that produced the phenomena of the heavens (cosmology), the earth's atmosphere (meteorology), and the earth itself (such as minerals, plants, and animals, including human beings). The two topics of plants and animals fell generally under the study of the soul, understood as that which distinguishes living from nonliving beings (see Blair, Chapter 17, this volume). Natural philosophy also addressed questions that would now be seen as metaphysical, such as the nature of space and time and the relation of God to creation (see Garber, Chapter 2, this volume).

Because natural philosophy sought the universal causes of phenomena, it was distinct from natural history, which described *naturalia* and their particular properties; insofar as this was an object of systematic study, rather than a tool for biblical exegesis or a reservoir for sermon examples and recreational art and literature, it fell under the purview of medicine because some minerals and animals, and many plants, were used in therapeutics. Alchemy had a rather separate existence, not being a university subject, though it was sometimes pursued by physicians because the chemical treatment of substances often aimed at the preparation of medications.

The *scientiae mediae* (or *mathematica media*, "mixed mathematics") differed from natural philosophy in that they dealt with matter considered solely from the standpoint of quantity, without respect to causes. In addition to the pure mathematical disciplines of arithmetic and geometry, mathematics included astronomy and astrology (the two terms were often used interchangeably), optics, harmonics, and mechanics.[7] These disciplines were in turn distinct from the "mechanical arts," which would have included practical applications of mathematical knowledge in fields such as architecture, navigation, clockmaking, and engineering (Figures 1.1 and 1.2).

Because all of these disciplines were conceived as separate pursuits, with their own methods, goals, and widely varying degrees of intellectual and social status, it would have been highly unusual, at least in the late fifteenth century, to find the same person involved in all or most of them. Natural philosophy was part of the university curriculum but was usually taught as

---

[6] Eileen Serene, "Demonstrative Science," in *The Cambridge History of Later Medieval Philosophy: From the Rediscovery of Aristotle to the Disintegration of Scholasticism, 1100–1600*, ed. Norman Kretzmann, Anthony Kenny, and Jan Pinborg (Cambridge: Cambridge University Press, 1982), pp. 496–517.
[7] William Wallace, "Traditional Natural Philosophy," in *The Cambridge History of Renaissance Philosophy*, ed. Charles B. Schmitt, Quentin Skinner, and Eckhard Kessler with Jill Kraye (Cambridge: Cambridge University Press, 1988), pp. 201–35.

Figure 1.2. *Horologia ferrea* (Iron clocks). Jan Galle after Joannes Stradanus (Jan van der Straet), ca. 1580, from *Nova reperta*. In *Speculum diuersarum imaginum speculatiuarum a varijs viris doctis adinuentarum, atq[ue] insignibus pictoribus ac sculptoribus delineatarum...* (Antwerp: Jan Galle, 1638). Reproduced by permission of the Print Collection, Miriam and Ira D. Wallach Division of Art, Prints and Photographs, The New York Public Library, Astor, Lenox and Tilden Foundations.

propadeutic to the higher faculty of medicine, at least at Italian universities, and often by medical men. The *quadrivium* of mathematical sciences (arithmetic, geometry, music, and astronomy) and the *trivium* of the verbal ones (grammar, logic, and rhetoric), which together constituted the seven "liberal arts," would have been taught with varying emphases in the university to prepare students for their studies in philosophy. University-trained physicians would have learned some astrology and some natural history – the latter as part of the study of materia medica – but apothecaries, who belonged to the ranks of merchants, would have been the experts in this area. Similarly, mixed mathematicians who consulted concerning fortifications, hydraulics, horology, mapmaking, and a host of other practical activities tended to work out of artisanal studios or as adjuncts to princely courts rather than as university professors.

Hence early modern career trajectories can often appear to modern eyes at once as dazzlingly diverse and oddly circumscribed: A Renaissance engineer such as Leonardo da Vinci painted, designed buildings and machines, drew maps, and built fortresses and canals. But (despite his curiosity about human anatomy) he would not have treated patients nor (despite his speculative ideas on the nature of water) would he have taught a university class in natural philosophy. The multifaceted "Renaissance man" is to some extent a trick of historical perspective, which creates polymathesis out of what was simply a different classification of knowledge and a different professional division of labor.

Similarly, because modern "science" maps so awkwardly onto early modern natural knowledge, there is some temptation to see the latter as a crazy quilt of mismatched parts seeking – finally – to merge into the new conglomerate recognized in the late seventeenth-century arrangement of books in the Tellier library (or even the nineteenth-century category of "science").[8] Yet the older classifications of knowledge and divisions of labor appeared just as coherent to those who lived them as the modern constellation of natural science does to twenty-first-century readers. The most generally accepted division of human knowledge in premodern Europe parsed it not primarily according to subject matter (e.g., nonliving versus living beings), nor according to methods used (e.g., experimenting in laboratories versus reading books in libraries or classrooms), but rather according to whether it served purposes that were "speculative" (i.e., theoretical), "practical" (i.e., related to leading a good and useful life), or "factive" (i.e., related to the production of things in the arts and trades).[9]

What makes the study of nature during the early modern period so difficult to describe, however, is not so much the gap between this period's classifications of knowledge and ours, nor the cumbersome lists (natural philosophy, natural history, medicine, mixed mathematics, mechanical arts) and coinages ("chymistry," "natural knowledge") that try to bridge that gap, but rather the fact that the gusher of novelty that flooded sixteenth- and seventeenth-century Europe also reconfigured knowledge and careers over the course of the early modern period itself. By the turn of the seventeenth century, there were university professors of medicine who not only wrote treatises on natural philosophy but also contributed to cutting-edge mathematics (Girolamo Cardano, 1501–1576), or who began by teaching mathematics but who moved on (and up) to courtly careers in natural philosophy and commissions in engineering (Galileo Galilei, 1564–1642). University-trained physicians turned to peasants and artisans for instruction (Theophrastus Bombastus von Hohenheim, known as Paracelsus, ca. 1493–1541); artisans themselves set forth natural philosophical theories in print (Bernard Palissy, ca. 1510–ca. 1590). What was studied (and in what combinations), how it was studied, where, and by whom were in remarkable flux during this period.

---

[8] Cunningham and Williams, "De-Centring the 'Big Picture'"; and Sydney Ross, "'Scientist': The Story of a Word," *Annals of Science*, 18 (1962), 65–86.
[9] See James A. Weisheipl, "The Classification of the Sciences in Medieval Thought," *Mediaeval Studies*, 27 (1965), 54–90.

Introduction: The Age of the New

These changes often meshed with the enormous political, religious, social, and economic transformations that characterized the early modern era, some of which are alluded to in the title page engraving of *Nova reperta*. The invention and diffusion of printing created new kinds of authors and readers (see Johns, Chapter 15, this volume). The religious movements of the Reformation and Counter-Reformation demanded adjustments in not only what was taught but how (see Feldhay, Chapter 29, this volume). Incessant wars of unprecedented length and scale fed demands for improved military technology (see DeVries, Chapter 14, this volume). These wars, together with frequent episodes of religious persecution, triggered waves of forced migration among scholars and skilled artisans, while competition among courts and wealthy cities opened up possibilities for social advancement to these and other practitioners of natural knowledge (see Moran, Chapter 11, this volume). European commerce expanded dramatically in scope and scale. The mineral wealth brought back from the New World reshaped the European economy, while shiploads of new flora and fauna arriving in European ports from exotic lands stimulated natural history and medicine (see the following chapters in this volume: Eamon, Chapter 8; Findlen, Chapter 19). The geography of changes in natural knowledge closely tracked that of religious, military, and economic developments, beginning in northern Italy in the early sixteenth century, spreading to the prosperous towns of Switzerland and southern Germany by the latter part of the century and subsequently to the Low Countries, and then, by the late seventeenth century, to France and England.[10]

In addition to these interlocking transformations, there were others specific to the learned realm. Perhaps the most far-reaching was the intellectual movement known as humanism: the study of Greek and Roman texts not as timeless contributions to a transhistorical intellectual enterprise, as the philosophical and logical works of Aristotle had been treated in medieval schools and universities, but as works of a particular time and place. Because these texts reflected the languages and cultures of the authors that produced them, in all their historical specificity, they needed to be read with those particularities in mind. Humanists' editions and translations of these texts – both those long known and those newly rediscovered – together with their erudite commentaries on them, dramatically expanded the body of works available to students of nature in the sixteenth and seventeenth centuries, making accessible a variety of philosophical and medical traditions in addition to the Aristotelian and Galenic: Platonism (and neo-Platonism), Stoicism, Skepticism, Epicureanism, and Hippocratism.[11]

---

[10] For some sense of the geographical distribution and varying tempos of these developments, see Roy Porter and Mikuláš Teich, eds., *The Renaissance in National Context* (Cambridge: Cambridge University Press, 1992); and Porter and Teich, eds., *The Scientific Revolution in National Context* (Cambridge: Cambridge University Press, 1992).

[11] Jill Kraye, "Philologists and Philosophers," in *The Cambridge Companion to Renaissance Humanism*, ed. Jill Kraye (Cambridge: Cambridge University Press, 1996), chap. 8; and Vivian Nutton,

This proliferation of information and possible approaches to the natural order and human cognition had a great impact on natural inquiry (see the following chapters in this volume: Blair, Chapter 17; Joy, Chapter 3; Garber, Chapter 2).[12] In some areas, the new scholarship led to heated debates with more traditional scholars about the value and interpretation of familiar texts – witness the flurry of attacks on and defenses of Pliny's *Natural History* in the 1490s (see Chapter 19, this volume). More generally, however, the broader range of books available – thanks in large part to printing – together with the humanists' cultivation of an elegant Latin style modeled on that of ancient authors, created new scholarly and literary sensibilities. For many sixteenth-century scholars, educated into such sensibilities, the works of medieval interpreters seemed not so much wrong as old-fashioned, poorly informed, and narrowly conceived. A few of these interpreters gained new life after the middle of the sixteenth century, particularly those, such as Thomas Aquinas, whom the Counter-Reformation Church proposed as the touchstones of philosophical and theological orthodoxy. For the most part, however, medieval commentaries, even standbys such as those of Paul of Venice in logic and philosophy or Jacopo da Forlì in medicine, simply ceased to be reprinted.

Thus, new early modern approaches to natural inquiry should not be seen in the first instance as an attack on the doctrines and methods contained in the works of Aristotle and his medieval Arabic and Latin commentators – an impressive intellectual edifice modern scholars often refer to by the shorthand term "scholasticism." Such attacks, although the stuff of popular historiographic legend – crystallized around heroic figures such as Galileo and Francis Bacon (1561–1626) – were less common than one might gather from the many textbooks on the history of early modern science that embrace, with varying degrees of enthusiasm, the premise of a "Scientific Revolution." More typically, as the chapters in Parts I and III of this volume demonstrate, the process of change was gradual and sporadic, shaped well into the first half of the seventeenth century by serious, widespread, and accepted efforts to accommodate ancient texts to newer methods and discoveries.[13] In this intellectual environment of accommodation rather than wholesale innovation, it comes as no surprise that van der Straet's *Nova reperta*, the initial designs

---

"Hippocrates in the Renaissance," in *Die Hippokratischen Epidemien: Theorie-Praxis-Tradition*, ed. Gerhard Baader and Rolf Winau (Sudhoffs Archiv, Beiheft 27) (Stuttgart: Franz Steiner Verlag, 1989), pp. 420–39.

[12] See Anthony Grafton, "The New Science and the Traditions of Humanism," in Kraye, ed., *Cambridge Companion*, chap. 11; and Anthony Grafton, with April Shelford and Nancy Siraisi, *New Worlds, Ancient Texts: The Power of Tradition and the Shock of Discovery* (Cambridge, Mass.: Belknap Press, 1992).

[13] See, for example, Christia Mercer, "The Vitality and Importance of Early Modern Aristotelianism," in *The Rise of Modern Philosophy: The Tension Between the New and Traditional Philosophies from Machiavelli to Leibniz*, ed. Tom Sorrell (Oxford: Clarendon Press, 1993); and Ian Maclean, *Logic, Signs, and Nature in the Renaissance: The Case of Learned Medicine* (New York: Cambridge University Press, 2002).

of which date to the 1580s, privileged as sites of dramatic innovation the mechanical arts rather than textual disciplines such as natural philosophy, theoretical medicine, or even natural history. It was only toward the middle of the seventeenth century that the weight of scholarly opinion – and even then there were many objectors – shifted from gradual, accommodationist strategies to calls for more fundamental change, as more and more voices argued that the old edifice of natural knowledge needed to be torn down and a new one constructed, however unclear the shape of that new edifice might be.

Given the vast transformations that characterized the history of early modern Europe, and the impact of those transformations on the organization of knowledge in both theory and practice, the chapters in this volume, especially those in Part III: "Dividing the Study of Nature," necessarily represent a compromise between early modern and modern categories. Although the aim of Part III is to acquaint readers with the substantive changes that occurred in natural knowledge, neither all of the chapter headings nor their arrangement would have been recognizable to early modern Europeans, even those most abreast of new developments. In order to have made them so, the chapters on "Astronomy" and "Astrology," for example, would have needed to be merged, as would indeed all the chapters relating to mixed mathematics: astronomy/astrology, optics, acoustics (or rather, music), mechanics, and parts of the mechanical arts. There would also have been good historical arguments for combining the chapters on "Medicine" and "Natural History," at least for the earlier part of the period. The title of Chapter 21, "From Alchemy to 'Chymistry'," epitomizes the historiographic problems of trying to fix a moving target – and one that emphatically does not become modern chemistry by the end of the period covered in this volume.[14] Quite apart from the difficulties of finding authors to write about branches of knowledge that have since been split up, with their splinters redistributed elsewhere, many readers would be ill-served by a work that presumed a detailed knowledge of the early modern ways of thinking it was supposed to explain. Hence, although each chapter strives to make clear the place of its topic in early modern schemes of knowledge, we have in some cases separated subjects that would have been combined in those schemes and have occasionally relabeled them.

We would therefore recommend that the chapters in Part III be read in tandem with those in Part II: "Personae and Sites of Natural Knowledge," which describe who was making knowledge where. Some of the scenes described in Part II will be familiar: the professor lecturing in the university lecture hall, or the virtuoso performing an experiment in a scientific academy (see the following chapters in this volume: Shapin, Chapter 6; Grafton, Chapter 10; Moran,

---

[14] William R. Newman and Lawrence Principe, "Alchemy versus Chemistry: The Etymological Origins of a Historiographic Mistake," *Early Science and Medicine*, 3 (1998), 32–65.

Chapter 11). But others will be less so: the tutor employed by an aristocratic family (see Chapter 6, this volume), the apothecary or herbwoman selling medicinal plant products, exotic or domestic (see Chapter 8, this volume), whole households practicing astronomy or natural history (see the following chapters in this volume: Schiebinger, Chapter 7; Cooper, Chapter 9), or military engineers computing the optimal angle of fortifications (see Chapter 14, this volume). No single rubric, modern or early modern, describes what kind of people they were (by gender, rank, confession, or profession) or what kind of knowledge they were forging. For the sake of convenience, we have tried to use the umbrella terms "students of nature" (or "naturalists" or "natural inquirers") and "natural knowledge," which have some seventeenth-century antecedents but were not recognized by most contemporaries as a comprehensive category for all of these varied activities.

Moreover, the relationship between the disciplines of Part III and the personae and sites of Part II was crosshatched and complex. For example, although a disparate crowd of physicians, engineers, alchemists, astronomers, and even natural philosophers might spend parts of their careers at court, the lecture hall was considerably less permeable. Scholars, master artisans, apprentices, and clients of various social ranks might meet in workshops, cannon foundries, or distilleries, as shown in the densely populated engravings of van der Straet's *Nova reperta* (e.g., the clockmaker's shop of Figure 1.2). Academicians and apothecaries might rub shoulders in the piazza or coffeehouse (see the following chapters in this volume: Eamon, Chapter 8; Findlen, Chapter 12; Johns, Chapter 15); correspondents in an epistolary network might never rub shoulders anywhere and for that reason might enjoy greater freedom to indulge in discussions and debates on specialized topics (see Harris, Chapter 16, this volume). Read side-by-side, the chapters in Parts II and III show that the new associations between fields of knowledge (e.g., between alchemy and natural philosophy, or between engineering and mathematics) were matched by new associations between people in new places: the botanical garden, the anatomy theater, and the metropolitan print shop and bookseller.

These associations were made possible in part by the mobility of many practitioners of early modern knowledge. For some, this mobility was voluntary, as in the case of the English astronomer Edmond Halley's (ca. 1656–1743) voyage to Saint Helena or the German naturalist Maria Sybilla Merian's (1647–1717) expedition to Surinam. For others, it was vocational, as for Jesuit missionaries to China or Peru, or the engineers who traveled from court to court offering their services to build fortifications or ornamental fountains. For still others it was involuntary, as when the Protestant astronomer Johannes Kepler (1571–1630) was forced to leave his teaching post in Catholic Graz or the Dutch natural philosopher Christiaan Huygens (1629–1695) gave up his position as president of the Paris Académie Royale des Sciences after the revocation of the Edict of Nantes in 1685. Whether willed or not, these travels enlarged the range of natural phenomena studied and thickened contacts among those who studied them. As one of the favorite biblical quotations of the era put it: "Many shall go to and fro, and knowledge shall be increased" (Daniel 12:4).

Knowledge was not only increased in some quantitative fashion during this period; it was also qualitatively transformed. The chapters in Part I: "The New Nature" address shifts in the foundations and sources of natural knowledge as well as in its characteristic forms of explanation and proof. To fuse natural philosophy with natural history, for example, or terrestrial with celestial mechanics, involved rethinking the nature of knowledge and even the nature of nature. Sometimes the problem was methodological: In traditional classifications of knowledge, where each discipline was held to have its own distinctive axioms and modes of argumentation, to mingle, for example, mathematical cosmology with physical astronomy, let alone with theology and biblical exegesis, was according to some authorities to commit an elementary category mistake.[15] There were also epistemological stumbling blocks: How could the particulars of experience, so variable and tied to local circumstance, ever yield reliable universal generalizations? Thus syllogisms with universal premises and conclusions gave way to other kinds of proof. New forms of experience, such as experiments and structured programs of observation, were adapted from practices in the workshop, sickroom, shipboard, and field, and articulated into new types of arguments that depended heavily on analogy, the credibility of testimony, and the consilience of evidence. Moreover, ways of knowing that were long deemed inferior by the learned were elevated to higher status, first within court culture and then among scholars, often by way of court-sponsored academies: *Historia*, the knowledge of particulars, was promoted to equal standing with *philosophia*, the knowledge of universals, and the know-how of peasants, mariners, and artisans was recognized in some quarters as genuine knowledge.

With new explanations, arguments, and modes of inquiry, ontology also shifted: An explanation of natural phenomena couched in terms of qualities observable to the unaided senses assumed a nature different from one that appealed to microscopic mechanisms, magical natures, or invisible forces. The furniture of the universe changed alongside standards of intelligible explanations.

The chapters in Part IV: "The Cultural Meanings of Natural Knowledge" describe how natural knowledge interacted with the symbols, values,

---

[15] See Aristotle, *Posterior Analytics*, 1.7 (75a38–b21); Robert S. Westman, "Proof, Poetics, and Patronage: Copernicus's Preface to *De revolutionibus*," in *Reappraisals of the Scientific Revolution*, ed. David C. Lindberg and Robert S. Westman (Cambridge: Cambridge University Press, 1990), esp. pp. 183–4. The interactions between mathematical and physical astronomy in the sixteenth century were complex; for a survey of the spectrum of positions, see N. Jardine, *The Birth of History and Philosophy of Science: Kepler's "A Defence of Tycho against Ursus" with Essays on its Provenance and Significance* (Cambridge: Cambridge University Press, 1988), pp. 225–57.

ambitions, and imaginary of early modern Europe. It would be misleading to describe these interactions in terms of the context of natural knowledge because in most cases no hard-and-fast boundary separated the topics under consideration from the production of natural knowledge itself. Hence headings of the form "Science and X," although perhaps helpful to orient modern readers, presume autonomous fields of activity that in many cases had yet to crystallize as such. This is particularly true with respect to the interactions of natural philosophy and theology, but some forms of early modern art and literature were also so tightly intertwined with coeval natural inquiry that it is more accurate to treat them as expressions of a common endeavor. So whether one describes the highly detailed reportage of natural and human phenomena common to authors of early novels and authors of articles in the *Philosophical Transactions of the Royal Society of London* as literary or as scientific realism seems a moot point; the same might be said about the techniques of mimesis used in Dutch genre painting and botanical illustration.

In the case of the chapters on "Gender" and "European Expansion and Self-Definition," other dynamics are explored. Moralists and philosophers had long invoked the natural order to shore up the political, social, and religious orders. Over the course of the early modern period, many of these hierarchies and arrangements were reshuffled. At the same time, Europeans faced the task of incorporating into older intellectual structures their relationships with the non-European peoples and civilizations they encountered in the course of voyages of trade, conquest, and mission. New forms of natural knowledge that developed over the course of the sixteenth and seventeenth centuries – together with the new forms of authority they attributed to nature – became important resources to these ends.[16]

Although the organization of this hefty volume into four parts will, we hope, make it more easily navigable for readers unlikely to read it cover to cover, we do want to draw attention to thematic connections that may not be obvious from part headings and chapter titles. If, for example, a chapter relates its topic explicitly to developments in medicine or mechanics, we assume the reader needs no further clues as to where to find out more. But if the link to other chapters in the volume is less apparent but still significant, we have inserted internal cross-references, a convention we have also followed in this introduction.

There are certainly omissions in this volume, some that we recognize all too clearly and others that will become visible only in the context of further scholarship. But the omission that is likely to arouse the most surprise is in the title itself: Where is the Scientific Revolution? Our avoidance of the phrase is intentional. The cumulative force of the scholarship since the 1980s has been to insert skeptical question marks after every word of this ringing three-word

---

[16] See Lorraine Daston and Fernando Vidal, eds., *The Moral Authority of Nature* (Chicago: University of Chicago Press, 2004).

phrase, including the definite article. It is no longer clear that there was any coherent enterprise in the early modern period that can be identified with modern science, or that the transformations in question were as explosive and discontinuous as the analogy with political revolution implies, or that those transformations were unique in intellectual magnitude and cultural significance.[17] Few professional historians of science embrace the more extravagant claims once made by historians of science such as E. A. Burtt, Alexandre Koyré, or Herbert Butterfield about the world-shaking significance of the Scientific Revolution as "the real origin both of the modern world and of the modern mentality."[18] Even the canonical texts of the Revolution's heroes – for example, Galileo, Bacon, or Isaac Newton (1642–1727) – appear modern only if read (as they often are) with the greatest selectivity.

Although traditional claims about the Scientific Revolution as the wellspring of modernity (or even of modern science) no longer convince, nothing has yet challenged contemporaries' own view of their epoch as drenched in novelty. On the contrary, historical research across a broad range of topics has confirmed their impression of pell-mell change at every level: the astounding growth in the number of plant species and mathematical curves identified, for example; the creation of whole new ways of conceiving the natural order, such as the idea of "natural law";[19] the deployment of natural philosophers as technical experts on the government payroll and of natural philosophy as the best argument for religion. The transformations that occurred between about 1490 and 1730 were huge, and hugely varied, as documented by the chapters in this volume.

It is, however, precisely the variety of these transformations that frustrates attempts to corral them into any single historical event, whether revolutionary or evolutionary, disciplined or dispersed. Narratives about changes in astronomy and cosmology, from Nicholas Copernicus (1473–1543) to Newton, have

---

[17] These points are cogently made in Steven Shapin, *The Scientific Revolution* (Chicago: University of Chicago Press, 1996), pp. 3–5; see also Margaret J. Osler, "The Canonical Imperative: Rethinking the Scientific Revolution," in *Rethinking the Scientific Revolution*, ed. Margaret J. Osler (Cambridge: Cambridge University Press, 2000), pp. 3–24. The essays in this latter volume, especially when read in conjunction with those in Lindberg and Westman, eds., *Reappraisals of the Scientific Revolution*, give some idea of major trends in specialist scholarship since the mid-1990s and their historiographic reverberations.

[18] Herbert Butterfield, *The Origins of Modern Science, 1300–1800*, rev. ed. (New York: Free Press, [1957] 1965), p. 8; cf. E. A. Burtt, *The Metaphysical Foundations of Modern Physical Science* (Garden City, N.Y.: Doubleday, [1924] 1954), pp. 15–24, and Alexandre Koyré, *From the Closed World to the Infinite Universe* [1957] (Baltimore: Johns Hopkins University Press, 1979), pp. 1–3.

[19] The term "law" was applied to natural phenomena by Seneca (*Naturales quaestiones*, VII. 25.3) in the context of comets, and was used occasionally in medieval Latin grammar, optics, and astronomy: Jane E. Ruby, "The Origins of Scientific Law," *Journal of the History of Ideas*, 47 (1986), 341–59. Only in the seventeenth century, however, did it become the predominant term for natural regularities. See Friedrich Steinle, "The Amalgamation of a Concept – Law of Nature in the New Sciences," in Friedel Weinert, ed., *Laws of Nature: Essays on the Philosophical, Scientific, and Historical Dimensions* (Berlin: De Gruyter, 1995), pp. 316–68; John R. Milton, "Laws of Nature," in *The Cambridge History of Seventeenth-Century Philosophy*, ed. Daniel Garber and Michael Ayers, 2 vols. (Cambridge: Cambridge University Press, 1998), 1: 680–701.

traditionally furnished the backbone of historical accounts of the Scientific Revolution. The changes in this field were unquestionably momentous, driven to a large extent by techniques and imperatives developed within a discipline that had already achieved a distinct intellectual identity in late antiquity. But the merging of natural history with natural philosophy was no less momentous a change, although it did not culminate in a dramatic synthesis or system, and depended on a far more motley ensemble of methods: field observation, experiment, collecting, travel, letter-writing, classification, and exchange. These were cobbled together from sites and practices foreign to both disciplines and to one another (e.g., the apothecary shop, humanist correspondence, travel diaries, alchemical stills, and cabinets of curiosities). The remarkable transformations of early modern anatomy and physiology – despite the coincidence of the publication date of Andreas Vesalius's (1514–1564) *De humani corporis fabrica* (On the Fabric of the Human Body, 1543) with Copernicus's *De revolutionibus orbium celestium* (On the Revolutions of the Heavenly Spheres, 1543) – were largely separate from both of the two preceding stories, bringing us into worlds of Christian ritual and absolutist spectacle. Does it really make sense to fit all of these varied developments into one Grand Change, whatever we choose to call it?[20]

It is of course no coincidence that so many remarkable changes, however disparate in substance, pace, and outcome, occurred in the same time span of about two hundred years. In some cases, the synergy between fields such as natural philosophy and the mechanical arts – remote from one another at the beginning of the period but neighbors in the classification of knowledge by its end – was powerful and fruitful. In other cases, however, the cross-fertilization took place less among various kinds of natural knowledge than between natural knowledge and some other major transformation in early modern European society: The dynamic expansion of natural history, for example, owed far less to natural philosophy, mixed mathematics, or even medicine than to the booming trade with the Far East and the Far West that flooded European markets with new commodities and naturalia, many of them previously unknown to learned Europeans.[21] In general, the key question is not whether the innovations and transformations of the early modern period interacted with one another – they undeniably did, in complex and consequential ways – but rather which interactions were strong and which weak, which sustained and which episodic, and why. It is debatable whether

---

[20] These examples are not meant to echo the contrast of "classical" and "Baconian" sciences in Thomas S. Kuhn, "Mathematical versus Experimental Traditions in the Development of Physical Science," in Kuhn, *The Essential Tension: Selected Studies in Scientific Tradition and Change* (Chicago: University of Chicago Press, 1977), pp. 31–65, although they second the spirit of that essay. The "conceptual transformations" (p. 45) in early modern natural history and anatomy do not seem minor to us, although they are of a different kind than those that occurred in astronomy.

[21] Pamela H. Smith and Paula Findlen, eds., *Merchants and Marvels: Commerce, Science, and Art in Early Modern Europe* (New York: Routledge, 2002).

Introduction: The Age of the New

the interactions between elements in the field somewhat anachronistically defined as natural knowledge were, for any given case, more significant than those between that element and some other area undergoing and precipitating rapid change during this period, such as printing or the elaboration of the culture of the early modern courts.[22]

Yet the story of the Scientific Revolution retains its hold, even on those scholars who have contributed to its unraveling. Part of the reluctance to relinquish the historical narrative is due to the brilliance with which it has been told and retold in books that are deservedly numbered among the classics of the history of science.[23] Its drama of worlds destroyed and reconstructed recruited many historians of early modern science to the discipline and still entrances students in introductory courses.[24] But the magnetism of the mythology of the Scientific Revolution radiates beyond the classroom, to the airwaves of the public broadcasting system and the pages of the *New York Times*. It is a genuine mythology, which means it expresses in condensed and sometimes emblematic form themes too deep to be unsettled by mere facts, however plentiful and persuasive. The Scientific Revolution is a myth about the inevitable rise to global domination of the West, whose cultural superiority is inferred from its cultivation of the values of inquiry that, unfettered by religion or tradition, allegedly produced the sixteenth- and seventeenth-century "breakthrough to modern science."[25] It is also a myth about the origins and nature of modernity, which holds both proponents and opponents in its thrall. Those who regret "the modern mentality" as the

---

[22] Adrian Johns, *The Nature of the Book: Print and Knowledge in the Making* (Chicago: University of Chicago Press, 1998). The literature on early modern European courts is enormous; see, for example, Ronald G. Asch and Adolf M. Birke, eds., *Prince, Patronage, and the Nobility: The Court at the Beginning of the Modern Age, c. 1450–1650* (Oxford: Oxford University Press, 1991). Lisa Jardine, *Ingenious Pursuits: Building The Scientific Revolution* (New York: Anchor Books, 1999), deftly interweaves various forms of seventeenth-century natural knowledge with coeval intellectual, economic, and cultural changes.

[23] In addition to the works mentioned in note 18, see E. J. Dijksterhuis, *The Mechanization of the World Picture*, trans. C. Dikshoorn (Princeton, N.J.: Princeton University Press, [1950] 1986); Thomas S. Kuhn, *The Copernican Revolution: Planetary Astronomy in the Development of Western Thought* (New York: Vintage, 1957); I. Bernard Cohen, *The Birth of a New Physics* (Garden City, N.Y.: Doubleday, 1960); Marie Boas Hall, *The Scientific Renaissance, 1450–1630* (New York: Dover, 1962); A. Rupert Hall, *The Revolution in Science, 1500–1750*, 2nd ed. (London: Longmans, [1962] 1983); and Richard S. Westfall, *The Construction of Modern Science: Mechanisms and Mechanics* [1971] (Cambridge: Cambridge University Press, 1977). For an overview of the historiography and extensive bibliography up to about 1985, see H. Floris Cohen, *The Scientific Revolution: A Historiographical Inquiry* (Chicago: University of Chicago Press, 1994).

[24] Most of the books written about the Scientific Revolution were and are intended as textbooks for introductory-level history of science courses, such as Shapin, *The Scientific Revolution*; John Henry, *The Scientific Revolution and the Origins of Modern Science* (New York: St. Martin's Press, 1997); James R. Jacob, *The Scientific Revolution: Aspirations and Achievements, 1500–1700* (Amherst, N.Y.: Humanity Books, 1998); and Peter Dear, *Revolutionizing the Sciences: European Knowledge and Its Ambitions, 1500–1700* (Princeton, N.J.: Princeton University Press, 2001).

[25] See, for example, Toby E. Huff, *The Rise of Early Modern Science: Islam, China, and the West* (Cambridge: Cambridge University Press, 1993), quotation at p. 12.

"disenchantment of the world" are as captivated as those who celebrate it as a liberation from obfuscation and tyranny.[26]

The need for such a myth overwhelms its incoherences: Natural knowledge circa 1730 was assuredly not the modern science that arose in name and in fact in the mid-nineteenth century as an integrated enterprise of institutionally sponsored research, technological invention, and industrial application.[27] Furthermore, it is unclear what either kind of knowledge had to do with that mist-shrouded entity known as "the modern mind," which has been variously equated with Cartesian rationalism, capitalist calculation, secularization, hard-headed materialism, imperialist expansion, the demise of anthropocentrism, and a certain skepticism about the existence of fairies.

The pessimistic conclusion that might be drawn from this account of the tenacity of the Scientific Revolution in the historiography of science is that it will last as long as the myth of modernity, of which it is part and parcel. But modernity itself has a history, myths and all. These began in the early modern period, with publications such as the *Nova reperta*, self-conscious reflections on the relative accomplishments of the Ancients versus the Moderns,[28] and the quickening tempo of innovation in almost every realm, from church to marketplace, library to laboratory. These novelties were by no means unanimously welcomed; indeed, many were criticized just because they were new. By the mid-seventeenth century, however, "new" was fast becoming a term of praise rather than opprobrium. Innovation itself was not new, but the self-confident insistence on it was. Instead of requiring disguise or justification as a revival of older customs or a return to purer ideas, novelty became its own justification. In his 1686 popularization of Copernican astronomy, the French natural philosopher Bernard le Bovier de Fontenelle promised "all the news [*nouvelles*] that I know about the heavens, and I believe that none are fresher."[29]

Astronomy had become as new as the "New" World, the subject of the first engraving in the *Nova reperta*, which sets the framework for the rest. It shows Amerigo Vespucci, holding a mariner's astrolabe and a banner surmounted by a cross, confronting America, personified as a naked woman (Figure 1.3). The image emphasizes the enormous cultural difference between the elegantly

---

[26] The evocative phrase originates with Max Weber, "Wissenschaft als Beruf [1917]," in *Max Weber Gesamtausgabe*, Abt. I: *Schriften und Reden*, ed. Wolfgang J. Mommsen and Wolfgang Schluchter, together with Birgitt Morgenbrod (Tübingen: J.C.B. Mohr, 1992), 17: 70–111, at p. 109.

[27] For an account of the Scientific Revolution that spans the seventeenth through the nineteenth centuries, see Margaret C. Jacob, *The Cultural Meaning of the Scientific Revolution* (New York: Alfred A. Knopf, 1988).

[28] Richard Foster Jones, *Ancients and Moderns: A Study of the Rise of the Scientific Movement in Seventeenth-Century England*, rev. ed. (New York: Dover, [1961], 1982); and Joseph M. Levine, *Between the Ancients and the Moderns: Baroque Culture in Restoration England* (New Haven, Conn.: Yale University Press, 1999).

[29] Bernard le Bovier de Fontenelle, *Entretiens sur la pluralité des mondes*, ed. François Bott (Paris: Editions de l'Aube, [1686], 1990), p. 133.

Introduction: The Age of the New

Americen Americus retexit, & Semel vocauit inde semper excitam.

Figure 1.3. *America*. Jan Galle after Joannes Stradanus (Jan van der Straet), ca. 1580, from *Nova reperta*. In *Speculum diuersarum imaginum speculatiuarum a varijs viris doctis adinuentarum, atq[ue] insignibus pictoribus ac sculptoribus delineatarum* ... (Antwerp: Jan Galle, 1638). Reproduced by permission of the Print Collection, Miriam and Ira D. Wallach Division of Art, Prints and Photographs, The New York Public Library, Astor, Lenox and Tilden Foundations.

clothed and technologically advanced Europeans and the culturally backward Americans, in a timeless rural landscape, who evoke simultaneously the primitive inhabitants of the "New" World and – in the context of the entire series – Europe's own primitive past. This is the early modern period's own myth of modernity – one at least as spellbinding as that created for it by latter-day historians.

Part I

THE NEW NATURE

2

PHYSICS AND FOUNDATIONS

Daniel Garber

In our times, the domain of the physical sciences is reasonably well defined. Although, at its edges, the less empirically grounded parts of the physical sciences may merge into philosophical speculation, it is no compliment to a scientist to characterize his or her work as "philosophical." In this respect, we have moved a considerable distance from the early modern period. For many European thinkers in the sixteenth and seventeenth centuries, an account of the world around them was radically incomplete without a larger background picture in which to embed it, a picture that often included elements such as the basic categories of existence and the relation of the natural world to God. Many shared the sense of the interconnectedness of knowledge and felt the need for what might be called a foundation for the science that treats the natural world.

The project did not have precise boundaries, nor is it easy to characterize what it is that we are talking about when we are talking about the foundations of our understanding of the physical world. In many ways, the enterprise of providing foundations for a view of the physical sciences was shaped by two traditions, the Aristotelian tradition in philosophy and the Christian tradition in theology. As I shall argue in more detail, the Aristotelian tradition was a common element in the intellectual background of every serious thinker of the period and provided a model for what a properly grounded science should look like. Even for many of those who would reject the Aristotelian tradition in favor of other ancient traditions (such as atomism or Hermeticism) or other views of the world not obviously connected with ancient philosophical traditions, the Aristotelian tradition was hard to escape. But the Aristotelianism at issue was one deeply imbued with the spirit of Christian theology. From the time that Aristotelianism was introduced to the Latin West in the late twelfth and early thirteenth centuries, Christian doctrines about creation, divine omnipotence, and divine freedom put serious constraints on how Aristotelian doctrines were received. These constraints continued to play a role in how Europeans thought about the natural world throughout the period of the sixteenth and seventeenth centuries, and very often (though not always) entered into the versions of other non-Aristotelian philosophies proposed and adopted. Furthermore, the Christian God often provided an important resource in understanding the foundations of the natural world; for example, serving as the ultimate ground of the laws of motion for Descartes or the ground of absolute space for Newton. In this way, Christian theology and Aristotelian philosophy wind their ways throughout the questions that I will take up in this chapter.

FOUNDATIONS

It is tempting to frame the question of foundations in terms of physics and its metaphysical foundations,[1] but the question is somewhat more complex than that simple formulation would suggest.

In its strict Aristotelian meaning, metaphysics was usually taken to be the science of being qua being, the science of being as such. In addition, metaphysics was often taken to include an account of God, separated (i.e., immaterial) substances, and substance in general. Physics, on the other hand, was taken to be the study of natural things, things with natures, where natures were understood to be internal principles of motion and rest. Although the view that physics depends in some substantive way on metaphysics was not completely unheard of among medieval Aristotelian schoolmen, physics was generally held to be a discipline largely independent of metaphysics, and as a more concrete discipline dealing with sensible things, it should be studied *before* the student took up metaphysics. Therefore, in this strict sense, for an Aristotelian, one could not properly talk about the metaphysical *foundations* of physics.[2]

---

[1] Historians who do include E. A. Burtt, *The Metaphysical Foundations of Modern Physical Science: A Historical and Critical Essay* (London: Routledge and Kegan Paul, 1932); E. W. Strong, *Procedures and Metaphysics: A Study of the Philosophy of Mathematical-Physical Science in the Sixteenth and Seventeenth Centuries* (Berkeley: University of California Press, 1936); Alexandre Koyré, *Metaphysics and Measurement: Essays in Scientific Revolution* (Cambridge, Mass.: Harvard University Press, 1968); Gerd Buchdahl, *Metaphysics and the Philosophy of Science: The Classical Origins, Descartes to Kant* (Cambridge, Mass.: MIT Press, 1969); and Gary Hatfield, "Metaphysics and the New Science," in *Reappraisals of the Scientific Revolution*, ed. David Lindberg and Robert Westman (Cambridge: Cambridge University Press, 1990), pp. 93–166.

[2] For a discussion of the meanings of the term "metaphysics" among medieval Aristotelians, see John Wippel, "Essence and Existence," in *The Cambridge History of Later Medieval Philosophy*, ed. Norman Kretzmann, Anthony Kenny, and Jan Pinborg (Cambridge: Cambridge University Press, 1982), pp. 385–410, esp. pp. 385–92. On the question of ordering knowledge in late scholastic thought, see Daniel Garber, *Descartes' Metaphysical Physics* (Chicago: University of Chicago Press, 1992), pp. 58–62; and Roger Ariew, "Descartes and the Late Scholastics on the 'Order of the Sciences'," in *Conversations with Aristotle*, ed. Constance Blackwell and Sachiko Kusukawa (London: Ashgate, 1999). It should be noted that the term "metaphysics" as it was first used did not designate any discipline or subject matter. It was originally coined simply to designate the somewhat heterogeneous group of treatises that followed Aristotle's physical treatises in the ordering given in the edition of his writings by Andronicus of Rhodes. See G. E. R. Lloyd, *Aristotle: The Growth and Structure of His Thought* (Cambridge: Cambridge University Press, 1968), pp. 13–14.

But the view that metaphysics provides a kind of foundation for physics did indeed appear in the seventeenth century, most famously in the metaphysical physics of René Descartes (1596–1650) and Gottfried Wilhelm Leibniz (1646–1716). As Descartes wrote in the preface to the 1647 French edition of his *Principia philosophiae* (Principles of Philosophy, 1644): "The whole of philosophy is like a tree. The roots are metaphysics, the trunk is physics, and the branches emerging from the trunk are all the other sciences, which may be reduced to three principal ones, namely medicine, mechanics and morals."[3]

In this case, it may therefore be proper to talk about the metaphysical foundations of physics. However, it is important to note that the conception of both metaphysics and physics at work here is somewhat idiosyncratic, very different from that found in the Aristotelian tradition or even in other contemporary writers. For Descartes, for example, the study of being qua being that is at the center of Aristotelian metaphysics had no place at all in his philosophy.[4] What his philosophy did contain, on the other hand, was an account of how we acquire knowledge of the physical world, something quite foreign to most other conceptions of metaphysics. Furthermore, because Descartes recognized no internal principles of motion and rest of the sort that define the subject matter of physics for the Aristotelian schoolmen, his conception of physics was very different from theirs.

For Leibniz, too, the world of mechanist physics was grounded ultimately both in metaphysical objects, simple substances or monads, and in metaphysical principles, the principles by virtue of which God chose to create this world.[5] Although Leibniz's conceptions of metaphysics and physics were, in a way, closer to the Aristotelian conceptions,[6] they were still distant enough from them (and from Descartes' conceptions of the domains) to make any general comparison of the relation between metaphysics and physics problematic and unilluminating.[7] Problems with characterizing our question in

---

[3] See René Descartes, *Oeuvres de Descartes*, ed. Charles Adam and Paul Tannery, new ed., 11 vols. (Paris: CNRS/J.Vrin, 1964–74), 9B: 14. In quoting Descartes, I will generally follow the translations in *The Philosophical Writings of Descartes*, ed. and trans. John Cottingham, Robert Stoothoff, Dugald Murdoch, and Anthony Kenny, 3 vols. (Cambridge: Cambridge University Press, 1984–91). Because this latter book is keyed to the Adam and Tannery edition, I will not give separate references to it.

[4] This has led Jean-Luc Marion to the bold (and somewhat paradoxical) conclusion that Descartes does not have a metaphysics. See Jean-Luc Marion, *On Descartes' Metaphysical Prism* (Chicago: University of Chicago Press, 1999), chap. 1. On Descartes' conception of metaphysics and physics and the order of knowledge, see Garber, *Descartes' Metaphysical Physics*, chap. 2.

[5] For a detailed development of this theme, see Daniel Garber, "Leibniz: Physics and Philosophy," in *The Cambridge Companion to Leibniz*, ed. Nicholas Jolley (Cambridge: Cambridge University Press, 1995), pp. 270–352.

[6] As I discuss later in this chapter, Leibniz did recognize a sense in which the schoolmen were right to say that bodies are composed of matter and form.

[7] Just how far the term "metaphysics" strayed from its earlier signification can be seen in the next century, where in his *Discours préliminaire* (1751), d'Alembert characterized it as "the experimental physics of the soul"! See Jean Le Rond d'Alembert, *Preliminary Discourse to the Encyclopedia of Diderot*, trans. R. N. Schwab and W. E. Rex (Chicago: University of Chicago Press, 1995), p. 84.

terms of the metaphysical foundations of physics are compounded further by the fact that for many seventeenth-century students of nature, the term metaphysics did not come up at all, or if it did, it was explicitly rejected. Both Thomas Hobbes and Pierre Gassendi, for example, rejected the enterprise of metaphysics, strictly speaking.[8] Yet, in a number of such cases, as we shall see, they would certainly have acknowledged having views about the foundations of the physical world.

There are other ways in which the question of foundations came up in the seventeenth-century study of nature. For example, within the context of the Aristotelian system, mechanics, a "middle science" or branch of mixed mathematics, was distinguished from physics by virtue of the fact that whereas physics studies bodies insofar as they are natural and governed by internal principles of motion and rest, mechanics studies bodies insofar as they are constrained and made to do things that, left to their own natures, they would not do. In this context, mechanics makes use of some physical principles, such as the principle that heavy bodies tend to fall toward the center of the earth (which coincides with the center of the world in the Aristotelian system).[9] In this sense, one might say that physics is foundational with respect to mechanics. Similar points could be made about astronomy, optics, and harmonics, which are also branches of mixed mathematics. Furthermore, a number of figures drew distinctions between first causes and hidden natures on the one hand and phenomenal effects, their causal consequences, on the other. In his *Essay Concerning Human Understanding* (1690), for example, John Locke (1632–1704) famously distinguished between the real essence and the nominal essence. The real essence was the corpuscular substructure, the causal nexus from which flow the properties that make a body the body that it is, whereas the nominal essence was the collection of phenomenal properties accessible to our senses that result from that real essence, and in terms of which we sort bodies into categories.[10] Although this distinction between

---

[8] Hobbes often spoke contemptuously of metaphysics; see especially Thomas Hobbes, *Leviathan; or, The matter, forme, & power of a commen-wealth ecclesiasticall and civill* (London: Andrew Crooke, 1651), chap. 46. However, in his own program for philosophy, following the logic, he does begin with what he called "first philosophy," which, for him, consisted of definitions. See Thomas Hobbes, *De corpore* (London: Andrew Crooke, 1655), pt. 2. Gassendi's posthumous *Syntagma philosophicum* in Pierre Gassendi, *Opera omnia*, 6 vols. (Lyon: Laurentius Anisson and Ioan. Baptista Devenet, 1658) also began with logic, but he moved directly from there into physics. Some of Descartes' followers also sidestepped their master's demand for metaphysical foundations and went directly into physics. See, for example, Henricus Regius, *Fundamenta physices* (Amsterdam: Ludivicus Elzevirius, 1646); and Jacques Rohault, *Traité de physique* (Paris: Charles Savreux, 1671).

[9] On the relation between mechanics and physics, see Domenico Bertoloni Meli, "Guidobaldo dal Monte and the Archimedean Revival," *Nuncius*, 7 (1992), 3–34; James G. Lennox, "Aristotle, Galileo, and 'Mixed Sciences'," in *Reinterpreting Galileo*, ed. William A. Wallace (Washington, D.C.: Catholic University of America Press, 1986), pp. 29–51; and Peter Dear, *Discipline and Experience: The Mathematical Way in the Scientific Revolution* (Chicago: University of Chicago Press, 1995).

[10] See John Locke, *An Essay Concerning Humane Understanding, in four books*, 3.6 (London: Printed by Eliz. Holt for Thomas Basset, 1690). One can find similar themes in other works of the period. See, for example, Robert Lenoble, *Mersenne ou La naissance du mécanisme*, 2nd ed. (Paris: J.Vrin, 1971), chap. 9; Tulio Gregory, *Scetticismo ed empirismo: Studio su Gassendi* (Bari: Laterza, 1961); Galileo

the phenomena and their underlying causes was usually drawn specifically in order to deny that we have any knowledge of those causes, it represented another way in which one could talk about the foundations of a science of the physical world. Also current, both in Aristotelian physics texts and in later non-Aristotelian texts, was a distinction between the general part of physics, which contained a general account of the contents of the physical world and the general principles that things follow, and the special part of physics, which treated the explanation of the behavior of specific kinds of bodies.[11] Again, this is another way of capturing the distinction between foundational questions and other questions in the science of body and in physics.[12]

For all these reasons, framing the question of foundations in terms of the metaphysical foundations of physics does not capture what is of interest. But although the question is difficult to formulate precisely, there is a real sense in which early modern practitioners of the sciences of body recognized and debated foundational questions related to the ground-level kinds of things that existed in the world, their natures, and their relations to God and spirit. In this chapter, I survey some sixteenth- and seventeenth-century conceptions of the foundations of the sciences of the physical world, understood in this broad and somewhat imprecise sense. I begin with an overview of the Aristotelian foundations and a brief survey of some of the alternatives to this conception of the world put forward by Renaissance thinkers. Then I discuss some foundational issues connected with the so-called mechanical philosophy that came to dominate the field by the end of the seventeenth century.

THE ARISTOTELIAN FRAMEWORK

Aristotle's philosophy, as developed by his medieval followers, was at the center of the school curriculum in the sixteenth century, as it was in the centuries before, and it remained central in the schools well into the seventeenth century. There were, of course, some significant variations between different schools and universities in different regions that corresponded to

---

Galilei, *Istoria e dimostrazioni intorno alle macchie solari* . . . (Rome: Giacomo Mascardi, 1613), translated in Stillman Drake, *Discoveries and Opinions of Galileo* (Garden City, N.Y.: Doubleday, 1957), pp. 123 ff.

[11] In Eustachius a Sancto Paulo's enormously popular and often reprinted Aristotelian textbook, the *Summa philosophiae quadripartita* (Paris: Carolus Chastellain, 1609), the physics (one of the four parts of the book) is organized in this way. (My references are to the edition published in Cambridge by Rogerus Daniel in 1648.) The first part of the *Physica* deals with the "natural body in general." Part II then deals with inanimate bodies (the heavens, the earth, the elements, etc.), and the third treats animate things. Descartes' *Principia philosophiae* is similarly organized, with Part II treating "the principles of material things," Part III treating "the visible world" (i.e., the heavens), and Part IV treating specific kinds of bodies on earth, such as the magnet. Descartes died before he could complete two additional books on living things. One can find similar principles of organization in both Hobbes and Gassendi.

[12] One has to be a bit careful here. It is "science of body" and not "science of matter"; as we shall see, for an Aristotelian, matter, strictly speaking, is only one constituent of body, which also includes form.

different academic traditions and different religious persuasions (see Blair, Chapter 17, this volume).[13] But virtually all teachers, whether Catholic or Protestant, Northern or Southern European, could agree with the Jesuit *Ratio studiorum* (Plan of Studies) of 1586, their manual of instruction, in holding that, at least in the classroom, "in logic, natural philosophy, morals and metaphysics, the doctrine of Aristotle is to be followed."[14] Because this formed the basis of the education of virtually every literate person in early modern Europe, the works of Aristotle and, even more so, the numerous textbooks that gave accessible treatments of the Aristotelian philosophy offered a common vocabulary and conceptual framework with which to view the natural world.[15]

Natural philosophy, or physics, was generally defined by the schoolmen as the science of natural bodies (see Chapter 17, this volume). And so, for example, physics dealt with the natural fall of earthy bodies as their natures carry them toward the center of the universe. It was contrasted with the sciences of the artificial, such as mechanics, which dealt with ways of accomplishing goals that are contrary to the natures of things, such as when we use a lever or a pulley to raise a heavy body some definite distance.[16] As treated in physics, bodies (substances) were comprehended in terms of primary matter, substantial form, and privation. Primary matter was that which underlies change and persists when a body changes from one kind of thing to another. Substantial form, on the other hand, was that which characterizes a thing as the kind of thing that it is; it was what changed when a body became a thing of a different kind. In living things, the form was known as a soul. Privation was not really distinct from matter; it was the lack of some particular property in matter that allows that matter to acquire some property at a later time. In the strict Thomistic tradition, matter was pure potentiality and form pure actuality, and the one could not exist without the other. Scotist

---

[13] There are a number of different scholastic traditions within Aristotelian thought, as well as different humanist traditions. On this, see Charles Schmitt, *Aristotle and the Renaissance* (Cambridge, Mass.: Harvard University Press, 1983); and Roger Ariew "Descartes and the Scotists," chap. 2 of his *Descartes and the Last Scholastics* (Ithaca, N.Y.: Cornell University Press, 1999).

[14] S. J. Ladislaus Lukás, ed., *Ratio atque institutio studiorum . . .* (Rome: Institutum Historicum Societatis Iesu, 1986), p. 98. For a detailed discussion of the differences between sixteenth- and early seventeenth-century universities, emphasizing the centrality of Aristotle, see Richard Tuck, "The Institutional Setting," in *The Cambridge History of Seventeenth-Century Philosophy*, ed. Daniel Garber and Michael Ayers, 2 vols. (Cambridge: Cambridge University Press, 1998), 1: 14–23.

[15] For discussions of the burgeoning Aristotelian literature in the sixteenth and seventeenth centuries, see William Wallace, "Traditional Natural Philosophy," in *The Cambridge History of Renaissance Philosophy*, ed. Charles B. Schmitt, Quentin Skinner, and Eckhard Kessler with Jill Kraye (Cambridge: Cambridge University Press, 1988), pp. 201–35, esp. pp. 225 ff.; Charles B. Schmitt, "The Rise of the Philosophical Textbook," in Schmitt and Skinner, eds., *The Cambridge History of Renaissance Philosophy*, pp. 792–804; and Patricia Rief, "The Textbook Tradition in Natural Philosophy, 1600–1650," *Journal of the History of Ideas*, 30 (1969), 17–32.

[16] See Franciscus Toletus, *Commentaria una cum quaestionibus in octo libros de physica auscultatione* (Venice: Apud Iuntas, 1589), fol. 4v et seq.; Eustachius, *Physica*, in *Summa philosophiae quadripartita*, pp. 112–13; pseudo-Aristotle, *Mechanics*, 847a10 ff.

and Ockhamist traditions, however, gave form and matter more capacity for independent existence.[17]

For Aristotle, space was so closely connected with the body that occupies it that he denied the existence of empty space.[18] He wrote in the *Physics*: "Now it [space or place] has three dimensions, length, breadth, depth, the dimensions by which all body is bounded. But the place cannot *be* body; for if it were there would be two bodies in the same place. . . . What in the world, then, are we to suppose place to be?"[19]

The answer to this question is, evidently, "nothing," or at least nothing independent of the body that occupies it. If there were empty space, "how then will the body of the cube differ from the void or place that is equal to it? And if there can be two such things, why cannot there be any number coinciding?"[20] As a consequence, Aristotle rejected the idea of empty space as incoherent. Aristotle also used a number of arguments from the supposed incoherence of motion in a vacuum to argue for the impossibility of vacua in nature. By the thirteenth century, scholastic writers were beginning to attribute to nature a *horror vacui*, a kind of force by which nature resists allowing a vacuum to form.[21] However, Aristotle's medieval followers had some trouble with his doctrine of space and vacuum. One consequence was that without space outside of the (finite) world, not even God would seem to be able to move the universe, if he chose to do so. This apparent consequence of Aristotelian doctrine was rejected in the famous condemnation of Aristotle by Étienne Tempier, the bishop of Paris, in 1277: "[We condemn the proposition] that God could not move the heavens with rectilinear motion; and the reason is that a vacuum would remain."[22] As a result, scholastic Aristotelians had the difficult task of introducing the possibility of some kind of empty space into the universe without violating the basic principles of the Aristotelian philosophy.[23]

---

[17] Aquinas gives a lucid account of these notions and their relations in his essay "De principiis naturae," in Thomas Aquinas, *Opuscula omnia*, ed. P. Mandonnet, 5 vols. (Paris: Lethielleux, 1927), 1: 8–18, trans. Robert P. Goodwin in Thomas Aquinas, *Selected Writings of St. Thomas Aquinas* (Indianapolis: Bobbs-Merrill, 1965), pp. 7–28. For a different exposition of these notions, influenced by the later thought of William of Ockham and John Duns Scotus, see the *Physica* of Eustachius in his *Summa philosophiae quadripartita*, 1.1–1.3.

[18] See Edward Grant, *Much Ado about Nothing: Theories of Space and Vacuum from the Middle Ages to the Scientific Revolution* (Cambridge: Cambridge University Press, 1981), chap. 1.

[19] Aristotle, *Physics*, 4.1 (209a 5–8, 14). Translations of Aristotle are taken from *The Complete Works of Aristotle*, ed. Jonathan Barnes, 2 vols. (Princeton, N.J.: Princeton University Press, 1984) 1: 355.

[20] Aristotle, *Physics*, 4.8 (216b 9–11), 1: 367.

[21] See Grant, *Much Ado about Nothing*, chap. 4, for a history of this notion.

[22] "Condemnation of 1277," para. 49, in Edward Grant, ed., *A Source Book in Medieval Science* (Cambridge, Mass.: Harvard University Press, 1979), p. 48. See also Grant, "The Condemnation of 1277, God's Absolute Power, and Physical Thought in the Late Middle Ages," *Viator*, 10 (1979), 211–44.

[23] See Grant, *Much Ado about Nothing*, chaps. 5–6; Pierre Duhem, *Medieval Cosmology: Theories of Infinity, Place, Time, Void, and the Plurality of Worlds*, ed. and trans. Roger Ariew (Chicago: University of Chicago Press, 1985), chaps. 5–6, 9–10.

These are the most general principles of the Aristotelian physical world. But also important was the Aristotelian doctrine of what specific bodies there are in the world. Within the sublunar world, the world below the sphere of the moon, there were four elements: earth, water, air, and fire. By virtue of the form it has, each of the elements had a characteristic array of what were generally called primary and motive qualities. The primary qualities were hot, cold, wet, and dry. Earth was cold and dry; water, cold and wet; air, hot and wet; and fire, hot and dry. In addition to the primary qualities, the elements had motive qualities, either heavy or light; earth and water, the heavy elements, had a tendency to fall downward toward the center of the world, and air and fire tended to rise and move away from the center of the world. Strictly speaking, however, these motive qualities derived from the fact that each of the elements had a proper place, with earth at the center, then water, air, and fire, respectively. When separated from that proper place, the elements had a tendency to move toward it.[24] In nature, however, the elements were rarely, if ever, found in their pure form. They were normally thought to be mixed together, giving rise to bodies that had properties different from those of the elements of which they were composed. The complex theory of mixtures gave rise to some of the most heated disputes in late medieval and early modern Aristotelianism (see Joy, Chapter 3, this volume).[25] Because things in the sublunar world were composed of different elements that were capable of separating, the sublunar world was a world of things in flux that were generated as the elements combined and corrupted as the elements separated.

Fundamentally distinct was the world of heavenly bodies. These bodies were made up not of the four elements but of a fifth element, the quintessence. Celestial physics was taken to be altogether different from terrestrial physics. Rather than moving in rectilinear paths, celestial bodies moved in perfect circles. Rather than a world of change, of generation and corruption, like the sublunar world, the celestial world was taken to be an unchanging world of physical perfection.[26]

Insofar as Aristotelianism represented orthodoxy, the overt rejection of this tradition constituted a touchstone of modernity; those who rejected the Aristotelian tradition were called "new philosophers" or "renovators" or "innovators" by their sixteenth- and seventeenth-century contemporaries. In the following sections, I survey a number of such figures and movements.

---

[24] Compare the account in Eustachius, *Physica*, pp. 206–11.
[25] See also Anneliese Maier, *On the Threshold of Exact Science* (Philadelphia: University of Pennsylvania Press, 1982), chap. 6.
[26] For an account of medieval Aristotelian cosmology, see Edward Grant, *Planets, Stars, and Orbs: The Medieval Cosmos, 1200–1687* (Cambridge: Cambridge University Press, 1994), esp. pt. 2.

RENAISSANCE ANTI-ARISTOTELIANISMS: CHYMICAL PHILOSOPHIES

Alchemy, chemistry, or, as some historians now prefer to refer to it, chymistry, goes back to ancient thought in one form or another (see Newman, Chapter 21, this volume).[27] But the sixteenth century was a time of particular interest in chymistry. The idea of chymistry meant many things to many people of the period, and it is very dangerous to generalize.[28] Chymistry was both theory and practice, involving both an account of at least a part of the natural world and an application of that understanding to the practical problems of transforming base metals into gold and silver. It also involved other aspects of what we might now call chemical engineering, as well as the problem of curing patients.[29] For some people, the theoretical part of chymistry dealt with only a part of nature, with mixtures or with metals.[30] But for others, chymistry was itself the whole of natural science, a genuine natural philosophy, and a conception of the foundations of natural science alternative to that offered by the Aristotelians insofar as chymical philosophers offered an alternative conception of the basic categories and principles of the physical world. In his popular and often reprinted *Traicté de la chymie* (Treatise on Chemistry, 1660), Nicaise Le Fèvre (1610–1669), for example, distinguished three sorts of chymistry: philosophical, medical, and pharmaceutical. But the first was for him the most important, the most basic. He wrote:

> [The first sort of chymistry is] wholly Scientifical and given to Contemplation, and may be very well termed *Philosophical*, having only its end in the knowledge of Nature, and of its effects; because it takes for object those on[l]y things which are constituted out of our power: So that this kinde of Chymical Philosophy, doth rest satisfied in the knowledge of the nature

---

[27] For a survey of early chymistry, see Allen G. Debus, *The Chemical Philosophy: Paracelsian Science and Medicine in the Sixteenth and Seventeenth Centuries*, 2 vols. (New York: Science History Publications, 1977), vol. 1, chap. 1; and William Newman, *Gehennical Fire: The Lives of George Starkey, an American Alchemist in the Scientific Revolution* (Cambridge, Mass.: Harvard University Press, 1994), chap. 3. Newman emphasizes especially the contributions of pseudo-Geber and Lull. Historiographical trends of the 1990s suggest that there is no substantive distinction between alchemy and chemistry in the period, and some have suggested using the archaic "chymistry" as a neutral term. I will follow that practice in this chapter. See Lawrence Principe, *The Aspiring Adept: Robert Boyle and His Alchemical Quest* (Princeton, N.J.: Princeton University Press, 1998), pp. 8–10; William Newman and Lawrence Principe, "Alchemy vs. Chemistry: The Etymological Origins of a Historiographic Mistake," *Early Science and Medicine*, 3 (1998), 32–65.

[28] This is a point emphasized by Principe in *The Aspiring Adept*, pp. 214 ff.

[29] For a study of some of the practical aspects of chymistry focused on one particular practitioner, Johann Joachim Becher (1635–1682), see Pamela H. Smith, *The Business of Alchemy: Science and Culture in the Holy Roman Empire* (Princeton, N.J.: Princeton University Press, 1994).

[30] For a discussion of the place of chymistry among the sciences, see, for example, Jean-Marc Mandosio, "Aspects de l'alchimie dans les classifications des sciences et des arts au XVIIe siècle," in *Aspects de la tradition alchimique au XVIIe siècle*, ed. Frank Greiner (Paris: S.É.H.A., and Milan: ARCHÉ, 1998), pp. 19–61.

of the Heavens and Starres, the source and original of the Elements, the cause of Meteors, original of Minerals, and the way by which Plants and Animals are propagated. . . . We say then, that Chymistry makes all natural things, extracted by the omnipotent hand of God, in the Creation, out of the Abysse of the Chaos, her proper and adæquate object. . . . To make it short, It's nothing else but Physick, or knowledge of Nature it self, reduced to operation, and examining all its Propositions by reasons grounded upon the evidence and testimony of the senses.[31]

As such, chymistry aimed to replace the natural philosophy of the Aristotelians as taught in the schools. Le Fèvre went on to contrast the empty abstractions of the school philosophers with the down-to-earth and concrete approach of the chymists:

If you ask from the School-Philosopher, What doth make the compound of a body? He will answer you, that it is not yet well determined in the Schools: That, to be a body, it ought to have quantity, and consequently be divisible; that a body ought to be composed of things divisible and indivisible, that is to say, of points and parts; but it cannot be composed of points. . . . [Le Fèvre continues with a long and somewhat comic rehearsal of the hesitations and uncertainties in the schoolman's answer.] You see then, that Chymistry doth reject such airy and notional Arguments, to stick close to visible and palpable things, as it will appear by the practice of this Art: For if we affirm, that such a body is compounded of an acid spirit, a bitter or pontick salt, and a sweet earth; we can make manifest by the touch, smell, taste, those parts which we extract, with all those conditions we do attribute unto them.[32]

Important to the chymical thought of the period was the work of Theophrastus Bombastus von Hohenheim, known as Paracelsus (1493–1541). Trained as a physician, he focused much of his writing on medical topics, where he opposed the authority of Galen and Aristotle in favor of an empirically based medicine that made extensive use of chymical remedies. But Paracelsus and his numerous followers were also associated with a more general intellectual reform, a philosophy of nature grounded in chymistry.[33]

As with other sixteenth-century reformers of natural philosophy, Paracelsus and his followers were motivated in good part by religious and theological

---

[31] Nicaise Le Fèvre [Nicasius le Febure], *A Compleat Body of Chymistry . . .* (London: Thomas Ratcliffe, 1664), pp. 7, 9. Although French, Le Fèvre moved to London and became a member of the Royal Society of London. The book was originally published in French in 1660 but appeared quickly in English translation (1662), "Rendered into English by P. D. C. Esq. one of the Gentlemen of his Majesties Privy Chamber." It then came out in numerous editions in both French and English, with at least one German edition (1676). A fifth French edition came out as late as 1751.

[32] Le Fèvre, *A Compleat Body of Chymistry*, p. 10.

[33] The standard scholarly edition of Paracelsus's chymical and medical writings is Paracelsus, *Sämtliche Werke*, ed. Karl Sudhoff and William Matthiessen, 14 vols. (Munich: R. Oldenbourg, O. W. Barth, 1922–33). Collections of Paracelsus's writings in English include *The Hermetic and Alchemical Writings of Paracelsus*, ed. A. E. Waite, 2 vols. (Berkeley: Shambhala, 1976), and *Selected Writings*, ed. Jolande Jacobi, trans. Norbert Guterman (Princeton, N.J.: Princeton University Press, 1995).

questions.³⁴ Aristotle and Galen, heathen philosophers, were to be replaced by a genuinely Christian philosophy. For reformers of this sort, philosophy began with a return to the ancient wisdom found in the sacred scriptures, particularly the Old Testament, which predates the works of the pagan philosophers. But, at the same time, their chymical philosophy also turned to God's second book, the book of nature, for knowledge of the world. Peter Severinus (1540–1602), a late sixteenth-century follower of Paracelsus, famously advised those who seek wisdom to sell everything they owned, travel the world to observe what it contains, and then to build furnaces to probe its secrets (see Smith, Chapter 13, this volume).³⁵

What emerged out of this study was a view of the world that was in some ways structurally similar to the Aristotelian world but in some ways radically different. According to Paracelsus, everything could be explained through three chymical principles, the *tria prima*: salt, sulphur, and mercury. (It is not altogether clear what the relation was between the *tria prima* and the Aristotelian four elements, nor what became of matter and form in the Paracelsian scheme.) For Paracelsus, everything was explicable chymically through combinations and transmutations of these principles. Indeed, even the creation story of Genesis could be interpreted chymically, as the successive separation of things from an initial *mysterium magnum* by way of chymical processes. In this way, the entire world was regarded as a vast chymical laboratory. Chymical transformations were driven by heat and fire, ultimately derived from the sun and from God himself. But the Paracelsian world was more than just chymistry. Also important to the chymical philosophy of Paracelsus were elaborate relations and harmonies among phenomena at all different levels, the macrocosm/microcosm analogy. In particular, Paracelsus held that the human being, the microcosm, is a representation of the universe as a whole, the macrocosm, and that there are thus systematic relations, reflections, and sympathies that hold between the two. This had important consequences for Paracelsian medicine and additionally for the practice of Paracelsian science. By virtue of these correspondences, the Paracelsian magus, through his own character and discipline, was capable of concentrating the celestial powers in himself and bringing about works. Hence, for the Paracelsian, science was not a neutral activity: The moral status of the philosopher had a central role to play in the enterprise. Furthermore, as with many other philosophies of the period, the world of Paracelsus's chymical philosophy was animated: Paracelsus saw the fire that was at the center of his philosophy as being, in some sense, equivalent to life itself.

---

³⁴ My account of Paracelsus's views is drawn from the following sources: Allen G. Debus, *The Chemical Philosophy*, esp. vol. 1, chaps. 1–2; Debus, *Man and Nature in the Renaissance* (Cambridge: Cambridge University Press, 1978), esp. chap 2; and Brian Copenhaver and Charles B. Schmitt, *Renaissance Philosophy* (Oxford: Oxford University Press, 1992), pp. 306 ff.
³⁵ Cited in Debus, *Man and Nature*, p. 21.

Numerous works in chymistry followed the Paracelsian revival. Although there was considerable disagreement on detail, all agreed in seeing a certain small number of chymical principles and their combinations as essential to the project, and most shared a chymical cosmology and an interest in applying chymical ideas to medicine. Also important here was the importation into more traditional chymical theories of corpuscular ideas, in the sense that chymical elements were taken to be divisible to some smallest parts that retain their natures as elements. Main figures in the later chymical tradition include Severinus, Thomas Erastus (1524–1583), Daniel Sennert (1572–1637), Robert Fludd (1574–1637), Oswald Crollius (1560–1609), George Starkey (1628–1665), and Johannes Baptista Van Helmont (1579–1644).[36] Even a number of figures usually associated with the mechanistic strains of thought to be discussed later, such as Robert Boyle (1627–1691) and Isaac Newton (1642–1727), had serious interests in chymistry.[37]

The intellectual center of chymistry in the sixteenth and early seventeenth centuries was probably Germany; it was out of Germany that the Rosicrucians came, making a kind of religion out of their chymical philosophy.[38] But chymistry was also widespread in other European countries.[39] Chymists occupied a wide range of roles in society. Some taught in universities, particularly in faculties of medicine, and some worked at courts, particularly in the German-speaking countries. Many practiced chymistry as

---

[36] Newman, *Gehennical Fire*, emphasizes the importance of corpuscular strains of seventeenth-century chymistry, which, he argues, derives from the thirteenth-century *Summa perfectionis* of pseudo-Geber. For a general survey of alchemy in the seventeenth century, see Debus, *Chemical Philosophy*, chaps. 3–7. For some studies of particular chymists of the period, see Newman, *Gehennical Fire* (a study of the American and English chymist George Starkey); Smith, *The Business of Alchemy*; Bruce Moran, *Chemical Pharmacy Enters the University: Johannes Hartmann and the Didactic Care of Chymiatria in the Early Seventeenth Century* (Madison, Wis.: American Institute of the History of Pharmacy, 1991); Bernard Joly, *Rationalité de l'alchemie au XVIIe siècle* (Paris: J. Vrin, 1992) (a study of Pierre-Jean Fabre); Hans Kangro, *Joachim Jungius' Experimente und Gedanken zur Begründung der Chemie als Wissenschaft* (Wiesbaden: Franz Steiner Verlag, 1968); Robert Halleux, "Helmontiana," *Academiae analectica, Koninklijke Academie, Klasse der Wetenschappen*, 45 (1983), 35–63; and Halleux, "Helmontiana II," *Academiae analectica*, 49 (1987), 19–36.

[37] For Boyle and chymistry, see Principe, *The Aspiring Adept*. For Newton, see Betty Jo Teeter Dobbs, *The Foundations of Newton's Alchemy; or, "The hunting of the greene lyon"* (Cambridge: Cambridge University Press, 1975); and Richard S. Westfall, "Newton and the Hermetic Tradition," in *Science, Medicine, and Society in the Renaissance*, ed. Allen G. Debus, 2 vols. (New York: Science History Publications, 1972), 2: 183–98.

[38] The classic work on this subject is Frances A. Yates, *The Rosicrucian Enlightenment* (London: Ark Paperbacks [Routledge and Kegan Paul], 1986; orig. publ. 1972).

[39] For accounts of the lively discussions over chymistry in seventeenth-century England and France, see Allen G. Debus, *The English Paracelsians* (New York: Watts, 1965); Allen G. Debus, *Science and Education in the Seventeenth Century* (New York: Science History Publications, 1970) (dealing with debates over chymistry in England); and Allen G. Debus, *The French Paracelsians* (Cambridge: Cambridge University Press, 1991). For discussions of chymistry in the Holy Roman Empire in the period, see Bruce Moran, *The Alchemical World of the German Court: Occult Philosophy and Chemical Medicine in the Circle of Moritz of Hessen, 1572–1632* (Sudhoffs Archiv, Beihefte 29) (Stuttgart: Franz Steiner Verlag, 1991).

a trade, either connected with medicine or with metallurgy and the like.[40] Chymistry remained, in one way or another, a part of the texture of much scientific thought throughout the early modern period.

RENAISSANCE ANTI-ARISTOTELIANISMS: THE ITALIAN NATURALISTS

Another group that set itself against Aristotle in the sixteenth century has come to be known as the Italian naturalists.[41] The rediscovery of Platonic texts in the fifteenth century presented European thinkers with a new way of looking at the world that was often at odds with the dominant Aristotelianism. The Latin translations of Plato by Marsilio Ficino (1433–1499), first published in 1484, were enormously popular. Included in Ficino's commentary on Plato's *Phaedrus* were translations of the neo-Platonist Proclus. Ficino's Latin translation of Plotinus appeared a few years later, in 1492.[42] The reintroduction of Plato and neo-Platonism into the intellectual world of the sixteenth century gave rise to a number of interesting new natural philosophies, including those of Girolamo Fracastoro (1470–1553), Bernardino Telesio (1509–1588), Girolamo Cardano (1501–1576), Francesco Patrizi (1529–1597), Giordano Bruno (1548–1600), and Tommaso Campanella (1568–1639).[43] These thinkers can also be construed as offering an alternative conception of the foundations of the physical world.

These natural philosophers shared a general scorn for Aristotelian natural philosophy, particularly its categories of matter and form.[44] At least three of these figures, Telesio in his *De rerum natura* (On the Nature of Things, 1563), Campanella in his *Universalis philosophiae, seu metaphysicarum rerum . . .*

---

[40] I am indebted to conversations and correspondence with Tara Nummedal for information on her work about the chymist's life in German countries in the period. See Tara E. Nummedal, "Adepts and Artisans: Alchemical Practice in the Holy Roman Empire, 1550–1620," Ph.D. dissertation, Stanford University, Stanford, Calif., 2001. For the case of a chymist hanged for counterfeiting in France, see Adrien Baillet, *La vie de M. Descartes*, 2 vols. (Paris: Daniel Horthemels, 1691), 1: 231, and *Le Mercure françois; ou, la suitte de l'histoire de la paix*, 25 vols. (Paris: Iean and Estienne Richer, 1612–; this vol., 1633), 17: 713–23.

[41] The figures discussed in this section are often referred to as Renaissance philosophers of nature. The term, however, is a modern designation and now generally thought to be inappropriate. See Paul O. Kristeller, *Eight Philosophers of the Italian Renaissance* (Stanford, Calif: Stanford University Press, 1964), pp. 94–6, 110–12. For a general overview, in addition to Kristeller, see Copenhaver and Schmitt, *Renaissance Philosophy*, chap. 5; and Alfonso Ingegno, "The New Philosophy of Nature," in *The Cambridge History of Renaissance Philosophy*, pp. 236–63. My own accounts of these thinkers draw heavily on these sources.

[42] For details on the transmission of Platonic texts in the Renaissance, see Anthony Grafton, "The Availability of Ancient Works," in Schmitt and Skinner, eds., *The Cambridge History of Renaissance Philosophy*, pp. 767–91.

[43] Not all scholars link these philosophers to the strict Platonic tradition. See, for example, Frances Yates, *Giordano Bruno and the Hermetic Tradition* (Chicago: University of Chicago Press, 1964), who links Bruno to the Hermetic tradition.

[44] See Copenhaver and Schmitt, *Renaissance Philosophy*, pp. 303 ff.

dogmata (Doctrines of the Universal Philosophy, that is, of Metaphysical Things, 1638), and Patrizi in his *Nova de universis philosophia* (New Philosophy of Everything, 1591), challenged Aristotelian conceptions of space and place and argued that space exists prior to everything and independent of body, an empty container that is, in part, filled by the physical world.[45] They also shared a view of the world as animate; as one study has eloquently characterized it, their world "was an enchanted world of ensouled objects linked together and joined to a higher realm of spirit and absolute being."[46] Writing in his *De sensu rerum et magia* (On the Sense of Things and on Magic, 1620), Campanella asserted that "the world is a feeling animal . . . [whose] parts partake in one and the same kind of life"; it posesses "a spirit . . . both active and passive in nature."[47]

However, in other respects, these natural philosophers differed considerably from one another. In his *De contagione* (On Contagion, 1546), Fracastoro saw attraction and sympathy, suitably interpreted in quasi-mechanistic and atomistic terms, as a basic phenomenon in nature.[48] For his part, Telesio rejected Aristotle's conception of body in terms of matter and form, replacing it with a conception of the world that is grounded in heat and cold, immaterial (but natural) agents that enter into lifeless matter and thereby animate it. According to Telesio, virtually everything that we see around us in the physical world is the result of a struggle between these two fundamental and immaterial agents, which oppose each other. Although Campanella began his career as a follower of Telesio,[49] in later years he came to think that Telesio's physical theory needed deeper grounding. He held that Telesio was wrong to think of hot and cold as natural agents and argued that their

---

[45] On conceptions of space and vacuum in sixteenth-century Italian thought, see Grant, *Much Ado about Nothing*, pp. 192–206. Although Telesio thought that a vacuum was possible and could be produced, he did not believe that it occurred naturally. See Charles B. Schmitt, "Experimental Arguments For and Against a Void: The Sixteenth-Century Arguments," *Isis*, 58 (1967), 352–66. More generally, on Telesio, see Copenhaver and Schmitt, *Renaissance Philosophy*, pp. 309–14; Schmitt and Skinner, eds., *The Cambridge History of Renaissance Philosophy*, pp. 250–2; and Kristeller, *Eight Philosophers*, chap. 6. On Campanella, see Copenhaver and Schmitt, *Renaissance Philosophy*, pp. 317–28; and Schmitt and Skinner, eds., *The Cambridge History of Renaissance Philosophy*, pp. 257–61, 294–5. On Patrizi, see Schmitt and Skinner, eds., *The Cambridge History of Renaissance Philosophy*, pp. 256–7; 292–3; and Kristeller, *Eight Philosophers*, chap. 7.

[46] Copenhaver and Schmitt, *Renaissance Philosophy*, p. 288. The passage continues: "A universal world-soul pervades all creatures and makes all creatures, even rocks and stones, alive and sentient in some degree. Stars and planets are mighty living divinities, so astrological bonds and forces of sympathy unify all things in the lower world under the rule of the higher; microcosm reflects macrocosm as man's lesser world mirrors the greater world of universal nature. Hidden symmetries and illegible signatures of correspondence energize and symbolize a world charged with organic sympathies and antipathies. The natural philosopher's job is to break these codes and uncover their secrets."

[47] Quoted in Brian Copenhaver, "Astrology and Magic," in Schmitt and Skinner, eds., *The Cambridge History of Renaissance Philosophy*, pp. 264–300, esp. p. 294.

[48] On Fracastoro, see Copenhaver and Schmitt, *Renaissance Philosophy*, pp. 305–6.

[49] In his *Philosophia sensibus demonstrata* (Philosophy Demonstrated through the Senses, 1591), Campanella, like Telesio, rejected the form and matter of the Aristotelians; Telesio argued that body (mass) is animated by the manifest principles of heat and cold.

efficacy is traced back to God and the world soul.[50] In contrast, light formed the foundation of Patrizi's conception of the world in his *Nova de universis philosophia*. The notion of light was quite complex for Patrizi, who distinguished between the incorporeal light that emanates from God and other spirits and the corporeal light found in the physical world. For Patrizi, light of one sort or another explained everything in the physical world: life, the structure of the heavens, and the nature of an extracorporeal region where eternal beings can be found. Ultimately, light was grounded in God and a neo-Platonic hierarchy of being, beginning with The One. God was present at every level, working through the incorporeal element of light.[51] The views of others in this group, particularly Cardano and Bruno, are more difficult to characterize in a few words. Although Bruno was not altogether consistent as a thinker, there are a number of clear themes in his dense and complex writings. Bruno rejected the Aristotelian conceptions of God, substance, matter, and form. In *De la causa, principio, et uno* (On Cause, Principle, and Unity, 1584), he held that God is the only substance, and all finite things are just aspects of God. Bruno did hold, in a sense, that the main principles of body are matter and form. However, he often treated them as coinciding with one another in a very non-Aristotelian way.[52] Cardano's *De subtilitate* (On Subtlety, 1550) was a jumble of largely anti-Aristotelian views challenging various elements of the Aristotelian foundations of physics but obscure about what should replace them.[53]

None of these natural philosophers formed a lasting school or posed any serious danger to the reigning Aristotelianism of the schools. Their quest for novelty and originality may have undermined any serious attempt to form real traditions in a stable natural philosophy; they seem to have shared little more than a more or less animistic conception of the universe and a general sense that Aristotle had gotten it all wrong. Also important here was the fact that this philosophy never seemed to have any real institutional or professional home. Ficino was linked to the Medici court; Telesio had his own institute, the Accademia Cosentina, in the town of Cosenza, to promote his brand of natural philosophy; Patrizi was bishop of Gaeta; Fracastoro and Cardano were both physicians and taught medicine for at least a part of their careers; and Bruno and Campanella, both Dominicans, lived colorful lives that involved wandering through Europe disseminating their teachings and trying (unsuccessfully) to avoid getting into trouble with the authorities.

---

[50] See his *De sensu rerum et magia* (1620) and his *Universalis philosophiae, seu metaphysicarum rerum . . . dogmata* (1638). On Campanella, see the references cited in note 45.
[51] On Patrizi, see the references cited in note 45.
[52] On Bruno, see Copenhaver and Schmitt, *Renaissance Philosophy*, pp. 314–17; and Hilary Gatti, *Giordano Bruno and Renaissance Science* (Ithaca, N.Y.: Cornell University Press, 1999).
[53] On Cardano, see Copenhaver and Schmitt, *Renaissance Philosophy*, pp. 308–9; *The Cambridge History of Renaissance Philosophy*, pp. 247–50; and Anthony Grafton, *Cardano's Cosmos: The Worlds and Works of a Renaissance Astrologer* (Cambridge, Mass.: Harvard University Press, 1999).

Their views were widely disseminated in Italy. But they were also well known in intellectual circles outside of Italy. Bruno's visit to England in 1583–5 had lasting effects; the influence of Italian philosophy can also be seen in the physics sketched out by Francis Bacon (1561–1626).[54] In France, Marin Mersenne (1588–1648) and Jean-Cecile Frey (ca. 1580–1631), defenders of the Aristotelian tradition in the 1620s, regularly listed Telesio, Bruno, and Campanella among their main opponents.[55] Pierre Gassendi (1592–1655), another anti-Aristotelian, seems to have borrowed from Patrizi's *Discussiones peripateticae* (Peripatetic Discussions, 1581) in his *Exercitationes paradoxicae adversus Aristoteleos* (Paradoxical Exercises against the Aristotelians, Part I, 1624, Part II published posthumously in 1658).[56] Later in the seventeenth century, these Italian neo-Platonists would constitute one of the important influences on the so-called Cambridge Platonists, including Henry More (1614–1687) and Ralph Cudworth (1617–1688).

RENAISSANCE ANTI-ARISTOTELIANISMS: MATHEMATICAL ORDER AND HARMONY

Behind many of the anti-Aristotelian views discussed in the last two sections lay another kind of foundational commitment, a commitment to the mathematical rationality and order of the world. In this view, which threads its way through chymical, Platonist, and other views, the world is governed by geometric and arithmetic structures. There are a number of different versions of this broadly Pythagorean view, which was concerned more with the large-scale structure of the cosmos than with the detailed analysis of matter. It is not surprising that this view became associated with music and the idea that nature is to be understood in terms of notions such as harmony. It must be remembered here that in the early seventeenth century, music was one of the middle sciences, along with astronomy, optics, and mechanics (see Andersen and Bos, Chapter 28, this volume). Traditional music theory dealt largely with numerical proportions, which were correlated with the notes of the scale and, in appropriate combinations, led to consonances. In this way, music was a science that dealt with harmony and order, both in the narrow

---

[54] See Graham Rees, "Bacon's Speculative Philosophy," in *The Cambridge Companion to Bacon*, ed. Markku Peltonen (Cambridge: Cambridge University Press, 1996), pp. 121–45.

[55] See, for example, the (unpaginated) preface to Mersenne's *Quaestiones . . . in Genesim* (Questions on Genesis, Paris: Sebastian Cramoisy, 1623). On Mersenne's relations with Italian naturalism, see Lenoble, *Mersenne*, chap. 3. Jean-Cécile Frey attacks them in his *Cribrum philosophorum qui Aristotelem superiore et hac aetate oppugnarunt* (A Sieve for Philosophers Who Oppose Aristotle Both in Earlier Times and in Our Own, 1628) in his posthumous *Opuscula varia* (Various Works, Paris: Petrus David, 1646), pp. 29–89. On Frey, see Ann Blair, "The Teaching of Natural Philosophy in Early Seventeenth-Century Paris: The Case of Jean Cécile Frey," *History of Universities*, 12 (1993), 95–158.

[56] On this, see pp. x–xi of Rochot's introduction to Gassendi, *Exercitationes paradoxicae adversus aristoteleos*, ed. and trans. [French] Bernard Rochot (Paris: J. Vrin, 1959).

sense of interest to practicing musicians and in a broader sense, in which it was of interest to natural philosophy.

For the English natural philosopher Robert Fludd, who was also very much a partisan of the chymical philosophies, a fundamental analogy for understanding the world was musical.[57] In one version, given in his *Utriusque cosmi maioris scilicet minoris metaphysica, physica atque technica historia* (The Physical, Metaphysical, and Technical History of Both Cosmoses, Namely the Greater and the Lesser, 1617–21),[58] Fludd's image of the world was based on the monochord, a string stretched between two bridges that was widely used in theoretical studies of music (see Figure 2.1). He pictured the cosmos as a monochord, with one end of the string anchored at the center of the Earth, and the other in the heavens. The sun is placed squarely at the middle of the string, dividing the string into two octaves. The notes of the scale (A, B, C, etc.) then mark out different regions of the cosmos, both subsolar and supersolar. Another more geometrical rendering of the same basic cosmology is given in Figure 2.2. This representation introduces two pyramids, which Fludd calls the material pyramid and the formal pyramid. The actual sounding music of the world results from an interaction between the two.[59]

For Athanasius Kircher (1601–1680), German by birth but a long-time professor at the Jesuit Collegio Romano in Rome, who also dabbled in chymistry, among many other pursuits, the cosmos was more like an organ[60] (see Figure 2.3). Instead of Fludd's one level of being, represented by the monochord, in his *Musurgia universalis* (Universal Harmony, 1650), Kircher recognized ten, which he likened to stops in an organ. The first six represented the results of the six days of creation; the remaining four dealt with other aspects of the world. When God, the divine organist, had pulled out all the stops, the world was then constituted. Each of these stops, of course, involved numerical proportions – harmonies – which blended together to produce the harmonies of the world as a whole. Within each rank, Kircher presented a vision of the harmonies at work. So, for example, at the level of cosmology, he argued for a conception of a harmony manifested in the relations each planet held with respect to the others, the whole relationship being governed by the sun.

---

[57] For accounts of Fludd's cosmology, see, for example, Robert Westman, "Nature, Art, and Psyche: Jung, Pauli, and the Kepler-Fludd polemic," in *Occult and Scientific Mentalities in the Renaissance*, ed. Brian Vickers (Cambridge: Cambridge University Press, 1984), pp. 177–229; and Eberhard Knobloch, "Harmony and Cosmos: Mathematics Serving a Teleological Understanding of the World," *Physis*, 32 (1995), 55–89. For an account of Fludd's chymical work, see Debus, *Chemical Philosophy*, chap. 4.
[58] Oppenheim and Frankfurt. "Technical" doesn't quite capture what Fludd has in mind here, which is the history with respect to its creation and construction.
[59] See Knobloch, "Harmony and Cosmos," p. 73.
[60] For an account of Kircher's views, see Knobloch, "Harmony and Cosmos," pp. 76–82. For a brief overview of Kircher's connection to chymistry, see Claus Priesner and Karin Figala, eds., *Alchemie: Lexikon einer hermetischen Wissenschaft* (Munich: C. H. Beck, 1998), pp. 196–8.

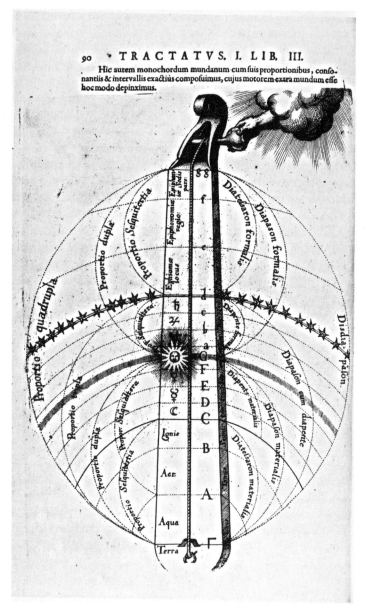

Figure 2.1. Representation of the cosmos in terms of a monochord. In Robert Fludd, *Utriusque cosmi maioris scilicet minoris metaphysica, physica atque technica historia*, 2 vols. (Oppenheim: Aere Johan-Theodori de Bry, typis Hieronymi Galleri, 1617–21), 1: 90. Reproduced by permission of the Rare Book Division, Department of Rare Books and Special Collections, Princeton University Library.

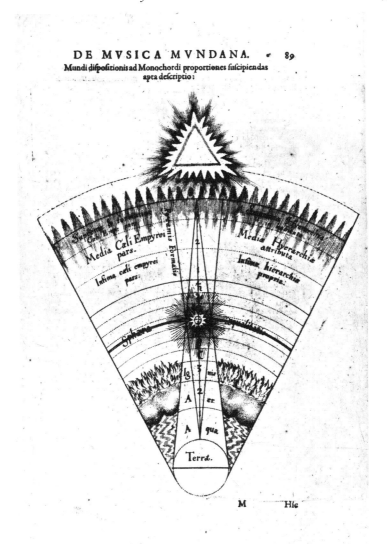

Figure 2.2. Alternative representation of the cosmos in terms of interpenetrating pyramids. In Robert Fludd, *Utriusque cosmi maioris scilicet minoris metaphysica, physica atque technica historia*, 2 vols. (Oppenheim: Aere Johan-Theodori de Bry, typis Hieronymi Galleri, 1617–21), 1: 90. Reproduced by permission of the Rare Book Division, Department of Rare Books and Special Collections, Princeton University Library.

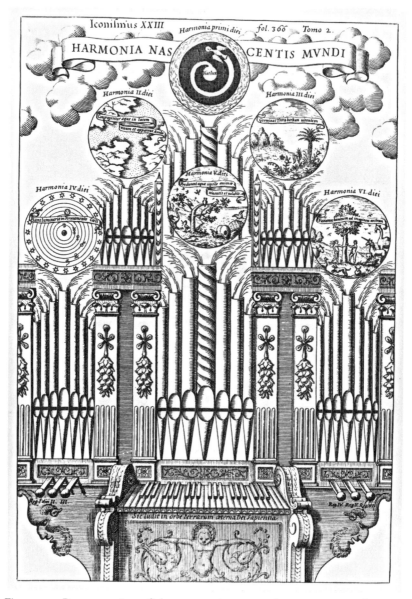

Figure 2.3. Representation of the cosmos in terms of an organ. In Athanasius Kircher, *Musurgia universalis, sive, Ars magna consoni et dissoni in X. libros digesta...*, 2 vols. (Rome: Haeredes Francisci Corbelletti, 1650), 2: 366. Reproduced by permission of the Rare Book Division, Department of Rare Books and Special Collections, Princeton University Library.

But the most interesting person in this group of Pythagoreans was the German astronomer and astrologer Johannes Kepler (1571–1630). Kepler was a technical astronomer well versed in the mathematical arcana of the subject, who knew how to construct an astronomical argument on the basis of observations. But just as interesting as the mathematical astronomy was a certain style of argument Kepler used that reveals an underlying view of the world that was in some ways similar to that of Fludd and Kircher.[61]

One of Kepler's best-known arguments was the explanation of why there are exactly six planets, including earth, and why they have the distances from one another that they do. In the *Mysterium cosmographicum* (The Mystery of the Universe, 1596; 2nd ed., with extensive notes, 1621), Kepler first argued that the distances among the planets, including earth, correspond to the distances one gets by nesting the five Platonic regular solids within one another: the tetrahedron (pyramid), cube, octahedron (formed by eight equilateral triangles), dodecahedron (12 pentagons), and icosahedron (20 equilateral triangles). Unfortunately, the world was not quite as simple as this model would suggest. Because the orbits of the planets turned out to be elliptical, as Kepler himself discovered, they did not fit this simple model, which implied circular orbits. However, Kepler was able to accommodate this within his model by regarding the elliptical orbit as a deviation from the circular orbit due to a magnetic attraction to or repulsion from the sun. For Kepler, this only showed an even greater rationality in the universe insofar as the deviations from the circular orbit give rise to pleasing celestial harmonies, literally a music of the spheres.[62]

Kepler also recognized harmonies in a broader sense – as correspondences among the different parts of the universe. For example, in arguing for Copernican cosmology in the *Epitome astronomiae copernicanae* (Epitome of Copernican Astronomy, 1618–21), Book IV, he compared the three regions of the Copernican cosmology – the central sun, the outer sphere of the fixed stars, and the intermediate region of the planets – with the Trinity. Kepler went on to compare the sun with the common sense in animals, located in the head, the globes that surround the sun with the sense organs, and the fixed stars with the sensible objects. He also compared the sun with the central fireplace and with the heart of the world, the seat of reason and life.[63] This is strongly reminiscent of the analogies drawn by Paracelsus and the chymical

---

[61] For a detailed discussion of this aspect of Kepler's thought, see Bruce Stephenson, *The Music of the Heavens: Kepler's Harmonic Astronomy* (Princeton, N.J.: Princeton University Press, 1994). I am deeply indebted to Rhonda Martens for her help in understanding Kepler's views.

[62] See Johannes Kepler, *Epitome astronomiae copernicanae*, in Johannes Kepler, *Gesammelte Werke*, ed. W. von Dyck and M. Caspar, 20 vols. to date (Munich: C. H. Beck, 1937–), 7: 275, translated in *Epitome of Copernican Astronomy IV*, in *Ptolemy, Copernicus, Kepler* (Great Books of the Western World), ed. Robert Maynard Hutchins, 54 vols. (Chicago: Encyclopaedia Britannica, 1952), 16: 845–960, esp. p. 871.

[63] See Kepler, *Gesammelte Werke*, 7: 258–60, translated in Hutchins, ed., *Ptolemy, Copernicus, Kepler*, pp. 853–6.

philosophers between the macrocosm and the microcosm, whereby the cosmos in its structure reflects the human being and the human being reflects the larger world.

Kepler was, first and foremost, an astronomer who based his astronomical models on observation; indeed, the best observations obtainable. Kepler, of course, famously struggled to use the unprecedentedly accurate data of Tycho Brahe (1546–1601) in formulating his theory of the orbit of Mars. We must appeal to observation in order to determine the real motions of planets. In response to Fludd's fanciful symbolic representations of the cosmos, Kepler replied: "I have demonstrated that the whole corpus of tempered Harmonics is to be found completely in the extreme, proper motions of the planets according to measurements which are certain and demonstrated in Astronomy. To [Fludd], the subject of World Harmony is his picture of the world; to me it is the universe itself or the real planetary movements."[64]

But, for Kepler, observation alone was not enough to fix the real structure of the world: For that, we need to know that the structures discovered by observation correspond to a geometrical archetype. The discovery that the resulting model derived from observation satisfies an elegant geometrical schema permits assertions about the way the world really is. Kepler wrote in Book I of the *Epitome*: "Astronomers should not be granted excessive licence to conceive anything they please without reason: on the contrary, it is also necessary for you to establish the probable causes of your Hypotheses which you recommend as the true causes of Appearances. Hence, you must first establish the principles of your Astronomy in a higher science, namely Physics or Metaphysics."[65]

Mathematical harmonies had their role to play for Kepler, but only in tandem with observation. In this emphasis on observation as grounds for the claims about harmony, Kepler separated himself both from what Fludd had done and from what Kircher was yet to do.[66]

In many ways, Kepler's view of the basic nature of the cosmos agreed with elements of the worldviews of his contemporaries. Like that of many of his contemporaries, his universe was, in a sense, animistic. Kepler freely compared the sun with the intelligence of the world and with the heart of the world, and he compared the world with an animal and argued that the sun has a soul and is, in a sense, a living being.[67] However, from time to time he also used another, very different analogy. In a letter to Herwart von Hohenberg dated 10 February 1605, Kepler wrote:

[64] Johannes Kepler, *Harmonices mundi libri V*, in *Gesammelte Werke*, 6: 376–7, quoted in Westman, "Nature, Art, and Psyche," p. 206.
[65] Kepler, *Gesammelte Werke*, 7: 25, quoted in Robert Westman, "Kepler's Theory of Hypotheses and the 'Realist Dilemma'," *Studies in History and Philosophy of Science*, 3 (1972), 233–64, esp. p. 261.
[66] On the controversy between Fludd and Kepler, see Westman, "Nature, Art, and Psyche"; Knobloch, "Harmony and Cosmos"; and Judith V. Field, "Kepler's Rejection of Numerology," in Vickers, ed., *Occult and Scientific Mentalities*, pp. 273–96.
[67] Kepler, *Gesammelte Werke*, 7: 259–60, 298 ff., translated in Hutchins, ed., *Ptolemy, Copernicus, Kepler*, pp. 855–6, 896 ff.

My goal is to show that the heavenly machine is not a kind of divine living being but similar to a clockwork insofar as almost all the manifold motions are taken care of by one single absolutely simple magnetic bodily force, as in a clockwork all motion is taken care of by a simple weight. And indeed I also show how this physical representation can be presented by calculation and geometrically.[68]

This analogy leads us in the direction of a conception of the foundations of the physical world that is very different from the one that we have been considering so far, which came to be called the mechanical philosophy.[69] In radical contrast with the Renaissance world, infused with soul, sentience, intelligence, and harmony, the mechanical philosophy took as central the image of the machine.

THE RISE OF THE MECHANICAL AND CORPUSCULAR PHILOSOPHY

Many of the trends discussed in the previous sections persisted well into the seventeenth century and beyond, though sometimes in rather altered versions. However, there is another extremely important trend that emerged sometime in the sixteenth century and came to flourish in the seventeenth century: the mechanical (or corpuscular) philosophy.[70] The English natural philosopher Robert Boyle gave a particularly concise and cogent account of this position in his important essay *The Origin of Forms and Qualities according to the Corpuscular Philosophy* (1666).

The mechanical philosophy, as Boyle presented it, replaced the explanation of the manifest properties of bodies in terms of the Aristotelian notions of form, matter, and privation, with a view in accordance with which those properties are "produced Mechanically, I mean by such Corporeall Agents, as do not appear, either to Work otherwise, then by vertue of the Motion, Size,

---

[68] Kepler, *Gesammelte Werke*, 15: 146, quoted in Max Caspar, *Kepler* (London: Abelard-Schuman, 1959), p. 136.
[69] Insofar as it involves the magnet, arguably it does not get us all the way to a genuine mechanical conception of the world, where everything happens through size, shape, motion, and the impact of bodies on one another.
[70] Among contemporaries, the two names are virtually synonymous. The *Oxford English Dictionary* (q.v. mechanical) cites John Harris's *Lexicon Technicum* (1704) on this question: "Mechanical Philosophy, is the same with the Corpuscular, which endeavours to explicate the Phænomena of Nature from Mechanical Principles." Robert Boyle seems to identify the two in his *Of the Excellency and Grounds of the Corpuscular or Mechanical Philosophy* (1674). Calling it "corpuscular" emphasizes that the manifest properties of bodies are to be explained in terms of their smaller parts, and calling it "mechanical" emphasizes that the principles used in explanation are broadly mechanical. For histories of seventeenth-century science that emphasize the mechanical philosophy, see E. J. Dijksterhuis, *The Mechanization of the World Picture*, trans. C. Dikshoorn (Oxford: Oxford University Press, 1961); Richard S. Westfall, *The Construction of Modern Science: Mechanisms and Mechanics* (New York: John Wiley, 1971); and Marie Boas Hall, "The Establishment of the Mechanical Philosophy," *Osiris*, 10 (1952), 412–541.

Figure and Contrivance of their own Parts."[71] Boyle explicated this view in a number of basic theses: (1) "there is one Catholick or Universal Matter common to all Bodies, by which I mean a Substance Extended, divisible and impenetrable"; (2) "to discriminate the Catholick Matter into variety of Natural Bodies, it must have Motion in some or all its designable Parts"; (3) "Matter must be actually divided into Parts, ... and each of the primitive Fragments ... must have two Attributes, its own Magnitude ... and its own *Figure* or *Shape*."[72] In this way, the mechanical or corpuscular philosophy rejected the explanation of physical phenomena in terms of Aristotelian forms and qualities, the innate tendencies of substances to behave in particular ways. It also sought to eliminate all sensible qualities from objects themselves; the Aristotelian's hot and cold, wet and dry, are eliminated as real qualities of things, as are sensible qualities such as color and taste. For the mechanical philosopher, everything, be it terrestrial or celestial, natural motion or constrained, must be explained in terms of the size, shape, and motion of the parts that make it up, just as the behavior of a machine is explained. As Descartes summarized the program:

> Men who are experienced in dealing with machinery can take a particular machine whose function they know and, but looking at some of its parts, easily form a conjecture about the design of the other parts, which they cannot see. In the same way I have attempted to consider the observable effects and parts of natural bodies and track down the imperceptible causes and particles which produce them.[73]

In this way, the image of the macrocosm and the microcosm, central to chymical philosophies and Renaissance naturalism, found its way into mechanism after a fashion. For the mechanical philosopher, as for the chymist and the Renaissance naturalist, what happens at one level reflects and is reflected by what happens at every other level.

Another important feature of the mechanist foundations of nature was laws of nature. The idea of natural law in the sense of moral laws governing human behavior decreed by God was founded long before the early modern period; it seems to be a direct extension of the notion of a law in the ordinary political sense.[74] But the idea that there are general laws that govern insentient and inanimate nature, mathematically formulable regularities that govern

---

[71] Robert Boyle, *The Works of Robert Boyle*, ed. Michael Hunter and Edward B. Davis, 14 vols. (London: Pickering and Chatto, 1999–2000), 5: 302.
[72] Boyle, *Works*, 5: 305–307.
[73] René Descartes, *Principia philosophiae* (Amsterdam: Ludovicus Elzevirius, 1644), 4.203. For a discussion of some of the epistemological implications of this view, see Larry Laudan, "The Clock Metaphor and Hypotheses: The Impact of Descartes on English Methodological Thought, 1650–1670," in his *Science and Hypothesis* (Dordrecht: Reidel, 1981), pp. 27–58.
[74] For an account of natural law theories in the seventeenth century, see Knud Haakonssen, "Divine/Natural Law Theories in Ethics," in Garber and Ayers, eds., *The Cambridge History of Seventeenth-Century Philosophy*, 2: 1317–57.

all bodies, was an apparently new feature of the mechanical philosophy of the seventeenth century; with the idea that there is one kind of matter in the whole of the universe came the idea that there is one set of laws that governs that matter. Although perhaps not the first to have such an idea, Descartes was responsible for its first appearance in print in a self-conscious and foundational context. In his *Principia philosophiae*, Descartes announced "certain rules or laws of nature, which are the secondary and particular causes of the various motions we see in particular bodies."[75] The laws of nature in question are three laws governing the motion of bodies, including two laws governing the persistence of motion and a law governing collision. Although his laws were considerably debated, and alternatives were proposed by Huygens, Leibniz, Newton, and others, after Descartes, the idea that the world is governed by precise mathematical laws seemed to become a central part of the mechanist foundations of the physical sciences.[76]

Galileo Galilei (1564–1642) (along with his Italian followers) is generally credited with being one of the founders of the mechanist program in the early part of the century.[77] In Northern Europe, an atomist mechanist program was initiated in the 1610s by Isaac Beeckman (1588–1637), a somewhat itinerant schoolmaster in the Netherlands who was known to Descartes, Mersenne, Gassendi, and many other thinkers of the period.[78] By the late 1620s, this program had made its way to France and was being pursued by Mersenne,

---

[75] Descartes, *Principia philosophiae*, 2.37. Descartes' laws were first announced as such in Chapter 7 of his *Traité de la lumière* (Treatise on Light, 1633), which remained unpublished until 1664, by which time the idea of laws of nature was firmly established. Galileo had presented what we would today call laws of motion, a version of the so-called law of inertia and the law of free fall, in his *Dialogo sopra i due massimi sistemi del mondo* (1632), in *Opere di Galileo Galilei*, ed. A. Favaro (Florence: Barbèra, 1890–1910), 7: 44–53, 173–5, translated in *Dialogue Concerning the Two Chief World Systems – Ptolemaic and Copernican*, trans. Stillman Drake (Berkeley: University of California Press, 1967), pp. 20–8, 147–9; and *Discorsi e dimostrazioni matematiche intorno a due nuove scienze* (Leiden, 1638), in *Opere di Galileo Galilei*, 8: 209–10, 243, translated with introduction and notes by Stillman Drake in *Two New Sciences: Including Centers of Gravity & Force of Percussion* (Madison: University of Wisconsin Press, 1974), pp. 166–7, 196–7. But aside from the problems of interpretation, particularly with respect to the so-called law of inertia, Galileo himself never characterizes these as "laws"; in his thought they have the character of regularities that govern heavy bodies in the vicinity of the centers toward which they are attracted. Francis Bacon talked about the forms that constitute particular qualities (heat, light, and weight, for example) as constituting laws in the sense that whenever the form or nature was present, the quality would be as well. See Bacon, *Novum Organum*, 1.17. But this seems to be a very different sense of law.

[76] For a general discussion of the idea of laws of nature in the seventeenth century, see J. R. Milton, "Laws of Nature," in Garber and Ayers, eds., *The Cambridge History of Seventeenth-Century Philosophy*, 1: 680–701.

[77] The literature on Galileo is enormous, and the main aspects of his career are well known. For a survey of some aspects of this question with respect to Galileo, see Peter Machamer, "Galileo's Machines, His Mathematics, and His Experiments," in *The Cambridge Companion to Galileo*, ed. Peter Machamer (Cambridge: Cambridge University Press, 1998), pp. 53–79.

[78] Beeckman's notebooks, which include records of his conversations with Descartes, for example, are published as *Journal tenu par Isaac Beeckman de 1604 à 1634*, ed. Cornelis de Waard, 4 vols. (The Hague: Martinus Nijhoff, 1939–53). For an account of his life and thought, see Klaas van Berkel, *Isaac Beeckman (1588–1637) en de Mechanisering van het Wereldbeeld* (with a summary in English) (Amsterdam: Rodopi, 1983).

Gassendi, Gilles Personne de Roberval (1602–1675), Thomas Hobbes (1588–1679), and Kenelm Digby (1603–1665), the last two visiting from England.[79] Descartes took his version of it to the Netherlands starting in the late 1620s.[80] Although he was not uncontroversial there, Descartes had many Dutch followers, including a number in the universities.[81] The program even had some success in Germany, though Germany was intellectually more conservative than Western Europe.[82] There was a tradition of atomism in England that went back to the early part of the century, but it was given new life with the introduction of Cartesian and Gassendist ideas at mid-century.[83] By the 1660s or 1670s, mechanist approaches to nature were found virtually throughout Europe and seem to have dominated intellectual discourse. By and large, the mechanical philosophy flourished outside the universities, first in salons and private academies, such as Mersenne's academy in Paris and the Montmort academy that followed it, and then in institutions such as the Royal Society of London and the Académie Royale des Sciences in Paris.[84] But the philosophy also found some success in the educational institutions in the Netherlands, France, and even Germany.[85]

---

[79] On Mersenne, see Robert Lenoble, *Mersenne ou La naissance du mecanisme* (Paris: J. Vrin, 1971). For the diffusion of Gassendi's thought in Europe, see *Gassendi et l'Europe*, ed. Sylvia Murr (Paris: J. Vrin, 1997), pt. II. On Hobbes, see F. Brandt, *Hobbes's Mechanical Conception of Nature* (Copenhagen: Levin and Munksgaard, 1928).

[80] For the diffusion of Cartesian thought, the best general reference is still Francisque Bouillier, *Histoire de la philosophie cartésienne*, 3rd ed., 2 vols. (Paris: Delagrave, 1868). On the reception of Cartesian ideas in Italy, see Giulia Belgioioso, *Cultura a Napoli e cartesianesimo* (Galatina: Congedo editore, 1992).

[81] See Theo Verbeek, *Descartes and the Dutch: Early Reactions to Cartesianism (1637–1650)* (Journal of the History of Philosophy Monograph Series) (Carbondale: Southern Illinois University Press, 1992).

[82] See Francesco Trevisani, *Descartes in Germania: La ricezione del cartesianesimo nella facoltà filosofica e medica di Duisberg (1652–1703)*, (Milan: Franco Angeli, 1992); and Christia Mercer, *Leibniz's Metaphysics: Its Origins and Development* (Cambridge: Cambridge University Press, 2001).

[83] On atomism in England, see Robert H. Kargon, *Atomism in England from Hariot to Newton* (Oxford: Oxford University Press, 1966). On Cartesianism in England, see Alan Gabbey, "Philosophia Cartesiana Triumphata: Henry More (1646–1671)," in *Problems of Cartesianism*, ed. T. M. Lennon, J. M. Nicholas, and J. W. Davis (Kingston and Montreal: McGill-Queens University Press, 1982), pp. 171–249.

[84] On the Royal Society of London, see, for example, Michael Hunter, *Establishing the New Science: The Experience of the Early Royal Society* (Woodbridge: Boydell Press, 1989). On the Mersenne circle, the Montmort academy, and the Académie Royale des Sciences, see Harcourt Brown, *Scientific Organizations in Seventeenth-Century France (1620–1680)* (Baltimore: Williams and Wilkins, 1934); Frances A. Yates, *The French Academies of the Sixteenth Century* (London: Routledge, 1988; orig. publ. 1947), chap. 12; Roger Hahn, *The Anatomy of a Scientific Institution: The Paris Academy of Sciences, 1666–1803* (Berkeley: University of California Press, 1971); Alice Stroup, *A Company of Scientists: Botany, Patronage, and Community at the Seventeenth-Century Parisian Royal Academy of Sciences* (Berkeley: University of California Press, 1990). On the Cartesian salons in Paris, see Erica Harth, *Cartesian Women* (Ithaca, N.Y.: Cornell University Press, 1992).

[85] See Verbeek, *Descartes and the Dutch*; Trevisani, *Descartes in Germania*; Mercer, *Leibniz's Metaphysics*; and Laurence Brockliss, "Les atomes et le vide dans les collèges de plein-exercice en France de 1640–1730," in *Gassendi et l'Europe*, ed. Sylvia Murr (Paris: J. Vrin, 1997), pp. 175–87. Interesting in this connection is a battle between the older Aristotelians and the younger Cartesians on the faculty of the Université d'Angers in the early 1670s. On this, see Roger Ariew, "Cartesians, Gassendists, and Censorship," chap. 9 of his *Descartes and the Last Scholastics*. Cartesianism seems to come somewhat

When Boyle introduced the general principles of the mechanical philosophy, he quite explicitly put aside differences among different sects, claiming to write "rather for the Corpuscularians in general, than any party of them."[86] But one can find among practitioners who identified themselves as mechanical philosophers or were identified by their contemporaries as mechanical philosophers a variety of different conceptions of the worldview that underlies the world of corpuscles in collision. In the sections that follow, I discuss some of the important variants of the mechanical philosophy.

THE MECHANICAL PHILOSOPHY: THEORIES OF MATTER

An important aspect of the foundations of physics was the conception of the nature of matter, the stuff of which the physical world is ultimately made. In the mechanical philosophy, one important strand of thinking about the nature of matter was the revival of ancient atomism.[87] When looking at atomism in the early seventeenth century, it is important to remember that there were a variety of atomisms in play, not all of which fit in with a mechanist or corpuscular philosophy. For example, among a number of chymists and Aristotelian natural philosophers there was the view that the elements can be divided into minimal parts that would lose their status as elements if divided further. Because these smallest parts are distinguished from one another by having different essences, this *minima naturalia* view fails to satisfy Boyle's definition of the mechanical philosophy.[88] But more influential was the revival of the atomism of Epicurus and Lucretius. There were a number of people involved in this revival, including Sebastian Basso (ca. 1560–ca. 1621), Nicholas Hill (ca. 1570–ca. 1610), David van Goorle (1591–1612), among others. But the key figure was Pierre Gassendi. Gassendi's project was more than just natural philosophy; his aim was to rehabilitate

---

later into Italy. On this, see Belgioioso, *Cultura a Napoli e cartesianesimo*; and Claudio Manzoni, *I cartesiani italiani (1660–1760)* (Udina: La Nuova Base, 1984).

[86] Boyle, *Works*, 3: 7.

[87] For general histories of atomism, see the still classic Kurd Lasswitz, *Geschichte der Atomistik vom Mittelalter bis Newton*, 2 vols. (Hamburg: L. Voss, 1890); Andrew Pyle, *Atomism and Its Critics from Democritus to Newton* (Bristol: Thoemmes Press, 1997); and Antonio Clericuzio, *Elements, Principles, and Corpuscles: A Study of Atomism and Chemistry in the Seventeenth Century* (Dordrecht: Kluwer, 2000). Kargon's *Atomism in England*, gives a good history of atomism in seventeenth-century England. For an account of the variety of atomisms available in the early seventeenth century, see Lynn Sumida Joy, *Gassendi the Atomist* (Cambridge: Cambridge University Press, 1987), chap. 5. For an account of the revival of Epicureanism, see Howard Jones, *The Epicurean Tradition* (London: Routledge, 1989). For a more general account of corpuscularianism, see Norma Emerton, *The Scientific Reinterpretation of Form* (Ithaca, N.Y.: Cornell University Press, 1984), chaps. 3–4.

[88] On this doctrine, see Pierre Duhem, *Système du monde*, 10 vols. (Paris: Hermann, 1958), 7: 42–54; Emerton, *The Scientific Reinterpretation of Form*, chaps. 3–4; Newman, *Gehennical Fire*, pp. 24 ff.; Roger Ariew, "Descartes, Basso, and Toletus: Three Kinds of Corpuscularians," chap. 6 of his *Descartes and the Last Scholastics*. The position can be found in the writings of pseudo-Geber (on which see Newman, *Gehennical Fire*, pp. 94 ff.), Julius Caesar Scaliger, and Johannes Baptista Van Helmont, among many others.

Epicurean philosophy as a whole and present a cleansed version acceptable to a Christian audience.[89] For Gassendi, as for Epicurus, the world was made up of two principles: atoms and the void. Atoms were taken to be the smallest parts of matter, possessed of size, shape, weight, and nothing else. Although finite in size, and thus having physical parts, atoms were taken to be indivisible. In this way, they constituted the smallest level of analysis for any body. Furthermore, all the manifest properties of bodies were to be explained in terms of the size, shape, and motion of these atoms.[90]

Descartes presented an alternative mechanist foundation for the physical world. The commitment to a metaphysical grounding for physics was basic to Descartes' thought. One of the central elements of his metaphysics was his doctrine of the essence of body and its distinction from mind. Body, for Descartes, was a substance whose essence is extension and extension alone. By that, Descartes meant to exclude all properties in bodies except for size, shape, and motion; in this sense, one can say that bodies, or material substances, are, for Descartes, the objects of geometry made concrete.

Because bodies are the objects of geometry made real, they are infinitely divisible, and there is no smallest part of matter. Just as any finite line can be divided into smaller parts, so can any finite body be divided into smaller parts. (Although he differed from Descartes in many respects, Hobbes agreed with him in holding that matter is infinitely divisible and that there are no smallest particles.) Furthermore, insofar as they are extended and extended alone, Cartesian bodies have no innate tendency to descend or to do anything else. Gravity, for Descartes, was something that had to be explained in terms of the interaction between the heavy body and the particles in the ether that surround it; it could not be a basic, inherent property of body as it was for the Aristotelians and would become for the Newtonians.[91]

---

[89] Epicurus faced the normal obstacles encountered by any pagan author attempting to enter the Christian intellectual world, and then some. In addition to the stigma of an ethics based on pleasure, Epicurus did his best to demystify the physical world by offering systematic naturalistic explanations of everything his contemporaries attributed to the gods. Epicurus furthermore argued that the gods themselves were made up of atoms and that they lived in places distant from the human realms and were uninterested in human affairs. On the Christianization of Epicurus's thought, see Margaret J. Osler, "Baptizing Epicurean Atomism: Pierre Gassendi on the Immortality of the Soul," in *Religion, Science, and Worldview*, ed. M. J. Osler and P. L. Farber (Cambridge: Cambridge University Press, 1985), pp. 163–83. It should be noted here that there are disagreements about whether Gassendi was a genuine believer or whether, in the end, he was a freethinker or even an atheist. The classic development of the view of Gassendi as a libertine is found in René Pintard, *Le libertinage érudit dans la première moitié du XVIIe siècle* (Paris: Boivin, 1943; Geneva: Slatkine Reprints, 1983). It is answered in Paul O. Kristeller, "The Myth of Renaissance Atheism and the French Tradition of Free Thought," *Journal of the History of Philosophy*, 6 (1968), 233–44.

[90] Gassendi's atomism is developed at some length in his posthumous *Syntagma philosophicum* (1658), in Gassendi, *Opera omnia*, 6 vols. (Lyon: Laurentius Anisson and Ioan. Baptista Devenet, 1658), 1: 256A ff. See also Bernard Rochot, *Les travaux de Gassendi sur Epicure et sur l'atomisme, 1619–1658* (Paris: J. Vrin, 1944).

[91] Descartes' physics is developed in the early *Le monde*, written in 1630–3 but first published in 1664 (Paris: Theodore Giraud, 1664), and in the *Principia philosophiae*, pt. 2. For discussion of Descartes' physics and its metaphysical foundations, see Daniel Garber, *Descartes' Metaphysical Physics* (Chicago: University of Chicago Press, 1992). The relation between these issues in Descartes and in the schoolmen is discussed in Dennis Des Chene, *Physiologia: Natural Philosophy in Late*

Descartes and Gassendi represented the two main poles in seventeenth-century theories of matter.[92] There is every reason to believe that it was these two positions that Boyle had in mind when he chose to put aside the differences among different groups of corpuscularians. Although they may have differed on the question of whether there is an ultimate level of analysis of body, or whether every body, no matter how small, is divisible into smaller parts, they agreed in rejecting Aristotelian form and matter and in holding that the manifest properties of bodies are to be explained in terms of their size, shape, and motion. But, in addition to these positions, other alternatives were available.

Although the theory of matter was not central in the thought of Galileo, he did seem to subscribe to a kind of corpuscularianism. In a celebrated passage from the *Il Saggiatore* (The Assayer, 1623), he asserted: "To excite in us tastes, odors, and sounds I believe that nothing is required in external bodies except shapes, numbers, and slow or rapid movements. I think that if ears, tongues, and noses were removed, shapes and numbers and motions would remain, but not odors or tastes or sounds."[93]

However, it is important to note that Galileo's ultimate particles seem not to have been the small but finite corpuscles Boyle had in mind, but "infinitely many unquantifiable atoms," suggesting an infinitesimal conception, though this idea was not worked out in great detail.[94] Coordinate with the infinitesimal particles were infinitesimal voids. The consistency of bodies, Galileo argued, is caused by these tiny voids, interspersed in bodies, together with "the repugnance nature has against allowing a void to exist."[95] Galileo was, of course, aware of the Aristotelian arguments against the void from the infinite speed that a body in motion would seem to have when moved in a vacuum, but he thought that these arguments could be answered.[96]

One of the most interesting attempts to ground the conception of body and matter in connection with the mechanical philosophy is found in the work of Leibniz. From his earliest youth, Leibniz was captivated by the

---

Aristotelian and Cartesian Thought (Ithaca, N.Y.: Cornell University Press, 1996). For an account of Cartesian physics in late seventeenth-century figures, see Paul Mouy, *Le développement de la physique cartésienne, 1646–1712* (Paris: J. Vrin, 1934). For Descartes' relation to atomism, see Sophie Roux, "Descartes Atomiste?" in *Atomismo e continuo nel XVII secolo*, ed. Egidio Festa and Romano Gatto (Naples: Vivarium, 2000), pp. 211–73.

92 On the relations between Cartesianism and Gassendism later in the century, see Thomas M. Lennon, *The Battle of the Gods and Giants: The Legacies of Descartes and Gassendi, 1655–1715* (Princeton, N.J.: Princeton University Press, 1993).

93 Galileo Galilei, *Il Saggiatore* (Rome: Giacomo Mascardi, 1623), in *Opere di Galileo Galilei*, 6: 350, translated in Drake, *Discoveries and Opinions of Galileo*, pp. 276–7. On Galileo's atomism, see William R. Shea, "Galileo's Atomic Hypothesis," *Ambix*, 17 (1970), 13–27; A. Mark Smith, "Galileo's Theory of Indivisibles: Revolution or Compromise," *Journal of the History of Ideas*, 27 (1976), 571–88; and Giancarlo Nonnoi, "Galileo Galilei: quale atomismo?" in *Atomismo e continuo nel XVII secolo*, ed. Egidio Festa and Romano Gatto, pp. 109–49.

94 Galileo Galilei, *Discorsi e dimostrazioni*, in *Opere di Galileo Galilei*, 8: 71–2, translated in Drake, *Two New Sciences*, p. 33.

95 Galileo, *Discorsi*, in *Opere di Galileo Galilei*, 8: 59, translated in Drake, *Two New Sciences*, p. 19.

96 Galileo, *Discorsi*, in *Opere di Galileo Galilei*, 8: 105–6, translated in Drake, *Two New Sciences*, p. 65.

mechanical philosophy. But Leibniz's mechanism was not uncritical.[97] He came to see a number of problems with the mechanist conception of body in both the Cartesian and the atomist versions. Against the Cartesian conception of body, a substance whose essence is extension, he argued that extension is not itself the kind of thing that can exist alone. Rather, he argued, it is a relative notion that presupposes some quality that is extended. Just as one cannot have a father without a child, one cannot have mere extension without there being some quality that is extended.[98] Elsewhere, Leibniz argued that because Cartesian bodies are divisible, indeed infinitely divisible, they lack the kind of genuine unity required for something to be a substance.[99] Leibniz had a number of arguments against the atomists as well. If there are parts of matter that are indivisible, then they must be infinitely hard because all elasticity comes from smaller parts that can move with respect to one another. But if atoms were infinitely hard, then in collision, their speeds would change instantaneously, which violates Leibniz's principle that nature makes no leaps (the Principle of Continuity). He also argued that atoms are impossible because there is no reason why God should stop the divisibility of a piece of matter in one place rather than another, in violation of his celebrated Principle of Sufficient Reason.[100]

Despite his criticism of the prevailing mechanist accounts of body, Leibniz continued throughout his life to hold that there is a sense in which everything can be explained in terms of size, shape, and motion. But behind the extended bodies of the mechanical philosophy, he argued, there must be something more real, which he called individual substances; in that sense, his position constitutes a kind of substantial atomism. Sometimes these individuals were conceived of based on the model of Cartesian living things – corporeal substances with souls attached to bodies, making those bodies both active and genuinely unified. But more often, particularly in his later writings, Leibniz appealed to his monads. Modeled on Cartesian souls (that is, incorporeal substances), monads were genuinely active and genuine individuals. The bodies of everyday experience were just the confused appearance presented

---

[97] See, for example, the intellectual biography Leibniz gives for his dealings with mechanism in his letter to Nicholas Remond, 10 January 1714, in Gottfried Wilhelm Leibniz, *Die philosophischen Schriften*, ed. C. I. Gerhardt, 7 vols. (Berlin: Weidmannsche Buchhandlung, 1875–90), 3: 606–7, translated in Leibniz, *Philosophical Papers and Letters*, ed. and trans. L. E. Loemker (Dordrecht: Reidel, 1969), pp. 654–5.

[98] This argument is found in an essay dated 1702, in Gottfried Wilhelm Leibniz, *Mathematische Schriften*, ed. C. I. Gerhardt, 7 vols. (Berlin and Halle: A. Asher et comp. and H. W. Schmidt, 1849–63), 6: 99–100, translated in Gottfried Wilhelm Leibniz, *Leibniz: Philosophical Essays*, ed. and trans. Roger Ariew and Daniel Garber (Indianapolis: Hackett, 1989), p. 251.

[99] See, for example, Leibniz's letter to Arnauld, 30 April 1686, in Leibniz, *Die philosophischen Schriften*, 2: 96, translated in Ariew and Garber, eds. and trans., *Leibniz: Philosophical Essays*, p. 85. For an account of this and other arguments against the Cartesian conception of body, see Daniel Garber, "Leibniz and the Foundations of Physics: The Middle Years," in *The Natural Philosophy of Leibniz*, ed. K. Okruhlik and J. R. Brown (Dordrecht: Reidel, 1985), pp. 27–130.

[100] For an exposition of Leibniz's arguments against atomism, see Garber, "Leibniz: Physics and Philosophy," pp. 321–5.

by these substances; both the bodies and the laws that they obey are ultimately grounded in the world of genuine substances. What was truly real, for Leibniz, were these substances. Mechanism for Leibniz was grounded in something not purely material, either corporeal substances, which involve an immaterial soul, or monads, which are themselves immaterial substances.

Mechanist corpuscularianism often presented itself as a replacement for an Aristotelian conception of body. But this was not always the case. As mentioned earlier, there was an atomistic and corpuscularian tradition separate from the Epicurean and mechanist tradition and quite consistent with an Aristotelian conception of body, the *minima naturalia* view on which elements that by their nature were distinct were divisible into smallest parts that are also by their nature distinct. There were, in addition, many who tried to render the full-blown mechanical philosophy consistent with the Aristotelian philosophy that many mechanists thought it was meant to replace. Digby's widely read *Two Treatises* (1644), one of the early works written from a mechanist point of view, evinced great respect for the Aristotelian point of view and tried to show its consistency with Digby's own system. In the second half of the seventeenth century, as the mechanist program was gaining serious momentum, there were numerous books with titles like Jean-Baptiste Du Hamel's *De consensu veteris et novae philosophiae* (On the Agreement of the Old and New Philosophy, Paris, 1663), Jacques Du Roure's *La physique expliquée suivant le sentiment des anciens et nouveaux philosophes; & principalement Descartes* (Physics Explained in accordance with the Opinions of the Old and the New Philosophers, and Especially that of Descartes, Paris, 1653), Johannes de Raey's *Clavis philosophiae naturalis sive Introductio ad contemplationem naturae aristotelico-cartesiana* (The Key to Natural Philosophy; or, Introduction to the Aristotelio-Cartesian Contemplation of Nature, Leiden, 1654), René Le Bossu's *Parallèle des principes de la physique d'Aristote & celle de René Des Cartes* (The Parallels between the Principles of the Physics of Aristotle and René Descartes, Paris, 1674). Some of these works were simply comparisons of the old and the new. But, in numerous cases, authors tried to render consistent the matter and form of the schools with the size, shape, and motion of the moderns.[101] One of the young Leibniz's earliest surviving writings is a letter he wrote to his teacher, Jakob Thomasius (1622–1684), on 20/30 April 1669 (published by him a year later, virtually unchanged), naming a number of the most prominent adherents of this position and outlining his own way of reconciling Aristotelianism and the mechanical philosophy.[102] The ideas there were rather naive; he argued that Aristotelian

---

[101] On this theme in seventeenth-century thought, see Christia Mercer, "The Vitality and Importance of Early Modern Aristotelianism," in *The Rise of Modern Philosophy*, ed. Tom Sorell (Oxford: Oxford University Press, 1993); and Mercer, *Leibniz's Metaphysics*.

[102] The letter can be found in Gottfried Wilhelm Leibniz, *Sämtliche Schriften und Briefe*, ed. Deutsche [before 1945, Preussische] Akademie der Wissenschaften (Berlin: Akademie Verlag, 1923–), 2.1: 15, translated in Loemker, ed. and trans., *Philosophical Papers and Letters*, pp. 93–103.

notions of matter, form, and change can be interpreted in mechanist terms, and that this is how Aristotle himself had understood them, a far cry from the much more sophisticated reconciliation one finds in Leibniz's mature writings. But, in a real sense, though the details change, the idea of grounding mechanistic physics on Aristotelian foundations remained with Leibniz for much of his life. Following Aristotelian practice, Leibniz often characterized his substances, both corporeal substances and monads or simple substances, in terms of matter and form, as I discuss in more detail. In this way, he could claim to have reconciled the new mechanical philosophy with the old scholastic Aristotelian philosophy. As Leibniz put it in the *Discours de métaphysique* (Discourse on Metaphysics, 1686), "the thoughts of the theologians and philosophers who are called scholastics are not entirely to be disdained."[103]

THE MECHANICAL PHILOSOPHY: SPACE, VOID, AND MOTION

Among the foundational issues, questions about space, place, and void were important to the Aristotelian philosophy of the schools and were widely discussed by some of the opponents to Aristotelianism discussed earlier. But the reintroduction of atomism by many mechanists brought with it a renewed interest in these questions and some new positions worth examining.

As discussed earlier, for Aristotle, empty space was impossible: All space was filled with body and could not be otherwise. Although he rejected Aristotle in many other respects, this was an issue on which Descartes agreed with him. For Descartes, as for Aristotle, space was not something over and above body. Because the nature of body is extension, and because every property (such as extension) requires something that instantiates that property, anything extended must be body. For Descartes, space was simply an abstract way of talking about extended bodies and their relations to one another, and the very idea of a vacuum was a conceptual impossibility. As a consequence, the world was full for Descartes, and there was no empty space, nor could there be. Because space was just a relation among bodies, place was defined in terms of the relations among bodies, as was motion for Descartes. Motion was a change of situation with respect to the bodies neighboring a given body. Although there was no fact of the matter whether a given body or

---

[103] Gottfried Wilhelm Leibniz, *Discours de métaphysique* (written in 1686, unpublished during Leibniz's lifetime), para. 11, in Leibniz, *Sämtliche Schriften und Briefe* 6.4: 1529–88. They are not entirely to be disdained, but not entirely to be followed either. For the schoolmen, form was to explain the details of the behavior of bodies: why some fall and some rise; why some are hot and others are cold. This was not so for Leibniz. For Leibniz, all explanation in physics was in terms of size, shape, and motion. Matter and form enter in only to ground the reality of body by providing unity, and the general laws of motion by providing force and activity. In this way, Leibniz argued "that the belief in substantial forms has some basis, but that these forms do not change anything in the phenomena and must not be used to explain particular effects," *Discours de métaphysique*, para. 10.

its neighborhood is really moving when the two are separating from one another, there was, for Descartes, a fact of the matter about whether they are separating. In this way, Descartes hoped to make a real distinction between motion and rest, and reject the evident relativism that his position would seem to entail.[104]

The plenist position characterized the later Cartesian school and quite naturally went with the view that body is divisible to infinity. If the world is filled with no empty spaces, then bodies must be divisible indefinitely in order to prevent empty spaces from being formed as larger bodies move. Indeed, there are some circumstances in which bodies must actually be divided to infinity in order to guarantee that there are no vacua.[105] However, Descartes' position on the nature of motion was not generally followed. Christiaan Huygens (1629–1695), in his youth a follower of Descartes, built a physics where motion is understood to be relative to an arbitrarily chosen resting point.[106]

Those who revived atomism in the seventeenth century tended to favor views of space that held it to be independent of body and capable of existing empty, without body. As already mentioned, Galileo had rejected Aristotle's ban on the vacuum. For Galileo, the consistency of bodies was explained at least in part by the interspersal of tiny vacua throughout matter.[107] Like Epicurus, Gassendi argued for the existence of void space from the fact that, without a void, motion would be impossible, either at the macroscopic or the microscopic level. Although others had opposed the Aristotelian ban on the vacuum, Gassendi took the argument one step further, arguing that space is something that must be conceived outside of the Aristotelian categories of substance and accident.[108] But it was probably Gassendi's espousal of this position that would influence later thinkers such as Locke. As Locke wrote in his *Essay Concerning Human Understanding* (1690): "If it be demanded (as it usually is) whether this *Space* void of *Body* be *Substance* or *Accident*, I shall

---

[104] This position is developed, for example, in Descartes' *Principia philosophiae*, 2.1–35. For a fuller discussion of the issues raised, see Garber, *Descartes' Metaphysical Physics*, chaps. 5–6.

[105] See Descartes, *Principia philosophiae*, 2.34–35. Descartes argues that in a specified region, for any body, however small, in that region, one can find a body smaller still. Because he wants to reserve the term "infinity" for God alone, Descartes calls this indefinite divisibility rather than infinite divisibility.

[106] The relativity of motion is central to Huygens's derivation of the laws of impact. By virtue of the doctrine of the relativity of motion, what appear as different physical situations in Descartes' derivation (*Principia philosophiae*, 2.40, 46–52) are identified with one another, allowing Huygens to present laws much more elegant than Descartes'. See Christiaan Huygens, *De motu corporum ex percussione* (1659), in Christiaan Huygens, *Oeuvres complètes*, ed. D. Bierans de Haan, J. Bosscha, D. J. Kortweg, and J. A. Vollgraff, 22 vols. (The Hague: Société Hollandaise des Sciences and Martinus Nijhoff, 1888–1950), 16: 30–168, trans. Richard J. Blackwell in "Christiaan Huygens's The Motion of Colliding Bodies," *Isis*, 68 (1977), 574–97. See also the discussion in Dijksterhuis, *The Mechanization of the World Picture*, pp. 373–80.

[107] See Galileo, *Discorsi e dimostrazioni*, in *Opere di Galileo Galilei*, 8: 71–2, translated in Drake, *Two New Sciences*, p. 33.

[108] Gassendi, *Opera*, 1: 182A. The position here is reminiscent of the one that Patrizi had taken some years earlier. On Patrizi's theory of space, see Grant, *Much Ado about Nothing*, pp. 204–5.

readily answer, I know not: nor shall be ashamed to own my Ignorance, till they that ask, shew me a clear distinct *Idea* of *Substance*."[109]

Locke also rejected with vigor the Cartesian identification of space with body.[110] As a result, he saw no problem with recognizing the possibility of empty space. He wrote: "Whatever Men shall think concerning the existence of a *Vacuum*, this is plain to me, That we have as clear an *Idea of Space distinct from Solidity*, as we have of Solidity distinct from Motion, or Motion from Space."[111]

Unlike Gassendi, Locke stopped short of saying that space definitely falls outside the categories of substance and accident, and he stopped short of asserting that space is a something that contains bodies, as opposed to a relation of sorts among bodies. But Locke was quite clear about rejecting the Cartesian identification of body and space and the consequent impossibility of the vacuum.

A similar position can be found in the writings of the Cambridge Platonist Henry More. Like Gassendi before him, More believed that space should be thought of as a container that contains all of the bodies in nature. But unlike Gassendi and Locke, More did not want to accommodate space by rejecting the categories of substance and accident. Although More agreed with Descartes that extension must be the property of something, he disagreed with Descartes in his claim that all extension must be body. Unlike Descartes, More argued that both body and soul are extended, the one extended and penetrable, the other extended and impenetrable. More argued that the appropriate substance to which to attribute the infinite extension of space is neither finite body nor finite spirit but God himself.[112]

Possibly related to More's view is one of Newton's, in his *Principia mathematica philosophiae naturalis* (Mathematical Principles of Natural Philosophy, 1687). There Newton presented an absolutist conception of space, which he contrasted with a relativist conception: "Absolute space, in its own nature, without relation to anything external, remains always similar and immovable. Relative space is some movable dimension or measure of the absolute spaces; which our senses determine by its position to bodies."[113]

It is with respect to the immobile framework of this absolute space that absolute (as opposed to relative) motion is to be measured: Absolute motion

---

[109] Locke, *Essay*, 2.13.17.
[110] Ibid., 2.13.11–17, 23–7.
[111] Ibid., 2.13.26.
[112] See Henry More, *An Antidote Against Atheism*, appendix, chap. 7, in his *A Collection of Several Philosophical Writings of Dr Henry More* . . . (London: Printed by James Flesher for W. Morden, 1662); and More, *Enchiridion metaphysicum* (London: Printed by James Flesher for W. Morden, 1671), chap. 8.
[113] Isaac Newton, *Philosophiae naturalis principia mathematica*, ed. Alexandre Koyré and I. Bernard Cohen, 2 vols. (Cambridge, Mass.: Harvard University Press, 1972), 1: 46, trans. Andrew Motte in Isaac Newton, *Mathematical Principles of Natural Philosophy*, revised by Florian Cajori, 2 vols. (Berkeley: University of California Press, 1934), 1: 6.

is simply motion with respect to this immobile framework.¹¹⁴ Newton gave a number of criteria by which one can tell whether one is in motion, absolutely speaking, including his famous bucket experiment.¹¹⁵ As More did, Newton seems to have identified space with God himself. In the General Scholium added to the second edition of the *Principia* (1713), Newton wrote that "He endures forever, and is everywhere present; and, by existing always and everywhere, he constitutes duration and space."¹¹⁶ Elsewhere, Newton talked about space as God's sensorium: God "is more able by his Will to move the Bodies within his boundless uniform Sensorium, and thereby to form and reform the Parts of the Universe, than we are by our Will to move the Parts of our own Bodies."¹¹⁷

An interesting kind of intermediate position between the Cartesian and the Gassendist is found in Leibniz. Against the conception of space found, for example, in an Epicurean atomist such as Gassendi, Leibniz offered a conception of space as relative:

> I hold space to be something merely relative. . . . I hold it to be an order of coexistences, as time is an order of successions. For space denotes, in terms of possibility, an order of things which exist at the same time, considered as existing together. . . . Space is nothing else but . . . order or relation, and is nothing at all without bodies but the possibility of placing them.¹¹⁸

Although Leibniz agreed with Descartes in rejecting the idea of space as something that exists independently of the bodies that fill it, he disagreed with Descartes' identification of body and space. But although it is conceivable for Leibniz that there could be empty space, a wise God would not leave any space unfilled. In this way, Leibniz shared the Cartesian commitment to the idea that all space is full of body (along with the idea that all body is divisible

---

[114] Although he agrees, in a sense, with Descartes in distinguishing motion and rest, his conception of the distinction is altogether different. See Newton's critique of Descartes' conception of motion in Isaac Newton, *De gravitatione* . . . , published in *Unpublished Scientific Papers of Isaac Newton*, ed. A. R. Hall and M. B. Hall (Cambridge: Cambridge University Press, 1962), pp. 89–156 (Latin original followed by English translation).

[115] In the bucket experiment, Newton imagines a bucket hung by a twisted cord and spun about so that the cord untwists. As the motion of the bucket communicates itself to the water, the surface of the water will become more and more concave as the water ascends the sides of the bucket. Newton writes: "The ascent of the water shows its endeavor to recede from the axis of its motion; and the true and absolute circular motion of the water, which is here directly contrary to the relative, becomes known, and may be measured by this endeavor." (Isaac Newton, *Principia mathematica* . . . , 1: 51, trans. Motte in Newton, *Mathematical Principles*, 1: 10.) The classic article on the question of Newton and absolute space and motion is Howard Stein, "Newtonian Space-Time," *Texas Quarterly*, 10 (1967), 174–200, reprinted in *The Annus Mirabilis of Sir Isaac Newton, 1666–1966*, ed. Robert Palter (Cambridge, Mass.: MIT Press, 1970).

[116] Newton, *Principia mathematica*, 2: 761, trans. Motte in Newton, *Mathematical Principles*, 2: 545.

[117] Question 31 in Isaac Newton, *Opticks; or, A Treatise of the Reflections, Refractions, Inflections & Colours of Light* (New York: Dover, 1952), p. 403; see also Question 28 in Newton, *Opticks*, p. 370.

[118] Leibniz to Clarke, 25 February 1716 (Leibniz's Third Paper), para. 4 in G. W. Leibniz and Samuel Clarke, *Correspondance Leibniz-Clarke*, ed. André Robinet (Paris: Presses Universitaires de France, 1957), p. 53; and G. W. Leibniz and Samuel Clarke, *The Leibniz-Clarke Correspondence*, ed. H. G. Alexander (Manchester: Manchester University Press, 1956), pp. 25–6.

to infinity) while sharing with the Gassendists the view that a vacuum is possible.[119] Interestingly enough, even though space was relative for Leibniz, motion was not. Leibniz held that in any situation in the physical world, one can designate any point as being immobile and the laws of physics will not be violated in that frame. But he also believed that at the metaphysical level of forces, there is a real distinction between motion and rest, and a fact of the matter about which bodies are really moving. Real motion, for Leibniz, involved real force: The bodies that are in motion are endowed with what he called living force (mass times velocity squared, mv^2).[120]

The question of absolute versus relative space gave rise to one of the most celebrated scientific disputes in the period, the debate between Leibniz and the Newtonians, as it unfolded in a series of letters between Leibniz and the English divine and friend of Newton's, Samuel Clarke (1675–1729).[121] There were many arguments on a number of issues, including the role of God in the universe and Leibniz's views on the relativity of space, time, and motion. A central consideration related to Leibniz's so-called Principle of Sufficient Reason, the claim that there must be a reason for everything. Leibniz pointed out that if there were absolute space, as Newton held, then one is forced to make distinctions without real differences. For example, if the world were to be moved five inches to the left, or if east and west were to be systematically reversed, the absolutist would have to hold that these worlds were really different. But if so, then there could be no reason for God to choose one of them over any of the others: Because the worlds are equally orderly and indistinguishable in all of their phenomena, God would violate the Principle of Sufficient Reason if he created any of them at all. This, for Leibniz, was a good reason for adopting a theory of space in which such worlds are not genuinely different. (This, of course, has the effect that, in the case at hand, because there is no difference between the starting place and the ending place, there is no motion either, properly speaking.) But Clarke was not satisfied. For Clarke, God was free to do what he liked: God's decision to create one possible universe over other possible and even indistinguishable universes is

---

[119] For a more detailed account of Leibniz on space, see Garber, "Leibniz: Physics and Philosophy," pp. 301 ff.

[120] See, for example, *Discours de métaphysique*, para. 18; and Leibniz to Huygens 12/22 June 1694, in Leibniz, *Mathematische Schriften*, 2: 184, translated in Ariew and Garber, eds. and trans., *Leibniz: Philosophical Essays*, p. 308. For a discussion of Leibnizian relativity, see Howard Stein, "Some Philosophical Prehistory of General Relativity," in *Foundations of Space-Time Theories*, ed. J. Earman, C. Glymour, and J. Stachel (Minnesota Studies in the Philosophy of Science, 8) (Minneapolis: University of Minnesota Press), pp. 3–49, esp. pp. 3–6, with notes and appendices; and Garber, "Leibniz: Physics and Philosophy," pp. 306 ff.

[121] For a close discussion of the exchange, see Ezio Vailati, *Leibniz and Clarke: A Study of Their Correspondence* (Oxford: Oxford University Press, 1997). Although it is clear that Newton played some role behind the scenes in Clarke's side of the correspondence, the exact extent is unclear. See Vailati, *Leibniz and Clarke*, pp. 4–5, and the references cited therein.

all the reason that is needed.[122] This exchange nicely illustrates the extent to which theological concerns were central to foundational debates about the nature of the physical world.

The issue of the nature of space and the possibility of a vacuum was one of the most important foundational issues in seventeenth-century physics. But even though it was foundational, aspects of the issue were thought to be amenable to empirical investigation, particularly the question of the real existence of the vacuum. In 1644, Evangelista Torricelli (1608–1647), a student of Galileo who worked in Florence, found that when one filled a tube that was closed on one side with mercury and then stood the tube up in a pool of mercury, if the tube was long enough, the mercury in the tube would fall and leave what appeared to be an empty space at the top.[123] This gave rise to considerable debate and discussion. The classic experiments were performed by Blaise Pascal (1623–1662) (see Dear, Chapter 4, this volume). There were two sets of experiments. The first were reported in Pascal's *Expériences nouvelles touchant le vide* (New Experiments on the Vacuum, 1647). There Pascal varied the experiments, using tubes of different widths, heights, and shapes. He used water and wine in addition to mercury in an attempt to show that the space at the top of the column was genuinely empty and filled neither with vapor from the liquid below nor with air that may have been in the liquid or seeped in through the pores in the tubes. He argued at that point that the column was held up by a limited "fear of the vacuum," a variant of the conception of the *horror vacui* common in Aristotelian science. Pascal's view changed in the *Récit de la grande expérience de l'équilibre des liqueurs* (Account of the Great Experiment on the Equilibrium of Fluids, 1648). There Pascal reported on the famous Puy de Dôme experiment, where his brother-in-law, Florin Périer, carried a barometer to the top of the Puy de Dôme, a high mountain in the Auvergne region of France, and compared the reading at the top with the reading of a similar apparatus at the bottom of the mountain. The fact that the column of mercury at the top was lower than the column of mercury at the bottom established, for Pascal, that it was the pressure of the air that kept the column at the level that it was; as one goes higher in the atmosphere, that air pressure decreases, causing the decrease in the length of the column. Pascal also concluded that nature does not abhor a vacuum and

---

[122] See, for example, Leibniz to Clarke (Leibniz's Third Paper), 25 February 1716, para. 5, and Clarke's reply, Clarke to Leibniz, 15 May 1716 (Clarke's Third Reply), paras. 2, 5. Interestingly enough, in his correspondence with Clarke, Leibniz does not discuss Newton's bucket experiment for distinguishing between absolute and relative motion. However, he discusses it elsewhere, and rejects it. See Leibniz to Huygens 4/14 September 1694, in Leibniz, *Mathematische Schriften*, 2: 199, translated in *Leibniz: Philosophical Essays*, p. 308–9.

[123] The classic account of this discovery and its consequences remains C. de Waard, *L'expérience barométrique: ses antécédents et ses explications* (Thouars [Deux-Sèvres]: Imprimerie Nouvelle, 1936).

that all of the phenomena that had been attributed to the supposed horror of the vacuum are caused by the pressure of the ambient air.[124]

Pascal's experiments were widely discussed, though not universally accepted as establishing what Pascal claimed they did. Descartes, of course, for whom extension and body were the same, could not accept Pascal's conclusion that the vacuum exists. Although he was perfectly prepared to agree with Pascal that it was air pressure that supported the column of mercury, Descartes believed that the apparently empty space at the top of the column was really subtle matter that had entered through the pores of the glass.[125] This position was developed in more detail in a series of letters that Étienne Noël (1581–1659) sent Pascal in autumn 1647. (Noël was a Jesuit and may possibly have been Descartes' philosophy teacher at the Jesuit Academy of La Flèche.) Noël argued that the fact that light passes through the vacuum shows that the glass must have pores in order to allow the particles of light to pass through. And if light can pass through, so could small particles from the atmosphere.[126] This consideration was trenchant enough that even some supporters of the vacuum, such as Gassendi and his English follower Walter Charleton (1620–1707), agreed that it cast doubt on Pascal's conclusion.[127] In the end, the problem was solved (as many metaphysical problems seem to be) by simply setting the issue aside.[128] In his *New Experiments Physico-Mechanical, touching the Spring of the Air* (1660), where he first reported his famous air-pump experiments, Boyle wrote: "The Controversie about a *Vacuum* [seems to be] rather a Metaphysical, then a Physiological Question; which therefore we shall here no longer debate, finding it very difficult either to satisfie Naturalists with this Cartesian Notion of a Body, or to manifest wherein it is erroneous, and substute a better in its stead."[129]

For Boyle, the foundational question that goes beyond the ability of the experimenter to determine is a question that should be left aside.

---

[124] The *Expériences nouvelles* can be found in Blaise Pascal, *Oeuvres complètes*, 7 vols. (Paris: Desclée de Brouwer, 1964–), 2: 493–513, translated in Blaise Pascal, *Provincial Letters, Pensées, Scientific Treatises*, trans. Thomas M'Crie (Great Books of the Western World), ed. Robert Maynard Hutchins, 54 vols. (Chicago: Encyclopaedia Britannica, 1952), 33: 359–81. The *Récit* can be found in Pascal, *Oeuvres complètes*, 2: 677–90, translated in Hutchins, ed., *Provincial Letters*, pp. 382–9. For accounts of the arguments, see, for example, P. Guenancia, *Du vide à Dieu: Essai sur la physique de Pascal* (Paris: Maspero, 1976); and Simone Mazauric, *Gassendi, Pascal et la querelle du vide* (Paris: Presses Universitaires de France, 1998).

[125] See Garber, *Descartes' Metaphysical Physics*, pp. 136–43.

[126] For Noël's correspondence with Pascal, see Pascal, *Oeuvres complètes*, 2: 513–40. For a survey of Noël's arguments, see Garber, *Descartes' Metaphysical Physics*, p. 143.

[127] See Gassendi, *Opera*, 1: 205A; and Walter Charleton, *Physiologia Epicuro-Gassendo-Charltoniana* (London: Printed by T. Newcomb for T. Heath, 1654), pp. 42–4.

[128] See Steven Shapin and Simon Schaffer, *Leviathan and the Air-Pump: Hobbes, Boyle, and the Experimental Life* (Princeton, N.J.: Princeton University Press, 1985), pp. 45 ff., 119 ff.

[129] Boyle, *Works*, 1: 198.

THE MECHANICAL PHILOSOPHY: SPIRIT, FORCE, AND ACTIVITY

In the orthodox mechanical philosophy, everything was to be explained in terms of size, shape, motion, and the collision of corpuscles with one another, all governed by the laws of nature. This would seem to exclude any intrusion of mentality or incorporeal substance into the physical world. Among the main figures, only Hobbes espoused a straightforwardly materialistic philosophy and eliminated mind altogether.[130] Descartes introduced mind as a thinking thing, in contrast with body, whose essence is extension alone. As a consequence of these conceptions, mind and body were completely distinct from one another, and the one could exist without the other. Because this entailed a rejection of the Aristotelian conception of a soul, the principle of life, Descartes was committed to explaining the phenomena of life – digestion, reproduction, involuntary motions, and so forth – in purely mechanistic terms. The mind, an incorporeal and nonextended substance, explained thought and reason. But insofar as some of our activities involve rational processes of thought and choice and voluntary motion (I reach out and choose a book rather than a pack of playing cards), the mental world did on some occasions intrude into the physical world for Descartes.[131]

Henry More took Descartes' position further still. In his earlier years, More corresponded with Descartes and did much to advocate the study of his thought in England.[132] But even though he was a great advocate of the mechanical philosophy in many ways, More was convinced that much that the mechanists claimed to be able to explain mechanistically could not be so explained and required an appeal to what he called the "spirit of nature." This incorporeal principle was taken to explain "what remands down a stone toward the Center of the Earth . . . keeps the Waters from swilling out of the Moon, curbs the matter of the sun into roundness of figure," among many other things.[133] More characterized this spirit of nature as "a substance

---

[130] There are some others whose views are associated with materialism. In his set of objections to Descartes' *Meditations*, Gassendi seems to adopt a materialist view against Descartes' famous dualism; see Descartes, *Oeuvres*, 7: 262–70, and his expansion of this in his *Disquisitio Metaphysica* (Amsterdam: Johannes Blaev, 1644), Gassendi, *Opera*, 3: 284B ff. However, in the *Syntagma*, he comes out quite clearly for the existence of incorporeal substance. See Gassendi, *Opera*, 2: 440A ff. Another character in the period often accused of materialism is Spinoza. Although his complex metaphysics does allow for the possibility of being interpreted in this way, insofar as the mind and body are, in a sense, identical, it can also be interpreted in other ways. See Benedict de Spinoza *Ethics*, in Spinoza, *Opera*, ed. Carl Gebhardt (Heidelberg: C. Winter, 1925), vol. 2, pp. 84–96, esp. Part 2, props. 1–13.

[131] For a development of this reading, see Daniel Garber, "Mind, Body, and the Laws of Nature in Descartes and Leibniz," in Garber, *Descartes Embodied* (Cambridge: Cambridge University Press, 2001), pp. 133–67.

[132] On More's role in the diffusion of Cartesianism, see Alan Gabbey, "Philosophia Cartesiana Triumphata: Henry More (1646–1671)."

[133] Henry More, *A Collection*, p. xv.

incorporeal, but without Sense and Animadversion, pervading the whole Matter of the Universe, and exercising a Plastical power therein ... raising such Phaenomena in the world, by directing the parts of Matter and their Motion, as cannot be resolved into mere Mechanic powers."[134] More's conception of the world extended to other kinds of spirits as well. Along with his friend, the English natural philosopher Joseph Glanvill (1636–1680), More proselytized for the recognition of disembodied spirits, ghosts, and witches, arguing that they should be accepted by the very standards of belief espoused by the Royal Society.[135]

Another mechanist view that granted a large role to incorporeal substance was Leibniz's, where the ultimate entities, corporeal substances or monads, are understood to be immaterial substances or at least endowed with immaterial substances. But, Leibniz held, though the mechanist world is grounded in something that goes beyond matter and motion, everything in the physical world can be explained in terms of size, shape, and motion. For Leibniz, the appeal to incorporeal substance was needed not to explain individual events in the physical world but rather the very existence and nature of laws that govern those events. For example, Leibniz argued that if bodies were mere extension, as the Cartesians held, and contained nothing immaterial, then one body could not resist another in a collision, and a body A in motion colliding with a body B at rest would put body B into motion without diminishing the speed of body A in any way. In this situation, various conservation laws, such as the conservation of momentum and the conservation of mv^2, would be violated. In this way, Leibniz took great pains to distance himself from views such as More's, which involved the direct intervention of incorporeal substance in the material world.[136]

Closely related to the question of incorporeal substance in natural philosophy is the question of the activity of bodies and the real existence of force in the physical world. If the essence of body is extension alone, then it would appear that there is no room in body for any activity at all. For that reason, Descartes held that the motion of bodies in the world derives directly from

---

[134] Henry More, *The Immortality of the Soul*, p. 193, in More, *A Collection*. A similar view is found in More's friend and colleague Ralph Cudworth. See Ralph Cudworth, *The True Intellectual System of the Universe* (London: Richard Royston, 1678). What corresponds in Cudworth's thought to More's Spirit of Nature is what he calls the plastic natures. Indeed, Cudworth goes so far as to argue that the purely materialistic (and atheistic) form in which atomism has come down to us is a perversion of the original, which before Democritus and Leucippus included incorporeal souls and an incorporeal deity in addition to atoms and the void (1.18, 41 ff.).

[135] See Daniel Garber, "Soul and Mind: Life and Thought in the Seventeenth Century," in Garber and Ayers, eds., *The Cambridge History of Seventeenth-Century Philosophy*, pp. 776 ff.

[136] See Part I of Gottfried Wilhelm Leibniz, *Specimen dynamicum* [1695], in Leibniz, *Mathematische Schriften*, 6: 242–3, translated in Ariew and Garber, eds. and trans., *Leibniz: Philosophical Essays*, pp. 125–6; and Gottfried Wilhelm von Leibniz, *De ipsa natura* [1698], para. 2, *Die philosophischen Schriften*, 4: 504–5, translated in Ariew and Garber, eds. and trans., *Leibniz: Philosophical Essays*, p. 156.

God himself or from the finite minds to which he gave the ability to move bodies. As he wrote to Henry More:

> The translation that I call motion, is not something with less being than figure has, that is, it is a mode in body. But the moving force can be that of God, conserving as much translation in matter, as he placed in it in the first moment of creation, of that of some other created substance, such as our mind, or some other thing [an angel, for example] to which he gave the force for moving a body. . . . I consider "matter left free and having no other impulse" as plainly at rest. Moreover, it is impelled by God, conserving as much motion or translation in it as he placed there in the beginning.[137]

In this way, all motion (at least, all motion that does not derive from finite minds) derives directly from God. Despite this feature of his account of body, Descartes made free use of the notion of force in his physics. But as I discuss later in this chapter, given Descartes' grounding of the laws of nature (in which the notion of force plays its role) in God, it is fair to interpret his appeal to force as an indirect appeal to God. For example, it is because God maintains the motion that a body has that it appears to resist being stopped or being deflected from its rectilinear path.[138]

A general trend within Cartesian metaphysical physics after Descartes' death was the development and ultimate dominance of the doctrine of occasionalism. Although Descartes allowed that minds can be the causes of motion as well, many of Descartes' later followers, including Gérauld de Cordemoy (1626–1684), Louis de La Forge (1632–ca. 1666), Johann Clauberg (1622–1665), and Nicolas Malebranche (1638–1715), took the doctrine one step further and argued that God is the *only* genuinely efficacious cause in the world, eliminating both bodies and minds as real causes. For a variety of reasons, they argued that what appear to be instances of body–body causality (one body collides with another) or mind–body causality (the mind wills to raise the arm of the body to which it is attached) are really caused by God, carrying out the effects in accordance with laws that he has ordained for himself. According to one popular argument, for example, God's conservation of the world from moment to moment, which underlies Descartes' view of the laws of motion, makes any causal relations between finite creatures, minds or bodies, otiose. Another central argument, due to Malebranche, eliminates finite causes by arguing that only in the case of God do we find the necessary connection between cause and effect required for a genuine causal relation.[139]

---

[137] Descartes, *Oeuvres*, 5: 403–4. The quotation in the passage is from More's letter to Descartes. There is a certain amount of controversy over whether the "some other thing" to which God gave the ability to move bodies is another body or another kind of spirit. On this, see Garber, *Descartes' Metaphysical Physics*, pp. 303–4.

[138] See Garber, *Descartes' Metaphysical Physics*, chap. 9.

[139] On occasionalism, see *Causation in Early Modern Philosophy*, ed. Steven Nadler (University Park: The Pennsylvania State University Press, 1993); and para. 10 of Nadler, "Doctrines of Explanation in Late Scholasticism and in the Mechanical Philosophy," in Garber and Ayers, eds., *The Cambridge History*

The atomist Gassendi would appear to be opposed to Descartes on this score. Gassendi did agree with Epicurus in holding that there is a sense in which bodies are genuinely active. Unlike Descartes, Gassendi held that God, in creating bodies, created them with genuine self-motion. Gassendi wrote in the *Syntagma philosophicum* (Treatise on Philosophy, 1658): "It seems that we must say . . . that the first moving cause in physical things is atoms; while they move through themselves and through the force which is continually received from the Author from the beginning, they give motion to all things. And therefore these atoms are the origin, principal, and cause of all motions which are in nature."[140]

But it is clear that for Gassendi, as for Descartes, the foundation of this activity was God: God was "the Author" who must continually sustain the force that he has given to bodies.

Leibniz seems to have taken Gassendi's views of the activity of bodies one step further by seeing force and activity not merely as properties of the basic stuff of the world but as, in a sense, definitive of the very notion of body. He wrote in an essay entitled "On the Correction of Metaphysics and the Concept of Substance" (1694):

> I say that this power of acting inheres in all substance, and that some action always arises from it, so that the corporeal substance itself does not, any more than spiritual substance, ever cease to act. This seems not to have been perceived clearly by those who have found the essence of bodies to be in extension, alone or together with the addition of impenetrability, and who seem to conceive of bodies as absolutely at rest.[141]

Given the close connection between activity and substantiality, it is not surprising that the notion of force entered into the very definition of substance for Leibniz. In his dynamics, Leibniz made two important distinctions with respect to force. First of all, there was the distinction between primitive and derivative forces, the distinction between the subject that is exerting the force (primitive) and the actual force exerted by the substance at a particular time (derivative). Derivative forces manifest themselves in motion and the resistance to motion at the level of observable bodies, governed by laws of motion that Leibniz proposes. Then there is the distinction between active and passive forces. Passive forces are exerted in reaction to other forces that act on the body; these forces include impenetrability and resistance. Active forces are exerted by the substance without being acted on; these include

---

of Seventeenth-Century Philosophy. On the argument for occasionalism from divine sustenance, see Daniel Garber, "How God Causes Motion: Descartes, Divine Sustenance, and Occasionalism," in Garber, *Descartes Embodied*, pp. 189–202. For the argument from necessary connection, see Nicolas Malebranche, *De la recherche de la verité* (Paris: A. Pralard, 1674–5), 6.2.3.

[140] Gassendi, *Opera*, 1: 337A; cf. 1: 279B, 1: 280A.
[141] Leibniz, *Die philosophischen Schriften*, 4: 468–70, translated in Loemker, ed. and trans., *Philosophical Papers and Letters*, p. 433.

living force (the force associated with motion) and dead force (the kind of force found in a stretched rubber band). Leibniz claimed that primitive active force is, properly speaking, the substantial form of a substance, whereas primitive passive force constitutes the primary matter.[142]

For Leibniz, force and activity were essential parts of substance and thus very different from the inert corporeal substances of the Cartesian tradition. But, despite that, they do not act independently of God. Leibniz wrote in the essay "De ipsa natura" ("On Nature Itself," 1698):

> The very substance of things consists in a force for acting and being acted upon. From this it follows that persisting things cannot be produced if no force lasting through time can be imprinted on them by the divine power. Were that so, it would follow that no created substance, no soul would remain numerically the same, and thus, nothing would be conserved by God, and consequently everything would merely be certain vanishing or unstable modifications and phantasms, so to speak, of one permanent divine substance.[143]

It is a subtle position that Leibniz was trying to outline here. Although God must continually conserve the world, for Leibniz as for many of his contemporaries, what he must conserve is a world of active substances that contain within themselves the grounds of their own activity.

THE MECHANICAL PHILOSOPHY: GOD AND FINAL CAUSES

It is evident from the preceding discussion that God had a large role to play in the mechanical philosophy. God was identified by some with the container space; he was appealed to in order to determine what is a rational choice and what is not in determining the structure of the world; and he was appealed to as the primary cause of motion in the world and as the ground of force and activity in the world. The mechanist's philosophy was infused with the divine spirit, in a sense. In addition to these uses of God in the mechanical philosophy, I would like to discuss two additional themes that relate to God and the mechanical philosophy: the controversies over final causes, and the use of God in the derivation of the laws of motion.

The world of Christian scholasticism was a world full of meaning: divine plans and divine designs. One of Descartes' most controversial positions was to put such considerations out of bounds for the physicist. He wrote: "When dealing with natural things we will, then, never derive any explanations from

---

[142] Gottfried Wilhelm Leibniz, *Specimen dynamicum*, pt. 1, in *Mathematische Schriften*, 6: 236 ff., translated in Ariew and Garber, eds. and trans., *Leibniz: Philosophical Essays*, p. 119 ff.
[143] "De ipsa natura" [1698], sec. 8, *Die philosophischen Schriften*, 4: 508, translated in Ariew and Garber, eds. and trans., *Leibniz: Philosophical Essays*, pp. 159–60.

the purposes which God or nature may have had in view when creating them [and we shall entirely banish from our philosophy the search for final causes]. For we should not be so arrogant as to suppose that we can share in God's plans."[144]

Benedict de Spinoza (1636–1677) took the argument one step further and denied not only that we could know final causes but that, strictly speaking, God had no intentions. The appendix to Part I of his posthumously published *Ethica* (1677) gave an elaborate argument for why it is wrong to think of God anthropomorphically, as if he acted with intentions.

Needless to say, this was not a position that was popular among most thinkers of the period. Boyle, for example, wrote an essay directly opposing Descartes, as well as those more radical than Descartes who eliminated final causes altogether, *A Disquisition about the Final Causes of Natural Things* (1688).[145] Although Boyle recommended that "a *Naturalist*, who would Deserve that Name, must not let the Search or Knowledge of *Final Causes* make him Neglect the Industrious Indagation of *Efficients*," he argued that "all Consideration of *Final Causes* is not to be Banish'd from Natural Philosophy: but *that* 'tis rather Allowable, and in some Cases Commendable, to Observe and Argue from the Manifest Uses of Things, that the Author of Nature Pre-ordain'd those ends and uses."[146] More generally, Boyle held that "by being addicted to *Experimental Philosophy*, a Man is rather Assisted than Indisposed, to be a *Good Christian*," as the subtitle to his *Christian Virtuoso* (1690–1) reads.[147]

Newton, too, embraced final causes. Writing in the celebrated General Scholium, added to the end of the second edition of the *Principia* in 1713, and referring to the order of the heavenly bodies, Newton noted that "It is not to be conceived that mere mechanical causes could give birth to so many regular motions. . . . This most beautiful system of the sun, planets, and comets, could only proceed from the counsel and dominion of an intelligent and powerful Being."[148] In this way, God is very much present to the world in ordering it and shaping it.

But the philosophically most sophisticated defense of final causes in the period was probably that of Leibniz. As a mechanist, Leibniz held that everything could be explained in terms of size, shape, and motion, in terms of efficient causes. But he also held that everything can be explained in terms

---

[144] *Principia philosophiae*, 1.28. The material in brackets is from the 1647 French translation. Before Descartes, Bacon had also rejected final causes in physics. See Francis Bacon, *Novum Organum* (London: Joannes Billius, 1620), 1.48 and 2.2; and Bacon, *De dignitate et augmentis scientiarum* (London: I. Haviland, 1623), 3.4.
[145] Boyle, *Works*, 11: 79–151.
[146] Ibid., 11: 151.
[147] Ibid., 11: 281.
[148] Newton, *Principia mathematica . . .* , 2: 760, translated in Newton, *Mathematical Principles*, 2: 544; cf. Query 31 of Newton, *Opticks*, p. 402. There Newton dismisses Descartes' attempt to derive the current state of the world from an initial chaos without appeal to final causes as "unphilosophical."

of God's intentions. As he wrote in the *Specimen dynamicum* (An Example from the Dynamics, 1695):

> In general, we must hold that everything in the world can be explained in two ways: through the *kingdom of power*, that is, through *efficient causes*, and through the *kingdom of wisdom*, that is, through *final causes*, through God, governing bodies for his glory, like an architect, governing them as machines that follow the *laws of size* or *mathematics*, governing them, indeed, for the use of souls. . . . These two kingdoms everywhere interpenetrate each other without confusing or disturbing their laws, so that the greatest obtains in the kingdom of power at the same time as the best in the kingdom of wisdom.[149]

Leibniz did not think that we should always appeal directly to final causes. He wrote in an essay from 1702: "[I]t is empty to resort to the first substance, God, in explaining the phenomena of his creatures, unless his means or ends are, at the same time, explained in detail, and the proximate efficient or even the pertinent final causes are correctly assigned, so that he shows himself through his power and wisdom."[150]

However, in some cases, particularly in optics, Leibniz thought that final causes could be very helpful in discovering things that are too difficult to discover using efficient causes, such as the sine law of refraction.[151]

This difference in attitude toward final causes is reflected in the very different ways in which Descartes and Leibniz derived the laws of motion from God. For Descartes, the laws of motion he proposed were justified by the claim that in sustaining the world from moment to moment, as he must do for it to remain in existence, God also preserves a certain quantity of motion in the world, and certain features of that motion, for example the tendency of a body in motion to remain in uniform rectilinear motion. In justification of his famous law of the conservation of quantity of motion (size times speed) in his *Principia philosophiae* (1644), Descartes wrote:

> For we understand that God's perfection involves not only his being immutable in himself, but also his operating in a manner that is always utterly constant and immutable. Now there are some changes whose occurrence is guaranteed either by our own plain experience or by divine revelation, and

---

[149] Gottfried Wilhelm Leibniz, *Specimen dynamicum*, pt. I, in Leibniz, *Mathematische Schriften*, 6: 243, translated in Ariew and Garber, eds. and trans., *Leibniz: Philosophical Essays*, pp. 126–7.

[150] Gottfried Wilhelm Leibniz, "On Body and Force, May 1702," in Leibniz, *Die philosophischen Schriften*, 4: 397–8, translated in Ariew and Garber, eds. and trans., *Leibniz: Philosophical Essays*, p. 254.

[151] See Leibniz, *Specimen dynamicum*, pt. I, in Leibniz, *Mathematische Schriften*, 6: 243, translated in Ariew and Garber, eds. and trans., *Leibniz: Philosophical Essays*, pp. 126–7; Gottfried Wilhelm Leibniz, "A Letter of Mr. Leibniz . . ." (July 1687), in Leibniz, *Die philosophischen Schriften*, 3: 51–2, translated in Loemker, ed. and trans., *Philosophical Papers and Letters*, p. 351. The sine law of refraction is discussed in Leibniz, *Discours de métaphysique*, para. 22. A specific example Leibniz refers to on a number of occasions is the "Unicum Opticae, Catoptricae, et Dioptricae Principium," *Acta eruditorum*, June 1682: 185–90, in Gottfried Wilhelm Leibniz, *Opera omnia*, ed. Louis Dutens (Geneva: Fratres de Tournes, 1768), 3: 145–51.

either our perception or our faith shows us that these take place without any change in the creator; but apart from these we should not suppose that any other changes occur in God's works, in case this suggests some inconstancy in God. Thus, God imparted various motions to the parts of matter when he first created them, and he now preserves all this matter in the same way, and by the same process by which he originally created it; and it follows from what we have said that this fact alone makes it most reasonable to think that God likewise always preserves the same quantity of motion in matter.[152]

Descartes suggested similar derivations for the three subsidiary laws of motion that he proposes. It is important to note here that Descartes was *not* appealing to God's intentions or God's choice. The laws he proposed derive directly from God's nature: It is because of his immutability that God must act in the way in which he does, and because he acts that way, bodies obey Descartes' laws of motion.

Leibniz rejected Descartes' incorrect laws and replaced them with a set of conservation laws very much like the ones now used in classical mechanics. However, Leibniz also rejected the way in which Descartes derived the laws from God.

[The laws of motion] do not derive entirely from the principle of necessity, but from the principle of perfection and order; they are an effect of the choice and the wisdom of God. I can demonstrate these laws in many ways, but it is always necessary to assume something which is not absolutely geometrically necessary. These beautiful laws are a marvelous proof of an intelligent and free being [God], against the system of absolute and brute necessity of Straton and Spinoza.[153]

In this way, the laws of nature, for Leibniz, derive from the free choice of a God who chooses the laws appropriate for this best of all possible worlds.

BEYOND THE MECHANICAL PHILOSOPHY: NEWTON

In many ways, Newton's world was the by then familiar mechanist/corpuscularian world of bodies governed by laws of motion. Although Newton eschewed any systematic statement of his theory of matter, it is reasonably clear that he rejected the Cartesian metaphysical physics and subscribed to a version of atomism in which he recognized both atoms and the

---

[152] Descartes, *Principia philosophiae*, 2.36.
[153] Gottfried Wilhelm Leibniz, *Theodicy*, 1.345, in Leibniz, *Die philosophischen Schriften*, 6: 319; see also, for example, Leibniz, *Discours de métaphysique*, para. 21. Also see Gottfried Wilhelm Leibniz, *Principes de la nature et de la grâce* (written in 1714, but unpublished during Leibniz's lifetime), para. 11, in Leibniz, *Die philosophischen Schriften*, 6: 598–606. Strato of Lampsacus (d. 270 B.C.E.) was an ancient follower of Aristotle who had the reputation of denying providence. None of his works survive.

void.¹⁵⁴ Perhaps most surprising to his contemporaries, and most disturbing as well, was the extent to which Newton was willing to add active powers to bodies. Again, in the thirty-first Query to his *Opticks*, Newton wrote concerning the atoms that make up bodies:

> It seems to me farther that these Particles have not only a *Vis inertiae* . . . , but also that they are moved by certain active Principles, such as is that of Gravity, and that which causes Fermentation, and the Cohesion of Bodies. These Principles I consider not as occult Qualities, supposed to result from the specifick Forms of Things, but general Laws of Nature, by which the things themselves are form'd.¹⁵⁵

Newton's world was thus an active world composed of bodies with active principles, including but not limited to gravitation, that are central to the formation of the world we see around us.¹⁵⁶ In adding these active forces, perhaps as a result of his chymical studies,¹⁵⁷ Newton departed from the strict Boylean mechanism that was the hallmark of the previous generation; he thus admitted that not everything can be explained by matter and motion alone, and that there is action that does not work by direct collision but at a distance. It was this to which Leibniz, for example, objected. Leibniz saw Newton's obscure forces as a step backward from the clarity and intelligibility of the mechanical philosophy, a reversion back to the scholastic philosophy that the mechanical philosophy was supposed to replace, a departure from the clarity of action by impact, and a return to the obscurity of influences and occult qualities. With Newton (and his followers) in mind, Leibniz complained bitterly of the people of his day who "have such a lust for variety that, in the midst of an abundance of fruits, it seems they want to revert to acorns"; rejecting the clear truths of the mechanical philosophy, they show their "love for difficult nonsense."¹⁵⁸

Leibniz did not live to see Newton's acorns grow into mighty oaks, or his nonsense transformed into the new common sense. Although Newton's conception of the world came to dominate European thought in the eighteenth

---

¹⁵⁴ Kargon, *Atomism in England*, chap. 9.
¹⁵⁵ Newton, *Opticks*, p. 402.
¹⁵⁶ See the discussion by Daniel Garber, John Henry, Lynn Joy, and Alan Gabbey, "New Doctrines of Body and Its Powers, Place, and Space," in Garber and Ayers, eds., *The Cambridge History of Seventeenth-Century Philosophy*, pp. 553–623, at pp. 602 ff. It should be noted that there is considerable disagreement about the status of gravitation in Newton: whether he really thought that gravitation was a basic force of nature, or whether he thought that it could be explained by more basic mechanical causes. However, at least some of his followers were willing to take the plunge and accept action at a distance. See, for example, Roger Cotes's preface to the second edition of Newton's *Principia* (1713), in *Principia mathematica*, 1: 19–35, esp. 27–8, translated in *Mathematical Principles*, 1: xx–xxxiii, esp. xxvii. On the status of gravitation, see Ernan McMullin, *Newton on Matter and Activity* (Notre Dame, Ind.: Notre Dame University Press, 1978), chap. 3.
¹⁵⁷ On Newton and chymistry, see Westfall, "Newton and the Hermetic Tradition," and Dobbs, *The Foundations of Newton's Alchemy*, chap. 6.
¹⁵⁸ Gottfried Wilhelm Leibniz, *Antibarbarus physicus*, in Leibniz, *Die philosophischen Schriften*, 7: 337, translated in Ariew and Garber, eds. and trans., *Leibniz: Philosophical Essays*, p. 31.

century and replaced the stricter mechanical philosophy of the seventeenth century, the particular foundations that Newton himself supplied were not always adopted along with the physics. There were attempts to ground Newton's physics in different metaphysics, including the idealistic metaphysics of Bishop Berkeley (1685–1753), the monadological metaphysics of Leibniz's German followers, the atoms of force of Rudjer Bošković (1711–1787), David Hume's (1711–1776) psychologistic foundations of causality, and the magisterial system of Immanuel Kant (1724–1804). But in contrast with the sixteenth and seventeenth centuries, when the foundational enterprise was closely linked with the scientific enterprise itself, later developments in technical physics seemed largely independent of the different attempts to provide it with appropriate foundations.

CONCLUSION: BEYOND FOUNDATIONS

The ultimate fate of the Newtonian system in the eighteenth century illustrates a fundamental shift in scientific thought with regard to foundational questions. In the beginning of the period examined in this chapter, the idea of foundations is quite central to the idea of the study of nature. By the end of the seventeenth century, this idea had not been altogether abandoned by any means but had changed its status in fundamental ways. By this time, I think it is fair to say that the enterprise of physics and the enterprise of grounding physics have largely separated from one another and become rather separate disciplines.

This separation had been prepared for some time before. Already in the works of Boyle, questions about the vacuum and the infinite divisibility of matter, questions that go beyond the ability of experiment to resolve, had become metaphysical in a pejorative sense and had been placed beyond the domain of the natural philosopher. By the end of the seventeenth century, even Leibniz, one of the heirs of the program for a metaphysical physics, had come to separate the domain of physics proper from its metaphysical foundations and argued that the physicist need not concern himself with that domain. Leibniz's grounding of his mechanist world in a conception of substance was very different from that of Descartes, involving the positing of incorporeal substances in nature and the way in which God enters into the metaphysical grounding of his conception of the natural world. But, Leibniz argued, metaphysics and theology should not be the concern of the physicist, properly speaking. Writing in his *Discourse on Metaphysics*, he noted:

> Just as a geometer does not need to burden his mind with the famous labyrinth of the composition of the continuum, there is no need for any moral philosopher and even less need for a jurist or statesman to trouble himself with the great difficulties involved in reconciling free will and God's providence, since the geometer can achieve all his demonstrations and the

statesman can complete all his deliberations without entering into these discussions, discussions that remain necessary and important in philosophy and theology. In the same way, a physicist can explain some experiments, at times using previous simpler experiments and at times using geometric and mechanical demonstrations, without needing general considerations from another sphere. And if he uses God's concourse, or else a soul, animating force [*archée*], or something else of this nature, he is raving just as much as the person who, in the course of an important practical deliberation, enters into a lofty discussion concerning the nature of destiny and the nature of our freedom.[159]

In this disciplinary separation of foundations from the science that it grounds are born both philosophy and science as we have come to know them.

---

[159] Leibniz, *Discours de métaphysique*, para. 10, in Leibniz, *Sämtliche Schriften und Briefe* 6.4: 1543–44, translated in Ariew and Garber, eds. and trans., *Leibniz: Philosophical Essays*, p. 43.

3

SCIENTIFIC EXPLANATION FROM FORMAL CAUSES TO LAWS OF NATURE

Lynn S. Joy

The story of the changing forms of explanation adopted in the early modern sciences is too often told as a story of the wholesale rejection of the systematic Aristotelian treatment of causal questions that flourished in medieval as well as ancient science. Narratives of this sort have ignored a promising alternative way of understanding the multifaceted transformation that occurred in early modern natural philosophers' beliefs about causality. By focusing instead on the Aristotelian tradition's contributions to the development of rival forms of explanation, it becomes possible to characterize these new sorts of explanations against a rich conceptual background. Of course, scientific innovators in the period 1500–1800 did widely reject Aristotle's account of the four kinds of causes as a source of acceptable theories in the specific sciences.[1] But a more tempered view of this rejection may better reveal how the new sorts of explanations were actually conceived by their originators.

THREE NOTABLE CHANGES IN EARLY MODERN SCIENTIFIC EXPLANATIONS

This chapter considers three notable changes in early modern scientific explanations. The first was a change in the overall purpose of scientific research that

---

[1] Historians of science and philosophy have assessed the contributions of Aristotelian thought to the growth of early modern science in strikingly different ways. Some have viewed the rejection of Aristotelian principles as crucial to the development of early modern science, whereas others have argued for the indispensability of some of these same principles in its development. Readers interested in interpretations tracing the rejection of Aristotle should consult, for example, Charles Coulston Gillispie, *The Edge of Objectivity* (Princeton, N.J.: Princeton University Press, 1960), pp. 11–16, 266–8, 285; and Carolyn Merchant, *The Death of Nature: Women, Ecology, and the Scientific Revolution* (New York: Harper and Row, 1980), pp. 99–126, esp. 112, 121–6. By contrast, interpretations that show the *indispensability* of Aristotelian ideas include: William A. Wallace, *Causality and Scientific Explanation*, 2 vols. (Ann Arbor: University of Michigan Press, 1972 and 1974), vol. 1; and Dennis Des Chene, *Physiologia: Natural Philosophy in Late Aristotelian and Cartesian Thought* (Ithaca, N.Y.: Cornell University Press, 1996), pp. 53–251.

was initiated by those critics of Aristotelianism who relinquished Aristotle's goal of understanding the form of each natural substance. Rather than trying to elucidate each substance's form, early modern innovators in the specific sciences, as well as natural philosophy, sought to determine the fundamental constituent parts – whether elements or atoms – of each kind of material body and also to identify the lawlike regularities exhibited in the organization and motions of these fundamental elements or atoms.[2] Such a redirection of the purpose of scientific research also produced new definitions of the metaphysical requirements that must be satisfied for something to count as a cause.

A second notable change consisted in the replacement of long-standing Aristotelian explanations of specific kinds of natural phenomena. In astronomy, the contested explanations dealt with well-established observations of planetary motion as well as newly discovered effects such as the appearance of a supernova or of an apparently new comet. For example, Tycho Brahe's description of the comet of 1577 – as located far above the sphere of the moon and orbiting the sun – and Galileo Galilei's subsequent endorsement of just the first part of this description in his *Istoria e dimostrazioni intorno alle macchie solari e loro accidenti* (Letters on Sunspots, 1613) set the stage for a bitter series of disputes between Galileo and the Jesuit Horatio Grassi regarding the explanation of three comets observed in 1618.[3] Both Galileo and Grassi had already rejected the usual Aristotelian view that comets are meteorological phenomena occurring beneath the sphere of the moon. Their disagreement concerned whether comets are the same kind of objects as planets and whether measurements of their parallaxes are reliable indicators of their distances from Earth. However, it also encompassed the two thinkers' disputes about a variety of other issues, including the nature of human sense perception, the reflection of sunlight by planets, and the heating of terrestrial bodies. Such wide-ranging contested explanations were to be found outside astronomy, too, in sciences that specialized in the study of terrestrial phenomena such as mechanics, natural history, alchemy, and medicine.[4]

---

[2] For a detailed analysis of this first notable change in early modern treatments of causality, see the sections of this chapter on "God as a Final Cause and the Emergence of Laws of Nature" and "Intrinsic versus Extrinsic Efficient Causes among the Corpuscular Physicists."

[3] Galileo Galilei, *Letters on Sunspots*, excerpts translated in *Discoveries and Opinions of Galileo*, ed. Stillman Drake (Garden City, N.Y.: Doubleday Anchor Books, 1957), p. 119. See also Drake's introduction to Galileo's *The Assayer* [1623] in *Discoveries and Opinions of Galileo*, esp. pp. 221–7, and Stillman Drake, *Galileo at Work* (Chicago: University of Chicago Press, 1978), pp. 264–73, for a summary of Mario Guiducci (and Galileo Galilei), *Discorso delle comete*... [1619].

[4] On changes in the science of natural history, see Lorraine Daston and Katharine Park, *Wonders and the Order of Nature, 1150–1750* (New York: Zone Books, 1998), pp. 217–31; also their "Unnatural Conceptions: The Study of Monsters in Sixteenth- and Seventeenth-Century France and England," *Past and Present*, 92 (1981), 20–54; and Lorraine Daston, "Baconian Facts, Academic Civility, and the Prehistory of Objectivity," *Annals of Scholarship*, 8 (1991), 337–63. On developments in the science of alchemy, see William R. Newman, *Gehennical Fire: The Lives of George Starkey, an American Alchemist in the Scientific Revolution* (Cambridge, Mass.: Harvard University Press, 1994),

Finally, a third notable change in early modern scientific explanations was signaled by natural philosophers' waning interest in metaphysical discussions of the nature of causality itself. Increasingly, they addressed epistemological questions that had little to do with Aristotelian investigations of an ontology of causes. Their epistemological inquiries explored, among other topics, how a hitherto unknown cause becomes known to human observers and what method of investigation enables the observer *to discover* the particular cause of an individual effect in a specific science.[5] Greater attention was paid as well to the task of trying *to produce* certain natural effects because some investigators were now convinced that if they could reproduce the relevant natural phenomena, they would learn how the same effects were brought about by nature.

How did these three changes occur? Were they perhaps made possible by the same concepts that were being called into question by modern critics of Aristotelianism? Did Aristotle's account of the four causes assist such critics in articulating new conceptions of scientific explanation despite the fact that they showed little interest in basing their explanations on his concept of substance? The gap between the new and old conceptions of scientific explanation was large indeed. Still, the old causal concepts continued to be applied in the new natural philosophies, although such applications often occurred in contexts where Aristotelian assumptions concerning substance and nature were supplanted by rival assumptions about material bodies and a mechanistic nature. The continuing usefulness of the account of the four causes stemmed in part from earlier revisions that medieval Islamic and Christian interpreters such as Avicenna (Ibn Sīnā), Albertus Magnus, and Thomas Aquinas had made in its scope. These medieval interpreters had extended their discussions to encompass the causal powers of a supernatural God and the occult powers of the stars, planets, and ordinary bodies on Earth – any of which could act together with Aristotle's four causes of a natural substance.[6] Even the seventeenth-century thinkers who rejected explanations in terms of the four causes seem to have profited from such revised versions

---

pp. 92–169. Concerning the juxtaposition of traditional and innovative explanations in medicine, see A. Wear, R. K. French, and I. M. Lonie, eds., *The Medical Renaissance of the Sixteenth Century* (Cambridge: Cambridge University Press, 1985); and Harold J. Cook, "The New Philosophy and Medicine in Seventeenth-Century England," in *Reappraisals of the Scientific Revolution*, ed. David C. Lindberg and Robert S. Westman (Cambridge: Cambridge University Press, 1990), pp. 397–436.

[5] See the last section of this chapter, "Active and Passive Principles as a Model for Cause and Effect," which considers two classic cases that together illustrate the third notable change in early modern treatments of causality.

[6] Besides the medieval Aristotelians who made such revisions, the Renaissance Platonists also contributed significantly to the elaboration of scientific explanations that invoked the occult qualities of terrestrial and celestial bodies. A good account of the relationship between the medieval Aristotelian and Renaissance Platonist treatments of occult qualities is given in Brian P. Copenhaver, "Natural Magic, Hermetism, and Occultism in Early Modern Science," in Lindberg and Westman, eds., *Reappraisals of the Scientific Revolution*, pp. 261–301.

of Aristotle's account because these versions helped them to articulate new types of explanations.

The narrative that follows does not try to survey every new type of explanation that emerged in early modern science. Rather, it describes a limited number of historical cases to formulate two theses about how laws of nature and material efficient causes became central features of modern scientific explanations and why formal causes were rejected. The first thesis argues that *Aristotle's conception of causal explanation – while in many ways incompatible with explanations based on laws of nature and material efficient causes – actually served as the source of certain definitive features of this modern conception of scientific explanation.*[7] The second thesis suggests that *the decline of explanations in terms of the four causes occurred not because the new conception of scientific explanation was shown to be rationally superior to Aristotle's conception but because the latter had been seriously weakened by the efforts of its early modern defenders to rehabilitate it.*[8]

CAUSALITY IN THE ARISTOTELIAN TRADITION

Aristotle's account of causality provided early modern thinkers who had learned it in their university training – usually from textbooks but sometimes from the original works of Aristotle or his Arabic and Latin commentators – with a philosophical vocabulary whose concepts guided their expectations about the kinds of causes that were appropriate to identify in scientific explanations.[9] According to his account, the four causes – matter, form, efficient

---

[7] This chapter thus takes issue with one of the claims made by John R. Milton in his "Laws of Nature," in *The Cambridge History of Seventeenth-Century Philosophy*, ed. Daniel Garber and Michael Ayers, 2 vols. (Cambridge: Cambridge University Press, 1998), 1: 680–701. The present study disagrees with Milton's assertion that (p. 684): "The fundamental reason why no clear well-defined notion of a law of nature had emerged by the end of the sixteenth century is that there was no room for any such idea within the inherited ... systems of Aristotelian physics and epicyclic astronomy, whether geocentric or heliocentric.... What was still lacking was a new kind of natural philosophy, which could serve as a satisfactory replacement for scholastic Aristotelianism." By contrast, I argue in what follows that the new natural philosophies of Boyle and Newton – including their respective treatments of laws of nature – crucially relied on several important Aristotelian precedents for their conceptualization. Therefore, although it is true that their notions of laws of nature were incompatible with Aristotelian substance theory and causal theory, their notions were actually conceived in particular terms borrowed from these two theories.

[8] See the section of this chapter on "Intrinsic versus Extrinsic Efficient Causes among the Aristotelian Reformers."

[9] Numerous published editions and translations of Aristotle's works, commentaries on them, and textbooks summarizing them were available for use in the early modern universities. For surveys of these various publications, see Charles B. Schmitt, *Aristotle and the Renaissance* (Cambridge, Mass.: Harvard University Press, 1983); and F. Edward Cranz, *A Bibliography of Aristotle Editions, 1501–1600* (Bibliotheca Bibliographica Aureliana, 38) (Baden-Baden: Verlag Valentin Koerner, 1971). Concerning the use of different kinds of Aristotelian texts in university education, see L. W. B. Brockliss, *French Higher Education in the Seventeenth and Eighteenth Centuries* (Oxford: Oxford University Press, 1987), esp. pp. 337–443; and Brian P. Copenhaver and Charles B. Schmitt, *Renaissance Philosophy* (Oxford: Oxford University Press, 1992), pp. 60–126. On the teaching of some of Aristotle's works in the

cause, and final cause – are the four kinds of causes whose combined presence or action is required for the coming-to-be of a substance and whose combined presence or action best explains the motions or changes that a substance undergoes. Aristotle had developed this account in his philosophical investigations of what is the basic unit of existence in the natural world, namely, what is a primary substance. In medieval and Renaissance textbooks, Aristotle's account was therefore commonly taught together with his definition of a primary substance. These textbooks defined a primary substance as an individual organism, such as a particular plant (e.g., this oak tree) or a particular animal (e.g., this man named "Peter").[10] They also taught that earth, fire, air, and water – the four elements – are primary substances. This definition of what counts as a primary, or individual, substance was crucial to Aristotelian natural philosophers because one of their aims was to identify and classify the basic units of existence in the natural world.[11] Once they had identified these substances, their other aim was to explain how each individual substance came to exist and to possess its characteristic properties. Here they utilized Aristotle's four causal concepts to provide scientific explanations, for instance, of how an acorn, from which an oak tree is observed to develop, comes to exist and why both the acorn and the oak tree possess characteristic observable properties that make them stages in the growth of a *single* organism, which counts as an individual substance. These natural philosophers also extended their inquiries to study the species consisting of *all* oak trees by observing and describing the common properties of the individual oak

---

secondary schools, see Paul F. Grendler, *Schooling in Renaissance Italy: Literacy and Learning, 1300–1600* (The Johns Hopkins University Studies in Historical and Political Science, 107) (Baltimore: Johns Hopkins University Press, 1989), pp. 203, 268–71.

[10] See, for instance, the textbook account of substance in Gregor Reisch, *Margarita philosophica*, 2.5, 3rd Basel ed., expanded by Oronce Finé (Basel: S. H. Petri, 1583), pp. 135–6. Despite the popularity of textbooks such as the *Margarita philosophica*, it is important to remember that university teachers and students also had access to highly sophisticated treatments of Aristotle's account of substance. One of the most important of these for seventeenth-century readers was Francisco Suárez, *Disputationes metaphysicae* (Salamanca: Joannes and Andreas Renault, 1597), esp. Disputations 32–4.

[11] Twentieth-century scholars of Aristotle's concept of *ousia*, or substance, point out that his definition of what is a primary substance underwent significant changes between his expositions of the concept in the *Categories* and in the *Physics* or *Metaphysics*. They usually agree that, in the earlier work, Aristotle referred to particular objects, such as a certain man or a certain horse, as primary substances. However, they disagree about the extent to which he revised this concept in the development of his physics and metaphysics. Michael J. Loux sees the *Categories* account of primary substance as already anticipating Aristotle's later treatment of it as the essence or universal instantiated in a particular object. But Sarah Waterlow and Jonathan Lear, respectively, interpret Aristotle as firmly holding that a primary substance is an individual object, such as a certain man or horse, in the *Categories*. They argue that he changed his definition of this concept only in the *Metaphysics*. See Michael J. Loux, *Primary Ousia* (Ithaca, N.Y.: Cornell University Press, 1991), pp. 2–17; Sarah Waterlow, *Nature, Change, and Agency in Aristotle's Physics* (Oxford: Oxford University Press, 1982), pp. 41–2, 48–54, 87–92; Jonathan Lear, *Aristotle: The Desire to Understand* (Cambridge: Cambridge University Press, 1988), pp. 257–9, 265–73. For a general introduction to Aristotle's thought, especially his natural philosophy, see G. E. R. Lloyd, *Aristotle: The Growth and Structure of His Thought* (Cambridge: Cambridge University Press, 1968).

Scientific Explanation 75

trees that compose the species.[12] They explained the characteristic properties of all members of the species oak tree in terms comparable to the four causes that explained each individual oak tree.

In his *Physics* and *Metaphysics*, Aristotle had defined the four causes using various examples drawn from art as well as nature. He had often referred to the better-known examples of human actions and artifacts (artificial substances such as statues and household utensils) to indicate the meaning of a causal concept when the relevant meaning might otherwise have been more obscure if he had used only natural substances to illustrate it. Hence a typical statement of Aristotle's definitions of matter, form, efficient cause, and final cause asserted:

> We call a cause (1) that from which (as immanent material) a thing comes into being, e.g. the bronze of the statue and the silver of the saucer, and the classes which include these. (2) The form or pattern, i.e. the formula of the essence, and the classes which include this (e.g. the ratio 2:1 and number in general are causes of the octave) and the parts of the formula. (3) That from which the change or the freedom from change first begins, e.g. the man who has deliberated is a cause, and the father a cause of the child, and in general the maker a cause of the thing made. . . . (4) The end, i.e. that for the sake of which a thing is, e.g. health is the cause of walking. For why does one walk? We say 'in order that one may be healthy,' and in speaking thus we think we have given the cause.[13]

The writers of medieval and Renaissance textbooks confronted a difficult task when they tried to paraphrase and clarify such passages from Aristotle for their student readers. But this same difficulty had also been experienced even by the sophisticated twelfth- and thirteenth-century Arabic and Latin commentators on his writings because the task of specifying the precise matter, form, efficient cause, and final cause of an individual substance was deceptively simple.[14] The matter of a statue is the bronze from which it is shaped. The form of a statue is its shape, which is produced by a human agent, the sculptor. The sculptor, in shaping the statue, acts as its efficient cause. The final cause, or end, of the statue is its completion in the finished form that was

---

[12] Reisch, *Margarita philosophica*, 2.5, pp. 135–6.
[13] Aristotle, *Metaphysics*, 5.2, trans. W. D. Ross, in *The Complete Works of Aristotle*, ed. Jonathan Barnes, 2 vols. (Princeton, N.J.: Princeton University Press, 1984), 2: 1600 (1013a24–35).
[14] Two commentators whose writings continued to shape Aristotelian discussions of causality during the period 1500–1700 were the twelfth-century Arabic commentator Averroes and the thirteenth-century Latin commentator Thomas Aquinas. Early modern Aristotelians who were also influential in defining the terms of these discussions included Francisco Suárez, Julius Pacius, Jacopo Zabarella, and various other teachers affiliated with either the Jesuit College in Coimbra or the University of Padua. See, for example, Collegium Conimbricense, *Commentarii . . . in octo libros Physicorum*, 2 vols. (Lyons: Buysson, 1594); Collegium Conimbricense, *Commentarii . . . in duos libros De generatione et corruptione* (Lyons: Buysson, 1600); Francisco Suárez, *Disputationes metaphysicae*; Julius Pacius, *Aristotelis Naturalis auscultationis libri VIII* [1596] (repr., Frankfurt: Minerva, 1964); *Aristotelis Peripateticorum principis organum* [1597] (repr., Hildesheim: Geory Olms, 1967). See also the last section of this chapter, where several works of Jacopo Zabarella are examined.

originally envisioned by the sculptor. To explain the causes of an individual *natural* substance, textbook writers and philosophical commentators alike usually transferred the specification of the four causes from the better-known artifact examples to the case of a natural substance such as an oak tree. However, the exact analogy between these cases could not easily be drawn. An oak tree is a composite of matter and form that together constitute an individual substance. Strictly speaking, neither the oak tree's matter nor its form can exist apart from the other prior to their union, which brings the oak tree into existence as an individual substance, and therefore it is difficult to describe the material cause and the formal cause independently of each other. Furthermore, according to Aristotle, the oak tree is a natural and not an artificial substance; hence it must possess within itself its own principle of motion or change. Because its form serves as this principle, it counts not only as the oak tree's formal cause but as its efficient cause, too. The form even does triple duty in an Aristotelian explanation of a natural substance because it also serves as the substance's final cause, or end. The purpose of the developing oak tree is to become a mature oak tree as defined by the form.

Aristotelian textbook writers and philosophical commentators thus struggled to interpret Aristotle's definitions of the four causes as well as his concept of nature, which attributed to each primary substance its own nature. Many simply quoted parts of Aristotle's statement of the concept:

> By nature the animals and their parts exist, and the plants and the simple bodies (earth, fire, air, water). . . . Each of them has within itself a principle of motion and of stationariness (in respect of place, or of growth and decrease, or by way of alteration). . . . Nature is a principle or cause of being moved and of being at rest in that to which it belongs primarily, in virtue of itself and not accidentally. . . . Things have a nature which have a principle of this kind. Each of them is a substance.[15]

Some commentators then noted Aristotle's observation that, for any *natural* substance, its form just is its principle of motion or rest, and as such its form does triple duty as the substance's formal cause, efficient cause, and final cause.[16] Of course, this last feature of the conception of the four causes, as applied to natural substances, led both puzzled schoolboys and professors to ask: Which aspects of the form are responsible for its several causal powers? What precisely is a causal power? Does causality itself have a nature or cause?

Reflections such as these, on the nature of causality itself, had stimulated investigations in both science and metaphysics from the time of Averroes (Ibn

---

[15] Aristotle, *Physics*, 2.1, trans. R. P. Hardie and R. K. Gaye, in Aristotle, *Complete Works*, ed. Barnes, 1: 329 (192b10–23, 192b33–4).

[16] See, for instance, Reisch, *Margarita philosophica*, 8.13, p. 638; and Thomas Aquinas, *De principiis naturae*, chap. 4, translated in *Aquinas on Matter and Form and the Elements*, ed. Joseph Bobik (Notre Dame, Ind.: University of Notre Dame Press, 1998), pp. 71–3. Aristotle's statements concerning the relationship among the form, the efficient cause, and the end – which gave rise to the view that these three kinds of causes coincide in natural substances – occur in his *Physics*. See Aristotle, *Complete Works*, ed. Barnes, 1: 330–1, 338 (193b6–18, 194a27–30, 198a22–30).

Rushd) and Aquinas in the twelfth and thirteenth centuries to the beginning of the seventeenth century. As natural philosophers, some Aristotelians had sought better to understand the forms of all natural substances in the belief that, if they could determine exactly the form of each substance, they would also discover its efficient cause and its final cause – thus achieving a complete causal understanding of each natural substance. Others had sharply distinguished between the form, on the one hand, and the efficient and final causes, on the other, and they had stressed the importance of identifying the respective powers of each of these kinds of causes.[17] Furthermore, as philosophers, every one of them had been committed to establishing the metaphysical requirements that must be satisfied for something to count as a cause of each kind. But despite their multifaceted research, Aristotelians by the second half of the sixteenth century had faced growing criticisms of their explanations in the specific sciences and challenges from rival metaphysical treatments of causality.

GOD AS A FINAL CAUSE AND THE EMERGENCE OF LAWS OF NATURE

The first change, which concerned the overall purpose of scientific research, is dramatically illustrated by the differences between the views of Gregor Reisch (1467–1525), author of a widely read sixteenth-century Aristotelian textbook, the *Margarita philosophica* (The Philosophical Pearl, 1503), and those of Robert Boyle (1627–1691), the well-known seventeenth-century advocate of the mechanical philosophy.[18] Reisch, who reiterated the Aristotelians' scientific goal of understanding the form of each natural substance, and Boyle, whose goal was to identify the corpuscular structures of material bodies and the lawlike regularities governing them, envisioned radically different aims for science. Still, they agreed on at least two crucial points: A natural philosopher must reserve an important place for God in the explanation of natural phenomena, and God's role in such explanations is that of an extrinsic final cause (see Garber, Chapter 2, this volume).

Reisch began from the assumption that a further distinction, besides those among the four kinds of causes, needed to be made between what he called "intrinsic causes" and "extrinsic causes."[19] Intrinsic causes are those causal

---

[17] A prominent exponent of this line of interpretation was Suárez. See Francisco Suárez, *On Efficient Causality: Metaphysical Disputations 17, 18, and 19*, trans. Alfred J. Freddoso (New Haven, Conn.: Yale University Press, 1994), 17.1, pp. 3–10.

[18] See note 10. All citations of Reisch's *Margarita philosophica* in this article refer to the 1583 Basel edition. Unlike some of the earlier Basel editions, this one contains printed page numbers, making citation easier. Although in it Reisch's text has been expanded by Oronce Finé, all the passages from it cited here remained unchanged from the 1517 Basel edition, which does not contain any additions by Finé. For an account of the various editions of Reisch's work, see John Ferguson, "The *Margarita Philosophica* of Gregorius Reisch: A Bibliography," *The Library*, 10 (1929), 194–216.

[19] Reisch, *Margarita philosophica*, 8.12–13, pp. 636–8.

powers that belong to a natural substance itself and can be exercised only by the substance. In Reisch's view, two of the four Aristotelian causes – form and matter – can only be intrinsic and not extrinsic because the union of form and matter defines an individual substance. But efficient causes and final causes can be either intrinsic or extrinsic, depending on whether the specific efficient cause or final cause is *a part of* the substance under study.[20] The case of God's action as a final cause of a natural substance would thus count as an example of an extrinsic final cause in that God's causal power does not belong to the affected substance itself yet does influence its behavior. The affected substance is of course still also influenced by its own intrinsic final cause, which is its form. Reisch portrayed God as similar to a human craftsman who possesses the idea of an artifact and whose idea then determines the goal to be achieved when the craftsman acts to produce the artifact. Citing precedents for this view in Augustine's writings, he described the forms of natural substances as divine ideas in the mind of God.[21] Each divine idea serves as the archetype of the form that intrinsically causes the characteristic properties of an individual substance. In other words, the extrinsic causal power of God concurs, or acts together, with the intrinsic causal power of the form of a substance to produce that substance.

Although Boyle rejected the sixteenth-century Aristotelians' claim that forms are the intrinsic final causes of natural substances, he did develop further their treatment of God as an extrinsic final cause. But his view of God's role was not the same as that taught by Reisch. The latter had characterized God as a final cause in that God possesses the archetype of the form of each individual substance and serves as the divine end that moves all natural substances – through their desires, inclinations, or other means – toward their individual ends.[22] Boyle, however, spoke of God as a final cause because he believed God to be the divine mind that conceives of the providential order of all natural bodies and the divine will that commands all material bodies to obey the laws of nature.[23] This account arose from his criticism of the Aristotelians for failing to address satisfactorily a question that should have been central to their causal explanations: Precisely how does a form direct a natural substance toward its individual end?[24] Still, Boyle worried that his own answer to a comparable question – how does a divine will command material bodies to obey the laws of nature? – might itself be viewed as inadequate. It might even, he feared, be conflated with the principles of the seventeenth-century Epicureans and Stoics. Indeed his description of

---

[20] Ibid.
[21] Ibid., p. 637.
[22] Ibid., pp. 637–8.
[23] Robert Boyle, *A Disquisition about the Final Causes of Natural Things* (London: Printed by H. C. for John Taylor, 1688), pp. 91–6; Boyle, *A Free Enquiry into the Vulgarly Receiv'd Notion of Nature* (London: Printed by H. Clark for John Taylor, 1686), pp. 40–3, 124–7.
[24] Boyle, *Final Causes*, pp. 87–90; Boyle, *Notion of Nature*, pp. 26–8, 44–7.

how the laws of nature regulate the motions of material bodies could easily have been mistaken for the Epicurean notion of nature, according to which mobile atoms serve as the efficient causes of the motions of ordinary bodies that are composed of these atoms. His reference to a providential natural order might, on the other hand, have been confused with the Stoics' conception of a God-like nature, which operates according to its own material and rational principles and does not rely on a Christian God for its existence.

Boyle's strategy was to steer a middle course between the modern revivers of Epicurean atomism and the modern revivers of Stoic natural philosophy.[25] His middle course avoided the extreme of ascribing too little order and purpose to the natural world, yet it also avoided the other extreme of regarding nature itself as a divinity that fashions an order so complete that there is no need for God to act as its final cause. The modern Epicureans, Boyle thought, ascribed too little order to nature by giving a dominant role to chance in their explanations of why material bodies possess certain characteristic properties.[26] Like their Hellenistic predecessors, they lacked an understanding of the overall design of the world and thus theorized about atoms as the efficient causes of natural phenomena without indicating the ends that these causes are supposed to achieve. This resulted in the inability of modern Epicureans to describe adequately either the design of individual bodies or the structure of nature as a whole. A few modern atomists, among them the Christian Epicurean Pierre Gassendi (1592–1655), did manage to avoid such inadequacies by replacing Epicurus's chance with the providence of a Christian God.[27] But this did not prevent Boyle from criticizing Gassendi's Hellenistic predecessors in his *A Disquisition about the Final Causes of Natural Things* (1688):

> There are some effects, that are so easy . . . to be produc'd, that they do not infer any knowledge or intention in their Causes; but there are others, that require such a number and concourse of conspiring Causes, and such a

---

[25] Boyle, *Final Causes*, pp. 3–4, 45–9, 100–1, 104–6; Boyle, *Notion of Nature*, pp. 64–5. In addition to steering a middle course between the Epicureans and Stoics, Boyle also adopted in some works a middle position between the Epicureans and the Cartesians. His *Disquisition about the Final Causes of Natural Things*, for example, reserved its lengthiest criticisms not for the followers of Epicurus, who held that the world's atomic constitution needed no deity to create it, but for the followers of Descartes, who refused to claim any knowledge of God's purposes. The latter group, Boyle thought, were so impressed by the gap between God's omniscience and the finite knowledge of human beings that they jeopardized their own belief in God's powers by limiting scientific explanations to efficient causes and by declining to speculate about the final causes of nature, which are God's purposes. See Boyle, *Final Causes*, pp. A3–A5; and René Descartes, *Principles of Philosophy*, 1.24–8, in *The Philosophical Writings of Descartes*, trans. John Cottingham, Robert Stoothoff, Dugald Murdoch, and Anthony Kenny, 3 vols. (Cambridge: Cambridge University Press, 1984–91), 1: 201–2.

[26] Boyle, *Final Causes*, pp. 3–4, 45–9, 160–1; Boyle, *Notion of Nature*, pp. 64–5.

[27] Two studies that analyze this and other aspects of Gassendi's revival of Epicureanism from different perspectives are: Lynn Sumida Joy, *Gassendi the Atomist, Advocate of History in an Age of Science* (Cambridge: Cambridge University Press, 1987); and Margaret J. Osler, *Divine Will and the Mechanical Philosophy: Gassendi and Descartes on Contingency and Necessity in the Created World* (Cambridge: Cambridge University Press, 1994).

continued *series* of motions or operations, that 'tis utterly improbable, they should be produced without the superintendency of a Rational Agent, Wise and Powerfull enough to range and dispose the several intervening Agent's and Instruments ... of such a remote effect. And therefore it will not follow, that if chance could produce slight contexture in a few parts of matter; we may safely conclude it able to produce so exquisit ... a Contrivance, as that of the Body of an Animal. ... There is incomparably more Art express'd in the structure of a Doggs foot, then in that of the famous Clock at *Strasburg*.[28]

Here, in comparing a dog's foot and the clock at Strasbourg, he drew an important parallel between the need for a divine artisan in the production of natural mechanisms and the need for human artisans in the construction of machines such as the famous clock. This comparison, Boyle believed, clearly revealed the inadequacy of those atomist explanations that refused to acknowledge the need for final causes in determining the structure and purpose of natural bodies.

By contrast, the modern Stoics were judged by Boyle to have made the opposite error of endorsing an overly speculative notion of nature. Their mistake was to hold that nature itself is a deity and to see divine purposes everywhere in the natural world.[29] As such, the Stoics' notion of nature left no room for the actions of a Christian God because God's causal powers would be superfluous in determining the ends of a natural order that can itself function as a unified, intelligent, and living being capable of determining its own ends. This criticism, however, did not apply unilaterally to all modern Stoics, for many of them were Christians. Perhaps the most influential was Justus Lipsius (1547–1606), the Flemish editor and popularizer of the writings of the Roman Stoic Seneca. Lipsius worked tirelessly to reconcile both Stoic physics and ethics with Christian theology, but when conflicts between them became irresolvable, he did not hesitate to uphold Christian doctrines over whatever physical or ethical principle had been taught by Seneca. As he reminded readers of his *Manuductio ad stoicam philosophiam* (Introduction to Stoic Philosophy, 1604), "No one should place the End or happiness in Nature, as the Stoics do; unless by the interpretation which I gave, namely in God."[30] Lipsius thus exploited certain readings of the ancient texts in order to enhance the credibility of Stoic views among Christian readers. A case in point was his attempt to show that no heresy had been committed when Seneca denied that God created matter. This denial was not heretical because the term "matter" did have two Stoic usages, one referring to universal, or primary, matter and another referring to particular, or secondary, matter from

---

[28] Boyle, *Final Causes*, pp. 45–7.
[29] Boyle, *Notion of Nature*, pp. 100–1, 104–6, 120–1.
[30] Justus Lipsius, *Manuductionis ad stoicam philosophiam libri iii* [1604], in Justus Lipsius, *Opera omnia*, 4 vols. (Wesel, 1675), 4: 617 ff., translated in Jason Lewis Saunders, *Justus Lipsius: The Philosophy of Renaissance Stoicism* (New York: The Liberal Arts Press, 1955), p. 55.

which individual finite bodies are formed. Lipsius argued that when Seneca denied God's creation of matter, he had been referring to primary matter, which – because it is the eternal substratum of everything that exists – is identical to God.[31] Thus, Seneca's denial of God's creation of primary matter was understandable because it would have made no sense for him to say that what is eternal requires creation or that God creates himself.

Lipsius's Stoicism and Gassendi's Epicureanism represented attempts to reconcile early modern Christian beliefs and Hellenistic pagan principles. In this they had something significant in common with Boyle's own efforts to articulate a natural theology. The difficulties Boyle encountered in doing so are instructive. His metaphor portraying the whole of nature as if it were, like a watch or clock, a single "Cosmical Mechanism," and his arguments from design, which extolled the skillful design of the human eye, the eye of a fly, or a dog's foot, could be interpreted in incompatible ways.[32] If nature were a single world machine, for instance, this might show that God – the divine artisan – had produced it for a transcendent purpose. But it could equally well show that the world is simply an eternal mechanism that needs no divine creator and has no purpose beyond the systematic functioning of its parts. Moreover, although the intricate designs of parts of animals such as the human eye might be construed as works of God, such natural designs might also be regarded as evidence that denies God's role as a final cause. The human eye, considered simply as a material mechanism, may have no purpose beyond the systematic functioning of its parts.

Not only Boyle but Lipsius and Gassendi, too, were liable to be asked: Does the fact that nature as a whole functions as a machine and that individual material bodies also function as machines require the existence of God? Would such a nature, encompassing the laws of motion as well as "the general Fabrick of the World, and the Contrivances of particular Bodies," be inconceivable without appealing to God's power as an extrinsic final cause?[33] These three thinkers clearly thought so because they believed that the very existence of laws of nature presupposed the existence of a necessary relationship between God and the laws of nature, and it was this relationship that enabled them to define *what is* a law of nature in the first place. Later in this chapter, we shall see how various attempts to define laws of nature crucially depended on the concept of God as an extrinsic final cause and on the concept of matter as an extrinsic efficient cause. But first we need to consider what several prominent sixteenth-century Aristotelians had to say in defense of the forms as *intrinsic* efficient causes. An examination of their work is relevant to our inquiry in that it suggests a compelling reason why Aristotle's account of the four causes declined in influence among the early moderns.

---

[31] Saunders, *Justus Lipsius*, pp. 166–7.
[32] Boyle, *Final Causes*, pp. 18, 44, 47–9; Boyle, *Notion of Nature*, p. 73.
[33] Boyle, *Notion of Nature*, p. 41.

INTRINSIC VERSUS EXTRINSIC EFFICIENT CAUSES AMONG THE ARISTOTELIAN REFORMERS

A century before Boyle played his part in articulating new laws of nature, the defenders of substantial forms had investigated a set of problems in the specific sciences that had forced them to recognize the difficulty of trying to explain all varieties of natural phenomena by means of intrinsic efficient causes. These Aristotelian reformers, including Agostino Nifo (ca. 1469–1538), Julius Caesar Scaliger (1484–1558), Jacopo Zabarella (1532–1589), and their successors, had kept alive certain medieval inquiries concerning the causal powers of the forms of substances (substantial forms) and the forms of the four elements (the substantial forms of earth, fire, air, and water). Their inquiries had combined the study of chemical compounds called "mixts" with philosophical analyses that were often borrowed from the medieval Arabic and Latin commentators on Aristotle's *On Generation and Corruption*.[34]

In one of its simpler versions, the problem consisted of how to explain the differences between a mixt, such as a metal alloy produced by combining two molten metals, and a mere mixture, such as a multigrain mixture composed of two sorts of grain.[35] The process of combination (or mixtion) created a single entity (a mixt) homogeneous in all its parts, although its components, such as the two metals, could still be separated from one another by further processes. By contrast, the process of composition produced, in the latter case, a multigrain entity (a mixture) that was not homogeneous in all its parts because each individual grain retained its original identity as one of the two sorts of grain composing the mixture. What explanation could be given for the generation of an apparently distinct natural substance in the first case but not in the second? Did the forms of the mixt's component metals change into a single new form, that of the metal alloy? The thirteenth-century philosopher and natural historian Albertus Magnus had investigated this and a variety of other possible mixts, which he had called "intermediates" because they shared certain properties of infusible stones as well as certain properties of fusible metals.[36] Among these intermediates were mineral salts, which he had described as naturally occurring mixts that are combined from stones and metals in the earth.

---

[34] An instructive example of this type of scientific treatise is Jacopo Zabarella's *Liber de mistione* (1590). This short work contains his survey of those medieval predecessors, including Avicenna, Averroes, Aquinas, and Scotus, whose explanations of mixts he evaluates together with Aristotle's own views on the subject. See *Liber de mistione*, in Jacopo Zabarella, *De rebus naturalibus libri XXX* [1607] (repr., Frankfurt: Minerva, 1966), cols. 451–80.

[35] Aristotle, *On Generation and Corruption*, 1.10 (327a30–328b25), discussed in Norma E. Emerton, *The Scientific Reinterpretation of Form* (Ithaca, N.Y.: Cornell University Press, 1984), p. 77. For an opposing view of medieval and early modern treatments of Aristotle's account of combination, see John E. Murdoch, "The Medieval and Renaissance Tradition of *Minima Naturalia*," in *Late Medieval and Early Modern Corpuscular Matter Theories*, ed. Christoph Lüthy, John E. Murdoch, and William R. Newman (Leiden: Brill, 2001), pp. 91–131.

[36] Emerton, *Scientific Reinterpretation of Form*, pp. 77–8.

Albertus Magnus and other medieval writers, including Avicenna, Averroes, Roger Bacon, Thomas Aquinas, John Duns Scotus, and Albert of Saxony, had also tackled philosophically more complex versions of the problem of mixts. Hence it was not surprising that a sixteenth-century defender of substantial forms such as Julius Caesar Scaliger interpreted the problem of mixts as one of how to describe the relationship between genuine substances and mixts of the four Aristotelian elements. He argued that a mixt somehow acquires a new form that differs from the forms of the elements constituting the mixt. In Scaliger's account, which owed its main idea to the Islamic philosopher Averroes (1126–1198), the new dominant form takes over the organization of the mixt from the weakened forms of the constituent elements:

> The nature of the elements [is understood] not only with respect to themselves but also with respect to their mixts. With respect to itself, it [each element] has a form which it gives up in order to obtain a nobler form [in the mixt]. Thus neither do the forms [of the elements combined in the mixt] remain, nor are the qualities deprived of their forms, but in a different way they are accommodated to the substance of the mixt. For a new generation it is necessary that the forms of the parts, subdued by one another's qualities, should have laid aside the original inflexibility of nature under the dominion of one [form] that is more powerful.[37]

Scaliger's contemporaries also studied Aristotle's use of forms to accomplish a second task in his physics, that of specifying the differences among the four kinds of change: generation or corruption, alteration, growth or diminution, and local motion. Aristotle had specified their differences by asking, for each kind of change, whether there is a single underlying substance that is the subject of the change.[38] Because he had presupposed that the basic unit of existence is the substance that results from generation, he had required that explanations of the three other kinds of natural change should refer first and foremost to the form of the individual substance underlying the change, not to the other substances whose proximity in time and place one might otherwise think would causally affect the changing individual substance. Nifo, Scaliger, and Zabarella tried to enhance this account of change by analyzing what exactly happens to the parts of a substance and the degrees or qualities of a form during the four kinds of change. Nifo, for instance, followed Averroes by treating such a question in terms of the minimal parts of a substance and the minimal degrees, or parts, of a form. In his exposition of Aristotle's *Physics*, he wrote:

---

[37] Julius Caesar Scaliger, *Exotericarum exercitationum liber XV de subtilitate ad Hieronymum Cardanum* [1557], ex. 16, pp. 34–5, translated in Emerton, *Scientific Reinterpretation of Form*, p. 83. I have added the second and fourth bracketed insertions to clarify this quotation.

[38] Aristotle, *On Generation and Corruption*, 1.4, trans. H. H. Joachim, in Aristotle, *Complete Works*, ed. Barnes, 1: 522–3 (319b8–320a1).

Averroes held that growth, generation, and alteration take place by means of minima.... He held that there are maximum and minimum degrees of any naturally intensible form.... The agent can alter the first minimum part of the subject by one degree of quality, then by means of the first it will alter the second to one degree; while it alters the second to one degree, it will push the first up to two degrees; [etc.].... By flux should be understood the reception by which the subject successively receives the form ... and by the flowing form Averroes understands the form which is received by this successive reception.[39]

Despite some apparent resemblances between Nifo's theory of the minimal parts of a substance and the corpuscular explanations advanced by later physicists, the sixteenth-century defenders of the forms were careful to qualify their statements about *minima*. They repeatedly pointed out that minimal parts of a substance and minimal degrees, or parts, of a form should not be confused with the atoms of Democritus or Epicurus. The minimal parts of a substance – unlike the atoms, whose overall structure and motions determine the properties of a material body – are always dependent, for their identity and existence, on the whole substance. Moreover, the minimal parts of a form always depend, for their causal powers, on the whole form, which alone can act as the intrinsic cause of the essential properties of a substance.

During the first half of the seventeenth century, however, a different group of defenders of forms explored just how far explanations based on forms could be accommodated to explanations based on the extrinsic causal powers of elements and atoms. Francis Bacon (1561–1626) and Daniel Sennert (1572–1637), two important members of this group, each pieced together a revised account of the forms from non-Aristotelian as well as Aristotelian sources. Sennert's *Tractatus de consensu et dissensu Galenicorum et Peripateticorum cum Chymicis* (Treatise on the Agreement and Disagreement of the Galenists and Aristotelians with the Chemists, 1619), as its title indicates, aimed to survey the agreements and disagreements concerning chemical phenomena among three important natural philosophical traditions. The principles of Galenic medicine, Aristotelian physics, and Paracelsian alchemy were selectively applied by Sennert to address the problem of mixts and various other problems involving chemical phenomena. Like Nifo and Scaliger, he affirmed the emergence in a mixt of a dominant form whose intrinsic causal power determines the essential properties of the mixt, and he denied that, in the absence of a dominant form, any configuration of

---

[39] Agostino Nifo, *Expositio super octo Aristotelis Stagiritae libros de physico auditu: Averrois ... in eosdem libros proemium ac commentaria* [1552], fols. 96v, 97v, 112r, 213r, translated in Emerton, *Scientific Reinterpretation of Form*, pp. 93, 101.

atoms or extrinsic efficient causes could produce the new substance of the mixt.[40]

Bacon's *Novum organum* (New Organon, 1620) was published just one year after Sennert's *Tractatus*, and in it he recommended a noteworthy revision of the concept of form:

> When I speak of Forms, I mean nothing more than those laws [*leges*] and determinations of absolute actuality [*actus puri*], which govern and constitute [*ordinant et constituunt*] any simple nature, as heat, light, weight, in every kind of matter and subject that is susceptible of them. Thus the Form of Heat and the Form of Light is the same thing as the Law of Heat or the Law of Light.[41]

Whether this recommendation was intended by Bacon to be as radical as it now seems is difficult to say because in making it he appears to have assumed that there is no serious inconsistency in claiming that a form, which he treated as an intrinsic efficient cause, can be redefined as a scientific law, which he treated as a regularity in the behavior of extrinsic efficient causes. A Baconian form produces a particular property in a body by means of the latent configuration and latent process of the parts of the body; hence the form's causal powers are derived from the structure and motions of the body's parts.[42] Yet, in Bacon's account, these parts only acquire the capacity to bring about latent configurations and latent processes from the forms themselves – of heat, light, and so forth – which are embodied in the parts. Thus, his concept of form confused the powers of extrinsic and intrinsic efficient causes.

Except for Bacon, defenders of forms, such as Nifo, Scaliger, Zabarella, and Sennert, did not usually compete with the early modern corpuscular physicists in explaining natural phenomena *by means of natural laws*. The corpuscular physicists were advocates of extrinsic efficient causes, whereas the Aristotelian reformers were trying to preserve the credibility of substantial forms as intrinsic efficient causes. Because the two groups were engaged for the most part in incompatible explanatory enterprises, it would be misleading to attribute the declining influence of substantial forms theories to the rational superiority of the corpuscular theories. A corpuscular theorist such as René Descartes (1596–1650), for instance, *began* by assuming that

---

[40] Daniel Sennert, *Tractatus de consensu et dissensu Galenicorum et Peripateticorum cum Chymicis*, in Daniel Sennert, *Opera omnia*, 4 vols. (Lyons: Hugetan and Ravaud, 1650), 3: 779–80, translated in Emerton, *Scientific Reinterpretation of Form*, p. 119.

[41] Francis Bacon, *Novum organum*, in Francis Bacon, *The Works of Francis Bacon*, ed. James Spedding, Robert L. Ellis, and Douglas D. Heath, 14 vols. (London: Longmans, 1857–74), 4: 146; 1: 257–8, cited in Antonio Pérez-Ramos, "Bacon's Forms and the Maker's Knowledge Tradition," in *The Cambridge Companion to Bacon*, ed. Markku Peltonen (Cambridge: Cambridge University Press, 1996), p. 107.

[42] Bacon, *Novum organum*, in Bacon, *Works*, 4: 122–6, 151–8; 1: 230–4.

his theory of matter and its constituent parts was preferable to explanations of a body's observable properties that claimed that these properties were brought about by the body's substantial form. He then devised an argument to show that a substantial form is simply a disposition constituted by the shapes, sizes, positions, and movements of the material parts of a body that enables the body to cause motions in the nerves of a human observer. Accordingly, a substantial form does not exist as anything above and beyond the constituent material parts of a body. But this argument against substantial forms presupposed the plausibility of Descartes' theory of matter and could only impugn the existence of substantial forms from a standpoint that had already excluded them from its account of bodies.[43] Why, then, did explanations in terms of intrinsic efficient causes suffer a decline during the period? Although this question needs further investigation, one important reason for the decline may have been the Aristotelians' growing awareness of a serious contradiction within their own substance theory.

The introduction of various concepts of *minima* put defenders of forms in the position of having to affirm contradictory claims about substantial forms. On the one hand, the whole substantial form itself was said *to determine completely* the identity and causal powers of the parts (or degrees, or qualities) of the form. This was implied by the form's status as an intrinsic efficient cause. On the other hand, it was clear that more sophisticated analyses of motion and change in terms of a form's parts were badly needed to supplement traditional Aristotelian explanations of even the simplest chemical phenomena. Merely invoking the intrinsic causal powers of a substantial form could not begin to explain, for example, how a metal alloy acquires its characteristic properties. The Aristotelians who theorized about the form's minimal parts thus treated the parts themselves as possessing causal powers that could act independently of the whole form when they argued that such minimal parts successively brought the whole form into existence. Therefore, these theorists were simultaneously committed to the view that the causal powers of the form's minimal parts *cumulatively determine* the identity and powers of the whole form and to the view that the whole form is what fully determines the identity and causal powers of its minimal parts. From first principles concerning substantial forms, two contradictory claims were being asserted. These contradictory claims revealed a serious weakness in sixteenth-century Aristotelian substance theory and its seventeenth-century extrapolations: the weakness that several of the most promising updated versions of the theory could not be further developed without jeopardizing its logical coherence.

---

[43] Further discussion of Descartes' argument is given in the next section of this chapter and in note 49.

INTRINSIC VERSUS EXTRINSIC EFFICIENT CAUSES AMONG THE CORPUSCULAR PHYSICISTS

The second notable change in early modern treatments of scientific explanations involved the replacement of long-standing Aristotelian explanations in the specific sciences by new explanations that described how the changes and motions exhibited by ordinary bodies are caused by the matter, elements, or atoms constituting those bodies. Such explanations invoking the causal powers of matter, elements, or atoms were often preferred for their simplicity and clarity by scientific innovators who objected to the incoherence or obscurity of the Aristotelian concept of form. But even these apparently simpler kinds of explanations required careful development because if matter, elements, or atoms can serve as efficient causes, then the character of their relationship to the laws of nature needs to be spelled out. How should one characterize the law-governed behavior itself of matter, elements, or atoms? Here is where the continuing reference to God as a final cause suggested a meaningful way to define exactly what a law of nature is and what matter does when it serves as an efficient cause.

God's role was not simply to create the providential natural order governed by laws of nature but also to design the constituent atoms or parts of each material body and to endow them with the relevant quantities of motion that would enable them to serve as the efficient causes of each body's changing states. An atom or part possessing a certain quantity of motion is an *extrinsic* efficient cause because it serves – through impact – as a cause of the motion of atoms or parts external to itself and because it is also a cause of the overall configuration of the other atoms or parts that, together with itself, compose a larger body. This overall configuration and the sum total of the motions of the constituent atoms or parts fully determine all the properties possessed by any material body. Hence there can be within such a composite body no intrinsic efficient cause such as an Aristotelian form because the body's constitution is fully determined not by the substantial form of the whole body but rather by its atoms or parts, which are extrinsic efficient causes.

This sort of explanation of the properties of ordinary bodies resembled the type of explanation advanced by the ancient Greek and Hellenistic atomists, who had also anticipated another seventeenth-century innovation, the distinction between the primary and secondary qualities of bodies.[44] Although the early atomists had not fully explored the epistemological issues raised by such a distinction, they had offered accounts of how the perceived qualities

---

[44] Discussions of a comparable distinction are found, for instance, in Epicurus, "Letter to Herodotus," in Diogenes Laertius, *Lives of Eminent Philosophers*, Loeb Classical Library nos. 184–5, trans. R. D. Hicks, 2 vols. (Cambridge, Mass.: Harvard University Press, 1925), 2: 10.48–55, esp. 54–5, pp. 576–85; and Titus Lucretius Carus, *On the Nature of the Universe [De rerum natura]*, trans. R. E. Latham and rev. John Godwin (London: Penguin Books, 1994), bk. 2, ll. 333–477, 730–990, pp. 46–9, 55–62; bk. 4, ll. 24–263, 523–718, pp. 95–101, 108–13.

of ordinary bodies are caused by the properties of the atoms that compose those bodies. Specifically, the sizes, shapes, and motions of the atoms cause human observers to perceive the particular color, taste, smell, touch, and sound of each ordinary body. Modern philosophers, including the rationalist Descartes and the empiricist John Locke (1632–1704), developed causal theories of sense perception similar in certain limited respects to the accounts of the early atomists.

Locke explicitly defined the terms "primary quality" and "secondary quality" in his *Essay Concerning Human Understanding* (1690), and the *Essay*'s multiple definitions of the primary–secondary quality distinction reflected the views of many of his scientific contemporaries. Although he was openly skeptical about the natural philosophers' ability to give satisfactory corpuscular explanations of what he called the "real essence" of each kind of body, he nonetheless endorsed their corpuscular explanations of the secondary qualities of bodies. According to Locke, primary qualities – solidity, extension, figure, number, and motion or rest – are those properties belonging to both atoms and composite bodies that can never be taken away from an atom or a composite body by any process of division or destruction.[45] By its very nature, any material body – whether it is a single atom or a body composed of many atoms – will always possess some solidity, extension, figure, number, and motion or rest. Locke further defined primary qualities as properties that exist in a body whether or not they are perceived by human observers.[46] Secondary qualities, by contrast, are those properties, such as color, taste, smell, touch, and sound, that are commonly thought to belong to composite bodies but do not really exist in those bodies independently of their being perceived by human observers. Such secondary qualities were regarded by him as mere sense perceptions caused by the interaction between "the Bulk, Figure, Texture, or Motion of some of the insensible parts" of a human observer's sense organs and the primary qualities of the atoms or parts of the observed body.[47] He speculated that God had originally endowed material bodies with the primary qualities that enabled them to exercise these causal powers.[48] The effects produced by a given body's causal powers could include not only changes in the primary qualities of other bodies but also a human observer's perception of the secondary qualities attributed to the given body.

Locke's predecessor and philosophical rival Descartes had offered an account of the properties of material substance that may seem to agree wholly with this distinction between primary and secondary qualities. Descartes, too, had focused on the causal relationships between, on the one hand, the qualities of light, color, smell, taste, sound, and heat and cold, and, on the

---

[45] John Locke, *An Essay Concerning Human Understanding*, 2.8.9 (4th rev. ed., London, 1700), ed. Peter H. Nidditch (Oxford: Oxford University Press, 1975), p. 135.
[46] Ibid., 2.8.23, p. 141.
[47] Ibid., 2.8.24, p. 141.
[48] Ibid., 2.8.23, p. 140; 2.23.12–13, pp. 302–4.

other hand, the parts of a material substance that produce the sensation of these qualities in human beings. He had even suggested that the existence of Aristotelian substantial forms could be undermined by equating them with such qualities, whose existence as anything beyond the mere effects of material particles in motion could be shown to be doubtful. Descartes' *Principia philosophiae* (Principles of Philosophy, 1644) thus combined a version of the primary–secondary quality distinction with an argument against substantial forms:

> Now we understand very well how the different size, shape and motion of the particles of one body can produce various local motions in another body. But there is no way of understanding how these same attributes (size, shape and motion) can produce something else whose nature is quite different from their own – like the substantial forms and real qualities which many <philosophers> suppose to inhere in things; and we cannot understand how these qualities or forms could have the power subsequently to produce local motions in other bodies. Not only is all this unintelligible, but we know that the nature of our soul is such that different local motions are quite sufficient to produce all the sensations in the soul. . . . In view of all this we have every reason to conclude that the properties in external objects to which we apply the terms light, colour, smell, taste, sound, heat and cold – as well as other tactile qualities and even what are called 'substantial forms' – are . . . simply various dispositions in those objects [in the shapes, sizes, positions and movements of their parts] which make them able to set up various kinds of motions in our nerves <which are required to produce all the various sensations in our soul>.[49]

Descartes' *Principia* had nonetheless tempered its account of the distinction between what Locke later called "primary and secondary qualities" by prominently featuring an additional Cartesian distinction among the attributes, modes, and qualities of a substance:

> We employ the term *mode* when we are thinking of a substance as being affected or modified; when the modification enables the substance to be designated as a substance of such and such a kind, we use the term *quality*;

---

[49] René Descartes, *Principles of Philosophy* (Cottingham trans.), 4.198, p. 285. See also René Descartes, *Oeuvres de Descartes*, ed. Charles Adam and Paul Tannery, 12 vols., rev. ed. (Paris: J. Vrin/CNRS, 1964–76), 8A (Latin text), 4.198, pp. 322–3. The argument that Descartes articulated in this passage is notable because he does attempt to show that if body is extension alone, then body must exclude the substantial forms and real qualities that the scholastic Aristotelians attributed to bodies. Substantial forms and real qualities are excluded because their existence in bodies is no more real than the existence of secondary qualities. Both a substantial form and a secondary quality are mere effects on the human observer caused by an observed body's modes of extension. In this argument, Descartes does make the sort of case against substantial forms that some historians of philosophy find lacking in his work. Concerning the absence of such arguments in his writings, see Daniel Garber, *Descartes' Metaphysical Physics* (Chicago: University of Chicago Press, 1992), p. 110.

and finally, when we are simply thinking in a more general way of what is in a substance, we use the term *attribute*.⁵⁰

Each substance has one principal property which constitutes its nature and essence, and to which all its other properties are referred. Thus extension in length, breadth and depth constitutes the nature of corporeal substance.... Everything else which can be attributed to body presupposes extension, and is merely a mode of an extended thing.⁵¹

In making this additional distinction, Descartes had endorsed a substance metaphysics that explained the nature of material substance *in general* by referring to what he held to be its one essential property, extension (see Chapter 2, this volume). His substance metaphysics further explained the nature of any *particular kind* of material body by referring to the modifications of extension – which he had called "qualities" – that typically occur in a particular kind of body. The additional distinction also showed that Cartesian explanations do not admit the existence of indivisible units of matter. Indeed, Descartes, in his *Principia*, had denied the possibility of atoms of matter!⁵² Hence he had described the light, color, smell, taste, sound, and heat and cold of a particular body not as effects caused by atoms but as effects of the modifications of extension occurring in the body and in the human observer who perceives the body. Such modifications – called "modes" – included "all shapes, the positions of parts and the motions of the parts" of a body.⁵³ Nevertheless, these shapes, sizes, positions, and motions of a body's parts were fundamentally different from Locke's primary qualities because Descartes had not treated them as properties themselves but rather as modifications of the one essential property of material substance, namely, extension.

Here a major question must be raised about the various early modern conceptions of scientific explanation that appealed to extrinsic efficient causes: Do the *laws of nature*, which not only govern ordinary bodies but also govern the constituent atoms of ordinary bodies and even the modes of Cartesian extension, really provide explanations of these bodies' changing states? Innovators in mechanics and the science of motion, from Galileo and Descartes to Christiaan Huygens (1629–1695) and Robert Hooke (1635–1703), saw nothing absurd about attributing to brute matter – that is, nonhuman and even nonliving matter – the *capacity to obey laws*.⁵⁴ The fact that brute matter

---

⁵⁰ Descartes, *Principles of Philosophy* (Cottingham trans.), 1.56, p. 211.
⁵¹ Ibid., 1.53, p. 210.
⁵² Ibid., 2.20, p. 231.
⁵³ Ibid., 1.65, p. 216.
⁵⁴ Galileo Galilei, *Dialogue Concerning the Two Chief World Systems – Ptolemaic and Copernican*, trans. Stillman Drake (Berkeley: University of California Press, 1953), pp. 20–1, 222–9; Galileo, *Two New Sciences*, trans. Stillman Drake (Madison: University of Wisconsin Press, 1974), pp. 225, 232–4; René Descartes, *Principles of Philosophy*, trans. Valentine Rodger Miller and Reese P. Miller (Dordrecht: Kluwer, 1991), 2.36–53, pp. 57–69. On Huygens, see E. J. Dijksterhuis, *The Mechanization of the World Picture*, trans. C. Dikshoorn (Princeton, N.J.: Princeton University Press, 1961), pp. 373–6,

possesses neither the intelligence to understand what a law means nor the will to obey it ceased to be, by the second half of the seventeenth century, a decisive reason for corpuscular theorists to doubt that matter can obey laws. Of course, a law of motion, when applied to material objects such as moving billiard balls, might still be interpreted by some physicists as a mere metaphor because billiard balls, unlike human beings, do not have the capacity to comprehend a superior's command to obey a law. In this metaphorical sense, a law of motion might serve as part of an explanation that defines the cause of a billiard ball's motion in terms of coordinated intentional actions performed by both God and the billiard ball: God conceives and commands a particular law, and subsequently the billiard ball is motivated to conform to the law.

However, in the physics of Descartes and Huygens, these metaphorical descriptions were usually accompanied by definitions that stipulated new literal meanings for terms such as "law of motion."[55] Henceforth, laws of motion could be defined *literally* as regularities in the movements and dispositions of bodies such as billiard balls, and these bodies were conceived as inanimate things whose constituent matter had been created by God such that their matter could be affected only by *extrinsic* efficient causes. Accordingly, matter's obedience to a law of motion could not involve its having the *intrinsic* motivations of a law-abiding, thinking being. Its obedience could only refer to the regularity it exhibits in its motions, which are produced by extrinsic efficient causes such as the impact of one billiard ball on another. The literal meaning of a scientific law was thus becoming – for many innovators in the study of motion and mechanics – preferable to its metaphorical meaning. Boyle aptly expressed this preference, although he did so while reminding readers of the conceptual ties between the two sorts of meanings:

> Each part of this great Engine, the *World*, should without either Intention or Knowledge, as regularly and constantly Act towards the attainment of the respective Ends which he [God] design'd them for, as if themselves really understood, and industriously prosecuted, those Ends. Just as in a well made Clock, the Spring, the Wheels, the Ballance, and the other parts, *tho'* each of them Act according to the Impulses it receives . . . by the other pieces of the Engine, without knowing what the Neighbouring Parts, or what themselves do . . . ; yet . . . they would not move more conveniently, nor better perform the Functions of a Clock, if they *knew* that they were to make the *Index* truly mark the Hours, and *intended* to make it do so.[56]

458–63. On Hooke, see I. Bernard Cohen, *The Birth of a New Physics*, 2nd ed. (New York: W. W. Norton, 1985), pp. 150–1, 218–21.

[55] Descartes, *Principles of Philosophy* (Miller trans.), 2.36–53, pp. 57–69; Dijksterhuis, *Mechanization of the World Picture*, pp. 373–6.

[56] Boyle, *Final Causes*, pp. 91–2. Boyle's italics.

Corpuscular physicists such as Boyle came to rely less and less on the assumption that natural laws primarily refer – or even, as some earlier thinkers had held, can only refer – to the intentional actions of law-abiding, thinking beings. Although many of them continued to emphasize God's relationship to matter when defining what counts as a law of nature, they now also defined such laws in terms of lawlike regularities, according to which the observable features of any ordinary body are explained as the effects of the organization and motions of the body's constituent atoms. Why did they take this decisive step that wholly changed the meaning of their concept of a law of nature? Their Aristotelian background supplied two important reasons for taking this step.

The first reason was that neither the seventeenth-century corpuscular physicists nor the Aristotelian reformers who had preceded them could have progressed much further in developing their respective kinds of scientific explanation if they had continued to accept, as a basic commitment, the distinction between intrinsic and extrinsic efficient causes. Explanations in the specific sciences increasingly required the operation of efficient causes that could no longer be clearly characterized as purely intrinsic or purely extrinsic but were instead questionable combinations of both. The corpuscular physicists' willingness to give up their basic commitment to the distinction between intrinsic and extrinsic causes therefore served as a critical precondition for their own redefining of what counts as a law of nature. It freed them from having to resolve the conflict between (a) their principle that brute matter moves only by impact as an extrinsic cause when it obeys a law of motion and (b) their belief that even an inanimate body, when it is regulated by God's law, may possess the intrinsic motivations of a law-abiding, thinking being. Simply by refusing to worry about how the same body can be governed by both extrinsic and intrinsic causes, the corpuscular physicists could now envision the possibility that the concept of a law of nature governing brute matter might encompass both notions.

Another important reason for changing the meaning of the concept of a law of nature grew out of the corpuscular physicists' development of the distinction between proper causes (causes *per se*) and accidental causes (causes *per accidens*) – two modes of causation that Aristotle himself had discussed when inquiring how the four kinds of causes could be better understood by considering whether they were proper or accidental causes. Medieval and sixteenth-century commentators, from Aquinas to Francisco Suárez (1548–1617), had reiterated Aristotle's statements about these modes of causation, and they had integrated their own distinction between intrinsic and extrinsic causes into an overall account of causality that also featured proper and accidental causes.[57] But when the former distinction became difficult to

---

[57] For Aristotle's exposition of the distinction between proper and accidental causes, see his *Physics*, 2.3, in *Complete Works of Aristotle*, ed. Barnes, 1: 333 (195a 27–195b 5). An important sixteenth-century

Scientific Explanation 93

uphold in the new physics or in the chemistry of mixts, early modern natural philosophers found that they could still rely on the latter distinction between proper and accidental causes. When reconceived as the difference between active and passive principles, the latter distinction would lead them to the brink of accepting radically different ideas of cause and effect.

ACTIVE AND PASSIVE PRINCIPLES AS A MODEL FOR CAUSE AND EFFECT

What was the Aristotelian distinction between proper and accidental causes? How did reconceiving it in terms of active and passive principles help to bring about the third notable change in early modern scientific explanations, a turning away from metaphysical analyses of causality in favor of epistemological and practical inquiries into the best methods of discovering the relations between causes and effects? Aquinas, in *De principiis naturae* (On the Principles of Nature, ca. 1252), had defined the distinction between proper and accidental causes by using examples of human action or artifacts to illustrate how the distinction works in cases of both natural and artificial change:

> A cause is said to be a cause *per se* when it, precisely as such, is a cause of something. For example, a builder [precisely as such, i.e., as a builder] is the cause of a house, and the wood [precisely as such, i.e., as wood] is the matter of the bench. A cause is said to be a cause *per accidens* when it happens to be conjoined to that which is a cause *per se*, as when we say that the grammarian builds. The grammarian is said to be the cause of the building *per accidens*, i.e., not inasmuch as the grammarian is a grammarian, but inasmuch as it happens to the builder that the builder is a grammarian.[58]

Aristotle's commentators did not always agree about whether the difference between causes *per se* and causes *per accidens* should parallel the difference between intrinsic and extrinsic causes. Aquinas, for instance, held that it did not, noting that all four causes – matter, form, the efficient cause, and the final cause – can count as causes *per se*, whereas only matter and form are said to be intrinsic, and only the efficient cause and final cause are said to be extrinsic. Yet, by the early sixteenth century, these distinctions were being qualified in a variety of ways, so much so that it was possible for Reisch, in

---

treatment of this distinction was given in Suárez, *Metaphysical Disputations 17, 18, and 19*, 17.2, pp. 11–16.

[58] Thomas Aquinas, *De principiis naturae*, chap. 5, translated in Bobik, ed., *On Matter and Form*, p. 82. Regarding the distinction between the difference between causes *per se* and causes *per accidens* and the difference between intrinsic and extrinsic causes, see Ibid., chap. 3, in Bobik, ed., *On Matter and Form*, pp. 39–40. See also Saint Thomas Aquinas, *De principiis naturae*, critical Latin text, ed. John J. Pauson (Textus Philosophici Friburgenses, 2) (Fribourg: Société Philosophique, 1950), chap. 5, pp. 99–100.

his *Margarita philosophica*, to speculate about how the distinction between proper and accidental causes coincided with what he called the difference between "active and passive principles."[59]

The distinction between proper and accidental causes would remain a feature of both Aristotelian and non-Aristotelian scientific explanations well into the eighteenth century. However, the methods for identifying such causes underwent a remarkable transformation. This transformation was epistemological in that it concerned how demonstratively certain knowledge of a cause *per se* could be demarcated from the mere experience of one or more associations between a cause *per accidens* and a particular effect – experience that might otherwise be mistaken for knowledge of a cause *per se*. A good way to understand what this involved is to consider the divergent accounts of a well-known method for identifying causes, the method of *regressus* (also known by the names of its subparts, the methods of resolution and composition), given by two representative thinkers at opposite ends of the transformation. Zabarella, the sixteenth-century Aristotelian reformer, and Isaac Newton (1642–1727), chief architect of the new physics, both articulated their notions of causality by discussing the methods of resolution and composition. But Zabarella believed that he could acquire knowledge of causes *per se* by means of these methods, whereas Newton used the methods to acquire a knowledge of what he called "active and passive principles."[60] Just how their respective claims to causal knowledge diverged from one another reveals that Newton's search for active and passive principles, although it differed from the Aristotelian reformer's causal inquiries, still preserved the notion of a cause *per se* because it aimed to achieve the knowledge of at least some causes

---

[59] Reisch, *Margarita philosophica*, 2.9–10, pp. 142–4; 8.11–12, pp. 635–6. Although Reisch's account of actions and passions in Book 2 is basically consistent with his discussion of active and passive principles in Book 8, it does nonetheless focus more narrowly on the *extrinsic relationship* between an action and a passion. The relations between actions and passions are defined by him in terms of such extrinsic relationships. However, according to his somewhat different distinction between active and passive principles in Book 8, certain kinds of *intrinsic* as well as extrinsic causes can count as active principles.

[60] Isaac Newton, *Opticks*, 3.31 (4th ed., London, 1730; repr. New York: Dover Publications, 1979), pp. 401–3. Philosophers and historians of science have offered various interpretations of the significance of Newton's active principles in his alchemy and his physics. My own account differs from those given by the following scholars because I emphasize the role of Newton's methods of analysis and synthesis in his acquisition of the knowledge of active and passive principles. Readers who wish to consider other accounts of his active and passive principles should examine: Richard S. Westfall, *Never at Rest: A Biography of Isaac Newton* (Cambridge: Cambridge University Press, 1980), esp. pp. 299–310; Betty Jo Teeter Dobbs, *The Janus Faces of Genius: The Role of Alchemy in Newton's Thought* (Cambridge: Cambridge University Press, 1991), esp. pp. 24–57, 94–6; J. E. McGuire, "Force, Active Principles, and Newton's Invisible Realm," *Ambix*, 15 (1968), 154–208; Ernan McMullin, *Newton on Matter and Activity* (Notre Dame, Ind.: University of Notre Dame Press, 1978), esp. pp. 43–56; and McMullin, "The Impact of Newton's *Principia* on the Philosophy of Science," *Philosophy of Science*, 68 (2001), 279–310. McGuire provides an analysis of the relationship between Zabarella's and Newton's views concerning active and passive principles that is different and perhaps incompatible with my account. See J. E. McGuire, "Natural Motion and Its Causes: Newton on the 'Vis Insita' of Bodies," in *Self Motion: From Aristotle to Newton*, ed. Mary Louise Gill and James G. Lennox (Princeton, N.J.: Princeton University Press, 1994), pp. 305–29.

per se. Newton's notions of active and passive principles were nonetheless a far cry from what Zabarella himself would have counted as knowledge of proper causes.

Zabarella employed the method of resolution to discover the existence of a cause *per se*, and he used the method of composition to confirm that the causal powers of such a cause necessarily produce the effect attributed to it. In works on logic and the methodology of the sciences, he thus extended and developed several of his predecessors' accounts of scientific knowledge, particularly those deriving from Aristotle's *Posterior Analytics*.[61] Among the issues Zabarella reviewed were: (1) how the method of resolution establishes as a fact the existence of a cause *per se*, as contrasted with a merely accidental cause; and (2) how resolution and composition together enable a natural philosopher to establish the existence of a cause *per se* and to confirm its causal powers even when the cause in question is wholly inaccessible to human sense perception. To address both of these issues, he recommended that the method of *regressus* be interpreted as having three parts, the second of which links the resolutive process to the compositive process. Zabarella's definition of these parts in *De regressu* (1578) summarizes how resolution and composition are supposed to work:

> The regress thus consists necessarily of three parts. The first is a 'demonstration that' (*quod*), by which we are led from a confused knowledge of the effect to a confused knowledge of the cause. The second is this 'mental consideration,' by which from a confused knowledge of the cause we acquire a distinct knowledge of it. The third is demonstration in the strictest sense (*potissima*), by which we are at length led from the cause distinctly known to the distinct knowledge of the effect.[62]

This three-step process may be applied to various kinds of examples. Suppose, for instance, a neighbor notices that the foundations of a new home have appeared across the street and she sees strange men walking around it. The neighbor can, by the method of resolution, be led from a confused knowledge of the effect (the rough outline of a new house) to a confused knowledge of its cause (the men who seem to be housebuilders). Next, through what Zabarella called a "mental consideration," the neighbor can acquire from her confused knowledge of the cause a more distinct knowledge of it. She acquires this perhaps by observing that the strange men are carrying saws and hammers as they walk around the foundation, and she compares these men and their equipment with her memory of housebuilders carrying similar

---

[61] See, for example, his *Libri quatuor de methodis* (1578), his *Liber de regressu* (1578), and his *Commentarii in duos Aristotelis libros posteriores analyticos* (1594) in Jacopo Zabarella, *Opera logica* [1597] (repr. Hildesheim: Georg Olms, 1966), cols. 275–334, 479–98, 615–1284.

[62] Jacopo Zabarella, *De regressu*, chap. 5, in Zabarella, *Opera logica*, col. 489, translated in John Herman Randall, Jr., *The School of Padua and the Emergence of Modern Science* (Padua: Editrice Antenore, 1961), p. 58.

equipment whom she has seen at other building sites. Now in possession of a distinct knowledge of the cause *per se* (the housebuilders) of the new house, the neighbor employs the method of composition to make an inference from the action of this cause *per se* to the effect (the new home) produced by this cause. She thus finishes a full *regressus*, having gone through the combined methods of resolution and composition.

Such an example, although it is not one of Zabarella's own, illustrates his understanding of how resolution together with a mental consideration establishes the existence of a cause *per se*, as contrasted with a merely accidental cause. The neighbor identifies the housebuilders as the proper cause of the new home, and she does *not*, for instance, identify a group of medical doctors as the proper cause. If the housebuilders happen to be a group of doctors who are constructing the home as part of a local charity's project for weekend volunteers, this would be an accidental feature because, considered simply as a group of doctors, the men on the building site would count merely as a cause *per accidens* of the new home. Zabarella's confidence that the method of resolution identifies a proper rather than an accidental cause rested on his assumption that terms such as "housebuilder" and "doctor" refer to kinds of human beings, each kind possessing certain essential properties that necessarily determine the conditions in which such a human being counts as a proper cause rather than an accidental one. He would have seen no difficulty, therefore, in the neighbor's going through a mental process of induction, at the end of which she became certain that the men on the building site – because of their resemblance to other observed housebuilders – *really are* the causes of the new home. Nor would Zabarella have entertained the skeptical doubt that the neighbor's association of these housebuilders with the new home might be just a coincidence – a coincidence that relates the new home to a regularly observed accidental cause that does not by itself possess the power to build a house. He did of course possess a sophisticated grasp of the Aristotelian tradition's various treatments of certain classic examples of the contrast between genuine causal claims (such as "the eclipse of the moon is caused by the earth's shadow") and statements describing a mere association of appearances (such as "all ravens are black").[63] But the assumption – in my example – that the term "housebuilder" refers to a distinct kind of human being, one of whose essential properties is to build houses, would have dispelled any such skeptical doubt in this case.

Zabarella's assumption that causal terms refer to distinct kinds of beings whose essential properties necessarily determine whether they are causes *per se* in a given situation also extended to his use of natural kind terms. Natural

---

[63] Aristotle, *Posterior Analytics*, 2.2, in Aristotle, *Complete Works*, 1: 148 (90a1–34); Zabarella, *Commentarii in duos Aristotelis libros posteriores analyticos*, 2.1, in Zabarella, *Opera logica*, cols. 1049–61; Zabarella, *Liber de speciebus demonstrationis*, chap. 10, in Zabarella, *Opera logica*, cols. 429–31; Zabarella, *Libri duo de propositionibus necessariis*, bk. 2, in Zabarella, *Opera logica*, cols. 407–12.

kind terms are those that scientific investigators employ in classifying natural substances, and for Zabarella these would have included terms such as "oak tree," "fire," and "planet." He treated such terms as referring to natural substances whose essential properties determine their causal powers. Hence, in Zabarella's view, a natural substance's essential properties guarantee that the conclusion reached by the method of resolution, when it is supplemented by the relevant mental consideration, must be a true statement.

What would happen to Zabarella's confidence in the combined methods of resolution and composition if these methods were applied to a case where the cause, whose existence is to be established, is wholly inaccessible to human sense perception? Moreover, what would happen to the interpretation of these methods if they were to be practiced by a scientist who did *not* assume that the world is composed of natural substances as conceived by Aristotle and who thus denied that the essential properties of a thing are necessarily related to each other by means of the form of the substance to which they belong? The fact that Zabarella explicitly raised the first question but not the second reveals a great deal about how his view of resolution and composition differed from Newton's view of these methods.[64] By contrasting Zabarella's treatment of the first question with Newton's answer to the second, one can begin to understand the remarkable transformation in methods for identifying causes *per se* that occurred among thinkers such as Newton, who no longer believed that substantial forms can comprise genuine causes.

Among his illustrations of a cause *per se* that is wholly inaccessible to human sense perception, Zabarella included the case of prime matter as it was characterized in Aristotle's *Physics*. Medieval commentators on this work had defined prime matter as "that matter . . . which is understood without any form and privation, but is subject to form and privation."[65] Nothing precedes prime matter in existence, but prime matter does not exist in the same way that a substance exists, for it is not a composite of form and matter. Zabarella thus regarded prime matter as inaccessible to human sense perception because he believed that only a substance that has both form and matter can be perceived. To apply his scientific method to this case, he introduced a greater flexibility in the middle step, or "mental consideration," of his *regressus*. During the middle step, the definition of a relevant term such as "prime matter" was now permitted into the scientist's

---

[64] Here I take issue with John Herman Randall, Jr.'s influential reading of Zabarella's distinction between using the resolutive method to discover principles that are known by induction and using the resolutive method to discover principles that are unknown *secundum naturam* and yet knowable through a demonstration *a signo*. Randall compares the former sort of discovery to the discovery of a Newtonian formal principle and the latter sort to the discovery of a Newtonian explanatory principle. However, I argue that Zabarella's and Newton's principles should not be compared in this way. See Randall, *School of Padua*, p. 53.

[65] Aquinas, *De principiis naturae*, chap. 2, translated in Bobik, ed., *On Matter and Form*, p. 25; Latin text in Aquinas, *De principiis naturae*, ed. Pauson, chap. 2, p. 85.

reasoning process.[66] Such a definition serves the purpose of clarifying the confused knowledge of an unobservable cause, which the scientist has already obtained through his reasoning from an observable effect to this cause. Once his confused knowledge is clarified, the definition also guarantees that there is a necessary connection between the cause and the effect. In this way, a scientist can acquire knowledge of the existence of a cause *per se* even if the cause in question is, like prime matter, wholly inaccessible to sense perception. Zabarella summarized the process by referring to those steps Aristotle had taken to identify prime matter as the cause of the generation of a natural body:

> From the generation of substances he shows that prime matter occurs: from a known effect an unknown cause. For generation is known to us by sense but the underlying matter is in the highest degree unknown. So after the proper subject, that is a perishable natural body, in which each is originally present, has been considered, it is demonstrated that there is present in it a cause, on account of which the effect is present in the same, and it is *demonstratio quod* which is thus formed: where there is generation there is underlying matter; but in a natural body there is generation; so in a natural body there is matter.[67]
>
> For this reason Aristotle, who wished to teach us a distinct, not merely a confused, knowledge of principles, . . . began to investigate the nature and conditions of the matter which he has discovered. . . . In its own nature matter must lack all forms and have the potentiality to receive all. . . . It readily becomes apparent to us that such matter is the cause of generation.[68]

It is perhaps tempting to think that a case such as this shows Zabarella's preoccupation with the modern scientist's problem of how to infer, from the observable phenomena, the existence of an unknown cause that is as yet unobserved but could be observed if the right experiments were carried out. However, this would seriously underestimate his interest in establishing the existence of causes that are *unobservable in principle* – causes that scientists today would describe as theoretical objects that can never be detected even by the most advanced experimental equipment. Zabarella, too, thought of himself as trying to solve a problem regarding causes that are unobservable in principle – the problem of how in particular to establish the existence of a cause *per se* that is neither a substance nor an essential property of a substance. In his example of prime matter, Zabarella could not appropriately use the resolutive method together with a mental process of induction to establish

---

[66] Zabarella, *De regressu*, chap. 5, in Zabarella, *Opera logica*, cols. 487–9.
[67] Zabarella, *De regressu*, 4.485, translated in Nicholas Jardine, "Epistemology of the Sciences," in *The Cambridge History of Renaissance Philosophy*, ed. Charles B. Schmitt, Quentin Skinner, Eckhard Kessler, and Jill Kraye (Cambridge: Cambridge University Press, 1988), pp. 691–2.
[68] Jacopo Zabarella, *De regressu*, chap. 5, in Zabarella, *Opera logica*, col. 488, translated in Jardine, "Epistemology," pp. 692–3.

a causal relationship between prime matter and the generation of a natural body. This was because prime matter cannot be an essential property of any substance because it is defined as what must be *prior to* the existence of any substance. As such, it also can never be observed, and hence it can never be perceived in any association – causal or otherwise – with an observable effect such as the generation of a natural body.

Zabarella's response to the problem was to extend the method of resolution to cover those special cases where an Aristotelian natural philosopher could not assume that the necessary relationship between a substance and its essential properties will determine what counts as a cause *per se*. What he seems never to have anticipated was that among his successors would be physicists, likewise studying cases where this assumption did not hold, who would nonetheless reject his own extension of the resolutive method. They would reject it because of its aim of discovering causes that are unobservable in principle. These new physicists would instead devote themselves to solving a quite different methodological problem: How does a physicist infer, from the observable phenomena, the existence of a cause *per se* that is *as yet* unobserved but could be observed if the right experiments were carried out?

Newton, whose scientific work often dealt with objects that exemplified this new methodological problem, employed a version of resolution and composition that he called "analysis and synthesis" (see Andersen and Bos, Chapter 28, this volume).[69] However, he did not share Zabarella's confidence that these methods would yield certain knowledge of every sort of cause *per se*, and thus he refrained from speculating about the nature of theoretical objects that are unobservable in principle. Even when discussing the nature of God in his *Philosophiae naturalis principia mathematica* (Mathematical Principles of Natural Philosophy, 1687), Newton exercised this restraint and spoke about God only insofar as certain aspects of God could be known through human sense experience. "For all discourse about God is derived through a certain similitude from things human, which while not perfect

---

[69] Newton was also familiar with the development of the methods of analysis and synthesis in Greek geometry by, among others, the fourth-century mathematician Pappus of Alexandria. As an early modern scholar, he had access to several distinct versions of the two methods bearing the names "analysis" (or "resolution") and "synthesis" (or "composition"). These versions had been formulated by the Greek geometers, the Aristotelians, Galen, Chalcidius, and other older sources. My treatment of Newton's methods of analysis and synthesis concentrates on how he conceived of their repeated use to acquire knowledge of more and more general causes. I also point out that Newton seemed to regard the laws of motion and even the law of gravity as themselves *effects* whose more general causes could be learned through the further application of the methods of analysis and synthesis. Thus, on these two significant points, my treatment differs from those given by the following authors: Andrea Croce Birch, "The Problem of Method in Newton's Natural Philosophy," in *Nature and Scientific Method*, ed. Daniel O. Dahlstrom (Studies in Philosophy and the History of Philosophy, 22) (Washington, D.C.: Catholic University of America Press, 1991), pp. 253–70; Henri Guerlac, "Newton and the Method of Analysis," in *Dictionary of the History of Ideas*, ed. Philip P. Wiener, 5 vols. (New York: Charles Scribners' Sons, 1973–74), 3: 378–91; and Niccolò Guicciardini, "Analysis and Synthesis in Newton's Mathematical Work," in *The Cambridge Companion to Newton*, ed. I. Bernard Cohen and George E. Smith (Cambridge: Cambridge University Press, 2002), pp. 308–28.

is nevertheless a similitude of some kind. . . . And to treat of God from phenomena is certainly a part of natural philosophy."[70]

In both the *Principia* and the *Opticks*, Newton's methods of analysis and synthesis were applied typically to cases that he thought could be studied through the inductive association of phenomena that either are observable at the present time or will be observable in the future through new experiments and better scientific instruments. He emphasized such a restriction when stipulating his "Rules for the Study of Natural Philosophy" in Book III of the *Principia*.[71] The first rule, which appeared in every edition of this work during his lifetime, stated: "No more causes of natural things should be admitted than are both true and sufficient to explain their phenomena." The second rule, also appearing in every edition, focused on the associations of similar phenomena: "Therefore, the causes assigned to natural effects of the same kind must be, so far as possible, the same." The fourth rule, added by Newton to the third edition of the *Principia* (1726), was even more explicit about basing one's claims *only on what can be observed*: "In experimental philosophy, propositions gathered from phenomena by induction should be considered either exactly or very nearly true notwithstanding any contrary hypotheses, until yet other phenomena make such propositions either more exact or liable to exceptions."

Newton also carefully distinguished between what he characterized as *active* relationships among the associated phenomena and *passive* relationships among the associated phenomena. His search for the active and passive principles governing these two kinds of relationships developed into a methodology for discovering weaker as well as stronger causal relations among different types of natural phenomena. His passive principles were lawlike regularities in the association between the observed states of one or more bodies. Newton's first law of motion, for example, described a basic regularity in the successive states of any given body: "Every body perseveres in its state of being at rest or of moving uniformly straight forward, except insofar as it is compelled to change its state by forces impressed." The law was also characterized, in his *Principia*, as involving an "inherent force of matter . . . by which every body, so far as it is able, perseveres in its state either of resting or of moving uniformly straight forward."[72] In the *Opticks*, Newton added that this *vis inertiae* (force of inertia) in particles of matter is "accompanied with such passive Laws of Motion as naturally result from that Force, but also that they [the particles of matter] are moved by certain

---

[70] Isaac Newton, *The Principia: Mathematical Principles of Natural Philosophy*, trans. I. Bernard Cohen and Anne Whitman (Berkeley: University of California Press, 1999), bk. 3, pp. 942–3.

[71] Newton, *Principia*, bk. 3, pp. 794–6. See also the Latin text of the third edition (London, 1726) in Isaac Newton, *Isaac Newton's Philosophiae naturalis principia mathematica*, ed. Alexandre Koyré and I. Bernard Cohen, with Anne Whitman, 2 vols. (Cambridge, Mass.: Harvard University Press, 1972), 2: 550–5.

[72] Newton, *Principia*, law 1, p. 416, and definition 3, p. 404.

active Principles, such as is that of Gravity."[73] These active principles consisted of laws of attraction or repulsion between material bodies, including the gravitational attraction among the stars and planets and among ordinary terrestrial bodies, but also including the short-range attractions and repulsions exhibited by magnetic, electrical, and chemical phenomena. Even the cause of fermentation, which keeps the heart and blood in perpetual motion and heated, was counted as one of Newton's active principles.[74] He had already anticipated some aspects of this theory of active principles when summarizing his law of gravity earlier in the *Principia*:

> *Gravity exists in all bodies universally and is proportional to the quantity of matter in each.* We have already proved that all planets are heavy [or gravitate] toward one another and also that the gravity toward any one planet, taken by itself, is inversely as the square of the distance of places from the center of the planet. And it follows . . . that the gravity toward all the planets is proportional to the matter in them.
>
> . . . Therefore the gravity toward the whole planet arises from and is compounded of the gravity toward the individual parts. We have examples of this in magnetic and electric attractions. For every attraction toward a whole arises from the attractions toward the individual parts.[75]

When Newton reflected on how such principles operate together in a unified natural world, he provided further clues concerning his general beliefs about causality. The methods of analysis and synthesis were employed by him to identify both the weaker, or passive, principles and the stronger, or active, principles. By analysis, he identified through inductive reasoning the lawlike regularities whose principles, taken together, constitute a hierarchical system of laws. The more general a cause is, the more active is its principle and the higher is its law's ranking in the unified system of laws. Conversely, by synthesis, Newton confirmed through deductive reasoning that the lower-level laws are deducible from the higher-level laws of the system. He then interpreted this logical relationship in causal terms. Passive principles are maintained in their operations by the stronger, active principles. Active principles approximate Aristotelian causes *per se* in that they are active powers resembling Zabarella's essential properties of a substance. Despite the fact that Newton rejected any scientific explanation based on the form or essential properties of an Aristotelian substance, he still expected that his discovery of the correct active principles would eventually culminate in his establishing the existence of at least some causes *per se*. As he progressed from a knowledge of the weaker, or passive, principles toward a knowledge of the stronger, or active, principles, he thereby sought to arrive at a knowledge of some genuine

---

[73] Newton, *Opticks*, 3.31, p. 401.
[74] Ibid., pp. 376, 399.
[75] Newton, *Principia*, 3.7.7, pp. 810–11.

causes. Yet he tried to do so without making what he regarded as the serious error of hypothesizing about causes that are unobservable in principle.

Newton's distinctive use of the methods of analysis and synthesis was nowhere more evident than in his long-standing deliberations about whether the force of gravity can be explained. How this use differed from Zabarella's employment of resolution and composition may be seen in Newton's attempts to show that gravity is not an occult, or hidden, cause. One such attempt occurs near the end of the *Opticks*, where he discussed the application of passive and active principles to the material corpuscles constituting ordinary bodies:

> These Particles have not only a *Vis inertiae*, accompanied with such passive Laws of Motion as naturally result from that Force, but also . . . they are moved by certain active Principles, such as is that of Gravity, and that which causes Fermentation, and the Cohesion of Bodies. These Principles I consider, not as occult Qualities, supposed to result from the specifick Forms of Things, but as general Laws of Nature, by which the Things themselves are form'd; their Truth appearing to us by Phaenomena, though their Causes be not yet discover'd. For these are manifest Qualities, and their Causes only are occult. And the *Aristotelians* gave the Name of occult Qualities, not to manifest Qualities, but to such Qualities only as they supposed to lie hid in Bodies, and to be the unknown Causes of manifest Effects: Such as would be the Causes of Gravity, and of magnetick and electrick Attractions, and of Fermentations, if we should suppose that these Forces or Actions arose from Qualities unknown to us, and uncapable of being discovered and made manifest. Such occult Qualities put a stop to the Improvement of natural Philosophy, and therefore of late Years have been rejected.[76]

Here Newton remarks that gravity would have been treated by thinkers like Zabarella as an occult quality – as a wholly unobservable essential property of a body that nonetheless produces observable effects, such as changes in the inertial states of other bodies. But he is anxious to correct this mistaken interpretation of his law of gravity, which he instead characterizes as an active principle that is entirely *manifest*, or observable. Gravity is not occult because its existence is established by the method of analysis (resolution) that relates observable phenomena by means of inductive reasoning. Gravity therefore consists, in part, of a lawlike regularity associating the observed states of two or more bodies. From the accelerated motion of a body free-falling toward the earth, for instance, the physicist reasons inductively that this body's motion resembles the accelerated motion of all other bodies that free-fall toward the earth. The physicist can then ask what *causes* the regularity of these accelerated motions. At that point, he performs some additional inductive reasoning according to the method of analysis. This enables him

---

[76] Newton, *Opticks*, 3.31, p. 401.

to recognize that the regularity of these accelerated motions resembles the attractions between magnets, although – as Newton noted in the *Principia* – the force of gravity differs in kind from magnetic force because the force of attraction between two magnets is not proportional to the quantity of matter in the magnets.[77] Having recognized this albeit limited resemblance between accelerated motions and magnetic attractions, the physicist can now redescribe the accelerated motions in question as *effects of the gravitational attraction* between falling bodies and the earth. Thus, through a second use of the method of analysis, which involves additional inductive reasoning, the physicist has discovered what Newton calls an "active principle."

Knowledge of an active principle such as the law of gravity brought Newton closer to an ideal knowledge of causes *per se* than did knowledge of a passive principle such as the law of inertia, he thought, for only an active principle can attribute *activity* to the bodies it associates. But how much further did he expect to progress in his approximation of such ideal knowledge? The closing pages of the *Opticks* contain an interesting prediction:

> And although the arguing from Experiments and Observations by Induction be no Demonstration of general Conclusions; yet it is the best way of arguing which the Nature of Things admits of, and may be looked upon as so much the stronger, by how much the Induction is more general. . . . By this way of Analysis we may proceed from Compounds to Ingredients, and from Motions to the Forces producing them; and in general, from Effects to their Causes, and from particular Causes to more general ones, till the Argument end in the most general.[78]

In predicting the eventual discovery of the most general cause or causes, Newton hinted that active principles such as the law of gravity could themselves be treated as *observable effects*. They, too, could be conceived as having causes, which are more general than they themselves are and are discoverable by means of inductive reasoning. Through further use of the method of analysis, therefore, physicists should be able to discover these more general causes, and eventually they should be able to understand the *most general* cause, which is the cause *per se* of all the active principles. The physicists' understanding of this most general cause, in Newton's view, would count as knowledge of the cause *per se* of gravity not only because it would represent an advance over his own knowledge of the law of gravity, but more importantly because it would signal that their scientific knowledge of material bodies had reached completion. Of what could such complete scientific knowledge possibly consist? Newton left readers of the *Opticks* with a final suggestion awaiting confirmation through future applications of the methods of analysis and synthesis. Such knowledge would consist of "the Wisdom and Skill of a

---

[77] Newton, *Principia*, 3.6.6.5, p. 810.
[78] Newton, *Opticks*, 3.31, p. 404.

powerful ever-living Agent, who being in all Places, is more able by his Will to move the Bodies within his boundless uniform Sensorium, and thereby to form and reform the Parts of the Universe, than we are by our Will to move the Parts of our own Bodies."[79]

Thus, in revising and extending the scope of inductive reasoning, Newton helped to create a new model of knowledge of cause and effect. This model has sometimes been characterized as a precursor of the eighteenth-century philosopher David Hume's probabilistic treatment of cause and effect in response to his own skeptical problem of induction.[80] But such a characterization fails to account adequately for Newton's belief in the necessary relationships among the active and passive principles of nature and his repeated use of analysis and synthesis for the purpose of acquiring a unified knowledge of the causal structure of the world. It also underestimates the role of Aristotelian causal concepts in the development of his new model of scientific explanation. Newton's causal claims deserve to be studied in their own right because they presupposed the indispensability of at least some causes *per se*. He still tried – in his studies of moving bodies, optical phenomena, and alchemical phenomena – to achieve a knowledge of this kind of cause.

Newton was neither the first nor the only early modern thinker to elaborate concepts of active and passive principles in his scientific explanations. Justus Lipsius and other revivers of the Stoic tradition earlier had written works discussing active and passive principles of nature and the active and passive qualities of the four elements. The Cambridge Platonists Henry More (1614–1687) and Ralph Cudworth (1617–1688) had also articulated, in their account of the world soul, certain spiritual principles that, they believed, guided the motions of passive matter. Most importantly, in the alchemical tradition, innovators from Paracelsus (Theophrastus Bombastus von Hohenheim, ca. 1493–1541) to Johannes Baptista van Helmont (1579–1644) had preceded Newton in both the experimental and theoretical investigations of active and passive chemical principles, such as the active principle of fermentation. Newton himself had studied many of these predecessors' writings and techniques at various points in his schooling and adult career.[81] However, what is especially instructive about his account of active and passive principles was its bridging of the gap between the sixteenth-century Aristotelians' theories of substantial forms and the early modern innovators' respective corpuscular and alchemical philosophies. The contrast between his methodology and Zabarella's epitomized not only the changing epistemology of early modern science but also the emergence of a different relationship between the epistemological aims of scientists such as Newton and their metaphysical commitments.

---

[79] Ibid., p. 403.
[80] David Hume, *A Treatise of Human Nature* [1739–40], 1.3.1–16, 1.4.1–2, ed. L. A. Selby-Bigge, rev. text by P. H. Nidditch (Oxford: Oxford University Press, 1978), pp. 69–218.
[81] Dobbs, *Janus Faces of Genius*, pp. 24–57, 94–6; Westfall, *Never at Rest*, pp. 299–310.

Newton shared with many of his contemporaries a continuing metaphysical commitment to the existence of at least some causes *per se* – a commitment that nevertheless excluded Aristotle's account of the four causes and hence the very principles on which the concept of a cause *per se* was based. Thus, although Aristotelian causal concepts were indispensable in defining several new kinds of scientific explanation, the history of how they made possible the transformation of early modern science is arguably the story of how these causal concepts became increasingly unrecognizable to the very thinkers who relied on them. If this is so, then there is every reason to think that a conceptual revolution of considerable magnitude did in fact occur in early modern beliefs about causality. Its hallmarks were the decline of formal causes and the rise of laws of nature in the explanation of natural phenomena. Yet such a revolution shaped far more than our theories about matter and motion. The three changes in conceptions of scientific explanation traced in this chapter continued to develop long after the sixteenth and seventeenth centuries. They structured later scientific thought about the unity of nature and the kinds of events or objects that can appropriately be described by natural laws. They also established important constraints that guided the formation of modern beliefs about human nature in the eighteenth and nineteenth centuries. Indeed, perhaps their most striking consequence is still with us today in our persistent reflections on whether to explain human actions and passions as law-governed natural effects or whether to dispense with laws of nature altogether when trying to explain ourselves.

4

THE MEANINGS OF EXPERIENCE

Peter Dear

The categories of "experience" and "experiment" lay at the heart of the conceptions of natural knowledge that dominated European learning at both the beginning and the end of the Scientific Revolution. The Latin words generally used to denote "experience" in both the medieval and early modern periods, *experientia* and *experimentum*, were generally interchangeable, with no systematic distinction between them except in particular contexts to be discussed; both are related to the word *peritus*, meaning skilled or experienced. Besides these terms and their vernacular cognates, another related Latin term, *periculum* ("trial" or "test"), began to be used in the late sixteenth century to designate the deliberate carrying out of an experiment (*periculum facere*), initially in the mathematical sciences. By the end of the seventeenth century, the construal of experience as "experiment" in this sense had acquired a wide and influential currency.

At the start of the sixteenth century, scholastic versions of Aristotelian natural philosophy dominated the approach to knowledge of nature that informed the official curricula of the universities (see the following chapters in this volume: Blair, Chapter 17; Garber, Chapter 2); Aristotle's writings stress repeatedly the importance of sense experience in the creation of reliable knowledge of the world. Nonetheless, during the seventeenth century, many of the proponents of what came to be called by some (rather obscurely) "the new science" criticized the earlier orthodoxy of what Aristotelian natural philosophy (or "physics") had become on the grounds that it paid insufficient attention to the lessons of experience. For example, Francis Bacon (1561–1626) wrote in his *New Organon* of 1620 that Aristotle "did not properly consult experience ...; after making his decisions arbitrarily, he parades experience around, distorted to suit his opinions, a captive."[1] Intellectual reformers such as Bacon commonly represented traditional Aristotelian philosophy as being

---

[1] Francis Bacon, *The New Organon*, 1.8, ed. and trans. Lisa Jardine and Michael Silverthorne (Cambridge: Cambridge University Press, 2000), p. 52.

obsessed with logic and wordplay rather than as attempting to come to grips with things themselves by means of the senses.

The so-called Aristotelian worldview[2] was, in its lowest common denominator, the standard framework of philosophical education in the universities and colleges of Europe in the early modern period. In practice, this means that the curricular structure of such institutions was coordinated with Aristotle's writings, together with commentaries on them. Thus, the teaching of natural philosophy used such works as Aristotle's *Physics*, *De anima* (On the Soul), and *De caelo* (On the Heavens), as well as aspects of his *Metaphysics*, together with other more minor Aristotelian texts. That situation, well-established at the beginning of the sixteenth century, continued at most universities through the seventeenth century, albeit with considerable shifts in emphasis and interpretation over time. Sporadic attempts to revise this standard curricular arrangement made little progress; a planned wholesale restructuring of the natural philosophical curriculum in the German Lutheran universities, instigated by Philip Melanchthon (1497–1560) with the intention of displacing Aristotle, soon fell flat.[3] Aristotle's approach to the philosophy of nature, then, was part of a pedagogical tradition based on the use of his texts. His philosophy thus inevitably shaped the categories of thought even of those who, increasingly in the seventeenth century, explicitly rejected his authority.

Bacon's criticism of Aristotle gives the impression that the Aristotelian approach subordinated experience to abstract reasoning, using experience only as a means of confirming preconceptions. This was indeed a common criticism in the seventeenth century. Nonetheless, scholastic philosophers who took their lead from Aristotle's texts stressed, following the master himself, that all knowledge had its origin in the senses: "There is nothing in the mind which was not first in the senses," ran a scholastic maxim.[4] This emphasis on the sensory origin of knowledge looks like a radical empiricism that makes direct experience paramount. Indeed, Aristotle himself had regarded even mathematics, apparently the intellectual field of knowledge furthest removed from the messiness of experience, as rooted in the senses: We gain our ideas of number from seeing collections of things in the world, and our ideas of geometrical figures from spatial experience.

---

[2] For ambiguities of the category "Aristotelianism," see Charles B. Schmitt, *Aristotle in the Renaissance* (Cambridge, Mass.: Harvard University Press, 1983); Edward Grant, "Ways to Interpret the Terms 'Aristotelian' and 'Aristotelianism' in Medieval and Renaissance Natural Philosophy," *History of Science*, 25 (1987), 335–58. For approaches to issues of experience and experiment in the study of the natural world among ancient Greeks themselves, see the classic essay by G. E. R. Lloyd, "Experiment in Early Greek Philosophy and Medicine," in G. E. R. Lloyd, *Methods and Problems in Greek Science* (Cambridge: Cambridge University Press, 1991), pp. 70–99.

[3] On Melanchthon and Pliny, see Sachiko Kusukawa, *The Transformation of Natural Philosophy: The Case of Philip Melanchthon* (Cambridge: Cambridge University Press, 1995), esp. pp. 51, 136–7. On changes in French curricula, see L. W. B. Brockliss, *French Higher Education in the Seventeenth and Eighteenth Centuries: A Cultural History* (Oxford: Clarendon Press, 1987).

[4] Paul Cranefield, "On the Origins of the Phrase *Nihil est in intellectu quod non prius fuerit in sensu*," *Journal of the History of Medicine*, 25 (1970), 77–80.

There is thus an apparent contradiction between Bacon's denial that there was an adequate place for experience in Aristotelian philosophy and the foundational role of sensory experience in the work of contemporary Aristotelian philosophers themselves. This contradiction may be explained by considering the ways in which experience was *used* in the making of knowledge during the Scientific Revolution.

EXPERIENCE AND THE NATURAL PHILOSOPHY OF ARISTOTLE IN EARLY MODERN EUROPE

There is nowadays nothing extraordinary in the idea that "experience" can be a category worthy of historical investigation. Rather than a fundamental, unproblematic means of acquiring knowledge, sensory experience as a form of knowledge – generally under the terminological guise of "observation" – has since the 1970s at least been regarded by philosophers of science as constituted and ordered through prior conceptual categories.[5] "Experience," in this view, depends on the expectations and presumptions of the observer. This thesis, currently accepted by practically all philosophers, is designated by the term "theory-ladenness of observation."[6] Philosopher of science Norwood Russell Hanson illustrated the idea by imagining the astronomers Johannes Kepler (1571–1630) and Tycho Brahe (1546–1601) on a hill at dawn. Each looks to the east to observe the sun, but do they, the one a Copernican who believes that the sun stands still in the center of the universe and the other a geocentrist who believes that the sun circles the Earth, see the same thing? Hanson said that in an important sense they do not: There is, he writes, "a difference between a physical state and a visual experience."[7]

In studying the meanings of experience in the early modern period, however, we find more at stake than just the interpretation of perceptions. There is another philosophical issue, namely the relationship between the experience of a single event and the perception of a truth that holds generally. Kepler, to borrow Hanson's example, did not, as he stood on the hill, simply experience the earth happening on that occasion to roll around to reveal the sun ever farther above the horizon. He saw an instance of a regular natural occurrence, reflecting the Copernican structure of the universe. In a sense, we have to do here with what, much later, came to be called the "problem of induction." But around 1600, for figures such as Kepler and Tycho, the issue was integrated closely with both the specifics of Aristotelian epistemology and Aristotle's view of nature.

---

[5] Although similar ideas can, of course, be traced back much further: see, for example, Michael Friedman, *Kant and the Exact Sciences* (Cambridge, Mass.: Harvard University Press, 1992).
[6] Norwood Russell Hanson, *Patterns of Discovery: An Inquiry into the Conceptual Foundations of Science* (Cambridge: Cambridge University Press, 1958); Thomas S. Kuhn, *The Structure of Scientific Revolutions* (Chicago: University of Chicago Press, 1962).
[7] Hanson, *Patterns of Discovery*, pp. 5–8, at p. 8.

The Meanings of Experience

For Aristotle, a science of the physical world should, ideally, take the form of a logical deductive structure derived from incontestable basic statements or premises. The model for this was the structure of classical Greek geometry as exemplified in Euclid's *Elements*, where the truth of unexpected conclusions can be demonstrated by deduction from a delimited set of prior, and supposedly obvious, accepted axioms (such as that "when equals are subtracted from equals, the remainders are equal"). In the case of sciences that concerned the natural world, however, such axioms could not be known by simple introspection. In those cases, the axioms had to be rooted in familiar and commonly accepted *experience*. Thus "the sun rises in the east" was unshakably and universally known to be true through experience, as was the doctrine that acorns grow into oak trees, or even the apparently more recondite principle that, in a homogeneous medium, vision (and hence perhaps light rays, depending on one's theory) occurs in straight lines – because everyone knows that it is impossible to see around corners. On the basis of such experiences, firm deductive sciences of astronomy, plants, and optics could be erected. To do this in practice was, of course, much more difficult than to lay it out as an ideal, but as an ideal it dominated scholastic thought well into the seventeenth century.

This kind of experience, therefore, was of universal behaviors rather than particulars: The sun *always* rises in the east; acorns *always* (barring accidents) grow into oak trees.[8] Singular experiences (such as the eruption of Vesuvius in 79 C.E. or the coronation of Pope Urban VIII) were more problematic because they could only subsequently be known by historical report, as something that had happened on a particular occasion. They were thus unfit to act as scientific axioms because they could not receive immediate free assent from all: Most people had not witnessed them. A science needed to be *certain*, whereas histories were matters of fallible record and testimony.[9] The difficulty was unavoidable; most, if not all, of an individual's knowledge of the world relies very heavily on things believed from the testimony of others.[10] We will later see how, those subscribing to an Aristotelian ideal of science of this kind developed a variety of techniques to "universalize" their own specialist empirical work.

---

[8] This kind of *ceteris paribus* assumption was justified in medieval and later scholasticism in the guise of so-called *ex suppositione* reasoning: If the oak tree actually does grow from the acorn, the explanation provided will constitute a necessary scientific demonstration of that process. See especially William A. Wallace, "Albertus Magnus on Suppositional Necessity in the Natural Sciences," in *Albertus Magnus and the Sciences: Commemorative Essays 1980* (Toronto: Pontifical Institute of Medieval Studies, 1980), ed. James A. Weisheipl, pp. 103–28; reprinted as Wallace, *Galileo, the Jesuits, and the Medieval Aristotle* (Aldershot: Variorum, 1991), chap. 9.

[9] On these matters, see Stephen Pumfrey, "The History of Science and the Renaissance Science of History," in *Science, Culture, and Popular Belief in Renaissance Europe*, ed. Stephen Pumfrey, Paolo L. Rossi, and Maurice Slawinski (Manchester: Manchester University Press, 1991), pp. 48–70.

[10] The role of trust is stressed in Steven Shapin, *A Social History of Truth: Gentility and Science in Seventeenth-Century England* (Chicago: University of Chicago Press, 1994).

Aristotle's natural philosophy was especially concerned with "final causes," the purposes or ends toward which processes tended or that explained the conformation and capacities of something (see Joy, Chapter 3, this volume). Living creatures were model instances: All the parts of an animal's body seem to be fitted to their particular functions, and by studying their behaviors passively one could find out what they were doing – that is, what they were *for*. Active interference, by setting up artificial conditions, would risk subverting the natural course of things, hence yielding misleading results; experimentation would be just such interference. Experiments in the inanimate world ran into the same problem: Using a balance with unequal arms to raise a heavy weight (resting on the shorter arm) by using a lighter weight (resting on the longer arm), for example, would misrepresent the relative tendencies of those weights to strive toward the center of the earth. To the extent that Aristotle's natural philosophy sought the final causes of things, and thereby to determine their natures, experimental science was therefore disallowed.

Beyond the confines of academic practice, "experience" had other connotations as well. In the sixteenth century, opponents of university learning, most prominently Paracelsus (Theophrastus Bombastus von Hohenheim, 1493–1541) in the 1530s and 1540s, held up untutored experience as an alternative to the elaborate epistemology of the Aristotelians. Paracelsus advocated a closer acquaintance with things themselves as the way to acquire knowledge of a practical, operational kind – in contrast with Aristotelian focus on philosophical *understanding*. The particular concern of Paracelsus was with healing, an unavoidably practical specialty. By stressing knowledge of the properties of things and how to make use of them, Paracelsus turned attention squarely onto the practical experience of the artisan, who was taken to have an intimate, almost mystical rapport with things themselves.[11] The burgeoning tradition of "natural magic," and the popular "books of secrets" of the same period, promoted similar attitudes.[12] Others subsequently in the sixteenth century, particularly (although by no means exclusively) in England, advocated a similar upgrading of artisanal knowledge, their most accomplished representative being Bacon. In the closing decade of the sixteenth century and the first quarter century or so thereafter, Bacon promoted a reformed "natural philosophy" directed toward ends different from that of

---

[11] On Paracelsus, see Walter Pagel, *Paracelsus: An Introduction to Philosophical Medicine in the Era of the Renaissance*, 2nd rev. ed. (Basel: Karger, 1982), and more broadly Andrew Weeks, *Paracelsus: Speculative Theory and the Crisis of the Early Reformation* (Albany: State University of New York Press, 1997).

[12] William Eamon, *Science and the Secrets of Nature: Books of Secrets in Medieval and Early Modern Culture* (Princeton, N.J.: Princeton University Press, 1994). These views go back at least to Roger Bacon in the thirteenth century – see Roger Bacon, *Opus Majus*, ed. John Henry Bridges (1897–1900; facsimile repr. Frankfurt am Main: Minerva, 1964) – and were also represented in the sixteenth and seventeenth centuries by the so-called Hermetic tradition. On the latter, see the classic argument by Frances A. Yates, "The Hermetic Tradition in Renaissance Science," in Yates, *Collected Essays*, vol. 3, *Ideas and Ideals in the North European Renaissance* (London: Routledge and Kegan Paul, 1984), pp. 227–46.

the schools, emphasizing the practical benefits to be derived from knowledge of nature and praising the craft knowledge of artisans. Bacon held up "experience" as the route to such knowledge, by which he meant the scrupulous examination and collection of facts regarding the properties and behaviors of physical phenomena (see Serjeantson, Chapter 5, this volume). These facts remained, however, generic: They concerned "how things behave" and took for granted the establishment of such general facts from singular instances, much like the Aristotelian kind.[13] The main exception was Bacon's concern with "monsters" and other pretergenerations, that is, individual cases where nature does *not* behave in its normal, regular way.[14] Bacon's well-known disdain for final causes in natural philosophy meant in addition that, unlike an orthodox Aristotelian natural philosopher, he had no epistemological difficulties in using artificial situations, such as experimental contrivances, in generating telling facts (quite apart from his moral objections to an art/nature division).[15]

However, even within the domain of scholastic orthodoxy, there were other sciences concerning the natural world besides natural philosophy that exposed differing concerns about final causes. For the mathematical sciences, as we shall see, the kind of knowledge sought was uncompromised by final causes and hence permitted experimental contrivance, with no worries about "monsters." By contrast, in the study of medicine and the human body, issues of regularity and variability played a crucial role in determining criteria of health.

EXPERIENCES OF LIFE AND HEALTH

The teaching of human anatomy formed an integral part of an early modern medical education in the universities, and, like other areas of the study of nature, it already had its established ways of doing things. In the sixteenth century, with frequent bows to the example set by the ancient Greco-Roman physician Galen, anatomists conceived of their enterprise as being above all one of disciplined *seeing*; and what they saw in the corpses that they dissected was taken by many, following the precedent of Galen's views, to be

---

[13] See, for example, the list of "Instances meeting in the nature of heat" in Francis Bacon, *New Organon*, 2.11, pp. 110–11.
[14] On the "monstrous" in this period, see Lorraine Daston and Katharine Park, *Wonders and the Order of Nature, 1150–1750* (New York: Zone Books, 1998), esp. chap. 5; and Zakiya Hanafi, *The Monster in the Machine: Magic, Medicine, and the Marvelous in the Time of the Scientific Revolution* (Durham, N.C.: Duke University Press, 2000). Bacon also discusses "unique instances," which are "wonders of species" (that is, unique *kinds* of beings), and distinguishes them from particular "errors of nature," which are "wonders of individuals," such as monsters, that do not form a collective species: see Francis Bacon, *New Organon*, 2.28–29, pp. 147–9.
[15] There were also precedents for the overcoming of an art/nature distinction in the alchemical tradition: see William R. Newman, "Art, Nature, and Experiment among some Aristotelian Alchemists," in *Texts and Contexts in Ancient and Medieval Science: Studies on the Occasion of John E. Murdoch's Seventieth Birthday*, ed. Edith Sylla and Michael McVaugh (Leiden: E. J. Brill, 1997), pp. 305–17.

representative of all human beings.¹⁶ Controversy over the details of this issue continued through the sixteenth century, with some, such as Realdo Colombo (ca. 1510–1559) at Padua, maintaining the strong uniformity of human anatomy and the rarity of anomalies, whereas Colombo's predecessor at Padua, Andreas Vesalius (1514–1564), frequently paid lip service to that ideal while in practice routinely noting variations found both among individuals and those systematically caused by age, sex, and regional or ethnic differences.[17]

The Padua-trained English physician William Harvey (1578–1657), writing in the 1620s, followed what was by then established anatomical practice by regarding his work on the circulation of the blood as fundamentally a matter of *looking* in the right way ("autoptic" experience).[18] Harvey displays once again the impact of broadly Aristotelian epistemological doctrines on understandings of active, interventionist experience of nature. Intervention by way of vivisection necessarily put the animal subject into an unnatural, traumatized condition and could accordingly be represented as an illegitimate way to obtain knowledge of natural functioning. This objection to such research procedures as Harvey himself employed carried considerable weight in the mid-seventeenth century, and Harvey was in no position to shrug them off. In his inquiry into the circulation of the blood, published in *De motu cordis* (On the Motion of the Heart, 1628), he had adopted the conventional stance of sixteenth-century anatomists, such as that of his own teacher at Padua, Girolamo Fabrici (ca. 1533–1619), whereby the investigator was understood to be acquiring unmediated ocular evidence of the way things stood in the body rather than to be testing hypotheses by means of artificial experiment. Thus, Harvey could see himself as *demonstrating* the circulation of the blood, in the literal sense of *showing* it; the universalization of his particular experiences was no more problematized than was the norm for anatomical knowledge in this period.[19] However, this approach was not sufficient to exempt him from methodological criticism. In his *Exercitatio anatomica* (Anatomical Exercise, 1649), Harvey responded to various critics, and the objection that appeared

---

[16] See articles in Andrew Wear, R. K. French, and I. M. Lonie, eds., *The Medical Renaissance of the Sixteenth Century* (Cambridge: Cambridge University Press, 1985), esp. Andrew Cunningham, "Fabricius and the 'Aristotle Project' in Anatomical Teaching and Research at Padua," pp. 195–222. Gabriele Baroncini, *Forme di esperienza e rivoluzione scientifica* (Bibliotheca di Nuncius, Studi e testi IX) (Florence: Leo S. Olschki, 1992), is a particularly useful discussion of ideas of experience in philosophy with special focus on medical and life-science authors.
[17] Nancy G. Siraisi, "Vesalius and Human Diversity in *De humani corporis fabrica*," *Journal of the Warburg and Courtauld Institutes*, 57 (1994), 60–88.
[18] Andrew Wear, "William Harvey and the 'Way of the Anatomists'," *History of Science*, 21 (1983), 223–49; see also Wear, "Epistemology and Learned Medicine in Early Modern England," in *Knowledge and the Scholarly Medical Traditions*, ed. Don Bates (Cambridge: Cambridge University Press, 1995), pp. 151–73. In general, see also Baroncini, *Forme de esperienza*, chap. 5: "Harvey e l'esperienza autoptica."
[19] Wear, "William Harvey"; see also Roger French, *William Harvey's Natural Philosophy* (Cambridge: Cambridge University Press, 1994), p. 316.

The Meanings of Experience 113

to give him the most trouble was the methodological denial of the legitimacy of vivisection experiments because of their unnaturalness.[20] He could do little more than reaffirm his conclusions on the basis of the specifics of his particular procedures and the inferences drawn from them:

> And lest anyone should have recourse to the statement that these things are so when Nature is upset and preternaturally disposed, but not, however, when she is left to herself and acts freely, since in an ill and preternatural disposition appearances are not the same as in a natural and healthy one – it must therefore be said and thought that although (with the vein divided) it may seem or be stated as preternatural for so much blood to get out of the far portion because Nature is upset, yet the dissection does not close the near part to prevent anything moving out or being pressed out, whether or not Nature is upset.[21]

That is, Harvey's interventions did not, he thought, interfere with nature in any relevant way because they did not "upset" those particular matters that were under investigation.

Methodological concerns in anatomy owed their greatest textual debts to Galen, but Harvey himself, as is well known, was something of an acolyte of Aristotle. The most striking example is Harvey's use of Aristotle at the beginning of his last major work, *De generatione animalium* (On the Generation of Animals, 1651). Apart from praising Aristotle's own zoological investigations (including Aristotle's treatise with the same Latin title as Harvey's own), Harvey also employed Aristotle's account of the proper structure of scientific argument as found in the latter's *Posterior Analytics*.[22] Galen's own pronouncements on these matters were themselves heavily indebted to Aristotle. Harvey's approach to such questions, intended to show the orthodoxy of his procedures in the face of objections from fellow anatomists, therefore involved him in subtle renegotiations of the proper interpretation of Aristotelian teachings – much as was done in the case of the mathematical sciences.

Indeed, Harvey explicitly invoked the original mathematical model of deductive argument from which Aristotle himself had apparently constructed his general account of scientific procedure.[23] Harvey's position on the place of sensory experience in the making of knowledge about nature is quite clear: "Whoever wishes to know what is in question (whether it is perceptible and visible, or not) must either see for himself or be credited with belief in the

---

[20] French, *William Harvey's Natural Philosophy*, p. 277.
[21] William Harvey, "A Second Essay to Jean Riolan," in Harvey, *The Circulation of the Blood and Other Writings*, trans. Kenneth J. Franklin (London: Dent/Everyman's Library, 1963), p. 155.
[22] Charles B. Schmitt, "William Harvey and Renaissance Aristotelianism: A Consideration of the Praefatio to *De generatione animalium* (1651)," in *Humanismus und Medizin*, ed. Rudolf Schmitz and Gundolf Keil (Weinheim: Acta Humaniora, 1984), pp. 117–38.
[23] See the discussion in G. E. R. Lloyd, *Magic, Reason, and Experience: Studies in the Origin and Development of Greek Science* (Cambridge: Cambridge University Press, 1979), chap. 2.

experts, and he will be unable to learn or be taught with greater certainty by any other means."[24] The reliability of sensory experience in making natural knowledge is itself attested by geometry: "If faith through sense were not extremely sure, and stabilized by reasoning (as geometers are wont to find in their constructions), we should certainly admit no science: for geometry is a reasonable demonstration about sensibles from non-sensibles. According to its example, things abstruse and remote from sense become better known from more obvious and more noteworthy appearances."[25]

The case history as a medical genre can be traced back to the Hippocratic writings (ca. 450–ca. 350 B.C.E.) of Greek antiquity.[26] Case histories recorded in detail the progression of a disease in a particular patient from onset to resolution (either death or a return to health). Their meaning was contested in antiquity itself, with different medical sects interpreting them as either particular instances of independently existing disease entities (a case of measles, for example) or as being wholly specific to the individual patient.[27] Through most of the sixteenth and seventeenth centuries, the Latin European academic approach to medicine (the one so violently opposed by the mid-sixteenth-century physician Paracelsus and his later followers) was derived from the writings of Galen; following his general theoretical approach, physicians usually treated case histories as means to determine the generic nature of the ailment (typically in terms of an imbalance of the four humors). During the late Middle Ages and continuing into the sixteenth century, the term *experimentum* was employed by many medical writers to designate a specific remedial recipe, thereby indicating the remedy's legitimate foundation

---

[24] Harvey, *Circulation*, p. 166. Harvey's necessary reliance on "the experts" is reflected also in his dedication of *De motu cordis* to the Royal College of Physicians (p. 5): "The booklet's appearance under your aegis, excellent Doctors, makes me more hopeful about the possibility of an unmarred and unscathed outcome for it. For from your number I can name very many reliable witnesses of almost all those observations which I use either to assemble the truth or to refute errors; you so instanced have seen my dissections and have been wont to be conspicuous in attendance upon, and in full agreement with, my ocular demonstrations of those things for the reasonable acceptance of which I here again most strongly press." For more on the common expression "ocular demonstration" as used here by Harvey, see Thomas L. Hankins and Robert J. Silverman, *Instruments and the Imagination* (Princeton, N.J.: Princeton University Press, 1995), esp. p. 39, and also Barbara J. Shapiro, *A Culture of Fact: England, 1550–1720* (Ithaca, N.Y.: Cornell University Press, 2000), p. 172, noting the expression's use in the context of English religious apologetics. The term, of course, refers to first-hand eyewitnessing.

[25] Harvey, *Circulation*, p. 167; see also French, *William Harvey's Natural Philosophy*, p. 278.

[26] For an excellent overview, see G. E. R. Lloyd, "Introduction," in *Hippocratic Writings*, ed. G. E. R. Lloyd (Harmondsworth: Penguin, 1978). On the complex relations between natural philosophy, medicine, and natural history in the seventeenth century, see Harold J. Cook, "The New Philosophy and Medicine in Seventeenth Century England," in *Reappraisals of the Scientific Revolution*, ed. David C. Lindberg and Robert S. Westman (Cambridge: Cambridge University Press, 1990), pp. 397–436; and Cook, "The Cutting Edge of a Revolution? Medicine and Natural History Near the Shores of the North Sea," in *Renaissance and Revolution: Humanists, Scholars, Craftsmen, and Natural Philosophers in Early Modern Europe*, ed. J. V. Field and Frank A. J. L. James (Cambridge: Cambridge University Press, 1993).

[27] John Scarborough, *Roman Medicine* (Ithaca, N.Y.: Cornell University Press, 1969), provides a convenient overview of the Hellenistic and Greco-Roman medical sects and writers.

in the writer's experience of the ailment and its treatment. In the sixteenth century, Girolamo Cardano (1501–1576) provided a notable example of such usage by someone explicitly aware of the ways in which medicine fell short of being a true, or "perfect," science because it did not strictly demonstrate from unquestionable principles.[28] There also lingered around medical uses of *experimentum* a certain aura of the occult (in the sense of "hidden" from normal cognitive comprehension), which resonated with the term's use by Roger Bacon in the thirteenth century.[29]

EXPERIENCE AND NATURAL HISTORY: INDIVIDUALS, SPECIES, AND TAXONOMY

At the turn of the eighteenth century, Étienne Chauvin's *Lexicon philosophicum* (1692 and 1713) described a terminological distinction that seems to have become commonplace during the preceding several decades. *Experientia*, according to Chauvin, holds a place among physical principles second only to reason, "for reason without experience is like a ship tossing about without a helmsman." Chauvin distinguished among three kinds of experience: the experience that is acquired unintentionally in the course of life; the kind gained from deliberate examination of something, but in the absence of any particular expectation of the eventuality; and the experience acquired with the purpose of determining the truth of a conjectured explanation (*ratio*).[30] Chauvin, employing an additional Latin word (*experimenta*), then proceeded to describe the nature of a properly philosophical experience: It should be based on "experiments" of varying kinds and of considerable number; these experiments properly encompass mechanical artifice as well as natural history.[31] A philosophical experience, therefore, is made from numerous experiments,[32] much as, in Aristotle's account, an experience was made from many memories of the same thing.[33] Unlike Aristotle, however, Chauvin did

---

[28] Nancy G. Siraisi, *The Clock and the Mirror: Girolamo Cardano and Renaissance Medicine* (Princeton, N.J.: Princeton University Press, 1997), chap. 3, esp. pp. 45, 59–60; Baroncini, *Forme di esperienza*, pp. 109–10. See also, for example, Francis Bacon's usage in Bacon, *Of the Proficiencie and Advancement of Learning* (London: Henrie Tomes, 1605), 2.8.

[29] Jole Agrimi and Chiara Crisciani, "Per una ricerca su *experimentum-experimenta*: Riflessione epistemologica e tradizione medica (secolo XIII–XV)," in *Presenza del lessico greco e latino nelle lingue contemporanee*, ed. Pietro Janni and Innocenza Mazzini (Macerata: Università degli Studi di Macerata, 1990), pp. 9–49.

[30] Étienne Chauvin, *Lexicon Philosophicum* (Leeuwarden, 1713; facsimile repr. Düsseldorf: Stern-Verlag Janssen, 1967), p. 229, col. 2.

[31] Ibid., p. 230, col. 1. Varying the kinds of experiments underpinning a philosophical claim was something that Francis Bacon had also advocated; he criticized William Gilbert for building an entire philosophy from nothing but magnetic experiments. See Francis Bacon, *New Organon*, 1.54.

[32] The Jesuit mathematician Christopher Scheiner had used the same terminological distinction at the beginning of the seventeenth century. See Peter Dear, *Discipline and Experience: The Mathematical Way in the Scientific Revolution* (Chicago: University of Chicago Press, 1995), pp. 55–7.

[33] Aristotle, *Metaphysics*, 6.2 (1026b 29–32).

not worry about the differences in philosophy between experience of artificial constructs (mechanical artifice, recalling those of René Descartes) and experience of natural processes – including what Chauvin called "natural history"

Indeed, natural history itself was an area of research, and a rubric, in rapid reconstruction during the seventeenth century. Francis Bacon had stressed the importance of a comprehensive natural history, or descriptive account of natural phenomena, as the prolegomenon to the construction of a true natural philosophy. Bacon meant not just the subject matters that are nowadays understood by "natural history" but all natural phenomena, animate and inanimate.[34] Natural history was principally to be distinguished from "civil history," comprised of accounts of human affairs; both kinds of histories were descriptive accounts, neither (supposedly) giving causal *explanations* of the matters that they addressed. A generally Baconian sense of natural history remained particularly important in English natural philosophy, including that of the early Royal Society, for the rest of the century.[35] But in those studies to which the term "natural history" later came to be restricted, chiefly botany and zoology, issues of singulars and universals arose similar to those just discussed.

In sixteenth-century Italy, Ulisse Aldrovandi (1522–1605) and other naturalists first began to collect actual specimens of plants rather than simply describing the plants as they appeared in situ.[36] This new practice was essential to the notion of natural historical knowledge as being centered on collections of specimens brought from many different locations. It was adopted into the new botanical gardens (usually associated, as in Italy, with universities) that began to be founded in France on the Italian model in the second half of the sixteenth century[37] (see Findlen, Chapter 19, this volume). The collection of specimens (which can be seen as, among other things, ancestors of the nineteenth-century type specimen in paleontology)[38] brought experience into the making of natural history in a way that effectively reinforced conceptual categories with tangible, visible exemplars. An aspect of the new approaches to botany in this period is the use of naturalistic drawings of plants, as represented, for example, by Otto Brunfels's *Herbarum vivae eicones* (1530), which exploited the new printing technologies.[39] As in the case of Vesalius, however, such use of visual representations was controversial: Vesalius was

---

[34] See Francis Bacon, *New Organon*, "The Great Renewal," pp. 20–1.
[35] See, for example, Shapiro, *A Culture of Fact*, pp. 114–16 and chap. 5.
[36] Paula Findlen, *Possessing Nature: Museums, Collecting, and Scientific Culture in Early Modern Italy* (Berkeley: University of California Press, 1994), p. 166.
[37] Karen Meier Reeds, *Botany in Medieval and Renaissance Universities* (New York: Garland, 1991), which includes a reprint of Reeds, "Renaissance Humanism and Botany," *Annals of Science*, 33 (1976), 519–42.
[38] Ronald Rainger, *The Understanding of the Fossil Past: Paleontology and Evolution Theory, 1850–1910* (Princeton, N.J.: Princeton University Press, 1982).
[39] Reeds, *Botany*, pp. 28–32.

obliged to defend himself, in the preface to *De humani corporis fabrica* (On the Fabric of the Human Body, 1543), against those who thought that the provision of a book purporting to show details of the human body would merely encourage medical students to rely on the book instead of looking for themselves – precisely the opposite of Vesalius's announced intention.[40] So, too, some sixteenth-century herbalists reiterated the arguments of ancient writers such as Pliny and Galen against providing pictures of plants, which were potentially deceptive and inferior to careful observations of the real thing.[41]

Apart from the principal pharmaceutical uses of plants, botanical taxonomy emerged as a serious issue in the sixteenth century in part because of the sheer number of new plants then reaching Europe for the first time. Andrea Cesalpino (1519–1603), at the University of Pisa, provided the most influential taxonomic model in his *De plantis libri xv* (1583). Its philosophical justification (and it is significant that Cesalpino felt the need for one) was derived from Aristotle: Taking reproduction as a fundamental function in the perpetuation of species, and thereby following Aristotle's general precepts, Cesalpino justified using the reproductive parts of plants (flowers and fruit) as possessing characters that would relate most fundamentally to the essential nature of the plant itself.[42] Thus, such characters were the proper ones to use as discriminatory criteria in classification. This general approach, despite differences in the details of taxonomic schemes, was followed by subsequent taxonomists throughout the seventeenth century.[43] It was a way of presenting the practical task of classification as more than just a cataloging system added to descriptive natural history; by privileging particular characters on theoretical grounds, natural history could also strive toward the higher status of natural *philosophy*.

The indeterminacy of sensory experience in such matters as the allocation of species to appropriate genera became increasingly clear to the English botanist John Ray (1627–1705) in the 1690s. Taxonomical practice had become a matter of deciding the significance of similarities and differences, a move, in effect, from experience (which revealed a specimen's relevant characters) to the knowledge of that specimen's essential nature (what kind of thing it really was). Ray, however, denied the possibility of such an inductive move from experience to knowledge of essences. In polemics conducted in the 1690s with the continental taxonomists Augustus Bachmann (in Germany) and Joseph Tournefort (in France), Ray argued that the categorization of

---

[40] See Vesalius's preface, in C. D. O'Malley, *Andreas Vesalius of Brussels, 1514–1564* (Berkeley: University of California Press, 1964), pp. 322–3.
[41] Reeds, *Botany*, pp. 31–2.
[42] See, for example, Findlen, *Possessing Nature*, p. 58.
[43] For its continuing importance for Linnaeus in the eighteenth century, see Sten Lindroth, "The Two Faces of Linnaeus," in *Linnaeus: The Man and His Work*, ed. Tore Frängsmyr (Canton, Mass.: Science History Publications, 1994), pp. 1–62, esp. pp. 35–6.

organisms into larger groups such as, for example, the aggregation of species into genera, could never achieve philosophical soundness. Creatures should properly be grouped together according to common essential characters; that is, characters expressive of the creature's essential nature. Thus, classification according to accidental characters – characters *not* expressive of the thing's essence – would not be a true, natural classification. But, Ray queried, how are we to know which characters of an organism are essential to it and which merely accidental? He adduced the case of whales, where, depending on our choice of characters, the animals could be grouped together with fish (if such matters as living exclusively in the water and possessing fins were taken as essential characters) or with warm-blooded land animals (if live births and air breathing were taken as essential).[44] Ray's skeptical stance was thus directly constitutive of his views regarding the relationship between experience in natural history and the expression of formalized knowledge of nature.[45] Experience meant much more than descriptive observation.

Such taxonomic concerns were ridiculed by Jonathan Swift's satire of the projects of the Royal Society of London in *Gulliver's Travels* (1726), cloaked as those of the fictitious academicians of Lagado. The latter wished to abolish the use of words and instead to communicate through the display of the things themselves – for "words were only names for things."[46] Swift's immediate target here would seem to have been (rather than John Locke) the universal language projects of the Royal Society in the 1660s, of which John Wilkins's *Essay Towards a Real Character and a Philosophical Language* (1668) was the most celebrated. Wilkins's book was an attempt to encompass all things in the world within a comprehensive language scheme built on taxonomical principles. The scheme was, at root, essentialist; that is, it assumed the possibility of identifying the true *kinds* of things found in the world in order to designate each of them with its own name.[47] The language schemes of Wilkins and others in England in the 1650s and 1660s were in this way structured on fundamentally Aristotelian principles. They assumed, just as scholastic Aristotelians had done, that sensory experience yielded concepts that could then be denoted by words: The deep psychological trick lay precisely in the extraction of concepts regarding the universal essences of kinds of things from the singulars of actual experience.[48] It was the possibility of this (Aristotelian) trick that Ray denied at the end of the seventeenth

---

[44] Phillip R. Sloan, "John Locke, John Ray, and the Problem of the Natural System," *Journal of the History of Biology*, 5 (1972), 1–53.
[45] On skepticism in early modern Europe, see especially Richard H. Popkin, *The History of Scepticism from Savonarola to Bayle* (New York: Oxford University Press, 2003).
[46] Jonathan Swift, *Gulliver's Travels*, 3.5 (Oxford: Oxford University Press, 1948), p. 223.
[47] A point argued by Hans Aarsleff, "Wilkins, John," *Dictionary of Scientific Biography*, 14, 361–81; reprinted in Aarsleff, *From Locke to Saussure* (Minneapolis: University of Minnesota Press, 1982).
[48] On the Aristotelian structure of such schemes, see Mary M. Slaughter, *Universal Languages and Scientific Taxonomy in the Seventeenth Century* (Cambridge: Cambridge University Press, 1982).

century. At the same time, however, the work of Ray's contemporary Isaac Newton was refashioning such issues in the context of the mathematical sciences.

EXPERIENCE AND THE MATHEMATICAL SCIENCES

The emergence of something resembling "experimental science" in this period occurred most evidently in the so-called mathematical sciences. Following the widely accepted Aristotelian view, these were frequently represented as branches of natural knowledge that concerned only the quantitative, measurable properties of things rather than questions having to do with what *kinds* of things they were. Those latter questions fell under the general disciplinary heading of "natural philosophy" but not "mathematics." Thus, such sciences as astronomy (studying the positions and movements in the sky of celestial objects) and geometrical optics (studying the quantitative behavior of geometrically construed light rays) were branches of "mathematics." They were also the sciences that made the greatest use of specialized instruments such as quadrants and astrolabes, and sometimes, especially in optics, custom-made experimental apparatus, to generate precise empirical results. This meant that they provided to their practitioners recondite knowledge that was, for that reason, hard to fit into the mold of a demonstrative science because it was not rooted in generally accepted experience.[49]

The disciplinary structure of sixteenth-century universities (including, in the latter part of that century, the influential new colleges of the Jesuits) reified a conceptual scheme that placed mathematical sciences in a category separate from that of natural philosophy.[50] The arts curriculum of medieval and early modern universities had derived from the late antique classification of the *trivium*, comprising the headings grammar, logic, and rhetoric, and the *quadrivium*, consisting of arithmetic, geometry, astronomy, and music.[51] The last two items stood for a slew of mathematical sciences of the physical world, including in addition such disciplines as geography, geometrical optics, and mechanics (statics). They were known in the early modern period

---

[49] Canonical examples are found in Ptolemy's *Almagest* (astronomy) and in Alhazen's optical text, usually known in its Latin version as *Perspectiva*, first printed in Federicus Risnerus, ed., *Opticae thesaurus: Alhazeni arabis libri septem, nunc primum editi* . . . (Basel: per Episcopios, 1572). There were many individual practitioners of such sciences in the sixteenth and seventeenth centuries who rejected such a sharp separation of natural philosophy from mathematical sciences, Kepler prominent among them; but doing so could expose such dissenters to sharp methodological critique, as discussed.

[50] For the official statement in the Jesuits' 1599 *Ratio studiorum* of the disciplinary and conceptual distinction between natural philosophy and mathematics, see Mario Salmone, ed., *Ratio atque institutio studiorum Societatis Jesu: L'ordinamento scolastico dei collegi dei Gesuiti* (Milan: Feltrinelli, 1979), p. 66.

[51] Essays on the quadrivial disciplines in the early Middle Ages may be found in David L. Wagner, *The Seven Liberal Arts in the Middle Ages* (Bloomington: Indiana University Press, 1983).

under various labels, such as "subordinate," "middle," or "mixed" sciences.[52] Aristotle had proposed a particular way to understand how they could be seen as true sciences.

According to Aristotle, a true science (*episteme*) should be founded on its own proper principles unique to that science. Subject matters were thus distributed into distinct sciences according to the content of their principles, so that the principles of a science would always be of the same genus as its subject matter. The requirement thus served to ensure the possibility of a formal deductive link between premises and conclusions. However, disciplines such as astronomy and music apparently violated this rule: They drew on the results of pure mathematics (arithmetic and geometry) to apply them to something other than pure quantity, in this case celestial motions and sounds. Consequently, Aristotle made a special accommodation for them by classifying them as sciences *subordinate* to higher disciplines.[53] Aristotle's solution to the problem was rather ad hoc; in the sixteenth century, it provoked scholastic discussions on whether demonstrations in a mixed mathematical science really did yield true scientific knowledge if the presupposed theorems of arithmetic or geometry were not actually proved alongside them.[54] These doubts were accompanied by suggestions that mathematical demonstrations did not conclude through arguments that specified the *causes* of the conclusions (that is, of the effects to be explained).[55] On those grounds, they were not to be placed on a par with philosophical demonstrations.

In the late sixteenth and early seventeenth centuries, however, the argument that mixed mathematics did not produce genuine scientific knowledge was fiercely contested by prominent Jesuit mathematicians, most important among them Christoph Clavius (1538–1612). In making his case, Clavius relied heavily on the authority of Aristotle and other ancient sources. Aristotle had not only made the parts of mixed mathematics into subordinate sciences, thereby implicitly affirming their scientific status, but had also explicitly included mathematics as a part of philosophy. Clavius used this scheme, citing Ptolemy as an additional witness, to suggest that mathematics was not only the equal of the qualitative and undoubtedly scientific natural philosophy but was in fact its superior: "For [Ptolemy] says that natural philosophy and metaphysics, if we consider their mode of demonstrating, are rather to

---

[52] See further discussion in Dear, *Discipline and Experience*, p. 39; for medieval and sixteenth-century background, see W. R. Laird, "The Scientiae mediae in Medieval Commentaries on Aristotle's Posterior Analytics," Ph.D. dissertation, University of Toronto, 1983, esp. chap. 8.

[53] Two central texts are: Aristotle, *Posterior Analytics*, 1.7; Aristotle, *Metaphysics*, 13.3 (esp. 1078a 14–17). See Richard D. McKirahan, *Principles and Proofs: Aristotle's Theory of Demonstrative Science* (Princeton, N.J.: Princeton University Press, 1992).

[54] See for documentation and further references William A. Wallace, *Galileo and His Sources: The Heritage of the Collegio Romano in Galileo's Science* (Princeton, N.J.: Princeton University Press, 1984), p. 134.

[55] A useful discussion is in Nicholas Jardine, "Epistemology of the Sciences," in *The Cambridge History of Renaissance Philosophy*, ed. Charles B. Schmitt, Quentin Skinner, and Eckhard Kessler with Jill Kraye (Cambridge: Cambridge University Press, 1988), pp. 685–711, at pp. 693–7.

The Meanings of Experience 121

be called conjectures than sciences, on account of the multitude and discrepancy of opinions."[56] These attitudes were by no means confined to Jesuit mathematicians (one can also point to figures such as the Englishman John Dee in the second half of the sixteenth century),[57] but Jesuit writers such as Clavius were especially influential in the seventeenth century because they were widely read and cited by other, often non-Jesuit (and non-Catholic), mathematicians.[58]

Mathematicians in the early seventeenth century, particularly among the Jesuits and those influenced by them, thus continued to look to Aristotelian and other classical sources as their disciplinary models. This gave them work to do if mixed mathematical sciences, which concerned the natural world and therefore rested largely on sensory evidence, were to remain scientifically valid in Aristotle's sense. In order to universalize experiential premises, such premises needed to command assent because they were evident, not because of particular events adduced in their support. What they produced therefore did not look like "experimental science" in the modern sense: From this Aristotelian perspective, statements of individual events are not evident and indubitable but rely on historical reports that are necessarily *fallible*. The Aristotelian model of a science thus took scientific knowledge to be fundamentally open and public insofar as scientific demonstration derived from principles that commanded universal assent. Singular experiences, experimental events, were not public because they were known directly only to those few who had actually witnessed them; such experiences were in consequence questionable elements of scientific discussion.

By way of compensation, therefore, mathematical scientists depended to some degree on their reputations as reliable truth-tellers, or at least (especially in the case of the Jesuits) on the reputations of their institutions. They did not need to rely exclusively on such vulnerable foundations, however. Contemporary astronomers, for example, were not in the habit of publishing raw astronomical data: Rather than presenting immediate observational results confirmed by their own testimony, astronomers typically used their data as a means of generating, via geometrical models of celestial motions, predictive tables of planetary, solar, or lunar positions. In other words, there was no formal methodological separation between the observational and the calculational parts of the enterprise. As the Jesuit Niccolò Cabeo wrote in his commentary on Aristotle's *Meteorology* (1646),[59] there was a necessary reliance

---

[56] Christoph Clavius, "In sphaeram Ioannis de Sacro Bosco commentarius," in Clavius, *Opera mathematica*, 5 vols. in 4 (Mainz: A. Hierat for R. Eltz, 1611–12), 3: 4. Cf. Ptolemy, *Almagest*, 1.1; see the English translation in *Ptolemy's Almagest*, trans. G. J. Toomer (London: Duckworth, 1984), p. 36.

[57] See especially John Dee, *The Mathematicall Preface to the Elements of Geometrie of Euclide of Megara*, intro. Allen G. Debus (1570; facsimile repr. New York: Science History Publications, 1975).

[58] See, for example, the English Protestant Isaac Barrow's use of Jesuit sources in the 1660s in Dear, *Discipline and Experience*, p. 223; such attention was commonplace.

[59] Niccolò Cabeo, *In quatuor libros Meteorologicorum Aristotelis commentaria* (Rome: Francisco Corbeletti, 1646), p. 399, col. 2.

in astronomy on testimony and human records. But Cabeo did not see this fact as methodologically disabling because the universalized "experiences" that result from the accumulated data provide astronomy with an apparently self-validating character: Cabeo noted that, as a result of the long process of astronomical endeavor since antiquity, there had emerged "from the power of those observations laws and canons of celestial motions which correspond best to things."[60] It did not matter that the observational data derived from nonevident historical testimony because data could *never* be evident; they were not in themselves universals. The acceptance of the principles on which astronomy was based depended instead on an ongoing familiarity with their verisimilitude as guides to current and future appearances.[61] The legitimacy of the knowledge claimed by astronomy depended on the discipline's continuing practice.

The Antwerp Jesuit François d'Aguilon, writing on optics in 1613, had also expressed the importance of transcending particulars, but in a slightly different way directly beholden to Aristotelian epistemology:

> For a single [sensory] act does not greatly aid in the establishment of sciences and the settlement of common notions, since error can exist which lies hidden for a single act. But if [the act] is repeated time and again, it strengthens the judgement of truth until finally [that judgement] passes into common assent; whence afterwards [the resulting common notions] are put together, through reasoning, as the first principles of a science.[62]

These remarks clearly appealed to Aristotle's definition of "experience" in his logical treatise *Posterior Analytics*: "From perception there comes memory . . . and from memory (when it occurs often in connection with the same thing), experience; for memories that are many in number form a single experience."[63] For Aguilon, repetition was essential to creating a properly scientific experience. Repetition combats deception by the fallible senses or by the unfortunate choice of an atypical instance, and hence ensures a reliable statement about how nature behaves "always or for the most part," as Aristotle had put it.[64] The result is experience adequate to establish the empirical "common notions" that form the basis of a science.

The Aristotelian kind of scientific experience held sway even among figures later regarded as opponents of Aristotle. In his famous account of fall along inclined planes, published in the *Discorsi e dimostrazioni matematiche intorno a due nuove scienze* (Discourses and Mathematical Demonstrations

---

[60] Ibid.
[61] Dear, *Discipline and Experience*, p. 95.
[62] Franciscus Aguilonius, *Opticorum libri sex* (Antwerp: Ex officina Plantiana, 1613), pp. 215–16.
[63] Aristotle, *Posterior Analytics*, 2.19 (100a 4–9), trans. Jonathan Barnes in Aristotle, *The Complete Works of Aristotle: The Revised Oxford Translation*, ed. Jonathan Barnes (Princeton, N.J.: Princeton University Press, 1984) pp. 165–6.
[64] Aristotle, *Metaphysics*, 6.2 (1026b 29–32).

Concerning Two New Sciences, 1638),[65] Galileo Galilei (1564–1642) tried to establish the authenticity of the experience that falling bodies in fact behave as he claimed they do by deriving it from the memory of many individual instances. He did not describe a specific experiment or set of experiments carried out at a particular time, together with a detailed quantitative record of the outcomes; instead, he just wrote that, with apparatus of a kind carefully specified, he had found that the results agreed exactly with his expectations, in trials repeated "a full hundred times." This last phrase (found frequently, in various forms, in contemporary scholastic writings) means "countless times."[66]

Galileo's approach was mirrored by that of many contemporary writers in the mixed mathematical sciences: Detailed accounts of experimental or observational apparatus were commonly followed by assertions of the unvarying results of their proper use.[67] In a search to win assent for their less than obvious empirical principles, such writers avoided the thorny issue of trust by refusing, in effect, to acknowledge distrust as a relevant option;[68] reputation and institutional credibility took the strain. René Descartes (1596–1650) adopted a comparable approach to the same problems: His famous attempt to provide a solid grounding for knowledge took its lead from deductive mathematical reasoning but also reserved a place for experience. Descartes, too, finessed the problem of trust by refusing to treat it as an issue. In the *Discours de la méthode* (Discourse on Method, 1637), he transparently invited others to assist in his work by furnishing him with "the cost of the experiences that he would need" because information received from other people would typically yield only prejudiced or confused accounts, or at least would oblige him to waste his own valuable time by repaying his informants with explanations and discussions. Descartes wanted to make the requisite experiences himself or pay artisans to do them – the incentive of financial gain ensuring that the latter would do exactly as they were instructed.[69] Descartes was intent only on satisfying himself, as if that should be enough for all.

---

[65] The standard English edition is Galileo Galilei, *Discourses and Demonstrations Concerning Two New Sciences*, trans. Stillman Drake (Madison: University of Wisconsin Press, 1974).

[66] Dear, *Discipline and Experience*, pp. 129–32; cf. Charles B. Schmitt, "Experience and Experiment: A Comparison of Zabarella's View with Galileo's in *De motu*," *Studies in the Renaissance*, 16 (1969), 80–138.

[67] See also Dear, *Discipline and Experience*, p. 80. A precedent for the description of apparatus appears in Ptolemy's second-century account of astronomical sighting instruments in the *Almagest*.

[68] See Shapin, *Social History of Truth*.

[69] René Descartes, *Discours de la méthode*, pt. 4, in Descartes, *Oeuvres de Descartes*, ed. Charles Adam and Paul Tannery, 8 vols. (Paris: J. Vrin, 1964–76), 6: 72–3. On Descartes and experiment, see Daniel Garber, "Descartes and Experiment in the *Discourse* and the *Essays*," in *Essays on the Philosophy and Science of René Descartes*, ed. Stephen Voss (New York: Oxford University Press, 1993), pp. 288–310; Garber, *Descartes' Metaphysical Physics* (Chicago: University of Chicago Press, 1992), chap. 2; Desmond Clarke, *Descartes' Philosophy of Science* (Manchester: Manchester University Press, 1982), pp. 22–3. On the contemporary work in France of Mersenne and Pascal, see Dear, *Discipline and Experience*, chaps. 5, 7; and Christian Licoppe, *La formation de la pratique scientifique: Le discours de l'expérience en France et en Angleterre (1630–1820)* (Paris: Éditions de la Découverte, 1996), chap. 1.

EVENT EXPERIMENTS AND "PHYSICO-MATHEMATICS"

In the study of the inanimate world, set-piece experiments seem first to have entered significantly into knowledge-making practices in the domain of the mathematical sciences. Here we first find regular use of historical reports of particular events to justify universal statements about how some aspect of nature behaves. Hints of this departure are found in Galileo's mathematical work on motion, but whereas Galileo tried to avoid placing the justification for his claims squarely on historical reports, other writers on the mathematical sciences were beginning to narrate particular, contrived events. Thus, Jesuit mixed mathematicians, including the astronomer Giambattista Riccioli (1598–1671), reported experiments that involved dropping weights from the tops of church towers to determine their rates of acceleration, and gave places, dates, and witnesses to underwrite their stories.[70] One of the most famous such instances in the seventeenth century took place in 1648. Blaise Pascal (1623–1662), in Paris, had asked his brother-in-law, Florin Périer, off in the French provinces, to take a mercury barometer up a nearby mountain, the Puy-de-Dôme, to determine whether the mercury's height in the glass tube would decrease with increasing altitude. Pascal expected that it would, and believed that such a result would confirm his conviction that the mercury column in the tube was sustained by the weight of the air – there being less atmospheric air to weigh down and thus counterweight the mercury at higher elevations than at lower ones.

Périer's report on the trial was quickly published by Pascal. It gives a detailed, circumstantial account of Périer's trip up the mountain and back one day in September, in the company of named witnesses, and the measurements that were made along the way. This was not an entirely unequivocal use of a recorded event as justifying evidence for a claim about nature because Pascal buttressed his brother-in-law's narrative by using its results to predict an analogous drop in the height of mercury as a result of carrying similar apparatus up church towers in Paris; he then asserted that actual (unspecified) trials bore out that prediction.[71] Nonetheless, Pascal's promotion of the Puy-de-Dôme trial indicates the role that contrived, set-piece experiments, historically reported, were beginning to play.

The general introduction of this kind of "experimental experience" from the mathematical sciences into the wider arena of natural philosophy may be

---

[70] Alexandre Koyré, "A Documentary History of the Problem of Fall from Kepler to Newton: De motu gravium naturaliter cadentium in hypothesi terrae motae," *Transactions of the American Philosophical Society*, n.s. 45 (1955), pt. 4. The further concept of the "virtual witness," one who experiences vicariously the empirical findings of others by means of reading a detailed circumstantial account of the proceedings, was first put forward in Steven Shapin, "Pump and Circumstance: Robert Boyle's Literary Technology," *Social Studies of Science*, 14 (1984), 481–520. See also Henry G. Van Leeuwen, *The Problem of Certainty in English Thought, 1630–1690* (The Hague: Martinus Nijhoff, 1963).

[71] Dear, *Discipline and Experience*, pp. 196–201.

The Meanings of Experience

traced by reference to the gradual emergence in the seventeenth century of a new term, "physico-mathematics." The idea that mathematics, particularly the mixed mathematical disciplines, could yield genuinely *causal* scientific knowledge of natural bodies and phenomena became a virtual commonplace during the first half of the seventeenth century.[72] The gradual introduction of the category "physico-mathematics" served in effect to elevate the status of mathematical sciences to the level of physics (natural philosophy) without formally violating the long-standing and well-entrenched Aristotelian disciplinary separation of the two.[73]

The new category made it easier for mathematical scientists to make philosophical claims that had previously been fiercely contested. Galileo's dispute over floating bodies in 1612 had taken the form of an assertion of the rights of mathematics over those of physics.[74] A process of disciplinary imperialism, whereby subject matter usually regarded as part of physics was taken over by mathematics, operated to upgrade the status and explanatory power of the mathematical sciences. The label "physico-mathematics" made the move explicit to all.

The term appears both in popular vernacular texts and in workaday academic settings during the 1620s, 1630s, and 1640s. This seems to have occurred along with a restructuring of what was demanded of physical knowledge itself. Thus the stress found in Clavius's writings on the demonstrative certainty of mathematics came to overshadow the Aristotelian physicists' fundamentally teleological causal-explanatory ambitions. The Jesuit-educated Marin Mersenne (1588–1648), for example, was familiar with the texts in which Clavius had paraded certainty as a mark of the superiority of mathematics over physics;[75] the appeal of mixed mathematics as an exemplar of a new philosophy of nature for Mersenne was of a piece with the widespread and growing acceptance of the idea of a true "physico-mathematics" that would combine mathematical demonstration with physical subject matter.

Cambridge mathematician Isaac Barrow (1630–1677) provided a useful picture of "physico-mathematics" in England in the 1660s, by which time the term had become firmly established in mathematical usage. In discussing mathematical terms and categories in his *Mathematical Lectures* (read in

---

[72] On conceptual aspects of the relationship between physics and mathematics, and the role of mixed mathematics as mediator, among Jesuit mathematicians in the early decades of the seventeenth century, see Ugo Baldini, *Legem impone subactis: Studi su filosofia e scienza dei Gesuiti in Italia, 1540–1632* (Rome: Bulzoni, 1992), chap. 1. On pressures promoting the revaluing of mathematical knowledge, see Mario Biagioli, "The Social Status of Italian Mathematicians, 1450–1600," *History of Science*, 27 (1989), 41–95.

[73] Dear, *Discipline and Experience*, chap. 6, sec. IV.

[74] See Mario Biagioli, *Galileo, Courtier: The Practice of Science in the Culture of Absolutism* (Chicago: University of Chicago Press, 1993), chap. 4.

[75] Peter Dear, *Mersenne and the Learning of the Schools* (Ithaca, N.Y.: Cornell University Press, 1988), pp. 37–9.

that decade but not published until 1683), Barrow made the usual distinction between "pure" and "mixed" mathematics. He noted that the latter dealt with physical accidents rather than with the nature of quantity in itself, and that some people were wont to call its divisions (in Latin) "Physico-Mathematicas."[76] Barrow insisted that physics and mathematics were strictly inseparable: All physics implicates quantity and hence mathematics.[77] Barrow's position accorded well with the famous title later used by his student and successor as Lucasian Professor of Mathematics at Cambridge, Isaac Newton (1642–1727). Newton's *Principia mathematica philosophiae naturalis* (Mathematical Principles of Natural Philosophy, 1687) possessed a title that would have been unthinkable by earlier – Aristotelian – standards because by definition natural philosophy could not have had *mathematical* principles.[78] The *Principia* sums up neatly the direction that arguments concerning the potential of the mixed mathematical sciences had taken during preceding decades and shows precisely the point at which experimental contrivance and historical reporting about it were by 1687 flooding into natural philosophical practice.

NEWTONIAN EXPERIENCE

It is with the Royal Society of London, founded in the early 1660s, and especially with the exemplary work of the English natural philosopher Robert Boyle (1627–1691), that concern with experimental reports became most clearly established as the foundation of a new natural philosophy.[79] This was not an uncontested victory, and the opposition to an experimental approach to natural philosophy was not restricted to Thomas Hobbes (1588–1679); he and many others, including Aristotelians, regarded experimental knowledge as nothing but descriptive natural history, unfit for grounding philosophical

---

[76] Isaac Barrow, *Lectiones mathematicae* (1683), reprinted in *The Mathematical Works of Isaac Barrow, D.D.*, ed. William Whewell (Cambridge, 1860; facsimile repr., Hildesheim: Georg Olms, 1973), p. 31 (lect. 1); see also p. 89 (lect. 5).
[77] Ibid., p. 41 (lect. 2); see Edwin Arthur Burtt, *The Metaphysical Foundations of Modern Science: A Historical and Critical Essay* (Garden City, N.Y.: Doubleday [Anchor Books], 1954), pp. 150–5, and additional discussion and references in Dear, *Discipline and Experience*, pp. 222–7.
[78] See Dear, *Discipline and Experience*, chap. 8.
[79] Steven Shapin and Simon Schaffer, *Leviathan and the Air-Pump: Hobbes, Boyle, and the Experimental Life* (Princeton, N.J.: Princeton University Press, 1985), chap. 2; Shapin, "Pump and Circumstance." See also Michael Ben-Chaim, "The Value of Facts in Boyle's Experimental Philosophy," *History of Science*, 38 (2000), 1–21, and more generally Lorraine Daston, "Baconian Facts, Academic Civility, and the Prehistory of Objectivity," *Annals of Scholarship*, 8 (1991), 337–63. Another discussion of the "factual" is Mary Poovey, *A History of the Modern Fact: Problems of Knowledge in the Sciences of Wealth and Society* (Chicago: University of Chicago Press, 1998), esp. chaps. 1–3, with much discussion of English material from the seventeenth and eighteenth centuries; see also Shapiro, *A Culture of Fact*. Daniel Garber, "Experiment, Community, and the Constitution of Nature in the Seventeenth Century," *Perspectives on Science*, 3 (1995), 173–205, discusses the differences between such explicit attention to communal fact-making and Descartes' insistence on the capacity of the solitary knower.

knowledge.[80] The sources of the Royal Society's predilection for historical reports as the core of its communal enterprise, however, are difficult to pin down. The early Fellows usually credited Francis Bacon with having inspired their enterprise, and indeed their professed concern with useful knowledge, and with empirical investigation as the means to its acquisition, resonated strongly with Bacon's work. Bacon's name was also invoked on the Continent, by luminaries of the Paris Académie Royale des Sciences (founded in 1666).[81] It is noteworthy, however, that the other famed assembly of experimenters at this time, the Florentine Accademia del Cimento (founded in 1657 but dissolved by 1667), whose published experiments resembled quite closely many of those by Boyle and others, made virtually no mention of Bacon at all.[82] In England, however, Isaac Newton represents an especially significant expression of the development of experimental reports; as a result of his work, the "experimental philosophy" promoted by Robert Boyle was wedded to the quasi-experimental practices of the mixed mathematical sciences to yield a new synthesis that became established in the eighteenth century as one of the many senses of "Newtonianism."[83] The methodological hallmark of Newtonianism came to be a characteristically agnostic stance toward fundamental causal claims regarding the inner natures, or essences, of the things being investigated.[84] Thus, Newton represented his ideas on light and colors as being solidly rooted in experience; they did not, he claimed, exceed the (conveniently) high degree of certainty that the mathematical science of optics traditionally afforded. Newton purported to be able to show by experiment that white light was a mixture of the colors. When refracted through a prism to produce a spectrum, white light was separated into its components; the refractive colors were not (as had formerly been thought) newly created as modifications of the white light. Newton denied that these claims relied in any way on a particular hypothesis regarding the true *nature* of light – a particle or a wave theory, for example.

Newton's optical work lay squarely within the tradition of geometrical optics, one of the mixed mathematical sciences. Newton's work, however, also needs to be understood in the specifically physico-mathematical terms of Barrow. Newton stepped beyond the boundaries of classical mixed mathematics when he began to address questions of natural philosophy from the basis of his *mathematical* conclusions: "These things being so, it can

---

[80] See above all Shapin and Schaffer, *Leviathan and the Air-Pump*.
[81] See in general Alice Stroup, *A Company of Scientists: Botany, Patronage, and Community at the Seventeenth-Century Parisian Royal Academy of Sciences* (Berkeley: University of California Press, 1990); and Roger Hahn, *The Anatomy of a Scientific Institution: The Paris Academy of Sciences, 1666–1803* (Berkeley: University of California Press, 1971), chap. 1.
[82] W. E. Knowles Middleton, *The Experimenters: A Study of the Accademia del Cimento* (Baltimore: Johns Hopkins University Press, 1971), pp. 331–2.
[83] Robert E. Schofield, "An Evolutionary Taxonomy of Eighteenth-Century Newtonianisms," *Studies in Eighteenth-Century Culture*, 7 (1978), 175–92.
[84] French, *William Harvey*, pp. 315–16, highlights an analogous attitude in Harvey's later work.

no longer be disputed, whether there be colours in the dark, nor whether they be the qualities of the objects we see, nor perhaps whether Light be a Body."[85] By taking over topics from natural philosophy, he had thus adopted the presumptions that had driven the increasing use of the label "physico-mathematics" throughout the century. Nonetheless, physical causation was still to be kept distinct from the characteristic concerns of the mathematical sciences, and claims to any degree of certainty were to be warranted through the secure exemplars of mathematics. Newton had adopted this line in his almost contemporaneous Latin lectures on optics:

> [T]he generation of colors includes so much geometry, and the understanding of colors is supported by so much evidence [*evidentiâ*: "evidentness"], that for their sake I can thus attempt to extend the bounds of mathematics somewhat, just as astronomy, geography, navigation, optics, and mechanics are truly considered mathematical sciences even if they deal with physical things: the heavens, earth, seas, light, and local motion. Thus although colors may belong to physics, the science of them must nevertheless be considered mathematical, insofar as they are treated by mathematical reasoning.[86]

Thus, according to Newton, "with the help of philosophical geometers and geometrical philosophers, instead of the conjectures and probabilities that are being blazoned about everywhere, we shall finally achieve a natural science supported by the greatest evidence."[87] This kind of evidence (i.e., "evidentness") is precisely that of the mathematician and incorporates the evidentness of sensory experience.

The controversies that followed the initial publication of Newton's optical ideas in 1672 concerned precisely these kinds of issues.[88] Some critics, such as Robert Hooke (1635–1702) of the Royal Society, granted all of Newton's empirical claims but nonetheless denied the inferences that Newton made from them. Others complained that the experiments did not yield the results that Newton claimed. Little in the way of straightforward, unproblematic confirmation of Newton's optical work, based on some ideal of experimental replication, was involved in the future of Newtonian optics.[89] Given Newton's

---

[85] Isaac Newton, "New Theory about *Light* and *Colours*," *Philosophical Transactions*, 6 (1672), 3075–87, at p. 3085.
[86] Isaac Newton, *The Optical Papers of Isaac Newton, Vol. I: The Optical Lectures, 1670–72*, ed. Alan E. Shapiro (Cambridge: Cambridge University Press, 1984), p. 439.
[87] Ibid.
[88] The most penetrating analysis along these lines is still Zev Bechler, "Newton's 1672 Optical Controversies: A Study in the Grammar of Scientific Dissent," in *The Interaction Between Science and Philosophy*, ed. Yehuda Elkana (Atlantic Highlands, N.J.: Humanities Press, 1974), pp. 115–42.
[89] Two different accounts are those of Simon Schaffer, "Glass Works: Newton's Prisms and the Uses of Experiment," in *The Uses of Experiment: Studies in the Natural Sciences*, ed. David Gooding, Trevor Pinch, and Simon Schaffer (Cambridge: Cambridge University Press, 1989), pp. 67–104, which concerns the social issues that determined the fortunes of Newtonian optics in England, and Alan E. Shapiro, "The Gradual Acceptance of Newton's Theory of Light and Color, 1672–1727," *Perspectives on Science: Historical, Philosophical, Social*, 4 (1996), 59–140, who stresses, contra Schaffer, the theoretical arguments involved in Newton's ultimate success.

concern with evident experience, the difficulties that he experienced in convincing others are an object lesson in the difficulties involved in arguing from experiments to conclusions.

Newton's most famous pronouncement on proper procedure in the sciences appears in Query 31 in the third edition of the *Opticks* (1717):

> As in Mathematicks, so in Natural Philosophy, the Investigation of difficult Things by the Method of Analysis, ought ever to precede the Method of Composition. This Analysis consists in making Experiments and Observations, and in drawing general Conclusions from them by Induction, and admitting of no Objections against the Conclusions, but such as are taken from Experiments, or other certain Truths. For Hypotheses are not to be regarded in experimental Philosophy. And although the arguing from Experiments and Observations by Induction be no Demonstration of general Conclusions; yet it is the best way of arguing which the Nature of Things admits of, and may be looked upon as so much the stronger, by how much the Induction is more general. And if no Exception occur from Phaenomena, the Conclusion may be pronounced generally.[90]

The mathematical prototype of "induction" in Newton's usage appears to relate to Isaac Barrow's views on the subject. Echoing Aristotle, Barrow had allowed that knowledge of a universal in geometry could be acquired through experience of a single example – as, for example, in the inspection of the properties of a single triangle.[91] Similarly, Newton allowed the "inductive" constitution of a universal truth from the outcome of a single experimental procedure.[92] The difficulty of attributing a philosophical (rather than merely historical) meaning to particular events had formerly left the Royal Society's enterprise at something of an impasse, which reproduced the basic problem of using singular knowledge-claims to warrant universal ones. But Newton's work provided a model for validating experimental particularities in natural philosophy in terms developed within the mathematical sciences. Whereas the experimental events recounted by Robert Boyle (1627–1691), including his famous accounts of air-pump trials,[93] had aimed at reporting the natural behavior of historical singulars (such that Boyle was hard-pressed to justify

---

[90] Isaac Newton, *Opticks* [4th ed., 1730] (New York: Dover, 1952), p. 404. The parts of this passage that first appeared in the 1717 edition are noted in Henry Guerlac, "Newton and the Method of Analysis," in *Dictionary of the History of Ideas*, ed. P. P. Wiener, 5 vols. (New York: Charles Scribner's Sons, 1973), 3: 378–91, at p. 379.

[91] See Dear, *Discipline and Experience*, chap. 8, sec. III.

[92] See Paul K. Feyerabend, "Classical Empiricism," in *The Methodological Heritage of Newton*, ed. Robert E. Butts and John W. Davis (Toronto: University of Toronto Press, 1970), pp. 150–70, at p. 162, n. 10; and Alan E. Shapiro, *Fits, Passions, and Paroxysms: Physics, Method, and Chemistry and Newton's Theories of Colored Bodies and Fits of Easy Reflection* (Cambridge: Cambridge University Press, 1993), pp. 34–5.

[93] Shapin and Schaffer, *Leviathan and the Air-Pump*.

extending their significance much beyond their sources),[94] Newton's importation of the experimental practices of the mathematical sciences gave event experiments a philosophical respectability that they had formerly lacked.

Nonetheless, in imposing a particular methodological model onto natural philosophical practice, Newton had already been compelled to alter the achievable goals of natural philosophy. Sensory experience, as constituted in mathematical sciences, was never able to observe causes qua causes. Thus, Newton's optics could never demonstrate the truth of any particular theory of the nature of light – a feature that Newton tried to turn to his advantage by contrasting the demonstrability of his assertions concerning optical phenomena. Similarly, in the case of inverse-square-law universal gravitation, Newton claimed to demonstrate (again, from experiment and observation) the existence of a force acting between all particles of ordinary matter, but he excused himself from any obligation to prove a theoretical cause of that force. Whatever the nature of the physical process manifested as gravitational attraction, the measurable *properties* of that force were as Newton demonstrated them to be (see Joy, Chapter 3, this volume).[95]

To his eighteenth-century readers, Newton's work represented a newly consolidated conception of scientific experience that departed from the scholastic model. Whereas for an Aristotelian philosopher "experience" was the source of one's knowledge of how the world was wont to behave, for a natural philosopher of the eighteenth century it had become a technique for interrogating nature (if necessary, "torturing" it, in Francis Bacon's phrase),[96] and one that yielded, above all, operational rather than essential knowledge. No longer a matter of "what everyone knows," the experimental approach to knowledge aimed at accumulating records of natural phenomena the truth of which would be accepted by others on the basis of personal and institutional authority or on the word of appropriate witnesses.

CONCLUSION

By the turn of the seventeenth century, the two most prominent forums for the pursuit of the sciences (natural philosophical and mathematical) were London's Royal Society and the Académie Royale des Sciences in Paris. Both prided themselves on conducting experimental investigations, and both put

---

[94] Shapin, *A Social History of Truth*, chap. 7, esp. pp. 347–9, discusses Boyle's views on the variation in physical properties of the "same" chemical substances obtained from different localities.

[95] For one among many treatments of Newton's attitude toward his demonstrations on gravity, see I. Bernard Cohen, *The Newtonian Revolution: With Illustrations of the Transformation of Scientific Ideas* (Cambridge: Cambridge University Press, 1980), chap. 3, esp. pp. 74–5.

[96] Julian Martin, *Francis Bacon, the State, and the Reform of Natural Philosophy* (Cambridge: Cambridge University Press, 1992), p. 166, elucidates the connection of this phrase with Bacon's experience in contemporary English legal procedure. See also, for a caveat on overreading the "torture" metaphor, Peter Pesic, "Wrestling with Proteus: Francis Bacon and the 'Torture' of Nature," *Isis*, 90 (1999), 81–94, which maintains that a better translation of Bacon's term is "vexation."

"experience" high on the list of cognitive desiderata. The practical convergence between them is striking. Although the Royal Society was much more emphatic in its rhetorical stress on what Boyle had dubbed "the experimental philosophy," activities by members of the Académie's "physical" section (devoted to nonmathematical, qualitative sciences such as natural history and chemistry) could easily have found their place alongside the empirical material published by the Royal Society's Fellows. From Edme Marriotte on the physiology of plants (as well as on his own version of "Boyle's Law"), to Guillaume Homberg on their chemical analysis, to Christiaan Huygens's stress in the 1660s on the importance of Baconian empiricism even in the conduct of the mathematical sciences, the members of the Académie pursued an investigative style that was becoming the norm in the new natural philosophy of the decades around 1700.[97]

The varieties of experience in the sciences of early modern Europe thus ran the gamut from mathematics through the traditional topics of natural philosophy to natural history. In each case, there was much room available for dispute and contestation of what experience was and how it should be used, and what kind of natural philosophy could be underpinned by experience. Everyone, however, including the sternest of skeptics (Descartes among them), agreed that experience was crucial to the achievement of natural knowledge. At the end of the period, much of the practical implementation of this rhetorical stress on experience had begun to take the form of stylized, set-piece investigations that established the lessons of specific experiences in a solid literary archive of accredited books and journals. Experience, a perennial topic of philosophical discussion, was now a practical ideological element of the scientific enterprise.

---

[97] Stroup, *Company of Scientists*, esp. pp. 134–7 and chap. 12; Frederic L. Holmes, *Eighteenth-Century Chemistry as an Investigative Enterprise* (Berkeley: Office for History of Science and Technology, University of California at Berkeley, 1989); Licoppe, *La formation de la pratique scientifique*, chap. 2; and Christian Licoppe, "The Crystallization of a New Narrative Form in Experimental Reports (1660–1690): Experimental Evidence as a Transaction Between Philosophical Knowledge and Aristocratic Power," *Science in Context*, 7 (1994), 205–44. Both John L. Heilbron, *Physics at the Royal Society During Newton's Presidency* (Los Angeles: William Andrews Clark Memorial Library, 1983), and Marie Boas Hall, *Promoting Experimental Learning: Experiment and the Royal Society, 1660–1727* (Cambridge: Cambridge University Press, 1991), carry the story of the Royal Society's experimental endeavors into the eighteenth century.

5

PROOF AND PERSUASION

R. W. Serjeantson

Questions of proof and persuasion are important in the history of the sciences of any period, but they are particularly pressing in the case of early modern Europe.[1] The sixteenth and seventeenth centuries saw more self-conscious theoretical reflection on how to discover and confirm the truths of nature than any period before or since; the same period also manifested a huge range of practical strategies by which investigators of the natural world set about demonstrating their findings and convincing their audiences of their claims. Studying these strategies of proof and persuasion has opened up vistas of opportunity for historians of the sciences in early modern Europe. In a range of disciplines, from the social history of medicine to the history of philosophy, historians of the period have argued for the ineradicable significance of forms of proof and persuasion in understanding their various

---

[1] It has even been argued that "credibility should not be referred to as a 'fundamental' or 'central' topic – from a pertinent point of view it is the *only* topic" (Steven Shapin, "Cordelia's Love: Credibility and the Social Studies of Science," *Annual Review of Sociology*, 3 (1995), 255–75, at pp. 257–8). For general studies of what is now often known as "the rhetoric of science," see John Schuster and Richard R. Yeo, eds., *The Politics and Rhetoric of Scientific Method: Historical Studies* (Dordrecht: Reidel, 1986); Andrew E. Benjamin, G. N. Cantor, and J. R. R. Christie, eds., *The Figural and the Literal: Problems of Language in the History of Science and Philosophy, 1630–1800* (Manchester: Manchester University Press, 1987); Charles Bazerman, *Shaping Written Knowledge: The Genre and Activity of the Experimental Article in Science* (Madison: University of Wisconsin Press, 1988); L. J. Prelli, *A Rhetoric of Science: Inventing Scientific Discourse* (Columbia: University of South Carolina Press, 1989); Alan G. Gross, *The Rhetoric of Science* (Cambridge, Mass.: Harvard University Press, 1990); Jan V. Golinski, "Language, Discourse, and Science," in *Companion to the History of Modern Science*, ed. R. C. Olby, G. N. Cantor, J. R. R. Christie, and M. J. S. Hodge (London: Routledge, 1990), pp. 110–23; Peter Dear, ed., *The Literary Structure of Scientific Argument: Historical Studies* (Philadelphia: University of Pennsylvania Press, 1991); Marcello Pera and William R. Shea, *Persuading Science: The Art of Scientific Rhetoric* (Canton, Mass.: Science History Publications, 1991); Alan G. Gross and William M. Keith, eds., *Rhetorical Hermeneutics: Invention and Interpretation in the Age of Science* (Albany: State University of New York Press, 1997); the essay review of the books by Gross, Prelli, and Dear by Trevor Melia, *Isis*, 83 (1992), 100–106; and the special issues of two journals: "Symposium on the Rhetoric of Science," *Rhetorica: A Journal of the History of Rhetoric*, 7, no. 1 (1989) and "The Literary Uses of the Rhetoric of Science," *Studies in the Literary Imagination*, 22, no. 1 (1989).

objects of inquiry.² The rhetorical form of texts and even objects has come to be seen as constitutive of their meaning, not separable from it. Furthermore, an increasing number of studies have shown how early modern physicians, mathematical practitioners, and natural philosophers all exploited the different and historically specific resources of proof and persuasion that they had at their disposal.

The study of proof and persuasion provides a further opportunity to the historian: It offers a means of bridging the gap between a text (or a practice) and its reception. As the reception, rather than the genesis, of developments in the sciences has become an increasingly important aspect of historiography, it has also become increasingly apparent that this reception history is often extremely difficult to reconstruct. The evidence for reading practices, or for the individual decisions that led to one account being accepted over another, is often much more sparse than the evidence that allows the reconstruction of the processes resulting in a particular theory or practice. It is here that the study of proof and persuasion can come in. The ways in which writers and practitioners set about persuading their audiences of the truth or utility of their arguments can also offer a yardstick against which their intentions can be judged. Additionally, the study of proof and persuasion provides a means of recovering the expectations with which arguments might have been received – expectations that can sometimes be set against evidence for actual instances of reception. To put it another way, the history of proof and persuasion brings together approaches to the history of the sciences that analyze conceptual, technical, and metaphysical developments with approaches that analyze the sciences' social functions and the roles or identities – or in early modern terms, the ethos – of their protagonists.³

In the sixteenth and seventeenth centuries, changes in the conception of nature and in the ways that nature was studied encouraged the proliferation of very different techniques of probation. Humanists took persuasion to be their greatest imperative; they revived and imitated ancient literary styles and forms by which to accomplish this goal. The scholastic commentary tradition of the sixteenth century mutated into the university textbook of the

---

² For medicine, see David Harley, "Rhetoric and the Social Construction of Sickness and Healing," *Social History of Medicine*, 12 (1999), 407–35. For philosophy, see Thomas M. Carr, Jr., *Descartes and the Resilience of Rhetoric* (Carbondale: Southern Illinois University Press, 1990); and Quentin Skinner, *Reason and Rhetoric in the Philosophy of Hobbes* (Cambridge: Cambridge University Press, 1996), esp. pp. 7–15.

³ On this issue, see Steven Shapin and Simon Schaffer, *Leviathan and the Air-Pump: Hobbes, Boyle, and the Experimental Life* (Princeton, N.J.: Princeton University Press, 1985), esp. pp. 13–15; Robert S. Westman, "Proof, Poetics, and Patronage: Copernicus' Preface to *De revolutionibus*," in *Reappraisals of the Scientific Revolution*, ed. David C. Lindberg and Robert S. Westman (Cambridge: Cambridge University Press, 1990), pp. 167–205; Nicholas Jardine, "Demonstration, Dialectic, and Rhetoric in Galileo's *Dialogue*," in *The Shapes of Knowledge from the Renaissance to the Enlightenment*, ed. Donald R. Kelley and Richard H. Popkin (Dordrecht: Kluwer, 1991), pp. 101–21, esp. pp. 115–16; and Peter Dear, *Discipline and Experience: The Mathematical Way in the Scientific Revolution* (Chicago: University of Chicago Press, 1995), p. 5.

seventeenth century. The prestige of mathematics and mathematical accounts of demonstration about the natural world rose dramatically and helped spur on the earliest investigations into mathematical probabilities. The new forms of natural history and experimental report that arose in the seventeenth century were founded on a notion of "fact" derived from the human sciences of history and law. Finally, the new kinds of institutions that were formed to study nature, from anatomy theaters to the royal academies, all brought with them different expectations about what constituted a plausible claim to truth. Nonetheless, there were also constants and continuities in the theory and practice of proof and persuasion in this period. These make it possible to trace a path through the competing claims for plausibility in early modern natural knowledge. In this chapter, I begin by considering the different conceptions of proof and persuasion that obtained in different disciplines. I then discuss how these conceptions were affected by developments in the study of nature and, in particular, by the incorporation of mathematics and experiment into the discipline of natural philosophy. The chapter closes by considering mechanisms of proof and persuasion in two distinct but overlapping areas: the printed book and institutions for the pursuit of natural knowledge.

DISCIPLINARY DECORUM

The learned culture that was transmitted through and beyond the universities of early modern Europe was structured in terms of distinct intellectual disciplines. Each of these disciplines possessed its own body of knowledge and practices, but there was also a great deal of shared knowledge in the form of commonplaces, *loci classici*, and maxims that operated across the range of arts and sciences.[4] In the context of the universities, there was also a marked degree of hierarchy within these disciplines, with the basic discipline, grammar, at the bottom, and the highest discipline, theology, at the top. It is true that Renaissance humanists challenged these scholastic notions of disciplinary hierarchy by reasserting the late-antique notion of the "encyclopedia" or circle of learning, prizing the arts of grammar, rhetoric, poetry, history, and moral philosophy over the mathematical sciences, natural philosophy, and metaphysics.[5] Nonetheless, in the course of the sixteenth century, the

---

[4] See Ian Maclean, *The Renaissance Notion of Woman: A Study in the Fortunes of Scholasticism and Medical Science in European Intellectual Life* (Cambridge: Cambridge University Press, 1980), pp. 4–5.
[5] Paul O. Kristeller, "The Modern System of the Arts," in *Renaissance Thought II* (New York: Harper, 1964), pp. 163–227; Donald R. Kelley, *Renaissance Humanism* (Boston: Twayne Publishers, 1991), p. 3; Maurice Lebel, "Le concept de l'encyclopaedia dans l'oeuvre de Guillaume Budé," in *Acta Conventus Neo-Latini Torontonensis: Proceedings of the Seventh International Congress of Neo-Latin Studies, Toronto, 8 August to 13 August 1988*, ed. Alexander Dalzell, Charles Fantazzi, and Richard J. Schoeck (Binghamton, N.Y.: Medieval and Renaissance Texts and Studies, 1991), pp. 3–24. See more generally Erika Rummel, *The Humanist–Scholastic Debate in the Renaissance and Reformation* (Cambridge, Mass.: Harvard University Press, 1995).

universities generally absorbed this challenge, and their basic structures, at least, were not fundamentally displaced.

If anything, Renaissance humanists encouraged a high degree of self-consciousness about questions of proof and persuasion because of the emphasis they laid on the three disciplines of the trivium: grammar, rhetoric, and dialectic. The richly elaborated investigations of late medieval scholastic logic may have become a common butt of humanist derision, but the proponents of the new learning were fascinated by the possibilities of the art of rhetoric for achieving the union of eloquence and wisdom.[6] This fascination encouraged the rise of the phenomenon of "humanist dialectic," a highly rhetoricized account of the argumentative process that extends in a tradition from Lorenzo Valla's (1407–1457) *Repastinatio* (Re-excavation), through Rudolph Agricola's (ca. 1443–1485) influential *De inventione dialectica* (On Dialectical Invention, 1479), to Petrus Ramus's (1515–1572) *Dialectic* in Latin and French (1555) and beyond.[7] The thriving Aristotelian tradition of the sixteenth century was also affected by these developments, elaborating more formal accounts of method and scientific demonstration in medicine and natural philosophy than most humanists could stomach.

Within the disciplinary structure of the late Renaissance arts and sciences, issues of proof and persuasion were formally addressed in the disciplines of logic and rhetoric, respectively. These disciplines therefore have a privileged place in the history of the subject. They had, nonetheless, rather different procedures and ends. Logic – the "art of arts and the science of sciences" in the oft-cited description of the medieval logician Peter of Spain – was concerned both with scientific (that is, certain) demonstration and (in the form of dialectic) with arguments that were merely probable. The art of rhetoric, in contrast, taught the theory and practice of persuasive argument. In its Ciceronian conception, this involved educating the orator to speak elegantly and copiously on any subject with direct application to a specific

---

[6] See Jerrold E. Seigel, *Rhetoric and Philosophy in Renaissance Humanism: The Union of Eloquence and Wisdom, Petrarch to Valla* (Princeton, N.J.: Princeton University Press, 1968).
[7] Lorenzo Valla, *Laurentii Valle repastinatio dialectice et philosophie*, ed. Gianni Zippel, 2 vols. (Padua: Antenore, 1982); Rudolf Agricola, *De inventione dialectica libri tres / Drei Bücher über die Inventio dialectica: Auf der Grundlage der Edition von Alardus von Amsterdam [1539]*, ed. Lothar Mundt (Tübingen: Max Niemeyer, 1992); Petrus Ramus, *Dialectique 1555: Un manifeste de la Pléiade*, ed. Nelly Bruyère (De Pétrarque à Descartes, 61) (Paris: J. Vrin, 1996). See further Cesare Vasoli, *La dialectica e la retorica dell'umanesimo: "Invenzione" e "metodo" nella cultura del XV e XVI secolo* (Milan: Feltrinelli, 1968); Lisa Jardine, "Lorenzo Valla and the Intellectual Origins of Humanist Dialectic," *Journal of the History of Philosophy*, 15 (1977), 143–64; John Monfasani, "Lorenzo Valla and Rudolph Agricola," *Journal of the History of Philosophy*, 28 (1990), 181–200; Peter Mack, *Renaissance Argument: Valla and Agricola in the Traditions of Rhetoric and Dialectic* (Brill's Studies in Intellectual History, 43) (Leiden: E. J. Brill, 1993); Lisa Jardine, "Distinctive Discipline: Rudolph Agricola's Influence on Methodical Thinking in the Humanities," in *Rodolphus Agricola Phrisius (1444–1485): Proceedings of the International Conference at the University of Groningen, 28–30 October 1985*, ed. F. Akkerman and A. J. Vanderjagt (Leiden: E. J. Brill, 1988), pp. 38–57; Walter J. Ong, S.J., *Ramus, Method, and the Decay of Dialogue: From the Art of Discourse to the Art of Reason* (Cambridge, Mass.: Harvard University Press, 1958); E. Jennifer Ashworth, "Logic in Late Sixteenth-Century England: Humanist Dialectic and the New Aristotelianism," *Studies in Philology*, 88 (1991), 224–36.

audience. In its Aristotelian form – less prominent in the earlier part of the period than the Ciceronian form – rhetoric used reasonable argument (logos) and drew upon the moral character (ethos) of the speaker in a bid to excite the passions (pathos) of the audience and thereby persuade them of the truth of the speaker's position. Both logic and rhetoric were very widely taught in the schools, colleges, and universities of early modern Europe, with significant continuities between the Protestant and Catholic worlds (see the following chapters in this volume: Blair, Chapter 17; Grafton, Chapter 10). (Protestants frequently used Catholic books for teaching and scholarship; because of the prohibition of the Inquisition, the reverse was less common.)

Nonetheless, one of the most characteristic aspects of the disciplinary structure of late Renaissance learned culture was the assumption that different standards of proof were applicable to different disciplines. This assumption was often given an Aristotelian justification from a text in the *Nicomachean Ethics* (1.3):

> It is the mark of an educated man to look for precision in each class of things just so far as the nature of the subject admits: it is evidently foolish to accept probable reasoning from a mathematician and to demand from a rhetorician demonstrative proofs.[8]

This doctrine of different standards of proof for different disciplines played out in various ways, according to different conceptions of disciplinary classification. One of the most pervasive distinctions, which also ultimately derived from Aristotle, was between the theoretical and the practical disciplines. Arithmetic, geometry, physics, astronomy, optics, and metaphysics were, for an Aristotelian such as the Spanish Jesuit Franciscus Toletus (1532–1596), theoretical or contemplative sciences. In contradistinction, moral philosophy, history, and, to an extent, medicine were considered as practical (or active) disciplines.[9] Other classifications drew upon Renaissance conceptions of the difference between the arts (conceived as bodies of practical precepts) and the sciences (conceived as bodies of theoretical knowledge).[10] Finally – although this was less often invoked – disciplines might be distinguished on

---

[8] Aristotle, *Nicomachean Ethics*, trans. W. D. Ross, revised by J. O. Urmson, in *The Complete Works of Aristotle: The Revised Oxford Translation*, ed. J. Barnes, 2 vols. (Bollingen Series, 71) (Princeton, N.J.: Princeton University Press, 1984), 2: 1729–1867, at p. 1730, 1.3 (1094b24–26). For an example, see Charles B. Schmitt, "Girolamo Borro's *Multae sunt nostrarum ignorationum causae* (Ms. Vat. Ross. 1009)," in *Philosophy and Humanism: Essays in Honor of Paul Oskar Kristeller*, ed. Edward P. Mahoney (Leiden: E. J. Brill, 1976), pp. 462–76, at p. 474. On the implications of Aristotle's text for developing notions of probability, see Lorraine Daston, "Probability and Evidence," in *The Cambridge History of Seventeenth-Century Philosophy*, ed. Daniel Garber and Michael Ayers, 2 vols. (Cambridge: Cambridge University Press, 1998), 2: 1108–44, at p. 1108.
[9] See William A. Wallace, "Traditional Natural Philosophy," in *The Cambridge History of Renaissance Philosophy*, ed. Charles B. Schmitt, Quentin Skinner, and Eckhard Kessler, with Jill Kraye (Cambridge: Cambridge University Press, 1988), pp. 201–35, at p. 210.
[10] Wilhelm Schmidt-Biggemann, "New Structures of Knowledge," in *A History of the University in Europe*, gen. ed. Walter Rüegg (Cambridge: Cambridge University Press, 1992–), vol. 2: *Universities in Early Modern Europe (1500–1800)*, ed. Hilde de Ridder-Symoens (1996), pp. 489–530, at pp. 497–8.

the basis of their object of study: The French jurist and natural philosopher Jean Bodin (1530–1596) distinguished in his *Methodus ad facilem historiarum cognitionem* (Method for the Easy Understanding of History, 1566) between *res humanae* (human affairs), which are dependent upon will (*voluntas*); *res naturales* (natural affairs), which operate through causes (*per causas*); and *res divinae* (divine affairs), which were the province of God.[11]

These disciplinary distinctions had important implications for conceptions of proof and persuasion. The techniques of rhetorical persuasion – including circumstantial arguments directed to specific audiences, figures of speech, and the appeal to trusted authorities – were considered particularly appropriate for the practical, human sciences of history and moral philosophy. In contrast, within the theoretical science of university natural philosophy – and sometimes, for polemical purposes, outside it – the use of rhetoric and argument from authority tended to be frowned upon in favor of formally correct syllogisms, unadorned arguments, and universal rather than particular conclusions. The reason for this was that, from the Aristotelian perspective, which remained institutionally dominant throughout the sixteenth century and in some places retained its dominance throughout the seventeenth century as well, natural philosophy was considered a science (*scientia*); that is, a body of knowledge potentially capable of certain demonstration.

Nonetheless, although the assumptions about proof and persuasion that derived from the *trivium* were pervasive, they were also malleable – and modified when applied to the higher university disciplines of medicine, law, and theology. The status of medicine was a question frequently debated by medical writers: Was it a science, like its junior partner natural philosophy, or an art? By the late Renaissance, writers on medicine and law were elaborating versions of logic in their respective disciplines that were notably distinct from that familiar from the arts course. Medical authors acknowledged that they used concepts such as "contrary," "similarity," and "sign" in a less rigorous way than they were applied in logic. Lawyers often reduced the standard four Aristotelian causes (material, efficient, formal, and final) to two (mischief and remedy) or even one (motive), whereas medical doctors added a further four causes (subjective, instrumental, necessary, and catalytic) to the Aristotelian quartet. The situation was similar with respect to the "circumstances" that writers in philosophy and the sciences used to classify the variable subject matter of their disciplines. Lawyers tended to work with the standard six circumstances derived from antique rhetorical theory (who, what, where,

---

[11] Jean Bodin, "Methodus ad facilem historiarum cognitionem," in *Oeuvres philosophiques*, ed. Pierre Mesnard (Paris: Presses Universitaires de France, 1951), pp. 99–269. Translated as Jean Bodin, *Method for the Easy Comprehension of History*, trans. Beatrice Reynolds (New York: Columbia University Press, 1945). See also Donald R. Kelley, "The Development and Context of Bodin's Method," in *Jean Bodin: Verhandlungen der internationalen Bodin-Tagung in München 1970*, ed. Horst Denzer (Munich: Beck, 1973); repr. in Donald R. Kelley, *History, Law, and the Human Sciences: Medieval and Renaissance Perspectives* (London: Variorum Reprints, 1984), chap. 8, pp. 123–50, esp. p. 148.

when, why, by what means), whereas medical doctors listed as many as twenty-two in their efforts to get at the variety of symptoms within the Galenic theory of human idiosyncrasy.[12]

As I will show, notions of proof and persuasion derived from the *trivium* came under increasing strain in the course of the sixteenth and seventeenth centuries, particularly as a result of developments in natural philosophy. The decline of the Aristotelian disciplinary structure meant that the Aristotelian prohibition of *metabasis* – the use of the methods appropriate to one discipline in a different one – increasingly lost its force.[13] Developments in mathematics, mechanics, probability theory, and conceptions of experience within natural philosophy all changed the forms of proof that were considered appropriate for different disciplines. "To me," wrote the English chymist Robert Boyle (1627–1691) in his *Disquisition on the Final Causes of Things* (1688), "'tis not very material, whether or no, in Physicks or any other Discipline, a thing be prov'd by the peculiar Principles of that Science or Discipline; provided it be firmly proved by the common grounds of Reason."[14] Finally, and perhaps most importantly in the domain of natural knowledge, the "new philosophy" of the seventeenth century was characterized by a vehement and sustained attack on the value of conventional logic and rhetoric for either discovering or communicating knowledge about the natural world.

THEORIES OF PROOF AND PERSUASION

What, then, did it mean to "prove" something in early modern Europe? According to the *Lexicon philosophicum* (Philosophical Lexicon, 1613) of the Marburg philosopher Rudolph Goclenius (1547–1628), "*to prove* generally means: to make known the truth of something; to confirm a matter in whatever way."[15] Early modern notions of both proof and persuasion had truth as their object: Like their Roman counterparts, sixteenth-century rhetoricians were reluctant to concede Plato's accusation that rhetoric sacrificed veracity in the cause of persuasion.[16] Goclenius's definition allows that things can be

---

[12] Ian Maclean, "Evidence, Logic, the Rule and the Exception in Renaissance Law and Medicine," *Early Science and Medicine*, 5 (2000), 227–57, at pp. 238 and 240.
[13] Amos Funkenstein, *Theology and the Scientific Imagination from the Middle Ages to the Seventeenth Century* (Princeton, N.J.: Princeton University Press, 1986), pp. 36, 296, 303–4.
[14] Robert Boyle, *A Disquisition about the Final Causes of Natural Things* [1688], in *The Works of Robert Boyle*, ed. Michael Hunter and Edward B. Davis, 14 vols. (London: Pickering and Chatto, 2000), 11: 79–151, at p. 91. The passage is discussed in Edward B. Davis, "'Parcere nominibus': Boyle, Hooke and the Rhetorical Interpretation of Descartes," in *Robert Boyle Reconsidered*, ed. Michael Hunter (Cambridge: Cambridge University Press, 1994), pp. 157–75, at p. 164.
[15] Rudolph Goclenius, *Lexicon philosophicum* (Frankfurt am Main: Petrus Musculus and Rupert Pistorius, 1613), p. 879 (s.v. "Probare"): "Probare . . . Generaliter significat declarare veritatem alicuius rei, rem confirmare quoquo modo."
[16] Charles Trinkaus, "The Question of Truth in Renaissance Rhetoric and Anthropology," in *Renaissance Eloquence: Studies in the Theory and Practice of Renaissance Rhetoric*, ed. James J. Murphy (Berkeley: University of California Press, 1983), pp. 207–20; Hugh M. Davidson, "Pascal's Arts of Persuasion," in *Renaissance Eloquence*, pp. 292–300, at pp. 292–3; and Wayne A. Rebhorn,

proved with different degrees of certainty ("confirmed") and by a variety of means ("in whatever way"). The purpose of proof, furthermore, is to "make something known" (*res declarare*). This was a constant object of theories of probation, but it also incorporated a recurrent theoretical tension: Should a proof proceed according to a method of discovery or a method of doctrine? That is to say, are things (*res*) best explained in terms of how they were found out or in terms that emphasize their organization for pedagogical purposes? This dilemma was bequeathed to early modern natural philosophers from antiquity and was at the heart of some of the most often-reprinted writings on method, such as Jacopo Acontius's (1492–ca. 1566) *De methodo, hoc est de recta investigandarum tradendarumque artium ac scientiarum ratione* (On Method; that is, on the Right Way of Investigating and Imparting the Arts and Sciences, 1558).[17] It was a dilemma that a number of seventeenth-century writers on methods of discovery resolved by, in effect, denying that they were concerned with problems of teaching at all.

It is helpful to consider early modern theories of proof and persuasion in terms of three broad categories suggested by the disciplinary structure of early modern learning: demonstration, probability, and persuasion. The first two categories were the province of logic, which was sometimes divided into demonstration, or the science of certain proof, and dialectic, the logic of probabilities.[18] The third category, persuasion, was the province of rhetoric. (An analogous, although not identical, threefold structure can be found in scholastic theories of cognition in the period, with early modern Thomists distinguishing human understanding according to the degree of certainty inhering in it. Thus certain knowledge (*scientia*), opinion (*opinio*), and faith (*fides*) all had their own forms of certainty: metaphysical, physical, and moral, respectively.)[19]

The different forms of proof – demonstration, probability, and persuasion (*demonstratio, probabilitas, persuasio*) – were extensively discussed in the thousands of works on logic and rhetoric that were written, taught, and published in the sixteenth and seventeenth centuries, and every natural philosopher educated to any level beyond that of rudimentary Latin grammar would have encountered them in some form or another. How far early modern

---

"Introduction," in *Renaissance Debates on Rhetoric*, ed. and trans. Wayne A. Rebhorn (Ithaca, N.Y.: Cornell University Press, 2000), pp. 1–13, at pp. 7–9.

[17] Jacobus Acontius, *De methodo . . . Über die Methode*, ed. Lutz Geldsetzer, trans. Alois von der Stein (Düsseldorf: Stern-Verlag/Janssen and Co., 1971).

[18] See Aristotle, *Topica*, 1.1; Pierre Gassendi, *Institutio logica* (1658), ed. and trans. Howard Jones (Assen: Van Gorcum, 1981), p. 64; E. Jennifer Ashworth, "Historical Introduction," in *Language and Logic in the Post-Medieval Period* (Dordrecht: Reidel, 1974), pp. 1–25, at p. 25; and Ashworth, "Editor's Introduction," in Robert Sanderson, *Logicae artis compendivm* (1618), ed. E. Jennifer Ashworth (Instrumenta Rationis, 2) (Bologna: Cooperativa Libraria Universitaria Editrice Bologna, 1985), pp. ix–lv, at pp. xxxv–xxxvii.

[19] See, for example, Roderigo de Arriaga, "Disputationes logicae," in *Cursus philosophicus* (Paris: Jacques Quesnel, 1639), pp. 33–212, at p. 200; and Peter Dear, "From Truth to Disinterestedness in the Seventeenth Century," *Social Studies of Science*, 22 (1992), 619–31, at pp. 621–2.

philosophers – and indeed learned writers more generally – applied the probative theories of the trivium to their own practices of investigation and composition is a further question. The disciplines of the trivium were sometimes regarded in early modern Europe as the intellectual equivalent of water wings: something to be cast off when the art had been thoroughly learned.[20] Their close association with the schools also sometimes made resort to them in any overly apparent way suspect in extrascholastic contexts. Nonetheless, all three forms of proof were deployed in early modern philosophy, both natural and moral. At the most basic level, the probative claims of a work might be signaled by its title: Christoph Hellwig's *De studii botanici nobilitate oratio* (Oration on the Nobility of the Study of Botany, 1666) indicates its aim to persuade its audience of the merits of a form of natural knowledge that had grown steadily in significance over the sixteenth and seventeenth centuries. The English physician William Gilbert's (1544–1603) book *De magnete* (On the Magnet, 1600) has as its subtitle "A new physics [*physiologia*], demonstrated with both arguments and experiments."[21] Galileo Galilei's (1564–1642) *Discorsi e dimostrazioni matematiche intorno a due nuove scienze* (Discourses and Mathematical Demonstrations Concerning the Two New Sciences, 1638) likewise emphasizes the solidity of his claims for mechanics and local motion and their mathematical foundation. Furthermore, it was not uncommon to use different kinds of proof at different points within the same work. This is illustrated by another work by Galileo, his *Dialogo sopra i due massimi sistemi del mondo* (Dialogue on the Two Chief World Systems, 1632), which at different points draws upon all three resources of demonstration, probable argument, and rhetorical persuasion.[22]

The most ambitious formal accounts of the probative process produced by early modern natural philosophers took the form of doctrines of "method." The sixteenth century saw an upsurge of interest in questions of method – that is, in theoretical accounts of how knowledge is obtained and demonstrated.[23] Medieval discussions of method focused upon scientific proof by means of the so-called demonstrative regress, or *regressus*. This involved finding a cause from its effect by induction and then demonstrating that effect back from its cause in order to obtain causal – and hence scientific – knowledge of a phenomenon. Accounts of *methodus* by Renaissance philosophers retained this preoccupation with causal demonstration while increasingly bringing philological discoveries to bear upon it. The basic context for accounts of

[20] See Samuel Butler, *Prose Observations*, ed. Hugh de Quehen (Oxford: Clarendon Press, 1979), p. 128: "A logician, Gramarian, and Rhetorician never come to understand the true end of their Arts, untill they have layed them by, as those that have learned to swim, give over their bladders that they learnd by."
[21] William Gilbert, *De magnete, magneticisque corporibus, et de magno magnete tellure; Physiologia nova, plurimis & argumentis, & experimentis demonstrata* (London: Peter Short, 1600; facsimile repr. Brussels: Culture and Civilisation, 1967).
[22] Nicholas Jardine, "Demonstration, Dialectic, and Rhetoric," pp. 101–21.
[23] Neal W. Gilbert, *Renaissance Concepts of Method* (New York: Columbia University Press, 1960).

demonstration in sixteenth century academic natural philosophy remained, however, Aristotelian logic, and specifically the account of scientific demonstration in Aristotle's *Posterior Analytics*, II.13. This text, commentaries upon it, and redactions of it in textbooks and lecture courses encouraged the widespread view among early modern Aristotelians that a proof qualified as "scientific" only if it was derived from premises that were universal. This was to be achieved by means of a syllogism, the middle term of which expressed the operative cause.[24] The purpose of this form of scientific demonstration was to acquire certain knowledge of phenomena through "absolute demonstration" (*demonstratio potissima*). This characteristically consisted of four stages: (1) observation, which provided "accidental" knowledge of an effect; (2) induction, which allowed demonstration of the cause from the effect (*demonstratio quia*); (3) *consideratio* (or *negotiatio* or *meditatio*), by means of which the mind came to grasp the necessary association of the proximate cause with the effect; and (4) demonstration of the effect from the cause (*demonstratio propter quid*), which finally provided certain knowledge (*scientia*) of the phenomenon. It was commonly stipulated that the argument should be in the first figure (*Barbara*) of the syllogism; that is, with a universal major premise and an affirmative minor one.[25]

Medieval accounts of method, such as that of Pietro d'Abano in his *Conciliator* of the early fourteenth century, had also often sought to reconcile medical and philosophical traditions. Sixteenth-century discussions of "method" continued to draw inspiration from medical theory, revitalized by philological interest in the original Greek texts of Galen. The humanist physician Niccolò Leoniceno's (1428–1524) discussion in his *De tribus doctrinis ordinatis secundum Galeni sententiam opus* (Treatise on the Three Types of Teaching, Ordered According to the Opinion of Galen, 1508) of Galen's use of the term *didaskalia* ("didactics") in the prologue to the *Ars medica* was particularly significant. In this work, Leoniceno argued that Galen was not primarily concerned with the method of scientific demonstration (*modus doctrinae*) but with the method of organizing a whole science for teaching (*ordo docendi*).[26] As suggested earlier, this distinction between discovery and doctrine was widely endorsed by sixteenth-century physicians and philosophers. It was developed in a particularly influential way in the *De methodis* (1578) of the Paduan philosopher Jacopo Zabarella (1533–1589). Zabarella's application

---

[24] See Dear, *Discipline and Experience*, p. 36.
[25] Nicholas Jardine, "Epistemology of the Sciences," in Schmitt, Skinner, and Kessler, eds., *The Cambridge History of Renaissance Philosophy*, pp. 685–711, esp. p. 687. See also William A. Wallace, *Galileo and his Sources: The Heritage of the Collegio Romano in Galileo's Science* (Princeton, N.J.: Princeton University Press, 1984), pp. 125–6.
[26] William F. Edwards, "Niccolò Leoniceno and the Origins of Humanist Discussion of Method," in *Philosophy and Humanism: Renaissance Essays in Honor of Paul Oskar Kristeller*, ed. Edward Mahoney (Leiden: E. J. Brill, 1976), pp. 283–305. On Leoniceno, see Vivian Nutton, "The Rise of Medical Humanism: Ferrara, 1464–1555," *Bulletin of the Society for Renaissance Studies*, 11 (1997), 2–19.

of the term *methodus* to issues of discovery and *ordo* to issues of doctrine and organization governed the terms of the debate for the subsequent fifty years.[27]

The sixteenth-century fascination with theories of scientific demonstration persisted throughout the seventeenth century. Accounts of method underwent constant modification but remained part of a recognizable generic tradition. There was progressively less interest in *regressus* theory properly speaking – although there is a clear continuity, for instance, between Thomas Hobbes's (1588–1679) account of *methodus* in the *De homine* (1658) and those of late Renaissance Paduan Aristotelians[28] – and a correspondingly greater interest in concepts of method derived not from logic but from geometry. In particular, Euclid's distinction in the *Elements* between analysis and synthesis was endowed with increased significance (see Andersen and Bos, Chapter 28, this volume). René Descartes (1596–1650) took up these terms for his account of scientific discovery in his significantly titled *Discours de la méthode* (Discourse on Method, 1637).[29] Isaac Newton (1642–1727) also drew upon geometrical terminology when he asserted in the third edition of the *Opticks* (1721) that "As in Mathematicks, so in Natural Philosophy, the Investigation of difficult Things by the Method of Analysis, ought ever to precede the Method of Composition."[30] In general, as seventeenth-century natural philosophers abandoned the search for essential properties in favor of a more phenomenological understanding of nature, they also lost interest in Aristotelian traditions of method that emphasized demonstrative certainty. Indeed, even within sixteenth-century Aristotelianism, objections were raised against *regressus* as the best account of demonstration in natural philosophy. It was accused of circularity. It was sometimes even suggested – for instance by the Italian philosopher Agostino Nifo (ca. 1469–1538) – that certain questions in natural philosophy were incapable of achieving demonstrative certainty

---

[27] Jacopo Zabarella, *De methodis libri quattuor. Liber de regressu*, ed. Cesare Vasoli (Bologna: Cooperativa Libraria Universitaria Editrice Bologna, 1985); John Herman Randall, *The School of Padua and the Emergence of Modern Science* (Padua: Editrice Antenore, 1961); Luigi Olivieri, ed., *Aristotelismo veneto e scienza moderna*, 2 vols. (Padua: Editrice Antenore, 1983); Nicholas Jardine, "Keeping Order in the School of Padua: Jacopo Zabarella and Francesco Piccolomini on the Offices of Philosophy," in *Method and Order in Renaissance Philosophy of Nature: The Aristotle Commentary Tradition*, ed. Daniel Di Liscia, Eckhard Kessler, and Charlotte Methuen (Aldershot: Ashgate, 1997), pp. 183–209; and Irena Backus, "The Teaching of Logic in Two Protestant Academies at the End of the 16th Century: The Reception of Zabarella in Strasbourg and Geneva," *Archiv für Reformationsgeschichte*, 80 (1989), 240–51.

[28] William F. Edwards, "Paduan Aristotelianism and the Origins of Modern Theories of Method," in *Aristotelismo veneto e scienza moderna*, ed. Olivieri, 1: 206–20.

[29] René Descartes, *Discours de la méthode* [1637], in *Oeuvres de Descartes*, ed. Charles Adam and Paul Tannery, 12 vols. (Paris: J. Vrin, 1964–76), vol. 6: *Discours de la méthode & Essais* (1973), pp. 1–151, at p. 17. See further Stephen Gaukroger, *Cartesian Logic: An Essay on Descartes's Conception of Inference* (Oxford: Clarendon Press, 1989), pp. 72–102; Benoît Timmermans, "The Originality of Descartes's Conception of Analysis as Discovery," *Journal of the History of Ideas*, 60 (1999), 433–48.

[30] Isaac Newton, *Opticks* (New York: Dover, 1979), p. 404.

because the cause would always remain hidden.³¹ In this case, logical proofs in natural philosophy left the realm of the demonstrative and entered the province of the probable.

The discipline that generated and policed probable argument was dialectic. Like the proofs of scientific demonstration, dialectical arguments were generally framed syllogistically. But they did not seek to generate the certainty of *scientia*. Dialectical conclusions remained probable either because the premises were not certain or because the inferential process was conjectural. In the first case, premises might be supplied by – as Aristotle had put it in a widely repeated formula – "reputable opinions" that were accepted by "everyone, or by the majority, or by the wise."³² In the second case, the basic inferential mechanism of dialectic was the so-called topical syllogism, in which the middle term was provided by a general "topic" or *locus* that helped shed light on the question at hand. These topics commonly included categories such as definition, genus, species, cause, effect, antecedent, consequent, greater, less, and argument from authority.³³ The probable arguments of dialectic might thus include argument from comparisons, analogies, and examples. By the sixteenth and seventeenth centuries, dialectical reasoning had also come to comprehend the issue, which in the ancient world had been a predominantly rhetorical one, of inference from signs.³⁴ This was a subject of some debate in sixteenth-century natural philosophy, and it was particularly important in early modern learned medicine, in which semiology comprised one of the five parts of medical studies (the others being physiology, aetiology, therapeutics, and hygiene).³⁵

---

31 Nicholas Jardine, "Galileo's Road to Truth and the Demonstrative Regress," *Studies in History and Philosophy of Science*, 7 (1976), 277–318, at pp. 290–91; N. Jardine, "Epistemology of the Sciences," esp. p. 689; N. Jardine, "Demonstration, Dialectic, and Rhetoric," p. 111.
32 Aristotle, *Topica*, I. 1.
33 See generally Michael C. Leff, "The Topics of Argumentative Invention in Latin Rhetorical Theory from Cicero to Boethius," *Rhetorica*, 1 (1983), 23–44; and F. Muller, "Le *De inventione dialectica* d'Agricola dans la tradition rhétorique d'Aristote à Port-Royal," in Akkerman and Vanderjagt, eds., *Rodolphus Agricola Phrisius (1444–1485)*, pp. 281–92. Despite their titles, both of these discussions have more to say about dialectic than rhetoric. See also Niels Jørgen Green-Pedersen, *The Tradition of the Topics in the Middle Ages: The Commentaries on Aristotle's and Boethius' "Topics"* (Munich: Philosophia Verlag, 1984). For analyses of the topics in use, see Jean Dietz Moss, "Aristotle's Four Causes: Forgotten *topos* of Renaissance Rhetoric," *Rhetoric Society Quarterly*, 17 (1987), 71–88; and Angus Gowland, "Rhetorical Structure and Function in the *Anatomy of Melancholy*," *Rhetorica*, 19 (2001), 1–48, at pp. 29–31. For remarks about the decline of the topical tradition, see Ann Moss, *Printed Commonplace-Books and the Structuring of Renaissance Thought* (Oxford: Clarendon Press, 1996), pp. 255–81. For a modern attempt to recruit "topical logic" into science studies, see Lawrence J. Prelli, *A Rhetoric of Science: Inventing Scientific Discourse* (Columbia: University of South Carolina Press, 1989).
34 Daniel Garber and Sandy Zabell, "On the Emergence of Probability," *Archive for the History of Exact Science*, 21 (1979), 33–53; and John Poinsot (John of St. Thomas), *Tractatus de signis: The semiotic of John Poinsot* [1632], ed. John Deely with Ralph Austin Powell (Berkeley: University of California Press, 1985). John Deely, *Early Modern Philosophy and Postmodern Thought* (Toronto Studies in Semiotics) (Toronto: University of Toronto Press, 1994) is sometimes suggestive but historically uneven.
35 Ian Maclean, "The Interpretation of Natural Signs: Cardano's *De subtilitate* versus Scaliger's *Exercitationes*," in *Occult and Scientific Mentalities in the Renaissance*, ed. Brian Vickers (Cambridge:

The manner in which dialectic functioned in practice in sixteenth-century natural philosophy can be illustrated by a treatise on sublunary aerial phenomena written by Marcus Frytschius, a citizen of the Lausitz *Sechsstädtebund*. Frytschius's *Meteorum* (Concerning Meteors), published in Nuremberg in 1563, is explicitly organized by the precepts of dialectic, and by the "topics" in particular. In his discussion of comets, for example, Frytschius sought to defend the standard position that they are earthly, not heavenly, phenomena.[36] He proved this by a number of arguments distinguishing comets (the predicate) from stars (the subject). He then went on to prove the same point with eight arguments drawn from the subject (stars). The eighth argument is taken from the proper nature of stars:

> No heavenly body or star has a tail.
> Comets have a tail
> Therefore a comet is not a star.[37]

The final argument Frytschius produced to prove that comets are not stars does not rely upon ratiocination but rather upon testimony: It is the argument from authority. "Seneca, who cites the author Epigenes, who says that the Chaldeans maintain it, also testifies that comets are not stars. And this is the common judgement of the learned."[38] Individually, these arguments were not demonstrative: They do not conform to the strict requirements of *regressus* theory. Taken together, however, they all tend to confirm the probability of the desired conclusion. Of course, the arguments depend upon tacit assumptions about the nature of comets – assumptions laid starkly bare by the syllogistic form in which they are framed. These assumptions changed over the course of the sixteenth and seventeenth centuries, as Pierre Bayle's notoriously debunking *Pensées diverses sur la comète* (Various Thoughts on the Comet, 1682) illustrates. But even in the case of Bayle's book, the strongly

---

Cambridge University Press, 1984), pp. 231–52; Brian K. Nance, "Determining the Patient's Temperament: An Excursion into Seventeenth-Century Medical Semiology," *Bulletin of the History of Medicine*, 67 (1993), 417–38; and Roger French, "Sign Conceptions in Medicine from the Renaissance to the Early Nineteenth Century," in *Semiotik: Ein Handbuch zu den zeichentheoretischen Grundlagen von Natur und Kultur*, ed. Roland Posner, Klaus Robering, and Thomas Albert Sebeok (Berlin: Walter de Gruyter, 1998).

[36] On cometary theory in the sixteenth century, see Peter Barker and Bernard R. Goldstein, "The Role of Comets in the Copernican Revolution," *Studies in History and Philosophy of Science*, 19 (1988), 299–319; and Tabitta van Nouhuys, *The Age of Two-Faced Janus: The Comets of 1577 and 1618 and the Decline of the Aristotelian World View in the Netherlands* (Brill's Studies in Intellectual History, 89) (Leiden: E. J. Brill, 1998).

[37] Marcus Frytschius, *Meteorum, hoc est, impressionum aerearum et mirabilium naturae operum, loci fere omnes, methodo dialectica conscripti, & singulari quadam cura diligentiaque in eum ordinem digesti ac distributi* (Nuremberg: Montanus and Neuber, 1563), fols. 99v–102r, at fol. 101v: "Octavum argumentum. A proprie stellarum natura. Sydus sive stella non habet comam. Cometa habet comam. Ergo Cometa non est stella."

[38] Ibid. fols. 101v–102r: "Cometas non esse stellas, testatur & Seneca, qui citat authorem Epigenem qui ait Chaldeos affirmare, quod Cometae non sint stellae. Et haec est usitata eruditorum sententia."

dialectical character of the arguments (if not their reduction to syllogistic form) remained.³⁹

From logic we turn to rhetoric. "The proofs and Demonstrations of *Logicke*, are toward all men indifferent, and the same," wrote the English philosopher and common lawyer Francis Bacon (1561–1626) in his *Advancement of Learning* (1605), "but the Proofes and perswasions of *Rhetoricke*, ought to differ according to the Auditors."⁴⁰ Sound logic, whether demonstrative or probable, was taken to persuade by virtue of its universally valid rationality. Effective rhetoric, by contrast, willingly took advantage of local knowledge. The "topics" of rhetorical theory were less abstract and more specific than those of logic; they might include considerations of where someone was born, their parentage, their loyalties, and their character. Insofar as the object of natural philosophy was taken to be the universal manifestation of nature, then its proofs would be logical. This was in pointed contradistinction with moral and political philosophy, which took human actions as their object and hence employed proofs more closely associated with the disciplines of rhetoric and history. In practice, however, natural philosophers in the early modern period were scarcely less aware of the need to appeal to specific audiences. Like their moral philosophical counterparts, they were concerned with effective techniques of persuasion. Scholarly studies of the rhetoric of science in the early modern period have approached the subject from a range of positions. Some have used a more or less anachronistic understanding of "rhetoric."⁴¹ Historically more successful studies, however, have drawn upon

---

³⁹ See Pierre Bayle, *Pensées diverses sur la comète* [1682], ed. A. Prat, rev. by Pierre Rétat, 2 vols. in 1 (Paris: Société des Textes Français Modernes, 1994), and also Rétat's "Avertissement de la deuxième edition," pp. 11, 21.

⁴⁰ Francis Bacon, *The Advancement of Learning* [1605], ed. Michael Kiernan (The Oxford Francis Bacon, 4) (Oxford: Clarendon Press, 2000), p. 129. See also Francis Bacon, *De augmentis scientiarum*, in *Works*, ed. James Spedding, Robert Leslie Ellis, and Douglas Denon Heath, 7 vols. (London: Longman, 1857), I: 413–837, at p. 673; bk. 6, chap. 3: "Siquidem probationes et demonstrationes Dialecticae universis hominibus sunt communes; at probationibus et suasiones Rhetoricae pro ratione auditorum variari debent."

⁴¹ Richard Foster Jones, "The Rhetoric of Science in England of the Mid-Seventeenth Century," in *Restoration and Eighteenth-Century Literature*, ed. Carroll Camden (Chicago: University of Chicago Press, 1963), pp. 5–24; James Stephens, "Rhetorical Problems in Renaissance Science," *Philosophy and Rhetoric*, 8 (1975), 213–29; John R. R. Christie, "Introduction: Rhetoric and Writing in Early Modern Philosophy and Science," in Benjamin et al., eds., *The Figural and the Literal*, pp. 1–9; Robert E. Stillman, "Assessing the Revolution: Ideology, Language and Rhetoric in the New Philosophy of Early Modern England," *The Eighteenth Century: Theory and Interpretation*, 35 (1994), 99–118; Michael Wintroub, "The Looking Glass of Facts: Collecting, Rhetoric and Citing the Self in the Experimental Natural Philosophy of Robert Boyle," *History of Science*, 35 (1997), 189–217; K. Neal, "The Rhetoric of Utility: Avoiding Occult Associations for Mathematics through Profitability and Pleasure," *History of Science*, 37 (1999), 151–78; and Maurice Slawinski, "Rhetoric and Science/Rhetoric of Science/Rhetoric as Science," in *Science, Culture, and Popular Belief in Renaissance Europe*, ed. Stephen Pumfrey, Paolo Rossi, and Maurice Slawinski (Manchester: Manchester University Press, 1991), pp. 71–99. See also James Stephens, *Francis Bacon and the Style of Science* (Chicago: University of Chicago Press, 1975); James P. Zappen, "Science and Rhetoric from Bacon to Hobbes: Responses to the Problem of Eloquence," in *Rhetoric 78, Proceedings of the Theory of Rhetoric: An Interdisciplinary Conference*, ed. Robert Brown, Jr. and Martin Steinman, Jr. (Minneapolis: University of Minnesota Center for Advanced Studies in Language, Style, and Literary Theory, 1979), pp. 399–419; Zappen,

early modern conceptions of rhetoric to explain aspects of the composition, arguments, and reception of works in early modern natural philosophy and medicine.[42] The writings of Galileo, in particular, have proved amenable to historical analysis through the categories of Renaissance rhetoric.[43]

From the early Renaissance onward, the art of rhetoric was lovingly cultivated as the supreme means of persuasion by writers, preachers, and politicians. The Renaissance revival of ancient learning brought with it a fascination with ancient eloquence. This fascination was stimulated by the Byzantine rhetorical tradition, by the rediscovery in 1416 of the full manuscript of Quintilian's *Institutio oratoria* (On the Education of the Orator) by the Italian humanist Poggio Bracciolini, and by the increasing impact of successive Latin translations of Aristotle's *Rhetoric* in the sixteenth century.[44] The medieval rhetorical tradition of the *ars dictaminis* was developed in many directions,[45] particularly in the areas of epistolography,[46] the *ars praedicandi*,[47] epideictic

---

"Francis Bacon and the Historiography of Scientific Rhetoric," *Rhetoric Review*, 8 (1989), 74–88; John C. Briggs, *Francis Bacon and the Rhetoric of Nature* (Cambridge, Mass.: Harvard University Press, 1989); and David Heckel, "Francis Bacon's New Science: Print and the Transformation of Rhetoric," in *Media, Consciousness, and Culture: Explorations of Walter Ong's Thought*, ed. Bruce E. Gronbeck, Thomas J. Farrell, and Paul A. Soukup (Newbury Park, Calif.: Sage, 1991), pp. 64–76.

[42] See, for example, Brian Vickers, "The Royal Society and English Prose Style: A Reassessment," in *Rhetoric and the Pursuit of Truth: Language Change in the Seventeenth and Eighteenth Centuries*, ed. Brian Vickers and Nancy Struever (Los Angeles: William Andrews Clark Memorial Library, 1985), pp. 1–76; Jean Dietz Moss, "The Interplay of Science and Rhetoric in Seventeenth-Century Italy," *Rhetorica*, 7 (1989), 23–4; Moss, *Novelties in the Heavens: Rhetoric and Science in the Copernican Controversy* (Chicago: University of Chicago Press, 1993); John T. Harwood, "Science Writing and Writing Science: Boyle and Rhetorical Theory," in *Robert Boyle Reconsidered*, pp. 37–56; and Gowland, "Rhetorical Structure and Function in the *Anatomy of Melancholy*."

[43] Brian Vickers, "Epideictic Rhetoric in Galileo's *Dialogo*," *Annali dell'Istituto e Museo di Storia della Scienza di Firenze*, 8 (1983), 69–102; Jean Dietz Moss, "Galileo's *Letter to Christina*: Some Rhetorical Considerations," *Renaissance Quarterly*, 36 (1983), 547–76; Moss, "Galileo's Rhetorical Strategies in Defense of Copernicanism," in *Novità celesti e crisi del sapere: atti del convegno internazionale di studi Galileiani*, ed. Paolo Galluzzi (Florence: Giunti Barbèra, 1984), pp. 95–103; Moss, "The Rhetoric of Proof in Galileo's Writings on the Copernican System," in *Reinterpreting Galileo*, ed. William A. Wallace (Washington, D.C.: Catholic University of America Press, 1986), pp. 179–204; A. C. Crombie and Adriano Carugo, "Galileo and the Art of Rhetoric," *Nouvelles de la république des lettres*, 2 (1988), 7–31, reprinted in Crombie, *Science, Art, and Nature in Medieval and Modern Thought* (London: Hambledon, 1996), pp. 231–55; Nicholas Jardine, "Demonstration, Dialectic, and Rhetoric"; and Moss, *Novelties in the Heavens*, pp. 75–300. See also Maurice A. Finocchiaro, *Galileo and the Art of Reasoning: Rhetorical Foundations of Logic and Scientific Method* (Boston Studies in the Philosophy of Science, 61) (Dordrecht: Reidel, 1980).

[44] See John Monfasani, "Humanism and Rhetoric," in *Renaissance Humanism: Foundations, Forms, and Legacy*, ed. Albert Rabil, 3 vols. (Philadelphia: University of Pennsylvania Press, 1988), vol. 3: *Humanism and the Disciplines*, pp. 171–235, esp. pp. 177–84.

[45] James J. Murphy, *Rhetoric in the Middle Ages: A History of Rhetorical Theory from St. Augustine to the Renaissance* (Berkeley: University of California Press, 1974); Ronald Witt, "Medieval *ars dictaminis* and the Beginnings of Humanism: A New Construction of the Problem," *Renaissance Quarterly*, 35 (1982), 1–35; Virginia Cox, "Ciceronian Rhetoric in Italy, 1260–1350," *Rhetorica*, 17 (1999), 239–80; and Judith Rice Henderson, "Valla's *Elegantiae* and the Humanist Attack on the *Ars dictaminis*," *Rhetorica*, 19 (2001), 249–68.

[46] Judith Rice Henderson, "Erasmus on the Art of Letter-writing," in Murphy, ed., *Renaissance Eloquence*, pp. 331–55; Henderson, "Erasmian Ciceronians: Reformation Teachers of Letter-writing," *Rhetorica*, 10 (1992), 273–302; Henderson, "On Reading the Rhetoric of the Renaissance Letter," in *Renaissance-Rhetorik*, ed. Heinrich F. Plett (Berlin: Walter de Gruyter, 1993).

[47] John W. O'Malley, *Praise and Blame in Renaissance Rome: Rhetoric, Doctrine, and Reform in the Sacred Orators of the Papal Court, c. 1450–1521* (Durham, N.C.: Duke University Press, 1979). See also Debora

(the rhetoric of praise and blame),[48] and *elocutio* (the study of the figures and tropes).[49] A very large body of theoretical treatises, covering all or some of the five parts (*inventio, dispositio, elocutio, memoria,* and *pronuntiatio*) and three *genera* (demonstrative or epideictic, deliberative, and forensic) of Ciceronian oratory,[50] were published to satisfy the voracious appetite of schoolteachers, university scholars, preachers, and courtiers for guidance in techniques of eloquence and persuasion.[51] This rhetorical culture also encouraged less formulaic reflections on the nature and function of oratory,[52] as well as innumerable orations, epistles, eulogies, sermons, addresses, defenses, attacks, and prefaces, almost all of which can be regarded as being informed in some way or another by the art of rhetoric.[53]

One term in particular was central to the rhetorical account of proof and persuasion: the notion of credit or belief (*fides*). In a formula widely taken up from Cicero's *De partitione oratoria* (On the Classification of Oratory), rhetorical argument was said to be "a plausible invention to generate belief" (*probabile inventum ad faciendam fidem*).[54] This "belief" was double-edged. In the first place, it was necessary that the orator be credible – that he (the orator in ancient and Renaissance rhetorical theory was assumed to be male) possess a good ethos. Recommended techniques for achieving this ethos included promising an audience novelty, emphasizing personal probity, speaking moderately and without partiality, and, if possible, without

---

Kuller Shuger, *Sacred Rhetoric: The Christian Grand Style in the English Renaissance* (Princeton, N.J.: Princeton University Press, 1988), which does not limit its discussion to English rhetoricians; Harry Caplan and H. H. King, "Latin Tractates on Preaching: A Booklist," *Harvard Theological Review*, 42 (1949), 185–206; and King, "Pulpit Eloquence: A List of Doctrinal and Historical Studies in English," *Speech Monographs*, 22 (1955), 1–159.

[48] O. B. Hardison, *The Enduring Monument: A Study of the Idea of Praise in Renaissance Literary Theory and Practice* (Chapel Hill: University of North Carolina Press, 1962); and John M. McManamon, *Funeral Oratory and the Cultural Ideals of Italian Humanism* (Chapel Hill: University of North Carolina Press, 1989).

[49] Brian Vickers, "Rhetorical and Anti-rhetorical Tropes: On Writing the History of *elocutio*," *Comparative Criticism*, 3 (1981), 105–32; and Richard A. Lanham, *A Handlist of Rhetorical Terms*, 2nd ed. (Berkeley: University of California Press, 1991).

[50] For discussion of the *partes* and *genera* of Ciceronian rhetoric, see Brian Vickers, *In Defence of Rhetoric*, rev. ed. (Oxford: Clarendon Press, 1989), pp. 52–82.

[51] James J. Murphy, "One Thousand Neglected Authors: The Scope and Importance of Renaissance Rhetoric," in Murphy, ed., *Renaissance Eloquence*, pp. 20–36.

[52] Rebhorn, ed. and trans., *Renaissance Debates on Rhetoric*, provides a useful anthology.

[53] For guidance into this mass of literature and some of the issues raised by it, see the essays in Murphy, ed., *Renaissance Eloquence*, and Peter Mack, ed., *Renaissance Rhetoric* (Basingstoke: Macmillan, 1994). See also Vickers, *In Defence of Rhetoric*. For bibliographies of Renaissance rhetoric, see James J. Murphy, *Renaissance Rhetoric: A Short-title Catalogue of Works on Rhetorical Theory from the Beginning of Printing to A.D. 1700* (New York: Garland, 1981); Paul D. Brandes, *A History of Aristotle's Rhetoric: With a Bibliography of Early Printings* (Metuchen, N.J.: Scarecrow Press, 1989); James J. Murphy and Martin Davis, "Rhetorical Incunabula: A Short-title Catalogue of Texts Printed to the Year 1500," *Rhetorica*, 15 (1997), 355–470; and Heinrich F. Plett, *English Renaissance Rhetoric and Poetics: A Systematic Bibliography of Primary and Secondary Sources* (Symbola et Emblemata, 6) (Leiden: E. J. Brill, 1995).

[54] Cicero, *De partitione oratoria*, 2.1, trans. E. W. Sutton and H. Rackham, in *De oratore III, De fato, Paradoxa stoicorum, De partitione oratoria* (Loeb Classical Library) (London: Heinemann, 1948), pp. 305–421, at p. 314.

impugning an adversary's character.⁵⁵ Rhetorical theory generally counseled establishing the speaker's ethos at the beginning of an oration, which is why such devices can so often be found in prefaces of early modern books. Bacon, for instance, consistently drew upon the modesty topos encouraged by these notions of ethos to advance his argument that knowledge would advance further through the contributions of many modest inquirers (such as himself) than through the proud individual systematizing of previous philosophers:

> And I have also followed the same humility in my teaching which I applied to discovering. For I do not try either by triumphant victories in argument, nor by calling antiquity to my aid, nor by any usurpation of authority, nor by a veil of obscurity either, to invest these my discoveries with any majesty, which might easily be done by anyone trying to bring lustre to his own name rather than light to the minds of others.⁵⁶

The second task of rhetorical *fides* was to instill belief not in the rhetorician himself but in what he had to say. In order to achieve this, it was necessary above all for the orator to find or discover (*invenire*) arguments – the province of the part of rhetoric known as *inventio*, and described by the anonymous author of the very widely read *Rhetorica ad Herennium* as "the most important and most difficult" part of rhetoric.⁵⁷ A number of techniques were available in Renaissance rhetorical theory for "discovering" credible (*probabile*) arguments. The orator might resort to the "topics" discussed earlier in respect to dialectic. Or he might draw upon the "commonplaces" (*loci communes*). These were set arguments that could be drawn upon whether one was attacking or defending a case: For instance, one might argue for witnesses against arguments, or vice versa.⁵⁸ This in turn emphasizes another important aspect of early modern rhetoric: its two-sidedness. Rhetorical theory taught the skill of arguing on both sides of the question (*in utramque partem*); the *locus classicus* for this was Lactantius's account in the *Institutiones divinae* (XV. 5) of

---

⁵⁵ See Skinner, *Reason and Rhetoric*, pp. 127–33.
⁵⁶ Francis Bacon, *Novum organum with Other Parts of the Great Instauration*, ed. Peter Urbach, trans. John Gibson (Chicago: Open Court, 1994), p. 14 (preface to the *Instauratio magna*), translating Francis Bacon, *Novum organum* [1620], ed. Thomas Fowler, 2nd ed. (Oxford: Clarendon Press, 1889), pp. 166–7: "Atque quam in inveniendo adhibemus humilitatem, eandem et in docendo sequuti sumus. Neque enim aut confutationum triumphis, aut antiquitatis advocationibus, aut authoritatis usurpatione quadam, aut etiam obscuritatis velo, aliquam his nostris inventis majestatem imponere aut conciliare conamur; qualia reperire non difficile esset ei, qui nomini suo non animis aliorum lumen affundere conaretur." See further James S. Tillman, "Bacon's *ethos*: The Modest Philosopher," *Renaissance Papers*, (1976), 11–19.
⁵⁷ *Rhetorica ad Herennium*, 2.i.1, trans. Harry Caplan (Loeb Classical Library) (London: Heinemann, 1954), p. 58: "De oratoris officiis quinque inventio et prima et difficillima est."
⁵⁸ On the theory of commonplaces in Renaissance rhetoric, see Quirinus Breen, "The Terms 'loci communes' and 'loci' in Melanchthon," *Church History*, 16 (1947), 197–209; Sister Marie Joan Lechner, *Renaissance Concepts of the Commonplaces* (New York: Pageant, 1962); Francis Goyet, *Le Sublime du lieu commun: L'Invention rhétorique dans l'Antiquité et à la Renaissance* (Paris: Honoré Champion, 1996); and Moss, *Printed Commonplace-Books*.

the skeptic Carneades, who argued equally persuasively for justice one day and against it the next.

Rhetorical and dialectical *inventio*, the art of finding plausible or probable arguments, was ipso facto also associated with the discovery of new truths. For this reason, *inventio* was the part of rhetoric and logic that impinged most significantly on theoretical accounts of the study of nature in early modern Europe.[59] The Italian natural philosopher Giambattista della Porta (1535–1615) drew on the semiotic theory found in the Aristotelian rhetorical tradition for his *De humana physiognomonia* (On Human Physiognomy, 1586).[60] The German Reformer Philip Melanchthon (1497–1560) used natural philosophical *loci* to structure his teaching of the subject.[61] Bacon was preoccupied by the process of discovery,[62] and in his late works, he frequently drew upon the Aristotelian rhetorical notion of "particular topics" to structure his investigations of natural phenomena.[63] ("Particular topics" were articles of inquiry appropriate to specific investigations; they were opposed to the "general topics" that were appropriate to inquiries in any discipline.)[64] Bacon saw these particular topics as "a sort of mixture of logic and of the proper material itself of individual sciences."[65] The German polymath Gottfried Wilhelm Leibniz (1646–1716) equated the art of invention with "la science generale [sic]."[66] As we shall see, however, the assumption that rhetoric and logic per se might help in discovering new truths about nature came under increasingly sustained attack in the course of the seventeenth century.

---

[59] See also Theodore Kisiel, "Ars inveniendi: A Classical Source for Contemporary Philosophy of Science," *Revue internationale de philosophie*, 34 (1980), 130–54.

[60] Cesare Vasoli, "L'analogia universale: La retorica come semiotica nell'opera del Della Porta," in *Giovan Battista della Porta nell'Europa del suo tempo* (Naples: Guida Editori, 1990), pp. 31–52. See also Giovanni Manetti, "Indizi e prove nella cultura greca: Forza epistemica e criteri di validità dell'inferenza semiotica," *Quaderni storici*, 85, no. 29 (1994), 19–42; Donald Morrison, "Philoponus and Simplicius on tekmeriodic Proof," in Di Liscia, Kessler, and Methuen, eds., *Method and Order in Renaissance Philosophy of Nature*, pp. 1–22.

[61] Sachiko Kusukawa, *The Transformation of Natural Philosophy: The Case of Philip Melanchthon* (Cambridge: Cambridge University Press, 1995), pp. 151–3.

[62] Lisa Jardine, *Francis Bacon: Discovery and the Art of Discourse* (Cambridge: Cambridge University Press, 1974); William A. Sessions, "Francis Bacon and the Classics: The Discovery of Discovery," in *Francis Bacon's Legacy of Texts: "The Art of Discovery Grows with Discovery,"* ed. William A. Sessions (New York: AMS, 1990), pp. 237–53.

[63] See Francis Bacon, *De augmentis scientiarum*, in *Works*, 1: 633–9; Bacon, *Historia ventorum* (London: M. Lownes, 1622); Bacon, *Historia vitae et mortis* (London: M. Lownes, 1623); and further, Paolo Rossi, *Francis Bacon: From Magic to Science*, trans. Sacha Rabinovitch (London: Routledge and Kegan Paul, 1968), pp. 157, 216–19.

[64] Aristotle, *The "Art" of Rhetoric*, trans. John Henry Freese (Loeb Classical Library) (London: Heinemann, 1926), pp. 30–3 (I.ii.21–2).

[65] Bacon, *De augmentis*, 5.3, in *Works*, 1: 635: "Illi autem mixturae quaedam sunt, ex Logica et Materia ipsa propria singularum scientiarum."

[66] Gottfried Wilhelm Leibniz, "Discours touchant la méthode de la certitude et l'art d'inventer," in *Philosophische Schriften*, ed. Carl Immanuel Gerhardt, 7 vols. (Berlin: Weidman, 1875–90), 7: 174–83, at p. 180.

DISCIPLINARY RECONFIGURATIONS

As the scope, content, and social setting of natural philosophy changed in the course of the late sixteenth and seventeenth centuries, so did techniques of proof and persuasion. The early modern period saw significant developments not just in the content of natural philosophy but also in its exposition. As with content and exposition, the impetus for the critique of the broadly Aristotelian traditional natural philosophy lay very largely in the late Renaissance revaluation of other schools of ancient philosophy besides Aristotle's. Neo-Stoicism, Ciceronian and Pyrrhonian skepticism, and, in the seventeenth century, Epicureanism all contributed to bringing established forms of proof and persuasion into doubt.[67] In terms of changing the content of natural philosophy in the early seventeenth century, Epicureanism had the greatest impact, as its doctrine of atomism helped to spawn corpuscularianism and the mechanical philosophy more generally.[68] In terms of casting doubt on received views about proof and persuasion, however, the Pyrrhonian skepticism that arose after the Latin translation of Sextus Empiricus's *Outlines of Pyrrhonism* in 1569 had the most impact. The Pyrrhonian assertion that nothing could be known with certainty was deeply threatening to conventional assumptions about the possibility of certain demonstration. This critique was particularly developed in the late sixteenth century by the medically trained author Francisco Sánchez (ca. 1550–1623) in his *Quod nihil scitur* (That Nothing Is Known, 1581) and the magistrate Michel de Montaigne (1533–1592) in his *Essais* (Essays, 1580, 1588, 1593). Sánchez's treatise elaborated a more thoroughgoing assault on philosophical claims to demonstrative *scientia* than that of the eclectic vernacular humanist Montaigne; Sánchez concluded his treatise with the explanation that "I was not anxious myself to perpetrate the fault I condemn in others, namely to prove my assertion with arguments that were far-fetched, excessively obscure, and perhaps more doubtful than the

---

[67] See Gerhard Oestreich, *Neostoicism and the Early Modern State*, ed. Brigitta Oestreich and H. G. Koenigsberger, trans. David McLintock (Cambridge: Cambridge University Press, 1982); and for its impact on natural philosophy, Peter Barker and Bernard R. Goldstein, "Is Seventeenth-Century Physics Indebted to the Stoics?" *Centaurus*, 27 (1984), 148–64; and Margaret J. Osler, ed., *Atoms, Pneuma, and Tranquillity: Epicurean and Stoic Themes in European Thought* (Cambridge: Cambridge University Press, 1991). On Ciceronian skepticism, see Charles B. Schmitt, *Cicero Scepticus: A Study of the Influence of the 'Academica' in the Renaissance* (Archives internationales d'histoire des idées, 52) (The Hague: Martinus Nijhoff, 1972). On the fortunes of the Pyrrhonian skepticism that developed after the publication of Sextus Empiricus's *Outlines of Pyrrhonism* in 1562 (and Latin translation in 1569), see Richard H. Popkin, *The History of Scepticism from Erasmus to Spinoza*, 2nd ed. (Berkeley: University of California Press, 1979). On skepticism in relation to natural knowledge, see Nicholas Jardine, "Scepticism in Renaissance Astronomy: A Preliminary Study," in *Scepticism from the Renaissance to the Enlightenment*, ed. Richard H. Popkin and C. B. Schmitt (Wiesbaden: Harrassowitz, 1987), pp. 83–102. On Epicureanism, see Howard Jones, *The Epicurean Tradition* (London: Routledge, 1989); and J. J. MacIntosh, "Robert Boyle on Epicurean Atheism and Atomism," in Osler, ed., *Atoms, Pneuma, and Tranquility*, pp. 197–219.

[68] Daniel Garber, "Apples, Oranges, and the Role of Gassendi's Atomism in Seventeenth-Century Science," *Perspectives on Science*, 3 (1995), 425–8.

very problem under investigation."⁶⁹ It was in response to this skeptical challenge that earlier seventeenth-century philosophers elaborated their theories, notably the French Minim Marin Mersenne (1588–1648) and Descartes, in their natural philosophy, and Hugo Grotius (1583–1645) and Edward, Lord Herbert of Cherbury (1583–1648), in their moral and metaphysical philosophies.⁷⁰

Whether mechanical, experimental, or natural historical, the new forms of natural philosophy that built upon the doctrines of ancient philosophical schools were self-consciously new. The proofs of rhetoric and, above all, logic, however, remained strongly associated with the older philosophy of the schools. Hence, as natural philosophers in the seventeenth century became increasingly critical of the intellectual and institutional constraints of the universities, they also criticized their methods of probation. Thus, a significant aspect of the novelty of the new philosophy consisted in a deep dissatisfaction – a dissatisfaction that amounted practically to crisis – with received techniques of proof and persuasion.

As we have seen, university natural philosophy in the Renaissance was conceived in the dominant Aristotelian tradition as a contemplative science founded upon certain demonstrations. These demonstrations were ideally composed of syllogisms. In this understanding, natural philosophy both proceeded logically and was underpinned by logical principles. One of the central aspects of the new forms of natural philosophy that developed from the late sixteenth century onward, however, was an attack on logic in general and the syllogism in particular as a means of making discoveries about nature.⁷¹ Thus, one of the central features of the dissolution of the Aristotelian tradition in natural philosophy was a systematic critique of received methods of proof and persuasion.

This willingness to criticize conventional forms of probation explains why Bacon preferred aphorisms to Aristotelian axioms and in the *Novum organum* (New Organon, 1620) repeatedly attacked syllogisms: "We reject proof by syllogism, because it operates in confusion and lets nature slip out of our

---

⁶⁹ Francisco Sánchez, *That Nothing is Known (Qvod nihil scitvr)*, ed. Elaine Limbrick, trans. Douglas F. S. Thomson (Cambridge: Cambridge University Press, 1988), p. 163, translation from pp. 289–90: "Nec enim quod in aliis ego damno, ipse committere volui: ut rationibus a longe petitis, obscurioribus, & magis forsan quaesito dubiis, intentum probarem." On the relations of Montaigne's *Essais* to the arts course, and particularly the arts of proof and persuasion, see Ian Maclean, *Montaigne philosophe* (Paris: Presses Universitaires de France, 1996), esp. pp. 39–53.

⁷⁰ See further Peter Dear, *Mersenne and the Learning of the Schools* (Ithaca, N.Y.: Cornell University Press, 1988), pp. 23–47; Richard Tuck, "The 'modern' Theory of Natural Law," in *The Languages of Political Theory in Early-Modern Europe*, ed. Anthony Pagden (Cambridge: Cambridge University Press, 1987), pp. 99–119; and R. W. Serjeantson, "Herbert of Cherbury Before Deism: The Early Reception of the *De veritate*," *The Seventeenth Century*, 16 (2001), 217–38, at p. 220.

⁷¹ See William Eamon, *Science and the Secrets of Nature: Books of Secrets in Medieval and Early Modern Culture* (Princeton, N.J.: Princeton University Press, 1996), pp. 292–6.

hands."[72] What was required was not a formal analysis of propositions but an investigation of the things from which those propositions were abstracted. The syllogism is "by no means equal to the subtlety of things"; it "compels assent without reference to things."[73] In an analogous vein, Descartes argued that syllogisms "are of less use for learning things than for explaining to others the things one already knows."[74] Boyle liked to "insist rather on Experiments than Syllogismes," comparing "those Dialectical subtleties, that the Schoolmen too often employ about Physiological Mysteries" to "the tricks of Jugglers" (i.e., conjurers).[75] The second secretary of the Royal Society, Robert Hooke (1635–1702), allowed some virtue to logic, but asserted that it was "wholly deficient" for "Inquiry into Natural Operations."[76] Numerous other writers also developed the *novatores*' attack on logic as the basis of proof in natural philosophy.[77]

Not all new philosophers, however, rejected the use of logic outright. Hobbes was scornful of the English Catholic philosopher Thomas White's (1593–1676) assertion that "Philosophy must not be treated logically."[78] Both Hobbes and Pierre Gassendi retained syllogistic as part of their philosophical systems.[79] Other authors, such as Hobbes's bitter opponent Seth Ward (1617–1689), Savilian Professor of Astronomy at Oxford, defended the universal subservience of logic to the "enquiry of all truths," and even the application of the syllogism to a newly mathematized "Physicks."[80] But throughout the seventeenth century, natural philosophers devoted intensive efforts to trying to establish probative procedures that would replace the increasingly

---

[72] Francis Bacon, *Advancement of Learning*, p. 124; Bacon, *The New Organon*, trans. Michael Silverthorne, ed. Lisa Jardine (Cambridge: Cambridge University Press, 2000), p. 16 (*Distributio operis*); see also p. 83 (bk. I, aphorism 104) and p. 98 (bk. I, aphorism 127).
[73] Francis Bacon, *New Organon*, p. 35 (bk. 1, aphorism 13). See further L. Jardine, *Francis Bacon*, esp. pp. 84–5, and L. Jardine, "Introduction" to Bacon, *New Organon*, vii–xxviii.
[74] René Descartes, "Discourse on the Method," trans. John Cottingham, Robert Stoothoff, and Dugald Murdoch, in *The Philosophical Writings of Descartes*, 3 vols. (Cambridge: Cambridge University Press, 1985–91), I: 111–51, at p. 119, translating Descartes, *Discours de la méthode*, p. 17: "pour la Logique, ses syllogismes & la pluspart de ses autres instructions seruent plutost a expliquer a autruy les choses qu'on sçait." See further Carr, *Descartes and the Resilience of Rhetoric*, pp. 41–2.
[75] Robert Boyle, *The Sceptical Chymist* [1661], in Boyle, *Works*, 2: 205–378, at p. 219. See further Jan V. Golinski, "Robert Boyle: Scepticism and Authority in Seventeenth-Century Chemical Discourse," in Benjamin et al., eds., *The Figural and the Literal*, pp. 58–82, at p. 67.
[76] Robert Hooke, "A General Scheme, or Idea of the Present State of Natural Philosophy," in Hooke, *Posthumous Works*, ed. Richard Waller (London: Samuel Smith and Benjamin Walford, 1705; facsimile repr. London: Frank Cass, 1971), pp. 1–70, at p. 6.
[77] Charles Webster, ed., *Samuel Hartlib and the Advancement of Learning* (Cambridge: Cambridge University Press, 1970), p. 77; John Webster, *Academiarum examen* (London: Giles Calvert, 1654; facsimile. repr. in Allen G. Debus, *Science and Education in the Seventeenth Century: The Webster-Ward Debate* (London: Macdonald, 1970), pp. 32–40. Eusebius Renaudot, ed., *A General Collection of Discourses of the Virtuosi of France* (London: Thomas Dring and John Starkey, 1664), sig. §4r–v.
[78] Thomas Hobbes, *Thomas White's "De Mundo" Examined*, trans. Harold Whitmore Jones (Bradford: Bradford University Press, 1976), p. 26, chap. 1, sec. 4.
[79] Gassendi, *Institutio logica*; Thomas Hobbes, *De corpore*, in *The English Works of Thomas Hobbes*, ed. Sir William Molesworth, 11 vols. (London: Bohn, 1839), vol. 1.
[80] [Seth Ward], *Vindiciae academiarum* (Oxford: Thomas Robinson, 1654; facsimile repr. in Debus, *Science and Education*), p. 25.

discredited ones of Aristotelian logic.[81] Some of the most famous treatises in the philosophy of early modern science exemplify this search: works such as Bacon's *Novum organum* (1620) – which advertised its ambition to replace Aristotle's *Organon* in its very title – and Descartes' *Discours de la méthode pour bien conduire sa raison et chercher la vérité dans les sciences* (Discourse on the Method for Conducting One's Reason Well and for Seeking Truth in the Sciences), which likewise emphasized its place in the tradition of writings on "method." Leibniz made numerous efforts to produce an "art of invention,"[82] and Hooke attempted to synthesize a "General Scheme, or Idea of the Present State of Natural Philosophy" that would allow for certainty of demonstration.[83]

The significance of comparable attacks on rhetoric is harder to assess. Indeed, characterizing the changing place of rhetoric in early modern natural philosophy is an extremely vexing matter, about which it is hard to make firm generalizations. Although rhetoric had always been taken to have a legitimate place in certain aspects of natural philosophy – notably in *parerga* such as dedications and prefaces – its legitimacy in arguments about nature per se was generally held to be doubtful.[84] In particular, the techniques of rhetorical *elocutio* were interdicted, most famously by the Royal Society's ecclesiastical hired pen Thomas Sprat (1635–1713) in his *History of the Royal-Society of London* (1667): "Who can behold, without indignation, how many mists and uncertainties, these specious *Tropes* and *Figures* have brought upon our Knowledge?"[85] This attack on figures of speech was licensed by the pervasive early modern dichotomy between "words" and "things" (*res et verba*): The rhetorical devices of metaphor, simile, and amplification belonged squarely

---

[81] For a case study of this phenomenon, see Stephen Clucas, "In Search of 'the true logick': Methodological Eclecticism among the 'Baconian Reformers'," in *Samuel Hartlib and Universal Reformation: Studies in Intellectual Communication*, ed. Mark Greengrass, Michael Leslie, and Timothy Raylor (Cambridge: Cambridge University Press, 1994), pp. 51–74.

[82] See, for example, Leibniz, "Disourse touchant la méthode de la certitude et l'art d'inventer"; and Louis Couturat, *La Logique de Leibniz* (Paris: Félix Alcan, 1901).

[83] Hooke, "A General Scheme." On Hooke's philosophy of science, see D. R. Oldroyd, "Robert Hooke's Methodology of Science as Exemplified in his 'Discourse of Earthquakes'," *British Journal for the History of Science*, 6 (1972), 109–30; Oldroyd, "Some 'Philosophical Scribbles' Attributed to Robert Hooke," *Notes and Records of the Royal Society*, 35 (1980), 17–32; Olroyd, "Some Writings of Robert Hooke on Procedures for the Prosecution of Scientific Inquiry, Including his 'Lectures of Things Requisite to a Natural History'," *Notes and Records of the Royal Society of London*, 41 (1987), 146–67; and Lotte Mulligan, "Robert Hooke and Certain Knowledge," *The Seventeenth Century*, 7 (1992), 151–69.

[84] See J. D. Moss, *Novelties in the Heavens*, p. 3; also Hobbes, *White's "De mundo" Examined*, p. 26, chap. 1, sec. 4: "Philosophy should therefore be treated logically, for the aim of its students is not to impress, but to know with certainty. So philosophy is not concerned with rhetoric."

[85] Thomas Sprat, *The History of the Royal-Society of London* (London: J. Martyn and J. Allestry, 1667; facsimile repr. London: Routledge and Kegan Paul, 1958), p. 112. On the significance of Sprat's comments on the Royal Society's putative "manner of discourse," see Vickers, "The Royal Society and English Prose Style"; Werner Hüllen, "Style and Utopia: Sprat's Demand for a Plain Style, Reconsidered," in *Papers in the History of Linguistics*, ed. Hans Aarsleff, Louis G. Kelly, and Hans-Josef Niederehe (Amsterdam: John Benjamins, 1987), pp. 247–62; and Hüllen, "*Their Manner of Discourse*": *Nachdenken uber Sprache im Umkreis der Royal Society* (Tübingen: Narr, 1989).

to the realm of *verba*. For this reason, the claim that you studied things whereas your opponent was merely studious of words was one of the more hackneyed charges in early modern controversy. This did not, however, lessen the force of the charge.[86] Experimental natural philosophers, in particular, liked to accord a probative force to things that words (on their account) could never possess: According to the Secretary of the Paris Académie Royale des Sciences, Bernard le Bovier de Fontenelle (1657–1757), "Physics holds the secret of shortening countless arguments that rhetoric makes infinite."[87]

If anything, the shift from the schools to the investigations of private individuals and academicians as sites of innovation in natural knowledge during the course of the sixteenth and seventeenth centuries may have led to a rise, rather than a decline, in the significance of rhetoric. The situation is comparable with the discovery by the earlier humanists of the polemical power of elegant and persuasive language in their attacks on the schools.[88] Almost all of the new vernacular natural philosophers were familiar with the textbooks and other productions of school philosophy, but they increasingly rejected both their language – Latin – and their more formulaic habits of expression. Perhaps the most significant changes in early modern techniques of proof and persuasion were brought about by two other concurrent developments in the study of nature. The first was the incorporation of considerations of continuous and discontinuous quantities – the mathematics of geometry and arithmetic – into the study of the natural world. The second was a reconfiguration of the way in which experience contributed to the knowledge of nature; that is to say, the incorporation of experiment into natural philosophy.[89]

MATHEMATICAL TRADITIONS

As the earlier quotation from Aristotle's *Nicomachean Ethics* suggested, mathematics – and, in particular, geometry – had a privileged place with respect to the certainty of its proofs. The nature of that certainty was a matter of debate. In his *Commentarium de certitudine mathematicarum* (Treatise on the Certainty of Mathematics, 1547), the Italian philosopher Alessandro Piccolomini (1508–1579) argued that mathematics did not owe its certainty to the

---

[86] See Wilber Samuel Howell, "*Res et verba*: Words and Things," *ELH: A Journal of English Literary History*, 13 (1946), 131–42; Ian Maclean and Eckhard Kessler, eds., *Res et Verba in der Renaissance* (Wiesbaden: Harrassowitz, 2002); and Roger Hahn, *The Anatomy of a Scientific Institution: The Paris Academy of Sciences, 1666–1803* (Berkeley: University of California Press, 1971), p. 7.
[87] Bernard le Bovier de Fontenelle, *Digression sur les Anciens et les Modernes* [1688], ed. Robert Shackleton (Oxford: Clarendon Press, 1955), p. 164.
[88] See, for example, Rummel, *Humanist–Scholastic Debate*, esp. p. 41.
[89] The significance of these two traditions in early modern natural philosophy was influentially developed by Thomas Kuhn, "Mathematical versus Experimental Traditions in the Development of Physical Science," *Journal of Interdisciplinary History*, 7 (1976), 1–31.

fact that its demonstrations conformed to Aristotelian criteria for *scientia*.[90] Several authors, most notably the Jesuit Benito Pereira (1535–1610), developed this position. They argued that mathematical demonstrations were not *demonstrationes potissimae* on the grounds that they did not provide an explanation in terms of the four causes of Aristotelian logic.[91] The challenge to the demonstrative status of mathematics did not go unmet. Two other Jesuit mathematicians, Christoph Clavius (1538–1612) and Christoph Scheiner (1573–1650), reasserted the scientific status of mathematics on the basis of its demonstration of conclusions "by axioms, definitions, postulates, and suppositions."[92] (In his *Algebra* of 1608, Clavius even attempted to describe mathematics in syllogistic terms.) These arguments were taken up by Mersenne.[93] The Italian mathematicians Francesco Barozzi (1537–1604) and Giuseppe Biancani (1566–1624), and the English mathematicians Isaac Barrow (1630–1677) and John Wallis (1616–1703), also defended the claim of mathematics to be a causal science. By the time of Barrow's celebrated *Lectiones mathematicae* (Mathematical Lectures, delivered in 1665), the most pressing challenge to the certainty of mathematics was no longer seen to come from writers such as Pereira but rather from the French natural philosopher Pierre Gassendi (1592–1655). In the second part of his *Exercitationes paradoxicae adversos Aristoteleos* (Paradoxical Exercises Against the Aristotelians, published posthumously in 1658), Gassendi had argued that no science, including mathematics, could be said to provide causal knowledge in Aristotle's terms.[94]

For natural philosophers, however, doubts about the status of mathematics were less important than questions about whether and how to incorporate quantity into the hitherto qualitative study of nature. In the seventeenth century, the increasingly widely held assumption that nature was mathematical

---

[90] N. Jardine, "Epistemology of the Sciences," p. 697.
[91] See Benito Pereira, *De communibus omnium rerum naturalium principiis et affectionibus libri quindecem* (Rome, 1576), p. 24; Paolo Mancosu, *Philosophy of Mathematics and Mathematical Practice in the Seventeenth Century* (Oxford: Oxford University Press, 1996), esp. p. 13; and Alistair Crombie, "Mathematics and Platonism in the Sixteenth-Century Italian Universities and in Jesuit Educational Policy," in *Prismata, Naturwissenschaftsgeschichtliche Studien*, ed. Y. Maeyama and W. G. Saltzer (Wiesbaden: Franz Steiner Verlag, 1977), pp. 63–94, at p. 67.
[92] Dear, *Discipline and Experience*, p. 41, quoting Christoph Scheiner, *Disquisitiones mathematicae* (1614).
[93] Peter Dear, *Mersenne and the Learning of the Schools* (Ithaca, N.Y.: Cornell University Press, 1988), p. 72.
[94] Paolo Mancosu, "Aristotelian Logic and Euclidean Mathematics: Seventeenth-Century Developments of the *Quaestio de certitudine mathematicarum*," *Studies in History and Philosophy of Science*, 23 (1992), 241–65. On Gassendi's attack on Aristotelianism, see Barry Brundell, *Pierre Gassendi: From Aristotelianism to a New Natural Philosophy* (Synthèse Historical Library: Texts and Studies in the History of Logic and Philosophy, 30) (Dordrecht: Reidel, 1987.) See also Wolfgang Detel, *Scientia rerum natura occultarum: Methodologische Studien zur Physik Pierre Gassendis* (Quellen und Studien zur Philosophie, 14) (Berlin: Walter de Gruyter, 1978); Lynn Sumida Joy, *Gassendi the Atomist: Advocate of History in an Age of Science* (Cambridge: Cambridge University Press, 1988); and Margaret J. Osler, *Divine Will and the Mechanical Philosophy: Gassendi and Descartes on Contingency and Necessity in the Created World* (Cambridge: Cambridge University Press, 1994).

in structure led natural philosophers from Galileo to Newton to the further assumption that the surest form of natural proof was mathematical demonstration. The mathematical tradition that Galileo helped legitimate for natural philosophy had been, for much of the sixteenth century, a craft tradition, whose practitioners employed their mechanical knowledge in architecture, fortification, navigation, and machinery (see the following chapters in this volume: Bertoloni Meli, Chapter 26; Bennett, Chapter 27). The incorporation of "mixed mathematics" into natural philosophy brought with it the assumption that the universe was causally deterministic. Appropriately rigorous demonstration could reveal this determinism.[95]

In large part because of their successful incorporation of mathematics, the more mechanical forms of natural philosophy survived the seventeenth century with their claims to certainty intact. The nature of that certainty, however, was no longer expressed in Aristotelian terms. Indeed, the prestige of geometry as the only truly demonstrative science flourished throughout the period, from Pietro Catena's *Oratio pro idea methodi* (Oration on the Idea of Method, 1563) and Petrus Ramus's *Proemium mathematicum* (Mathematical Introduction, 1567) to the efforts of seventeenth-century philosophers to extend its methods into realms beyond that of geometry properly speaking. Hobbes called geometry "the onely Science it hath pleased God hitherto to bestow upon mankind."[96] In his "De l'Esprit géometrique et de l'art de persuader" ("The Geometric Spirit and the Art of Persuasion"), Blaise Pascal (1623–1662) said of geometry that it was "almost the only human science that produces demonstrations infallibly" because it defines all of its terms and proves all of its propositions.[97] Geometry, and more specifically its axiomatic method, was widely taken up as a model in the human sciences as well. The natural-law theories of the early Grotius, in his *De iure praedae* (On the Law of Plunder, 1604–5), and of Hobbes were also strongly inflected by the search for quasi-geometrical proofs.[98] Most famously of all, perhaps,

---

[95] Lorraine Daston, "The Doctrine of Chances without Chance: Determinism, Mathematical Probability, and Quantification in the Seventeenth Century," in *The Invention of Physical Science: Intersections of Mathematics, Theology and Natural Philosophy Since the Seventeenth Century: Essays in Honor of Erwin N. Hiebert*, ed. Mary Jo Nye, Joan L. Richards, and Roger H. Stuewer (Boston Studies in the Philosophy of Science, 139) (Dordrecht: Kluwer, 1992), pp. 27–50, esp. pp. 34 and 47.

[96] Thomas Hobbes, *Leviathan* [1651], ed. Richard Tuck, rev. ed. (Cambridge: Cambridge University Press, 1996), chap. 4, p. 28.

[97] Blaise Pascal, "De l'esprit géometrique et de l'art de persuader," in *Oeuvres complètes*, ed. L. Lafuma (Paris: Éditions du Seuil, 1963), pp. 348–56, at p. 349: "presque la seule des sciences humaines qui en produise d'infaillibles, parce qu'elle seule observe la véritable méthode, au lieu que toutes les autres sont par une nécessité naturelle dans quelque sorte de confusion que les seuls géomètres savent extrêmement reconnaître."

[98] Wolfgang Röd, *Geometrischer Geist und Naturrecht: Methodengeschichtliche Untersuchungen zur Staatsophilosophie im 17. und 18. Jahrhundert* (Munich: Verlag der Bayerischen Akademie der Wissenschaften, 1970); Ben Vermeulen, "Simon Stevin and the Geometrical method in *De jure praedae*," *Grotiana*, n.s. 4 (1983), 63–6; and Tuck, "The 'Modern' Theory of Natural Law."

Benedict (Baruch) de Spinoza's (1632–1677) *Ethics* (completed in 1675) was also "demonstrated in the geometrical manner."[99]

EXPERIMENT

The second principal development within natural philosophy that had a decisive impact on techniques of proof and persuasion was the experiment (see Dear, Chapter 4, this volume). In the course of the seventeenth century, natural philosophers increasingly appealed to the results of specific experiments rather than, as previously, to a philosophical consensus about what happens "all or most of the time." This new notion of experiment had several consequences. First, syllogistic forms of argument fell out of favor. Second, experimental reports tended to take on a "historical" or narrative form, with the consequence that their readers became what have been called "virtual witnesses."[100] Furthermore, for reasons that I will explain, experimental reports also appealed to actual witnesses to a much greater extent than before, emphasizing their skill, social standing, or philosophical reputation.

The new and paradoxical discipline of "experimental natural philosophy" came to prominence in the second half of the seventeenth century. But it by no means commanded universal assent, and controversies over its findings provide a valuable insight into its claims for proof and its capacity for persuasion. One of the most celebrated quarrels over the function of experiment in natural philosophy occurred in the 1660s between Boyle and Hobbes. Hobbes challenged the experimentalists' claims to proof on several grounds. He pointed out that their meetings, and hence the matters of fact they endeavored to demonstrate, were not open to public witness. He further denied that the phenomena the experimentalists described counted as philosophical in any case because they neither demonstrated effects from causes nor inferred causes from effects. For Hobbes, observations or experiments did not prove phenomena; they illustrated conclusions already arrived at by properly philosophical procedures.[101] Spinoza, too, questioned Boyle's conclusions. He thought that because Boyle did "not put forward his proofs as mathematical" when he tried in his *Certain Physiological Essays* (1661) to show that all tactile qualities depend on mechanical states, "there will be no need to inquire whether they are altogether convincing."[102]

---

[99] Benedict (Baruch) de Spinoza's *Ethica ordine geometrico demonstrata* was first published in his *Opera posthuma* ([Amsterdam?], 1677).
[100] Shapin and Schaffer, *Leviathan and the Air-Pump*, pp. 60–5. See also Golinski, "Robert Boyle," esp. p. 68.
[101] Shapin and Schaffer, *Leviathan and the Air-Pump*, pp. 111–54.
[102] Spinoza to Oldenburg, April 1662, in A. Rupert Hall and Marie Boas Hall, eds., *The Correspondence of Henry Oldenburg*, 13 vols. (Madison/London: University of Wisconsin Press/Mansell/Taylor and Francis, 1965–86), 1: 452–3 (text), 462 (translation). See further Shapin and Schaffer, *Leviathan and*

Thus, changing conceptions of natural philosophy in the seventeenth century, and experimentalism in particular, brought with them new forms of proof. Perhaps the most important of these new forms was the fact. The concept of "fact" (*fait, Tatsache*) is the most important conceptual link between the natural and human sciences of the early modern period.[103] Facts originated in legal discourse; in particular, in the distinction between questions of fact and questions of law (the *de facto* and the *de jure*). The etymological root of fact is in "deed" (Latin *factum*), and in early usages the term retains suggestions of "event" or "action" even in spheres outside the law. The rise to prominence of the fact in natural science seems to have occurred concurrently with the increasing methodological importance ascribed to natural history. "Matter of fact" was originally the concern of history and law, disciplines that had as their object of inquiry volitional human actions.[104] Gradually, however, a term that had previously connoted human action exclusively began to be applied to natural events and objects of natural inquiry.

The Baconian emphasis on natural history as the necessary basis for any subsequent theoretical elaboration was undoubtedly important in this process. Bacon's writings had their greatest impact in England but were also influential in the Low Countries and, by the early eighteenth century, in Enlightenment France.[105] In this respect, it is perhaps no accident that Bacon trained and practiced professionally as a lawyer for most of his adult life.[106] Nonetheless, in his most sustained theoretical account of how to investigate the world, the Latin treatise *Novum Organum*, Bacon wrote more frequently in characteristic sixteenth-century terms of *res ipsae* ("things themselves") rather than of "matter of fact."[107] In this respect, the rise of the fact should

---

the Air-Pump, p. 253; and Golinski, "Robert Boyle," p. 75. Spinoza made a similar comment on Boyle's claim that "it would scarce be believ'd, how much the smallnesse of parts [of bodies] may facilitate their being easily put into motion, and kept in it, if we were not able to confirme it by Chymical Experiments." See Robert Boyle, *Certain Physiological Essays* [1661, 1669], in Boyle, *Works*, 2: 3–203, at p. 122. "One will never be able to prove this by chemical or other experiments," he wrote to Oldenburg, "but only by reason and calculation" (*nunquam chymicis neque aliis experimentis, nisi mera ratione et calculo aliquis id comprobare poterit*). Spinoza to Oldenburg, April 1662, in Oldenburg, *Correspondence*, 1: 454 (text), 463 (translation).

[103] For the case of England, see Barbara J. Shapiro, *A Culture of Fact: England, 1550–1720* (Ithaca, N.Y.: Cornell University Press, 2000).

[104] Lorraine Daston, "Strange Facts, Plain Facts, and the Texture of Scientific Experience in the Enlightenment," in *Proof and Persuasion: Essays on Authority, Objectivity, and Evidence*, ed. Suzanne Marchand and Elizabeth Lunbeck (Turnhout: Brepols, 1996), pp. 42–59.

[105] On the reception of Bacon's works, see Antonio Pérez-Ramos, *Francis Bacon's Idea of Science and the Maker's Knowledge Tradition* (Oxford: Clarendon Press, 1988), pp. 7–31; Pérez-Ramos, "Bacon's Legacy," in *The Cambridge Companion to Bacon*, ed. Markku Peltonen (Cambridge: Cambridge University Press, 1996), pp. 311–34; Alberto Elena, "Baconianism in the Seventeenth-Century Netherlands: A Preliminary Survey," *Nuncius*, 6 (1991), 33–47; Michel Malherbe, "Bacon, l'Encyclopédie et la Révolution," *Les études philosophiques*, 3 (1985), 387–404; and H. Dieckmann, "The Influence of Francis Bacon on Diderot's *Interprétation de la nature*," *Romanic Review*, 24 (1943), 303–30.

[106] See Julian Martin, *Francis Bacon, the State and the Reform of Natural Philosophy* (Cambridge: Cambridge University Press, 1992).

[107] Daston, "Strange Facts," pp. 42–3 and n. 3.

perhaps also be associated with the rapidly increasing tendency in the seventeenth century to write about natural philosophy in the vernacular and thereby escape the expectations about philosophical terminology and argument generated by the Latin of the schools.[108]

One of the most significant aspects of this new discourse of "fact" was that it conflicted with the characteristic scholastic assumptions about proof and persuasion already discussed. Experimental reports of matters of fact were about temporally and spatially specific particulars, and hence were not universal. "Matters of fact" therefore fell outside the scope of logical demonstration because they lacked the criterion of universality required for this in the Aristotelian tradition. For the most significant late sixteenth-century theorist of *methodus*, Jacopo Zabarella, history and the matter of fact it contained were incompatible with philosophical *scientia*: "History is the bare narration of past deeds, which lacks all artifice – except possibly that of eloquence."[109] From this perspective then, or from the perspective of some of the more rigorous philosophies that succeeded it, "facts" had a low standing because they could not easily be incorporated into universal causal demonstrations.[110] Nonetheless, many of the new experimental natural philosophers of the seventeenth century found this vernacular escape from the Latin methodological assumptions of the schools an advantage, and their successors quickly came to take the new language for granted. For an experimentalist writer such as Boyle, keen to disparage the claims of peripatetic natural philosophy, natural "facts" provided an invaluable argumentative ally. They helped supply him with a new "literary technology" of virtual witnessing: Matters of fact allowed Boyle to validate experiments and induce belief in his reports of them.[111]

The disjunction between the newer "historical" traditions of natural philosophy and the legacy of Aristotelian conceptions of the discipline helps explain the differences between the circumstantial, historical, and individual experiments reported in 1650s and 1660s England and those of other experimenters – such as Pascal – who reported their experiences in more universal

---

[108] On this point, see also Geoffrey Cantor, "The Rhetoric of Experiment," in *The Uses of Experiment: Studies in the Natural Sciences*, ed. David Gooding, Trevor Pinch, and Simon Schaffer (Cambridge: Cambridge University Press, 1989), pp. 159–80, at p. 170.

[109] Jacopo Zabarella, "De natura logica," in *Opera logica* (Basel: Conrad Waldkirchius, 1594), col. 100 (bk. 2, chap. 24): "At Historia [...] est nuda gestorum narratio, quae omni artificio caret, praeterque fortasse elocutionis." See further Anthony Grafton, *Commerce with the Classics: Ancient Books and Renaissance Readers* (Jerome Lectures, 20) (Ann Arbor: University of Michigan Press, 1997), p. 13 and n. 16. For Zabarella's account of "art," see Heikki Mikkeli, *An Aristotelian Response to Renaissance Humanism: Jacopo Zabarella on the Nature of Arts and Sciences* (Societas Historica Finlandiae Studia Historica, 41) (Helsinki: SHS, 1992), esp. pp. 29 and 107–10.

[110] Daston, "Strange Facts," p. 45, citing Jean Domat, *Les Loix civiles dans leur ordre naturel*, 2nd ed., 3 vols. (Paris: Jean Baptiste Coignard, 1691–97), 2: 346–7. See also Lorraine Daston, "Baconian Facts, Academic Civility, and the Prehistory of Objectivity," *Annals of Scholarship*, 8 (1991), 337–63, at p. 345.

[111] Steven Shapin, "Pump and Circumstance: Robert Boyle's Literary Technology," *Social Studies of Science*, 14 (1984), 481–520; Shapin and Schaffer, *Leviathan and the Air-Pump*, p. 60.

terms.[112] These differences should also alert us to the different conceptions of fact that obtained in different languages: English "facts" of the 1660s and 1670s seem to have been philosophically firmer than French *faits* of the same period.[113] The discourse of fact provided a new way of talking about the marvels, heteroclites, and pretergenerations of nature that absorbed so many contributors to the *Philosophical Transactions* or the *Journal des Savants* in the late seventeenth century. Early modern facts were not transparent expressions of the phenomena but constituted particular forms of experience, articulated in words. A *fait* in the *Mémoires* of the Académie Royale des Sciences was more than simply a *phénomène* or *observation*. Nonetheless, the reason late seventeenth-century natural philosophers prized facts was that they took them to offer a way of presenting experience without being committed to a preexisting explanatory framework. Modern scholars have found this kind of claim philosophically suspect, and it also had its contemporary critics.[114]

The incorporation of "matters of fact" into natural philosophy indicates a fundamental change in standards of proof in the discipline.[115] In scholastic terms, facts could not provide "metaphysical" or "mathematical certainty" (*scientia*) because they were particular, not universal. Nor did they even pertain, strictly speaking, to the realm of opinion (*opinio*), with its corresponding degree of "physical certainty." Instead, because facts depended upon testimony, they belonged to the realm of *fides* and hence possessed only "moral certainty."[116] This hierarchy of certainty explains why Descartes was at pains at the end of his *Principia philosophiae* (Principles of Philosophy, 1644) to assert that his explanations possessed more than moral certainty and to remind his readers that "there are some matters, even in relation to the things in nature, which we regard as absolutely, and more than just morally, certain."[117] These scholastic distinctions between different degrees of certainty were by their very nature predicated on the existence of different probative standards in different disciplines. For this very reason, however, they help illustrate one of the most significant developments in seventeenth-century

---

[112] Peter Dear, "Jesuit Mathematical Science and the Reconstitution of Experience in the Early Seventeenth Century," *Studies in History and Philosophy of Science*, 18 (1987), 133–75.

[113] Shapin and Schaffer, *Leviathan and the Air-Pump*, esp. pp. 22–6 and 315–16; and Daston, "Strange Facts," esp. p. 46.

[114] Daston, "Baconian Facts," pp. 342, 346, 347, and 355; Lorraine Daston and Katharine Park, *Wonders and the Order of Nature, 1150–1750* (New York: Zone Books, 1998), pp. 231–40; Daston, "Strange Facts," p. 47; and Descartes, *Discours de la méthode*, p. 73.

[115] Daston, "Baconian Facts," p. 346.

[116] On the notion of "moral certainty," see Dear, "From Truth to Disinterestedness"; and Barbara J. Shapiro, *Probability and Certainty in Seventeenth-Century England: A Study of the Relationships between Natural Science, Religion, History, Law, and Literature* (Princeton, N.J.: Princeton University Press, 1983), pp. 31–3.

[117] René Descartes, *Principia philosophiae*, in *Oeuvres*, 8: 328 (4.206): "Praeterea quaedam sunt, etiam in rebus naturalibus, quae absolute ac plusquam moraliter certa existimamus." Translation from René Descartes, *Principles of Philosophy* [1644], trans. John Cottingham, Robert Stoothoff, and Dugald Murdoch, in *Philosophical Writings of Descartes*, 1: 177–291, at p. 290.

natural philosophy: the incorporation of forms of proofs derived from the human sciences into the study of nature.

The increased philosophical status of the "fact" brought about a further decisive change in conceptions of proof and persuasion in natural philosophy: the philosophical rehabilitation of human testimony. Precisely because of their uniqueness, their specificity, and their historical situation, matters of fact depended upon the reports of human testimony. This presented a profound challenge to traditional accounts of probation. Argument from testimony had hitherto been regarded as a weak weapon in the argumentative armamentarium of the sciences. Testimony was strongly identified with argument from authority. In the realm of demonstrative science, however, argument from authority had no place whatsoever because what was being sought was not authoritative opinion, still less "matters of fact," but rather causal knowledge of the thing itself. Even in the probable reasoning of dialectic, argument from authority was regarded as the last and indeed the least of the "topics," most appropriate for confirming conclusions that had already been arrived at. Argument from authority was considered to be principally useful for persuasion, not proof; furthermore, it was regarded as having a greater role in the moral and political than in the natural sciences.[118]

The new emphasis on "matter of fact" changed all this, however. The need to draw upon human testimony in natural history and experiment forced an ongoing reappraisal of its status. Testimony was a vital form of proof in law courts, and natural philosophers began increasingly to draw upon legal theory and practice with respect to its use.[119] (This was also the period that saw the appearance of the expert witness in the courtroom.)[120] The "new philosophy" of the seventeenth century often characterized itself as having finally banished the principle of authority in natural inquiry. It portrayed the more traditional natural philosophy of the sixteenth century, by extension, in terms of the slavish adherence to authority that *novatores* such as Bacon and Descartes so effectively repudiated. Both theoretically and practically, however, this picture is mistaken, for at least in more natural-historically oriented natural philosophy, the development in the seventeenth century

---

[118] R. W. Serjeantson, "Testimony and Proof in Early-Modern England," *Studies in History and Philosophy of Science*, 30 (1999), 195–236.

[119] Barbara J. Shapiro, "The Concept 'Fact': Legal Origins and Cultural Diffusion," *Albion*, 26 (1994), 227–52.

[120] Catherine Crawford, "Legalizing Medicine: Early Modern Legal Systems and the Growth of Medico-Legal Knowledge," in *Legal Medicine in History*, ed. Catherine Crawford and Michael Clark (Cambridge: Cambridge University Press, 1994), pp. 89–116; Carol A. G. Jones, *Expert Witnesses: Science, Medicine, and the Practice of Law* (Oxford: Clarendon Press, 1994), esp. pp. 17–34; Nancy Struever, "Lionardo Di Capoa's *Parere* (1681): A Legal Opinion on the Use of Aristotle in Medicine," in *Philosophy in the Sixteenth and Seventeenth Centuries: Conversations with Aristotle*, ed. Sachiko Kusukawa and Constance Blackwell (Aldershot: Ashgate, 1999), pp. 322–36; Stephen Landsman, "One Hundred Years of Rectitude: Medical Witnesses at the Old Bailey, 1717–1817," *Law and History Review*, 16 (1986), 445–95; and Robert Kargon, "Expert Testimony in Historical Perspective," *Law and Human Behaviour*, 10 (1986), 15–20.

was very largely in the opposite direction. Trust in human testimony became more, not less, significant over the course of the sixteenth and seventeenth centuries.[121]

PROBABILITY AND CERTAINTY

From the middle of the seventeenth century onward, mathematics and "matters of fact" joined forces to provide a genuinely novel addition to the early modern repertoire of proof and persuasion: mathematical probability. The new probabilists began to theorize about how a posteriori knowledge of the natural and moral world might be able to generate an a priori expectation of future events.[122] Predicting future events had already preoccupied a range of sixteenth-century students of nature. Astrologers drew upon genitures and theories of astral influence to predict the longevity and political or social accomplishments of individuals. Medical astrologers applied these techniques to questions of health and disease, and learned physicians used Hippocratic notions of the course of a disease and syndromes of symptoms to establish medical prognoses.[123] The origin of theories of mathematical probability, however, is more usually taken to lie in questions about expected returns in games of chance. The Italian physician and polymath Girolamo Cardano (1501–1576) offered some suggestions in his *Liber de ludo aleae* (A Book on the Game of Dice), written circa 1520 but not published until 1663. He calculated odds successfully but looked unsuccessfully for a calculation that would hold for any single throw rather than an average run of throws; capricious *fortuna* dominates his account.[124] Similar questions about the equitable return in an interrupted game of chance were the spur for the earliest calculations of mathematical probability by Pascal, by Pierre de Fermat (1601–1665), and by the Dutch natural philosopher Christiaan Huygens (1629–1695).[125] As we have already seen, a concern with degrees of certainty was a common preoccupation of writers on logic, the soul, and – increasingly in the seventeenth century – the theory of historical knowledge.[126] It was in the latter realm that

---

[121] See Steven Shapin, *A Social History of Truth: Civility and Science in Seventeenth-Century England* (Chicago: University of Chicago Press, 1994).
[122] Antoine Arnauld and Pierre Nicole, *La logique ou l'art de penser* [1662–83], ed. Pierre Clair and François Girbal, 2nd ed. (Paris: J. Vrin, 1981), pp. 351–4 (pt. IV, chap. 16). See also Ian Hacking, *The Emergence of Probability: A Philosophical Study of Early Ideas about Probability, Induction, and Statistical Inference* (Cambridge: Cambridge University Press, 1975), pp. 73–101; and Daniel Garber and Sandy Zabell, "On the Emergence of Probability," *Archive for the History of Exact Science*, 21 (1979), 33–53.
[123] Maclean, "Evidence, Logic, the Rule and the Exception," pp. 250–51.
[124] Daston, "The Doctrine of Chances without Chance," pp. 38–40. See also Hacking, *Emergence of Probability*, pp. 54–6.
[125] Pascal, *Oevres complètes*, pp. 46–9; and Christiaan Huygens, *De ratione in ludo aleae* (1657). See further Hacking, *Emergence of Probability*, pp. 57–62; and Daston, "Probability and Evidence," pp. 1124–5.
[126] Carlo Borghero, *La certezza e la storia: Cartesianismo, pirronismo e conoscenza storica* (Milan: F. Angeli, 1983); Markus Völkel, "*Pyrrhonismus historicus*" und "*fides historica*": Die Entwicklung der

the new idea that one might be able to quantify certainty (rather than just qualify it) was most eagerly applied. In their *Logique de Port-Royal* (1662), Antoine Arnauld (1612–1694) and Pierre Nicole (1625–1695) applied nascent statistical techniques to a highly contentious question in ecclesiastical history – whether the Emperor Constantine had been baptized at Rome – and also to the (hypothetical) case of a falsely dated contract.[127] The new mathematical probability gave a great impetus to the growing seventeenth-century tendency to admit the less than certain into philosophy. Nonetheless, late seventeenth-century mathematical and philosophical probabilism, as it culminated in the writings of the mathematician Jakob Bernoulli (1655–1705), was deterministic.[128] It did not measure chance; it measured human uncertainty. The Aristotelian distinction between "things better known to us" and "things better known to nature" was transformed into an account that saw the probability calculus as a way of approaching the "objective certainty" possessed by events in the natural world.[129]

Thus, the impact of mathematical probability on the understanding of the natural world in the seventeenth century was slender in comparison with its influence in the nineteenth century.[130] Its broader intellectual impact, however, was more significant. The new probability theory was rapidly applied to a whole range of areas.[131] A treatise such as the English mathematician John Craig's (1662–1731) *Theologiae Christianae principia mathematica* (*Mathematical Principles of Christian Theology*, 1699) testifies to the widespread desire to apply the forms of proof of the new natural philosophy as a means of persuasion in fields far removed from it – in Craig's case, to make an argument about the necessary *terminus ante quem* of the second coming.[132] Mathematical probability, it was hoped, might allow for the quantification of witness testimony, as well as of mortality rates.[133]

The seventeenth century thus saw a radical revaluation of probable knowledge.[134] It would be misguided, however, to suggest that the quest for certainty

---

deutschen historischen Methodologie unter dem Gesichtspunkt der historischen Skepsis (Europäische Hochschulschriften, Reihe 3: Geschichte und ihre Hilfswissenschaften, 313) (Frankfurt am Main: Peter Lang, 1987).

[127] Antoine Arnauld and Pierre Nicole, *La logique ou l'art de penser* [1662], ed. Pierre Clair and François Girbal, 2nd ed. (Paris: J. Vrin, 1981), pp. 340–41, 348–9.
[128] Daston, "Probability and Evidence," esp. pp. 1137–8.
[129] Daston, "The Doctrine of Chances without Chance," pp. 28–9.
[130] Ian Hacking, *The Taming of Chance* (Cambridge: Cambridge University Press, 1990).
[131] Lorraine Daston, *Classical Probability in the Enlightenment* (Princeton, N.J.: Princeton University Press, 1988).
[132] Richard Nash, *John Craige's Mathematical Principles of Christianity* (Carbondale: Southern Illinois University Press, 1991).
[133] [George Hooper], "A Calculation of the Credibility of Human Testimony," *Philosophical Transactions of the Royal Society of London*, 21 (1699), 359–65; Jakob Bernoulli, *Ars conjectandi* [1713], in *Werke*, ed. B. L. van der Waerden, 3 vols. (Basel: Birkhäuser, 1969–75), 3: 107–259, esp. pp. 241–7. See further Daston, *Classical Probability in the Enlightenment*, pp. 306–42; Daston, "The Doctrine of Chances without Chance," p. 37; and Daston, "Probability and Evidence," pp. 1125–6.
[134] Shapiro, *Probability and Certainty in Seventeenth-Century England*; Daston, "Probability and Evidence"; Aant Elzinger, "Christiaan Huygens' Theory of Research," *Janus*, 67 (1980), 281–300.

about the natural world was entirely abandoned. The desire for demonstrative proof remained strong throughout the seventeenth century in all forms of philosophy, including natural philosophy. The paradigm of demonstrative certainty increasingly became mathematics and, in particular, Euclidean geometrical analysis. The successes of "mixed mathematics" in natural philosophy help explain why Leibniz, writing in 1685, thought that it was "our own century which has gone in for demonstrations on a large scale." Leibniz cited authors as diverse as Galileo – who "broke the ice" – and the Altdorf mathematician Abdias Trew (1597–1669), "who has reduced to demonstrative form the eight books of Aristotle's *Physics*."[135] By the end of the seventeenth century, and in particular because of the rapidly acquired authority of Newton's *Principia mathematica philosophiae naturalis* (Mathematical Principles of Natural Philosophy, 1687), it became a commonplace that the principles underpinning natural philosophy were mathematical. This intellectual hegemony was sometimes resented: The English essayist Samuel Parker noted in 1700 that "the Domain of Number and Magnitude" was undoubtedly "very large" but went on to ask pointedly, "Must they therefore devour all Relations and Properties whatsoever?"[136] These natural-historically inspired doubts notwithstanding, the probative virtues of numbers were increasingly proclaimed to be superior to – in the words of the political arithmetician William Petty (1623–1687) – the persuasions of "only comparative and superlative Words, and intellectual Arguments."[137] Words were of uncertain value and too easily manipulated; everyone, however, knew what was meant by a number.[138] By 1700, the powerful Renaissance fascination with the arts of verbal argument was drawing to a close.

PROOF AND PERSUASION IN THE PRINTED BOOK

One object in particular integrates much scholarship on early modern proof and persuasion: the printed book. Books were one of the principal means by which natural philosophers communicated their findings to their

---

[135] Gottfried Wilhelm Leibniz, "Projet et Essais pour arriver à quelque certitude pour finir une bonne partie des disputes et pour advancer l'art de inventer," in *Opuscules et fragments inédits de Leibniz*, ed. Louis Couturat (Paris: Presses universitaires de France, 1903), pp. 175–82. See Leibniz "[Préceptes pour advancer les sciences]," in *Philosophische Schriften*, 7: 157–73, at p. 166: "Abdias Trew, habile Mathematicien d'Altdorf, a reduit la physique d'Aristote en forme de demonstration."
[136] Samuel Parker, *Six Philosophical Essays upon Several Subjects* (London: Thomas Newborough, 1700), sig. A3r. See further Mordechai Feingold, "Mathematicians and Naturalists: Sir Isaac Newton and the Royal Society," in *Isaac Newton's Natural Philosophy*, ed. Jed Z. Buchwald and I. Bernard Cohen (Cambridge, Mass.: MIT Press, 2000), pp. 77–102.
[137] Sir William Petty, *Political Arithmetick; Or, a discourse concerning the extent and value of lands, people, . . . &c.* (London: Robert Clavel and Henry Mortlock, 1690), p. 9. On quantification as a significant aspect of the moral economy of science, see Lorraine Daston, "The Moral Economy of Science," *Osiris*, 10 (1995), 2–24, at pp. 8–12.
[138] Quentin Skinner, "Moral Ambiguity and the Art of Persuasion in the Renaissance," in *Proof and Persuasion*, pp. 25–41; and Daston, "Moral Economy," p. 9.

contemporaries and were the source most often drawn upon by historians of the sciences in early modern Europe. The format and presentation of early printed books – and of related media such as pamphlets and journals – played a significant role in persuading their readers of the veracity of their contents (see the following chapters in this volume: Grafton, Chapter 10; Johns, Chapter 15). These readers brought to them expectations about what constituted plausibility that printers and publishers conformed to and sometimes knowingly exploited.[139] Genre, format, *mise-en-page*, illustrations, paper, title-page information, and personalization of individual copies all contributed to the persuasive power of the printed book.

The issue of genre – or, more broadly, of literary form – is particularly significant for questions of proof and persuasion. Different modes of argument were associated with, and encouraged, different forms of exposition. The sixteenth and seventeenth centuries saw a proliferation in the generic forms in which natural philosophy was presented. The dominant form at the beginning of the period was the commentary. University teaching in the early years of the sixteenth century tended to involve the study of authoritative texts – such as Galen's *Ars medica* or Aristotle's *libri naturales* – and commentaries upon them.[140] Over the course of the century, the commentary tradition declined, to be gradually replaced by the textbook (the *cursus*, *systema*, or *compendium*). The explanation for this development is complex. It lies partly in the growing dissatisfaction with Aristotelian philosophy. The development of subjects – such as astronomy, optics, or botany – beyond the traditional ones of the *libri naturales* was also an important stimulus to the production of new syntheses. But insofar as the rise of the textbook was also brought about by a dissatisfaction with the expository mode of authoritative texts, and indeed also with a dissatisfaction with the principle of authority itself, it is also related to changing conceptions of proof and persuasion.[141] Whereas a commentary followed the preoccupations and arguments of its source text, a textbook could cover an entire discipline, or one area of a discipline, in a systematic manner. Alternatively, arguments in natural philosophical textbooks might now follow the structure of a disputation, with physical opinions being

---

[139] Adrian Johns, *The Nature of the Book: Print and Knowledge in the Making* (Chicago: University of Chicago Press, 1998), esp. pp. 28–40.

[140] Per-Gunnar Ottoson, *Scholastic Medicine and Philosophy: A Study of Commentaries on Galen's Tegni* (Uppsala: Institutionen för Idé-och Lärdomhistoria, Uppsala University, 1982); R. K. French, "Berengario da Carpi and the use of Commentary in Anatomical Teaching," in *The Medical Renaissance of the Sixteenth Century*, ed. A. Wear, R. K. French, and I. Lonie (Cambridge: Cambridge University Press, 1985), pp. 42–74.

[141] Patricia Reif, "Natural Philosophy in Some Early Seventeenth-Century Scholastic Textbooks," Ph.D. dissertation, St. Louis University, St. Louis, Mo., 1962; Reif, "The Textbook Tradition in Natural Philosophy, 1600–1650," *Journal of the History of Ideas*, 30 (1969), 17–32; Charles B. Schmitt, "Galileo and the Seventeenth-Century Text-Book Tradition," in *Novità celesti e crisi del sapere: atti del convegno internazionale di studi Galileiani*, ed. Paolo Galluzzi (Florence: Giunti Barbèra, 1984), pp. 217–28; and Schmitt, "The Rise of the Philosophical Textbook," in *Cambridge History of Renaissance Philosophy*, pp. 792–804.

proposed, objected to, and resolved, all in logical form, with the stages of the argument sometimes identified in the margin.[142] Beyond the university, there was even greater generic freedom. Natural philosophy was a significant component of early modern encyclopaedic works: Cardano's *De subtilitate* (On Subtlety, 1550) is a case in point. This work in turn was argued against point-by-point in the form of *exercitationes* by the humanist Julius Caesar Scaliger (1484–1558) in his *Exotericae exercitationes de subtilitate* (Popular Exercises on Subtlety, 1557). This work in its turn became much used as a textbook in the many universities of the German-speaking lands.[143]

In the late sixteenth and seventeenth centuries, the dialogue emerged as a particularly significant genre for transmitting natural philosophy. This form had its origins in the rhetorical emphasis on being able to argue on both sides of the question. In other respects, however, the dialogue belonged, as one might expect, to the probable realm of dialectic; the Italian theorist Sperone Speroni (1500–1588) (in an echo of the Thomist distinction) considered the serious dialogue as belonging, in terms of its "certainty," to the middle place of *opinione*, between the *scienza* of the demonstrative syllogism and the "persuasions" of rhetoric.[144] Thus dialectic was also significant for the dialogue form. In its sixteenth-century heyday, the dialogue was primarily deployed on moral and political subjects, whether in imitation of Cicero or Plato. In natural philosophy, however, the form came into its own in the seventeenth century, with significant contributions from Jean Bodin (1530–1596) in his *Universae naturae theatrum* (Theater of Universal Nature, 1596), Galileo in his *Dialogo sopra i due massimi sistemi del mondo* (1632), and Boyle in *The Sceptical Chymist* (1661).[145]

The emergence of experiment was also instrumental in encouraging the development of new literary forms for natural philosophy.[146] Some were co-opted from other fields. The essay was another genre that was also originally moral and political in nature but became a significant vehicle for the new philosophy. Inaugurated by Montaigne on the ancient model of, in particular,

---

[142] See, for example, Eustachius a Sancto Paulo, *Summa philosophiae quadripartita* (Paris, 1609), pt. 3: *De rebus physicis*. This work was widely used as a textbook in both Catholic and Protestant Europe. See Charles H. Lohr, *Latin Aristotle Commentaries*, 2 vols. (Florence: Leo S. Olschki, 1988), vol. 2: *Renaissance Authors*, s.v. "Eustacius."

[143] See Gabriel Naudé, *Instructions Concerning Erecting of a Library*, trans. John Evelyn (London: G. Bedle, T. Collins, and J. Crook, 1661; first published as *Advis pour dresser une bibliothèque*, 1644), p. 27: "*Scaliger*, who has so fortunately oppos'd *Cardan*, as that he is at present in some parts of *Germany* more followed then *Aristotle* himself."

[144] Sperone Speroni, *Apologia dei dialoghi* (1574–5), as discussed in Virginia Cox, *The Renaissance Dialogue: Literary Dialogue in Its Social and Political Contexts, Castiglione to Galileo* (Cambridge Studies in Renaissance Literature and Culture, 2) (Cambridge: Cambridge University Press, 1992), pp. 72–3 and p. 176, n. 13.

[145] For Bodin, see Ann Blair, *The Theater of Nature: Jean Bodin and Renaissance Science* (Princeton, N.J.: Princeton University Press, 1997). For Boyle, see Golinski, "Robert Boyle," p. 61. For Galileo, see Cox, *Renaissance Dialogue*, esp. pp. 32, 77, and 113.

[146] Geoffrey Cantor, "The Rhetoric of Experiment," pp. 162–3.

Plutarch, the essay quickly became associated with notions of trial (*Versuch*) and investigation. It was employed for this purpose by Descartes, in the *Essais* that followed the *Discours de la méthode* (1637), and Boyle, in his early *Certain Physiological Essays* (1661).[147] This did not prevent some of Boyle's readers, such as Leibniz, from wishing he would write in a more systematic form and provide "some kind of system of chymistry" (*corpus quoddam Chymicum*).[148] Some of these newer or co-opted literary forms did not last. Bacon advocated the aphorism as a means of delivering knowledge.[149] The English magus John Dee (1527–1608) had earlier transmitted his astronomical work by aphorisms, but Bacon's enthusiasm for the form was not widely followed.[150]

In contrast, journals became an important forum of enduring importance for reports of experimental and natural-historical "matters of fact" (particularly prodigious matters of fact). Several experimental societies produced a journal (or journals) to publish reports that would not make a book. The Royal Society had its *Philosophical Transactions* (from 1665) and, briefly, the *Philosophical Collections* (1679–1682). The medically inclined Academiae Naturae Curiosorum of Schweinfurt (founded in 1652) published the *Miscellanea curiosa*. In the beginning, journals such as these often owed their continuing existence to the efforts of a single individual: in the case of the Royal Society, to Henry Oldenburg (ca. 1618–1677) and Robert Hooke, respectively.[151] Other journals, such as the *Journal des savants* and the *Acta eruditorum* (which were even less exclusively natural philosophical than the *Philosophical Transactions*), thrived without institutional support.[152] Some submissions to

---

[147] Robert Boyle, *Certain Physiological Essays*, in Boyle, *Works*, 2: 3–203. On the significance of Boyle's use of the essay form, see Golinski, "Robert Boyle," pp. 62–3, 68.

[148] Leibniz to Oldenburg, 5 July 1674, in Oldenburg, *Correspondence*, 11: 43 (text), 46 (translation [modified]). See also Leibniz to Oldenburg, 10 May 1675, in Oldenburg, *Correspondence*, 11: 303 (text), 306 (translation [modified]): "I hope [. . .] that he will perfect philosophical Chymistry [. . .]. I beg you to urge him to some time vehemently at least to write distinctly and openly what his opinions on that subject are." See also Golinski, "Robert Boyle," pp. 75–6.

[149] Francis Bacon, *Advancement of Learning*, p. 124. On Bacon's preference for aphorism, see Sister Scholastica Mandeville, "The Rhetorical Tradition of the Sententia, with a Study of its Influence on the Prose of Sir Francis Bacon and of Sir Thomas Browne," Ph.D. dissertation, St. Louis University, St. Louis, Mo., 1960; James Stephens, "Science and the Aphorism: Bacon's Theory of the Philosophical Style," *Speech Monographs*, 37 (1970), 157–71; Margaret L. Wiley, "Francis Bacon: Induction and/or Rhetoric," *Studies in the Literary Imagination*, 4 (1971), 65–80; L. Jardine, *Francis Bacon*, pp. 176–8; Alvin Snider, "Francis Bacon and the Authority of Aphorism," *Prose Studies: History, Theory, Criticism*, 11 (1988), 60–71; Stephen Clucas, "'A Knowledge Broken': Francis Bacon's Aphoristic Style and the Crisis of Scholastic and Humanist Knowledge-Systems," in *English Renaissance Prose: History, Language, and Politics*, ed. Neil Rhodes (Medieval and Renaissance Texts and Studies, 164) (Tempe, Ariz.: Medieval and Renaissance Texts and Studies, 1997), pp. 147–72; and L. Jardine, "Introduction," in Francis Bacon, *New Organon*, pp. xvii–xxi.

[150] Wayne Shumaker, ed. and trans., *John Dee on Astronomy: Propaedeutica Aphoristica (1558 and 1568), Latin and English* (Berkeley: University of California Press, 1978).

[151] See esp. Michael Hunter and Paul B. Wood, "Towards Solomon's House: Rival Strategies for Reforming the Early Royal Society," *History of Science*, 24 (1986), 49–108, at pp. 59–60.

[152] See Augustinus Hubertus Laeven, *The Acta Eruditorum under the Editorship of Otto Mencke: The History of an International Learned Journal between 1682 and 1707*, trans. Lynne Richards (Amsterdam: APA–North Holland University Press, 1990).

these journals nonetheless remained influenced by the epistolary conventions of the rhetorical tradition.[153]

Two further forms of proof and persuasion in and beyond the printed book should be mentioned in concluding this section. The study of the persuasive power of illustrations and diagrams is a field that is still in its infancy, but it is one that has significant potential to develop the implications of Leibniz's comment that geometrical diagrams were "the most useful of characters" for recognizing, discovering, or proving that kind of truth.[154] Finally, there is the important matter of the significance of philosophical instruments as a means of proof and persuasion.[155]

PROOF, PERSUASION, AND SOCIAL INSTITUTIONS

Beyond the printed book, there is a wide range of cultural contexts in which techniques of proof and persuasion should be situated. Historians of early modern Europe have considered them in a range of ways: in terms of the "places" in which these techniques functioned;[156] the social roles of their authors (see Shapin, Chapter 6, this volume);[157] the professional disciplines of the late Renaissance university;[158] the non- or antischolastic ambitions of the experimental academies of the seventeenth century; the political constitution

---

[153] Jean Dietz Moss, "Newton and the Jesuits in the *Philosophical Transactions*," in *Newton and the New Direction in Science: Proceedings of the Cracow Conference 25 to 28 May 1987*, ed. G. V. Coyne, M. Heller, and J. Zycinski (Vatican City: Vatican Observatory, 1988), pp. 117–34.

[154] Gottfried Wilhelm Leibniz, "Dialogue on the connection between things and words [1677]," in *Selections*, ed. Philip P. Wiener (New York: Charles Scribner's Sons, 1951), pp. 6–11, at p. 9, translating "Dialogus, August, 1677," in Leibniz, *Philosophische Schriften*, 7: 190–4. For suggestions of future directions of research, see Shapin and Schaffer, *Leviathan and the Air-Pump*, p. 146; John T. Harwood, "Rhetoric and Graphics in *Micrographia*," in *Robert Hooke: New Studies*, ed. Michael Hunter and Simon Schaffer (Woodbridge: Boydell Press, 1989), pp. 119–47; Johns, *Nature of the Book*, pp. 22–3; Dennis L. Sepper, "Figuring Things Out: Figurate Problem-Solving in the Early Descartes," in *Descartes' Natural Philosophy*, ed. Stephen Gaukroger, John Schuster, and John Sutton (London: Routledge, 2000), pp. 228–48.

[155] See further Michael Aaron Dennis, "Graphic Understanding: Instruments and Interpretation in Robert Hooke's *Micrographia*," *Science in Context*, 3 (1989), 309–64; W. D. Hackmann, "Scientific Instruments: Models of Brass and Aids to Discovery," in *Uses of Experiment*, pp. 31–65, esp. pp. 33–4; and Stephen Johnston, "Mathematical Practitioners and Instruments in Elizabethan England," *Annals of Science*, 48 (1991), 319–44, esp. p. 329.

[156] Nicholas Jardine, "The Places of Astronomy in Early Modern Culture," *Journal for the History of Astronomy*, 29 (1998), 49–68.

[157] Robert S. Westman, "The Astronomer's Role in the Sixteenth Century: A Preliminary Study," *History of Science*, 18 (1980), 105–47; Steven Shapin, "'A scholar and a gentleman': The Problematic Identity of the Scientific Practitioner in Early Modern England," *History of Science*, 29 (1991), 279–327; and Adrian Johns, "Prudence and Pedantry in Early Modern Cosmology: The Trade of Al Ross," *History of Science*, 35 (1997), 23–59.

[158] Maclean, "Evidence, Logic, the Rule and the Exception"; see also Maclean, *Interpretation and Meaning in the Renaissance: The Case of Law* (Cambridge: Cambridge University Press, 1992), pp. 77, 102, 105, 167 n. 279.

Proof and Persuasion 169

of the society that produced such academies;[159] and the incorporation of ideals of civility and etiquette into natural philosophy.[160]

In institutional terms, the most significant development of the early modern period was the rise of philosophical academies, a development that Fontenelle thought was a necessary consequence of the "renewal of the true philosophy" that he attributed to the seventeenth century.[161] Explicitly and implicitly, these academies defined themselves against the universities – even as they denied that they presented any threat to established modes of education.[162] Several studies since the 1970s have emphasized that the role of the universities in the early modern study of nature was not as negligible or even negative as has sometimes been assumed.[163] Nonetheless, the new philosophical academies allowed the development of new forms of authentication and encouraged the rejection of older ones – a process helped by the studied neglect of the traditional disciplines of proof and persuasion, rhetoric, and logic, which went along with the academies' desire to avoid questions of politics and religion.[164]

One of the most significant manifestations of these new forms of proof concerned how experimental reports were published. In this, however, as in most other matters, not all experimental academies followed the same pattern. A number of the secrets of nature exposed in della Porta's *Magia naturalis*

---

[159] Shapin and Schaffer, *Leviathan and the Air-Pump*; and Mario Biagioli, "Scientific Revolution, Social Bricolage, and Etiquette," in *The Scientific Revolution in National Context*, ed. Roy Porter and Mikuláš Teich (Cambridge: Cambridge University Press, 1992), pp. 11–54.

[160] Daston, "Baconian Facts."

[161] Bernard le Bovier de Fontenelle, ed., *Histoire de l'Académie Royale des Sciences*, 9 vols. (Paris: Gabriel Martin, 1729–33), 1: 5: "le renouvellement de la vraye Philosophie a rendu les Académies de Mathematique & de Phisique . . . necessaires." See further Hahn, *Anatomy*, p. 1.

[162] Mordechai Feingold, "Tradition versus Novelty: Universities and Scientific Societies in the Early Modern Period," in *Revolution and Continuity: Essays in the History and Philosophy of Early Modern Science*, ed. P. Barker and R. Ariew (Washington, D.C.: Catholic University of America Press, 1991), pp. 45–59; Michael Hunter, *Establishing the New Science: The Experience of the Early Royal Society* (Woodbridge: Boydell Press, 1989), pp. 2–3; and Hunter, *Science and Society in Restoration England* (Cambridge: Cambridge University Press, 1981), pp. 145–7.

[163] John Gascoigne, "A Reappraisal of the Role of the Universities in the Scientific Revolution," in *Reappraisals of the Scientific Revolution*, pp. 207–60; Charles Schmitt, "Philosophy and Science in Sixteenth-Century Italian Universities," in *The Renaissance: Essays in Interpretation*, ed. André Chastel, Cecil Grayson, Marie Boas Hall, Denys Hay, Paul Oskar Kristeller, Nicolai Rubinstein, Charles B. Schmitt, Charles Trinkhaus, and Walter Ullmann, (London: Methuen, 1982), pp. 297–336; David A. Lines, "University Natural Philosophy in Renaissance Italy: The Decline of Aristotelianism?" in *The Dynamics of Natural Philosophy in the Aristotelian Tradition (and Beyond): Doctrinal and Institutional Perspectives*, ed. Cees Leijenhorst, Christoph Lüthy, and Johannes M. M. H. Thijssen (Leiden: E. J. Brill, 2002); Mordechai Feingold, *The Mathematicians' Apprenticeship: Science, Universities, and Society in England, 1560–1640* (Cambridge: Cambridge University Press, 1984); John Gascoigne, "The Universities and the Scientific Revolution: The Case of Newton and Restoration Cambridge," *History of Science*, 23 (1985), 391–434; and Christine Shepherd, "Philosophy and Science in the Arts Curriculum of the Scottish Universities in the Seventeenth Century," Ph.D. dissertation, University of Edinburgh, Edinburgh, Scotland, 1975.

[164] See, for example, the proposal for a Compagnie des Sciences et des Arts sent to Christiaan Huygens in about 1663, *Oeuvres complètes de Huygens*, 22 vols. (The Hague: Martinus Nijhoff, 1888–1950), 4: 325–9.

(1558; revised and expanded edition, 1589) probably owe their presence to his membership in the Accademia dei Segreti, but the book, not unreasonably, appeared as della Porta's own.[165] However, as we have seen, several of the earlier experimental societies began to produce volumes of "collections" (*recueils*), "essays" (*saggi*), *ephemerides*, or *Transactions*. In the case of the Parisian Bureau d'Adresse, centered around Théophraste Renaudot (1586–1653), these took the form of short discussions of questions on all manner of different subjects, both moral and natural. Although the form of the "question" could be seen as scholastic hangover, the manner of the discussions was not.[166] Renaudot's questions were debated anonymously. The same anonymity obtained in the *Saggi di naturali esperienze* that appeared in 1667 from the posthumously christened Accademia del Cimento, which had been founded in 1657 and was defunct by the time its proceedings were published.[167] The collective voice of this publication, composed by the virtuoso Count Lorenzo Magalotti, and prominently authorized on its title page by its patron, Prince Leopold of Tuscany, both precluded any persuasive appeal to the credibility of an individual experimenter and ironed out the disagreements that can be found in the academicians' private correspondence.[168] In partial contrast, the early publications of the Académie Royale des Sciences were not anonymous in any consistent sense, but their *Mémoires* on the natural history of plants and animals or their *Recueil* of mathematical treatises emphasized that responsibility for the contents lay as much with the Académie as an institution as with the individual named academicians.[169]

The early experimental academies of the seventeenth century gave the publications they sponsored or lent their name to something that universities had also (but much less systematically) provided: an imprimatur. Both the Royal Society and the Académie Royale des Sciences published books under their own imprints. Some, such as Hooke's *Micrographia* (1665), were intellectual and financial successes; other sponsored publications might be

---

[165] On della Porta, see the essays collected in *Giovan Battista della Porta nell'Europa del suo tempo* and Eamon, *Science and the Secrets of Nature*, pp. 194–232.

[166] Théophraste Renaudot, *Recueil general des questions traictees és conferences du Bureau d'Adresse* (Paris: G. Loyson, 1655-6). On the question as a characteristic form of scholastic investigation, see Brian Lawn, *The Rise and Decline of the Scholastic "Quaestio Disputata": With Special Emphasis on Its Use in the Teaching of Medicine and Science* (Leiden: E. J. Brill, 1993).

[167] Accademia del Cimento, *Saggi di naturali esperienze fatte nell'Accademia del Cimento sotto la protezione del Serenissimo Principe Leopoldo di Toscana e descritte dal Segretario di essa Accademia* (Florence: Giuseppe Cocchini, 1667). See also the subsequently compiled collection of experiments edited by Giovanni Targioni Tozzetti, *Atti e memorie inedite dell'Accademia del Cimento*, 3 vols. (Florence, 1780). On the Accademia del Cimento, see W. E. K. Middleton, *The Experimenters: A Study of the Accademia del Cimento* (Baltimore: Johns Hopkins University Press, 1971); and M. L. R. Bonelli and Albert Van Helden, *Divini and Campani: A Forgotten Chapter in the History of the Accademia del Cimento* (Florence: Istituto e Museo di Storia della Scienza, 1981).

[168] Biagioli, "Scientific Revolution," pp. 27–31.

[169] Hahn, *Anatomy*, p. 26.

failures in one or both respects.¹⁷⁰ In the case of the Académie Royale, there evolved a quasi-legal procedure of lending the credit of the society to certain publications by allowing authors to add the phrase "apprové par l'Académie" to the censor's approbation at the front of their works. Whereas in England any author who was a fellow of the Royal Society might advertise that fact on his title page – and many did – in Paris, only works examined by the Académie as a whole might carry the designation of "Academician."¹⁷¹

Perhaps the most significant development in the natural philosophy of the experimental societies, however, was in respect to manners. Most of the new private academies founded in the late sixteenth and seventeenth centuries included instructions on etiquette.¹⁷² In itself, this perhaps says little – early modern university statutes were, after all, overwhelmingly concerned with issues of behavior and discipline. Nonetheless, the ethos of the princely humanist academies of late Renaissance Italy was self-consciously one of civility, conversation, and consensus, and this ethos was taken up by the larger, ultimately more stable, and more exclusively natural-philosophically inclined northern European academies of the late seventeenth century. The formal disputations that were an integral component of university pedagogy were often explicitly condemned – even if the quarrels that replaced them sometimes appeared little better. Most importantly, as disputation was devalued, so too were the formalized procedures of proof and persuasion that had underpinned it. These were replaced by less stereotyped techniques and procedures that owed more to the conditions obtaining in the academies and in the wider society, techniques that were derived from legal practice, from courtesy manuals, or from epistolary convention.

When it suited them, the experimental royal academies made a virtue of the publicity of their activities, in tacit contrast with the purportedly solitary pursuit of university learning.¹⁷³ In his *History of the Royal Society*, Sprat asked "all *sober men*" whether "they will not think, they are fairly dealt withal,

---

[170] The officially sponsored publication of Thomas Sprat's *History of the Royal-Society* (1667) was arguably counterproductive. See Paul B. Wood, "Methodology and Apologetics: Thomas Sprat's *History of the Royal Society*," *British Journal for the History of Science*, 13 (1980), 1–26; Hunter, *Science and Society*, esp. p. 148; and Hunter, "Latitudinarianism and the 'Ideology' of the Early Royal Society: Thomas Sprat's *History of the Royal Society* (1667) Reconsidered," in Hunter, *Establishing the New Science*, pp. 45–71. Supporting the publication of John Ray's posthumous *Historia piscium* almost bankrupted the Society; see Sachiko Kusukawa, "The *Historia piscium* (1686)," *Notes and Records of the Royal Society of London*, 54 (2000), 179–97.

[171] Hahn, *Anatomy*, pp. 22–9. Biagioli, "Scientific Revolution," p. 37, argues that "experimental philosophers ... could be legitimate individual authors only in so far as they were members of a gentlemanly corporation (like the Royal Society)" (his emphasis).

[172] Daston, "Baconian Facts," p. 351.

[173] See, for example, Steven Shapin, "'The Mind Is Its Own Place': Science and Solitude in Seventeenth-Century England," *Science in Context*, 4 (1990), 191–218. Nonetheless, early modern universities offered much in the way of public exhibition of their activities. See Giovanna Ferrari, "Public Anatomy Lessons and the Carnival: The Anatomy Theatre of Bologna," *Past and Present*, 117 (1987), 50–106; and Kristine Louise Haugen, "Imagined Universities: Public Insult and the *Terrae filius* in Early Modern Oxford," *History of Universities*, 16 (2000), 1–31.

in what concerns their *Knowledg*, if they have the concurring Testimonies of *threescore or an hundred*?"[174] Appeals to consensus and credit replaced formalized eristic and the expression of *opiniones*. In fact, however, none of these societies were public in the way that the teaching of natural philosophy became public in the eighteenth century. Membership in them was restricted, whether statutorily or informally. Furthermore, several societies had strong tendencies toward secrecy. In the case of the Accademia del Cimento, this was a function of Prince Leopold's desire not to compromise his social position and to control disputes among the academicians – who were not permitted to identify themselves as such. In the case of the Royal Society of London, the urge toward secrecy stemmed from a desire to persuade members to divulge discoveries and from Hooke's personal concern with properly establishing intellectual priority.[175]

A further question frequently encountered in the early experimental academies was the role of principles of explanation. Should experiments simply demonstrate "matters of fact," or should they be placed within an explanatory philosophical framework? The first statutes of the Royal Society commanded that "[i]n all Reports of Experiments to be brought into the Society, the matter of fact shall be barely stated, without any prefaces, apologies, or rhetorical flourishes; and entered so in the Register-book." If Fellows wanted to conjecture a causal explanation for the phenomena they delivered, then they had to do so separately from the account of the experiment.[176] Likewise, Fontenelle emphasized that in the Académie Royale des Sciences, "we do not fail to hazard conjectures about causes – but they are only conjectures."[177] There were a wide variety of early attempts to guide or reform the Royal Society of London in the first forty years of its existence. These position papers quickly ceased to consider the place of philosophical authorities, but they did turn on the relative weight to be accorded to observation, experiment, cause, hypothesis, and (in what is perhaps a Cartesian echo) the "principle[s] of philosophy."[178]

If experiments were to be placed in a philosophical framework, however, which one was it to be? The competing legacies of Aristotle, Bacon, Descartes, and Gassendi overshadowed much experimental natural philosophy in the late seventeenth-century academies and societies. Some groups, such as the Cartesian one coordinated by Jacques Rohault (1618–1672), openly professed a single philosophical authority. The Accademia del Cimento, in contrast,

---

[174] Sprat, *History of the Royal-Society*, p. 100.
[175] Shapin and Schaffer, *Leviathan and the Air-Pump*, p. 113; Biagioli, "Scientific Revolution," pp. 27–8; and Hunter and Wood, "Towards Solomon's House," pp. 74–5.
[176] *The Record of the Royal Society for the Promotion of Natural Knowledge*, 4th ed. (London: The Royal Society, 1940), p. 290; discussed by Hunter, *Establishing the New Science*, pp. 24–5.
[177] Bernard le Bovier de Fontenelle, "Preface" to the *Histoire de l'Académie royale des sciences. Année M.DC.XX* (Paris: Jean Boudot, 1702), sig. ĩ2r: "On ne laisse pas de hasarder des conjectures sur les causes, mais ce sont des conjectures." See further Hahn, *Anatomy*, pp. 33–4.
[178] Hunter and Wood, "Towards Solomon's House," p. 66.

set out to test various tenets of Aristotle's natural philosophy. The larger societies, however, tended to eschew individual philosophical authority.[179] Samuel Sorbière (1615–1670), a guiding spirit of the Académie Montmort, claimed that the early members of the Royal Society were divided in their allegiance between Descartes (favored by the mathematicians) and Gassendi (favored by the "men of General Learning"). Sprat denied that this division existed but emphasized (perhaps somewhat misleadingly) the Society's Baconian inspiration.[180] The Society of Jesus, meanwhile, maintained its adherence to the authority of Thomist-Aristotelianism throughout the seventeenth century.[181]

Perhaps the most significant explanation for the changing practices of proof and persuasion fostered by the new philosophical and experimental societies is that they very rarely included pedagogy as part of their brief.[182] In the middle years of the seventeenth century, numerous schemes were proposed for new educational institutions that would teach the experimental natural philosophy for which the universities had at that point found little room.[183] But the education of the young was a task that the generally wellborn men who constituted the membership of the early societies kept at arm's length. Nonetheless, even if the experimenters largely managed to avoid wielding the early modern pedagogue's whip, they did not succeed in avoiding more symbolic forms of violence. Although the Republic of Letters and its associated institutions certainly liked to conceive of themselves as the most civil of civil societies, their ideals of etiquette and decorum were fundamentally fragile. Sixteenth-century disputes over natural knowledge attained extraordinary levels of bitterness and vituperation.[184] Despite the

[179] Hahn, *Anatomy*, p. 31. For a revision of one aspect of Hahn's account of the early Académie, see Robin Briggs, "The Académie Royale des Sciences and the Pursuit of Utility," *Past and Present*, 131 (1991), 38–87.

[180] Thomas Sprat, *Observations on Monsieur de Sorbier's Voyage into England* (London: John Martyn and James Allestry, 1668), p. 144. On the notion of "general learning," see Meric Casaubon, *Generall Learning: A Seventeenth-Century Treatise on the Formation of the General Scholar*, ed. Richard Serjeantson (Renaissance Texts from Manuscript, 2) (Cambridge: RTM, 1999).

[181] Marcus Hellyer, "'Because the authority of my superiors commands': Censorship, Physics, and the German Jesuits," *Early Science and Medicine*, 1 (1995), 319–54.

[182] William Petty's suggestion that the Royal Society offer courses in natural philosophy at a charge of £1 per month had to await the appearance of experimental lecturers such as William Whiston in the early eighteenth century. See Hunter, *Establishing the New Science*, pp. 2, 202; and S. D. Snobelen, "William Whiston: Natural Philosopher, Prophet, Primitive Christian," Ph.D. dissertation, University of Cambridge, 2001.

[183] Abraham Cowley, *A Proposition for the Advancement of Experimental Philosophy* (London: Printed by J. M. for Henry Herringman, 1661; facsimile repr. as *The Advancement of Experimental Philosophy*, Menston: Scolar Press, 1969); John Evelyn to Robert Boyle, 3 September 1659, describing his plan for a "College," printed in *The Correspondence of Robert Boyle, 1636–91*, ed. Michael Hunter, Antonio Clericuzio, and Lawrence M. Principe, 6 vols. (London: Pickering and Chatto, 2001), 1: 365–9; Hunter, *Establishing the New Science*, pp. 157, 181–4; and Webster, *Great Instauration*, pp. 88–99.

[184] The dispute between Julius Caesar Scaliger and Girolamo Cardano is a case in point. See Anthony Grafton, *Cardano's Cosmos: The Worlds and Works of a Renaissance Magician* (Cambridge, Mass.: Harvard University Press, 1999), p. 4; and Maclean, "The Interpretation of Natural Signs," pp. 231–52.

injunctions of the civilizing process, the new natural philosophical etiquette was arguably no more successful at controlling controversy than its more self-consciously disputatious predecessor.[185]

CONCLUSION

Issues of proof and persuasion in early modern Europe cannot be separated from the theoretical accounts that were formulated about them at the time. It is not a straightforward matter to claim that an argument proves something conclusively when it failed to prove it to its original audiences.[186] Claims for demonstration must be understood within the context of contemporary procedures of proof and persuasion. Although these provide a necessary starting point, contemporary accounts of how proof and persuasion function cannot simply be used to explain all practical manifestations of natural argumentation in the period in which they appear. Other factors – contingencies of publication, language, illustration, and distribution – necessarily come into play. More obviously, social, political, and institutional commitments also affected to a profound degree how and why particular arguments were accepted.[187]

Detailed examination of the multifarious ways in which such local commitments affected questions of proof and persuasion is beyond the scope of this study. Longer-term and wider-scale developments, however, can be identified more clearly. The most important factor within these developments is education. Questions of proof and persuasion in early modern Europe were closely associated with teaching, for pedagogy was the principal arena in which probation and persuasion occurred. As we have seen, the fundamental assumptions about proof and persuasion were imparted by the training in logic and rhetoric in the early modern schools and universities. The teaching of these disciplines remained a constant throughout the sixteenth and seventeenth centuries, and there were thus significant continuities in practices of proof and persuasion throughout the period. The scope of the application of rhetoric and logic, however, changed dramatically. Their applicability to natural philosophy came under intense pressure in the form of challenges from skepticism, mathematical techniques, and new conceptions of experiment.

Furthermore, study of the natural world was increasingly undertaken by individuals who had little or no connection with the universities and who

---

[185] Daston, "Baconian Facts," p. 353; Anne Goldgar, *Impolite Learning: Conduct and Community in the Republic of Letters, 1680–1750* (New Haven, Conn.: Yale University Press, 1995); Anthony Grafton, "Jean Hardouin: The Antiquary as Pariah," *Journal of the Warburg and Courtauld Institutes*, 62 (1999), 241–67; and Hahn, *Anatomy*, pp. 30–1.

[186] See R. H. Naylor, "Galileo's Experimental Discourse," in *Uses of Experiment*, pp. 117–34, at p. 130.

[187] See, for example, Nicholas Jardine, "Keeping Order in the School of Padua: Jacopo Zabarella and Francesco Piccolomini on the Offices of Philosophy," in Di Liscia, Kessler, and Methuen, eds., *Method and Order in Renaissance Philosophy of Nature*, pp. 183–209.

were indeed frequently markedly hostile toward them. The freedom of these individuals and of the institutions they formed from the imperative to impart their investigations to the young systematically was perhaps the most significant factor freeing them from the probative habits of the schools and allowed for the period's striking proliferation of techniques, methods, and forms of presentation. Once sixteenth- and seventeenth-century investigators into the natural world – whether natural philosophical, mathematical and astronomical, or medical – freed themselves from the imperative to teach, they also freed themselves from traditions of proof and persuasion dictated, often literally, by the schools.

Inquiry into the early modern natural world, then, was inextricably bound up with the ways in which it was presented. Forms of proof and persuasion cannot be dissociated from the content of natural knowledge in the sixteenth and seventeenth centuries; changes in this content in turn had a significant impact on forms of proof and persuasion. These changing conceptions of probation may well also have had profound implications for early modern notions of "science."[188] For an Aristotelian of the sixteenth century, *scientia* precisely consisted in being able to demonstrate with certainty the causes of an observed effect. The new mathematical and experimental strands of natural philosophy, however, cast that presupposition into doubt. As the task of natural philosophy changed in the course of the sixteenth and seventeenth centuries from explanation to description,[189] claims to a "scientific" knowledge of the natural world became problematic. At the end of the seventeenth century, the English philosopher John Locke (1632–1704) manifested an acute consciousness of the implications of the new experimental natural philosophy for the older conception of "science." In his *Essay Concerning Human Understanding* (1690), Locke noted that the getting and improving of knowledge about natural substances was "*only by Experience* and History." But this, he went on, "makes me suspect, that natural Philosophy is not capable of being made a Science."[190] For better or worse, however, Locke's successors did not take him at his word. By the twentieth century, natural philosophy had become simply "science," a discipline whose persuasive power was greater than that of natural philosophy had ever been.

---

[188] On this subject generally, see Ernan McMullin, "Conceptions of Science in the Scientific Revolution," in *Reappraisals of the Scientific Revolution*, pp. 27–92.

[189] On this development, see Peter Dear, *Revolutionizing the Sciences: European Knowledge and Its Ambitions, 1500–1700* (Princeton, N.J.: Princeton University Press, 2001), esp. pp. 3–4, 13–15, 44, 65, and 170.

[190] John Locke, *An Essay Concerning Human Understanding* [1690], ed. Peter H. Nidditch (Oxford: Clarendon Press, 1975), p. 645 (4.12.10). See further Margaret J. Osler, "John Locke and the Changing Ideal of Scientific Knowledge," *Journal of the History of Ideas*, 31 (1970), 3–16, at p. 15; and McMullin, "Conceptions of Science," pp. 75–6. Compare the similar remark in John Locke, *Some Thoughts Concerning Education*, ed. John W. Yolton and Jean S. Yolton (The Clarendon Edition of the Works of John Locke) (Oxford: Clarendon Press, 1989), p. 245.

Part II

PERSONAE AND SITES OF NATURAL KNOWLEDGE

6

THE MAN OF SCIENCE

Steven Shapin

It is difficult to refer to the early modern man of science in other than negative terms. He was not a "scientist": The English word did not exist until the nineteenth century, and the equivalent French term – *un scientifique* – was not in common use until the twentieth century. Nor did the defined social and cultural position now picked out by "the scientist's role" exist in the early modern period. The man of science did not occupy a single distinct and coherent role in early modern culture. There was no one social basis for the support of his work. Even the minimal organizing principle for any treatment of the man of science – that he was someone engaged in the investigation of nature – is, on reflection, highly problematic. What *conceptions* of nature, and of natural knowledge, were implicated in varying cultural practices? The social circumstances in which, for example, natural philosophy, natural history, mathematics, chemistry, astronomy, and geography were pursued differed significantly.

The *man* of science was, however, almost always male, and to use anything but this gendered language to designate the pertinent early modern role or roles would be historically jarring. The system of exclusions that kept out the vast numbers of the unlettered also kept out all but a very few women. And although it is important to recover information about those few female participants, it would distort such a brief survey to devote major attention to the issue of gender[1] (see the following chapters in this volume: Schiebinger, Chapter 7; Cooper, Chapter 9; Outram, Chapter 32).

Any historically responsible treatment of the early modern man of science has to embrace a splitting impulse and resist temptations toward facile

---

[1] Women do become rather more substantial philosophical presences in the salons of the Enlightenment; see, for example, Dena Goodman, "Enlightenment Salons: The Convergence of Female and Philosophic Ambitions," *Eighteenth-Century Studies*, 22 (1989), 329–50.

This chapter was substantially written while the author was a Fellow of the Center for Advanced Study in the Behavioral Sciences, Stanford, California. He thanks the Center and the Andrew W. Mellon Foundation for their support.

generalization.² The diversity of past patterns needs to be insisted upon, and not as a matter of mere pedantry. Even those historical actors concerned with bringing into being a more coherent and dedicated role for some version of the man of science were well aware of contemporary diversities. Francis Bacon (1561–1626) noted that "natural philosophy, even among those who have attended to it, has scarcely ever possessed, especially in these later times, a disengaged and whole man . . . , but that it has been made merely a passage and bridge to something else."³

So the man of science was not a "natural" feature of the early modern cultural and social landscape: One uses the term faute de mieux, aware of its impropriety in principle, yet confident that no mortal historical sins inhere in the term itself. Although it is a proper historical question to ask "how we got from there to here," one should at the same time be wary about transporting into the distant past the coherences of present-day social roles. Despite the legitimacy of asking how the relatively stable professionalized role of the modern scientist emerged from diverse sixteenth- and seventeenth-century arrangements, it would be misleading to mold historical inquiry solely to fit the contours of present-day interest in "origins stories" or to construe historical inquiry solely as a search for traces of present arrangements.⁴

Early modern scientific work – of whatever version – was pursued within a range of traditionally established social roles. One has to appreciate the expectations, conventions, and ascribed attributes of those existing roles, as well as the changes they were undergoing and their mutual relations, in order to understand the social identities of men of science in the period. Yet, vital as it is to insist on the heterogeneity of existing roles in which natural knowledge was harbored and extended in the early modern period, a brief survey such as this one can treat just a few of the more consequential roles – and here I have elected to focus on the university scholar or professor, the medical man, and the gentleman.

---

² For justification of such splitting sensibilities, see, for example, Thomas S. Kuhn, "Mathematical versus Experimental Traditions in the Development of Physical Science," in *The Essential Tension: Selected Studies in Scientific Tradition and Change*, ed. Thomas S. Kuhn (Chicago: University of Chicago Press, 1977), pp. 31–65, and the archaeology of disciplines and roles mooted in Robert S. Westman, "The Astronomer's Role in the Sixteenth Century: A Preliminary Study," *History of Science*, 18 (1980), 105–47.

³ Francis Bacon, *The New Organon* [1620], bk. 1, aphorism 80, ed. Fulton H. Anderson (Indianapolis: Bobbs-Merrill, 1960), p. 77.

⁴ A well-known essay on "the emergence and development of the social role of the scientist," strongly shaped by the assumptions of structural-functionalist sociology and by the so-called professionalization model, is Joseph Ben-David, *The Scientist's Role in Society* [1971] (Chicago: University of Chicago Press, 1984), esp. chaps. 4–5 (for early modern topics). Note that the negative claims of this and the preceding paragraph are direct contradictions of Ben-David's assertion (p. 45; cf. p. 56 n. 20) that it was in the seventeenth century that "certain men . . . view[ed] themselves for the first time as scientists and [saw] the scientific role as one with unique and special obligations and possibilities." For well-judged criticism of ahistorical assumptions in Ben-David's account, see Thomas S. Kuhn, "Scientific Growth: Reflections on Ben-David's 'Scientific Role'," *Minerva*, 10 (1972), 166–78; cf. Roy Porter, "Gentlemen and Geology: The Emergence of a Scientific Career, 1660–1920," *The Historical Journal*, 21 (1978), 809–36, at pp. 809–13.

A more complete survey would be able to treat a whole range of other contemporary roles and their importance for the conduct of natural knowledge. The clerical role, for example, overlapped significantly, but only partially, with that of the university scholar, and a number of key figures spent the whole, or very considerable portions, of their working lives within religious institutions or sustained by clerical positions: among many examples, Nicholas Copernicus (1473–1543) in his Ermland chapter house, Marin Mersenne (1588–1648) in the order of Minims in Paris, and Pierre Gassendi (1592–1655), whose canonry at Digne assured his financial independence. The significance of the priestly role for contemporary appreciations of the proper relationship between natural knowledge and religion cannot be overemphasized. When some seventeenth-century practitioners circulated a conception of natural philosophers as "priests of nature," they meant to display the theological equivalence of the Books of Nature and Scripture and also to imbue scientific work with the aura surrounding the formally religious role.[5]

Still other major scientific and philosophical figures spent much of their careers as amanuenses, clerks, tutors, or domestic servants of various kinds to members of the gentry and aristocracy, a common career pattern for Renaissance humanist intellectuals in several countries. Thomas Hobbes (1588–1679) functioned in a variety of domestic service roles to the Cavendish family for almost the whole of his adult life, and one of John Locke's (1632–1704) first positions was as private physician, and later as general secretary, to the Earl of Shaftesbury. Relationships binding the practice of science to the patronage of princes and wealthy gentlemen were pervasive and consequential: The significance of the Tuscan court's patronage for Galileo Galilei's "socioprofessional identity" and for the direction of his scientific work has been vigorously asserted, and the importance of patronage and clientage relations for the careers and authority of very many other notable early modern men of science – and for the authority of the knowledge they produced – merits much fuller study.[6] Finally, a more extensive account of the early modern man of science would treat a whole range of less exalted figures – mathematical practitioners, instrument makers, lens grinders, and various types of "superior artisans" – whose significance both for the practical conduct of scientific research and for the development of empirical methods was much insisted upon by the Marxist historiography of the 1930s and 1940s and as vigorously denied by idealist historians.[7]

---

[5] See, for example, Harold Fisch, "The Scientist as Priest: A Note on Robert Boyle's Natural Theology," *Isis*, 44 (1953), 252–65; and Simon Schaffer, "Godly Men and Mechanical Philosophers: Souls and Spirits in Restoration Natural Philosophy," *Science in Context*, 1 (1987), 55–85.

[6] Mario Biagioli, *Galileo, Courtier: The Practice of Science in the Culture of Absolutism* (Chicago: University of Chicago Press, 1993); see also Bruce T. Moran, ed., *Patronage and Institutions: Science, Technology, and Medicine at the European Court, 1500–1750* (Woodbridge: Boydell Press, 1991).

[7] For classic stress on the crucial significance of craft roles in the emergence of modern science, see Edgar Zilsel, "The Sociological Roots of Science," *American Journal of Sociology*, 47 (1942), 544–62. For Alexandre Koyré–inspired rejection of any such idea, see A. Rupert Hall, "The Scholar

THE UNIVERSITY SCHOLAR

The man of science, and almost all specific versions thereof, represented a subset of the early modern learned classes. By construing the investigation of nature as an act within learned culture, one is immediately marking out a massively important social division in early modern Europe, that between those who were literate and those who were not, between those who had passed through formal schooling and those who had not. European cultures did differ in the extent to which their populations were schooled, and therefore literate, but, in general, the fraction of the literate was very small and that of the learned even smaller.[8] What was understood about the characters of the learned elite was, mutatis mutandis, understood of the learned man of science as well.

By no means all noteworthy early modern men of science were systematically shaped by university training. Among those who did not formally attend university at all were Blaise Pascal (1623–1662), Robert Boyle (1627–1691), and René Descartes (1596–1650), though Descartes' training at the Jesuit school of La Flèche was considerably more significant to his intellectual development than was Boyle's time at Eton College. At both ends of the social scale, the future man of science might escape university training – those being bred to artisanal or mercantile work, such as the potter and natural historian Bernard Palissy (1510–1590) or the merchant and microscopist Antonie van Leeuwenhoek (1632–1723), because they lacked the means or current interest,[9] and the aristocrat (e.g., Boyle) because private resources might be preferred and because there was no professional or material inducement to secure formal

---

and the Craftsman in the Scientific Revolution," in *Critical Problems in the History of Science*, ed. Marshall Clagett (Madison: University of Wisconsin Press, 1959), pp. 3–23. For revived interest in the role and standing of mathematical practitioners, see, for example, Mordechai Feingold, *The Mathematicians' Apprenticeship: Science, Universities, and Society in England, 1560–1640* (Cambridge: Cambridge University Press, 1984); J. A. Bennett, "The Mechanics' Philosophy and the Mechanical Philosophy," *History of Science*, 24 (1986), 1–28; Bennett, "The Challenge of Practical Mathematics," in *Science, Culture, and Popular Belief in Renaissance Europe*, ed. Stephen Pumfrey, Paolo L. Rossi, and Maurice Slawinski (Manchester: Manchester University Press, 1991), pp. 176–90; Mario Biagioli, "The Social Status of Italian Mathematicians, 1450–1600," *History of Science*, 27 (1989), 41–95; Richard W. Hadden, *On the Shoulders of Merchants: Exchange and the Mathematical Conception of Nature in Early Modern Europe* (Albany: State University of New York Press, 1994); Frances Willmoth, *Sir Jonas Moore: Practical Mathematics and Restoration Science* (Woodbridge: Boydell Press, 1993); Amir Alexander, "The Imperialist Space of Elizabethan Mathematics," *Studies in History and Philosophy of Science*, 26 (1995), 559–91; Stephen Johnston, "Mathematical Practitioners and Instruments in Elizabethan England," *Annals of Science*, 48 (1991), 319–44; and Katherine Hill, "'Juglers or Schollers?': Negotiating the Role of a Mathematical Practitioner," *British Journal for the History of Science*, 31 (1998), 253–74.

[8] For treatment of changing relations between elite and lay cultures in the early modern period, see Peter Burke, *Popular Culture in Early Modern Europe* (London: Temple Smith, 1978), esp. chaps. 2 and 9; see also Paul J. Bagley, "On the Practice of Esotericism," *Journal of the History of Ideas*, 53 (1992), 231–47; and Carlo Ginzburg, "High and Low: The Theme of Forbidden Knowledge in the Sixteenth and Seventeenth Centuries," *Past and Present*, 73 (1976), 28–41.

[9] The experimentalist Robert Hooke was at Christ Church, Oxford, as a chorister, and it is unclear whether he ever availed himself of formal university instruction.

training. For a larger number of other men of science, university education was part of a background preparation for roles in civic life, and the acquisition of scientific expertise, or at least of that expertise for which they became known, occurred elsewhere. The mathematician Pierre de Fermat (1601–1665) and the astronomer Johannes Hevelius (1611–1687) studied law at a university, as did many other future men of science; William Gilbert (1544–1603), author of *De magnete* (On the Magnet, 1600), and the mathematician and physicist Isaac Beeckman (1588–1637) studied medicine; and Johannes Kepler (1571–1630) studied mainly theology.

In their mature careers, however, many scientific practitioners in the sixteenth and seventeenth centuries were professionally engaged by universities or related institutions of higher learning, though the proportion of these among the great figures making up the canon of early modern science can be overestimated.[10] Andreas Vesalius (1514–1564), Galileo, and Isaac Newton (1642–1727) were professors (for at least part of their careers), whereas Copernicus, Kepler, Bacon, Descartes, Mersenne, Pascal, Boyle, Tycho Brahe (1546–1601), and Christiaan Huygens (1629–1695) were not. Moreover, the professorial role was by no means a stable one. Although for late twentieth-century scientists a permanent university appointment generally represents a natural career culmination, this was not necessarily the case for the early modern man of science. Occupying a university chair or fellowship might be just an episode in a career that included a variety of other social roles. There was indeed an early modern pattern of using university employment as a stepping stone to more desirable positions directly supported by court patronage. A figure such as the mathematician and astronomer Christoph Clavius (1538–1612) was arguably exceptional in remaining at his professorial position (in the Jesuits' Collegio Romano) for almost the whole of his adult life. Both Isaac Barrow (1630–1677) and his successor in the Cambridge Lucasian Chair of Mathematics, Isaac Newton, abandoned their university appointments while they were relatively young men – Barrow for brighter prospects as a royal chaplain (returning to Cambridge later as Master of Trinity and University Vice Chancellor), and Newton (after health problems) to become an official of the Royal Mint. Their contemporary Seth Ward (1617–1689), the Savilian Professor of Astronomy at Oxford, gave up his professorial career in early middle age, accepting several church livings and ultimately becoming bishop of Exeter.

Thomas Willis (1621–1675) vacated the Sedleian Chair of Natural Philosophy at Oxford for a lucrative medical practice in London. Vesalius left his teaching at the University of Padua in mid-career for medical service in the

---

[10] This brief survey does not aim at a prosopography of early modern men of science and their institutional affiliations. Such an exercise would first have to establish social and intellectual criteria for identifying who *was* a man of science, whereas a major purpose of this chapter is to draw attention to the problematic nature of *any* coherent set of criteria the present-day historian might draw up.

imperial household; the astronomer Gian Domenico Cassini I (1625–1712) combined duties as a professor at the University of Bologna with engineering work for the pope before abandoning both for a stipend as a member of the new Académie Royale des Sciences in Paris; the French Huguenot inventor Denis Papin (1647–1712) had no compunction about leaving his chair of mathematics at the university of Marburg because of its miserable salary and heavy teaching load, and the Danish astronomer Ole Römer (1644–1710) equally understandably quit his chair of mathematics at the University of Copenhagen to become a powerful officeholder – first mayor and then state councillor. Hence the identification of scientific work with the professorial career was significant but tenuous and patchy during the early modern period. If you were, for example, a cleric-professor, or a physician-professor, then it needs no special explanation that you gave up your chair – and even gave up your scientific research – when better-paid or more prestigious ecclesiastical or medical opportunities presented themselves.

Professional affiliation with institutions of higher education and the stewardship of learning meant three things above all. First, it signaled links with organized forms of Christian religion. Throughout the early modern period, universities outside Italy were widely under church control – the Reformation splitting the institutional nature of that control but not, with some important exceptions, diluting it. The universities had as one of their major purposes the training of individuals for clerical roles, and membership in the clergy, or formal subscription to church doctrines, were very general conditions for matriculation, graduation, or entry to the fellowship and professoriate.

Second, the university combined curatorial and culturally reproductive roles, and its professors' activities and identities were primarily understood in those lights. Universities signified both responsible custodianship of the knowledge inherited from the past and its reliable transmission to future generations, and, although a significant number of professors took it upon themselves to engage in research that challenged orthodox beliefs, nowhere in early modern Europe was such a conception of the professorial role standard. Original research was not, so to speak, a role requirement.

Third, affiliation with the university associated the man of science with specific hierarchical social forms: The master was understood to be a master of knowledge traditionally accumulated and traditionally vouched for, and his institutional purpose was to transmit that mastery to future generations. The value placed on these hierarchical forms implicated the value placed on traditional forms of knowledge. The "modern" assault on school-knowledge proceeded importantly by way of criticisms of the schools' hierarchical social forms and the role of the professor in those forms. The university setting vouched for expertise, authenticity, and orthodoxy, and those ascribed characteristics spoke in favor of the knowledge housed there. But to those of a mind to criticize university arrangements, the same site and role were associated with authoritarianism, dogmatism, pedantry, disputatiousness, and melancholic sequestration from the civic and material worlds.

Indeed, some of the new scientific societies that began to emerge in the mid-seventeenth century developed in self-conscious opposition to the universities: A peaceable and useful community of inquiring equals was juxtaposed with bastions of school-mastery, divisiveness, and inconsequentiality.[11] The Royal Society of London was a notable site in which such sentiments were expressed, whereas in Germany Gottfried Wilhelm Leibniz's (1646–1716) plans for a state-supported scientific academy stressed the importance of selecting persons who were not only knowledgeable but who were "also endowed with a unique goodness of mind; in whom rivalry and jealousy are wanting; who will not use despicable devices to appropriate for themselves the labors of others; who are not factious and have no wish to be regarded as the founders of sects; who labor for love of learning and not for ambition or sordid pay."[12] In such venues, disapproving assessments of the professorial character precipitated by negation, as it were, the developing identity of the free academic member of the Republic of Science. Yet, apart from a very general commitment to a harmoniously collaborative – or at least collective – pursuit of natural knowledge, there is no single coherent pattern to be discerned in the establishment or structure of seventeenth-century scientific societies. Members of the Académie Royale des Sciences in Paris enjoyed substantial Crown pensions and devoted themselves effectively to the extension of state power through reformed natural knowledge and technology, but, although fellows of the Royal Society of London intermittently expressed their desire to realize the imperializing dreams of the utopian research institute described in Bacon's *New Atlantis* (1627), the English Crown offered no stipends and little financial support. Charles II laughed at them for wasting their time on intellectual trivialities.[13]

---

[11] Some of these issues are treated for the English setting in Allen G. Debus, *Science and Education in the Seventeenth Century: The Webster–Ward Debate* (London: Macdonald, 1970); Michael R. G. Spiller, *"Concerning Natural Experimental Philosophie": Meric Casaubon and the Royal Society* (The Hague: Martinus Nijhoff, 1980); and James R. Jacob, *Henry Stubbe, Radical Protestantism and the Early Enlightenment* (Cambridge: Cambridge University Press, 1983), esp. chap. 5. The relations between the Royal Society of London and gentlemanly conventions are briefly treated later in this chapter. For a general sketch of the academic institutional form as it developed in Europe beginning in the mid-fifteenth century, see Ben-David, *The Scientist's Role*, pp. 59–66.

[12] Gottfried Wilhelm Leibniz, "On the Elements of Natural Science," in Leibniz, *Philosophical Papers and Letters* [ca. 1682–4], ed. and trans. Leroy E. Loemker, 2nd ed. (Dordrecht: Reidel, 1969), pp. 277–90, at p. 282. For the context and outcome of Leibniz's plans for establishing scientific societies, see Ayval Ramati, "Harmony at a Distance: Leibniz's Scientific Academies," *Isis*, 87 (1996), 430–52.

[13] There is a very large secondary literature on particular seventeenth-century scientific societies, as well as some attempt to identify their collective significance: see, for example, Sir Henry Lyons, *The Royal Society, 1660–1940: A History of Its Administration under Its Charters* (Cambridge: Cambridge University Press, 1944), chaps. 1–4; Dorothy Stimson, *Scientists and Amateurs: A History of the Royal Society* (New York: Henry Schuman, 1948); Sir Harold Hartley, ed., *The Royal Society: Its Origins and Founders* (London: The Royal Society, 1960); Margery Purver, *The Royal Society: Concept and Creation* (Cambridge, Mass.: MIT Press, 1967); Michael Hunter, *Establishing the New Science: The Experience of the Early Royal Society* (Woodbridge: Boydell Press, 1989); Hunter, *The Royal Society*

Membership in a scientific society or academy therefore had no one stable significance for the identity of the seventeenth-century man of science, though eighteenth-century developments, and especially patterns emerging in France, did eventually make the academic role increasingly important for scientific identity.[14] The role of the seventeenth-century scientific academician might be recognized as a modified form of long-standing social roles – the court bureaucrat or the recipient of Crown patronage – or, where the ties between scientific societies and the state were weaker, patterns of gentlemanly conversation and virtuosity might be more central to his identity. In the former case, the contribution of academic membership to the recognized role of the man of science could be substantial; in the latter, the significance of such membership might be subsumed in the gentlemanly role.

THE MEDICAL MAN

The profession of medicine also associated the pursuit of natural knowledge with recognized and authoritative early modern social roles, and many medical men pursued scientific investigations within the rubric of a professorial role, such as Vesalius (at Padua) and Marcello Malpighi (1628–1694) (at Bologna). Established colleges of physicians and surgeons might also offer quasi-academic roles, such as the lectureship on surgery held for many years by William Harvey (1578–1657) at the London Royal College of Physicians. Nevertheless, the medical role was one that in principle provided for the authoritative pursuit of natural knowledge outside the rubric of the

---

and Its Fellows, 1660–1700: The Morphology of an Early Scientific Institution (British Society for the History of Science Monographs, 4) (Chalfont St. Giles: British Society for the History of Science, 1982); Roger Hahn, *The Anatomy of a Scientific Institution: The Paris Academy of Sciences, 1666–1803* (Berkeley: University of California Press, 1971); Claire Salomon-Bayet, *L'Institution de la science et l'expérience du vivant: Méthode et expérience à l'Académie Royale des Sciences, 1666–1793* (Paris: Flammarion, 1978); Alice Stroup, *A Company of Scientists: Botany, Patronage, and Community at the Seventeenth-Century Parisian Royal Academy of Sciences* (Berkeley: University of California Press, 1990); W. E. Knowles Middleton, *The Experimenters: A Study of the Accademia del Cimento* (Baltimore: Johns Hopkins University Press, 1971); Knowles Middleton, "Science in Rome, 1675–1700, and the Accademia Fisicomathematica of Giovanni Giustino Ciampiani," *British Journal for the History of Science*, 8 (1975), 138–54; David S. Lux, *Patronage and Royal Science in Seventeenth-Century France: The Académie de Physique in Caen* (Ithaca, N.Y.: Cornell University Press, 1989); Daniel Roche, *Le siècle des lumières en province: Académies et académiciens provinciaux, 1680–1789*, 2 vols. (Paris: Mouton, 1978); K. Theodore Hoppen, *The Common Scientist in the Eighteenth Century: A Study of the Dublin Philosophical Society, 1683–1708* (London: Routledge and Kegan Paul, 1970); Harcourt Brown, *Scientific Organizations in Seventeenth Century France (1620–1680)* (Baltimore: Williams and Wilkins, 1934); Martha Ornstein, *The Role of Scientific Societies in the Seventeenth Century* (Chicago: University of Chicago Press, 1928); R. J. W. Evans, "Learned Societies in Germany in the Seventeenth Century," *European Studies Review*, 7 (1977), 129–51; and James E. McClellan III, *Science Reorganized: Scientific Societies in the Eighteenth Century* (New York: Columbia University Press, 1985), chaps. 1–2. See also many of the works cited in notes 17–24.

[14] Eighteenth-century developments are treated in Steven Shapin, "The Image of the Man of Science," in *The Cambridge History of Science*, vol. 4: *Eighteenth-Century Science*, ed. Roy Porter (Cambridge: Cambridge University Press, 2003), pp. 159–83. See also works cited in note 24.

universities or, indeed, of incorporated learning. To become a physician, of course, one had to pass through the institutions of higher learning – sometimes only quite nominally – but once one had done so, one could occupy that role, and be active in scientific inquiry, without necessarily being a member of any university or in the pay of any medical corporation.[15]

Unlike the role of the university scholar in general, the social role of the medical man strongly linked natural knowledge with practical interventions. No matter how much the physician's role – though not the surgeon's or apothecary's – was argued to belong to the world of polite and pure learning, the value of the physician's knowledge was nevertheless vouched for by its ability both to explain the real vicissitudes of human bodies and, where possible, to guide those practices that maintained health and alleviated disease.[16] Although physicians were commonly mocked for what were seen as their illegitimate therapeutic pretensions, the very existence of the role testified to the overall esteem in which formal medical knowledge was held and the overall efficacy attributed to that knowledge. Medicine was therefore one important domain within which natural knowledge enjoyed well-established social authority and credibility.

Moreover, medical roles – unlike those of the professoriate generally – were centrally concerned with the description, explanation, and management of natural bodies. And however much many early modern philosophers insisted upon the dual nature of human beings – spiritual and material – the medical role tended to focus its interventions on human beings in their material aspects. For these reasons, it was common for medical men to pursue those scientific subjects most closely linked with the form and functioning of the human body. The medical role therefore "naturally" propelled some of its members toward the study of anatomy and physiology, including among very many examples Harvey, Malpighi, Willis, Santorio Santorio (1561–1636), Olof Rudbeck (1630–1702), Richard Lower (1631–1691), Francesco Redi (1626–ca. 1697), and Regnier de Graaf (1641–1673). Similar professional concerns attracted others to natural history, such as Conrad Gessner (1516–1565), Jan Swammerdam (1637–1680), and Nehemiah Grew (1641–1712), or chemistry, in the cases of Georgius Agricola (1494–1555) and John Mayow (1641–1679).[17]

---

[15] Training in natural philosophy and natural history was a key preparatory requirement for a medical degree at many medieval and early modern universities. That is one reason why so many men trained in natural philosophy and natural history were physicians, and also why membership of early scientific societies was so heavily weighted toward medical men.

[16] The cultural and social boundaries that reserved "professional" standing to bookishly trained physicians and that relegated surgeons and apothecaries to trade or craft status were hard to enforce. In England, at any rate, more liberal and inclusive notions of "the medical profession" were emerging by the late seventeenth and early eighteenth centuries, with interesting consequences for relations between medicine and the culture of science; see, for example, Geoffrey Holmes, *Augustan England: Professions, State, and Society, 1680–1730* (London: Allen and Unwin, 1982), chaps. 6–7.

[17] See, for example, Harold J. Cook, "Physicians and Natural History," in *Cultures of Natural History*, ed. Nicholas Jardine, James A. Secord, and Emma C. Spary (Cambridge: Cambridge University Press, 1996), pp. 91–105. Cook notes how materia medica provided a substantive link between natural

However, the participation of medical men was not confined to subjects strictly related to medical practice; see, for example, the work of such physicians as Gilbert (in magnetism), Nicolaus Steno (1638–1686) (in geology), and Henry Power (1623–1668) (in experimental natural philosophy). John Locke earned a medical degree before establishing his reputation in mental and political philosophy, and it might be said that Thomas Sydenham's (1624–1684) key achievement was a methodology of quite wide scientific applicability. Nor was substantial interest in medical subjects restricted to those occupying the social role of physician or surgeon: Bacon, Descartes, and Boyle lacked professional qualifications but either theorized on medical subjects or dabbled in medical therapeutics and dietetics.

THE GENTLEMAN

Like the roles of the scholar and the medical man, the gentlemanly role offered both problems and opportunities for changing conceptions of what it was to make natural knowledge. On the one hand, the traditional gentlemanly role was not, of course, primarily defined around the acquisition and pursuit of formal knowledge, though humanist writers argued strenuously through the sixteenth and early seventeenth centuries that virtuous and polite knowledge *ought to be* central to legitimate conceptions of gentility. Although there were important overlaps between the gentle and the learned classes, gentlemanly culture was uncomfortable – in England more than in Italy or France – with the idea that the wellborn should make the pursuit of formal knowledge a professional activity, either in a remunerative sense or in the sense of the pursuit being fundamental to one's social identity. Scholars might in many cases be genuinely respected by gentle society, but that society importantly distinguished the roles of the gentleman and the professional scholar or pointed to features of the scholar's "character" that handicapped his ability to take part in gentlemanly conversation. Particular targets of criticism were the scholar's traditional isolation, his "morose" or "melancholic" complexion, his tendency toward disputation, and his pedantry.[18]

On the other hand, the gentle classes were widely literate, sometimes well educated, and, especially on the Continent, often disposed to act as patrons to men of science – in the case of the "mixed" mathematical sciences because of their acknowledged utility to the arts of war, wealth-getting, and political

---

history, chemistry, and medical therapeutic concerns; see also Paula Findlen, *Possessing Nature: Museums, Collecting, and Scientific Culture in Early Modern Italy* (Berkeley: University of California Press, 1994), chap. 6.

[18] Steven Shapin, "'A Scholar and a Gentleman': The Problematic Identity of the Scientific Practitioner in Early Modern England," *History of Science*, 29 (1991), 279–327; Shapin, *A Social History of Truth: Civility and Science in Seventeenth-Century England* (Chicago: University of Chicago Press, 1994), chaps. 2–4; Adrian Johns, "Prudence and Pedantry in Early Modern Cosmology: The Trade of Al Ross," *History of Science*, 35 (1997), 23–59.

control, and, in the case of other scientific practices, such as astronomy or natural history, because they lent luster to the patron and sparkle to civil conversation.[19] The gentry, aristocracy, and nobility therefore controlled an enormously important pool of resources for supporting the work of men of science, while cultural and social attitudes placed obstacles between patronage or amateurism, on the one hand, and the professional pursuit of, or systematic identification with, scientific practice on the other. In the sixteenth and early seventeenth centuries, those obstacles could in principle be set aside – there were some very notable aristocratic men of science – but contemporary culture possessed few resources for appreciating and approving a substantive merger between the role of the professionally learned and the role of the gentle.

Those cultural resources soon began to be available, with potential consequences for changing notions of the social role of the man of science and of scientific knowledge itself. Beginning in the late sixteenth century, Francis Bacon – English aristocrat and Lord Chancellor – argued strenuously for methodological and organizational reforms in natural knowledge that would at once make that knowledge an effective arm of state power and render it a pursuit suitable for civically engaged gentlemen. Natural knowledge was to be hauled out of the privacy of the traditional scholar's study – which made science disputatious, wordy, and barren – and into the bright light of real-world phenomena and practical civic concerns.[20] The reformed man of science was supposed to live a *vita activa*, and reformed science was to be done in public places.[21]

Bacon's vision of a civically pertinent science practiced by civically situated scholars was further developed in England starting in the 1660s by the new Royal Society of London. Here such publicists as Henry Oldenburg (1618–1677), Thomas Sprat (1635–1713), and Joseph Glanvill (1636–1680) announced that the Royal Society had turned traditionally deductive natural philosophical practice upside down, and, placing particular facts before causal and metaphysical systems, had cured science of its disputatiousness, pedantry, individualism, authoritarianism, and aridity. And when the social and intellectual virtues of the new practice were embodied in the person of the Honourable Robert Boyle – a great Anglo-Irish aristocrat – the Royal

---

[19] See, for example, Biagioli, *Galileo, Courtier*; Mario Biagioli, "Le prince et les savants: La civilité scientifique au 17ᵉ siècle," *Annales: Histoire, Sciences Sociales*, 50 (1995), 1417–53; Biagioli, "Etiquette, Interdependence, and Sociability in Seventeenth-Century Science," *Critical Inquiry*, 22 (1996), 193–238; Willmoth, *Sir Jonas Moore*; Findlen, *Possessing Nature*; Moran, ed., *Patronage and Institutions*; Stroup, *A Company of Scientists*; and Pamela H. Smith, *The Business of Alchemy: Science and Culture in the Holy Roman Empire* (Princeton, N.J.: Princeton University Press, 1994).

[20] See Julian Martin, *Francis Bacon, the State, and the Reform of Natural Philosophy* (Cambridge: Cambridge University Press, 1992).

[21] For early modern debates over whether the scientific life should be "active" or "contemplative," see Owen Hannaway, "Laboratory Design and the Aim of Science: Andreas Libavius versus Tycho Brahe," *Isis*, 77 (1986), 585–610; and Steven Shapin, "The House of Experiment in Seventeenth-Century England," *Isis*, 79 (1988), 373–404.

Society declared that it had realized Bacon's dream of joining a new science to a new social role for the man of science: not a professional scholar, not a schoolman, not a slave to a philosophical system, not a professional cleric, and not a professional physician, but a free, independent, modest, and virtuous seeker of truth about God's nature. Science, the Society said, had been remade into both a polite and a useful practice, fit for gentlemanly participation and equipped to secure and extend state power.[22]

It is the gentlemanly pattern of changing conceptions of the social role of the man of science that poses the greatest challenge to the traditional "professionalization model." Historians and sociologists working within that model searched the historical record for traces of modern arrangements, particularly for emerging appreciations of the distinctiveness and autonomy of science and for a remunerative basis for the conduct of scientific research. Yet gentle culture tended to be suspicious of intellectual specialization and scholarly isolation, and, again especially in England, those who offered their intellectual labor in exchange for pay were sometimes considered to have sacrificed that freedom of action and integrity considered vital to making reliable knowledge.[23] Where the pursuit of natural knowledge was not specifically sustained by resources attached to such other social roles as that of the university scholar, the cleric, and the physician, that pursuit – like most other early modern learned activities – was supported and made possible by accumulated capital. Inherited independent means overwhelmingly provided the practical resources to seek natural knowledge, while such independence might be pointed to as a powerful symbolic guarantee of the integrity and disinterestedness of the authentic amateur, he who pursued knowledge for love rather than for lucre.

The gentlemanly conception of a new social role for the man of science was important in new practitioners' self-conceptions and in justifications of new intellectual practices. Yet its wider cultural legitimacy was circumscribed, both in England and on the Continent. In England, influential wits and courtiers poked fun at the utilitarian pretensions of the Royal Society and recognized no substantial differences between the new social forms and the old pedantry and dispute. In the Royal Society itself, Boylean patterns of modest

---

[22] The significance of particular patterns of gentility associated with some Continental men of science has been addressed by Stephen Gaukroger, *Descartes: An Intellectual Biography* (Oxford: Oxford University Press, 1995), esp. pp. 28–67; Peter Dear, "A Mechanical Microcosm: Bodily Passions, Good Manners, and Cartesian Mechanism," in *Science Incarnate: Historical Embodiments of Natural Knowledge*, ed. Christopher Lawrence and Steven Shapin (Chicago: University of Chicago Press, 1998), chap. 2; Albert Van Helden, "Contrasting Careers in Astronomy: Huygens and Cassini," *De zeventiende eeuw*, 12 (1996), 96–105; and Victor E. Thoren, *The Lord of Uraniborg: A Biography of Tycho Brahe* (Cambridge: Cambridge University Press, 1990).

[23] Studies of Hooke and Boyle that have treated these aspects of remunerated science include Stephen Pumfrey, "Ideas above His Station: A Social Study of Hooke's Curatorship of Experiments," *History of Science*, 29 (1991) 1–44; Steven Shapin, "Who Was Robert Hooke?," in *Robert Hooke: New Studies*, ed. Michael Hunter and Simon Schaffer (Woodbridge: Boydell Press, 1989), pp. 253–85; and Shapin, *A Social History of Truth*, chap. 8.

empiricism and polite probabilism were soon challenged by a Newtonian persona and a Newtonian natural philosophical program that suggested to many a revival of older conceptions of scholarly isolation and philosophical authority. Early Royal Society rhetoric about the proper conduct of inquiry and the proper role of the man of science was widely applauded on the Continent, but the grip of corresponding social patterns was never very secure in France, Italy, and the German states. Everywhere the social role of the man of science remained heterogeneous, the pursuit of natural knowledge adventitiously attached in all sorts of ways to the preexisting social roles of the professional scholar, the medical man, the gentleman, and to as many other roles as figured in the production of learned culture generally.[24]

[24] The early to mid-eighteenth century developed much more elaborate cultures of both politeness and utility, and more contested notions of the role of the man of science within those cultures. On politeness, see Anne Goldgar, *Impolite Learning: Conduct and Community in the Republic of Letters, 1680–1750* (New Haven, Conn.: Yale University Press, 1995); Geoffrey V. Sutton, *Science for a Polite Society: Gender, Culture, and the Demonstration of Enlightenment* (Boulder, Colo.: Westview Press, 1995); and Alice N. Walters, "Conversation Pieces: Science and Politeness in Eighteenth-Century England," *History of Science*, 35 (1997), 121–54. For science and utility, see Larry Stewart, *The Rise of Public Science: Rhetoric, Technology, and Natural Philosophy in Newtonian Britain, 1660–1750* (Cambridge: Cambridge University Press, 1992); Jan Golinksi, *Science as Public Culture: Chemistry and Enlightenment in Britain, 1760–1820* (Cambridge: Cambridge University Press, 1992), esp. chap. 4; and also Shapin, "The Image of the Man of Science."

7

WOMEN OF NATURAL KNOWLEDGE

Londa Schiebinger

"L'esprit n'a point de sexe" ("the mind has no sex"), declared François Poullain de la Barre (1647–1723) in 1673 in an effort to level what he considered "the most remarkable of all prejudices": the inequality of the sexes.[1] An ardent Cartesian, he set out to demonstrate that the mind – distinct from the body – has no sex. New attitudes toward women, such as those voiced by Poullain and others, raised questions about female participation in natural knowledge, itself a novel enterprise struggling for recognition within established hierarchies. In the sixteenth and seventeenth centuries, the relation of natural inquiry to church, king, households (grand and humble), princely coffers, and global and local marketplaces was in a state of flux. Important questions remained to be answered about natural knowledge – its ideals and methods, its proper limits, and who should mold them.[2] The looser institutional organization and openings in attitudes allowed women to enter into natural inquiry through a number of informal arrangements and, in some cases, make important contributions to natural knowledge.

At a time when participation in natural inquiry was regulated to a large extent by social standing, men and women seeking to understand nature came primarily from two distinct social groups: learned elites and artisans (see Shapin, Chapter 6, this volume). The humanistic literati mixed in

---

[1] François Poullain de la Barre, *De l'égalité des deux sexes: Discours physique et moral* (Paris: Jean du Puis 1673), preface. Materials in this chapter are drawn in part from Londa Schiebinger, *The Mind Has No Sex? Women in the Origins of Modern Science* (Cambridge, Mass.: Harvard University Press, 1989), pp. 1–101.

[2] Alexandre Koyré, *From the Closed World to the Infinite Universe* (Baltimore: Johns Hopkins University Press, 1957); Robert Merton, *Science, Technology, and Society in Seventeenth Century England* [1938] (New York: H. Fertig, 1970); A. Rupert Hall, *The Revolution in Science, 1500–1750* (New York: Longmans, 1983); H. Floris Cohen, *The Scientific Revolution: A Historiographical Inquiry* (Chicago: University of Chicago Press, 1994); S. A. Jayawardene, *The Scientific Revolution: An Annotated Bibliography* (West Cornwall: Locust Hill Press, 1996). The notion that universities stood in the way of the new sciences has been challenged in Mordechai Feingold, *The Mathematicians' Apprenticeship: Science, Universities, and Society in England, 1560–1640* (Cambridge: Cambridge University Press, 1984).

courtly circles, scientific academies, and salons, while skilled craftsmen and craftswomen fashioned telescopes and astrolabes, made maps, and refined techniques for capturing with exactitude the minutest details of natural phenomena. In addition to these two groups, European peasants, fishermen, women who gathered medicinal herbs, and others served as informants to naturalists. William Eamon (Chapter 8, this volume) discusses how Ulisse Aldrovandi (1522–1605) visited fish markets to learn the names, habits, and unique characteristics of fish. Harold Cook has argued against a historiography that emphasizes too stringent a separation of head and hand, suggesting that especially in the Dutch Republic (and one might add the German lands) precisely the marriage of book learning and craft skills produced that ferment in knowledge still sometimes instructively referred to as the Scientific Revolution.[3] Nonetheless it is useful to highlight the nonacademic training offered within artisanal workshops that worked to the advantage of women and men of lower estates.

This chapter investigates the shifting institutional foundations of natural knowledge during the revolutions that marked its origins in the sixteenth and seventeenth centuries, and the changing fortunes of women within those institutions. We look first at the world of learned elites: universities, princely courts, informal humanist circles, scientific academies, and Parisian salons. These networks of literati are contrasted with the workshops of the skilled craftsmen and craftswomen. The chapter closes with a look outward from Europe, investigating the naturalists who undertook long and arduous journeys during the expansive voyages of scientific discovery.

LEARNED ELITES

Without proper training, access to libraries, instruments, and networks of communication, it is difficult for anyone – man or woman, highborn or lowborn, European or non-European – to make significant contributions to knowledge. Historically, women have not fared well in European institutions of learning. From their origins in the twelfth century, universities were, in principle, closed to women. Unlike religious houses, which had been centers of learning for both men and women, universities provided formal training in theology, law, and medicine aimed at preparing young men for careers in the church, government, or teaching. Women, barred from these learned professions, were not expected to enter the university.[4]

---

[3] Harold Cook, "The New Philosophy in the Low Countries," in *The Scientific Revolution in National Context*, ed. Roy Porter and Mikuláš Teich (Cambridge: Cambridge University Press, 1992), pp. 115–49. For a critique of the notion of a "scientific revolution," see Steven Shapin, *The Scientific Revolution* (Chicago: University of Chicago Press, 1996).

[4] Paul Kristeller, "Learned Women of Early Modern Italy: Humanists and University Scholars," in *Beyond Their Sex: Learned Women of the European Past*, ed. Patricia Labalme (New York: New York

Although today it would be difficult for anyone prohibited from entering universities to work in science, this was not the case in the early modern period. At this time, as Steven Shapin discusses (Chapter 6, this volume), "men of science" cultivated natural knowledge in a variety of settings: Galileo Galilei (1564–1642) was a resident astronomer at the court of Cosimo de' Medici; Francis Bacon (1561–1626) and Gottfried Wilhelm Leibniz (1646–1716) were government ministers as well as men of letters; and René Descartes (1596–1650), Christiaan Huygens (1629–1695), and Robert Boyle (1627–1691) were men of independent means.

In the absence of clearly established prerequisites of education and certification, participation in natural knowledge was regulated largely by networks of princely, aristocratic, and ecclesiastical patronage. The key to courtly and private patronage was power – not raw military might, but rather a highly ritualized exchange of gifts and status. A prince's courtiers, some of whom, such as Galileo, were mathematicians and philosophers, added to the luxurious ostentation of a court where displays of self-glorification affirmed the prince's title and power. In their turn, courtiers basked in the reflected glory of their patrons. Such an exchange is portrayed in the frontispiece to Johannes Kepler's (1571–1630) *Tabulae Rudolphinae* (Rudolphine Tables, 1627); here Emperor Rudolf II's imperial eagle drops talers from its beak and spreads protective wings over Kepler's "temple of astronomy."[5] The development of informal intellectual circles worked to the advantage of wellborn women whose high social standing allowed them to wield influence in the learned world, as it did in other domains of culture. Genteel women insinuated themselves into networks of learned men by exchanging patronage or public recognition for discourse with men of lesser rank but of significant intellectual stature.

Women in princely courts and the informal scientific circles that emerged from them served as important patrons, interlocutors, hostesses, and ready consumers of natural knowledge and curiosities – matters of import in an age when patronage often structured a naturalist's identity and career.[6] In the exchange characteristic of this system, Christina (1626–1689), queen of Sweden, invited Descartes to her court in the 1640s to serve as her tutor in natural philosophy and mathematics and to draw up regulations for her scientific academy. In the 1690s, Sophie Charlotte (1668–1705), electress of Brandenburg and later queen of Prussia, supported Leibniz in founding the

---

University Press, 1984), pp. 117–28; David Noble, *A World without Women: The Christian Clerical Culture of Western Science* (Oxford: Oxford University Press, 1993).

[5] I. Bernard Cohen, *Album of Science: From Leonardo to Lavoisier, 1450–1800* (New York: Scribner, 1980), p. 53, n. 68; Bruce Moran, ed., *Patronage and Institutions: Science, Technology, and Medicine at the European Courts, 1500–1750* (Rochester: Boydell Press, 1991); and Mario Biagioli, *Galileo, Courtier: The Practice of Science in the Culture of Absolutism* (Chicago: University of Chicago Press, 1993).

[6] On the creation of identities, see Stephen Greenblatt, *Renaissance Self-Fashioning: From More to Shakespeare* (Chicago: University of Chicago Press, 1980); on the economy of discourse characteristic of this period, see Anne Goldgar, *Impolite Learning: Conduct and Community in the Republic of Letters, 1680–1750* (New Haven, Conn.: Yale University Press, 1995), pp. 12–53.

Societas Regia Scientiarum, with its new astronomical observatory, in Berlin.[7] To my knowledge, however, no woman served as court philosopher; there was, in other words, no female Galileo, a client of princely patronage whose charge it was to plumb the depths of natural philosophy.[8] Although a few wellborn women, such as the Princess Elisabeth (1618–1680) of Bohemia, proved themselves acute natural and moral philosophers (as Elisabeth did in her correspondence with Descartes), most served as patrons rather than as producers of natural knowledge.

In the late seventeenth century, the scepter of learning passed from courtly circles to learned academies. Historians of science have identified the founding of Europe's scientific academies – the Accademia dei Lincei in Rome, the Accademia del Cimento in Florence, the Royal Society in London, and the Académie Royale des Sciences in Paris – as key steps in the emergence of modern natural knowledge.[9] These princely academies provided social prestige and often religious and political protection for the fledgling natural knowledge. State recognition of natural knowledge also coincided with a more stringent exclusion of women from scientific institutions.[10] This exclusion of women, however, was not a foregone conclusion and requires explanation.

The seventeenth-century scientific academies had their roots in two distinct traditions – the medieval university and the Renaissance court. Insofar as academies were rooted in universities, an explanation for women's exclusion is easily found in the traditions of those all-male institutions. It is also possible, however, to see scientific societies as descendants of courtly circles and the informal intellectual gatherings that emerged alongside them.[11] If we emphasize the continuities between scientific academies and Renaissance courtly culture – where women were active participants – it becomes more difficult to explain the exclusion of women from these academies.

Take the case of the Parisian Académie Royale des Sciences. Women joined in the informal *réunions*, salons, and scientific circles that flourished in late sixteenth- and early seventeenth-century Paris.[12] They gathered among the curious every Monday at Hermeticist Théophraste Renaudot's (1586–1653) Maison du Grand Coq on the Ile de la Cité in Paris to observe his

---

[7] Adolf von Harnack, *Geschichte der Königlich Preussischen Akademie der Wissenschaften zu Berlin* [1900], 3 vols. (Hildesheim: Georg Olms, 1970), 1: 124.

[8] At the French court, Christine de Pizan (ca. 1363–ca. 1431) wrote several commissioned works in the fifteenth century.

[9] David Lux, *Patronage and Royal Science in Seventeenth-Century France: The Académie de Physique in Caen* (Ithaca, N.Y.: Cornell University Press, 1989); and Alice Stroup, *A Company of Scientists: Botany, Patronage, and Community at the Seventeenth-Century Parisian Royal Academy of Sciences* (Berkeley: University of California Press, 1990).

[10] Joan Landes, ed., *Feminism, the Public and the Private* (Oxford: Oxford University Press, 1998).

[11] Frances Yates, *The French Academies of the Sixteenth Century* (London: Warburg Institute, 1947), p. 1; and Martha Ornstein, *The Role of Scientific Societies in the Seventeenth Century* (Chicago: University of Chicago Press, 1928). On women as cultural ambassadors, see Susan Groag Bell, "Medieval Women Book Owners: Arbiters of Lay Piety and Ambassadors of Culture," *Signs*, 7 (1982), 742–68.

[12] G. Bigourdan, "Les premières sociétés scientifiques de Paris au XVIIe siècle," *Comptes rendues de l'Académie des Sciences*, 163 (1916), 937–8.

experiments. Women were also present among the Cartesians, persons of "all ages, both sexes, and all professions," who gathered every Wednesday at Jacques Rohault's (1620–1675) home to watch him attempt to give an experimental base to Descartes' physics.[13] In the years preceding the founding of the Académie Royale des Sciences, women attended the Palais Précieux pour les Beaux Esprits des Deux Sexes and flocked to the salons of the Marquise de Sévigné (1626–1696) and the Duchess of Maine (1676–1753). The number of women attending informal academies grew at such a rate that Pierre Richelet (1626–1698) added the word *académicienne* to his famous dictionary in the 1680s, explaining that this was a new word signifying a person of the fair sex belonging to an academy of *gens de lettres*, coined on the occasion of the election of Madame des Houlières (1638–1694) to the Académie Royale d'Arles.[14]

Despite their prominence in informal scientific circles, women were not to become members of the Académie Royale des Sciences. Why not? Certain aspects of the French academic system could have encouraged the election of gentlewomen. Seventeenth-century academies perpetuated Renaissance traditions where learning mixed with elegance, adding grace to life and beauty to the soul. The Académie retained a conviviality in its program, with rules of etiquette and a routine of dinners and musical entertainment, all of which tended to blur the boundaries that would later separate the academies from the salons.[15] This was an atmosphere in which wellborn women might have flourished. At the same time, the Académie was monarchical and hierarchical. At its head sat twelve honorary nobles whose presence was largely ornamental; working naturalists – the new aristocracy of talent – found themselves on a lowlier rung. Yet noble birth was not enough to secure even an honorary place for women. The closed and formal character of the academy discouraged the election of women. Membership in the academy was a public, salaried position with royal protection and privileges.[16] Although a salaried position in itself might not preclude women – the illustrious Marie le Jars de Gournay (1565–1645), for example, received a modest *pension* from Richelieu until her death in 1645 – in the case of the Académie, with the membership limited to forty, the election of a woman would have displaced a man.

---

[13] Claude Clerselier, ed., *Lettres de Mr. Descartes* [1659], 6 vols. (Paris: Charles Angot, 1724), 2: preface. On Renaudot's gatherings, see Howard Solomon, *Public Welfare, Science, and Propaganda in Seventeenth Century France: The Innovations of Théophraste Renaudot* (Princeton, N.J.: Princeton University Press, 1972).
[14] Pierre Richelet, *Dictionnaire de la langue françoise, ancienne et moderne*, 3 vols. (Lyon, 1759), 1: 21.
[15] Harcourt Brown, *Scientific Organizations in Seventeenth-Century France, 1620–1680* (Baltimore: Williams and Wilkins, 1934); and Roger Hahn, *The Anatomy of a Scientific Institution: The Paris Academy of Science, 1666–1803* (Berkeley: University of California Press, 1971).
[16] Members supplemented the modest salary of 2,000 livres per year with private funds. Charles Gillispie, *Science and Polity in France at the End of the Old Regime* (Princeton, N.J.: Princeton University Press, 1980), pp. 81–2.

Women fared no better in England with the founding of the Royal Society of London in 1662. The Royal Society was open – at least ideologically – to a wide range of people. Thomas Sprat (1635–1713), the first historian of the society, emphasized that valuable contributions were to come from both learned and vulgar hands: "from the Shops of *Mechanicks*; from the Voyages of *Merchants*; from the Ploughs of *Husbandmen*; from the Sports, the Fishponds, the Parks, the Gardens of *Gentlemen*."[17] In fact the Royal Society never made good its claim to welcome men of all classes; the entrance fees and weekly dues alone discouraged those of humble means. Merchants and tradesmen comprised only four percent of the society's membership; the vast majority of the members (at least fifty percent in the 1660s) came from the ranks of gentlemen virtuosi, or wellborn connoisseurs of the new natural knowledge.[18] Considering that the Society relied for its monies on dues paid by members, the absence of noblewomen from the ranks of enthusiastic patrons is difficult to explain.

One woman in particular, Margaret Cavendish (1623–1673), Duchess of Newcastle, was a qualified candidate, having written some eight books on natural philosophy. Fellows of noble birth bestowed prestige upon the new Society; men above the rank of baron could become members without scientific qualifications. However, when Cavendish – a duchess – asked for nothing more than a visit, her request aroused great controversy. Her now famous visit took place in 1667. Robert Boyle prepared his "experiments of . . . weighing of air in an exhausted receiver; [and] . . . dissolving of flesh with a certain liquor."[19] The duchess, accompanied by her ladies, was much impressed and left (according to one observer) "full of admiration."[20] She did not, however, when asked, contribute funds to the Royal Society.[21]

Margaret Cavendish's one fleeting encounter with the men of London's Royal Society indeed appears to have set a precedent – a negative one: no woman was elected to full membership until 1945. This pattern did not hold uniformly across Europe. The Académie Royale des Sciences did not admit

---

[17] Thomas Sprat, *History of the Royal Society of London* (London: Printed by T. R. for J. Martyn and J. Allestry, 1667), pp. 62–3, 72, 435.

[18] The society required new members to pay an admittance fee of 10, and later 20, shillings. (Peers were required to pay £5.) Fellows were expected to pay a weekly subscription of 1 shilling. See Michael Hunter, *The Royal Society and Its Fellows, 1660–1700: The Morphology of an Early Scientific Institution* (Chalfont St. Giles: British Society for the History of Science, 1982), pp. 15, 24, tables 5–7.

[19] Thomas Birch, *History of the Royal Society*, 4 vols. (London: Printed for A. Millar, 1756–7), 2: 175.

[20] Samuel Pepys, *The Diary of Samuel Pepys*, ed. Robert Latham and William Matthews, 11 vols. (London: Bell, 1970–83), 8: 243. See also Samuel Mintz, "The Duchess of Newcastle's Visit to the Royal Society," *Journal of English and Germanic Philology*, 51 (1952), 168–76; Douglas Grant, *Margaret the First: A Biography of Margaret Cavendish, Duchess of Newcastle, 1623–1673* (London: Hart-Davis, 1957); Kathleen Jones, *A Glorious Fame: The Life of Margaret Cavendish, Duchess of Newcastle, 1623–1673* (London: Bloomsbury, 1988). For other women, see Lynette Hunter, "Sisters of the Royal Society: The Circle of Katherine Jones, Lady Ranelagh," in *Women, Science, and Medicine, 1500–1700*, ed. Lynette Hunter and Sarah Hutton (Gloucestershire: Sutton, 1997), pp. 178–97.

[21] Michael Hunter, *Establishing the New Science: The Experience of the Early Royal Society* (Woodbridge: Boydell Press, 1989), pp. 167, 171.

women until the 1970s, but Italian academies in Bologna, Padua, Rome, and elsewhere did admit a few accomplished women, such as Madeleine de Scudéry (1607–1701) in the seventeenth century, and Laura Bassi (1711–1778) and Emilie du Châtelet (1706–1749) in the eighteenth century. The Académie Royale des Sciences et Belles-Lettres in Berlin (as it was styled in the eighteenth century) also admitted honorary luminaries, including Catherine the Great of Russia (1729–1796) and Duchess Juliane Giovane, a poet and woman of letters.[22]

The focus of historians on academies has drawn attention away from another legitimate heir of courtly circles – the salons. In contrast with the massive public receptions of the Italian *saloni*, the French salons offer a unique example of intellectual institutions run by women. Featuring intimate intellectual gatherings in the sitting rooms of socially prominent women, these elegant gatherings of diverse character competed with academies for the attention of the learned. Like the French academies, the salons created cohesion among intellectual elites; Bernard de Fontenelle (1657–1757), for example, longtime secretary of the Académie Royale des Sciences, became *président* of Madame Lambert's (1647–1733) salon. They also played a crucial role in assimilating the rich and talented into the French aristocracy.[23] The discussion of natural knowledge – examination of the exact characteristics of the two chameleons sent to Scudéry by the consul of Alexandria in 1672, for example – was fashionable in Scudéry's salon as well as in the salons of Madame Rochefoucauld and Madame Tencin (1685–1749).[24] Despite their informal and private character, salons wielded influence in public matters: Women, such as Madame Lambert, served as intellectual power brokers at a time when natural knowledge was organized through highly personalized patronage systems.

---

[22] Kathleen Lonsdale and Marjory Stephenson were elected to the Royal Society in 1945 (*Notes and Records of the Royal Society of London*, 4 [1946], 39–40). See also Joan Mason, "The Admission of the First Women to the Royal Society of London," *Notes and Records of the Royal Society of London*, 46 (1992), 279–300. On du Châtelet, see Mary Terrall, "Emilie du Châtelet and the Gendering of Science," *History of Science*, 33 (1995), 283–310; and Terrall, "Gendered Spaces, Gendered Audiences: Inside and Outside the Paris Academy of Sciences," *Configurations*, 3 (1995), 207–32. On Bassi, see Paula Findlen, "Science as a Career in Enlightenment Italy: The Strategies of Laura Bassi," *Isis*, 84 (1993), 441–69; and Beate Ceranski, *"Und Sie Fürchtet sich vor Niemandem": Die Physikerin Laura Bassi, 1711–1778* (Frankfurt: Campus Verlag, 1996). See also Paula Findlen, "A Forgotten Newtonian: Women and Science in the Italian Provinces," in *The Sciences in Enlightened Europe*, ed. William Clark, Jan Golinski, and Simon Schaffer (Chicago: University of Chicago Press, 1999), pp. 313–49.

[23] Carolyn Lougee, *Le paradis des femmes: Women, Salons, and Social Stratification in Seventeenth Century France* (Princeton, N.J.: Princeton University Press, 1976), pp. 41–53; and Dena Goodman, *The Republic of Letters: A Cultural History of the French Enlightenment* (Ithaca, N.Y.: Cornell University Press, 1994), chap. 3.

[24] Madeleine de Scudéry wrote her *Histoire de deux caméléons* as a rebuttal to Claude Perrault's *Description anatomique d'un caméléon*. Her paper was eventually published in the Académie's *Mémoires pour servir à l'histoire naturelle des animaux* (Papers for a Natural History of Animals, 1671–6). See Erica Harth, *Cartesian Women: Versions and Subversions of Rational Discourse in the Old Regime* (Ithaca, N.Y.: Cornell University Press, 1992), pp. 98–110; Gillispie, *Science and Polity in France*, pp. 7, 94.

Salonnières experienced the same limits to their power as other highborn women in this period: They maneuvered to ensure the election of favored male candidates to prestigious posts, but not women. Because women were barred from the centers of scientific culture, such as the Royal Society of London and the Académie Royale des Sciences in Paris, their relationship to knowledge was inevitably mediated by a man, whether that man was their husband, companion, or tutor.

Some historians have taken the case of women as consumers of natural knowledge as the paradigmatic example of women's participation in natural inquiry. Yet relegating women to the status of hostess or amateur diminishes the contributions that women such as Maria Sibylla Merian (1647–1717) made to natural knowledge. Not all natural inquiry in early modern Europe was transacted within elite social settings. In the workaday world of artisanal workshops, women's contributions (like men's) depended less on learned discourse and more on practical innovations in illustrating, calculating, or observing.

ARTISANS

Sociologist Edgar Zilsel was among the first to point to the skills of "artist-engineers" as being central to the development of modern natural knowledge.[25] It has become commonplace to malign scholarship on artisans' contributions as the product of Marxist historiography (as indeed it was in the 1930s and 1940s). One might today, however, join scholarship in this area to laboratory studies (see Smith, Chapter 13, this volume). To be sure, gentlewomen such as Mary Sidney Herbert (1561–1621), Countess of Pembroke, built elaborate laboratories in their private residences and employed men of humbler origins, such as Adrian Gilbert, half-brother to Sir Walter Raleigh (1552–1618), as her "Laborator" to assist her in compounding household medicines, such as "Adrian Gilbert's Cordiall Water."[26] By the same token, princes welcomed court engineers and architects – men unskilled in learned discourse but with considerable technical expertise – to construct ostentatious gardens and waterworks and fabulous facades, and undertake other feats of artistic and technical virtuosity in improving fortifications and ballistics.[27] Independent craftsmen and women, who employed keen observational skills within household workshops, also secured an empirical base for fields such as astronomy and natural history. Women were at best bystanders in gentlemen's laboratories (even when present among the spectators,

---

[25] Edgar Zilsel, "The Sociological Roots of Science," *American Journal of Sociology*, 47 (1942), 545–6; Arthur Clegg, "Craftsmen and the Origin of Science," *Science and Society*, 43 (1979), 186–201.
[26] Margaret Hannay, "'How I These Studies Prize': The Countess of Pembroke and Elizabethan Science," in Hunter and Hutton, eds., *Women, Science, and Medicine*, pp. 108–21.
[27] William Eamon, "Court, Academy, and Printing House: Patronage and Scientific Careers in Late-Renaissance Italy," in Moran, ed., *Patronage and Institutions*, pp. 25–50, esp. pp. 31–2.

they – like the humble male "laborants" or "operators" – rarely featured among the "modest witnesses" whose signatures validated experiments in early modern England). Nonetheless they were prominent within artisanal workshops, especially on the Continent (see Cooper, Chapter 9, this volume).[28]

The new value attached to the traditional skills of artisans in this period helps explain the success women enjoyed as astronomers in this period. Between 1650 and 1710, some fourteen percent of astronomers in German lands were women (a higher percentage even than is true in Germany today).[29] Astronomy was never officially an organized guild, yet craft traditions that molded much of working life in early modern Europe were very much alive in the practices of astronomy, especially in Germany, the Low Countries, and parts of Poland. Astronomers, for example, derived income from artisanal activities, such as preparing popular almanacs and calendars – what Leibniz called "libraries for the common man." By choosing astronomers known for their calendar making and establishing a monopoly on the sale of calendars, the Royal Society of Sciences in Berlin hoped to capture this income for itself.[30]

Women's exclusion from universities set limits on their participation in astronomy; for instance, Maria Margarethe Winckelmann's (1670–1720) sighting of an important comet was attributed to her husband in part because she was not educated in Latin and could not easily publish her finding in the *Acta eruditorum*, then the leading journal for natural knowledge in German lands.[31] The actual work of observing the heavens, however, took place in this period largely outside the universities and was commonly learned under the watchful eye of a master. Gottfried Kirch (1639–1710), one of Germany's leading astronomers, for example, studied at Johannes Hevelius's (1611–1687) private observatory, built across the roofs of three adjoining houses in Danzig in 1640; this was as important for his astronomical career as his study of mathematics at the University of Jena.

Whereas men's work in the trades was typically regulated by their occupational status (apprentice, journeymen, master), women's was more commonly governed by their familial and marital status.[32] Trained by her father (or occasionally by her mother), a woman moved, in typical guild fashion,

---

[28] Steven Shapin, *A Social History of Truth: Civility and Science in Seventeenth-Century England* (Chicago: University of Chicago Press, 1994); Donna Haraway, *Modest_Witness@Second_Millennium* (New York: Routledge, 1997), pp. 29–32.
[29] This estimate is drawn from Joachim von Sandrart, *Teutsche Academie der edlen Bau-, Bild- und Mahlerey-Künste* (Frankfurt: J. P. Miltenberger 1675); Friedrich Lucae, *Schlesische Fürsten-Kron oder eigentliche, wahrhaffte Beschreibung Ober- und Nieder-Schlesiens* (Frankfurt am Main: Knoch 1685); Frederick Weidler, *Historia astronomiae* (Wittenberg: Gottlieb Heinrich Schwartz, 1741).
[30] Harnack, *Geschichte der Königlich Preussischen Akademie der Wissenschaften*, 1: 48–9.
[31] Schiebinger, *The Mind Has No Sex?* pp. 82–98.
[32] Margaret Wensky, *Die Stellung der Frau in der stadtkölnischen Wirtschaft im Spätmittelalter* (Cologne: Bohlau, 1981); and Merry Wiesner, *Working Women in Renaissance Germany* (New Brunswick, N.J.: Rutgers University Press, 1986).

from being an assistant to her father to becoming an assistant to her husband. Elisabetha Koopman of Danzig (1647–1693), like other women in this period, wed with care to ensure her place in astronomy. In 1663, she married a leading astronomer Hevelius, a man thirty-six years her elder. Hevelius, a brewer by trade, took over the lucrative family beer business in 1641. His first wife, Catherina Rebeschke (1613–1662), had managed the household brewery, leaving him free to serve in city government and to pursue his avocation, astronomy. When she died in 1662, Hevelius married Koopman, who had been interested in astronomy for many years. In appropriate guild fashion, Elisabetha Hevelius served as chief assistant to her husband in both the family business and the family observatory. In her pathbreaking work, Margaret Rossiter has described "women's work" in nineteenth- and twentieth-century science (and especially in astronomy) as typically involving tedious computation, lifelong service as an assistant, and the like – all of which are a legacy of the guild wife.[33] The role of the guild wives, however, cannot be collapsed into that of a mere assistant; wives were of such import to production that every guild master, at least in Germany, was required by law to have one.[34] The very different structure of the workplace in the early modern period allowed the wife a more comprehensive role. For twenty-seven years, Elisabetha Hevelius collaborated with her husband, observing the heavens in the cold of night by his side.[35]

COLONIAL CONNECTIONS

Historians have lavished attention on universities, princely courts, scientific academies, salons, and even artisanal workshops as loci of intellectual ferment in early modern Europe. Today, new attention is being brought to bear on another aspect of early modern natural knowledge – overseas exploration. In this context, domestic and colonial botanical gardens (and later menageries and natural history museums) served as displays of princely élan, experimental stations for economic and medicinal horticulture, collection points for voyagers, and, last but not least, innovative institutions of the new natural history.[36] One could argue that the opening of the Jardin Royal des Plantes

---

[33] See Margaret Rossiter, "'Women's Work' in Science, 1880–1910," *Isis*, 71 (1980), 381–98; and Rossiter, *Women Scientists in America: Struggles and Strategies to 1940* (Baltimore: Johns Hopkins University Press, 1982), pp. 51–72.
[34] Merry Wiesner, "Women's Work in the Changing City Economy, 1500–1650," in *Connecting Spheres: Women in the Western World, 1500 to the Present*, ed. Marilyn Boxer and Jean Quataert (New York: Oxford University Press, 1987), pp. 64–74, esp. p. 66.
[35] After her husband's death, Elisabetha Hevelius edited and published their joint works: *Catalogus stellarum fixarum* (1687); *Firmamentum Sobiescianum* (1690), containing fifty-six star maps; and *Prodromus astronomiae* (1690), a catalogue of 1,564 stars and their positions.
[36] Lucile Brockway, *Science and Colonial Expansion: The Role of the British Royal Botanic Gardens* (New York: Academic Press, 1979); Alfred Crosby, *Ecological Imperialism: The Biological Expansion of Europe, 900–1900* (New York: Cambridge University Press, 1986); Nicholas Jardine, James A. Secord,

Médicinales (Jardin des Plantes) in Paris in 1635 was as important to the new natural knowledge as the founding of the much celebrated Académie Royale des Sciences. Plants were shipped from abroad to gardens in Paris, Pisa, Leiden, Montpellier, Heidelberg, and elsewhere in attempts to create a microcosm of the world's flora for the purposes of acclimatizing useful medical herbs, identifying profitable woods and agricultural plants, satisfying popular demand for ornamental exotics, and developing classification schemes on a global basis.

Europeans making forays into nature and foreign scientific traditions in the sixteenth and seventeenth centuries came from varied backgrounds. Jesuit missionaries served as major conduits for scientific knowledge into Europe (though Protestants were often suspicious of knowledge so transmitted, as was the case with quinine, originally known as "Jesuits' Bark").[37] Physicians such as Paul Hermann (1640–1695) collected as they served in various parts of the world for the various India companies; Hermann later became a professor of botany at the university in Leiden. Even merchants, such as Jakob Breyne (1637–1697), occasionally joined the frenzied exchange of exotic plant and animal stuffs characteristic of this period.

Female naturalists, however, rarely figured in Europe's rush to know exotic lands. Moral and bodily imperatives discouraged women from voyaging to unknown lands; physicians warned that white women taken to very warm climates succumbed to "copious menstruation, which almost always ends, in a short space of time, in fatal hemorrhages of the uterus."[38] There was also the often-expressed fear that women giving birth in the tropics would deliver children resembling the native peoples of those areas.[39]

The German-born Maria Sibylla Merian was one of the few women who undertook her own course of study (of insects) and traveled independently

---

and Emma C. Spary, eds., *Cultures of Natural History* (Cambridge: Cambridge University Press, 1996); David Miller and Peter Reill, eds., *Visions of Empire: Voyages, Botany, and Representations of Nature* (Cambridge: Cambridge University Press, 1996); Marie-Noëlle Bourguet and Christophe Bonneuils, eds., *De l'inventaire du monde à la mise en valeur du globe: Botanique et colonisation*, special issue, *Revue française d'histoire d'Outre-Mer*, 86 (1999); Tony Rice, *Voyages: Three Centuries of Natural History Exploration* (London: Museum of Natural History, 2000); Emma C. Spary, *Utopia's Garden: French National History from the Old Regime to Revolution* (Chicago: University of Chicago Press, 2000); Richard Drayton, *Nature's Government: Science, Imperial Britain, and the "Improvement" of the World* (New Haven, Conn.: Yale University Press, 2000); Roy MacLeod, ed., *Nature and Empire: Science and the Colonial Enterprise*, special issue, *Osiris*, 15 (2000); Pamela Smith and Paula Findlen, eds., *Merchants and Marvels: Commerce, Science, and Art in Early Modern Europe* (New York: Routledge, 2002); and Londa Schiebinger and Claudia Swan, eds., *Colonial Botany: Science, Commerce, and Politics in the Early Modern World* (Philadelphia: University of Pennsylvania Press, 2005).

[37] Cromwell considered Peruvian bark a "Popish remedy." Saul Jarcho, *Quinine's Predecessor: Francesco Torti and the Early History of Cinchona* (Baltimore: Johns Hopkins University Press, 1993), p. 46.

[38] Johann Blumenbach, *The Natural Varieties of Mankind* [1795], trans. Thomas Bendyshe [1865] (New York: Bergman, 1969), p. 212, n. 2. Blumenbach codified notions long current in Europe.

[39] Marie Helene Huet, *Monstrous Imagination* (Cambridge, Mass.: Harvard University Press, 1993); Londa Schiebinger, *Nature's Body: Gender in the Making of Modern Science* (Boston: Beacon Press, 1993).

in pursuit of natural history in this period. The daughter of the well-known artist Matthäus Merian the elder, Merian had been trained from an early age in the workshop of her stepfather (her own father died shortly after her birth) in the arts of illustration and copper-plate engraving.[40] In 1665, she married Johann Andreas Graff, one of her stepfather's favorite pupils. The couple set up their own household workshop in Nuremberg, where her husband published Maria Sibylla (now) Graffin's *Blumenbuch* (Book of Flowers, 1675–1680), a collection of illustrations to be used as patterns for artists and embroiderers.

In 1699, having left her husband and reclaimed her father's famous name, Merian set sail for Surinam, then a Dutch colony. She had some connections to Surinam through her merchant son-in-law and the Labadists, an experimental religious community with holdings in both the Netherlands and its colonies. She was not, however, schooled, as the great Joseph Pitton de Tournefort (1656–1708) had been, to be sent into the field, nor had she been commissioned to make the journey by a trading company, scientific society, or Crown as were many of the naturalists in this period. Her interest was self-generated and largely self-supported, part of her lifelong quest to find another variety of caterpillar as economically significant as the silkworm. For two years, she collected, studied, and drew the insects and plants of the region.[41]

Despite her rarity as a female naturalist, Merian's practices in the field were by and large similar to those of her male colleagues. Like Hans Sloane (1660–1753), physician to the English governor in Jamaica from 1687 to 1689 and future president of London's Royal Society, she was keen to collect from the local inhabitants the best information concerning the exotic plants and insects she encountered.[42] Like the German astronomer Peter Kolb (1675–1726), who wrote an early ethnology of the Africans at the Cape of Good Hope, Merian developed deep friendships with several Amerindians and displaced Africans in Surinam who served as her guides to desirable specimens and provided access to dangerous, often impassible regions.[43] Like the men, Merian had

---

[40] Women had long been active as illustrators; nuns had illuminated manuscripts, and other women were active members of painters' guilds. See Ann Sutherland Harris and Linda Nochlin, *Women Artists, 1550–1950* (Los Angeles: Los Angeles County Museum of Art, 1976); Madeleine Pinault, *The Painter as Naturalist: From Dürer to Redouté*, trans. Philip Sturgess (Paris: Flammarion, 1991), pp. 43–6.

[41] On Merian, see Elisabeth Rücker, *Maria Sibylla Merian, 1647–1717* (Nuremberg: Germanisches Nationalmuseum, 1967); Margarete Pfister-Burkhalter, *Maria Sibylla Merian: Leben und Werk, 1647–1717* (Basel: GS-Verlag, 1980); Natalie Zemon Davis, *Women on the Margins: Three Seventeenth-Century Lives* (Cambridge, Mass.: Harvard University Press, 1995); Helmut Kaiser, *Maria Sibylla Merian: Eine Biographie* (Düsseldorf: Artemis und Winkler, 1997); and Kurt Wettengl, ed., *Maria Sibylla Merian, 1647–1717: Artist and Naturalist*, trans. John Southard (Ostfildern: G. Hatje, 1998).

[42] Hans Sloane, *A Voyage to the Islands Madera, Barbadoes, Nieves, St. Christophers, and Jamaica; with the Natural History . . .* , 2 vols. (London: Printed by B. M. for the author, 1707–25); and Maria Sibylla Merian, *Metamorphosis insectorum Surinamensium* [1705], ed. Helmut Decker (Leipzig: Insel-Verlag A. Kippenberg, 1975), introduction, p. 38.

[43] Peter Kolb, *The Present State of the Cape of Good Hope*, trans. Guido Medley (London: W. Innys, 1731).

assistants: her twenty-one-year-old daughter, whom she had trained, and her slaves, who served as her guides and hacked paths for her through dense "thorns and thistles."[44] Merian also followed the practice common up to that time of retaining native names and recording much else that native peoples told her about the plants and animals she studied. In the introduction to her celebrated *Metamorphosis insectorum Surinamensium* (Metamorphosis of the Insects of Surinam, 1705), which she advertised as the "first and strangest work done in America," she wrote: "The names of the plants I have kept as they were given by the natives and Indians in America."[45]

Although Merian's homespun enterprise was similar in many respects to those of a number of male naturalists, such as Sloane and others working in the Caribbean, it contrasts sharply with that of Hendrick van Reede tot Drakenstein (1636–1691), a military man and colonial administrator, whose interest in botany was driven by the need to protect his troops from beriberi, dysentery, cholera, jaundice, malaria, and other tropical diseases. (Merian and van Reede are linked for posterity through Carolus Linnaeus's, 1707–1778, contempt of the botanical nomenclatures of both.)[46] As governor of Malabar for the Dutch East India Company from 1670 to 1677, van Reede produced a magisterial twelve-volume opus, *Hortus Malabaricus* (Flora of Malabar, 1678–1693), describing 740 plants of the region. To compile his complex text, van Reede coordinated the efforts of at least twenty-five men from many distinct cultures, castes, and classes, and two continents.[47] Only an administrator of van Reede's stature could command the necessary resources, contacts, and personnel to mount a project of this magnitude.

The negotiation between European and exotic natural knowledge traditions is a complicated story that remains to be told. In many instances, indigenous informants included unlettered women who passed along hard-won knowledge to lettered naturalists who, by systematizing it, were able to make previously local knowledge more universally available. Historian Richard Grove has claimed that some of the collecting and much of the cataloguing for Garcia da Orta's (1500–1568) well-known *Coloquios dos simples e drogas . . . da India* (Colloquies on the Simples and Drugs of India, 1563), for example, was done by a Konkani slave girl known only as Antonia.[48] Charles Clusius (1526–1609; also da Orta's translator) praised country "women root

---

[44] Merian, *Metamorphosis*, commentary to plate no. 36.
[45] Merian, *Metamorphosis*, introduction, p. 38. See also Londa Schiebinger, *Plants and Empire: Colonial Bioprospecting in the Atlantic World* (Cambridge, Mass.: Harvard University Press, 2004).
[46] Carolus Linnaeus, *Critica botanica* (Leiden: Conrad Wisshoff, 1737), no. 218.
[47] Hendrik Adriaan van Reede, *Hortus Malabaricus* (Amsterdam: Johan van Someren and Johan van Dyck, 1678–93); van Reede provided an extensive description of how the text was compiled in vol. 3, pp. iii–xviii. See also J. Heniger, *Hendrik Adriaan van Reede tot Drakenstein and Hortus Malabaricus* (Rotterdam: Balkema, 1986); and K. S. Manilal, ed., *Botany and History of Hortus Malabaricus* (Rotterdam: Balkema, 1980).
[48] Richard Grove, *Green Imperialism: Colonial Expansion, Tropical Island Edens, and the Origins of Environmentalism, 1600–1860* (Cambridge: Cambridge University Press, 1995), p. 81.

cutters" (*rhizotomae mulierculae*), who supplied him with information about the medical properties of plants indigenous to his own country.[49] Herman Busschof (1625–1672), working for the Dutch East India Company in Batavia, wrote a treatise on an "Indian Doctress" who provided an ingenious cure for his troublesome gout.[50] Women's role in the voyages of scientific exploration is an area where research remains to be done.

The more fluid state of scientific culture in early modern Europe left room for innovation. New institutions and new calls for equality provided openings in intellectual culture that allowed a few women to contribute to the making of natural knowledge. Although women did not fare well in traditional institutions of learning, such as universities, they had a foothold (however tenuous) in courtly circles, learned salons, artisanal workshops, and other settings fostering the emergence of modern science. The sixteenth and seventeenth centuries saw a number of women studying the medicinal qualities of plants, collecting exotic insects, and studying the movements of the heavens. In many instances, their efforts were supported by natural philosophers – Descartes, Poullain de la Barre, and Leibniz among them. Sustained negotiations over sites and boundaries in this period set the stage for women to work at the margins of Enlightenment science – before the twentieth century, one of the high tides of women's contributions to natural knowledge.

---

[49] Charles de l'Ecluse, *Rariorum aliquot stirpium, per Pannoniam, Austriam, et vicinas . . . historia* (Antwerp: C. Plantin, 1583), p. 345. See also Jerry Stannard, "Classici and Rustici in Clusius' *Stirp. Pannon. Hist.* (1583)," *Festschrift anlässlich der 400 jährigen Wiederkehr der wissenschaftlichen Tätigkeit von Carolus Clusius (Charles de l'Escluse) im pannonischen Raum*, ed. Stefan Aumüller (Burgenländische Forschungen herausgegeben vom Burgenländischen Landesarchiv, Sonderheft 5) (Eisenstadt: Amt der Bürgenländischen Landesregierung, Landesarchiv, 1973), pp. 253–69. I thank Claudia Swan for this reference.

[50] Herman Busschof, *Two Treatises* (London: Printed by H. C. and are to be sold by Moses Pitts, 1676). I thank Roberta Bivins for this reference.

8

MARKETS, PIAZZAS, AND VILLAGES

William Eamon

Long before natural objects became subjects for experimental study in the laboratory, they had been commodities traded in the marketplace. In the early modern period, as European merchant vessels ventured far beyond the Mediterranean, this marketplace expanded rapidly, thereby increasing the variety and geographical diversity of the commodities traded therein.[1] These changes were vividly reflected in the stockpiling of goods in warehouses for wholesale trade and in the accumulation of exotic natural and artificial objects in museums and cabinets of curiosities. From the gigantic warehouses of Amsterdam and the Hague to the bustling ports of Marseille and Venice, early modern collectors busily gathered specimens of exotic flora and fauna, shells, coral, and other objects from distant parts of the world.

The dramatic increase in the pace of trade, population growth, and the rise of credit all led to an expansion of the distribution network: in particular, to a rise in the number and variety of shops. In 1606, Lope de Vega wrote of Madrid, "*Todo se ha vuelto tiendas*" ("Everything has turned into shops"), while Daniel Defoe lamented that shops in seventeenth-century London had spread "monstrously."[2] The boom in shopkeeping not only increased the diversity of items available to consumers but also created spaces for conversation and for gaining information about natural and manufactured goods. In the early modern period, craftsmen's shops were also workshops and were thus important sources of natural and technological information.

Echoing humanist educational ideals, the young Gargantua of Rabelais's *La vie très horrifique du grand Gargantua, père de Pantagruel* (The Most Horrific Life of the Great Gargantua, Father of Pantagruel, 1534) visited jewelers, goldsmiths, alchemists, weavers, dyers, instrument makers, and other

---

[1] Fernand Braudel, *The Wheels of Commerce*, trans. Siân Reynolds (New York: Harper and Row, 1982), chap. 1. See in general Pamela H. Smith and Paula Findlen, eds., *Merchants and Marvels: Commerce, Science, and Art in Early Modern Europe* (New York: Routledge, 2002).
[2] For the boom in shopkeeping, see Braudel, *Wheels of Commerce*, pp. 68–75.

craftsmen to learn about the properties of things. To study the nature of herbs, he "visited the druggists' shops, the herbalists, and the apothecaries, and carefully examined the fruit, roots, leaves, gums, seeds, and foreign ointments."[3]

The growing recognition of the marketplace as a site of natural knowledge signaled important shifts in the definition of knowledge and of who might qualify as natural knowers. It also raised fundamental questions, never fully resolved in the early modern period, about whose knowledge was considered valid and authoritative. Some sixteenth-century humanists and natural philosophers maintained that a truer understanding of nature might be gained in markets and workshops than from books.[4] In the early seventeenth century, Francis Bacon (1561–1626) made that claim a central tenet of his philosophical program, which was known as the *Instauratio Magna* (Great Renewal). Despite the popularity of Bacon's philosophy in mid-seventeenth-century England, during the latter part of the century natural philosophers as a whole became less receptive to craft knowledge as an avenue to knowing nature. The failure of the history of trades project of the Royal Society of London signaled the abandonment of its original Baconian vision of uniting natural philosophy and the arts. Nevertheless, the assimilation of artisanal knowledge into natural philosophy left a lasting legacy in the impetus it gave to the census of natural objects and to the experimental investigation of their properties.

MARKETS AND SHOPS

The global enterprises of early modern merchant capitalists stimulated interest in the natural history of Asia, Africa, the Middle East, and the New World. Venice's unique position as an entrepôt of trade with Constantinople, Syria, and Egypt made the city an unrivaled center of pharmaceutical research. In the 1540s, the botanist Pietro Andrea Mattioli urged the Venetian Senate to procure from everywhere its galleys sailed the herbs, liquors, and minerals needed to prepare classical drugs.[5] The search for the ingredients of the ancient drug theriac, the universal antidote first described by Galen, is characteristic of the manner in which capitalism, together with the new humanist focus on ancient sources, contributed to the reform of botany

---

[3] François Rabelais, *The Histories of Gargantua and Pantagruel*, trans. J. M. Cohen (London: Penguin Books, 1955), p. 93.

[4] Paolo Rossi, *Philosophy, Technology, and the Arts in the Early Modern Era*, trans. S. Attanasio (New York: Harper and Row, 1970), p. 3 ff.; Edgar Zilsel, "The Sociological Roots of Science," *American Journal of Sociology*, 47 (1941/2), 544–62. See also the essays in Smith and Findlen, eds., *Merchants and Marvels*.

[5] Pier Andrea Mattioli, *I discorsi nei sei libri di Pedacio Dioscoride* (Venice: Vincenzo Valgrisi, 1559), introductory epistle.

and materia medica (see Findlen, Chapter 19, this volume).⁶ Also, the brisk trade in exotic naturalia made available to the medical community an entire new body of medicinal drugs.⁷ With the new materia medica, physicians and apothecaries even claimed to be able to compose medicines that would succeed where traditional drugs failed. In Germany, the Fugger commercial house established a flourishing trade in guaiacum, a popular treatment for syphilis that was made from *Guaiacum officinale*, a New World tree. Because of the phenomenal demand for guaiacum, as well as the Fuggers' virtual monopoly on the drug, the "Holy Wood" was said to have sold for as much as seven gold crowns per pound.⁸

In exploiting the possibilities of the global economy, the European states facilitated the introduction of new natural knowledge. Philip II of Spain, motivated by a desire to realize the economic potential of his vast American empire, solicited information about the geography and natural history of the New World. In 1571, he appointed Juan López de Velasco to the newly created post of cosmographer-chronicler of the Indies, instructing him to compile maps, cosmographic tables, records of tides and eclipses, and a natural history of the Indies. To obtain this information, López de Velasco drew up a questionnaire that was circulated in 1577 to the local councils in New Spain. The "Relaciones Geográficas" ("Geographic Reports") gave Philip detailed information about a vast empire that the emperor himself was unable to see. Such projects to "make visible" distant, invisible worlds, although motivated by political and economic interests, coincided with the new philosophy's aim of bringing to light invisible secrets.⁹ In the mid-seventeenth century, the English virtuoso Joseph Glanvill (1636–1680) proclaimed that opening up an "America of secrets and an unknown Peru of nature" was "the noble end of true Philosophy."¹⁰

---

⁶ Giuseppe Olmi, "Farmacopea antica e medicina moderna: La disputà sulla teriaca nel Cinquecento bolognese," *Physis*, 19 (1977), 197–245. In addition, see Richard Palmer, "Medical Botany in Northern Italy in the Renaissance," *Journal of the Royal Society of Medicine*, 78 (1985), 149–57; Karen Reeds, *Botany in Medieval and Renaissance Universities* (New York: Garland Press, 1991).

⁷ José María Lopez-Piñero and José Pardo Tomás, *La influencia de Francisco Hernández (1515–1587) en la constitución de la botánica y la materia médica modernas* (Cuadernos Valencianos de Historia de la Medicina y de la Ciencia, 51) (Valencia: University of Valencia/C.S.I.C., 1996); José Pardo Tomás and María Luz López Terrada, *Las primera noticias sobre plantas americanas en las relaciones de viajes y crónicas de Indias (1493–1553)* (Cuadernos Valencianos de Historia de la Medicina y de la Ciencia, 40) (Valencia: University of Valencia/C. S. I.C., 1996); and Simon Varey, ed., *The Mexican Treasury: The Writings of Dr. Francisco Hernández* (Stanford, Calif.: Stanford University Press, 2000).

⁸ Robert S. Munger, "Guaiacum, the Holy Wood from the New World," *Journal of the History of Medicine and Allied Sciences*, 4 (1949), 196–229.

⁹ David C. Goodman, *Power and Penury: Government, Technology and Science in Philip II's Spain* (Cambridge: Cambridge University Press, 1988); and Barbara E. Mundy, *The Mapping of New Spain: Indigenous Cartography and the Maps of the Relaciones Geográficas* (Chicago: University of Chicago Press, 1996).

¹⁰ Joseph Glanvill, *The Vanity of Dogmatizing* (London: Printed by H. C. for Henry Eversden, 1661), p. 178.

To take advantage of anticipated future demand, merchant capitalists stockpiled goods in gigantic warehouses.[11] Historians have argued that the commercial practice of accumulating material objects facilitated the growth of natural history collections. Some evidence supports this claim. The Augsburg merchant and collector Philipp Hainhofer bought shells for his collection from Dutch merchants at Frankfurt fairs, and the port of Marseille supplied collectors with coral, shells, and art objects from the East and West Indies.[12] It was no coincidence that the accumulation of naturalia in museums and cabinets of curiosities became fashionable just as the market economy began to flourish in Western Europe. The curiosity cabinets so typical of the Mannerist age testify as much to the acquisitiveness as to the curiosity of the early modern period.[13] Although the tulip mania of 1636–7 in The Netherlands was an extreme example of speculation in exotic naturalia, it illustrates how the new economy brought about a "culture of collecting."[14]

Among nature's more astute observers were the craftsmen, shopkeepers, and vendors who processed and traded in natural objects in workshops and market squares. Prior to the sixteenth century, natural philosophers rarely sought out the expertise of such "experienced" persons. To the medieval scholastics, as Peter Dear has noted (Chapter 4, this volume), experience meant common knowledge that depended on the senses, but its truth did not rest on particular instances. By definition, singular events did not reveal how nature behaves but instead were considered as anomalies or even as "marvels."[15] In the early modern period, however, the meaning of experience changed. Instead of denoting general empirical statements (e.g., "Heavy bodies fall"), experience tended to signify specific, often unique descriptions of

---

[11] Braudel, *The Wheels of Commerce*, pp. 94–7.
[12] Lorraine Daston and Katharine Park, *Wonders and the Order of Nature, 1150–1750* (Cambridge, Mass.: Zone Press, 1998), p. 265; Hans-Olof Böstrom, "Philipp Hainhofer and Gustavus Adolphus' Kunstschrank," in *The Origins of Museums: The Cabinet of Curiosities in Sixteenth- and Seventeenth-Century Europe*, ed. Oliver Impey and Arthur MacGregor (Oxford: Clarendon Press, 1985), pp. 90–101; Antoine Schnapper, *Le géant, la licorne et la tulipe: Collections et collectionneurs dans la France du XVIIe siècle* (Paris: Flammarion, 1988), pp. 220–1.
[13] Harold J. Cook, "The Moral Economy of Natural History and Medicine in the Dutch Golden Age," in *Contemporary Explorations in the Culture of the Low Countries*, ed. William Z. Shetter and Inge Van der Cruysse (Publications of the American Association of Netherlandic Studies, 9) (Lanham Md.: American Association of Netherlands Studies, 1996), pp. 39–47; and Pamela Smith, *The Business of Alchemy: Science and Culture in the Holy Roman Empire* (Princeton, N.J.: Princeton University Press, 1994). On collecting, see Paula Findlen, *Possessing Nature: Museums, Collecting, and Scientific Culture in Early Modern Italy* (Berkeley: University of California Press, 1994); Findlen, "The Economy of Scientific Exchange in Early Modern Italy," in *Patronage and Institutions: Science, Technology, and Medicine at the European Court, 1500–1750*, ed. Bruce T. Moran (Bury St. Edmunds: Boydell Press, 1991), pp. 5–24; Impey and MacGregor, eds., *The Origins of Museums*; and Daston and Park, *Wonders and the Order of Nature*, chaps. 4 and 7.
[14] Simon Schama, *The Embarrassment of Riches: An Interpretation of Dutch Culture in the Golden Age* (Berkeley: University of California Press, 1988), pp. 350–66.
[15] Lorraine Daston, "Baconian Facts, Academic Civility, and the Prehistory of Objectivity," *Annals of Scholarship*, 8 (1991), 337–63.

natural phenomena.[16] Although he followed Aristotle in maintaining that there could be no certain knowledge of particulars, the humanist Jean Bodin (1530–1596) urged naturalists to "examine the treasures of the singular things of nature."[17] In early modern natural philosophy, knowledge of experiential "facts," in the sense of "nuggets of experience detached from theory," took on heightened significance.[18]

Like nature itself, the marketplace was a repository of empirical knowledge. In his 1586 autobiography, the Bolognese naturalist Ulisse Aldrovandi reported that a visit to the fish markets of Rome in 1549–50 was the critical event that sparked his interest in natural history. Just as René Descartes observed and dissected animals in the Paris butcher shops a century later, Aldrovandi went to fishmongers to learn the names, habits, and characteristics of fish.[19] Similarly, early modern natural philosophers looked to the workshops of perfumers, metallurgists, jewelers, dyers, and other craftsmen for trade secrets that might lead to an understanding of the properties of matter. Vannoccio Biringuccio, a Sienese mine supervisor, wrote in 1540 that goldsmiths possessed important alchemical secrets.[20] William Gilbert, the author of an important treatise, *De magnete* (On the Magnet, 1600), acknowledged his debt to artisans and instrument makers, and Robert Boyle (1627–1691) credited dyers for the information that led him to the discovery of chemical color indicators.[21]

The shops of apothecaries and distillers were also sites of natural knowledge. In the second half of the sixteenth century, the pharmacies of Francesco Calzolari in Verona and Ferrante Imperato in Naples had natural history museums where the virtuosi gathered to view and discuss rarities.[22] Pharmacists also operated distilleries. Giorgio Melichio ran one at the Struzzo pharmacy in Venice, and the surgeon Leonardo Fioravanti concocted distilled drugs in the Orso pharmacy in Campo Santa Maria Formosa. Physicians relied upon pharmacists and distillers for information about medicinal herbs.[23] The Dominicans operated a distillery in the Campo Frari

---

[16] Peter Dear, *Discipline and Experience: The Mathematical Way in the Scientific Revolution* (Chicago: University of Chicago Press, 1995), pp. 13–14.
[17] Jean Bodin, *Universae naturae theatrum* [1596], quoted in Ann Blair, *The Theater of Nature: Jean Bodin and Renaissance Science* (Princeton, N.J.: Princeton University Press, 1997), p. 99.
[18] Lorraine Daston, "The Factual Sensibility," *Isis*, 79 (1988), 452–67, at p. 465.
[19] René Descartes, *Treatise of Man*, trans. Thomas S. Hall (Cambridge, Mass.: Harvard University Press, 1972), pp. xii–xiii; and Findlen, *Possessing Nature*, p. 175.
[20] Vannoccio Biringuccio, *Pirotechnia*, trans. Cyril Stanley Smith and Martha Teach Gnudi (Cambridge, Mass.: MIT Press, 1959; orig. publ. 1942), p. 367. More generally on artisans as observers of nature, see Pamela H. Smith, *The Body of the Artisan: Art and Experience in the Scientific Revolution* (Chicago: University of Chicago Press, 2004).
[21] Boyle published his discovery in his *Experiments and Considerations Touching Colours* (1664); see William Eamon, "New Light on Robert Boyle and the Discovery of Colour Indicators," *Ambix*, 27 (1980), 204–9. On Gilbert, see Edgar Zilsel, "The Origins of William Gilbert's Scientific Method," *Journal of the History of Ideas*, 2 (1941), 1–32.
[22] Paula Findlen, *Possessing Nature*, pp. 31–2, 245–6.
[23] Richard Palmer, "Pharmacy in the Republic of Venice in the Sixteenth Century," in *The Medical Renaissance of the Sixteenth Century*, ed. Andrew Wear, R. K. French, and Ian M. Lonie (Cambridge: Cambridge University Press, 1985), pp. 100–17.

in Venice that was a bustling center of medical discussion. Lay people also took up the trade. In Germany, the aquavit distilleries (*Wasserbrennereien*) began as home industries run chiefly by women. Despite attempts by some city governments to curb the trade, aquavit women continued to make products in simple kitchen stills.[24] Whereas university scholars felt the constraints of theological condemnations of alchemy, artisans used alchemical techniques such as distillation freely, unconcerned with philosophical justifications or apologies (see Newman, Chapter 21, this volume). Indeed, the sixteenth-century humanist and naturalist Conrad Gessner reported that empirics were more receptive than the physicians to distilling drugs.[25]

Competition transformed the medical economy no less than the broader economy. As competition intensified among the providers of medical services, the demand for certain kinds of services accelerated, particularly for "specifics" (remedies for particular illnesses) as opposed to the complex regimens of diet, conduct, and hygiene traditionally prescribed by university-trained physicians. Changing fashions created a demand for new drugs. "Paracelsian" remedies, distilled products, and medical "secrets" advertised in broadsides and popular medical tracts became increasingly fashionable, and empirics and charlatans invaded the piazzas to vend their nostrums. Fioravanti exploited the market for medical fashions by giving his "secrets" catchy trade names such as "angelic electuary" (*elettuario angelico*), "blessed oil" (*olio benedetto*), and *dia aromatica*, the "fragrant goddess" he prescribed as the first course of action against almost every ailment he encountered.[26] As empirics proliferated, the official medical community grew more vigilant in regulating them.[27] In theory, the conflict between the learned physicians and the empirics involved two diametrically opposed medical ontologies and healing strategies. The "high" medical tradition regarded disease as an imbalance of bodily humors, the remedy for which was to preserve and restore a proper humoral balance. For empirics, on the other hand, disease was a sickness or evil (*male*); healing was an active intervention with specific remedies to attack the disease and drive it out.[28] In practice, however, it was not

---

[24] Activities at the Frari distillery are revealed in depositions relating to the trial of Fra Antonio Volpe, Archivio di Stato, Venice, Sant'Uffizio, busta 23. For Germany, see Robert James Forbes, *A Short History of the Art of Distillation* (Leiden: E. J. Brill, 1948), pp. 90–1, 102–3.

[25] Conrad Gessner, *Schatz* [1582], quoted by Joachim Telle in *Bibliotheca Palatina: Katalog zur Ausstellung vom 8. Juli bis 2. November 1986, Heiliggeistkirche Heidelberg*, ed. Elmar Mittlar, 2 vols. (Heidelberg: Braus, 1986), 1: 337. In addition, see Joachim Telle, ed., *Pharmazie und der gemeine Mann: Hausarznei und Apotheke in deutschen Schriften der frühen Neuzeit* (Austellungskataloge der Herzog August Bibliothek, 36) (Braunschweig: Waisenhaus, 1982).

[26] William Eamon, "'With the Rules of Life and an Enema': Leonardo Fioravanti's Medical Primitivism," in *Renaissance and Revolution: Humanists, Scholars, Craftsmen, and Natural Philosophers in Early Modern Europe*, ed. J. V. Field and F. A. J. L. James (Cambridge: Cambridge University Press, 1993), pp. 29–44.

[27] David Gentilcore, "'Charlatans, Mountebanks, and Other Similar People': The Regulation and Role of Itinerant Practitioners in Early Modern Italy," *Social History*, 20 (1995), 297–314.

[28] David Gentilcore, *Healers and Healing in Early Modern Italy* (Manchester: Manchester University Press, 1998), p. 182. In addition, see Gianna Pomata, *Contracting a Cure: Patients, Healers, and*

so much the treatments that empirics used but the healer's inner character that allegedly separated physicians from the rest. Thus, in 1606 the English physician Eleazar Dunk wrote that the difference between learned doctors and empirics was that the latter were "ignorant," "boastful," and hasty in promising cures before they had ascertained the cause of the disease.[29] Lacking in well-tempered judgments, according to the physicians, the large and growing numbers of empirics were a public danger and should be regulated accordingly.

The market economy also created new sites of formal scientific instruction. The *maestri del abbaco*, or arithmetic teachers, taught practical mathematics to children of the merchant class from their shops in the piazzas. By the sixteenth century, private abbacists had set up shop in most major Italian cities. The Florentine abbacists were generally lower-middle-class professionals. Most lived and had their shops in the working-class Oltrarno district, although at least one, Giovanni di Bartolo, had his school in the more fashionable Piazza Santa Trinità. Using the instructional system elaborated by Leonardo Fibonacci in the early thirteenth century, the abbacists disseminated the Hindu-Arabic numeral system not only in Italy but throughout Europe. Between 1530 and 1586, at least six *Rechenmeister* (masters of calculation) were residing in Strasbourg, where they taught arithmetic, geometry, and accounting. By 1613, Nuremberg could boast forty-eight such schools.[30] Although arithmetic was taught at the universities as part of the seven liberal arts, its content was theoretical and devoid of practical applications. Hence merchants and civil servants preferred to send their children to the abbacists. Many students of the early reckoning schools, for example Niccolo Tartaglia (1499–1557), went on to become distinguished mathematicians (see Andersen and Bos, Chapter 28, this volume).

In their call to move away from armchair philosophy, Renaissance naturalists praised the practical empiricism of the workshop. Bacon extolled "maker's knowledge" as superior to speculation and urged philosophers to

---

the Law in Early Modern Bologna, translated by the author with the assistance of R. Foy and A. Taraboletti-Segre (Baltimore: Johns Hopkins University Press, 1998), pp. 129–36.

[29] Eleazar Dunk, *The Copy of a Letter written by E. D. Doctour of Physicke to a Gentleman, by whom it was published* (London, 1606), quoted in Harold J. Cook, "Good Advice and Little Medicine: The Professional Authority of Early Modern Physicians," *Journal of British Studies*, 33 (1994), 1–31, at p. 19. In addition, see Cook, *The Decline of the Old Medical Regime in Stuart London* (Ithaca, N.Y.: Cornell University Press, 1986).

[30] Paul F. Grendler, *Schooling in Renaissance Italy: Literacy and Learning, 1300–1600* (Baltimore: Johns Hopkins University Press, 1989), pp. 22–3, 306–23; Warren Van Egmond, "The Commercial Revolution and the Beginnings of Western Mathematics in Renaissance Florence, 1300–1500", Ph.D. dissertation, Indiana University, Bloomington, Ind., 1976; and Miriam Chrisman, *Lay Culture, Learned Culture: Books and Social Change in Strasbourg, 1480–1599* (New Haven, Conn.: Yale University Press, 1982), p. 183. See also Frank J. Swetz, *Capitalism and Arithmetic: The New Math of the 15th Century*. (La Salle, Ill.: Open Court, 1987), which contains an English translation of the Treviso arithmetic of 1478.

compile histories of the mechanical arts.[31] Yet admitting this knowledge into natural philosophy was problematic, both socially and epistemologically. Galileo Galilei (1564–1642), although praising the Venetian Arsenal as a font of empirical information, left little doubt that geometrical demonstrations were more certain than mechanics' skills.[32] Although an experience could just as well occur in the workshop as in the laboratory, such information did not count as knowledge until some recognized form of authority confirmed it. When the Delft draper Antonie van Leeuwenhoek communicated his first microscopic observations to the Royal Society of London in 1673, he sent them under a cover letter by the physician Reginald de Graaf that introduced Leeuwenhoek as "a certain very ingenious person."[33] One medium through which craft information entered letters was the "books of secrets," which represented trade secrets as "experiments." The "professors of secrets" who compiled these works created an image of natural philosophy as a hunt for "secrets of nature," in contrast with the scholastic view of natural philosophy as logical demonstrations of ordinary phenomena.[34] Another such medium was technological treatises in the vernacular by craftsmen such as Bernard Palissy (1510–1590), a French potter, or by humanists such as Georgius Agricola (1494–1555), who wrote a Latin treatise on mining.[35] Either way, from the point of view of the official scientific and medical disciplines, the ability to read and write – whether in Latin or the vernacular – marked the essential division between those who provided empirical information and those who created scientific knowledge (see Shapin, Chapter 6, this volume).

NATURAL KNOWLEDGE IN THE PIAZZA

Entrance into the workshops and commercial houses of early modern Europe was never completely open. The hefty fees charged by abbacists, for example, restricted their services to the sons of well-to-do merchants and civil servants, excluding lower-class pupils. Craftsmen notoriously guarded their trade secrets in closed workshops. Calzolari's museum, located above his house, adjacent to his pharmacy, was open only to the virtuosi to whom he wanted to display his collection.

---

[31] Antonio Pérez-Ramos, *Francis Bacon's Idea of Science and the Maker's Knowledge Tradition* (Oxford: Clarendon Press, 1988).
[32] Galileo Galilei, *Two New Sciences*, trans. Stillman Drake (Madison: University of Wisconsin Press, 1974), pp. 11–12
[33] Steven Shapin, *A Social History of Truth: Civility and Science in Seventeenth-Century England* (Chicago: University of Chicago Press, 1994), p. 304.
[34] William Eamon, *Science and the Secrets of Nature: Books of Secrets in Medieval and Early Modern Culture* (Princeton, N.J.: Princeton University Press, 1994), chap. 8; Eamon, "Science as a Hunt," *Physis*, 31 (1994), 393–432.
[35] Rossi, *Philosophy, Technology, and the Arts in the Early Modern Era*.

The piazza itself, on the other hand, was an open and public area, bounded only by the buildings whose street-level spaces housed the shops of merchants and craftsmen.[36] It was not a single homogeneous space but a network of separate and interconnected niches, each carved out by the entertainers and vendors who set up the temporary stages and benches from which they sold their wares. The piazza was also the scene of public performances of natural knowledge that virtually anyone could view. It was there that *ciarlatani* and mountebanks put on their comedies, displayed marvels, and demonstrated nature's wonders in order to attract the crowds of buyers for their nostrums and pamphlets.

Theatrical elements were essential in the marketplace defined by the piazza. In order to attract the throngs of people that became the buyers of their nostrums, mountebanks performed comic routines, using the stock characters of what would later be called (in politer circles) *commedia dell'arte*.[37] Ridicule and laughter were the charlatan's stock in trade, for the merrier the crowd, the sooner it would part with its money. The physician was the usual butt of the mountebank's joke; but the mountebank's success depended upon making himself the fool as well. An English quack's harangue described the German doctor Waltho Van Claturbank as a "chymist and dentifricator" who had attended twelve universities and offered cures for paralitick paraxysms, illiac passion, hen-pox, hog-pox, and whores-pox. He sold a styptic for "the restoration of maidenhood" along with his "Carthamophra of the Triple Kingdom," two drops of which would restore health to any who may "chance to have his brains beat out, or his head chop'd off."[38] Was the mountebank primarily an entertainer or a seller of medical nostrums? Obviously he was both. His success depended less upon the efficacy of his cures than upon the complicity of his audience and its willingness to suspend disbelief.[39]

The piazzas were also the sites of displays of exotic rarities and demonstrations of nature's wonders. A sixteenth-century *ciarlatano* who called himself "il Persiano" claimed that he possessed "marvelous occult secrets of nature" from Persia, and a Venetian distiller advertised a cabinet of curiosities that included "ten very stupendous monsters, marvelous to see, among which there are seven newborn animals, six alive and one dead, and three imbalmed female infants." The fascination with exotica and rarities, which

---

[36] On the architecture of the early modern Italian piazza, see Donatella Calabi, ed., *Fabbriche, piazze, mercati: La città italiana nel Rinascimento* (Rome: Officina Edizioni, 1997).

[37] K. M. Lea, *Italian Popular Comedy: A Study in the Commedia dell'Arte, 1560–1620*, 2 vols. (New York: Russell and Russell, 1962; orig. publ. 1934), esp. 1: 17–128. See also Gentilcore, *Healers and Healing in Early Modern Italy*, chap. 4.

[38] Roger King, "Curing Toothache on the Stage? The Importance of Reading Pictures in Context," *History of Science*, 33 (1995), 396–416, at p. 407.

[39] Alison Klairmont Lingo, "Empirics and Charlatans in Early Modern France: The Genesis of the Classification of the 'Other' in Medical Practice," *Journal of Society History*, 19 (1986), 583–603; and Roy Porter, "The Language of Quackery in England, 1660–1800," in *The Social History of Language*, ed. Peter Burke and Roy Porter (Cambridge: Cambridge University Press, 1987), pp. 73–103.

historians have noted was an important aspect of the sensibility of early modern science, was as much a part of the culture of the piazza as of the court.⁴⁰

Piazzas were socially leveling spaces; it was not just the "common people" who gathered there. People of all social classes witnessed and participated in the cultural performances that took place in the city squares. Like many English travelers to Italy, the English virtuoso John Evelyn (1620–1706) observed the *ciarlatani* with fascination. A phosphorus experiment performed in the Royal Society in 1671 reminded him of a mountebank he had seen in the Piazza Navona in Rome doing tricks with a phosphorescent ring to draw an audience, "and having by this surprising trick, gotten Company about him, he fell to prating for the vending of his pretended Remedies."⁴¹ Never before or since, Evelyn reported, had he seen such a brilliant phosphor. He always regretted not having purchased the recipe.

Rubbing elbows with the *ciarlatani* in town squares were prophets dressed in sackcloth declaiming their tales of catastrophic and prodigious happenings. Interpreting the meanings of natural and celestial events, or of monstrous births, they foretold death, famine, and war, and urged the populace to repentance.⁴² Both anomalous prodigies and normal astrological events were objects of fascination, packed with hidden meanings, which popular prophets and almanac makers promised to clarify. With the advent of the new technology of printing, news of such portents as the Ravenna monster (a misbirth that reportedly occurred in 1512) spread rapidly through pamphlets and broadsides, feeding the voracious popular appetite for prodigies.⁴³ Almanacs and astrological predictions were also widely published in popular editions.⁴⁴

---

⁴⁰ Benedetto, detto il Persiano, *I maravigliosi, et occulti secreti naturali* (Rome, 1613); Giulielmo Germerio Tolosano, *Gioia preciosa . . . Opera à chi brama la sanità utilissima & necessaria* (Venice, 1604); Tomaso Garzoni, *La piazza universale di tutte le professioni del mondo*, ed. Paolo Cherchi and Bieatrice Collina, 2 vols. (Turin: Einaudi, 1997), 2: 1188–97; Eamon, *Science and the Secrets of Nature*, chap. 7; and Lorraine Daston, "Marvelous Facts and Miraculous Evidence in Early Modern Europe," *Critical Inquiry*, 18 (1991), 93–124. On the fascination with wonders in the courts, see Daston and Park, *Wonders and the Order of Nature*, pp. 100–8.

⁴¹ John Evelyn, *The Diary of John Evelyn*, ed. E. S. De Beer, 6 vols. (Oxford: Oxford University Press, 1955), 4: 253. On the phosphorus experiments in the Royal Society, see Jan Golinski, "A Noble Spectacle: Phosphorus and the Public Culture of Science in the Early Royal Society," *Isis*, 80 (1989), 11–39.

⁴² Ottavia Niccoli, *Prophecy and the People in Renaissance Italy*, trans. Lydia G. Cochrane (Princeton, N.J.: Princeton University Press, 1990); and Sara Schechner Genuth, *Comets, Popular Culture, and the Birth of Modern Cosmology* (Princeton, N.J.: Princeton University Press, 1997).

⁴³ On monsters and their meaning in early modern Europe, see Daston and Park, *Wonders and the Order of Nature*, chap. 5. On the Ravenna monster and popular prophecies, see Niccoli, *Prophecy and the People*, pp. 35–46.

⁴⁴ Bernard Capp, *English Almanacs, 1500–1800: Astrology and the Popular Press* (Ithaca, N.Y.: Cornell University Press, 1979); Patrick Curry, *Prophecy and Power: Astrology in Early Modern England* (Princeton, N.J.: Princeton University Press, 1989); Paola Zambelli, "Fino del mondo o inizio della propaganda? Astrologia, filosofia della storia e propaganda politico-religiosa nel dibattito sulla congiunzione del 1524," in *Scienze, credenze occulte, livelli di cultura* (Florence: Leo S. Olschki, 1982), pp. 291–368; and Zambelli, ed., *'Astrologi hallucinati': Stars and the End of the World in Luther's Time* (Berlin: Walter de Gruyter, 1986).

However, it is unclear how seriously the urban population took the science of astrology. Just as the *ciarlatani* lampooned the doctors and imitated the "professors of secrets" by selling their own cheap pamphlets of "secrets," ballad singers satirized the astrologers in carnival songs such as the Italian one by "Doctor Master Pegasus Neptune" that predicted "conjunctions of cheese and lasagna" and "a flood of Dalmatian wine," followed by "horrendous winds shot off like bombards sending off stupendous stenches."[45]

Among the most dramatic exhibitions performed in the piazzas were executions and punishments of criminals. The executioner's craft required considerable knowledge of anatomy. He had to judge the physical condition of prisoners when deciding which torture to apply and had to nurse prisoners back to health after interrogations. Because the executioners were the principal suppliers of cadavers for public anatomy demonstrations, they had to know how to execute criminals in a manner that caused little damage to the body.[46] Because of their hands-on anatomical knowledge, executioners, as well as butchers and surgeons, were sometimes called upon to perform autopsies under the supervision of physicians. Research has revealed that, ironically, executioners also acted as healers. In early modern Germany, people consulted them for treatments of broken bones, sprains, and other injuries. Executioners prescribed medicines composed of human body parts, including human fat and blood. In the execution ritual, the body of the executed sinner acquired a kind of sacred power, which the executioner was able to tap as he handled the cadaver.[47]

The use of human body parts as medicaments had roots deep in folklore. In German folk medicine, human skull, baked and ground into a powder, was prescribed as a remedy for epilepsy, and women in labor wore straps of tanned human skin as belts to ease the birthing process. Human blood was reportedly so valued as a cure for the falling sickness that epileptics waited at the scaffold as a beheading took place in order to drink the "poor-sinner's blood" while it was fresh and warm.[48] Yet the therapeutic use of human body parts was not limited to folk medicine. The *Pharmacopoeia Londinensis* (London Pharmacopoeia) of 1618 recommended powdered human skull for epilepsy.[49]

---

[45] *Prognostico: over diluvio consolatorio composto per lo eximio Dottore Maestro Pegaso Neptunio: el qual dechiara de giorno in giorno que che sarà nel mese de febraro; Cosa belissima & molto da ridere* (n.p., n.d. [Venice?, 1524]), quoted in Niccoli, *Prophecy and the People*, p. 158.

[46] See Andrea Carlino, *La fabbrica del corpo: Libri e dissezione nel Rinascimento* (Turin: Einaudi, 1994), chap. 2; and Katharine Park, "The Criminal and the Saintly Body: Autopsy and Dissection in Renaissance Italy," *Renaissance Quarterly*, 47 (1994), 1–33.

[47] Kathy Stuart, "The Executioner's Healing Touch: Health and Honor in Early Modern German Medical Practice," in *Infinite Boundaries: Order, Disorder, and Reorder in Early Modern German Culture*, ed. Max Reinhart (Sixteenth Century Essays and Studies, 40) (Kirksville, Mo.: Thomas Jefferson University Press, 1998), pp. 349–79.

[48] Stuart, "Executioner's Healing Touch," p. 360.

[49] William Brockbank, "Sovereign Remedies: A Critical Depreciation of the London Pharmacopoeia," *Medical History*, 8 (1964), 1–14, at p. 3.

In 1662, the physician Johann Joachim Becher urged that apothecaries keep at least twenty-four types of human material in stock.[50]

Natural knowledge from the piazzas raised fundamental questions about how experimental knowledge might be controlled. As the case of the phosphorus demonstrations suggests, even in the relatively closed environs of the Royal Society of London, experiments could easily degenerate into theatrical spectacles. To prevent this, some virtuosi urged that the Society clarify the boundaries separating experimental "matters of fact" from the flamboyant spectacles of the mountebanks. As one Fellow of the Society put it, "an Artist or Experimenter, is not to be taken for a maker of gimbals, nor an observer of Nature for a wonder-monger."[51] Although demonstrations of rarities and unusual phenomena were important resources in expanding the public culture for science, the danger that experiments might bedazzle onlookers instead of enlighten them was always present.[52]

NATURAL KNOWLEDGE IN THE COUNTRYSIDE AND VILLAGES

A sixteenth-century commonplace held that the common people possessed "secrets" making up a body of natural knowledge unknown to the savants. Leonardo Fioravanti investigated the "rules of life" observed by the peasants of Calabria, and the Danish Paracelsian Petrus Severinus exhorted naturalists in 1571 to "study the astronomy and terrestrial philosophy of the peasantry."[53] The seventeenth-century French jurist René Choppin echoed a familiar trope when he wondered, "how many savants in Medicine have been outdone by a simple old peasant woman, who with a single plant or herb has found a remedy for illnesses despaired of by physicians."[54] What would intellectuals have discovered had they studied the "natural philosophy" of the countryside and villages?

Gardeners, orchard keepers, farmers, and beekeepers accumulated vast amounts of empirical information about plants and animals, but only in

---

[50] Stuart, "Executioner's Healing Touch," p. 360.
[51] Michael Hunter and Paul B. Wood, "Towards Solomon's House: Rival Strategies for Reforming the Early Royal Society," *History of Science*, 24 (1986), 49–108, at p. 81; Steven Shapin and Simon Schaffer, *Leviathan and the Air-Pump: Hobbes, Boyle, and the Experimental Life* (Princeton, N.J.: Princeton University Press, 1985), pp. 112–15; and Steven Shapin, "The House of Experiment in Seventeenth-Century England," *Isis*, 79 (1988), 373–404, at pp. 388–90.
[52] Golinski, "A Noble Spectacle," pp. 38–9.
[53] Leonardo Fioravanti, *Capricci medicinali* (Venice: Lodovico Avanzo, 1561), pp. 53–4; Petrus Severinus (Peder Sørensen), *Idea medicinae philosophicae* (The Hague: Adrianus Ulacq, 1660), p. 39, quoted in Allen G. Debus, *The English Paracelsians* (New York: Franklin Watts, 1965), p. 20.
[54] Quoted by Natalie Zemon Davis, "Proverbial Wisdom and Popular Errors," in Davis, *Society and Culture in Early Modern France* (Stanford, Calif.: Stanford University Press, 1984), p. 261.

exceptional instances did they record such data in writing.[55] Classical agricultural writers continued to be reprinted, but early modern naturalists looked upon these works with growing skepticism. The seventeenth-century English virtuoso Ralph Austen contended that the ancient sources were "full of dangerous and hurtfull instructions, and things notoriously untrue." Following Bacon's advice, seventeenth-century reformers such as Gabriel Plattes and Samuel Hartlib believed that it was necessary to replace traditional agricultural writings with compilations of current practices that had been subjected to experimental tests.[56] In order to put this principle into practice, the Royal Society of London appointed a "Georgical Committee." However, the committee was dissolved after only a few years of activity.[57]

Medical practice in the countryside, as in the city, was characterized by a combination of naturalistic, religious, and magical healing. Some village "wise women" were respected for their empirical skills in identifying plants and their healing properties: Even physicians copied herb women's recipes into their medical formularies. The botanist Pietro Andrea Mattioli (1501–1577) observed that shepherds, peasants, and herb women, through their vast experience, had botanical knowledge of which the physicians were ignorant. The Portuguese botanist Amato Lusitano, who was in Ferrara in the mid-sixteenth century, recalled that the herbalists who collected plants for Duke Borso d'Este's herbarium frequently benefited from consultations with herb women.[58]

Village healers rarely distinguished among physical, magical, and religious remedies. Diamante di Bisa, a sixty-six-year-old widow interrogated by the Inquisition of Modena in 1599, explained that to treat newborn infants:

> I take walnut oil, rue, feverfew, saliva, and wild thyme, and boil them together in a clay pot, and with this I anoint the sick child on the umbilical cord, the throat, the kidneys and the pulse, making the sign of the cross with my hand, saying, "In nomine patris et filii et spiritus sancti, Amen." Then I say

---

[55] An important exception is the fourteenth-century tract on grafting and orchard keeping by Gottfried of Franconia, edited by Gerhard Eis, *Gottfrieds Pelzbuch: Studien zur Reichweite und Dauer der Wirkung des mittelhochdeutschen Fachschriftums* (Südosteuropäische Arbeiten, 38) (Munich: Callwey, 1944), fifteenth-century English trans. ed. Willy Braekman, *Geoffrey of Franconia's Book of Trees and Wine* (Scripta, 24) (Brussels: UFSAL, 1989).

[56] Charles Webster, *The Great Instauration: Science, Medicine, and Reform, 1626–1660* (New York: Holmes and Meier, 1976), pp. 470–1.

[57] Reginald Lennard, "English Agriculture under Charles II: The Evidence of the Royal Society's 'Enquiries'," *Economic History Review*, 4 (1932), 23–45.

[58] Albano Biondi, "La signora delle erbe e la magia della vegetazione," in *Cultura popolare nell'Emilia Romagna: Medicina erbe e magia* (Milan: Silvana Editoriale, 1981), pp. 185–203; Katharine Park, "Medicine and Magic: The Healing Arts," in *Gender and Society in Renaissance Italy*, ed. Judith Brown and Robert Davis (London: Longmans, 1997); Jole Agrimi and Chiara Crisciani, "Savoir médical et anthropologie religieuse: Les représentations et les fonctions de la *vetula* (XIIIe–XVe siècle)," *Annales: Economies, Sociétés, Civilisations*, 48 (1993), 1281–1333.

three Pater Nosters and three Ave Marias, asking God and the Holy Trinity that he should return to health.[59]

The Inquisition vigorously persecuted "superstitious" practices such as "signing" illnesses (as this procedure was called), which the Church regarded as unlawful encroachments on its spiritual territory.[60] Although Diamante prescribed herbs routinely used by physicians to treat infants and women in childbirth, the medicinal use of prayers and religious rituals by a lay person worried the religious authorities, particularly in the Counter-Reformation period.

The physicians also distanced themselves from healers such as Diamante di Bisa, condemning them for their supposed ignorance, superstition, and malice. Yet the distinction between naturalistic and magical healing was blurry, even in the eyes of the doctors. Not only did they refuse to rule out the efficacy of magical cures, but the revival of learned magic made them increasingly curious about the therapeutic uses of occult forces.[61] So great was the medical community's enthusiasm for amulets, for example, that as the seventeenth century drew to a close, the German physician Jacob Wolff wrote a 400-page catalog of diseases deemed treatable with the devices.[62] The condemnation of village healers was not primarily an attack on magical healing but an attempt by the physicians to control competition and, as the physician Antonio Guaineri had already put it in the fifteenth century, to "establish a distance between [themselves] and vulgar practitioners."[63]

Yet if the philosophical and juridical divide between village healers and learned physicians was wide, they had much in common in terms of actual practice. The general physiology that governed medical practice, both popular and learned, concerned maintaining the proper flow of fluids in the body. Accordingly, therapeutics typically focused on evacuating superfluous fluids through bleeding, vomiting, sweating, purging, and the like.[64] Physicians and village healers used similar techniques to promote conception and to influence or predict the sex of infants. A sixteenth-century physician's

---

[59] Mary O'Neil, "Discerning Superstition: Popular Errors and Orthodox Response in Late Sixteenth Century Italy," Ph.D. dissertation, Stanford University, Stanford, Calif., 1981, p. 75 (slightly modified). For similar examples from Venice, see Guido Ruggiero, *Binding Passions: Tales of Magic, Marriage, and Power at the End of the Renaissance* (Oxford: Oxford University Press, 1993), chap. 3; and Marisa Milani, ed., *Antiche pratiche di medicina popolare nei processi del S. Uffizio (Venezia, 1572–1591)* (Padua: Centrostampa Palazzo Maldura, 1986).

[60] Mary O'Neil, "Magical Healing: Love Magic and the Inquisition in Late Sixteenth-Century Modena," in *Inquisition and Society in Early Modern Europe*, ed. Stephen Haliczer (Totowa, N.J.: Barnes and Noble, 1967), pp. 88–114.

[61] Nancy G. Siraisi, *Medieval and Early Renaissance Medicine: An Introduction to Knowledge and Practice* (Chicago: University of Chicago Press, 1990), p. 152.

[62] Jacob Wolff, *Curiosus amuletorum scrutator* (Frankfurt: F. Groschuffius, 1692). For the medical debate over amulets, see Martha Baldwin, "Toads and Plague: Amulet Therapy in Seventeenth-Century Medicine," *Bulletin of the History of Medicine*, 67 (1993), 227–47. In addition, see Keith Thomas, *Religion and the Decline of Magic* (New York: Charles Scribner's Sons, 1971), pp. 177–92.

[63] Park, "Medicine and Magic," p. 9.

[64] Pomata, *Contracting a Cure*, pp. 129–32; Eamon, "'With the Rules of Life and an Enema,'" pp. 29–34.

concoctions to promote pregnancy (one including egg yolks, goat's milk, and the right testicles from a ram and a pig) do not seem very far removed from the remedies uncovered by the Inquisition in its interrogation of rural wise women – except that the latter often had illicit prayers adjoined to them.[65] Such similarities led the Neapolitan naturalist and magus Giambattista della Porta, in his unpublished *Criptologia* (ca. 1604), to conclude that popular magic held great truths, albeit truths distorted by popular superstitions. In reality, he maintained, the effects produced by witches and cunning men were accomplished by natural forces, and the various ceremonies, spells, and prayers connected with them were of no value.[66]

Was village magic a debased form of learned magic, or was the latter derived from the former? Inquisitorial interrogations reveal that some wise women and cunning men owned popular handbooks on magic, divination, and physiognomy. Other evidence, however, suggests that the information exchange may have been in the opposite direction. The seventeenth-century English antiquarian John Aubrey recorded folk practices in the belief that they might yield useful information. He thought that shepherds' weather prognostications were worth examining, and folk customs were "relique[s] of Naturall Magick." Della Porta carried out experiments on witchcraft practices in order to "unmask [their] frauds" and to reveal the natural causes underlying them.[67] Like many contemporaries, della Porta accepted magical happenings as real effects of natural causes masked in the fraudulent trappings of magic.

The Venetian career of Bartholomeo Riccio, a snakehandler from Puglia, sheds additional light on the information exchange that occurred when rural medical traditions entered the urban marketplace. Riccio was one of the shamanistic *sanpaolari*, healers who claimed descent from St. Paul.[68] In order to obtain a license to sell his *gratia di San Paolo* (St. Paul's grace) in Venice, Riccio had to prove the worth of his cure to the health board. Thus, in 1580, he appeared before the board, serpents in hand, and proceeded with his "demonstration," allowing himself to be bitten on the chest by his snakes. Although the bites swelled and turned black, Riccio applied his Maltese earth ointment and, to the physicians' astonishment, was immediately healed. As a snakehandler who collected vipers to sell to pharmacists as ingredients to make theriac, Riccio must have been quite a sight on the Piazza San Marco. He could catch and kill vipers barehanded, without injury to himself, "to the amazement of everyone." Without much difficulty, Riccio convinced the physicians of the efficacy of his remedy and received a license to exercise

---

[65] Giovanni Marinello, *Le medicine partenenti alle infermità delle donne* (Venice: Francesco de'Franceschi, 1563), p. 70.
[66] Giambattista della Porta, *Criptologia*, ed. Gabriella Belloni (Edizione Nazionale dei Classici del Pensiero Italiano, ser. 2, 37) (Rome: Centro Internazionale di Studi Umanistici, 1982), p. 158.
[67] John Aubrey, "Remaines of Gentilisme and Judaisme," in *Three Prose Works*, ed. John Buchanan-Brown (Carbondale: Southern Illinois University Press, 1972), p. 132; della Porta, *Criptologia*, p. 158; and Thomas, *Religion and the Decline of Magic*, pp. 228–9.
[68] Garzoni, *La piazza universale*, 2: 1195–6.

his profession: that of snakehandling as an entertainment to attract buyers for his remedies.⁶⁹ The Venetian health board's determination reflected the prevailing views of the established medical community. Academic physicians did not contest the *efficacy* of St. Paul's earth as a poison antidote. Instead, they attacked the *sanpaolari* for dissimulation, counterfeiting Maltese earth, and deceiving the people with various tricks to protect themselves against venomous bites.⁷⁰

If some intellectuals were curious about folk customs, the learned community in general grew increasingly critical of popular culture. In a series of works on "popular errors," physicians and intellectuals lashed out against "superstitions," reserving special venom for wise women and midwives.⁷¹ The English physician Sir Thomas Browne, in his *Pseudodoxia epidemica* of 1646, condemned not only popular medical practices but popular knowledge in its entirety. To Browne, popular culture teemed with superstition, ignorance, and perversion.⁷²

CONCLUSION: POPULAR CULTURE AND THE NEW PHILOSOPHY

The tension between the idea of "going to the people" for natural knowledge and intellectuals' contempt for popular culture surfaced in the 1660s in the Royal Society's collaborative "history of trades" project.⁷³ Inspired by Bacon's idea that the workshop was a sort of laboratory that "takes off the mask and veil from natural objects," the history of trades aimed both to improve technology and to furnish natural philosophy with experiments.⁷⁴ A similar effort was undertaken in the Paris Académie Royale des Sciences when in 1675

---

⁶⁹ The relevant documents are in Archivio di Stato, Venice, Provveditori alla Sanità, Reg. 734, c. 177v (1580) and Reg. 735, c. 135v (1583).
⁷⁰ For example, Mattioli and Martin Del Rio. See Brizio Montinaro, *San Paolo dei serpenti: Analisi di una tradizione* (Palermo: Sellerio, 1996), pp. 78–80; Angelo Turchini, *Morso, morbo, morte: La tarantola fra cultura medica e terapia popolare* (Milan: Franco Angeli, 1987), pp. 152–3; Thomas Freller, "'Lingue di serpi', 'Natternzungen' und 'Glossopetrae': Streiflichter auf die Geschichte einer populären 'kultischen' Medizin der frühen Neuzeit," *Sudhoffs Archiv*, 81 (1997), 63–83. To prevent counterfeiting Maltese earth, some physicians recommended that the Order of Malta certify with a seal the authenticity of the earth.
⁷¹ Peter Burke, *Popular Culture in Early Modern Europe* (New York: Harper and Row, 1978), chap. 8. Examples of books on "popular errors" include Laurent Joubert, *Popular Errors*, trans. Gregory David de Rocher (Tuscaloosa: University of Alabama Press, 1989); and Scipione Mercurio, *De gli errori popolari d'Italia* (Verona: Rossi, 1645). In addition, see Eamon, *Science and the Secrets of Nature*, pp. 259–66.
⁷² Thomas Browne, *Pseudodoxia Epidemica*, in *The Works of Sir Thomas Browne*, ed. Geoffrey Keynes, 4 vols. (Chicago: University of Chicago Press, 1964), vol. 2.
⁷³ Walter E. Houghton, Jr., "The History of Trades: Its Relation to Seventeenth Century Thought," *Journal of the History of Ideas*, 3 (1942), 51–73, 190–219; and Kathleen H. Ochs, "The Royal Society of London's History of Trades Programme: An Early Episode in Applied Science," *Notes and Records of the Royal Society of London*, 39 (1985), 129–58.
⁷⁴ Francis Bacon, *Parasceve*, in *The Works of Francis Bacon, Baron of Verulam, Viscount of St. Alban, and Lord Chancellor of England*, [1857–74], ed. James Spedding, Robert Leslie Ellis, and Douglas Denon Heath, 14 vols. (New York: Garrett Press, 1968), 4: 257.

the Crown's chief minister, Jean-Baptiste Colbert, instructed the academy to begin a comprehensive description of the mechanical arts.[75]

From the outset, however, the virtuosi faced numerous obstacles in constructing histories of the trades. Craftsmen were naturally reluctant to reveal their secrets because their livelihood depended upon maintaining a monopoly over special techniques. In addition, the scope of the history of trades project was simply too vast to be accomplished even by men fired by Baconian zeal. After all, its promoters were virtuosi, men of a thousand interests who naively thought they could quickly master any craft well enough to be the artisan's instructor. But the history of trades was no task for dilettantes; it required the prolonged, concentrated effort of generations. Moreover, the history of trades project required gentlemen to go to places they were loath to frequent. Despite his ardent support of the project, John Evelyn confined his efforts mainly to "aristocratic" arts such as engraving, oil painting, miniature painting, annealing, enamel, and marble paper. Yet even these arts proved to be too debasing for Evelyn, who abandoned his part of the project altogether because of the necessity of "conversing with mechanical and capricious persons."[76]

Learned suspicion of popular culture was another obstacle in the way of constructing histories of the trades.[77] In the Paris Académie, the exchange of information between artisans and scientists was nothing like what Bacon imagined. Instead of viewing the crafts as sources of natural knowledge, the Paris academicians aimed to impose "scientific" standards upon the mechanical arts.[78] Such elitist attitudes also surfaced in the Royal Society. In his outline for a history of trades, Evelyn preserved a hierarchical ranking of the crafts, beginning with the "Usefull and purely Mechanic," ascending to the "Polite and More Liberall," then to the "Curious," and ending finally with "Exotick, and very rare Seacretts."[79] Eventually, Evelyn opted against publishing his findings because he feared it might "debase much of their esteem by prostituting them to the vulgar."[80]

Finally, the membership of the early modern scientific societies grew increasingly more elite. Despite the Royal Society's claim that it champi-

---

[75] Roger Hahn, *The Anatomy of a Scientific Institution: The Paris Academy of Sciences, 1666–1803* (Berkeley: University of California Press, 1971), p. 68.
[76] John Evelyn, *The Diary and Correspondence of John Evelyn*, ed. William Bray (London: Charles Scribner's Sons, 1903), p. 590.
[77] Daston and Park, *Wonders and the Order of Nature*, chap. 9; and Eamon, *Science and the Secrets of Nature*, pp. 259–66.
[78] Hahn, *Anatomy of a Scientific Institution*, pp. 185–94. The resentment among artisans caused by these measures swelled to a chorus of protest against academic "despotism" during the French Revolution.
[79] Royal Society, Classified Papers, III(i), fol. 1. Compare Robert Hooke's more "Baconian" outline in *The Posthumous Works of Robert Hooke*, ed. Richard Waller (New York: Johnson Reprint Corp., 1969), pp. 24–6. On these proposals, see Michael Hunter, *Science and Society in Restoration England* (Cambridge: Cambridge University Press, 1981), pp. 99–101.
[80] Evelyn to Boyle, 9 August 1659, in Robert Boyle, *The Works of the Honourable Robert Boyle* [1772], ed. Thomas Birch, 6 vols. (repr. Hildesheim: Georg Olms, 1966), 6: 287–8; and Evelyn, *Diary and Correspondence*, p. 578.

oned "useful knowledge," tradesmen constituted only about three percent of its membership in 1672.[81] Robert Hooke (1635–1702), the Society's curator of experiments, envisioned it as a small, highly disciplined corps dedicated to the pursuit of natural knowledge. Comparing it with the conquistadors who took Mexico under Cortés's command, he wrote, "This newfound land [nature] must be conquerd by a Cortesian army well Disciplined and regulated though their number be but small."[82] Toward the end of the seventeenth century, with the society's original Baconian vision on the wane, the history of trades project was silently set aside.

The argument for the social utility of science did not die out, but increasingly it was framed within the context of the growing split between elite and popular cultures. Hooke's voice was but one of many signaling the rise of the expert. To Galileo, Copernicanism presumed the need for a deeper, more subtle understanding of Scripture that only natural philosophers could make.[83] Despite its emphasis upon "matters of fact," the new philosophy's validity rested on the claim that "naive" empirical knowledge was inherently unreliable. In the final analysis, the new philosophers asserted, the mysteries of the universe were beyond the capacities of the vulgar. By redefining the sites of science and by invalidating popular testimony, the new philosophy disqualified the people from the arenas where natural knowledge was produced.

---

[81] Michael Hunter, *The Royal Society and Its Fellows, 1660–1700: The Morphology of an Early Scientific Society* (Chalfont St. Giles: British Society for the History of Science, 1982), p. 40.

[82] "Proposalls for the Good of the Royal Society," Royal Society, Classified Papers, xx.50, fols. 92–4, quoted in Hunter and Wood, "Toward Solomon's House," p. 87.

[83] Galileo Galilei, *Letter to the Grand Duchess Christina*, in Galileo, *Discoveries and Opinions of Galileo*, trans. Stillman Drake (New York: Anchor Books, 1957), pp. 181–2.

9

HOMES AND HOUSEHOLDS

Alix Cooper

Where did early modern natural inquiry take place? Research by historians of science has begun to suggest that many of the activities crucial to the Scientific Revolution took place not only in such recognizably new and innovative sites as botanical gardens, anatomy theaters, laboratories, and the quarters of scientific societies but also – and often simultaneously – within the seemingly humble and prosaic spaces of natural inquirers' own homes and households. These domestic spaces in fact saw the production of natural knowledge of all kinds, as their occupants used them as places not just to sleep but also to think, write, calculate, observe, and experiment on natural phenomena. Furthermore, while doing so, they frequently ended up enlisting household members in these projects. In this way, homes and households became crucial sites for the pursuit of natural knowledge in early modern Europe.

Few historians of science have paid attention to these kinds of "private" spaces. One of the main reasons for this is almost certainly the way in which, over the past several centuries, scientific work has gradually come to be conceptualized as occurring primarily *outside* the home. This particular assumption is itself a historical artifact, stemming from modern changes in the organization of work more broadly. During the nineteenth century in particular, as more and more people abandoned home-based workshops and began to travel to new places of employment, newly labeled "scientists" likewise increasingly came to work outside the home in institutional spaces that were perceived as religiously and emotionally neutral. In the process, considerable ideological boundaries were erected between work and family, and between public and private realms, which have continued to shape modern thinking.[1]

---

[1] See, for example, Dorinda Outram, "Before Objectivity: Wives, Patronage, and Cultural Reproduction in Early Nineteenth-Century French Science," in *Uneasy Careers and Intimate Lives: Women in Science, 1789–1979*, ed. Pnina G. Abir-Am and Dorinda Outram (New Brunswick, N.J.: Rutgers

If we wish to understand how early modern natural inquiry was actually practiced, however, it is necessary to put aside modern preconceptions and enter the world of the early modern home, for in early modern Europe and even beyond, that was indeed where a considerable amount of all production, craft and otherwise, took place, including – as this chapter will show – the production of knowledge about the natural world. Only by examining this crucial setting is it possible to recover some sense of the wide range of people actually involved in projects of natural inquiry in early modern Europe. As a glimpse of the early modern scientific household reveals, the study of nature engaged not just learned and professional men but also a wide array of unacknowledged and (to our modern eyes) seemingly invisible collaborators to be found at home, from wives and children through domestic servants. The pursuit of natural knowledge was thus not only an individual enterprise – for "great men" only – but a collaborative and in many cases a collective one. Although individual contributions can be difficult to document – many women and servants, for example, had not been taught more than a rudimentary literacy and thus did not leave much of a paper trail, and early modern literary conventions tended to preclude the mentioning of household members in published work – enough manuscript evidence has survived in the form of handwritten laboratory notes, household recipe books, and the like to give us a window into their participation in early modern natural inquiry, though much research still needs to be done.

This chapter will examine some of the various ways in which home and household came to provide important frameworks for the gathering of natural knowledge in early modern Europe. As I will show, numerous scientific activities were performed either within the home itself (that is to say, literally within the spatial confines of a residence) or, more broadly, by members of a household, which might include not only a paterfamilias but also wife, sons, daughters, other relatives, and domestic servants. Natural inquiry in early modern Europe thus often constituted a *family* project to which a variety of household members would contribute, providing crucial support and continuity for scientific activities at a time when formal institutional support was often lacking. Indeed, the household model for natural inquiry was to demonstrate its staying power by enduring well into the nineteenth century. During the crucial years of the Scientific Revolution, however, it proved particularly important as a model for the pursuit of natural knowledge.

University Press, 1987), pp. 1–30; and Londa Schiebinger, *The Mind Has No Sex? Women in the Origins of Modern Science* (Cambridge, Mass.: Harvard University Press, 1989). It is important to note that much of what we now know about households and homes in early modern science results from the work of historians who have investigated the careers of women in science and discovered that their family status – as wives, daughters, or widows – significantly shaped these careers.

DOMESTIC SPACES

To examine some of the opportunities for science that the early modern home provided, let us take a brief tour through the interior architecture not, perhaps, of a rural or peasant home, which would typically have consisted of a single room for working, eating, and sleeping, but rather that of a larger, more prosperous urban residence. Here could be found all sorts of places where activities that might today be called "scientific" were avidly pursued. The study, or *studio*, for example, was one such place. Usually adjoining the bedroom, it provided virtuosi with, on the one hand, a private refuge for solitary contemplation and, on the other, a semipublic space where they could introduce distinguished visitors to the collections of books, globes, mathematical instruments, and curiosities both artificial and natural that often lined its walls (and even ceiling). French polymath Pierre Borel (ca. 1620–1671) termed his a "world within the home"[2] (see Findlen, Chapter 12, this volume). The study, which also came to be labeled a *museum* (abode of the Muses), was thus a liminal space with multiple uses that reflected and enabled the intellectual aspirations of its occupants, whether surgeons such as Ambroise Paré (ca. 1510–1590), who filled his study with monstrous specimens to illustrate his book *On Monsters*, or mathematicians such as John Dee (1527–1608), who retreated to his "private study" behind double doors to cast horoscopes and commune with angels.[3]

Science did not remain cloistered in the study, however, but overflowed into the rest of the house. The Renaissance anatomist Andreas Vesalius (1514–1564) was notorious for dissecting human cadavers in his own chambers, sometimes keeping them there for weeks on end.[4] Nor was this practice, apparently, that unusual; in 1519, Italian medical student Ippolito of Montereale had already reported with delight on an animal dissection he had observed at his teacher Giovanni Lorenzo's home, "so we could see the inner

---

[2] Quoted in Paula Findlen, "Masculine Prerogatives: Gender, Space, and Knowledge in the Early Modern Museum," in *The Architecture of Science*, ed. Peter Galison and Emily Thompson (Cambridge, Mass.: MIT Press, 1999), p. 36. On the organization and ideals of the study, see also Dora Thornton, *The Scholar in His Study: Ownership and Experience in Renaissance Italy* (New Haven, Conn.: Yale University Press, 1997); and Steven Shapin, "'The Mind Is Its Own Place': Science and Solitude in Seventeenth-Century England," *Science in Context*, 4 (1990), 191–218. The layout of residences differed, of course, from place to place and period to period, in accordance with such other factors as wealth, social status, and occupation. For an introduction to the development of house interiors during this period, see Witold Rybczynski, *Home: A Short History of an Idea* (London: Penguin, 1986), pp. 11–75.

[3] See Ambroise Paré, *On Monsters and Marvels*, trans. Janis L. Pallister (Chicago: University of Chicago Press, 1982), pp. 49, 52, 134, 141, 150; and Deborah Harkness, "Managing an Experimental Household: The Dees of Mortlake and the Practice of Natural Philosophy," *Isis*, 88 (1997), 259.

[4] C. D. O'Malley, *Andreas Vesalius of Brussels, 1514–1564* (Berkeley: University of California Press, 1964), pp. 64, 112. See also pp. 44–5 on the difficulties encountered by Renaissance anatomists seeking suitable places to carry out their dissections.

parts and the origin of the nerves."[5] Those who wished to study living rather than dead bodies, however, repaired to the homes of others, paying visits to the sick in their bedrooms. Here doctors, midwives, and other medical practitioners consulted with patients and prescribed elaborate remedies for their ills. Although hospitals, with their never-ending supply of poor patients with a wide variety of conditions (and little authority to direct their own care), were increasingly becoming the principal locus for clinical research and high-level training, physicians and surgeons nonetheless treated most of their clients at home.

Meanwhile, in the shop or workshop, which in the houses of artisanal families usually adjoined the living quarters, illustrations were drawn, apothecaries' remedies compounded, and scientific instruments designed and perfected.[6] Kitchens and basements or root cellars formed improvised laboratories for women to tinker with and write down medical recipes, whether of the more herbally based Galenic or chemically based Paracelsian kind. It was popular for English women of some means to have stills and alembics in their kitchens for making "essences"; some, such as Lady Grace Mildmay (1552–1620), turned entire rooms into still-rooms and effectively ran pharmaceutical dispensaries from their homes, leading English virtuoso John Evelyn to comment of the gentlewomen of his youth that "their recreations were in the distillatorie."[7] Even more well-to-do experimenters such as Robert Boyle (1627–1691) set up not just one but a series of rooms specially furnished with stills and other necessary equipment to conduct their "trials" and "assays."[8]

Natural inquiry could also be, and was, avidly pursued outside. In kitchen gardens, medicinal simples were cultivated and all manner of "experiments" performed on the vegetable world, while backyards served as "theaters" to investigate local flora and fauna.[9] Even the rooftops of a house might be put to use if necessary. The astronomer Johannes Hevelius (1611–1687) built first a small watchtower and then a large platform on his roof in Danzig upon which to store his telescopes, quadrants, and sextants and from which to gaze at the stars. As he proudly informed the readers of his *Machinae*

---

[5] Dorothy M. Schullian, "An Anatomical Demonstration by Giovanni Lorenzo of Sassoferrato, 19 November 1519," in *Miscellanea di scritti di bibliografia ed erudizione in memoria di Luigi Ferrari* (Florence: Leo S. Olschki, 1952), pp. 489, 494.

[6] Schiebinger, *The Mind Has No Sex?* pp. 66–118; see also Pamela H. Smith, *The Body of the Artisan: Art and Experience in the Scientific Revolution* (Chicago: University of Chicago Press, 2004), pp. 95–6.

[7] Lynette Hunter, "Women and Domestic Medicine: Lady Experimenters, 1570–1620," in *Women, Science and Medicine, 1500–1700: Mothers and Sisters of the Royal Society*, ed. Lynette Hunter and Sarah Hutton (Stroud: Sutton, 1997), pp. 89–107, esp. pp. 95–6; Linda Pollock, *With Faith and Physic: The Life of Tudor Gentlewoman Lady Grace Mildmay, 1552–1620* (London: Collins and Brown, 1993), pp. 98–102; and Leonard Guthrie, "The Lady Sedley's Receipt Book, 1686, and other Seventeenth-Century Receipt Books," *Proceedings of the Royal Society of Medicine*, 6 (1913), 165.

[8] Steven Shapin, "The House of Experiment in Seventeenth-Century England," *Isis*, 79 (1988), 373–404.

[9] Alix Cooper, "Inventing the Indigenous: Local Knowledge and Natural History in the Early Modern German Territories," Ph.D. dissertation, Harvard University, Cambridge, Mass., 1998.

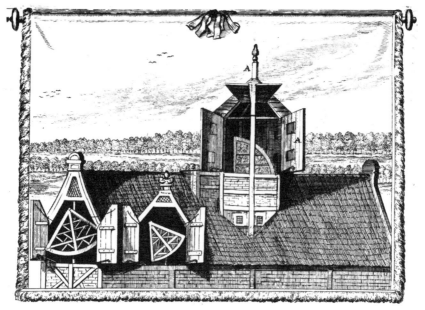

Figure 9.1. Hevelius's house in Danzig. In Johannes Hevelius, *Machinae coelestis pars prior* (Danzig: Simon Reiniger, 1673). Reproduced by permission of the Department of Printing and Graphic Arts, Houghton Library, Harvard College Library. Typ 620.73.451F.

coelestis (1673), these various jury-rigged observatories were all conveniently "contained within the limits of my house, so you don't even need to leave the house, or cross the street . . . to get to another observatory" (Figure 9.1). Noting further that his study was handily located just down the stairs, and that his print shop, with its engraving equipment, was even closer, on the second floor, he triumphantly concluded that his multiple observatories, despite or perhaps even because of their convenient setting right on top of his home, were lacking in nothing that he might need to make "any kind of observations whatsoever."[10]

It must be stressed, however, that natural inquiry was not confined solely to prosperous urban households. On the lowest rungs of the social ladder, peasant homes held carefully gathered herbs, and though learned physicians repeatedly expressed their scorn for home remedies, unofficial healers occasionally fired back with statements such as that of one Ann Windsor, in the sixteenth century, that "kitchen physic I believe is more proper . . . than the Dr's filthy physic."[11] Meanwhile, on the social ladder's highest rungs, kings' and princes' households or courts often served to stage especially massive and complex ventures into natural inquiry, bolstered by their patrons' much

---

[10] Johannes Hevelius, *Machinae coelestis* (Danzig: Simon Reiniger, 1673), pp. 446–7.
[11] Quoted in Pollock, *With Faith and Physic*, p. 94.

more substantial resources (see Moran, Chapter 11, this volume).[12] On the Danish island of Hven, for example, the noble-born astronomer Tycho Brahe (1546–1601) masterfully designed an entire palace, the famous Uraniborg, to serve not only as family residence but as his astronomical observatory and alchemical laboratory as well, on a scale far upstaging that of any other protoscientist of the time.[13] Even on these grander scales, however, the study of the natural world was influenced by similar patterns: of familial interaction, the structuring of space, the division of labor, and the management of household affairs.

NATURAL INQUIRY AS A FAMILY PROJECT

To understand the full significance of the early modern home as a site for early modern science, it is necessary to look beyond the mere physical spaces provided by the home as a dwelling – its rooms and chambers – and to contemplate the household itself as an institution. Social historians have long emphasized the centrality of the family as a unit of economic production and inheritance in early modern Europe. In cultures in which the distinction between "public" and "private" had not yet coalesced in its modern form, and the workplace had not yet been relocated away from the home, the extended household was responsible both for its members' material maintenance and for cultural reproduction more generally – for the transmission of customs and practices from one generation to the next.[14]

The family, furthermore, had long been seen as a model for social relations more generally, guiding the roles of older and younger, male and female, superior and subordinate. Aristotle (384–322 B.C.E.) had declared the household (*oikos*) the foundation of social order. Thus it came to serve, often quite explicitly, as a model for politics and government in early modern Europe.

---

[12] See Mario Biagioli, *Galileo, Courtier: The Practice of Science in the Culture of Absolutism* (Chicago: University of Chicago Press, 1993); and Bruce T. Moran, ed., *Patronage and Institutions: Science, Technology, and Medicine at the European Court, 1500–1750* (Woodbridge: Boydell Press, 1991).

[13] On the design of Uraniborg, see Owen Hannaway, "Laboratory Design and the Aim of Science: Andreas Libavius versus Tycho Brahe," *Isis*, 77 (1986), 585–610. But see also Jole Shackelford, "Tycho Brahe, Laboratory Design and the Aim of Science: Reading Plans in Context," *Isis*, 84 (1993), 211–30; and William R. Newman, "Alchemical Symbolism and Concealment: The Chemical House of Libavius," in Galison and Thompson, eds., *The Architecture of Science*, pp. 59–77.

[14] The literature on this topic is vast and controversies numerous; for a historiographical introduction, see Michael Anderson, *Approaches to the History of the Western Family, 1500–1914* (New York: Cambridge University Press, 1980). General surveys of the field, from a variety of methodological and national perspectives, include Steven Ozment, *When Fathers Ruled: Family Life in Reformation Europe* (Cambridge, Mass.: Harvard University Press, 1983); Edmund Shorter, *The Making of the Modern Family* (New York: Basic Books, 1975); Lawrence Stone, *The Family, Sex, and Marriage in England, 1500–1800* (New York: Harper and Row, 1977); Jean-Louis Flandrin, *Families in Former Times*, trans. Richard Southern (Cambridge: Cambridge University Press, 1979); and Michael Mitterauer and Reinhard Sieder, *The European Family: Patriarchy and Partnership from the Middle Ages to the Present*, trans. Karla Oosterveen and Manfred Hörzinger (Chicago: University of Chicago Press, 1982). For an important early discussion of "cultural reproduction" as applied to the history of science, see Outram, "Before Objectivity."

Patterns of authority within the family, it was believed, formed the basis for relations between ruler and subjects, with the monarch or prince as paternal and benevolent head not only of his own household or court but of the social body on a broader scale.[15] Likewise, the model of the household anchored many early modern conceptions of economic activity, especially with the rise of the economic philosophy of cameralism, which saw the state as a household, requiring proper management to ensure its prosperity and self-sufficiency.[16]

The intellectual sphere, including many of the more formal institutions of early modern science, likewise reflected this family model. This is perhaps most obvious in the case of the princely court, which functioned as a household writ large and saw competition for the favor of the paterfamilias – in this case, the prince – generate considerable interest in the pursuit of nature's more spectacular forms[17] (see Chapter 11, this volume). Famous physicist and astronomer Galileo Galilei (1564–1642), for example, parlayed his eye-catching telescopic accomplishments into a successful bid for the patronage of the Medici court, thus enabling him to exchange his own resource-poor household for a greater one when he moved to Florence as philosopher and mathematician to Cosimo II.[18]

The dominance of the family model can also be seen in early modern university training as images of the solitary scholar, derived from clerical and monastic ideals of celibacy, yielded to a new vision of the scholar as married and participating fully in society as paterfamilias in his own right.[19] Professors in the early modern university often fulfilled the paternal role by taking

---

[15] See, for example, Jean Bethke Elshtain, ed., *The Family in Political Thought* (Amherst: University of Massachusetts Press, 1982); Ernst H. Kantorowicz, *The King's Two Bodies: A Study in Medieval Political Theology* (Princeton, N.J.: Princeton University Press, 1957); Joan B. Landes, *Women and the Public Sphere in the Age of the French Revolution* (Ithaca, N.Y.: Cornell University Press, 1988), pp. 17–22; Simon Schama, *The Embarrassment of Riches* (Berkeley: University of California Press, 1988), pp. 375–98; and Lynn Hunt, *The Family Romance of the French Revolution* (Berkeley: University of California Press, 1992).

[16] See Albion W. Small, *The Cameralists: The Pioneers of German Social Policy* (Chicago: University of Chicago Press, 1909); Erhard Dittrich, *Die deutschen und österreichischen Kameralisten* (Darmstadt: Wissenschaftliche Buchgesellschaft, 1973); Kurt Zielenziger, *Die alten deutschen Kameralisten* (Jena: Gustav Fischer, 1914); and Keith Tribe, "Cameralism and the Science of Government," *Journal of Modern History*, 56 (1984), 263–84. For the intersection of cameralism and science, see Pamela H. Smith, *The Business of Alchemy: Science and Culture in the Holy Roman Empire* (Princeton, N.J.: Princeton University Press, 1994); R. Andre Wakefield, "The Apostles of Good Police: Science, Cameralism, and the Culture of Administration in Central Europe, 1656–1800," PhD diss., University of Chicago, 1999; and Alix Cooper, "'The Possibilities of the Land': The Inventory of 'Natural Riches' in the Early Modern German Territories," in *Oeconomies in the Age of Newton*, ed. Margaret Schabas and Neil DeMarchi (Durham, N.C.: Duke University Press, 2003), pp. 129–53.

[17] See note 13.

[18] Biagioli, *Galileo, Courtier*.

[19] Gadi Algazi, "Scholars in Households: Reconfiguring the Learned Habitus, 1400–1600," *Science in Context*, 16 (2003), 9–42. See also A. A. MacDonald, "The Renaissance Household as Centre of Learning," in *Centres of Learning: Learning and Location in Pre-Modern Europe and the Near East*, ed. Jan Willem Drijvers and A. A. MacDonald (Leiden: E. J. Brill, 1995), pp. 289–98. I would like to thank Dr. Algazi for alerting me to this reference.

in students as boarders; it was thus common for students to lodge with their professor or *Doktorvater* and eat dinner at his table, assuming the role of sons[20] (see Grafton, Chapter 10, this volume). In addition to dissecting sheep at the home of his teacher Giovanni Lorenzo in Perugia, Ippolito of Montereale lived with him, and Galileo, before he was fortunate enough to obtain Medici patronage, had to take in student boarders to supplement his income.[21] Even the scientific academies that came to be formed over the course of the seventeenth century can themselves be seen as following a family model, as members of the Royal Society under the presidency of Isaac Newton (1642–1727), for example, sometimes mirrored the behavior of squabbling siblings, to be publicly rebuked from the head of the table.[22] The household, in short, served in early modern Europe as a general pattern – social, emotional or affective, and physical – for many other kinds of "fictive families" or ersatz households, including but not limited to those of the court, university, and scientific academy, with which it coexisted and overlapped.

This model proved highly suited to the production of natural knowledge in many ways. One of the most important was by enabling activities that could not be carried out entirely by a single individual but rather required cooperative work and support, as was the case for so many of the new empirical sciences, such as natural history and observational astronomy. Structuring the division of labor among household members, the household also ensured the continuity of knowledge and skills and their transmission into the next generation. When Prussian physician and botanist Christian Mentzel (1622–1701) decided it was time to teach his son botany, for example, he "imposed on" him as an "exercise" the time-consuming task of constructing a global multilingual index of plants; his confidence that his son's "juvenile age" would make him "apter for work" paid off, as the boy produced an extremely thorough index, which his father was then able to publish in the confidence that he had also contributed to passing down his own skills.[23]

This transmission of scientific projects from one generation to another often also took place on what could be termed a material as well as an intellectual plane. Sons and daughters inherited not only a close familiarity with the activities of their parents, and the skills and networks of social connections necessary to continue practicing them – what might be termed the "intellectual capital" of a family project – but also its physical capital. Workshops,

---

[20] William Clark, "From the Medieval Universitas Scholarium to the German Research University: A Sociogenesis of the German Academic," Ph.D. dissertation, University of California, Los Angeles, 1986, p. 257; Rainer Müller, "Student Education, Student Life," in *Universities in Early Modern Europe, 1500–1800*, ed. Hilde de Ridder-Symoens (Cambridge: Cambridge University Press, 1996), pp. 345–6. For the example of Linnaeus and his own flock of students, see Lisbet Koerner, *Linnaeus: Nature and Nation* (Cambridge, Mass.: Harvard University Press, 1999).

[21] Dava Sobel, *Galileo's Daughter* (New York: Penguin, 2000), p. 23. On Ippolito, see note 5.

[22] See, for example, John Heilbron, *Physics at the Royal Society during Newton's Presidency* (Los Angeles: William Andrews Clark Memorial Library, 1983), pp. 16, 35–6.

[23] Christian Mentzel, *Pinax botanonymos polyglottos katholikos* (Berlin: Runge, 1682), sig. (a).

tools, scientific instruments, collections of naturalia and scientific curiosities, and, last but not least, book collections were usually private property in societies where lending libraries and public museums only became common well into the eighteenth-century Enlightenment; before then, few universities, courts, or scientific academies could count on well-stocked libraries, let alone the proper facilities and equipment with which to conduct science (see Grafton, Chapter 10, this volume). During the early modern period, individual practitioners of natural philosophy or natural history therefore often found themselves forced to draw upon their own family resources, both intellectual and material, unless they managed to persuade a patron to share with them some of the resources of his or her own household.[24]

The sheer number and prominence of families involved in the early modern study of nature testifies to their centrality to the enterprise. In astronomy, for example, the Cassini family at the Paris Observatory initiated a quasi-dynasty, with successive generations of Cassinis reigning over astronomical observation in France from the late seventeenth century until the fall of the Bastille in 1789;[25] and in early eighteenth-century Prussia, astronomy likewise became a "family business" for Gottfried Kirch (1639–1710), his wife, Maria Winkelmann (1670–1720), their son Christfried (1694–1740), and their daughters Christine (ca. 1696–1782) and Margaretha (dates unknown).[26] The contemporary literature of natural history is likewise rich with scientifically oriented households, such as the Camerarius and Volckamer families in the Holy Roman Empire, the Bauhins in Switzerland, and, perhaps most notably, the household of the renowned Swedish naturalist Carolus Linnaeus (1707–1778), whose daughter published an independent observation on the luminescence of nasturtiums.[27] Medical vocations also tended strongly to "run in

---

[24] Yet again Galileo Galilei (1564–1642) is a case in point. Before he succeeded in attracting the patronage of the Medici family (see Biagioli, *Galileo, Courtier*), he literally turned his own household into a workshop in a number of ways: drawing on his father's mathematical training to develop his own talents; self-publishing a book touting a geometric and military compass he had invented, with its place of publication listed as "In the Author's House"; and hiring a full-time live-in instrument maker to produce these compasses under his own roof. See Sobel, *Galileo's Daughter*, pp. 18, 26, and 27.

[25] So many Cassinis rose to prominence that, to clear up the potential confusion, authors sometimes resort to giving them the dynastic labels of Cassini I, II, III, and IV; see, for example, the *Dictionary of Scientific Biography*, ed. Charles Coulton Gillispie (New York: Scribner, 1981), 3: 100–9. It is perhaps worth noting that the Cassinis themselves intermarried with another prominent astronomical family, the Maraldis, resulting in yet another intergenerational collaboration (see Gillispie, ed., *Dictionary of Scientific Biography*, 9: 89–91).

[26] Schiebinger, *The Mind Has No Sex?* pp. 82–99. This pattern continued well into the nineteenth century, as witnessed by the well-known astronomical contributions of William Herschel (1738–1822), his sister Caroline Herschel (1750–1848), lauded for her observations of comets, and William's son John Frederick William Herschel (1792–1871). See Rob Iliffe and Frances Willmoth, "Astronomy and the Domestic Sphere: Margaret Flamsteed and Caroline Herschel as Assistant-Astronomers," in Hunter and Hutton , eds., *Women, Science, and Medicine, 1500–1700*, pp. 235–65; in the Herschel household, as Caroline Herschel noted, William Herschel had had "almost every room turned into a workshop" (p. 258).

[27] See Ann B. Shteir, *Cultivating Women, Cultivating Science: Flora's Daughters and Botany in England, 1760 to 1860* (Baltimore: Johns Hopkins University Press, 1996), p. 51.

the family," as seen, for example, in the Platter dynasty in sixteenth-century Basel.²⁸

This may have been partially because of the increasing tendency of university professors, especially from the seventeenth century on, to form families closely linked by intermarriage, with professorships and other posts often handed down from fathers to sons or, more indirectly, to sons-in-law.²⁹ This formed part of a more general pattern of the traditional inheritance of both occupations and avocations, which was not confined to the learned elite but flourished in artisanal and craft families more generally, such as those of the Musschenbroeks in Leiden, who spent several generations manufacturing air-pumps and microscopes before finally breaking into the physics professoriate.³⁰ In the family-structured world of early modern Europe, what might look like nepotism to modern eyes was rather viewed as a legitimate transmission of valuable traditions and skills; and, as the examples cited show, some of the most well-known figures of the era passed on their knowledge not just through the impersonal means of institutions and written work but in this most "personal" way.

DIVIDING LABOR IN THE SCIENTIFIC HOUSEHOLD

How then did the early modern household function to enable natural inquiry? To explore this further requires examination of the different roles that members of the household played at different times. An early modern household often embraced not only a "nuclear family" of parents and children but also a range of other possible members. At any point in time, these might include close relatives and other kin and, depending on the wealth and status of the family, other individuals of various kinds, from lodgers, boarders, guests,

---

[28] See Emmanuel Le Roy Ladurie, *The Beggar and the Professor: A Sixteenth-Century Family Saga*, trans. Arthur Goldhammer (Chicago: University of Chicago Press, 1997), esp. pp. 48–9, 114–7, 342, and 344–6. Although Thomas Platter, Sr., began his career as an illiterate peasant boy, his sons Felix and Thomas, Jr., fulfilled their father's medical aspirations, with the former becoming one of the most renowned physicians of sixteenth-century Basel. Each of the three left behind a journal; see, for example, Sean Jennett, trans., *Beloved Son Felix: The Journal of Felix Platter, a Medical Student in Montpellier in the Sixteenth Century* (London: Muller, 1961).

[29] See Friedrich W. Euler, "Entstehung und Entwicklung deutscher Gelehrtengeschlechter," in *Universität und Gelehrtenstand, 1400–1800*, ed. Helmuth Rössler and Günther Franz (Limburg: C. A. Starke Verlag, 1970), pp. 183–232; Clark, "From the Medieval Universitas Scholarium to the German Research University," pp. 372–3; and Algazi, "Scholars in Households," p. 25.

[30] Maurice Daumas, *Scientific Instruments of the Seventeenth and Eighteenth Centuries*, trans. and ed. Mary Holbrook (New York: Praeger, 1972), pp. 84–5. For some further examples of multigenerational families of mathematical practitioners, scientific instrument makers, botanical illustrators, and cartographers, respectively, see E. G. R. Taylor, *The Mathematical Practitioners of Tudor and Stuart England* (Cambridge: Cambridge University Press, 1954), pp. 166–7, 169, 171, 173, 176, 177, 185, 192–3, 199, 200, 201, 203–4, 207; Daumas, *Scientific Instruments of the Seventeenth and Eighteenth Centuries*, pp. 64, 65, 67–8, 68–9, 70, 73, 75–6, 77–8, 83, 84, 85, 87; Wilfrid Blunt and William T. Stearn, *The Art of Botanical Illustration* (Kew: Royal Botanic Garden, 1994), pp. 94, 108–12, 128, 145, 151–3; and Norman J. W. Thrower, *Maps and Civilization: Cartography in Culture and Society*, 2nd ed. (Chicago: University of Chicago Press, 1999), pp. 120, 279, n. 45.

acquaintances, and clients to domestic servants such as cooks, farmhands, chambermaids, stable boys, gardeners, manservants, apprentices, clerks, and personal secretaries.[31] Domestic servants were not generally seen as independent "employees" in the modern sense; rather, living in the household, they were regarded as part of it, subject to the authority of its common head, and were often given quasi-familial status.[32] In their capacity as low-ranking household members, they were assigned a variety of tasks, often menial or manual, and some of these assistants, hired for their mechanical or other useful skills, became the "invisible technicians" whose labor was indispensable in an emerging culture of observation and experiment.[33] At a time when few universities or scientific academies could boast of extensive (or indeed any) official laboratory space, a few wealthy natural philosophers such as Boyle built laboratories in their homes and staffed them with "operators," manservants chosen specifically for their ability to carry out the kinds of manual work (such as experiment) their masters felt would be inappropriate for "gentlemen" (see Smith, Chapter 13, this volume). In the experiments that Boyle and others conducted in their home laboratories, their chambers were far from private; gentlemanly "witnesses" were invited to view experiments, but generally only *after* servants had already perfected their skills in carrying them out.[34] Thus the home was not just an innocuous substitute for floor space not available elsewhere; experiments conducted in the home reflected the resources of the householder, with the "invisibility" of the technicians a direct result of their position within the household not as significant individuals in their own right but as contributors to the family project.

Wives and other female relatives, such as sisters and daughters, likewise performed crucial roles in the early modern scientific household that have often been invisible to modern historians (see Schiebinger, Chapter 7, this volume). Wives did not necessarily distance themselves from their husbands' work, as in the later Victorian ideology of separate public and private spheres; rather, each was expected to serve as her husband's "helpmeet" or companion, helping him accomplish what needed to be done.[35] In this capacity, wives

---

[31] Demographers still debate the currency of the "nuclear family" relative to other forms of families, such as the larger "stem family," in early modern Europe. It is not disputed, however, that the early modern family, especially in prosperous households, might include considerably more individuals than the family of today. For a discussion of this issue, see Anderson, *Approaches to the History of the Western Family*, pp. 4–24.

[32] On the lives and roles of domestic servants in early modern Europe, see, for example, Marjorie McIntosh, "Servants and the Household Unit in an Elizabethan English Community," *Journal of Family History*, 9 (1984), 3–23; Cissie Fairchilds, *Domestic Enemies: Servants and Their Masters in Old Regime France* (Baltimore: Johns Hopkins University Press, 1984); and Ann Kussmaul, *Servants in Husbandry in Early Modern England* (Cambridge: Cambridge University Press, 1981).

[33] See Steven Shapin, "The Invisible Technician," *American Scientist*, 77 (1989), 554–63, and, for a further development of these ideas, his *A Social History of Truth: Civility and Science in Seventeenth-Century England* (Chicago: University of Chicago Press, 1994), pp. 355–407.

[34] See Shapin, "The House of Experiment."

[35] Considerable debate exists concerning the role of the wife in the early modern household. Many histories of family change in Europe have argued that the emergence of the modern world (variously dated) also saw the rise of the "companionate marriage" and a shift from the family as a place

often played active roles in family projects, generally in accordance with a gendered division of labor. One of the most important ways they did so was by "managing" the household. It has been shown, for example, how Jane Dee, wife of the sixteenth-century British astrologer and communer with angels John Dee (1527–1608), worked to ensure his professional success by managing the entrance of visitors and potential patrons into the rooms where he worked and by coping with the assortment of peculiar and unreliable assistants he brought into their household.[36] The *salons* or social gatherings that elite seventeenth- and eighteenth-century French women directed in their drawing rooms can be seen as continuing in this tradition, enabling wives to garner patronage for their husbands' careers while creating intellectual spaces in the home.[37]

Women contributed to family projects in other ways as well. In craft settings, masters' wives and daughters were expected to take part in common tasks.[38] Here, too, gendered divisions of labor manifested themselves. In natural history, for example, wives, daughters, and other female members of the household were often trained to paint or otherwise illustrate plants or other specimens rather than formally "describing" them in Latin, a task allocated to their fathers and brothers. In Danzig, on the shores of the Baltic Sea, the early eighteenth-century physician and naturalist Johann Philip Breyne (1680–1764), himself the son of a naturalist father, had his daughters illustrate the exotic specimens he collected.[39] Meanwhile, across the Atlantic Ocean, Jane Colden (1724–1766) used her artistic training to produce one of the first local floras in North America, with the support of her father.[40] In astronomy, tasks were less obviously gendered during this period, and the activity of astronomical observation seems, in itself, to have been one regarded as suitable for women. Scholars have noticed that many of the observations written down in the notebooks of the English astronomer John Flamsteed (1646–1719), for example, are in the handwriting of his wife, Margaret; many similar cases have been found.[41] Alternatively, wives might contribute to

of economic production to a place for love, affection, and "sentiment"; see, for example, Shorter, *The Making of the Modern Family*; Stone, *The Family, Sex and Marriage in England, 1500–1800*; and Flandrin, *Families in Former Times*. But see also Ozment, *When Fathers Ruled*, for a challenge to this view, with his argument that both companionate marriage and signs of affection are visible even in the earlier forms of the "patriarchal" family.

[36] Harkness, "Managing an Experimental Household."
[37] Dena Goodman, *The Republic of Letters: A Cultural History of the French Enlightenment* (Ithaca, N.Y.: Cornell University Press, 1994); and Outram, "Before Objectivity."
[38] Merry E. Wiesner, *Working Women in Renaissance Germany* (New Brunswick, N.J.: Rutgers University Press, 1986), pp. 152–7.
[39] On wives and daughters as illustrators, see Shteir, *Cultivating Women, Cultivating Science*, pp. 178–82. On women's botanical painting and drawing more generally, see Madeleine Pinault, *The Painter as Naturalist*, trans. Philip Sturgess (Paris: Flammarion, 1991), pp. 43–6. Shteir notes that in the "botanical dialogues" that women began to publish in the eighteenth and early nineteenth centuries, they usually set their fictive conversations at home in the parlor or breakfast room (see pp. 81–3, 110, 174).
[40] Shteir, *Cultivating Women, Cultivating Science*, p. 52.
[41] Lesley Murdin, *Under Newton's Shadow: Astronomical Practices in the Seventeenth Century* (Bristol: Adam Hilger, 1985), p. 64; Iliffe and Willmoth, "Astronomy and the Domestic Sphere," pp. 244–57.

the maintenance of the household by practicing various professions of their own, such as those of midwifery and other medical specialties; such women often handed down their roles from mother to daughter.[42] If a woman's husband died, leaving her widowed, she often carried on the family craft or business (for example, printing or the apothecary trade), sometimes with resistance from guild officials but also with a degree of independence from male control that was almost impossible in early modern Europe for women from the artisanal classes to achieve in any other way[43] (see Schiebinger, Chapter 7, this volume).

Finally, sons had roles of their own to play in the workings of the scientific household. As has already been mentioned, they had a strong tendency to "inherit" the occupations of their fathers, not only in the university but also in craft or guild settings. This was reflected in their education, both formal and informal; sons were often exposed to their fathers' work and from a very early age were trained in the necessary skills. At the beginning of the early modern period, for example, Jacopo Berengario of Carpi (ca. 1460–ca. 1530) worked with his father as an apprentice surgeon before becoming a renowned anatomist at the University of Bologna, and at the end of it, the renowned Swiss physician Johann Jakob Scheuchzer (1672–1733) shared numerous botanizing field trips with his father and grandfather (both physicians) and was also included in many of their daily rounds.[44] Although a father might take on an apprentice or other students, in many cases his son would be his primary student and would be expected to learn to support the family and to carry on the family name after the father's death. To ensure that this process would occur smoothly, sons would gradually be exposed to various aspects of their fathers' work, and, in many cases, ended up helping

---

See also the discussion of Elisabetha Koopman, wife of the astronomer Johannes Hevelius, by Londa Schiebinger in Chapter 7 of this volume, and her portrayal of Maria Winckelmann in *The Mind Has No Sex?* pp. 82–99. For the case of Sophie Brahe, who helped her older brother Tycho observe a lunar eclipse in 1573, see John R. Christianson, *On Tycho's Island: Tycho Brahe and His Assistants, 1570–1601* (Cambridge: Cambridge University Press, 2000), pp. 57, 258–64.

[42] Wiesner, *Working Women in Renaissance Germany*, pp. 37–73, discusses women in the healing professions. She notes, for example, that when summoned before authorities to defend their medical practice, women cited their "feminine skills" (p. 54); in a further example of the division of medical labor, Jewish women enjoyed particular success as "eye-doctors," or oculists, in southern German cities before they were ousted by barber-surgeons (p. 50).

[43] See Olwen Hufton, "Women Without Men: Widows and Spinsters in Britain and France in the Eighteenth Century," *Journal of Family History*, 9 (1984), 355–76, and her *The Prospect Before Her: A History of Women in Western Europe, 1500–1800* (New York: Alfred A. Knopf, 1995), pp. 221–54; see also Wiesner, *Working Women in Renaissance Germany*, pp. 157–63. Although single or separated women were stigmatized in early modern Europe, they, too, might end up with similar arrangements. For the case of Maria Sibylla Merian, see Natalie Zemon Davis, *Women on the Margins: Three Seventeenth-Century Lives* (Cambridge, Mass: Harvard University Press, 1995); and Schiebinger, Chapter 7, this volume.

[44] Vittorio Putti, *Berengario da Carpi: Saggio biografico e bibliografico seguito dalla traduzione del "De fractura calvae sive cranei"* (Bologna: L. Cappelli, 1937); Hans Fischer, *Johann Jakob Scheuchzer (2. August 1672 – 23. Juni 1733), Naturforscher und Arzt* (Zürich: Leemann, 1973), pp. 14–15; and Rudolf Steiger, *Johann Jakob Scheuchzer (1672–1733). 1. Werdezeit (bis 1699)* (Zürich: Leemann, 1927), p. 21.

with it, before or after leaving the family home to pursue further education or apprenticeships elsewhere. Like servants, children (including daughters) might be called upon to perform especially manual or menial work; Felix and Ursula Platter prepared and folded paper for their father's print shop "till their fingers bled."[45]

In a final gesture, sons might be called in to complete projects left unfinished by their father's deaths. In natural history, for example, it was all too common for the publication of local floras, herbals, and other encyclopedic publications to be delayed indefinitely as more and more information was assembled, and upon the illness or death of the prime compiler, his son would be an obvious choice to finish the job and thereby ensure the project's long-delayed entry into the public world of natural knowledge. Thus, in seventeenth-century Königsberg, when physician and naturalist Johann Loesel (1607–1655) fell sick and was unable to publish his work on the local flora of the region, he had his son (also called Johann) publish the book in his stead; a year later, the elder Loesel died.[46] This kind of arrangement ensured that a life's precious work would not be lost but carried on into the next generation.

Early modern homes and households thus served to provide an important element of continuity in an age in which support for scientific activities tended to be inconstant, financially meager, and unevenly distributed. Only with the full support of the household, and in particular with the participation of family members, could many of the laborious, "Baconian" tasks of early modern science, which tended to require extensive information gathering and many years of labor, be brought to fruition. With the rise of scientific academies and other such institutions in the second half of the seventeenth century, the domestic model came gradually to be eclipsed by other, more visible sites for the production of natural knowledge in specialized research facilities. This process was a very slow one, however, and even after middle-class ideologies of the nineteenth century proclaimed science a creation of the public sphere, separate from the private sphere of home and household, family settings continued to offer useful, often crucial resources for the pursuit of science.[47]

---

[45] Le Roy Ladurie, *The Beggar and the Professor*, p. 133.
[46] Johann Loesel, *Plantas in Borussia sponte nascentes e manuscriptis Parentis mei divulgo* (Königsberg: Mensenius, 1654), dedication. See also Alix Cooper, "The Death of the Naturalist: The Labor of Posthumous Publication in Early Modern Natural History," paper presented at the History of Science Society annual meeting, Pittsburgh, Pennsylvania, November 1999.
[47] On the modern persistence of the family in science, see Abir-Am and Outram, eds., *Uneasy Careers and Intimate Lives*; and Helena M. Pycior, Nancy G. Slack, and Pnina G. Abir-Am, eds., *Creative Couples in the Sciences* (New Brunswick, N.J.: Rutgers University Press, 1996). Science was later, of course, brought back into the "private sphere" of the home both through the late nineteenth- and twentieth-century "domestic science" movements, aiming to instruct women on the principles of cookery, housekeeping, and other feminine disciplines, and, even earlier, through the popularization of science for women and children. On the latter, see James A. Secord, "Newton in the Nursery: Tom Telescope and the Philosophy of Tops and Balls," *History of Science*, 23 (1985), 127–51.

10

LIBRARIES AND LECTURE HALLS

Anthony Grafton

Classrooms and libraries called up radically different images in the minds of sixteenth- and seventeenth-century writers. The ideal classroom, as described by teachers such as Desiderius Erasmus (1465–1536) and embodied in public rooms in universities and colleges, professors' teaching rooms in their own houses, and tutors' rooms in palaces and noble villas, was a space of moderate size, designed and equipped as systematically as one of Henry Ford's factories to produce one sort of product: an educated Christian gentleman. A high pulpit, surrounded by desks with benches, dramatized the central role of the teacher and the knowledge he provided. Axioms in Greek and Latin and pictures of plants, animals, and ancient heroes, the latter equipped with moralizing captions, helped students both to memorize and to internalize their teacher's lessons. The only voice to be heard, in theory, was that of the teacher, explicating an assigned text. And the only knowledge transmitted was that presented in the texts – ancient knowledge authenticated by its patina of age and cultural authority, and presented in the true, moral light by an informed and upright teacher.[1]

The ideal library, by contrast, offered a radically different vision of knowledge. As Ioannes Meursius portrayed it in his 1625 celebration of Leiden University, a good library was housed in a spacious room, illuminated by tall windows (Figure 10.1). Its books, arranged in bookcases organized by subject matter, covered the intellectual waterfront: They dealt with modern history, mathematics, and astronomy, as well as classical literature and history. The equipment of the ideal library included more than books. Figure 10.1 depicts portraits of the princes of Orange, globes, and a locked bookcase stuffed with Joseph Scaliger's precious collection of Oriental manuscripts; a massive view of the city of Constantinople suggested that many roads led to the kingdom of useful knowledge. The Figure also shows grave and bearded gentlemen, most

---

[1] See Desiderius Erasmus, *The Education of a Christian Prince*, trans. Michael Cheshire and Michael Heath; *Collected Works of Erasmus*, 86 vols. (Toronto: University of Toronto Press, 1986), 27: 210.

Libraries and Lecture Halls

BIBLIOTHECA PUBLICA.

Figure 10.1. *Bibliotheca publica* in Leiden. In Johannes van Meurs, *Athenae Batavæ, Sive, De urbe Leidensi, & Academia, virisque claris*... (Leiden: Apud A. Cloucquium et Elsevirios, 1625), p. 36. Reproduced by permission of the Department of Rare Books and Special Collections, Princeton University Library.

of them wearing hats and one accompanied by his dog, stalking the aisles and engaging in excited discussions. Meursius, in other words, envisioned the library as a public theater for erudite and civil conversation between equals, where many voices, some virtual and some real, were to be heard at once.[2]

Erasmus and Meursius both sketched idealized visions of a messy and diverse reality. Yet both of them also provide genuinely vivid glimpses into sites of learning that could be found, in multiple forms, in every province of the learned landscape of sixteenth- and seventeenth-century Europe. These visions – and some of the dull, everyday facts and practices that underpinned them – have their place in a history of early modern science. Intellectual and cultural histories of this period have rightly emphasized other sites of learning treated elsewhere in this volume: the anatomy theater, the garden of simples, and the laboratory (see Findlen, Chapter 12; Smith, Chapter 13). In doing so, they accepted contemporaries' categories. Early modern reformers of learning often dramatized themselves as bold rebels out to overthrow the tyranny of book learning. Yet most influential students of nature mastered the elements of knowledge – human and natural – and many learned vital

---

[2] Ioannes Meursius, *Athenae Batavae* (Leiden: Elzevier, 1625). On this work and its intentions, see Anthony Grafton, *Athenae Batavae: The Research Imperative at Leiden, 1575–1650* (Scaliger Lectures, 1) (Leiden: Primavera Pers, 2003).

practices that they would apply as adults to the study of the natural world in these two bookish but contrasting milieus.

THE CLASSROOM

When the German medical reformer Theophrastus Bombastus von Hohenheim, known as Paracelsus (1493–1541), wanted to proclaim the full radicalism of his new approach to knowledge of the human body and its diseases, he challenged the ways of teaching that had been identified for centuries with the college and university. In 1527, he publicly burned the *Canon* of Avicenna (Ibn Sina), a central text in the medical curriculum. In more than one of his writings, moreover, he insisted that doctors needed to leave "that bare knowledge which their schools teach" and "learn of old Women, Egyptians and such-like persons, for they have greater experience in such things than all the Academicians."[3] The real doctor, according to Paracelsus, must turn to reading not the books of men but the larger book of nature.[4] Paracelsus was a self-proclaimed radical, who actually sympathized with the German peasants who rebelled against their lords in 1525. Yet others who totally rejected his social views accepted his critique of traditional learning. Martin Luther, for example, hoped to see Aristotle completely eliminated from the new Protestant curriculum in favor of Pliny,[5] and Andreas Vesalius (1514–1564), professor of anatomy at the University of Padua – for all his commitment to the study of Galen, in Greek – used the title page of *De humani corporis fabrica* (On the Fabric of the Human Body, 1543) to make clear that he based his claim to expert knowledge of the human skeleton and muscles on his own dissections.[6] Some of the seventeenth century's most influential writers on education – notably Tommaso Campanella (1568–1639) and John Amos Comenius (1592–1670) – continued this polemic against the preeminence of the word. They argued that a system of education based on direct knowledge of pictures and specimens, conducted in cities or schools that amounted to massive, accessible collections (*Kunst- und Wunderkammern*), would free humanity from subjection to the dead hand of past authority.[7]

---

[3] Paracelsus, *Of the Supreme Mysteries of Nature*, trans. R. Turner (London, 1655), as cited in Allen G. Debus, *The English Paracelsians* (New York: Franklin Watts, 1965), p. 22.
[4] See James Bono, *The Word of God and the Languages of Man*, vol. 1: *Ficino to Descartes* (Madison: University of Wisconsin Press, 1995).
[5] Sachiko Kusukawa, *The Transformation of Natural Philosophy: The Case of Philip Melanchthon* (Cambridge: Cambridge University Press, 1995).
[6] Andrea Carlino, *Books of the Body: Anatomical Ritual and Renaissance Learning*, trans. John Tedeschi and Anne Tedeschi (Chicago: University of Chicago Press, 1999).
[7] Tommaso Campanella, *The City of the Sun: A Poetical Dialogue*, ed. and trans. Daniel Donno (Berkeley: University of California Press, 1981); and Charles Webster, *The Great Instauration: Science, Medicine, and Reform, 1626–1660* (London: Duckworth, 1975).

A half-century of scholarship has shown that few of these prophecies were borne out in practice. Practically no teacher found it possible to dispense, in practice, with the central curriculum texts of the past. When Philipp Melanchthon (1497–1560) set out to create Protestant secondary school and university curricula that were not laden with the intellectual and theological sins of the past, he found it necessary to center the gymnasium on direct study of the Greek and Latin classics, and the university on that of Aristotle – even though he combined humanistic with scholastic approaches to the texts in a novel and highly influential fashion.[8] At the end of the sixteenth century, the faculty of the most innovative university in Europe, the Calvinist academy in Leiden, where Simon Stevin (1548–1620) taught practical mathematics in Dutch, still held fast to Aristotle in philosophy and offered Latin lectures on his works.[9]

Over time, the classical texts gained new companions in the classroom. Modern textbooks treated core subjects such as dialectic and rhetoric – Melanchthon himself composed groundbreaking treatments of rhetoric and theology for use in instruction – and soon branched out into subjects as varied as the best way to read history and the nature of the cosmos.[10] The new mathematics and the new astronomy, the new Machiavellian politics and the new Tacitean history, gradually invaded even the most traditionalist lecture halls, in the teachers' asides if not in the assigned texts or the formal lectures.[11] New forms of academic exercise also transformed teaching. In the course of the sixteenth and seventeenth centuries, students at the Illustrious Academy of Altdorf found themselves composing lush orations, inspired by medals stamped with emblematic images, at their graduation exercises.[12] Catholic Students in Jesuit colleges from Kiev to Coimbra, and their Protestant counterparts in Oxford and Cambridge, found themselves producing and performing plays, ancient and modern, to appreciative audiences that often included members of royal families.[13]

---

[8] Kusukawa, *Transformation of Natural Philosophy*.

[9] See Th. H. Lunsingh Scheurleer and G. H. M. Posthumus Meyjes, eds., *Leiden University in the Seventeenth Century: An Exchange of Learning* (Leiden: E. J. Brill, 1975); Anthony Grafton, "Civic Humanism and Scientific Scholarship at Leiden," in *The University and the City: From Medieval Origins to the Present*, ed. Thomas Bender (New York: Oxford University Press, 1988); and W. Otterspeer, *Groepsportret met Dame*, vol. 1: *Het bolwerk van de vrijheid: de Leidse universiteit, 1575–1672* (Amsterdam: Bert Bakker, 2000).

[10] See, for example, Mary Suzanne Kelly, *The De mundo of William Gilbert* (Amsterdam: Hertzberger, 1965); and Patricia Reif, "The Textbook Tradition in Natural Philosophy, 1600–1650," *Journal of the History of Ideas*, 30 (1969), 17–32.

[11] Mordechai Feingold, "The Humanities," in *The History of the University of Oxford*, vol. 4: *Seventeenth-Century Oxford*, ed. Nicholas Tyacke (Oxford: Clarendon Press, 1997), pp. 211–357; and "The Mathematical Sciences and New Philosophies," in ibid., pp. 359–448.

[12] F. J. Stopp, *The Emblems of the Altdorf Academy: Medals and Medal Orations, 1577–1626* (London: Modern Humanities Research Association, 1977).

[13] François de Dainville, *L'Éducation des jésuites: XVIe–XVIIIe siècles*, ed. Marie-Madeleine Compère (Paris: Éditions de Minuit, 1978).

Yet anyone who wished to be counted as a learned man had to study texts, word by word and line by line, and one learned to do this in the lecture hall. The teacher literally read the text aloud to his students. He then dictated a series of explanations, often on different levels. Often he began by introducing the text, saying something brief about its author and its genre. Then he would paraphrase it, word by word, in simple Latin, turning the complex word order of poetry into prose and clarifying demanding prose writers such as Tacitus or Livy. Only then, in most classrooms, would he identify and explain the difficult passages in the text, clearing up allusions to mysterious mythical figures and historical events and solving apparent puzzles and contradictions. Lectures moved slowly, often covering no more than thirty or forty lines an hour. But the rich coating of materials with which the teacher overlaid his text turned every course into a small encyclopedia.[14]

Students sat on their benches, entering these hierarchically ordered bits of information into printed copies of the ancient text or modern textbook in question, copies often prepared by the printer for this exercise. Bars inserted between the lines of type left interlinear white space where the student could enter the teacher's prose paraphrase of the text. Wide margins and interleaved sheets, or a separate notebook, enabled the student to record at least a sampling of the teacher's glosses to myths and metaphors. These records of instruction, often written so neatly as to reveal that they were compiled with the help of tutors, became memory palaces, stuffed with recondite and varied information keyed to the memorable ancient text that had stimulated the teacher's remarks.[15]

Bookish though the classroom was, it offered students a surprising amount of information about the natural, as well as the historical and moral, world. Courses on a number of classical authors – not only Pliny, whose encyclopedia played a central role in early modern understanding of the natural world, but poets such as Ovid, Lucretius, and Manilius – turned naturally into lessons on natural history and cosmology. When the late fifteenth-century humanist Paolo Marsi lectured on Ovid's *Fasti* at the University of Rome, he had to identify for his students the "Cilician *spica*" that the Romans had burned on 1 January. He explained in detail that it was not saffron, as others held, but nard, and he showed his listeners a *spica* of nard that he had picked in Cilicia.[16] When he had to discuss the health-giving spring of the nymph Iuturna in the Forum, he backed up his identification of it

---

[14] See in general Anthony Grafton and Lisa Jardine, *From Humanism to the Humanities: Education and the Liberal Arts in Fifteenth- and Sixteenth-Century Europe* (London: Duckworth, 1986), chaps. 1 and 7; and Kristine Haugen, "A French Jesuit's Lectures on Vergil, 1582–1583: Jacques Sirmond between Literature, History, and Myth," *Sixteenth Century Journal*, 30 (1999), 967–85.
[15] Ann Blair, "*Ovidius methodizatus*: The *Metamorphoses* of Ovid in a Sixteenth-Century Parisian College," *History of Universities*, 9 (1990), 73–118.
[16] Ovid, *Fasti*, ed. Pietro Marsi (Venice, 1482), [sig. a viii r].

by explaining that he had taken a student there and cured his dermatitis.[17] A century later, Louis Godebert, a regent master at the Parisian Collège de Lisieux, still turned his lectures on Ovid's *Metamorphoses* into a highly detailed, if completely traditional, description of the climatic zones (frigid, habitable, and torrid), the constellations, the causes of earthquakes, and the mechanisms that had produced the universal flood.[18] The nature that students met in humanistic classrooms was old-fashioned, but the information that they internalized, and the habits of mind that they made their own, stuck with them in adulthood, even as they went about the task of revising the curriculum and its philosophical foundations. Forms of instruction rooted in textual tradition left powerful residues in such prototypically modern thinkers as René Descartes (1596–1650), who developed his notion of "clear and distinct" ideas in dialogue with the ancient treatise on rhetoric by Quintilian that he had mastered at the Jesuit college of La Flèche.[19]

More importantly, what look at first sight like continuously practiced, uniform modes of instruction actually metamorphosed over time, and close study of texts proved capable of accommodating many different sets of interests. In the middle decades of the sixteenth century, for example, Pierre de la Ramée, or Ramus (1515–1572), professor of mathematics at the collège Royal in Paris, powerfully challenged the authority of central ancient authors such as Aristotle, whose views he legendarily dismissed as completely false. Yet Ramus, for all his iconoclasm in the realm of the text, completely accepted the central notion that formal education should rest first on the study of books. He simply insisted that each classical text had a dialectical core of argument, which could best be brought out by a summary or a diagram, as well as a rhetorical husk of allusions, metaphors, and figures of speech, which had to be identified and explicated. And Ramus reinforced the central position of classical teaching in the study of nature by insisting that ancient poets such as Manilius and Virgil offered rigorous and valid information.[20] Influential teachers at universities as far apart as Basel, Leiden, and Cambridge applied his methods to the letter. By doing so, they transformed classroom instruction on Latin verse and prose into a sort of training in formal argument and gave the authority of ancient writers on subjects such as agriculture and astronomy a new lease of authority.

At the highest level of specialized study, scholars also found ways to transform the teaching of texts into a training in new methods of research that could be applied as readily in the natural sciences as in the humanities – and even transform the classroom itself into a kind of seminar that stimulated independent work. No two masters of the world of late humanism detested

---

[17] Ibid., [sig. e vi r].
[18] Blair, "*Ovidius methodizatus*."
[19] Stephen Gaukroger, *Descartes: An Intellectual Biography* (Oxford: Clarendon Press, 1995).
[20] J. J. Verdonk, *Petrus Ramus en de wiskunde* (Assen: Van Gorcum, 1966); and Nelly Bruyère, *Méthode et dialectique dans l'oeuvre de La Ramée* (Paris: J. Vrin, 1984).

one another more than the Oxford scholar Henry Savile (1549–1622) and the Leiden scholar Joseph Scaliger (1540–1609), and with reason. Savile was just the sort of mathematician Scaliger loathed, one who believed he could interpret ancient texts on mathematics and astronomy more proficiently than Scaliger. And Scaliger was just the sort of humanist Savile despised, one who thought he could prove that all the astronomers of his time had misunderstood the problem of precession and all mathematicians had misunderstood the quadrature of the circle. Savile, of course, was right on the technical points at issue. But both men coincided in more substantial ways than they differed. Savile made his lectures on geometry at Oxford into a sophisticated inquiry – philosophical, philological, and scientific – into the actual development of mathematics in the Greek world.[21] Scaliger, for his part, refused to lecture at all. Yet he allowed young students to board with him and to frequent his table, and ended up giving no less a scholar than Hugo Grotius (1583–1645) private lessons in the study of ancient chronology and astronomy. Eventually, Grotius, while still a teenager working under Scaliger's supervision, prepared his own editions of ancient scientific texts, which he obtained by frequenting his teacher's house.[22] For all their adherence to a formidable tradition and all their insistence on the primacy of the text, the cases of these and other influential teachers make clear that text-based instruction was not a mere relic of an older information regime. Throughout the early modern period, its practitioners did their best to keep pace with the development of disciplines, and succeeded, in more than a few isolated cases, in showing that the commentary on a text could accommodate changes in method as well as new information about nature and society.

THE LIBRARY

Like the lecture hall, the library attracted some formidable attacks in the sixteenth and seventeenth centuries. For all his command of bookish culture, the English philosophical reformer Francis Bacon (1561–1626) considered libraries to be the repositories of an older and less powerful form of learning than those he preferred to pursue. It was true that books provided true "images of the minds" of great writers and that reading, as Bacon noted in his essays, made a full man. In the end, however, he saw libraries as the polar opposite of the new spaces of learning that he liked to conjure up in visions of the future of knowledge – such spaces, for example, as the imaginary laboratories, houses of deceit of the senses, and galleries of invention that he

---

[21] Mordechai Feingold, *The Mathematicians' Apprenticeship: Science, Universities, and Society in England, 1560–1640* (Cambridge: Cambridge University Press, 1984); Robert Goulding, "Sir Henry Savile and the Quadrature of the Circle," doctoral dissertation, Warburg Institute, University of London, 1998.

[22] Anthony Grafton, *Joseph Scaliger: A Study in the History of Classical Scholarship*, 2 vols. (Oxford: Clarendon Press, 1983–93), vol. 2.

envisioned for the inhabitants of Bensalem in his *New Atlantis* (1627). In these forcing-houses of knowledge, Bacon argued, men could detach themselves from inherited truths and force nature to yield its secrets. By comparison, Bacon compared libraries with shrines full of relics of the "ancient saints" – places where authorities were revered, not criticized.[23]

Bacon's contemporaries certainly considered him a critic of traditional forms of knowledge. In 1605, he sent the *Advancement of Learning* to Thomas Bodley (1545–1613), who had made himself a kind of patron saint of librarians when he endowed Oxford University with a splendid working library to replace the long-scattered collection given the university by Duke Humfrey of Gloucester. In an accompanying letter, Bacon praised Bodley for having "built an ark to save learning from deluge." But when he later asked Bodley's opinion of the manuscript of his *Cogitata et visa* (Things Thought and Seen, 1612), Oxford's great benefactor had to confess that they belonged to different parties: "I am one of that crew that saye there is and we possesse a farr greater holdfast of Certainteie in your Sciences, then you by your discourse will seeme to acknowledge." "Like a Caryors horse," Bodley admitted, he could not "bawke the beaten way in which I have bene trayned."[24]

This confrontation between two great English Protestant intellectuals seems to embody something larger: a disagreement of principle on the best sort of learning. Bacon, convinced that "antiquity in the order of time was the youth of the world,"[25] demanded that his contemporaries abandon or transform their traditional ways of obtaining knowledge about the world. Inventors, he argued, should replace authors as the objects of admiration, and notebooks filled with new observations and maxims drawn from them should ultimately take the place of the notebooks filled with excerpts from earlier texts that preserved older traditions of knowledge. Bodley, for his part, assured his pious librarian that he wanted to provide the teachers and students of the University of Oxford with "the greatest part of our Protestant writers," but he set out to gather a vast range of other texts as well, from the best editions of the Greek and Roman classics to the most recondite manuscripts from the Near and Far East. He also set great store by the creation of exact and detailed printed catalogues, inventories of the already known.[26] Where Bacon emphasized discovery, Bodley insisted on transmission; where Bacon idolized the new, Bodley stood by the old ways. So, at least, it seems, when

---

[23] Francis Bacon, *De augmentis scientiarum*, in *Works*, ed. James Spedding, Robert Ellis, and Douglas Heath, 7 vols. (London: Longmans, 1857–9), 1: 483, 486–7, as cited in Paul Nelles, "The Library as an Instrument of Discovery: Gabriel Naudé and the Uses of History," in *History and the Disciplines: The Reclassification of Knowledge in Early Modern Europe*, ed. Donald Kelley (Rochester, N.Y.: University of Rochester Press, 1997), p. 43.

[24] Ian Philip, *The Bodleian Library in the Seventeenth and Eighteenth Centuries* (Oxford: Clarendon Press, 1983), p. 3.

[25] Francis Bacon, *The Advancement of Learning*, 1, in *Works*, 2 vols. (London: William Ball, 1838), 1: 11.

[26] Philip, *The Bodleian Library*, p. 2; and Ian Philip, *The First Printed Catalogue of the Bodleian Library, 1605: A Facsimile* (Oxford: Clarendon Press, 1987).

one examines their two regimes of knowledge – and when one compares these with such new spaces of learning and sociability as the museums of Ulisse Aldrovandi (1522–1605) in Bologna or Elias Ashmole (1617–1692) in Oxford.[27]

Libraries, after all, were repositories, replete with the traditional genres and transmitted knowledge of the ancients. In the course of the fifteenth century, to be sure, they took on a new form and prominence, as great families founded libraries of a new kind: secular institutions, housed in large, narrow rooms lit by tall windows, filled with books in uniform, handsomely stamped bindings and designed not for the slow weaving of quotations into florilegia but for rapid erudite research and disputation. In fifteenth-century Florence, for example, the Medici family supported the creation of two purpose-built libraries: that of San Marco, where they deposited the vast collection of the passionate bibliophile Niccolò Niccoli (1363–1437), which that irascible but warm-hearted scholar had already made available to friends and colleagues with magnificent generosity; and the Laurenziana.[28] In sixteenth-century Italy and Northern Europe alike, kings and noblemen competed to equip their palaces and their universities with similar collections.[29]

Yet many of the new public libraries ended up parading the wealth, power, and culture of the rulers who had caused them to be assembled more effectively than they served the needs of scholars. Many of the greatest Italian libraries, from the Marciana to the Vatican, were notoriously hard to enter and harder still to work in. A ferocious librarian barred the magnificent library of the kings of France to scholars for decades. Only after this scary pedant fell out of his chair and burned to death in the fireplace could the English scholar Isaac Casaubon (1558–1614) manage to extract a vital manuscript of a Byzantine world chronicle from the collection and send it to his friend Joseph Scaliger in Leiden. Yet even Casaubon, who delighted in the riches of the Paris collection, never tried to catalogue it – although the lack of a catalogue regularly impeded his efforts to find precious manuscripts and lend them to erudite correspondents across Europe.[30]

University libraries did not always prove more accessible. Claude de Saumaise, a staggeringly erudite philologist, saw himself – though not all his colleagues agreed – as Joseph Scaliger's chosen successor in Leiden. Planning to follow up Scaliger's interests in ancient and modern Near Eastern astronomy and astrology, he asked for and was promised keys both to the library and to the locked bookcases that held its special collections, such as

[27] See Paula Findlen, *Possessing Nature: Museums, Collecting, and Scientific Culture in Early Modern Italy* (Berkeley: University of California Press, 1994); and Lorraine Daston and Katharine Park, *Wonders and the Order of Nature, 1150–1750* (New York: Zone Books, 1998).
[28] Anthony Hobson, *Great Libraries* (New York: Putnam, 1970); B. L. Ullman and Philip Stadter, *The Public Library of Renaissance Florence* (Padua: Antenore, 1972); and Guglielmo Cavallo, ed., *Le biblioteche nel mondo antico e medievale* (Rome: Laterza, 1988).
[29] André Vernet gen. ed., *Histoire des bibliothèques françaises*, 4 vols. (Paris: Promodis, 1988), vol. 2: *Les bibliothèques sous l'Ancien Régime, 1530–1789*, ed. Claude Jolly.
[30] Mark Pattison, *Isaac Casaubon, 1559–1614* (London: Longmans, Green, 1875), pp. 194–208.

Scaliger's Oriental manuscripts. Better still, the university curators gave him a house that opened into the cloister that housed the library, so that he could do research at any time, and in his dressing gown. Yet the library proved an inaccessible paradise. The librarian, Daniel Heinsius, was Saumaise's deadly enemy, and he fought an imaginative and effective rearguard action to make his foe – as Saumaise bitterly complained – the only scholar in Leiden who did not in fact have a key to the library. As its doors opened to the university public only twice a week, for a couple of hours each time, Saumaise found himself genuinely excluded. Thus Meursius's image of the Leiden library (Figure 10.1) misrepresented what was really more a cosmopolitan museum, kept to enhance the city's and the university's prestige by displaying the material results of their investment in culture, than a working scholarly institution.[31] The temptation to dismiss libraries as vestiges of an outworn information order, sanctioned only by tradition, is very strong. On the whole, students and even masters depended on their personal libraries for the books they used most intensively.[32]

Yet in the sixteenth and seventeenth centuries, libraries were more than prominent sites on tourists' mental maps of an imaginary learned Europe. Kings and princes, town councils and university curators, religious orders and wealthy scholars, established, enlarged, and sometimes endowed what they saw as potentially permanent, publicly accessible collections of books. From the first, moreover, they did so as part of what they saw as a larger campaign against ignorance and untruth – one in which libraries were meant to play a special intellectual role. The Florentine library of San Marco, for example, became more than a collection. In the late fifteenth century, it served as a central meeting place for the learned men who worked on the cutting edge of Laurentian culture, such as Pico della Mirandola and Marsilio Ficino, and the deeply erudite texture of their work reflected their access to rare materials, in some cases specially gathered for them. Ficino's *De vita* (On Life, 1489), an immensely learned work on the astrological and medical care of the self, and Pico's *Disputationes contra astrologiam divinatricem* (Disputations against Divinatory Astrology, 1496), a massive and searing criticism of classical astrology, contradicted one another on many points, but both of them shaped the debates of doctors, astrologers, and medical humanists for generations to come. Both works were studded with passages drawn from recondite manuscript materials and exemplified the fruitful ways in which natural philosophy and humanist scholarship could collide in the right library.[33]

---

[31] E. Hulshoff Pol, "The Library," in *Leiden University in the Seventeenth Century*, ed. Lunsingh Scheurleer and Meyjes, pp. 395–459.

[32] E. S. Leedham-Green, *Books in Cambridge Inventories*, 2 vols. (Cambridge: Cambridge University Press, 1986).

[33] Eugenio Garin, *La biblioteca di San Marco* (Florence: Le Lettere, 1999); Paola Zambelli, *L'Ambigua natura della magia*, 2nd ed. (Venice: Marsilio, 1996); Anthony Grafton, *Commerce with the Classics: Ancient Books and Renaissance Readers* (Ann Arbor: University of Michigan Press, 1997), chap. 2; and

Even Johannes Kepler (1571–1630), who transformed classical astronomy into a genuinely new science and resented the time he had to spend on what he called "philological tasks," knew these texts and employed their methods throughout his career.[34]

These practices of fifteenth-century Italian scholars were physically transplanted, in the decades around 1500, to Northern Europe. Johannes Trithemius (1462–1516), the learned Benedictine abbot of Sponheim and Würzburg, made the two libraries he built up into centers of research into both history (especially the literary history of the German world) and nature (especially as understood by earlier writers on science and magic). His immense bibliography of magical works, the *Antipalus maleficiorum* (The Enemy of Witchcraft, 1605), became a standard guide to the literature of magic, learned and illicit alike, in Europe's great age of demonological scholarship.[35] His example and his writings inspired the English mathematical scholar John Dee (1527–1608), whose magnificent library at Mortlake, also designed as an instrument of historical and magical learning, contained a number of copies of Trithemius's own books, their margins spiderwebbed with Dee's precise, eloquent marginal notes.[36] In Dee's house, the study of nature rested on the largest collection of manuscripts and printed books in the United Kingdom, as well as on a vast range of instruments and specimens.[37] At Dee's Mortlake, as at Aldrovandi's Bologna, collections of books and collections of naturalia occupied adjoining spaces, and natural historians and philosophers conducted their inquiries by scrutinizing the manuscripts that weighted down their shelves as well as by pulling on the threads of international networks of colleagues in botany and zoology and ransacking local markets for striking specimens.

If Galileo Galilei (1564–1642) gave the model for a new style in natural philosophy, one that really did find new conventions for presenting evidence and drawing arguments from it, Kepler and Savile, Pierre Gassendi (1592–1655) and Marin Mersenne (1588–1648), Athanasius Kircher (1602–1680), and Gottfried Wilhelm Leibniz (1646–1716) continued to find the pursuit of erudition productive throughout the seventeenth century. The erudite working habits inculcated in libraries do much to explain the persistence of humanistic forms of inquiry, as in Aldrovandi's "emblematic natural history"

Darrel Rutkin, "Astrology, Natural Philosophy, and the History of Science, c. 1250–1700," Ph.D. dissertation, Indiana University, Bloomington, 2002.

[34] Grafton, *Commerce with the Classics*, chap. 5.

[35] Johannes Trithemius, *Antipalus maleficiorum* (Mainz: Lippius, 1605). See Klaus Arnold, *Johannes Trithemius (1462–1516)*, 2nd ed. (Würzburg: Schöningh, 1991).

[36] For Dee's notes, see *John Dee's Library Catalogue*, ed. Julian Roberts and Andrew G. Watson (London: Bibliographical Society, 1990), and the microfilm collection *Renaissance Man: The Reconstructed Libraries of Renaissance Scholars, 1450–1700*, ser. I: *The Books and Manuscripts of John Dee, 1527–1608* (Marlborough: Adam Matthew, 1991–).

[37] William Sherman, *John Dee: The Politics of Reading and Writing in the English Renaissance* (Amherst: University of Massachusetts Press, 1995); and Deborah Harkness, *John Dee's Conversations with Angels: Cabala, Alchemy, and the End of Nature* (Cambridge: Cambridge University Press, 1999).

and Bacon's efforts to read ancient myths as allegorical accounts of natural processes.[38]

In a deeper sense, too, the forms of erudition that the library promoted shaped visions and practices for the study of nature. In the sixteenth century, libraries became weapons in a new form of confessional warfare – one in which the archive of early Christianity was the chief realm of struggle. In Protestant Magdeburg and Oxford, and then in Rome and Milan, the warring churches built up not only systematic collections for church history but also collaborative research teams. Younger scholars did the humble work of collating and excerpting texts and monuments.[39] More experienced ones turned the resulting collections of material into narrative prose, which still others verified and revised. In his *New Atlantis*, Bacon – whose own practices as a researcher involved much excerpting, the results of which moved seamlessly into his collections of empirical observations – took the Magdeburg team as the model for his own grand design for collaborative research, Salomon's House, with its serried ranks of specialized knowledge workers. Bacon dramatized the need for patient teamwork and specialization of intellectual functions not because he had seen visions and dreamed dreams of the Cavendish Laboratory but because he saw the impact of similar practices on the work of historical scholars in his own day.[40] Meanwhile, Bacon's own views began to shape the collecting practices of librarians from Gabriel Naudé (1600–1653) to Leibniz. Learned libraries became the sites where the preeminently Baconian discipline of *historia litteraria* – systematic inquiry into the history of the disciplines and the reasons for their success or failure – could best be pursued.[41] In this and other respects, Bacon's own practices remained anchored to the erudite library and its ways to an extent that would no doubt have surprised his friend and critic Bodley.

---

[38] William Ashworth, "Natural History and the Emblematic World View," in *Reappraisals of the Scientific Revolution*, ed. David C. Lindberg and Robert Westman (New York: Cambridge University Press, 1990), pp. 303–32; Ashworth, "Emblematic Natural History of the Renaissance," in *Cultures of Natural History*, ed. Nicholas Jardine, James A. Secord, and Emma C. Spary (Cambridge: Cambridge University Press, 1995), pp. 17–37; Findlen, *Possessing Nature*; Charles Lemmi, *The Classic Deities in Bacon: A Study in Mythological Symbolism* (Baltimore: Johns Hopkins University Press, 1933); and Paolo Rossi, *Francis Bacon: From Magic to Science*, trans. Sacha Rabinovitch (Chicago: University of Chicago Press, 1968).

[39] Pamela Jones, *Federico Borromeo and the Ambrosiana: Art, Patronage, and Reform in Seventeenth-Century Milan* (Cambridge: Cambridge University Press, 1993); Simon Ditchfield, *Liturgy, Sanctity, and History in Tridentine Italy: Pietro Maria Campi and the Preservation of the Particular* (Cambridge: Cambridge University Press, 1995); and Gregory Lyon, "Baudouin, Flacius and the Plan for the Magdeburg Centuries," *Journal of the History of Ideas*, 64 (2003), 253–72.

[40] See Anthony Grafton, "Where was Salomon's House? Ecclesiastical History and the Intellectual Origins of Bacon's *New Atlantis*," in *Die europäische Gelehrtenrepublik im Zeitalter des Konfessionalismus*, ed. Herbert Jaumann (Wiesbaden: Harrassowitz, 2001), pp. 21–38. For an example of these practices in Bacon's larger milieu, see George William Wheeler, ed., *Letters of Sir Thomas Bodley to Thomas James* (Oxford: Clarendon Press, 1926).

[41] Wilhelm Schmidt-Biggemann, *Topica Universalis* (Hamburg: Meiner, 1983); and Martin Gierl, *Pietismus und Aufklärung: Theologische Polemik und die Kommunikationsreform der Wissenschaft am Ende des 17. Jahrhunderts* (Göttingen: Vandenhoeck and Ruprecht, 1997).

Thus, humanist schools and massive libraries were not the central stages on which the dramas of the new natural philosophy were acted. But they continued, until 1700 and after, to serve vital functions in the economy of knowledge and instruction. Moreover, and more unexpectedly, they inspired and supported the new natural philosophy of the sixteenth and seventeenth centuries, in some ways more effectively and consistently than they supported traditional forms of instruction in Aristotelian natural philosophy or Galenic medicine.

11

COURTS AND ACADEMIES

Bruce T. Moran

An English courtier of the twelfth century lamented that "In the court I exist and of the court I speak, what the court is, God knows. I know not."[1] The same difficulty affects court studies; no one definition of a courtly "site" can stand equally well for all periods, places, and historical circumstances. In the early modern era, political patronage and clientage networks functioned as effective means of government administration;[2] this made the court a "point of contact" in the ongoing exchange and political maneuvering between a ruler and those seeking to influence the direction of royal or princely power, rather than a physical location. Some members of the court resided at a distance from the ruler himself, maintaining a more remote presence as part of a courtly circle. A court was thus more than a household, more than buildings, and more than ritualistic events based in legal custom or ceremonial-administrative protocols. It was also an "abstract totality," a society of individuals in service to, but not necessarily in immediate attendance upon, a sovereign.[3]

The court was an "ethos" as well as an institution,[4] and particular courts gave rise to particular sorts of cultures, each with its own attitudes and habits, its own system for judging merit and value, and its own social and symbolic mechanisms for directing the behavior of its members. Courts also varied according to size and relative number within specific linguistic regions. In politically fragmented areas, courts were larger in number but

---

[1] Quoted in Ralph A. Griffiths, "The King's Court during the War of the Roses: Continuities in an Age of Discontinuities," in *Princes, Patronage, and the Nobility: The Court at the Beginning of the Modern Age, c. 1450–1650*, ed. Ronald G. Asch and Adolf Birke (Oxford: Oxford University Press, 1991), p. 67.

[2] Sharon Kettering, *Patrons, Brokers and Clients in Seventeenth Century France* (New York: Oxford University Press, 1986).

[3] Ronald G. Asch, "Introduction: Court and Household from the Fifteenth to the Seventeenth Centuries," in Asch and Birke, eds., *Princes, Patronage, and the Nobility*, pp. 1–38.

[4] See R. J. W. Evans, "The Court: A Protean Institution and an Elusive Subject," in Asch and Birke, eds., *Princes, Patronage, and the Nobility*, pp. 481–91.

smaller in territories of jurisdiction. Ducal courts predominated in German-speaking areas, and wealthier Italian courts, including the papal court in Rome, were able to bestow status and authority in a proportion far exceeding regional power. In more centralized states such as England, France, and Spain, larger, royal courts overshadowed lesser aristocratic households. In 1522, the immediate household of the French king Francis I (1494–1547) comprised 540 officials divided into sixty categories.[5] By contrast, the territorial courts of German princes were far less imposing (the total number of those attending the Bavarian court in 1500 was around 160)[6] and relied to a larger extent upon the services of indentured retainers (*Diener und Räte von Haus aus*).

Whether a court was large or small, the personality and interests of its ruler directed court life and organized its vitality as a cultural site. In this regard, Renaissance and early modern courts shared much in common with their medieval predecessors. For example, the Valois king Charles V (ruled 1364–80) and his brother, the Duc de Berry, were well known for their artistic and literary interests. Charles was particularly fond of the occult arts, establishing a College of Astrology and Medicine at the University of Paris in 1371 and leaving a library of over a thousand books at his death, many of which related to the arts of astrology, geomancy, chiromancy, and necromancy, with a further seventy volumes given over to astronomy.[7] Magic, astrology, and alchemy depended upon the mastery of method and the application of specific procedures, and princes often supported them from economic and political motives and sought to derive social utility from the occult sciences as forms of court technology. The same motives, combined with the desire for cultural advantage over other courts, ensured attention to esoteric traditions within the category of applied arts well into the early modern era. At the same time, patronage of the applied arts themselves increased, as princes championed efforts in a variety of other areas, encouraging artisanal ventures as well as supporting creativity in military engineering, precision mechanics, observational astronomy, and medicine (see DeVries, Chapter 14, this volume).

Where there were courts, there were also courtiers, and descriptions of the latter ranged from cultured advisers aspiring to virtue by reason of their noble lineage to "base sycophants" and "crumb-catching parasites." In his *L'Arte aulica* (The Courtly Art, 1601), Lorenzo Ducci, secretary to Cardinal Giovan Francesco Biandrate at Ferrara, distinguished "honour," which he

---

[5] R. J. Knecht, "Francis I: Prince and Patron of the Northern Renaissance," in *The Courts of Europe: Politics, Patronage, and Royalty, 1400–1800*, ed. A. G. Dickens (New York: McGraw-Hill, 1977), pp. 99–120.

[6] Maximilian Lanzinner, *Fürst, Räte und Landstände: Die Entstehung der Zentralbehörden in Bayern, 1511–1598* (Veröffentlichungen des Max-Planck-Instituts für Geschichte, 61) (Göttingen: Vandenhoeck und Ruprecht 1980); and Dieter Stievermann, "Southern German Courts around 1500," in Asch and Birke, eds., *Princes, Patronage, and the Nobility*, pp. 157–72.

[7] Hilary M. Carey, "Astrology at the English Court in the Later Middle Ages," in *Astrology, Science, and Society: Historical Essays*, ed. Patrick Curry (Bury St. Edmunds: Boydell Press, 1987), pp. 41–56.

defined as "the opinion held of anothers vertue," from "the honours which are the courtiers end," which he described as "degrees, dignities, power, wealth, and the reputation which spring from them";[8] he observed that just as a tailor knew his cloth and the physician knew the functioning of the human body, courtiers studied the nature of their prince so as to "gently wrest into the prince's mind a love and liking of him" for the sake of personal advancement.[9] The courtier described by the Italian nobleman, writer, and court diplomat Baldassare Castiglione (1478–1529), on the other hand, was a man skilled in both arms and letters. As Castiglione emphasized in *Il cortegiano* (The Courtier, 1528), the courtier's eloquence was not a manipulative tool but the expression of real knowledge gained through classical education; the good courtier deployed this knowledge with grace (*grazia*) and a nonchalance (*sprezzatura*) intended to disguise the difficulty of a particular action. Castiglione assumed noble birth for his courtier-virtuoso, yet the unity of action, refinement, and contemplation that he described also moved non-aristocrats to strive, on the basis of intellectual merit and virtuosity, to enhance the reputation of their prince. As Shakespeare noted about the order of nature, court societies also consistently observed "degree, priority, and place";[10] the resulting dynamics of competition, ambition, dependence, and rivalry fostered its fair share of flatterers and dissemblers. The unhappy courtier "Misaulus," brought to literary life by the humanist, poet, and imperial knight Ulrich von Hutten (1488–1523), complained of imprisonment at court, where the golden chains he bore around his neck were signs of servitude and captivity.[11]

SCIENCE AT COURT

However, the same social forces that made a competitive tournament of court life selectively focused and encouraged individual talents, dignified innovation, and sometimes shaped new topics and directions of natural inquiry. In this way, the court served as an important social site for introducing novel views and technologies and for criticizing older ideas.[12] It also provided a

[8] Lorenzo Ducci, *Arte aulica di Lorenzo Ducci, nella quale s'insegna il modo, che deve tenere il cortigiano per devenir possessore della gratia del suo principe* (Ferrara: Lorenzo Baldini, 1601); an English translation appeared in 1607, translated by Edward Blount, as *Ars Aulica; or, the Courtiers Arte* (London: Melchior Bradwood for Edward Blount, 1607), p. 17.
[9] Ducci, *Arte aulica di Lorenzo Ducci*, p. 100. See Sydney Anglo, "The Courtier: The Renaissance and Changing Ideals," in Dickens, ed., *The Courts of Europe*, pp. 51–2.
[10] William Shakespeare, *Troilus and Cressida*, I.iii.85–86.
[11] Ulrich von Hutten, *Misaulus qui et dicitur Aula Dialogus*, in *Des teutschen Ritters Ulrich von Hutten sämmtliche Werke*, ed. Ernst Hermann Joseph Münch, 3 vols. (Berlin: G. Reimer Verlag, 1823), 3: 18.
[12] See the essays in *Patronage and Institutions: Science, Technology and Medicine at the European Court, 1500–1750*, ed. Bruce T. Moran (Rochester, N.Y.: Boydell Press, 1991), esp. Paula Findlen, "The Economy of Scientific Exchange in Early Modern Italy," pp. 5–24; William Eamon, "Court, Academy, and Printing House: Patronage and Scientific Careers in Late-Renaissance Italy," pp. 25–50; Lesley B. Cormack, "Twisting the Lion's Tail: Practice and Theory at the Court of Henry Prince of

degree of freedom from the intellectual constraints of other social institutions, especially universities, and allowed participation by those excluded from traditional sites and learned professions. Although sometimes viewed as being concerned mostly with amusements and recreations, individual courts encouraged technical ingenuity and theoretical speculation in the very spectacles that were designed as entertainment and offered, on these and other occasions, opportunities for the cross-fertilization of skills between learned and lay protégés. Automata, precision clockwork, illusionistic imitations of the natural world, manufactures from porcelain, hard stone, or rock crystal, theatrical machinery, pyrotechnic displays, and the romance of collecting associated with the *Kunstkammer* (cabinet of art) all combined aspects of art, nature, and science (sometimes as exhibitions of mastery, sometimes as part of the rhetoric of rivalry, or *paragone*) in the ceremonies, festivals, and splendid protocols of court life.[13]

Patronage and clientelism were the most important tools of the court and, outside of the male-dominated papal court and the households of ecclesiastical officials, the use of those tools extended to women as well as men (see Schiebinger, Chapter 7, this volume). Isabella d'Este, marchesa of Mantua, had no trouble constraining the artist Perugino to accept her particular requirements for birds, trees, and specific backgrounds in the paintings she commissioned from him in the early sixteenth century, although this may not have improved the final result.[14] In England and France, aristocratic women in the early modern era proved especially eager to use noble networks to pursue questions of natural philosophy, mathematics, and medicine. Where natural philosophy still remained part of elite literary culture, privilege and patronage became the levers whereby women such as Christina of Sweden (1626–1689), Anna of Denmark (1574–1619), mother of the Stuart navigator prince, Henry, and Margaret Cavendish, Duchess of Newcastle (1623–1673), established themselves on the margins of scientific communities and took part in scientific debate.[15] Yet, even within such aristocratic settings, there

---

Wales," pp. 67–83; Harold Cook, "Living in Revolutionary Times: Medical Change under William and Mary," pp. 111–35; Bruce T. Moran, "Patronage and Institutions: Courts, Universities, and Academies in Germany: An Overview, 1450–1700," pp. 169–83; Pamela H. Smith, "Curing the Body Politic: Chemistry and Commerce at Court, 1664–70," pp. 195–209; and Alice Stroup, "The Political Theory and Practice of Technology under Louis XIV," pp. 211–34.

[13] Thomas DaCosta Kaufmann, ed., *The Mastery of Nature: Aspects of Art, Science, and Humanism in the Renaissance* (Princeton, N.J.: Princeton University Press, 1993).

[14] Charles Hope, "Artists, Patrons, and Advisers in the Italian Renaissance," in *Patronage in the Renaissance*, ed. Guy Fitch Lytle and Stephen Orgel (Princeton, N.J.: Princeton University Press, 1981), pp. 293–343, esp. pp. 307 ff.

[15] Leeds Barroll, *Anna of Denmark, Queen of England: A Cultural Biography* (Philadelphia: University of Pennsylvania Press, 2002); Susanna Åkerman, *Queen Christiana of Sweden and Her Circle: The Transformation of a Seventeenth-Century Philosophical Libertine* (Leiden: E. J. Brill, 1991); and Londa Schiebinger, *The Mind Has No Sex? Women in the Origins of Modern Science* (Cambridge, Mass.: Harvard University Press, 1989), pp. 44 ff.

was a certain degree of "status dissonance,"[16] where abilities and attainments remained unacknowledged and unrewarded. Whereas men above the rank of baron could become members of the Royal Society without much scrutiny, noblewomen were absent from the Society's rolls, and Cavendish herself unleashed a tidal wave of controversy when she asked to visit one of its working sessions.[17]

Styles of patronage, whether in the form of subsidies, appointments, or gift-like "gratifications," not only mirrored the tastes, interests, and political situations of individual benefactors but also reflected the degree to which patrons themselves had become personally involved in particular endeavors. In the case of the Medici family in Florence, as well as that of the papal court in Rome, it has been argued that great patrons elected to eschew signs of expertise in order to avoid being seen as technicians and thus lose social status and power.[18] Yet patronage dynamics of a different sort animated other courts and allowed some rulers, such as the German Landgrave Wilhelm IV of Hesse-Kassel (1532–1592), the French king Francis I (1494–1547), and the Hapsburg Holy Roman Emperor Rudolf II (1552–1612), to indulge their own learning and to pursue projects as practitioners, collectors, and savants. In such settings, those serving the court were often led to accept a combination of social and professional roles. The collaborative efforts that sometimes resulted brought together scholars and artisans (e.g., mathematicians and instrument makers) and conferred social prestige, in the form of aristocratic legitimation, to labors associated with chemical and mechanical workshops.

As professional identities conformed to court expectations, new occupational patterns emerged. Freed from the restrictions of teaching traditional textual canons and set loose from the intellectual constraints of university curricula, astronomers who chose to work at court began to involve themselves more directly in the practical and empirical aspects of studying the heavens, taking part in observational programs at the court's expense, building instruments, and applying themselves to the critical discussion of natural philosophy.[19] At the Prague court of the emperor Rudolf II (ruled 1576–1612), science, art, humanism, and technology intertwined, thanks in large part to the heterogeneity of the interests and backgrounds of court members. Other interests merged there as well. A famous scene created by Rudolf's personal artist Aegidius Sadeler (1570–1629) in 1607 of the imperial reception hall of Hradschin Castle depicted commercial, artistic, and social interaction inside a courtly space (Figure 11.1). Within such a

[16] The term is used by Werner Gundersheimer, "Patronage in the Renaissance: An Exploratory Approach," in Lytle and Orgel, eds., *Patronage in the Renaissance*, pp. 3–23, at p. 18.
[17] Schiebinger, *The Mind Has No Sex?* p. 25.
[18] Mario Biagioli, *Galileo, Courtier: The Practice of Science in the Culture of Absolutism* (Chicago: University of Chicago Press, 1993), pp. 73 ff.
[19] Robert S. Westman, "The Astronomer's Role in the Sixteenth Century: A Preliminary Study," *History of Science*, 18 (1980), 105–47.

Figure 11.1. Vladislav Hall, Hradschin Castle, Prague. Aegidius Sadeler, 1607, engraving. Reproduced by permission of The Metropolitan Museum of Art, Harris Brisbane Dick Fund, 1953. [53.601.10(1)].

space, a variety of noble and non-noble visitors (note the Persian contingent at the center-left) could take advantage of the opportunity for the exchange of knowledge or for fashioning informal ties that might result in future projects.

At Prague, some natural investigators became salaried employees, and others gained appointments that allowed them to continue their own intellectual interests within a practical courtly milieu.[20] In this way, the botanist Charles de L'Écluse (1526–1609) came to supervise the imperial gardens in Vienna during the 1570s. Through his numerous contacts, especially Ogier Ghiselin de Busbecq, rare seeds and bulbs, including tulips (some of the first in Europe), found their way to the garden of Maximillian II (1527–1576),[21] and L'Écluse became one of several botanists, including Pier Andrea Matthioli, Rembert Dodonaeus, Hugo Blotius, and Oliver Busbeck, who helped make up what has been called a "court academy" linked to the imperial court. L'Écluse was particularly well known, however, and his portrait, as well as

[20] R. J. W. Evans, *Rudolf II and His World: A Study in Intellectual History, 1576–1612* (Oxford: Oxford University Press, 1973); and Erich Trunz, *Wissenschaft und Kunst im Kreise Rudolfs II, 1576–1612* (Kieler Studien zur deutschen Literaturgeschichte, 18) (Neumünster: Wachholtz, 1992).
[21] Anna Pavord, *The Tulip* (London: Bloomsbury, 1999), pp. 57–60.

the portraits of other botanists, made up part of a gallery adjoining the garden at Pisa described in the 1640s by the English diarist and early promoter of the Royal Society John Evelyn (1620–1706). Here, as at other Italian gardens, arrangements of plants were set in close proximity to the display of natural rarities – stones, gems, shells, and other precious materials – connecting the garden, with its embedded sculptures, grottoes, and antiquities, to cabinets of natural curiosities.[22] At the imperial court as well, the artistry of the garden, containing rare plants and supervised by some of the most experienced naturalists of the time, stimulated debates about artistic creativity and the control of nature while providing yet another opportunity for the portrayal of the emperor's own political and cultural aspirations.

Another of the imperial "learned celebrities," Paulus Fabritius (ca. 1519–1589), also accepted multiple roles at court, receiving an appointment as imperial *mathematicus* while serving the emperors Ferdinand, Maximilian II, and Rudolf II as personal physician. Fabritius's involvement in the construction of two triumphal arches to welcome the emperor Rudolf II into Vienna in 1577 made use of professional and cultural interchange within the courtly site to turn a court spectacle into a public demonstration of an important technical aspect of the Copernican theory. One arch was festooned with poems and contained a representation of Europe in the figure of a woman constructed so that she knelt before the emperor. Celestial and terrestrial globes made of stone appeared beneath statues of Maximilian and Rudolf, respectively, and each rotated on its axis as the emperor passed by. The turning terrestrial globe revealed the words: "from the opinions of Heraclides of Pontus, Ekphantes the Pythagorean, and Nicolas Copernicus."[23] With others in the imperial circle, Fabritius helped make astronomical observations, composed astronomical works, and discussed novel theories. The tradition of exploring innovative astronomical ideas was thus well established by the time Tycho Brahe (1546–1601) and Johannes Kepler (1571–1630) arrived at the Rudolphine court.

From Prague, Tycho bestowed his own aristocratic status upon activities associated with the study of astronomy and offered hospitality to others, including Kepler, whose talents he found useful in supporting his own cosmological views. Although he succeeded Tycho as court mathematician-astronomer, Kepler's ambition in coming to Prague was to acquire details of Tycho's systematic observations of the planets. It would take Kepler years to gather these together, and their publication, under the title of the *Tabulae Rudolphinae* (Rudolphine Tables) in honor of the emperor, would not occur until 1627, long after the death of both Tycho and Rudolf. Gaining access

---

[22] John Dixon Hunt, "'*Curiosities* to Adorn *Cabinets* and *Gardens*,'" in *The Origins of Museums: The Cabinet of Curiosities in Sixteenth- and Seventeenth-Century Europe*, ed. Oliver Impey and Arthur MacGregor (Oxford: Clarendon Press, 1985), pp. 193–203.

[23] Thomas DaCosta Kaufmann, "Astronomy, Technology, Humanism, and Art at the Entry of Rudolf II into Vienna, 1577," in Kaufmann, ed., *The Mastery of Nature*, chap. 5, esp. pp. 140–4.

to the observations was a complicated court affair and obliged Kepler in the beginning to defend Tycho's claims in a priority dispute with another imperial mathematician, Nicholas Reymers Baer, called Ursus (d. 1600).[24] Following Tycho's death, Kepler gained a court appointment and dedicated both his *Astronomiae pars optica* (*Optical Part of Astronomy*, 1604) and the *Astronomia nova* (New Astronomy, 1609) to the emperor. The latter work bore the marks of imperial publication on its title page, where it was described as appearing "by order and munificence of Rudolf II Emperor of the Romans, etc. worked out at Prague in a tenacious study lasting many years by His Holy Imperial Majesty's mathematician Johannes Kepler."[25] Distribution of the book was technically reserved to Rudolf to be vended privately; even though Kepler eventually sold the edition to the printer to recover funds promised but not paid, there is nonetheless an important sense in which one of the most influential books in the history of astronomy may be considered a courtly production.

Dedicating books to the emperor or to other courtly patrons could bring substantial social rewards. The quest for imperial patronage had earlier led the Brussels-born physician Andreas Vesalius (1514–1564) to dedicate his celebrated anatomical text *De humani corporis fabrica* (On the Fabric of the Human Body, 1543) to the emperor Charles V and its companion volume, the *Epitome*, to Charles's son, Philip II. In this case, the dedications served to reinstate the status of an entire family at court. Although several of Vesalius's forebears had served as imperial physicians, the illegitimacy of Andreas's father allowed him to advance only to the post of imperial apothecary. Andreas reclaimed the family's traditional position in his great work, raising high above the *Fabrica*'s famous frontispiece the heraldic device (three weasels) that had been granted by the emperor to his great-grandfather. The dedications had the desired effect. Whereas the first edition of the *Fabrica* referred to Vesalius as "Professor of the School of Physicians at Padua," the second edition (1555) described him as "Physician of the Most Invincible Emperor Charles V."[26]

Many physicians were also mathematicians, and one important job of court mathematicians was the construction of horoscopes. Astrology related to medical practice as a form of "astronomical engineering." Especially if the

---

[24] Nicholas Jardine, *The Birth of History and Philosophy of Science: Kepler's 'A Defence of Tycho against Ursus' with Essays on its Provenance and Significance* (Cambridge: Cambridge University Press, 1984); Owen Gingerich and Robert Westman, *The Wittich Connection: Conflict and Priority in Late Sixteenth-Century Cosmology* (Transactions of the American Philosophical Society, 78, p. 7) (Philadelphia: American Philosophical Society, 1988), esp. pp. 42–76.

[25] Johannes Kepler, *Gesammelte Werke*, vol. 3: *Astronomia Nova*, ed. Max Caspar (Munich: C. H. Beck, 1937), title page; see also Johannes Kepler, *New Astronomy*, trans. William H. Donahue (Cambridge: Cambridge University Press, 1992).

[26] C. D. O'Malley, *Andreas Vesalius of Brussels, 1514–1564* (Berkeley: University of California Press, 1964); and Andrew Cunningham and Tamara Hug, *Focus on the Frontispiece of the Fabrica of Vesalius, 1543* (Cambridge: Cambridge Wellcome Unit for the History of Medicine, 1994).

patient was a member of a royal family, a high clergyman, or a member of the upper nobility, casting an accurate nativity became an important part of medical diagnosis and treatment, affecting the physician's choice of diet and drugs and determining the timetable for bleeding, crises, and critical days.[27] As medical philosophies developed that emphasized the integration of the individual within the larger world, astrology, like alchemy, gained even more relevance as a technology of courtly life.[28] At the Prague court, Rudolf II became a devotee of astrology and supported as well an array of cabbalists, alchemists, and self-proclaimed magicians. But astrological prognostications had long interested many other kings and princes and had become one of the common features of court culture.[29] In Italy, the libraries of the Gonzaga, Visconti, and Sforza families, as well as those of the Dukes of Urbino and Ferrara, suggest a strong interest in astrology during the fourteenth and fifteenth centuries,[30] and Philip II of Spain (the dedicatee of Vesalius's *Epitome*) encouraged work in astrology and alchemy in addition to promoting practical ventures in medicine, architecture, navigation, and military technology.[31] Astrology related to dynastic ambitions, and participation in the literary and emblematic discourse surrounding such dynastic concerns was one way to secure the attention and support of a powerful patron.

Making his own discoveries fit the dynastic rhetoric of the Medici court became, it has been argued, one of the primary means by which Galileo Galilei (1564–1642) pursued a program of social self-fashioning and legitimation.[32] In this view, Galileo was mathematician, natural philosopher, and court strategist all at once. Although attentive to the possibilities of social advancement by means of dramatic intellectual spectacle, he also recognized in the court a way to advance his discoveries and perhaps also saw there a vehicle for promoting the credibility of his own theoretical claims. In referring to the four moons that he had discovered orbiting Jupiter as the "Medicean Stars" (after the Grand Duke Cosimo and his three brothers), Galileo transformed a matter of science into a "matter of state." Tuscan ambassadors in Prague, Paris, London, and Madrid were promised copies of Galileo's *Sidereus Nuncius* (Starry Messenger, 1610) and held out hope for the arrival of telescopes

---

[27] Lynn White, Jr., "Medical Astrologers and Late Medieval Technology," *Viator*, 6 (1975), 295–308, esp. p. 296; Nancy G. Siraisi, *Medieval and Early Renaissance Medicine: An Introduction to Knowledge and Practice* (Chicago: University of Chicago Press, 1990), esp. pp. 128 ff.
[28] William Newman, "Technology and Alchemical Debate in the Late Middle Ages," *Isis*, 80 (1989), 423–45.
[29] Hilary M. Carey, *Courting Disaster: Astrology and the English Court and University in the Later Middle Ages* (New York: St. Martin's Press, 1992).
[30] Pearl Kibre, "The Intellectual Interests Reflected in Libraries of the Fourteenth and Fifteenth Centuries," *Journal of the History of Ideas*, 7 (1946), 257–97, esp. pp. 285–7; and Hilary Carey, "Astrology at the English Court," in Curry, ed., *Astrology, Science, and Society*, p. 47.
[31] David C. Goodman, *Power and Penury: Government, Technology, and Science in Philip II's Spain* (Cambridge: Cambridge University Press, 1988).
[32] Biagioli, *Galileo, Courtier*, chap. 1.

constructed by Galileo but paid for from the Medici court treasury.[33] In this case, the social network afforded by court connections and ambassadorial channels produced a powerful tool of observational verification and also stimulated the discussion of cosmological issues.

In evaluating courts as sites of natural knowledge in the early modern era, it is important not to limit discussion solely to the best-known personalities and most powerful rulers. Projects cultivated at smaller courts also helped to establish new social conditions that allowed naturalists to interact with and describe nature in novel ways. Especially where the consolidation of regional power and claims to legal jurisdiction linked political and economic ambitions to the patronage of projects involving practical mathematics, the making of precision instruments (including navigational devices, proportional compasses, triangulation instruments, and surveying tools) acquired courtly status. In this regard, a style of patronage that has been called "utilitarian" predominated in Elizabethan and Jacobean England. Prominent within this courtly environment was the family of William Cecil (Lord Burghly, 1520–1598), which, according to historians Stephen Pumfrey and Frances Dawbarn, "formed the vital centre of a network of cultural, artistic, and intellectual patronage unequalled in England in the second half of the sixteenth and early seventeenth centuries." Burghly's patronage extended to projects in agriculture and the mechanical arts, as well as alchemical schemes, including participation in a short-lived society for the art of making copper and quicksilver by means of transmutation. Consistent with the utilitarian goals of patronage in early modern England, another favorite of the Elizabethan court, Robert Dudley (Earl of Leicester, 1532–1588), offered support to the well-known astronomer Thomas Digges (ca. 1545–1595). It was not innovative astronomy that Dudley desired, however. The support of Digges stemmed from more pragmatic mathematical concerns, especially Digges's talents as a military engineer and surveyor.[34]

Technical feats of precision engineering made possible the production of mechanical automata as well as clockwork-driven celestial globes and astronomical clocks. Some of the best examples of such self-automated celestial automata were to be found at the court of the German Landgrave Wilhelm IV of Hesse-Kassel (1532–1592) at the end of the sixteenth century, where they arose as part of a courtly aesthetic that emphasized technical precision and mirrored the prince's zeal for reform of astronomical measurement through the creation of more exact observational methods and instruments.[35]

---

[33] Concluding remarks by Albert van Helden in his translation of Galileo Galilei, *Sidereus Nuncius, or The Sidereal Messenger* (Chicago: University of Chicago Press, 1989), p. 100.
[34] Stephen Pumfrey and Frances Dawbarn, "Science and Patronage in England, 1570-1625: A Preliminary Study," *History of Science*, 42(2004), 137–88.
[35] Bruce T. Moran, "German Prince-Practitioners: Aspects in the Development of Courtly Science, Technology, and Procedures in the Renaissance," *Technology and Culture*, 22 (1981), 253–74.

Wilhelm was a true prince-astronomer, and his court at Kassel became a prominent site for serious projects of observational astronomy. Those who contributed to the efforts in Kassel – for example, the mechanician Jost Bürgi (1552–1632) and the astronomer-mathematician Christoph Rothmann (died after 1597) – did so as serious collaborators in projects often chosen by the prince. Correspondence between courts provided routes for the exchange of observations and opinions; such avenues were especially important to Rothmann, whose correspondence with Tycho Brahe in the late 1580s included debates concerning the substance of the heavens and arguments in favor of the Copernican hypothesis.[36] At the court of Wilhelm's son, Moritz (1572–1632), the prince's patronage of alchemy and occult philosophy extended across institutional sites to the University of Marburg, where Moritz created the new professorship of chemical medicine (*chymiatria*). For the post, the prince chose Johannes Hartmann (1568–1631), a professor of mathematics at Marburg, whose own attempt at career building led him to turn his attention to medicine and chemistry, in line with the Kassel prince's patronage interests.[37]

Although some who sought court vocations bent their own talents to the interests of patrons, others attempted to shape those interests by influencing patronage decisions or helping to alter previous patronage patterns. At the Kassel court of Moritz of Hesse, the court physician Jacob Mosanus (1564–1616) argued for a change in the prince's patronage of chrysopoeia (gold making) in favor of endeavors leading to the preparation of chemical medicines.[38] In the well-chronicled case of the German court physician and mathematician Johann Joachim Becher (1635–1682), a client of the Bavarian court sought to shift the attention of his prince from alchemical and chemical projects toward more certain commercial and technical ventures.[39]

At many courts, particularly in Italy and Northern Europe, alchemical interests combined with interests in magic and the occult arts and led to the support of nontraditional medical ideas and procedures. Paracelsian physicians in particular often seem to have relied upon court positions to establish

---

[36] Bernard R. Goldstein and Peter Barker, "The Role of Rothmann in the Dissolution of the Celestial Spheres," *British Journal for the History of Science*, 28 (1995), 385–403.
[37] Bruce T. Moran, *The Alchemical World of the German Court: Occult Philosophy and Chemical Medicine in the Circle of Moritz of Hessen (1572–1632)* (Stuttgart: Franz Steiner Verlag, 1991); Moran, *Chemical Pharmacy Enters the University: Johannes Hartmann and the Didactic Care of Chymiatria in the Early Seventeenth Century* (Madison, Wis.: American Institute of the History of Pharmacy, 1991); Heiner Borggrefe, Vera Lüpkes, and Hans Ottomeyer, eds., *Moritz der Gelehrte, Ein Renaissancefürst in Europa* (Eurasburg: Edition Minerva, 1997); Heiner Borggrefe, "Das alchemistische Laboratorium Moritz des Gelehrten im Kasseler Lusthaus," in *Landgraf Moritz der Gelehrte: Ein Calvinist zwischen Politik und Wissenschaft*, ed. Gerhard Menk (Marburg: Trautvetter und Fischer, 2000), pp. 229–52; Hartmut Broszinski, "Die alchemistische Bibliothek des Landgrafen Moritz: Der Landgraf und die Bücher," in Menk, ed., *Landgraf Moritz der Gelehrte*, pp. 253–62.
[38] Moran, *The Alchemical World of the German Court*, pp. 70 ff.
[39] Pamela H. Smith, *The Business of Alchemy: Science and Culture in the Holy Roman Empire* (Princeton, N.J.: Princeton University Press, 1994).

the credibility of their medical theories and practices, as was the case in the courts of Cosimo II and Don Antonio de' Medici in Florence.[40] Members of the English court proved helpful in advancing the claims of those seeking to organize themselves into a society of chemical physicians in the mid-1660s.[41] Earlier, the French Paracelsian physician Joseph Duchesne (Quercetanus) (ca. 1544–1609) was able to compose a list of princes who were sympathetic to chemical medicine. These included the emperor and the king of Poland, as well as the archbishop of Cologne, the Duke of Saxony, the Landgrave of Hesse, the Margrave of Brandenburg, the dukes of Braunschweig and Bavaria, and the princes of Anhalt.[42] One of the first publishers of Paracelsus's (Theophrastus Bombastus von Hohenheim, 1493–1541) works, Adam von Bodenstein, was court physician to the German Duke of Neuburg, who later became Elector of the Palatinate, Ottheinrich. Later, The Danish professor of medicine and royal physician Petrus Severinus (1542–1602) helped to "deradicalize" Paracelsus's ideas while making them more acceptable to scholarly communities in a book dedicated to his king, Frederik II.[43] The chemical investigations of Oswald Croll (ca. 1560–1608), which would lead to the creation of one of the most significant expositions of Paracelsian remedies, the *Basilica chymica* (1609), took shape with the financial support of the Calvinist prince Christian I of Anhalt-Bernburg. Croll, who was named *medicus ordinarius* (chief medical representative) by the prince and who acknowledged his debt to Christian in his preface to the *Basilica*, was also a political agent of the court, representing Anhalt in confessional and political dealings with Protestants in Bohemia.[44]

The medical controversies aroused by court physicians not only reaffirmed the potential of the court as a locus of innovation but sometimes inspired further refinement of intellectual positions among interested onlookers who were themselves removed from the courtly "site." The most notable disputes raged at Paris. There three Paracelsian physicians associated with the

---

[40] Paolo Galluzzi, "Motivi paracelsiani nella Toscana di Cosimo II e di Don Antonio dei Medici: Alchimia, medicina 'chimica' e riforma del sapere," in *Scienze, credenze occulte, livelli di cultura: Convegno internazionale di studi (Firenze, 26–30 giugno 1980)* (Florence: Leo S. Olschki, 1982), pp. 31–62.

[41] Harold J. Cook, *The Decline of the Old Medical Regime in Stuart London* (Ithaca, N.Y.: Cornell University Press, 1986), pp. 145 ff.

[42] Hugh Trevor-Roper, "The Court Physicians and Paracelsianism," in *Medicine at the Courts of Europe, 1500–1837*, ed. Vivian Nutton (London: Routledge, 1990), pp. 79–94, at p. 89.

[43] Jole Shackelford, "Paracelsianism and Patronage in Early Modern Denmark," in Moran, ed., *Patronage and Institutions*, pp. 85–109; Shackelford, "Early Reception of Paracelsian Theory: Severinus and Erastus," *Sixteenth Century Journal*, 26 (1995), 123–36; Ole Peter Grell, "The Acceptable Face of Paracelsianism: The Legacy of *Idea Medicinae* and the Introduction of Paracelsianism into Early Modern Denmark," in *Paracelsus: The Man and His Reputation, His Ideas and Their Transformation*, ed. Ole Peter Grell (Leiden: E. J. Brill, 1998), pp. 245–67.

[44] Owen Hannaway, *The Chemists and the Word: The Origins of Didactic Chemistry* (Baltimore: Johns Hopkins University Press, 1975), p. 2 ff.; and Wilhelm Kühlmann and Joachim Telle, eds., *Oswald Crollius, De signaturis internis rerum: Die lateinische Editio princeps (1609) und die deutsche Erstübersetzung (1623)* (Stuttgart: Franz Steiner Verlag, 1996), Einleitung, pp. 6 ff.

court of the French king Henry IV – Jean Ribit, sieur de la Rivière (ca. 1571–1605), Duchesne, and Theodore de Mayerne (1573–1655) – contended with the faculty of medicine;[45] controversies concerning the use of chemical medicines continued to involve court personalities (specifically the Franciscan Gabriel Castagne, who dedicated a work on potable gold to Marie de' Medici and named himself councilor and chaplain to the king, and the physician and advocate of chemical and metallic remedies Nicolas Abraham de la Framboisier) during the reign of Louis XIII (ruled 1610–1643).[46]

The interest at court in chemical medicines was partly ideological – the retinue of Henry IV (1553–1610) contained several Huguenot physicians – and partly practical, based on the appeal of novel procedures promising quicker cures with fewer unpleasant side effects to endure. In particular, the dispute between Duchesne and the faculty of medicine at Paris, led by its censor Jean Riolan (the elder, 1539–1606), attracted the attention of the German chemist and schoolmaster Andreas Libavius (ca. 1555–1616). Libavius worried that the condemnation of court doctors was also a censure of practical alchemy. At the same time, he chafed at the possibility that Paracelsian physicians might use the power of the court to redefine the conditions of linguistic authority in medicine and replace traditional Greek sources of medical terminology with their own formulations.[47] In the end, however, the more pressing need to defend the utility of chemistry in medicine placed Libavius on the side of court doctors. In this instance, a debate focused on the court, complicated by the interweaving of political, religious, and medical intricacies, helped to clarify the personal position of a courtly outsider among Galenic, Hippocratic, and Hermetic medical philosophies.

CABINETS AND WORKSHOPS

Courts also became sites of early museums or cabinets of curiosities, where works of art, antiquities, objects of nature, and mechanical marvels were collected together and offered yet another type of aulic spectacle while drawing attention to princely wealth and power. Whereas earlier medieval collections

---

[45] See Hugh Trevor-Roper, "The Paracelsian Movement," in *Renaissance Essays*, ed. Hugh Trevor-Roper [1961] (Chicago: University of Chicago Press, 1985), pp. 149–99; Allen G. Debus, *The French Paracelsians: The Chemical Challenge to Medical and Scientific Tradition in Early Modern France* (Cambridge: Cambridge University Press, 1991), pp. 46–65; and Didier Kahn, "Inceste, assassinat, persécutions et alchimie en France et à Genève (1576–1596): Joseph Du Chesne et Mlle. de Martinville," *Bibliothèque d'humanisme et renaissance*, 63 (2001), 227–59.

[46] Stephen Bamforth, "Paracelsisme et médecine chimique à la cour de Louis XIII," in *Paracelsus und seine internationale Rezeption in der Frühen Neuzeit*, ed. Heinz Schott and Ilana Zinguer (Leiden: E. J. Brill, 1998), pp. 222–37.

[47] Bruce T. Moran, "Libavius the Paracelsian? Monstrous Novelties, Institutions, and the Norms of Social Virtue," in *Reading the Book of Nature: The Other Side of the Scientific Revolution*, ed. Allen G. Debus and Michael T. Walton (Kirksville, Mo.: Sixteenth Century Journal Publishers, 1998), pp. 67–79; Moran, "Medicine, Alchemy, and the Control of Language: Andreas Libavius Versus the Neoparacelsians," in Grell, ed., *Paracelsus: The Man and His Reputation*, pp. 135–49.

of wondrous and precious objects tended to be associated with dynasties and religious institutions, the fourteenth and fifteenth centuries witnessed the construction of collections of a more purely natural sort by individuals, especially by members of the higher nobility and by wealthy participants of the urban elite.[48] The most famous sixteenth- and early seventeenth-century cabinets were to be found in the palaces of the Medici in Florence and the Gonzaga in Mantua, and included as well the collection at the castle Ambras of the Hapsburg archduke Ferdinand II and those of the emperors Maximilian II in Vienna and Rudolf II in Prague.[49] Smaller courts could also assemble significant collections, and many took shape as a result of the patronage of German princes, especially August of Saxony, the Bavarian dukes Wilhelm IV and Albrecht V, August of Braunschweig-Lüneburg, and the princes of Hesse-Kassel.[50] The Berlin *Kunst- und Naturalienkammer* of Friedrich-Wilhelm and Frederick III of Brandenburg and, in France, the *Cabinet du Roi* begun by Louis XIII, became especially well known in the eighteenth century. Sometimes entire private collections of natural objects ended up in princely hands, where their arrangements were guided by artistic and historical criteria rather than by principles of natural philosophy. Such was the fate of a large collection of natural and artificial objects brought together by the Danish physician, university professor, and court adviser Olaus Worm (1588–1654), whose collection was incorporated into the Copenhagen *Kunstkammer* of the Danish king Frederik III in 1650.[51]

---

[48] Lorraine Daston and Katharine Park, *Wonders and the Order of Nature, 1150–1750* (New York: Zone Books, 1998), pp. 86 ff.

[49] Giuseppe Olmi, "Science – Honor – Metaphor: Italian Cabinets of the Sixteenth and Seventeenth Centuries," pp. 5–16; Elisabeth Scheicher, "The Collection of the Archduke Ferdinand II at Schloss Ambras: Its Purpose, Composition, and Evolution," pp. 29–38; Rudolf Distelberger, "The Hapsburg Collections in Vienna during the Seventeenth Century," pp. 39–46; and Eliska Fucíková, "The Collection of Rudolf II at Prague: Cabinet of Curiosities or Scientific Museum?" pp. 47–53; all in Impey and MacGregor, eds., *The Origins of Museums*. Also Thomas DaCosta Kaufmann, *The School of Prague: Painting at the Court of Rudolf II* (Chicago: University of Chicago Press, 1988); Paula Findlen, *Possessing Nature: Museums, Collecting, and Scientific Culture in Early Modern Italy* (Berkeley: University of California Press, 1994); and Daston and Park, *Wonders and the Order of Nature*, pp. 135–72.

[50] Julius von Schlosser, *Die Kunst- und Wunderkammern der Spätrenaissance: Ein Beitrag zur Geschichte des Sammelwesens* (Leipzig: Klinkhardt und Biermann, 1908); Gerhard Händler, *Fürstliche Mäzene und Sammler in Deutschland von 1500–1620* (Strassburg: Heitz und Cie, 1933); Werner Arnold, ed., *Sammler, Fürst, Gelehrter: Herzog August zu Braunschweig und Lüneburg, 1579–1666* (Wolfenbüttel: Herzog August Bibliothek, 1979); Joachim Menzhausen, "Elector Augustus's *Kunstkammer*: An Analysis of the Inventory of 1587," pp. 69–75; Lorenz Seelig, "The Munich *Kunstkammer*, 1565–1807," pp. 76–89; Franz Adrian Dreier, "The *Kunstkammer* of the Hessian Landgraves in Kassel," pp. 102–9; and Christian Theuerkauff, "The Brandenburg *Kunstkammer* in Berlin," pp. 110–14; all in Impey and MacGregor, eds., *The Origins of Museums*; and Thomas DaCosta Kaufmann, *Court, Cloister, and City: The Art and Culture of Central Europe, 1450–1700* (Chicago: University of Chicago Press, 1995).

[51] H. D. Schepelern, "Natural Philosophers and Princely Collectors: Worm, Paludanus, and the Gottorp and Copenhagen Collections," in Impey and MacGregor, eds., *The Origins of Museums*, pp. 121–7; and Jole Shackelford, "Documenting the Factual and the Artifactual: Ole Worm and Public Knowledge," *Endeavour*, 23 (1999), 65–71.

Nevertheless, in many such princely or royal collections, examples of human technical virtuosity were given a place beside exhibits that emphasized the complexity of nature. Collections in princely hands accentuated foreign novelties and the wondrous and marvelous aspects of a world in which mystical and practical artistry resided together in a common workshop. In these collections, natural wonders combined with the marvels of virtuoso performance and encompassed in a single site specimens of both craft mysteries and the secrets of nature.[52]

The workshop was closely related to the curiosity cabinet, and some objects on display combined fictive and natural elements in such a way as to communicate dynastic or personal messages when works of nature were marvelously turned into works of art. When, in 1588, Cardinal Grand Duke Ferdinando I appointed the Roman nobleman Emilio de' Cavalieri as superintendent of artistic productions at the Florentine court, the *patente* required him to supervise all "jewelers, carvers of any type [*intagliatori di qual si vogla sorte*], cosmographers, goldsmiths, the makers of miniatures, gardeners of the gallery, turners, confectioners, clock makers, artisans of porcelain, distillers, sculptors, painters, and makers of artificial gems."[53] This combination of skills created collaborations between *artefici* and recalled the mix of experts that had worked for Grand Duke Francesco I (1541–1587) at the Casino Mediceo, where alchemists and medicinally inclined *stillatori* (distillers) labored in close proximity and sometimes were the same person. Francesco liked to visit court workshops and became fascinated by technical expertise and the "books of secrets" tradition. The French essayist Michel de Montaigne (1533–1592) noted in his travel journal: "The same day we saw the palace of the duke, where he himself takes pleasure in working at counterfeiting oriental stones and cutting crystals: for he is a prince somewhat interested in alchemy and the mechanical arts."[54]

For Francesco's successor, Ferdinando, a personal interest in creations from hard stone, such as porphyry, provided the aesthetic basis for another sort of collaboration that brought together the arts of metallurgy, botany, and distillation as distillers created herbal *tempering media* with which to harden the steel tools needed to create stone inlays. In this case, courtly value was placed not just on the collection and display of minerals and stones, which indicated access to costly and exotic things, but also on the ability to act upon them, using specially tempered tools to cut hard stones, changing rock crystal into

---

[52] William Eamon, *Science and the Secrets of Nature: Books of Secrets in Medieval and Early Modern Culture* (Princeton, N.J.: Princeton University Press, 1994), pp. 221 ff.

[53] Quoted in Suzanne B. Butters, "'Una pietra eppure non una pietra': Pietre dure e botteghe medicee nella Firenze del Cinquecento," in *Arti fiorentine: La grande storia dell'artigianato*, vol 3: *Il Cinquecento*, ed. Franco Franceschi and Gloria Fossi (Florence: Giunti Gruppo Editoriale, 2000), p. 144.

[54] Quoted in Paolo Rossi, "*Sprezzatura*, Patronage, and Fate: Benvenuto Cellini and the World of Words," in *Vasari's Florence: Artists and Literati at the Medicean Court*, ed. Philip Jacks (Cambridge: Cambridge University Press, 1998), pp. 55–69, at p. 64.

glass and refashioning the mass into other shapes, or embroidering white marble with other materials. Technology and collaboration within court workshops created powerful images of the Medici-*medicus*, guarantor of public health. In defying the nature of hard stones, the collaboration produced metaphors of personal hardness, moral transparency, and spiritual control.[55]

In some instances, collections drew attention toward reconsidering local natural phenomena, and in others toward acknowledging as part of nature's perfection the place of the monstrous, exotic, and rare. In either case, observers were drawn to consider the particulars of nature and were often brought face to face with the discontinuities of a presumed natural order. Those who witnessed such novelties could lay claim to a type of authority based on experience rather than texts and would insist, in debates about natural philosophy, on the value of empirically derived arguments as opposed to reasoning based on assertion.[56] Collecting was also pleasurable, and the desire for what is pleasing cannot be discarded as either a motive for or a consequence of the scientific and technological projects organized within the courtly site. Emotions became means of cognition within the spaces where the exotic and curious forms of nature nurtured a passion for wonder. Taking pleasure in the right sorts of things had long been associated with the habituation of a virtuous person. Within the courtly context, emotions connected to wonder, pleasure, and even horror helped form and alter beliefs about nature and contained just enough momentum to make the assessment of the natural order more than a purely reflective maneuver.[57]

Merchants and missionaries as well as natural philosophers attempted to garner favor by pleasing the court with gifts of natural curiosities. Some were among the most well-known collectors of naturalia in the early modern era. Although he was seldom in Florence, the Medici court naturalist, Ulisse Aldrovandi (1522–1605), became a favorite of the Grand Duke Francesco, who, in return for expressions of praise and loyalty, intervened on Aldrovandi's behalf in seeking financial concessions from the Senate of Bologna.[58] Another successful court client, the Jesuit Athanasius Kircher (1602–1680), used collecting and publishing as a means of participating in court debates.[59] His controversy with the court physician, naturalist, and superintendent of the ducal pharmacy Francesco Redi (1626–1698), concerning the efficacy of the so-called snakestone (one of many curiosities collected

---

[55] Suzanne B. Butters, *The Triumph of Vulcan: Sculptors' Tools, Porphyry, and the Prince in Ducal Florence*, 2 vols. (Florence: Leo S. Olschki, 1996), 1: 215–77, 333–50; and Butters, "'Una pietra eppure non una pietra,'" in Franceschi and Fossi, eds., *Arti fiorentine*, 3: 144–63.
[56] Lorraine Daston, "The Factual Sensibility," *Isis*, 79 (1988), 452–67.
[57] Daston and Park, *Wonders and the Order of Nature*, pp. 144 ff. Regarding emotions and rationality, see Jon Elster, *Alchemies of the Mind: Rationality and the Emotions* (Cambridge: Cambridge University Press, 1999); and John M. Cooper, *Reason and Emotion: Essays on Ancient Moral Psychology and Ethical Theory* (Princeton, N.J.: Princeton University Press, 1999).
[58] Findlen, *Possessing Nature*, pp. 352–75, esp. pp. 359 ff.
[59] Ibid., pp. 78 ff., 217 ff., 346 ff.

at the Tuscan court of Ferdinand II) became part of an intense competition for the patronage of the Medici in Florence. The debate centered in part on the trustworthiness of experiment and testimony, and brought into question the authority of courtier witnesses as opposed to the statements of distant (Jesuit) observers.[60] Indeed, the pharmacy, or *spezieria*, of the Tuscan court was a space in which the exploration of the larger relevance of natural objects was actively pursued. Within this particular subsite of the court, personal experience sometimes merged with other cultural functions of a literary, historical, and poetical nature, as in Redi's attempts to discover the lethal powers of vipers and the means by which the Egyptian queen Cleopatra might have committed suicide.[61]

FROM COURT TO ACADEMY

Scientific ideals such as the recognition of the value to natural inquiry of precision observation, collaboration, practical experience, and technical expertise found fertile ground within courtly contexts, and rulers helped to advance the claims and discoveries of their protégés. Nevertheless, the emergence of new scientific organizations and academies, especially in the late seventeenth century, encouraged a shift in claims to authority from the invocation of personal relationships and privileged social status to membership within collective institutions. There, credibility emerged as a result of corporate effort and combined with experimental practices and the communal determination of "matters of fact."[62] The shift from "the flesh and blood patron" to the "*persona ficta* of the corporation" was not abrupt, however.[63] Members of court aristocracies shaped and influenced early scientific academies, and participation by the nobility enhanced their respectability. Courtly ties also allowed some, such as the early seventeenth-century French scholar and patron Nicolas-Claude Fabri de Peiresc (1580–1637), to form dyadic alliances (relationships based on friendship and loyalty) to advance the prospects of individuals involved in the investigation of nature – among whom, in the case of de Peiresc, were Tommaso Campanella (1568–1639), Marin Mersenne (1588–1648), Galileo, and Pierre Gassendi (1592–1655).[64]

---

[60] Martha Baldwin, "The Snakestone Experiments: An Early Modern Medical Debate," *Isis*, 86 (1995), 394–418; Paula Findlen, "Controlling the Experiment: Rhetoric, Court Patronage, and the Experimental Method of Francesco Redi," *History of Science*, 31 (1993), 35–64.

[61] Jay Tribby, "Cooking (with) Clio and Cleo: Eloquence and Experiment in Seventeenth Century Florence," *Journal of the History of Ideas*, 52 (1991), 417–39.

[62] Steven Shapin and Simon Schaffer, *Leviathan and the Air-Pump: Hobbes, Boyle, and the Experimental Life: Including a Translation of Thomas Hobbes, Dialogus physicus de natura aeris by Simon Schaffer* (Princeton, N.J.: Princeton University Press, 1985); Mario Biagioli, "Scientific Revolution, Social Bricolage, and Etiquette," in *The Scientific Revolution in National Context*, ed. Roy Porter and Mikulas Teich (Cambridge: Cambridge University Press, 1992), pp. 11–54.

[63] Biagioli, *Galileo, Courtier*, epilogue.

[64] Lisa T. Sarasohn, "Nicolas-Claude Fabri de Peiresc and the Patronage of the New Science in the Seventeenth Century," *Isis*, 84 (1993), 70–90; Sarasohn, "Thomas Hobbes and the Duke of Newcastle:

As with individual patronage styles, the involvement of princes in the founding of academies was based on special interest and personal identity. Sometimes personalities at the same court reached opposite opinions about the significance and purpose of inquiring into nature. The young Marchese di Monticello, Federico Cesi (1585–1630), founded the Accademia dei Lincei (Academy of the Lynxes) at Rome in 1603 but saw the academy disbanded by his father, who had grown suspicious of what he thought to be his son's magical projects. Only later, following the death of the elder duke and Federico's own ascension to the dukedom of Aquasparta, was the academy once again revived and set on a course of corporate endeavor.[65]

Another early scientific academy, the Accademia del Cimento (Academy of Experiment, established in 1657) depended directly upon the organization and financial support of Leopoldo de' Medici. The academy was oriented toward experiment, but its members served the prince as participants within a private company rooted in patronage and still based their social credentials on princely authority.[66] Leopoldo, who supervised the work of the academy and designated the times when it would meet, nevertheless stepped aside when members disagreed, allowing his academician-protégés the freedom to make their own decisions when confronted by contending claims. Later academicians prided themselves on remaining detached from the bitter personal squabbling that characterized scholastic disputation. However, private ambitions still prompted conflict within corporate sites. Rivalries formed when patrons were absent or became tired of the organizations they had founded. And yet, at least when assembled, academicians seem to have accepted the necessity of civil behavior, as well as a degree of theoretical and factual impartiality, as the best means for constructing a knowledge of nature on a collective basis.[67]

To a considerable degree, emerging scientific academies reflected court sentiments. Courtly etiquette influenced their protocols,[68] and the new academies themselves sometimes exhibited an unmistakable courtly bearing. Although concerned with maintaining decorum and concentrating upon depersonalized facts, they still continued to be troubled by differences in social status, kept their business secret, and remained fundamentally suspicious of

---

A Study in the Mutuality of Patronage before the Establishment of the Royal Society," *Isis*, 90 (1999), 715–37.

[65] Giuseppe Gabrieli, "Federico Cesi Linceo," *Nuova antologia*, ser. 7, 277 (1930), 352–69.

[66] W. E. Knowles Middleton, *The Experimenters* (Baltimore: Johns Hopkins University Press, 1971); Paulo Galluzzi, "L'Accademia del Cimento: 'Gusti' del principe, filosofia e ideologia dell'esperimento," *Quaderni storici*, 16 (1981), 788–844; and Michael Segre, *In the Wake of Galileo* (New Brunswick, N.J.: Rutgers University Press, 1991), pp. 127–40.

[67] Lorraine Daston, "Baconian Facts, Academic Civility, and the Prehistory of Objectivity," *Annals of Scholarship*, 8 (1991), 337–63.

[68] Mario Biagioli, "Etiquette, Interdependence, and Sociability in Seventeenth-Century Science," *Critical Inquiry*, 22 (1996), 193–238.

outsiders.[69] Moreover, scientific academies were just one example of learned societies that were formed in the early modern era. Philology, literature, and theology had preceded natural inquiry at court as magnets for patronage, and learned societies that promoted these areas of cultural endeavor existed long before the formation of academies devoted to science. In this regard, we still do not know the degree to which new sixteenth- and seventeenth-century developments in natural inquiry actually produced the changes reflected in organizations such as della Porta's very private Accademia Secretorum Naturae (Academy of the Secrets of Nature, ca. 1560) and Cesi's Accademia dei Lincei, or rather fitted themselves into social forms already well established. Indeed, some historians now reject the view that one form of social organization (i.e., academies) simply replaced another (courts) and suggest that, for instance, the founding in Paris of the Académie Royale des Sciences (1666) followed not as a result of the failure of private patronage but developed with a shift of patronage styles from one organized around individuals to one focused on more resourceful corporate institutions.[70]

What is clear is that lines once thought to have divided the court from other scientific "sites" now seem to be far less distinct. Even as princes became more distant from the work of academicians, a courtly ethos still prevailed. A good example is the German academy of those devoted to nature, the Academia or Collegium Naturae Curiosorum, founded in 1652 in Schweinfurt, a society that later became known as the Leopoldina. Although initially a society of physicians interested primarily in the medicinal properties of various parts of nature, the Academia obtained state sponsorship after revising its statutes and adding experiment to curiosity as part of its official program. The president and editor of the academy's journal were thereafter appointed *Pfalzgrafen* and given, by virtue of their rank and status, legal powers of legitimation within a wide social spectrum. They possessed the right to appoint notaries and judges, legitimize illegitimate children, bestow titles upon lower nobility, acknowledge adoptions, free slaves, restore honor, confer coats of arms, crown poets, and grant degrees to doctors, licentiates, masters, and baccalaureates in the faculties of philosophy, law, and medicine. The legitimizing power of the state and the granting of official recognition came to the academy by virtue of conferred title.[71]

---

[69] Alice Stroup, *A Company of Scientists: Botany, Patronage, and Community at the Seventeenth-Century Parisian Royal Academy of Sciences* (Berkeley: University of California Press, 1990), pp. 199 ff.
[70] David S. Lux, *Patronage and Royal Science in Seventeenth-Century France: The Académie de Physique in Caen* (Ithaca, N.Y.: Cornell University Press, 1989); and Stroup, *A Company of Scientists*.
[71] Rolf Winau, "Zur Frühgeschichte der Academia Naturae Curiosorum," in *Der Akademiededanke im 17. und 18. Jahrhundert* (Wolfenbütteler Forschungen, 3), ed. Fritz Hartmann and Rudolf Vierhaus (Bremen: Jacobi Verlag, 1977), pp. 117–38; "Das Kaiserliche Privileg der Leopoldina vom 7. August 1687," in *Das Kaiserliche Privileg der Leopoldina vom 7. August 1687*, trans. Siegfried Kratzsch (Acta Historica Leopoldina, 17) (Halle: Deutsche Akademie der Natur Forscher Leopoldina 1987), pp. 57–67; and Georg Uschmann, "Ein Wendepunkt in der Geschichte der Leopoldina," in *Das Kaiserliche Privileg*, pp. 7–13.

Long after the Royal Society and the Académie Royale des Sciences made their appearances, the German physician, poet, and Göttingen professor Albrecht von Haller (1708–1777) continued to insist upon a close relationship between prince and academy. In his view, the sovereign was a moral witness to the work of academies and could provide the financial means for obtaining the space and machinery of experimental science. Most importantly, Haller argued, the image of the sovereign at the head of a scientific society helped to maintain order and decorum among academy members.[72] Haller's remarks may have been partially motivated by personal ambition, yet he nonetheless underscored the social value of the prince's allegorical presence in imparting intellectual prestige and, to some degree, social legitimacy upon scientific claims.

The Royal Society and the Académie Royale des Sciences represented the emergence of new kinds of institutions oriented toward the collective and systematic exploration of specific subjects, each ultimately with its own methodological preferences and with its own means of reporting conclusions and discoveries through official publications. Although humanist models of a courtly rather than scholastic sensibility promoted the ideal of civil discourse in the new organizations, their formation, membership, purpose, and degree of independence in choosing projects still resonated, in individual cases, with the background noise of state sponsorship. Yet here there were also great differences. The Paris Académie, financed by the state, received not only an annual budget but a list of projects to be carried out by its members. The Royal Society, on the other hand, remained relatively impoverished even with its royal charter of 1662. The result was that the Society soon settled into the likeness of a gentleman's club as opposed to the promised "powerhouse of dynamic utilitarianism." In this case, aristocratic attention turned toward the society once its members were able to popularize the new learning as a fitting and necessary component of learned judgment.[73]

Elsewhere, a more direct connection between science and the state was forged by the German philosopher, mathematician, and theological rationalist Gottfried Wilhelm Leibniz (1646–1716). In his design for the Vienna society, Leibniz looked to the unification of *sapientia* (wisdom) and *potentia* (power) in the service of social reform;[74] he saw learned societies as providing

---

[72] Otto Sonntag, "Albrecht von Haller on Academies and the Advancement of Science: The Case of Göttingen," *Annals of Science*, 32 (1975), 379–91.

[73] Michael Hunter, "The Crown, the Public and the New Science, 1689–1702," in *Science and the Shape of Orthodoxy: Intellectual Change in Late Seventeeth-Century Britain*, ed. Michael Hunter (Rochester, N.Y.: Boydell Press, 1995), pp. 151–66, esp. p. 153; Hunter, *Establishing the New Science: The Experience of the Early Royal Society* (Rochester, N.Y.: Boydell Press, 1989). On popularization, language, and the aristocracy at the Royal Society, see also Steven Shapin, *A Social History of Truth: Civility and Science in Seventeenth-Century England* (Chicago: University of Chicago Press, 1994); Jan V. Golinski, "A Noble Spectacle: Phosphorus and the Public Cultures of Science in the Early Royal Society," *Isis*, 80 (1989), 11–39; and Peter Dear, "*Totius in Verba*: Rhetoric and Authority in the Early Royal Society," *Isis*, 76 (1985), 145–61.

[74] Richard Meister, *Geschichte der Akademie der Wissenschaften in Wien* (Vienna: Adolf Holzhausens, 1947); Werner Schneiders, "Gottesreich und gelehrte Gesellschaft: Zwei politische Modelle bei

the rulers of political societies with the knowledge needed to bring about rational social change. In another design, for the Berlin Academy (founded as the Societas Regia Scientiarum in 1700), science became the servant of the state, with Leibniz as its first president. Especially during the reign of Frederick the Great of Prussia (ruled 1740–1786), who helped direct some of the projects of the academy, academicians performed their part in the grand scheme of *raison d'état*, wherein the activities of each organ of society contributed to the security and well-being of an autonomous, meta-personal "state" being.[75] Piety, prestige, and pleasure, three of the most important motives of earlier Renaissance patronage,[76] were now digested within the ideological viscera and moral self-consciousness of a Leviathan-like body politic.

Court patronage stimulated innovation and curiosity, and the courtly site contributed constructive elements to scientific inquiry. Technical expertise, novel procedures, and the critique of ancient traditions stand out prominently within court environments, in contrast with universities, as means appropriate for acquiring knowledge of nature. Court spaces sometimes impinged upon other institutional settings, such as academies, schools, and guilds, and produced important social and intellectual results. Artists connected with courts and freed from the artisan conditions imposed by guilds, it has been argued, were also liberated from expectations of service to the city and could become imaginative aesthetic outsiders capable of more critical introspection.[77] The same ideological transformation, which brought the civility of intellectual pursuits to the *mechanical arts* (perhaps straining in the process the traditional patronage hierarchy that had served the literati so well), can be considered a leitmotif of court culture in the late Renaissance. There were also pedagogical consequences of the euphoria for creativity at court. As in the case of the chemical-medical interests of the German prince Moritz of Hessen, courtly patronage dynamics sometimes extended to universities and led to important disciplinary changes. Even when conciliar bodies (i.e., academies) with legal and administrative authority functioned mainly outside the court, the sovereign power of the ruler, now in more symbolic corporate attire, continued to affect claims to authority and to influence the protocol and organization of academies committed to experimental science.

G. W. Leibniz," in Hartmann and Vierhaus, eds., *Der Akademiegedanke im 17. und 18. Jahrhundert*, pp. 47–62.

[75] Adolf Harnack, *Geschichte der Königlich Preussischen Akademie der Wissenschaften zu Berlin* (Berlin, 1900; repr., Hildesheim: Georg Olms, 1970), pp. 317–94; Ronald S. Calinger, "Frederick the Great and the Berlin Academy of Sciences (1740–1766)," *Annals of Science*, 24 (1968), 239–49; and Hans Aarsleff, "The Berlin Academy under Frederick the Great," *History of the Human Sciences*, 2 (1989), 193–206.

[76] Peter Burke, *Culture and Society in Renaissance Italy, 1420–1540* (New York: Charles Scribner's Sons, 1972), chap. 4.

[77] Martin Warnke, *The Court Artist: On the Ancestry of the Modern Artist*, trans. David McLintock (Cambridge: Cambridge University Press, 1993).

12

ANATOMY THEATERS, BOTANICAL GARDENS, AND NATURAL HISTORY COLLECTIONS

Paula Findlen

At the end of the sixteenth century, the English lawyer and natural philosopher Francis Bacon (1561–1626) began to fantasize about the locations for knowledge. The *Gesta Grayorum* (1594), a court revel performed before Queen Elizabeth I and attributed to Bacon, described an imaginary research facility containing "a most perfect and general library" and "a spacious, wonderful garden" filled with wild and cultivated plants and surrounded by a menagerie, aviary, freshwater lake, and saltwater lake. Spaces for living nature were complemented by a museum of science, art, and technology – "a goodly huge cabinet" housing artifacts ("whatsoever the hand of man by exquisite art or engine has made rare in stuff"), natural oddities ("whatsoever singularity, chance, and the shuffle of things hath produced"), and gems, minerals, and fossils ("whatsoever Nature has wrought in things that want life and may be kept"). The fourth and final component was a space in which to test nature, "a still-house, so furnished with mills, instruments, furnaces, and vessels as may be a palace fit for a philosopher's stone." The totality of these facilities, Bacon concluded, would be "a model of the universal nature made private."[1] This statement suggested a new idea of empiricism that privileged human invention and demonstration over pure observation and celebrated the communal aspects of observing nature over the heroic efforts of the lone observer. Nature had to be reconstructed within a microcosm, creating an artificial world of knowledge in which scholars prodded, dissected, and experimented with nature in order to know it better.

Some thirty years later, the continued fantasy of a society organized around knowledge led Bacon to write his famous utopia, the *New Atlantis* (published posthumously in 1627), in order to demonstrate how an empirical worldview could transform an entire society. The nucleus of Bacon's utopian society, Bensalem, was a structure called Salomon's House, the knowledge-making

---

[1] Francis Bacon, *Gesta Grayorum*, in John Nichols, *The Progresses and Public Processions of Queen Elizabeth*, 3 vols. (London: John Nichols and Son, 1823), 3: 290.

center of the realm. Surrounded by artificial mines, lakes, a botanical garden, and a menagerie, and made of "high towers, . . . great and spacious houses, . . . [and] certain chambers," it represented a full elaboration of science as an activity that removed nature *from* nature in order to study it better. Bacon's remarkable array of unique spaces for science mirrored the variety of possible experiences that one could have of nature, isolating all natural objects and processes. The inhabitants of Bensalem proudly told their English visitors that, in doing this, they had made natural things "by art greater much than their nature."[2] They not only knew nature but used their knowledge to improve upon it. This statement epitomized Bacon's definition of good science as an invention of the human mind in contemplation of nature.

Bacon's fascination with the special sites in which to gain experience of nature did not emerge ex nihilo. Like many aspects of his natural philosophy, it was based on a keen understanding of developments in European science in the preceding half-century. Between the 1530s and the 1590s, anatomy theaters, botanical gardens, and cabinets of curiosities became regular features of the pursuit of scientific knowledge.[3] All of these structures shared the common goal of creating purpose-built spaces in which scholars could use the best intellectual, instrumental, and manual techniques of science to gain knowledge of the natural world. In effect, they acted in ways similar to Bacon's utopian vision of science; to differing degrees, they removed natural artifacts from their original locations, placing them inside new spaces for the specific purpose of studying them in order to improve natural knowledge. The proliferation of anatomy theaters, botanical gardens, and museums reflected the ways in which interpreting nature had become tied to ambitious empirical projects of investigating nature in toto, with all the attendant difficulties of gathering and storing materials, while at the same time encouraging smaller experiential projects that sought to understand unique aspects of nature by creating artificial conditions in which to experiment (see Dear, Chapter 4, this volume).

Bacon could not have sketched his famous portrait of Salomon's House as a teeming beehive of empirical activity without the work of observing nature that had occurred in the preceding half-century. During the Renaissance, the idea of experiencing nature firsthand had become an increasingly important part of medical education (see Cook, Chapter 18, this volume). Physicians, who had opened bodies occasionally throughout the late Middle Ages, reinvigorated their interest in the manual art of dissection, rubbing elbows with surgeons whose cutting abilities made them artisans rather than philosophers

---

[2] Francis Bacon, *The Great Instauration and New Atlantis*, ed. Jerry Weinberger (Wheeling, Ill.: Harland Davidson, 1980), pp. 72–4.
[3] Libraries, observatories, and laboratories also were purpose-built spaces in which knowledge could be gained (see Grafton, Chapter 10, and Smith, Chapter 13, this volume).

of nature.⁴ They also renewed their interest in the natural material out of which medicines were made, collaborating and occasionally clashing with apothecaries in their efforts to gain practical knowledge of plants.⁵ Bacon was correct in stating that the initial goals behind the desire for experience were somewhat narrow, reflecting the expanded scope of the physician's competency in all realms of medicine. Some university-educated physicians had become encyclopedists, studying everything and anything related to the microcosm of man, but it was not yet clear that they had developed a full appreciation of the need to study nature on its own terms and not just for the sake of medicine.

The anatomy theater, the botanical garden, and the natural history museum were all a direct result of the medical fascination with experience in the early sixteenth century. All found their nascent formulation during the 1530s in European cities that had strong traditions of medical education. Their gradual institutionalization across the sixteenth and early seventeenth centuries offers an important means for understanding how early modern scholars integrated the study of the material world of nature into their definition of science. Anatomizing, botanizing, and collecting were not a routine part of natural philosophy in 1500. A century later, studying nature without using some of these techniques of investigation was no longer possible. Many of the great naturalists of the sixteenth and seventeenth centuries, from Konrad Gesner in the 1550s to John Ray in the 1690s, constructed a new science of nature based on extensive field research, collecting, and collating of specimens. They could not have done these things without defining new locations for natural inquiry. Thus, the new purpose-built spaces gave the study of nature a new direction and intensity in addition to offering defined locations, both inside and outside universities, in which to observe specimens. They were indeed houses of knowledge.

ANATOMIZING

The idea of creating a special, enclosed space in which to study nature emerged at a very early stage in the realm of human anatomy. During the late Middle

---

⁴ On the revival of dissecting practices and their relation to the medical idea of experience, see especially Vivian Nutton, "Humanistic Surgery," in *The Medical Renaissance of the Sixteenth Century*, ed. Andrew Wear, Roger French, and I. M. Lonie (Cambridge: Cambridge University Press, 1985), pp. 75–99; Andrea Carlino, *Books of the Body: Anatomical Ritual and Renaissance Learning*, trans. John Tedeschi and Anne C. Tedeschi (Chicago: University of Chicago Press, 1999); Giovanna Ferrari, *L'Esperienza del passato: Alessandro Benedetto filologo e medico umanista* (Florence: Leo S. Olschki, 1996); and Andrew Cunningham, *The Anatomical Renaissance: The Resurrection of the Anatomical Projects of the Ancients* (Brookfield: Scolar, 1997).

⁵ On the botanical idea of experience, see Agnes Arber, *Herbals, Their Origin and Evolution: A Chapter in the History of Botany, 1470–1670*, 3rd ed. (Cambridge: Cambridge University Press, 1986); Karl H. Dannenfeldt, *Leonhard Rauwolf: Sixteenth-Century Physician, Botanist, and Traveler* (Cambridge, Mass.: Harvard University Press, 1968); Karen Reeds, *Botany in Medieval and Renaissance Universities* (New York: Garland, 1991); and Reeds, "Renaissance Humanism and Botany," *Annals of Science*, 33 (1976), 519–42.

Ages, occasional dissections of human cadavers had become a standard part of medical education in universities such as those at Bologna, Padua, and Montpellier, where surgery was part of the medical curriculum. The first documented formal dissection of this sort was recorded in Padua in 1341, though it is clear that the practice was considerably older; the *Anatomia* (1316) of Mondino de' Liuzzi, professor of medicine at the University of Bologna, was certainly composed on the basis of actual dissections.[6] Such practices accelerated in the second half of the fifteenth century. The increased circulation of ancient and medieval anatomical treatises, which was a direct result of the invention of printing, and a growing fascination with the postmortem as a means of understanding the causes of disease led medical professors and their students to demand more frequent dissections.[7]

In response to these changes in medical training and practice, a curious new structure appeared during the winter months in various European cities: the temporary anatomy theater. Usually built of wood, it was an ephemeral structure, not unlike a theatrical set, designed to accommodate the occasional dissection that the early Renaissance university demanded. These temporary theaters could be built in preexisting spaces – lecture halls or, better yet, churches and public piazzas that were already designed to accommodate audiences in the hundreds. (Before then, small audiences of students seem simply to have stood around a table observing the body during a lecture.) The Italian physician Alessandro Benedetti, who taught at Bologna in the late fifteenth century, provided the earliest and most elaborate description of the new structures in his *Anatomice: sive, de historia corporis humani libri quinque* (Anatomy: or, Five Books on the History of the Human Body, 1502) when he wrote:

> A temporary theater should be built at a large and well-ventilated place, with seats arranged in a circle, as in the Colosseum in Rome and the Arena in Verona, sufficiently large to accommodate a great number of spectators in such a manner that the teacher would not be inconvenienced by the crowd.... The corpse has to be put on a table in the center of the theater in an elevated and clear place easily accessible to the dissector.[8]

Over the next few decades, temporary theaters became a popular feature of medical instruction. By the 1520s, even less well-known medical faculties, such as those at Pisa and Pavia, supplied a structure in which to dissect, and by the 1540s, the idea of the anatomy theater had been so well integrated into medical training that the French physician Charles Estienne

---

[6] Andrea Carlino, "The Book, the Body, the Scalpel," *RES*, 16 (1988), 31.
[7] Katharine Park, "The Criminal and Saintly Body: Autopsy and Dissection in Renaissance Italy," *Renaissance Quarterly*, 47 (1994), 1–33; Carlino, *Books of the Body*; and Cunningham, *Anatomical Renaissance*.
[8] In Arturo Castiglione, "The Origin and Development of the Anatomical Theater to the End of the Renaissance," *Ciba Symposia*, (3 May 1941), 831. See also Ferrari, *L'esperienza del passato*, esp. pp. 166–73.

(ca. 1505–1564) insisted that anatomy could not be taught properly without a *locus anatomicus*, a place in which to anatomize.[9] He compared the human body to "anything that is exhibited in a theater in order to be viewed."[10] The structure of the anatomy theater played an important role in drawing attention to the visual aspects of the new anatomy. It made dissection a theatrical and often highly public event for medical students, physicians, surgeons, and a general public curious about the secrets of the body.

Accounts of actual dissections correspond remarkably well to ideal descriptions of the anatomy theater. The famous Flemish anatomist Andreas Vesalius (1514–1564), who performed numerous dissections throughout Europe between the 1530s and 1543, usually worked in temporary wooden theaters. His controversial anatomy lectures in Bologna, in the winter of 1540, occurred in a temporary wooden amphitheater erected in the church of San Francesco. The German medical student Baldasar Heseler, who attended the lectures, described the building as holding almost two hundred spectators – medical students, university professors, and the general public – on four wooden benches.[11] Vesalius exploited contemporary theatrical techniques, diminishing the distance between the lecturer and the audience by allowing the audience to handle the organs as he removed them from the body. He emphasized the resulting shared experience: "[S]urely, lords, he said, you can learn only little from a mere demonstration, if you yourselves have not handled the objects with your hands."[12] Dissections in the Dutch anatomy theaters of the seventeenth century continued the tradition begun by Vesalius, passing body parts among the spectators in the highest rows in order to offer everyone a tactile and immediate visual experience of the body.[13]

In the decades following the publication of Vesalius's *De humani corporis fabrica* (On the Fabric of the Human Body, 1543), dissections became a more regular feature of Renaissance medical education. In October 1554, Felix Platter, a young medical student from Basel, described his enthusiasm for Guillaume Rondelet's anatomies at Montpellier, many of which occurred in the new quarters built to accommodate the current passion for anatomy: "I never miss[ed] the dissections of men and animals that took place in the

---

[9] E. Ashworth Underwood, "The Early Teaching of Anatomy at Padua, with Special Reference to a Model of the Padua Anatomical Theatre," *Annals of Science*, 19 (1963), 1–26, at p. 7. The most comprehensive account of the rise of the anatomy theater remains Gottfried Richter, *Das anatomische Theater* (Berlin: Ebering, 1936).

[10] Charles Estienne, *De dissectione partium corporis humani libri tres* (Paris: S. Colinaeum, 1545), quoted in Giovanna Ferrari, "Public Anatomy Lessons and the Carnival: The Anatomy Theater of Bologna," *Past and Present*, 117 (1987), 50–106, at p. 85.

[11] Baldasar Heseler, *Andreas Vesalius' First Public Anatomy at Bologna, 1540*, ed. Ruben Eriksson (Uppsala: Almquist and Wiksells, 1959), p. 85.

[12] Ibid., p. 291. I have modified the translation slightly. The standard biography of Vesalius remains Charles D. O'Malley, *Andreas Vesalius of Brussels, 1514–1564* (Berkeley: University of California Press, 1964).

[13] Jan C. C. Rupp, "Michel Foucault, Body Politics and the Rise and Expansion of Modern Anatomy," *Journal of Historical Sociology*, 5 (1992), 31–60, at p. 47.

Sites of Anatomy, Botany, and Natural History

Table 12.1. *Anatomy Theaters*

1556	Montpellier	1617	Paris
1557	London	1619	Amsterdam
1588	Ferrara	1623	Oxford
1589	Basel	1642	Rotterdam
1593	Leiden	1643	Copenhagen
1594	Padua	1654	Groningen
1595	Bologna	1662	Uppsala
1614	Delft		

College...."[14] Platter surely observed these things in the wooden dissecting room built between 1554 and 1556. Like the dissecting rooms of the Company of Barber-Surgeons and the Royal College of Physicians in London, used to instruct medical practitioners outside the orbit of the university, the Montpellier theater represented an early stage in the evolution of the anatomy theater from a temporary to a permanent structure.

By the 1590s, the anatomy theater had achieved a new level of legitimacy in the university training of physicians. During this period, all of the leading medical faculties in Europe built permanent anatomy theaters (Table 12.1). By 1595, a stone theater had replaced the earlier wooden one in Montpellier.[15] In 1589, Platter, by then a distinguished professor of medicine at the University of Basel, persuaded his institution to buy a building with a plot of land so that the newly appointed professor of anatomy and botany, Caspar Bauhin (1560–1624), and his students might dissect in the winter and botanize in the summer all in one place. The statutes describing Bauhin's position underscore its close relationship to the new buildings of science. "He should teach not so much by precepts, but more by ocular demonstrations," the medical faculty of Basel declared.[16] Unfortunately, such high sentiments were not backed up in perpetuity by the university. By the 1620s, after the generation of physicians who had installed the theater and garden retired, the facility had fallen into disrepair.

In the ancient university towns of Italy, however, principally Padua and Bologna, a new commitment to the place of anatomy in medical education led to the building of anatomical theaters that exist to this day. The young English medical student William Harvey (1578–1657), who came from London, where an early dissecting theater had been built not by a university but by the

---

[14] Felix Platter, *Beloved Son Felix: The Journal of Felix, Platter, a Medical Student at Montpellier in the Sixteenth Century*, trans. Seán Jennett (London: Frederick Muller, 1961), p. 88.

[15] Thomas Platter, *Journal of a Younger Brother: The Life of Thomas Platter as a Medical Student at Montpellier at the Close of the Sixteenth Century*, trans. Seán Jennett (London: Frederick Muller, 1963), p. 36: "There is a special dissecting room in the College, built of dressed stone in the form of an amphitheatre (*Theatrum anatomicum*) and designed to allow the greatest possible number of persons to see the operations."

[16] Reeds, *Botany*, pp. 95, 111 (quotation), 116, 130.

medical corporations of the city, who wished to train physicians and surgeons, learned to dissect by observing the work of his professor, Girolamo Fabrici (ca. 1533–1619), who designed and promoted Padua's elliptical anatomy theater (Figure 12.1).[17] Fabrici shared the sentiments of his colleagues in Bologna, who argued in 1595 that they wanted to dissect "without having to erect a new theater every year and tear it down after the dissection was completed."[18] They built a rectangular dissecting room, imitating the idea but not the form of the Paduan theater, because they felt that a rectangular room offered a more open use of space, in contrast with the cramped experience of standing in the tight wooden pews of the Paduan theater that spiraled upward to the heavens.

The success of Padua's model traveled not only to the neighboring city of Bologna but north to the new Dutch university of Leiden (founded 1575), whose faculty pioneered many of the new approaches to medicine and natural philosophy in the seventeenth century. Another student of Fabrici, Pieter Paaw, conceived the idea of a splendid anatomy theater for his university after becoming professor of anatomy in 1589. His theater, like many of the Dutch dissecting theaters, was built in a former church (which had been vacated by Catholics after the wars of religion) that also held the university library.[19] In contrast with their Italian counterparts, the Dutch theaters engaged more actively with the religious connotations of dissection as an art that inquired into the secrets of life through the observation of death. Paaw decorated the theater with articulated skeletons of humans and animals on which he placed Latin mottos conveying the transience of human life and moralizing the deaths of many of the criminals whose bodies ended up on the dissecting table and whose skeletons populated the theater. His successors quickly transformed the Leiden theater into a cabinet of curiosities, filling it with Chinese scrolls, porcelain teapots, Egyptian idols, and exotic plants.[20] Such decorations fulfilled the idea of the anatomy theater as a civic institution not simply for the use of medical professors and students but more generally a public site in which to display the curiosities of nature and art.

---

[17] Jerome Bylebyl, "The School of Padua: Humanistic Medicine in the Sixteenth Century," in *Health, Medicine and Mortality in the Sixteenth Century*, ed. Charles Webster and Margaret Pelling (Cambridge: Cambridge University Press, 1979), pp. 335–70; Andrew Cunningham, "Fabricius and the 'Aristotle Project' in Anatomical Teaching and Research at Padua," in *The Medical Renaissance of the Sixteenth Century*, ed. Andrew Wear, Roger French, and Ian Lonie (Cambridge: Cambridge University Press, 1985), pp. 195–222; and Roger French, *William Harvey's Natural Philosophy* (Cambridge: Cambridge University Press, 1994), pp. 59–68.

[18] Castiglione, "Origin," p. 842. On the Bologna theater, see Ferrari, "Public Anatomy."

[19] Th. H. Lunsingh Scheurleer, "Un amphithéâtre d'anatomie moralisé," in *Leiden University in the Seventeenth Century*, ed. Th. H. Lunsingh Scheurleer and G. H. M. Posthumus Meyjes (Leiden: E. J. Brill, 1975), pp. 217–77. The Delft anatomy theater built by the surgeons' guild was in the former convent of St. Mary Magdalene.

[20] William Schupbach, "Some Cabinets of Curiosities in European Academic Institutions," in *The Origins of Museums: The Cabinet of Curiosities in Sixteenth- and Seventeenth-Century Europe*, ed. Oliver Impey and Arthur MacGregor (Oxford: Clarendon Press, 1983), pp. 170–1.

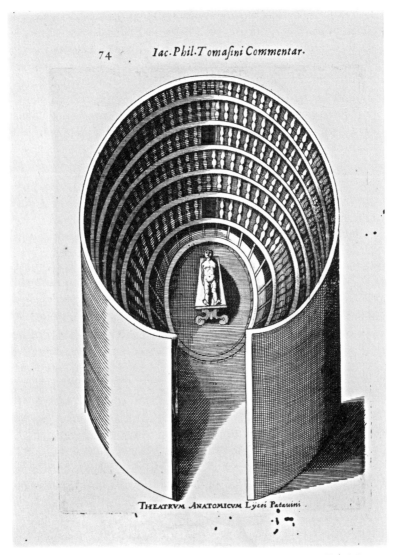

Figure 12.1. The Padua anatomy theater designed by Hieronymus Fabricius, 1595. In Giacomo Filippo Tomasini, *Gymnasium Patavinum* (Udine: Nicolaus Schirattus, 1654). Reproduced by permission of the Houghton Library, Harvard University.

As Leiden became the center of medical education in Northern Europe, it soon surpassed Padua as the model to emulate. Protestant anatomists in Germany, England, and the Netherlands contributed to the proliferation of dissecting theaters in universities, colleges of physicians, and surgeons' guilds. The appearance of anatomy theaters in medical corporations further underscores the practical appeal of anatomy not only as an important feature of university medical education but also as an activity that defined the professional lives of physicians and surgeons in the early modern period. The great anatomists of Scandinavia, such as Olaus Rudbeck (1630–1702), who identified the lymphatic vessels and designed Uppsala's anatomy theater in the Gustavianum in 1662 after studying in Leiden in 1653–4, participated in the Protestant exaltation of the pious physician who saw the dissecting theater as a religious temple.[21] What Harvey called the "ocular testimony" of the body emanated distinctively from the anatomy theater by the middle of the seventeenth century.[22] But it was a kind of experience that continued to have highly symbolic and metaphysical associations.

BOTANIZING

With the exception of surgeon's theaters, which had a more narrowly professional function, most anatomy theaters appeared in tandem with university botanical gardens. Although the botanical garden did not precede the permanent anatomy theater, it more quickly became part of the institutional culture of science in Renaissance Europe. Private botanical gardens flourished in the early sixteenth century not only as "physick gardens" filled with medicinal plants but also as pleasure gardens of the nobility and urban elite. By the 1530s, medical professors and their students botanized regularly during summer vacations. The city of Ferrara, an early center for the revival of natural history, had a ducal garden that university professors and students used for study.[23]

A steady stream of published herbals in the 1530s and 1540s, all lamenting the imperfection of botanical knowledge, made it clear how much remained to be known about plants. Yet the profusion of nature made it difficult to see all but the tiniest fraction of the plant world. One solution to this problem lay in the creation of public botanical gardens, associated primarily with universities and occasionally with princely courts, that functioned as living repositories of nature. On 29 June 1545, the Republic of Venice authorized the foundation of a botanical garden at the University of Padua so that "scholars and other gentlemen can come to the garden at all hours in the summer, retiring in the shade with their books to discuss plants learnedly, and investigating their nature peripatetically while walking."[24] The Grand Duke

---

[21] Gunnar Eriksson, *The Atlantic Vision: Olaus Rudbeck and Baroque Science* (Canton, Mass.: Science History Publications, 1994), pp. 1, 2, 10; and Karin Johannisson, *A Life of Learning: Uppsala University during Five Centuries* (Uppsala: Uppsala University Press, 1989), pp. 31–2. For the Danish equivalent, see V. Maar, "The Domus Anatomica at the Time of Thomas Bartholinus," *Janus*, 21 (1916), 339–49.

[22] Andrew Wear, "William Harvey and the 'Way of Anatomists,'" *History of Science*, 21 (1983), 223–49, at p. 230.

[23] Vivian Nutton, "The Rise of Medical Humanism: Ferrara, 1464–1555," *Renaissance Studies*, 11 (1997), 2–19, at p. 18.

[24] Marco Guazzo, *Historie . . . di tutti i fatti degni di memoria nel mondo* (Venice: Gabriele Giolito, 1546), quoted in Margherita Azzi Visentini, *L'Orto botanico di Padova e il giardino del Rinascimento* (Milan: Edizioni il Polifilo, 1984), p. 37.

of Tuscany, Cosimo I, concluded negotiations for a garden at the University of Pisa in July, founding another at the convent of San Marco in Florence in December.[25] By 1555, the Spanish royal physician Andrés Laguna felt that he could use the precedent of Italy as an argument for persuading Philip II to fund a royal physic garden at Aranjuez. "All the princes and universities of Italy take pride in having many excellent gardens, adorned with all kinds of plants found throughout the world," he wrote in his translation of the ancient Greek physician Dioscorides' *De materia medica*, "and so it is most proper that Your Majesty provide and order that we have at least one in Spain, sustained with royal stipends."[26]

By the end of the sixteenth century, most universities with strong medical faculties promoting this early modern program of learning, and a number of cities with strong colleges of physicians, had botanical gardens (Table 12.2).[27] These gardens, filled with New World plants as well as European varietals, claimed to contain the natural world in microcosm. Sunflowers from Peru, tulips from the Levant, and corn, potatoes, tomatoes, tobacco, and hundreds of other plants from the "Indies" transformed the botanical garden into another Eden, filled not only with the medicinal herbs of the ancient Near East that had been described in Greek and Roman pharmacopeias but also with the wonders of a newly discovered nature that came from the Americas.[28] Reflecting on the significance of the garden, Ulisse Aldrovandi (1522–1605), professor of natural history and founder of Bologna's botanical garden in 1568, wrote: "These public and private gardens, with the lectures [that accompany them], are the reason that natural things are elucidated, joined together with the New World that we are still discovering."[29]

Botanical gardens served several important functions. Physicians occasionally described them as public repositories of medicines in an age of plague, though one wonders how realistic it was to expect a single garden to halt a pandemic. More importantly, they were sites in which a new kind of medical

---

[25] Lionella Scazzosi, "Alle radici dei musei naturalistici all'aperto: Orti botanici, giardini, zoologici, parchi e riserve naturali," in *Stanze della meraviglia: I musei della natura tra storia e progetto*, ed. Luca Basso Peressut (Bologna: Cooperativa Libraria Universitaria Editrice Bologna, 1997), pp. 91–3. See also Fabio Garbari, Lucia Tongiorgi Tomasi, and Alessandro Tosi, *Giardino dei semplici: L'Orto botanico di Pisa dal XVI al XIX secolo* (Pisa: Cassa di Risparmio di Pisa, 1991); Alessandro Minelli, ed., *The Botanical Garden of Padua, 1545–1995* (Venice: Marsilio, 1995); and Else M. Terwen-Dionisius, "Date and Design of the Botanical Garden of Padua," *Journal of Garden History*, 14 (1994), 213–35.

[26] Andrés Laguna, *Pedacio Dioscórides Anazarbeo acerca de la materia medicinal y de los venenos mortiferos* [1555], quoted in José M. López Piñero, "The Pomar Codex (ca. 1590): Plants and Animals of the Old World and from the Hernandez Expedition to America," *Nuncius*, 7 (1992), 35–52, at p. 38.

[27] Andrew Cunningham, "The Culture of Gardens," in *Cultures of Natural History*, ed. Nicholas Jardine, James A. Secord, and Emma C. Spary (Cambridge: Cambridge University Press, 1996), pp. 38–56.

[28] John Prest, *The Garden of Eden: The Botanical Garden and the Re-Creation of Paradise* (New Haven, Conn.: Yale University Press, 1981).

[29] Biblioteca Universitaria, Bologna, *Aldrovandi*, MS 70, fol. 62r. See Antonio Baldacci, "Ulisse Aldrovandi e l'orto botanico di Bologna," in *Intorno alla vita e alle opere di Ulisse Aldrovandi* (Bologna: L. Beltrami, 1907), pp. 161–72.

Table 12.2. *Botanical Gardens*

1545	Padua	1589	Basel
1545	Pisa	1593	Montpellier
1545	Florence	1597	Heidelberg
1550s	Aranjuez	1623	Oxford
1563	Rome	1638	Messina
1567	Valencia	1641	Paris
1568	Bologna	1650s	Uppsala
1568	Kassell	1670s	Edinburgh
1577	Leiden	1673	Chelsea
1580	Leipzig		

professor, the professor of botany (or "medicinal simples," as it was often called), demonstrated the nature and virtues of plants to students. Finally, they became botanical research facilities in which scholars who sought to understand the plant as a natural rather than medical object did their earliest work on morphology and classification. The Italian physician Andrea Cesalpino (1519–1603) wrote his fundamental *De plantis* (1583) while teaching at the University of Pisa in proximity to its well-stocked garden. Bauhin, the great Swiss naturalist, traveled to the Padua and Bologna gardens in 1577–8 before becoming a teacher of botany in Basel. He wrote his *Pinax theatri botanici* (Index of a Botanical Theater, 1623), one of the earliest works to attempt a comprehensive cross-referencing of plant names and to refine plant classification, as the culmination of decades of work with plants in European botanical gardens.[30]

The public botanical garden exhibited key institutional characteristics that distinguished it from the private noble garden. Stern rules specified appropriate garden behavior, warning visitors that they could look at and smell but not pick or trample plants or attempt to take home branches, flowers, seeds, bulbs, and roots without the express permission of the custodian.[31] Botanical professors readily exchanged plants with other learned botanists, physicians, and apothecaries in order to keep their gardens full and varied and to please princely patrons and overseas merchants, who were the other important source of new plants. The goal, in all instances, was to maintain and improve the diversity and utility of nature that the garden revealed.

As the botanical garden became an important scientific research facility, one of the pressing questions concerned how it organized knowledge. The initial design of the Paduan garden, for example, emphasized an aesthetic and highly symbolic arrangement of plants on the outer edges of the garden and a

---

[30] Reeds, *Botany*, pp. 110–30; and Scott Atran, *Cognitive Foundations of Natural History: Towards an Anthropology of Science* (Cambridge: Cambridge University Press, 1990), pp. 135–42.
[31] The 1601 Leiden regulations appear in F. W. T. Hunger, *Charles de l'Escluse*, 2 vols. (The Hague: Martinus Nijhoff, 1927), 1: 249. For a similar set of regulations for Padua, see Minelli, *Botanical Garden*, p. 48.

more practical arrangement of those in the interior.[32] In the case of the former, the design of the garden outweighed any practical considerations of how to provide a plot of land in which plants could grow best; in the case of the latter, function won out over form, making long, rectangular flower beds a key feature of the botanical garden (Figure 12.2). Initially, the first-century Greek physician Dioscorides' *De materia medica*, the standard botanical textbook at most universities, defined which specimens should appear in the garden. Yet ancient botanical classifications could not contain all the Northern European, American, and Asian plants that were not indigenous to the ancient Mediterranean and thus were not described by Dioscorides. New ways of thinking about nature affected the structure of the garden itself. Increasingly, the most practical solution was to organize the garden as a microcosm of the world, dividing it geographically on the grounds that any alternative organization might be rendered problematic by the appearance of a new specimen.

By the 1590s, botanical gardens emphasized these practical configurations. The Leiden garden, founded in 1577, underwent a complete reorganization under the directorship of Carolus Clusius and especially Pieter Paaw.[33] Clusius simplified the design, creating four quadrants to represent the four continents (Europe, Asia, Africa, and America), each divided into sixteen beds. He organized plants by species rather than by medicinal use, reflecting the changing status of botany as a field worthy of independent study rather than a branch of medicine. The Montpellier Jardin du Roi, a royal garden founded just beyond the city walls by the professor of anatomy and botany Pierre Richer de Belleval, also favored a basic geometric design, clustering plants according to their natural habitats.[34] Such models indicate the direction of most seventeenth-century botanical gardens, whose creators increasingly viewed plants in scientific and commercial rather than symbolic terms, unlike the initial creators of the Italian Renaissance gardens. The botanical garden, like the anatomy theater, had become a standard means for experiencing and understanding nature.

COLLECTING

Visitors to the botanical gardens of Padua, Pisa, and Leiden in the 1590s and early 1600s discovered, to their delight, that the natural history museum, or "cabinet of curiosities" as it was commonly called at the time, had become an important feature of the garden. In his 1591 description of the Paduan garden, Girolamo Porro discussed the research facilities then under construction at Padua in a manner that strongly anticipated Bacon's ideal of a truly integrated

---

[32] Andrea Ubrizsy Savoia, "The Botanical Garden in Guilandino's Day," in Minelli, *Botanical Garden*, pp. 173, 181; and Terwen-Dionisius, "Date and Design," p. 220.
[33] Hunger, *Charles de l'Escluse*, I: 217–49; and W. K. H. Karstens and H. Kleibrink, *De Leidse Hortus, een Botanische Erfenis* (Zwolle: Uitgeverij Waanders, 1982).
[34] Reeds, *Botany*, pp. 80–90.

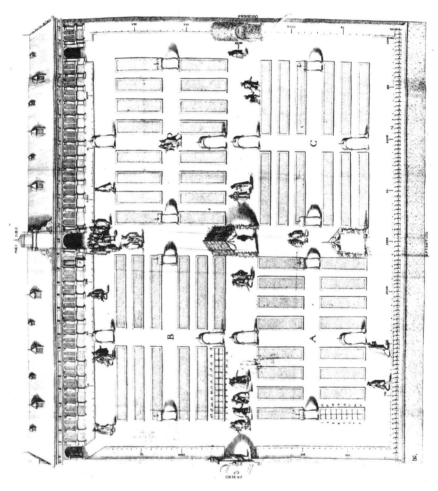

Figure 12.2. The botanical garden at the University of Leiden. Jacques de Ghein II, 1601. Reproduced by permission of the Nationaal Herbarium Nederland, Leiden.

facility for science. He described a series of rooms built on the edge of the garden. Some would be "employed for various medicinal operations as well as... foundries, distilleries, and so on." Subsequent rooms displayed minerals, marine life, stuffed terrestrial animals, and birds. Porro remarked, "This rich and varied array of things will form a wonderful and beautiful museum for the delight and education of scholars of this rare profession."[35]

As we have already seen, many university gardens were part of a research and teaching complex that housed an anatomy theater and various scientific collections accumulated by the medical faculty. In 1595, the University of Pisa added a distillery and foundry to its garden so that scholars might test nature in the laboratory – a reminder that the materials and tools of alchemy could also be useful in the realm of natural history. A gallery to house natural objects followed shortly thereafter, as also occurred at the University of Pisa, where the cabinet of curiosities occupied the upper floor of a gallery at the entrance to the garden.[36] The interrelationship among these different ways of examining nature suggests that it was experience of the material world in general that early modern scholars sought, above and beyond any single way of understanding nature. The animal skins, stuffed anteater, and Nile crocodile adorning the Leiden anatomy theater made it a cabinet of curiosities when dissections were not under way.[37] For similar reasons, it often made sense to hire a single individual who would hold the professorships of both anatomy and botany on the presumption that the winter skill of dissection translated smoothly into the summer skill of botanical demonstration.[38]

Yet subjecting nature to art, as Bacon put it, was by no means a uniform process. If an artificial nature was made in a microcosm, it was rendered artificial in many different ways to address diverse questions of knowledge. The anatomy theater, for example, celebrated the normal body. Vesalius had recommended that one choose male bodies that were as "normal as possible" for public dissections, contrasting them with private dissections in which "any body" was potentially interesting for understanding disease. Although he and a number of his contemporaries delighted in the study of

---

[35] Girolamo Porro, *L'Horto de i semplici di Padova* [1592]. I have used the translation in Vittorio Dal Piaz and Maurizio Rippa Bonati, "The Design and Form of the Paduan *Horto Medicinale*," in Minelli, *Botanical Garden*, pp. 42–3.

[36] Lionella Scazzosi, "Alle radici dei musei naturalistici all'aperto," in Peressut, ed., *Stanze della meraviglia*, p. 102. See Lucia Tongiorgi Tomasi, "Il giardino dei semplici dello studio pisano: Collezionismo, scienza e immagine tra Cinque e Seicento," in *Livorno e Pisa: Due città e un territorio nella politica dei Medici* (Pisa: Nistri-Lischi e Pacini, 1980), pp. 514–26; and Tongiorgi Tomasi, "Inventari della galleria e attività iconografica dell'orto dei semplici dello Studio pisano tra Cinque e Seicento," *Annali dell'Istituto e Museo di Storia della Scienza*, 4 (1979), 21–7.

[37] Jan C. C. Rupp, "Matters of Life and Death: The Social and Cultural Conditions of the Rise of Anatomical Theatres, with Special Reference to Seventeenth Century Holland," *History of Science*, 28 (1990), 272.

[38] This was certainly the case in Montpellier, Valencia, Basel, Leiden, and Uppsala. In part, the decision rested on the number of overall chairs in the medical faculty. Universities such as those at Bologna, Padua, and Pisa, with larger faculties, could afford to be more specialized in their appointments.

human diversity, they used the anatomy theater primarily to create that most artificial of humans – the typical male – a Greek statue peeled open like an onion and accompanied by a normative female who was defined exclusively by her reproductive organs.[39] By contrast, the botanical garden claimed to be a universal portrait of nature – an artificial paradise divested of much of its symbolic meaning as it strove to accommodate the ever-increasing number of plants. The cabinet of curiosities subscribed to neither of these paradigms of what an artificial nature might be. It offered a highly idiosyncratic image of nature. At times, collectors presented their cabinets as a true microcosm of the world and, at times, described them as selective accumulations of objects. Overwhelmingly, however, collectors followed a philosophy best articulated by the sixteenth-century Milanese physician Girolamo Cardano (1501–1576), who emphasized the subtlety and variety of nature. By bringing together all the strange, beautiful, and costly things of the world, the cabinet became a room of wonder in ways that the anatomy theater and botanical garden could never be.[40]

In the late sixteenth century, collecting became an important means of understanding nature.[41] In contrast with anatomy theaters and botanical gardens, which were institutional sites funded by civic governments and municipal corporations, the museum (the modern institution that emerged from the cabinet) arose primarily because of the efforts of individual physicians and naturalists who advocated an empirical approach to nature that made natural objects as important as books in the search for knowledge. Physicians' records are replete with accounts of collections such as the one Thomas Platter saw as a student in Montpellier in 1596 in the home of the recently deceased chancellor of the Faculty of Medicine, the distinguished physician Laurent Joubert. Platter noted exotic animals such as the pelican, chameleon, crocodile, and the fabled remora, but also "some remarkable freaks": an eight-footed pig, a two-headed goat, and large stones ejected from the bodies of Joubert's patients. On the ground floor was a whale's skeleton.[42] The overwhelming fascination with the wonders of nature typified the Renaissance collection.[43]

---

[39] Andreas Vesalius, *De humani corporis fabrica* [1543], in O'Malley, *Andreas Vesalius*, p. 343; and Nancy Siraisi, "Vesalius and Human Diversity in 'De humani corporis fabrica,'" *Journal of the Warburg and Courtauld Institutes*, 57 (1994), 60–88.

[40] Lorraine Daston and Katharine Park, *Wonders and the Order of Nature, 1150–1750* (New York: Zone Books, 1998).

[41] Krzysztof Pomian, *Collectors and Curiosities: Paris and Venice, 1500–1800*, trans. Elizabeth Wiles-Portier (London: Polity, 1990); Antoine Schnapper, *La géant, la licorne, la tulipe: Collections françaises au XVIIe siècle*, vol. I: *Histoire et histoire naturelle* (Paris: Flammarion, 1988); Giuseppe Olmi, *L'Inventario del mondo: Catalogazione della natura e luoghi del sapere nella prima età moderna* (Bologna: Il Mulino, 1992); Paula Findlen, *Possessing Nature: Museums, Collecting, and Scientific Culture in Early Modern Italy* (Berkeley: University of California Press, 1994); and Horst Bredekamp, *The Lure of Antiquity and the Cult of the Machine*, trans. Allison Brown (Princeton, N.J.: Markus Weiner, 1995).

[42] Thomas Platter, *Journal*, pp. 105–8.

[43] On the subject of wonder, see Joy Kenseth, ed., *The Age of the Marvelous* (Hanover, N.H.: Hood Museum of Art, 1991); and Daston and Park, *Wonders and the Order of Nature*.

The majority of collectors of natural objects were physicians, apothecaries, and natural philosophers, though virtually anyone with some education, some experience of travel, or some access to the networks by which scholars routinely traded objects could lay claim to being a collector (see Moran, Chapter 11, this volume). Quite typically, the scholars who played a prominent role in advocating for anatomy theaters and botanical gardens, and who taught in these settings, owned private collections. The Bologna physician Aldrovandi is a case in point. By the 1560s, Aldrovandi was well known as a collector of natural objects. In the next few decades, his home became one of the important research centers for natural history. Through a wide network of friends, colleagues, and patrons, he transformed his private collection into such an important public resource that in 1603, two years before his death, he persuaded the Senate of Bologna to maintain it as a civic museum.[44]

Aldrovandi's museum became the first public science museum when it opened in 1617, preceding even the Ashmolean Museum at Oxford, which opened with a 1683 bequest by Elias Ashmole, an avid alchemist, experimenter, and member of the early Royal Society. Visitors to anatomy theaters, botanical gardens, private collections, and princely treasuries were already accustomed to looking at curiosities as part of observing nature. They examined fossils in the Vatican mineralogical collection in Rome, wondered at the Holy Roman Emperor Rudolf II's collection of New World fauna in Prague, traipsed through the royal gardener John Tradescant's museum in Lambeth, and gazed at the Nile crocodiles that hung on the walls and ceilings of the apothecary Ferrante Imperato's famous museum in Naples, as shown in a woodcut from his *Dell'historia naturale* (Natural History, 1599) (Figure 12.3). Museums forced scholars to think of nature as a group of objects whose material specificity mattered very much in understanding them. Rather than contemplating nature as an abstract universal, collectors reveled in its particularities.[45] Such things were the facts born of experience.

The particulars that collectors found most appealing often had a direct connection with the commercial value of nature in the early modern period. It is not surprising that apothecaries such as Imperato played a prominent role in collecting culture. They collected nature to make a living from it. Exotic collectibles, such as true balsam from the East and New World cinnamon, which Columbus identified in 1492 as a potentially valuable commodity for his Spanish patrons to export back to Europe, were also important ingredients in the early modern pharmacopeia; apothecaries displayed them in rooms above their shops on the ground floor as a means of further reassuring customers that their medicines derived from a profound knowledge of the natural world. The expansion of collecting activities occurred in direct

---

[44] Findlen, *Possessing Nature*, pp. 24–31; and Cristiana Scappini and Maria Pia Torricelli, *Lo Studio Aldrovandi in Palazzo Pubblico (1617–1742)*, ed. Sandra Tugnoli Pattaro (Bologna: Cooperativa Libraria Universitaria Editrice Bologna, 1993).

[45] Lorraine J. Daston, "The Factual Sensibility," *Isis*, 79 (1988), 452–70; and Daston, "Baconian Facts, Academic Civility, and the Prehistory of Objectivity," *Annals of Scholarship*, 8 (1991), 337–63.

Figure 12.3. Ferrante Imperato's natural history museum in late sixteenth-century Naples. In Ferrante Imperato, *Dell'historia naturale* (Naples: Constantino Vitali, 1599). Reproduced by permission of the Rare Books Division, The New York Public Library, Astor, Lenox and Tilden Foundations.

proportion to the proliferation of new trade networks between Europe and the Americas, in conjunction with the old trading routes that linked Europe to the Levant (see Harris, Chapter 16, this volume). Although collectors rarely bought and sold specimens, preferring to exchange objects in kind, they nonetheless had a fairly precise understanding of the value of different parts of nature in a world shaped by commerce and trade.

With the rise of scientific academies in seventeenth-century Europe, the museum became part of the new institutional culture of science. In 1681, the naturalist Nehemiah Grew published a catalogue of the Royal Society of London's repository, begun shortly after its founding in 1660, in which he stated that the Society collected "not only Things strange and rare, but the most known and common amongst us."[46] Such an approach signaled a new goal for collecting, which soon came to emphasize the importance of collecting the whole of nature rather than its most unusual and rarest elements. The Académie Royale des Sciences in Paris also emphasized the

---

[46] Nehemiah Grew, *Musaeum Regalis Societatis, or a Catalogue & Description of the Natural and Artificial Rarities Belonging to the Royal Society and Preserved at Gresham College* (London: W. Rawlins, 1681), preface. See Arthur MacGregor, "The Cabinet of Curiosities in Seventeenth Century Britain," and Michael Hunter, "The Cabinet Institutionalized: The Royal Society's 'Repository' and Its Background," both in MacGregor and Impey, eds., *Origins of Museums*, pp. 147–58, 159–68.

desirability of collecting all parts of nature, signaling a new interest in the ordinary aspects of what it meant to experience nature, from the lowliest frond to the largest mammal.

In the ensuing century, as scientific academies appeared throughout Europe, the idea of the science museum grew in stature so that it was no longer a "cabinet of curiosities" but a series of repositories that housed instruments, such as telescopes, microscopes, air-pumps, barometers, and ultimately the machines of experimental physics, as well as artifacts. Each of these instruments furthered the idea of an artificial nature by making the scientific instrument a location in which to put nature to the test in ways that could not be done in nature (see Bennett, Chapter 27, this volume). Academies such as the Royal Society of London, which became famous for its repository of natural objects as well as its instruments, explicitly referred to their community as the realization of Salomon's House. By the early eighteenth century, it was commonly assumed that the purpose-built spaces for scientific inquiry that best resembled those described in the *New Atlantis* could be found in the rooms of such institutions as the Académie Royale des Sciences in Paris and the Instituto delle Scienze in Bologna. For a time, the academy, more than the university, became the place in which to coordinate all of the different ideas about building a house of knowledge.[47]

Bacon's ideal of a multidimensional research facility provided an important foundation upon which later generations could build by insisting that scientific training and work needed to occur in locations suited to the specific ends of science. It provided a philosophy that unified the disparate and often ad hoc activities that emerged in Renaissance spaces for natural inquiry. If understanding nature was a kind of art, then all of the ways in which nature could be rendered artificial became important in developing a complete understanding of the natural world. This was a lesson that experimental philosophers such as Robert Boyle (1627–1691) and Robert Hooke (1635–1702), two of the leading members of the early Royal Society, thoroughly absorbed in their quest to explore the less visible aspects of nature inside the machines, rooms, and repositories that made their community scientific.

---

[47] Paula Findlen, "Building the House of Knowledge: The Structures of Thought in Late Renaissance Europe," in Töre Frangsmyr, ed., *The Structure of Knowledge: Classifications of Science and Learning since the Renaissance* (Berkeley: Office for History of Science and Technology, University of California, 2001), pp. 5–51.

13

LABORATORIES

Pamela H. Smith

In 1603, after six years of construction, Count Wolfgang II von Hohenlohe put the finishing touches on a new two-story laboratory in his residence Schloss Weikersheim.[1] Many of the basic elements of his laboratory can be seen in the frontispiece from a work of theosophical alchemy, *Amphitheatrum sapientiae aeternae* (Amphitheater of Eternal Wisdom, 1609), by the physician and alchemist Heinrich Khunrath (1560–1605); see Figure 13.1. Although this frontispiece foregrounds the spiritual dimension of alchemy (for example, in the kneeling figure of the alchemist), it also illustrates the practical tools of the alchemical laboratory that Khunrath would have known from his work with Central European princes and alchemists.

As in the frontispiece, Wolfgang II's roomy laboratory had large, bright windows with extra-deep sills where vessels could be placed, as well as smaller window vents to allow smoke and steam to escape. One corner was occupied by a raised flat stone hearth or forge (like those used by blacksmiths), and, looming over it, a smoke hood, like the one shown in the engraving, to draw away vapors. (This did not, however, protect the laboratory workers from the many poisonous fumes that often billowed up from operations to fill the room.)[2] A large set of fixed bellows mounted at the side of the hearth fanned the coals in the forge and heated the smaller furnaces that were probably contained within it. Connected to the main chimney of the hearth in the Weikersheim laboratory were four brick furnaces, including one called a *Faule Heinz*, or Lazy Harry, on which many distillations could be carried out simultaneously; an assaying furnace in which refined gold and silver were assayed to determine their purity and ores were tested for metal content; and, probably, a sublimation furnace in which substances were heated until they

---

[1] The following description is drawn from Jost Weyer, *Graf Wolfgang II von Hohenlohe und die Alchemie: Alchemistische Studien in Schloß Weikersheim, 1587–1610* (Sigmaringen: J. Thorbeke, 1992), pp. 94–103.
[2] For a description of conditions in a metalworking shop, see Robert Barclay, *The Art of the Trumpet-Maker: The Materials, Tools, and Techniques of the Seventeenth and Eighteenth Centuries in Nuremberg* (Oxford: Clarendon Press, 1992), pp. 100–1.

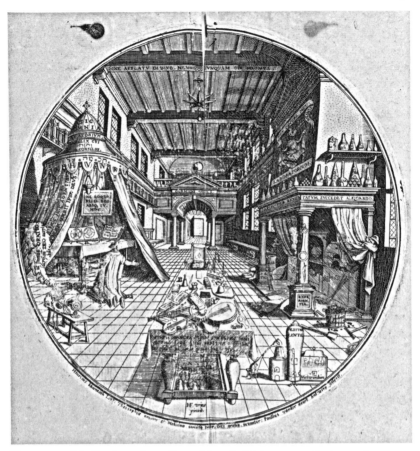

Figure 13.1. Idealized version of an alchemical laboratory. Paul van der Doort after Jan Vredeman de Vries, engraving. In Heinrich Khunrath, *Amphitheatrum sapientiae aeternae*, 2nd ed. (Hanau: G. Antonius, 1609). Reproduced by permission of the Research Library, The Getty Research Institute, Los Angeles.

vaporized and then condensed back to solidity by rapid cooling. Distillation and assay furnaces appear in the right foreground and under the smoke hood in Figure 13.1. Wolfgang II's laboratory was further outfitted with small, portable copper furnaces and, almost certainly, with a "philosophical furnace" for seeking the philosophers' stone (the substance that transmuted base metals into gold and silver). Built into the walls were at least four large closets, complete with shelves and drawers that held glass distillation vessels, often made to order;[3] metal, stone, and ceramic smelting vessels, retorts, mortars and pestles, and laboratory utensils, including funnels, stirring implements, tongs, and balances; and chemical substances, including metals, mineral ores,

---

[3] Weyer, *Graf Wolfgang II von Hohenlohe*, p. 153.

salts, and acids purchased from nearby apothecaries and merchants. As in Khunrath's image, additional vessels and instruments were stored elsewhere in the room, on shelves and probably on tables and the hearth as well. Wolfgang II's laboratory was also equipped with two footwarmers (used in the cold winters and for some small distillations), as well as a bed where the laboratory worker could rest while keeping the fire alive to carry out long chemical operations (some of which required forty days of slow-burning fire). Finally, a large drain ran through the laboratory to empty waste water into the moat around the castle.

Such a laboratory was staffed by servants of all types, who performed the heavy work (pulverizing ores, hauling wood and charcoal, and cleaning utensils and vessels), as well as by more specialized workers, such as distillers and apothecaries, who produced alcohol and acids, and various other miscellaneous persons; in Weikersheim, even a notary was sometimes used in laboratory work. Count Wolfgang used a single long-term laboratory worker, or *Laborant*, who appears to have held the major responsibility for the laboratory over a sixteen-year period. His tasks included carrying out its operations, by day or night, maintaining order, and, in particular, holding its activities in deepest secrecy, as well as guarding the laboratory key on his person at all times. Whereas the Weikersheim laboratory had only a few workers, the laboratory at the court of one of Count Wolfgang's close correspondents in alchemical matters, Duke Frederick of Württemberg in Stuttgart, employed ten *Laboranten* in 1608, and a total of thirty-three over the period from 1593 to 1608.[4] Count Wolfgang himself also worked in the laboratory, apparently sometimes with his wife, Magdalena,[5] attempting to carry out the recipes for both the philosophers' stone and simple medicaments contained in the numerous alchemical books and manuscripts in the palace library.[6]

Although Count Wolfgang called his early seventeenth-century workspace a *Laboratorium*, it resembled the workshops of apothecaries, metalworkers, and pigment makers, among many other types of artisans, and it shared many of its components with the workplaces of metal smelters, glassmakers, and potters. By the end of the seventeenth century, in contrast, the laboratory had become one of the hallmarks of the new science – the site where theories and hypotheses were purportedly tested by experiment and from which new discoveries and useful knowledge emerged. When and how the laboratory made the transition from artisanal workshop to its place as a central "site of science" tells us much about how a new active epistemology in natural

---

[4] On laboratory workers, see Weyer, *Graf Wolfgang II von Hohenlohe*, pp. 186–99.
[5] Ibid., p. 395.
[6] Ibid., pp. 200 ff.

philosophy emerged in the early modern period to define a new kind of science.[7]

THEORY AND PRACTICE

In antiquity, the work of the hands was generally delegated to slaves and therefore had a low epistemological status. In Greek thought, manual labor could not lead to "scientific" or theoretical knowledge (see Bennett, Chapter 27, this volume). In addition, manual work was associated with the production of goods for commerce or making a living, and, for the Greek philosophers, this disbarred it from partaking in an impartial, unsullied search for truth. Moreover, craftsmanship and other forms of *technē* involved sensation rather than reason and thus could never be separated from the world of changing objects. Thus, although for Plato and Aristotle manual crafts were part of knowledge in a broad sense,[8] they did not assist in the pursuit of virtue and were not a suitable activity for the independent and leisured free man.[9]

In the Aristotelian scheme of knowledge that shaped the attitudes of sixteenth-century intellectuals, theory was separate from practice. Whereas theory (*theôria*) involved abstract speculation on the causes of things, practice (*praxis*) could be of two kinds – things done and things made. "Things done" was the subject of disciplines such as history, politics, ethics, and economics. It was studied in the particular (by collection of experiences) and could not be formed into a deductive system; thus it was not as certain as theory. The other type of practice, *technē*, had nothing to do with certainty but instead was the base knowledge of how to make things or produce effects and was practiced by animals, slaves, and craftspeople. It was the only area of knowledge that was productive.

A new attitude toward practice, which differed sharply from Aristotle's concern with theory, appeared first in the Roman concern with political life; this practical form of knowledge (of "things done") was necessary for a life in the service of the republic. At the same time, early Christianity,

---

[7] Since the late seventeenth century, the practical activity of the laboratory has almost always involved some form of *labor* that implicated the human body, and the end product of the laboratory has come to be a written text that makes experience public and allows other scientists to replicate the experiments performed within it. See Bruno Latour, *Science in Action: How to Follow Scientists and Engineers through Society* (Cambridge, Mass.: Harvard University Press, 1987); and Peter Dear, "Narratives, Anecdotes, and Experiments: Turning Experience into Science in the Seventeenth Century," in *The Literary Structure of a Scientific Argument: Historical Studies*, ed. Peter Dear (Philadelphia: University of Pennsylvania Press, 1991), pp. 135–63.

[8] Carl Mitcham, "Philosophy and the History of Technology," in *The History and Philosophy of Technology*, ed. George Bugliarello and Dean B. Doner (Urbana: University of Illinois Press, 1979), pp. 163–201, at pp. 173–7.

[9] Elsbeth Whitney, *Paradise Restored: The Mechanical Arts from Antiquity through the Thirteenth Century* (Transactions of the American Philosophical Society, 80) (Philadelphia: American Philosophical Society, 1990), pp. 23–32.

which was the heir to this classical thought but also the genitor of a new and distinctive culture, came to regard manual labor as the mark of mortal life after the Fall as well as a necessary part of the redemption of human beings. This ambivalent stance toward labor was especially clear, for example, in the works of Augustine of Hippo (354–430), who characterized the arts as the way in which human beings could create a "second nature" for themselves but at the same time as a diversion from heavenly things.[10]

Reflecting Augustine's ambivalence, early medieval monastic movements viewed labor as penitential. The goal of such labor was, however, not the production of objects or effects but the sustenance of the monastic community and the subjugation of the will in the continual struggle between spirit and flesh. The leisured man (*otiosus*) of classical antiquity had thus become a hindrance to the work of human salvation. In the *Rule* of St. Benedict of Nursia (ca. 480–547), the monastery functioned as a workshop of the soul, and manual labor was placed at the center of the *opus Dei* (work of God).[11] Thus, monastic buildings included *scriptoria, oratoria, dormitoria, lectoria, refectoria*, and sometimes *laboratoria*, where distilling was done and medicaments prepared. This view of the redemptive potential of bodily labor continued to play an important role in European thought up through the seventeenth century.

A different appreciation of the arts and the labor they required entered Europe as a result of new familiarity with the works of Arabic authors and their redaction of ancient texts after the turn of the first millennium. Within the Arabic classification of the sciences, unlike that of Aristotle, the arts and crafts had a major place. In the Islamic world, many scholars were also instrument makers, and work with the hands did not disqualify a person from higher status. Contact with Arabic culture was particularly important for the art that came to be most frequently associated with the laboratory: alchemy. Most of the alchemical writings that entered Europe from the East were works attributed to the eighth-century alchemical writer Jābir ibn Ḥayyān (latinized as Geber), as well as the *Book of Secrets* of the ninth-century Persian author al-Rāzī (Rhazes).[12] Arabic alchemy included an exoteric practice that produced medicines, distillations, and other kinds of products, and an esoteric tradition that sought the divine spirit in material bodies. The alchemical laboratory, filled with the instruments of separation and analysis by fire – furnaces, distilling vessels, and crucibles – was the space in which these two

---

[10] This tendency was reinforced by the development in late antique encyclopedias and medieval universities of a course of education that focused on the systematization of the liberal arts and excluded the mechanical arts as being incapable of similar ordering; see, in general, A. Geoghegan, *The Attitude toward Labor in Christianity and Ancient Culture* (Washington, D.C.: The Catholic University of America Press, 1945); and Whitney, *Paradise Restored*.

[11] George Ovitt, *The Restoration of Perfection: Labor and Technology in Medieval Culture* (New Brunswick, N.J.: Rutgers University Press, 1987), pp. 104–5.

[12] William R. Newman, *The "Summa perfectionis" of pseudo-Geber* (Leiden: E. J. Brill, 1991) shows that Jābir ibn Ḥayyān was not identical with the Latin Geber, as had previously been believed.

searches merged. This alchemical tradition was novel to Europe because it encompassed both a manual and textual tradition and thus unified theory and practice in a way that few other areas of study in medieval Europe did. Alchemical activity would thus become one of the principal models for the new relationship between practice and theory in early modern natural philosophy.

TOWARD A NEW EPISTEMOLOGY

Although scholars such as Albertus Magnus (1206–1279) and Roger Bacon (ca. 1220–1292) read the works of Arabic writers with great interest and advocated the practice of a *scientia experimentalis*, and alchemists debated about the status of practical work,[13] a new epistemology, based on productive bodily labor, did not develop until the late Renaissance. One of the most important stimuli for this new attitude was the fifteenth-century introduction into Western Europe of the Hermetic Corpus, with its doctrine of the human being as a "little God" who could effect changes in nature by means of "natural magic." Because scholars viewed this body of texts as contemporary with Moses and thus as another source of divine revelation, its effect on western ideas about the relationship of human beings to nature was far-reaching. Whereas Aristotle had seen human art as imitating nature, Hermetic writings encouraged the image of human art as not just emulating but even rivaling nature. In both art theory and alchemy, art came to be seen as capable of perfecting and even superseding nature.[14] In the seventeenth century, this idea would be extended still further by individuals such as Francis Bacon (1561–1626) and Robert Boyle (1627–1691) to encompass the view that art could "master" and "try" nature in the disciplined space of the laboratory (see the following chapters in this volume: Dear, Chapter 4; Bennett, Chapter 27).

An equally important source for the new attitude toward human art (Aristotle's *technē*) was the organization of craftspeople into guilds and the increasing economic importance and political power of these guilds during

---

[13] William R. Newman, "Technology and Alchemical Debate in the Late Middle Ages," *Isis*, 80 (1989), 423–45. The *Libellus de alchimia*, ascribed to Albertus Magnus, states that the second precept of the art of alchemy (after the injunction that the alchemical worker must be silent and secretive) was that the worker should have "a place and a special house, hidden from men, in which there are two or three rooms" in which to carry out alchemical operations. The word used is *camerae*, not *laboratorium*. See *Libellus de alchemia*, trans. Sister Virginia Heines (Berkeley: University of California Press, 1958), p. 12.

[14] William R. Newman, "Technology and Alchemical Debate," p. 424; and Newman, *The "Summa perfectionis."* See, for example, Giambattista della Porta, *Magiae naturalis, sive de miraculis rerum naturalium* (Naples, 1558). Leon Battista Alberti's *Della pittura* (comp. 1435–6), as well as Leonardo da Vinci, Michelangelo Buonarotti, and Giorgio Vasari, expressed such views. See also Jan Bialostocki, "The Renaissance Concept of Nature and Antiquity," in *The Renaissance and Mannerism: Studies in Western Art* (Acts of the Twentieth International Congress of the History of Art, 2) (Princeton, N.J.: Princeton University Press, 1963), pp. 19–20, 23–30; and R. Hooykaas, "La Nature et l'art," in *Selected Studies in History of Science* (Coimbra: Acta Universitatis Conimbricensis, 1983), pp. 192–213, esp. pp. 197–206.

the late Middle Ages. As a result, the status of artisans and their labor in the workshop increased greatly. The humanists' reform of knowledge, and their reaction against the life of contemplation they saw medieval universities and monasteries as exemplifying, renewed the Roman perception of the value of the active life.[15] At the same time, the practical knowledge of "things made" had become a source of power and a subject of great interest for those concerned with city and state government. Scholars and humanists entreated their fellow reformers to go into the workshops of artisans to learn their practices. In 1531, for example, the pedagogical reformer Juan Luis Vives (1492–1540) encouraged scholars not to "be ashamed to enter into shops and factories, and to ask questions from craftsmen, and to get to know about the details of their work."[16] Men such as Vives regarded such contact with artisans as one part of a general reform of learning and as an opportunity to see practice embodied for the good of the community.[17]

Most importantly for the development of a new epistemology in natural philosophy, however, such men understood the observation of artisanal practices as a means for gaining new "scientific" knowledge of nature because they believed that the familiarity with matter and natural materials that underpinned artisans' productive capabilities could be incorporated into a theoretical framework for the reform of natural philosophy. The humanist Cipriano Piccolpasso, for example, wrote *I tre libri del'arte del vasaio* (The Three Books of the Potters' Art, ca. 1558) to allow the potters' practices to move from "persons of small account, [to] circulate in courts, among lofty spirits and speculative minds."[18] At the same time, artisan-artists such as Leonardo da Vinci (1452–1519), Albrecht Dürer (1471–1528), Benvenuto Cellini (1500–1571), and Bernard Palissy (ca. 1510–1590), among many others, carried on a lively exchange with humanist scholars and considered themselves thereby to have risen above the common guild artisan. This interaction between scholarly and artisanal cultures during the Renaissance is the most important source for the transformation of values that led to the legitimation of bodily labor in a specially designed space as a means of producing scientific knowledge.[19]

---

[15] For a general discussion, see, for example, Anthony Grafton and Lisa Jardine, *From Humanism to the Humanities: Education and the Liberal Arts in Fifteenth- and Sixteenth-Century Europe* (Cambridge, Mass.: Harvard University Press, 1986), introduction. An example of what *praxis* could mean can be found in Lisa Jardine and Anthony Grafton, "'Studied for Action': How Gabriel Harvey Read His Livy," *Past and Present*, 129 (1990), 30–78.

[16] Juan Luis Vives, *De tradendis disciplinis* [1531], trans. Foster Watson (Totowa, N.J.: Rowman and Littlefield, 1971), p. 209.

[17] Eugenio Garin, *Science and Civic Life in the Italian Renaissance*, trans. Peter Munz (New York: Doubleday, 1969); R. Hooykaas, *Humanisme, science et Réforme: Pierre de la Ramée (1515–1572)* (Leiden: E. J. Brill, 1958); and Paolo Rossi, *Philosophy, Technology, and the Arts in the Early Modern Era*, trans. Salvator Attanasio (New York: Harper and Row, 1970).

[18] Cipriano Piccolpasso, *The Three Books of the Potter's Art* [ca. 1558], trans. Ronald Lightbown and Alan Caiger-Smith, 2 vols. (London: Scholar Press, 1980), 2: 6–7.

[19] On this subject, see in general Antonio Pérez-Ramos, *Francis Bacon's Idea of Science and the Maker's Knowledge Tradition* (Oxford: Clarendon Press, 1988); Pamela O. Long, "The Contribution of

A new impetus for this exchange between scholars and practitioners emerged with the growing power of territorial rulers all over Europe and the increasing fragmentation of the political scene. Nobles and princes took a new interest in the arts – especially those connected with warfare, building, the production of territorial income, self-representation, and display – and they employed and celebrated many artisans (see the following chapters in this volume: Bennett, Chapter 27; DeVries, Chapter 14). New opportunities of this sort encouraged artisans of all kinds (but especially gunners, fortification builders, architectural writers, machine builders, and alchemists) to write about their techniques, in part to advertise their know-how.[20] The *Feuerwerkbuch* of 1420, written by an anonymous gunner seeking a German patron in need of techniques of gunpowder making and the use of guns, missiles, and pyrotechnics, was just one of many such treatises that began to appear around 1400.[21] Other craftsmen, such as the goldsmiths Lorenzo Ghiberti (ca. 1378–1455) and Antonio Averlino (ca. 1400–ca. 1462, called Filarete), wrote treatises in which they both advertised their own skills and attempted to provide a theoretical elaboration of their craft experience. Although many of these practitioners often published only some of their "secrets," in order not to jeopardize their livelihood, their literary activities created a new link between texts and technique, inplying that manual techniques could be learned from texts as well as by individual apprenticeship and imitation of a master. The publication of replicable processes and techniques would become a hallmark of laboratory activity by the mid-seventeenth century (although the ideal of free and open communication even in the modern scientific laboratory has remained in a state of tension with a practical need for secrecy in order to establish priority and compete for research funding).

In the early sixteenth century, both religious and intellectual reformers increasingly advocated direct engagement with nature as a way to gain knowledge. According to Martin Luther (1483–1546), for example:

> We are now again beginning to have the knowledge of the creatures which we lost in Adam's fall. Now we observe the creatures rightly, more than popery does.... We, however, by God's grace are beginning to see his glorious work

---

Architectural Writers to a 'Scientific' Outlook in the Fifteenth and Sixteenth Centuries," *Journal of Medieval and Renaissance Studies*, 15 (1985), 265–98; William Eamon, *Science and the Secrets of Nature: Books of Secrets in Medieval and Early Modern Culture* (Princeton, N.J.: Princeton University Press, 1994); J. V. Field and Frank A. J. L. James, eds., *Renaissance and Revolution: Humanists, Scholars, Craftsmen and Natural Philosophers in Early Modern Europe* (Cambridge: Cambridge University Press, 1993); and Pamela H. Smith, *The Body of the Artisan: Art and Experience in the Scientific Revolution* (Chicago: University of Chicago Press, 2004).

[20] Pamela O. Long, "Power, Patronage, and the Authorship of *Ars*," *Isis*, 88 (1997), 1–41; and Bert S. Hall, "Der Meister sol auch kennen schreiben und lesen: Writings about Technology ca. 1400–ca. 1600 A.D. and their Cultural Implications," in *Early Technologies*, ed. Denise Schmandt-Besserat (Malibu, Calif.: Undena Publications, 1979), pp. 47–58.

[21] Long, "Power, Patronage," p. 18.

and wonder even in the flower. . . . We see the power of his word in the creatures. As he spoke, then it was, even in a peach-stone.[22]

There was no better place to come into contact with the inner workings of nature than in the alchemical laboratory because it was there that natural processes were recreated. Indeed, Luther addressed the rewards of alchemy directly:

> The science of alchymy I like very well, and indeed, 'tis the philosophy of the ancients. I like it not only for the profits it brings in melting metals, in decocting, preparing, extracting and distilling herbs, roots; I like it also for the sake of the allegory and secret signification, which is exceedingly fine, touching the resurrection of the dead at the last day.[23]

Interest in the particulars of nature – and in the practice and experience connected to the particular and ephemeral, rather than the general and demonstrable – grew in part out of the nominalist ideas that fueled Martin Luther's belief in the utter impotence of human beings to know God in a comprehensive or certain manner, and partly from a Pauline emphasis on knowing God through His Creation, a trend that became particularly strong in Calvinism.[24]

Theophrastus Bombastus von Hohenheim, called Paracelsus (1493–1541), exemplifies the path of those who put practice and the laboratory at the center of religious reform. This medical reformer took the methods of the practitioner and artisan to be the ideal mode of proceeding in the acquisition of all knowledge because the artisan worked directly with the objects of nature. Paracelsus considered this unmediated labor in and experience of nature as elevating the artisan above the scholar both spiritually and intellectually. Because the artisan imitated the processes of nature in his creation of works of art from the material of nature, he had a better understanding of nature, and because he was in contact with God's Creation, he also possessed a better knowledge of God's revelation.[25] Paracelsus called for the joining of theory and practice; he believed experience was the crucial link in obtaining knowledge. Experience was gained not through a process of reasoning – in

---

[22] Martin Luther, *Tischreden*, vol. 1, item 1160, quoted in James J. Bono, *The Word of God and the Languages of Man: Interpreting Nature in Early Modern Science and Medicine* (Madison: University of Wisconsin Press, 1995), p. 71.

[23] Martin Luther, *Table Talk*, trans. William Hazlitt (London: Bell, 1902), p. 326.

[24] For examples of Calvinist practitioners who expressed this belief, see Keith Cameron, introduction to his edition of Bernard Palissy, *Recepte veritable* (Geneva: Librairie Droz, 1988), pp. 15–16; and Catharine Randall, *Building Codes: The Aesthetics of Calvinism in Early Modern Europe* (Philadelphia: University of Pennsylvania Press, 1999).

[25] This is stated by Paracelsus in many works, but clearly in *Astronomia Magna: oder die gantze Philosophia sagax der grossen und kleinen Welt/ des von Gott hocherleuchten/ erfahrnen/ und bewerten teutschen Philosophi und Medici* [finished 1537–8, orig. publ. 1571], in Paracelsus, *Sämtliche Werke: Medizinische, naturwissenschaftliche und philosophische Schriften*, ed. Karl Sudhoff and William Matthiessen, 14 vols. (Munich and Berlin: R. Oldenbourg and O. W. Barth, 1922–33), vol. 12 (Munich: Oldenbourg, 1929), p. 11. See also Kurt Goldammer, *Paracelsus: Natur und Offenbarung* (Hanover: Theodor Oppermann Verlag, 1953).

this he echoed Luther's belief that salvation came not through reasoning but through faith – but by a union of the divine powers of mind and body with the divine spirit in matter,[26] a notion derived in part from Hermetic texts. Knowledge of nature, gained through experience and manual labor, was thus a form of worship, yielding understanding of God's Creation and bringing about redemption of the world and humankind.[27]

The primary model for all such redemptive knowledge was alchemy. Alchemical principles formed the basis for Paracelsus's medical-chemical theories, and he viewed the operations of alchemy in the laboratory as a search for the generative spirit in nature and the means to imitate it by art.[28] The alchemical processes of generation and maturation carried out in the microcosm of the laboratory imitated God's macrocosmic Creation, and the refining techniques of alchemy imitated, on a small scale, human redemption after the Fall. Work in the laboratory was thus devoted to imitating nature and was regarded as the basis of all knowledge about nature. Once Paracelsus's texts began to be published in the 1560s, his influence became profound. His ideas inspired medical practitioners and religious radicals of all stripes to regard alchemy and the "anatomy" of nature in the laboratory as the key both to the innermost secrets of the natural world and to its reformation.[29] Although Paracelsus had both his fervent supporters, such as Oswald Croll (ca. 1560–1609), and his detractors, such as Andreas Libavius (ca. 1560–1616), who feared the religious enthusiasm as well as the social and intellectual challenge inherent in Paracelsus's work,[30] both Croll and Libavius saw the laboratory as a space for the discovery of new and valuable knowledge about nature as well as a place that in one way or another contributed to general reform of the world.

By the mid-sixteenth century, the laboratory was referred to formally as *officina* or *laboratorium*, and numerous images that illustrated actual and ideal laboratories were published, both in technical treatises dealing with mining and smelting[31] and in works of spiritual alchemy, such as that of Heinrich Khunrath, in which the alchemist prays for divine revelation in a laboratory

---

[26] Paracelsus, *Das Buch Labyrinthus medicorum genant* [1538, orig. publ. 1553], in *Sämtliche Werke*, vol. 11 (Munich: R. Oldenbourg, 1928), p. 192.
[27] Walter Pagel, *Paracelsus: An Introduction to Philosophical Medicine in the Era of the Renaissance*, 2nd ed. (Basel: Karger, 1982), esp. pp. 50–65.
[28] In *Die neun Bücher de natura rerum* [1537], in *Sämtliche Werke*, vol. 11, Paracelsus equated alchemy and generation, and he regarded all principles and processes of nature as alchemical.
[29] B. J. T. Dobbs, *The Foundations of Newton's Alchemy; or, "The hunting of the greene lyon"* (Cambridge: Cambridge University Press, 1975); and Charles Webster, *The Great Instauration: Science, Medicine, and Reform, 1626–1660* (London: Gerald Duckworth, 1975).
[30] Owen Hannaway, *The Chemists and the Word: The Didactic Origins of Chemistry* (Baltimore: Johns Hopkins University Press, 1975).
[31] For example, Georg Agricola, *De re metallica* (Basel: H. Froben and N. Episcopus, 1556); Alvaro Alonso Barba, *El arte de los metales* (Madrid, 1640); Vannoccio Biringuccio, *De la pirotechnia* (Venice: Venturino Rossinello, 1540); Lazarus Ercker, *Beschreibung aller fürnemsten mineralischen Ertzt unnd Bergwercksarten* (Prague: G. Schwartz, 1574).

that spans both theory and practice (Figure 13.1). By this time, laboratories could be found in the homes of scholars and physicians (see Cooper, Chapter 9, this volume), noble residences, monasteries and convents, and artisanal workshops.[32]

EVOLUTION OF LABORATORY SPACES

Paracelsian ideas about the alchemical reform of the world impelled many nobles, particularly in the German territories, to found alchemical laboratories. The Medici, Spanish and Austrian Hapsburgs, Hohenzollerns, and many less powerful nobles maintained alchemists, established laboratories, and even practiced in the laboratories themselves.[33] In some courts, these laboratories were not distinguishable from the workshops of apothecaries, artisans, and instrument makers, whereas in others, gold was sought with avidity. In those of Emperor Rudolf II (1576–1612) and Landgrave Moritz of Hessen-Kassel (1572–1632), however, alchemical work in the laboratory was believed to be capable of bringing about a general material and intellectual reform that could reconcile the schisms within European Christendom. As part of the religious and intellectual reforms of his territory, Moritz of Hessen-Kassel founded a chair of *chymiatria* at the University of Marburg in 1609 and appointed to the position a Paracelsian, Johannes Hartmann (1568–1631), who remained professor until the chair was disbanded in 1621. Hartmann included laboratory teaching in the curriculum.[34] A surviving laboratory notebook from 1615–16 indicates that students learned laboratory techniques and medicinal preparations, and that Hartmann himself used the laboratory to attempt to reproduce some of Paracelsus's more arcane recipes. Although many accounts of laboratory experiments were written as synopses

---

[32] See, for example: Owen Hannaway, "Laboratory Design and the Aim of Science: Andreas Libavius versus Tycho Brahe," *Isis*, 77 (1986), 585–610; Weyer, *Graf Wolfgang II von Hohenlohe*; R. W. Soukup, S. von Osten, and H. Mayer, "Alembics, Curcurbits, Phials, Crucibles: A 16th-Century Docimastic Laboratory Excavated in Austria," *Ambix*, 40 (1993), 25; and their fuller investigation in Rudolf Werner Soukup and Helmut Mayer, *Alchemistisches Gold: Paracelsistische Pharmaka: Laboratoriumstechnik im 16. Jahrhundert. Chemiegeschichtliche und archäometrische Untersuchungen am Inventar des Laboratoriums von Oberstockstall/Kirchberg am Wagram*, ed. Helmuth Grössing, Karl Kadletz, and Marianne Klemun (Perspektiven der Wissenschaftsgeschichte, 10) (Vienna: Böhlau Verlag, 1997).

[33] Giulio Lensi Orlandi, *Cosimo e Francesco de' Medici alchimisti* (Florence: Nardini Editore, 1978). On Rudolf II, see R. J. W. Evans, *Rudolf II and His World: A Study in Intellectual History, 1576–1612* (Oxford: Oxford University Press, 1984). On other prince-practitioners, see A. Bauer, *Chemie und Alchymie in Österreich bis zum beginnenden XIX Jahrhundert* (Vienna: R. Lechner, 1883); Cristoph Meinel, ed., *Die Alchemie in der europäischen Kultur- und Wissenschaftsgeschichte* (Wiesbaden: O. Harrassowitz, 1986); Bruce T. Moran, *The Alchemical World of the German Court: Occult Philosophy and Chemical Medicine in the Circle of Moritz of Hessen (1572–1632)* (Sudhoffs Archiv, Beiheft 29) (Stuttgart: Franz Steiner Verlag, 1991); Bruce T. Moran, ed., *Patronage and Institutions: Science, Technology, and Medicine at the European Court, 1500–1750* (Rochester, N.Y.: Boydell Press, 1991); Heiner Borggrefe, Thomas Fusenig, and Anne Schunicht-Rawe, eds., *Moritz der Gelehrte: Ein Renaissancefürst in Europa* (Eurasburg: Edition Minerva, 1997); and Weyer, *Graf Wolfgang II von Hohenlohe*.

[34] Moran, *The Alchemical World of the German Court*, pp. 57–67.

of laboratory practice rather than as an explicit test of theory, they often implicitly tested hypotheses about the nature of matter.[35]

Although religious reform might have been paramount in the founding of the laboratories of some German princes, it took a back seat to the reform of learning in the laboratories established by some of the many learned academies that came into existence in the sixteenth and seventeenth centuries. The groups that established, or at least planned, laboratories included two in sixteenth-century Naples: the Accademia Segreta of Girolamo Ruscelli and the Accademia dei Secreti of Giambattista della Porta (1535–1615).[36] The noble Frederico Cesi's Accademia dei Lincei was established in the first years of the seventeenth century, and many more academies followed within the century, the most important being the Academia Naturae Curiosorum (1652) of the German territories, the Accademia del Cimento (1657) of Florence, the Royal Society of London (1660), and the Académie Royale des Sciences (1666) of Louis XIV in Paris. All these societies adhered to the new active method of philosophizing, which, although not spelled out or executed rigorously, was based upon active practice, including trial and experiment in a laboratory, to test hypotheses. Thus, by the middle of the seventeenth century, Aristotle's division of knowledge into theory, praxis, and *technē* had been turned inside out. The production of effects and tangible objects, above all in the laboratory, came to prove the certainty of theory.

These latter societies were inspired in part by the works of early seventeenth-century writers, particularly those of Francis Bacon (1561–1626), who advocated a "New Philosophy; or Active Science" (*The Great Instauration*, 1620). In the words of the English physician and natural philosopher William Gilbert (1544–1603), this "new sort of Philosophizing" will be produced by "true philosophizers, honest men, who seek knowledge not from books only but from things themselves."[37] In Bacon's posthumous utopia, *New Atlantis* (1627), as well as the utopias of his contemporaries, Tommaso Campanella's *City of the Sun* (written in 1602 and first published in 1623) and Johann Valentin Andreae's *Christianopolis* (1619), laboratories were the center of a reformed society in which humans had regained dominion over the natural world. These authors saw the laboratory as the place in which to reproduce in miniature the processes of nature, thereby revealing the causes of things and gaining power over nature. Many other writers developed these ideas in the seventeenth century, among them John Amos Comenius (1592–1670)

---

[35] On the contrasts between the goals of laboratory practice in the sixteenth and seventeenth centuries, see Ursula Klein, "The Chemical Workshop Tradition and the Experimental Practice: Discontinuities within Continuities," *Science in Context*, 9 (1996), 251–87. Newman, *The "Summa perfectionis,"* and William R. Newman and Lawrence M. Principe, *Alchemy Tried in the Fire: Starkey, Boyle, and the Fate of Helmontian Chymistry* (Chicago: University of Chicago Press, 2002), argue that alchemists' recipes incorporate the testing of hypotheses.
[36] Eamon, *Science and the Secrets of Nature*, pp. 148–51, 199–200.
[37] William Gilbert, *De magnete* (London: Peter Short, 1600), preface.

and Samuel Hartlib (d. 1662), who believed strongly in the central place of practice and the laboratory in religious and intellectual reform. Théophraste Renaudot (1584–1653) included laboratories in his Parisian Bureau d'Adresse (established in 1633) to make public valuable processes and medical preparations.[38] In the writings of "chemical philosophers" of the second half of the century, such as Johann Joachim Becher (1635–1682), Johann Rudolf Glauber (1604–1670), and John Webster (1610–1682), only resonances of this overt spiritual reform are visible, as the chemical laboratory came to be seen as relevant especially to the *material* reform of the world. It was also associated with the production of material things useful to the common good and often to the emerging nation-states in the form of trade and luxury commodities.[39] In 1668, for example, Joseph Glanvill wrote in *Plus Ultra* that, "Nature being known, it may be master'd, managed, and used in the services of human life."[40]

EXPERIMENT IN THE LABORATORY

In the second half of the seventeenth century, a new conception of replicable experimentation in the controlled space of the laboratory began to be constructed. Whereas the experiential knowledge gathered in the laboratory – like that collected in astronomical observatories, botanical gardens, curiosity cabinets, anatomy theaters, and physicians' case histories – had previously been framed within a number of already existing discourses, including Aristotelianism,[41] it began to take on a new form in the practices of members of some institutions, such as chemists in the Parisian Jardin Royal des Plantes (established in 1626),[42] and of scientific societies, such as the Académie Royale des Sciences and the Royal Society of London. "Demonstration" acquired its modern meaning of proving through an appeal to the senses rather than through an appeal to reason by deductive proof. In members'

---

[38] Howard M. Solomon, *Public Welfare, Science, and Propaganda in Seventeenth Century France* (Princeton, N.J.: Princeton University Press, 1972), pp. 56, 74.
[39] Allen G. Debus, *The Chemical Philosophy*, 2 vols. (New York: Science History Publications, 1977); Pamela H. Smith, *The Business of Alchemy: Science and Culture in the Holy Roman Empire* (Princeton, N.J.: Princeton University Press, 1994); and Smith, "Vital Spirits: Alchemy, Redemption, and Artisanship in Early Modern Europe," in *Rethinking the Scientific Revolution*, ed. Margaret J. Osler (Cambridge: Cambridge University Press, 2000), pp. 119–35.
[40] Joseph Glanvill, *Plus Ultra* (London: James Collins, 1668), p. 87. Although laboratories came to have multiple uses in the course of the seventeenth century, the pedagogic view of the laboratory as a place to discipline and morally edify students would reemerge in slightly transmuted form (to some extent replacing philological research as a component of *Bildung*) as scientific research in the laboratory was institutionalized in German, North American, and Japanese universities in the nineteenth century. See Larry Owens, "Pure and Sound Government: Laboratories, Playing Fields, and Gymnasia in the Nineteenth-Century Search for Order," *Isis*, 76 (1985), 182–94.
[41] Peter Dear, *Discipline and Experience: The Mathematical Way in the Scientific Revolution* (Chicago: University of Chicago Press, 1995).
[42] Ursula Klein, *Verbindung und Affinität: Die Grundlegung der neuzeitlichen Chemie an der Wende vom 17. zum 18. Jahrhundert* (Basel: Birkhauser, 1994).

private laboratories, as well as those established by learned societies, experiment came to mean more than simply carrying out the processes described in a recipe; instead, following Francis Bacon's dicta, it came to involve the "vexing" and "driving" of nature, often with the aid of fire, acids, and instruments such as the air-pump (which produced a vacuum), to draw out its hidden secrets.[43] The apologia for the Royal Society published by Thomas Sprat in 1667, as well as the works of Robert Boyle (1627–1691) and Robert Hooke (1635–1703), clearly developed the idea of the laboratory as a place where "matters of Fact" were judged and resolved upon by demonstration and discourse in a public (but carefully disciplined) space before reliable witnesses.[44] Boyle and Hooke, in particular, attempted to provide their readers with the possibility of "virtually witnessing" and even replicating their experiments in the laboratory.[45]

These matters of fact were supposed to form the stuff of eventual induction to theoretical principles, although the new philosophers were hesitant to embark on what they called "speculative" theory. The proceedings of the Accademia del Cimento, published in 1667,[46] also set out the academicians' experiments in plain prose and clear illustrations. These illustrations show some of their instruments, such as the vacuum pump, barometer, thermometer, telescope, and microscope, as well as the furnaces, crucibles, and distilling vessels of chymistry – the new term signaling the wider scope of laboratory practices in this period (see Newman, Chapter 21, this volume) – that had become standard components of the laboratory. In the Cimento's account of laboratory work, like others published in the seventeenth century, the relation of experiments to theory was assumed, although not always explicated.

In general, although by the late seventeenth century the laboratory had come to be regarded as a central site for the production of natural knowledge, it was in practice often a place of pedagogy, entertainment, and the production of goods as well.[47] For example, Nicolas Lemery (1645–1715) offered a course of chemistry in Paris that was organized around the laboratory, in which he claimed to demonstrate both theory and practice, although

---

[43] Peter Pesic, "Wrestling with Proteus: Francis Bacon and the 'Torture' of Nature," *Isis*, 90 (1999), 81–94.
[44] Thomas Sprat, *The History of the Royal Society* (London: Printed by T. R. for J. Martyn, 1667), pp. 99–100; Steven Shapin, "The House of Experiment in Seventeenth-Century England," *Isis*, 79 (1988), 373–404, at pp. 378, 390, and 399.
[45] Steven Shapin and Simon Schaffer, *Leviathan and the Air-Pump: Hobbes, Boyle, and the Experimental Life* (Princeton, N.J.: Princeton University Press, 1985), chap. 2; Steven Shapin, "Pump and Circumstance: Robert Boyle's Literary Technology," *Social Studies of Science*, 14 (1984), 481–520. See also Shapin, *A Social History of Truth: Civility and Science in Seventeenth-Century England* (Chicago: University of Chicago Press, 1994).
[46] Accademia del Cimento, *Saggi di naturali esperienze fatte nell'Accademia del Cimento sotto la protezione del serenissimo principe Leopoldo di Toscana e descritte dal segretario di essa accademia* (Florence: Giuseppe Cocchini, 1667).
[47] J. L. Heilbron, *Electricity in the 17th and 18th Centuries* (Berkeley: University of California Press, 1979), esp. pp. 140–52.

his classes consisted in making all kinds of chemical preparations that he afterward put on sale.[48] Lémery's course was preceded in Paris by the early seventeenth-century lectures of Jean Beguin (ca. 1550–ca. 1620) and by courses for Parisian apothecaries, focusing on preparations in the laboratory, which had existed from at least the sixteenth century.[49] Some natural philosophers, such as Becher and Hooke,[50] viewed the laboratory as a place where artisans' techniques and recipes could be tested and made available to the commonwealth; indeed, this became one of the explicit aims of the Académie Royale des Sciences, a part of the extension of royal power over autonomous bodies, including guilds.

ACADEMIC INSTITUTIONALIZATION OF THE LABORATORY

The new epistemology of natural knowledge involving use of the laboratory was institutionalized in only very few universities by the end of the seventeenth century, the most notable example being the University of Leiden. Established in 1575 during the Eighty Years' War in the struggle for independence from Spain, this illustrates the confluence of the many different streams that led to the new science and the central role of the laboratory in it. The University of Leiden was conceived as a place to train the citizens of the United Provinces in a proper Calvinist atmosphere, with an eye toward the common good of the republic. Its founding documents established the study of the ancients, particularly Aristotle, as the basis of teaching, but also emphasized the importance of practice, as it was meant to prepare its students for an active life in the service of the republic. This can be seen in a 1575 proposal for the curriculum in the medical faculty, which prescribed not only lectures and disputations but also practical training in "the inspection, dissection, dissolution and transmutation of living bodies, plants and metals."[51] As other medical faculties in Italy and Germany already had done, Leiden opened a botanical garden and anatomy theater in the early 1590s, and in 1636 began clinical teaching (which had some years before entered the curriculum at the University in Padua). But Leiden went further than these other universities in establishing a chemical laboratory in 1669.[52] (Before this, professors held tutorials in their private laboratories.)

---

[48] Charles Bedel, "Les cabinets de chimie," in *Enseignement et diffusion des sciences en France au XVIIIe siècle*, ed. René Taton (Paris: Hermann, 1964), p. 42.
[49] Charles Bedel, "L'Enseignement des sciences pharmaceutiques," in *Enseignement et diffusion*, p. 238.
[50] Michael Hunter, *Establishing the New Science: The Experience of the Early Royal Society* (Woodbridge: Boydell Press, 1989), p. 213.
[51] Professor of Theology, Gulielmus Feugueraeus, quoted in Harm Beukers, "Clinical Teaching in Leiden from Its Beginning until the End of the Eighteenth Century," *Clio Medica*, 21 (1987–8), 139–52, at p. 139.
[52] Harold J. Cook, "The New Philosophy in the Low Countries," in *The Scientific Revolution in National Context*, ed. Roy Porter and Mikuláš Teich (Cambridge: Cambridge University Press, 1992), pp. 115–149; and Harm Beukers, "Het Laboratorium van Sylvius," *Tijdschrift voor de geschiedenis der geneeskunde, natuurwetenschappen, wiskunde en techniek*, 3 (1980), 28–36.

It is a testament to the central place of medical practitioners in developing and institutionalizing the new philosophy that all of these sites for the practice of the new epistemology were at first connected with the medical faculty (see Findlen, Chapter 12, this volume). Only in 1674, when Burchardus de Volder began teaching *physica experimentalis* to support his lectures in *physica theoretica* at the University of Leiden, was the laboratory institutionalized in a faculty outside medicine. The physical laboratory, demonstrating the principles of Newtonian mechanics and developed in the context of teaching students, was established in some university curricula by the eighteenth century and outside universities in the demonstrations of popular lecturers, many of whom maintained their own experimental "cabinets."[53] The practical orientation of the university at Leiden, which was continued above all in the teaching of Herman Boerhaave (1668–1738), influenced the founding ideals of eighteenth-century universities, including those at Halle, Göttingen, and Edinburgh. Thus, by the end of the eighteenth century, the laboratory was firmly established not only in the meeting places of learned societies and the residences of nobles and princes (in which settings it began to have significance for national agendas), but had come to be institutionalized as a central part of the project of both elite and popular Enlightenment.[54]

The laboratory was in no way a site of classical science; rather, it lay at the heart of a new active mode of doing philosophy that formed in the early modern period. Developed above all in the context of alchemy, later chymistry, its orientation, emphasis, and goals were rooted in the practice and productivity of workshops. The story of its emergence as a site of modern science thus illuminates particularly well the distinctive elements of the new epistemology and the new science, which looked toward a direct engagement with nature as a source of "scientific," or certain, knowledge. This knowledge was obtained by means of bodily and sensory interactions with nature in the laboratory. Such active engagement with the natural world yielded an ever-increasing body of knowledge through which the hidden workings of nature and nature's laws would be revealed and which would result in useful products and processes that helped realize the aim of human dominion over nature.

---

[53] On physical cabinets of experiment, see Gerard l'Estrange Turner, "The Cabinet of Experimental Philosophy," in *The Origins of Museums: The Cabinet of Curiosities in Sixteenth- and Seventeenth-Century Europe*, ed. Oliver Impey and Arthur Macgregor (Oxford: Clarendon Press, 1985), pp. 214–22, at p. 222. See also Simon Schaffer, "Natural Philosophy and Public Spectacle in the Eighteenth Century," *History of Science*, 21 (1983), 1–43.

[54] On the beginnings of a connection between scientific research and national agendas, see Solomon, *Public Welfare*; Debus, *The Chemical Philosophy*; and Smith, *The Business of Alchemy*. On laboratory research as a part of Enlightenment, see Jan Golinski, *Science as Public Culture: Chemistry and Enlightenment in Britain, 1760–1820* (Cambridge: Cambridge University Press, 1992); Christoph Meinel, "*Artibus academicis inserenda*: Chemistry's Place in Eighteenth and Early Nineteenth Century Universities," *History of Universities*, 7 (1988), 89–115; Lisa Rosner, *Medical Education in the Age of Improvement: Edinburgh Students and Apprentices, 1760–1826* (Edinburgh: Edinburgh University Press, 1991); and Larry Stewart, *The Rise of Public Science: Rhetoric, Technology, and Natural Philosophy in Newtonian Britain, 1660–1750* (Cambridge: Cambridge University Press, 1992).

14

SITES OF MILITARY SCIENCE AND TECHNOLOGY

Kelly DeVries

It is a sad historical fact that during the Renaissance, one of the richest and most dynamic intellectual periods in Europe's history, there was also almost continuous warfare across the entire continent. The English and the Scots fought, the English and the French fought, the English fought with the Burgundians and the Spanish, and the English fought among themselves (the War of the Roses); the French also fought among themselves, the French and the Burgundians fought, the French and the Italians fought, the French and the Germans fought, and the French even fought with the Portuguese; the Burgundians and the Swiss fought, the Burgundians fought with the Germans, and the Burgundians fought on numerous occasions with their Low Country subjects; the Germans fought with the Italians, and the Germans fought among themselves; the Italians fought among themselves; the various Iberian kingdoms fought among themselves and against Spanish Muslims; the Danes fought with the Swedes, and both fought with the Norwegians; the Teutonic Knights fought with the Livonians and the Russians; and everyone tried (in vain) to fight against the Ottoman Turks.[1] These wars would continue sporadically throughout the early modern period until they culminated in another period of continuous European warfare from 1688 to 1815, dates that correspond to another rich and dynamic intellectual period, the Enlightenment.[2]

Unsurprisingly, this state of almost continuous warfare bore important fruits in the theory and the practice of the military arts, encouraged by both political and military leaders. Indeed, the cultivation of these fields was necessary for the survival of these leaders in an age some historians

---

[1] Kelly DeVries, "Was There a Renaissance in Warfare? Humanism and Technological Determinism, 1300–1559," forthcoming.
[2] See John A. Lynn, "International Rivalry and Warfare," in *The Eighteenth Century: Europe, 1688–1815*, ed. T. C. W. Blanning (Oxford: Oxford University Press, 2000), pp. 178–217.

have called "The Military Revolution."³ The wars and warriors of the period stimulated many important technical discoveries and innovations, altering the way warfare itself was waged and encouraging the development of new groups of practitioners or operators who could work with these discoveries and innovations. Many of these operators might in a later age have been called scientists or engineers: Not only did they develop new techniques and master new practical skills in fields as diverse as gunnery, fortification, gunpowder manufacture, and military surgery – skills that required long and disciplined work experimenting with materials, machines, and even human bodies – but a number of them produced systematic discussions of this work in the form of technical treatises. Some of these writings were purely practical, whereas others offered more general accounts of various aspects of military technology, articulated in mathematical terms. Indeed, the battlefield supplied the new sciences of the early modern period with some of their most challenging and fruitful problems, most notably that of projectile motion. In this sense, early modern warfare offered not only a kind of laboratory for the experimental development of a variety of new military machines and technical skills but also a spur to the mathematical analysis of motion (see Bennett, Chapter 27, this volume).

OFFENSIVE TECHNOLOGIES: GUNPOWDER AND GUNS

By the middle of the thirteenth century, gunpowder had made its way to Western Europe from Asia – most likely from China – by intermediaries yet to be ascertained. Incendiary materials of this sort were part of the more general realm of alchemy, which in this period focused in large part on the production of useful substances, from dyestuffs to drugs. The first surviving recipe for gunpowder appears in the *Epistola de secretis operibus artis et naturae, et de nullitate magiae* (Letter on the Secret Workings of Art and Nature, and on the Vanity of Magic) of the English natural philosopher Roger Bacon (ca.

---

³ The Military Revolution thesis originated with Michael Roberts's 1955 lecture at Queen's University and was printed the next year as a pamphlet under the title *The Military Revolution, 1560–1660* (Belfast: Marjory Boyd, 1956). Later, Geoffrey Parker extended this thesis with *The Military Revolution: Military Innovation and the Rise of the West, 1500–1800* (Cambridge: Cambridge University Press, 1988), now in a second edition (1995). There is also a small group of historians, myself among them, who oppose the notion of a technologically determined Military Revolution. See, for example, Kelly DeVries, "Debate – Catapults Are Not Atomic Bombs: Towards a Redefinition of 'Effectiveness' in Premodern Military Technology," *War in History*, 4 (1997), 454–70; DeVries, "Gunpowder Weaponry and the Rise of the Early Modern State," *War in History*, 5 (1998), 127–45; and Bert S. Hall, *Weapons and Warfare in Renaissance Europe: Gunpowder, Technology, and Tactics* (Baltimore: Johns Hopkins University Press, 1997). For a bibliography of the works, pro and con, inspired by the Military Revolution thesis, see Kelly DeVries, *A Cumulative Bibliography of Medieval Military History and Technology* (Leiden: E. J. Brill, 2002), pp. 578–87.

1219–ca.1292), written sometime between 1248 and 1267.[4] Similar recipes appeared in the *De mirabilibus mundi* (Concerning the Wonders of the World, ca. 1275), attributed to the German natural philosopher Albertus Magnus (ca. 1193–1280), as well as in the pseudonymous Marcus Graecus's *Liber ignium ad comburendos hostes* (Book of Fires for the Burning of Enemies, ca. 1300). These gunpowder recipes varied slightly, but all combined saltpeter, sulfur, and charcoal in a mixture that, when lit, combusted with a forceful explosion. Although Roger Bacon and Marcus Graecus both indicated their ideas on how to use this compound – both suggesting that if "an instrument" could be enclosed on one end, the explosion of gunpowder inside would produce "flying fire" – neither they nor the treatise attributed to Albertus Magnus described a weapon that could use their gunpowder mixture.[5] The weapon eventually invented in Europe was a tube-shaped affair that used gunpowder to discharge a missile, and it quickly became known as a "cannon," from the French *canon* (derived from the word for "tube"), or "gun," from the English *gynne* or *gunne* (derived from the word for "engine").[6]

It is difficult to determine when the first cannon was made in Europe. Written evidence from the early fourteenth century is scarce and often disputed. Less controversial are two artistic sources: one illumination in a manuscript of *De notabilibus, sapientiis et prudentiis regum* (Concerning the Majesty, Wisdom and Prudence of Kings) by the English political theorist Walter de Milemete (fl. 1326), and another in a copy of the pseudo-Aristotelian *De secretis secretorum* (The Secrets of Secrets), both produced in London circa 1326.[7] From this time through the fourteenth and fifteenth centuries, the number of references to gunpowder weapons increases slowly and steadily, as does evidence of their effectiveness in European warfare. By 1377, siege guns of indeterminate size had been successfully used by Philip the Good, Duke of Burgundy, to breach the walls of a fortification, Odruik, near Calais, causing

---

[4] The *Epistola de secretis artis et naturae, et de nullitate magiae* can be found in Appendix I of Roger Bacon, *Opera*, ed. J. S. Brewer (London: Rolls Series, 1859), pp. 550–1. Bacon's gunpowder recipes also appear in his later *Opus majus* (ca. 1267), edited in *The Opus Majus of Roger Bacon*, ed. J. H. Bridges, 2 vols. (London: Rolls Series, 1900), 2: 217–18; and Roger Bacon, *Opus tertium* (ca. 1268), in A. G. Little, ed., "Part of the Opus tertium," *British Society of Franciscan Studies*, 4 (1912), 51, reprinted in *Studies in English Franciscan History* (Manchester: Manchester University Press, 1917), p. 206.

[5] I have translated these recipes and discussed them more extensively in "Gunpowder and Early Gunpowder Weapons," in *Gunpowder: The History of an International Technology*, ed. Brenda Buchanan (Bath: University of Bath Press, 1996), pp. 123–5. There seems little doubt that Marcus Graecus is a pseudonym, although for whom is unknown.

[6] On the nomenclature of early gunpowder weapons, see Hall, *Weapons and Warfare in Renaissance Europe*, pp. 43–5; and Henry Brackenbury, "Ancient Cannon in Europe, Part I: From Their First Employment to A.D. 1350," *Proceedings of the Royal Artillery Institution*, 4 (1865), 288–9.

[7] Christ Church College, Oxford, MS no. 92, fol. 70v (Walter de Milemete); British Library, London, Additional Manuscript 47680, fol. 44v (Pseudo-Aristotle). See Kelly DeVries, "A Reassessment of the Gun Illustrated in the Walter de Milemete and Pseudo-Aristotle Manuscripts," *Journal of the Ordnance Society*, 15 (2003), 5–17.

its surrender;[8] and by 1382, gunpowder weapons used by the Ghentenaars proved forceful enough on a battlefield, Bevershoutsveld, outside the walls of medieval Bruges, to determine the defeat of their Brugeois enemies.[9]

Technological changes were also frequent during this early period of gunpowder weapon development. Guns, which according to the two illuminations mentioned measured one to one-and-a-half meters in length, were made both larger and smaller, eventually becoming hand-held. The methods of manufacture also changed, as guns were more frequently made at the foundry than on the forge. Transportation methods, metallurgy, and the composition of powder were tested and eventually improved.

These gunpowder weapons also required operators trained in their loading and firing, as well as smiths, carpenters, masons, founders, farriers, carters, and other men who worked with the guns, their mounts, and projectiles before and after their use in military engagements.[10] Inexperienced operators of early gunpowder weapons seem often to have lacked confidence in their machines. Errors in use were frequent, as many commanders, trained in the old tactics and strategies of warfare, appear to have been unable to recognize the capabilities of their new weapons and thus use them effectively. Such a case occurred even as late as 1472, when the Burgundian army, perhaps the best equipped with guns as well as the most progressive in initiating technological changes of these weapons, besieged Beauvais. As Burgundian chronicler Philippe de Commynes (ca. 1447–1511) reported:

> My lord of Cordes [one of the Burgundian leaders] . . . had two cannons which were fired only twice through the gate and made a large hole in it. If he had had more stones to continue firing he would have certainly taken the town. However, he had not come with the intention of performing such an exploit and was therefore not well provided.[11]

Misuse of the guns by inexperienced (and chemically ignorant) cannoneers also caused problems. Often, these early gunpowder weapons exploded, and sometimes more bizarre accidents occurred. At the siege of the Flemish castle of Gavere in July 1453, Flemish soldiers coming to relieve the castle

---

[8] An original account of this siege is found in Jean Froissart, *Chroniques*, ed. S. Luce, 10 vols. in 12 (Paris: Mme. Ve. J. Renouard, 1869–97), 8: 249.

[9] On this battle and its role in the history of early gunpowder weapons, see Kelly DeVries, "The Forgotten Battle of Bevershoutsveld, May 3, 1382: Technological Innovation and Military Significance," in *Armies, Chivalry, and Warfare in Medieval Britain and France* (Harlaxton Medieval Studies, 7), ed. Matthew Strickland (Stamford, Lincolnshire: Paul Watkins Publishing, 1998), pp. 289–303, reprinted in Kelly DeVries, *Guns and Men in Medieval Europe, 1200–1500: Studies in Military History and Technology* (Variorum Collected Studies) (Aldershot: Ashgate, 2002), essay VIII.

[10] These gunpowder artillery trains could actually be quite large. See Robert D. Smith, "Good and Bold: A Late 15th-Century Artillery Train," *The Royal Armouries Yearbook*, 6 (2001), 98–105; and Robert D. Smith and Kelly DeVries, *The Artillery of the Dukes of Burgundy, 1363–1477* (Woodbridge: Boydell Press, 2005).

[11] Philippe de Commynes, *The Memoirs of Philippe de Commynes*, ed. Samuel Kinser, trans. Isabelle Cazeaux, 2 vols. (Columbia: University of South Carolina Press, 1969), 1: 236.

broke ranks and fled because one of their cannoneers inadvertently allowed a spark from his igniter to fly into an open sack of gunpowder, which burst into flames. All the nearby cannoneers panicked and ran; when the rest of the army saw this, it also took flight.[12]

Still, the initial problems of using this technology did not dampen the innovative spirits of the individuals who were responsible for supplying these weapons to their various armies and apparently also for improving them. This was largely done in anonymity and perhaps in some secrecy as well; at least there are not many records that discuss the process.[13] Nonetheless, it is clear that people constantly experimented with both guns and gunpowder over the course of the early modern period. This primarily involved experimenting with the relative percentages of the mixtures of the three principal ingredients of gunpowder – saltpeter, sulfur, and charcoal. Roger Bacon's initial recipe called for seven parts of saltpeter to five parts of sulfur and five parts of charcoal, a compound that would certainly combust but was difficult to ignite. Later recipes made the mixture more ignitable by increasing the amount of saltpeter while decreasing the amounts of sulfur and charcoal. The percentages of the three substances varied often throughout the following centuries, with gunpowder chemists not standardizing the most effective mixture (75% saltpeter, 12.5% sulfur, and 12.5% charcoal) until the eighteenth century.[14]

There was also some experimentation with the constituent substances themselves. Imported and domestic sulfur were often called for in different recipes, sometimes for contradictory reasons, whereas some recipes indicated a preference for linden, hazelwood, or willow charcoal.[15] It was saltpeter, however, that seems to have received the most attention. The problem was that most saltpeter is highly soluble, which means that any gunpowder stored in a damp location or at sea would soon become unusable. To reduce this problem, experiments with various compounds of saltpeter (such as potassium nitrate, sodium nitrate, and calcium nitrate) and other techniques of keeping the saltpeter dry were also attempted, ranging from transporting it separately from the charcoal and sulfur to mixing wood ashes, which are rich in potassium carbonate potash, into the saltpeter.[16]

---

[12] Richard Vaughan, *Philip the Good: The Apogee of Burgundy* (London: Longmans, 1970), p. 34. See also Smith and DeVries, *Artillery of the Dukes of Burgundy*, for a more lengthy discussion of this siege.

[13] Although some historians suggest that technological innovations have been shrouded in secrecy since the beginning of time, evidence for this is obviously difficult to find and analyze. See Pamela O. Long, *Openness, Secrecy, Authorship: Technical Arts and the Culture of Knowledge from Antiquity to the Renaissance* (Baltimore: Johns Hopkins University Press, 2001).

[14] See Philippe Contamine, *War in the Middle Ages*, trans. Michael Jones (Oxford: Blackwell, 1984), p. 196; and Hall, *Weapons and Warfare in Renaissance Europe*, pp. 67–75.

[15] DeVries, "Gunpowder and Gunpowder Weapons," pp. 121–36, reprinted in DeVries, *Guns and Men in Medieval Europe*, essay XI.

[16] DeVries, "Gunpowder and Early Gunpowder Weapons"; and Bert S. Hall, "The Corning of Gunpowder and the Development of Firearms in the Renaissance," in Buchanan, ed., *Gunpowder*, pp. 87–120.

Despite the importance of saltpeter in the formation of early gunpowder, it is difficult to know where the earliest supplies of the substance came from or how it was produced. At least initially, all lands seem to have produced their own saltpeter, which was probably grown and harvested to the satisfaction of their military needs. By the turn of the fifteenth century, however, as the increasing use of gunpowder weapons required more and more saltpeter, imported saltpeter began to supplement domestic production. At the same time, technical descriptions of saltpeter production began to be written down, although it is questionable whether the producers actually understood how it actually functioned in gunpowder. The scale of these production sites is also unknown, as it was not until the middle of the sixteenth century that "nitre-beds" or "saltpeter plantations" – where the decomposition of animal and human wastes and vegetable materials could be managed – are mentioned in historical sources.[17] Early modern saltpeter production does seem to have been a very profitable industry, which undoubtedly encouraged numerous producers to enter it. However, the military needs of some states quickly led to its standardization and, ultimately, the regulation of production and ownership.[18]

Finally, attempts were made to engrain or "corn" the powder. Although the use of potassium nitrate and the addition of wood ashes were both meant to keep the saltpeter as dry as possible, the gunpowder made in this way (later called serpentine powder) remained soluble. In addition, the size of dry powder grains meant that the powder burned slowly, as there was little oxygen between the grains to accelerate burning. The solution to both problems was to increase the size of the gunpowder grains. Initially, this seems to have been done by introducing a liquid to the gunpowder – early recipes called for brandy, vinegar, or the urine of a "wine-drinking man" – molding the powder into cakes, and then crumbling it when ready for use. By the middle of the sixteenth century, however, this had proven to produce too many gunpowder grains that were too large for effective propulsion. Thus, a sifting "corning" process – using various sizes of sieves to standardize the grain size – replaced the crumbling of the powder. Still, throughout the early modern period, dry serpentine powder, large corned powder – akin to the crumbled powder – and fine corned powder existed side by side for use in cannons of various sizes and for different purposes.[19]

---

[17] Hall, *Weapons and Warfare in Renaissance Europe*, pp. 74–5. See also Alan R. Williams, "The Production of Saltpetre in the Middle Ages," *Ambix*, 22 (1975), 125–33.

[18] No comprehensive study of late medieval and early modern saltpeter production has been written. Local studies include: Stephen Bull, "Pearls from the Dungheap: English Saltpetre Production, 1590–1640," *Journal of the Ordnance Society*, 2 (1990), 5–10; Surirey de Saint-Rémy, "The Manufacture of Gunpowder in France (1702), Part I: Saltpetre, Sulphur, and Charcoal," trans. D. H. Roberts, *Journal of the Ordnance Society*, 5 (1993), 47–55; and Thomas Kaiserfeld, "Chemistry in the War Machine: Saltpeter Production in 18th Century Sweden," in *The Heirs of Archimedes: Technology, Science and Warfare, 1350–1800*, ed. Brett Steele (Cambridge, Mass.: MIT Press, 2005), pp. 275–92. See also Hall, *Weapons and Warfare in Renaissance Europe*, pp. 75–9.

[19] Hall, "The Corning of Gunpowder," pp. 94–107; and Hall, *Weapons and Warfare in Renaissance Europe*.

Most of the early modern people responsible for these innovations remain anonymous and their processes of invention and experimentation unrecorded and unknown. Indeed, the task of supplying gunpowder itself, at least initially, seems to have been assigned to a variety of occupations. In the fourteenth century, apothecaries seem to have supplied gunpowder to the armies; in the fifteenth and sixteenth centuries, it was supplied by masters of artillery, although whether they actually made the gunpowder or procured it from a subcontractor is unclear.[20] Only in the late sixteenth and early seventeenth centuries were ordnance offices set up in various European lands; these offices took care of the procurement, standardization, and secrecy of state gunpowder holdings.[21]

Most of the engineers and metallurgists responsible for contemporary innovations in gunpowder weapons also remain anonymous. These innovations were so numerous that they cannot be discussed here; it suffices to mention that this period saw the building of various sizes of gunpowder weapons, the development of hand-held gunpowder weaponry, the construction of gunpowder weapons specifically designed for use on board ships (with more easily interchangeable parts), the introduction and continued use of removable chambers (which allowed for quicker and safer firing),[22] the creation of new types of projectiles (bolts, stone balls, metal balls, incendiary balls, and, finally, various odd styles of shot, such as canister or bar shot), and the development of different mounts, aiming devices, and transportation methods (culminating in the invention of individual wheeled carriages that combined all three).[23]

The operators of these machines, again largely anonymous, became far more experienced and expert in their use in battles and sieges – as both attackers and defenders. Mistakes and danger persisted with gunpowder weapons more than with traditional arms, which is undoubtedly why cannoneers were paid a higher wage than other soldiers – indeed the same rate as the socially superior cavalryman. However, over the course of the sixteenth

---

[20] DeVries, "Gunpowder and Early Gunpowder Weapons," pp. 125–7. The expertise of these masters seems to have been gained either because they made the weapons or gunpowder or gained experience in using them in military situations. Although the names of several masters are known, only two are discussed widely in contemporary sources, the French brothers Jean and Gaspard Bureau, who are credited with the reorganization of the administration and tactical use of the French gunpowder artillery train during the latter stages of the Hundred Years War. Much of what the standard article (that of H. Dubled) on the Bureau brothers claims is difficult to ascertain. See H. Dubled, "L'Artillerie royale Française à l'époque de Charles VII et au début du règne de Louis XI (1437–1469): Les frères Bureau," *Memorial de l'artillerie française*, 50 (1976), 555–637.

[21] Richard Winship Stewart, *The English Ordnance Office, 1585–1625: A Case Study in Bureaucracy* (London: Royal Historical Society, 1996).

[22] See Kelly DeVries and Robert D. Smith, "Removable Powder Chambers in Early Gunpowder Weapons," in *Gunpowder, Explosives, and the State: A Technological History*, ed. Brenda Buchanan (Aldershot: Ashgate, 2005).

[23] Kelly DeVries, *Medieval Military Technology* (Peterborough: Broadview Press, 1992), pp. 157–8; and Robert D. Smith, "Casting Shot in the 16th Century," *Journal of the Ordnance Society*, 12 (2000), 88–92. For a more complete discussion of these matters, see also Smith and DeVries, *Artillery of the Dukes of Burgundy*.

Sites of Military Science and Technology

and seventeenth centuries, problems with gunpowder weapons became less severe. Operators of gunpowder weapons also began to receive more training in the discharge and tactical use of their weapons. Eventually, even this problem was solved, by the publication of illustrated vernacular manuals such as Dutch drillmaster Jacob de Gheyn's (1565–1629) *Wapenhandlingen van roers, musquetten end spiessen* (Arms Drill with Arquebus, Musket, and Pike). Originally published in 1607 and continually reprinted and translated for the next two centuries, this manual used both text and images to detail techniques of maneuvering, loading, and discharging firearms for the literate and illiterate alike.[24]

DEFENSIVE TECHNOLOGIES: ARMOR AND FORTIFICATION

The changes in warfare brought about by gunpowder weapons forced transformations in other military occupations as well. Some of the more traditional arms manufacturers, especially armorers, tried to keep pace with these gunpowder-related changes. Armorers did this first by adding metal plates to the traditional chain armor worn by most soldiers and then, in the middle of the fifteenth century, they began to form a suit of armor that provided complete protection for its wearer.[25] The construction of this type of armor, known generically as plate armor, required the most highly skilled metallurgy.[26]

It has often been thought that the use of gunpowder weapons caused the ultimate decline of armor.[27] Yet experiments undertaken in 1989 by curators at the Landeszeughaus in Graz, Austria, using original early modern gunpowder weapons against original armor breastplates, have shown that armor was protective against almost all forms of gunpowder weapons through the eighteenth century.[28] This might mean that the decline of early modern armor

---

[24] Jacob de Gheyn, *Wapenhandlinghe van roers, musquetten ende spiessen, achtervolgende de ordre van Sijn Excellentie Maurits, Prince van Orangie, Graue van Nassau, etc., Gouverneur ende Capiteijn Generael ouer Gelderlant, Hollant, Zeelant, Vtrecht, Overijssel, etc.* (The Hague, 1607); for an English edition, see J. B. Kist, ed., *The Exercise of Armes: All 117 Engravings from the Classic 17th-Century Military Manual* (Mineola, N.Y.: Dover 1999). See also Geoffrey Parker, *The Military Revolution: Military Innovation and the Rise of the West, 1500–1800* (Cambridge: Cambridge University Press, 1988), pp. 20–1.

[25] On the history of armor, the best work remains Claude Blair, *European Armour: circa 1066 to circa 1700* (London: B.T. Batsford, 1958).

[26] Matthias Pfaffenbichler, *Medieval Craftsmen: Armourers* (Toronto: University of Toronto Press, 1992).

[27] There are too many works on the decline of the armor industry to list here. Those interested should begin their study with the relevant chapters in Blair, *European Armour*, and Ewart Oakeshott, *European Weapons and Armour: From the Renaissance to the Industrial Revolution* [1980] (Woodbridge: Boydell Press, 2000); and, as a case study, Stuart Pyhrr, José-A. Godoy, and Silvio Leydi, *Heroic Armor of the Italian Renaissance: Filippo Negroli and His Contemporaries* (New York: The Metropolitan Museum of Art, 1998).

[28] The original findings of these experiments are recorded in Peter Krenn, ed., *Von alten Handfeuerwaffen: Entwicklung, Technik, Leistung* (Graz: Landeszeughaus, 1989), summarized in Peter Krenn, "Was leisteten die alten Handfeuerwaffen? Ergebnisse einer Ausstellung des

perhaps had a nontechnological cause, such as the declining availability of skilled craftsmen or metallurgical resources, or the high cost of such armor, that ultimately made it obsolete in the face of the larger number of men in arms during the same period. Evidence for this latter possibility comes from the fact that even before the end of the fifteenth century, many armorers, especially in Italy, had begun to nurture a luxury market, which combined their artisanal skills and their artistic talents to produce both defensible and beautiful armors.[29]

Fortification engineers, too, had to take into account the new destructive power of gunpowder weapons being used against traditional medieval fortresses. Initially, this was done simply by altering the existing structures, first with gunports constructed at vulnerable spots in fortification walls or gates. Later, boulevards and artillery towers were added to medieval fortifications. These were large structures of different sizes and shapes filled with gunpowder weapons meant to provide defensive flanking firepower in an effort to ward off attacks by offensive guns.[30] However, at some locations an entirely new fortification system was constructed that was designed completely with an eye toward defense against gunpowder weapons. This was an evolutionary process that began with the construction of multitowered structures and then led to the development of the low-walled, sloping, angular bastion, and finally to the construction of the *trace italienne* fortification, which transformed the angular bastion design into an entirely new and innovative fortress. This process was also expensive, with such fortification systems costing much more than their medieval precursors. Still, the *trace italienne* would remain the prominent fortification design of the early modern era, albeit frequently altered to suit geographical strengths or limitations or to meet new technological threats.[31] In general, the geometrical problems posed

---

Landeszeughauses in Graz," *Waffen- und Kostümkunde*, 32 (1990), 35–52; and Peter Krenn, Paul Kalaus, and Bert Hall, "Material Culture and Military History: Test-Firing Early Modern Small Arms," *Material History Review*, 42 (Fall 1995), 101–9. Surprisingly, it is the small hand-held gunpowder weapons, the pistols, that proved to have had the most penetrative ballistic force against armor.

[29] For a case study of this, see Pyhrr, Godoy, and Leydi, *Heroic Armor of the Italian Renaissance*.

[30] The literature on early modern fortifications is extensive. For the early changes, see Kelly DeVries, "The Impact of Gunpowder Weaponry on Siege Warfare in the Hundred Years War," in *The Medieval City under Siege*, ed. Ivy A. Corfis and Michael Wolfe (Woodbridge: Boydell Press, 1995), pp. 227–44, reprinted in DeVries, *Guns and Men in Medieval Europe*, essay XIII; and Kelly DeVries, "Facing the New Military Technology: Non-*Trace Italienne* Anti-Gunpowder Weaponry Defenses, 1350–1550," in Steele, ed., *The Heirs of Archimedes*, pp. 37–71. One should be cautious in using Parker, *The Military Revolution*, as he discredits medieval fortifications and their early anti-gunpowder weaponry defenses while giving far more defensive credit to *trace italienne* fortifications than is supported by contemporary evidence.

[31] The standard work on the development of the *trace italienne* fortifications remains John R. Hale, "The Early Development of the Bastion: An Italian Chronology," in *Europe in the Late Middle Ages*, ed. J. R. Hale (Evanston, Ill.: Northwestern University Press, 1965), pp. 466–94; see also his other study: *Renaissance Fortification: Art or Engineering?* (London: Thames and Hudson Press, 1977). For an excellent work on how much this type of fortification cost and how well it stood up against gunpowder weapons, see Simon Pepper and Nicholas Adams, *Firearms and Fortifications: Military*

by analysis and design of fortifications were not trivial, and they stimulated the development of synthetic geometry. This bore fruit in the new schools of military engineering founded in the eighteenth century, such as the French École Polytechnique, which became a center of study in this field.[32]

Large defensive structures of this sort, although a huge drain on the public coffers, were constructed to protect the general populace and enjoyed a high degree of public recognition as a result. It is perhaps for this reason that we often know the names of fortification designers, builders, and engineers, unlike those of the anonymous innovators who experimented with the composition of gunpowder and the design of guns. Almost all who worked on fortifications were well trained in the liberal arts and were literate in both Latin and several vernaculars. They were also highly respected as architects, on a par with the designers of religious and state buildings, and a number wrote treatises on fortifications and other types of construction. Especially famous were those who worked in Italy during the fifteenth and sixteenth centuries: Leon Battista Alberti (1404–1472), Antonio Francesco Averlino, also known as Filarete (ca. 1400–1469), Francesco di Giorgio Martini (1439–1502), Buonaccorso Ghiberti (1451–1516), Leonardo da Vinci (1452–1519), Antonio da Sangallo (1484–1546), and Giambattista Aleotti (1546–1636).[33] Indeed, the seventeeth-century Sébastien Le Prestre de Vauban (1633–1707), who was counted on by King Louis XIV both to build French fortifications

---

Architecture and Siege Warfare in Sixteenth-Century Siena (Chicago: University of Chicago Press, 1986).

[32] Ken Alder, *Engineering the Revolution: Arms and Enlightenment in France, 1763–1815* (Princeton, N.J.: Princeton University Press, 1997).

[33] General studies of these fortification designers and engineers can be found in Horst de La Croix, "The Literature on Fortification in Renaissance Italy," *Technology and Culture*, 4 (1963), 30–50; and Hale, *Renaissance Fortification*. For readily accessible treatises of Alberti, Filarete, Martini, and Ghiberti, see: Leon Battista Alberti, *L'Architettura (De re aedificatoria) Leon Battista Alberti*, ed. and trans. Giovanni Orlandi, 2 vols. (Milan: Edizioni Il Polifilo, 1966), translated as *On the Art of Building in Ten Books*, trans. J. Rykwert, Neil Leach, and Robert Tavernor (Cambridge, Mass.: MIT Press, 1988); Filarete, *Trattato di architettura*, ed. Anna Maria Finoli and Liliana Grassi, 2 vols. (Milan: Il Polifilo, 1972), translated as *Treatise on Architecture, Being the Treatise by Antonio di Piero Averlino, known as Filarete*, trans. John R. Spencer, 2 vols. (New Haven, Conn.: Yale University Press, 1965); Francesco di Giorgio Martini, *Trattati di architettura ingegneria e arte militare*, 2 vols. (Milan: Electa, 1967); and Buonaccorso Ghiberti, *A Translation of Vitruvius and Copies of Late Antique Drawings in Buonaccorso Ghiberti's Zibaldone*, trans. Gustina Scaglia (Philadelphia: American Philosophical Society, 1979). Treatises of Leonardo da Vinci, Sangallo, and Aleotti must be accessed through secondary studies devoted to their military architecture. See Renzo Manetti, "Antonio da Sangallo: Arte fortificatoria e simbolismo neoplatonico nella fortezza di Firenze," in *Architettura militare nell'Europa del XVI secolo: Atti del Convegno di Studi, Firenze, 25–28 Novembre 1986*, ed. Carlo Cresti, Amelio Fara, and Daniela Lamberini (Siena: Edizioni Periccioli, 1988), pp. 111–20; Pietro C. Marani, "L'Architettura militare di Leonardo da Vinci fra tradizione, rinnovamento e ripensamento," in Cresti, Fara, and Lamberini, eds., *Architettura militare*, pp. 49–59; Simon M. Pepper, "Planning Versus Fortification: Sangallo's Plan for the Defence of Rome," *Architectural Review*, 159 (1976), 163–9; and Paolo Zermani, "'Delle fortificazioni': 'Offensa' e 'difesa' nella teoria architettonica di Giovan Battista Aleotti," in Cresti, Fara, and Lamberini, eds., *Architettura militare*, pp. 337–50.

and to knock down those of opponents, is buried with highest honors in Les Invalides next to Napoleon Bonaparte.[34]

Another occupation that changed because of the increased use of gunpowder weapons was that of the military surgeon. As long as there had been armies, surgeons had accompanied them. Although these medical men could do little more than cleanse and suture wounds caused by bladed weapons or remove projectiles and patch wounds caused by archers and slingers, they seem to have been quite successful in these tasks, as is clear from previously healed war wounds in the torsos, limbs, and skulls of battlefield casualties found in excavations at Visby (1361) and Towton (1461).[35] With the increasing use of gunpowder-propelled projectiles, however, new, more gruesome wounds occurred, that required special attention. Initially, surgeons seem to have treated these wounds no differently from non-gunpowder-caused war wounds, cleansing and then either suturing or supperating the wound.[36] However, by the end of the fifteenth century, some developed the practice of treating gunpowder wounds with hot oil (and without cleansing them or removing debris). This new practice seems to have been driven by a novel belief that wounds caused by gunpowder weapons were poisoned and that cauterization was the best means for treating them, as argued by the papal surgeon Giovanni da Vigo (ca. 1450–1525) in his popular military surgical treatise *Practica in arte chirugica copiosa* (The Extensive Practice of Surgical Arts).[37] Nor did French military surgeon Ambroise Paré's (ca. 1510–1590) well-known criticism of the technique at the siege of Turin in 1536 invalidate this method of wound treatment, which unfortunately for its victims continued to be used to treat gunshot wounds until well into the eighteenth century, when it was replaced by amputation.[38] Surgeons were also forced to take care

---

[34] Vauban's treatises on military architecture have been translated into English as *A Manual of Siegecraft and Fortification*, trans. George A. Rothrock (Ann Arbor: University of Michigan Press, 1968). On fortification engineering in general during the late early modern period, see Christopher Duffy, *The Fortress in the Age of Vauban and Frederick the Great, 1660–1789* (London: Routledge and Kegan Paul, 1985); and Duffy, *Fire and Stone: The Science of Fortress Warfare, 1660–1860* (Newton Abbot: David and Charles, 1975).

[35] For the battle of Visby, see Bengt Thordemann, *Armour from the Battle of Wisby, 1361* [1939] ([Highland City, Tx.]: Chivalry Bookshelf, 2001). For the battle of Towton, see Veronica Fiorato, Anthea Boylston, and Christopher Knüsel, *Blood Red Roses: The Archaeology of a Mass Grave from the Battle of Towton, A.D. 1461* (Oxford: Oxbow Books, 2000).

[36] Kelly DeVries, "Military Surgical Practice and the Advent of Gunpowder Weaponry," *Canadian Bulletin of Medical History*, 7 (1990), 131–46, reprinted in DeVries, *Guns and Men in Medieval Europe, 1200–1500*, essay XVII.

[37] There is no modern edition of Giovanni da Vigo's surgical text, but see the reprint of the 1543 English translation, *Joannes da Vigo: The Most Excellent Workes of Chirurgerye* [1543] (Amsterdam: Da Capo Press Theatrum Orbis Terrarum, 1968). The popularity of this work is indicated by the number of editions and translations published during the end of the fourteenth and beginning of the fifteenth centuries. See DeVries, "Military Surgical Practice and the Advent of Gunpowder Weaponry," pp. 140–2.

[38] Ambroise Paré, *The Apologie and Treatise*, ed. and trans. Geoffrey Keynes (Chicago: University of Chicago Press, 1952), p. 138. Paré's treatment for gunshot wounds is discussed more extensively in his later work, *Ten Books of Surgery with the Magazine of the Instruments Necessary for It*, ed. and trans. Robert White Linker and Nathan Womack (Athens: University of Georgia Press, 1969), pp. 1–47.

of other gunpowder-related injuries, the most notable of these being powder burns acquired by gunners from the strong back-blast of their hand-held guns. So numerous were these injuries, in fact, that one commentator on the fifteenth-century Burgundian army, Olivier de la Marche, claimed that Duke Charles the Bold had to increase the number of his military surgeons just to treat the powder burns and other injuries of his gunners.[39]

COURTLY ENGINEERS AND GENTLEMAN PRACTITIONERS

With the exception of the designers and builders of fortifications, as I have already indicated, the names of most artisans and operators of these technologies are unknown. The socioeconomic evidence that does exist suggests that they came mostly from the artisanal or lower classes and were largely formally uneducated. Their discoveries and innovations undoubtedly brought some of them prominence and wealth, but, like other artisans of the early modern period, they appear in few historical records. However, one group of intellectuals began to build on and tie their quest for patronage to the advent and proliferation of gunpowder weapons. The earliest of these intellectuals, from the fifteenth and early sixteenth centuries, have most frequently been called "courtly engineers" by modern historians (see the following chapters in this volume: Bertoloni Meli, Chapter 26; Bennett, Chapter 27). Because their search for patronage required self-advertisement, their names are frequently known: Conrad Kyeser (1366–ca. 1405), Mariano Taccola (1381–ca. 1458), Roberto Valturio (1405–1475), and Francesco di Giorgio Martini (1439–1502) in the fifteenth century; and Leonardo da Vinci, Agostino Ramelli (1531–ca. 1600), and Jacques Besson in the sixteenth century (fl. 1550–1580). Unlike other intellectuals, such as Miguel de Cervantes de Saavedra (1547–1616) and François Rabelais (1490–ca. 1553), who spurned the new military technology as both unchivalrous and demonic, these men embraced gunpowder weapons.

With a few exceptions, such as Leonardo da Vinci, most military writers from the sixteenth century on took advantage of the newly invented printing press to publish their works. At times, these publications gave credibility to what were essentially erroneous ideas, such as those of the military surgeon Giovanni da Vigo, whom I have already mentioned, or publicized antiquated reactions to new military technologies. Among the latter were the treatises of the Florentine writer Niccolò Machiavelli (1469–1527), whose popular works

---

For a discussion of Paré's influence on later surgical history, see J. F. Malgaigne, *Surgery and Ambroise Paré*, trans. Wallace B. Hamby (Norman: University of Oklahoma Press, 1965); and the references in DeVries, "Military Surgical Practice and the Advent of Gunpowder Weaponry," p. 143, n. 3.

[39] Olivier de la Marche, *Estat de la maison de Charles le Hardi*, in *Collection complète des mémoires relatifs à l'histoire de France*, vols. 9 and 10, ed. M. Petitot (Paris: Foucault, 1820), 10: 493.

Arte della guerra (The Art of War, 1521), *Discorsi sulla prima deca di Tito Livio* (The Discourses on the First Ten Books of Livy, 1531), and *Il principe* (The Prince, 1532), which cautioned against the military use of gunpowder weapons, gained a greater audience and popularity throughout Europe because they were printed than they would have had they been written earlier.[40] Most often, however, the printing press was effective in disseminating many of the innovations in strategy, tactics, fortification construction, military surgery, military offensive and defensive technologies, and the use of and defense against gunpowder weapons that ultimately directed military theory and practice well into the nineteenth century.[41]

Although some of their scientific and technological designs and ideas were flawed, and even fantastic, as in the case of the flying machines of Leonardo da Vinci, the courtly engineers not only remained influential throughout the sixteenth and seventeenth centuries but even won a new audience and following. Some of their intellectual and military descendants patterned their works on those of their predecessors, writing treatises to gain employment or patronage. But others had no such goals when they devoted themselves to these scientific and technological military interests. As members of the social and economic elite, these men did not need this type of employment and saw the intellectualization of warfare, and of the technology behind it, as a means of gaining status and participating in the warfare of the day from the safety of their desks. Other writers on military topics were military men who sought elevation in status to the gentlemanly class, which their military skills alone had not brought them. The books of, for example, Thomas Harriot (1560–1621), Peter Whitehorne (fl. 1588), William Bourne (d. 1583), Leonard Digges (1588–1635), and Thomas Digges (d. 1595), to name only some of the English writers, testified to their scientific and technological interests as well as their pedagogical abilities. They discussed and debated the construction

---

[40] For references to da Vigo's works, see note 36. Machiavelli's military works – and I count both *The Discourses* and *The Prince* among them, as each devotes so much space to a discussion of Italian warfare and gunpowder weapons – spread throughout Italy and Europe in large numbers and translations. There are many versions of *The Discourses* and *The Prince* available today; *The Art of War* is most easily accessed in Niccolò Machiavelli, *The Art of War*, ed. and trans. Neal Wood (New York: Da Capo Press, 1965). Among the many studies of Machiavelli's military writings, see M. Hobohm, *Machiavellis Renaissance der Kriegkunst*, 2 vols. (Berlin: Curtius, 1913); and Sydney Anglo, "Machiavelli as a Military Authority: Some Early Sources," in *Florence and Italy: Renaissance Studies in Honour of Nicolai Rubinstein*, ed. P. Denley and C. Elam (London: Westfield College, University of London, Committee for Medieval Studies, 1988), pp. 321–4. For a discussion of Machiavelli's views on gunpowder weapons, see Ben Cassidy, "Machiavelli and the Ideology of the Offensive: Gunpowder Weapons in *The Art of War*," *Journal of Military History*, 67 (2003), 381–404.

[41] To date, no complete study of the relationship between printing and the proliferation of military arts and sciences has been undertaken, although several case studies, both geographical and authorial, point to what might be an extremely rich topic. See, for example, John R. Hale, "Printing and the Military Culture of Renaissance Venice," in Hale, *Renaissance War Studies* (London: Hambledon, 1983), pp. 429–70; Hale, "Gunpowder and the Renaissance: An Essay in the History of Ideas," in *From Renaissance to Counter-Reformation: Essays in Honour of Garrett Mattingly*, ed. Charles H. Carter (London: Cape, 1966), pp. 113–44; and Henry J. Webb, *Elizabethan Military Science: The Books and the Practice* (Madison: University of Wisconsin Press, 1965).

of gunpowder weapons, the mixture of gunpowder, the ballistics of shot, gunnery practice, and the logistics and provisioning of armies. They not only used traditional chemical and metallurgical terminology and philosophy in their discussions but also added lengthy discussions of tactics, drilling, training, and gunfire employing mathematics and geometry.[42]

Many of the works of these men became immensely successful and were used as instruction manuals for training soldiers and, especially, officers. Throughout the late sixteenth, seventeenth, and eighteenth centuries, officers received a training that taught not only tactics and strategy but also the technical details of warfare, as well as the general principles that informed these practices and techniques. This education not only created a new military class system, which was disrupted only with the French Revolution and would not be completely eliminated until the twentieth century, but also wedded science and technology to the military arts in a union that has yet to be broken.[43]

In addition to its effects on military training and practice, the booming market for military skills produced by the almost constant warfare of early modern Europe, together with the dramatic innovations that resulted from the introduction of guns and gunpowder, had important implications for natural inquiry. Gunnery, the building of war machines, and the design of fortifications were prominent among what were known as the "mechanical arts," which included the practical disciplines that involved working with machines. Over the course of the late sixteenth and seventeenth centuries, writers on the arts of warfare, whether courtly engineers, gentleman amateurs, or more humble practitioners who gained a professional status by discussing, debating, and writing on these military subjects, worked to develop more systematic conceptual foundations for these military practices, both to improve effectiveness and to promote their own interests and careers. Increasingly, they theorized foundations in terms of projective geometry and practical mathematics (see Bennett, Chapter 27, this volume), which had important effects on the discipline of mechanics itself.

---

[42] Currently, the most comprehensive work on the subject is Webb, *Elizabethan Military Science*, but this will be updated by Steven A. Walton, *The Scientific Military Gentleman: Scientific Knowledge, Military Technology, and Society in Early Modern England* (forthcoming). See also A. Rupert Hall, *Ballistics in the Seventeenth Century* (Oxford: Oxford University Press, 1965).

[43] John R. Hale, "The Military Education of the Officer Class in Early Modern Europe," in *Cultural Aspects of the Italian Renaissance: Essays in Honor of Paul Oskar Kristeller*, ed. C. H. Clough (Manchester: Manchester University Press, 1976), pp. 440–61, and in Hale, *Renaissance War Studies*, pp. 225–46; and Hale, "Military Academies on the Venetian Terrafirma in the Early Seventeenth Century," *Studi veneziani*, 15 (1973), 273–95, also in Hale, *Renaissance War Studies*, pp. 285–308. On changes in the engineering profession during the French Revolution, see Alder, *Engineering the Revolution*.

15

COFFEEHOUSES AND PRINT SHOPS

Adrian Johns

Experimental philosophy came to prominence on a wave of coffee. In the mid-1650s, a group of aspiring Oxford scholars met regularly at a new kind of place. Something like an alcohol-free tavern, it was presided over by an apothecary named Arthur Tillyard. There, spurred by liberal doses of thick, black liquid, they debated the new ideas transforming natural philosophy and the mathematical sciences. Peter Staehl, Robert Boyle's (1627–1691) German chymist, mounted experimental displays at the same location. This club of scholars moved to London in 1660 and reemerged as what would soon be called the Royal Society. At about the same time, the novel setting in which these rendezvous had originally occurred – the coffeehouse – also moved to London and began an extraordinary surge in popularity there. Together, the coffeehouse and the Royal Society would become perhaps the two most distinctive social spaces of Restoration England. The implications of their advent would stretch beyond England itself. What began in Oxford and London would grow to affect the fortunes of the sciences throughout Western Europe.

Experiment was not the only controversial investigatory practice fostered by the rise of the coffeehouse. In 1659, political philosopher James Harrington (1611–1677) – whose *Oceana* (1656) had founded republicanism on a natural philosophy of circulating particles – organized regular debates at Miles' Coffeehouse, in Westminster. Here soldiers, political figures, and ordinary citizens participated in exhilarating exchanges that ranged broadly over the history and philosophy of government. The Rota, as Harrington called this forum, both generated new knowledge and, far more importantly, exemplified a new way of proposing, debating, and resolving claims in general. It soon disintegrated, falling victim to the growing inevitability of the restoration of a Stuart monarchy. But the kind of sociability it had exemplified proved more resilient. Harrington's social experiment had disclosed the coming of a new and vital form of civil society that would come to redefine argument itself.[1]

---

[1] James Harrington, *Political Works*, ed. John Pocock (Cambridge: Cambridge University Press, 1977), pp. 112–17. For Harrington and natural philosophy, see J. H. Scott, "The Rapture of Motion: James

If coffeehouses and their clientele had been restricted to Oxford and London, their role in this new form of society would have been less significant. Before long, however, they were appearing in larger towns across England. In Continental Europe, too, they had taken root and were flourishing. Paris coffeehouses were reputedly premises of rare elegance, and those of Amsterdam rivaled London's as centers for gossip and conversation. Everywhere they proved popular. By 1700, England, France, and the Netherlands combined were importing some three million pounds of coffee per year, a huge amount for the time, indicating that the coffeehouses themselves had become the endpoints of a large international trade network. It was in London that they became particularly prevalent, however, and where they came to stand for a new form of sociability – the first coffeehouse to open in colonial America even called itself "the London" to evoke the association. By that time, anyone wanting to find out about politics, hear about the latest initiatives of the virtuosi, plan a publication, or sample a new medicine was well-advised to resort to a coffeehouse. These modest institutions were making possible a reconstitution of conversation itself, providing the nucleus for a public sphere.[2]

Yet coffee alone could scarcely have had such a pronounced effect. In the coffeehouses of Amsterdam, Paris, and London, it met with another thick, black liquid: ink. Each was culturally volatile, and their mixture proved both potent and dangerous. Coffeehouses were soon providing the economic foundation for novel forms of communication. They welcomed discussion fuel of all kinds, especially pamphlets and periodicals that could be consumed at a sitting. The coffeehouse provided a chance to read such materials, either silently to oneself or aloud to others, and an opportunity for conversation about what one read. Reading of this kind could transform the meanings even of traditional works. ("What Havock they made of *Bodin*, *Machiavel*, & *Plato*," exclaimed one observer.) Such reading, widely recognized as unprecedented, soon became an inescapable constant of cultural life. Sir Roger L'Estrange (1616–1704), the English Crown's chief policeman of the press, might bemoan the pamphlet press's inherent insubordination, but he was forced to do so in

---

Harrington's Republicanism," in *Political Discourse in Early Modern Britain*, ed. Nicholas Phillipson and Quentin Skinner (Cambridge: Cambridge University Press, 1993), pp. 139–63.

[2] On the coffee trade, coffeehouses, and the origins of the public sphere, see Fernand Braudel, *Civilization and Capitalism, 15th–18th Century*, vol. 1: *The Structures of Everyday Life*, trans. S. Reynolds (London: Fontana, 1985), pp. 256–60 (for the development of the trade); B. Cowan, *The Social Life of Coffee: The Emergence of the British Coffeehouse* (New Haven, Ct: Yale University Press, 2005); J. Leclant, "Le café et les cafés à Paris (1644–1693)," *Annales Économies, sociétés, civilisations*, 6 (1951), 1–14; Daniel Roche, *France in the Enlightenment*, trans. Arthur Goldhammer (Cambridge, Mass.: Harvard University Press, 1998), pp. 627–30 (for Paris); Margaret C. Jacob, *The Radical Enlightenment: Pantheists, Freemasons and Republicans* (London: Allen and Unwin, 1981), p. 183 (for The Hague); David S. Shields, *Civil Tongues and Polite Letters in British America* (Chapel Hill: University of North Carolina Press, 1997), pp. 18–22, 55–7 (for Philadelphia, a reference for which I am grateful to Margaret Meredith); and Deric Regin, *Traders, Artists, Burghers: A Cultural History of Amsterdam in the Seventeenth Century* (Amsterdam: Van Gorcum, 1976), pp. 137–8 (for Amsterdam).

pamphlets and journals that adopted exactly the forms suited to coffeehouse perusal. Indeed, L'Estrange's own office was financed by a legal monopoly on "news" – the principal epistemic novelty to issue from the coffeehouse and the printing press.[3]

The alliance of coffee and print transformed authorship, communication, and conversation. By the same token, the alliance also reshaped the materials by which historians now find out about such matters. The implications were by no means limited to political themes. Coffeehouse reading promised – or threatened – to transform *all* kinds of creative discussion, whatever the subject or occasion. It was in London in particular that its effects on natural philosophy came to be most strongly felt, thanks largely to the inauguration of experimental philosophy at the Royal Society. No other metropolis fostered the intersection of coffee, print, and experiment to quite the same extent or to such potent effect. By the beginning of the eighteenth century, the conjunction of coffee and print had established the foundations for a different age – an age characterized not by court masques, royal prerogative, and sectarian intransigence but by Newtonianism, public reason, and rotten boroughs.

PRINT

By the time coffee arrived in London, print had been there for almost two hundred years. There were at least 150 bookshops in the capital, served by forty to sixty printing houses. Their proprietors were in principle united as members of the Stationers' Company, a craft organization chartered in the mid-sixteenth century. In practice, not all participated: An unknowable number of illegal "private" presses existed, and many booksellers were not members of the Stationers' Company. Nonetheless, the Company's existence mattered greatly. It had successfully forged a social, moral, and economic order for the book trade and was legally and politically responsible for maintaining that order. Its monthly tribunal could deploy substantial powers in order to restore harmony when disputes among members seemed likely to

---

[3] Thomas St. Serfe, *Tarugo's Wiles; Or, The Coffee-House* (London: H. Herringman, 1668), p. 20. The literature on coffeehouses is now extensive; their presence in London over a long period is systematically surveyed in Bryant Lillywhite, *London Coffee Houses: A Reference Book of Coffee Houses of the Seventeenth, Eighteenth and Nineteenth Centuries* (London: Allen and Unwin, 1963), and introduced in Aytoun Ellis, *The Penny Universities: A History of the Coffee-Houses* (London: Secker and Warburg, 1956). For their importance to Robert Hooke, see especially Rob Iliffe, "Material Doubts: Hooke, Artisan Culture and the Exchange of Information in 1670s London," *British Journal for the History of Science*, 28 (1995), 285–318. For the novelty of news, see Joad Raymond, *The Invention of the Newspaper: English Newsbooks, 1641–1649* (Oxford: Clarendon Press, 1996) (which places the newspaper just earlier than the first coffeehouses); James Sutherland, *The Restoration Newspaper and its Development* (Cambridge: Cambridge University Press, 1986); and Charles John Sommerville, *The News Revolution in England: Cultural Dynamics of Daily Information* (New York: Oxford University Press, 1996), esp. pp. 75–84. For L'Estrange's monopoly on news, see George Kitchin, *Sir Roger L'Estrange: A Contribution to the History of the Press in the Seventeenth Century* (London: Kegan Paul, 1913), pp. 139–46.

disrupt the community. Even nonmembers were inclined to follow its conventions, and members and nonmembers alike were in consequence routinely identified as "Stationers." On these protocols, as much as on the press itself, depended the very existence of anything that could be called a *culture* of print.[4]

The Stationers' position in London resembled that of other printing and bookselling communities in Continental cities. Theirs was a prosperous and expanding business in need of regulation. A mere generation after the invention of the press in the mid-fifteenth century, printers had started to work in cities throughout Germany, the Netherlands, England, France, and the Italian states (Figure 15.1). By the mid-sixteenth century, print was already transforming the character of the book itself. A pair of workers using one press could produce perhaps 1,000 sheets of printed paper every day, and in larger printing houses there would be several presses at work simultaneously. Hence, the number of printed books that these operations made available was unprecedented. One of the many arresting statistics provided by historian of the book Henri-Jean Martin reveals that by around 1500 there was already a printed book in existence for every five living Europeans. At the great book fairs held annually in Cologne, Leipzig, and, especially, Frankfurt, vast numbers of volumes were traded – some 75,000 titles between 1564 and 1649, amounting to perhaps 60–100 million copies (depending on print run and allowing for smaller books that were often left out of the printed records). In 1564–9, 2,000 titles were advertised; in 1610–19, the years when the fairs reached their peak, the total was over 15,500. The learned made sure to keep abreast of current notions by scanning the Frankfurt catalogues.[5]

The "printing revolution" that ensued, however, was to be at least as important in qualitative and practical as in sheer quantitative terms. The press was put to use in innovative ways that coincided with and reinforced transformations in the life of the scholar and craftsman alike.[6] Protagonists of natural

---

[4] Adrian Johns, *The Nature of the Book: Print and Knowledge in the Making* (Chicago: University of Chicago Press, 1998), pp. 187–265.
[5] Lucien Febvre and Henri-Jean Martin, *The Coming of the Book: The Impact of Printing, 1450–1800*, trans. David Gerard (London: Verso, 1984), pp. 248–9. For Frankfurt and other fairs, see A. H. Laeven, "The Frankfurt and Leipzig Book Fairs and the History of the Dutch Book Trade in the Seventeenth and Eighteenth Centuries," in *Le magasin de l'univers: The Dutch Republic as the Centre of the European Book Trade*, ed. Christiane Berkvens-Stevelinck, H. Bots, P. G. Hoftijzer, and O. S. Lankhorst (Leiden: E. J. Brill, 1992), pp. 185–97. (I calculated the total of 75,000 from the table on p. 191.) For figures and analysis, see James Westfall Thompson, ed., *The Frankfurt Book Fair: The Francofordiense Emporium of Henri Estienne* (New York: Franklin, 1968); A. Dietz, *Zur Geschichte der Frankfurter Büchermesse, 1462–1792* (Frankfurt: 1921); H.-J. Martin, *Print, Power, and People in Seventeenth-Century France*, trans. David Gerard (Metuchen, N.J.: Scarecrow Press, 1993), pp. 198–200; and Martin, *The History and Power of Writing*, trans. Lydia G. Cochrane (Chicago: University of Chicago Press, 1994), pp. 247–51.
[6] For the advent and impact of works in the "mathematicals" (as enterprises such as surveying and navigation were called), see T. R. Adams, "The Beginning of Maritime Publishing in England, 1528–1640," *The Library*, 6th ser., 14 (1992), 207–20, and more generally S. Johnston, "Mathematical Practitioners and Instruments in Elizabethan England," *Annals of Science*, 48 (1991), 319–44.

Figure 15.1. A printing house in Holland. In Samuel Ampzing, *Beschryvinge ende lof der Stad Haerlem in Holland* (Haarlem: Adriaen Rooman, 1628). Reproduced by permission of the Bodleian Library. Douce A 219.

history, medicine, and the mathematical sciences had perceived the potential of the press from its early days; German astronomer Johannes Regiomontanus's (1436–1476) remarkably audacious proposal to publish a corpus of ancient mathematics actually predated the activities of the great humanist scholar-printers that are better-known today. Regiomontanus's proposal came to nothing, but in the sixteenth century such practitioners used print more successfully to address new audiences, to advance new claims, and even to reshape the identity of the investigator of nature.[7]

Natural philosophy in particular was moving away from its university setting, first to the environs of the royal court and later to semiautonomous or autonomous institutions, such as Giambattista della Porta's (1535–1615) Accademia dei Secreti (in Naples), Federico Cesi's (1585–1630) Accademia dei Lincei (Rome), the Accademia del Cimento (Florence), Théophraste Renaudot's (1586–1653) Bureau d'Adresse and the later Académie Royale des Sciences (Paris), and the Royal Society in London itself. With these changes in locale came transformations in the conduct and even content of natural knowledge.[8] In the hands of their protagonists, print facilitated the advent of conventions of openness, intellectual property, authorship, and collaboration. Della Porta's group had been secretive to the point of notoriety, but Cesi's academy planned collective action to communicate natural knowledge among its various projected houses and to publish it more broadly by means of its own printing press. Its *Gesta Lynceorum* were the earliest printed reports of such an academy. The Accademia del Cimento intended to issue reports of its experiments from the outset, and eventually did so, with the authorship of contributors subsumed under that of the prince. Renaudot went further and printed what he claimed to be extracts of the participants' actual conversations, though still shorn of their names. Here the aim was to maintain modesty and decorum. Any member making a claim, because his name was not attached to it, "was no-wise *interested* to maintain his *Sentence* upon any Point; but being once produc'd, it was as a thing expos'd to the company, and no more accounted any man's *Property*, then *Truth* it self." A generation later, the Royal Society did attach authorial identities to the claims it circulated, although almost always with disarming apologies for the

---

[7] See, in general, Elizabeth L. Eisenstein, *The Printing Press as an Agent of Change: Communications and Cultural Transformations in Early Modern Europe* (Cambridge: Cambridge University Press, 1979), pp. 520–635. For Regiomontanus, see Noel M. Swerdlow, "Science and Humanism in the Renaissance: Regiomontanus's Oration on the Dignity and Utility of the Mathematical Sciences," in *World Changes: Thomas Kuhn and the Nature of Science*, ed. Paul Horwich (Cambridge, Mass.: MIT Press, 1993), pp. 131–68; and Lisa Jardine, *Worldly Goods: A New History of the Renaissance* (London: Macmillan, 1996), pp. 348–50. For the innovative use of print by "professors of secrets," see William Eamon, *Science and the Secrets of Nature: Books of Secrets in Medieval and Early Modern Culture* (Princeton, N.J.: Princeton University Press, 1994), pp. 234–66. For the mathematical sciences, see also the introductory article by G. J. Whitrow, "Why did Mathematics begin to take off in the Sixteenth Century?" in *Mathematics from Manuscript to Print, 1300–1600*, ed. Cynthia Hay (Oxford: Clarendon Press, 1988), pp. 264–9.

[8] Mario Biagioli, "Scientific Revolution, Social Bricolage, and Etiquette," in *The Scientific Revolution in National Context*, ed. Roy Porter and Mikuláš Teich (Cambridge: Cambridge University Press, 1992), pp. 11–54. For della Porta, see Eamon, *Science and the Secrets of Nature*, p. 151. For Renaudot, see Howard M. Solomon, *Public Welfare, Science, and Propaganda in Seventeenth Century France: The Innovations of Théophraste Renaudot* (Princeton, N.J.: Princeton University Press, 1972), pp. 60–99; Geoffrey V. Sutton, *Science for a Polite Society: Gender, Culture, and the Demonstration of Enlightenment* (Boulder, Colo.: Westview, 1995), pp. 19–52; and Jeffrey K. Sawyer, *Printed Poison: Pamphlet Propaganda, Faction Politics, and the Public Sphere in Early Seventeenth-Century France* (Berkeley: University of California Press, 1990), pp. 136–7.

presumption implicit in the act of authorship.[9] In all of these ways, natural philosophers experimented with the use of printed materials for recording and communicating what would become scientific knowledge, just as they tried out various social practices for attaining that knowledge in the first place. There was nothing preordained about their eventual success, and it did not come about straightforwardly as a result of the sheer quantity of printed books. The press was rich in potential for many different ends, so in putting it to use as they did, the natural philosophers, physicians, and mathematicians of early modern Europe did more than advance individual items of new knowledge. They also helped to remake print culture itself.

The best bookshops and printing houses tended to cluster in discrete locations in major cities, such as St. Paul's Churchyard in London or the Rue S. Jacques in Paris.[10] But others could be found scattered throughout any metropolis. Upon entering such a place, one would expect to find a domestic as well as a commercial scene. In London, especially, premises were small: Many printers had only one or two presses. Without modern credit facilities, such operations required a regular turnover in order to survive. As a result, a printer's life was likely to be orientated around small-scale, local projects such as pamphlets and, increasingly, periodicals, which were pamphlets that were produced repeatedly on a regular schedule. More elaborate publications such as bibles might be undertaken, but they were always risky. Bibles were at least likely to find a substantial market; works such as atlases or mathematical treatises lacked even that security. Stationers were correspondingly reluctant to commit to their publication, and they might well lay them aside in midimpression to produce the ephemera on which their daily bread depended. What made such decisions even more likely was the threat of unauthorized printing, increasingly being termed "piracy" in the seventeenth century. Literary property did not exist in law at this point, except by virtue of privileges granted by the monarch. Craft conventions did proscribe comprinting, but privileges and conventions alike had no force (and very little weight) beyond state borders. The threat of piracy further compromised the prospects for printing learned works,[11] although it could sometimes be turned to a good use. Not a few scholars invoked the practice explicitly to excuse the apparent

---

[9] Martha Ornstein, *The Role of Scientific Societies in the Seventeenth Century* (Hamden: Archon, 1963), pp. 74–6; William Edgar Knowles Middleton, *The Experimenters: A Study of the Accademia del Cimento* (Baltimore: Johns Hopkins University Press, 1971), pp. 65–6; Théophraste Renaudot, *A Question*, series of three pamphlets (London: J. Emery, 1640); Renaudot, *A General Collection of Discourse of the Virtuosi of France* (London: T. Dring and J. Starkey, 1664), esp. sig. §4r–v; Renaudot, *Another Collection of Philosophical Conferences* (London: T. Dring and J. Starkey, 1665); Steven Shapin, *A Social History of Truth: Civility and Science in Seventeenth-Century England* (Chicago: University of Chicago Press, 1994), pp. 177–80.

[10] Martin, *Print, Power, and People*, pp. 219–26. For the French trade in general, see the definitive treatment in Roger Chartier and Henri-Jean Martin, eds., *Histoire de l'édition française*, 2nd ed., 4 vols. (Paris: Fayard, 1989–91), vols. 1–2.

[11] For London, see Johns, *The Nature of the Book*, pp. 58–186. For Paris, see Martin, *Print, Power, and People*, pp. 193–238. For Amsterdam, see Berkvens-Stevelinck, Bots, Hoftijzer, and Lankhorst, *Le*

immodesty of authorship, alleging either that they had reluctantly suffered an unauthorized edition or that they had published their own work solely in order to preempt one. Either way, however, this meant not so much beating the printers' and booksellers' culture as joining it.[12]

The print culture predominating in early modern Europe rested on apprehensions such as these. As a result, it was quite unlike its modern equivalent. Booksellers and printers, rather than authors, were its defining participants. These were not neutral intermediaries, with no more influence on communication than to provide its conditions of possibility. At the very least, it was said, they actively encouraged confrontational, *ad hominem* debate in order to secure sales. Finding that "the Spirit of contradiction prove[d] saleable," booksellers and printers reputedly created a new breed of hack writer instructed to spare "neither Bacons, Harveys, Digbys, Brownes, or any the like." The English puritan Francis Osborne confirmed that they would "*incourage*, if not *hire . . . Blasphemers* against the *Spirit* of *Knowledge*" whose books sold solely by virtue of the "*names* they pretend in their *Title-page* to *confute*."[13]

Printers' notorious neglect of worthy tomes in favor of pamphlets only made matters worse because they thereby at least delayed the publication of learned ideas and quite possibly prevented it altogether. This also had the subtler effect of demeaning the medium of print in general, thereby damaging the credibility of those works that *were* published. Members of the book trade allegedly denied to authors the profits generated by their efforts, and, because by craft convention they held perpetual property rights to the titles they published, they regarded themselves as authorized to intervene in the content of a work. In short, their practices affected both the reputations of would-be authors and the value attached to their claims. Learned writers faced an alien realm in which the preferred idiom was the pamphlet war. In such circumstances, it was difficult to know whether authorship was advisable, or even possible. No wonder "scribal publication" – the organized distribution of texts reproduced solely by hand – remained a viable and respected alternative to print well into the seventeenth century, and in some enterprises as late as 1700.[14]

Yet the potential of printed publication to inform and unite a scholarly community remained appealing. What to do? In general, it was best to build

---

magasin de l'univers. Dutch piracy is described in P. G. Hoftijzer, "'A Sickle unto thy Neighbour's Corn': Book Piracy in the Dutch Republic," *Quaerendo*, 27 (1997), 3–18.

[12] D. F. McKenzie, "Speech–Manuscript–Print," in *New Directions in Textual Studies*, ed. Dave Oliphant and Robin Bradford (Austin, Tex.: Harry Ransom Humanities Research Center, 1990), pp. 87–109.

[13] Richard Whitlocke, *Zootomia; or, Observations on the Present Manners of the English* (London: Printed by T. Roycroft, 1654), sig. Q4v; Francis Osborne, *A Miscellany of Sundry Essayes, Paradoxes, and Problematicall Discourses* (London: Printed by J. Grismond for R. Royston, 1659), sig. (a)4r–v.

[14] Harold Love, *Scribal Publication in Seventeenth-Century England* (Oxford: Oxford University Press, 1993). For pamphleteering in Holland, see Craig E. Harline, *Pamphlets, Printing, and Political Culture in the Early Dutch Republic* (Dordrecht: Martinus Nijhoff, 1987); for France, see Sawyer, *Printed Poison*.

an alliance with a printer or bookseller of known reputation, and one with access to the great fairs at Frankfurt, Leipzig, and Cologne. As a result, getting into print often involved negotiating with master printers far from one's own town. Andreas Vesalius (1514–1564), professor of anatomy in Padua, sought out Johannes Herbst (Oporinus) in Basel to print his *De Fabrica*; Oporinus's remarkable printing house also hosted Theophrastus Bombastus von Hohenheim, known as Paracelsus (1493–1541), and the Marian exile John Foxe (1516–1587), who worked there as a corrector. Paracelsus's own works were printed after his death, in Strasbourg, Frankfurt, and Cologne; the advent of Paracelsianism as a movement owed as much to the printers in these towns as it did to Paracelsus himself. Johannes Petreius produced Nicholas Copernicus's (1473–1543) *De revolutionibus orbium coelestium* (On the Revolutions of the Celestial Spheres, 1543) in Nuremburg, under the oversight of Joachim Rheticus (1514–1576), while Copernicus himself was on his deathbed in distant Frauenburg. Later, Johannes Kepler (1571–1630) found himself bereft of such assistance when he sought to publish his *Rudolfine Tables* (1627), and he spent months traveling through war-torn Europe in search of a qualified printer. Physician Robert Fludd (1574–1637) and, following on his heels, William Harvey (1578–1657), were still seeking this kind of help in the late 1620s when they sought out William Pfitzer in Frankfurt.[15]

In any case, the use of a distant printer brought its own problems. Piracy, as always, was one. Moreover, given the creative responsibilities conventionally considered appropriate to the master printer, personal attendance at the press was widely recognized as essential to ensure a correct printed text. Because this was clearly impossible in many cases, a cohort of local intermediaries came into being who were employed to address the problem. Rheticus was a notable early example.[16] When Kepler discovered that the defensive preface to Copernicus's *De revolutionibus* had actually been composed by Lutheran pastor Andreas Osiander (1498–1552), he was therefore looking at a kind of textual intervention that this realm of print had made entirely predictable.[17]

Eventually, some began to see a way to challenge the cultural conditions produced by the book trade at a more fundamental level. Helmontian physician William Rand told educational reformer and intelligencer Samuel Hartlib (ca. 1600–1662) of their ideas. "I have often meditated how learned

---

[15] Eisenstein, *Printing Press*, pp. 489–90, 535, 569–70 (for Oporinus); Roger French, *William Harvey's Natural Philosophy* (Cambridge: Cambridge University Press, 1994), p. 227; Hugh R. Trevor-Roper, *Renaissance Essays* (Chicago: University of Chicago Press, 1985), pp. 152–3; and Max Caspar, *Kepler*, ed. Owen Gingerich (New York: Dover, 1993), pp. 311–12.

[16] Anne Goldgar, *Impolite Learning: Conduct and Community in the Republic of Letters, 1680–1750* (New Haven, Conn.: Yale University Press, 1995), pp. 30–53.

[17] Kepler's revelation is reported in Johannes Kepler, *New Astronomy*, trans. William H. Donahue (Cambridge: Cambridge University Press, 1992), p. 28; for the preface itself, see Robert Westman, "Proof, Poetics and Patronage: Copernicus' Preface to *De revolutionibus*," in *Reappraisals of the Scientific Revolution*, ed. Robert Westman and David C. Lindberg (Cambridge: Cambridge University Press, 1990), pp. 167–205.

men might avoid being rook't as they are by stationers," Rand began. He proceeded to suggest that "Learned men in a body or Collegiate association, with leave of the Magistrate might well print their owne Inventions (which were but a modest request) in such a Manner & with such reputation of correctednes &c. that hardly any others would print them." Rand's proposal was for collective action by natural philosophers to reform a print culture that was itself a collective creation.[18]

Such schemes bore fruit in institutions such as Royal Society. The Society in particular is renowned for its pioneering role in experimental philosophy. Yet that effort depended centrally on a strategy for engaging with the world of print that is virtually unknown today. It partly involved implementing a plan similar to Rand's. The Society secured royal authority to intervene in the Stationers' realm by means of a privilege to appoint its own "printers" (a term that it interpreted to include booksellers). More broadly, however, the Society labored to develop polite conventions governing every element of communication, from authorial composition, through manuscript circulation, registration, and reading, to printing, publication, distribution, and response.

There were partial precedents on which to draw, both in scholarly society and in the book trade. In France, Renaudot's Bureau d'Adresse had striven to use print to expand its influence and had fought the effects of piracy; some of its conversations were translated and printed in London in the mid-1660s as the Royal Society honed its own conventions. Closer to home, the Stationers' Company itself had long maintained a Register Book by which to regulate claims of literary propriety at its monthly meetings.[19] The Royal Society virtuosi followed suit by inaugurating their own register, but to a large extent the Society was striking out on a new path. Its conventions soon came to permeate virtually everything it did. In fact, from its defining combats with philosopher Thomas Hobbes (1588–1679) to the presidency of Isaac Newton (1642–1727), the Royal Society's early history was constituted by a series of transactions that emerged out of either the conventions themselves or their claimed violation.[20]

Fellows of the Royal Society were expected to contribute regularly to the Society's activities. They did this by submitting machines, experiments, or, most commonly, reports. Such a report would then either be "perused" by two referees, or, occasionally, "read" aloud before the assembled company – a

---

[18] W. Rand to S. Hartlib, 14 February 1652, Sheffield University, Hartlib Papers, 62/17/1A–2B.
[19] George Havers, trans., *A General Collection of Discourse of the Virtuosi of France* (London: Printed for T. Dring and J. Starkey, 1663); and G. Havers and J. Davies, trans., *Another Collection of Philosophical Conferences of the French Virtuosi* (London: Printed for T. Dring and J. Starkey, 1665). For the Stationers' register, see Johns, *The Nature of the Book*, pp. 213–30.
[20] Hobbes claimed that his *Dialogus physicus* had been provoked by the Society's misuse of its registration regime. See Steven Shapin and Simon J. Schaffer, *Leviathan and the Air-Pump: Hobbes, Boyle, and the Experimental Life* (Princeton, N.J.: Princeton University Press, 1985), p. 348.

deceptively simple term for what was a closely constrained ritual. Either way, the reading was then certified by "registration." The paper would be dated and copied into a register book. This register (which in fact rapidly became a number of separate thematic volumes) was to be a decisive record of the Society's act of reading the pieces entered. No less significantly, it represented a secure archive of authorship – something not otherwise available because registration in the book trade preserved not authors' but booksellers' interests. Entries would then provide consensual grounds for further experimentation, conversation, and writing. A circulation of printed and manuscript materials thus formed the basis for the Royal Society's continuing vitality.

By establishing the basis for continuing conversation in the Society, however, the register system also defined the character of disagreement. When argument and confrontation occurred, it would usually be on the basis of authorship. The most characteristic form of debate in the experimental community, then – in Paris, as in London, and quite possibly elsewhere in Europe – was the priority dispute. Mechanical philosopher Robert Hooke (1635–1703) has gained a reputation for being peculiarly prone to launching such attacks, but in fact he was merely the most prominent complainant. Almost all prominent virtuosi were involved in such complaints, including Boyle, Christopher Wren (1632–1723), Christiaan Huygens (1629–1695), and, of course, Newton himself.[21]

The Society's conventions did not end with registration and the internal debates it generated. Registered papers might then be recommended for publication, either by the author concerned or, starting in 1665, in the *Philosophical Transactions*. Any correspondence produced by the resulting work would then provide yet more grist for the Society's weekly mill.[22] It is worth pausing over the role of the *Transactions* in particular.[23] The journal was invented by

---

[21] For Boyle, see Johns, *The Nature of the Book*, pp. 504–10. For Wren, see Thomas Sprat, *The History of the Royal-Society of London, For the Improving of Natural Knowledge* (London: printed by T. R. for J. Martyn and J. Allestry, 1667), pp. 311–19; and Christopher Wren, *Parentalia; or, Memoirs of the Family of the Wrens*, ed. S. Wren (London: printed for T. Osborn and R. Dodsley, 1750), p. 199. For Huygens, see Rob C. Iliffe, "'In the Warehouse': Privacy, Property and Priority in the Early Royal Society," *History of Science*, 30 (1992), 29–68. For Newton, see Alfred Rupert Hall, *Philosophers at War: The Quarrel Between Newton and Leibniz* (Cambridge: Cambridge University Press, 1980). For the importance of registration and priority at the Parisian Académie Royale des Sciences, see Roger Hahn, *The Anatomy of a Scientific Institution: The Paris Academy of Sciences, 1666–1803* (Berkeley: University of California Press, 1971), pp. 21–4, 27–8.

[22] The records of these proceedings are manifest throughout Thomas Birch, ed., *The History of the Royal Society*, 4 vols. (London: Printed for A. Millar, 1756–7). For registration, see also Shapin, *Social History of Truth*, pp. 302–4. For the conversational aspects of such inquiries, see Jay Tribby, "Cooking (with) Clio and Cleo: Eloquence and Experiment in Seventeenth-Century Florence," *Journal of the History of Ideas*, 52 (1991), 417–39.

[23] For Oldenburg's importance, see John Henry, "The Origins of Modern Science: Henry Oldenburg's Contribution," *British Journal for the History of Science*, 21 (1988), 103–10; and Michael Hunter, "Promoting the New Science: Henry Oldenburg and the Early Royal Society," *History of Science*, 26 (1988), 165–81. For the *Philosophical Transactions*, see E. N. da C. Andrade, "The Birth and Early Days of the *Philosophical Transactions*," *Notes and Records of the Royal Society*, 20 (1965), 9–22; and D. A. Kronick, "Notes on the Printing History of the Early *Philosophical Transactions*," *Libraries*

the Society's secretary, Henry Oldenburg (1618–1677). Oldenburg intended the periodical to provide him with a secure income and to extend the Society's civility into the broader republic of letters. In the latter aim, at least, it rapidly found success – even though Oldenburg failed in his original plan to produce a Latin version for Continental distribution. Although it was not financially secure for at least a generation, the *Transactions* took on unrivaled epistemic authority. It became one of the most widely imitated models for the learned periodical (see Serjeantson, Chapter 5, this volume). Such productions soon multiplied: the *Journal des sçavans*, the model for the *Transactions* itself, was launched in Paris in 1665, the *Acta eruditorum* in Leipzig in 1682, the *Histoire de l'Académie Royale des Sciences* in Paris in 1666, and the *Nouvelles de la République des Lettres* – perhaps the most successful of all – by philosopher and historian Pierre Bayle (1647–1706) in Amsterdam in 1684.[24] In this context, the success of the *Transactions*, fragile as it was, was the first major indication that the Society's intervention in the domain of the Stationers could pay off. In effect, the Society was proposing itself as an archetype for a new culture of the book.

The Society's right to license printers and grant imprimaturs – a privilege shared by no similar body – meant that it could reasonably hope to secure its fellows' authorship at the point of publication. Bishop Thomas Sprat (1635–1713) underscored the point in his apologetic *History of the Royal Society* (1667), mining the register to laud the achievements of Wren in particular. But the greatest exemplar was yet to come. Newton's *Principia mathematica naturalis philosophiae* (Mathematical Principles of Natural Philosophy, 1687) was to be by far the most important of all the works issued under the Society's aegis. Fittingly, the process of its publication fully reflected the uncertainties and compromises that still plagued the virtuosi in their use of the printed book.

The Society did not publish the *Principia* itself. A year earlier, in 1686, it had underwritten the printing of naturalist Francis Willughby's (1635–1672) posthumous *Historia piscium* (History of Fish, 1686), an expensive, richly illustrated folio that was to be the only book actually published by the Royal Society in the seventeenth century. At first, the virtuosi hoped that its production would mark a new initiative in the Society's mastery of publication techniques. But the *Historia piscium* proved to be a disappointment in terms of its natural history and a near disaster in terms of its economics. Its finances ravaged by the failure, the Society probably could not have funded the printing of a substantial new work even if doing so had been its usual

and Culture, 25 (1990), 243–68. For the general history of journals, see Arthur J. Meadows, ed., *Development of Science Publishing in Europe* (Amsterdam: Elsevier, 1980); and David A. Kronick, *A History of Scientific and Technical Periodicals: The Origins and Development of the Scientific and Technical Press, 1665–1790* (New York: Scarecrow Press, 1962).

[24] The world of these enterprises is excellently reconstructed in Goldgar, *Impolite Learning*, pp. 54–114. For the Latin *Philosophical Transactions*, see Adrian Johns, "Miscellaneous Methods: Authors, Societies and Journals in Early Modern England," *British Journal for the History of Science*, 33 (2000), 159–86.

policy. In the event, it reverted to its existing practice and prevailed upon astronomer Edmond Halley (1656–1742) to undertake the book. This Halley did, nudging Newton to supply the complete text and negotiating with the booksellers for the best possible terms. At first, Halley hoped to publish the book himself, employing printers directly. He planned to oversee their work in person and reduce expenses. Newton acceded to this. He told Halley that he was prepared to accept the title *Mathematical Principles of Natural Philosophy* (one that otherwise struck him as grandiose) in order to maximize sales, "which I ought not to diminish now 'tis yours." But Halley was eventually forced to use a bookseller after all. "I am contented to lett them go halves with me," he told Newton," rather than have your excellent work smothered by their combinations." The book was then printed at several houses, as was normal, under the oversight of Joseph Streater – a master printer rather notorious at the time for his piracies and his production of pornography. The undertaker was the international dealer Samuel Smith. Through Smith's trade contacts on the Continent, Newton's *Principia* was distributed to European centers and rapidly came to be accepted as the greatest work of natural philosophy of its time.[25]

Yet success in Europe might not mean success in the streets of London. One organization could scarcely transform overnight a print culture that had evolved over two hundred years, as the insecure fortunes of the *Transactions* itself indicated. The Society could pioneer a new scientific civility, but it could not so easily reach beyond the walls of Gresham College to transform the reading of its works. This was the preserve of the coffeehouses.

COFFEE

In Restoration England, there were three places where the whole political nation was supposed to coalesce. Parliament was one: Its conjunction of king, lords, and commons reputedly represented the country's respectable essence. The Royal Society was the second: Its supporters implied that it was modeled on Parliament, with meetings supposedly standing in for the nation at large (something that was not true of academies established in more absolutist realms, such as the Académie Royale des Sciences in Paris).[26] The

---

[25] I. Bernard Cohen, *Introduction to Newton's Principia* (Cambridge: Cambridge University Press, 1971), pp. 132–3, 136–7; A. N. L. Munby, "The Distribution of the first edition of Newton's *Principia*," *Notes and Records of the Royal Society*, 10 (1952), 28–39; Bodleian Library, Oxford, Smith correspondence, MS Rawl. Letters 114. The second edition of Newton's work was printed at Cambridge, for which remarkably full records have survived; for analysis, see Donald F. McKenzie, *The Cambridge University Press, 1696–1712: A Bibliographical Study*, 2 vols. (Cambridge: Cambridge University Press, 1966), 1: 73, 77, 100, 313–4, 330–6; 2: 24.

[26] For the notion of the Society as representing the political nation (that is, all enfranchised gentlemen), see Michael Hunter and P. B. Wood, "Towards Solomon's House: Rival Strategies for Reforming the Early Royal Society," *History of Science*, 24 (1986), 49–108, esp. pp. 68, 83. It did not represent the *people* – a very different group indeed, and a far larger one.

third, and in some ways the most genuinely representative, was the coffeehouse. Here, contemporaries said, could be found "the perfect Epitome of Mankind."[27] Unlike both Parliament and the Royal Society, a town's coffeehouses contained exemplars of all ranks and both genders. High-churchmen and dissenters, gentlemen, retailers, and mechanics – in a coffeehouse, all flocked together.

Or rather, not quite together. The interior of a coffeehouse typically contained several tables around which customers gathered (Figure 15.2). Before long, these tables became identified with different topics of conversation, or different "parties." Tories thus learned to deride their Whig neighbors gathered around the "treason table"; wits recorded tables devoted to conversations about painting, scholasticism ("the Table of *Salamanca*"), and the new philosophy of nature.[28] Despite the commonplace representation of indiscriminate mixing, the interior of the actual coffeehouse does seem to have provided for cultural differentiation – and along novel lines. Moreover, it is unlikely that many premises really attracted all sorts. Any given coffeehouse would have what locals called a "character." It would be known as attracting primarily a particular kind of customer. In London, for example, physicians went to Garraway's, whereas it was difficult to miss Halley and his friends at Child's, "for there some do lead their lives almost." Gangs would periodically descend from Tory coffeehouses upon Whig ones to jeer and foment violence. In short, the notion of coffeehouse society as leveling and representative referred not to particular sites but to coffeehouses in general. They implied the new notion of a *distributed* assembly, convening in multiple locations from Westminster to the Tower – and, as coffeehouses soon started to appear in towns and cities all over the land, far beyond.[29]

Three things bound that assembly together: circulating objects, the shared practices applied to them, and the resulting products. The objects were printed and manuscript texts, the practices those of reading and conversation, and the products the numinous but consequential entities known as "intelligence" and "news." Customers went to coffeehouses, obviously, to drink coffee, but they also went to read and debate. These were the prime "Rendezvous" where printed claims met their audiences. In fact, early coffeehouses had often abutted onto booksellers' shops, and well into the eighteenth century the economic power of coffeehouse audiences remained strongly influential on the output of their community. The result was a tribunal certain to affect at least the form of any given debate, and not unlikely to determine its outcome.

---

[27] *The Humours of a Coffee-House*, 1 (25 June 1707).
[28] St. Serfe, *Tarugo's Wiles*, pp. 21–4.
[29] Pierre Bayle, ed., *A General Dictionary*, 10 vols. (London: J. Bettenham, 1734–41), 7: 608–9n; Pierre Bayle, *The Two Sosias* (London: J. Bettenham, 1719), p. 9; and John Byrom, *Private Journal and Literary Remains*, ed. Richard Parkinson, 2 vols. in 4 (Manchester: Chetham Society, 1854–7), 1: 42.

Figure 15.2. A coffeehouse in London. In E. Ward, *The Fourth part of Vulgus Britannicus; or, British Hudibras* (London: James Woodward and John Morphew, 1710), frontispiece. Reproduced by permission of The Huntington Library, San Marino, California.

As befitted its place of origin, this tribunal displayed contradictory characteristics. One day it could credit astrological prophecies and tales of Jesuits massing by night to overthrow the monarchy. The next it could manifest a laceratingly critical wit, whose apparently Hobbesian exponents ridiculed talk of spirits and charged that anyone crediting immaterial essences was "unfit for Conversation." Coffeehouses thus became locations for both the building and destruction of belief. Auctions, for example, originated there.

Bidding for cloth, books, land, buildings, antiquities, and paintings (good ones, too, by Anthony Van Dyck and Peter Paul Rubens), participants ventured substantial sums of money on what they saw of the objects and their handlers in the coffeehouse. Subscription enterprises extended the principle. One could subscribe to the latest maritime insurance corporation at one coffeehouse or to the latest scholarly publication at another. In many cases, the economic viability of learned publications depended on coffeehouse customers being tempted. Projectors used coffeehouses to promulgate their plans for definitive encyclopedias, unsinkable ships, animal hospitals, and solutions to the longitude; wits used the same rooms to ridicule those plans.[30] Likewise, the supposed rarities long collected by noblemen in their cabinets were now made accessible to coffeehouse "cits." One could witness 250 of them at Don Saltero's in Chelsea; Saltero, whose real name was James Salter, had once been a servant to naturalist Hans Sloane (1660–1753) and had retained some of Sloane's surplus curiosities. Some of these marvels were genuine and important, others more dubious: Don Saltero proudly advertised Robinson Crusoe's shirt and Pontius Pilate's wife's chambermaid's sister's sister's hat. In general, the absurd hobnobbed with the authentic, the courtly and magical with the commercial and materialistic. The coffeehouse clientele struggled to produce viable distinctions by which to judge such displays. But if the effort gave drinkers a headache, no matter; they could always go to another coffeehouse and cure themselves with "The Honourable *R. Boyl*'s most excellent Lozenges."[31]

Although such examples may make them seem somewhat gullible, coffeehouse verdicts were more usually characterized as witheringly skeptical. Those coffeehouse tribunals that emulated Harrington's Rota and took on institutional airs tended to support the representation. Richard Leigh's "*Coffe-Academy*," for example, devoted its "*Gazett Philosophy*" to demolishing John Dryden. But it was not always so, and the greatest single example of a coffeehouse academy was creative to a fault. This was Stationer John Dunton's (1659–1733) so-called Athenian Society. With its periodical, the *Athenian Mercury*, outselling all others, Dunton's society epitomized the alliance between print and coffeehouse. It actually comprised three or four men meeting upstairs at Smith's, but it represented itself as a second Royal Society and announced its intention to produce a "new Systeme of experimental *Philosophy*." The *Mercury* called for readers to contribute experiments in order to

---

[30] Guildhall Library, London: Broadsides 8.147, 11.49, 13.54, 20.36; Fo. Pam. 5121; Andrew Yarranton, *England's Improvement*, 2 vols. (London: T. Parkhurst, 1698); *England's Improvements Justified* (n.p., n.d.); *A Continuation of the Coffee-House Dialogue, between Captain Y. and a young Baronet* (n.p., n.d.); *The Humours of a Coffee-House*, 4 (16 July 1707), 5 (23 July 1707), 6 (30 July 1707); and *The Weekly Comedy*, 19 (19 December 1707), 20 (26 December 1707).

[31] *A Catalogue of the Rarities to be seen at Don Saltero's Coffee-House* (London: T. Edlin, 1729), pp. 9, 15; and *The Weekly Comedy*, 18 (12 December 1707).

further its aim of becoming "*Boyl* reviv'd." Today the misrepresentation seems transparent. Yet it managed to fool Jonathan Swift, for one, and Swift never forgave Dunton for the deception. In all, the Athenian Society succeeded by epitomizing as no other body could the "sublime Spirit of Coffee."[32]

That the conjunction of coffeehouse and printing house was capable of producing such a creature as the Athenian Society hinted at both promise and danger, and indeed, the relationship between the real Royal Society and the coffeehouses proved an uneasy one. Individual fellows patronized coffeehouses frequently, and as a group they would regularly adjourn to one after their weekly meetings. Hooke was probably their greatest aficionado, discussing experiments, printing and bookselling practices, and his latest allegations against Newton before avid listeners at Jonathan's. Collectively, however, the coffeehouses represented a source of skepticism and of a criticism ready to be directed even to such specifics as Boyle's theories of the air.[33] Negotiating the transition between Gresham College and Garraway's or Jonathan's consequently became a delicate matter. Thomas St. Serfe's parody of a coffeehouse dispute over blood transfusion was uncomfortably apposite. In St. Serfe's version, a drinker claimed to have witnessed a bawdily farcical transfusion experiment at the Royal Society, yet still voiced the foolish virtuoso's hope that the process might enable him to "perpetuate himself to Eternity." In reality, when the Society attempted to discern the effects of its actual transfusion experiments on Arthur Coga, it found its almost equally grandiose expectations frustrated as soon as Coga left the room. When he next returned, any results of the experiment had been drowned out, Coga having consumed a large quantity of wine. As Philip Skippon remarked, "the Coffee-houses [had] endeavored to debauch the Fellow, and so consequently discredit the Royal Society, and make the Experiment ridiculous." They succeeded, and transfusion trials soon came to a halt.[34]

The same transition could also be put to use by Society members themselves. Hooke, for one, employed it against the Astronomer Royal, John Flamsteed (1646–1719). In 1681, Hooke confronted Flamsteed at Garraway's with a puzzle about telescope lenses. Flamsteed floundered, and Hooke humiliated him by proclaiming his invention of a machine to solve the problem mechanically. Hooke then took the issue back to the Royal Society and triumphed there, too, by actually showing such a machine. Flamsteed complained bitterly of Hooke's conduct. He routinely issued captious challenges

---

[32] Richard Leigh, *The Censure of the Rota* (Oxford: F. Oxlad, 1673), p. 1; Elkanah Settle, *The New Athenian Comedy* (London: Printed for Campanella Restio, 1693), p. 2; Charles Gildon, *The History of the Athenian Society* (London: Printed for J. Dowley, n.d. [1691]), pp. 32–3; and G. D. McEwen, *The Oracle of the Coffee House: John Dunton's Athenian Mercury* (San Marino, Calif.: Huntington Library, 1972), pp. 33–5.

[33] [Anon.], *An Excerpt out of a Book, Shewing, That Fluids rise not in the Pump* (n.p., n.d.): BL 536.d.19(6).

[34] St. Serfe, *Tarugo's Wiles*, pp. 18–20; and William Derham, ed., *Philosophical Letters between the Late Learned Mr. Ray and several of his Ingenious Correspondents* (London: W. and J. Innys, 1718), pp. 27–8.

in the coffeehouse, Flamsteed charged, in order to use skilled practitioners' arguments as his own in later Society meetings. And in the coffeehouse, at least, his vaunted machines never seemed to appear. Flamsteed had a point: In this instance, as in others, Hooke's device failed to work even at the Society. Yet Hooke won anyway because he had perfectly mastered two strategies of contrast: boastful concealment in the coffeehouse combined with modest display before the virtuosi. He put this mastery to work in order to further discussion, confirm his mathematical claim, and bolster his standing.[35]

What was it about coffee that caused these effects? Contemporaries were interested enough in the question to speculate that the drink might have some special quality. They believed it assuaged a wide range of medical conditions, from the king's evil (scrofula) to pimples. Reports that Harvey had "frequently" drunk coffee helped to establish its medicinal worth. (The reports were true, and on his death Harvey actually bequeathed some coffee to his physician peers.) It was also good for scribblers. "Clear your Eyes with the steam of the *Coffee*," one hack advised another; "'tis good for your *Brain*." The fluid's virtues clearly facilitated authorship, sociability, and sheer thinking. Wits who ridiculed such beliefs merely manifested the cynical humor of the excessively coffee-drenched – and the laugh was on them because such excess consumption invariably, it was believed, caused sexual impotence. Arguments like these exemplified links between sensibility, dietetics, and intellectual work that were prevalent in the natural philosophy of the time,[36] so it was fitting that John Houghton, apothecary, dealer in tea and chocolate, and, in alliance with the coffeemen, pioneer of printed advertising, discoursed at the Royal Society on these same subjects in the dying days of the seventeenth century (Figure 15.3). Like wine, Houghton concluded, coffee operated by a "potential heat," invigorating "heavy parts in the fermentative juices" and aiding the body's circulations. A circle had been closed: The effect of coffee on experimental philosophy was now explained by an experimental philosophy of coffee.[37]

---

[35] Cambridge University Library, Cambridge, Royal Greenwich Observatory Archive, Ms. RGO 1.50.K, fols. 251r–255v; Thomas Birch, ed., *The History of the Royal Society of London*, 4 vols. (London: A. Millar, 1756–7), 4: 100–1; Adrian Johns, "Flamsteed's Optics and the Identity of the Astronomical Observer," in *Flamsteed's Stars*, ed. F. Willmoth (Woodbridge: Boydell and Brewer, 1997), pp. 77–106.

[36] For example, Adrian Johns, "The Physiology of Reading in Restoration England," in *The Practice and Representation of Reading in England*, ed. James Raven, Helen Small, and Naomi Tadmor (Cambridge: Cambridge University Press, 1996), pp. 138–61; Steven Shapin, "The Philosopher and the Chicken: On the Dietetics of Disembodied Knowledge"; and Peter Dear, "A Mechanical Microcosm: Bodily Passions, Good Manners, and Cartesian Mechanism," in *Science Incarnate: Historical Embodiments of Natural Knowledge*, ed. Christopher Lawrence and Steven Shapin (Chicago: University of Chicago Press, 1998), pp. 21–50 and 51–82, respectively.

[37] Richard Bradley, *The Virtue and Use of Coffee* (London: E. Matthews and W. Mears, 1721), pp. 23–5; Sheffield University, Sheffield Hartlib Papers 42/4/4A; *A Dialogue between Tom and Dick* (1680), p. 2; St. Serfe, *Tarugo's Wiles*, p. 17; *The Women's Petition against Coffee* (London, 1674), pp. 3–4; J. Houghton, "A Discourse of Coffee," *Philosophical Transactions*, 256 (1699), 311–17. For more on early modern accounts of the physiological effects of coffee, see W. Schivelbusch, *Tastes of Paradise: A Social History of Spices, Stimulants, and Intoxicants* (New York: Pantheon, 1992), pp. 34–49.

Figure 15.3. Coffee plant. In Hans Sloane, "An Account of a Prodigiously Large Feather of the Bird *Cuntur*, brought from *Chili*, and Supposed to be a Kind of Vultur; and of the *Coffee-Shrub*," *Philosophical Transactions of the Royal Society*, 208 (February 1693/4), pp. 61–64. Reproduced by permission of The Huntington Library, San Marino, California.

AUDIENCES AND ARGUMENTS

Coffeehouses, Houghton concluded, "improve Arts, and Merchandize, and all other Knowledge." Many would agree. The distributed forum of adjudication created by coffeehouse reading and conversation marked a decisive break with previous social practices, and one properly identified with the modernizing elements of European society at the end of the seventeenth century. Today, historians associate coffeehouses and periodicals not with absolutism, predestinarian theology, and witchcraft – the preoccupations of an earlier age – but with stockjobbing, experimental philosophy, and the novel. They marked the origin, as philosopher Jürgen Habermas has famously argued, of "public reason" itself.[38]

There should be no doubt that the conjunction of print and coffee did foster major changes in political and intellectual culture. It is worth taking care, however, to specify the character of those changes. Too hasty an acceptance of the rhetoric of public reason might mask what were in fact highly consequential distinctions. For example, the harmony between the new philosophy, coffeehouse readerships, and the realm of the printers and booksellers was by no means automatic. Nor, when it did come into being, was it self-sustaining. Least of all was it unqualified. Moreover, the examples of Hooke and Coga imply that we should see public reason as at least as much an outcome as a cause and look to the opportunities offered by manipulation of cultural rifts to understand its emergence. After all, Hooke's perceived need for aggressive self-promotion before a Garraway's audience might be related historically to the grandiose pronouncements of earlier astronomers who were expected to supply regular discoveries for the prince (see Moran, Chapter 11, this volume). Certainly, Flamsteed was inclined to see flattery and vainglory in courtiers and coffee-drinkers alike, arguing that both were compelled to sacrifice personal vocation to the hunger of patrons for immediate spectacle. Tories, too, were not slow to use terms such as "arbitrary power" to describe the critical judgments of the coffeehouse. In this light, Hooke might be regarded as occupying all too comprehensible a position, fitting neatly into a historical trajectory connecting mathematical duelists such as Girolamo Cardano (1501–1576) in the Renaissance to "mechanick" projectors in the Restoration, and to the first Cambridge University Tripos mathematicians in the eighteenth century.[39] There clearly was a break

---

[38] J. Habermas, *The Structural Transformation of the Public Sphere*, trans. T. Burger and F. Lawrence (Cambridge: Polity, 1989). The specific location of the origin of public reason in the Restoration coffeehouse has been maintained in Steven Pincus, "'Coffee Politicians Does Create': Coffeehouses and Restoration Political Culture," *Journal of Modern History*, 67 (1995), 807–34.

[39] See the comparisons advanced by Molyneux and Flamsteed in Bayle, *General Dictionary*, 7, pp. 607–9nn. For mathematical duels in the Renaissance, see Oystein Ore, *Cardano: The Gambling Scholar* (Princeton, N.J.: Princeton University Press, 1953), and for the ability of early Tripos students to challenge each other to similar duels, see J. Gascoigne, "Mathematics and Meritocracy: The Emergence of the Cambridge Mathematical Tripos," *Social Studies of Science*, 14 (1984), 547–84.

with past scholarly practices: The collective audience of coffeehouse readers replaced the unique figure of the prince, and, however skillful he might be, Hooke could not make himself into a second Galileo Galilei (1564–1642). But the change was not so radical that contemporaries could not detect inheritances in coffeehouse custom, nor so fundamental that they were forced to conclude that the coffeehouse was incommensurable with the court.

This has implications for how modern historians should represent the culture of public reason that emerged at the end of the seventeenth century. Constructive philosophizing and destructive skepticism coexisted in coffeehouses along with high-church toryism, radical whiggery, and hard-headed proto-capitalism. Prophecies of the millennium competed for a hearing with the results of actuarial calculations. In terms of the subjects of debate, and indeed of its conduct, it is difficult to posit a clear division. The problem becomes one of interpreting a place in which such apparent signifiers of the premodern and the modern could be so closely juxtaposed and yet of which the accepted collective representation could be one of harmonious *reason*. Public reason was supposedly the result of a peaceful, transnational discussion carried on by a community scornful of confessional and political divisions and defined instead by its reading of printed periodicals. How could that be?

One answer, of course, is that this representation was not merely descriptive but was articulated for a purpose. The virtuosi sought not only to exploit public reason, and not only to defend themselves against the critics who exercised it, but to create, sustain, and reform it in its own right. Representations of public reason therefore served to construct as much as to describe. It was not print itself that guaranteed the public sphere, but print put to use in particular ways and in particular settings. The *and* in the title of this chapter is, then, just as important as the other terms. It was the conjunction of reading works in print and coffeehouse conversation that proved so consequential. Scientific authorship and modern print culture – even, perhaps, modern modes of argument themselves – may have originated in that conjunction.

16

NETWORKS OF TRAVEL, CORRESPONDENCE, AND EXCHANGE

Steven J. Harris

There can be little argument with the claim that early modern science and technology developed in large measure within a limited set of localized sites that ranged from state-supported scientific academies and observatories to botanical gardens, aristocratic collections, and apothecary shops. Nor can there be any question of the importance of local, face-to-face interactions in the creation of the "forms of life" that were characteristic of these social microenvironments. Yet it is equally true that local knowledge was very often embedded in geographically extended networks of communication and exchange. These multiple, often overlapping networks directly facilitated the gathering of information and natural objects as well as the dissemination of the natural knowledge produced at those sites. Indeed, between the fifteenth and eighteenth centuries, European knowledge of the natural world depended increasingly upon expert practitioners who were entrusted with providing reliable information and authentic natural specimens while traversing ever larger and ever more remote geographical tracts. Broadly stated, the early modern period witnessed unprecedented growth in the scale of scientific practice in regard to both the number of formally trained practitioners and the geographical range over which they traveled.

THE EXPANDING HORIZON OF SCIENTIFIC ENGAGEMENT

Although networks of travel and correspondence grew extensively during the late Middle Ages, they were almost without exception confined to the lands of Europe and the coasts of the Mediterranean Sea. Moreover, even within these geographical limits, networks of exchange lacked the density they were to achieve later. Transnational postal services connected only the major capitals and were primarily limited to the administrative and diplomatic correspondence of states (especially the Holy Roman Empire) and the papacy.¹ Although there was frequent exchange of manuscripts and correspondence among medieval universities and a constant circulation of professors and students, there were never more than eighty universities in all of Europe before 1500, and barely a third of that number could be counted as major intellectual centers.² Pilgrimages and crusades to the Holy Lands extended travel routes beyond the European coast but scarcely beyond the Mediterranean,³ and the long-distance travels to the Far East of thirteenth-century merchants such as the Venetian traveler Marco Polo (ca. 1254–1324) or missionaries such as the Franciscan friar William of Rubruck (ca. 1215–1270) failed to develop into regular, European-controlled routes.⁴

Scientific practices in the medieval period were similarly restricted by the circumscribed infrastructure of travel and communications. This is perhaps most easily seen in a comparison of medieval and early modern charting practices. Practical maps – that is, maps intended to guide and record actual movements through or observations of physical space – developed reciprocally with travel. Reliable maps were both a precondition for and a sign of travel. Medieval itinerate maps channeled pilgrims to holy sites such as Rome, Santiago de Compostela, and Jerusalem along well-beaten paths with little or no attention to surrounding territory. Fourteenth- and early fifteenth-century portolans, though rich in coastal detail and often remarkably accurate, were restricted in scope to the Mediterranean Sea and the Atlantic coast south of the British Isles.⁵ Medieval *mappae mundi* (maps of the world), even including exemplars as late as the one completed by the Venetian Fra Mauro in 1459, were primarily visual encyclopedias that drew upon literary sources rather than upon direct observations of lands beyond the Atlantic and Mediterranean coasts.⁶ The late sixteenth-century atlases

---

¹ Fernand Braudel, *The Wheels of Commerce* [1973] (New York: Harper and Row, 1979), pp. 409–10.
² Charles Homer Haskins, *The Rise of Universities* (Ithaca, N.Y.: Cornell University Press, 1957), pp. 29–31.
³ Annie Shaver-Crandell and Paula Gerson, eds., *The Pilgrim's Guide to Santiago de Compostela: A Gazetteer* (London: Harvey Miller Publishers, 1995).
⁴ See Christopher Dawson, *Mission to Asia* (Toronto: University of Toronto Press, 1995), pp. vii–xviii; and Ronald Latham, *The Travels of Marco Polo* (New York: Penguin, 1958), pp. 7–15. On the restricted range of medieval portolans, see Michel Mollat du Jourdin, *Europe and the Sea* (Oxford: Blackwell, 1993), pp. 24–38. For an overview of the limits of the medieval geographic horizon, see Seymour Phillips, "The Outer World of the European Middle Ages," in *Implicit Understanding: Observing, Reporting, and Reflecting on the Encounters between Europeans and Other Peoples in the Early Modern Era*, ed. Stuart B. Schwartz (Cambridge: Cambridge University Press, 1994), pp. 23–63, as well as his exhaustive treatment of the subject, J. R. S. Phillips, *The Medieval Expansion of Europe*, 2nd ed., (New York: Clarendon Press, 1998).
⁵ J. B. Harley and David Woodward, eds., *Cartography in Prehistoric, Ancient, and Medieval Europe and the Mediterranean* (Chicago: University of Chicago Press, 1987), p. 503: "[I]n the Middle Ages the production of both portolan charts and local and regional maps was concentrated within relatively few areas." Beacause their primary function was to aid navigation, portolans typically contained very little first-hand information about interior lands. See Michel Mollat du Jourdin and Monique de la Roncière, eds., *Sea Charts of the Early Explorers, 13th to 17th Centuries*, trans. L. Le R. Dethan (New York: Thames and Hudson, 1984), pp. 11–17.
⁶ Peter Whitfield, *The Image of the World* (London: The British Library, 1994), p. 32.

of Flemish geographer Gerardus Mercator (Gerhard Kremer, 1512–1594) or Dutch cartographer Abraham Ortelius (1527–1598), on the other hand, provided abundantly detailed surveys of European territories as well as world maps with reasonably accurate coastlines for the world's major landmasses (Australia and Antarctica excepted).[7]

Similarly, the most ambitious astronomical project of the Christian Middle Ages was the Alfonsine Tables, an ephemeris compiled under the patronage of Castilian king Alfonso the Wise (1221–1284) and completed around 1270. Even if we credit the claim that Alfonso established an astronomical congress or college, the number of Jewish, Muslim, and Christian scholars involved was no more than a dozen. The Tables themselves depended upon only a handful of original observations, all of which were evidently made in Toledo.[8] By comparison, there were several dozen new ephemerides produced in the sixteenth and seventeenth centuries, and the planetary and satellite tables compiled by astronomer Gian Domenico Cassini I (1625–1717) in the 1660s and 1670s drew on observations made in France, Denmark, Egypt, Cayenne, and the Antilles. His planisphere of 1696 required an even larger network of expert observers.[9]

The expansion in scales of practice that distinguishes the early modern period from the medieval was largely a concomitant of the sustained mastery of overseas navigation by a plurality of loosely interconnected, state-supported corporations. The principal engine driving overseas navigation was of course long-distance trade – what historian Fernand Braudel called "the real big business" of the era.[10] But in addition to the overseas trading companies, there were also colonial administrative bureaus and Catholic religious orders engaged in the overseas missions. Regardless of an organization's mandate, overseas travel entailed a number of interrelated problems that not only included the technology of ship and sail and the science of instrumentation and navigation but also depended upon the social science of administration, communications, and the training and disciplining of agents sent to remote locations. The social or organizational knowledge necessary

---

[7] Gerardus Mercator, *Atlas, or Geographicke Description of the World* [1636], facsimile edition of English translation by Henry Hexham (Amsterdam: Theatris Orbis Terrarum, 1968); and Abraham Ortelius, *Theatrum orbis terrarum* [1570], facsimile edition with introduction by R. A. Skelton (Amsterdam: N. Israel, 1964).

[8] Pierre Duhem, *Le système du monde*, 10 vols. (Paris: Librairie Scientifique Hermann, 1954), 2: 259–66; Charles Homer Haskins, *Studies in the History of Mediaeval Science* (London: Ungar, 1955), pp. 16–18; and J. L. E. Dreyer, *A History of Astronomy from Thales to Kepler* (New York: Dover, 1953), pp. 247–8. The only other notable observational program of the period was Jean de Linières's measurement of the positions of 47 stars. See A. C. Crombie, *The History of Science from Augustine to Galileo* (New York: Dover, 1995), p. 109.

[9] Dreyer, *History of Astronomy*, pp. 345, 404; René Taton, "Gian Domenico Cassini," *Dictionary of Scientific Biography*, vol. 2, ed. Charles Gillispie (New York: Scribners, 1970), pp. 101–3; and Lloyd A. Brown, *Jean Dominique Cassini and His World Map of 1696* (Ann Arbor: University of Michigan Press, 1941).

[10] Braudel, *Wheels of Commerce*, p. 403.

to operate long-distance networks was most immediately felt in the sciences that were directly associated with navigation, such as geography and cartography, as pilots returned with charts of distant coastlines. But regular cycles of overseas travel also influenced the practices of botany and natural history by expanding the ability of European collectors to gather and transport plant and animal specimens as well as natural curiosities of all sorts.[11] By the end of the seventeenth century, the English naturalist John Ray (1627–1705) implicitly acknowledged the worldwide scope of his field when he offered his readers an inventory of all known animal and plant species: 150 different quadrupeds; 500 species of birds; 1,000 fishes; 6,000 plants; and 10,000 insects.[12]

In astronomy, the mobility of expertise involved instrument-assisted astronomical observations, which made possible novel solutions to old problems. For example, the simultaneous observation of the parallax of Mars by Cassini in Paris and Jean Richer (1630–1696) in Cayenne in 1672 yielded the first rigorous measurement of the distance of the Earth from the sun.[13] The charting of southern constellations and the thematic mapping of winds, currents, and magnetic variation and declination also depended upon extended scales of practice in which the expert use of instruments played a central role.[14] Hence, many branches of early modern science were distinguished from their ancient and medieval predecessors by their much-expanded horizon of engagement with the natural world.

THE METRICS OF SCIENTIFIC PRACTICE

That "horizon," and scales of practice in general, may be defined along three orthogonal axes: (1) the spatial extent or geographical reach involved in collecting or observing; (2) the temporal extent or duration of a given program of observation, collection, or description; and (3) the number of people directly engaged in a program and their level of expertise or specialized training. How these dimensions played out in actual practice depended upon the field and project. Yet collectively they indicate the critical elements

---

[11] For a discussion of the expanding commerce in natural curiosities, see Paula Findlen, "Pilgrimages of Science," in Findlen, *Possessing Nature: Museums, Collecting, and Scientific Culture in Early Modern Italy* (Berkeley: University of California Press, 1994), pp. 155–92.

[12] John Ray, *The Wisdom of God Manifested in the Works of the Creation* (London: Samuel Smith, 1691), pp. 5–8.

[13] See Taton, "Gian Domenico Cassini," p. 103.

[14] For examples of charting southern constellations, see Alan Cook, *Edmond Halley: Charting the Heavens and the Seas* (Oxford: Clarendon Press, 1998), pp. 72–9, 439–41; and Deborah J. Warner, *The Sky Explored: Celestial Cartography, 1500–1800* (New York: Alan R. Liss, 1979), pp. 26–7, 52–5, 100–1, 194, 196–7, 255. For examples of maps of magnetic declination, see Cook, *Halley*, pp. 256–91 and Plate IX. On Athanasius Kircher's anticipation, by several decades, of Halley's desire to produce a map of ocean currents and his attempts to enlist his fellow Jesuits in the construction of a global map of magnetic declination, see Michael John Gorman, "The Scientific Counter-Reformation: Mathematics, Natural Philosophy and Experimentalism in Jesuit Culture, 1580–c.1670," Ph.D. dissertation, European University Institute, Florence, 1998, pp. 106–15.

in what may be called the metrics of scientific practice. The paradigmatic example of large-scale practices across all three dimensions was cartography. The accurate mapping of the coast of South America required many cycles of voyages to distant shores and a complex set of skills distributed among mathematicians, mapmakers, pilots, shipwrights, and even common sailors without whom ships did not sail. Stabilizing the cartographic shape of that continent took several generations.[15] By way of contrast, several of the most celebrated episodes of early modern science appear to have been small-scale practices performed in small spaces, over short periods, and by a few participants. Andreas Vesalius's (1514–1564) anatomical work, Galileo Galilei's (1564–1642) telescopic discoveries, and Robert Boyle's (1627–1691) air-pump experiments, for example, transpired largely within the confines of room-sized spaces over the course of a few weeks or months, and the respective skills as anatomist, instrument maker, and experimentalist were concentrated almost literally in their own hands. Of course these discoveries were not isolated, solipsistic events. The ongoing programs of creating an anatomical atlas of the human body, detailing the features and movements of celestial objects, or systematically assaying the properties of the "Torricellian space" were large-scale in that they required work in many lands, much time, and multiple types of expertise.

Francis Bacon (1561–1626) captured – but by no means invented – the spirit of large-scale scientific practice in his posthumously published utopian tract *New Atlantis* (1627). In describing the activities of "Salomon's House," a sort of research station with the aim of seeking out "the knowledge of Causes . . . and the enlarging of the bounds of Human Empire," Bacon assigned the "Merchants of Light" the task of gathering "the books, and abstracts [or summaries], and patterns of experiments" from around the globe. Upon receiving the literary treasures from the Merchants of Light, the "Depradators" who worked within Salomon's House "collect[ed] the experiments which are found in all [these] books" while the "Compilers" drew "the experiments of the former . . . into titles and tables."[16] Elsewhere Bacon called for the enlargement of human understanding by "going to and fro to remote and heterogeneous instances," and he insisted on the importance of overseas voyages for the laying open of the "intellectual globe."[17]

Along with growth in the scale of practice, there was also a general increase in the density of practice, or the concentration of communal attention on

---

[15] See, for example, Uta Lindgren, "Trial and Error in the Mapping of America during the Early Modern Period," in *America: Early Maps of the New World*, ed. Hans Wolf (Munich: Prestel-Verlag, 1992), pp. 145–60; and Eviatar Zerubavel, "The Mental Discovery of America," in *Terra Cognita: The Mental Discovery of America* (New Brunswick, N.J.: Rutgers University Press, 1992), pp. 36–66.
[16] Francis Bacon, *New Atlantis*, in *New Atlantis and The Great Instauration*, ed. Jerry Weinberger (Wheeling, Ill.: Harlan Davidson, 1989), pp. 71, 81.
[17] Bacon, *The New Organon*, aphorisms 47, 84, in *The New Organon and Related Writings*, ed. Fulton H. Anderson (Indianapolis: Bobbs-Merrill, 1960), pp. 51, 81.

a particular problem or project. Karl Marx once observed that a sparsely populated country with a dense communications system was virtually more densely populated than a populous country with poorly developed communications.[18] As a corollary, we may think of a scientific field as being more or less densely "populated" depending not only upon the number of its practitioners but also the effectiveness of its communications. A number of factors stand in the way of any simple tally of the actual population of practitioners. It is anachronistic to speak in terms either of a "scientific community" as a coherent group or of "scientist" as a professional designation during the early modern period (see Shapin, Chapter 6, this volume). Fields such as meteorology and natural history, though certainly considered part of natural philosophy broadly conceived, lacked institutional and professional autonomy. Mathematics, cosmography, astronomy, and medicine (including botany), on the other hand, did have the institutional support of universities, and those who were formally trained in these fields enjoyed enhanced social status.[19] The term "scientific practitioner" is here used as an umbrella term to group together everyone with formal education or training in the natural and mathematical sciences who was engaged in the observation, description, or manipulation of the natural world and who recorded such work on paper. It includes guild members, such as apothecaries, pilots, and chart makers, as well as university-educated elites, such as physicians and mathematicians. It excludes the illiterate peasant, miner, midwife, or sailor, who surely manipulated natural forces in their struggles to survive but could not, or did not, record their experiences. As a term of convenience, then, scientific practitioners were simply those with some degree of expertise who attempted to record in symbols, pictorially or textually, their knowledge of nature.

However one might define practitioners, we still lack an exhaustive demography of early modern science.[20] Nevertheless, the partial evidence currently available suggests a general rise in the absolute number of practitioners in the sixteenth and seventeenth centuries. The growth of vocations in the natural sciences that the sociologist Robert Merton identified in his study of seventeenth-century English Puritans seems to have been a Europe-wide phenomenon.[21] At least a quantitative survey of members of the Society of Jesus, a Catholic and largely Continental religious organization, shows similar upward trends across a range of scientific fields.[22] The number of publications

---

[18] Karl Marx, *Capital: A Critique of Political Economy*, trans. Eden and Cedar Paul, 2 vols. (London: J. M. Dent and Sons, 1930), 1: 372.
[19] See Mario Biagioli, "The Social Status of Italian Mathematicians, 1450–1600," *History of Science*, 27 (1989), 41–95.
[20] For an outline of what such a quantitative survey of early modern science might look like, see John L. Heilbron, "Science in the Church," *Science in Context*, 3 (1989), 3–28, at pp. 12–15.
[21] Robert K. Merton, *Science, Technology & Society in Seventeenth Century England* (Brighton, England: Harvester Press, 1970), pp. 40–54.
[22] For the Society of Jesus, see Steven J. Harris, "Transposing the Merton Thesis: Apostolic Spirituality and the Establishment of the Jesuit Scientific Tradition," *Science in Context*, 3 (1989), 29–65, at

in the mathematical and natural sciences passing through the Frankfurt Book Fair increased steadily over the same period, suggesting an increasing number of authors able to commit their knowledge of nature to the printed page.[23] Further indirect evidence comes from the growth of various types of scientific institutions. The emergence of anatomy theaters, botanical gardens, natural history cabinets, university chairs of mathematics,[24] and astronomical observatories[25] in the sixteenth century and their steadily increasing numbers in the seventeenth century demonstrate an expanding institutional base for various forms of natural inquiry. Presumably, there was also an increase in the number of anatomists, naturalists, mathematicians, and astronomers needed to staff these institutions, as well as the ancillary collaborators upon whom they relied for data and specimens.

There is also evidence in support of the "virtual," or communicative, aspect of Marx's demographic aphorism. Certain groups within this expanding population succeeded in coordinating their efforts as students and manipulators of nature by developing efficient modes of communication, thereby effectively amplifying their numbers. Although a headcount of this "virtual population" does not exist, it is possible to identify those groups that successfully exploited existing possibilities of communication. Before making such identifications, however, a brief review of the nature of the changes in the available modes of communication is in order.

CORRESPONDENCE NETWORKS, LONG-DISTANCE TRAVEL, AND PRINTING

Broadly speaking, the fundamental changes in communications in the early modern period depended on innovations in postal services, overseas travel, and printing. Although independent from one another in origin and largely technological in character, interactions among these three developments were of critical importance for the intensification of scientific practices. Singly and

---

pp. 39–45. A broadly similar pattern of growth is evident in Richard S. Westfall's survey of members of the European scientific community as reflected in the *Dictionary of Scientific Biography*. See Richard S. Westfall, "Science and Technology during the Scientific Revolution: An Empirical Approach," in *Renaissance and Revolution: Humanists, Scholars, Craftsmen and Natural Philosophers in Early Modern Europe*, ed. Judith V. Field and Frank A. J. L. James (Cambridge: Cambridge University Press, 1993), pp. 63–72.

[23] Franz Zarncke, "Erläuterung der graphischen Tafeln zur Statistik der deutschen Buchhandels in den Jahren 1564 bis 1765," appendix to Friedrich Kapp, *Geschichte der deutschen Buchhandels* (Leipzig: Verlag des Berufvereins der Deutschen Buchhandler, 1886), pp. 786–809.

[24] For a discussion of chairs of mathematics in English universities, see Mordechai Feingold, *The Mathematicians' Apprenticeship: Science, Universities and Society in England, 1560–1640* (Cambridge: Cambridge University Press, 1984), esp. pp. 445–85. For chairs of mathematics in Jesuit universities, see Steven J. Harris, "Les chaires de mathématiques," in *Les jésuites à la Renaissance: Système éducatif et production du savoir*, ed. Luce Giard (Paris: Presses Universitaires de France, 1995), pp. 239–62.

[25] Derek Howse, "The Greenwich List of Observatories: A World List of Astronomical Observatories, Instruments, and Clocks, 1670–1850," *Journal of the History of Astronomy*, 25 (1994), 207–18.

collectively, they offered a rich set of resources to individual practitioners and especially to groups interested in understanding or controlling the natural world. As the critical elements in the infrastructure of early modern communications, they made possible – but did not compel – the emergence of the Republic of Letters.

The most mundane element of the Republic of Letters was also its sine qua non; that is, the regular and reliable circulation of correspondence. The growth of papal and state diplomatic corps and their need for the speedy and secure exchange of correspondence is one part of the story. Another is the emergence of regular courier services either within international merchant and trading houses, such as that of the Fuggers in Augsburg, or as an independent postal service, as in the southern German principality of Thurn und Taxis.[26] The volume and range of correspondence among merchants increased rapidly during the Renaissance, and the demand for mercantile intelligence is inseparable from the emergence of the earliest newsletters.[27] Yet another component of the Republic of Letters is the administrative correspondence of the major religious orders, especially the Jesuits, whose "edifying reports" from the overseas missions circulated well beyond the membership of the order itself.[28] Even though networks of diplomatic, commercial, and missionary correspondence were independently organized and designed primarily to fulfill the interests of the host institution, they often overlapped. The contents of the Fugger newsletter, for example, benefited from the close alliance between the House of Fugger and the Hapsburgs, on the one hand, and Fugger patronage of the Jesuits on the other.[29] Expansion of overseas trade, colonial administration, and missionary work thus laid the foundation for a truly global network of epistolary exchange and, at the same time, an increase in the density of exchange among European commercial and governmental centers.[30]

---

[26] Victor von Klarwill, *The Fugger News-Letter* (London: John Lane, 1924), pp. xviii–xxix. Although the Fugger *Neue Zeitungen* predate the invention of printing, they flourished especially after 1500. By 1600, the Fugger firm had branches in towns from Madrid, London, and Antwerp to Danzig, Venice, and numerous overseas outposts. See Wolfgang Behringer, *Thurn und Taxis: Die Geschichte ihrer Post und ihrer Unternehmen* (Munich: Piper, 1990).

[27] Braudel, *Wheels of Commerce*, pp. 409–12; and Jürgen Habermas, *The Structural Transformation of the Public Sphere*, trans. Thomas Burger (Cambridge, Mass.: MIT Press, 1991; orig. publ. 1962), pp. 16–18.

[28] Steven J. Harris, "Confession-Building, Long-Distance Networks, and the Organization of Jesuit Science," *Early Science and Medicine*, 1 (1996), 287–318, esp. pp. 303–8.

[29] Klarwill, *Fugger News-Letters*, pp. xiv–xv. The surviving Fugger "news-letters" comprise about 35,000 pages.

[30] The emergence of *Zeitunger*, or writers of newsletters, was by no means unique to the German Fuggers; there were the *scrittori d'avisi*, who composed the *Notizie scritte* issued by the Venetian government, the *gazettani* and the Jesuit *hijuela* and *hebdomadarius* in Rome, the *nouvellistes* in Paris, the *Nouvellanten* in Augsburg, and the writers of letters in London. See Jürgen Habermas, *Public Sphere*, p. 253, n. 32; Klarwill, *Fugger News-Letters*, pp. xiv, xviii; and Harris, "Confession-Building," pp. 299–301.

Initially, it was chiefly Italian humanists who were able to integrate their scholarly and literary concerns with the mundane possibilities offered by state and commercial postal services to create the Republic of Letters.[31] Part utopian fiction, part political ideology, and always highly adaptable, the Republic functioned as an elite market of intellectual exchange throughout the early modern period, though with increasing effectiveness after about 1600. The classical and literary interests of the humanists quickly grew to embrace a wide range of intellectual topics, including the study of nature and even rather technical discussions in mathematics and experimental philosophy. In the sixteenth century, naturalists such as Conrad Gessner (1516–1565) and Ulisse Aldrovandi (1522–1605) found correspondence to be the most useful medium for the exchange of information on plants and animals, as well as actual specimens and books relating to the natural world (see Findlen, Chapter 19, this volume). In the second quarter of the seventeenth century, the Minim friar Marin Mersenne (1588–1648) cultivated a network of more than seventy correspondents, including eminent French mathematicians such as René Descartes (1596–1650), Gilles de Roberval (1602–1675), Pierre de Fermat (1601–1665), and Blaise Pascal (1623–1662), among whom he circulated challenging problems in geometry and analysis.[32] By the middle of the century, the Jesuit Athanasius Kircher (1602–1680), stationed in Rome, had built up an enormous network of some 760 correspondents largely by exploiting contacts with his confreres in the Society's schools and overseas missions. Many of his exchanges centered on selected scientific themes such as magnetic declination, astronomical observations, and the medicinal properties of plant and animal remedies.[33] By the end of the century, German polymath Gottfried Wilhelm Leibniz (1646–1716) took it as a matter of course that his network of some 400 correspondents would include Frenchmen, Italians, Englishmen, and even Chinese mandarins (mediated by Jesuit missionaries). The rhetorical justification of the Republic of Letters during its golden age in the eighteenth century stressed the universality of discourse and the high-minded cultural goals of its "citizens."[34] Be that as it may, the day-to-day

---

[31] Dena Goodman, *The Republic of Letters: A Cultural History of the French Enlightenment* (Ithaca, N.Y.: Cornell University Press, 1994), pp. 15–23. See also Lorraine Daston, "The Ideal and the Reality of the Republic of Letters in the Enlightenment," *Science in Context*, 4 (1991), 367–86.

[32] On Mersenne's correspondence, see Bernard Rochet, *La correspondance scientifique du Père Mersenne* (Paris: Palais de la Découverte, 1966).

[33] More than two thousand letters that Kircher received have survived and are now stored in the Vatican Library in Rome. Digital images of the entire collection are now available on the Web site of the Institute and Museum of History of Science in Florence. (The site's URL is http://archimede.imss.fi.it/kircher.) See also John E. Fletcher, "A Brief Survey of the Unpublished Correspondence of Athanasius Kircher, S.J. (1602–1680)," *Manuscripta*, 13 (1969), 150–60; Fletcher, "Astronomy in the Life and Correspondence of Athanasius Kircher," *Isis*, 61 (1970), 52–67; and Fletcher, "Medical Men and Medicine in the Correspondence of Athanasius Kircher," *Janus*, 56 (1970), 259–77.

[34] Daston, "Republic of Letters," pp. 367–9, 377. See also the essays by Françoise Waquet, "L'Espace de la République des Lettres" and Willem Frijhoff, "La circulation des hommes de savoir: Pôles, institutions, flux, volumes," in *Commercium litterarum: Forms of Communication in the Republic*

operation of this "state" rested ultimately on the mundane infrastructure of commercial and administrative transactions.

At the same time that correspondence networks proliferated, travel extended beyond European shores to those of Africa, Asia, and the New World.[35] Although the initial motivations for overseas travel were overwhelmingly trade and territorial conquest, the prosecution of those ends required a range of scientific knowledge. First and foremost were the fields of geography and cartography because reliable sea charts were of immediate concern. Both the Portuguese and Spanish crowns established royal chartrooms as integral parts of their many *Casas*, or houses of exploration and trade.[36] In order to sustain cycles of travel, it was also necessary to observe and record information about prevailing winds and currents. To retain overseas outposts and colonies, Europeans needed also to become students of climates, vegetation, landscapes, and peoples of distant regions. In most places and often within a few decades of initial contact, the transient adventures of explorers, conquistadors, and opportunistic traders gave way to the more or less sedentary work of colonial administrators, missionaries, and factors (mercantile agents).

With regular cycles of commercial transport to and from the colonial periphery and a burgeoning epistolary commerce at home, it was a relatively easy matter to incorporate reports and objects from overseas into domestic circulation. Jesuit missionaries in India, for example, were initially encouraged by their order's founder, Ignatius Loyola, to provide anecdotal reports of natural phenomena.[37] By the seventeenth century, they were sending regular

---

of Letters, 1600–1750, ed. Hans Bots and Françoise Waquet (Amsterdam: APA-Holland University Press, 1994).

[35] Fernand Braudel, *The Structures of Everyday Life* [1973] (New York: Harper and Row, 1979), pp. 402–15.

[36] As early as 1415, the Portuguese had established in Lisbon the Casa da Ceuta (House of Ceuta, Ceuta being a port on the African side of the Strait of Gibraltar) to oversee trade with North Africa. By the middle of the fifteenth century, Prince Henry had created a sister organization, the Casa da Guiné in Lagos, which was later relocated to Lisbon. By the close of the century, both houses had been combined and renamed the Casa da India. See Donald F. Lach, *Asia in the Making of Europe*, 3 vols. (Chicago: University of Chicago Press, 1965), 1: 92–3; and Francisco P. Mendes da Luz, *O Conselho da India* (Lisbon: Divisaão de Publicaçoes e Biblioteca, Agência Geral do Utramar, 1952), pp. 30–9. The Spanish houses of trade are discussed later in this chapter. For the cartographic activity of both the Portuguese and Spanish *casas*, see Mollat du Jourdin and de la Roncière, eds., *Sea Charts of the Early Explorers*.

[37] As early as 1547, we find Ignatius Loyola urging missionaries in India to send information about "such things as the climate, diet, customs and character of the natives and of the peoples of India." His motive became clear in his instructions to the Jesuit superior in Goa in 1554: "Some leading figures who in this city [Rome] read with much edification for themselves the letters from India, are wont to desire, and they request me repeatedly, that something should be written regarding the cosmography of those regions where ours [i.e., members of the Society of Jesus] live. They want to know, for instance, how long are the days of summer and of winter; when summer begins; whether the shadows move towards the left or towards the right. Finally, if there are things that may seem extraordinary, let them be noted, for instance, details about animals and plants that are either not known at all, or not of such a size, etc. And this news – sauce for the taste of a certain curiosity that is not evil and is wont to be found among men – may come in the same letters or in other letters separately." See Harris, "Confession-Building," pp. 304–5.

shipments of such exotic – and often therapeutic – natural wonders as guaiacum, bezoar, and snakestone.[38]

Indeed, it was no accident that the growth of the Republic of Letters and curiosity cabinets coincided with the expansion of European travel because each fed the other. The curiosity of scholars and the thirst of administrators of overseas enterprises for hard information led to a demand for "news from the Indies." Colonial bureaucrats, commercial agents, and missionaries – themselves often the products of a humanist education – could easily meet that demand in the reports written in the course of their duties. The Fugger *Neue Zeitungen* (literally, "new journal" or "newspaper") occasionally carried reports of natural wonders and geographical notices; for example, they circulated Christopher Columbus's (ca. 1446–1506) letter of 1493.[39] The Spanish Consejo Real y Supremo de las Indias (Council of the Indies), based in Seville, sent questionnaires to colonial bureaucrats in New Spain that requested information about everything from topography to botany. Much of that information found its way into the published *Chronicles of the Indies*.[40] In addition to confidential administrative reports, the Society of Jesus also gathered, edited, and circulated "edifying news" intended to boost morale within the order and advertise the Society's good works to outsiders. Such reports were the basis of the Society's *Annuae litterae* (Annual Letters), published irregularly from 1550 onward, and the various journals its members edited in the eighteenth century: the *Journal de Trévoux* (1701–62), *Lettres édifiantes et curieuses* (Edifying and Curious Letters, 1702–76), and *Neuer Welt-Bott* (New World-Messenger, 1726–58).[41] Thus, as the range of natural phenomena known to Europeans grew, so did the means of communicating about them.

The third, and in many respects the most pervasive, change in early modern modes of communication resulted from the printing press. Although much could be said regarding the role of printing in the growth of natural knowledge, the specific question here is the role it played in expanding the scale and density of scientific practices. Most obviously, the printed page served as both a storehouse and a clearinghouse for information. The increased reliability of tables of numbers (e.g., for ephemerides, geographical positions, trigonometry, and logarithms) and their increased availability in comparison

---

[38] Martha Baldwin, "The Snakestone Experiments: An Early Modern Medical Debate," *Isis*, 86 (1995), 394–418, esp. pp. 396–8.
[39] Klarwill, *Fugger News-Letters*, p. XX.
[40] Antonio de Herrera y Tordesillas, *Historia general de los hechos de los castellanos en las Islas de Tierra Firme del mar océano* (Madrid: Emplenta Real 1601–15). Moreover, it appears likely that José de Acosta's *Historia natural y moral de las Indias* (Seville: Juan de Leon, 1590) also benefited from the *Consejo's* questionnaires; see Gonzalo Menendez-Pidal, *Imagen del mundo hacia 1570: Segun noticias del Conselo de Indias y de los tratadistas españoles* (Madrid: Gráficas Ultra, 1944), pp. 15–16. For detailed accounts of the Spanish questionnaires, see Clinton R. Edwards, "Mapping by Questionnaire: An Early Spanish Attempt to Determine New World Geographical Positions," *Imago Mundi*, 23 (1969), 17–28; and Barbara E. Mundy, *The Mapping of New Spain: Indigenous Cartography and the Maps of the Relaciones Geográficas* (Chicago: University of Chicago Press, 1996), pp. 18–19.
[41] Harris, "Confession-Building," pp. 303–8.

with manuscript tables facilitated their use both domestically and overseas.[42] The portability of printed books in general made possible the mobility of libraries or, more precisely, the ability to assemble libraries almost anywhere in the world where Europeans settled – something that could scarcely have been achieved in a scribal culture. Also, the standardization of texts contributed toward – though by no means guaranteed – the standardization of technique. This is perhaps most noticeable in the use of basic geometric propositions in fields as disparate as surveying, astronomy, military architecture, and cartography – branches of mixed mathematics that gained passage on almost every overseas voyage.[43] For example, Jesuit libraries in Beijing, Lima, and Japan had copies of Nicholas Copernicus's (1473–1543) *De revolutionibus orbium coelestium* (On the Revolutions of the Celestial Spheres, 1543), as well as works by Christoph Clavius (1537–1612), Descartes, and Isaac Newton (1642–1727).[44] The Dutch brought Euclid's *Elements* to Japan, and the English instructed servants of Indian rajas in the science of surveying.[45]

The printed book was a storehouse not only by virtue of the mass-produced word. By conjoining image with text, the "picture book" shared in the benefits – and occasional liabilities – of being standardized, accessible, and portable.[46] The proliferation of illustrated anatomies,[47] botanical and

---

[42] See, for example, Elizabeth Eisenstein, *The Printing Press as an Agent of Change: Communications and Cultural Transformations in Early Modern Europe* (Cambridge: Cambridge University Press, 1979), pp. 580–8. Mechanical reproduction alone, however, did not guarantee reliability. Vigilance on the part of the author was crucial in order to prevent the mass production of errors on the part of the printer and proofreader. See Adrian Johns, *The Nature of the Book: Print and Knowledge in the Making* (Chicago: University of Chicago Press, 1998), pp. 4–5, 358.

[43] See J. L. Heilbron, *Geometry Civilized: History, Culture, and Technology* (Oxford: Clarendon Press, 1998), pp. 18–24; and Alfred Crosby, *The Measure of Reality: Quantification and Western Society, 1250–1600* (Cambridge: Cambridge University Press, 1997), pp. 95–108.

[44] On the richness of scientific treatises in Jesuit mission libraries, see Nathan Sivin, "Copernicus in China," *Studia Copernicana*, 6 (1973), 63–122, esp. pp. 63–6; J. Laures, S.J., "Die alte Missionsbibliothek im Pei-t'ang zu Peking," *Monumenta Nipponica* (Tokyo), 2 (1939), 124–39, esp. pp. 128–31; Shigeru Nakayama, *A History of Japanese Astronomy: Chinese Background and Western Impact* (Cambridge, Mass.: Harvard University Press, 1969), pp. 82–5; and Luis Martín, S.J., *The Intellectual Conquest of Peru: The Jesuit College of San Pablo, 1567–1767* (New York: Fordham University Press, 1968), pp. 95–7.

[45] On "Dutch learning" in Japan, see Nakayama, *A History of Japanese Astronomy*, pp. 165–87. On the introduction of contemporary European science into India by the English, see S. N. Sen, "The Introduction of Western Science in India during the 18th and the 19th Century," in *Science, Technology, and Culture in India*, ed. Surajit Sinha (New Delhi: Research Council for Cultural Studies Indic International, 1970), pp. 14–43, esp. pp. 14–17.

[46] Walter Ong, S.J., *Orality and Literacy: The Technologizing of the Word* (London: Methuen, 1982), pp. 117–34. See especially Eisenstein, *The Printing Press as an Agent of Change*, pp. 108–13, 266–70, 566–74. As in the case of textual reliability, woodcut and engraving technologies by themselves did not guarantee the integrity of the representation. Unlike woodcuts, engravings were printed separately from the text, and only careful control by the author could avoid the confusion of misplaced or mislabeled illustrations. Poorly made or imperfect plates, image reversal (or inversion), poor craftsmanship, and outright plagiarism (most notably in the case of "fugitive sheets" from anatomical works) were just a few of the hazards that could degrade representational fidelity. See Johns, *Nature of the Book*, pp. 434–41.

[47] K. B. Roberts and J. D. W. Tomlinson, *The Fabric of the Body: European Traditions of Anatomical Illustration* (Oxford: Clarendon Press, 1992), pp. 7–10, 34–43, 69–96, 125–346; and K. B. Roberts,

zoological handbooks,[48] and geographical[49] and celestial[50] atlases – all of which had their origins in the sixteenth century – not only demonstrates the synergy between text and image but also points to the growth of discipline-specific publications. The best-known exemplars from each of these categories are Vesalius's *De humani corporis fabrica* (On the Fabric of the Human Body, 1543 and 1555); Gessner's *Catalogus plantarum* (Catalog of Plants, 1542); and Joan Blaeu's (ca. 1599–1673) *Atlas maior* (Great Atlas, 1662). In these cases, the richness and quality of the illustrations combined with systematic arrangement of image and text to produce truly remarkable compendia of their respective fields.

The public world of the printing press intersected with the private world of personal correspondence in multiple ways. As noted, topographical and ethnographical information gathered via questionnaires sent to New Spain by the "House of the Indies" eventually found its way into the multivolume *Historia general* (1601–15) of Antonio de Herrera (1549–1625), and letters from Jesuit missionaries provided the raw materials for both the Society's published *Annuae litterae* and its eighteenth-century periodical literature.[51] Kircher distinguished himself not only as the most prolific of Jesuit intelligencers, but also as one of the most-published authors of the Society of Jesus. In most of the more than twenty volumes he published in the natural and mathematical sciences, Kircher drew extensively upon his missionary correspondents for the kaleidoscope of "remote and heterogeneous" reports that his encyclopedic works comprise.[52]

These transitory links between correspondence and publication finally gained a measure of institutional permanence in the 1660s with the founding of state-supported scientific academies and the invention of learned journals devoted to scientific topics.[53] The career of Henry Oldenburg (ca. 1620–1677) nicely illustrates this point. Before the foundation of the Royal Society of London in 1660, he was "private intelligencer" – that is, corresponding as private individual and not as an officer of a formal institution – much like his mentor Samuel Hartlib (ca. 1600–1662) in the previous generation. After the

---

"The Contexts of Anatomical Illustration," in *The Ingenious Machine of Nature: Four Centuries of Art and Anatomy*, ed. Mimi Cazort, Monique Kornell, and K. B. Roberts (Ottawa: National Gallery of Canada, 1996), pp. 71–103.

[48] Agnes Arber, *Herbals* (Cambridge: Cambridge University Press, 1938); and Gill Saunders, *Picturing Plants: An Analytical History of Botanical Illustration* (Berkeley: University of California Press, 1995).

[49] See Adolf Erik Nordenskiöld, *Facsimile-Atlas to the Early History of Cartography with Reproductions of the Most Important Maps Printed in the XV and XVI Centuries* (New York: Dover, 1973); and Mollat and de la Ronciére, eds., *Sea Charts of the Early Explorers*.

[50] Deborah J. Warner, *The Sky Explored: Celestial Cartography, 1500–1800* (New York: Alan R. Liss, 1979).

[51] See notes 39 and 40.

[52] Paula Findlen, "Scientific Spectacle in Baroque Rome: Athanasius Kircher and the Roman College Museum," *Roma moderna e contemporanea*, 3 (1995), 625–65.

[53] David A. Kronick, *A History of Scientific and Technical Periodicals: The Origins and Development of the Scientific and Technical Press, 1665–1790* (Metuchen, N.J.: Scarecrow Press, 1976), pp. 77–112.

inauguration of the *Philosophical Transactions* in 1665, Oldenburg served very effectively as its first editor. Now private correspondence could pass regularly into the public domain – that is, after it first passed through the hands of gate-keeper editors such as Oldenburg – and achieve the permanence and visibility of other published works. Conversely, readers' subscriptions to journals billed as "published correspondence" often acted as a spur to further correspondence, either private or published, thus strengthening the circulation of ideas.[54]

To summarize, the multiple overlapping and geographically expanding networks of travel and communication that developed over the course of the early modern period rank among its most distinctive features. As the German political philosopher Jürgen Habermas and others have argued, "the traffic in commodities and news created by early capitalist long-distance trade" was part and parcel of "a new social order."[55] A concomitant of that new order was the growth of a "public sphere" in which the exchange of information, ideas, and opinions became part of the fabric of urban life. Crisscrossing the emerging public sphere and binding its parts together were the quasi-public correspondence of intelligencers and bureaus of address and the various genres of published material, ranging from proto-newspapers and pamphlets to broadsides and learned journals. These novel modes of communication provided the infrastructure of the Republic of Letters, which may be understood simply as that region of the public sphere inhabited by the largely nonproductive – in the strict economic sense, though of course not in the sense of "knowledge production" – and educated elites. Their celebration of "public knowledge," "public good," and "detachment" from the menial business of getting and spending should not obscure their dependence upon mundane realities of long-distance travel and the acquisition and exchange of information. Indeed, segments of the Republic of Letters, perhaps best exemplified in Bishop Thomas Sprat's (1635–1713) *apologia* of 1667 for the newly founded Royal Society, happily proclaimed the advantages that the "new science" would have for commerce and manufacture.

Within the fields of activity concerned with the production and use of natural knowledge, the opportunities of long-distance travel and communication also presented special challenges. Put simply, how could "local" practices be extended to distant places?[56] In order for an expansion in the

---

[54] A. Rupert Hall and Marie Boas Hall, eds., *The Correspondence of Henry Oldenburg*, 13 vols., vols. 1–9 (Madison: University of Wisconsin Press, 1965–73), vols. 10–11 (London: Mansell, 1975–6), vols. 12–13 (London: Taylor and Francis, 1986). See also Michael Hunter, "Promoting the New Science: Henry Oldenburg and the Early Royal Society," *History of Science*, 26 (1988), 165–81.

[55] Habermas, *Public Sphere*, pp. 14–15. Compare this with Braudel's claim that "long-distance trade undoubtedly played a leading role in the genesis of merchant capitalism and was for a long time its backbone," *Wheels of Commerce*, p. 403.

[56] For a discussion of the problems associated with "making science travel," see Steven Shapin, "Here and Everywhere: Sociology of Scientific Knowledge," *Annual Review of Sociology*, 21 (1995), 307.

scale of scientific practice to take place, it was first necessary to project social and cognitive conventions beyond the well-ordered social space of the study, laboratory, chartroom, or observatory and to create the virtual space of large-scale enterprises. Expanding scales of practice, then, required finding ways to guarantee – or at least enhance – the mobility of expertise and criteria for evaluating testimony from remote observers.

VIRTUAL SPACES AND THEIR EXTENSION

A number of studies have focused in some detail on the social codes attached to particular architectural spaces in which various forms of natural knowledge were negotiated.[57] The expansion in the scale of scientific practices poses the question as to how these social relations could be stretched and reworked to embrace widely scattered practitioners. In the absence of a concrete architectural space as a stage for the enactment of social codes through face-to-face interactions, how could the social contract of knowledge making – sociability, obligation, trust, reliability, and expertise – be maintained? In other words, how were sites devoted to the making of natural knowledge able to achieve sociability at a distance?

As historians of science David Lux and Harold Cook have argued, the increase in private travel, in the form of *peregrinatio academica* – or the "Grand Tours" of European capitals by well-to-do young gentlemen and their tutors – greatly enlarged the circle of personal contacts and thereby facilitated the circulation of reports among trustworthy acquaintances.[58] In addition to establishing friendships, such contacts honed the skills of both host and traveler as judges of character. Personal interactions allowed each party to observe clothing, personality, demeanor, etiquette, and conversational ability and thereby to evaluate the character of the other. Indeed, evaluation of character could commence even before the first face-to-face encounter, because travelers were frequently preceded – or accompanied – by letters of introduction either from friends of the host or indirect

---

[57] See, for example, Steven Shapin and Simon Schaffer, *Leviathan and the Air-Pump: Hobbes, Boyle, and the Experimental Life* (Princeton, N.J.: Princeton University Press, 1985), for social and experimental spaces associated with Boyle's laboratory, and Shapin's discussion of the social space of the Royal Society in his "The House of Experiment in Seventeenth-Century England," *Isis*, 79 (1988), 373–404. For Galileo's movement through the social space of the Tuscan court, see Mario Biagioli, *Galileo, Courtier: The Practice of Science in the Culture of Absolutism* (Chicago: University of Chicago Press, 1993); and Biagioli, "Galileo the Emblem Maker," *Isis*, 81 (1990), 230–58. For a discussion of the mathematical and experimental spaces of the Jesuit Collegio Romano, see Gorman, "The Scientific Counter-Reformation," pp. 160–97. For studies on the interface between social and architectural space, see Peter Galison and Emily Thompson, eds., *The Architecture of Science* (Cambridge, Mass.: MIT Press, 1999).

[58] David S. Lux and Harold J. Cook, "Closed Circles or Open Networks? Communicating at a Distance During the Scientific Revolution," *History of Science*, 43 (1998), 180–211, esp. pp. 184–5. See also Justin Stagel, *A History of Curiosity: The Theory of Travel, 1550–1800* (Chur: Harwood, 1995).

acquaintances of proven integrity.[59] In this way, circles of trust could be extended to two and three degrees of social consanguinity.

As important as these informal lines of contact were in establishing networks of trustworthy informants, institutional and corporate entities played an even larger role. Here it will be useful to distinguish between geographical and virtual spaces. The early voyages of discovery and the subsequent cycles of trade and colonial travel vastly increased the geographical horizon of European activity. Virtual space, on the other hand, is an extension of the metaphor of social space or "spatial formations."[60] The term is used here to mean the social relations that make possible the controlled circulation of people, reports, and objects in physical space and especially the ability to "traverse" that space vicariously. The test of the integrity of the virtual space attached to a particular geographic site, say a chartroom in Seville or an observatory in Paris, is whether stay-at-home practitioners could obtain reports or objects from a distant place "as if" they were there. Such virtual participation in distance practices depended principally upon the following elements: sustained cycles of long-distance travel and communications; the training and deployment of agents deemed reliable and trustworthy; and the standardization of methods of observation and conventions of representation. Extending the scale of scientific practices was a matter of extending the virtual spaces attached to localized sites of knowledge production.

The most promising – though by no means the only – places to look for such configurations are the chartered corporations engaged in overseas activities beacause it was here that expertise, social order, and travel came together to form robust long-distance networks.[61] Falling into this category are a number of organizations: overseas trading companies such as the Dutch and English East India companies; colonial administrative bureaus; religious orders engaged in overseas missions; and (eventually) the large state-supported scientific academies, which from the time of their foundations entertained visions of overseas scientific expeditions.[62]

Disparate as these corporations were in regard to their mandates and modes of organization, they all shared a number of structural features. Each was a legally constituted corporation established by the sovereign authority of

---

[59] Lux and Cook, "Closed Circles," pp. 186–7.
[60] Nigel Thrift, "On the Determination of Social Action in Space and Time," in *Spatial Formations* (London: Sage, 1996), pp. 63–95.
[61] Steven J. Harris, "Long-Distance Corporations, Big Sciences, and the Geography of Knowledge," *Configurations*, 6 (1998), pp. 271, 276–9.
[62] In the same year as its foundation, 1660, the Royal Society issued instructions for merchants traveling to the Canary Islands to replicate the Torricellian experiment on the mountain of Tenerife; the plan appears not to have been executed. See Thomas Birch, *The History of the Royal Society of London*, 4 vols. (New York: Johnson Reprint Corp., 1968), 1: 1–2. In January 1667, just three weeks after its first meeting, the Académie Royale des Sciences reviewed Adrien Auzout's plan for an astronomical expedition to Madagascar; war with the Netherlands scuttled the project. See John W. Olmstead, "The Scientific Expedition of Jean Richer to Cayenne," *Isis*, 34 (1942), 118–19. Plans for later expeditions were more successful.

Crown, Parliament, or papacy. Each enjoyed legal jurisdiction over a prescribed domain of activity and was thus entitled to enact by-laws for its internal governance and statutes to regulate the actions of its members.[63] Although the division and flow of power varied markedly among these corporations, most were hierarchically organized, with a centralized administrative apparatus and fixed headquarters. Corporate membership was explicit; members were bound by both written and unwritten rules of conduct, and corporate leaders had at their disposal mechanisms for the social and cognitive training and disciplining of members.

Of course, methods of recruitment, levels of competence, and disciplinary mechanisms varied widely from corporation to corporation. The Society of Jesus enjoyed perhaps the highest level of reliability among its members because of its highly selective recruiting practices and long period of education. Yet, compared with the Dutch East India Company or the Spanish House of Trade, it was the most impoverished and the least able to control its lines of transport because it owned no ships and could only book passage for its missionaries and material goods from "external" sources. Such differences in recruitment, training, discipline, and material resources suggest that there were a number of different ways to master the operation of long-distance networks, and the plurality of more or less successful long-distance strategies attests in turn to the flexibility of the social technology necessary to define collective goals while delimiting individual behavior.

What distinguished overseas trading companies, colonial offices, and missionary orders from other early modern corporations were charters that obliged them to engage in geographically and temporally extended practices and therefore to master the operation of long-distance networks.[64] Put differently, if a long-distance corporation was to fulfill its mandate, it had to define and inculcate its "ways of proceeding" at the center and extend those ways to its periphery of action. Concretely, that meant that corporate leaders had to recruit and train reliable agents, send them to remote regions to undertake corporate business, and maintain cycles of correspondence that consisted primarily of directives from headquarters and intelligence reports from the field. Here again, strategies varied greatly from corporation to corporation. Some, like the Society of Jesus, excelled at instilling in their members a deep sense of commitment to corporate causes. Others, such as the Dutch East India Company, used wealth – or the promise of wealth – as an incentive for compliance among its upper officials and (occasionally) physical coercion among its common sailors.[65] No strategy functioned perfectly, and all suffered at one

---

[63] Toby Huff, *The Rise of Early Modern Science: Islam, China, and the West* (Cambridge: Cambridge University Press, 1993), pp. 136–7.
[64] For discussions of long-distance networks as sociological models, see John Law, "On the Methods of Long-Distance Control: Vessels, Navigation and the Portuguese Route to India," *Sociological Review Monographs*, 32 (1986), 234–63; and Latour, *Science in Action*, pp. 223–32.
[65] C. R. Boxer, *The Dutch Seaborne Empire, 1600–1800* (New York: Penguin, 1973), p. 79.

time or another from breaches of trust, breakdowns in communication, or divergence of goals. The overall effectiveness of these corporations, however, cannot be doubted. Collectively, they were largely successful in achieving the intertwined goals of overseas commerce, colonization, and conversion.

Although the disinterested pursuit of scientific knowledge was never a primary goal of these corporations, the operation of long-distance networks of any sort required knowledge of certain parts of the natural world. Corporations engaged in the carrying trade needed to gain knowledge of wind, currents, and coastlines, as the rich tradition of Spanish and Dutch cartography demonstrates. Both generals of missionary orders and heads of colonial bureaus administering the distant lands of the crown had to understand something about climates, landscapes, and indigenous peoples – what contemporaries usually lumped together under the headings of "natural and moral histories."[66] Because of the need to have healthy agents in the field, all also had to concern themselves with diets, diseases, and possible indigenous cures.[67]

The list of scientific practices implicated in the activities of long-distance corporations is a long one, and so by way of illustration let us consider just three examples. In 1503, the Spanish Crown established the Casa de la Contratación in Seville as one of its earliest bureaus responsible for overseas transport. The Casa trained and licensed pilots, drafted and enforced laws pertaining to navigation, and maintained the *Padrón Real*, or master map, of overseas voyages.[68] Outward-bound pilots made copies of the master map and were given explicit instructions regarding the recording of latitudes and landfalls on their copies and in their logbooks. The chief cosmographer conferred with returning pilots before entering any corrections or additions to the master map. Thus, some of the best sea charts of the early sixteenth century were produced under the authority and direction of the Casa.

By the seventeenth century, it was the chartrooms of the Dutch East India Company (Verenigde Oostindische Compagnie, or VOC) that gathered and collated the cartographic information from its pilots for maps of southern Africa, India, Indonesia, and Japan. As masters of the seventeenth-century carrying trade, the VOC also became one of the primary channels for the transport of medically and commercially valuable plant materials and natural exotica of all types. A number of VOC physicians, hired by the directors to attend to the health of the company's agents, assembled pharmaceutical recipes, herbaria, and gardens while also composing some of the best

---

[66] Both Herrera's "Chronicles" and da Acosta's "Moral and Natural History" fit under this category. See note 40.
[67] Harris, "Long-Distance Corporations," pp. 285–94.
[68] Edward L. Stevenson, "The Geographical Activities of the *Casa de la Contratacion*," *Annals of the Association of American Geographers*, 17 (1927), 39–59. See also Menendez-Pidal, *Imagen del Mundo*, pp. 4–5.

botanical handbooks of the century.[69] Not surprisingly, the botanical gardens of Amsterdam and Leiden became rich repositories for plant specimens and plant-related knowledge obtained by company officers working in the remote outposts of the VOC.

Among the Catholic religious orders engaged in the foreign missions, it was the Society of Jesus that excelled in the long-distance sciences. Cinchona was known to contemporaries as "Jesuit bark" in acknowledgment of the role of the Society's pharmacists in first bringing the antifebrile to the attention of European physicians – and for reputedly possessing a near monopoly on its early distribution.[70] Jesuit pharmacists working the South American missions were quick to exploit the medical knowledge of indigenous peoples and, by the eighteenth century, had developed a supply network that stretched from Mexico to Argentina.[71] The Society also lent its name to a number of other botanical commodities, including Jesuits' drops (a concoction of garlic, Peruvian balsam, and sarsaparilla), Jesuit nut (the seed of the Chinese water chestnut), Jesuit tea (yerba maté), and the Ignatian bean (a name applied to both the seed of the poisonous *Strychnos ignatii* and the seed of the medicinally useful *Fevillea trilobata*). As missionaries traveling the interiors of distant lands in search of new peoples to proselytize, Jesuits were frequently obliged to play the parts of explorer, naturalist, and cartographer. The request from superiors in Rome to find a land route from India to China resulted in the treks of Antonio de Andrade (1580–1634) and Bento de Goes (1562–1607) across the Himalayas.[72] Similar directives from Rome to seek out the elusive "Prester John" (i.e., the legendary Christian King of Africa) inspired Pedro Paez (1564–1622) to travel through Ethiopia and as far as the Upper Nile.[73] Jacques Marquette's (1636–1675) partial exploration of the Mississippi River,[74] Eusebius Kino's (1644–1711) exploration of northwest New Spain,[75] and Samuel Fritz's (1654–1728) journey down the Amazon and

---

[69] For a list of botanical and pharmaceutical publications from the physicians of the VOC, see Peter van der Krogt, ed., *VOC: A Bibliography of Publications Relating to the Dutch East India Company, 1602–1800* (Utrecht: HES Publishers, 1991), pp. 364–80.

[70] For the Society's implication in the cinchona trade, see Saul Jarcho, *Quinine's Predecessor: Francesco Torti and the Early History of Cinchona* (Baltimore: Johns Hopkins University Press, 1993), pp. 4–5, 14–17; and Jaime Jaramillo-Arango, "A Critical Review of the Basic Facts in the History of Cinchona," *Journal of the Linnaean Society (London)*, 53 (1949), 272–309.

[71] For an overview of Jesuit missionary-apothecaries in South America, see Renée Gicklhorn, *Missionsapotheker: Deutsche Pharmazeuten im Lateinamerika des 17. und 18. Jahrhunderts* (Stuttgart: Wissenschaftliche Verlagsgesellschaft, 1973), pp. 33–88.

[72] Cornelius Wessels, S.J., *Early Jesuit Travelers in Central Asia, 1603–1721* (The Hague: Martinus Nijhoff, 1924; repr. New Delhi: Asian Educational Services, 1992), pp. 1–68.

[73] Philip Caraman, *The Lost Empire: The Story of Jesuits in Ethiopia, 1535–1634* (London: Sidgwick and Jackson, 1985).

[74] Joseph P. Donnelly, S.J., *Jacques Marquette, S.J., 1637–1675* (Chicago: Loyola University Press, 1985), pp. 204–29.

[75] Herbert Bolton, *The Rim of Christendom: A Biography of Euschio F. Kino* (New York: Macmillan, 1936).

Orinoco rivers[76] were all similarly motivated. All resulted in manuscript maps sent either to Rome or to Jesuit universities, and all were eventually published as part of Jesuit-authored geography texts.[77]

CONCLUSION

The claim here is not that all or most of the science of the early modern period issued from the activities of long-distance corporations. Rather, it is that long-distance corporations adequately mastered the social, administrative, and technological challenges necessary to extend social and cognitive conventions beyond local settings. In doing so, they greatly facilitated the expansion in the scales of scientific practices. The increase in the density of practice, however, was primarily – though not exclusively – a consequence of the successful exploitation of new modes of communication, especially the circulation of correspondence in the Republic of Letters and the use of the printing press for scientific publication. No great divide separated domestic correspondence, publication, and overseas travel. Indeed, the administration of long-distance corporations depended upon the frequent exchange of internal, private, and sometimes secret correspondence. Although the various genres of administrative correspondence did not generally figure into the epistolary commerce of the Republic of Letters, more accessible forms of correspondence (for example, reports of natural curiosities such as exotic materia medica) often flowed along corporate channels and into outlets that fed the Republic.

The critical difference between the two was one of social structure. The "Republic" was less a state than a loose confederation of weak social associations, most often held together by the energy and dedication of a single intelligencer – and therefore likely to collapse upon his or her death. Participation in a correspondence circle was voluntary, informal, and occasional. Moreover, the stated goals of the citizens of the Republic were, at least in principle, largely intellectual and cultural. However important corporation-mediated knowledge of the natural world may have been, that knowledge was a by-product of commercial, territorial, and clerical agendas. Correspondence circles lacked explicit criteria for membership, and they possessed no mechanisms for compliance beyond social suasion and civility. As imperfect as discipline and compliance may have been within long-distance corporations, they greatly exceeded what one could hope for from the loose webs of

---

[76] Josef Gicklhorn and Renée Gicklhorn, *Im Kampf um den Amazonas: Forscherschicksal des P. Samuel Fritz* (Prague: Noebe, 1943).

[77] Paez's travels to the source of the Blue Nile, the first ever by a European, were reported in Athanasius Kircher's *Oedipus aegyptiacus* (Rome: Vitalis Mascardi, 1652–4), and Martini's exploration of China found its way into Kircher's *China monumentis . . . illustrata* (Amsterdam: Jacob à Meurs, 1667). Kino's geographical reconnaissance of northwest Mexico appeared in Heinrich Scherer's *Geographia naturalis* (Munich: Rauchin, 1703).

correspondence that connected the Republic of Letters. Yet, in contrast with the more rigid framework of corporations, such informal arrangements could be a source of strength: Correspondence networks could shrink and grow, topics could be raised and dropped, requests could be eagerly answered or politely ignored, and the noble (if vague) rhetoric of the Republic of Letters could sustain the loyalty of some of the best literary and scientific minds of the period.[78] The Republic of Letters may have been a sparsely populated country, yet it attained a remarkable degree of influence in the intellectual affairs of Europe by virtue of the density of its communications. In combination, long-distance corporations and the consortium of overlapping correspondence networks functioned as two powerful and interlinked engines: the former binding together scholars and experts from various countries and disciplines, the latter drawing an ever-widening horizon of engagement into that domestic network of exchange.

The cognitive consequences of the expanded scales of scientific practice were multiple, but only two of the broadest trends will be sketched here. In cosmography, the transition from portolans based on rhumb lines (an array of compass bearings connecting coastal points over restricted regions) to sea and world charts based on Ptolemaic coordinates (lines of latitude and longitude encircling the globe) encouraged a reconceptualization of space and ways of fixing locations in space. The "tunnel vision" characteristic of pilgrim maps and routers, both of which depicted only narrow pathways across land and sea, gave way to an expanded "peripheral vision" that embraced the entire surface of the Earth – including its unexplored regions, hence the seemingly paradoxical possibility of locating terra incognita on a map. Once the Earth's surface had been conceived of as a mappable space and conventions for measurement and coordinates had been stabilized (not, in practice, a trivial matter), landmasses and place names could be added indefinitely while still preserving cartographic notions of position and distance. Moreover, it was no coincidence that the systematic mapping of visible stars – and not just those associated astrologically with the signs of the zodiac – on star charts and celestial globes occurred more or less simultaneously with the mapping of the Earth's surface. Extending "peripheral vision" to the heavens was contingent upon travel to the southern hemisphere and upon the establishment of conventions of instrumentation, measurement, and mapping. The celestial sphere, like the terraqueous sphere, became a mappable space capable of holding an indefinite number of new objects, all of which could be located and their (angular) distances from one another measured.

In natural history, the proliferation of new objects from overseas – plant, animal, and mineral specimens, medicinal simples and compounds, and artifacts produced by distant peoples – contributed to the breakdown of traditional schemes of classification. In what the historian of science William

---

[78] Daston, "Republic of Letters," pp. 375–81.

Ashworth has called "the emblematic view of nature" of the Renaissance, most domestic or familiar plants and animals were embedded in a rich literary tradition that drew upon historical, mythological, scriptural, and moral allusions as well as a host of presumed natural affinities, similitudes, and signatures that linked them to each other, to humans (i.e., to parts of the human body or aspects of human temperaments), and even to the stars. Ashworth has argued convincingly that European naturalists' confrontation with exotic nature compelled them to look upon natural objects unadorned – that is, devoid of the sort of emblematic meanings that attended their European counterparts.[79] The transition from an organizational scheme based on literary and moral associations to one based on careful description of the "thing-in-itself," exhaustive cataloging, and eventual systematic classification (as in the works of the eighteenth-century Swedish naturalist Carolus Linnaeus) was therefore driven in large part by the burgeoning inventory of natural objects, and that inventory was made available by expanding scales of practice in natural history.

[79] William Ashworth, "Natural History and the Emblematic World View," in *Reappraisals of the Scientific Revolution*, ed. David C. Lindberg and Robert S. Westman (Cambridge: Cambridge University Press, 1990), pp. 303–33, 318–19, 322–4.

Part III

DIVIDING THE STUDY OF NATURE

17

NATURAL PHILOSOPHY

Ann Blair

"Natural philosophy" is often used by historians of science as an umbrella term to designate the study of nature before it could easily be identified with what we call "science" today. This is done to avoid the modern and potentially anachronistic connotations of the term "science." But "natural philosophy" (and its equivalents in different languages) was also an actor's category, a term commonly used throughout the early modern period and typically defined quite broadly as the study of natural bodies. As the central discipline dedicated to laying out the principles and causes of natural phenomena, natural philosophy underwent tremendous transformations during the early modern period. From its medieval form as a bookish Aristotelian discipline institutionalized in the universities, natural philosophy became increasingly associated during the sixteenth and seventeenth centuries with new authorities, new practices, and new institutions, as is clear from the emergence of new expressions such as the "experimental natural philosophy" of Robert Boyle (1627–1691) and the Royal Society of London or the *Philosophiae naturalis principia mathematica* (Mathematical Principles of Natural Philosophy, 1687) of Isaac Newton (1642–1727).[1]

Traditional natural philosophy (that is, of the bookish, largely Aristotelian variety) continued to prevail in university teaching through much of the seventeenth century (see Grafton, Chapter 10, this volume), but it, too, was transformed by the innovations of the period, which prompted attempts at adaptation as well as staunch resistance. By 1700, it had yielded definitively in all but the most conservative contexts to the mechanical, mathematized natural philosophies of Cartesianism and Newtonianism.[2] Nonetheless, the

---

[1] See Robert Boyle, *Some Considerations Concerning the Usefulnesse of Experimental Natural Philosophy* (Oxford: Printed by Henry Hall for Richard Davis, 1664).

[2] The basic principles and strategies of Aristotelian natural philosophy have also survived in restricted, confessional circles down to the twentieth century. See, for example, Charles Frank, S.J., *Philosophia naturalis* (Freiburg: Herder, 1949).

I am grateful to Roger Ariew, Laurence Brockliss, Mordechai Feingold, Anthony Grafton, James Hankins, and the editors of this volume for helpful comments on earlier drafts of this chapter.

term "natural philosophy" continued to be current (notably in English) through the eighteenth century, its broad scope left intact by the transitions to new methods and explanatory principles. The concept and the term were replaced starting in the early nineteenth century by the emergence and professionalization of specialized scientific disciplines with which we are familiar today, from biology and zoology to chemistry and physics.[3]

THE UNIVERSITY CONTEXT OF NATURAL PHILOSOPHY

"Philosophia naturalis" served as a translation of Aristotle's *physikē epistēmē* and was also called "physica" or "physice" (a shortened version of the same expression).[4] It originally designated one of the three branches of speculative philosophy delineated by Aristotle, alongside mathematics and metaphysics.[5] As institutionalized in the universities of medieval Christendom, starting in the thirteenth century, natural philosophy consisted in the study of and commentary on Aristotle's *libri naturales*. These comprised (as in the regulations of Paris of 1255, equivalents of which prevailed in universities throughout Europe) Aristotle's *Physics, On the Heavens, Meteorology, On the Soul, On Generation and Corruption,* the *History,* and *Parts of Animals,* the shorter works known collectively as the *parva naturalia* – including *On Sleep and Waking, On Memory and Remembering,* and *On Life and Death* – and two tracts now considered of doubtful authenticity, *On Causes* and *On Plants*.[6] But, given the special emphasis on logic in the medieval curriculum, natural philosophy was generally reduced in practice to the study of the *Physics* on the

---

[3] For one account of the demise of natural philosophy as a concept after 1800, see Simon Schaffer, "Scientific Discoveries and the End of Natural Philosophy," *Social Studies of Science*, 16 (1986), 387–420. On the emergence of new terms, see Sydney Ross, "*Scientist*: The Story of a Word," *Annals of Science*, 18 (1962), 65–85. In different languages, "natural philosophy" had different connotations. In French, "philosophie naturelle" was not as widely used as the English equivalent – "physique" was preferred by both traditionalists such as Scipion Dupleix and innovators such as the Cartesians. When used in titles, "philosophie naturelle" tended to signify books with Hermetic or alchemical interests, as in Jean d'Espagnet's *Philosophie naturelle restituée* (Latin 1623, French translation 1651), or Pierre Arnaud, compiler, *Trois traitez de philosophie naturelle . . . ascavoir le secret livre du tres ancien philosophe Artephius* (1612) or the anonymous *Flambeau de la philosophie naturelle*. In German, where Latin persisted longer as the language of science, "Naturphilosophie" first appeared in book titles at the end of the eighteenth century and blossomed with the philosophies of Schelling, Goethe, and others who looked for unifying organic forces in nature.

[4] "Physiologia," literally the explanation (*logos*) of nature (*physis*), is another near synonym of these terms; for some examples and discussion, see Roger Ariew and Alan Gabbey, "The Scholastic Background," in *The Cambridge History of Seventeenth-Century Philosophy*, ed. Daniel Garber and Michael Ayers, 2 vols. (Cambridge: Cambridge University Press, 1998), 1: 425–53, esp. pp. 427–8.

[5] See Aristotle, *Metaphysics*, 6.1.(1026a). Speculative philosophy was contrasted with practical philosophy, which embraced ethics and the mechanical arts.

[6] *Chartularium universitatis Parisiensis*, vol. 1, no. 246, as cited and discussed in Pearl Kibre and Nancy Siraisi, "The Institutional Setting: The Universities," in *Science in the Middle Ages*, ed. David Lindberg (Chicago: University of Chicago Press, 1978), p. 131. On the medieval universities more generally, see *Universities in the Middle Ages*, ed. Hilde de Ridder-Symoens (Cambridge: Cambridge University Press, 1992).

one hand (with some consideration of *On the Heavens* and *Meteorology*) and *On the Soul* on the other (with some reference to the *parva naturalia*).[7] At the University of Paris, for example, once Aristotle had become the centerpiece of the curriculum in the mid-thirteenth century, a candidate for a bachelor's degree took only a minimum of natural philosophy and focused primarily on logic. Natural philosophy featured mostly in the two years of additional course work for the master's degree, which was required in order to teach or to continue on to a higher faculty (i.e., law, medicine, or theology).[8] Despite variations between institutions, some of which offered more instruction at the undergraduate level in the *quadrivium* (the mathematical disciplines of arithmetic, geometry, astronomy, and music), this basic pattern remained the norm in Europe until the end of the seventeenth century.

Broadly speaking, the institutional structures of the medieval universities remained in place throughout the early modern period, but the rapid expansion in higher education starting around 1500 and the new technology of printing fostered new pedagogical developments. Throughout Europe, students attended universities in greater numbers in the sixteenth century; the dates at which attendance curves peaked vary from place to place, from around 1590 in Castile to 1660 in Louvain.[9] About one hundred new universities were founded between 1500 and 1650 (whereas ten existing universities were abolished, transferred, or merged in the same period). The new foundations were often associated with a religious offensive. In the first half of the sixteenth century, they clustered in Spain, affirming the effects of the *reconquista* – that is, the "reconquest" of Spain from its Muslim and Jewish inhabitants. After the peace of Augsburg (1555) established the principle of religious territoriality (*cujus regio, eius religio*), new universities multiplied in the principalities of Central and Eastern Europe, as each region needed schools appropriate to its ruler's religious choice.[10]

The growth of state bureaucracies also required more educated elites to fill them, which prompted the formation of new educational institutions. These included the *collèges de plein exercice* at the University of Paris and the Jesuit colleges founded across Europe, which offered instruction independent of the faculty of arts that combined a secondary education in Latin

---

[7] For the strand of natural philosophy focused on the soul and what came to be called "psychologia" in the late sixteenth century (which I will not discuss much), see the chapters by Katharine Park and Eckhard Kessler in The *Cambridge History of Renaissance Philosophy*, ed. Charles B. Schmitt, Quentin Skinner, and Eckhard Kessler, with Jill Kraye (Cambridge: Cambridge University Press, 1988), pp. 453–534.
[8] Edward Grant, *Planets, Stars, and Orbs: The Medieval Cosmos, 1200–1687* (Cambridge: Cambridge University Press, 1994), p. 20.
[9] Maria Rosa di Simone, "Admission," in *A History of the University in Europe*, vol. 2: *Universities in Early Modern Europe (1500–1800)*, ed. Hilde de Ridder-Symoens (Cambridge: Cambridge University Press, 1996), pp. 285–325, esp. p. 299.
[10] Willem Frijhoff, "Patterns," in de Ridder-Symoens, ed., *Universities in Early Modern Europe*, pp. 43–110, esp. p. 71. On German foundations, see *Beiträge zu Problemen deutscher Universitätsgründungen der frühen Neuzeit*, ed. Peter Baumgart and Notker Hammerstein (Nendeln: KTO Press, 1978).

and elementary Greek grammar and rhetoric with two or more years of university-level work devoted to philosophy. Students could attend such colleges alongside or instead of university courses for the B.A., although degrees could only be conferred by the universities near which these colleges were often located.[11] Special schools also catered to the sons of the nobility (*Ritterakademien* in the Holy Roman Empire; *collegi dei nobili* in Italy; *gymnasia illustria* in the United Provinces of the northern Netherlands; and academies such as that of Pluvinel in France).[12] Finally, the various religious orders and the secular clergy ran monastic schools and seminaries to train their members.

The general trend across Europe during the sixteenth century, under confessional and administrative pressures to educate more students faster (notably to serve as preachers and bureaucrats), was to compress subjects previously reserved for the later into the earlier years of study.[13] As a result, more students were exposed to instruction in natural philosophy, notably for the B.A. This trend, combined with the spread of printing, fueled a great increase in the number and kinds of books on natural philosophy, particularly of the pedagogical variety.[14] For the professors, there were numerous editions, translations, commentaries, and specialized treatises, whether of the traditional scholastic or the newer humanist variety.[15] Humanist editions and translations strove to strip away the legacy of the medieval Arabic transmission in favor of a translation from the Greek original into elegant Ciceronian Latin. Humanists delved into the newly recovered Greek commentaries on Aristotle from late antiquity, for example those of Themistius (first published in Latin in 1481), Alexander of Aphrodisias (first published in Latin in 1495), or Simplicius (first published in Greek in 1499), but the medieval commentary of Averroes (Ibn Rushd), a scholastic favorite, remained standard for many university professors.[16]

---

[11] Laurence Brockliss, *French Higher Education in the Seventeenth and Eighteenth Centuries* (Oxford: Clarendon Press, 1987), pp. 19–26.

[12] Hilde de Ridder-Symoens, "Mobility," in de Ridder-Symoens, ed., *Universities in Early Modern Europe*, pp. 416–48, at p. 432. On Pluvinel, see Ellery Schalk, *From Valor to Pedigree: Ideas of Nobility in France in the Sixteenth and Seventeenth Centuries* (Princeton, N.J.: Princeton University Press, 1986), chap. 8.

[13] Richard Tuck discusses the different forms of this trend in different universities in "The Institutional Setting," in Garber and Ayers, eds., *The Cambridge History of Seventeenth-Century Philosophy*, pp. 9–32, esp. pp. 16–17.

[14] For a more detailed account of the different genres of Aristotelica in the Renaissance, see Charles B. Schmitt, *Aristotle and the Renaissance* (Cambridge, Mass.: Harvard University Press, 1983), chap. 2.

[15] I follow Grant's definition of a scholastic as one who was trained at a European university and probably also taught at one for some time, and who commented on the natural books of Aristotle. I also follow Grant on the related broad definition of an Aristotelian as one who commented on Aristotle and whose commentary was not solely designed to refute him. See Grant, *Planets, Stars, and Orbs*, pp. 21–3. One can distinguish, as I try to here at various points, between medieval and early modern scholasticism; I reserve "neoscholastic" to describe nineteenth- and twentieth-century movements of Thomist revival.

[16] Jill Kraye, "The Philosophy of the Italian Renaissance," in *The Renaissance and Seventeenth-Century Rationalism*, ed. G. H. R. Parkinson (Routledge History of Philosophy, 4) (London: Routledge, 1993), pp. 16–69, at pp. 24–5.

For students, aids to the acquisition of Aristotelian natural philosophy included Latin editions shorn of cumbersome commentaries but instead enhanced with such trappings as summaries, dichotomous tables, and indexes. The genre of the philosophical textbook, which offered a succinct compendium or manual of natural philosophy, flourished in the sixteenth century.[17] Catholic textbooks were often structured around the traditional medieval *quaestio*, a question in "whether" (e.g., whether the world is eternal) around which one gathered arguments, objections, and responses to objections in favor of alternative solutions before reaching a conclusion.[18] Protestant textbooks, on the other hand, straying more readily from medieval practice and in imitation of Philip Melanchthon (1497–1560), who was the first to include Aristotle in the Lutheran curriculum, tended to pose simplified questions ("What is the world?") that called for definitions and descriptions rather than subtle argumentation and might be answered by a series of numbered propositions.[19]

Most notorious for their pedagogical reductions of complex material were the Calvinist pedagogues who followed the French educational reformer Petrus Ramus (Pierre de La Ramée, 1515–1572). They favored the use of dichotomous tables, from the disposition of which the student would supposedly be able to master any topic. For example, a textbook by Wilhelm Scribonius, already in its fourth edition in 1600, presented a vast topic such as animals by providing the proper subdivisions of it without any descriptions or explanations. Scribonius divided animals into rational and irrational, the latter into those living in water and those on land, the land animals into reptiles and quadrupeds, the quadrupeds into oviparous and viviparous, the viviparous into those with cleft hooves and those with solid hooves, and so on. Textbooks of these various kinds ensured among students a broad diffusion of the basic elements of Aristotelian physics.[20]

Although the flow of university texts, from theses and textbooks to commentaries and treatises, continued exclusively in Latin into the eighteenth century, the first vernacular textbooks of Aristotelian natural philosophy,

---

[17] Charles Schmitt, "The Rise of the Philosophical Textbook," in Schmitt, Skinner, and Kraye, eds., *The Cambridge History of Renaissance Philosophy*, pp. 792–804, esp. pp. 796 ff. For a survey and bibliography of the genre, see Mary Richards Reif, "Natural Philosophy in Some Early Seventeenth-Century Scholastic Textbooks," Ph.D. dissertation, Saint Louis University, St. Louis, Mo., 1962.
[18] For examples of this format, see Domingo de Soto, *Super octo libros physicorum Aristotelis* (Venice: Franciscus Zilettus, 1582), or Franciscus Piccolomini, *Librorum ad scientiam de natura attinentium partes quinque* (Frankfurt: Wecheli haeredes, 1597).
[19] Philip Melanchthon, *Doctrinae physicae elementa* (Lyon: Jean de Tournes and Gul. Gazeius, 1552). On this use of "loci" to structure his textbooks, see Sachiko Kusukawa, *The Transformation of Natural Philosophy: The Case of Philip Melanchthon* (Cambridge: Cambridge University Press, 1995), p. 174.
[20] Guilelmus Adolphus Scribonius, *Physica et sphaerica doctrina*, 4th ed., with annotations by the English doctor Thomas Bright (Frankfurt: Palthenius, 1600), pp. 188–9. Scribonius taught at the school of Corbach from 1576 to 1583. On this context see Joseph Freedman, "Aristotle and the Content of Philosophy Instruction at the Central European Schools and Universities during the Reformation Era (1500–1650)," *Proceedings of the American Philosophical Society*, 137 (1993), 213–53. On the diffusion of Ramism, see Joseph Freedman, "The Diffusion of the Writings of Petrus Ramus in Central Europe, c. 1570–1630," *Renaissance Quarterly*, 46 (1993), 98–152.

starting in 1595, testified to the broadening of the audience seeking a university-style education. These books probably appealed to privately tutored noblemen, to students so weak in Latin that they needed a vernacular crib, to intellectually ambitious barber-surgeons or artisans (such as the potter Bernard Palissy, ca. 1510–1590), and to women, as one dedication suggests.[21] The authors of these works complained of the difficulty of their task, which required coining new vernacular terms to match technical Latin ones, but they were no doubt proud, as one voluble French translator was, to satisfy the desires of "those very studious in French books . . . who had often begged [him] to give them some book in French to attain knowledge of the secrets of nature" and in so doing to "enrich, embellish and adorn our language after the example of the ancients."[22] None of Aristotle's actual writings about nature were translated into vernaculars, however. A set of problems offering questions and answers about the human body and health circulated in the sixteenth and seventeenth centuries as the *Problemata Aristotelis* (Problems of Aristotle), in Latin and in German, French, and English translations, but this text was composed in the Middle Ages and bore no relation to the ancient *Problems* now identified as pseudo-Aristotelian.[23] This work is representative of the most popular extension of Aristotelianism, alongside collections of sayings or short excerpts attributed to Aristotle that were also available in Latin and in the vernacular, such as Jacques Bouchereau's *Flores Aristotelis* (Flowers of Aristotle, first published in 1560) or William Baldwin's *Sayings of the Wise* (first published in 1547).[24]

The extent to which formal changes in the transmission of natural philosophy at the Renaissance universities made the discipline more open to new ideas is debatable. The medieval *quaestio*, after all, lent itself perfectly well to departures from Aristotle's original concerns or arguments, although medieval authors tended to mask their innovations rather than point them out.[25] Renaissance commentaries certainly gave their authors a wide berth for innovation, allowing for digressive discussions that could stray from the

---

[21] The first textbook of Aristotelian natural philosophy in French was dedicated to a woman, Jacquete de Mombrom, lady of a number of viscounties and baronies; see Jean de Champaignac, *Physique françoise* (Bordeaux: S. Millanges, 1595). Champaignac's work was soon followed by others in the early seventeenth century, including those of Scipion Dupleix (1603), Théophraste Bouju (1614), and René de Ceriziers (1643); see Ann Blair, "La persistance du latin comme langue de science à la fin de la Renaissance," in *Sciences et langues en Europe*, ed. Roger Chartier and Pietro Corsi (Paris: Ecole des Hautes Etudes en Sciences Sociales, 1996), pp. 21–42, at p. 40.

[22] François de Fougerolles, *Le théâtre de la nature universelle* (Lyon: Pillehotte, 1597), sig. ++3r, ++1v, as discussed in Ann Blair, *The Theater of Nature: Jean Bodin and Renaissance Science* (Princeton, N.J.: Princeton University Press, 1997), pp. 205–6.

[23] See Ann Blair, "Authorship in the Popular 'Problemata Aristotelis,'" *Early Science and Medicine*, 4 (1999), 1–39; and Blair, "The *Problemata* as a Natural Philosophical Genre," in *Natural Particulars: Nature and the Disciplines in Early Modern Europe*, ed. Anthony Grafton and Nancy Siraisi (Cambridge, Mass.: MIT Press, 1999), pp. 171–204.

[24] Schmitt, *Aristotle and the Renaissance*, p. 54.

[25] See Edward Grant, "Medieval Departures from Aristotelian Natural Philosophy," in *Studies in Medieval Natural Philosophy*, ed. Stefano Caroti (Florence: Leo S. Olschki, 1989), pp. 237–56, at p. 255.

initial passage or opinion at issue.[26] Textbooks, in which the author constructed a systematic presentation of his own, albeit within an Aristotelian framework, have been hailed as the "pedagogical expression of a serious revolution, that which gave birth to Descartes."[27] Each of these forms offered opportunities for modifying the tradition even as they transmitted it.

Rather than singling out the Renaissance as a time of decadent or eclectic Aristotelianism, scholars, beginning in the 1970s, have emphasized the vitality and variety of Aristotelian philosophy throughout the nearly 500 years of its dominance (ca. 1200–1690). Indeed there never was a period characterized by the spread or imposition of a monolithic interpretation of Aristotle. Medieval Aristotelianism had always embraced a wide range of positions, from Averroism to positions tinged with Platonism, such as those of Thomas Aquinas on the soul, to the nominalist probes of the limits of reason in Scotism and Ockhamism; the *quaestio* itself as a form encouraged awareness of the multiplicity of possible arguments and solutions. In the analysis of the historian of science Edward Grant, flexibility was a central feature of Aristotelianism as a philosophical system and the key to its long survival. Aristotle's own obscurities and ambiguities precluded agreement on any one interpretation so that variety of interpretation was perforce the norm. At the same time, Aristotelian principles, with their nearly universal applicability, could be used to generate new theories and respond to new concerns (as in medieval theology, for example). Furthermore, the fact that natural philosophy was fragmented into hundreds of separate *quaestiones* (e.g., concerning Aristotle's *Physics*, book IV: Is place immobile? Is the concave surface of the moon the natural place of fire? Is every being in a place? Is the existence of a vacuum possible? Is a resisting medium required in the motion of bodies?) masked the inconsistencies generated by that flexibility and discouraged debate about the system as a whole.[28]

Aristotelian natural philosophy faced a number of challenges in the Renaissance that stemmed from a new awareness of alternative ancient philosophies, the resurgence of religious objections, and recent empirical observations and discoveries, as I will describe. But the result was hardly a turn away from Aristotle. As Charles Lohr has pointed out, "The number of

---

[26] For a study of the range of Renaissance commentary on Avicenna, see Nancy Siraisi, *Avicenna in Renaissance Italy: The Canon and Medical Teaching in Italian Universities after 1500* (Princeton, N.J.: Princeton University Press, 1987), p. 177. Biblical commentary could also serve as the opportunity to discuss natural philosophical issues, as in Francisco Vallès, *De iis quae scripta sunt physice in libris sacris, sive de sacra philosophia* (Turin: haeredes Nicolae Bevilacquae, 1587). On the genre of the commentary more generally, see Jean Céard, "Les transformations du genre du commentaire," in *L'Automne de la Renaissance, 1580–1630*, ed. Jean Lafond and André Stegmann (Paris: J. Vrin, 1981), pp. 101–16.

[27] François de Dainville, *La géographie des humanistes* (Paris: Beauchesne, 1940), pp. 222–3.

[28] Edward Grant, "Aristotelianism and the Longevity of the Medieval World View," *History of Science*, 16 (1978), 93–106. The examples of *quaestiones* are taken from Albert of Saxony (ca. 1316–1390), "Questions on the Eight Books of Aristotle's *Physics*," trans. Edward Grant, in *A Source Book in Medieval Science*, ed. Edward Grant (Cambridge, Mass.: Harvard University Press, 1974), pp. 199–203, at p. 201.

Latin Aristotle commentaries [in all fields] composed between 1500 and 1650 exceeds that of the entire millennium from Boethius to Pomponazzi."[29] Of all the areas to which commentaries on Aristotle could be devoted, natural philosophy was second only to logic in the number of commentaries produced; at least one-third of all Aristotle commentators wrote on one or more aspects of natural philosophy – more than those who wrote on metaphysics, ethics, rhetoric, or politics combined.[30] Printing and the expansion of higher education no doubt account for the explosive nature of this growth of Aristotelica,[31] but these figures are eloquent testimony to the fact that Aristotle was still *the* Philosopher to print, teach, and study.

Aristotle alone came complete with interpretive formulations finely honed over centuries of debate and reflection, which adapted his philosophy to the needs and concerns of Christian orthodoxy. Only for Aristotle did there already exist a vast arsenal of pedagogical presentations and tools suitable for students at various levels on which professors could build without having to start from scratch. Finally, given its flexibility, Aristotelianism had the resources with which to respond to many of the new challenges. As a result, these challenges generated more interpretations and adaptations rather than a decline in Aristotelianism. Institutional and intellectual factors together can account for the continued vitality and increased productivity of Aristotelian natural philosophy through the first half of the seventeenth century. Aristotelianism remained the common philosophical ground of the Renaissance, the point of reference in relation to which every new philosophy had to prove its tenability.[32]

ARISTOTELIANISM AND THE INNOVATIONS OF THE RENAISSANCE

Charles Schmitt has outlined two different kinds of eclecticism, or openness to innovation, evident in Aristotelian natural philosophy. The first, already present in the Middle Ages, was an openness to new developments

---

[29] Charles Lohr, "Renaissance Latin Aristotle Commentaries: Authors A–B," *Studies in the Renaissance*, 21 (1974), 228–89, at p. 228.

[30] This estimate is based on a tally of the authors identified in Charles Lohr, "Renaissance Latin Aristotle Commentaries: Authors A–B," and its sequels in *Renaissance Quarterly*, 28 (1975), 689–741 ("C"); 29 (1976), 714–45 ("D–F"); 30 (1977), 681–741 ("G–K"); 31 (1978), 532–603 ("L–M"); 32 (1979), 529–80 ("N–Ph"); 33 (1980), 623–734 ("Pi–Sm"); and 35 (1982), 164–256 ("So–Z"). Or see Lohr, *Latin Aristotle Commentaries* (Florence: Olschki, 1998), vols. 2 and 3.

[31] Charles Schmitt estimates that 3,000–4,000 editions of Aristotelica appeared in print before 1600, in contrast with approximately 500 editions of Platonica. See Schmitt, *Aristotle and the Renaissance*, p. 14. For a bibliography of Platonica, see James Hankins, *Plato in the Renaissance*, 2 vols. (Leiden: E. J. Brill, 1990), 2: 669–796.

[32] Eckhard Kessler, "The Transformation of Aristotelianism," in *New Perspectives on Renaissance Thought: Essays in the History of Science, Education, and Philosophy*, ed. John Henry and Sarah Hutton (London: Duckworth, 1990), pp. 137–47, at p. 146.

that emerged within the tradition. The second involved a willingness to draw on sources outside that tradition and was a specific characteristic of Aristotelianism in the early modern period.[33] In the first instance, the universities of the Renaissance inherited the full range of Aristotelian positions found in the Middle Ages, displaying plenty of internal eclecticism: Thomists and Scotists were widespread throughout Europe; Italian universities were known for their Averroists; in Germany there were also Albertists; at the University of Krakow in the sixteenth century, Aristotle was "still read with the eyes of John Buridan"; and in early sixteenth-century Paris, Spanish scholars such as Juan de Celaya (1490–1558) and Luis Coronel (d. 1531) followed the calculatory tradition of fourteenth-century Oxford.[34] Not only were there disagreements between these scholastic "sects" but there were equally important disagreements within them because of the variety of ways of being a "Thomist" or an "Averroist."[35]

In addition, during the Renaissance, Aristotelian natural philosophers faced a number of new challenges, which originated outside the universities and outside the Aristotelian tradition – from the humanists and the newly recovered ancient sources they made available, from the Protestant and Catholic Reformations and their concern with making philosophy better serve religion, and from the emergence of new empirical observations and mathematical methods. The responses of Aristotelian natural philosophers ranged from the selective adoption of certain innovations to conservative defenses of received opinion.

The humanists fostered the study of a "new Aristotle" based on new, more elegant Latin translations, such as those by Leonardo Bruni (1369–1444) and Theodore Gaza (1400–1476), a new emphasis on Aristotle's ethical and political writings, and newly recovered ancient commentaries (by Themistius or Simplicius, for example). Italian humanists also revived a number of other ancient philosophical authorities, including Plato and Hermes Trismegistus, the legendary Egyptian priest; Epicurus; and the skeptic Sextus Empiricus. Although various works of Plato, including the *Timaeus*, with its account of the origins of the world, had been available in Latin in the Middle Ages, the arrival of Byzantine émigrés in fifteenth-century Italy gave a new seriousness to the study of Plato as a philosopher. Georgios Gemistos Pletho (ca. 1360–1454) was exceptional in promoting Plato with the idea of rebuilding the polytheistic paganism of ancient Greece.[36] Most humanists valued Plato

---

[33] Schmitt, *Aristotle and the Renaissance*, p. 92.
[34] See John Murdoch, "From the Medieval to the Renaissance Aristotle," *New Perspectives on Renaissance Thought*, pp. 163–76, at p. 167.
[35] Christia Mercer, "The Vitality and Importance of Early Modern Aristotelianism," in *The Rise of Modern Philosophy: The Tension between the New and Traditional Philosophies from Machiavelli to Leibniz*, ed. Tom Sorell (Oxford: Clarendon Press, 1993), pp. 33–67, at pp. 45–6.
[36] This paragraph is indebted to Jill Kraye, "The Philosophy of the Italian Renaissance," pp. 26–37. See, more generally, Hankins, *Plato in the Renaissance*.

instead as a buttress to Christianity and adduced in support of this interpretation the writings of the neo-Platonists, especially Plotinus (ca. 205–269) and Proclus (ca. 410–485). Early proponents of Plato did not necessarily attack Aristotle. Although George of Trebizond framed his *Comparatio Platonis et Aristotelis* (Comparison of Plato and Aristotle, 1458) as a preemptive defense of Aristotle against the Platonists, other humanists attempted to reconcile the two by following the Byzantine position that the two philosophers were fundamentally in agreement.[37]

The Florentine philosopher Marsilio Ficino (1433–1499) was the first to develop Platonism into a system complete enough to rival Aristotle's. Ficino composed voluminous translations of and commentaries on Plato and the Hermetic texts and offered his own synthesis of Christianity and Platonism in his *Theologia platonica* (composed around 1474 and published in 1482). He contrasted this "pious philosophy" with what he considered the impieties of scholastic Aristotelianism.[38] Defenders of Plato maintained that Plato's belief in individual immortality and in the creation of the world by a divine Demiurge made his philosophy more easily reconciled with Christianity, but critics noted the difficulties posed by Plato's belief in the transmigration of souls and by the fact that the creation described in the *Timaeus* was not a creation ex nihilo but rather from preexisting matter. Platonism found support here and there throughout the early modern period, for example among German mystics from Nicholas of Cusa (1401–1464) to Jakob Boehme (1575–1624) or from isolated individuals such as Symphorien Champier (ca. 1470–1539) in France and Leo Ebreo (ca. 1460–1523) in Portugal, down to Henry More (1614–1687) and Ralph Cudworth (1617–1688), who as fellows at Cambridge used Platonism to combat materialist interpretations of the new mechanical philosophy.[39] Only in Italian universities were a few professorships created for the teaching of Platonism alongside the usual Aristotelianism – in Pisa (1576), Ferrara (1578), and Rome (1592), the latter two having been created for Francesco Patrizi (1529–1597) in particular.[40]

The impact of Renaissance Platonism and Hermeticism on scientific developments has been much debated. Frances Yates argued that the neo-Platonist emphasis on the successive emanations from a perfect being to progressively lower orders of existence helped inspire enthusiasm for heliocentrism,

---

[37] Notably Theodore Gaza (1400–1476) and Cardinal Bessarion (1403–1472).
[38] James Hankins, "Marsilio Ficino as a Critic of Scholasticism," *Vivens Homo*, 5 (1994), 325–34.
[39] See Brian Copenhaver, *Symphorien Champier and the Reception of the Occultist Tradition in Renaissance France* (The Hague: Mouton, 1978). There is no synthetic work on the reception of Plato in the seventeenth century, but see Christia Mercer, "Humanist Platonism in 17th Century Germany," in *Humanism and Early Modern Philosophy*, ed. Jill Kraye and M. W. F. Stone (London: Routledge, 2000), pp. 238–58. Note also Mercer's critique of the term "Neoplatonist," pp. 251–2. On Henry More's natural philosophy, see Alan Gabbey, "Philosophia Cartesiana Triumphata: Henry More (1646–71)," in *Problems of Cartesianism*, ed. Thomas Lennon, John M. Nicholas, and John W. Davis (Kingston: McGill-Queen's University Press, 1982), pp. 171–249.
[40] Kraye, "The Philosophy of the Italian Renaissance," p. 47.

Natural Philosophy 375

which placed the sun at the center of vital emanations of heat and light[41] (see Copenhaver, Chapter 22, this volume). But most thinkers inspired by Platonism or Hermeticism remained hostile to Copernicanism.[42] The notable exceptions, Giordano Bruno (1548–1600) and Johannes Kepler (1571–1630), each also had other motivations for their choice of heliocentrism. Bruno embraced Copernicanism in the context of an infinitist cosmology, which he justified as a tribute to divine omnipotence, free from the standard cosmological and physical assumptions.[43] Johannes Kepler hailed Copernicanism as mathematically superior because it established the order of and distance between the planets and aided him in his goal of elucidating the geometrical or musical harmonies inherent in these relationships.[44] The impact of Platonism on Galileo has long been a matter of debate; recent work has emphasized the need to consider, in this controversy, how Platonism was understood in the Renaissance.[45] At this point, Platonism can plausibly be credited with fostering a renewed interest in geometrical-mathematical methods, which a few Italian professors of philosophy hailed as a replacement for the dry logicism of Aristotle.[46] In addition, Platonism offered one of the first viable alternatives to Aristotelian natural philosophy and helped to challenge a number of its specific assumptions, including, for example, the Aristotelian notion of the quintessence, a fifth element peculiar to the superlunary world that distinguished it from the sublunary.[47]

Other philosophical alternatives to both Aristotle and Plato were brought to light by humanist discoveries of long-lost manuscripts. Ancient atomism, for example, was first revived with Poggio Bracciolini's discovery in 1417 of a manuscript of Lucretius's *De natura rerum* (On the Nature of Things) in

---

[41] See in particular Frances Yates, *Giordano Bruno and the Hermetic Tradition* (Chicago: University of Chicago Press, 1964).
[42] This point is made forcefully in Robert Westman, "Magical Reform and Astronomical Reform: The Yates Thesis Reconsidered," in Robert Westman and J. E. McGuire, *Hermeticism and the Scientific Revolution* (Los Angeles: The William Andrews Clark Memorial Library, 1977), pp. 3–91. On the difficulties of defining "Hermeticism," see Westman, p. 70, and more generally Brian Copenhaver, "Natural Magic, Hermetism, and Occultism in Early Modern Science," in *Reappraisals of the Scientific Revolution*, ed. David Lindberg and Robert Westman (Cambridge: Cambridge University Press, 1990), pp. 261–302.
[43] See Westman, "Magical Reform," p. 30.
[44] See Judith V. Field, *Kepler's Geometrical Cosmology* (Chicago: University of Chicago Press, 1988); and Bruce Stephenson, *The Music of the Heavens: Kepler's Harmonic Astronomy* (Princeton, N.J.: Princeton University Press, 1994). Recent work on Kepler has tended to emphasize the religious origins of his ideas; see Job Kozhamthadam, *The Discovery of Kepler's Laws: The Interaction of Science, Philosophy and Religion* (Notre Dame, Ind.: University of Notre Dame Press, 1994). On Kepler's university context which was devoid of Platonism, see Charlotte Methuen, *Kepler's Tübingen: Stimulus to a Theological Mathematics* (Aldershot: Ashgate, 1998), p. 222.
[45] For an assessment, see James Hankins, "Galileo, Ficino, and Renaissance Platonism," in Kraye and Stone, eds., *Humanism and Early Modern Philosophy*, pp. 209–37.
[46] Notably Giuseppe Biancani (1566–1624) and Jacopo Mazzoni (1548–1598); see Paolo Galluzzi, "Il 'Platonismo' del tardo Cinquecento e la filosofia di Galileo," in *Ricerche sulla cultura dell'Italia moderna*, ed. Paola Zambelli (Bari: Laterza, 1973), pp. 39–79.
[47] James Hankins, "Platonism, Renaissance," in *Routledge Encyclopedia of Philosophy*, ed. Edward Craig, 10 vols. (London: Routledge, 1998), 7: 439–47, at p. 446.

the library of a Swiss monastery. The translation by Ambrogio Traversari of Diogenes Laertius's *Lives of Eminent Philosophers* (first published in 1533) gave a new currency to the opinions of many ancient figures, including Epicurus, who had long been dismissed as a mere libertine.[48] The Stoics, too, were proposed, notably by Justus Lipsius (1547–1606), as offering an alternative and more pious natural and moral philosophy than Aristotelianism.[49] The Pre-Socratics and the Pythagoreans also found appeal, especially to philosophers in the Platonic vein.

Hostility to Aristotle was especially widespread among a group of late sixteenth-century Italian philosophers often called "nature philosophers" because of their emphasis on natural philosophy (see Garber, Chapter 2, this volume).[50] Although their philosophies were innovative, they remained speculative, without empirical or mathematical components, and were stymied by the Counter-Reformation Church, which exhibited a preference for Aristotle and the Thomist synthesis after the Council of Trent (1545–1563).[51] Among the earliest of these critics of Aristotle was the Italian physician and polymath Girolamo Cardano (1501–1576), who, for example, reduced Aristotle's four elements to three by eliminating fire as an element. Despite incurring an accusation of heresy for casting a horoscope of Christ in 1570 and a scathing attack by Julius Caesar Scaliger in defense of Aristotle, Cardano acquired an international reputation for his books on natural philosophy as well as his practice of medicine and astrology.[52] Francesco Patrizi developed a more systematic new philosophy to replace Aristotelianism in his *Nova de universis philosophia* (New Philosophy of the Universe, 1591), which relied on Platonic

---

[48] See Charles L. Stinger, *Humanism and the Church Fathers: Ambrogio Traversari (1386–1439) and Christian Antiquity in the Italian Renaissance* (Albany: State University of New York Press, 1977), p. 79; see also *Ambrogio Traversari nel VI centenario della nascita*, ed. Gian Carlo Garfagnini (Florence: Leo S. Olschki, 1988).

[49] Peter Barker, "Stoic Contributions to Early Modern Science," in *Atoms, Pneuma, and Tranquility: Epicurean and Stoic Themes in European Thought*, ed. Margaret Osler (Cambridge: Cambridge University Press, 1991), pp. 135–54.

[50] On each of these authors, see Alfonso Ingegno, "The New Philosophy of Nature," in Schmitt, Skinner, and Kraye, eds., *The Cambridge History of Renaissance Philosophy*, pp. 236–63, esp. pp. 247–63; and Paul O. Kristeller, *Eight Philosophers of the Italian Renaissance* (Stanford, Calif.: Stanford University Press, 1964). More generally, on the different types of philosophers in this period, see Paul Richard Blum, *Philosophenphilosophie: und Schulphilosophie: Typen des Philosophierens in der Neuzeit* (Stuttgart: Franz Steiner Verlag, 1998).

[51] Stuart Brown, "Renaissance Philosophy Outside Italy," in Parkinson, ed., *The Renaissance and Seventeenth-Century Rationalism*, pp. 70–103, at p. 75.

[52] On the conflict with Scaliger, see especially Ian Maclean, "The Interpretation of Natural Signs: Cardano's *De Subtilitate* vs. Scaliger's *Exercitationes*," in *Occult and Scientific Mentalities in the Renaissance*, ed. Brian Vickers (Cambridge: Cambridge University Press, 1984), pp. 231–52. On Cardano's reception, see Kristian Jensen, "Cardanus and His Readers in the Sixteenth Century," and Ian Maclean, "Cardano and His Publishers, 1534–1663," in *Girolamo Cardano: Philosoph, Naturforscher, Arzt*, ed. Eckhard Kessler (Wiesbaden: Harrossowitz, 1994). On Cardano's work in astrology and medicine, see, respectively, Anthony Grafton, *Cardano's Cosmos: The Worlds and Works of a Renaissance Astrologer* (Cambridge, Mass.: Harvard University Press, 1999), and Nancy G. Siraisi, *The Clock and the Mirror: Girolamo Cardano and Renaissance Medicine* (Princeton, N.J.: Princeton University Press, 1997).

sources to portray God as an incorporeal, intellectual light, who pours forth light and heat to create the world, generating successively lower levels of being. Despite initially receiving the approval of Pope Clement VIII, the book was placed on the Index of Forbidden Books in 1594; Patrizi continued to teach Platonic philosophy, first in Ferrara and then in Rome, until his death in 1597, but at that point the papal theologian Robert Bellarmine, who as cardinal would later take a stern line against Galileo, concluded that Platonism was more dangerous to Christianity than Aristotelianism and recommended that Patrizi's chair of Platonic philosophy be suppressed.[53]

Bernardino Telesio (1509–1588) offered yet another alternative to received philosophy, rejecting Aristotelianism on the grounds that it was in conflict with the senses and with Scripture and instead explaining the natural world as the interaction between the two principles of hot and cold. In order to Christianize this revival of pre-Socratic naturalism, he introduced a universal spirit (also reminiscent of Stoic *pneuma*) that infused the world and from which he drew new definitions of time and space. Telesio's works were condemned posthumously in 1593.[54]

Tommaso Campanella (1568–1639), a disciple of Telesio, carried the idea of the world-spirit to the extreme of envisioning the whole universe as a living animal in which God was omnipresent and immanent (a position called "pansensism"). Nature was full of correspondences and divine messages that the natural philosopher could interpret, especially through astrology. Campanella was imprisoned in 1599 for fomenting rebellion in Calabria against Spanish domination there, and spent most of the next thirty years in jails; he was released in 1629 by Pope Urban VIII and practiced astral magic with him to ward off evil celestial influences. When Spain threatened to have him extradited, Campanella fled to France in 1634. He had the support there of a circle of "libertine" philosophers, who increasingly became disillusioned with his querulous demands for greater recognition.[55] Finally, Giordano Bruno (1548–1600), drawing on a wide range of sources, that included the atomist Lucretius, his own contemporary Telesio, and

---

[53] See Luigi Firpo, "The Flowering and Withering of Speculative Philosophy – Italian Philosophy and the Counter Reformation: The Condemnation of Francesco Patrizi," in *The Late Italian Renaissance, 1525–1630*, ed. Eric Cochrane (London: Macmillan, 1970), pp. 266–86, at p. 278. For more recent work on Patrizi, see Luc Deitz, "Space, Light, and Soul in Francesco Patrizi's *Nova de universis philosophia* (1591)," in Grafton and Siraisi, eds., *Natural Particulars: Nature and the Disciplines in Renaissance Europe*, pp. 139–69 and the literature cited therein.

[54] See also D. P. Walker, *Spiritual and Demonic Magic* (Notre Dame, Ind.: University of Notre Dame Press, 1975), pp. 189–92, and, for the most recent treatment, Martin Mulsow, *Frühneuzeitliche Selbsthaltung: Telesio und die Naturphilosophie der Renaissance* (Tübingen: Max Niemeyer Verlag, 1998).

[55] See John M. Headley, *Tommaso Campanella and the Transformation of the World* (Princeton, N.J.: Princeton University Press, 1997), pp. 162–3, 104 ff. On Campanella's practice of astral magic with the pope, see Walker, *Spiritual and Demonic Magic*, chap. 7. On the French reception of Campanella, see Michel-Pierre Lerner, *Tommaso Campanella en France au XVIIe siècle* (Naples: Bibliopolis, 1995).

neo-Platonists such as Plotinus and Nicholas of Cusa, suggested that all matter is infused by soul.[56] Rather than proposing a pious philosophy like that of Ficino, however, Bruno's solution was to subsume religion under a rationalistic worldview, and it was probably this naturalism rather than any particular aspect of his theories (such as Copernicanism) that led to his being burned at the stake for heresy in 1600.[57]

Although these Italian nature philosophers did not succeed in unseating Aristotle from his position of philosophical dominance and, given the persecution they faced, did not garner many followers, they did leave their contemporaries and successors with an increased awareness of the possibility of developing viable philosophical alternatives to Aristotelianism. Criticism of Aristotle on specific issues for his obscurity and internal inconsistencies became increasingly common.[58] Whereas some tried to develop an entire philosophy based on an ancient authority other than Aristotle, others combined Aristotelianism with positions borrowed from a mix of the different thinkers that had recently been rediscovered.[59] Philosophical diversity also prompted two new kinds of responses: syncretism on the one hand and skepticism on the other.

Giovanni Pico della Mirandola (1463–1494) set the standard for the syncretic position in gathering 900 theses drawn from a wide range of philosophical traditions, from the medieval Arabs to the Hermetic texts, with the idea of showing that each philosophical tradition was an incomplete manifestation of a single (Christian) truth. Although this work (the *Conclusiones*, 1486) was condemned by Pope Innocent VIII in 1488, it was widely read and cited in the Renaissance, in part for its doxography (i.e., its collection of philosophical opinions) and in part for its syncretic approach, which was perpetuated by Francesco Giorgi (1460–1540) and Agostino Steuco (1497–1548), among others.[60] By contrast, Giovanni Pico's nephew Gianfrancesco Pico

---

[56] For the most recent work, see Hilary Gatti, *Giordano Bruno and Renaissance Science* (Ithaca, N.Y.: Cornell University Press, 1999), esp. chaps. 6–8.

[57] Kraye, "The Philosophy of the Italian Renaissance," pp. 49–50. On Bruno's trial, see Luigi Firpo, *Il processo di Giordano Bruno*, ed. Diego Quaglioni (Rome: Salerno, 1993). John Bossy argues rather speculatively that Bruno actively worked to undermine the Catholic missions in England while he was there in the 1580s and wrote virulently against the papacy, but these activities, performed under a pseudonym, were not known to the Inquisitors and did not play a role in his trial; see John Bossy, *Giordano Bruno and the Embassy Affair* (New Haven, Conn.: Yale University Press, 1991), esp. pp. 179–80. For an overview of the controversies surrounding Bruno, see Michele Ciliberto, "Giordano Bruno tra mito e storia," *I Tatti Studies*, 7 (1997), 175–90.

[58] See Charles B. Schmitt, "Aristotle as a Cuttlefish: The Origin and Development of a Renaissance Image," *Studies in the Renaissance*, 12 (1965), 60–72.

[59] See, for example, the eclecticism of Jean Bodin, as discussed in Blair, *The Theater of Nature*, pp. 107–15.

[60] Replacing earlier problematic editions, there is now *Syncretism in the West: Pico's 900 Theses (1486): The Evolution of Traditional Religious and Philosophical Systems*, ed. and trans. S. A. Farmer (Tempe, Ariz.: Medieval and Renaissance Texts and Studies, 1998). See also Charles B. Schmitt, "Perennial Philosophy: From Agostino Steuco to Leibniz," *Journal of the History of Ideas*, 27 (1966), 505–32; and Wilhelm Schmidt-Biggemann, *Philosophia perennis: Historische Umrisse abendländischer Spiritualität in Antike, Mittelalter und Früher Neuzeit* (Frankfurt: Suhrkamp, 1998).

della Mirandola (1469–1533) concluded from the same variety of philosophical opinion that all philosophy is false and that only the Christian faith offers certainty. The persistent appeal of this skeptical, fideist position in the sixteenth century to authors ranging from Henricus Cornelius Agrippa von Nettesheim (1486–1535) to Michel de Montaigne (1533–1592) and Francisco Sanchez (ca. 1550–1623)[61] prompted René Descartes (1596–1650), Marin Mersenne (1588–1648), and Francis Bacon (1561–1626), among others in the early seventeenth century, to look for a more solid foundation than philosophical authority on which to ground natural knowledge.

THE IMPACT OF THE REFORMATIONS AND RELIGIOUS CONCERNS

A second challenge to received Aristotelianism stemmed from the renewal of moral and religious objections. Francesco Petrarca, or Petrarch (1304–1374), was one of the first to mock Aristotelianism as sterile and irrelevant to the more important ethical and religious concerns of life. Petrarch complained bitterly of those who attacked him because he refused to worship Aristotle as they did and instead pointed out the limitations of philosophical knowledge when compared with the rewards of religious contemplation:

> Thus we come back to what Macrobius says.... "It seems to me that there was nothing this great man [Aristotle] could not know." Just the opposite seems to me true. I would not admit that any man had knowledge of all things through human study. This is why I was torn to pieces, and ... this is what is claimed to be the reason: I do not adore Aristotle. But I have another whom to adore. He does not promise me empty and frivolous conjectures of deceitful things which are of use for nothing and not supported by any foundation. He promises me the knowledge of Himself.[62]

Petrarch thus raised the classic Christian objections to Aristotle, which had motivated the condemnation of Aristotelianism when it was first introduced in the universities in the thirteenth century.[63]

---

[61] See Richard H. Popkin, *The History of Scepticism from Erasmus to Spinoza*, rev. ed. (Berkeley: University of California Press, 1979); José R. Maia Neto, "Academic Skepticism in Early Modern Philosophy," *Journal of the History of Ideas*, 58 (1997), 199–220; Zachary Sayre Schiffman, *On the Threshold of Modernity: Relativism in the French Renaissance* (Baltimore: Johns Hopkins University Press, 1991); Francisco Sanchez, *That Nothing Is Known*, ed. Elaine Limbrick, trans. Douglas Thomson (Cambridge: Cambridge University Press, 1984); and Agrippa von Nettesheim, *Of the Vanitie and Uncertaintie of Artes and Sciences* [first published in Latin, 1526], ed. Catherine Dunn (Northridge: California State University Press, 1974).

[62] Petrarch, "On His Own Ignorance and That of Many Others," trans. Hans Nachod, in *The Renaissance Philosophy of Man*, ed. Ernst Cassirer, Paul Oskar Kristeller, and John Herman Randall, Jr., (Chicago: University of Chicago Press, 1948), p. 101.

[63] The most extensive condemnations were those of Etienne Tempier, bishop of Paris, in 1277. See J. M. M. H. Thijssen, "What Really Happened on 7 March 1277? Bishop Tempier's Condemnation and Its Institutional Context," in *Texts and Contexts in Ancient and Medieval Science: Studies on the*

Although Aristotle had so rapidly and effectively been Christianized (through the work of Thomas Aquinas, among others) that by 1325 he had become the standard philosophical authority in universities, religious objections to Aristotle became once again a powerful line of attack against his authority in the Renaissance.[64] In particular, Aristotle's discussions of the eternity of the world, the necessity of natural law, and the immortality of the soul were obviously not in agreement with Christian doctrines about the creation of the world, the possibility of miraculous exceptions to the laws of nature, and the survival and judgment of the individual soul after death. Throughout the early modern period, natural philosophers had to show how Aristotelianism or any other philosophical system they would prefer to it could be reconciled with Christian doctrines on these issues. As a result, the eternity of the world and the immortality of the soul were often standard topics in early modern natural philosophy.[65]

At the same time as the humanists leveled these strictures against scholasticism, the Church also became increasingly hostile to the scholastic separation between philosophy and theology that gave philosophers in the faculty of arts a degree of institutional and intellectual independence. Instead, at the Fifth Lateran Council (1512–1517), the Church called on philosophy to play an active role in supporting religious doctrines and launched an offensive in particular against the Averroist strand of Aristotelianism represented in many Italian universities. The Council mandated that philosophers demonstrate the immortality of the soul, whereas a number of scholastic philosophers had long since concluded that this question could not be resolved on philosophical grounds alone.

Pietro Pomponazzi (1462–1525), professor at Padua, in defending the independence of philosophy from such religious mandates, flouted the decree of the Lateran Council in his *De immortalitate animae* (On the Immortality of the Soul, 1516) in which he concluded that the soul could be shown on purely rational grounds to be mortal rather than immortal. After a papal condemnation in 1518, Pomponazzi published a *Defensorium* including orthodox proofs of the immortality of the soul and refrained from publishing his other

---

Occasion of John E. Murdoch's Seventieth Birthday, ed. Edith Sylla and Michael McVaugh (Leiden: E. J. Brill, 1997), pp. 84–105.

[64] The role of religious objections to Aristotle has been well studied for the medieval period but much less so for the early modern one. For an account of Renaissance philosophy that brings this later religious anti-Aristotelianism to the fore, see Stephen Menn, "The Intellectual Setting," in Garber and Ayers, eds., *The Cambridge History of Seventeenth-Century Philosophy*, 1: 33–86.

[65] John Murdoch concludes of the late medieval period that "in a very important way natural philosophy was not about nature." The observation applies equally well to most traditional natural philosophy in the early modern period, which perpetuated the speculative discussions typical of medieval science. See John Murdoch, "The Analytic Character of Late Medieval Learning: Natural Philosophy without Nature," in *Approaches to Nature in the Middle Ages*, ed. Lawrence D. Roberts (Binghamton, N.Y.: Center for Medieval and Renaissance Studies, 1982), pp. 171–213, at p. 174.

highly naturalistic treatments of fate and miracles.⁶⁶ Nonetheless, Paduan Aristotelians continued to be known for their commitment to naturalistic Aristotelianism. Cesare Cremonini (1550–1631), for example, did not attempt to Christianize his interpretation of Aristotle's position on the eternity of the world and denied the intervention of God in the sublunary realm; for this he was investigated by the Inquisition, though he retained his high-paying position at the University of Padua.⁶⁷ But Cremonini remained the exception. Over the course of the sixteenth century, most Aristotelian natural philosophers conformed to religious tenets or avoided questions with theological implications, leaving them to metaphysics.⁶⁸

More generally, the new awareness of the shortcomings of Aristotle even among Aristotelians led them to think of themselves as increasingly independent philosophers. For example, the German professor of philosophy Bartholomaeus Keckermann (1571–1609) distinguished the "bad Peripatetics," who were concerned only with what Aristotle said, from the good ones, like himself or the Paduan philosopher Jacopo Zabarella (1533–1589), who pursued the truth beyond what Aristotle had established.⁶⁹ Indeed, Zabarella described his goal as the pursuit of reason rather than Aristotelian authority. In his treatises on logic and method, he drew on the full range of sources available in his day, including medieval and the newly recovered ancient commentaries as well as sources outside the Aristotelian tradition.⁷⁰ Many late Aristotelians justified taking liberties with their chosen authority by reiterating in various forms a dictum first coined by Aristotle himself to explain his own independent search for truth: "amicus Plato, sed magis amica veritas" ("Plato is my friend, but truth is a greater friend").⁷¹ For example, the general of the Dominican order, Thomas de Vio, known as Cardinal Cajetan (1468–1534), preferred Thomas Aquinas as an authoritative philosopher to

---

⁶⁶ See Martin L. Pine, *Pietro Pomponazzi: Radical Philosopher of the Renaissance* (Padua: Editrice Antenore, 1986).
⁶⁷ Kraye, "The Philosophy of the Italian Renaissance," p. 42. For new documents concerning a 1604 denunciation of Cremonini on the grounds that he did not teach the immortality of the soul, see Antonino Poppi, ed., *Cremonini e Galilei inquisiti a Padova nel 1604: Nuovi documenti d'archivio* (Padua: Editrice Antenore, 1992). On his thought more generally, see Heinrich C. Kuhn, *Venetischer Aristotelismus am Ende der aristotelischen Welt: Aspekte der Welt und des Denkens des Cesare Cremonini (1550–1631)* (Frankfurt: Peter Lang, 1996); and *Cesare Cremonini (1550–1631): Il suo pensiero e il suo tempo* (Documenti e Studi, 7) (Cento: Baraldi, 1990).
⁶⁸ On the role of metaphysics in supplying philosophical demonstrations of religious tenets, see Charles Lohr, "Metaphysics," in Schmitt, Skinner, and Kraye, eds., *The Cambridge History of Renaissance Philosophy*, pp. 537–638, esp. pp. 614 ff.
⁶⁹ As discussed in Mercer, "The Vitality and Importance of Early Modern Aristotelianism," pp. 41–2.
⁷⁰ Schmitt, *Aristotle and the Renaissance*, p. 11; and, more generally, Heikki Mikkeli, *An Aristotelian Response to Renaissance Humanism: Jacopo Zabarella on the Nature of Arts and Sciences* (Helsinki: SHS, 1992). For early modern treatments of method, see Daniel A. di Liscia, Eckhard Kessler, and Charlotte Methuen, eds., *Method and Order in Renaissance Philosophy of Nature: The Aristotle Commentary Tradition* (Aldershot: Ashgate, 1997) and Neal Ward Gilbert, *Renaissance Concepts of Method* (New York: Columbia University Press, 1960).
⁷¹ See Leonardo Taran, "Amicus Plato sed magis amica veritas: From Plato to Aristotle to Cervantes," *Antike und Abendland*, 30 (1984), 93–124; and Henry Guerlac, "Amicus Plato and Other Friends," *Journal of the History of Ideas*, 39 (1978), 627–33.

the Aristotle whom Aquinas was supposedly interpreting. In response to Pomponazzi's irreligious interpretation of Aristotle, Cajetan concluded that Aristotle had deviated from the true principles of philosophy, notably on the question of the immortality of the soul.[72]

Among Protestants, the desire to be rid of the medieval legacy of scholasticism led to an initial contempt for Aristotle, most notably by Luther. After an early attempt to use lectures on Pliny and natural history as an introduction to natural philosophy, Philip Melanchthon (1497–1560) returned to Aristotelian categories and scholastic methods in devising the Lutheran curriculum.[73] Among Calvinists, there was some attempt, notably by the French theologian Lambert Daneau (1530–1595), to devise a "Christian physics" based primarily on the Bible.[74] But even Daneau strove to reconcile Aristotelian opinion with the biblical statements.[75] In the main, the Calvinist professors of philosophy at the new German universities composed Aristotelian textbooks where the usual topics were "reduced" according to Ramist principles. Lutheran and Calvinist commentators on Aristotle readily relied on and cited Catholic authorities such as Suarez or Zabarella.[76] Although the reverse was less often the case (presumably because of Catholic censorship), this cross-confessional contact is evidence of the fundamental similarities between Catholic and Protestant Aristotelianisms.

The Reformations, both Protestant and Catholic, also had an impact on the justifications of natural philosophy in bringing back to the fore a concern for Christian (and not specifically denominational) piety. Textbooks of all confessions framed natural philosophy as a pious exercise. In what he boasted was the first work of its kind, a *Compendium naturalis philosophiae* (1542), the Franciscan Frans Titelmans began with a three-page prose "psalm to the Creator, the one and triune Lord," and each of the twelve books into which

---

[72] Charles H. Lohr, "The Sixteenth-Century Transformation of the Aristotelian Natural Philosophy," in *Aristotelismus und Renaissance: In memoriam Charles Schmitt*, ed. Eckhard Kessler, Charles H. Lohr, and Walter Sparn (Wiesbaden: Harrassowitz, 1988), pp. 89–99, at pp. 90–1.

[73] See Kusukawa, *The Transformation of Natural Philosophy*, p. 175; and Charlotte Methuen, "The Teaching of Aristotle in Late Sixteenth-Century Tübingen," in *Philosophy in the Sixteenth and Seventeenth Centuries: Conversations with Aristotle*, ed. Constance Blackwell and Sachiko Kusukawa (Aldershot: Ashgate, 1999), pp. 189–205.

[74] Lambert Daneau, *Physica Christiana* (Geneva: Petrus Santandreanus, 1576); on this agenda more generally, see Ann Blair, "Mosaic Physics and the Search for a Pious Natural Philosophy in the Late Renaissance," *Isis*, 91 (2000), 32–58.

[75] On Daneau, see Olivier Fatio, *Méthode et théologie: Lambert Daneau et les débuts de la scolastique réformée* (Geneva: Droz, 1976); and Max Engammare, "Tonnerre de Dieu et 'courses d'exhalations encloses es nuées': Controverses autour de la foudre et du tonnerre au soir de la Renaissance," in *Sciences et religions de Copernic à Galilée (1540–1610)* (Rome: École Française de Rome, 1999), pp. 161–81. More generally, the latter volume contains a rich sampling of current work on science and religion in the sixteenth and early seventeenth centuries.

[76] For a detailed study of the citation patterns of a Calvinist professor, see Joseph Freedman, *European Academic Philosophy in the Late Sixteenth and Seventeenth Centuries: The Life, Significance, and Philosophy of Clemens Timpler (1563/4–1624)*, 2 vols. (Hildesheim: Georg Olms, 1988). Zabarella is the modern author most cited in Timpler's natural philosophy; see Freedman, *European Academic Philosophy*, 1: 276 and n. 180 at 2: 642.

his 400-page work was divided closed with similar psalms. This intermingling of psalmic piety with a pedagogical exposition of Aristotle would not become a lasting feature of the textbook genre, but it reveals the uneasiness of the author in presenting Aristotle "straight up," especially to the broad and inexperienced readership targeted by an introductory textbook. Through the psalms, Titelmans meant to give concrete expression to his objectives, which remained the refrain of natural philosophy textbooks and treatises for well over a century:

> I saw that the discipline of physics, if it was treated rightly and according to its dignity, was of the greatest importance to sacred Theology and for the fuller knowledge of God; and led in an admirable fashion not only to the knowledge of God, but also to excite the love of God: which two things (that is, the knowledge and the love of God) must be the final and principal end of all honorable studies.[77]

Similarly, Protestant textbooks, following the lead of Melanchthon, praised natural philosophy as an incitement to piety for revealing the benevolent providence of God. The actual practice of natural philosophy was not much affected by these reiterations, but they gave renewed prominence to natural theological arguments from design that defended the existence and worship of God against what contemporaries perceived as a threat from the rise of atheism.[78]

Given its general natural theological usefulness, natural philosophy elicited considerable agreement across confessional lines not only within Christianity but also among the Jewish minorities concentrated in Italian cities and in Central and Eastern Europe.[79] Although Jews were not often included in the natural philosophical discussions among Christians in the early modern period, the late Renaissance (ca. 1550–1620) was a period of relative openness of Jewish thinkers to Christian scientific developments.[80] In particular, David Gans (1541–1613), who lived in Prague and maintained contacts at the court of Rudolf II, notably with Tycho Brahe (1546–1601) and Johannes Kepler, tried to promote natural philosophy among his Jewish contemporaries in the

---

[77] "Videbam physicam disciplinam, si recte pro sua dignitate tractaretur, ad sacram Theologiam plurimum omnino habere momenti, et ad Dei pleniorem cognitionem: neque ad Dei tantum cognitionem, verumetiam ad Dei excitandum amorem, mirum in modum conducere: quae duo (nempe Dei cognitio et amor) omnium debent esse honestorum studiorum ultimus atque praecipuus finis." Frans Titelmans, *Compendium philosophiae naturalis* [1542] (Paris: Michael de Roigny, 1582), p. 4. On this work, see Schmitt, "The Rise of the Philosophical Textbook," pp. 795–6.

[78] For a discussion of the vexing question of the existence of atheists in this period, see Michael Hunter and David Wootton, eds., *Atheism from the Reformation to the Enlightenment* (Oxford: Clarendon Press, 1992).

[79] For a brief discussion of the irenic role of natural philosophy, a topic that warrants further study, see Blair, *Theater of Nature*, pp. 26 and 147–8.

[80] For a lucid presentation of the historiography and history of Jewish attitudes toward natural philosophy in this period, see David B. Ruderman, *Jewish Thought and Scientific Discovery in Early Modern Europe* (New Haven, Conn.: Yale University Press, 1995), esp. chap. 2 and pp. 370–1.

hope of enhancing relations between Jews and Christians. He saw in natural philosophy a theologically neutral area by the study of which Jews could improve their standing among Christians.[81] Although Gans's works were not published in his day and scientific study remained peripheral in Jewish education, rabbis such as Moses Isserles in Cracow and the Maharal (Judah Loew ben Bezalel) in Prague encouraged naturalistic pursuits and recognized natural philosophy as a legitimate sphere of knowledge separate from the sacred. In addition, the number of Jews studying medicine at Padua rose steadily from the sixteenth to the eighteenth centuries, ensuring the diffusion of a secular medical training to Jews who returned to practice medicine in their towns of origin.[82] Nevertheless, the attractiveness of the kabbalah with its very different mode of thought on the one hand and the pressures of the well-established pattern of cultural isolation in which most Jews lived on the other kept in check a wide acceptance of natural philosophy in Jewish circles.[83]

NEW OBSERVATIONS AND PRACTICES

Aristotelian natural philosophers responded to new scientific observations and practices in such areas as astronomy, natural history, or magnetism, which originated outside the universities. The development during the Renaissance of new sites of scientific practice, such as observatories, laboratories, princely courts, foreign travel, or technical schools providing instruction in navigation and other mathematical arts, generated new approaches to nature that were quite foreign to the bookish and disputatious methods of Aristotelian natural philosophers.[84] Bypassing the university's once solid monopoly on scientific discourse, in the sixteenth century autodidacts and artisans could, thanks to printing, both learn from and contribute to widely diffused discussions about nature (see Bennett, Chapter 27, this volume). For example, Niccolò Tartaglia (ca. 1499–1557), the son of a post-rider, who taught himself mathematics from the alphabet to the solution to third-degree equations, worked as a teacher of mathematics in Venice; in what was likely a bid for patronage, he dedicated to the Duke of Urbino, Francesco Maria della Rovere, a study of ballistics in which he determined the angle at which a cannon should

---

[81] Noah Efron, "Irenism and Natural Philosophy in Rudolfine Prague: The Case of David Gans," *Science in Context*, 10 (1997), 627–49. See also André Neher, *Jewish Thought and the Scientific Revolution of the Sixteenth Century: David Gans (1541–1613) and His Times*, trans. David Maisel (Oxford: Oxford University Press, 1986).

[82] Ruderman, *Jewish Thought and Scientific Discovery*, chap. 3.

[83] For an example of the interactions of kabbalah and natural philosophy, see David B. Ruderman, *Kabbalah, Magic, and Science: The Cultural Universe of a Sixteenth-Century Jewish Physician* (Cambridge, Mass.: Harvard University Press, 1988). On seventeenth-century Jewish thought more generally, see Isadore Twersky and Bernard Septimus, eds., *Jewish Thought in the Seventeenth Century* (Cambridge, Mass.: Harvard University Press, 1987); and David Ruderman and Giuseppe Veltri, eds., *Cultural Intermediaries: Jewish Intellectuals in Early Modern Italy* (Philadelphia: University Press, 2004).

[84] See, for example, E. G. R. Taylor, *The Mathematical Practitioners of Tudor and Stuart England* (Cambridge: Cambridge University Press, 1967).

be pointed to maximize its range.[85] Bernard Palissy (1510–1590), a potter employed by the French queen Catherine de Medici, articulated his pride in his artisanal knowledge of the interactions of water and clay in a vernacular dialogue in which empirically minded "pratique" consistently mocked and defeated the learned pretensions of "theorique."[86] Authors in these new modes of natural philosophical inquiry worked independently of, and often with hostility toward, the universities. Nonetheless, a few innovations developed outside the universities were selectively incorporated into university teaching.

Certainly one of the great challenges to Aristotelian natural philosophy stemmed from the accumulation of theoretical and observational innovations in astronomy. Copernicanism was discussed but almost universally dismissed in universities prior to 1640; the principal exception was a circle of scholars at the University of Wittenberg who were willing to entertain Copernicanism as a useful hypothesis in the 1560s and 1570s.[87] From the early seventeenth century on, however, the theory was gradually given more careful consideration.[88] At the University of Paris and other Catholic institutions, the Tychonic system was generally preferred until the acceptance of Cartesianism (in the 1690s at the University of Paris) or Newtonianism (e.g., in the liberalized climate in Rome in the 1740s). The papal ban on works expounding heliocentrism was finally lifted in 1757.[89] Although immune to the papal condemnation of Galileo (1633), Protestants, too, raised objections to Copernicus on physical and biblical grounds. For example, Christian Wurstisen (1544–1588) was forbidden from teaching Copernicanism at the University of Basel

---

[85] See the translation of his *Nova scientia* [1537] in *Mechanics in Sixteenth-Century Italy*, ed. and trans. Stillman Drake and I. E. Drabkin (Madison: University of Wisconsin Press, 1969).

[86] Bernard Palissy, *Discours admirable des eaux et fontaines* (Paris: M. le Jeune, 1580). On Palissy in particular and on this theme more generally, see Paolo Rossi, *Philosophy, Technology, and the Arts in the Early Modern Era*, ed. Benjmain Nelson, trans. Salvator Attanasio (New York: Harper and Row, 1970), pp. 1–4.

[87] See Robert Westman, "The Melanchthon Circle, Rheticus, and the Wittenberg Interpretation of the Copernican Theory," *Isis*, 66 (1975), 165–93.

[88] For example, one professor at Paris who had rapidly dismissed Copernicanism in a physics course of 1618–19 gave more careful consideration to the objections and possible responses to them in a later course of 1628; see Ann Blair, "The Teaching of Natural Philosophy in Early Seventeenth-Century Paris: The Case of Jean Cécile Frey," *History of Universities*, 12 (1993), 95–158, at p. 126. One can find a similar discussion of arguments and counter-arguments with no final conclusion in a physics lecture by Caspar Barlaeus at the University of Amsterdam in 1636; see Paul Dibon, *La philosophie néerlandaise au siècle d'or*, vol. 1: *L'Enseignement philosophique dans les universités à l'époque précartésienne (1575–1650)* (Amsterdam: Elsevier, 1954), p. 234.

[89] For the case of France and especially Paris, see Laurence Brockliss, "Copernicus in the University: The French Experience," in *New Perspectives on Renaissance Thought: Essays in the History of Science, Education, and Philosophy*, ed. John Henry and Sarah Hutton (London: Duckworth, 1990), pp. 190–213, esp. pp. 191–7. On other Catholic contexts, including Southern France, Italy, Spain, and Portugal, see W. G. L. Randles, *The Unmaking of the Medieval Christian Cosmos, 1500–1760: From Solid Heavens to Boundless Aether* (Aldershot: Ashgate, 1999), chaps. 7–8. On the lifting of the ban, see Pierre-Noël Mayaud, *La condamnation des livres coperniciens et sa révocation à la lumière de documents inédits* (Rome: Editrice Pontificia, 1997).

after he had begun to do so while he taught mathematics there from 1564 to 1586.⁹⁰

For Protestants and Catholics alike, to accept heliocentrism required jettisoning many fundamental tenets of Aristotelian physics and opening oneself to considerable religious objections. In particular, Aristotelian physics dictated that the earth, as the heaviest of the elements, naturally rested at the center of the universe, whereas only the celestial bodies made of the perfect fifth element could revolve in eternal circular motion. Biblical passages such as Joshua 10:12, in which Joshua asked the sun to "stand still over Gibeon" to give him more time to finish a battle, seemed a powerful objection to many – from Catholics such as Cardinal Bellarmine, who saw in Galileo's arguments no grounds for replacing the traditional interpretation of the Church Fathers, to Protestants such as Tycho Brahe, who felt that such biblical statements about philosophy should be acknowledged as authoritative and unambiguous.⁹¹

There were nonetheless innovations in astronomy that were less radical than heliocentrism itself and to which Aristotelian natural philosophy proved more permeable. These included the discovery by Tycho Brahe, from his well-equipped observatory on the Danish island of Hven, that there was no observable parallax for the comet of 1577. Brahe concluded that the comet had appeared in the highest regions of the heavens, above the sphere of the moon. Like the new star of 1572, which he had already described, the comet therefore constituted an example of change in the part of the heavens that was immutable according to the Aristotelian cosmology. Reaction among natural philosophers to this specific challenge to the Aristotelian theory of the heavens was varied. At the University of Paris, for example, one professor rejected Brahe's parallax measurement (although it certainly was the best available); another discussed comets and the arguments for and against their superlunary nature without concluding one way or the other; another allowed that there were two kinds of comets – some were sublunary, as Aristotle described, and others superlunary, such as that observed by Brahe, and of supernatural origin; and still another simply abandoned the traditional sub- and superlunary distinction in favor of a fluid heaven, following Brahe and the Stoics (see Donahue, Chapter 24, this volume).⁹²

---

⁹⁰ Wolfgang Rother, "Zur Geschichte der Basler Universitätsphilosophie im 17. Jahrhundert," *History of Universities*, 2 (1982), 153–191, at p. 169.
⁹¹ On the problem of biblical interpretation posed by Copernicanism, see Richard Blackwell, *Galileo, Bellarmine, and the Bible* (Notre Dame, Ind.: University of Notre Dame Press, 1991); and Gary B. Deason, "John Wilkins and Galileo Galilei: Copernicanism and Biblical Interpretation in the Protestant and Catholic Traditions," in *Probing the Reformed Tradition: Historical Essays in Honor of Edward A. Dowey, Jr.*, ed. Elsie Anne McKee and Brian G. Armstrong (Louisville, Ky.: Westminster/John Knox Press, 1989). See also Ann Blair, "Tycho Brahe's Critique of Copernicus and the Copernican System," *Journal of the History of Ideas*, 51 (1990), 355–77.
⁹² See Roger Ariew, "The Theory of Comets at Paris, 1600–50," *Journal of the History of Ideas*, 53 (1992), 355–72.

In various ways, Aristotelian natural philosophers could absorb the observation into their philosophical scheme without any threat to their Aristotelian allegiance. The same was true of the sunspots and the irregularities of the moon observed through the telescope by Galileo, which also violated the Aristotelian principle of the immutability of the superlunary world.[93] Thus a number of Aristotelian natural philosophers were aware of and willing to accept some recent astronomical innovations.

The Renaissance also witnessed an explosion of natural-historical knowledge prompted by voyages to the New World and by increased documentation of the flora and fauna of regions both exotic and familiar. Although Aristotle himself was a keen observer of natural particulars and composed a number of natural-historical works, natural history did not get much attention in the standard cursus of Aristotelian natural philosophy. It was rather the purview of medical doctors seeking to catalog remedies in mineral, vegetable, and animal substances. One university professor of philosophy explained in the early seventeenth century that natural history was rarely taught because its topics were not demonstrative or difficult enough to require a teacher and because there was not enough time to fit them into the philosophy curriculum.[94] Textbooks on natural philosophy generally simply enumerated the large categories of natural history (birds, quadrupeds, fish, snakes, and insects, for example) without paying any attention to the particular features of each species that would detract from the universal quality of the *scientia* of natural philosophy.[95] Authors working outside the universities, free from the time constraints of a set curriculum and generally more open to a broader range of recent work in natural history or travel accounts, often devoted more attention to natural particulars, as in Girolamo Cardano's *De subtilitate rerum* (On the Subtlety of Things, 1550) and *De rerum varietate* (On the Variety of Things, 1557) or Jean Bodin's *Universae naturae theatrum* (Theater of Universal Nature, 1596).

Nonetheless, observations from the New World and other places entered Aristotelian natural philosophy at the universities in various ways (see the following chapters in this volume: Findlen, Chapter 19; Vogel, Chapter 20). For example, all commentators acknowledged that recent experience had disproved the ancient notion that the torrid zone was uninhabitable. The Jesuit commentators at the University of Coimbra in Portugal (active 1592–8), for example, debated in their frequently reprinted commentaries on Aristotle the number of continents and the proportion of sea and land, and adduced a

---

[93] See Roger Ariew, "Galileo's Lunar Observations in the Context of Medieval Lunar Theory," *Studies in History and Philosophy of Science*, 15 (1984), 213–26.

[94] Gilbert Jacchaeus, *Institutiones physicae*, 1.4, 4th ed. (Schleusingen: Petrus Schmit, 1635), p. 13, as quoted in Reif, "Natural Philosophy in Some Early Seventeenth-Century Scholastic Textbooks," p. 66.

[95] For an example, see Wilhelm Scribonius, *Physica et sphaerica doctrina* (Frankfurt: Palthenius, 1600), p. 189 ff.

mix of ancient, medieval, and modern authors, explicitly noting the priority of experience over received authority on these issues.[96] The Jesuits were also well known for their courses on geography, which integrated the reports of faraway missionaries and tutored the future ruling elites of different nations in local geography and hydrography – hardly Aristotelian topics.[97] Perhaps in conscious emulation of the Jesuits or perhaps in response to the interests of their students, who were also destined to be officers of the new bureaucracies, university professors could also include natural historical and geographical topics that ranged well beyond the prescribed Aristotelian texts. For example, student notes extant in manuscript and published form show how one professor at Paris in the 1620s, Jean-Cecile Frey, discussed the New World in a physics course of 1618 after offering more standard commentaries on *On the Heavens* and *On Generation and Corruption*. Professors could also introduce a broad range of topics in extracurricular instruction, which was especially common in the residential colleges of Oxford or Paris. This was the most likely locus for Frey's more unusual courses – on druidic philosophy and the "admirable things of the Gauls" (covering the noteworthy natural and human features of contemporary France) or on "curious propositions about the universe," which contained a motley selection of travel lore.[98]

The Jesuits were particularly noted among Aristotelian natural philosophers for their openness to new empirical and mathematical methods. Although they did not practice experiments or the observation of specific, punctual events, the Jesuits are credited with incorporating the evidence of common experience in theorizing about natural philosophy.[99] The Jesuits harbored magneticians such as Niccolò Cabeo (1586–1650) and Athanasius Kircher (1602–1680), who adopted the experimentalism of William Gilbert (1544–1603), although the Society formally banned some of Gilbert's propositions in 1651.[100] Especially at the Collegio Romano, which trained the elite of the Jesuit intellectuals, professors such as Christoph Clavius (1537–1612) kept abreast of new developments in the mathematization of physical phenomena such as motion, which seemed an impossible crossing of disciplinary boundaries to most traditional Aristotelians.[101] (See the following chapters in this volume: Bennett, Chapter 27; Bertoloni Meli, Chapter 26.) Throughout the seventeenth century, the Jesuits, despite their continued allegiance to Aristotle and the Tychonic system, included prominent astronomers noted for their observational feats.[102]

---

[96] Dainville, *La géographie des humanistes*, pp. 25–35.
[97] Ibid., pp. 343–74.
[98] See Blair, "The Teaching of Natural Philosophy," esp. pp. 124–7.
[99] See Peter Dear, *Discipline and Experience: The Mathematical Way in the Scientific Revolution* (Chicago: University of Chicago Press, 1995), chap. 2.
[100] Stephen Pumfrey, "Neo-Aristotelianism and the Magnetic Philosophy," in Henry and Hutton, eds., *New Perspectives on Renaissance Thought*, pp. 177–89, esp. p. 184.
[101] Dear, *Discipline and Experience*, pp. 33–42.
[102] Recent historiography on the Jesuits stresses the diversity of Jesuit involvement in science, notably in *The Jesuits: Cultures, Sciences, and the Arts, 1540–1773*, ed. John W. O'Malley, Gauvin Alexander

Even at the universities, there was some penetration of the new methods. Starting in the late sixteenth century, universities throughout Europe increasingly featured botanical gardens and anatomy theaters, which were generally associated with the medical faculties, and (in the seventeenth century) observatories and chemical laboratories.[103] At Oxford and Cambridge, chairs were established in the mathematical disciplines. The Savilian chairs of geometry and astronomy were founded at Oxford in 1619 and 1621 and the Lucasian chair of mathematics at Cambridge in 1663.[104] Students' notebooks and book ownership records provide evidence of both formal and informal instruction in geography at Oxford and Cambridge.[105] Students could also engage in extracurricular scientific activities in the laboratories that friends or tutors kept in their rooms.[106] Although many a new philosopher complained that his years of study were wasted,[107] instruction in early modern universities could be quite wide-ranging and could integrate new elements of theory and practice. Official curricula, like most university statutes or the Jesuit *ratio studiorum*, generally mentioned only Aristotelian works and in some cases called explicitly for allegiance to them.[108] But official curricula do not give us a full picture of the teaching to which a student was actually exposed. Professorial treatises and student notes and commonplace books, many more of which deserve to be studied, reveal the diversity of topics that students encountered, from the private laboratories in some college rooms

---

Bailey, and Steven J. Harris (Toronto: University of Toronto Press, 1999), particularly the articles by Rivka Feldhay, Michael John Gorman, Steven J. Harris, Florence Hsia, and Marcus Hellyer. See also Steven J. Harris, "Confession-Building, Long-Distance Networks, and the Organization of Jesuit Science," *Early Science and Medicine*, 1 (1996), 287–318, and the rest of this special issue on "Jesuits and the Knowledge of Nature"; and Rivka Feldhay, "Knowledge and Salvation in Jesuit Culture," *Science in Context*, 1 (1987), 195–213.

[103] See, for example, the case of Leiden as discussed in *Leiden University in the Seventeenth Century*, ed. Th. H. Lunsing Scheurleer and G. H. M. Posthumus Meyjes (Leiden: E. J. Brill, 1975). For that of Altdorf, see Olaf Pedersen, "Tradition and Innovation," in de Ridder-Symoens, ed., *A History of the University in Europe*, pp. 451–88, at p. 473. On botanical gardens at Italian universities, see Paula Findlen, *Possessing Nature: Museums, Collecting, and Scientific Culture in Early Modern Italy* (Berkeley: University of California Press, 1994), pp. 256–61.

[104] A convenient list of English professorships in the sciences can be found in Robert Merton, *Science, Technology, and Society in Seventeenth Century England* (Atlantic Highlands, N.J.: Humanities Press, 1970), p. 29.

[105] Lesley B. Cormack, *Charting an Empire: Geography at the English Universities, 1580–1620* (Chicago: University of Chicago Press, 1997), pp. 27 ff.

[106] For a detailed study of the student experience at Oxford and Cambridge, see Mordechai Feingold, *The Mathematicians' Apprenticeship: Science, Universities, and Society in England, 1560–1640* (Cambridge: Cambridge University Press, 1984); for a general reassessment of the role of universities in the Scientific Revolution, see John Gascoigne, *Science, Politics, and Universities in Europe, 1600–1800* (Aldershot: Ashgate, 1998).

[107] See, for example, the complaints discussed in Mercer, "The Vitality and Importance of Early Modern Aristotelianism," pp. 34–8.

[108] For the text of the Jesuit injunction to follow Aristotle in natural philosophy, among other fields, see Roger Ariew, "Descartes and Scholasticism: The Intellectual Background to Descartes' Thought," in *The Cambridge Companion to Descartes*, ed. John Cottingham (Cambridge: Cambridge University Press, 1992), pp. 58–90, at pp. 64–5. For a French royal decree of 1671 calling for no other doctrine to be taught than "the one decreed by the rules and statutes of the university," see *Descartes' Meditations: Background Source Materials*, ed. Roger Ariew, John Cottingham, and Tom Sorell (Cambridge: Cambridge University Press, 1998), p. 256.

in Cambridge to the druids in Frey's extracurricular Paris instruction or the Pre-Socratics praised in the teaching of one Paduan professor in the 1640s.[109] Nonetheless, exposure to new methods and topics remained an optional extra and never took on the dominant or obligatory character of the more traditional parts of the curriculum.

RESISTANCE TO RADICAL INNOVATION

Given the diversity of opinion it embraced, early modern Aristotelian natural philosophy cannot easily be defined as a set of philosophical positions.[110] One scholar has concluded that, "probably not one of Aristotle's doctrines was held by all early modern scholastics."[111] Certainly most Aristotelian philosophers adhered to a set of core beliefs. The three principles of form, matter, and privation constitute the bedrock of Aristotle's theory of substance and change (called hylemorphism, from *hyle*, meaning matter, and *morphe*, meaning form) (see Joy, Chapter 3, this volume). Matter is passive but has the potential to become a substance when it is informed by a substantial form; form is the active principle that gives qualities to the substance and experiences change. Privation, or the absence of form, is necessary to explain the state of matter before it becomes substance, but its importance to late Aristotelians was on the wane.[112]

Most famously, Aristotelians adhered to the notion that the sublunary world consisted of four elements – air, earth, water, and fire; but late Aristotelians did not unanimously hold even this central tenet. For example, one Théophraste Bouju, royal counselor and almoner, who claimed for his vernacular coverage of quadripartite philosophy in 1614 the authority of Aristotle, nonetheless rejected fire as an element.[113] This was also the position of the Italian physician Girolamo Cardano, who in his mid-sixteenth-century works of philosophy and medicine prided himself on rejecting the received authorities, respectively Aristotle and Galen.[114] The main difference

---

[109] On Claude Bérigard (d. ca. 1663), who taught at Pisa from 1627 to 1639, and then at Padua until 1663, and other examples of Italian professors who strayed from the official curriculum, see Brendan Dooley, "Social Control and the Italian Universities: From Renaissance to Illuminismo," *Journal of Modern History*, 61 (1989), 205–39, for example, p. 229.

[110] Edward Grant, "Ways to Interpret the Terms 'Aristotelian' and 'Aristotelianism' in Medieval and Renaissance Natural Philosophy," *History of Science*, 25 (1987), 335–58.

[111] Roger Ariew, "Aristotelianism in the Seventeenth Century," in *Routledge Encyclopedia of Philosophy*, ed. Edward Craig, vol. 1 (London: Routledge, 1998), pp. 386–93, at p. 386.

[112] As discussed in more detail in Roger Ariew and Alan Gabbey, "The Scholastic Background," in Garber and Ayers, eds., *The Cambridge History of Seventeenth-Century Philosophy*, pp. 429–32.

[113] Theophraste Bouju, *Corps de toute la philosophie divisé en deux parties*, vol. 1 (Paris: Charles Chastellain, 1614), ch. 18, pp. 405–8, as cited in Roger Ariew, "Theory of Comets at Paris, 1600–1650," *Journal of the History of Ideas*, 53 (1992), 355–72, at p. 360. For Bouju's opinions on the void and place, see Ariew and Gabbey, "The Scholastic Background," pp. 436, 438. As a commentator on Aristotle who was active outside the universities, Bouju was particularly free from traditional interpretive constraints.

[114] See Siraisi, *The Clock and the Mirror*, pp. 56–8 and 138–42.

Natural Philosophy

between Bouju and Cardano resides not in their actual position rejecting fire as an element but in the way in which they couched it: Whereas Bouju proclaimed himself an Aristotelian, Cardano thought of himself and was thought of by contemporaries as an anti-Aristotelian innovator ("novator") – a term freighted, unlike today, with mostly negative connotations. Given the doctrinal flexibility of "Aristotelianism," self-definition is a key factor for the historian to consider in distinguishing an eclectic Aristotelian from a critic of Aristotle, because both might share some of the same positions despite being in opposing camps.[115]

Although Aristotelian natural philosophers in the early modern period boasted of novelties of their own and took liberties with received Aristotelian philosophy, they bristled at explicit attacks against Aristotle. From Theophrastus Bombastus von Hohenheim, called Paracelsus (1493–1541), who called for bonfires of authoritative texts at the University of Basel in 1527, to the three young philosophers at the University of Paris who advertised in 1624 a public defense of fourteen atomist theses "against Aristotle, Paracelsus, and the Cabbalists," those who publicly attacked Aristotle were rapidly condemned. Paracelsus was drummed out of Basel, and the 1624 disputation, forbidden by the Sorbonne in a ban enforced by the Parlement of Paris, never took place.[116] In both cases, the attacks on Aristotle were perceived to threaten the stability of the institutional university hierarchy and by extension of society itself. The three young philosophers provoked such a strong reaction at a time when in less formal, private venues similar challenges to Aristotle were readily discussed because they were perceived not as disinterested seekers after truth but rather as arrogant troublemakers, intentionally attracting large crowds to hear their scandalous attacks on their elders' orthodox doctrines. University and civil authorities cracked down hard, particularly because the bloody consequences of doctrinal disputes during the recent wars of religion (1562–98) were still vividly remembered.[117]

Explicit anti-Aristotelianism, especially when it threatened to strike within the university, triggered reiterations of the commitment to Aristotle that

---

[115] This is also the conclusion I reached in confronting the statements made about Aristotle with the actual reliance on Aristotelian categories in the cases of the explicitly anti-Aristotelian, but nonetheless traditionalist, Jean Bodin and the explicitly Aristotelian, but innovative, Jean-Cécile Frey; see Ann Blair, "Tradition and Innovation in Early Modern Natural Philosophy: Jean Bodin and Jean-Cécile Frey," *Perspectives on Science*, 2 (1994), 428–54.

[116] On the reception of Paracelsus, see Heinz Schott and Ilana Zinguer, eds., *Paracelsus und seine internationale Rezeption in der Frühen Neuzeit: Beiträge zur Geschichte des Paracelsismus* (Leiden: E. J. Brill, 1998).

[117] See Daniel Garber, "Defending Aristotle/Defending Society in Early Seventeenth-Century Paris," in *Wissenideale and Wissenkulturen in der Frühen Neuzeit*, ed. Claus Zittel and Wolfgang Detel (Berlin: Akademie-Verlag, 2002), pp. 135–60. Jean-Baptiste Morin vividly describes the scene of the 1624 thesis defense before setting out to refute the theses himself in *Refutation des thèses erronées d'Anthoine Villon dit le soldat philosophe et Estienne de Claves medecin chimiste* (Paris: Printed by the author, 1624).

one finds in official documents.[118] University professors were not the only ones to defend Aristotle when a wave of anti-Aristotelian works appeared in the 1620s. The Minim Marin Mersenne (1588–1648), who maintained a large international network of correspondents and both convened and attended the kinds of informal gatherings that were especially interested in new philosophies, nonetheless judged quite harshly those who wrote against Aristotle:

> [Aristotle] transcends all that is sensible and imaginable, and the others crawl on the earth like little worms: Aristotle is an Eagle in Philosophy, the others are only like chicks who want to fly before having wings.[119]

Late Aristotelian natural philosophers may have become increasingly eclectic in the positions they embraced and thought of themselves more as independent philosophers than as commentators on Aristotle, but they also became increasingly strident in their explicit allegiance to and defense of Aristotle against his detractors. This explicit allegiance was what they most unambiguously had in common.

Despite a certain presence of new empirical and mathematical methods at the universities, Aristotelian natural philosophers could not accept the call to reject received philosophy and ancient authorities as mere opinion and to build certain knowledge instead on mathematical and empirical foundations. Aristotelian natural philosophy was defined as the search for *scientia*, or certain knowledge, to be acquired through deductive causal explanation rather than empirical or mathematical description; the new philosophies of the seventeenth century would share this goal of causal explanation but proposed methods very different from the bookish cycle of philosophical discussion practiced by the Aristotelians. The philosophies of the sixteenth century, by contrast, which relied on replacing Aristotle with an alternative ancient philosopher as their champion (e.g., Plato, Epicurus, or the Stoics) by and large perpetuated the methods of Aristotelian natural philosophy. These traditional natural philosophies were overwhelmingly bookish in their sources, drawing the problems they wished to explain from authoritative texts rather than from observations of nature or experiments. Their explanations relied on dialectical argumentation rather than mathematical demonstration, and their motives were entirely speculative, with no concern for the possibility

---

[118] See, for example, the refutation of criticisms of Aristotle on points large and small in Jean-Cécile Frey's *Cribrum philosophorum* [1628] as discussed in Blair, "Teaching Natural Philosophy in Early Seventeenth-Century Paris," pp. 117–20.

[119] Marin Mersenne, *La vérité des sciences*, 1 [1625] (facsimile Stuttgart-Bad Cannstatt: Friedrich Frommann Verlag, 1969), chap. 5, pp. 109–10. On Mersenne, see Peter Dear, *Mersenne and the Learning of the Schools* (Ithaca, N. Y.: Cornell University Press, 1988). On the rather austere order of Minims, which nonetheless produced a number of important French philosophers in the seventeenth century, see P. J. S. Whitmore, *The Order of Minims in 17th-Century France* (The Hague: Martinus Nijhoff, 1967).

of practical applications. By contrast, the mechanical philosophers, who prevailed by the end of the seventeenth century and called for experiments to acquire data or confirm theory, strove for mathematical laws as the ideal expression of natural phenomena and promised (albeit often on the thinnest of grounds) practical applications for the future (see Dear, Chapter 4, this volume). After successfully weathering the threat of alternative traditional natural philosophies through the mid-seventeenth century, and issuing strident condemnations of the mechanical philosophers as late as the 1670s, Aristotelianism finally succumbed first to mechanical philosophy and then to Newtonianism.

The mechanical philosophy was flexible enough to embrace both experimentation and mathematization and a radical enough departure from Aristotelianism that attempts to reconcile the two (in philosophies dubbed "nov-antiqua") did not have broad success. Cartesianism first entered the French universities, one of the last major bastions of Aristotelianism, in the 1690s, although near Barcelona students were still producing under dictation courses in the old style, commenting on the *Physics* and *On the Heavens* through the eighteenth century.[120] After 1668, no new Latin editions of the works of Aristotle were issued until the activities of classical scholars in the nineteenth century.[121]

FORCES FOR CHANGE IN THE SEVENTEENTH CENTURY

Whereas the Aristotelians controlled the universities, the "new philosophers" relied on new kinds of institutions to develop their ideas and gain a following. These more or less formal gatherings ranged from "academies" with princely patronage to informal meetings in individual homes. Often formed on the model of literary societies, the groups that focused on scientific questions operated in the vernacular. The first may well have been the group of *curiosi* that Giambattista della Porta (1535–1615) gathered around him in Naples in the 1560s as the Accademia dei Secreti; membership was reserved for those who could contribute a new observation. The Roman Accademia dei Lincei, founded by the nobleman Federico Cesi in 1603 and famous for including Galileo among its members, and the Florentine Accademia del

---

[120] See, for example, the student notebooks that have survived from the Academia Cervariensis (or University of Cervera, which replaced the University of Barcelona from 1714 to 1821) and are preserved at the Biblioteca de Catalunya, Barcelona, such as those of Jaume Puig, "Tractatus in octo libros physicos Aristotelis," MS 1647 (1741–2); Joseph Vallesca, "Cursus aristotelicus," MS 2521 (18th century); and Josephus Osset, "Philosophiae novo-antiquae institutiones," MS 602 (1779), in which, despite the title, one still finds a defense of Aristotle, notably against the Stoics and Plato. This very late Aristotelianism, which may well also be found elsewhere, would certainly be worth charting systematically.

[121] Lohr, "Renaissance Latin Aristotle Commentaries, A–B," p. 230.

Cimento, founded in 1657, were both especially oriented toward the collection of observations and the performance of experiments.[122]

In the Holy Roman Empire, scientific societies appeared later, starting in the mid-seventeenth century, and focused especially on attempts to form a pansophic philosophy to counteract the religious and political splintering of the empire as consolidated by the Thirty Years War. The short-lived Societas Ereunetica, the Academia Naturae Curiosorum founded in Schweinfurt in 1652 (reorganized in 1677 as the Academia Leopoldina under the auspices of the emperor but with no fixed location), and the Collegium Experimentale founded in Altdorf in the 1670s promoted rosicrucianism, alchemy, and the study of mirabilia, respectively; these emphases accentuated the growing cultural divergence between Eastern and Western Europe.[123] More successful in gaining a Europe-wide audience were the *Acta eruditorum*, a learned journal founded in Leipzig in 1682, and the plan for a "Societas scientiarum" conceived by Gottfried Wilhelm Leibniz (1646–1716) among many other projects for implementing his utopian visions of international scientific and philosophical collaboration. The plan called for a society under the patronage of the elector of Brandenburg comprising members based in Berlin and correspondents reporting from elsewhere, divided into departments of physics, mathematics, German languages, and literature. Although the plan was adopted in 1700 with Leibniz as the Society's president, the Berliner Sozietät der Wissenschaften was not inaugurated until 1711 because of difficulties in securing sufficient revenues.[124] In England and France, a series of informal gatherings starting in the 1630s culminated in the foundation of the Royal Society in 1662 and the Académie Royale des Sciences in 1666, both of which emphasized the utilitarian goals of science following the ideals of Francis Bacon.[125]

Amid these various gatherings of the early seventeenth century, the multiplication of attacks on Aristotle and the sense that skepticism was a dangerous threat to be countered created an atmosphere in which it seemed that anyone could offer a "new philosophy." For example, the Bureau d'Adresse of Théophraste Renaudot held weekly discussions on philosophy from 1633

---

[122] On Della Porta, see William Eamon, *Science and the Secrets of Nature: Books of Secrets in Medieval and Early Modern Culture* (Princeton, N.J.: Princeton University Press, 1994), chap. 6. On Cesi's project, see Brendan Dooley, ed. and trans., *Italy in the Baroque: Selected Readings* (New York: Garland, 1995), pp. 23–37.

[123] For the German societies, see R. W. J. Evans, "Learned Societies in Germany in the 17th Century," *European Studies Review*, 7 (1977), 129–52; *Der Akademiegedanke im 17. und 18. Jahrhundert*, ed. Fritz Hartmann and Rudolf Vierhaus (Bremen: Jacobi Verlag, 1977), particularly R. Winau, "Zur Frühgeschichte der Academia Curiosorum," pp. 117–38; and Pedersen, "Tradition and Innovation," p. 484.

[124] See Ayval Ramati, "Harmony at a Distance: Leibniz's Scientific Academies," *Isis*, 87 (1996), 430–52, esp. pp. 449–51; and Hans-Stephan Brather, ed., *Leibniz und seine Akademie: Ausgewählte Quellen zur Geschichte der Berliner Sozietät der Wissenschaften, 1697–1716* (Berlin: Akademie Verlag, 1993).

[125] On the antecedents to these two societies, see respectively Charles Webster, *The Great Instauration: Science, Medicine, and Reform, 1626–60* (New York: Holmes and Meier, 1975); and Harcourt Brown, *Scientific Organizations in Seventeenth-Century France, 1620–80* (New York: Russell and Russell, 1967).

to 1642 in which the public was invited to participate according to rules that called for reasoned and amiable interchange on any philosophical topic, excluding politics and religion.[126] Judging from the printed record of these sessions (in which the participants remain anonymous), a wide range of questions were debated, in French, around Aristotle and the new philosophies – from traditional questions about the origins of motion, vapors, or thunder, or whether one can demonstrate the immortality of the soul, to questions of more recent origin concerning the merits of such novelties as heliocentrism and the mechanical and chemical philosophies. Richelieu, whose support made the existence of the Bureau possible, may well have initiated discussion of practical questions about navigation and how to determine longitude.[127] At a more select and less formal venue in Paris, a lecture at the home of the papal nuncio in Paris in 1628, where one Sieur de Chandoux was touting his philosophical system, Descartes rose to refute him and so impressed Cardinal Bérulle that he later enjoined Descartes to carry on the search for a new philosophy.[128] Descartes' brief from this leading figure of the French Counter-Reformation was to combat skepticism by devising a new philosophy that would be both certain, to counter the skeptics, and pious, to counter the impieties proposed in place of Aristotle.

THE ORIGINS OF THE MECHANICAL PHILOSOPHY

In the 1620s, a Europe-wide spate of anti-Aristotelian works appeared, notably by the Frenchman Sebastian Basso (fl. ca. 1560–1621), the French Oratorian Pierre Gassendi (1592–1655), the Dutchman David van Goorl (b. 1591), and the Englishman Nicholas Hill (ca. 1570–1610), all of them atomists.[129] Rather than a single philosophy, atomism loosely designates a number of different theories premised on the idea that matter is constituted of the coalescence of indivisible atoms.[130] Some atomists, such as Daniel Sennert (1572–1637), professor at the University of Wittenberg, were keen to derive the notion from Aristotle and to do so relied on passages in Averroes' commentary

---

[126] Garber, "On the Front-Lines of the Scientific Counter-Revolution."
[127] Howard M. Solomon, *Public Welfare, Science, and Propaganda in Seventeenth-Century France: The Innovations of Théophraste Renaudot* (Princeton, N.J.: Princeton University Press, 1972), pp. 72–3; and Simone Mazauric, *Savoirs et philosophie à Paris dans la première moitié du XVIIe siècle: Les Conférences du Bureau d'Adresse de Théophraste Renaudot (1633–42)* (Paris: Publications de la Sorbonne, 1997).
[128] As described in Adrien Baillet, *La vie de Monsieur Descartes* [1691], 1.14 (New York: Garland, 1987), pp. 160–5.
[129] Pintard, *Le libertinage érudit*, pp. 42–3. On Basso, see Christoph Luethy, "Thoughts and Circumstances of Sébastien Basson: Analysis, Micro-History, Questions," in *Early Science and Medicine*, 2 (1997), 1–73.
[130] See Daniel Garber, John Henry, Lynn Joy, and Alan Gabbey, "New Doctrines of Body and Its Powers, Place, and Space," in *The Cambridge History of Seventeenth-Century Philosophy*, 1: 553–623; and Stephen Clucas, "The Atomism of the Cavendish Circle: A Reappraisal," *The Seventeenth Century*, 9 (1994), 247–73, which questions the notion of "atomism" as a unified doctrine.

that discussed the existence of the smallest units of a substance, or *minima naturalia*;[131] this was the kind of atomism most often found in university contexts.[132] Others couched their theories as refutations of Aristotelian physics, grounded in an alchemical notion of "seeds" of matter (e.g. Paracelsus and Johannes Baptista van Helmont) or in Epicureanism.[133]

Gassendi proposed a full-scale revival of Epicureanism, an ancient philosophy long reviled as irreligious because of its explanations based on the chance encounters of atoms. To make Epicureanism compatible with Christianity (and even more "pious" than Aristotelianism, Gassendi claimed), he rejected the Epicurean notion of eternal uncreated atoms. Instead, he maintained that atoms were divinely created and endowed with motion by God, and he introduced into the strictly naturalistic system of Epicurus immaterial beings, including angels and rational souls, that did not jeopardize the atomic structure of material ones.[134] Gassendi directed his system against the Aristotelians, but by mid-century it became clear that the main opposition to Aristotelianism would come from another innovator, René Descartes (1596–1650). Whereas Gassendi's works were never translated from Latin, Descartes' theories were more broadly popularized, notably in French.[135] Furthermore, Descartes' followers proved skillful at adapting his original philosophy in response to objections, easing its spread into the universities.[136]

Descartes' philosophy can be seen as a kind of atomism, although Descartes differed from Gassendi on infinite divisibility, which Gassendi denied, and on the existence of the void, which Descartes denied. In the *Discours de la méthode* (Discoure on Method, 1637), Descartes described how through systematic doubt he eliminated all previous philosophical commitments as mere opinion and started from scratch to build a solid philosophy based

---

[131] E. J. Dijksterhuis, *The Mechanization of the World Picture, Pythagoras to Newton* (Princeton, N.J.: Princeton University Press, 1961), 3.5.C: "The defection from Aristotelianism," p. 282.
[132] See the discussion of Harvard University in William R. Newman, *Gehennical Fire: The Lives of George Starkey, an American Alchemist in the Scientific Revolution* (Cambridge, Mass.: Harvard University Press, 1994), esp. pp. 20–32.
[133] See Clucas, "The Atomism of the Cavendish Circle," pp. 251–2. On Paracelsus, see Walter Pagel, *Paracelsus*, 2nd ed. (Basel: Karger, 1982), p. 85.
[134] Garber, Henry, Joy, and Gabbey, "New Doctrines of Body," pp. 569–73. See Pierre Gassendi, *Exercitationes paradoxicae adversus Aristoteleos* [1624], ed. and trans. Bernard Rochot (Paris: J. Vrin, 1959). See also Margaret Osler, "Baptizing Epicurean Atomism: Pierre Gassendi on the Immortality of the Soul," in *Religion, Science, and Worldview: Essays in Honor of Richard S. Westfall*, ed. Margaret J. Osler and Paul Lawrence Farber (Cambridge: Cambridge University Press, 1985), pp. 163–84; and, more generally, Lynn Joy, *Gassendi the Atomist: Advocate of History in an Age of Science* (Cambridge: Cambridge University Press, 1987).
[135] On Descartes' authorial strategies, see Jean-Pierre Cavaillé, "'Le plus éloquent philosophe des derniers temps': Les stratégies d'auteur de René Descartes," *Annales: Histoire, Sciences Sociales*, 2 (1994), 349–67.
[136] For a more detailed comparison of the reception of the two thinkers, see Thomas Lennon, *The Battle of Gods and Giants: The Legacies of Descartes and Gassendi, 1655–1715* (Princeton, N.J.: Princeton University Press, 1993); and Laurence Brockliss, "Descartes, Gassendi, and the Reception of the Mechanical Philosophy in the French Colleges de Plein Exercice, 1640–1730," *Perspectives on Science*, 3 (1995), 450–79, and more generally this entire journal issue entitled *Descartes versus Gassendi*.

only on "clear and distinct" ideas.[137] From the existence of the thinking self (*cogito ergo sum*) Descartes established the existence of God, guarantor of the truth of the clear and distinct ideas, and, by further rational deduction, the building blocks of an entire cosmology. From the basic principles that matter is extension, that all phenomena can be explained as matter in motion, and that secondary qualities can be reduced to the primary qualities of size, shape, and motion, Descartes envisioned the world as a plenum of particles of matter of various sizes set in motion by God and self-perpetuating since then. The interaction of the particles according to various rules of impact had generated all natural phenomena – from the planets and their movement in circular vortices to the sensations of taste or smell in the body. Descartes' philosophy was designed as a complete overhaul of existing philosophies, Aristotelian and atomist. Because of his commitment to heliocentrism, he feared that his work would be placed on the Index of Forbidden Books following the condemnation of Galileo in 1633, and as a result he left his cosmological treatise *Le monde* (The World) unpublished during his lifetime.

It is not easy to explain the success of Cartesianism.[138] Like other new theories, it caught the fancy of the young especially, among whom it generated unusual enthusiasm. Christiaan Huygens (1629–1695), for example, described with some bemusement in later years how he was enthralled, at the age of fifteen or sixteen, by the novel and pleasing aspect of the vortices of particles that constituted Descartes' cosmology.[139] The enthusiastic support of the younger and more reckless of contemporary philosophers certainly did not enhance the appeal of Cartesianism to the rest of the philosophical community. At the then recently founded University of Utrecht, the bold teachings of Henricus Regius (1589–1679), which were never condoned by Descartes, prompted an official condemnation of Cartesianism in 1641.[140] Professors at Utrecht were forbidden to teach Cartesianism on the grounds that it undermined the foundations of traditional philosophy and the acquisition of the technical terms commonly used by traditional authors and because "various false and absurd opinions either follow from the new philosophy or can rashly be

---

[137] These claims are inevitably exaggerated. For a study of the legacy of scholasticism in Descartes, see Etienne Gilson, *Études sur le rôle de la pensée médiévale dans la formation du système cartésien* (Paris: J. Vrin, 1930); and Roger Ariew, *Descartes and the Last Scholastics* (Ithaca, N.Y.: Cornell University Press, 1999).

[138] The term was coined in 1662 as a pejorative by the Cambridge Platonist Henry More; see Brian Copenhaver, "The Occultist Tradition and Its Critics," in Garber and Ayers, eds., *The Cambridge History of Seventeenth-Century Philosophy*, 1: 485.

[139] Christiaan Huygens, appendix to the letter of 26 February 1693 to Bayle, in *Oeuvres*, 10.403, as quoted in Dijksterhuis, *The Mechanization of the World Picture*, p. 408.

[140] Regius maintained, for example, that a human being is not a substantial unity but only an *ens per accidens*; Descartes commented that "you could hardly say anything more offensive." See Descartes, *Oeuvres de Descartes*, ed. Charles Adam and Paul Tannery, 12 vols (Paris: Le Cerf, 1897–1910), 2: 460, as quoted in Geneviève Rodis–Lewis, "Descartes' Life and the Development of His Philosophy," in Cottingham, ed., *The Cambridge Companion to Descartes*, pp. 43 and 55. Recent scholarship has emphasized the independence of Regius from Descartes; see Paul Dibon, "Der Cartesianismus in den Niederlanden," in *Die Philosophie des 17. Jahrhunderts*, ed. Jean-Pierre Schobinger, 4 vols. (Basel: Schwabe, 1992), 2: 357–8.

deduced by the young."[141] Although Cartesianism was banned in Leiden, too, the Low Countries also harbored some of the earliest university interest in Descartes, notably at the new institution of Groningen.[142]

In France, Cartesianism was condemned by the king and the university (in 1671) after Cartesian attempts to account for the Eucharistic transformation had been condemned by the pope in 1663.[143] But outside the universities, Cartesianism inspired the friendly critique of the Jansenist Antoine Arnauld (1612–1694), which marked the beginning of the association many contemporaries saw between Cartesianism and Jansenism, an oppositional religious and political faction.[144] Descartes also inspired the occasionalism of the Oratorian Nicolas Malebranche (1638–1715) and, most effectively, the popularization efforts of Jacques Rohault (1618–1672). Rohault gave weekly public lectures in Paris expounding Descartes' physics, which even included experiments. His *Traité de physique* (1671) became the standard textbook of Cartesian physics and was used, in Latin and then in English translation, across the Continent as well as at Oxford and Cambridge.[145] Rohault's success, like that of Robert Chouet, who introduced Cartesian physics at the Academy of Geneva in 1669 without provoking controversy, rested on the strategy of minimizing the differences between Cartesianism and Aristotelian natural philosophy.[146]

The eclecticism of Cartesians willing to compromise on the points that most irked Aristotelians (among them Descartes' rejection of hylemorphism, his heliocentric cosmology, and his mechanistic interpretation of animals) certainly contributed to their success. For example, Cartesians lecturing at the university were often willing to put their views in an Aristotelian framework by organizing their discussions under such scholastic headings as "matter" and "form"; some even tried to read Cartesian views into Aristotle, claiming that earlier commentators had misunderstood him.[147] They

---

[141] As quoted in Nicholas Jolley, "The Reception of Descartes' philosophy," in *The Cambridge Companion to Descartes*, p. 395. On this "quarrel of Utrecht," see Theo Verbeek, ed., *La querelle d'Utrecht: René Descartes et Martin Schoock* (Paris: Impressions Nouvelles, 1988).

[142] Theo Verbeek, *Descartes and the Dutch: Early Reactions to Cartesian Philosophy, 1637–1650* (Carbondale: Southern Illinois University Press, 1992), pp. 85–6. For a rich history of the reception of Cartesianism and Copernicanism in the Low Countries, see Rienk Vermij, *The Calvinist Copernicans: The Reception of the New Astronomy in the Dutch Republic 1575–1750* (Amsterdam: Koninklijke Nederlanse Akademie van Wetenschappen, 2002).

[143] See Roger Ariew, "Damned If You Do: Cartesians and Censorship, 1663–1706," *Perspectives on Science*, 2 (1994), 255–74.

[144] See Tad Schmalz, "What Has Cartesianism to Do with Jansenism?" *Journal of the History of Ideas*, 60 (1999), 37–56; and Brockliss, "Descartes, Gassendi, and the Reception of the Mechanical Philosophy," p. 473.

[145] On Rohault, his successor Pierre-Sylvain Régis, and Malebranche, see P. Mouy, *Le développement de la physique cartésienne, 1646–1712* (Paris: J. Vrin, 1934), esp. pp. 108–16; on his textbook see G. Vanpaemel, "Rohault's 'Traité de physique' and the Teaching of Cartesian Physics," *Janus*, 71 (1984), 31–40.

[146] See Michael Heyd, *Between Orthodoxy and the Enlightenment: Jean-Robert Chouet and the Introduction of Cartesian Science in the Academy of Geneva* (The Hague: Martinus Nijhoff, 1982), pp. 116–17.

[147] See Roger Ariew and Marjorie Grene, "The Cartesian Destiny of Form and Matter," *Early Science and Medicine*, 2 (1997), 302–25, at pp. 321–2.

also avoided claiming heliocentrism as an unimpeachable fact and proposed it merely as a hypothesis; dropping Descartes' metaphysical underpinnings of his physics, they limited Descartes' physics of matter in motion to the inorganic world, thereby skirting the delicate question of sentient beings.[148] Furthermore, a strong institutional separation between philosophy and theology, such as existed in Geneva, enabled Cartesian physics to be adopted without provoking fears of irreligious consequences. In Germany, where there was little tradition of a separation of philosophy and theology, Cartesianism was slow to spread, despite the inroads made by Johann Clauberg (1622–1665) in Duisburg.[149] In Italy, Cartesianism appeared as part of a "mechanist syncretism" starting in Naples in the 1660s.[150]

Although Colbert rejected Cartesians as members in the early Académie Royale des Sciences on the grounds that they were excessively dogmatic, after his death in 1683 the Académie became more closely associated with Cartesianism, even more so after Malebranche and Bernard le Bovier de Fontenelle (1657–1757) became members in 1699 (and the latter became secretary of the Académie). The University of Paris followed suit, and Cartesianism became the norm there in the 1690s. The Jesuits, forbidden from teaching the new philosophy, soon found themselves a laughingstock and their classes deserted in the early eighteenth century.[151] Ironically, Aristotelianism yielded to Cartesianism in France just at the time when Descartes' cosmology had been debunked by the work of Huygens and Newton; but the French natural philosophers, loath to abandon their national champion, only cast off Descartes for Newton some fifty years later, in the 1740s.[152]

THE TRANSFORMATION OF NATURAL PHILOSOPHY BY EMPIRICAL AND MATHEMATICAL METHODS

In England, the spread of the mechanical philosophy was enhanced by contact with the philosophies of Gassendi and Descartes not only through

---

[148] Brockliss, "Descartes, Gassendi, and the Reception of the Mechanical Philosophy," p. 469. More generally, see Dennis Des Chene, *Physiologia: Natural Philosophy in Late Aristotelian and Cartesian Thought* (Ithaca, N.Y.: Cornell University Press, 1996).

[149] See Francesco Trevisani, *Descartes in Germania: La ricezione del cartesianesimo nella facoltà filosofica e medica di Duisburg (1652–1703)* (Milan: Franco Angeli, 1992), p. 13.

[150] On Cartesianism in Italy, see Claudio Manzoni, *I cartesiani italiani (1660–1760)* (Udine: La Nuova Base, 1984); Mario Agrimi, "Descartes nella Napoli di fine Seicento," in *Descartes: Il Metodo e i Saggi (Atti del Convegno per il 350 anniversario della pubblicazione del Discours de la méthode e degli Essais)*, ed. Giulia Belgioioso, Guido Cimino, Pierre Costabel and Giovanni Papuli, 2 vols. (Florence: Istituto della Enciclopedia Italiana, 1990), 2: 545–86; Giulia Belgioioso, *Cultura a Napoli e cartesianesimo: Scritti su G. Gimma, P. M. Doria, C. Cominale* (Galatina: Congedo Editore, 1992); "Philosophie aristotélicienne et mécanisme cartésien à Naples à la fin du XVIIe siècle," *Nouvelles de la République des Lettres*, 1 (1995), 19–47; and the special issue of the *Giornale critico della filosofia italiana*, 16 (1996).

[151] Brockliss, "Descartes, Gassendi and the Reception of the Mechanical Philosophy," p. 464.

[152] These transitions are described in Brockliss, *French Higher Education*.

print but also through the travel to France of émigrés during the civil war of the 1640s. Thomas Hobbes and William Cavendish, for example, returned to England enthusiastic about mechanical philosophy.[153] Although Hobbes (1588–1679) favored the rationalist methods of Descartes, most English mechanical philosophers grafted onto the basic principles of matter-in-motion practices of observation and experimentation inspired by Francis Bacon.[154] Bacon developed no philosophical system to replace Aristotle's and was never successful during his lifetime in gaining the support he sought for a reform of society through a reform of natural philosophy.[155] Nonetheless, after his death in 1626 (fittingly, so the contemporary story went, from pneumonia contracted while observing a chicken frozen in winter[156]), his work inspired natural philosophers, especially in England but also on the Continent, well into the eighteenth century. Bacon called for a collaborative pursuit of natural knowledge through the systematic observation of nature, both in its natural state and "on the rack," that is, in artificial experiments contrived to highlight otherwise hidden features. In his *Novum Organum* (New Organon, 1620), designed to replace the logical *Organon* of Aristotle, Bacon described a method for the careful derivation of generalizations from the accumulation of natural historical particulars.[157]

English mechanical philosophers such as Robert Boyle (1627–1691) and Robert Hooke (1635–1703) adhered to the principle that everything could be explained by matter in motion. But they shunned what they perceived to be the dogmatism of Descartes, with his a priori rationalist assumptions, such as his denial of the possibility of a void. Instead they favored a new experimental method that differed from the concept of "experience" current among both eclectic Aristotelians and the new philosophers of the Continent, including Galileo, Descartes, and Blaise Pascal (1623–1662). Rather than invoking "experience" unproblematically as what was commonly known to happen in nature and using it as a quick justification to arrive at general principles, the English experimentalists described with precision specific events that actually happened in nature, produced by experimental conditions designed to elicit unusual phenomena (such as the air-pump), and they were cautious

---

[153] Robert Hugh Kargon, *Atomism in England from Hariot to Newton* (Oxford: Clarendon Press, 1966), pp. 63 and 69.

[154] On the opposition of Hobbes to Boyle, see Steven Shapin and Simon Schaffer, *Leviathan and the Air-Pump: Hobbes, Boyle, and the Experimental Life* (Princeton, N.J.: Princeton University Press, 1985).

[155] See Julian Martin, *Francis Bacon, the State, and the Reform of Natural Philosophy* (Cambridge: Cambridge University Press, 1991).

[156] For the contemporary accounts and a reinterpretation of Bacon's death, see Lisa Jardine and Alan Steward, *Hostage to Fortune: The Troubled Life of Francis Bacon* (London: Victor Gollancz, 1998), pp. 502–11.

[157] For an entry into the vast literature on Bacon, see Antonio Pérez-Ramos, *Francis Bacon's Idea of Science and the Maker's Knowledge Tradition* (Oxford: Clarendon Press, 1988); and Markku Peltonen, ed., *The Cambridge Companion to Bacon* (Cambridge: Cambridge University Press, 1996).

about offering causal explanations of the observed phenomena.[158] Rejecting explanations that attributed moral qualities to nature (such as the "fear of the void" associated with Aristotle), Boyle introduced qualities that he attributed to the particles of matter, such as the springiness of air particles (later interpreted as the discovery of "Boyle's law"), although he could not explain springiness itself in terms of the shape and size of the particles.[159] Boyle remained mindful of the limits of human ability to understand all the reasons of nature and was satisfied, like most English experimentalists, with what he considered to be probable rather than certain knowledge resulting from experimental investigations.[160]

In a parallel, more mathematical tradition, Continental natural philosophers such as Galileo and his followers pursued mathematics as the key to certainty in natural philosophy. Although Galileo probably did perform inclined-plane experiments, he often idealized "experience" as what would happen in nature under perfect conditions (e.g., in freefall without air resistance).[161] The new physics of motion expressed in mathematical laws that he developed inspired the modifications by Christiaan Huygens and was a prerequisite to Newton's synthesis of the new physics with the new astronomy.[162] It also marked the end of the traditional distinction between physics as the science of real bodies and mathematics as the study of abstract and unreal entities. This separation had already been eroded in some circles by the study of "mixed mathematical" disciplines such as optics or astronomy. In a separate strand of mathematization, Kepler had discovered three laws of mathematical correlations in the planetary motions; his method was grounded in the conviction that God had created the universe according to "number, weight and measure" and therefore according to mathematical laws, and in the painstaking attention to empirical precision with which he manipulated the data collected by Tycho Brahe.[163]

Both Galileo and Kepler carried out much of their innovative mathematical and observational work under the auspices of princely patronage.

---

[158] For a useful rapid presentation of the problem of "experience," see Steven Shapin, *The Scientific Revolution* (Chicago: University of Chicago Press, 1996), pp. 80–96.

[159] On the gap between mechanical explanations and experimental results, see Christoph Meinel, "Early Seventeenth-Century Atomism: Theory, Epistemology, and the Insufficiency of Experiment," *Isis*, 79 (1988), 68–103.

[160] See Jan Wojcik, *Robert Boyle and the Limits of Reason* (Cambridge: Cambridge University Press, 1997); Rose-Mary Sargent, *The Diffident Naturalist: Robert Boyle and the Philosophy of Experiment* (Chicago: University of Chicago Press, 1995); Steven Shapin, *A Social History of Truth: Civility and Science in Seventeenth-Century England* (Chicago: University of Chicago Press, 1994), chaps. 4 and 7.

[161] For an introduction to the large historiography on Galileo's methods, see Stillman Drake, *Galileo at Work: His Scientific Biography* (Chicago: University of Chicago Press, 1978); and William Shea, *Galileo's Intellectual Revolution* (London: Macmillan, 1972). On Galileo's followers, see Michael Segre, *In the Wake of Galileo* (New Brunswick, N.J.: Rutgers University Press, 1991).

[162] See I. B. Cohen, *The Birth of a New Physics*, rev. ed. (New York: W. W. Norton, 1985).

[163] See references cited in note 44 and Bruce Stephenson, *Kepler's Physical Astronomy* (New York: Springer, 1987).

They began their careers teaching mathematics at the university or equivalent institutions, Galileo first at Pisa and then at Padua from 1592 to 1610, and Kepler at the Protestant seminary in Graz from 1594 to 1600, In 1600, Kepler began as Tycho Brahe's assistant at the court of the Holy Roman Emperor Rudolf II in Prague and became imperial mathematician upon Tycho's death in 1601.[164] In 1610, Galileo was named court mathematician to Cosimo II de' Medici, Grand Duke of Tuscany in Florence.[165] In moving to positions at court Galileo and Kepler were freed from the constraints of often low-paid teaching and the traditional notions of what should be taught. Both worked to support heliocentrism at a time when Copernicus's theory was considered by others as at best a useful computational tool; they challenged the traditional distinctions and hierarchy between the disciplines in using mathematics to address physical questions about the nature of motion or the cosmos.[166] Similarly, those who contributed most to the development of mechanical philosophy relied mostly on new institutions such as the Royal Society. Although he settled in Oxford in 1656, Robert Boyle was an independently wealthy gentleman with no connection to the University. Isaac Newton held the Lucasian professorship of mathematics from 1669 to 1701, but even before he left Cambridge in 1696 for an appointment as warden of the Mint in London, his teaching elicited almost no notice from students or contemporaries.[167] Instead, Newton sent his first major piece of work, the reflecting telescope, to the Royal Society in 1671; he was elected a Fellow in 1672 and then President of the Royal Society in 1703. Despite contentious relations with various other Fellows, most notably the curator of experiments, Robert Hooke (1635–1703), the Royal Society constituted his primary scientific audience.

Newton puzzled many contemporaries by offering mathematical laws but no causal explanations in his *Principia mathematica philosophiae naturalis* (Mathematical Principles of Natural Philosophy, 1687); he counted on the certainty of mathematics to forestall the disputatiousness that he so disliked among natural philosophers. But this strategy nonetheless embroiled him in controversy: Leibniz, among others, accused him of reintroducing "occult qualities" (see Copenhaver, Chapter 22, this volume) because, although his theory of gravitation provided a single powerful explanation for the tides, the

---

[164] On this environment, see R. W. J. Evans, *Rudolf II and His World: A Study in Intellectual History, 1576–1612* (Oxford: Clarendon Press, 1973).

[165] See Mario Biagioli, *Galileo, Courtier: The Practice of Science in the Culture of Absolutism* (Chicago: University of Chicago Press, 1993).

[166] On the significance of this shift, see Robert Westman, "The Astronomer's Role in the Sixteenth Century: A Preliminary Study," *History of Science*, 18 (1980), 105–47.

[167] See Richard Westfall, *Never at Rest: A Biography of Isaac Newton* (Cambridge: Cambridge University Press, 1980), p. 210. For a general introduction to Newton studies, see John Fauvel, Raymond Flood, Michael Shortland, and Robin Wilson, eds., *Let Newton Be!* (Oxford: Oxford University Press, 1988).

motions of the moon and the planets, and projectile motion, Newton gave no causal account for gravitation itself, concluding, in his "General scholium":

> I frame no hypothesis; for whatever is not deduced from the phenomena is to be called an hypothesis; and hypotheses, whether metaphysical or physical, whether of occult qualities or mechanical, have no place in experimental philosophy.[168]

Newton moderated this stance somewhat in editions of his *Opticks* after the addition in 1706 of queries 25–31, which contained speculations about the nature of light and attraction, among other topics.

In addition to his publications on mathematical and physical subjects, Newton remained concerned with a full range of traditional topics, as is evident from his abundant theological and alchemical writings left in manuscript.[169] Although he definitively transformed physics into its modern form as a technical mathematical discipline, Newton has been described as the last of the Renaissance natural philosophers. His diverse interests were all part of a quest to understand the workings of God in the world – for example, in nature through the motions of the planets that God regulates and sustains, and in history through the fulfillment of biblical prophecies.[170] One of the ways in which early modern natural philosophy differs from the various "sciences" that later replaced it is that natural philosophy was unified by its search for a better understanding of God – of divine creation (in natural historical disciplines) and divine laws (in the mathematized disciplines).[171]

THE SOCIAL CONVENTIONS OF THE NEW NATURAL PHILOSOPHY

By the late seventeenth century, the Royal Society of London and the Paris Académie Royale des Sciences played leading roles in defining the practices of natural philosophy that were increasingly imitated throughout Europe and in reforming the universities. The Baconian ideal influenced both institutions, as they pursued in different ways a collaborative model of natural philosophy with utilitarian ambitions.[172] For lack of the royal patronage it

---

[168] Isaac Newton, "General Scholium," in *Newton: Texts, Backgrounds, Commentaries*, ed. I. Bernard Cohen and Richard Westfall (New York: W. W. Norton, 1995), pp. 339–42, at p. 342.
[169] See Betty Jo Teeter Dobbs, *The Janus Faces of Genius: The Role of Alchemy in Newton's Thought* (Cambridge: Cambridge University Press, 1991); and James E. Force and Richard Popkin, eds., *Essays on the Context, Nature, and Influence of Isaac Newton's Theology* (Dordrecht: Kluwer, 1990).
[170] Betty Jo Dobbs, "Newton as Final Cause and First Mover," *Isis*, 85 (1994), 633–43.
[171] See Andrew Cunningham, "How the *Principia* Got Its Name; or, Taking Natural Philosophy Seriously," *History of Science*, 29 (1991), 377–92, at p. 384.
[172] See Robin Briggs, "The Académie Royale des Sciences and the Pursuit of Utility," *Past and Present*, 131 (1991), 38–88; on Bacon's influence in France earlier in the century, see Michèle LeDoeuff, "Bacon chez les grands au siècle de Louis XIII," in *Francis Bacon: Terminologia e fortuna nel XVII*

had hoped for, the Royal Society was financed by its members, who paid an annual subscription and actively recruited the eminent to enhance its standing. Far-flung members who never attended meetings could contribute observations by correspondence, but the day-to-day activities of the Society were dominated by a core group of less than twenty Fellows.[173] The Académie Royale des Sciences was more tightly hierarchized (into descending ranks of *honoraires, pensionnaires, associés,* and *élèves*) and at its core comprised an elite of twenty-two members selected first by Colbert and later by the members in session; they received an annual stipend as officers of the king and performed various specific tasks, such as the administration of patents and prizes.[174] Despite these different formats, both the Royal Society and the Académie Royale hoped to mobilize natural philosophers to undertake collective natural histories in order to promote the material welfare of society.

Colbert assigned the Académie task of drawing up an inventory of machines in the country, and although models of machines and volumes of careful illustrations were collected, the project was never completed. The Académie then undertook a vast history of plants, instigated by Claude Perrault (1613–1688) and directed by various members in turn; but, although some results were published, the project was never realized according to its initial ambitions because of a lack of funding and personal rivalries as well as intellectual disagreements, notably concerning the appropriate balance between description and illustration on the one hand and causal explanation and chemical analysis on the other.[175] The sessions were closed, but the proceedings were published in the *Mémoires*. At the Royal Society, an active core of members attended its meetings and discussed the results of experiments performed by the curator of instruments. Without a specific agenda, the collaborative accumulation of results was realized in the wide range of material covered in the *Philosophical Transactions*; these developed a distinctive rhetoric to describe experiments to the members who could not attend and enlist their support as "virtual witnesses" to the phenomena.[176] The model of the natural philosopher as a gentleman, as epitomized for example by Robert Boyle, emphasized civility of conversation over passionate debate as the ideal form of interaction and encouraged members of the Royal Society to present

secolo, ed. Marta Fattori (Seminario Internazionale, Rome, 11–13 March 1984) (Rome: Edizioni dell'Ateneo, 1984), pp. 155–78.

[173] Michael Hunter, *The Royal Society and Its Fellows, 1660–1700: The Morphology of an Early Scientific Institution* (Oxford: British Society for the History of Science, 1994), chap. 1.

[174] On the hierarchical structure, see Pedersen, "Tradition and Innovation," p. 484. See Roger Hahn, *The Anatomy of a Scientific Institution: The Paris Academy of Sciences, 1666–1803* (Berkeley: University of California Press, 1971); and David J. Sturdy, *Science and Social Status: The Members of the Académie des Sciences, 1666–1750* (Woodbridge: Boydell Press, 1995).

[175] Alice Stroup, *A Company of Scientists: Botany, Patronage, and Community at the Seventeenth-Century Parisian Royal Academy of Sciences* (Berkeley: University of California Press, 1990).

[176] Peter Dear, "*Totius in Verba*: Rhetoric and Authority in the Early Royal Society," *Isis,* 76 (1985), 145–61.

their findings as modest observations of fact with only cautious references to theoretical claims.[177]

Both the Royal Society and the Académie Royale des Sciences explicitly banned religious and political discussions and dogmatism of any kind. (Jesuits and Cartesians were both barred from the Académie by Colbert for that reason.) The Royal Society and the Académie Royale conferred on natural philosophy a new institutional and intellectual autonomy. In these settings, the review and agreement of respected peers constituted the criterion of acceptability instead of adherence to preestablished conclusions set by church or state.[178] Furthermore, the disputatiousness for which traditional natural philosophy had become notorious was considered a vice best avoided in the new environment of the academies. Although results fell short of expectations and perhaps the first functionally useful item to stem from the Baconian research program was Benjamin Franklin's lightning rod (1750), both the Royal Society and the Académie Royale des Sciences successfully propagated the idea that science could be useful to state and society.[179]

CONCLUSION

The evolution of natural philosophy between 1500 and 1700 can be traced in a nutshell in encyclopedic reference works. Gregor Reisch's *Margarita philosophica* (1503), two short books on the principles and the origins of natural things, summarized Aristotle's *Physics* and sketched his *Meteorology* and natural histories (with additional material drawn from Pliny). Natural philosophy appeared as a largely static field covered by ancient authorities. A century later, the *Encyclopedia* (1630) of Johann Heinrich Alsted (1588–1638) crystallized many of the developments of the Renaissance. The eight parts in which physics was divided featured Aristotelian notions (principles, elements, and meteorological theories), enhanced with new, often modern, and even anti-Aristotelian authorities. Alsted coined new terms and lent credence to new subfields; most were traditional topics (mictologia, phythologia, empsychologia, and therologia[180]), but they also included "physiognomia," which incorporated Paracelsian signatures and neo-Platonic correspondences.

---

[177] Shapin, *A Social History of Truth*, pp. 308–9.
[178] These peer groups were formed differently, however. Whereas the Fellows of the Royal Society elected their own new members, the Paris Académie remained administered, and its members selected, by one of the ministers of the king of France. Contrast Hunter, *The Royal Society and Its Fellows*, p. 10, and Hahn, *The Anatomy of a Scientific Institution*, p. 59.
[179] For the lightning rod, see I. Bernard Cohen, *Revolution in Science* (Cambridge, Mass.: Belknap Press, 1985), p. 325. More generally, on the legitimation of science after Newton, see Larry Stewart, *The Rise of Public Science: Rhetoric, Technology, and Natural Philosophy in Newtonian Britain, 1660–1750* (Cambridge: Cambridge University Press, 1992).
[180] Mictologia is the science of metals; phytologia of plants; empsychologia of soul; and therologia of wild animals. See Johann Heinrich Alsted, *Encyclopedia*, vol. 2 (Herborn, 1630; facsimile Stuttgart-Bad Cannstatt: Frommann-Holzboog, 1989).

Each part concluded with a peroration vaunting the contribution of that field to piety and the greater glory of God. Aristotle still set the framework for physics, but new authorities, a new conception of independent philosophizing, and a renewed concern for religious piety motivated a work of synthesis that was so eclectic and inclusive as to verge on incoherence.[181]

Less than a century after Alsted, John Harris's *Lexicon technicum* (1708–10) still used the same terms and elements of definition: "Physicks or natural philosophy is the speculative knowledge of all natural bodies . . . and of their proper natures, constitutions, powers and operations." But the means of achieving the understanding of nature no longer bore any relation to Aristotle's physics. Instead, electricity, effluvia, elasticity, magnetism, and light were the recurring themes; the authorities cited centered around Newton, Edmund Halley, Nehemiah Grew, and Boyle. Traditional philosophies had not disappeared completely from memory but were assigned a place in a historical/hierarchical classification that made clear the superiority of the new mechanical natural philosophy. First came the Pythagoreans and Platonists, who relied on symbols, and next the Peripatetics with their toolbox of principles, qualities, and attractions, whose "physicks is a kind of metaphysics." The experimental philosophers, dominated by the chemists, made many discoveries but fell into theories and hypotheses. The last group were "the mechanical philosophers who explicate all the phenomena of nature by matter and motion . . . by effluvia and subtle particles etc. . . . by the known and established laws of motion and mechanicks: And these are, in conjunction with the [experimental philosophers] the only true philosophers."[182] In Harris's time natural philosophy remained a largely speculative search for a causal understanding of the regularities of nature, as Aristotle had defined it, but the forces for change, which accelerated the transformation of Aristotelianism during the Renaissance, had unleashed in the seventeenth century a radical restructuring of the discipline around new premises, new practices, and new institutions.

---

[181] On Alsted's eclecticism and in particular his tendency to juxtapose without integrating conflicting views, see Howard Hotson, *Johann Heinrich Alsted (1588–1638): Between Renaissance, Reformation, and Universal Reform* (Oxford: Clarendon Press, 2000), pp. 223–4.

[182] John Harris, *Lexicon technicum* (London: Dan Brown [and 9 others], 1704), "physiology." Harris acknowledges that his source here is John Keill, *Introductio ad veram physicam* (Oxford: Sheldonian Theatre, at the expense of Thomas Bennet, 1702).

18

MEDICINE

Harold J. Cook

If one looks at changes in perceptions of nature through the eyes of physicians, several fundamental themes stand out when considering the sixteenth and seventeenth centuries. Physicians were a highly literate group who expressed themselves on paper while also exhibiting great sensitivity to changes in both the science and the art of their discipline. They were educated in one of the three higher university faculties that awarded a doctorate (the others being law and theology). When those holding the medical doctorate (M.D.) referred to themselves as physicians, they were associating themselves with the study of nature, because the word for nature is Latinized from the Greek word for nature, *physis*, like our modern "physics" (see Blair, Chapter 17, this volume). Most universities had therefore accepted "physic" as one of the three higher faculties because of the argument that the science of physic was worthy of academic study even if the art of medicine was not.[1] As Aristotle had put it, insofar as physic was a science, it "does not theorize about the individual but the class of phenomena."[2] Moreover, as Aristotle had also made clear, the rigorous generalizations of science were related to causal reasoning: That is, in its scientific aspects, physic offered not only generalizable but also causal explanations. It was the certainty of causal natural explanation that made physic a science.

By the end of the seventeenth century, however, the science of physic had been fundamentally altered. When the eminent physician Samuel Garth (1661–1719) addressed his colleagues in honor of the famous William Harvey (1578–1657) and spoke about their common profession of "physick," he revealed a view quite different from Aristotle's. Garth began not with a discussion of the causes of things but with notable examples of new cures that

---

[1] Faye Marie Getz, "The Faculty of Medicine before 1500," in *The History of the University of Oxford* (Oxford: Clarendon Press, 1984–2000), vol. 2: *Late Medieval Oxford* (1992), ed. J. I. Catto and Ralph Evans, pp. 373–405.
[2] Artistotle, *Rhetoric*, 1.2 (1356b).

physicians could perform, whether they understood the reasons or not. When he came to broaden his discourse to include the places where certainty could be had in the knowledge of nature, he spoke not of generalizable reasoning but of investigation into particulars: He spoke of botany (the "shapes and tastes of an infinite number of plants"), mineralogy, and all other aspects of the description of nature. Physic "pursues nature through a thousand windings and meanders; the very Center of the Earth escapes it not, neither is there any thing in the Ocean hid from it." This was a godly art. It encompassed not only a complete knowledge of nature but the active hunt for new and deeper knowledge of it.[3] The physicians often tried to apply their knowledge to the preservation of health and remediation of disease, but physic itself rested on an active investigation of nature and nature's secrets. The natural philosophical foundations of physic had become less like a learned debate about the causes of things, as in the early sixteenth century, and more like an active and close description of the material world. The seat of the godly art had become not a mountainous height but an oceanic depth. Not only did this affect physic and physicians, physicians helped to effect the change.

THE SCIENCE OF PHYSIC

At the beginning of the sixteenth century, the science of physic consisted of both *theoria* and *practica*. Both rested on the exercise of reason rather than the art of treating disease. So said the great eleventh-century *Canon* of Avicenna (Ibn Sina), which remained an especially fundamental and widely taught summary of physic.[4] Avicenna explained that the science of physic had both theoretical and practical parts in the same way philosophy did. But because few people considered the parallels with philosophy, they often got the wrong idea about the "practice" of physic:

> When we say that practice proceeds from theory, we do not mean that there is one division of medicine by which we know, and another, distinct therefrom, by which we act – as many examining this problem suppose. We mean instead that these two aspects are both sciences – but one dealing with the basic problems of knowledge, the other with the mode of operation of these principles. The former is theory; the latter is practice.[5]

In this important passage, Avicenna continued to argue (in line with common philosophical opinion) that physic's *theoria* offers certainty because it is based on fully accepted general principles, whereas *practica*, which deals with

---

[3] Frank H. Ellis, "Garth's Harveian Oration," *Journal of the History of Medicine*, 18 (1963), 8–19.
[4] Nancy Siraisi, *Avicenna in Renaissance Italy* (Princeton, N.J.: Princeton University Press, 1987).
[5] Avicenna, *Canon*, trans. O. Cameron Gruner, modified and annotated by Michael McVaugh, in *A Source Book in Medieval Science*, ed. Edward Grant (Cambridge, Mass.: Harvard University Press, 1974), p. 716.

particulars, offers only opinion or judgment. Both are nevertheless based on reason rather than action. The practice of physic, then, concerned the ability to move intellectually from knowledge rooted in certainty to opinion based upon that certainty: to move from the universal to the particular. The physician did this on the basis of his reasoned knowledge of nature rather than his clinical experience, in the same way that philosophers dealt with theory and practice. The physician could "practice" his science even if he never treated patients.

The academic study that Avicenna had in mind made its first goal that of giving learned advice on maintaining health and prolonging life. Ideally, the physician advised on how to regulate a person's life so that his or her unique constitution or temperament remained in harmony with the world. The principles of physic were therefore intended for use by the healthy as well as the sick.[6] As a consequence, learned physic was associated with "regimen" (the Latin word for control, guidance, rule, or governance) and throughout the early modern period was commonly referred to as "preventive" or "dietetic" medicine (after the Greek word for a way of life, *diaita*). Thus, the study of physic entailed far more than knowing how to treat disease. It allowed the physician to weigh the unique individuality of any patient's temperament and circumstances against the regularities of nature, to give a causal explanation of what was right or wrong that was tailored to the individual and couched in the language of eminent authorities, to prognosticate the future course of a person's health, and to preserve and prolong life by recommending a proper manner of living.

Although many texts might be used to study and teach the theory and practice of physic, by the early modern period it had become common to formulate overarching summaries presented in five parts (which by the seventeenth century were called the medical "institutes," like the institutes of Roman law and Christian religion). First came the basic principles of nature: the elements, temperaments, humors, spirits, parts of the body, faculties, and actions. A very elementary outline of this first part of the medical institutes might begin with the common knowledge that one could divide the causes of things into four kinds: material, formal, efficient, and final (see Joy, Chapter 3, this volume). A bowl, for example, might be discussed in terms of its material composition, its shape, its maker, and its purpose. Although all bowls might share certain characteristics, each one was different. To analyze human physiology, which was far more complicated, one began with the particularities of each person. The physician could conclude with a discussion of the patient's

---

[6] Vivian Nutton, "Les exercices et la santé: Hieronymus Mercurialis et la gymnastique médicale," in *Le corps à la Renaissance: Actes du XXXe Colloque de Tours 1987*, ed. Jean Céard, Marie-Madeleine Fontane, and Jean-Claude Marselu (Paris: Aux amateurs de livres, 1990), pp. 295–308; and Gianna Pomata, *Contracting a Cure: Patients, Healers, and the Law in Early Modern Bologna*, trans. Gianna Pomata, with the assistance of Rosemarie Foy and Anna Taraboletti-Segre (Baltimore: Johns Hopkins University Press, 1998), esp. pp. 135–9.

ancestors and purpose in life but would begin with a judgment about the material and formal causes as combined in his or her temperament or constitution. The body would be analyzed in terms of the "naturals" (the parts that made up the body, each of which in turn had its purposes and the vital abilities, or faculties, to carry out its purposes) and the contranaturals (the things that worked against the ordinary function of the body and its parts, such as inherited defects). Among the naturals were the four humors, each of which combined two of the four qualities: Yellow bile was composed of the dry and hot, black bile of the dry and cold, phlegm of the wet and cold, and blood of the wet and hot. Although a learned account of one's temperament would have to account for all of one's makeup, people often oversimplified by merely referring to the predominant humor: One was choleric, melancholic, phlegmatic, or sanguine. Hair color, body type, dominant behavior, and mood all showed one's inner nature to the outward world. Students of physic also studied astrology, together with the relevant mathematics, to help them determine someone's nature, and basic works such as Aristotle's *Meteorologica*, which described the processes of natural cycles, including putrefaction, were heavily used for understanding causes.[7]

Also in this first part of the institutes would be an account of the six nonnaturals, things outside the body that could affect it, tipping it into illness or keeping it healthy: air, food and drink, exertion and rest, sleep and waking, retentions and evacuations, and passions of the mind.[8] If the proper course of life were followed for anyone's particular temperament, disease would be averted and health maintained; if one of the nonnaturals caused one's natural temperament to become unnatural (such as by causing a naturally sanguine person to become too cold or dry through improper food or drink), then illness would surely ensue.

After a discussion of the elements of nature and how they combined to make up human physiology, the second part of the institutes added diagnosis (the causes and semiotics of diseases), the third prognosis, and the fourth hygiene (the preservation of health and prevention of disease). Last came treatment (how to restore the body to health). Except for the last of the five, which was often difficult to reason out as a consequence of the four others, mastering the five parts of physic required knowledge of nature and its causes.

---

[7] See, for example, Michael MacDonald, *Mystical Bedlam: Madness, Anxiety, and Healing in Seventeenth-Century England* (Cambridge: Cambridge University Press, 1981); Sachiko Kusukawa, "*Aspecto divinorum operum*: Melanchthon and Astrology for Lutheran Medics," in *Medicine and the Reformation*, ed. Ole Grell and Andrew Cunningham (London: Routledge, 1993), pp. 33–56.

[8] The term "nonnaturals" (*res non naturales*) first appeared during the period of Latin translation from Arabic medical texts (roughly the thirteenth century), although the concept can be found in Galen's *Technē iatrikē*. See L. J. Rather, "The 'Six Things Non-Natural': A Note on the Origin and Fate of a Doctrine and a Phrase," *Clio Medica*, 3 (1968), 337–47. Also see Jerome J. Bylebyl, "Galen on the Non-Natural Causes of Variation in the Pulse," *Bulletin of the History of Medicine*, 45 (1971), 482–5; P. H. Niebyl, "The Non-Naturals," *Bulletin of the History of Medicine*, 45 (1971), 486–92; and Nancy G. Siraisi, *The Clock and the Mirror: Girolamo Cardano and Renaissance Medicine* (Princeton, N.J.: Princeton University Press, 1997), pp. 70–90.

It should therefore come as no surprise that at the beginning of the sixteenth century, the fundamentals of physic were everywhere taught according to the same methods as all other academic subjects: by commentary on authoritative texts and by disputation (see the following chapters in this volume: Blair, Chapter 17; Grafton, Chapter 10). Students might have a knowledge of medical practice – evidence of this knowledge might even be required before they could earn the M.D. – but they learned how to treat patients in the usual way, by apprenticing with a practitioner or in a hospital, which was a personal arrangement not ordinarily provided by the university. The science of physic developed the academic knowledge and reason of the students, yielding certainty in the investigation of the causes of things. And because professors taught by commentary and disputation, no one could master the science of medicine by reading books alone: One had to attend a medical faculty and become familiar with the customs of verbal inquiry and debate. As a matter of course, the rules of university discourse and disputation also affected the structure of medical texts.[9] The few who completed the medical doctorate had marked themselves as learned men who were able to explain the true causes of things as well as to give advice and treatments based upon that understanding.

By the beginning of the sixteenth century, the emergence of medical humanism added additional methods of investigation to those of reading, lecturing, and disputing. Like many of their contemporaries, medical humanists were convinced that by restoring the purity of ancient texts through the difficult practice of philology, true and potent wisdom could be recaptured. Unlike earlier humanists, who wished to revive Roman culture, medical humanists tried to recover the most authentic records of the sources of all philosophical medicine, especially those of Greece. These Hellenists took an active part in the collecting, editing, and printing of texts. They edited and translated Aristotle,[10] Plato, and other philosophers. But a flood of new and often far better editions of medical authors such as Dioscorides and Galen, and soon Hippocratic texts, were also among those flowing from famous humanist presses such as that of the Venetian Aldus Manutius.[11]

---

[9] Roger K. French, "Berengario da Carpi and the Use of Commentary in Anatomical Teaching," in *The Medical Renaissance of the Sixteenth Century*, ed. Andrew Wear, Roger K. French, and Ian M. Lonie (Cambridge: Cambridge University Press, 1985), pp. 42–74, 296–8; and Roger French, *William Harvey's Natural Philosophy* (Cambridge: Cambridge University Press, 1994), pp. 58–70.

[10] Charles B. Schmitt, "Aristotle Among the Physicians," in Wear et al., eds., *The Medical Renaissance of the Sixteenth Century*, pp. 1–15, 271–9.

[11] R. J. Durling, "A Chronological Census of Renaissance Editions and Translations of Galen," *Journal of the Warburg and Courtauld Institutes*, 24 (1961), 230–305; Durling, "Leonhard Fuchs and His Commentaries on Galen," *Medizinhistorisches Journal*, 24 (1989), 42–7; Francis Maddison, Margaret Pelling, and Charles Webster, eds., *Essays on the Life and Works of Thomas Linacre, c. 1460–1524* (Oxford: Clarendon Press, 1977); Ian M. Lonie, "The 'Paris Hippocratic': Teaching and Research in Paris in the Second Half of the Sixteenth Century," in Wear et al., eds., *The Medical Renaissance of the Sixteenth Century*, pp. 155–74, 318–26; and Vivian Nutton, "Hippocrates in the Renaissance,"

Medical Hellenists studied many subjects related to the foundations of the science of physic.[12] Some, such as the famous Hermetic and neo-Platonic philosopher and physician Marsilio Ficino (1433–1499), embarked upon a search for the keys to unlock the secrets of nature by using techniques of natural magic to absorb and control the spiritual realms. Ficino and others struggled to recover a preclassical knowledge of astrology, amulets, talismans, rituals, and music in order to heal the soul and body as well as interpret the orphic characters in which a true knowledge of the universe was contained. Others adopted a more mathematical stance. For instance, Reiner Gemma Frisius (1508–1555) of Louvain published a new edition of Peter Apian's *Cosmography* in 1529, defended the Copernican system, and trained a generation of outstanding astronomers, mathematicians, surveyors, and mapmakers. Within three years of receiving his medical doctorate in 1536, he had been appointed to a chair of medicine, which he held until his death. There he continued to teach the new cosmography and astronomy as well as medicine.[13] His Italian contemporary, Girolamo Cardano (1501–1576), also became famous for contributing to a similar range of subjects.[14] Still others grappled with discrepancies among texts that began to appear with the proliferation of new editions of ancient writings. For example, Niccolò Leoniceno (1428–1524) became embroiled in a controversy about Pliny's *Historia naturalis*, arguing that even the best editions of that encyclopedia of nature contained medical errors attributable to Pliny himself, which one could see by comparing it with Dioscorides' materia medica.[15]

When it came to recommendations for treatment, the recovery of classical and Hellenistic texts also made it clear that the "Arab" medicine adapted by the medieval Latins differed from Hellenic recommendations. For instance, Pierre Brissot (1478–1522), working in Paris during an epidemic of pleurisy in 1514, rejected contemporary doctrines about venesection in favor of what he considered to be Galen's methods, bleeding patients copiously and without

---

in *Die Hippokratischen Epidemien: Theorie – Praxis – Tradition*, ed. Gerhard Baader and Rolf Winau (*Sudhoffs Archiv*, Beiheft 27) (Stuttgart: Franz Steiner Verlag, 1989), pp. 420–39.

[12] D. P. Walker, *Spiritual and Demonic Magic from Ficino to Campanella* (London: Warburg Institute, 1956); James J. Bono, *The Word of God and the Languages of Man: Interpreting Nature in Early Modern Science and Medicine*, vol. 1: *Ficino to Descartes* (Madison: University of Wisconsin Press, 1995); and William Eamon, *Science and the Secrets of Nature: Books of Secrets in Medieval and Early Modern Culture* (Princeton, N.J.: Princeton University Press, 1994), pp. 91–266.

[13] George Kish, *Medicina, Mensura, Mathematica: The Life and Works of Gemma Frisius, 1508–1555* (Minneapolis: Publication of the Associates of the James Ford Bell Collection, 1967).

[14] Siraisi, *The Clock and the Mirror*.

[15] Vivian Nutton, "The Perils of Patriotism: Pliny and Roman Medicine," in *Science in the Early Roman Empire: Pliny the Elder, His Sources and Influence*, ed. R. French and F. Greenaway (Totowa, N.J.: Barnes and Noble, 1986), pp. 30–58; Roger K. French, "Pliny and Renaissance Medicine," in French and Greenaway, eds., *Science in the Early Roman Empire*, pp. 252–81; Steven Eardley, "Italian Miscellanies and the Refabrication of Rome: Humanist Collections and the Cult of Antiquity," Ph.D. dissertation, University of Wisconsin–Madison, 1998, pp. 194–209; and Anthony Grafton, *Bring Out Your Dead: The Past as Revelation* (Cambridge, Mass.: Harvard University Press, 2001), pp. 2–10.

regard to whether the phlebotomy was performed on the left or right side. This caused a stir among learned physicians, who were roused still further when Brissot took his methods to Portugal and in 1518 again had success with his treatments, provoking polemical attacks. When Brissot's reply gained posthumous publication in 1525, the issue was joined, helping to divide humanists from "Arabists."[16] Physicians also searched ancient texts for authentic and potent remedies known to the Greeks but since forgotten or corrupted. At first centered on the close study of newly recovered texts, the search for classical simples (individual, uncompounded substances, mostly herbal) soon extended to garden and field as well. For instance, the powerful preservative and antidote known as "theriac" had been recommended by Galen and others, but some of the ancient recipes for theriac compounded up to eighty-one simples, not all of which were clearly identified. The intensive search for the authentic ancient remedies yielded results that helped to bring about "a quiet revolution" in medical botany, as will be discussed further (see Findlen, Chapter 19, this volume).[17]

By the 1530s, more than one medical humanist was suggesting that although restoring the purity of ancient medical texts was necessary and solved many pressing questions, perhaps not only Pliny but even the wisest Greek physicians had sometimes erred. Among the most famous subjects about which doubts began to be expressed regarding the knowledge of the ancients was anatomy. Professors of physic agreed with those among the ancients who had argued that the understanding of health and disease was rooted in an understanding of the ordinary course of nature in the body (physiology) and that this in turn could best be understood in relation to a study of the parts of the body (anatomy). Anatomy literally meant a "cutting open" of the body, which clearly required a form of manual operation. From the late Middle Ages onward, beginning in Italy (perhaps because of the prevailing view that the soul leaves the body quickly after death), public anatomies of human bodies took place to demonstrate the parts of the body to magistrates, students of physic and theology, and interested onlookers.[18] It seems

---

[16] John B. de C. M. Saunders and Charles Donald O'Malley, eds., *Andreas Vesalius Bruxellensis, The Bloodletting Letter of 1539: An Annotated Translation and Study of the Evolution of Vesalius's Scientific Development* (New York: Henry Schuman, 1947).

[17] Quotation from Richard Palmer, "Pharmacy in the Republic of Venice," in Wear et al., eds., *The Medical Renaissance of the Sixteenth Century*, p. 110; also see Palmer, "Medical Botany in Northern Italy in the Renaissance," *Journal of the Royal Society of Medicine*, 78 (1985), 149–57; Gilbert Watson, *Theriac and Mithridatium: A Study in Therapeutics* (London: Wellcome Historical Medical Library, 1966); Jay Tribby, "Cooking (with) Clio and Cleo: Eloquence and Experiment in Seventeenth-Century Florence," *Journal of the History of Ideas*, 52 (1991), 417–39; and Paula Findlen, *Possessing Nature: Museums, Collecting, and Scientific Culture in Early Modern Italy* (Berkeley: University of California Press, 1994), pp. 241–56. For an example of the search for one authentic ingredient of theriac, see Clifford M. Foust, *Rhubarb: The Wondrous Drug* (Princeton, N.J.: Princeton University Press, 1992).

[18] Katharine Park, "The Criminal and the Saintly Body: Autopsy and Dissection in Renaissance Italy," *Renaissance Quarterly*, 47 (1994), 1–33; Park, "The Life of the Corpse: Division and Dissection in

that ordinarily on these grand occasions a professor read and commented on an anatomical text while someone else, such as a surgeon, actually did the cutting. This hardly challenged the view that better anatomical knowledge might be gathered from acquiring the philological skills necessary for decoding Greek and other ancient texts. But matching words to the parts of the body was no easy task. To solve many riddles, then, professors and students – as well as artists such as Leonardo da Vinci (1452–1519) – surreptitiously undertook private anatomies in which they put their own hands to the work. When Andreas Vesalius (1514–1564) explained how to dissect bodies, he wrote not only about how to perform a public anatomy but about the importance of private dissections, "which are undertaken very frequently."[19]

Vesalius was the most notable of the many who were widening their search for anatomical knowledge and in the process questioning whether authorities such as Galen had been entirely correct. Having studied at Louvain (where he first began to investigate anatomy on his own) and Paris (where he continued his anatomical work among some of the foremost medical Hellenists of the day), he took up a position as a teacher of anatomy and surgery at Padua in 1537. There he developed the studies that led to the publication of his famous *De humani corporis fabrica* (On the Fabric of the Human Body) of 1543. Vesalius's book presented evidence that some parts of the human body, such as the net-like covering of the brain, were imaginary, and that others, such as the septum in the heart, had been described incorrectly. Galen had done his dissections on animals rather than human bodies, Vesalius concluded, and therefore had gotten many things wrong. Vesalius criticized texts on the basis of his own experience, and he shared his studies with others by innovative methods of presentation: The *De fabrica* is famous for its close integration of text and illustration, the latter executed by some of the most skilled artists and woodblock cutters of the time. It also detailed the steps by which Vesalius performed his anatomies, and it described and illustrated the instruments he used so that others could do the same.[20]

Vesalius claimed greater authority than close readers of ancient texts or even the antique authors themselves, all because he had seen with his own eyes. The connection between this appeal to the eyes and anatomy is still found in modern languages, as in the English "autopsy," which had become commonly applied to postmortem anatomies by the mid-seventeenth century.[21]

Late Medieval Europe," *Journal of the History of Medicine*, 50 (1995), 111–32; and Roger K. French, *Dissection and Vivisection in the European Renaissance* (Aldershot: Ashgate, 1999).

[19] Charles D. O'Malley, *Andreas Vesalius of Brussels, 1514–1564* (Berkeley: University of California Press, 1964), p. 343; and Andrea Carlino, *Books of the Body: Anatomical Ritual and Renaissance Learning* (Chicago: University of Chicago Press, 1999).

[20] O'Malley, *Andreas Vesalius*; Vivian Nutton, "'Prisci dissectionum professores': Greek Texts and Renaissance Anatomists," in *The Uses of Greek and Latin: Historical Essays*, ed. A. C. Dionisotti, Anthony Grafton, and Jill Kraye (London: Warburg Institute, 1988), pp. 111–26; and Glenn Harcourt, "Andreas Vesalius and the Anatomy of Antique Sculpture," *Representations*, 17 (1987), 28–61.

[21] *Oxford English Dictionary*, 2nd ed., rev. (Oxford: Oxford University Press, 1991), s.v. "autopsy."

The Latin *autopsia* was derived from the Greek word for an eyewitness, *autoptēs*, and applied to a category of rhetoric in which one appealed to authority based on being present at an event. The appeal inevitably used the first person singular: "I saw" or a similar construction. The "autoptic imagination" that has been noticed from the mid-sixteenth century onward in the works of those who wrote of the New World had been present even longer among those who wrote of their medical experiences and observations.[22] It was, moreover, in the works of people such as the physician, mathematician, and natural philosopher Cardano that the boundary between "impersonal academic or scientific discussion and personal history" was breached "regularly and persistently."[23] Cardano also famously published one of the first autobiographies, the *Liber de vita propria* (Book of My Own Life), in 1575.

For Vesalius and other physicians, then, anatomy was one of the deepest expressions of their search for knowledge of self through witnessing the body. The famous orphic dictum to "know thyself" (*nosce te ipsum*) had long been closely associated with physic as well as moral philosophy. In the act of opening bodies of the dead, the anatomist transgressed many common sensibilities in order to gain knowledge of the human frame – his own as well as that of others. At the same time that they conveyed a sense of wonder about the magnificent plan of God and nature in granting humans such a body, anatomy lessons also conveyed potent messages about mortality and death. Public anatomies were therefore often attended by important dignitaries as well as students of religion and medicine.[24] When not in use, anatomy theaters might become places for the contemplation of this world and the next by other means. Decorated with human skins and skeletons, often posed to symbolize vices and virtues or famous people (such as Adam and Eve) or holding banners with Latin mottos that reminded the reader to prepare for the return to dust, displaying the instruments of dissection and hung with pictures about the world and the heavens, anatomy theaters such as the University of Leiden's conveyed cold lessons even in summer.[25]

Attempts to ground moral and physical knowledge on the close study of natural phenomena rather than first principles became ever more common over the course of the sixteenth century. The medical faculties at the universities of Padua, Montpellier, and Leiden became especially renowned for giving greater attention to the particulars of nature in their teaching than had been usual. In addition to Aristotelians such as the radical natural

---

[22] Anthony Pagden, *European Encounters with the New World* (New Haven, Conn.: Yale University Press, 1993), pp. 51–87.
[23] Siraisi, *The Clock and the Mirror*, p. 9.
[24] Giovanna Ferrari, "Public Anatomy Lessons and the Carnival: The Anatomy Theatre of Bologna," *Past and Present*, 117 (1987), 50–106; and Jonathan Sawday, *The Body Emblazoned: Dissection and the Human Body in Renaissance Culture* (London: Routledge, 1995).
[25] Th. H. Lunsingh Scheurleer, "Un amphithéâtre d'anatomie moralisée," in *Leiden University in the Seventeenth Century: An Exchange of Learning*, ed. Th. H. Lunsingh Scheurleer and G. H. M. Posthumus Meyjes (Leiden: Universitaire Pers Leiden/E. J. Brill, 1975), pp. 217–77.

philosopher and physician Pietro Pomponazzi (1462–1525) and anatomists such as Vesalius, Padua employed Giambattista da Monte (1498–1552), who is famous for being the first to teach clinical medicine in a university setting, beginning about 1543. Padua and Pisa also established physic gardens for their medical faculties in the mid-1540s, with other universities following suit.[26] The first professor of simples at Bologna and first director of the botanical garden at Pisa, Luca Ghini, developed methods for pressing and drying plants on paper for the purposes of identification (the making of herbaria), making botanical study something that could be done all year long.[27] When the University of Leiden was established in the late 1570s, it was explicitly modeled after Padua, with a botanical garden, an anatomy theater, and even, a few decades later, an arrangement with a local hospital to allow bedside clinical teaching. The earliest members of the Leiden medical faculty were educated at Padua and Montpellier, with many becoming renowned naturalists in their own right. Not surprisingly, then, a great many of the most important botanists and natural historians of the sixteenth century were physicians. For example, Guillaume Rondelet (1507–1566), a regius professor of medicine at Montpellier from 1545 until his death, was an intensely active anatomist and botanist, an author of several medical works, and famous for his massive book on fish, which remained authoritative for over one hundred years. He inspired his students and other scholars to take up natural history investigations as an essential part of medical study.

In works of academic physicians, there is thus clear evidence of learned authors increasingly resorting to close investigations of nature in which establishing precise physical details became of central importance.[28] The classical and medieval presumption that the science of physic flowed from the certainty of first principles was being undermined in favor of the description of particulars.

NEW WORLDS, NEW DISEASES, NEW REMEDIES

But academic physic had to adapt to changes arising from outside the halls of the universities as well as from within. The activity of putting learning into practice, whether for prevention or treatment, brought learned physicians into close and sometimes strained conversation with many other people. Although physicians practiced among and upon the poor as well as others,

---

[26] Jerome J. Bylebyl, "The School of Padua: Humanistic Medicine in the Sixteenth Century," in *Health, Medicine, and Mortality in the Sixteenth Century*, ed. Charles Webster (Cambridge: Cambridge University Press, 1979), pp. 335–70.
[27] Karen Meier Reeds, *Botany in Medieval and Renaissance Universities* (New York: Garland, 1991), esp. pp. 35–6.
[28] Andrew Wear, "William Harvey and the 'Way of the Anatomists,'" *History of Science*, 21 (1983), 223–49.

they lived mainly among educated urbanites: elite and well-to-do craftsmen, merchants, lawyers, magistrates, gentry, nobility, courtiers, high-ranking military and naval officers, princes of the church, lawyers and judges, and their wives, daughters, and female companions. The upper reaches of this civil society in which the physicians moved demanded intense sociability and assumed dominion. Servants, the occasional domestic slave, laborers, shopkeepers, and other ordinary people did as ordered and stood aside offering marks of respect upon their passing – or were supposed to. But physicians circulated among the various social strata of this civic world more than most.

Like many other urbane and educated people, physicians and other well-placed urban practitioners had complicated but mainly disparaging attitudes toward the popular medical views of their day, rooted as they were in the life of the mass of the population. In the face-to-face world of small rural villages and households, most knowledge was acquired from family and neighbors, an often poorly educated minister and schoolmaster, or a lord, lady, or retainer (see Eamon, Chapter 8, this volume). Rural people of all ranks often shared pastimes and outlooks. Plants, animals, and places were thought to have special powers, or to be inhabited by particular spirits, which could do good or ill. Matter associated with the bodies of executed criminals, childbirth, bodily excrements, or insects and vermin might also hold great power; so did objects and persons associated with the Christian sacraments. Particular people (herb-wives, seventh sons, or posthumous children, for example) might know more than others about how to use or control such powers, producing events such as welcome rain or harmful hail, personal health and the love of others, or ill will and even death. The rural world therefore was not only full of helpful people and resources but also contained a plenitude of spiritual and bodily dangers. The urban, educated, and well-to-do tended to see these "dark corners of the land" as being populated by ignorant bumpkins, superstitious peasants, or worse. The clergymen, judicial inquisitors, and physicians who investigated rural doings sometimes interpreted them as originating in dealings with the devil, who was often thought to play an active role in nature.[29] Yet physicians were also among the first to attribute witchcraft to delusional illnesses.[30]

---

[29] See, for example, Keith Thomas, *Religion and the Decline of Magic* (New York: Charles Scribner's Sons, 1971); Peter Burke, *Popular Culture in Early Modern Europe* (London: Temple Smith, 1978); Carlo Ginzburg, *The Night Battles: Witchcraft and Agrarian Cults in the Sixteenth and Seventeenth Centuries*, trans. John Tedeschi and Anne Tedeschi (Baltimore: Johns Hopkins University Press, 1983); and Brian P. Levack, *The Witch-Hunt in Early Modern Europe* (London: Longmans, 1987).

[30] See Johann Weyer's famous *De praestigiis daemonum* [1563], trans. John Shea (Binghamton, N.Y.: Medieval and Renaissance Texts and Studies, 1991), in which Weyer argued that most cases of witchcraft were due to natural causes, but in those cases where it was caused by demonic forces, the physician ought to stand aside for the minister. For an account of Cesalpino's more forceful argument that the physician and natural causes ought to take precedence over clerics and the supernatural, see Mark Edward Clark and Kirk M. Summers, "Hippocratic Medicine and Aristotelian Science in the *Daemonum investigatio peripatetica* of Andrea Cesalpino," *Bulletin of the History of Medicine*, 69 (1995), 527–41.

But if both dangers and benefits for the urbane lurked in the rural undergrowth, so, too, could they be found in the thickets of urban milieux. Many itinerant practitioners moved through neighborhoods and markets, whether cataract couchers, hernia repairers, bonesetters, or the makers of secret medicines, selling their skills or commodities for a fee. Given the increasing availability of printers and the larger number of literate urbanites, many practitioners made themselves known not only directly, through oral pitches to crowds, but indirectly, through broadsides and handbills and, after their mid-seventeenth-century development, newspaper advertisements. Such people – often called mountebanks or quacksalvers, or simply "quacks" – were quite common and could travel from Hungary to the British Isles and back on their medical peregrinations. Quacks were familiar enough to figure often in the drama and literature of the period. For instance, the main character in H. J. C. von Grimmelshausen's picaresque novel *Simplicius Simplicissimus* from about 1668 is a boy pulled away from his rural home by the horrors of the Thirty Years War who grows up having various adventures; this Simplicissimus for a time turned quack, tricking simple townspeople into thinking that his electuaries and other remedies had great and healthful powers. By this means, he was able to buy food and a horse, earning "much money on the way, and so [he] came safely to the German border."[31]

Some of the remedies of these quacks may have been more honest and sincere than those of Simplicissimus. The noted French surgeon Ambroise Paré (1510–1590), for example, obtained the recipe for a salve for gunshot wounds from a surgeon of Turin who had become known for his healing balm.[32] Many recipes both new and old turned up in the "Sachliteratur" (handbooks of useful information) and so-called books of secrets.[33] The Platter family of Basel, which produced several generations of noted medical professors, retained close connections with its rural relatives, who commonly used traditional remedies.[34] Members of civil society themselves, both men and women, kept medical recipe books and traded information about remedies by letter. Robert Boyle (1627–1691) was only one of the most notable to publish parts of his collection as a help to ordinary people.[35]

---

[31] H. J. C. von Grimmelshausen, *The Adventurous Simplicissimus*, trans. A. T. S. Goodrick (Lincoln: University of Nebraska, 1962), esp. pp. 252–4.

[32] Ambroise Paré, *The Apologie and Treatise*, ed. and trans. Geoffrey Keynes [1951] (New York: Dover, 1968), p. 24.

[33] Eamon, *Science and the Secrets of Nature*.

[34] Emmanuel Le Roy Ladurie, *The Beggar and the Professor: A Sixteenth-Century Family Saga*, trans. Arthur Goldhammer (Chicago: University of Chicago Press, 1997).

[35] Robert Boyle, *Of the Reconcileableness of Specifick Medicines to the Corpuscular Philosophy* (London: Printed for Samuel Smith, 1685); and Boyle, *Medicinal Experiments* (London: Samuel Smith, 1692–3). Also see Barbara Beigun Kaplan, *'Divulging of Useful Truths in Physick': The Medical Agenda of Robert Boyle* (Baltimore: Johns Hopkins University Press, 1993).

Learned physicians often objected strenuously to the practices of the unlearned, or "empirics."[36] But more challenging to learned physic than the secret remedies of the Simplicissimuses of the world were the surgeons, apothecaries, and other literate practitioners without medical doctorates who might be found among the urban elite. Many large towns and cities in Europe were governed in part or wholly by members of the guilds or other corporations, and as members of guilds, apothecaries and surgeons were included among the politically empowered. Some became quite wealthy as well. Apothecaries had begun as wholesale importers of spices and other merchantable wares, and might therefore be found not only in their own guilds but among the members of grocers or other trading companies. As retailing became more common in sixteenth- and seventeenth-century towns, many of the apothecaries kept shops from which they dispensed medicines of all kinds. Despite prohibitions to the contrary, in places where they were numerous they often gave out medical advice as well, even going into homes to do so.[37]

Surgeons were already breaking off from companies of barber-surgeons by the sixteenth century. The barber-surgeons used instruments to cut hair, shave beards, or open veins for the purpose of bleeding. Surgeons proper, often elite members of the barber-surgeon guilds, not only opened veins but treated broken bones, hernias, cataracts, ulcerated or pustulous skin, and diseases that produced noticeable troubles to the exterior of the body. They might also amputate limbs, remove cancerous growths, cut for the removal of stones in the bladder, repair fistulas, replace noses with skin grafts, or make prostheses. They, too, often treated illnesses more generally. The surgeons and their apprentices also supplied most of the medical assistance aboard ship and in the military, serving state and commerce. In Italy and a few other places, such as Montpellier and Leiden, surgery was included among the academic subjects taught, and throughout Europe surgeons were increasingly involved in learned medicine, even in promoting medical humanism.[38]

The diversity of urban medical practitioners and practices often upset physicians. They not only wished to protect themselves from having their patients poached but also argued that public health and safety demanded order. Legally chartered bodies of physicians therefore frequently attempted to prohibit or regulate the practices of others. They did so in cooperation with the civil authorities, who were mostly municipal magistrates but sometimes princely or even royal officers. Physicians often gained the juridical

---

[36] The contemporary literature was voluminous; for an early vernacular example, see Laurent Joubert, *Erreurs populaires au fait de la médecine et régime de santé* (Bordeaux: S. Millanges, 1579).
[37] In England, the "General Practitioners" arose from the surgeon-apothecaries, although they were more closely controlled elsewhere. See Irvine Loudon, *Medical Care and the General Practitioner, 1750–1850* (Oxford: Oxford University Press, 1986).
[38] Vivian Nutton, "Humanist Surgery," in Wear et al., eds., *The Medical Renaissance of the Sixteenth Century*, pp. 75–99, 298–303; and Mary C. Erler, "The First English Printing of Galen: The Formation of the Company of Barber-Surgeons," *Huntington Library Quarterly*, 48 (1985), 159–71.

power to inspect apothecary shops, examine the apprentices of surgeons or apothecaries before they could become freemen of their guilds, prohibit the practices of those who did not have their license, and even punish for bad practice. In return for assisting the physicians, the civil authorities received the cooperation of the physicians in developing plague orders, caring for the poor, and policing the health conditions of the cities.[39]

For their part, the many opponents of the physicians claimed that physic was an often useless academic exercise that was of no value when it came to curing people of diseases. Even if physicians knew something about the diseases described by the ancients, their opponents argued, new and powerful diseases had come into being. An often-cited case in point was syphilis: Many came to believe that it was a new disease introduced from New Spain by the crew of Columbus.[40] But other new diseases were present as well, such as the sweating sickness that attacked the regions around the Baltic and North Sea in the mid-sixteenth century.[41] The accompanying argument that new kinds of practices and practitioners should be allowed to deal with the new diseases carried some weight. For example, in the case of syphilis, lesions appeared on the skin, which caused surgeons to claim a right to treat the disease.

More powerful, perhaps, was the common opinion that God had created a remedy for each disease, if only humankind had the ingenuity to find it. The doctrine of signatures was one example of this belief: The liver-shaped leaves of liverwort, for example, indicated its efficacy in cases of liver disease. In a more general way, the idea that syphilis originated in the New World reinforced the belief in the usefulness of a remedy, guaiac wood, said to be used by the local people to counter its effects. Guaiac wood became an important commodity imported from New Spain.[42] Other new remedies, too, soon gained strong footholds in the European market. Among the most renowned were tobacco, chocolate, sarsaparilla, sassafras, and chinchona bark from the New World, and coffee, tea, camphor, opium, and others from the East

---

[39] Andrew W. Russell, ed., *The Town and State Physician in Europe from the Middle Ages to the Enlightenment* (Wolfenbüttel: Herzog August Bibliothek, 1981); Katharine Park, *Doctors and Medicine in Early Renaissance Florence* (Princeton, N.J.: Princeton University Press, 1985); John T. Lanning, *The Royal Protomedico: The Regulation of the Medical Profession in the Spanish Empire*, ed. John J. TePaske (Durham, N.C.: Duke University Press, 1985); Harold J. Cook, "Policing the Health of London: The College of Physicians and the Early Stuart Monarchy," *Social History of Medicine*, 2 (1989), 1–33; and Frank Huisman, "Itinerant Medical Practitioners in the Dutch Republic: The Case of Groningen," *Tractrix*, 1 (1989), 63–83.

[40] F. Guerra, "The Dispute Over Syphilis, Europe Versus America," *Clio Medica*, 13 (1978), 39–61; Frank B. Livingstone, "On the Origin of Syphilis: An Alternative Hypothesis," *Current Anthropology*, 32 (1991), 587–90; and Jon Arrizabalaga, John Henderson, and Roger French, *The Great Pox: The French Disease in Renaissance Europe* (New Haven, Conn.: Yale University Press, 1997).

[41] John A. H. Wylie and Leslie H. Collier, "The English Sweating Sickness (Sudor Anglicus): A Reappraisal," *Journal of the History of Medicine*, 36 (1981), 425–45.

[42] Sigrid C. Jacobs, "Guaiacum: History of a Drug; a Critico-Analytical Treatise," Ph.D. dissertation, University of Denver, 1974.

Indies.⁴³ Patients who wanted the latest medicines often pushed reluctant physicians into prescribing imported (and often expensive) drugs; on the other hand, physicians who promoted new remedies were often embraced by the apothecaries.

Many of the physicians' fears about folk medicine, secret remedies, new medicines, unorthodox theory, and unregulated claims to expertise found a focus in Paracelsus and his followers. Theophrastus Bombastus von Hohenheim (ca. 1493–1541), better known as Paracelsus ("better than Celsus," the Roman medical encyclopedist), became one of the most notorious itinerant practitioners of the early sixteenth century; in German-speaking lands, his name can still provoke passions of reverence or disdain. He has come to symbolize the beginnings of the incorporation of alchemy into medicine: Although there were earlier precedents, after him work in this arena tended to be colored by Paracelsus's reputation. Alchemy was partly a craft, partly a philosophy trying to explain phenomena that did not fit well with the four-element theory, and partly a mystical religion. But the connections of alchemy to mining and industries such as bleaching and dying also became important avenues by which practical methods of analysis made their way into medicine. Distillation, for example, was crucial to the development of early modern alchemical processes: By the thirteenth century, the distillation apparatus had become powerful enough to extract "aqua vitae" (the water of life) from wine. By the sixteenth century, the essences of many things could be distilled (as "liqueurs"), and metallurgy had developed improved methods to extract silver and gold from ore via methods that used either heat or mercury.

Using these and other means, alchemists often sought the universal substrate of nature, called the philosophers' stone, which could in turn be used to transform base metals into noble ones or to make a medicine that would prolong life and cure all diseases. The explanation for many of the processes and results of alchemical work involved notions of the microcosm and macrocosm (in which the material and spiritual things of the human body were a reflection in miniature of the things found in the larger universe) and of the operations of sympathy and antipathy (love and hate, attraction and repulsion, male and female). Paracelsus fit such views together with a vigorous popular Christianity and folklore.⁴⁴ To some, especially in Germanic-speaking

---

⁴³ See, for instance, Saul Jarcho, *Quinine's Predecessor: Francesco Torti and the Early History of Cinchona* (Baltimore: Johns Hopkins University Press, 1993); Sophie D. Coe and Michael D. Coe, *The True History of Chocolate* (London: Thames and Hudson, 1996); and M. N. Pearson, ed., *Spices in the Indian Ocean World* (Aldershot: Variorum, 1996).

⁴⁴ See, for example, Walter Pagel, *Paracelsus: An Introduction to Philosophical Medicine in the Era of the Renaissance* (New York: S. Karger, 1958); Allen G. Debus, *The Chemical Philosophy: Paracelsian Science and Medicine in the Sixteenth and Seventeenth Centuries* (New York: Science History Publications, 1977); Charles Webster, *From Paracelsus to Newton: Magic and the Making of Modern Science* (Cambridge: Cambridge University Press, 1982); and Z. R. W. M. von Martels, ed., *Alchemy Revisited* (Leiden: E. J. Brill, 1990).

regions, Paracelsian writings held the promise of a deep new wisdom. The Danish physician Petrus Severinus (1540–1602) made his views more widely known and academically respectable by epitomizing and explaining them in Latin.[45] By the end of the sixteenth century, even well-educated physicians might refer to Paracelsus with praise, although such people remained outside the orthodox Catholic, Anglican, Lutheran, and Calvinist mainstreams.

By the same period, medical chemists (sometimes called "iatrochemists") had become a powerful force throughout Europe more generally, whether or not they agreed with the basic principles that Paracelsus had promoted. Like some of their learned contemporaries, the chemists sharply criticized logic and dialectic as working against the physician and obscuring the "light of nature." They argued that their experimental approaches would lead to a reformed medicine through the discovery of new and powerful cures. Paracelsians also advocated three immaterial "principles" – rather than four elements – as the root causes of material change: salt (the principle of solidity), sulfur (that of inflammability), and mercury (that of fluidity, including smokiness). They also opposed the dissection of dead bodies, believing in the superiority of their own "chemical anatomy" of the living microcosm and macrocosm. Many also adopted the Paracelsian idea that each disease had an ontological cause: an "archeus" or spirit that had entered into the body and set itself against the local archei necessary for the proper functioning of the body. Using military metaphors, iatrochemists depicted the foreign archei as invaders to be countered by the administration of "specific" drugs, which aided the defense of the body's own archei. Many of the specifics were made with heavy metals (including mercury and antimony) detoxified by washing them in water or alcohol or by oxidation. Followers of Paracelsus also tried to develop a "panacea" or universal cure-all in the form of the liquor alcahest, a universal solvent that attacked the solids (tartar) that blocked the vessels in the body. Medical chemists came to constitute a sometimes loosely knit body of practitioners who published vigorously, and often in the vernacular, gaining converts to their views among other practitioners, the public, and royalty. By the second decade of the seventeenth century, few learned physicians opposed the use of chemical methods of analysis and treatment in toto, although most sharply criticized the chemists for single-mindedness.[46]

---

[45] Jole Shackelford, "Paracelsianism in Denmark and Norway in the 16th and 17th Centuries," Ph.D. dissertation, University of Wisconsin–Madison, 1989, pp. 1–186.
[46] See, for instance, Allen G. Debus, *The English Paracelsians* (New York: Franklin Watts, 1966); Debus, *The French Paracelsians: The Chemical Challenge to Medical and Scientific Tradition in Early Modern France* (Cambridge: Cambridge University Press, 1991); Bruce T. Moran, *The Alchemical World of the German Court: Occult Philosophy and Chemical Medicine in the Circle of Moritz of Hessen (1572–1632)* (*Sudhoffs Archiv*, Beiheft 29) (Stuttgart: Franz Steiner Verlag, 1991); Mar Rey Bueno, *Los señores del fuego: Destiladores y espagiricos en la corte de los Austrias* (Madrid: Ediciones Corona Borealis, 2002); and Miguel López Pérez, *Asclepio renovado: Alquimia y medicina en la España moderna (1500–1700)* (Madrid: Ediciones Corona Borealis, 2003).

One of the most influential medical chemists of the seventeenth century was Johannes Baptista van Helmont (1579–1644). Van Helmont was a well-educated aristocrat from Flanders who took up medical alchemy after being frustrated with the nature of academic medical teaching. In 1621, he fell into a bitter dispute with the Jesuits and Catholic Inquisition of the Spanish Netherlands with the publication of his first book, on the weapon salve. The salve he promoted cured by sympathy, being applied not to a wound but to the weapon that made the wound. To the authorities, this smacked of magic, even heresy. Van Helmont's views were denounced by the medical faculty at Louvain in 1623, and after interrogations by the Inquisition in 1627, he was forced to acknowledge his errors. He was again compelled to confess errors in 1630, and in 1634 he was arrested and put under house arrest for two years (a process not unlike that experienced by Galileo). Most of his works were published posthumously by his son. Van Helmont advocated chemical remedies, many of which were based on minerals, but he did not accept the Paracelsian three principles. He also became known as a firm advocate of experimentalism. One of his most famous experiments was that of the willow tree. He planted a small sapling in a barrel, weighed it, and added only water over the following months. In the end, the tree had grown, adding great weight, with the application of nothing but water. Van Helmont thought this confirmed his argument about the elementary nature of water. Van Helmont is also credited with arguments about the "wild spirits" that are sometimes given off by a thing when it is forced to give up its fixed state – these are now called "gases."[47] By the middle of the seventeenth century, as his works became available, the attraction of Helmontianism could be felt in many places.

Challenges to a traditional understanding of physic therefore came not only from within, via medical Hellenists, but from without, via apothecaries, surgeons, chemists, and even quacks and folk-practitioners. Physicians had to understand, and to dismiss or adapt, the views of others. To do so while at the same time maintaining their claim to oversee all aspects of medicine, they had to assimilate or reject both theoretical and empirical claims made by their rivals, especially the latter. As one of the most eminent learned French physicians put it: "The knowledge, collection, choice, culling, preservation, preparation, correction, and task of mixing of simples all pertain to pharmacists; yet it is especially necessary for the physician to be expert and skilled in these things. If, in fact, he wishes to maintain and safeguard his dignity and authority among the servants of the art, he should teach *them* these things."[48] A generation or two later, the author of this claim, Jean Fernel (ca. 1497–1558), would have had to add chemistry as well.

---

[47] Charles Webster, "Water as the Ultimate Principle in Nature: The Background to Boyle's *Skeptical Chymist*," *Ambix*, 13 (1966), 96–107; and Walter Pagel, *Jean Baptista Van Helmont: Reformer of Science and Medicine* (Cambridge: Cambridge University Press, 1982).

[48] Jean Fernel, *Methodo medendi*, quoted in Reeds, *Botany in Medieval and Renaissance Universities*, pp. 25–6.

TOWARD MATERIALISM

Although physic was centering more and more on the accumulation and close examination of physical details, many physicians were reluctant to give up causal explanation. In accordance with classical principles, it was in anatomy – and the physiology based it – that physicians continued to seek most explanations for individual health and disease. They did this both by more and more precise investigation into the matter and operations of the body and by debate about the consequences of the new findings. But as the axiomatic certainty of first principles such as the four elements and humors, and formal and final causes weakened, a close description of physical details alone threatened to become the sum of natural knowledge.

The authoritative texts on physic written by Jean Fernel underline the significance of observation in mid-sixteenth-century learned physic. But Fernel's books remained works that strove to find meaning through exegesis of texts and things in order to uncover hidden causes that could not be had through a knowledge of *res* (things) alone. His was a pious quest, like that of most other learned contemporaries, and consequently he firmly rejected the heretical – or at least heterodox – doctrines of the Paracelsians and others. For instance, in the process of developing his arguments, Fernel placed great weight on the immaterial breath of life and reason (*spiritus*), imbuing the body with abilities it would not have had if it had been framed from the lifeless elements alone.[49] Fernel's dualistic view, in which a semi-divine *spiritus* cooperates with the material body, was not the only possible physiological theory, but most orthodox ones continued to assume a dualist framework. Attempts to put forth monist theories appeared to transform the body either into an immaterial assembly of potent energies (as in Paracelsianism) or into nothing but a finely tuned physical machine. All kinds of religious enthusiasts and heretics were associated with the notion that the world was composed of divine or demonic energies. They clearly threatened the traditional establishment of church and state and were frequently suppressed. The threat of materialism, on the other hand, was associated with cynics and atheists such as the infamous Machiavelli.

Philosophical materialism was also to be found among Italian philosophers who took up Averroist themes, such as Cesare Cremonini and Pietro Pomponazzi (who argued that even a rational soul needed to be material in order to think). Padua in particular became known as a place where Aristotelian and Averroist teachings were publicly debated and supported, and where detailed investigations of the body continued. The professor of medicine Santorio Santorio (or Sanctorius, 1561–1636) invented many instruments, such as a clinical thermometer, to measure or manipulate the physical body

---

[49] James J. Bono, "Reform and the Languages of Renaissance Theoretical Medicine: Harvey Versus Fernel," *Journal of the History of Biology*, 23 (1990), 345–64.

and studied the "insensible respiration" of the body by carefully and frequently weighing himself, all that he ate and drank, and all that he excreted, finding that some weight disappeared in the process.

It was a student of the Paduans who most clearly and explicitly rejected much of the learned language about spirits and other immaterial bodily agents, William Harvey (1578–1657), who was credited with the discovery of the circulation of the blood. First educated in Kent and Cambridge, Harvey was one of several English students who studied in Padua between 1590 and 1604 during a period when a peace treaty with Spain made such travel possible. From 1599 to 1602, when he received his M.D., he worked with professors such as Sanctorius and the famous Girolamo Fabrici de Aquapendente (1537–1619), a student of Gabriele Falloppia, who in turn had been a student of Vesalius. Harvey also developed an Aristotelian outlook.[50] At Padua, he learned not only to improve his ability in philosophy, disputation, and human dissection – which he had already studied in England[51] – but learned to use other experimental techniques, such as vivisection and comparative anatomy. Upon returning to England, Harvey joined the London College of Physicians, and in 1615 took up the position of Lumleian Lecturer in anatomy.

During preparations for these annual public lectures, Harvey developed his ideas about the circulation of the blood, which he announced in his *De motu cordis et sanguinis* (On the Motion of the Heart and Blood) of 1628. He may have begun by studying the relationships between the movement of the heart and lungs, both of which were described by Galen in terms of the faculties of nutrition and vivification. By examining animal subjects, both living and dead, Harvey built on the idea of the "minor circulation," known to the medieval Ibn al-Nafis and to some sixteenth-century Europeans such as Michael Servetus (1509–1553) and Andrea Cesalpino (1519–1603). This idea held that the heart sends blood into the aorta and pulmonary artery, and hence into the lungs, and receives it back from the lungs via the vena cava and pulmonary vein. Because perfect motion is circular, according to Aristotle and many other ancients, Harvey may have seen his view as a demonstration of the perfection of God's creation.

Harvey depended not only on classical doctrines but also on close empirical investigations, which demonstrated to him at least two basic facts: The valves in the veins prevent the blood from moving in any direction except toward the heart; and the heart's movement of diastole (its squeezing) rather than the systole (its engorgement) was the active motion of the heart, causing the blood to rush outward. If one concluded (as Harvey did) from this second

---

[50] Walter Pagel, *William Harvey's Biological Ideas: Selected Aspects and Historical Background* (New York: Hafner, 1967).
[51] Peter Murray Jones, "Thomas Lorkyn's Dissections, 1564/5 and 1566/7," *Transactions of the Cambridge Bibliographical Society*, 9 (1988), 209–29.

movement that, in addition to the minor circulation, the heart forced the blood into the arteries that fed the rest of the body, and estimated (erring on the low side) the amount of blood pushed into the arteries by the heart in a period of one hour, the amount of blood pumped out by the heart would be seen to weigh more than the human body itself. This would be impossible to sustain unless the blood returned to the heart from the arteries via the veins, with their unidirectional flow.[52]

The publication of *De motu* led to many controversies throughout Europe, although Harvey seems to have quickly persuaded most of his English colleagues of his point of view; within a generation, few learned physicians anywhere remained holdouts against the theory of circulation. But even if one adopted Harvey's theory, its full implications were to be accepted much more slowly. The circulation of the blood challenged fundamental physiological principles. For instance, medieval and early modern dietetic physic was founded on the notion that what one ingests is concocted in various organs and flows from them to all parts of the body via the veins, arteries, and nerves, to be taken up as needed, just as irrigation ditches feed growing crops. If one now decided that, for instance, the veins did not originate in the liver and go outward to all parts of the body but moved the blood from all parts toward the heart, then how did nutrition occur? What was the purpose of the liver or any other organ or organ system? The number of unanswered questions that followed from Harvey's discovery was almost endless.

An even more fundamental set of questions emerged. In a letter, Harvey had used the analogy of a "pump" to describe the action of the heart. Could one stop there, and not discuss the powers of life that caused it to beat? A pump was built and powered by people; what causes lay behind the "pumping" of the heart? When one looked for other causal silences in Harvey's description, it was immediately apparent that in Harvey's view it was not necessary to speak about the concoction of the venous blood into arterial blood and its infusion with animal spirits in the heart, as ancient philosophy would have it. Did immaterial faculties reside in the organs and enable them to carry out their proper physiological responsibilities? In defending his view of circulation against the Parisian physician Jean Riolan, and in further investigations, published as *De generatione animalium* (On the Generation of Animals) in 1651, Harvey explicitly rejected previous discussions of *spiritus*, arguing that the vital forces of the body were contained in its material blood.[53] What had

---

[52] See, especially, Geoffrey Keynes, *The Life of William Harvey* (Oxford: Clarendon Press, 1966); Gweneth Whitteridge, *William Harvey and the Circulation of the Blood* (London: Macdonald, 1971); Jerome J. Bylebyl, "The Growth of Harvey's *De Motu Cordis*," *Bulletin of the History of Medicine*, 47 (1973), 427–70; Robert G. Frank, Jr., *Harvey and the Oxford Physiologists: A Study of Scientific Ideas and Social Interaction* (Berkeley: University of California Press, 1980); and French, *William Harvey's Natural Philosophy*, pp. 94–113.

[53] James J. Bono, "Reform and the Languages of Renaissance Theoretical Medicine"; Thomas Fuchs, *The Mechanization of the Heart: Harvey and Descartes*, trans. Marjorie Grene (Rochester, N.Y.: University of Rochester Press, 2001).

once been an organized set of principles about how the body functioned had been broken to bits by Harvey, whether he intended it or not. In the process, he threatened the dualist doctrines of ordinary Christianity as well. (Harvey was a firm defender of learned physic against those who challenged it, took the side of his king in the English civil wars, and was likely a Laudian in religion; that is, he was far from revolutionary in other parts of his life.) Many immediate research questions followed: Each organ deserved much closer examination, and the purposes of blood, of circulation itself, and of respiration all required new answers.[54]

Harvey's monist and vitalist views may have influenced political theorists such as Thomas Harrington.[55] Perhaps Harvey also had an effect on his friend Thomas Hobbes (1588–1679),[56] one of the most notorious monist materialists of the age, who wrote of Harvey with the greatest admiration. Hobbes had turned to a study of atomism and corpuscularianism during his visit to France in 1634–7, where he became a member of the Mersenne circle and gained an acquaintance with the work of Descartes and Gassendi, among others. For the next few years, he rigorously developed his ideas about how "whatever we experience, whether in sleep or waking, or at the hands of a malicious demon, has been caused by some material object or objects impinging upon us."[57] By the time of the publication of his *Leviathan* (1651), Hobbes was able to set out a theory of civil society that was grounded on the material nature of the person as a self-moving and self-directing being concerned with self-preservation and driven by the passions. Hobbes rooted this view of human nature in the new philosophical idea that all could be reduced to matter and motion: "Life is but a motion of Limbs," he declared in the second sentence of *Leviathan*.[58]

It was, however, the work of René Descartes (1596–1650) that made the threat of materialist monism in physiology plainest. Ironically, this was despite Descartes' professed dualism. The place in his system that he reserved for the rational soul (which also contained volition) was so small as to be virtually unnecessary to human life. Descartes relied heavily on his knowledge of the new physiology in developing his philosophical system.[59] He had embarked upon his intellectual quest, he declared in the *Discours de la méthode*

---

[54] Frank, *Harvey and the Oxford Physiologists*.
[55] I. Bernard Cohen, "Harrington and Harvey: A Theory of State Based on the New Physiology," *Journal of the History of Ideas*, 55 (1994), 187–210.
[56] According to Aubrey, Harvey left ten pounds to Hobbes in his will.
[57] Richard Tuck, "Hobbes and Descartes," in *Perspectives on Thomas Hobbes*, ed. G. A. J. Rogers and Alan Ryan (Oxford: Clarendon Press, 1988), pp. 11–41, quotation at p. 40. For a reply to Tuck, see Perez Zagorin, "Hobbes's Early Philosophical Development," *Journal of the History of Ideas*, 54 (1993), 505–18.
[58] Thomas Hobbes, *Leviathan*, ed. Richard Tuck (Cambridge: Cambridge University Press, 1991), p. 9.
[59] On Descartes' lifelong medical interests, see Gary Hatfield, "Descartes' Physiology and its Relation to his Psychology," in *The Cambridge Companion to Descartes*, ed. John Cottingham (New York: Cambridge University Press, 1992); Thomas S. Hall, "Descartes' Physiological Method: Position, Principle, Examples," *Journal of the History of Biology*, 3 (1970), 53–79; G. A. Lindeboom, *Descartes*

(Discourse on Method, 1637), in a search for useful knowledge, and "the maintenance of health is undoubtedly the chief good and the foundation of all other goods in this life."[60] In 1645, Descartes repeated to the Marquess of Newcastle that "the preservation of health has always been the principal end of my studies";[61] and in 1646 he wrote to Hector-Pierre Chanut that, because of this, "I have spent much more time" on medical topics than on moral philosophy and physics (although he had to concede that he had not yet found sure ways to preserve life).[62] The studies of his Dutch acquaintances set him on his path. The mathematician and physician Isaac Beeckman (1588–1637), one of the most important promoters of early modern atomism, apparently introduced the young Descartes to corpuscularian theories.[63] (It seems to have been Beeckman's reading of the polemics of Fernel against atomism that led Beeckman to work out a detailed defense of it.[64]) Descartes also spent time studying with François de le Boë Sylvius (1614–1672) when in 1639 and 1640 Sylvius gave private lectures in anatomy and physiology in Leiden, being among the first on the Continent to demonstrate the circulation of the blood.[65] (Sylvius and his students remained generally pro-Cartesian afterward; although they did not accept all of its doctrines, they were inclined toward a Cartesian-like physiological mechanism, with chemical reactions providing the energy to drive many processes.) Descartes seems to have periodically returned to dissecting during his many years of residence in the Netherlands.

It was, however, especially during the mid- to late 1640s, when he gave medical advice to the Princess Elisabeth of Bohemia, that Descartes reconsidered his physiological views with energy and attention. In the process, he minimized the distinction he had earlier (and famously) made between the rational soul and physical being, commenting that it was when one refrained from philosophy that one understood the union of soul and body most

---

and Medicine (Amsterdam: Rodopi, 1979); and Richard B. Carter, *Descartes' Medical Philosophy: The Organic Solution to the Mind–Body Problem* (Baltimore: Johns Hopkins University Press, 1983).

[60] John Cottingham, Robert Stoothoff, and Dugald Murdoch, eds. and trans., *The Philosophical Writings of Descartes*, 3 vols. (Cambridge: Cambridge University Press, 1985–91), 1: 142–3. Anthony Levi, *French Moralists: The Theory of the Passions, 1585 to 1649* (Oxford: Clarendon Press, 1964), p. 248, notes that "in his *Discours*, Descartes seems . . . to envisage the spiritual perfection of man as a function of medicine, a practical application of the exact deductive physics."

[61] Letter of October 1645, in *The Philosophical Writings of Descartes*, 3: 275.

[62] Letter of 15 June 1646, in *The Philosophical Writings of Descartes*, 3: 289.

[63] Jean Bernhardt, "Le rôle des conceptions d'Isaac Beeckman dans la formation de Thomas Hobbes et dans l'élaboration de son 'Short Tract,'" *Revue d'histoire des sciences*, 40 (1987), 203–15; H. H. Kubbinga, "Les premières théories 'moléculaires': Isaac Beeckman (1620) et Sébastien Basson (1621): Le concept d' 'individu substantiel' et d' 'espèce substantielle,' " *Revue d'histoire des sciences*, 37 (1984), 215–33; and Kubbinga, "La première spécification dite 'moléculaire' de l'atomisme épicurien: Isaac Beeckman (1620) et le concept d'individu substantiel," *Lias*, 11 (1984), 287–306.

[64] Beeckman's biography has been given definitive treatment in Karel van Berkel, *Isaac Beeckman (1588–1637) en de mechanisering van het wereldbeeld* (Amsterdam: Rodopi, 1983).

[65] On the reception of Harvey's theory in the Netherlands, see M. J. van Lieburg, "Zacharias Sylvius (1608–1664), Author of the 'Praefatio' to the First Rotterdam Edition (1648) of Harvey's 'De motu cordis,'" *Janus*, 65 (1978), 241–57; and van Lieburg, "Isaac Beeckman and His Diary-notes on William Harvey's Theory on Blood Circulation," *Janus*, 69 (1982), 161–83.

clearly: "It does not seem to me that the human mind is capable of forming a very distinct conception of both the distinction between the soul and the body and their union" at the same time.[66] Beginning in the mid-1640s, stimulated by Elisabeth's questions and her illnesses (which he traced to the passions loosed on her by the ill fortunes of her family[67]), Descartes began to develop powerful arguments about how to treat the body and its passions materialistically. At the same time, a bitter debate over Cartesianism had erupted at the University of Utrecht, where the medical professor Regius and his students introduced Descartes' views in a strongly materialist manner.[68] In his last two books, *Les passions de l'âme* (The Passions of the Soul, 1649) and *De homine* (On Humankind, 1662) – the latter published posthumously from a manuscript and notes by the Leiden medical professor Florentius Schuyl[69] – he continued to reserve a small place for the rational soul but otherwise set out a materialist view of human physiology.

Descartes famously proposed that all the functions of the animal body, including all the abilities of the soul except for reason and volition, could be described in terms of matter in motion. When it came to the circulation of the blood, he differed with Harvey's vitalist explanation, setting out an explanation based on inert matter. When blood entered the chambers of the heart, Descartes declared, the heat of the organ caused the blood to ferment quickly, causing a kind of contained explosion that drove the blood into the arteries. All other necessary activities could be explained, Descartes thought, without resorting to propositions that placed any vitality in matter itself. He also argued that the body could be controlled from the pineal gland – a recently discovered formation of the brain that lay in its center. This was a more radical materialism than that of Harvey, who had at least placed vital energies in the blood, although Descartes' materialism was not quite as thoroughgoing as Hobbes's.

Investigations into the structures of the body and their functions rapidly piled up. For instance, the lacteal vessels were discovered by Gasparo Aselli in the early 1620s, several other investigations were published between 1651 and 1653 (by Jean Pecquet, Olof Rudbeck, and Thomas Bartholin) that corrected and clarified their course in the body, and their valves were discovered in 1665 by Frederik Ruysch. Many investigators used the microscope as an

---

[66] Letter of 28 June 1643, in *The Philosophical Writings of Descartes*, 3: 226–9, esp. p. 227.
[67] He generally believed that the passions greatly affected the actions of the heart and other organs, thereby causing putrefactions in the blood, which gave rise to fevers. See Theo Verbeek, "Les passions et la fièvre: L'Idée de la maladie chez Descartes et quelques Cartésiens néerlandais," *Tractrix*, 1 (1989), 45–61.
[68] Theo Verbeek, ed. and trans., *La querelle d'Utrecht: René Descartes et Martin Schoock* (Paris: Les impressions nouvelles, 1988); Verbeek, "Descartes and the Problem of Atheism: The Utrecht Crisis," *Nederlands archief voor kerkgeschiedenis*, 71 (1991), 211–23; Verbeek, *Descartes and the Dutch: Early Reactions to Cartesian Philosophy, 1637–1650* (Carbondale: Southern Illinois University Press, 1992); Verbeek, ed., *Descartes et Regius* (Amsterdam: Rodopi, 1994).
[69] G. A. Lindeboom, *Florentius Schuyl (1619–1669) en zijn betekenis voor het Cartesianisme in de geneeskunde* (The Hague: Martinus Nijhoff, 1974).

aid in their close anatomical work after its brilliant use by Robert Hooke (1635–1703) of London, Antonie van Leeuwenhoek (1632–1723) of Delft, and Marcello Malpighi (1628–1694) of Bologna. The pancreatic duct was identified by Georg Wirsung in 1642; the capillaries by Malpighi in 1661; the fine structure of the kidneys by Lorenzo Bellini in 1662; the parotid duct by Nicholaus Steno in 1662; the fine detail of the human uterus and follicles by Johann Swammerdam and Reinier de Graaf at Leiden in the early 1670s; the lyphoid follicles of the small intestine by Johann Conrad Peyer in 1677; the salivary glands and tear ducts by Anton Nuck in the 1680s; the fine anatomy of the nervous system by Raymond Vieussens (published in 1685); the "bodies" (arachnoidal granulations) by Antonio Pacchioni in 1697; and so on. Many discoverers have left their mark in anatomical terms: To mention only English investigators, there is the antrum of Nathaniel Highmore (1651), the capsule of Francis Glisson (1654), the duct of Thomas Wharton (1656), the circle of Thomas Willis (1664), the canals of Clopton Havers (1691), and the glands of William Cowper (1694). Most of these anatomical discoveries, and many more, were made by physicians, although by the end of the century surgeons were beginning to supplant physicians in the writing of anatomical texts. By the end of the seventeenth century, most up-to-date physicians had also come to see the body as a closely knit structure of fine vessels and glands through which fluids moved (no longer the classical humors), being transformed into one substance or another through chemical processes.[70]

Moreover, by the mid-seventeenth century, chemistry had been incorporated into academic medicine at many schools. The medical faculty at Leiden, for instance, constructed a chemical laboratory in a corner of the botanical garden in 1669, making chemistry a subject that no longer had to be studied outside the regular curriculum as it long had been. The informal curriculum in chemistry had been available in Paris, too, for many years, following the establishment of a chemical laboratory in the Jardin des Plantes by Gui de la Brosse in the 1630s under the patronage of Cardinal Richelieu.[71] Montpellier had developed a flourishing study in chemistry, and one of its graduates, Théophraste Renaudot (1583–1653), also with the patronage of Richelieu, established an assembly in Paris in which chemical medicine was publicly debated.[72] In the 1650s, chemical teaching became common

---

[70] Edward G. Ruestow, "The Rise of the Doctrine of Vascular Secretion in the Netherlands," *Journal of the History of Medicine*, 35 (1980), 265–87.

[71] Rio Howard, "Guy de La Brosse: Botanique et chimie au début de la Révolution Scientifique," *Revue d'histoire des sciences*, 31 (1978), 301–26; Howard, "Medical Politics and the Founding of the Jardin des Plantes in Paris," *Journal of the Society for the Bibliography of Natural History*, 9 (1980), 395–402; Howard, "Guy de la Brosse and the Jardin des Plantes in Paris," in *The Analytic Spirit*, ed. Harry Woolf (Ithaca, N.Y.: Cornell University Press, 1981), pp. 195–224.

[72] Howard Solomon, *Public Welfare, Science, and Propaganda in Seventeenth Century France: The Innovations of Théophraste Renaudot* (Princeton, N.J.: Princeton University Press, 1972); and Kathleen Wellman, *Making Science Social: The Conferences of Théophraste Renaudot* (Norman: University of Oklahoma Press, 2003).

even at Oxford.⁷³ The supposition of some historians that physicians of the late seventeenth century can be divided into two groups, those who used chemical explanations (iatrochemists) and those who used mechanical ones (iatromechanists or iatromathematicians), is therefore mostly a product of our own categories rather than a reflection of the common integration of both in the work of contemporary physicians. When physicians wrote about the overall functions of the body, both chemical and anatomical findings were incorporated into their descriptions.⁷⁴

It was nevertheless a generally materialist (usually described as "mechanist") chemistry and physiology that was conveyed. Thomas Willis (1621–1675) used the concept of fermentation to explain many of the material functions of the body and disease. And even though he reserved a role for a soul in his physiological scheme, he localized the site of thinking in the material structures of the brain rather than in the open spaces of the ventricles, where most medical writers had for ages placed the immaterial processes of thought.⁷⁵ Other often-cited – and appropriate – examples are the views of Giovanni Alfonso Borelli (1608–1679) and Giorgio Baglivi (1668–1706). Borelli had studied with Galileo and worked closely with Malpighi.⁷⁶ In his *De motu animalium* (On Animal Motion, 1680–1), Borelli attempted to apply mathematical descriptions to the motions of the body, coupling them with the common chemical theory of fermentation to account for muscular action (arguing that a fluid discharged by the nerves into muscles almost instantaneously fermented and so inflated the muscles). Malpighi's pupil Baglivi also described the actions of the body as a series of materialist interactions, likening them to machines such as bellows, flasks, and sieves.⁷⁷ Despite the explicit words of Willis and many others, the new appreciation of the body therefore threatened to renew the old slander that "where there are three physicians there are two atheists."

---

⁷³ Guy Meynell, "Locke, Boyle and Peter Stahl," *Notes and Records of the Royal Society*, 49 (1995), 185–92.
⁷⁴ Harm Beukers, in his "Mechanistiche principes bij Franciscus dele Boë, Sylvius," *Tijschrift voor de geschiedenis der geneeskunde, natuurwetenschappen, wiskunde en teckniek*, 5 (1982), 6–15, emphasizes that the terms "iatrochemical" and "iatromechanical" philosophies were coined after the seventeenth century, so that trying to sort out the medical ideas of those such as Sylvius into one camp or the other presupposes a false dichotomy.
⁷⁵ Robert G. Frank, Jr., "Thomas Willis and His Circle: Brain and Mind in 17th-Century Medicine," in *The Languages of Psyche: Mind and Body in Enlightenment Thought*, ed. G. S. Rousseau (Berkeley: University of California Press, 1990), pp. 107–46; John P. Wright, "Locke, Willis, and the Seventeenth-Century Epicurean Soul," in *Atoms, Pneuma, and Tranquillity: Epicurean and Stoic Themes in European Thought*, ed. Margaret J. Osler (Cambridge: Cambridge University Press, 1991), pp. 239–58; and Robert L. Martensen, "'Habit of Reason': Anatomy and Anglicanism in Restoration England," *Bulletin of the History of Medicine*, 66 (1992), 511–35.
⁷⁶ See Domenico Bertoloni Meli, ed., *Marcello Malpighi: Anatomist and Physician* (Florence: Leo S. Olschki, 1997); and Bertoloni Meli, "Shadows and Deception: From Borelli's *Theoricae* to the *Saggi* of the Cimento," *British Journal for the History of Science*, 31 (1998), 383–402.
⁷⁷ Maria Pia Donato, "L'Onere della prova: Il Sant'Uffizio, l'atomismo e i medici romani," *Nuncius*, 18 (2003), 69–87.

CONCLUSION

By the end of the seventeenth century, what had been for Avicenna a less than precise bodily "matter" expressing itself through flows of urine, feces, blood, and phlegm had become materialized with precision. The fine structures of the body had been distinguished and their movements, growth, and changes were being charted. At the same time, the causal analyses that Avicenna had used had been fundamentally challenged. In the process, much had changed: Humors had disappeared, whereas bodily fluids, even lymph, had grabbed attention; it no longer mattered whether someone's temperament was sanguine or choleric, for her physiology was the same as her neighbor's. Not only had the basic principles of physic been undermined, opening new questions to study, but certainty had come to rest on knowledge of material detail. What might truly be called a research tradition had grown up in the medical ranks. Moreover, the hope that a new understanding of nature rooted in details might lead to the betterment of the human condition had clearly come to take pride of place in medicine.

The famous *Institutiones Medicae* (Medical Institutes, orig. ed. 1706) of Hermann Boerhaave (1668–1738) illustrates nicely how certainty now lay in material things rather than causal first principles. At Leiden, Boerhaave taught a new generation of clinically and chemically oriented medical professors and physicians from all over Europe and Britain. He held that the truths of physic could be discovered only by observation supplemented with reason. This meant that all true physical knowledge was built on sense experience. Physic therefore could account only for those things "which are purely material in the human Body, with mechanical and physical Experiments." First causes "are neither possible, useful, or necessary to be investigated by a Physician." Looking back to his hero Hippocrates, Boerhaave explained that "the Art of Physic was anciently established by a faithful Collection of Facts observed, whose Effects were afterwards explained, and their Causes assigned by the Assistance of Reason; the first carried Conviction along with it, and is indisputable; nothing being more certain than Demonstration from Experience, but the latter is more dubious and uncertain."[78]

Boerhaave's contemporary, Baglivi, agreed entirely: Certainty in the knowledge of physic remained the end, "For the Art is made up of such things as are fully Survey'd, and plainly Understood, and of such perceptions as are not under the controul of Opinion. It gives certain Reasons which are plac'd in due Order, and chalks out certain Paths, to keep its Sons from going astray. Now what is more uncertain than the Hypotheses?" As long as "Observation is the Thread to which Reason must point," all will be well. But "'tis manifest, that not only the Original of Medicine, but whatever solid

---

[78] [Herman Boerhaave], *Dr. Boerhaave's Academical Lectures on the Theory of Physic* (London: W. Innys, 1743–51), quotations at pp. 63, 71, and 42.

Knowledge 'tis entituled to, is chiefly deriv'd from Experience."[79] That certainty now resided in observation and experience of things was also clearly expressed by Boerhaave's and Baglivi's English contemporary Hans Sloane (1660–1753), who would later become President of both the Royal Society of London and the London College of Physicians. Sloane, too, wrote about how knowledge was no longer established on first principles but on physical observation:

> Knowledge based on Observations of Matters of Fact, is more certain than most Others, and in my slender Opinion, less subject to Mistakes than *Reasonings, Hypotheses*, and *Deductions* are. . . . These are things we are sure of, so far as our Senses are not fallible; and which, in probability, have been ever since the Creation, and will remain to the End of the World, in the same Condition we now find them.[80]

Consequently, the relationship between theory and practice had also changed. The textbook of François van den Zype (or Zypaeus), for instance, had its origin in the curriculum of the University of Louvain, which still required teaching to be based on Avicenna.[81] But Zypaeus's *Fundamenta medicinae physico-anatomica* (Physico-Anatomical Fundamentals of Medicine, 1683) differed from the *Canon* of Avicenna in several important ways.[82] Like many contemporary textbook writers, Zypaeus jettisoned what Avicenna had called *theoria* (the description of the elements, qualities, four causes, form and matter, naturals, nonnaturals, and contranaturals) after a few general remarks. He noted that previous doctrines had been changed by a revival of "the Doctrine of Hippocrates . . . in the Academies of France, [and] by the Experiments of the Chymists." Physic was further "improved with the greatest Pains, by Observations made in Mechanics, Natural Philosophy, and Chymistry, without Regard to any particular Sect." This eclectic view meant that no particular theory on the frame of nature was offered. Instead, the text immediately moved from Avicenna's science to what he had relegated to art. It remarked that the art of physic is acquired by means of "Observation and Reasoning." Observation must be of "all Things in the human Body, either

---

[79] [Giorgio Baglivi], *The Practice of Physick, Reduc'd to the Ancient Way of Observations* (London: Andr. Bell, Ral. Smith, Dan. Midwinter, Tho. Leigh, Will. Hawes, Will. Davis, Geo. Strahan, Bern. Lintott, Ja. Round, and Jeff. Wale, 1704), pp. 5, 9, 15, originally published as Giorgio Baglivi, *De praxi medica: Ad priscam observandi rationem revocanda* (Rome: Typis Dominici Antonii Herculis, 1696).

[80] Hans Sloane, *A Voyage To the Islands Madera, Barbados, Nieves, S. Christophers and Jamaica, with the Natural History of the Herbs and Trees, Four-footed Beasts, Fishes, Birds, Insects, Reptiles, &c.* (London: Printed by B. M. for the author, 1707), sig. Bv. Also see G. R. De Beer, *Sir Hans Sloane and the British Museum* (London: Oxford University Press, 1953).

[81] The statutes set down in the sixteenth century continued to require medical teaching of the five medical institutes "iuxta seriem doctrinarum Avicennae." See Léon van der Essen, *L'Université de Louvain (1425–1940)* (Brussels: Éditions Universitaires, 1945), pp. 253–4.

[82] The *Fundamenta* was republished at Brussels in 1687 and 1693, went through a fourth edition at Lyon in 1692, and yet a fifth (at Brussels) in 1737.

well, sick, dying, or dead." Reasoning itself had become only "an accurate Observation" seeing "those Things which pass in the human Body, unobservable by the Senses."[83] Moreover, where Avicenna had placed *practica*, Zypeaus described various therapies (which Avicenna had not considered part of the "science" but the "art" of physic). In other words, in Zypeaus's textbook, the principles by which one could preserve health (the old *practica*) had become the new *theoria*, and mere empirical details of therapy had become the new *practica*.

By the beginning of the eighteenth century, the eternal and certain could thus be found in nothing less than the close study of natural things – of "matters of fact." Certainty about causal reasonings had declined enormously, meaning that physic no longer possessed the qualities necessary for a science in the Aristotelian sense. Despite arguments about theory, no single outlook replaced the synthesis of Galen or Avicenna. Rather, "Hippocrates" became the model. This Hippocrates was not the man who first introduced concepts such as the four humors; the early modern Hippocrates stood for close observation and study of signs and things.[84] People were proud to refer to the Netherlands' new Hippocrates (Boerhaave), or to England's (Thomas Sydenham, 1624–1689). When they did so, they esteemed them not for their grand theory but for their close synthesis of medical knowledge derived from detailed observations and investigations and for their concern for utility in fighting disease. The classical definition of medicine was "to preserve health and prolong life," or sometimes "to preserve health and restore it if lost."[85] For the Moderns, however, the fight against disease came first. From the physicians' point of view, there had indeed been a revolution in science.

---

[83] English translation of the *Fundamenta* (2nd ed.) by Johannes Groenevelt, *The Rudiments of Physick Clearly and Accurately Describ'd and Explain'd* (Sherborne: R. Goadby, and London: W. Owen, 1753), pp. 22–3.
[84] Wesley D. Smith, *The Hippocratic Tradition* (Ithaca, N.Y.: Cornell University Press, 1979).
[85] "Finis remotus Medicinae est corporis humani sanitas, quae sit praesens, per medicinam est conservanda, si absens, restauranda": Françoise Zypaeus, *Fundamenta Medicinae* (Brussels: Aegidium T'Serstevens, 1687), p. 3.

19

NATURAL HISTORY

Paula Findlen

In the midst of his great *Historia animalium* (History of Animals, 1551–8), the Swiss-German naturalist Conrad Gessner (1516–1565) offered the following reflection on the process of creating knowledge. "Reason and experience are the two pillars of scientific work," he affirmed. "Reason comes to us from God; experience depends on the will of man. Science is born from the collaboration of the two."[1] Gessner's experience gathering materials for a new history of nature in the mid-sixteenth century gave him direct insight into the problems of combining reason and experience. The more material he uncovered, the more difficult it was to organize the natural world into distinctly logical patterns. By placing great emphasis on experience, Gessner had amassed enough material to write four hefty volumes that far surpassed what anyone had known before about animals. But he confessed that experience alone was an undisciplined kind of knowledge. It was reason that allowed him to give some semblance of order to nature and to interpret the similarities and differences he saw among the natural things of the world.

Gessner's methodological lessons in the midst of his Renaissance zoology remind us that the natural sciences were an important arena in which new definitions of knowledge arose from an increased emphasis on experience. In the early modern period, natural history was an important, controversial, and much discussed kind of knowledge. Natural history was a truly encyclopedic science in which broad sectors of society participated, although not, at this point, as a unified group. Learned scholars delighted in the questions of terminology that allowed them to use their formidable linguistic erudition, developing a more precise vocabulary for the natural world that conformed to their experience of it. Philosophers immersed themselves in paradoxes of classification. Travelers provided fresh and plentiful observations that conveyed the vast expanse of nature to an eager audience at home. Physicians

---

[1] Conrad Gessner, *Historia animalium*, quoted in Lucien Braun, *Conrad Gessner* (Geneva: Editions Slatkine, 1990), p. 15. Braun does not indicate from which edition he takes this quotation.

and apothecaries transformed their professional acumen about the medicinal uses of plants into full-scale observation of nature. Civil servants in nascent European empires asked themselves how natural commodities might benefit the state. Princes and patricians delighted in the cultivation and display of curiosities in their gardens and homes. Artists worked closely with physicians and philosophers to determine how their skills might convey accurately the fruits of observation and classification, and used natural objects in their own compositions. It is little wonder that natural history became the "big" science of the late seventeenth and eighteenth centuries, when the proliferation of European overseas empires further enlarged its scope.[2] It was a vast collective enterprise with intellectual, political, and economic implications.

Until the early 1990s, historians of science placed little emphasis on developments in natural history as part of the transformation in knowledge that occurred during the sixteenth and seventeenth centuries, preferring to concentrate on physics and astronomy.[3] Natural history did not fit well with the model of science that had been developed in the study of these other subjects. It produced no singular moment of discovery, no dramatic transformation of mind that we might associate with one individual. It yielded no heady confrontation with religious authority in the manner of Copernican astronomy, nor did tales of gathering Alpine flora and Mediterranean fauna have the same illicit appeal as Andreas Vesalius's exaggerated accounts of nocturnal body snatching to supply cadavers for dissection.[4]

---

[2] For the exponential growth of natural history in the eighteenth century, the following works provide a good introduction: Tore Frängsmyr, ed., *Linnaeus: The Man and His Work* (Canton, Mass.: Science History Publications, 1994; orig. publ. 1983); Mary Pratt, *Imperial Eyes: Travel Writing and Transfiguration* (London: Routledge, 1992); Londa Schiebinger, *Nature's Body: Gender in the Making of Modern Science* (Boston: Beacon Press, 1993); Richard H. Grove, *Green Imperialism: Colonial Expansion, Tropical Island Edens and the Origins of Environmentalism, 1600–1860* (Cambridge: Cambridge University Press, 1995); and Nicholas Jardine, James A. Secord, and Emma C. Spary, eds., *Cultures of Natural History* (Cambridge: Cambridge University Press, 1996), pp. 127–245.

[3] Several early exceptions to this approach include Karen Reeds, "Renaissance Humanism and Botany," *Annals of Science*, 33 (1976), 519–42; Barbara Shapiro, "History and Natural History in Sixteenth- and Seventeenth-Century England," in Barbara Shapiro and Robert G. Frank, Jr., *English Scientific Virtuosi in the Sixteenth and Seventeenth Centuries* (Los Angeles: William Andrews Clark Memorial Library, 1979); and Joseph M. Levine, "Natural History and the Scientific Revolution," *Clio*, 13 (1983), 57–73. An overview can be found in Jardine, Secord, and Spary, eds., *Cultures of Natural History*. The status of natural history within accounts of the Scientific Revolution has been raised in Harold J. Cook, "The Cutting Edge of a Revolution? Medicine and Natural History Near the Shores of the North Sea," in *Renaissance and Revolution: Humanists, Scholars, Craftsmen and Natural Philosophers in Early Modern Europe*, ed. J. V. Field and Frank A. J. L. James (Cambridge: Cambridge University Press, 1993), pp. 45–61.

[4] Needless to say, historians of these other sciences have found this older approach also unsatisfactory in explaining the development of the fields of physics, mathematics, astronomy, and anatomy. The current significance of natural history in accounts of the Scientific Revolution reflects this widespread shift in historical interpretation, exemplified well in such general works as Steven Shapin, *The Scientific Revolution* (Chicago: University of Chicago Press, 1996). See, for example, Brian Ogilvie, "Observation and Experience in Early Modern Natural History," Ph.D. dissertation, University of Chicago, 1997; Alix Cooper, "Inventing the Indigenous: Local Knowledge and Natural History in Early Modern Germany," Ph.D. dissertation, Harvard University, Cambridge, Mass., 1998; and

Rather than concluding that natural history was a less important or less innovative kind of science, as earlier historians of science occasionally suggested, we might consider instead how the significance of natural history in the early modern period reflected a different intellectual model of the growth of scientific thought. Natural history described a kind of incremental (as opposed to revolutionary) knowledge that emerged directly out of ancient and medieval encyclopedism. It also encompassed some of the most fundamental developments in Renaissance medicine, most notably the growing emphasis on the evidence of the senses and a deep conviction that knowledge of the natural world provided a crucial foundation for understanding the human body because the body was a microcosm of all that the terrestrial world contained. Finally, natural history was the site in which scholars struggled most openly with the problem of new knowledge. Catalogues of new stars seemed insignificant in their size and scope in comparison with the flood of reports from all corners of the globe regarding the novelties one could find in terrestrial nature. "What a great abundance of the rarest things are found in the newly discovered lands," exclaimed the Italian naturalist Ulisse Aldrovandi (1522–1605) in 1573.[5] The combination of these different factors shaped natural history in the sixteenth and seventeenth centuries.

THE REVIVAL OF AN ANCIENT TRADITION

Natural history was an ancient form of scientific knowledge, most closely associated with the writings of the Roman encyclopedist Pliny the Elder (ca. 22–78). His loquacious and witty *Historia naturalis* offered an expansive definition of this subject. Pliny's natural history broadly described all entities found in nature, or derived from nature, that could be seen in the Roman world and read about in its books; art, artifacts, and peoples as well as animals, plants, and minerals were included in his project.[6] Thus, Pliny's definition of "nature" included everything natural as well as artificial, and the idea of "history" underscored the important role of description in understanding nature rather than any specific sense of the past.[7] Pliny imagined the unit of natural history to be something he called a *factum*, not by any means our

---

Antonio Barrera, "Science and the State: Nature and Empire in Sixteenth-Century Spain," Ph.D. dissertation, University of California, Davis, 1999.
[5] Ulisse Aldrovandi, *Discorso naturale*, in Sandra Tugnoli Pattaro, *Metodo e sistema delle scienze nel pensiero di Ulisse Aldrovandi* (Bologna: Cooperativa Libraria Universitaria Editrice Bologna, 1981), p. 205.
[6] Roger French and Frank Greenaway, eds., *Science in the Early Roman Empire: Pliny the Elder, His Sources and Influence* (London: Croom Helms, 1986); and Mary Beagon, *Roman Nature: The Thought of Pliny the Elder* (Oxford: Clarendon Press, 1992).
[7] In its original usage, *historia* connoted an ahistorical idea of description rather than a chronological ordering of events, which explains why disciplines such as geology and paleontology did not emerge until ideas about history changed beginning in the late seventeenth century.

modern sense of the fact but rather an early term for information gathered through a variety of techniques considered reliable by the standards of the time (which might include reliable hearsay, the words of authorities, and other forms of indirect evidence). He compiled twenty thousand of these singular pieces of information in his work, through personal observation, the reports of others, and the writings of one hundred authors.[8]

Pliny did not see himself as the creator of an entirely new enterprise. Rather, he compiled a comprehensive and well-organized guide to the overwhelming amount of information about the natural world that was already available in antiquity. The intellectual genealogy that Pliny offered for natural history traced its origins to the work of his Greek predecessors, most notably Aristotle (384–322 B.C.E.) and his disciple Theophrastus. Their treatises on animals and plants represented the earliest surviving efforts to describe and classify the natural world in any detail and to offer general principles regarding the anatomy, physiology, reproduction, and habits of living things.[9] In this respect, the Aristotelian approach to nature differed noticeably from that of Pliny. It helped to establish a causal rather than descriptive foundation for the study of nature, downplaying the wonders of nature that Pliny highlighted in favor of describing the structure and function of specimens that typified the general rule of nature. All of these ancient investigators attempted to form basic conclusions about the nature of nature itself.

Pliny's contemporary, the Greek physician Dioscorides (ca. 40–80), offered yet another model of what natural history could be. *De materia medica* (On Medical Material), one of the most successful and enduring herbals of antiquity, emphasized the importance of understanding the natural world in light of its medicinal efficacy.[10] Dioscorides' treatise described approximately 550 plants and provided succinct descriptions of the virtues of Mediterranean plants in curing various ailments. Implicitly, it suggested that any description of nature should always be in the service of medicine, making greater knowledge of nature a precondition to the improvement of health. In the next century, the prolific Roman physician Galen further underscored this

---

[8] Pliny, *Natural History*, preface 17–18, trans. H. Rackham, 10 vols. (Cambridge, Mass.: Harvard University Press, 1938–63), 1: 13. For a discussion of the fact, see Lorraine J. Daston, "Baconian Facts, Academic Civility, and the Prehistory of Objectivity," *Annals of Scholarship*, 8 (1991), 337–63.

[9] G. E. R. Lloyd, *Science, Folklore, and Ideology: Studies in the Life Sciences in Ancient Greece* (Cambridge: Cambridge University Press, 1983). For a general overview, see Roger French, *Ancient Natural History* (London: Routledge, 1994). The most important works in this category are Aristotle's *History of Animals, Generation of Animals*, and *Parts of Animals*, as well as Theophrastus's *Enquiry into Plants*.

[10] John Riddle, *Dioscorides on Pharmacy and Medicine* (Austin: University of Texas Press, 1985). For the later fortunes of the work of Dioscorides, see Riddle, "Dioscorides," in *Catalogus translationum et commentariorum: Mediaeval and Renaissance Latin Translations and Commentaries*, 8 vols. (Washington, D.C.: Catholic University Press, 1960–), vol. 4 (1980), ed. Paul O. Kristeller and F. Edward Cranz, pp. 1–143; and Jerry Stannard, "Dioscorides and Renaissance Materia Medica," in *Materia Medica in the XVIth Century*, ed. M. Florkin (Analecta Medico-Historica, 1) (Oxford: Pergamon, 1966), pp. 1–21.

idea by writing copiously about pharmacology.[11] The ancient tradition of writing about nature's medicinal uses continued through the Middle Ages, when Christian fascination with the symbolic properties of animals, and occasionally plants, gave rise to new genres of writing about nature such as the medieval bestiary.

Thus, early modern scholars who perused the works of the ancients had a multiplicity of approaches to nature to consider. Like Dioscorides, they could devote themselves entirely to the improvement of medicine, as the majority did in the late fifteenth and early sixteenth centuries. They could expand and critique the Aristotelian project to comprehend causally and classify the natural world, or they could take up Pliny's idea of understanding the world through its natural description.

First and foremost, however, aspiring naturalists had to contend with a world of books. Natural history easily made the transition from the written to the printed word with the appearance of the printing press in the mid-fifteenth century largely because of its strength in the manuscript culture of medieval and early Renaissance science.[12] The first printed edition of Pliny's *Natural History* appeared in 1469. By 1600, no less than fifty-five editions had rolled off the presses.[13] During that same period, Aristotle's zoological writings began to appear in the original Greek rather than through the Latin translations of medieval Arabic commentaries. Manuscript editions of Theophrastus's botanical writings, known only indirectly prior to the fifteenth century, arrived in the Rome of Pope Nicholas V. A great patron of learning, Nicholas V commissioned between the 1440s and 1470s not just one but two translations of a good Greek manuscript of Aristotle's three books on animals, desiring Latin editions that were "no less elegant and correct than that in which they are possessed among the Greeks."[14] A Latin translation of Theophrastus's *De plantis* (On Plants) and *De causis plantarum* (On the Causes of Plants), completed by 1454 for the same pope by Theodore Gaza,

---

[11] For a discussion of Galen's contributions to medicine and natural philosophy, see Owsei Temkin, *Galenism: The Rise and Decline of a Medical Philosophy* (Baltimore: Johns Hopkins University Press, 1973).

[12] On medieval natural history, see Jerry Stannard, "Natural History," in *Science in the Middle Ages*, ed. David C. Lindberg (Chicago: University of Chicago Press, 1976), pp. 429–66; and David C. Lindberg, "Natural History," in Lindberg, *The Beginnings of Western Science* (Chicago: University of Chicago Press, 1992), pp. 348–53.

[13] Albert Labarre, "Diffusion de l'*Historia naturalis* de Pline au temps de la Renaissance," in *Festschrift für Claus Nissen* (Wiesbaden: Guido Pressler, 1973), p. 451. See also Martin Davies, "Making Sense of Pliny in the Quattrocento," *Renaissance Studies*, 9 (1995), 240–57; and John Monfasani, "The First Call for Press Censorship: Nicolò Perotti, Giovanni Andrea Bussi, Antonio Moreto, and the Editing of Pliny's *Natural History*," *Renaissance Quarterly*, 41 (1988), 1–31.

[14] Nancy G. Siraisi, "Life Sciences and Medicine in the Renaissance World," in *Rome Reborn: The Vatican Library and Renaissance Culture*, ed. Anthony Grafton (Washington, D.C.: Library of Congress, 1993), p. 174. This quote comes from the preface of George Trebizond's 1449–50 Latin translation of Aristotle's *Historia animalium*, *De partibus animalium*, and *De generatione animalium*. The second translation was completed by Theodore Gaza in 1473 or 1474. On the fortunes of Aristotle in this period, see Charles Schmitt, *Aristotle in the Renaissance* (Cambridge, Mass.: Harvard University Press, 1983).

engendered even more excitement because the works of Theophrastus had not been read directly by anyone in Western Europe since late antiquity.[15] They, too, found their way quickly into print in the 1490s. A printed Greek edition of Dioscorides appeared in 1499, preceded by a Latin edition in 1478.[16]

The profusion of published natural histories had two immediate effects. It increased the accessibility of ancient accounts of nature in Western Europe, and it allowed scholars to compare these texts with one another. When Aristotle's translator George Trebizond invoked the criterion of a "correct" text around 1450, he suggested the growing awareness of the ways in which problems of translation made it difficult to know what the ancients had really said about nature; Greek words frequently had been mangled by Latin and Arabic translators and further misinterpreted as a result of the errors that inevitably occurred with repeated copying of the same texts over many centuries. The world of words, like nature itself, was infinite and subtle in its variations. The more early Renaissance scholars studied the writings of the ancients, the more frustrating the task of recuperating the original meaning seemed.

The ancients themselves, as much as their medieval and early Renaissance translators, proved to be a source of frustration. Take the case of Pliny, who based his *Natural History* on extensive readings of Greek and Roman authors. By the 1490s, discerning readers of Pliny had noted a disturbing fact about their favorite encyclopedist: His command of Greek sources was notoriously unreliable. Very little of what he said about plants, for instance, could be correlated with the remarks of Dioscorides. A debate ensued that seemed to have very little to do with the actual stuff of nature and everything to do with language. Were the mistakes Pliny's or did the fault lie with later copyists and editors? Did he know the difference, for example, between a strawberry bush and a variety of similarly leafy plants that did not yield such delectable fruit?

The first salvo was launched by the Italian physician Niccolò Leoniceno (1428–1524), who taught medicine in Ferrara, then a center for the kind of humanistic learning that was deeply engaged in debates about textual criticism and accurate philology. His *De Plinii et plurium aliorum medicorum in medicina erroribus* (On the Errors in Medicine of Pliny and Many Other Medical Practitioners, 1492) reported numerous errors in Pliny's twenty thousand facts, many of them a result of mistranslation from his Greek sources.[17]

---

[15] Charles Schmitt, "Theophrastus," in Kristeller and Cranz, eds., *Catalogus translationum et commentariorum*, vol. 2 (1971), pp. 239–322.

[16] The Venetian printer Aldus Manutius published a Greek edition of Theophrastus's works in 1497 in conjunction with his edition of all the known works of Aristotle that appeared between 1495 and 1498. The 1499 edition of the works of Dioscorides also came from the Aldine press, which suggests how important this one publisher was to the dissemination of ancient scientific texts in the first fifty years of printing.

[17] On the debates about Pliny, see Lynn Thorndike, "The Attack on Pliny," in Thorndike, *A History of Magic and Experimental Science*, 8 vols. (New York: Columbia University Press, 1923–58), 4: 593–

Had Pliny only known Greek, Leoniceno suggested, he would have known the strawberry.

Respondents to Leoniceno's attack on Pliny initially came from the local intellectual community in northern Italy, then the leading center for medical education and humanistic study. As the debate continued into the sixteenth century, it attracted the attention of Northern European scholars.[18] The diversity of occupations of those involved reflected the widespread appeal of natural history. Participants in this public debate in no way defined a professional community of naturalists. Pandolfo Collenuccio, who responded to Leoniceno's treatise with his *Pliniana defensio* (Plinian Defense, 1493), was a lawyer and humanist. Ermolao Barbaro (1454–1493), whose *Castigationes plinianae* (Plinian Castigations, 1492) claimed to have uncovered some five thousand errors in the *Natural History* – none of them Pliny's, he felt, but all the fault of bad copyists of early manuscripts and of misinformation in the Greek sources – belonged to one of the leading families of Venice and was an important humanist. Alessandro Benedetti taught medicine in Bologna. Despite their diversity of occupations, all of them felt qualified to comment upon the problems Leoniceno had raised about Pliny's veracity because no one group could claim a monopoly on the kinds of expertise needed to assess natural history.

The result of this debate was a growing public recognition that the ancient texts of natural history were far from perfect. A great deal of scholarly work needed to be done examining different editions of key authors to determine what they really had said. Barbaro, for instance, read Pliny's work in light of the writings of Theophrastus and Dioscorides. Comparing books, however, quickly became an imperfect exercise in knowledge. Collenuccio argued that it was not enough to "read authors, look at plant pictures, and peer into Greek vocabularies."[19] Observation offered the potential for greater certainty than words alone. On this important point, virtually all of the participants in the debate on Pliny agreed, even as they argued vociferously about everything else. Leoniceno had argued from the start for a natural history written "not

---

610; Arturo Castiglioni, "The School of Ferrara and the Controversy on Pliny," in *Science, Medicine, and History*, ed. E. Ashworth Underwood, vol. 1 (Oxford: Oxford University Press, 1953), pp. 269–79; Charles G. Nauert, "Humanists, Scientists, and Pliny: Changing Approaches to a Classical Author," *American Historical Review*, 84 (1979), 72–85; Giovanna Ferrari, "Gli errori di Plinio: Fonti classiche e medicina nel conflitto tra Alessandro Benedetti e Niccolò Leoniceno," in *Sapere e/è potere*, ed. A. Cristiani, 2 vols. (Bologna: Istitute per la Storia di Bologna 1990), 2: 173–204; and Ogilvie, "Observation and Experience," pp. 89–112. For a general account of the intellectual climate in which these debates occurred, see Vivian Nutton, "The Rise of Medical Humanism: Ferrara, 1464–1555," *Renaissance Studies*, 11 (1997), 2–19; and Daniela Mugnai Carrara, *La biblioteca di Niccolò Leoniceno* (Florence: Leo S. Olschki, 1991).

[18] Peter Dilg, "Die botanische Kommentarliteratur Italiens um 1500 und ihr Einfluss auf Deutschland," in *Der Kommentar in der Renaissance*, ed. August Buck and Otto Herding (Bonn: Harald Boldt, 1975), pp. 225–52.

[19] Pandolfo Collenuccio, *Pliniana defensio*. As quoted in Edward Lee Greene, *Landmarks of Botanical History*, ed. Frank N. Egerton, 2 vols. (Stanford, Calif.: Stanford University Press, 1983), 2: 551.

from words, but from things."²⁰ Neither he nor any of his critics trusted words alone by the end of their dissection of Pliny.

Natural history offers us an ironic lesson about the growth of empirical practice at the beginning of the sixteenth century. In many instances, scholars observed nature more closely because they had decided to read ancient works of science more comprehensively and carefully. Whether one examined animals and plants to prove the ancients right or to demonstrate their fallibility ultimately did not matter. Initially the goal was to correct Pliny's words through observation of nature rather than to observe per se. In contrast, by the end of the sixteenth century, commentators on Pliny could more accurately be described as naturalists. In 1572, the German naturalist Melchior Wieland based his criticisms of Pliny on years of travel in the Near East as well as on his work as the curator of the University of Padua's famous botanical garden.[21] Although he possessed the same skills in Greek that had produced Leoniceno's testy criticisms eighty years earlier, Wieland could no longer be accused by his critics of lacking knowledge of nature. He reflected the new image of natural history as an observational practice.

WORDS AND THINGS

It was one thing for a handful of humanist commentators to suggest that examining nature would resolve their questions about an ancient text. This provided a specific reason for a limited group of scholars to gain experience of nature but said little about the means by which observation was to be generally put into practice or what its other sources of inspiration might have been. During the 1530s and 1540s, observation began to play a systematic role in the study of the natural world. Practitioners of medicinal botany, the most established part of natural history because of its close association with medicine, were at the forefront of this transformation. They saw their work as a revival of an ancient science. The German herbalist Otto Brunfels wrote in his *Herbarum vivae eicones* (Living Images of Plants, 1530) that he hoped "to bring back to life a science almost extinct."[22] Similarly, the University of Ferrara's first professor of medical botany, Gaspare Gabrieli, commented in his inaugural lecture of 1543 that "herbal medicine is despised and neglected by everyone."[23] Both comments reflected the growing perception that new information and new techniques of conveying knowledge, most notably the use of

---

[20] Niccolò Leoniceno, *De Plinii et plurium aliorum medicorum in medicina erroribus* (Basel: Henricus Petrus, 1529), p. 215.
[21] Nauert, "Humanists," pp. 84–5; and Anthony Grafton, "Rhetoric, Philology and Egyptomania in the 1570s: J. J. Scaliger's Invective Against M. Guilandinus's *Papyrus*," *Journal of the Warburg and Courtauld Institutes*, 42 (1979), 167–94.
[22] Greene, *Landmarks of Botanical History*, 1: 244. Greene quotes from the 1532 edition.
[23] Gaspare Gabrieli, *Oratio habita Ferrariae in principio lectionum de simplicium medicamentorum facultatibus anno MDXLII per me Gasparem Gabrielum*. A modern version of this text is found in

printing to circulate words and images more widely, would recuperate botany, and by extension natural history, as a subject that learned men should study.

Given the number of words spilled over Pliny's errors, and the amount of energy spent creating splendid new editions of Aristotle and Theophrastus, the image of natural history as a neglected field of study seemed unsustainable by the mid-sixteenth century. Yet it was precisely the well-known problems with these texts that made it possible to present the revival of natural history as a matter of urgent necessity. Naturalists had the fruits of several decades of renewed observation of the natural world upon which to build. Although their knowledge of non-European (in fact, non-Mediterranean) nature was still sparse, new accounts of European nature that paid greater attention to Northern Europe, coupled with tantalizing glimpses of New World nature, provided just enough information for them to imagine how limited the ancient geography of nature actually was. If the ancients had described less than one-hundredth of the plants in the world, as one physician estimated in 1536, then naturalists had a great deal of new knowledge to contribute to society.[24]

The new influx of natural knowledge came from two distinct sources: changes in medical education, and the conquest and exploration of the Americas and the East Indies. In the first instance, a vocal sector of university-trained physicians argued strongly for a new kind of medical education in the universities that made medicinal botany a precondition of medical expertise. In the introduction to *De historia stirpium commentarii insignes* (Remarkable Commentaries on the History of Plants, 1542), one of the first botanical texts to make illustrations an important part of natural description, the German physician Leonhart Fuchs (1501–1566) lamented that few physicians knew plants well. "They appear to think that this kind of information does not belong to their profession," he observed.[25] As a professor of medicine at the newly created University of Tübingen, Fuchs was in a position to change medical pedagogy. He belonged to a generation that introduced the teaching of Dioscorides' *De materia medica* into the curriculum. By the late 1540s, one could hear lectures on this text not only in the most venerable medical faculties such as Montpellier, where the French physician Guillaume Rondelet began to teach Dioscorides in 1545, but also in new universities such as Wittenberg, which formally added Dioscorides to its curriculum in 1546.[26]

Felice Gioelli, "Gaspare Gabrieli: Primo lettore dei semplici nello studio di Ferrara (1543)," *Atti e memorie della Deputazione Provinciale Ferrarese di Storia Patria*, ser. 3, vol. 10 (1970), 31.

[24] The physician in question, Antonio Musa Brasavola, taught medicine and medicinal botany at Ferrara until 1541. His *Examen omnium simplicium medicamentorum* (1536) is an early example of an attempt to teach botany as a dialogue among observers studying nature in the field.

[25] Leonhart Fuchs, *De historia stirpium* [1542], "Epistola nuncupatoria," p. v. I have used the translation in Greene, *Landmarks of Botanical History*, 1: 276.

[26] Karen Reeds, *Botany in Medieval and Renaissance Universities* (New York: Garland, 1991), p. 57; and Karl H. Dannenfeldt, "Wittenberg Botanists During the Sixteenth Century," in *The Social History*

The Italian universities played a leading role in defining the new institutional culture of natural history, which was a direct result of their strength in medical education (see Grafton, Chapter 10, this volume). In 1533, the University of Padua established the first permanent chair in "medicinal simples," which primarily covered botanical knowledge.[27] One year later, the University of Bologna appointed Luca Ghini to a similar position, where he taught the writings of Theophrastus. By 1545, both Pisa and Padua had botanical gardens in which professors demonstrated plants to students when they were not taking students on botanical trips during the summer months.[28] For the rest of the sixteenth century, scholars interested in the study of nature felt obliged to travel and study in Italy not only because it contained many of the Mediterranean plants described by the ancients but also because its universities promoted a new kind of natural history that emphasized evidence of the senses.

Observation gradually assumed an important place in pedagogy, leading some naturalists to redefine natural history as "sensory natural history" in order to distinguish the textual tradition of nature from its early modern counterpart.[29] Professors who taught medicinal botany in this fashion, such as Fuchs and Ghini, received lucrative offers from other universities, which hoped to increase the prestige of their medical faculties through the addition of a famous natural history professor. The majority of these positions appeared in Italy. Important medical centers in Europe followed their example in the 1570s through 1660s as natural history became a more integrated part of the medical curriculum. Basel created a professorship in 1589; Montpellier formalized the place of medicinal botany in its curriculum in 1593. Newer

---

of the Reformation, ed. Lawrence P. Buck and Jonathan W. Zophy (Columbus: Ohio State University Press, 1972), p. 226. Wittenberg was also the site of an interesting but ultimately unsuccessful attempt to introduce Pliny into the curriculum in 1543; for these and other curricular reforms, see Sachiko Kusukawa, *The Transformation of Natural Philosophy: The Case of Philip Melanchthon* (Cambridge: Cambridge University Press, 1995).

[27] A temporary chair was established at the University of Rome in 1513 but did not last for more than a few years.

[28] For a more detailed account of these developments, see Charles Schmitt, "Science in the Italian Universities in the Sixteenth and Seventeenth Centuries," in *The Emergence of Science in Western Europe*, ed. Maurice Crosland (New York: Macmillan, 1975), pp. 35–56; Schmitt, "Philosophy and Science in Sixteenth Century Italian Universities," in Schmitt, *The Aristotelian Tradition and the Renaissance Universities* (London: Variorum, 1984), chap. 15, pp. 297–336; Paula Findlen, "The Formation of a Scientific Community: Natural History in Sixteenth-Century Italy," in *Natural Particulars: Nature and the Disciplines in Renaissance Europe*, ed. Anthony Grafton and Nancy Siraisi (Cambridge, Mass.: MIT Press, 1999), pp. 369–400; and Findlen, *Possessing Nature: Museums, Collecting, and Scientific Culture in Early Modern Italy* (Berkeley: University of California Press, 1994). The rise of the botanical garden is discussed in Margherita Azzi Visentini, *L'Orto botanico di Padova e il giardino del Rinascimento* (Milan: Edizioni il Polifilo, 1984); Alessandro Minelli, ed., *The Botanical Garden of Padua, 1545–1995* (Venice: Marsilio, 1995); and Fabio Garbari, Lucia Tongiorgi Tomasi, and Alessandro Tosi, *Giardino dei Semplici: L'Orto botanico di Pisa dal XVI al XIX secolo* (Pisa: Cassa di Risparmio, 1991).

[29] This phrase comes from the pen of Ulisse Aldrovandi, one of the most vocal advocates of the history of sensible, or sensory, things in the second half of the sixteenth century. Biblioteca Universitaria, Bologna, *Aldrovandi* MS. 21, vol. IV, c. 36.

universities such as Leiden (founded 1575) actively recruited distinguished naturalists, persuading a reluctant Carolus Clusius (1526–1609, also known as Charles de L'Écluse) to return home in 1594 to help develop its botanical garden after he had spent a good part of his career as a court naturalist to the Holy Roman Emperor Maximilian II in Vienna.[30] By the seventeenth century, most medical students enjoyed some basic training in botany and comparative anatomy prior to graduation.

The new emphasis on observation did not displace books. Upon the death of the French naturalist Guillaume Rondelet (1507–1566), his loyal students joked that no one "wore out Dioscorides with so much use."[31] Reading and discussing ancient texts continued to generate new ways of looking at things. The German naturalist Valerius Cordus (1515–1544) traveled to Italy in order to create new descriptions of the plants that Dioscorides had identified from living specimens.[32] Cordus participated in the invention of the field trip as a fundamental part of natural history – yet another innovation that owed a great deal to the desires of Renaissance naturalists to find new ways to use ancient texts. Within a few decades, the field trip became a source of authoritative knowledge in itself, though books still played an important role in its formulation. These books included pocket editions of standard botanies in which travelers might record their own field observations. "I have made many pilgrimages to various lands," declared Luigi Anguillara, one of the directors of the Paduan botanical garden, in 1561, informing his readers that he was no armchair naturalist.[33] Increasingly, the image of the naturalist emphasized his role as an active observer of the world. Obviously, the distinction between the naturalist who participated in an occasional summer excursion in the Alps, primarily collecting specimens at home, and one who spent years in distant lands was significant in terms of what experience of nature actually meant. Yet both activities were crucial in building a new encyclopedia of nature.

The field trip became one of the important markers of communal activity among Renaissance naturalists. Its pedagogical role in training medical students complemented its important function in defining natural history as a

---

[30] Reeds, *Botany*, pp. 83, 111. On Clusius, also known by his Flemish name, Charles de L'Écluse, see F. W. T. Hunger, *Charles de l'Escluse*, 2 vols. ('s-Gravenhage: Martinus Nijhoff, 1927).

[31] Joannes Posthius, *De obitu D. Guillelmi Rondeletii*, in Reeds, *Botany*, p. 66.

[32] Greene, *Landmarks of Botanical History*, I: 375–6; and A. G. Morton, *History of Botanical Science* (London: Academic Press, 1981), p. 126. For a discussion of the early development of such practices and their relationship to Renaissance humanism, see Peter Dilg, "Studia humanitatis et res herbaria: Euricius Cordus als Humanist und Botaniker," *Rete*, 1 (1971), 71–85. This last essay discusses the work of Valerius's father, Euricius.

[33] Luigi Anguillara, *Semplici dell'Eccellente M. Luigi Anguillara, liquali in piu pareri a diversi nobili huomini scritti appaiono* (Venice: Vincenzo Valgrisi, 1561), p. 15. This approach to nature is discussed in Findlen, *Possessing Nature*, pp. 155–92; and in Brian Ogilvie, "Travel and Natural History in the Sixteenth Century," in Brian Ogilvie, Anke te Heesen, and Martin Gierl, *Sammeln in der Frühen Neuzeit* (Max-Planck-Institut für Wissenschaftsgeschichte, Preprint 50) (Berlin: Max-Planck-Institut für Wissenschaftsgeschichte, 1996), pp. 3–28.

science of collective observation and description. New books of nature were created from information painstakingly gathered firsthand and compared with old descriptions and eventually with old specimens. In time, the importance of the specimen complicated the role of the learned word because there were plenty of new words to be written as a result of looking at nature. Perhaps it did not matter that the French ichthyologist and ornithologist Pierre Belon couldn't read "two lines of Pliny," as his contemporaries alleged.[34] He had seen more of the world than most of them when he traveled in the Near East from 1546 until 1549, writing about his journey in a rich French vernacular that spoke directly to his contemporaries of his experiences. By the 1560s, naturalists spoke enviously of their colleagues who had had the opportunity to travel outside of Europe and thus obtain a direct knowledge of those parts of the natural world that most scholars experienced only in their studies, museums, and botanical gardens. "If I had had the luck to find a patron, or if my fortune had not been so restricted," wrote Gessner in his *Historia animalium*, "I would have traveled to the most distant lands, propelled by my strong passion to know."[35]

The thousands of unpublished pages of observational notes left behind by Renaissance naturalists suggest how important this activity was to their definition of science. Even when confined to travel within Europe, they found plenty to say about nature that was novel and illuminating. Valerius Cordus, for example, eulogized by the community of naturalists because of his untimely death while botanizing in Italy in 1544, bequeathed to posterity notes filled with precise descriptions of plants that set a new standard when they appeared posthumously in 1561 under the careful editing of Gessner. Cordus attempted to write a complete description of a plant that took into account not only surface details such as the appearance of leaves and flowers but also more subtle characteristics such as the nature and timing of fructification and the appearance of the seeds, roots, and loculi.[36] This was a level of specificity that no ancient natural history offered.

University professors who taught medicinal botany, and later natural history, formalized the role of observation by taking students on field trips in the summer months. The Flemish naturalist Clusius's careful study of the plants of Languedoc was a direct result of his education in Montpellier under Rondelet, who was so famous for his field trips that François

---

[34] Pierre Belon, *L'Histoire de la nature des oyseaux*, ed. Philippe Glardon (Geneva: Librairie Droz, 1997), p. xxiii. For an interesting discussion of Belon's world, see George Huppert, *The Style of Paris: Renaissance Origins of the French Enlightenment* (Bloomington: Indiana University Press, 1999).

[35] Conrad Gessner, *Historia animalium*, 1, in Braun, *Conrad Gessner*, p. 60. Karl H. Dannenfeldt's *Leonhard Rauwolf: Sixteenth-Century Physician, Botanist, and Traveler* (Cambridge, Mass.: Harvard University Press, 1968) is an excellent account of one of the few naturalists to travel extensively in the late sixteenth century.

[36] Morton, *History of Botanical Science*, p. 126.

Rabelais gently satirized him in his *Gargantua* (1534).[37] In 1595, the Swiss medical student Thomas Platter noted that the professor of anatomy and botany at Montpellier was officially required to take students on summer excursions.[38] By the end of the century, a great deal was known about the plants in the vicinities of the leading medical universities in Europe such as Bologna, Padua, Basel, and Montpellier. Similarly famous were Mount Pilatus, near Lucerne, and Mount Baldo, in the vicinity of Verona – the two mountains described by the Zurich physician Gesner and the Veronese apothecary Francesco Calzolari (1521–ca. 1600), both of whom routinely led groups to their summits, demonstrating plants as they went.[39] Each created a local laboratory for the investigation of the natural world.

The more naturalists observed nature in situ, the more they realized that limited contact with specimens did not yield enough knowledge to describe and compare medicinal herbs. They needed to take nature home. By the 1540s, the herbarium (a collection of dried plants) played an important role in natural history, its use advocated most forcefully by Luca Ghini, who had been among the first naturalists to initiate the field trip. Rondelet probably learned the technique of drying plants from Ghini during a trip to Italy in 1549 and then passed it on to many of his students. Felix Platter, for instance, described how he "collected plants, and arranged them properly on paper" as a medical student in Montpellier in 1554.[40] When Michel de Montaigne visited him in Basel in October 1580, he was amazed to see twenty-year-old specimens glued into nine volumes in Platter's study.[41]

The herbarium provided naturalists with a convenient tool with which to organize specimens, modifying ancient theories of classification with modern examples and testing general theories about the nature of plants against specific examples. It facilitated an important new project for natural history: to make new books of nature out of modern ingredients that supplemented and eventually replaced those of the ancients. Similarly, the new passion for

---

[37] F. David Hoeniger, "How Plants and Animals Were Studied in the Mid-Sixteenth Century," in *Science and the Arts in the Renaissance*, ed. John W. Shirley and F. David Hoeniger (Washington, D.C.: Folger Shakespeare Library, 1985), p. 139. On the influence of Rondelet's pedagogy on many northern naturalists, see Reeds, *Botany*, esp. pp. 55–72.

[38] Thomas Platter, *Journal of a Younger Brother: The Life of Thomas Platter as a Medical Student at Montpellier at the Close of the Sixteenth Century*, trans. Seán Jennett (London: Frederick Muller, 1963), p. 36.

[39] Conrad Gessner, "Descriptio Montis Fracti sive Montis Pilati, ut vulgo nominant, iuxta Lucernam in Helvetia," in Gesner, *De raris et admirandis herbis, quae sive quod noctu luceant, sive alias ob causas, Lunariae nominantur, Commentariolus* (Zurich: Andreas Gesner and Jakob Gesner, 1555); and Francesco Calzolari, *Il viaggio di Monte Baldo* (Venice: Vincenzo Valgrisi, 1566).

[40] Felix Platter, *Beloved Son Felix: The Journal of Felix Platter, a Medical Student in Montpellier in the Sixteenth Century*, trans. Seán Jennett (London: Frederick Muller, 1961), p. 88. See also Walther Rytz, "Das Herbarium Felix Platters: Ein Beitrag zur Geschichte der Botanik des XVI. Jahrhunderts," *Verhandlungen der Naturforschenden Gesellschaft in Basel*, 44 (1932–33), 1–222. This subject has received an interesting treatment in Ogilvie, "Observation and Experience," pp. 200–71.

[41] Michel de Montaigne, *Montaigne's Travels*, trans. Donald M. Frame (San Francisco: North Point Press, 1983), p. 14.

collecting natural objects, as seen particularly in the case of Ulisse Aldrovandi (1522–1605) in Bologna, whose museum was among the largest and most visited in Europe by the 1570s, created rich repositories of artifacts from which to write new zoologies and mineralogies.[42] Gradually, ancient description paled in comparison with the flood of words that poured from the pens of eager observers.

THINGS WITHOUT NAMES

One of the fundamental reasons for writing new natural histories related to the impact of the Americas, and long-distance travel in general, on thought about the natural world.[43] The larger the world became, in written description and in actual experience, the more limited Aristotle's, Dioscorides', and Pliny's image of nature seemed. What had the ancients really known? The Mediterranean surely, and parts of the Near East and North Africa. They knew little of Northern Europe and nothing of the Americas and Asia. As accounts of the Indies flooded Europe in the wake of Columbus's landing on San Salvador – which quite appropriately occurred in the same year that the debates on Pliny began – Renaissance naturalists found themselves awash in a sea of uncertain claims and unverified but intoxicating new facts about nature.

Initially, accounts of the New World emphasized its marvelous qualities. Columbus, who had read Pliny with some care, measured what he saw against his expectations of a natural world that, in true Plinian fashion, allegedly became more extraordinary and surprising the farther one traveled away from the center of the known world. He did not expect American nature to be ordinary, reporting in 1493 his surprise that he had seen no human monsters. Other aspects of American nature fulfilled his expectations, leaving him initially speechless at the sight of animals and plants that had no corollary in Europe. Of the trees of Hispaniola, Columbus wrote: "It grieves me extremely that I cannot identify them, for I am quite certain that they are all valuable and I am bringing samples of them and of the plants also."[44] When words were

---

[42] On Aldrovandi's collecting, see Giuseppe Olmi, *L'Inventario del mondo: Catalogazione della natura e luoghi del sapere nella prima età moderna* (Bologna: Il Mulino, 1992); and Findlen, *Possessing Nature*. For an interesting discussion of the evolution of a kind of natural history publication specific to the museum – the museum catalogue – see Alix Cooper, "The Museum and the Book: The *Metallotheca* and the History of an Encyclopaedic Natural History in Early Modern Italy," *Journal of the History of Collections*, 7 (1995), 1–23.

[43] Henry Lowood, "The New World and the European Catalog of Nature," in *America in European Consciousness*, ed. Karen Ordahl Kupperman (Chapel Hill: University of North Carolina Press, 1995), pp. 295–323.

[44] J. M. Cohen, ed. and trans., *The Four Voyages of Christopher Columbus* (London: The Cresset Library, 1969), pp. 69–70. These entries in Columbus's logbook are from 19 and 21 October 1492, respectively. The following works are especially useful in studying Columbus's encounter with nature: Antonello Gerbi, *Nature in the New World*, trans. Jeremy Moyle (Pittsburgh, Pa.: University of Pittsburgh

inadequate, things themselves became more important. They demonstrated their own existence, challenging those who saw them to capture their reality. In time, Columbus and those who followed him to the Indies found words to describe the unknown, familiarizing and eventually transforming many marvels into part of the ordinary fabric of nature.

Columbus's initial response to American nature contained elements of a procedure that became a common feature of early modern natural history. He did not see the point of describing all aspects of nature – only those that were relevant to his most immediate concerns. Collecting and describing nature became two of the most fundamental scientific activities associated with the New World.[45] Already in the 1490s, ships returned home laden with parrots, monkeys, iguanas, macaws, and other curiosities.[46] The arrival of these specimens imposed new pressures on traditional accounts of nature, making even more apparent the importance of observation as a source of information. Many aspects of American nature, from the exotic armadillo to the lowly potato, lacked a textual presence in the European natural history tradition. They demanded new space in the crowded pages of natural history writing, offering up the kind of information that a commentary on a writer such as Aristotle or Pliny could not comfortably accommodate. Very quickly, these aspects of American nature demanded their own histories.[47]

Such histories posed unique challenges because they lacked the standard ingredients of an authoritative description of a natural specimen, namely a long list of authorities whose words had solidified the meaning of an object over centuries. Take the case of corn, which was called "Turkish grain" (*turcicum frumentum*) by many naturalists when it first appeared in the pages of European natural histories in the 1540s; modern vernacular words for corn, such as *granturco* in Italian, still bear the traces of this misperception. Readers of New World reports were perfectly aware that it did not come from Turkey, but others exhibited an all too typical confusion between the old

---

Press, 1985), pp. 12–26; Mary B. Campbell, *The Witness and the Other World: Exotic European Travel Writing, 400–1600* (Ithaca, N.Y.: Cornell University Press, 1988), pp. 165–209; Stephen Greenblatt, *Marvelous Possessions: The Wonder of the New World* (Chicago: University of Chicago Press, 1991), pp. 52–85. For a general discussion of science and the metaphors of discovery, see Paula Findlen, "Il nuovo Colombo: Conoscenza e ignoto nell'Europa del Rinascimento," in *La rappresentazione letteraria dell'alterità nel Cinquecento*, ed. Lina Bolzoni and Sergio Zatti (Lucca: Pacini-Fazzi, 1997), pp. 219–44.

[45] The disintegration of the early model of discussing American nature in relation to European nature is described well in Richard White, "Discovering Nature in North America," *Journal of American History*, 79 (1992), 874–91; and Raquel Álvarez Peláez, *La conquista de la naturaleza americana* (Madrid: Consejo Superior de Investigaciones Científicas, 1993).

[46] Wilma George, "Sources and Background to Discoveries of New Animals in the Sixteenth and Seventeenth Centuries," *History of Science*, 18 (1980), 79–104.

[47] The consequences of this shift can be seen in William J. Ashworth, "Natural History and the Emblematic World View," in *Reappraisals of the Scientific Revolution*, ed. David C. Lindberg and Robert S. Westman (Cambridge: Cambridge University Press, 1990), pp. 303–32; and Ashworth, "The Persistent Beast: Recurring Images in Zoological Illustration," in *The Natural Sciences and the Arts* (Uppsala: S. Academiae Ubsaliensis, 1985), pp. 46–66.

Indies to the East and the new ones that had appeared in the West. By 1570, the Italian naturalist Pier Andrea Mattioli (1500–1577) pointed out this mistake, though other naturalists continued to disagree with him throughout the sixteenth century.[48] The problem was that too few naturalists had seen American maize to offer a reliable account of it. As long as corn was described as an extra-European product, its novelty – though not its exact provenance – was conveyed.

The age of conquest and discovery created a distinctive genre of natural history: the natural history of the Indies, both East and West. In an age of great curiosity about exotic nature, writing travel accounts and natural histories became a lucrative enterprise. Works such as the notary Gonzalo Fernández de Oviedo's *Historia general y natural de las Indias* (General and Natural History of the Indies, 1535–49), the physician Garcia da Orta's *Coloquios dos simples e drogas he cousas mediçinais da India* (Colloquies on Simples and Medicinal Drugs from India, 1563), and the Jesuit José Acosta's *Historia natural y moral de las Indias* (Natural and Moral History of the Indies, 1590) fed the European appetite for information about American and Asian nature. They anticipated a renewed fascination with locality as one of the conceptual parameters for studying nature, reviving the ancient Hippocratic interest in places as an important way of understanding nature. The "Indies" became one of the most strongly defined geographic entities in the study of nature.[49] It was an ideal site, yielding hundreds of novel flora and fauna on which naturalists could hone their newly developed observational skills.

Oviedo (1478–1557), who had first arrived in the Americas in 1514 as a gold mine inspector, proudly proclaimed his departure from the tradition of writing natural history as a commentary on the ancients when he declared: "My intention is not to relate the things . . . that have been written by other authors, but the notable things that come to my attention in these Indies of ours."[50] Many of his descriptions – of tobacco plants, rubber trees, and a myriad of other novelties – offered the first account of such things for European readers. Oviedo's *Historia general y natural de las Indias*, and many similar works that followed, went into multiple editions in many languages over the next century.

One of the crucial issues that Oviedo confronted concerned the credibility of knowledge. Much as the critics of Pliny had turned to observation as a means of correcting a text, Oviedo grappled with the obverse of this coin: how to use words to make things believable. Drawing inspiration from Pliny, who had briefly raised the problem of credible knowledge in his own fact-gathering, Oviedo used his skills as a notary to emphasize the importance

---

[48] Lowood, "New World," in Kupperman, ed., *America in European Consciousness*, p. 300.
[49] The reaction to the emphasis on exotic nature can be found in the development of self-conscious traditions of local natural history within Europe, especially in Germany and England; see Cooper, "Inventing the Indigenous".
[50] Gonzalo Fernández de Oviedo, *Historia*, 5.3, in Gerbi, *Nature in the New World*, p. 226.

of good witnesses.⁵¹ Whenever possible, he observed nature firsthand. When describing things that he himself had not seen, he preferred to rely on multiple corroborative accounts of the same phenomenon rather than privileging the individual report. Such techniques did not yield reports of absolute certainty, and early modern naturalists expended a good deal of ink correcting each other's exaggerations and misidentifications. These techniques went a long way, however, toward persuading readers that Oviedo was not simply another purveyor of tall tales about the marvels of the East but rather an observer whose personal experience of the Americas and whose careful management of information made him a trustworthy source of knowledge.⁵² He was a worthy precursor to Francis Bacon in the development of important ideas about the relationship between the quality of good testimony and the reliability of knowledge.

Very few naturalists made it to the Americas. Instead, they relied on examination of New World specimens in European collections and the reading of New World accounts to enlarge their portrait of nature. Gessner, for example, owned and annotated works such as André Thevet's *Les Singularitez de la France antarctique* (Singularities of Antarctic France, 1558), writing on its frontispiece: "I noted and drew [its] animals and plants."⁵³ He included corn in a 1542 plant catalogue and was growing tobacco and tomatoes in his garden by the 1550s. Gessner's cautious, probing attitude toward reports of a new nature led him to include only one American animal, the opossum, in the first volume of his *Historia animalium* (1551). By the time he completed this work in 1558, other examples of the bounty of the New World had found a place in his encyclopedia, though they by no means overwhelmed its Old World content.⁵⁴

The next generation of naturalists engaged more passionately and systematically with American nature. Aldrovandi, for instance, fancied himself a new Columbus and repeatedly attempted to interest various rulers in financing a trip to the Indies.⁵⁵ Where he failed, others, working for monarchs with overseas empires, succeeded. In 1570, Philip II of Spain ordered his royal

---

[51] The role of witnessing in early modern science has been most famously discussed in Steven Shapin and Simon Schaffer, *Leviathan and the Air-Pump: Hobbes, Boyle and the Experimental Life* (Princeton, N.J.: Princeton University Press, 1985), with attention to the development of an experimental community in mid-seventeenth-century England. The case of Oviedo suggests how the problems of long-distance communication in early accounts of the Indies already raised the issues of credibility and certainty (see Harris, Chapter 16, this volume).

[52] For the marvels of the East, see Lorraine Daston and Katharine Park, *Wonders and the Order of Nature, 1150–1750* (New York: Zone Books, 1998), pp. 21–66.

[53] Urs B. Leu, "Konrad Gesner und die Neue Welt," *Gesnerus*, 49 (1992), 285. Subsequent information about Gessner's interest in the new world is derived from this article.

[54] George, "Sources," pp. 81–3, 87. She estimates that the New World accounted for 9% of the total content of Gessner's zoology.

[55] Mario Cermenati, "Ulisse Aldrovandi e l'America," *Annali di botanica*, 4 (1906), 3–56; Olmi, "'Magnus campus': I naturalisti italiani di fronte all'America nel secolo XVI," in Olmi, *L'Inventario del mondo*, pp. 211–52; and Findlen, "Il Nuovo Colombo."

physician Francisco Hernández (1517–1587) to sail for New Spain, charging him with the mission of writing its natural history. Hernández remained in New Spain for seven years, primarily exploring the flora and fauna of Mexico. During that period, he and his European assistants worked extensively with native artists and informants to create a catalogue of Mexican nature that would bring together the best of European and American knowledge. Hernández not only learned Nahuatl to discuss what he observed with local informants but also translated his work into Nahuatl so that, appearing simultaneously in Spanish and in the most prominent indigenous language, it would be useful to all of New Spain's inhabitants.[56] He returned to Spain in February 1577 with a ship full of seeds and roots, an extensive herbarium, and thirty-eight volumes of notes and illustrations. Unfortunately, Philip's interest in the natural history of New Spain had waned, so very little of Hernández's work saw its way into print. What did appear was financed primarily by a Roman scientific academy, the Accademia dei Lincei (1603–30), which understood the value of his work. The original manuscripts burned in a fire at the Escorial in 1671.[57] Despite the intrinsic value of Hernández's work, it was too dependent on royal patronage to proceed without the continued interest of the Spanish monarch. In his case, the ingredients that had made such an ambitious project possible were also the reasons why it was unable to come to fruition.

A similar fate befell the Raleigh expedition to Virginia. In 1584, the artist John White accompanied Sir Walter Raleigh, the English astronomer Thomas Harriot (1560–1621), and other colonists to Roanoke. Harriot described and White drew. Although they intended to publish a natural history of Virginia, the project floundered, much like the colony. Erratic interest on the part of the Crown, funding problems, and the difficulties of settling the colony all contributed to its demise. Eventually Harriot published *A Briefe and True Report of the New Found Land of Virginia* (1590), in which a few engravings based on White's drawings appeared. Others became the basis for drawings of insects in Thomas Penny's *Theatrum insectorum* (Theater of

[56] The full history of this fascinating expedition remains to be written, but there has been a resurgence of interest in Hernández and his world. See German Somolinos d'Ardois, *El Doctor Francisco Hernández: La primera expedicion cientifica en America* (Mexico City: Secretaría del Educación Pública, 1971); Jacqueline de Durand-Forest and E. J. de Durand, "À la découverte de l'histoire naturelle en Nouvelle Espagne," *Histoire, Economie et Société*, 7 (1988), 295–311; José M. López Piñero, "The Pomar Codex (ca. 1590): Plants and Animals of the Old World and from the Hernández Expedition to America," *Nuncius*, 7 (1992), 35–52; Simon Varey and Rafael Chabráu, "Medical Natural History in the Renaissance: The Strange Case of Francisco Hernández," *Huntington Library Quarterly*, 57 (1994), 125–51. For a broader account of the relations between Spain's empire and its vision of nature, see Barrera, *Science and the State*.

[57] An editorial and translation project published in 2000 will make parts of the Hernández corpus more widely available. See Simon Varey, ed., Rafael Chabrán, Cynthia Chamberlin, and Simon Varey, trans., *The Mexican Treasury: The Writings of Dr. Francisco Hernández* (Stanford, Calif.: Stanford University Press, 2000); and Simon Varey, Rafael Chabrán, and Dora B. Weiner, eds., *Searching for the Secrets of Nature: The Life and Works of Dr. Francisco Hernández* (Stanford, Calif.: Stanford University Press, 2000).

Insects), edited and published posthumously by Thomas Moffet in 1634.[58] White's animals found their way into other accounts of English expeditions to the New World. But no complete natural history ever appeared. By contrast, the natural history of Brazil was a successful collaboration that resulted in Georg Markgraf's *Historia naturalis Brasiliae* (Natural History of Brazil), edited by Willem Pies and published in 1648.[59]

The most systematic effort to integrate descriptions of non-European nature into natural history occurred at the hands of Clusius. In 1564, during a trip to Spain and Portugal to observe Iberian plants, Clusius found himself in a bookstore in Lisbon, looking at a copy of Orta's *Colloquies*. He purchased it and, recognizing that such important information would never reach a wide audience if it remained in Portuguese, he began to translate it into Latin, abridging it so that it might be more directly informative.[60] Successively, he performed the same operation on the writings of Nicolas Monardes and Cristobal Acosta.[61]

At the same time, Clusius carefully amassed a collection of New World flora and fauna that would have been the envy of any naturalist at the end of the sixteenth century. He profited from his time at the imperial court in Vienna and, more briefly, in Prague, where the Holy Roman Emperors Maximilian II and Rudolf II delighted in collecting American artifacts as talismans of their ability to command the length and breadth of the globe (see Vogel, Chapter 33, this volume). Once again, royal patronage of natural history played an important role in encouraging certain directions in the study and representation of nature. Rudolf II's sizable menagerie and considerable library on the New World helps to explain why a number of naturalists in Prague, far from trading centers such as Seville, Venice, and Amsterdam, where New World artifacts were in plentiful supply, seem to have had a great deal of knowledge about the Americas.[62] Upon returning to the Netherlands, Clusius tried to exploit the proximity of the Dutch East India Company by attempting to give its physicians and apothecaries instructions on what to collect in travel. He complained, however, that they did not

---

[58] Paul Hulton, *America 1585: The Complete Drawings of John White* (Chapel Hill: University of North Carolina Press, 1984).

[59] E. van den Boogaart, with H. R. Hoetink and P. J. P. Whitehead, eds., *Johan Maurits van Nassau-Siegen, 1604–1679: A Humanist Prince in Europe and Brazil* (The Hague: The Johan Maurits van Nassau Stichting, 1979); and P. J. P. Whitehead and M. Boeseman, *A Portrait of Dutch 17th Century Brazil: Animals, Plants and People by the Artists of Johan Maurits of Nassau* (Amsterdam: North Holland, 1989).

[60] Garcia da Orta, *Aromatum et simplicium aliquot medicamentorum apud Indos nascentium historia*, ed. and trans. Carolus Clusius (Antwerp: Plantinus, 1567). This episode is discussed in Grove, *Green Imperialism*, pp. 77–81; Hunger, *Charles de l'Escluse*, vol. 1; and Ogilvie, "Observation and Experience," pp. 372–8.

[61] Nicolas Monardes, *Simplicium medicamentorum ex Novo Orbe delatorum*, ed. and trans. Carolus Clusius (Antwerp: Plantinus, 1579); and Cristobal Acosta, *Aromatum & medicamentorum in orientali India nascentium liber*, ed. and trans. Carolus Clusius (Antwerp: Plantinus, 1582).

[62] Eliska Fucíková, *Rudolf II and Prague: The Court and the City* (New York: Thames and Hudson, 1997).

seem to appreciate sufficiently the value of his enterprise, though some animals and plants captured in trade did make their way into his collection.[63] Dutch collectors such as the physician Bernard Paludanus (1550–1633), whose museum was not as large as Aldrovandi's but surely richer in American and Asian artifacts, seem to have fared better, judging by the quantity of objects that came directly to his museum through the trade routes.

Clusius planted what marvels of the Indies he managed to obtain in the Leiden botanical garden, and consulted extensively with friends and fellow naturalists, reminding them in his letters to send gifts to augment his project. He also benefited from the more workaday knowledge of the men he employed as gardeners – whose knowledge of plants derived not from books and learned debates about nature but from their experience as practicing horticulturalists who tended the gardens of Europe as artisans rather than as scholars of nature.[64] Clusius's *Exoticorum libri decem* (Ten Books of Exotic Things, 1605) represented the summation of all that he had read, seen, and discussed about nature beyond Europe.[65] It was a supreme testimonial to the quest to find words for things that had not been seen before 1492.

SHARING INFORMATION

By the mid-sixteenth century, naturalists shared information regularly. The scientific letter became the most important tool of communication in the development of natural history (see Harris, Chapter 16, this volume).[66] Techniques of comparing and collecting specimens worked better when discussing plants rather than animals, because plants were easy to transport and preserve. Yet all information was potentially transportable as long as words and images sufficed. Letters traveled hundreds of miles, often folded around seeds or accompanying a precious piece of a friend's herbarium. Clusius, for example, obtained some of the first tulip bulbs ever seen in the Netherlands in 1569 by writing a friend in Vienna for a sample from the Habsburg ambassador to the Ottoman sultan, who had brought them back from Istanbul.[67] Such singular exchanges accumulated over the years to create entire botanical gardens, natural history collections, and ultimately publishable natural

---

[63] Ogilvie, "Observation and Experience," pp. 390–2.
[64] Claudia Swan, *The Clutius Botanical Watercolors: Plants and Flowers of the Renaissance* (New York: Abrams, 1998).
[65] The full title is: *Exoticorum libri decem: quibus animalium, plantarum, aromatum, aliorumque peregrinorum fructuum historiae describuntur*.
[66] On the subject of exchange, see Paula Findlen, "The Economy of Scientific Exchange in Early Modern Europe," in *Patronage and Institutions: Science, Technology, and Medicine at the European Courts, 1500–1750*, ed. Bruce Moran (Woodbridge: Boydell Press, 1990), pp. 5–24; Giuseppe Olmi, "'Molti amici in varii luoghi.' Studio della natura e rapporti epistolari nel secolo XVI," *Nuncius*, 6 (1991), 3–31; and Ogilvie, "Observation and Experience," pp. 166–70.
[67] Ernest Roze, *Charles de l'Escluse d'Arras, le propagateur de la pomme de terre au XVIe siècle* (Paris: J. Rothschild, 1899), p. 53.

histories. Most naturalists felt so indebted to their patrons and colleagues, given the enormous expense and effort required to produce an illustrated natural history, that they profusely thanked them when they reached the point of publication. Aldrovandi, perhaps, took this to an extreme when he signed the title page of all his books *Ulyssis Aldrovandi et amicorum* – Ulisse Aldrovandi and friends.[68] Because it had taken him most of a lifetime to publish even the first volume of his natural history, which appeared in print in 1599 when Aldrovandi had achieved the ripe old age of seventy-seven, he had many debts to acknowledge.

The regular exchange of words and things suggests how important collaboration was in the pursuit of natural history. It was also a means for younger members of the community to introduce themselves to prominent naturalists. As a medical student studying in Montpellier in 1563, Jean Bauhin the Younger (1541–1612) sent the revered Gessner some of his best dried herbs accompanied by his own descriptions, knowing that Gessner had returned to the study of botany after completing his magisterial *Historia animalium*, the most extensive zoology since the works of Aristotle and his great medieval commentator Albertus Magnus. Months later, Bauhin's herbarium languished with Gessner in Zurich, though Gessner attempted to console his young friend over the temporary loss of his specimens with the promise of naming a plant after him. Finally, in July 1565, Gessner wrote, "Now I have your dried herbs in hand and soon, with the help of God, I will end my study of them to return them to you, as you ask."[69] Surely only Gessner was surprised to discover that Bauhin ceased to be so generous with his senior colleague after this experience, a lesson that many young naturalists learned in dealing with their elders, who did not always repay their gifts with generosity.[70]

What did a naturalist such as Gessner do with the materials scholars sent him? "I have more notes in my head, perhaps, than on slips of paper," he confessed to Fuchs in 1556.[71] Yet those slips of paper were crucial to Gessner's system of recording information. Like many Renaissance scholars, he reorganized all the information he received – from books, letters, and observations – into a format that allowed him gradually to create descriptions

---

[68] See, for example, Ulisse Aldrovandi, *Ornithologiae, hoc est de avibus historiae libri* (Bologna: Franciscus de Franciscis Senensis, 1599–1603). On Aldrovandi, see Giuseppe Olmi, *Ulisse Aldrovandi: Scienza e natura nel secondo Cinquecento* (Trent: Libera Università degli Studi di Trento, 1976); Sandra Tugnoli Pattaro, *Metodo e sistema delle scienze nel pensiero di Ulisse Aldrovandi* (Bologna: Cooperativa Libraria Universitaria Editrice Bologna, 1981); and Findlen, *Possessing Nature*.
[69] Conrad Gessner, *Vingt lettres à Jean Bauhin fils (1563–65)*, trans. Augustin Sabot (Saint-Étienne: Publications de l'Université de Saint-Étienne, 1976), p. 88.
[70] On this general subject, see Findlen, "The Formation of a Scientific Community."
[71] John L. Heller and Frederick G. Meyer, "Conrad Gesner to Leonhart Fuchs, October 18, 1556," *Huntia*, 5 (1983), 61. I have modified the translation slightly. On Gesner, see Hans Wellisch, "Conrad Gesner: A Bio-Bibliography," *Journal of the Society for the Bibliography of Natural History*, 7 (1975), 151–247; Braun, *Conrad Gessner*; and Udo Friedrich, *Naturgeschichte zwischen artes liberales und frühneuzeitlicher Wissenschaft: Conrad Gesners "Historia animalium" und ihre volkssprachliche Rezeption* (Tübingen: Max Niemeyer, 1995).

of individual objects that were as complete as the information then available allowed. "I have already noted the names of each of those [plants] that you describe, and I have sorted them in my notebooks, so that when I am ready [to write] the history of each one, I examine your descriptions in detail," he told Bauhin.[72] He did the same thing with information on English nature provided by William Turner and John Caius.[73] The letter, in short, was a form of communication that facilitated the encyclopedic process of gathering and sorting information. It allowed a town physician such as Gessner, who rarely left Zurich after 1546 (except to ascend to the Alps to botanize), to command the world. It exemplified well the mixing of words and things in the study of nature.

The size, expense, and relative fragility of animals in comparison with plants meant that they circulated in more finite quantities than portable plants and inanimate stone. Although some letters occasionally accompanied crates of skins, bones, and stuffed bodies, they more often described animals with words and images. Both Gessner and Aldrovandi, the two sixteenth-century naturalists most enamored of the idea of writing an entire history of animals, published descriptions of many animals that they had not personally seen, though they provided more information on those they were able to dissect as well as read about. When Clusius wished to gain access to the first birds of paradise to arrive in Europe with their feet – for the first two-thirds of the sixteenth century, it was believed that they did not have feet because native hunters cut them off when preparing them for European traders[74] – he was annoyed to discover that all examples had gone to Prague at the request of Rudolf II.[75] Field mice, snakes, and similarly ordinary fauna were in plentiful supply, but more exotic animals often could be seen alive only in the menageries of princes or seen stuffed in the cabinets of curiosities owned by many scholars, nobles, and patricians. When Charles IX's toucan died after the lengthy voyage from Brazil to Paris, a typical fate for most New World animals, the king gave the body to his royal surgeon, Ambroise Paré, to embalm and display.[76] Even though it began to mold, Paré continued to

---

[72] Gessner, *Vingt lettres*, p. 81. For a more detailed discussion of the use of commonplace books in Renaissance science, see Ann Blair, "Humanist Methods in Natural Philosophy: The Commonplace Book," *Journal of the History of Ideas*, 53 (1992), 541–51.

[73] Vivian Nutton, "Conrad Gesner and the English Naturalists," *Medical History*, 29 (1985), 93–7.

[74] Wolfgang Harms, "On Natural History and Emblematics in the Sixteenth Century," in *The Natural Sciences and the Arts: Aspects of Interaction from the Renaissance to the Twentieth Century* (Acta Universitaria Upsaliensis, nova ser., 22) (Uppsala: Almquist and Wiksell, 1985), pp. 67–83.

[75] Ogilvie, "Observation and Experience," p. 383. On the role of nature in the court of Rudolf II, see Thomas DaCosta Kaufmann, *The Mastery of Nature: Aspects of Art, Science, and Humanism in the Renaissance* (Princeton, N.J.: Princeton University Press, 1993), pp. 100–28, 174–94; and Fučíková, *Rudolf II and Prague*.

[76] Ambroise Paré, *On Monsters and Marvels*, trans. Janis L. Pallister (Chicago: University of Chicago Press, 1982), p. 139. For more on the collecting culture of this period, see Kryzsztof Pomian, *Collectors and Curiosities: Paris and Venice, 1500–1800*, trans. Elizabeth Wiles-Portier (London: Polity, 1990); Joy Kenseth, ed., *The Age of the Marvelous* (Hanover, N.H.: Hood Museum of Art, 1992); Olmi, *L'inventario del mondo*; Findlen, *Possessing Nature*; Horst Bredekamp, *The Lure of Antiquity and*

show it to visitors because it was rare to have this kind of specimen, however imperfect.

The difficulties of acquiring and preserving animals made illustrations even more important to zoology than to botany. They were often the only source of visual information and frequently accompanied the letters naturalists and their patrons exchanged. The printed image was usually created from an artist's or a naturalist's drawing. In the wake of Otto Brunfels's advertisement in 1530 of his herbal as a repository of "living images" – representations drawn from the living object – the idea of the image as a means of conveying crucial information about nature took shape. Fuchs declared in 1542 that "nature was fashioned in such a way that everything may be grasped by us in a picture."[77] He argued strongly for the particularity of each image, criticizing previous illustrators for making images generic rather than particular. Coordinating a well-drawn image based on observation with the catalogue of names associated with a particular plant or animal became a powerful shorthand for the object itself. In the second half of the sixteenth century, illustrations grew in number and importance. Gessner's *Historia animalium* was an encyclopedia of images as much as words, a total of approximately 1,200 woodcuts that included illustrations published in earlier natural histories and cosmographies as well as images made specifically for Gessner's zoology. Aldrovandi had a team of artists and engravers create approximately 3,000 wood blocks from his archive of natural images.[78]

By the 1550s, most naturalists advertised their publications as being filled with "portraits drawn from nature."[79] They championed the image as a guarantor of words because agreement between the two constituted a double form of proof. "If the ancients had drawn and painted all the things that

---

the *Cult of the Machine*, trans. Allison Brown (Princeton, N.J.: Markus Weiner, 1995); Luca Basso Peressut, ed., *Stanze delle meraviglie: I musei della natura tra storia e progetto* (Bologna: Cooperativa Libraria Universitaria Editrice Bologna, 1997); and Daston and Park, *Wonders*, pp. 135–72.

[77] Leonhart Fuchs, *De historia stirpium commentarii insignes* [1542], sig. B1r, in Sachiko Kusukawa, "Leonhart Fuchs on the Importance of Pictures," *Journal of the History of Ideas*, 58 (1997), 411. My additional comments on Fuchs are based on this article.

[78] William J. Ashworth, "Emblematic Natural History in the Renaissance," in Jardine, Secord, and Spary, eds., *Cultures of Natural History*, pp. 18, 27–9; Braun, *Conrad Gessner*, p. 68; and Giuseppe Olmi, "La bottega artistica di Ulisse Aldrovandi," in *De piscibus: La bottega artistica di Ulisse Aldrovandi*, ed. Enzo Crea (Rome: Edizioni dell'Elefante, 1993), p. 19. On natural history illustration, see also Olmi, *L'Inventario del mondo*, pp. 21–117; *Immagine e natura: L'Immagine naturalistica nei codici e libri a stampa delle Biblioteche Estense e Universitaria, Secoli XV–XVII* (Modena: Mucchi, 1984); and Brian Ogilvie, "Image and Text in Natural History, 1500–1700," in *The Power of Images in Early Modern Science*, ed. Wolfgang Lefèvre, Jürgen Renn, and Urs Schöpflin (Basel: Birkhäuser, 2003), p. 141–66.

[79] This phrase is from Pierre Belon's *L'Histoire de la nature des oyseaux, avec leurs descriptions, & naïfs portraicts retirez du naturel* (Paris: Guillaume Cauellat, 1555). Other examples include Guillaume Rondelet's *Liber de piscibus marinis in quibus verae piscium effigies expressae sunt* (Lyon: Matthew Bonhomme, 1554) and his *Universae aquatilium historiae pars altera cum veris ipsorum imaginibus* (Lyon: Matthew Bonhomme, 1555). Lucia Tongiorgi Tomasi also points to the frequency of such phrases as *vivae eicones*, *verae effigies*, and *vrais portraits* in her "Ulisse Aldrovandi e l'immagine naturalistica," in Crea, ed., *De piscibus*, p. 38.

they described," wrote Aldrovandi, "one would not find so many doubts and endless errors among writers."[80] Such an optimistic statement masked a number of complications that Aldrovandi himself knew well from his extensive efforts to create a visual archive of nature. Although many images were the product of personal observation, any naturalist who aspired to write a universal history of nature had to rely on the reports of others in order to create an adequate supply of information. Ascertaining the reliability of such material was a difficult task, which is why Pliny advised his readers not to trust images at all. Illustrations, however helpful, did not necessarily record the veracity of nature, though they did a great deal to record its variety, because conveying knowledge from the eye to the hand produced its own kind of errors. A drawing, let alone a print made from a drawing, yielded no simple truth of nature (see Niekrasz and Swan, Chapter 31, this volume).

Knowing how difficult it was to pass final judgment on all the information that was out there, naturalists hoped that the communal project of natural history ultimately would yield a common truth. "[I]f every person offers his observations in the public good," wrote Gessner in 1556, "there is hope that at some time it will come about that from them a single, perfectly complete work will be produced by someone who will add the final touch."[81] By the time of Gessner's death in 1565, the project of writing a new natural history was well under way but far from complete. Several decades later, Francis Bacon (1561–1626) would return to the idea of natural history as a collective enterprise, writing copiously about the need to realize the kind of project that Gesner had imagined decades earlier.[82]

The constant flow of materials among European naturalists by the late sixteenth century underscores their passionate belief that Gessner's dream of completing the project of natural history was a goal to be taken seriously.[83] If scientific letters at the dawn of the seventeenth century began to betray a certain skepticism about the plausibility of this project, they nonetheless maintained the structures of gathering and sharing information that had been put into place in the service of an encyclopedic ideal (see Harris, Chapter 16, this volume). "Since so many famous men are exhausting themselves in this work, it is to be hoped that the study of the knowledge of plants cannot help

---

[80] Ulisse Aldrovandi, "Avvertimenti del Dottor Aldrovandi," in *Trattati d'arte del Cinquecento*, ed. Paola Barocchi, 3 vols. (Bari: Laterza, 1960–2), 2: 513.
[81] Heller and Meyer, "Conrad Gesner," p. 67.
[82] See Francis Bacon's *Sylva sylvarum* as an example of the early seventeenth-century efforts to reprise this project.
[83] The image of natural history as an encyclopedic project that could nonetheless be completed was shared by many Renaissance naturalists. When Gessner died in 1565, Aldrovandi essentially replaced him as the naturalist most committed to writing a universal history of nature in imitation of the ancients. Later naturalists such as Clusius and Bauhin confined themselves to more restricted projects, though we should also think of the late seventeenth-century naturalist John Ray and his eighteenth-century successor, Carolus Linnaeus, as worthy successors to Gessner and Aldrovandi in the goal of studying all nature.

but progress," wrote Clusius to one of his correspondents in 1566.[84] This prediction turned out to be an accurate assessment of the kind of science that natural history promoted – knowledge whose gradual accumulation eventually allowed naturalists to derive general principles from it. When Caspar Bauhin's *Pinax Theatri Botanici* (Index of a Botanical Theater), appeared in 1623, the most important attempt to create a unified language and structure for all known plants prior to the work of John Ray and Carolus Linnaeus, it described and classified over 6,000 species.[85] Very few of these discoveries – a tenfold increase from what had been known a century earlier – were Bauhin's. Like the animals in Gessner's and Aldrovandi's large Latin folios, they represented the collective knowledge of a scientific community.

THE EMERGENCE OF THE NATURALIST

But who belonged to that community? What, for that matter, defined it?[86] The identity of the naturalist emerged from many diverse ingredients. He – and, with the exception of women such as seventeenth-century German artist and entomologist Maria Sybilla Merian (1647–1717), there seem to have been few women before the eighteenth century who actively collected, described, and drew nature, except as part of a family project (see Cooper, Chapter 9, this volume), let alone published the results of their research under their own name[87] – was most likely a physician by training, but not necessarily a practitioner of the healing arts. Felix Platter, for instance, wrote of his classmate Clusius that he "never practiced medicine."[88] The same was also true of Aldrovandi, who infuriated some of Bologna's physicians and apothecaries by claiming that his knowledge of nature made him a

---

[84] Roze, *Charles de l'Escluse*, p. 40.
[85] André Cailleux, "Progression du nombre d'espèces des plantes décrites de 1500 à nos jours," *Revue d'Histoire des Sciences*, 6 (1953), 44. The development of taxonomy in natural history has received the most attention in the realm of botany. Its broadest treatment can be found in Scott Atran, *Cognitive Foundations of Natural History: An Anthropology of Science* (Cambridge: Cambridge University Press, 1990). This work is an explicit response to Michel Foucault's account of the rise of classification and argues against Foucault's interpretation of classification as a reaction to the Renaissance doctrine of signatures in his *The Order of Things: An Archaeology of the Human Sciences* (New York: Random House, 1970). On Bauhin's contributions, see Reeds, *Botany*; Reeds, "Renaissance Humanism"; and Brian Ogilvie, "Encyclopedism in Renaissance Botany: From *Historia* to *Pinax*," in *Pre-Modern Encyclopedic Texts: Proceedings of the Second COMERS Congress, Groningen, 1–4 July 1996* (Leiden: E. J. Brill, 1997), pp. 89–99.
[86] This question has been most specifically addressed in Findlen, "The Formation of a Scientific Community"; and Ogilvie, "Observation and Experience," pp. 131–70.
[87] Kurt Wettengl, ed., *Maria Sibylla Merian: Artist and Naturalist, 1647–1717* (Frankfurt am Main: Gert Hatje, 1998); and Natalie Zemon Davis, *Women on the Margins: Three Seventeenth-Century Lives* (Cambridge, Mass.: Belknap Press, 1995), pp. 140–202. Women also appear in other contexts as illustrators of natural histories. See Agnes Arber, "The Colouring of Sixteenth-Century Herbals," reprinted in her *Herbals*, pp. 317–18; and J. D. Woodley, "Anne Lister, Illustrator of Martin Lister's *Historiae conchyliorum*," *Annals of Natural History*, 21 (1994), 225–9.
[88] Platter, *Beloved Son Felix*, p. 74.

better judge of medicines than they, despite his having never seen a patient. Medical education, in short, provided the skills but did not necessarily define a vocation for many of Europe's leading naturalists.[89]

The growing popularity of natural history created a variety of opportunities for aspiring naturalists to make a living studying nature. A number, of course, already had established professional identities as physicians and apothecaries before they embarked on the study of nature.[90] Others were drawn to the study of medicine because of their interest in nature. The case of Gesner – who taught Greek and Latin grammar to Zurich schoolboys and then taught mathematics, natural philosophy, and ethics at a local academy before becoming a town physician in 1554 – exemplifies this other tendency.[91] Yet another group abandoned the study of medicine as a result of their encounter with a dynamic generation of university professors who encouraged their interests in zoology, botany, and mineralogy.

University-educated naturalists enjoyed a growing number of professional opportunities by the late sixteenth century. They were leading candidates for positions in medical faculties as professors of medicinal botany and as demonstrators in university botanical gardens. (The latter category included a number of naturalists who did not have university degrees.)[92] Many also received court appointments in such cities as Madrid, Florence, Mantua, Ferrara, Prague, and Vienna – wherever rulers prized the study of nature, collected exotic marvels, and promoted the improvement of medicine.[93] Prior to the appearance of the 1554 Latin edition of his commentary on Dioscorides' *De materia medica*, Pier Andrea Mattioli (1501–1577) was a town physician in northern Italy with no strong network of patrons. After its initial success, he became an imperial physician, with the resources of the Habsburgs at his disposal to make his work even more magnificent. Imperial artists and

---

[89] In other instances, teaching medicinal botany became a means to a professorship but was clearly secondary to the desire to practice medicine. The case of the anatomist Gabriele Falloppia, who chastised Aldrovandi for spending too much time on subjects that he considered beneath a physician's dignity, exemplifies this other pattern; see Findlen, *Possessing Nature*, p. 255. The Bauhin family, which produced several generations of physicians who studied nature, also offers examples of physicians who felt that their work as naturalists at times conflicted with their medical practice; see Reeds, *Botany*. Even Gesner, who spent most of his life writing and publishing natural histories, complained that his need to earn a living from his publications conflicted with his work as a physician; see Wellisch, "Conrad Gesner," p. 164.

[90] Very few apothecary-naturalists published works of natural history, though many contributed to the exchange of information. Ferrante Imperato in Naples and Francesco Calzolari in Verona were important exceptions in this regard.

[91] Wellisch, "Conrad Gesner," pp. 153–63.

[92] Azzi Visentini, *Orto*; Tongiorgi Tomasi and Tosi, *Giardino*, pp. 27–114, passim, and L. Tjon Sie Fat and E. de Jong, eds., *The Authentic Garden: A Symposium on Gardens* (Leiden: Clusius Foundation, 1991), pp. 3, 37–69, offer good case studies of the personnel employed in botanical gardens.

[93] The courtly world in which natural history thrived is outlined in Moran, *Patronage and Institutions*; Dario A. Franchini, Renzo Magonari, Giuseppe Olmi, Rodolfo Signorini, Attilio Zanca, and Chiara Tellini Perina, *La scienza a corte: Collezionismo eclettico natura e immagine a Mantova fra Rinascimento e Manierismo* (Rome: Bulzoni, 1979); Kaufmann, *Mastery of Nature*; and Fucíková, *Rudolf II and Prague*.

engravers in Vienna worked with Mattioli to enhance his commentary, which contained approximately 1,200 images by the time of his death.[94] The Holy Roman Emperor Maximilian II rewarded Mattioli lavishly, ennobling his entire family.

As we have already seen in a number of cases, however, some important naturalists had no medical training at all. Oviedo occupied a variety of different posts in the nascent Spanish empire. He was a literate civil servant who used his abilities to write and assess evidence to create the natural history of the Indies. His appointment as "Official Chronicler of the Indies" in 1532 represented a logical culmination of his efforts to distinguish himself from the many other bureaucrats employed in the Americas by becoming the Spanish empire's Pliny. In his instance, it was his proven ability to write history for the state that led Charles V and his councilors to create a formal position that defined the work he was already doing.

The study of law, like the profession of the notary, also proved to be fertile ground in which naturalists could flourish. Most famous is the case of Francis Bacon, whose legal training and courtroom experience markedly shaped his ideas about evidence.[95] But he was not alone in finding in law and natural history compatible ways to exercise the mind. In late sixteenth-century Naples, one of the greatest naturalists in a city filled with interesting scholars and philosophers was the lawyer and nobleman Fabio Colonna (1567–1650). By the 1590s, Colonna was a judge whose duties led him to travel widely in southern Italy. In between cases, he observed the plants and fossils of the region. His *Phytobasanos* (Touchstone of Plants, 1592) identified and carefully described living examples of twenty-six plants found in Dioscorides' *De materia medica*. It also included six plants of southern Italy that no one had described previously and introduced the term "petal" as a standard descriptor in botanical terminology.[96]

Colonna's later works described over two hundred plant species using the same principles of "legal testimony" that he had advertised in his *Phytobasanos*. He advocated a fairly radical empiricism that emphasized direct

---

[94] Findlen, "The Formation of a Scientific Community."
[95] On Bacon and natural history, the literature continues to grow. See especially Julian Martin, *Francis Bacon, the State, and the Reform of Natural Philosophy* (Cambridge: Cambridge University Press, 1992); Antonio Pérez-Ramos, *Francis Bacon's Ideal of Science and the Maker's Knowledge Tradition* (Oxford: Clarendon Press, 1988); and Findlen, "Francis Bacon and the Reform of Natural History in the Seventeenth Century," in *History and the Disciplines: The Reclassification of Knowledge in Early Modern Europe*, ed. Donald Kelley (Rochester, N.Y.: University of Rochester Press, 1997), pp. 239–60. On the legal and cultural environment of Bacon's ideas about evidence, see Daston, "Baconian Facts"; Barbara Shapiro, "The Concept 'Fact': Legal Origins and Cultural Diffusion," *Albion*, 26 (1994), 1–26; and Shapiro, *Beyond "Reasonable Doubt" and "Probable Cause": The Anglo-American Law of Evidence* (Berkeley: University of California Press, 1991).
[96] Colonna has not yet received the full study that he deserves. See Greene, *Landmarks of Botanical History*, 2: 835–46; Nicoletta Morello, *La nascita della paleontologia nel Seicento: Colonna, Stenone e Scilla* (Milan: Franco Angeli, 1979); and Martin J. Rudwick, *The Meaning of Fossils: Episodes in the History of Paleontology*, 2nd ed. (Chicago: University of Chicago Press, 1985), pp. 42–4.

evidence of the senses over other forms of knowledge. "The observable thing perfectly gives a method," he wrote in 1618.[97] By observing animals and animal-like fossils, Colonna came to believe that fossils were animal remains or impressions, organic in origin, in contrast with the prevailing explanation that they were mere images of animals made in stone by a playful nature.[98] Others came to similar conclusions in the 1660s and 1670s by different means, and by the eighteenth century this new theory prevailed. Yet it was a lawyer who first argued this view by offering sustained evidence, even drawing and engraving his own images in order to unify every stage of his presentation. In doing this, he emulated the activities of artisan-naturalists such as Leonardo da Vinci (1452–1519) and the sixteenth-century French ceramicist Bernard Palissy (ca. 1510–1590), both of whom advocated a similar view of fossils and drew what they saw.[99]

Behind the divergent educational training and professional itineraries of many naturalists lay some core intellectual principles that only gradually eroded. To a certain degree, the identity of the naturalist continued to be defined by his relationship to the writings of the ancients. Many Renaissance naturalists proudly bore the epithet of the "new Aristotle" or the "new Pliny." At the beginning of the first volume of his *Ornithologia* (1599), Aldrovandi had the engraver record the following words beneath his portrait: "This is not you, Aristotle, but an image of Ulysses: though the faces are dissimilar, nonetheless the genius is the same."[100] The past continued to be a productive metaphor for the present. Oviedo and Hernández identified deeply with Pliny, undoubtedly recognizing themselves as heirs to Pliny's project of describing the nature of an empire – to such a degree in the latter's case that he produced one of the greatest translations of and commentaries on Pliny in the late sixteenth century.[101]

As the medical model of natural history, dominated by medicinal botany, waned at the end of the sixteenth century, the encyclopedic premise of natural

---

[97] Fabio Colonna, *La sambuca lincea* (Naples: C. Vitale, 1618). This passage is quoted in Giuseppe Olmi, "La colonia lincea di Napoli," in *Galileo e Napoli*, ed. F. Lomonaco and M. Torrini (Naples: Guida, 1987), p. 54, which places Colonna's work in the context of the Accademia dei Lincei's colony in Naples.

[98] On this symbolic way of looking at nature, against which Colonna reacted, see Foucault, *Order of Things*, pp. 17–45; Ashworth, "Natural History"; and Findlen, "Jokes of Nature and Jokes of Knowledge: The Playfulness of Scientific Discourse in Early Modern Europe," *Renaissance Quarterly*, 43 (1990), 292–331.

[99] Stephen Jay Gould, *Leonardo's Mountain of Clams and the Diet of Worms* (New York: Harmony, 1998), pp. 17–44; and Bernard Palissy, *Admirable Discourses*, trans. Aurèle La Rocque (Urbana: University of Illinois Press, 1957).

[100] Ulisse Aldrovandi, *Ornithologiae* (Bologna: Franciscus de Franciscis Senensis, 1599), n.p. The Sienese physician Pier Andrea Mattioli, who wrote one of the most popular commentaries on Dioscorides in the sixteenth century, also expressed a similar relationship to his ancient author; see Findlen, "The Formation of a Scientific Community"; and Jerry Stannard, "P. A. Mattioli: Sixteenth Century Commentator on Dioscorides," *Bibliographic Contributions, University of Kansas Libraries*, 1 (1969), 59–81.

[101] Gerbi, *Nature in the New World*, pp. 386–7; and Varey and Chabráu, "Medical Natural History."

history took on greater importance. "I take a strong interest in all of nature's curiosities," Clusius proclaimed in 1566, even though he concentrated primarily on the study of plants.[102] Aldrovandi wrote emphatically that he studied nature, "not as a physician, according to that more common practice, but as a philosopher."[103] In such statements, he underscored his distinctive contribution to natural history, which entailed its elevation to the status of natural philosophy (an Aristotelian ideal with which Bacon partly agreed when he made natural history the foundation of all natural philosophy by the 1620s) and the expansion of natural history from exclusively medical subjects to a more comprehensive account of nature. He, after all, was the only professor of natural history to occupy a position that had been specifically redefined from *lectura de simplicibus* (readership in simples) in 1543 to *lectura philosophiae naturalis ordinaria de fossilibus, plantis et animalibus* (ordinary readership in the natural philosophy of fossils, plants, and animals) in 1559.[104] The shift in title reflected an active attempt to make natural history a discipline autonomous from medicine. "Yet in these matters I am less a doctor than a natural philosopher," wrote the great English naturalist John Ray a century later, echoing the heritage of Aldrovandi, Clusius, and Bauhin.[105] Natural history would continue to be closely associated with medicine through the eighteenth century. But increasingly its leading practitioners studied nature apart from medicine.

Specialization was not the hallmark of the sixteenth- and early seventeenth-century naturalist. Gesner and Aldrovandi, imitating ancient writers such as Aristotle and Pliny, established the image of the naturalist as someone who aspired to describe a universal nature. Although many naturalists disagreed on what the best starting point might be – animals or plants? fossils or insects? – and how to arrange and interpret the details, they did not question the encyclopedism of their subject. The popularity of such works as Francis Bacon's *Novum organum* (New Organon) and *Instauratio magna* (Great Instauration), both of which appeared in 1620 and were read by the scholarly community in the second half of the seventeenth century, served to further the image of the naturalist as a collector, experimenter, and system-builder who was concerned with general inquiry into nature. "[B]estow on the world a general history of nature," wrote one friend to John Ray (1627–1705) in 1684, capturing the vastness of the enterprise in this pithy phrase.[106]

---

[102] Roze, *Charles de l'Escluse*, p. 43. Clusius's identity as a naturalist is discussed extensively in Ogilvie, "Observation and Experience."
[103] Ulisse Aldrovandi, *Lettere a Costanzo Felici*, ed. Giorgio Nonni (Urbino: Quattro Venti, 1982), p. 79.
[104] Findlen, *Possessing Nature*, pp. 253–6.
[105] Charles E. Raven, *John Ray, Naturalist: His Life and Works*, 2nd ed. (Cambridge: Cambridge University Press, 1986; orig. publ. 1950), p. 157. Of course, such efforts did not sever the possible uses of natural history for medicine, and many early modern naturalists actively continued to cultivate such ties, contra Aldrovandi and Clusius. See Harold J. Cook, "Natural History and Seventeenth-Century Dutch and English Medicine," in *The Task of Healing: Medicine, Religion, and Gender in England and the Netherlands, 1450–1800* (Rotterdam: Erasmus, 1996), pp. 253–70.
[106] Raven, *John Ray, Naturalist*, p. 212.

By the late seventeenth century, as Ray's numerous publications joined the works of many other scholars in redefining the natural world with a greater level of precision and detail than Gesner could ever have imagined, the context of natural history had changed dramatically. Even as the idea of a universal history remained important, divisions emerged among naturalists with regard to the best means of achieving this goal. If the sixteenth century was characterized by its simultaneous fascination with old books and new objects, the seventeenth century can be described as an age in which natural history became more conscious of the place of history in studying nature, more open to the role of instruments and experimentation, and more reliant on European overseas empires to provide materials for study. When Ray published his *Historia plantarum* (History of Plants) in 1686, he felt no need to reiterate the entire history of botany that preceded him because, he observed, "the Bauhin brothers had done this thoroughly."[107] The historical consciousness to which Ray alluded shaped natural history in at least two distinct ways. First, it conjured up the image of natural history as a cumulative discipline in which successive generations tried valiantly to tame unruly masses of information by developing better classification schemes and methodologies to identify the crucial characteristics of animals, plants, and minerals, such as the nascent work to classify plants according to their reproductive characteristics rather than the shape of their leaves. By Ray's time, Caspar Bauhin's *Pinax* was the starting point for any subsequent work in botany because it provided such a masterful synthesis of all that was known about plants before 1623, cross-referencing names and characteristics.

Although he admired the utility of such classificatory works and participated in their perfection, Ray was already less sanguine about the naturalist's holy grail: the quest for a perfect method by which all nature could be reduced to a simple set of characteristics. In the seventeenth century, the fascination with universal and artificial languages closely connected natural history to various intellectual schemes to reduce knowledge to a set of simple unifying principles. In his own work, Ray advocated the utility of classification for beginning naturalists but felt that it ultimately had its limits in accounting for all of nature's variety.[108] He did not want naturalists to oversimplify nature in the search for the crucial characteristics of different species, addressing evident mistakes of classification, such as placing whales among the fishes – an error Ray caught as he began to transform such categories as "quadruped," which classified animals according to the nature and arrangement of their limbs, into Linnaeus's idea of a "mammal."[109] History, in its root sense, reminded naturalists that resolving problems of nomenclature did not count as understanding nature.

---

[107] Ibid., p. 219.
[108] See John Ray, *Methodus plantarum* (London: Henry Faithhorne and John Kersey, 1682); and M. M. Slaughter, *Universal and Artificial Languages in the Seventeenth Century* (Cambridge: Cambridge University Press, 1982).
[109] On this episode, see Schiebinger, *Nature's Body*, pp. 40–74.

The second aspect of the naturalist's engagement with history regarded a transformation in the understanding of "natural history" as a historical as well as descriptive enterprise. The general portrait of nature prior to 1660 described an invariate rather than dynamic nature. All species existed across time, if not space. Such ideas found validation not only in Aristotle but also in the Bible, neither of which suggested that nature had changed since the moment of Creation. By the 1660s, the problem of understanding fossil formation seemed on the verge of undermining this ancient principle. Naturalists in many different parts of Europe – from Robert Hooke in England to Nicolaus Steno, a Danish scholar transplanted to Italy – argued that fossils were remains of creatures that had once existed.[110] As knowledge of fossils expanded, it became apparent that not every fossil corresponded to a living animal or plant. The fossil record was a historical record that eventually would revise the scientific understanding of human history in relation to nature's history. Naturalists in this early generation of fossil hunters were indeed historians – to such a degree that they often wrote about the natural and human history of a location simultaneously.[111] They struggled valiantly to contain their findings within biblical time, developing elaborate arguments about the role of the Flood in shaping the fossil record. By the mid-eighteenth century, naturalists such as Georges Leclerc, Comte de Buffon, began to suggest that nature's history significantly predated the human record.

The 1660s also represent an important watershed in the development of new techniques with which to observe nature. New instruments and a nascent culture of experimentation profoundly transformed the meaning of observation. The microscope, an instrument known to members of the Accademia dei Lincei in Rome as a modified version of Galileo's telescope, first gained popular currency with the publication of Robert Hooke's *Micrographia* (1665), which closely associated the compound microscope with the early Royal Society. In the hands of patient observers such as Marcello Malpighi, Jan Swammerdam, and Antonie van Leeuwenhoek, both handheld and compound microscopes yielded a myriad of surprising and fundamentally disturbing facts about nature.[112] Hooke, for instance, invited his readers to contemplate the wonders of an insect eye, the complexity of a flea, the patterns in a segment of cork, and the similarities between fossils and their living analogues when magnified. More controversially, Leeuwenhoek's sperm seemed to validate the idea of preformation when he claimed to see perfectly formed creatures inside them – and produced equally strong counterarguments from

[110] See Rhoda Rappaport, *When Geologists Were Historians, 1665–1750* (Ithaca, N.Y.: Cornell University Press, 1997), which contains an excellent bibliography.
[111] A good example of this genre is Robert Plot's *Natural History of Oxford-shire* (Oxford: Printed at the Theatre, 1677).
[112] Catherine Wilson, *The Invisible World: Early Modern Philosophy and the Invention of the Microscope* (Princeton, N.J.: Princeton University Press, 1995); Edward Ruestow, *The Microscope in the Dutch Republic: The Shaping of Discovery* (Cambridge: Cambridge University Press, 1996); and Marion Fournier, *The Fabric of Life: Microscopy in the Seventeenth Century* (Baltimore: Johns Hopkins University Press, 1996).

those who believed that the egg contained everything, making the microscope at best an indecisive tool for solving some of nature's most pressing mysteries.[113] After the initial enthusiasm for the microscope subsided, it was unclear in its first century of existence whether it would become anything more than a pleasing toy that introduced natural history to a general audience.[114]

By contrast, the experimental studies of insects in the late seventeenth century offered an excellent example of the development of observation from simple description to repeated observation and testing of diverse natural phenomena. Works such as the Tuscan physician Francesco Redi's (1626–1687) celebrated *Esperienze intorno alla generazione degl'insetti* (Experiments on the Generation of Insects, 1668) and Jan Swammerdam's *Historia insectorum generalis* (General History of Insects, 1669) used naked-eye observation and simple magnifying glasses more than microscopy to argue against the ancient idea that insects generated spontaneously. Redi, for instance, created a series of controlled experiments to demonstrate that maggots formed only when rotting meat was uncovered and larvae deposited, and never spontaneously from the animal itself.[115] No doubt inspired by reports of the work of Redi and Swammerdam, Maria Sibylla Merian recalled that she had begun her study of caterpillars in 1670 at the tender age of thirteen. By 1679, her *Raupenbuch* (Book of Caterpillars) was complete – the first of several publications illustrating and describing insect metamorphoses. So committed was Merian to the study of living nature that she described a gift of dead insects in 1672 as being "useless" to her work.[116] No inert specimen could provide the kind of observation that she and her contemporaries needed to study the caterpillar's life cycle.

Merian's dedication to the study of living nature led her to spend two years in the Dutch colony of Surinam between 1699 and 1701 with her younger daughter, Dorothea Maria, so that she might see for herself the insects that travelers had brought to Amsterdam and described in their travel accounts. Her *Metamorphosis insectorum surinamensium* (Metamorphosis of the Insects of Surinam, 1705) offers an excellent example of the ways in which European colonization helped to shape the culture of natural history by pushing it more insistently outside the boundaries of Europe itself (see Vogel, Chapter 33, this volume). Although ridiculed by the Dutch colonists for her interest in anything beyond the sugar cane that was the mainstay of the local economy, Merian nonetheless found herself in possession of a rich repository of natural lore that came directly from the African and Indian slaves who worked in

---

[113] Clara Pinto Correia, *The Ovary of Eve: Egg and Sperm and Preformation* (Chicago: University of Chicago Press, 1997).
[114] The same thing might also be said of Robert Boyle's air pump, also a product of the 1660s. It became an important site in which to examine nature in an artificial state, bringing living creatures to the brink of death with every drop of air that an experimenter removed, and demonstrating the significance of air for life.
[115] For an example of this kind of work, see Francesco Redi, *Experiments on the Generation of Insects*, trans. Mab Bigelow (New York: Kraus Reprint, 1969).
[116] Wettengl, *Maria Sibylla Merian*, p. 21.

the colony. Like the Spanish physician Hernández in the preceding century, Merian quickly understood that increasing natural knowledge through communication with native informants and making this information available to a European audience was one of the important scientific outcomes of the early age of imperialism (see Schiebinger, Chapter 7, this volume).

By the end of the seventeenth century, communicating natural knowledge had evolved well beyond random travelers' reports to become a sophisticated economy of information that linked mercantile and scientific activities. Just as Merian benefited from the Dutch presence in the New World, English naturalists began to see their own nation's overseas activities as a source of invaluable information. In 1661, the Royal Society appointed a committee that drew up a list of important questions about remote places that they might ask of merchants, sailors, and travelers.[117] In the next few decades, individual naturalists such as the physician John Woodward produced pamphlets instructing travelers how to observe and collect a nature that he would never see in person. Woodward's *Brief Tract for Making Observations in All Parts of the World* (1696) exemplifies the more entrepreneurial culture of late seventeenth-century natural history, in which naturalists encouraged travelers to become observers by offering them a share in the profit of collecting exotic natural objects for their wealthy patrons. Nature had indeed become a global commodity.

From the 1660s through the 1690s, the Baconian image of natural history as a common enterprise that defined the entire scientific community seemed on the verge of success. The nascent French and British overseas empires created a political and economic framework that built upon the earlier activities of the Portuguese, Spanish, and Dutch in making natural history the science of empire (see Chapters 16 and 33, this volume).[118] They provided an infrastructure in which to gather information to a degree that was unimaginable in 1492. At the same time, institutional developments in the realm of science gave natural history greater prominence. Early scientific societies, from the Accademia dei Lincei (1603–30) in Rome to the Royal Society (founded 1660) in London and the Académie Royale des Sciences (founded 1666) in Paris, all proposed natural history as their communal project. In many instances, Bacon's writings as well as the labors of his predecessors inspired these groups.[119]

By the time a Swedish pastor's son, Carolus Linnaeus (1707–1778), began his study of medicine at the University of Uppsala in 1728, in a facility that

---

[117] Daniel Carey, "Compiling Nature's History: Travellers and Travel Narratives in the Early Royal Society," *Annals of Science*, 54 (1997), 274–5.

[118] The most important theoretical formulation of this process remains Pratt, *Imperial Eyes*, which focuses on the eighteenth century. See also Grove, *Green Imperialism*; and Barrera, "Science and the State."

[119] The best case study of natural history within a scientific society is Alice Stroup, *A Company of Scientists: Botany, Patronage, and Community at the Seventeenth-Century Parisian Royal Academy of Sciences* (Berkeley: University of California Press, 1990).

already boasted a thriving botanical garden, he was heir to a lengthy and well-defined tradition of studying nature that had been in place for almost two centuries.[120] He guided visitors through the garden, developing the initial ideas for his *Systema naturae* (System of Nature, 1735), which laid out the basic principles of binomial nomenclature as a means of classifying not only those things that were known but all future things to be known. Linnaeus traveled abroad, especially to the Netherlands, where he worked for several years in the garden of a Dutch banker, returning home to become the most prominent member of the Uppsala medical faculty and a favorite of the Swedish nobility, who botanized with him on the weekends. He sent his disciples all over the world to collect specimens for his natural histories and obsessively edited and reedited his most important works until they swelled beyond recognition – fat tomes laden with information and erudition that seemed only to replicate some of the problems of the earlier generations. Linnaeus, in other words, pursued natural history on a scale that his predecessors could only dream of, but, on the level of practice, it was still fundamentally early modern, even as he returned to the questions of classification that naturalists such as Ray found unsatisfactory.

Had natural history changed very much by then? It certainly had routinized the process of gathering information on a global scale that had been such a thorny problem in the early sixteenth century. It had begun to grapple with problems of nomenclature in direct relation to the descriptive characteristics of animals and plants – a good example of how analysis of nature ultimately could not be divorced from experience. Issues of terminology and description were only beginning to enjoy the kind of prominence they would have by the mid-eighteenth century, and there was still much to know and describe of the natural world, as Linnaeus knew well when he trained his best students to travel far and wide in search of the rare and unknown specimen.

Natural history offered no simple or universally sustainable account of how nature worked as a whole. Its success lay with practitioners who privileged experience of nature as the foundation of all other aspects of science, which is undoubtedly why a methodical thinker such as Descartes found the naturalists' fascination with the material world so distracting – it cluttered the mind with more than it could ever possibly hope to know and with no end in sight to what else there was to be known.

---

[120] Frängsmyr, *Linnaeus*; James Larson, *Reason and Experience: Representation and the Natural Order in Carl von Linné* (Berkeley: University of California Press, 1971); Wilfrid Blunt, *The Compleat Naturalist: A Life of Linnaeus* (London: Collins, 1971); Schiebinger, *Nature's Body*; and Lisbet Koerner, *Linnaeus: Nature and Nation* (Cambridge, Mass.: Harvard University Press, 1999).

20

COSMOGRAPHY

Klaus A. Vogel
Translated by *Alisha Rankin*

In the early sixteenth century, "geography" was not yet a well-established science. Thus, it was quite remarkable that Erasmus of Rotterdam (ca. 1469–1536), one of the era's leading theologians and humanists, introduced the first Greek edition of the *Geography* of Ptolemy (ca. 100–170), published in Basel in 1533, by claiming that "hardly any other of the mathematical disciplines is more attractive or more necessary."[1] Erasmus called attention to the changing status of this newly emerging area of study and emphasized its importance. Only recently, he argued, had traditional limits of knowledge been overcome and scholastic speculations transformed into a clear new view of the earth:

> Earlier, there were more difficulties, since it was unclear if the heavens had a spherical form; since some believed that the world swam in the ocean as a ball swims in water, with only its tip showing and the rest covered with water; and since the men who spread this art in their writings also erred in many other things. Now that the thread has been laid by many others, but especially by Ptolemy, with whose guidance every man can easily find his way out of this labyrinth, the path is paved for you to reach the pinnacle of this art quickly and without deviation. Those who disregard it must frequently speculate hopelessly, in the interpretation of respected authors.[2]

As Erasmus composed this introduction, the first spectacular European overseas discoveries were still fresh in memory. Since the last decade of the fifteenth century, European seamen had circumnavigated Africa and discovered large islands and the American mainland in the western and southern oceans. These discoveries had triggered a revolution in both the content and the contours of the broad field known as cosmography. By 1533, it was obvious to learned and interested Europeans that the entire ocean could be navigated,

---

[1] [Ptolemy], *Claudii Ptolemaei Alexandrini philosophi . . . De Geographia libri octo* (Basel: Io. Froben, 1533), fol. 3r.
[2] Ibid.

that all parts of the earth were habitable, and that earth and water together formed one single globe.

These new ideas reflected a reorganization of the discipline of cosmography, a term generally taken to refer to the study of the entire universe, which included the central spheres of the four elements (earth, water, air, and fire) as well as the peripheral sphere of the planets and stars. Medieval cosmographers had employed an orderly network of meridians and parallels to map this universe, projecting these onto the earth, the innermost cosmic element and the location of life. They discussed the evidence for the spherical form of the elements; the borders of the habitable world; the locations of climate zones; the relationship between the spheres of earth and water; and the existence of *antipodes*, people dwelling on the opposite side of the earth.

Over the course of the fifteenth and sixteenth centuries, however, cosmographers began to focus increasingly on one aspect of their field: the emerging discipline that Ptolemy had referred to as "geography," which was confined to the systematic description of what the Greeks had known as the *oikumene*, the known, inhabited earth. In the course of the voyages of discovery, the terrestrial globe, understood as a sphere of earth and water (*globus terraquaeus*), replaced the traditional image of separate elementary spheres and became geography's fundamental model and object. But the relationship between cosmography and geography was initially unclear, and the two terms were often used indiscriminately; thus, until the end of the sixteenth century, we find Ptolemy's famous work referred to as both *Cosmography* and *Geography*. Yet the field itself was becoming increasingly differentiated, as is clear from the attempt by the German astronomer Peter Apian (Peter Bienewitz, 1501–1552) to introduce more precise definitions in his *Cosmographicus liber* (1524). There he explained the three terms "Cosmography," "geography," and "chorography" as follows: "Cosmography" refers to the system of spheres and the projection of the starry sky onto the earth's surface, which delineates the earth with the help of celestial coordinates; "geography" describes the terraquaeus sphere on the basis of larger features such as mountains, seas, and rivers; and "chorography" (or "topography") offers detailed depictions of individual, isolated places such as cities, towns, and harbors.[3]

Over the course of the sixteenth century, the relationships among the various fields associated with cosmography changed. Geography gained in importance and independence, and astronomy and geography diverged. By the end of this period, the all-encompassing term "cosmography" was in decline, and geography and astronomy were understood as independent research fields, largely distinct from one another, standing side-by-side as equals. Some scholars also recognized a separate discipline of "hydrography," the study of the seas. It is worth noting, however, that the body of men who

---

[3] [Peter Apian], *Cosmographicus liber a Petro Apiano mathematico studiose collectus* (Landshut: Johann Weißenburger, 1524), fol. 5r.

cultivated these various fields continued to be known generally as "cosmographers" – a usage I will retain for the purposes of this chapter.

As the field of cosmography expanded and ramified, so did its practitioners. In the fifteenth century, cosmographers had been mostly university-educated scholars who concerned themselves with maps of the world, geographical descriptions, and astronomical observations. During the maritime expansion in the late fifteenth and sixteenth centuries, cosmographers began to engage in a greater variety of cosmographic activities, some more artisanal and others more scholarly. Although there was no clearly demarcated profession or typical career, it is possible to describe some common attributes of early modern cosmographers. Practically oriented cosmographers, most of whom had experience at sea, flourished in the coastal regions of Europe, especially Portugal and Spain. They often came from sailor or merchant families and worked on ships or in ministries of navigation, such as the Almazém de Guiné e India (Storehouse of Guinea and India) in Lisbon and the Casa de Contratación (House of Trade) in Seville. They varied in their knowledge of Latin and mathematics, but they could determine the geographical latitude with simple methods and orient themselves using maps and globes.

In contrast, learned cosmographers, who created maps and spheres and wrote cosmographic texts, almost always possessed a university education. They knew Latin (and possibly some Greek) and had studied mathematics as part of the liberal arts. In the fifteenth century, many of them had been theologians – monks or even cardinals, or university scholars – who took up cosmographical activities on the side. During the sixteenth and seventeenth centuries, however, cosmographers increasingly came from the ranks of mathematicians, natural philosophers, and physicians. Whereas only a few important European courts employed cosmographers in the fifteenth century, from the early sixteenth century on they could also be found at minor courts, in larger trading companies, and at universities and academies, although some remained self-employed. Like printers, they needed technical expertise in addition to scholarly learning and the implements to manufacture maps, globes, and instruments. They sometimes handed down their knowledge, tools, scholarly contacts, and business partnerships to subsequent generations, creating dynasties of cosmographers such as the Apian, Mercator, Blaeu, and Cassini families (see the following chapters in this volume: Donahue, Chapter 24; Bennett, Chapter 27).

These sixteenth- and seventeenth-century cosmographers (later also called "geographers") were representatives of a young, emerging science. Cosmography united the natural philosophical conceptions of learned scholars, the experience of seamen and travelers, and cartographic handiwork. It included a strong practical element as well as the production of maps, globes, and descriptive narratives, in which beauty was valued as highly as practical utility. It profited from theology, history, and the study of classical literature as much as mathematics, astronomy, and navigation. Furthermore, geographical knowledge was indispensable for the development of trade and the measurement of new territories. Spheres and maps became symbols of a rising discipline, esteemed by clerics, princes, merchants, scholars, and the public, and old and new elites showed off the shimmer of globes and other cosmographic objects. Amalgamating theory, empirical method, and artisanal handiwork, the emergence of geography was trendsetting and exemplary for early modern natural knowledge.

COSMOGRAPHY BEFORE 1490

In the fifteenth century, Italy was the principal center of learned cosmography.[4] In Florence, Jacopo d'Angelo completed the first Latin translation of Ptolemy's *Geography* in 1406, the year in which his city won its long-desired access to the Mediterranean with the capture of Pisa, and a humanist circle met in the Florentine *Convento degli Angeli* to discuss geographical questions. Members of this circle included Palla Strozzi (ca. 1373–1462), the city's richest citizen; Leonardo Bruni (ca. 1369–1444), chancellor and city historian; Giorgio Antonio Vespucci, uncle of the future explorer Amerigo Vespucci; and Paolo dal Pozzo Toscanelli (1397–1482), the most important cosmographer of his day.[5] During the same period, the Camaldolite monk Fra Mauro (d. 1459) established his famous school of cartography in a monastery near Venice.[6]

Whereas Italian cosmography was dominated by scholars and merchants, Portuguese cosmographers focused on the practical problem of making the Atlantic coast accessible, cultivating their connections with Italy in pursuit of this goal.[7] In 1428, Prince Pedro, brother-in-law of the Portuguese Prince Henry the Navigator, traveled to Florence and Venice to obtain maps and documents for the academy in Sagres, and in 1458, the Portuguese King Alfonso V ordered a map of the world from Fra Mauro, which was delivered to him by the Venetian patrician Stefano Trevisan. Conversely, the most important Portuguese discoveries were made known to friendly Italian princes and merchants, and popes.

---

[4] Still valuable is O. Peschel and Sophus Ruge, *Geschichte der Erdkunde bis auf Alexander von Humboldt und Carl Ritter*, 2nd ed. (Munich: R. Oldenbourg, 1877). For a good survey, see Numa Broc, *La géographie de la Renaissance*, 2nd ed. (Paris: Éditions du Comité des Travaux historiques et scientifiques, 1986), with bibliography. A complete representation of cosmography in fifteenth-century Italy is lacking. For more references, see the important study by Patrick Gautier Dalché, "L'Oeuvre géographique du Cardinal Fillastre (m. 1428): Représentation du monde et perception de la carte à l'aube des découvertes," *Archives d'histoire doctrinale et littéraire du Moyen Age*, 59 (1992), 319–83.
[5] Roberto Almagià, "Il primato di Firenze negli studi geografici durante i secoli XV e XVI," *Atti della Società per il Progresso delle Scienze*, 18 (1930), 60–80; and Broc, *La géographie de la Renaissance*, p. 189.
[6] Placido Zurla, *Il mappamondo di Fra Mauro Camaldolese* (Venice, 1806).
[7] Broc, *La géographie de la Renaissance*, p. 192; and Peter Russell, *Prince Henry the Navigator* (New Haven, Conn.: Yale University Press, 2000).

Fifteenth-century cosmography comprised three distinct areas of knowledge and tradition: the *cartographic representation* of the *oikumene*, the *literary description* of certain places and regions, and *natural philosophical commentaries* on spherical cosmography. These three elements were only loosely connected with one another because they were bound up in different areas of practice, and they did not become fully integrated until after 1500.

In the field of cartography, cosmographers had traditionally created portolan charts for the Mediterranean coast; these combined measurements of sea distances with compass bearings, achieving a notable accuracy in the more frequently plotted coastal regions, although this diminished over longer distances. To address this problem, Ptolemy had advocated the use of astronomical observation to monitor outlying areas and as many locations on the map as possible.[8] He used a grid of meridians and parallels, which made possible both the design of regional maps and the representation of the entire *oikumene*. The manuscripts and early printed editions of the *Geography* included a list of Ptolemy's measurements of latitude and longitude, as well as basic maps derived from his specifications. Leading fifteenth-century cosmographers analyzed these data and corrected them using contemporary observations.

Following this tradition, Fra Mauro, the leading cartographer of the fifteenth century, also chose to chart the *oikumene* on a circular map, which his workshop produced in 1459–60 and which still resides in the Biblioteca Marciana in Venice. This map shows the Earth's inhabited hemisphere, made up of the three continents – Europe, Asia, and Africa – surrounded by a small fringe of the ocean. Fra Mauro's depiction of the earth surpassed those in contemporary editions of Ptolemy. Fra Mauro illustrated Scandinavia, Africa, and East Asia in their entirety, incorporated recent information on Portuguese voyages to West Africa, and corrected Ptolemy's portrayal of the Indian Ocean as completely enclosed by land to the south. Pointing to ancient accounts as well as reports from a reliable source who claimed to have sailed from India to far beyond Sofala (in modern Mozambique), he insisted that "without any doubt" one could sail around Africa.[9]

The second focus of traditional cosmographic knowledge was the study of classical and contemporary literary descriptions of the world. This body of literature was relatively small but quite diverse, as we can see from the section called "Cosmographers and Geographers" (*Cosmographi et geographi*) in a 1498 manuscript index of books owned by Hartmann Schedel (1440–1514), a Nuremberg physician and humanist who had studied in Italy and

---

[8] Hans von Mzik, ed. and trans., *Des Klaudios Ptolemäus Einführung in die darstellende Erdkunde: Erster Teil* (Vienna: Gerold and Co., 1938).
[9] Zurla, *Il mappamondo*; for the complete edition, see Tullia G. Leporace, ed., *Il mappamondo di Fra Mauro* (Venice: Istituto Poligrafico dello Stato, 1954).

written a well-known chronicle of the world (1493). His list contains eleven entries: two editions of Ptolemy (*Cosmographica Ptolemei*),[10] three geographical classics of late antiquity, the geography of Strabo (44 B.C.E–23 C.E.), the cosmography of Pomponius Mela (dated 44 C.E.), and the scholarly poem *De situ orbis* (The Places of the World, ca. 43 C.E.) by Dionysius Periegetes;[11] two contemporary geographical works, the (uncompleted) *Asia* and *Europa* of Aeneas Sylvius Piccolomini (1405–1464), who had been elected pope as Pius II; and three descriptions of Palestine, a vellum manuscript credited to the English mark, Venerable Bede (ca. 673–735) and two recently printed German travel narratives, by Johannes Tucher (1428–1491) and by Bernhard von Breidenbach (ca. 1440–1497).

Finally, learned cosmographers studied texts dealing with cosmography's natural philosophical basis, such as Johannes de Sacrobosco's *Tractatus de sphaera* (Treatise of the Sphere, ca. 1220) and commentaries and *quaestiones* on Aristotle's *Meteorology* and *On the Heavens*.[12] In this context, they debated the spatial configuration of the elements, the habitability of the sphere of earth, and the volume, placement, and dimensions of the sphere of water.

Most medieval Muslim and Christian scholars had argued that a smaller sphere of earth rested lightly and eccentrically inside the sphere of water (see Figure 20.1), surfacing above it only in the realm of the inhabited *oikumene*; as a result, they denied the existence of antipodes opposite to the known world. There were two common explanations given for this arrangement. One of these was laid out by the famous Parisian natural philosopher Jean Buridan (1300–ca. 1358) in a *quaestio* on the four books of Aristotle's *De caelo et mundo* (On the Heaven and the Earth). There, Buridan differentiated between the center of gravity and the center of volume in the spherical Earth. Once the sphere of earth emerged from the sphere of water he argued, the earth's center of gravity was displaced in the direction of the submerged parts. The visible surface of the earth, however, was kept dry and less dense because of the heat of the sun and other influences. In effect, the earth's center of gravity was permanently displaced from its center of volume, in contrast with the water's center of gravity, which always coincided with its center of volume because water is a fluid of uniform density. However, the center of gravity of the

---

[10] Klaus A. Vogel, "Neue Horizonte der Kosmographie: Die kosmographischen Bücherlisten Hartmann Schedels (um 1498) und Konrad Peutingers (1523)," *Anzeiger des Germanischen Nationalmuseums*, 67 (1991), 77–85.

[11] Not included under this heading were Pliny's *Historia naturalis* (Natural History, ca. 44–71) and the geography of Solinus, which he entered in another place.

[12] Klaus A. Vogel, "Das Problem der relativen Lage von Erd- und Wassersphäre im Mittelalter und die kosmographische Revolution," *Mitteilungen der Österreichischen Gesellschaft für Wissenschaftsgeschichte*, 13 (1993), 103–43, at pp. 109 ff.; and William G. L. Randles, "Classical Models of World Geography and Their Transformation Following the Discovery of America," in *The Classical Tradition and the Americas*, ed. Wolfgang Haase and Meyer Reinhold (Berlin: Walter de Gruyter, 1994), pp. 5–76; see also Klaus A. Vogel, "Sphaera terrae: Das mittelalterliche Bild der Erde und die kosmographische Revolution," Ph.D. dissertation, Universität Göttingen, 1995.

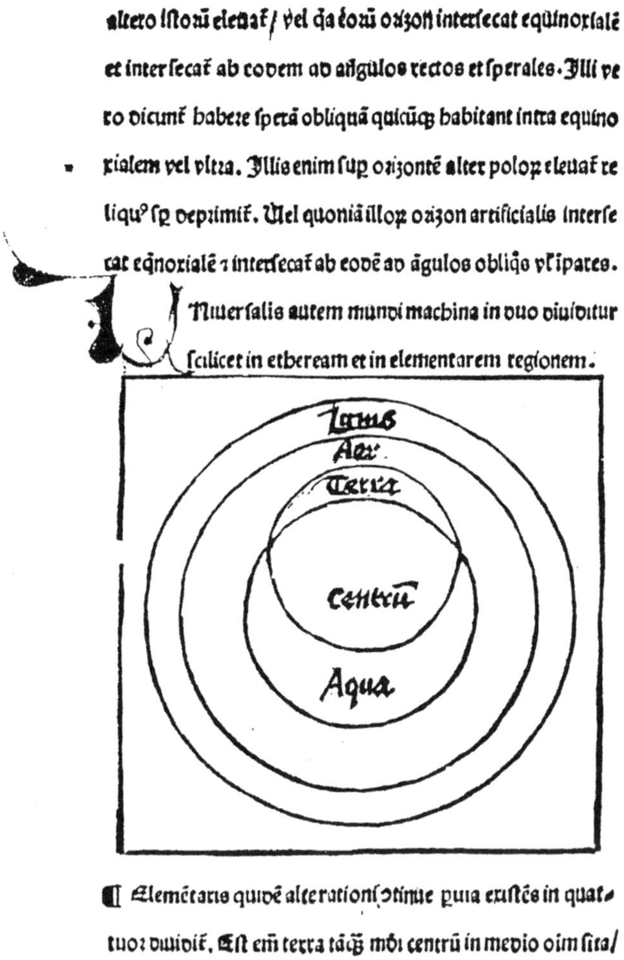

altero istorū elenaf/ vel q̃a eorū orizon intersecat equīnoxialē
et intersecaī ab eodem ao añgulos rectos et sperales. Illi ve
ro dicunt habere sperā obliquā quicūq̃ habitant intra equino
xialem vel vltra. Illis enim sup orizontē alter polop eleuaī re
liquꝰ sp deprimiī. Vel quoniā illop orizon artificialis interse
cat eqnorialē 1 intersecaī ab eodē ao āgulos obliqs vt ipates.
Uniuersalis autem mundi machina in duo diuiditur
scilicet in etheream et in elementarem regionem.

The spheres labeled: Ignis, Aer, Terra, Centrū, Aqua.

¶ Elemētaris quidē alterationi ꝯtinue puia existēs in quattuor diuidiſ. Est em̄ terra tāq̃ mōi centrū in medio oim sita/

Figure 20.1. The spheres of earth and water. In Johannes de Sacrobosco, *Spericum opusculum* (Leipzig: Martin Landsberg, ca. 1489). Reproduced by permission of The Herzog-August-Bibliothek Wolfenbüttel. HAB. 101.14 Qu (6). The common center (*Centrum*) of both spheres is in the lower part of the sphere of earth, whereas the sphere of water is concentric with the spheres of air (*Aer*) and fire (*Ignis*).

earth and the center of gravity of the water both corresponded to the center of the spherical universe as a whole. Due to the continual dryness of the upper hemisphere that showed above the surface of the water, this condition remained stable. Not even the erosion of the mountains changed the situation: When rivers eroded solid pieces of the upper hemisphere, the subsequent displacement of the earth's center of gravity effected a corresponding upward movement of the entire sphere until that center of gravity once again rested at the center of the cosmos (a process that incidentally explained the formation of mountain ranges). In this way, the sphere of earth would be continually rising out of the sphere of water.[13]

In the early fifteenth century, the Spanish Bishop Paul de Santa Maria (Paul of Burgos, 1351–1435) provided a complementary theological rationale for the emergence of the sphere of earth from the sphere of water in his 1481 addendum to Nicholas of Lyra's *Postilla* (1322–31), a widely used biblical commentary.[14] Citing Genesis and the Psalms, Paul explained the separation of the elements and the emergence of the dry surface of the earth as a consequence of God's intervention on the third day of creation; God had displaced the sphere of water in perpetuity away from the center of the cosmos, counteracting it only once, during the Flood.

Many of the most famous fifteenth-century maps of the world adopted these ideas. On his map of 1460, Fra Mauro integrated both explanations, as is clear from an inscription in the upper right-hand corner of his map. There he noted that God had distributed weight unevenly within the spheres of earth and water. The resulting emergence of the inhabited hemisphere of earth from the sphere of water made it providentially possible for living beings to inhabit the land. In the eyes of this cosmographically educated Camaldolite monk, the confinement of the *oikumene* to one hemisphere resulted simultaneously from God's will and the laws of celestial physics.

Until the last decade of the fifteenth century, almost all European scholars were convinced that the habitable region of the earth was confined to one hemisphere and surrounded by ocean. Piccolomini expressed this idea in *De mundo in universo* (On the World as a Whole, ca. 1450), a brief introduction to his *On Asia*: "Almost all agree that the form of the world is round. And they think the same of the earth, which is grounded in the middle of things, pulls all heavy things to itself, and submerges for the most part in water."[15]

GLOBUS MUNDI: DISCOVERIES AT SEA AND THE COSMOGRAPHIC REVOLUTION (1490–1510)

Learned cosmography was changed dramatically by the overseas discoveries of the European explorers. First and foremost, these discoveries altered the understanding and perception of the ocean, which ancient and medieval scholars had understood as encircling the inhabited *oikumene*. They viewed the ocean's water as a foreign element – a limit surmountable only in speculation. Over the course of the fourteenth and fifteenth centuries, however,

---

[13] Ernest A. Moody, "John Buridan on the Habitability of the Earth," *Speculum*, 16 (1941), 415–25; and Vogel, "Erd- und Wassersphäre im Mittelalter," pp. 114 ff.
[14] Nicolaus of Lyra, *Postilla super totam Bibliam* [1492], 4 vols. (Frankfurt am Main, 1971); and Vogel, "Erd- und Wassersphäre im Mittelalter," pp. 119 ff.
[15] Piccolomini, *Historia rerum ubique gestarum*, fol. a1v.

voyages by Portuguese and Spanish seamen had expanded the limits of the world accessible to Europeans, thus casting doubt on this idea. In the second decade of the fifteenth century, the Portuguese, under Prince Henry the Navigator, began systematically to make the western coast of Africa accessible.[16] In the 1440s, Dinis Dias sailed around Cape Verde and Nuno Tristâo reached the estuary of the river Gambia, and in 1460, Pedro de Sina sailed around the mountains of Sierra Leone. Just north of the equator, explorers found green, fruitful lands inhabited by dark-skinned people, overturning old theories of an uninhabitable "burned" or "torrid zone" (*zona torrida*). Shortly afterward, in 1474 or 1475, Lope Gonçalves and Rui de Sequeira crossed the equator, and in 1486 Bartholomeo Diaz returned to Lisbon, having navigated through a dangerous western storm around the southern cape of Africa, later named the Cape of Good Hope.

Although the Portuguese seamen, who had at first felt their way along the African shoreline like coastal explorers, became more experienced and more daring on the open seas, the ocean had yet to be conquered. Seafarers and contemporary scholars still clung to a cosmography consistent with the notion of one inhabited hemisphere. From this perspective, the prospect of a westward voyage into the unknown ocean to reach the mainland of Asia was revolutionary. Since the early 1480s, such a project had been proposed by Christopher Columbus – an experienced ship's captain from Genoa who had married a Portuguese noblewoman – against the resistance of his better-educated contemporaries. Their objections, which made perfect sense in the context of contemporary cosmographical knowledge, were summarized by a scholarly commission led by Fernando di Talavera, archbishop of Granada: (1) A western trip to Asia would take three years; (2) the western ocean was boundless and quite likely impossible to sail; (3) the antipodes did not exist because most of the sphere of earth was submerged in water; and (4) it was unlikely that anyone would discover unknown land of any worth so many centuries after creation.[17]

The debate about Columbus's proposal concerned not only the distance between the western and eastern edges of the *oikumene* but also the more wide-ranging problem of the relative size and position of the spheres of earth and water.[18] Against the claim, generally accepted by contemporary

---

[16] On the Portuguese discoveries, see Charles R. Boxer, *The Portuguese Seaborne Empire, 1415–1825*, 2 vols. (London: Alfred A. Knopf, 1969); J. H. Parry, *The Age of Reconnaissance: Discovery, Exploration, and Settlement, 1450 to 1650* (Berkeley: University of California Press, 1981); and Bailey W. Diffie and George D. Winius, *Foundations of the Portuguese Empire, 1415–1580* (Europe and the World in the Age of Expansion, 1) (Minneapolis: University of Minnesota Press, 1977).

[17] Samuel Eliot Morison, *Admiral of the Ocean Sea: A Life of Christopher Columbus* (Boston: Little, Brown, 1951), pp. 95 ff.; Paolo Emilio Taviani, *Cristoforo Colombo: La genesi della grande scoperta*, 2 vols. (Novara: Istituto Geografico de Agostini, 1974), pp. 205 ff.; and Felipe Fernandez-Armesto, *Columbus* (Oxford: Oxford University Press, 1991), pp. 23 ff.

[18] See the important article by William G. L. Randles, "The Evaluation of Columbus' 'India' Project by Portuguese and Spanish Cosmographers in the Light of the Geographical Science of the Period," *Imago Mundi*, 42 (1990), 50–64, esp. pp. 51 ff.

cosmographers, that a smaller sphere of earth floated within a larger sphere of water – an idea that excluded any western voyage – Columbus advocated a much simpler view of the earth. As he noted in a copy of the French theologian and cosmographer Pierre d'Ailly's (1350–1420) *Imago mundi* (Image of the World, ca. 1410), "Earth and water together form one round body."[19] Neither discussion nor debate could determine which of these two competing conceptions of the earth's form was correct. In this context, Columbus's voyage can be regarded as the first great experiment in the history of early modern science.

This experiment almost went awry on numerous occasions, but in the end it produced unexpected results because Columbus did not reach "India" (as he believed until his dying day); that is, China and the Far East. Instead he found new, unknown islands in the west, which were referred to as the "western antipodes" by the royal Spanish chronicler Peter Martyr of Anghiera (1457–1526), an Italian who became the official narrator of the Spanish voyages of discovery.[20] After the initial euphoria, reactions to Columbus's first western voyage remained subdued, both because Columbus's own account, published in the form of a letter in 1493 and disseminated throughout Europe, was not specific enough to clarify the geographical position of his newfound islands,[21] and because his voyage had not yet affected the foundations of cosmography.

Nevertheless, the return of Columbus gave the overseas voyages of exploration a new impetus. In 1499, the Portuguese seaman Vasco da Gama (ca. 1460–1524) returned to Lisbon, having rounded the Cape of Good Hope and reached Calicut (now in Kerala) on the west coast of India, thus opening the sea route to India. From that time on, the Portuguese played a substantial role in commerce with India, which had previously been monopolized by the Venetians. Cosmographically, however, the circumnavigation of Africa, although notable, was no sensation; the Roman encyclopedist Pliny the Elder (23–79) had already recounted the journeys of Hanno and Eudoxus, who had sailed the same route. The turning point came in 1500, when thirteen ships heading to India under the leadership of Pedro Alvarez Cabral (ca. 1460–1526), and manned by over 1,200 crew members, accidentally touched the present-day coast of Brazil. The following year another expedition of three ships was sent to explore the newly discovered coastline and sailed farther south along the coast of present-day Brazil and Argentina. This voyage was

---

[19] Edmund Buron, ed. and trans., *Ymago mundi de Pierre d'Ailly* (Paris: Maisonneuve frères, 1930), at p. 184, marg. 8: "aqua et terra simul facit corpus rotundum."
[20] Petrus Martyr d'Anghiera, *Opus epistolarum* [1530], no. 130, quoted in Guglielmo Berchet, ed., *Fonti italiane per la storia della scoperta del Nuovo Mondo* (Raccolta Colombiana), vol. 3, pt. 2 (Rome: Ministero della Pubblica Istruzione, 1893), p. 39.
[21] *De Insulis nuper inventis* (Basel: Bermann de Olpe, 1494), fol. dd6r; and Christopher Columbus, *Textos y documentos completos*, 3rd ed., ed. Consuelo Varela (Madrid: Alianza, 1989).

documented by the Florentine Amerigo Vespucci (1451–1512), a learned cosmographer who accompanied the expedition.

Vespucci's report, printed under the title *Mundus novus* (The New World, 1503), caused a sensation. Cosmographically precise, it did not hesitate to criticize the ancients, offering evidence that the earth was inhabitable outside of the known *oikumene*. As Vespucci noted, most classical authors had asserted

> that there is no continent beyond and south of the equator, but merely that sea which they called the Atlantic; furthermore, if any of them did affirm that a continent was there, they gave many arguments to deny that it was a habitable land. But this last voyage of mine has demonstrated that this opinion of theirs is false and contradicts all truth.[22]

In a few sentences, Vespucci made his point clear: The southern antipodes had been discovered, overturning all theories that restricted the inhabited world to Europe, Africa, and Asia and differentiated between an "upper" hemisphere (not covered by water) and a "lower" hemisphere (submerged). This was a cosmographic revolution. The discovery of the antipodes supported the simple round, integrated image of the earth advocated by Columbus since the 1480s: the modern terrestrial globe, composed of both earth and water.

Between 1503 and 1506, Vespucci's account appeared in many languages and dozens of printed editions. When the humanist Mathias Ringmann (ca. 1481–ca. 1511) and cosmographer Martin Waldseemüller (ca. 1470–ca. 1518), from St. Dié in Lorraine, published their *Cosmographiae introductio* (Introduction to Cosmography) in 1507, they portrayed Vespucci as a modern counterpart to Ptolemy and named the newly discovered land in the southwestern Atlantic "America" in his honor.[23]

The discovery of the antipodes and the evidence for the terraquaeus sphere still left the earth fixed and immobile at the center of the universe and the order of the planets untouched. But the Aristotelian principle of gravity was no longer as comprehensive as had been assumed. Because earth and water together formed a single sphere, it became easier to imagine this sphere in motion, displaced from the center of the universe. This was an important starting point for Nicholas Copernicus (1473–1543), who had studied in Cracow and Italy and who lived as a cleric in Frauenburg, in the kingdom of Poland, from 1503 on. Copernicus's famous *De revolutionibus orbium*

---

[22] *Albericus Vespuccius Laurentio Petri Francisci de Medicis* . . . (Paris: F. Baligault/Jehan Lambert, 1503/04), fol. a2r; Amerigo Vespucci, "Mundus Novus," in *Letters from a New World: Amerigo Vespucci's Discovery of America*, ed. Luciano Formisano, trans. David Jacobson (New York: Marsilio, 1992), p. 45.

[23] [Martin Waldseemüller], *Cosmographiae introductio, cum quibusdam geometriae ac astronomiae principiis ad eam rem necessariis* (St. Dié: G. and N. Lud, 1507), fol. a2v–a3r; Waldseemüller, *The Cosmographiae Introductio of Martin Waldseemüller*, ed. Joseph Fischer, Franz von Wieser, and Charles George Herbermann, trans. Edward Burke (New York: United States Catholic Historical Society, 1907); and Albert Ronsin, ed., *La fortune d'un nom: AMERICA, Le baptême du Nouveau Monde à Saint-Dié-des Vosges*, trans. Pierre Monat (Grenoble: Jérôme Millon, 1991).

coelestium (On the Revolutions of the Heavenly Spheres, 1543) proposed a heliocentric model of the universe, which he had formulated for the first time between 1507 and 1514 (see Donahue, Chapter 24, this volume). One of its first chapters, "How Earth Forms One Single Sphere with Water" (*Quomodo terra cum aqua unum globum perficiat*), rejected the Aristotelian model of the spheres of earth and water and the principle of heaviness that Aristotle had asserted as a universal cosmic force. "We should not heed certain Peripatetics who . . . assert that the earth emerges from the water," he wrote, "because its weight is not equally distributed due to its cavities, its center of gravity being different from its center of magnitude."[24]

Copernicus stressed that the existence of a single terraquaeus sphere had been proven by the most recent geographical discoveries, specifically mentioning Vespucci:

> This will be more clear when we add the islands discovered in our time under the kings of Spain and Portugal, and especially America, named after their finder, a ship's captain. On account of its still undisclosed magnitude this is thought to be another inhabited world, and there are also many other islands, heretofore unknown. So we should wonder even less that the Antipodes or Antichthones exist.[25]

Thus the evidence for the terraquaeus sphere provided by the overseas discoveries was crucial, even indispensable, for Copernicus's cosmological innovation (see Figure 20.2). Copernicus had not only studied astronomical classics and observed the heavens carefully but also responded attentively, competently, and creatively to the cosmographic revolution.[26]

COSMOGRAPHIA UNIVERSALIS: COSMOGRAPHY AS A LEADING SCIENCE (1510–1600)

With the transition to a new, simpler conception of the earth in the first decade of the sixteenth century, cosmography began its ascent. It won great recognition from popes and European monarchs, as well as from merchant families operating on an international scale. Prominent scholars occupied themselves with cosmography, both in seaports and inland, and its influence made its way into political literature and court painting. English humanist and statesman Thomas More (1478–1535) referred to Vespucci and the Portuguese accounts of the new lands in his *Utopia* (1516), which described

---

[24] Nicholas Copernicus, *Nicolai Copernici Torinensis De revolutionibus orbium coelestium libri VI* (Nürnberg: Io. Petreius, 1543), fol. a1r.
[25] Ibid., sig. a2r.
[26] Vogel, "Erd- und Wassersphäre im Mittelalter," pp. 135–8, with bibliography at p. 109, n. 17.

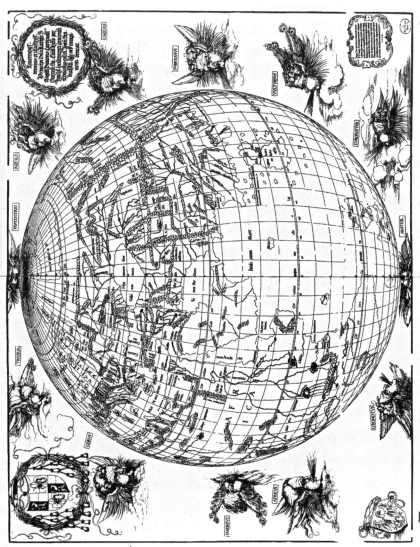

Figure 20.2. The Globe of the Old World. Johannes Sabius and Albrecht Dürer, 1515. Reproduced by permission of Staatsbibliothek zu Berlin Preussische Kulturbesitz. Kart. U 3530. This illustration, which does not include the most recent geographical discoveries of the day, shows the use of the image of the globe by two of the leading Continental cosmographers and artists of their time.

Figure 20.3. *The Ambassadors*. Hans Holbein the Younger, 1533, oil on wood. Reproduced by permission of the National Gallery, London. Holbein's famous painting shows cosmographical objects as symbols of commerce, politics, and modernity.

a fictitious island off the coast of America in the southern hemisphere.[27] In Hans Holbein's painting *The Ambassadors* of 1533 (see Figure 20.3), the sphere of the earth featured as a status symbol signifying the knowledge, modernity, and power of the political elite.

Meanwhile, the discoveries and conquests overseas continued.[28] The Portuguese had wrested rule of the Indian Ocean from the Arabs, and the Spanish captured the islands and coasts of the Caribbean and advanced through Panama to the Pacific Ocean (Balboa, 1513). A few years later, Ferdinand Magellan (ca. 1480–1521) led the first expedition to circumnavigate the Earth (1519–21). In the second half of the century, France, England,

---

[27] Klaus A. Vogel, "Neue Welt Nirgendwo? Geographische und geschichtliche Horizonte der 'Utopia' des Thomas Morus," in Klaus Vogel, ed., *Denkhorizonte und Handlungsspielräume: Historische Studien für Rudolf Vierhaus zum 70. Geburtstag* (Göttingen: Wallstein, 1992), pp. 9–32.

[28] See note 17.

and the Netherlands joined the movement of exploration and colonization. (see Harris, Chapter 16, this volume). France attempted to colonize Quebec in 1541–3, and England founded the Cathay Colony in 1577 and the colonies in Newfoundland in 1583. The Netherlands established the Foreign Trade Company in 1594 and the Dutch East India Company (Verenigde Oostindische Compagnie, VOC) in 1602.

These enterprises furthered the development of the new science of cosmography, beginning in the colonial centers of Lisbon and Seville. From the end of the fifteenth century on, the Almazém de Guiné e India in Lisbon operated a hydrographical bureau led by a nautical and cosmographic expert who held the title *almoxarife* (receiver or keeper of stores) and whose responsibilities included not only the management and administration of the Almazém, but also the selection and examination of the captains of royal ships.[29] (One person who held this office was Bartholomeo Diaz (ca. 1450–1500), who had sailed around Africa and watched over the building and armament of the ships for da Gama's fleet.) In 1547, the position of the *cosmógrapho-mor*, the principal royal cosmographer, was created – a post whose duties included managing the official nautical charts (*padrões*), examining and authorizing official manufacturers of sea charts and nautical instruments, and instructing and examining navigation officers (pilots, junior pilots, and masters) and recording their names in an official register. The first *cosmógrapho-mor*, Pedro Nuñez (1492–1577), had been a professor of mathematics in Coimbra since 1537 and was an influential scholar, who described the use of spherical geometry for navigational calculations of distance. Nuñez was succeeded in 1577 by Thomas d'Orta, who was succeeded in turn by João Baptista Lavanha in 1591.

Following the Portuguese example, the Casa de la Contratación was founded in Seville in 1503 as a center of administration for colonial trade and the dispatch of Spanish fleets to the Indies, under the leadership of Juan de Fonseca (1451–1524), bishop of Badajoz and Córdoba.[30] The Casa de la Contratación became a center for cosmographical expertise, to which all captains were required to report their observations upon return. In 1508, Spain began to employ four royal pilots, with Vespucci as *piloto mayor* (chief pilot), to search for the western passage to India. The *piloto mayor* was responsible for the examination and licensing of ships' captains as well as for the official royal sea chart, the *padrón real*. After Vespucci's death in 1512, this position

---

[29] A. Teixeira da Mota, "Some Notes on the Organization of Hydrographical Services in Portugal before the Beginning of the Nineteenth Century," *Imago Mundi*, 28 (1978), 51–60; and Kees Zandvliet, *Mapping for Money: Maps, Plans and Topographic Paintings and their Role in Dutch Overseas Expansion During the 16th and 17th Centuries* (Amsterdam: Batavian Lion International, 1998), pp. 16–21: "Map Production in Portugal."

[30] Ernesto Schäfer, *El Consejo Real y Supremo de las Indias: Su historia, organización y labor administrativa hasta la terminación de la Casa de Austria*, 2 vols. (Sevilla, 1935, 1947); and Ursula Lamb, *Cosmographers and Pilots of the Spanish Maritime Empire* (Variorum Collected Studies Series, 499) (Aldershot: Ashgate, 1995).

was held by Diaz de Solis (until 1515), Francisco Coto (until 1518), Sebastian Caboto (until 1522), Alonso de Chavez (until 1586), and Rodrigo Zamorano (until 1612).[31]

At the royal court in Madrid, the position of *cosmógrafo-cronista mayor de las Indias* (chief cosmographer-chronicler of the Indies) was set up under the auspices of the Consejo Real y Supremo de las Indias (Royal and Supreme Council of the Indies) and held by Juan López de Velasco (1530–1603) until 1588.[32] His duties were to draw up exact maps and to compose an extensive historical, geographical, and ethnographical description of Spanish America. In order to obtain exact longitudinal measurements, Velasco sent a briefing with detailed instructions for the observation of lunar eclipses in the colonies. In 1577 and again in 1584, he distributed an elaborate printed questionnaire to the colonies. This survey contained fifty questions about the geography, history, and living conditions of the inhabitants, natural resources, coastal conditions, and so on, and requested the completion of simple maps. The surviving responses, 208 *Relaciónes geograficas* and 92 maps made by Spaniards and natives between 1578 and 1586, are a unique source for contemporary life and geographical knowledge in the colonies.[33]

Interest in the course of overseas discoveries also remained high in Italy, where leading members of the Roman curia and merchants and scholars in commercial towns were kept abreast of cosmographic novelties through letters and personal accounts. Venice, Florence, and Rome were centers for the production of cosmographic works in this period;[34] the first substantial collection of travel narratives was printed in 1507, in Vicenza, under the title *Mondo Novo e paesi novamente ritrovati da Alberico Vespuzio fiorentino* (New World and Lands Newly Found by Amerigo Vespucci of Florence), edited by the humanist and university professor Fracanzano da Montalboddo and later published in Latin, German, and French.[35] In the following decades, Italians produced a steady stream of cosmographic titles, including the *Cosmographia in tres dialogos distincta* (Three Dialogues on Cosmography, 1543) of Francesco Maurolico (1494–1575), a Benedictine monk and mathematician teaching in Messina,[36] and *Delle navigazione e viaggi* (On Navigation and Voyages), by

---

[31] Zandvliet, *Mapping for Money*, pp. 23–5: "Map Production in Spain."
[32] Schäfer, *El Consejo Real*, pp. 147 ff.
[33] Harold F. Cline, "The Relaciones Geográficas of the Spanish Indies, 1577–1648," in *Handbook of Middle American Indians*, vol. 12 (Guide to Ethnohistorical Sources, 1), ed. Harold F. Cline (Austin: University of Texas, 1972), pp. 183–242; Donald Robertson, "The Pinturas (Maps) of the Relaciones Geográficas with a Catalog," in *Handbook of Middle American Indians*, pp. 243–78; and Barbara E. Mundy, *Indigenous Cartography and the Maps of the Relaciones Geográficas* (Chicago: University of Chicago Press, 1996).
[34] There is no synthetic work on Italian cosmography/geography in the sixteenth century; on geography in Rome, see Broc, *La géographie de la Renaissance*, pp. 199–204.
[35] Broc, *La géographie de la Renaissance*, p. 27.
[36] Francesco Maurolico, *Dialoghi tre della Cosmographia* (ms. Rome, 1536), ed. Giovanni Cioffarelli (forthcoming); Maurolico, *Cosmographia . . . in tres dialogos distincta* (Venice: Heirs of Lucantonio Giunta, 1543).

the Venetian diplomat Giambattista Ramusio (1485–1557), of which the first volume was published in 1550.[37] From the middle of the sixteenth century on, Rome once again became a center of cosmographic knowledge as the Jesuits began to play a leading role in the collection of geographical information and in cosmographic education. Among the most important figures in this connection were Christoph Clavius (1537–1612), a native of Bamberg, Germany, who taught at the Jesuit Collegio Romano,[38] and his pupil Matteo Ricci (1552–1610), who brought European cosmography to China (see Vogel, Chapter 33, this volume). Both the holdings of the Vatican Library and the opulent geographical wall maps in the Vatican galleries, painted between 1559 and 1583 by Pirro Ligorio and Ignatio Danti, bear witness to this late sixteenth-century flowering of Roman cosmography.[39]

Sixteenth-century cosmography also flourished north of the Alps, initially primarily in the Holy Roman Empire. The University of Vienna had long been a significant center for the mathematical sciences, including cosmography.[40] Mathematician and astronomer Georg Peurbach (1423–1461), author of the influential textbook *Theorica planetarum* (Theory of the Planets), had taught there in the fifteenth century, and another leading Viennese mathematician and astronomer, Georg Tannstetter (1482–1535), called Collimitius, mentioned several cosmographers by name in his catalogue *Viri mathematici* (Men of Mathematics), which he published in 1514.[41] A few of Collimitius's pupils, among them Peter Apian, were themselves later active as cosmographers. Scholars at the University of Vienna composed important commentaries on the geographical classics, including Vadian on Pomponius Mela (1518) and Camers on Solinus (1520).[42]

Among the most illustrious sixteenth-century cartographers of the Holy Roman Empire was Peter Apian, a member of what soon became a successful family of cosmographers.[43] Apian studied mathematics with Tannstetter in Vienna, where he received his baccalaureate degree in 1521. One year earlier, he had printed a modern map of the world, *Typus orbis universalis* (Prototype of

---

[37] Broc, *La géographie de la Renaissance*, pp. 37 ff.
[38] Ugo Baldini, ed., *Christoph Clavius e l'attività scientifica dei Gesuiti nell'età di Galileo* (Rome: Bulzoni, 1995).
[39] Broc, *La géographie de la Renaissance*, p. 202.
[40] Lucien Gallois, *Les géographes allemands de la Renaissance* (Paris: E. Leroux, 1890); Klaus A. Vogel, "Amerigo Vespucci und die Humanisten in Wien: Die Rezeption der geographischen Entdeckungen und der Streit zwischen Joachim Vadian und Johannes Camers über die Irrtümer der Klassiker," *Pirckheimer-Jahrbuch*, 7 (1992), 53–104, esp. pp. 67 ff.
[41] Franz Graf-Stuhlhofer, *Humanismus zwischen Hof und Universität: Georg Tannstetter (Collimitius) und sein wissenschaftliches Umfeld im Wien des frühen 16. Jahrhunderts* (Schriftenreihe des Universitätsarchivs, 8) (Vienna: WUV-Universitätsverlag, 1996), pp. 156–71.
[42] Joachim Vadian, *Pomponii Melae Hispani, libri de situ orbis tres, adiectis Ioachimi Vadiani in eosdem Scholiis*... (Vienna: Io. Singrenius, 1518); Johannes Camers, *In C. Iulii Solini Polyistora Enarrationes* (Vienna: Io. Singrenius, 1520); and Vogel, "Amerigo Vespucci und die Humanisten in Wien," pp. 77 ff.
[43] Hans Wolff, ed., *Philipp Apian und die Kartographie der Renaissance* (Bayerische Staatsbibliothek, Ausstellungskataloge, 50) (Munich: Anton H. Konrad, 1989).

the Universal Sphere), based on Waldseemüller's map, which depicted America. In the following years, he fabricated maps and globes, set up a workshop together with his brother Georg, and in 1524 published the first edition of his *Cosmographicus liber*. Beginning in 1526, he taught at the University of Ingolstadt, and in 1537 the son of Duke Albrecht of Bavaria was entrusted to him for lessons in cosmography, geography, and mathematics. Following the Diet of Augsburg in 1530, Apian stayed in contact with Emperor Charles V, who, after the publication of Apian's *Astronomicum caesareum* (Imperial Astronomy) in 1541, granted him knighthood and the noble title of Pfalzgrave and named him a court mathematician. In the same period, Peter's son Philip (1531–1589), the most gifted of his fourteen children, matriculated at the University of Ingolstadt and later inherited his father's life's work, taking over his father's teaching position at the university and directing a new land survey for Duke Albrecht that culminated in an enormous wall map made from twenty-four plates (*Chorographia Bavariae*, 1563).

During this period, Basel became the leading city for printing cosmographic works. In 1532, Basel theologian and humanist Simon Grynaeus (1493–1541) published his volume *Novus orbis regionum ac insularum veteribus incognitarum* (New World of Regions and Islands Unknown by the Old), the most complete travel narrative to date.[44] The *Cosmographia* of Sebastian Münster (1488–1552), a Hebrew scholar and cosmographer who lived in Basel from 1529, was printed there in 1544 and quickly became the most popular cosmographic work of the sixteenth century, running through 36 editions between 1544 and 1628.[45]

Word of the overseas discoveries had first come to scholars in the Holy Roman Empire via Antwerp, which had close ties to Iberian trade cities.[46] As in the southern European seaports, the cartographical workshops of Antwerp were important centers for the production and sale of portolan charts.[47] In the second half of the sixteenth century, they developed their own cosmographic practices, which were later partially adopted in Amsterdam. The cosmographer Gerardus Mercator (1512–1594) was born nearby, in Flemish Rupelmonde,[48] and learned cosmography and the construction of globes in Louvain from Reiner Gemma Frisius (1508–1555), a student of Apian, who later became a professor of medicine and Royal Cosmographer at the court of Emperor Charles V in Brussels. Antwerp also produced the learned merchant

---

[44] Broc, *La géographie de la Renaissance*, pp. 29–31.
[45] Karl Heinz Burmeister, *Sebastian Münster: Versuch eines biographischen Gesamtbildes* (Basel: Helbing and Lichtenhahn, 1969); Karl Heinz Burmeister, ed. and trans., *Briefe Sebastian Münsters, lat.-dt.* (Frankfurt am Main: Insel, 1964); and Broc, *La géographie de la Renaissance*, pp. 77–84.
[46] Letter from Johann Kollauer to Konrad Celtis, Antwerp, 4 May 1503, in *Der Briefwechsel des Konrad Celtis*, ed. Hans Rupprich (Humanistenbriefe, 3) (Munich: C. H. Beck, 1934), pp. 530 ff., no. 295; and Vogel, "Amerigo Vespucci und die Humanisten in Wien," pp. 63 ff., n. 30.
[47] Broc, *La géographie de la Renaissance*, pp. 196–8.
[48] Marcel Watelet, ed., *Gérard Mercator cosmographe, le temps et l'espace* (Mariakerke: Vanmelle, 1994); Broc, *La géographie de la Renaissance*, pp. 173–86.

Abraham Ortelius (1527–1598), author of the famous *Theatrum orbis terrarum* (Theater of the World, 1570), who was named Royal Cosmographer to Philip II of Spain in 1575.[49]

Cosmography developed later in France, in part because of the conservatism of Paris, a center of scholastic theology and natural philosophy.[50] Jacques Lefèvre d'Etaples (ca. 1455–ca. 1536), who wrote commentaries on Aristotle and translated the Bible, treated several cosmographic questions but never published a general overview. The situation changed with Orontius Finaeus (1494–1555), the most influential scholarly Parisian cosmographer, who became a professor of mathematics at the Collège Royal in 1531. Finaeus authored many important maps as well as a cosmography, *De mundi sphaera sive cosmographia* (On the Sphere of the World, or Cosmography, 1542), and educated some of the Jesuits who would later teach at the Collegio Romano.[51] Noteworthy on the literary scene was André Thevet (1516–1592), who became known as a traveler to America and the author of the *Cosmographie de Levant* (Cosmography of the Levant, 1554) and *Les singularitez de la France Antarctique* (The Singularities of Antarctic France, 1557). In 1559, Thevet was called to court, and shortly thereafter he was named Royal Cosmographer to Henry III.[52]

During the first half of the sixteenth century, English activity in this area was confined to the translation of important cosmographic works.[53] Thus, *The Cosmographical Glass* (1559), by William Cunningham (b. 1521), drew largely from the cosmographies of Apian and Finaeus.[54] Even later, English cosmography remained primarily oriented toward practice. Following the Spanish and Portuguese examples, Stephen Borough (1525–1584) was named as the first "Cheyffe Pylott of this our realme of Englande" in 1564.[55] John Dee (1527–1608), who boasted an outstanding mathematical education, composed the most significant work on navigation of his time, the four-volume *The British Complement of the Perfect Art of Navigation*, of which only volume 1 (1577) has survived. Dee's teachers and advisors included the five most important cosmographers of Europe: Nuñez, Gemma Frisius, Mercator, Ortelius, and Finaeus.[56] Finally, the most significant collection of travel narratives of

---

[49] R. W. Karrow, Jr., ed., *Abraham Ortelius (1527–1598), cartographe et humaniste* (Turnhout: Brepols, 1998).
[50] Broc, *La géographie de la Renaissance*, p. 123; François de Dainville, *La géographie des humanistes* (Paris: Beauchesne et ses fils, 1940; repr. Geneva: Slatkine, 1969), pp. 7 ff.
[51] Dainville, *La géographie des humanistes*, p. 36.
[52] Frank Lestringant, *L'Atélier du cosmographe* (Paris: Albin Michel, 1991), translated as *Mapping the Renaissance World*, trans. David Fausett (Cambridge: Polity, 1994).
[53] Eva G. R. Taylor, *Tudor Geography, 1485–1583* (London: Methuen, 1930); Taylor, *Late Tudor and Early Stuart Geography, 1583–1650* (London: Methuen, 1934); Taylor, *The Mathematical Practitioners of Tudor and Stuart England* (London: Methuen, 1954).
[54] Taylor, *Tudor Geography*, p. 26.
[55] David W. Waters, *The Art of Navigation in England in Elizabethan and Early Stuart Times* (London: Hollis and Carter, 1958), p. 515.
[56] Taylor, *Tudor Geography*, p. 75 ff.

the century was the work of Richard Hakluyt (1553–1616), a cleric and a member of the English embassy in Paris, whose three-volume *Principal Navigations, Voyages, and Discoveries of the English Nation* was printed in 1589. This collection contained over two hundred accounts of English land and sea voyages, from King Arthur to the latest trips to China.[57] In 1625, Samuel Purchas (ca. 1577–1626) published a new, expanded edition of Hakluyt's volume with the title *Hakluytus Posthumus, or Purchas his Pilgrimes*, which also included reports from England's rivals Spain and the Netherlands.[58]

Throughout Europe, sixteenth-century cosmographers produced a wide variety of objects and texts. These included celestial and terrestrial globes,[59] world and regional maps, commentaries on Johannes de Sacrobosco's *Tractatus de sphaera* (Treatise on the Sphere, 1220), cosmographic treatises, and historical-geographical descriptions of the world, as well as travel narratives and collections. Cosmographers also continued to edit and comment on ancient geographical texts. The most important of these (by Ptolemy, Strabo, Pomponius Mela, Solinus, and Dionysius Afer) had been edited in the fifteenth century, but the first editions of Ptolemy to include new maps of the world appeared in 1507 and 1508 in Rome. Waldseemüller and Ringmann included a more complete representation of the earth in their seminal 1513 edition of Ptolemy's *Geographia*, dedicated to the Emperor Maximilian. In addition to the twenty-seven ancient maps included in this work, their volume incorporated twenty modern charts, including a new map of the world (*Charta Marina*), ten land maps of Europe, four chorographical maps (of Switzerland, Lorraine, the Rhineland, and Crete), and five new maps of Africa and Asia.

One of the hallmarks of cosmography's success was the formalization of university instruction in geography in the second half of the sixteenth century. The Jesuits led on this front, in large part because of their global presence and their empirical understanding of the sciences (see Harris, Chapter 16, this volume). At the Collegio Romano, geography and astronomy were taught starting in 1550, following the example of Messina, using texts by authors such as Finaeus, Peurbach, and Johannes Stöffler. In their first and second years, students took courses on the "Spheres," on "Geography," and on the "Astrolabe," moving on to "Planetary Theory" in their third year; especially gifted students could also read Maurolico's *Three Dialogues on Cosmography* (1543). This curriculum was taught not only in Italy but also in Coimbra, Ingolstadt, Vienna, Cologne, Würzburg, Paris, Lyon, and Avignon.[60]

---

[57] Taylor, *Late Tudor and Early Stuart Geography*, pp. 1 ff.; Broc, *La géographie de la Renaissance*, pp. 38 ff.

[58] Taylor, *Late Tudor and Early Stuart Geography*, pp. 53 ff.; Broc, *La géographie de la Renaissance*, pp. 39 ff.

[59] On globes in the sixteenth and seventeenth centuries, see Elly Dekker and Peter van der Krogt, *Globes from the Western World* (London: Zwemmer, 1993).

[60] Ibid., pp. 36, 42.

A detailed program for the teaching of geography was also developed by the Italian Antonio Possevino (1533–1611), one of the authors of the Jesuit *Ratio studiorum* (Plan of Study, 1599). His two-volume *Bibliotheca Selecta* (Select Library, 1593), which traced the educational program of the Jesuits, included a treatise with the title *Methodus ad geographiam tradendam* (Method for Teaching Geography), which was later reprinted separately in Latin and Italian (in Venice and Rome in 1597).[61] Possevino first discussed Ptolemy and the geographers of antiquity, before moving on to the overseas discoveries and the work of modern geographers (especially in Germany and Italy). He stressed the uses of geography in the related sciences of physics, moral philosophy, and theology – a context important to the Jesuits. At the same time, Possevino saw geography as being closely connected with the writing of history. He thus placed himself in the tradition of the *Methodus ad facilem historiarum cognitionem* (Method for the Easy Comprehension of History, 1566) by the famous French jurist and scholar Jean Bodin (1530–1596).[62]

Cosmography became a leading science by confronting the learned – not only mathematicians but also natural philosophers and theologians – with new knowledge. At the same time, it acquired nobility and prestige, generating a host of works dedicated to popes, emperors, kings, aristocrats, and merchants. One of the most striking emblems of cosmography's new status and universal ambitions was Münster's *Cosmographia* (1544) (see Figure 20.4).[63] This work included a short history of creation, a treatise on the elements, a summary of mathematical cosmography, and, finally, a thorough representation of all countries, organized by continent. As far as possible, Münster used up-to-date descriptions and had new woodcuts made. The volume's concept was ambitious; directed at a broad audience, it became a notable publishing success, appearing in Latin and German (with a total circulation of 60,000 exemplars), as well as in French, Czech, Italian, and English translations, and inspiring several imitators and successors.

But it was Mercator who formulated a theoretical justification for the prominent role of cosmography among the sciences. Perhaps the most intellectual and influential cosmographer of his time, Mercator completed maps and globes for Charles V and the Duke of Cleves and composed the famous *Atlas sive cosmographicae meditationes de fabrica mundi* (Atlas; or, Cosmographical Meditations on the Fabric of the World, 1595).[64] In the preface to his *Chronologia* (1569), Mercator noted that cosmography was justly called the beginning and end of all natural philosophy, as it joined the heavens with

---

[61] Ibid., pp. 47–54.
[62] [Jean Bodin], *I. Bodini methodus ad facilem historiarum cognitionem* [1566] (Amsterdam: Jo. Ravesteiny, 1650; repr. Aalen: Scientia, 1967).
[63] See note 46.
[64] See note 49.

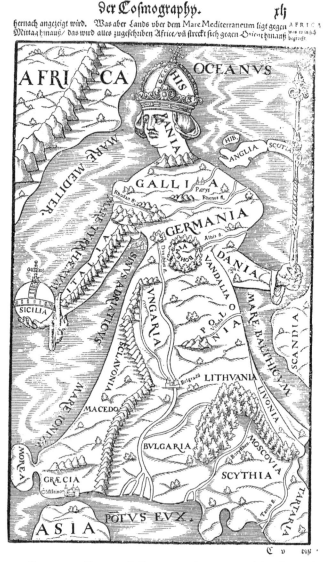

Figure 20.4. Europe as Queen of Cosmography, crowned by Spain (*Hispania*) and ruling over the continents of Africa and Asia. In Sebastian Münster, *Cosmographey* (Basel: Henricpetri, 1588; repr. Munich: Köbl, 1977), p. 41. The first known appearance of this iconography is in a woodcut by Johannes Putsch, Paris, 1537.

the earth. Cosmography dealt with the first and most important part of the world, on which the development, substance, and nature of all other things depended. Because chronology organized the historical-empirical data of human development, Mercator concluded that cosmography and chronology together constituted the all-encompassing basis of an empirical natural and human science.[65]

GEOGRAPHIA GENERALIS: TOWARD A SCIENCE OF DESCRIPTION AND MEASUREMENT (1600–1700)

In the last decades of the sixteenth century, the pace of overseas discoveries began to slow, eventually coming to a halt in the mid-seventeenth century. During the same period, scholarly interest, which had previously focused on the revolution in cosmography, began to turn to astronomy, which increasingly diverged from geography. The term "cosmography" came to be used less frequently or was equated with geography, as in the *Anatomia ingeniorum et scientiarum* (Anatomy of the Wits and the Sciences, 1615) of Bishop Antonio Zara from Aquileia. Geography moved from being an explosively expanding field of inquiry to an established one. At the same time, as classical texts and the methods of humanistic interpretation became less important, the role of practical considerations increased. Increasingly, the principal goal of the science of geography was to obtain more precise measurements of the surface of the earth, to describe it, and to discuss its properties.

In the ongoing divergence of geography and astronomy, Galileo's *Trattato della sfera ovvero cosmographia* (Treatise on the Sphere of Cosmography, ca. 1605–6) straddles the divide.[66] Composed for students at the University of Padua, this early text of Galileo was ptolemean, continuing the teaching on the "spheres." Although the style was fresh and the language vernacular Italian, the order of subjects and cosmographic reasoning remained traditional. There was little reasoning about Copernicus. The function of this text was clearly limited to basic university instruction. As far as geography was concerned, the facts were undisputed. The existence of the globe of earth and water had become common knowledge, and Galileo and other leading scholars were increasingly questioning the sky.[67]

The seventeenth century saw the continuing development of the practical branches of geography – cartography, navigation, and surveying. In the Netherlands, the cartographer Petrus Plancius (1552–1622) had made Portuguese cartographic knowledge accessible to the Dutch. The center of Dutch cartography moved from Antwerp and Louvain to Amsterdam with the founding of the Dutch East India Company in 1602,[68] and Hessel Gerritsz

---

[65] Gerardus Mercator, *Chronologia, hoc est Temporum demonstratio exactissima, ab initio mundi usque ad annum domini M.D.LXVIII* . . . (1st ed. 1569; Cologne: A. Birckmannus, 1579), fol. 2r: "Praefatio ad Lectorem."
[66] Galileo Galilei, "Trattato della sfera ovvero Cosmografia," in *Le opere di Galileo Galilei*, ed. Giuseppe Saragat, 20 vols. (Florence: G. Barbèra, 1968), 2: 203–55.
[67] Dainville, *La géographie des humanistes*, pp. 19 ff.
[68] Zandvliet, *Mapping for Money*, pp. 86 ff.

led its newly founded cartographic agency beginning in 1617. In 1620, a "cartographic depot" was established in Batavia, the capital city of the Dutch colonies. In 1633, Willem Blaeu (1571–1638) took over the cartographic office in Amsterdam; his descendants Joan Blaeu, Sr. and Joan Blaeu, Jr. dominated the Dutch cartographic business until 1705.

Meanwhile, navigational skills and knowledge continued to evolve in coastal regions.[69] Whereas the method of determining latitude by measuring pole heights or calculating solar culminations was sufficient for practical purposes, the determination of longitude remained a problem. Sixteenth-century attempts to calculate longitude by using clocks on ships instead of astrolabes – notably by Gemma Frisius in Louvain (from 1522) and William Borough – had failed because pendulum clocks on board ships simply did not work regularly enough (see the following chapters in this volume: Bertoloni Meli, Chapter 26; Bennett, Chapter 27). In the seventeenth century, the Sieur de St. Pierre suggested determining longitude by measuring lunar distances – a method that had already been considered in the late Middle Ages. This required exact observations of the moon's movement relative to the other stars and planets and led to the foundation in 1675 of the Royal Observatory in Greenwich by King Charles II. It was believed that only precise lunar observations could yield an exact longitude, prompting improvements in both the calculation of lunar motion and the measurement of celestial angles. Still, the method remained too imprecise. It was not until the beginning of the eighteenth century that the problem of determining longitude at sea was solved with the development of an exact chronometer in a competition between England and France.[70]

During the same period, geography made its name as an established science, with syntheses and textbooks.[71] In Amsterdam, Bernhardus Varenius (1622–1650) published his *Geographia generalis* (1650), which was studied and disseminated by Isaac Newton (1642–1727). Varenius provided an outstanding synthesis of developments in geography, incorporating the ideas of contemporary natural philosophers such as René Descartes (1596–1650). Only two years later, the French Jesuit Jean François published *La science de la géographie* (The Science of Geography, 1652). In it, he made an effort to legitimate geography as a genuine science and to present it systematically through its fundamental principles. In contrast, the work by the Italian Jesuit Giambattista Riccioli, *Geographia et hydrographia reformata* (Improved Geography and Hydrography, 1661), emphasized practical developments. This work, in which Riccioli compared the geographical observations made

---

[69] Eva G. R. Taylor, *The Haven-Finding Art: A History of Navigation from Odysseus to Captain Cook* (New York: Abelard-Schuman, 1957); and Waters, *Art of Navigation in England*.
[70] Taylor, *The Haven-Finding Art*, pp. 245 ff.: "The Longitude Solved."
[71] Broc, *La géographie de la Renaissance*, pp. 93 ff.; and Dainville, *La géographie des humanistes*, pp. 276 ff. (on Varenius and François), 445 ff. (on Riccioli).

by Jesuits around the globe, was highly regarded by the astronomer and geographer Gian Domenico Cassini I (1625–1712), who knew him personally.

Although interest in humanist geography, with its interpretation and critical expansion of classic texts, continued to decline in the second half of the seventeenth century,[72] cosmographers increasingly focused on improving measurement.[73] Cassini was one of the leading practitioners in this area.[74] After teaching geometry at the University of Bologna, where he also worked as an engineer, he relocated to Paris at the behest of French comptroller of Finances Jean-Baptiste Colbert (1619–1683), where he became a member of the Académie Royale des Sciences in Paris and in 1669 was named director of the Observatoire Royale de Paris. In addition to making significant astronomical discoveries, Cassini continued the measurements of the earth begun by Jean Picard and assumed the leadership of the Académie's most important projects: the delineation of the Parisian meridian line to the south and the exact calculation of the earth's diameter. He was later assisted by his son, Jacques Cassini II (1677–1756), who took over his activities and his membership in the Académie in 1712 and described the completion of the project started by his father in his *Traité de la grandeur et de la figure de la Terre* (Treatise on the Size and Shape of the Earth, 1723).

Gian Domenico Cassini had obtained some of his data by using the satellites of Jupiter to calculate exact longitude – a method that was to prove erroneous in the following century[75] (see Donahue, Chapter 24, this volume). His observations were complicated by the discovery that a pendulum oscillated more slowly at the equator than at the poles, as noted by Jean Richer (1630–1696) in 1672 on the island of Cayenne and by Edmond Halley (ca. 1656–1743) in 1675 on the island of Hebre. Christiaan Huygens (1629–1695) and Newton concluded from this that the earth was flattened at the poles – a conjecture confirmed in 1691 when Cassini's observations of Jupiter showed the same flattening.[76] Attempts to measure the Earth continued after Cassini's death in 1712. Most cosmographers were convinced that exact measurements and globally coordinated observations starting at the Parisian meridian would correct "dangerous errors" in land maps and sea charts, thus "perfecting geography and navigation," in accordance with the motto of the Académie.[77]

---

[72] Dainville, *La géographie des humanistes*, pp. 398 ff.: "La crise de l'humanisme en géographie, 1660–1700."

[73] Oscar Peschel and Sophus Ruge, in their *Geschichte der Erdkunde* (Munich: J. G. Cotta, 1865), outline their interpretation as follows: The Era of the Scholastics (until 1400), The Era of the Great Discoveries (1400–1650), The Era of Measurement (from 1650).

[74] Josef W. Konvitz, *Cartography in France, 1660–1848: Science, Engineering, and Statecraft* (Chicago: University of Chicago Press, 1987), pp. 4 ff.

[75] Taylor, *The Haven-Finding Art*, p. 248.

[76] Immanuel Kant, *Immanuel Kant's Physische Geographie*, 4 vols. (Mainz and Hamburg: G. Vollmer, 1801), 1: 29–35: "Abweichung der Figur der Erde von der Kugelgestalt" (on Richer, Cassini, Huygens, and Newton).

[77] Jacques Cassini, *Traité de la grandeur et de la figure de la Terre, par M. Cassini de l'Académie Royale des Sciences* (Amsterdam: Pierre de Coup, 1723), p. 1.

EXPERIENCE AND PROGRESS: CONTEMPORARY VIEWS OF THE EMERGENCE OF GEOGRAPHY

Although learned Europeans clearly recognized the sixteenth-century revolution in the experience and perception of the earth and the emergence of geography as a leading science, their assessment of these developments varied. Only a few years after the discovery of the New World became known, scholars began to reflect on the relationship between the old knowledge and the new and to argue about the errors of the classics.[78] The debate between Joachim von Watt (ca. 1484–1551), called Vadian, and Joannes Camers (ca. 1450–1546) exemplifies this new development. In 1515, Vadian, a celebrated poet laureate and a medical scholar at the University of Vienna, published a short tract in which he argued emphatically for the existence of the Antipodes and stressed the errors of ancient and medieval authors in this area.[79] Three years later, in his extensive edition of Pomponius Mela's *Geography* (1518), he gave the first critical analysis of this late antique text and seized upon the newly reopened question of the Antipodes.[80] On the other hand, Camers, a theologian and humanist originally from Italy and one of Vadian's older university colleagues, took up the opposite position. In his learned and equally extensive edition of another late antique classic, the *Geography* of Solinus (1520), he offered a philological analysis of the geographical text, without criticizing its content, and explicitly abstained from referring to modern ideas.[81] The ensuing methodological argument between the two scholars, infused with nationalist and reformist passions and set forth (though not resolved) in two publications from 1522,[82] may represent the beginning of the *querelle des anciens et modernes* (argument of the Ancients and Moderns), in which a scientific-critical engagement with the ancients confronted a philological-reconstructive one. The fact that this much broader argument, which was to extend well into the seventeenth century, had its roots in a cosmographical problem – how to assess the newest overseas discoveries and whether to

---

[78] Vogel, "Amerigo Vespucci und die Humanisten in Wien," pp. 77–104: "Der Streit zwischen Joachim Vadian und Johannes Camers über die Irrtümer der Klassiker."
[79] *Habes lector: hoc libello Rudolphi Agricolae Iunioris Rheti, ad Ioachimum Vadianum Helvetium Poetam Laureatam, Epistola, qua de locorum non nullorum obscuritate quaestio fit et percontatio* ... (Vienna: Io. Singrenius, 1515); and Vogel, "Amerigo Vespucci und die Humanisten in Wien," pp. 85–8.
[80] Vogel, "Amerigo Vespucci und die Humanisten in Wien," pp. 90 ff.
[81] Ibid., pp. 91–4.
[82] Joachim Vadian, "Loca aliquot ex Pomponianis commentariis repetita, indicataque, in quibus censendis, et aestimandis Ioanni Camerti Theologo Minoritano, viro doctissimo, suis in Solinum enarrationibus, cum Ioachimo Vadiano non admodum convenit," in Pomponius Mela, *Pomponii Melae De orbis situ libri tres, acuratissime emendati, una cum Commentariis Ioachimi Vadiani Helvetii castigatoribus* ... (Basel: A. Cratander, 1522); [Joannes Camers], *Io. Camertis ordinis Minorum, sacrae Theologiae professoris Antilogia, idest, locorum quorundam apud Iulium Solinum a Ioachimo Vadiano Helvetio confutatorum, amica defensio* (Vienna: Io. Singrenius, 1522); and Vogel, "Amerigo Vespucci und die Humanisten in Wien," pp. 95–102.

criticize long-standing geographical tradition – testifies to the importance of cosmography in the intellectual world of early modern Europe.[83]

The dispute between Vadian and Camers shows that the new cosmography had already overstepped the bounds of mathematics and won independent scientific status by the end of the second decade of the sixteenth century. In the course of the following two hundred years, the staggering consequences of the new knowledge of the Earth, both theoretical and practical, were formulated and reformulated many times. The French humanist Louis Le Roy (ca. 1510–1577), translator of Aristotle and Plato, emphasized this in *De la vicissitude ou variété des choses en l'univers* (On the Vicissitude or Variety of Things in the Universe, 1577):

> In our time, the Castilians have navigated beyond the Canary Islands, and have sailed in a westerly direction up to our Perieces [*Perioeciens*], whom they have subjected to the rule of Spain, with many cities and great lands, filled with gold and other goods discovered. And the Portuguese, proceeding in a southerly direction past the Tropic of Capricorn, reached our Anteces [*Antisciens*], showing, against the opinion of Aristotle and the poets of antiquity, that the entire middle Zone is inhabited, that is, the whole area between the two tropics. After that, they traveled across to India and reached our Antipodes, winning domination over them. . . . Thus we can truly affirm that the world today is completely manifest and all of humanity is known. All mortals can now exchange commodities with one another and provide for each other's dearth, like residents of one city and one republic of the world.[84]

Four decades later, in his *Instauratio magna* (Great Instauration, 1620), Francis Bacon (1561–1626) proposed the process of cosmographic discovery as exemplary for the reform of natural philosophy. Emphasizing the complementary roles of theory and experience, he argued that the human spirit must be prepared for a "voyage into the open sea":

> No doubt the ancients proved themselves in everything that turns on wit and abstract meditation, wonderful men. But as in former ages when men sailed only by observation of the stars, they could indeed coast along the shores of the old continent or cross a few small and Mediterranean seas; but before the ocean could be traversed and the new world discovered, the use of the mariner's needle, as a more faithful and certain guide, had to be found out; in like manner the discoveries which have been hitherto made in the arts

---

[83] Hans Baron, "The Querelle of the Ancients and the Moderns as a Problem for Present Renaissance Scholarship," *Journal of the History of Ideas*, 20 (1959), 3–22," slightly expanded in Baron, *In Search of Florentine Civic Humanism: Essays on the Transition from Medieval to Modern Thought*, vol. 2 (Princeton, N.J.: Princeton University Press, 1988), pp. 72–100, gives no reference to overseas discoveries and the emergence of geography.

[84] Louis Le Roy, *De la vicissitude ov variete des choses en l'vnivers . . . Par Loys Le Roy, dict Regivs. Av Treschrestien Roy de France, et de Poloigne Henry III. du nom* (Paris: Chez Pierre l'Huillier, 1577), bk. II, fol. 110 v.

and sciences are such as might be made by practice, meditation, observation, argumentation . . . but before we can reach the remoter and more hidden parts of nature, it is necessary that a more perfect use and application of the human mind and intellect be introduced.[85]

In his *Parallele des Anciens et Modernes en ce qui regarde les arts et les sciences* (Parallel of the Ancients and Moderns Regarding the Arts and Sciences, 1688), Charles Perrault (1628–1703), too, emphasized the superiority of the moderns in "Astronomy, Geography, Navigation, Physics, Chemistry, Mechanics."[86] For him, as for most sixteenth- and seventeenth-century natural philosophers, there was no doubt that the overseas explorations had caused a breakthrough in knowledge of the natural and human worlds. Seen from a later perspective, the emergence of geography during the sixteenth century still appears paradigmatic for the progress of the early modern sciences in general. As the Dutch historian Reijer Hooykaas has rightly stated, "Henry the Navigator, who organized the first great voyages of discovery, was no scientist, and he had no scientific aims. But it was his initiative that triggered off a movement which, growing into the avalanche of upheaval in sixteenth-century geography, opened the way for the reform, sooner or later, of all other scientific disciplines."[87] The influence of the overseas discoveries on the rebirth of geography and the reform of the sciences was not only a long-term process but also worked concurrently with the discoveries themselves. From the beginning, early modern European scholars reacted quickly and consequentially to the latest information from Portugal and Spain. The first example is Copernicus.[88] As we have seen, his theories, formulated shortly after 1503, were closely entwined with the overseas discoveries and the "cosmographic revolution": the emergence of the modern globe.

---

[85] [Francis Bacon], *Francisci de Verulamio, Summi Angliae Cancellarii, Instauratio magna* (London: Jo. Billius, 1620); translation cited from Francis Bacon, *The Great Instauration* [1620], in *New Atlantis and The Great Instauration*, ed. Jerry Weinberger (Wheeling, Ill.: Harlan Davidson, 1989), pp. 13–14.

[86] Charles Perrault, *Paralelle des Anciens et des Modernes en ce qui regarde les arts et les sciences. Dialogues*, 4 vols. (Paris: Jean Baptiste Coignard, 1688–96), vol. 1, 2nd ed.: fol. e4r.

[87] Reijer Hooykaas, "The Rise of Modern Science: When and Why?," *British Journal for the History of Science*, 20 (1987), 453–73, at p. 473.

[88] See notes 25–27.

21

FROM ALCHEMY TO "CHYMISTRY"

William R. Newman

Between the High Middle Ages and the end of the seventeenth century, the discipline of alchemy underwent a succession of remarkable changes, both in its internal configuration and in its outward dispersion. In a word, alchemy moved from a rather marginal position as a discipline concerned mainly with mineralogy, metallurgy, and the products of chemical technology to the center of the European stage, where it became the basis for a comprehensive theory of matter and the justification of a heterodox new medicine, occupying the best minds of the age. All the same, alchemy retained a striking continuity between its medieval and early modern incarnations. Up to the beginning of the Enlightenment, the writers of the popular new genre of "chymical textbooks" were paying tribute to Hermes Trismegistus, an ancient and numinous figure who supposedly founded the art of alchemy (see Copenhaver, Chapter 22, this volume).[1] Until the last quarter of the seventeenth century, these textbook authors made no strict demarcation between "alchemy" and "chemistry," and despite a misconception popular among historians, they did not normally disavow the transmutation of metals.[2]

The modern distinction between alchemy and chemistry, wherein the former refers exclusively to the transmutation of base metals into gold, is a caricature popularized above all by the *philosophes* of the French Enlightenment. In the Middle Ages, alchemy was commonly viewed as a subordinate and artisanal branch of physics, a sort of "applied science" based on general

---

[1] Hermes Trismegistus, or "thrice-greatest Hermes," was the supposed author of the *Corpus Hermeticum*, a collection of late antique dialogues written in Greek and passing down the supposed wisdom of ancient Egypt. See A.-J. Festugière, *La révélation d'Hermès Trismegiste*, 4 vols. (Paris: Lecoffre, 1944–54). Even Nicolas Lemery, whose *Cours de chimie* was reprinted in expanding editions into the eighteenth century, refers to Hermes as one of the fathers of "chimie." See Lemery, *Cours de chimie* (Paris: [Lemery], 1675), p. 2.

[2] For a sustained defense of the early modern identity of "alchemy" and "chemistry," see William R. Newman and Lawrence Principe, "Alchemy versus Chemistry: The Etymological Origins of a Historiographical Mistake," *Early Science and Medicine*, 3 (1998), 32–65.

principles supplied by natural philosophy.[3] It was classed within the field of "meteorology," that is, the study of matter below the sphere of the moon.[4] As its field of inquiry, medieval alchemy normally laid claim to subterranean (and hence sublunar) matter. As the heir to this tradition, sixteenth- and seventeenth-century alchemy already engaged in research and application over a broad spectrum of pursuits beyond the traditional attempt to transmute metals. These activities included metallurgical assaying, refining of salts, dye and pigment manufacture, the improvement of glass and ceramic formulas, the making of artificial gemstones, research on incendiary weapons, the study of chemical luminescence, techniques for the improvement of brewing, and a host of medical pursuits, such as the analysis and purification of existing drugs and the discovery and manufacture of entirely new pharmaceuticals. Although many of these concerns were already present in medieval alchemy, they acquired a new status in the sixteenth century, when they came to be widely viewed as applications of a unified alchemical theory with strong cosmological and religious overtones.

At the same time, alchemy changed its institutional home. Despite the similarity that scholastic authors often saw between alchemy and medicine, there was no "faculty of alchemy" in medieval universities, nor was the subject an official part of medieval curricula. Documentary and archaeological evidence suggest that the major domicile of medieval alchemy was the monastery, although it is clear that alchemists were also operating in courtly and medical circles.[5] The latter two venues acquired a vastly greater significance in the sixteenth century, and by the seventeenth, "chymistry," to use the early modern English orthography, had officially entered the university. The increasing popularity of alchemy in both learned and unlettered circles can be linked to four major movements – first, the topic was taken up by the burgeoning neo-Platonism of the Renaissance, with its focus on occult philosophy and magic; second, it formed the centerpiece of a comprehensive and controversial medical reform beginning in the sixteenth century; third, its technological and industrial potential began to be widely realized at a time when manual arts were gaining prestige over a broad area; and fourth, it provided a natural locus for unorthodox religious speculation at the height of the massive confessional upheaval of early modern Europe. In

---

[3] Thomas Aquinas, *Super Boetium de trinitate*, in *Sancti Thomae de Aquino opera omnia*, Leonine edition (Rome: Commissio Leonina, 1992), pp. 140–1.

[4] Petrus Bonus, *Margarita pretiosa*, in *Theatrum chemicum*, ed. Lazarus Zetzner, vol. 5 (Strasbourg: Eberhard Zetzner, 1660), pp. 511–713, esp. pp. 508, 513, and 528. For a concise description of the theory of *subalternatio*, see Steven J. Livesey, *Theology and Science in the Fourteenth Century* (Leiden: E. J. Brill, 1989), pp. 20–53. See also Dominicus Gundissalinus, *De divisione philosophiae*, ed. Ludwig Baur (Münster: Aschendorff, 1903), pp. 20–4.

[5] The many thirteenth- and fourteenth-century prohibitions against monastic practice of alchemy are good evidence for interest among the religious orders; see Robert Halleux, *Les textes alchimiques* (Turnhout: Brepols, 1979), p. 127. For archaeological evidence, see Stephen Moorhouse, "Medieval Distilling-Apparatus of Glass and Pottery," *Medieval Archaeology*, 16 (1972), 79–121; Isabelle Rouaze, "Un atelier de distillation du moyen âge," *Bulletin archéologique du Comité des Travaux Historiques et Scientifiques*, n.s. 22 (1989), 159–271.

acknowledgment of the increased disciplinary and practical scope of the early modern discipline when compared with medieval alchemy, I will henceforth use the inclusive seventeenth-century term "chymistry" to mean the totality of chemical/alchemical technology and theory as it existed in early modern Europe, especially after the onslaught of Paracelsianism.

THE EARLY SIXTEENTH CENTURY

The situation at the beginning of the sixteenth century was not one to suggest that alchemy would soon become the idée fixe of the learned. As Lynn Thorndike pointed out many years ago, alchemical incunabula are extremely rare, and the first half of the sixteenth century saw relatively few printed books on the subject.[6] This situation becomes comprehensible when one considers the disdain with which most early humanist authors, in particular, viewed alchemical pursuits. Already in the late fourteenth century the Tuscan poet Francesco Petrarca had ridiculed the attempt to transmute metals, and his satirical tone was amplified in the *Encomium Moriae* (Praise of Folly, 1511) of Desiderius Erasmus. The fifteenth-century Italian humanists Maffeo Vegio, Pandolfo Collenuccio, Tito Vespasiano Strozzi, and Ermolao Barbaro all stressed the inutility and low status of alchemy, and the famous polymath Leon Battista Alberti added that the transmutation of metals is "forbidden to mortals."[7] The denigration of alchemy by these writers combined a moralistic dislike of the venality implicit in gold making with an empirical observation that attempts to transmute base metals usually came to grief. A number of humanist authors also found support for their conviction that alchemy was fraudulent in a medieval anti-alchemical pronouncement, wrongly attributed to Aristotle, containing the claim that "the species of metals cannot be transmuted."[8]

Despite the censorious tone of numerous humanists, alchemy found a friendlier environment within the neo-Platonic magic of the court philosopher of Lorenzo de' Medici, Marsilio Ficino (1433–1499), who is famous for translating the Greek of Plato and Hermes Trismegistus into Latin. Ficino began an immensely influential reorientation of alchemy that was carried on by his acolytes, above all the "archimagus" of the sixteenth century, Heinrich

---

[6] Lynn Thorndike, *A History of Magic and Experimental Science*, vol. 5 (New York: Columbia University Press, 1941), pp. 532–49. See also Rudolf Hirsch, "The Invention of Printing and the Diffusion of Alchemical and Chemical Knowledge," *Chymia*, 3 (1950), 115–41.

[7] Sylvain Matton, "L'influence de l'humanisme sur la tradition alchimique," *Micrologus*, 3 (1995), 279–345.

[8] Matton, "L'influence," pp. 280–1, 285, 292–3, and 297. Although Matton makes no mention of it, the passages cited from Sebastian Brandt, Erasmus, and Alberti all reflect the influence of the Pseudo-Aristotelian phrase (actually by the Persian philosopher Ibn Sīnā, or Avicenna), which is usually identified by its opening words – *sciant artifices*. For the influence of the *sciant*, see William R. Newman, *Promethean Ambitions: Alchemy and the Quest to Perfect Nature* (Chicago: University of Chicago Press, 2004), pp. 34–114.

Cornelius Agrippa von Nettesheim (1486–1535) (see Copenhaver, Chapter 22, this volume). Several major changes in the standard view of alchemy can be traced directly to Ficino and Agrippa. First, in his *De vita coelitus comparanda* (On Making Your Life Agree with the Heavens), the third book of his *De vita* (On Life, 1489), Ficino explicitly linked the vital spirit of the cosmos with the alchemical quintessence, a physical substance that could be extracted by means of distillation and other techniques. Ficino also employed the term "elixir" (from the Arabic *al-iksīr*) for the quintessence: This word had been employed by Arabic alchemists as a synonym for the "philosophers' stone," a marvelous substance that could transmute base metals into noble ones.[9] Ficino's appropriation of these alchemical terms for the "spirit of the world" (*spiritus mundi*) would eventually lead to a new role for alchemy as the art by which the vital principle of the cosmos could be isolated. As Ficino himself wrote:

> Between the tangible and partly transient body of the world and its very soul, whose nature is very far from its body, there exists everywhere a spirit, just as there is between the soul and body in us, assuming that life everywhere is always communicated by a soul to a grosser body.... When this spirit is rightly separated and, once separated, is conserved, it is able like the power of seed to generate a thing like itself, if only it is employed on a material of the same kind. Diligent natural philosophers, when they separate this sort of spirit of gold by sublimation over fire, will employ it on any of the metals and will make it gold. This spirit rightly drawn from gold or something else and preserved, the Arab astrologers call Elixir. But let us return to the spirit of the world. The world generates everything through it (since, indeed, all things generate through their own spirit); and we can call it both "the heavens" and "quintessence."[10]

Ficino's brief reference to alchemy in *De vita* had an influence out of all proportion to its length.[11] The association that the prominent neo-Platonist drew between the alchemical quintessence and the spirit of the world gave alchemy a cosmic character that it had for the most part lacked in the Middle Ages, when it was seen primarily as a pursuit devoted to metals, minerals, and items of chemical technology, such as pigments and pharmaceuticals.[12]

---

[9] Paul Kraus, *Jābir ibn Hayyān: Contribution à l'histoire des idées scientifiques dans l'Islam*, 2 vols. (Cairo: Institut Français d'Archéologie Orientale, 1942), 2: 4.

[10] Marsilio Ficino, *Three Books on Life*, ed. and trans. Carol V. Kaske and John R. Clark (Binghamton, N.Y.: Medieval and Renaissance Texts and Studies, 1989), pp. 255–7. Kaske and Clark identify Ficino's source for the term "elixir" as the *Picatrix*, a famous work of Arabic magic that was circulating in Latin translation in Ficino's time.

[11] Sylvain Matton, "Marsile Ficin et l'alchimie: sa position, son influence," in Jean-Claude Margolin and Sylvain Matton, *Alchimie et philosophie à la Renaissance* (Paris: J. Vrin, 1993), pp. 123–92.

[12] An important exception is Roger Bacon, who thought that alchemists had discovered a way to isolate the *prima materia* of the world. See William R. Newman, "The Philosophers' Egg: Theory and Practice in the Alchemy of Roger Bacon," *Micrologus*, 3 (1995), 75–101.

Ficino's claim that alchemy could isolate the vital principle of the world was taken up eagerly by subsequent alchemists. Among the earliest of these was Giovanni Augurelli (1456–ca. 1524), an acquaintance of Poliziano and a friend of Ficino, whose *Chrysopoeiae libri tres* (Three Books on Making Gold) first appeared in 1515.[13] Dedicated to Pope Leo X, Augurelli's work is written in the form of a lengthy poem, in which such myths as the Golden Fleece, the labors of Hercules, and the love affairs of Venus are given alchemical interpretations.[14] Augurelli explicitly adopted Ficino's theory that the *spiritus mundi* is identical to the alchemical quintessence, thereby integrating this idea into the mainstream of alchemical literature.

Ficino's linkage of the alchemical quintessence to the *spiritus mundi* was also appropriated and expanded by his follower Agrippa von Nettesheim, in the latter's famous work *De occulta philosophia* (On Occult Philosophy, 1531–3).[15] But Agrippa added another important feature to the new understanding of alchemy in the form of an alchemically colored treatment of the four elements, which, like Ficino's association of the quintessence with the *spiritus mundi*, served to strengthen the cosmic significance of the discipline. Relying on his teacher Johannes Trithemius, abbot of Sponheim (1462–1516), Agrippa argued that each of the four elements was actually "three-fold" and contained its purer and simpler cognates within itself. Agrippa stated clearly that without direct knowledge of these simpler elements, one could not obtain success in natural magic.[16] In a later passage, Agrippa even suggested that elemental earth, to which he gave special prominence as the "center, foundation and mother" of all sublunar things, be purified and extracted from common humus by means of a process involving burning and washing.[17] The cosmic significance of alchemy was further asserted by Agrippa in the second book of *De occulta philosophia*, where he supplied a list of correspondences to the number one, or *monas*. In the elementary world, Agrippa argued that the number one is represented by "the philosophers' stone – the one subject and instrument of all natural and transnatural virtues."[18] In the metallurgical alchemy of the Middle Ages, the philosophers' stone had been a desideratum

---

[13] Giovanni Augurelli, *Ioannis Aurelii Augurelli P. Ariminensis chrysopoeiae libri III* (Basel: Johann Froben, 1518).

[14] The alchemical interpretation of ancient mythology was already present in medieval Greek sources, such as the eleventh-century *Suda*, and to a small extent in medieval Latin works. See Robert Halleux, "La controverse sur les origines de la chimie, de Paracelse à Borrichius," in *Acta Conventus Neo-latini Turonensis, Troisième Congrès International d'Études Neo-latines*, ed. Jean Margolin, 2 vols. (Paris: J. Vrin, 1980), 2: 807–19. See also Sylvain Matton, "L'interprétation alchimique de la mythologie," *Dix-huitième siècle*, 27 (1995), 73–87.

[15] Cornelius Agrippa, *De occulta philosophia libri tres*, ed. V. Perrone Compagni (Leiden: E. J. Brill, 1992), pp. 113–14.

[16] Agrippa, *De occulta philosophia*, p. 91. For a treatment of this passage, see William R. Newman, *Gehennical Fire: The Lives of George Starkey, an American Alchemist in the Scientific Revolution* (Cambridge, Mass.: Harvard University Press, 1994), pp. 213–21.

[17] Agrippa, *De occulta philosophia*, p. 93; Newman, *Gehennical Fire*, p. 214.

[18] Agrippa, *De occulta philosophia*, p. 257.

for transmuting metals. Now it had become the means by which natural magic worked as well. This statement is supported further by the accompanying text:

> There is one thing created by God, the subject of all the wonderfulness on earth or in the heavens: this thing is itself animal, vegetable, and mineral *in actu*, it is found everywhere yet known to very few, mentioned by none under its proper name but veiled in innumerable *figurae* and enigmas, without which neither alchemy, nor natural magic can attain its complete goal.[19]

Agrippa's description of the "one thing," replete with language traditionally used for the philosophers' stone, reinforced the new status that Ficino had conferred upon alchemy. It also made it possible to interpret one of the hallowed texts of the alchemical tradition, the *Tabula smaragdina*, or *Emerald Tablet*, of "Hermes Trismegistus" in a new light, albeit one inherited from Trithemius.

The *Emerald Tablet* is first found in the *Kitāb sirr al-khalīqa* (Book of the Secret of Creation), attributed to "Balinas" (Pseudo-Apollonius of Tyana, ca. eighth century).[20] Hermes says there that "that which is above is the same as that which is below," and follows this cryptic utterance with still more obscure material about the conversion of the "one thing" into earth by means of fire and a descent from heaven into earth. Medieval authors usually saw an encoded alchemical recipe in these lines, but in the sixteenth century the *Emerald Tablet* served as one basis for the comprehensive unification of alchemy and the neo-platonizing cosmology under discussion. Trithemius, whom Agrippa follows, took it literally as a cosmological statement concerning the soul of the world.

PARACELSUS

The new status of alchemy as a fundamental science deeply linked to natural magic and other occult pursuits was eagerly adopted by the Swiss iconoclast and medical writer Theophrastus von Hohenheim, or Paracelsus (1493–1541), whose massive output exercised a profound influence on the field. Paracelsus is known above all for his reorientation of alchemy away from the transmutation of metals and toward the pharmaceutical application of alchemical techniques. Although "alchemical" techniques such as distillation had been employed by physicians since at least the twelfth century, Paracelsus went much further in treating the body as a "chemical system."[21] Not only did he

---

[19] Ibid., p. 256: "una res est a Deo creata, subjectum omnis mirabilitatis, quae in terris et in coelis est: ipsa est actu animalis, vegetalis et mineralis, ubique reperta, a paucissimis cognita, a nullis suo proprio nomine expressa, sed innumeris figuris et aenigmatibus velata, sine qua neque alchymia neque naturalis magia suum completum possunt attingere finem."
[20] Julius Ruska, *Tabula smaragdina* (Heidelberg: Winter, 1926), pp. 6–38.
[21] Robert Halleux, *Textes alchimiques*, pp. 130–2.

Figure 21.1. The cosmology of the *Emerald Tablet* of Hermes. In *Musaeum hermeticum reformatum et amplificatum* (Frankfurt; H. à Sande, 1678).

view the organs of the body in terms of alchemical apparatus, as so many stills, filters, casks, and the like, but basing himself on the time-honored parallelism between the macrocosm (the cosmos at large) and the microcosm (the human body), Paracelsus argued that the minerals of the exterior world were to be found in a different form in the body. Because he generally championed a form of homeopathic therapy ("like cures like"), Paracelsus therefore stressed the importance of pharmaceuticals derived from minerals. His success in this endeavor was so striking that his followers coined a new term for his form of medical practice – *chymiatria*, or iatrochemistry. By no means did Paracelsus and his heirs forget the traditional goals of alchemy, however. "Chrysopoeia" and "argyropoeia," the making of gold and silver, remained as substantial components within the "chymistry" of the sixteenth and seventeenth centuries.[22]

Despite the much-lauded novelty of his *chymiatria*, Paracelsus owed a striking debt to his late medieval forebears. *De consideratione quintae essentiae omnium rerum* (On the Consideration of the Quintessence of all Things), written by the Franciscan John of Rupescissa (Jean de Roquetaillade, ca. 1300–ca. 1365), was known to Paracelsus. This work, probably composed in 1351 or 1352, gives comprehensive recipes for distilling and purifying ethyl alcohol, which John equates with the "quintessence" of Aristotelian cosmology.[23] In addition, John explains how the quintessence can be extracted from a variety of substances, including minerals, and ingested for medicinal purposes. This fantastically popular text was reworked and incorporated into the *Liber de secretis naturae* (Book of the Secrets of Nature) of Pseudo–Ramon Lull, where it acquired further dispersion. Paracelsus, who knew both the Rupescissan and Lullian traditions quite well, incorporates much of this material into his early *Archidoxis*, the first systematic treatment of his reworked alchemy. Even if Paracelsus was generally negative in his appraisal of medieval alchemy, it is clear that his own ideas evolved from that tradition, though in some instances mediated by contemporary works on distillation and strong waters.[24]

A further point of contact with medieval alchemy lies in the matter theory of Paracelsus. The medieval theory that metals originated from sulfur and mercury within the earth, ultimately deriving from Aristotle's *Meteorology*, 378a–b, was his direct inspiration. But Paracelsus added a third principle, salt, to the preexisting two, and argued that all things, not just metals and minerals, were composed from these three components. In this Paracelsus was motivated partly by theological concerns, for he argued that the *fiat lux* by which God created the world was necessarily tripartite, being the expression of the Father, Son, and Holy Spirit, and that the cosmos itself should therefore

---

[22] Newman and Principe, "Alchemy versus Chemistry," pp. 32–65.
[23] Robert Halleux, "Les ouvrages alchimiques de Jean de Rupescissa, Jean de Roquetaillade," in *Histoire littéraire de la France*, vol. 41 (Paris, 1981), pp. 241–84.
[24] Udo Benzenhoefer, *Johannes' de Rupescissa Liber de consideratione quintae essentiae omnium rerum deutsch* (Wiesbaden: Franz Steiner Verlag, 1989), pp. 57–82.

be tripartite.²⁵ Paracelsus did not think of his three principles as being unique, isolable pure substances that could be extracted from multiple compounds in the way that a modern chemist might say that the same sulfur can be obtained by decomposing either iron sulfide or copper sulfate. Instead, he argued that there were many sulfurs, mercuries, and salts, which differed from one another. Yet the material world bore the stamp of its creator in that each substance was composed of these three.²⁶

The choice of "salt" as the third Paracelsian principle surely reflects the increasingly important role of various salts in early modern technology. As Paracelsus was well aware in an age of increasing reliance on gunnery, saltpeter, or potassium nitrate was a key ingredient of gunpowder²⁷ (see DeVries, Chapter 14, this volume). Common salt and saltpeter also had remarkable powers of preserving dead flesh, which suggested profound medicinal properties, and saltpeter's vivifying power was corroborated by the fact that it could be used as a fertilizer.²⁸ Paracelsus may also have wished to incorporate the technology of the mineral acids into his theory of three principles. The mineral acids had been discovered in fairly pure form by the early fourteenth century and were associated in large measure with the "spirits" or volatile products of various substances considered "salts," notably vitriol (iron or copper sulfate), saltpeter, and common salt. The relationship in Paracelsus's mind between his three principles and the traditional four elements, however, is not clear. In his meteorological works, he refers to the four elements as "mothers" and argues that they exude the material from which the world is made. In his pharmacological *Archidoxis*, on the other hand, he speaks of elemental fractions made by distillation, and the three principles go unmentioned. Until the chronological development of the Paracelsian corpus is better sorted out, the reasons for this inconsistency will remain mysterious.²⁹

A final point of contact with medieval alchemy lies in Paracelsus's concept that the fundamental alchemical process is "Scheidung," or separation. Paracelsus envisioned processes ranging from the digestive system's separation of nutrient from excrement to the creative act of God Himself in terms of distillation and the removal of slag during the refining of metals. The notion

---

[25] Paracelsus, *Liber meteororum*, in *Sämtliche Werke*, ed. Karl Sudhoff, ser. 1, 14 vols. (Munich: Barth, Oldenbourg, 1922–33), 13: 135. See Andrew Weeks, *Paracelsus* (Albany: State University of New York Press, 1997), pp. 101–28; and Walter Pagel, *Paracelsus: An Introduction to Philosophical Medicine in the Era of the Renaissance* (Basel: Karger, 1958), pp. 100–5.

[26] Paracelsus, *Das Buch de mineralibus*, in *Sämtliche Werke*, 3: 41–3; and Reijer Hooykaas, *The Concept of Element: Its Historical-Philosophical Development*, trans. H. H. Kubbinga (privately printed without date or place), pp. 79–80.

[27] Bert Hall, *Weapons and Warfare in Renaissance Europe* (Baltimore: Johns Hopkins University Press, 1997).

[28] Henry Guerlac, "John Mayow and the Aerial Nitre," *Actes du VIIe Congrès International d'Histoire des Sciences* (Paris: Académie Internationale d'Histoire des Sciences, 1953), pp. 332–49; and Guerlac, "The Poet's Nitre," *Isis*, 45 (1954), 243–55. See also Allen G. Debus, "The Paracelsian Aerial Nitre," *Isis*, 55 (1964), 43–61.

[29] Weeks, *Paracelsus*, pp. 36–47.

that Gen. 1 refers to something like the fractional distillation in an alchemical vessel had already been intimated by Pseudo–Ramon Lull in the influential *Testmentum*, possibly composed as early as 1332.[30] Like Paracelsus, Pseudo-Lull stresses that the "purer part" of matter is all that will survive the conflagration at the end of time predicted by biblical tradition.[31] Paracelsus was fond of expressing this by means of a homey analogy – after the "Scheidung" enacted by the conflagration, the world will be clear and pure, like the egg white within an egg.[32]

REACTION TO AND INFLUENCE OF PARACELSUS

The reaction to Paracelsus was not one of immediate appreciation. Few of his books were printed during his lifetime, and much of his reputation was based on hearsay. During the 1560s and 1570s, however, the German followers of Paracelsus, such as Adam von Bodenstein, Michael Toxites, and Gerhard Dorn, began to publish his works.[33] This publishing effort reached a climax in the 1590s with the monumental collection of Paracelsus's works edited by Johann Huser, entitled *Bücher und Schrifften* (Books and Writings). The followers of Paracelsus were aided in this venture by a number of German princes, the most successful of whom was Ernst von Bayern (1554–1612), archbishop of Cologne and patron of Huser.[34] Ernst's interest in Paracelsianism and alchemy more generally was widely shared in the courts of Europe, especially those where German was spoken. The elector Palatine Ottheinrich (1502–1559) was already deeply interested in Paracelsus, and such later figures as Emperor Rudolph II, Wolfgang II von Hohenlohe, and Moritz of Hessen were also important patrons of Paracelsian alchemy.[35]

Despite his influence in courtly circles, the extreme iconoclasm of Paracelsus, reflected in such acts as the public burning of medical texts at the University of Basel in 1527, encouraged the demonization of his character by

---

[30] Michela Pereira and Barbara Spaggiari, *Il "Testamentum" alchemico attribuito a Raimondo Lullo* (Tavarnuzze: SISMEL, 1999), esp. pp. 12–24, 248–56. On the corpus of Pseudo-Lull, see Michela Pereira, *The Alchemical Corpus Attributed to Ramon Lull* (London: Warburg Institute, 1989).
[31] Pereira and Spaggiari, *Il "Testamentum,"* pp. 14–15, 170.
[32] Paracelsus, *Astronomia magna*, in *Sämtliche Werke*, 12: 322.
[33] Karl Sudhoff, *Bibliographia Paracelsica* (Berlin: Georg Reimer, 1898; repr. Graz: Akademische Druck, 1958), pp. 60–365.
[34] Joachim Telle, "Johann Huser in seinen Briefen," in *Parerga Paracelsica* (Stuttgart: Franz Steiner Verlag, 1991), pp. 159–248.
[35] Joachim Telle, "Kurfürst Ottheinrich, Hans Kilian und Paracelsus: Zum pfälzischen Paracelsismus im 16. Jahrhundert," in *Von Paracelsus zu Goethe und Wilhelm von Humboldt* (Salzburger Beiträge zur Paracelsus Forschung, 22) (Vienna: Verband der Wissenschaftlichen Gesellschaften Oesterreichs, 1981), pp. 130–46; R. J. W. Evans, *Rudolph II and His World: A Study in Intellectual History, 1576–1612* (Oxford: Clarendon Press, 1973); Jost Weyer, *Graf Wolfgang II von Hohenlohe und die Alchemie* (Sigmaringen: J. Thorbecke, 1992); and Bruce Moran, *The Alchemical World of the German Court: Occult Philosophy and Chemical Medicine in the Circle of Moritz of Hessen, 1572–1632* (Stuttgart: Franz Steiner Verlag, 1991).

academic opponents. One of the earliest of these was Thomas Erastus (1524–1583), a medical professor at the University of Heidelberg who is well known to modern historians as the eponymous founder of Erastianism, the doctrine that church must be subordinated to state.[36] Erastus's *Disputationes de nova Philippi Paracelsi medicina* (Disputations Concerning the New Medicine of Philip Paracelsus, 1571–3) is a sustained attack on Paracelsus that focuses on the supposedly sacrilegious, demonic, and dishonest elements in his work.[37] Erastus did not restrict his criticism to the person of Paracelsus but made his own preference for traditional botanical medicine clear as well. Yet this was not always the case with Paracelsus's critics: The Swiss physician and natural historian Conrad Gessner (1516–1565) despised the boasting and obscurity of his compatriot but incorporated many of the same chymical techniques into his own *Thesaurus Euonymi* (Treasure of Euonymous, 1552).[38] The Rothenburg schoolmaster Andreas Libavius (1540–1616) was equally hostile to Paracelsus's magic and his unorthodox religious ideas but came to be known as a champion of iatrochemistry, and his famous *Alchemia* of 1597 is often lauded for its textbook-like treatment of the same. Finally, Daniel Sennert (1572–1637), the influential medical professor of Wittenberg, wrote a synthetic work *De chymicorum cum Aristotelicis et Galenicis consensu ac dissensu* (On the Agreement and Disagreement of the Chymists with the Aristotelians and Galenists, 1619) that vilifies Paracelsus in terms like those of Erastus and yet defends chymistry at length.[39] One common thread uniting these authors, especially Libavius and Sennert, is their knowledge of medieval alchemy, including its pharmaceutical dimension. Hence it was possible to argue, as Libavius repeatedly did, that the valuable parts of iatrochemistry were already present in medieval writers such as Pseudo-Lull, and that Paracelsianism was the vulgar debasement of that tradition.[40] In reality, this charge remains unanswered to the present day, as the originality of Paracelsus's contribution is even now a matter of debate.[41]

Other opponents of Paracelsianism were less concerned with Paracelsus's disagreeable character than with the debunking of chymistry as a whole. A

---

[36] Charles D. Gunnoe, Jr., "Thomas Erastus and His Circle of Anti-Paracelsians," in *Analecta paracelsica*, ed. Joachim Telle (Stuttgart: Franz Steiner Verlag, 1994), pp. 127–48; and Lynn Thorndike, *A History of Magic and Experimental Science*, 8 vols. (New York: Columbia University Press, 1941), 5: 652–67.
[37] Thomas Erastus, *Disputationum de medicina nova Philippi Paracelsi* . . . (Basel: Petrus Perna, 1572).
[38] Conrad Gessner, *De secretis remediis liber aut potius thesaurus* . . . (Zurich: Andreas Gesner, 1552).
[39] Daniel Sennert, *De chymicorum cum aristotelicis et galenicis consensu ac dissensu* . . . (Wittenberg: Zacharias Schurer, 1619).
[40] See Newman, "Alchemical Symbolism and Concealment: The Chemical House of Libavius," in *The Architecture of Science*, ed. Peter Galison and Emily Thompson (Cambridge, Mass.: MIT Press, 1999), pp. 59–77.
[41] Gundolf Keil, "Paracelsus und die neuen Krankheiten," and "Mittelalterliche Konzepte in der Medizin des Paracelsus," in *Paracelsus – Das Werk – Die Rezeption*, ed. Volker Zimmermann (Stuttgart: Franz Steiner Verlag, 1995), pp. 17–46, 173–93. It must be admitted, however, that Keil's negative position toward Paracelsus is overly contentious and based on an inadequate reading of Paracelsus's works.

scurrilous and far-reaching debate erupted in France at the beginning of the seventeenth century between the proponents of chymistry and the Parisian medical faculty.[42] Joseph Du Chesne (ca. 1544–1609), or Quercetanus, physician to the French king Henry IV, published his work *De materia verae medicinae philosophorum priscorum* (On the Matter of the True Medicine of the Ancient Philosophers) in 1603, in which he claimed the authority of Hippocrates for the Paracelsian three principles.[43] His work was condemned by the Parisian medical faculty in the same year, which led to a free-for-all between Du Chesne's supporters and detractors. Among the latter, Jean Riolan the elder (1539–1606), censor of the medical faculty, and his son, Jean Riolan the younger (1577–1657), figured prominently. Their attacks on Du Chesne and his supporters were met with a blistering response by Libavius, attached to his *Alchymia* of 1606 (a reprint of the 1597 *Alchemia* with much new accompanying matter). The wounds of this debate continued to fester, with the Parisian medical faculty issuing periodic pronouncements against various chymical medicines – especially antimony – until the 1660s.

At the same time that the Parisian medical faculty was attacking chymical medicine, some opponents of Paracelsus were also assaulting the transmutation of metals. One of the most influential of these was Nicolas Guibert (ca. 1547–ca. 1620), whose *Alchymia ratione et experientia . . . impugnata* (Alchemy Impugned by Reason and Experience) of 1603 refers to Paracelsus as "the most foul and absolute great prince of liars who ever was, is, or will be, if you except the devil."[44] Guibert's attack, like that of the younger Riolan, was met by Libavius, who devoted his own *Defensio et declaratio perspicua alchymiae transmutatoriae* (Perspicuous Defense and Exposition of Transmutatory Alchemy) of 1604 to its discomfiture.[45] Indeed, Libavius was an eager champion of metallic transmutation: His *Alchymia* of 1606 contains four *Commentationes* (Treatises) specifically defending the traditional claims of chrysopoeia.[46]

One of the most significant results of the Paracelsian movement was the inauguration of the chymical textbook tradition and the incorporation of chymistry into university curricula. Shortly after the censure of Du Chesne by the Paris medical faculty, the king's almoner, Jean Beguin (ca. 1550–ca. 1620),

---

[42] Allen G. Debus, *The French Paracelsians* (Cambridge: Cambridge University Press, 1991), esp. pp. 46–101. See also Thorndike, *A History of Magic and Experimental Science*, 6: 247–53.

[43] Joseph Du Chesne, *Liber de priscorum philosophorum verae medicinae materia . . .* (Geneva: Haeredes Eustathii Vignon, 1603).

[44] N. Guibert, *Alchymia ratione et experientia ita demum viriliter impugnata . . .* (Strasbourg: Lazarus Zetzner, 1603), p. 77.

[45] Andreas Libavius, *Defensio et declaratio perspicua alchymiae transmutatoriae . . .* (St.-Ursel: Petrus Kopffius, 1604).

[46] Libavius's perennial defense of chrysopoeia must be stressed in the face of the rather whiggish picture of him given by Owen Hannaway, "Laboratory Design and the Aim of Science: Andreas Libavius versus Tycho Brahe," *Isis*, 77 (1986), pp. 585–610; and Hannaway, *The Chemists and the Word: The Didactic Origins of Chemistry* (Baltimore: Johns Hopkins University Press, 1975). For criticism of Hannaway's depiction of Libavius, see Newman, "Alchemical Symbolism and Concealment."

wrote a *Tyrocinium chymicum* (Chymical Training) (1610, 1612).[47] This work, which was to become one of the most popular books on chymistry of the seventeenth century, immediately displays its debt to Paracelsus by opening with a passage from the pseudo-Paracelsian work *De tinctura physicorum* (On the Philosophers' Tincture).[48] Beguin then defined chymistry in the following terms:

> The word Chymia is Greek: to the Latins it means the same thing as "art making liquor," or "dissolving solid things into liquor." Thus it is said that Chymia teaches par excellence to dissolve (which is more difficult) and to coagulate. If anyone should call it Alchymia, he denotes the excellency of it, in the manner of the Arabs. If he should call it Spagyria, he denotes its principal operations, namely *synkrisis* [association] and *diakrisis* [dissociation]. If he should call it "the Hermetic art," he refers to its inventor and its antiquity. If he should call it the distillatory art, he refers to its most excellent and easily its principal function.[49]

By explaining the significance of the various synonyms for *chymia*, Beguin managed to highlight the different connotations of the art. What is perhaps most striking is his emphasis on the twin operations of dissolution and coagulation: These are the primary tools of chymistry. Beguin reiterated this claim by asserting that the Paracelsian neologism *spagyria*, a synonym for chymistry, itself implied *synkrisis* and *diakrisis*, or "association and dissociation." Beguin's sources, such as Libavius, had already made this claim, interpreting *spagyria* as a fusion of the two Greek terms *span* ("to pull apart") and *ageirein* ("to bring together").[50] A few lines later, Beguin distinguished chymistry from physics, saying that chymistry has the "mixed and concrete body" as its object insofar as it is soluble and coagulable, whereas physics considers it qua mobile. In other words, chymistry studied the resolution and composition of bodies; physics dealt with their "motion," an Aristotelian category that included mainly alteration, augmentation or diminution, and local motion.

Beguin's definition of chymistry as the art of dissolving and coagulating bodies, itself an elaboration of the Paracelsian *spagyria*, would have

---

[47] The classic study of Beguin is that of T. S. Patterson, "Jean Beguin and His *Tyrocinium chymicum*," *Annals of Science*, 2 (1937), 243–98. Beguin's *Tyrocinium chymicum* exists in two early redactions (1610 and 1612), the first of which was published by his students. In the preface to the 1612 version, he complains that the 1610 printing was published without his authorization.

[48] Jean Beguin, *Tyrocinium chymicum recognitum et auctum* . . . (Paris: Matheus le Maistre, 1612), sig. [aiiv].

[49] Beguin, *Tyrocinium chymicum*, pp. 1–2: "Chymiae vocabulum Graecum est: Latinis idem, quod ars liquorem faciens: aut res solidas in liquorem solvens: dicta ita *kat' exochēn*, quod Chymia solvere (id quod difficilius) & coagulare doceat. Alchymiam si quis nuncuparit; Arabum more praestantiam eius: si Spagyriam; praecipua officia, *synkrisin* nempe & *diakrisin*: si artem Hermeticam; autorem & antiquitatem; si Destillatoriam; functionem eius praeclaram & facile principem insinuet."

[50] Andreas Libavius, *Commentariorum alchymiae* . . . *pars prima*, in Libavius, *Alchymia* (Frankfurt: Joannes Saurius, 1606), p. 77.

far-reaching consequences. It laid the grounds for defining chemistry as the science of analysis and synthesis, a definition that would last throughout the eighteenth century and still remain popular long after the Chemical Revolution inaugurated by Antoine-Laurent Lavoisier (1743–1794).[51] The more modern popularity of this definition probably had much to do with the benefit bestowed by giving chemistry a disciplinary space distinct from that of physics or natural philosophy. In the seventeenth century, however, chymistry was still in search of an entrée into the university, which it found only under the protectorship of medicine. The first known position of academic chymistry was conferred upon Johann Hartmann (1568–1631) in 1609 by the great alchemical Maecenas, Landgrave Moritz of Hessen. Hartmann himself came up with the idea of starting a *collegium chymicum*, or "chymical college," at the University of Marburg. The position that evolved out of Moritz's enthusiastic response was explicitly iatrochemical – *professor publicus chymiatriae*.[52] Hartmann's subsequent work was closely linked to that of Beguin: The German iatrochemist brought out his own edition of the *Tyrocinium chymicum* (written in 1618 and published in 1634), which he stuffed with his own chymical secrets and recipes. After Hartmann's appointment, other positions of chymistry arose. Zacharias Brendel, for example, began a public lecture course at the University of Jena around 1630, which was continued by Werner Rolfinck upon Brendel's death in 1638.[53] Other German universities soon followed, and by the eighteenth century, positions of chemistry within the medical faculties of European universities were commonplace.[54]

TRANSMUTATION AND MATTER THEORY

Despite the emphasis placed on pharmacy by the textbook tradition and early academic positions in chymistry, one should not be deluded into thinking that *chymiatria* (chymical medicine) had divorced itself from chrysopoeia (transmutatory alchemy) in the early seventeenth century. The quest for the philosophers' stone, the agent of metallic transmutation, could itself assume

---

[51] Examples of this definition are legion. Consider Georg Stahl, *Fundamenta chymiae dogmatico-rationalis experimentalis* (Nuremberg: Wolfgang: Mauritii Endteri Filiae, 1732), pt. 1, p. 1: "Chymistry in itself is the art of dissolving natural, inanimate, mixed, and composite bodies, and of transferring them into a new mixture or composition [*In se autem Chymia est ars, corpora naturalia inanimata, mixta, & composita dissolvendi, et in novam mixtionem vel compositionem transferendi*]." This may be compared with any number of nineteenth-century chemical textbooks, such as J. L. Comstock, *Elements of Chemistry* (New York: Robinson, Pratt, 1838), p. 9: "All chemical knowledge is founded on analysis and synthesis, that is, the decomposition of bodies, or the separation of compounds into their simple elements, or the recomposition of simple bodies into their compounds."
[52] Bruce T. Moran, *Chemical Pharmacy Enters the University* (Madison, Wis.: American Institute of the History of Pharmacy, 1991), pp. 15–16.
[53] Moran, *Chemical Pharmacy*, p. 10.
[54] Christoph Meinel, "Artibus academicis inserenda: Chemistry's Place in Eighteenth and Early Nineteenth Century Universities," *History of Universities*, 7 (1988), 89–115.

a medical dimension because it was widely thought that the stone was a cure not only for defective metals but for the ailing human body. Beguin himself edited the *Novum lumen chemicum* (New Chymical Light) of the famous transmutational alchemist Michael Sendivogius (1566–1636) in 1608, and some editions of the *Tyrocinium chymicum* begin with epistles describing Beguin's chrysopoetic efforts.[55] Hartmann was also deeply involved in the transmutation of metals before and after his appointment as *professor publicus chymiatriae*.[56] Even Robert Boyle (1627–1691) upheld the possibility of transmutation, and his personal papers reveal that the quest to find the philosophers' stone occupied him for some forty years.[57] Although there had been some earlier iatrochemists who disavowed transmutation, such as Guibert and Brendel's successor at Jena, Werner Rolfinck, the real divorce of "alchemy" from "chemistry" was officiated in the late seventeenth century when Nicolas Lemery (1645–1715) consciously excised *alchimie* – which he applied exclusively to metallic transmutation – from the third edition of his fabulously popular *Cours de Chimie* (Course of Chymistry, 1679).[58] The echoes of Lemery's derisory treatment of traditional alchemy resound in numerous influential writings, among them the bitter attack made on alchemy by Bernard Le Bovier de Fontenelle (1657–1757) in the 1722 *Histoire de l'Academie Royale des Sciences* (History of the Royal Academy of Sciences).[59] Fontenelle, in a brief introduction to Etienne François Geoffroy's *Des supercheries concernant la pierre philosophale* (Some Frauds Concerning the Philosophers' Stone), disparaged all alchemists as frauds and cheats. This is also the position of the important *Lexicon Technicum* (1704) of John Harris, which relies heavily on Lemery's rejection of alchemy.

Although the development of Lemery's motives remains unclear, it is at any rate certain that they had little or nothing to do with the eclectic corpuscular matter theory that he espoused. The claim of Hélène Metzger that seventeenth-century "corpuscular philosophy was radically different from the conception of the alchemists" and that the latter could not withstand "the assaults of the corpuscular and mechanical philosophy" was made in ignorance of the fact that the chymistry of the seventeenth century inherited a

---

[55] Michael Sendivogius, *Novum lumen chymicum* (Paris: Renatus Ruellius, 1608), sigs. aiir–[aiiiv] consist of a laudatory letter by Beguin, in which he says the following of Sendivogius: "Nor would I judge any of the philosophers up to now to have written more clearly and briefly about the power of art and nature." ("*Nec ullum hactenus Philosophorum clarius & brevius de artis & naturae potestate scripsisse iudicarem.*") See also Patterson, "Jean Beguin," pp. 245–7 and 296.
[56] Moran, *Chemical Pharmacy*, p. 14. Moran, *The Alchemical World of the German Court*, pp. 50–67.
[57] See Lawrence Principe, *The Aspiring Adept: Robert Boyle and His Alchemical Quest* (Princeton, N.J.: Princeton University Press, 1998). See also Principe, "Boyle's Alchemical Pursuits," in *Robert Boyle Reconsidered*, ed. Michael Hunter (Cambridge: Cambridge University Press, 1994), pp. 91–105. For Boyle's involvement with the chymist George Starkey, see Newman, *Gehennical Fire*, pp. 54–91, 170–5; and Newman and Principe, *Alchemy Tried in the Fire: Starkey, Boyle, and the Fate of Helmontian Chymistry* (Chicago: University of Chicago Press, 2002).
[58] Newman and Principe, "Alchemy versus Chemistry," pp. 59–63.
[59] *Mémoires de l'Académie Royale des Sciences, année MDCCXXII*, vol. 1, pp. 68–72.

well-developed corpuscular theory from medieval alchemy (see Joy, Chapter 3, this volume).[60] This theory had already appeared in the *Summa perfectionis* (Summa of Perfection), attributed to Geber, a classic text written around the end of the thirteenth century and still widely read in the seventeenth century. The essential points of Geber's theory are as follows. The four elements, viewed in corpuscular terms, combine in a "very strong composition" to make bigger particles of sulfur and mercury, which in turn bond to make various metals. The complete sublimability of mercury and sulfur without loss of their substance is explained on the assumption that they are composed of a homogeneous mass of very tiny particles – *subtiles partes*. Less volatile substances are made up of *grossae partes*, which are not as readily driven up by the fire. The same use of a size gradient at the micro level explains such laboratory processes as distillation, calcination, and the assaying procedures of cupellation and cementation: Wherever the separation of substances is effected by means of heat, it is caused by a lack of homogeneity between larger and smaller corpuscles. Similarly, the loose packing and tight packing of these corpuscles explains the variation in specific weight among different metals.[61] In all these instances, the striking interplay of theory with laboratory practice justifies the expression "experimental corpuscularism" in describing Geber's views about matter.

The corpuscular theory attributed to Geber was to have a vigorous afterlife in the seventeenth century. It supplied the primary inspiration for Daniel Sennert's atomism, which explained the material world in terms of minute corpuscles that combine and separate in processes of *synkrisis* and *diakrisis*. Sennert's theory was in turn appropriated by the young Robert Boyle, who turned it to his own purposes in his juvenile *Of the Atomicall Philosophy* (written in the mid-1650s). Traces of the theory can still be found in Boyle's mature works, such as *The Sceptical Chymist* (1661) and *The Origin of Forms and Qualities* (1666).[62] Geberian corpuscular theory also resurfaced in the writings of the Belgian chymist Johannes Baptista van Helmont (1579–1644). Van Helmont, the founder of an iatrochemical movement that was influential in the mid-seventeenth century, distanced himself from several key features of the Paracelsian system, in some instances returning to the ideas of pre-Paracelsian alchemists. He rejected the great emphasis that Paracelsus placed on the macrocosm–microcosm analogy and denied the irreducible status of

---

[60] Hélène Metzger, *Les doctrines chimiques en France du début du XVIIe à la fin du XVIIIe siècle*, vol. 1 (Paris: Les Presses Universitaires de France, 1923), p. 133. See also p. 27.
[61] William R. Newman, *The "Summa perfectionis" of Pseudo-Geber* (Leiden: E. J. Brill, 1991), esp. pp. 143–92. For an overview of alchemical corpuscular theory, see Newman, "Experimental Corpuscular Theory in Aristotelian Alchemy: From Geber to Sennert," in *Late Medieval and Early Modern Corpuscular Matter Theory*, ed. Christoph Lüthy, John E. Murdoch, and William R. Newman (Leiden: E. J. Brill, 2001), pp. 291–329.
[62] William R. Newman, "The Alchemical Sources of Robert Boyle's Corpuscular Philosophy," *Annals of Science*, 53 (1996), 567–85; and Newman, "Boyle's Debt to Corpuscular Alchemy," in Hunter, ed., *Robert Boyle Reconsidered*, pp. 107–18.

the three Paracelsian principles, mercury, sulfur, and salt. Like Boyle, whom he greatly influenced, van Helmont argued that in many instances the three principles were artifacts of combustion. In addition, van Helmont is known for his attempt to find a more "subtle" analytical agent than fire, an agent that would not combine with other substances but would merely "dissect" them into their constituents and then recede. His thoughts on this subject, he confessed, were derived from his research on mercury, and he openly admitted that one of his sources for this was Geber.[63] The goal of van Helmont's research, the so-called alkahest, was supposed to be a liquid made up of "the smallest atoms possible in nature." Because of their extreme minuteness, the atoms of the alkahest could penetrate to the depths of other substances and cut them apart. And just as in the case of Geber's explanation of mercury's separation from nonvolatile substances by means of sublimation, the particles would then recede because of their great homogeneity. Van Helmont's theory of the alkahest was in fact a hybrid based on the mercurial principle of the medieval alchemists and on the Paracelsian concept of a *sal circulatum* that could reduce substances to their *primum ens*, or first matter.[64]

SCHOOLS OF THOUGHT IN EARLY MODERN CHYMISTRY

The importance of mercury in alchemical theory is underscored by the fact that one of the major schools of alchemical thought in the seventeenth century believed that the philosophers' stone should be made from that substance. The famous eighteenth-century chemist Georg Stahl identified three "famous orders" or schools of alchemy – those who hoped to produce the philosophers' stone from "vitriol" (mainly copper or iron sulfate), those who used "niter" (saltpeter) as their starting point, and those who "expect to find the Secret in running Mercury."[65] According to Stahl, the leader of this final school was the "candid and ingenuous Author Philaletha," by whom he meant Eirenaeus Philalethes, perhaps the most popular alchemical writer of the late seventeenth century.[66] In reality, Eirenaeus Philalethes was George Starkey (1628–1665), who was born in Bermuda and educated at Harvard College, and who wrote chrysopoetic works under the Philalethes pseudonym and composed Helmontian iatrochemical texts under his Christian name. Starkey concocted a "sophic mercury" from antimony, silver, and mercury itself; this was supposed to mature into the philosophers' stone after long, gentle heating. The antimony was added to purge the quicksilver of its impurities,

---

[63] Newman, *Gehennical Fire*, p. 146.
[64] Ibid., pp. 146–8.
[65] Georg Stahl, *Philosophical Principles of Universal Chemistry* . . . , trans. Peter Shaw (London: Osborn and Longman, 1730), p. 395.
[66] Ibid., p. 402.

and the silver helped to amalgamate the mercury to the antimony. When gold was placed in this amalgam, it was supposed to "vegetate" and grow into a tree, which would ultimately give birth to the philosophers' stone. Modern laboratory replication has revealed that Starkey's recipe does produce a lovely crystalline tree, though not the agent of metallic transmutation.[67] Starkey's recipe provided the basis for Boyle's work *Of the Incalescence of Quick-silver with Gold* and informed much of Boyle's chymical practice.[68] It also had a significant influence on Isaac Newton (1642–1727): It is now known that the chrysopoetic *Clavis* once thought to be by Newton is actually a work of Starkey.[69] Traces of Starkey's Helmontian and Geberian matter theory are also found in Newton's work, as in Query 31 of *The Opticks*.[70]

The most serious competitor of the mercurial school of seventeenth-century alchemy was another widespread theory, also alluded to by Stahl, which claimed that the philosophers' stone should be derived from a "philosophical saltpeter" (*sal nitrum philosophicum*). The origins of the *sal nitrum* theory ultimately lie in the alchemically colored natural magic of Ficino and Agrippa as elaborated by Paracelsus. While Ficino had argued that the alchemists could isolate the universal spirit of life, Agrippa had pinpointed its material matrix – elemental earth. Paracelsus, for his part, had made "salt" a fundamental component of matter as one of the three principles. When later alchemical authors observed the efflorescence of saltpeter from rich humus, the inference was at hand that this substance, whose mysterious activity was evident in its use in gunpowder and nitric acid, in its action as a fertilizer, and in its property of releasing a "vital spirit" (oxygen) upon heating, was the proper source of Agrippa's "one thing created by God" – the philosophers' stone – that made transmutation a possibility. Because experimentation with ordinary potassium nitrate did not yield up the philosophers' stone, however, alchemists concluded that their *sal nitrum* must be sought in a more "general" state. Hence they attempted to derive a "nitrous spirit" directly from the atmosphere, from dew, from excrement, and from other substances where it should be present as a generalized principle of life. The riddling sentences of the *Emerald Tablet* seemed then to have found a solution – the descent of the "one thing" from heaven to earth referred to the vital, nitrous spirit, that should serve as the raw material of the philosophers' stone.

Hence many chymists of the seventeenth century, such as Michael Sendivogius and Thomas Vaughan, argued that the source of the philosophers'

---

[67] Newman and Principe, *Alchemy Tried in the Fire*, p. 185.
[68] Lawrence Principe, "Apparatus and Reproducibility in Alchemy," in *Instruments and Experimentation in the History of Chemistry*, ed. Trevor Levere and Frederic L. Holmes (Cambridge, Mass.: MIT Press, 2000), pp. 55–74. See also Principe, *Aspiring Adept*, p. 161, and Principe, "Robert Boyle's Alchemical Pursuits," pp. 96–7.
[69] William R. Newman, "Newton's Clavis as Starkey's Key," *Isis*, 78 (1987), 564–74.
[70] William R. Newman, "The Background to Newton's Chymistry," in *The Cambridge Companion to Newton*, ed. I. Bernard Cohen and George Smith (Cambridge: Cambridge University Press, 2002), pp. 358–69.

stone should be sought in substances rich in the philosophical saltpeter. Although there was considerable disagreement as to whether the precise source ought to be humus, dew, dung, or some other substance, the upholders of the *sal nitrum* theory universally decried the attempt to make the philosophers' stone with common quicksilver. In this they were directly opposed to the mercurial tradition represented by Starkey and his followers.[71]

The formation of distinct chrysopoetic schools in the sixteenth and seventeenth centuries, like the inauguration of the chymical textbook tradition, bears witness to the increasing divergence of traditions within the domain of early modern chymistry. Other subtraditions came into being as well, such as the genre of distillation books and the early sixteenth-century practical mining and assaying texts that openly display their borrowings from medieval alchemy.[72] Although it was possible for a mid-sixteenth-century metallurgical writer such as Georg Agricola to be cool toward alchemy, this was only the case because earlier German authors in the *Kunstbüchlein* (literally "art booklet") tradition had already sifted alchemical treatises for their metallurgical content and appropriated much that was directly useful to them.[73] The same pattern of appropriation and repudiation is strikingly evident in the attitude of scientific societies at the end of the seventeenth century. While writers such as Lemery and his colleagues at the French Académie Royale des Sciences were attacking alchemy as obscurantism and charlatanry, they were simultaneously employing apparatus, practices, and skills developed by alchemists in trying to analyze plants and minerals into their constituent Paracelsian principles, and in some instances borrowing the same corpuscular theory pioneered in alchemical texts.[74]

---

[71] Newman, *Gehennical Fire*, pp. 209–27.
[72] Paul Walden, *Mass, Zahl und Gewicht in der Chemie der Vergangenheit*, in *Sammlung chemischer und chemisch-technischer Vorträge begründet von F. B. Ahrens*, ed. H. Grossmann (Neue Folge, Heft 8) (Stuttgart: Ferdinand Enke, 1931), p. 3. For a short overview of the early modern genre of German artisanal booklets, see Ernst Darmstaedter, "Berg-, Probir- und Kunstbüchlein," in *Münchener Beiträge zur Geschichte und Literatur der Naturwissenschaften und Medizin*, 2/3 (1926), 101–206.
[73] Georgius Agricola, *De re metallica*, trans. Herbert Clark Hoover and Lou Henry Hoover (New York: Dover, 1950), pp. xxvii–ix and 607–15. See also William Eamon, *Science and the Secrets of Nature: Books of Secrets in Medieval and Early Modern Culture* (Princeton, N.J.: Princeton University Press, 1994), pp. 112–20.
[74] Frederic L. Holmes, *Eighteenth-Century Chemistry as an Investigative Enterprise* (Berkeley: University of California Press, 1989). Some editions of Lemery's *Cours de chymie* describe the formation of metals beneath the earth in a way that is strikingly similar to the *Summa perfectionis* of Geber and is surely borrowed from the tradition of alchemical corpuscular theory. See Nicolas Lemery, *Cours de chymie*, 10th ed. (Paris: Delespine, 1713), pp. 70–1: "The hardest, most compact, and heaviest metals are those in whose composition fermentation has made the most effective separation of the gross particles [*parties grossières*], so that what coagulates – being an assemblage of very subtle, divided bodies [*corps extrêmement subtils & divisez*] – undergoes a very strict union [*une union très-étroite*] that leaves only very tiny pores." Here we encounter the *grossae partes*, *subtiles partes*, strong composition, and interparticular pores of traditional corpuscular alchemy. Needless to say, the alchemical tradition does not account for Lemery's corpuscularism in toto because he incorporates complex ruminations about the shape and figure of his particles – a type of speculation that is notably absent from earlier alchemical texts.

In addition, it is clear that the seventeenth-century focus on experimentation as a whole – especially evident in the scientific societies – owes a serious debt to the alchemical tradition (see the following chapters in this volume: Dear, Chapter 4; Smith, Chapter 13). The earliest sustained series of experiments carried out by the Royal Society consists of an attempt made in 1664–5 to analyze May-dew and to extract from it the Sendivogian *sal nitrum*.[75] This is not perhaps surprising, given the prominence of the *sal nitrum* theory in Restoration England.[76] At a more fundamental level, however, alchemy guided the experimental ethos by stressing the need to subject matter to the artificial constraints of the laboratory, an idea later popularized by Francis Bacon (1561–1626) and his followers.[77] The identity of artificial and natural products had important implications in the realm of experiment because it meant that alchemists were not forced to view the experimental situation as producing an "unnatural" and hence invalid result (see Bennett, Chapter 27, this volume).[78] This had been evident since the Middle Ages, when a large apologetic literature grew up in defense of the "natural" (that is, genuine) character of alchemical gold and other products that had been fabricated by means of laboratory procedures using man-made apparatus. The works of Bacon and Boyle both draw on traditional defenses of alchemy when arguing for the identity of natural and artificial products, and the theme remained alive in the alchemical treatises of Isaac Newton.[79]

At the end of the seventeenth century, then, alchemy had diversified into a number of different traditions and influenced still others. On the one hand, transmutatory alchemy was now divided into clearly defined schools of practice, as in the example quoted from Stahl. On the other hand, the field was about to experience a decisive break between chrysopoetic chymists and those who disavowed transmutation, such as Fontenelle and Geoffroy. A number of quasi-industrial pursuits that had been fertilized if not spawned by chymistry were also crystallizing out of the mix – distillers of "strong waters," perfumers, metallurgists, and apothecaries were among the most obvious beneficiaries of

---

[75] Alan B. H. Taylor, "An Episode with May-Dew," *History of Science*, 22 (1994), 163–84.
[76] Robert G. Frank, Jr., *Harvey and the Oxford Physiologists* (Berkeley: University of California Press, 1980), pp. 115–39 and 221–45.
[77] For this theme, see Newman, *Promethean Ambitions*, chap. 5. See also Newman, "Alchemy, Domination, and Gender," in *A House Built on Sand*, ed. Noretta Koertge (Oxford: Oxford University Press, 1998), pp. 216–26.
[78] William R. Newman, "Art, Nature, and Experiment among Some Aristotelian Alchemists," in *Texts and Contexts in Ancient and Medieval Science*, ed. Edith Sylla and Michael McVaugh (Leiden: E. J. Brill, 1997), pp. 305–17. See also Newman, "Corpuscular Alchemy and the Tradition of Aristotle's *Meteorology*, with Special Reference to Daniel Sennert," *International Studies in the Philosophy of Science*, 15 (2001), 145–53.
[79] William R. Newman, "Alchemical and Baconian Views on the Art/Nature Division," in *Reading the Book of Nature*, ed. Allen G. Debus and Michael T. Walton (Kirksville, Mo.: Sixteenth Century Journal Publishers, 1998), pp. 81–90. For Newton's concern with the art/nature distinction, see Betty Jo Teeter Dobbs, *The Janus Faces of Genius: The Role of Alchemy in Newton's Thought* (Cambridge: Cambridge University Press, 1991), p. 267.

what had once been an exotic and secretive art.[80] Additionally, alchemy had significantly influenced the growth of experimental corpuscular theory and the new emphasis on experiment more generally. Although the relationship among these varied traditions is complex and difficult to unravel, one thing at least is quite evident. The once common claim that alchemy was an irrational delusion inimical to the main themes of the Scientific Revolution and restricted to the margins of European culture can no longer be maintained.[81] To the contrary, it is now clear that alchemy not only contributed in important ways to the seventeenth-century emphasis on corpuscular matter theory and experiment but also placed a new stress on mass balance (the identity of input and output weights in a chemical reaction) that would have major repercussions in the Chemical Revolution of the late eighteenth century.[82] In its altered guise as chymistry, the venerable discipline had captured the imagination of learned and popular culture alike, and it had become a mainstay in the early modern reformation of science and medicine.

---

[80] For the industrial implications of alchemy more generally, see Pamela H. Smith, *The Business of Alchemy: Science and Culture in the Holy Roman Empire* (Princeton, N.J.: Princeton University Press, 1994); and Tara Nummedal, "Practical Alchemy and Commercial Exchange in the Holy Roman Empire," in *Merchants and Marvels: Commerce, Science, and Art in Early Modern Europe*, ed. Pamela H. Smith and Paula Findlen (New York: Routledge, 2002), pp. 201–22.

[81] See Lawrence Principe and William R. Newman, "Some Problems with the Historiography of Alchemy," in *Secrets of Nature: Astrology and Alchemy in Early Modern Europe*, ed. William R. Newman and Anthony Grafton (Cambridge, Mass.: MIT Press, 2001), pp. 385–431.

[82] Newman and Principe, *Alchemy Tried in the Fire*, pp. 35–155 and 273–314.

22

MAGIC

Brian P. Copenhaver

The Middle Ages took magic seriously, though it was not a key issue for that period of European history, as it had been in late antiquity. Many medieval theologians treated magic with fear or loathing, in fact, and philosophers were often indifferent to it. But in the late fifteenth century, magic enjoyed a remarkable rebirth, acquiring the energy that kept it at the center of cultural attention for nearly two hundred years, as great philosophers and prominent naturalists tried to understand or confirm or reject it. After Marsilio Ficino (1433–1499) took the first steps in the renaissance of magic, prominent figures from all over Europe followed his lead, including Giovanni Pico della Mirandola (1463–1494), Johann Reuchlin (1455–1522), Pietro Pomponazzi (1462–1525), Paracelsus (Theophrastus Bombastus von Hohenheim, ca. 1493–1541), Girolamo Cardano (1501–1576), John Dee (1527–1608), Giordano Bruno (1548–1600), Giambattista della Porta (1535–1615), Tommaso Campanella (1568–1639), Johannes Baptista van Helmont (1579–1644), Henry More (1614–1687), and others of equal stature. Eventually, however, as Europe's most creative thinkers lost confidence in it, magic became even more disreputable than it had been before Ficino revived it. Around 1600, some reformers of natural knowledge had hoped that magic might yield a grand new system of learning, but within a century it became a synonym for the outdated remains of an obsolete worldview.[1] Before examining

---

[1] D. P. Walker, *Spiritual and Demonic Magic from Ficino to Campanella* [1958] (University Park: Pennsylvania State University Press, 2000); Frances Yates, *Giordano Bruno and the Hermetic Tradition* (London: Routledge and Kegan Paul, 1964); Brian P. Copenhaver, "Astrology and Magic," in *The Cambridge History of Renaissance Philosophy*, ed. Charles Schmitt, Quentin Skinner, and Eckhard Kessler, with Jill Kraye (Cambridge: Cambridge University Press, 1987), pp. 264–300; Copenhaver, "Natural Magic, Hermetism, and Occultism in Early Modern Science," in *Reappraisals of the Scientific Revolution*, ed. David C. Lindberg and Robert S. Westman (Cambridge: Cambridge University Press, 1990), pp. 261–301; Copenhaver, "Did Science Have a Renaissance?," *Isis*, 83 (1992), 387–407; and Copenhaver, "The Occultist Tradition and Its Critics in Seventeenth Century Philosophy," in *The Cambridge History of Seventeenth-Century Philosophy*, ed. Daniel Garber and Michael Ayers, 2 vols. (Cambridge: Cambridge University Press, 1998), 1: 454–512.

its extraordinary rise and fall in post-medieval Europe, we can begin with magic as described by one of its most voluble advocates, Heinrich Cornelius Agrippa von Nettesheim (1486–1535), a German physician and philosopher.

AGRIPPA'S MAGIC MANUAL

No one knew the risks and rewards of magic better than Agrippa. His notorious handbook, *De occulta philosophia* (On Occult Philosophy), had been circulated in manuscript by 1510, though it was printed only in 1533, over the complaints of Dominican inquisitors. Meanwhile, he had written another famous book, *De incertitudine et vanitate scientiarum* (On the Uncertainty and Vanity of the Sciences, 1526), wherein he recanted magic for religious reasons that had become urgent in the early years of the Reformation. Agrippa's change of heart – not really a change of mind – did nothing to diminish the enormous influence of the *Occult Philosophy*.

Agrippa's book was of great importance for natural philosophy because of its account of natural magic, which he described as

> ... the pinnacle of natural philosophy and its most complete achievement.... With the help of natural virtues, from their mutual and timely application, it produces works of incomprehensible wonder.... Observing the powers of all things natural and celestial, probing the sympathy of these same powers in painstaking inquiry, it brings powers stored away and lying hidden in nature into the open. Using lower things as a kind of bait, it links the resources of higher things to them ... so that astonishing wonders often occur, not so much by art as by nature.[2]

The plan of Agrippa's book reflects the triple hierarchy of his cosmos, where causality runs from above to below, from ideas in God's mind down through spiritual intelligences and heavenly bodies to animals, plants, and stones beneath the moon. Humans can ascend the magical channels that carry divine energies down to earth. Magicians can attract powers from on high by manipulating qualities, quantities, and minds: qualities of objects made of earthly matter in the lowest elementary world; quantities (figures and shapes as well as numbers) in these same lowly things and in the more sublime objects made of celestial matter in the middle world of heavenly spheres; and immaterial angelic minds, stationed in the highest intellectual world and free

---

[2] Heinrich Cornelius Agrippa von Nettesheim, *Opera quaecumque hactenus vel in lucem prodierunt vel inveniri potuerunt omnia* ... , 2 vols. (Lyon: Beringi fratres, ca. 1600), I: 526a (cited henceforth as *Opera*); Charles G. Nauert, *Agrippa and the Crisis of Renaissance Thought* (Urbana: University of Illinois Press, 1965), pp. 30–3, 59–60, 98–9, 106–15, 194–214, 335–8; and Walker, *Spiritual and Demonic Magic*, pp. 90–6.

of bodily quality or quantity. These three realms correspond to the three parts of Agrippa's occult philosophy: natural, mathematical, and ritual.[3]

Currents of power fuse the three realms in Agrippa's ambitious theory of magic. Just as forms flowing from God's mind reach down to the lowest material objects, so elements and qualities of matter extend upward, ever more refined, suffusing the whole hierarchy. Binding the whole together is the tenuous substance called spirit (*spiritus, pneuma*), not quite matter and not quite mind, the vehicle for exchanges of power between bodiless and embodied things. While in one sense the whole is embodied, through sympathies and similitudes, in another sense and through the same forces it is ensouled. A world soul mirrors not only human souls but also those of angels and demons, unencumbered by bodies and therefore very powerful. To energize links among minds, souls, spirits, and bodies, the magus starts with the natural magic of objects here on earth and moves up through mathematical, spiritual, and psychological magic, working on the self and on others and entering the middle world of figures and celestial influences, where the power of human imagination resonates with great effect.[4]

Up to this point, the occult philosophy might be acceptable for a pious Christian; it does not yet involve the spiritual persons – angels and demons – with whom the Church forbade dealings outside its own institutions. But Satan and his minions are cunning: With all the best intentions, a magus who starts with natural objects may end with illicit rites and evil spirits, inviting condemnation by the Church. Witches use both types of magic, natural and demonic, for their harmful spells (*maleficia*), which is where popular and learned magic merge most destructively in Agrippa's system.[5]

By the time Agrippa wrote, pagans and Christians had been testing the boundary between natural and demonic magic for two millennia. Agrippa knew the dangers, which explains why he came to disavow magic so passionately. Nonetheless, his arguments on behalf of a learned, philosophical magic are more compelling than his declamations against it. His occult philosophy is systematic, comprehensive, and grounded in authority and evidence, but it is not original. It is a vulgarization of the ancient magic revived in the fifteenth century by Ficino, which was summarized in his *De vita libri tres* (Three Books on Life, 1489) and then developed by Pico, Reuchlin, and others – including Pietro Pomponazzi, an Aristotelian natural philosopher whose work on the causes of magical effects was written (but not printed) before Agrippa's book was published in 1533.[6]

---

[3] Agrippa, *Opera*, 1: 1–4, 153–6, 310–11; 2: 90–1 (for quoted passage).
[4] Ibid., 1: 5–6, 18–19, 25–36, 40, 43–5, 68–70, 90–2, 128–38.
[5] Ibid., 1: 18–19, 40, 69–70, 90–2, 137–8, 268, 276, 361, 436–9.
[6] Marsilio Ficino, *Three Books on Life: A Critical Edition and Translation with Introduction and Notes*, ed. and trans. Carol Kaske and John Clarke (Binghamton, N.Y.: Medieval and Renaissance Texts and Studies, 1989); Brian P. Copenhaver, "Number, Shape, and Meaning in Pico's Christian Cabala: The Upright *Tsade*, the Closed *Mem*, and the Gaping Jaws of Azazel," in *Natural Particulars: Nature*

Ficino's sources included Greek manuscripts brought to Italy by Byzantine scholars, some arriving even before 1400, others driven west after Constantinople fell to the Turks in 1453. Texts of this provenance, assembled during the Middle Ages and now called the *Corpus Hermeticum*, had long been attributed to the Egyptian god Thoth, whom the Greeks named Hermes Trismegistus. This is the ancestry of the Hermetic writings and of the "Hermeticism" that has been contentious among historians of science ever since Frances Yates claimed that Renaissance magic was Hermetic and that the origins of modern science were to be found in that arcane wisdom.[7]

Scholars have challenged the Yates thesis since it was first proposed. One of their points, recognized long before by Byzantine scribes, is that the Hermetic writings are of two types, now called *technical* and *theoretical*. The major theoretical works are the Latin *Asclepius* and the Greek treatises that Ficino put into Latin as the *Pimander*. Their content is spirituality – pious speculation and exhortation about God, the cosmos, and the human condition. But these theoretical *Hermetica*, made famous in the Anglophone world by Yates, are not about *magical* theory or practice, which falsifies a large part of her claim that modern science grew out of Hermetic magic.[8]

Other texts attributed to Hermes have been called technical. These include dozens of works on alchemy, astrology, astronomy, botany, magic, medicine, pharmacy, and other practical topics that circulated in the Mediterranean region since antiquity in various languages, including Latin and Arabic. Unlike the theoretical treatises, some were known in the medieval West, disseminating technical information about magic and authenticating it with the name of Hermes. This Hermes, who presided over medieval guides to practical magic, was not Ficino's Hermes, a divine theologian and spiritual

---

and the Disciplines in Renaissance Europe, ed. Anthony Grafton and Nancy Siraisi (Cambridge, Mass.: MIT Press, 1999), pp. 25–76; Copenhaver, "The Secret of Pico's *Oration*: Cabala and Renaissance Philosophy," *Midwest Studies in Philosophy*, 26 (2002), 56–81; Copenhaver, "Astrology and Magic," pp. 267–85; and Copenhaver, "Did Science Have a Renaissance?," pp. 387–402.

[7] A. D. Nock and A.-J. Festugière, *Corpus Hermeticum*, 3rd ed., 4 vols. (Paris: Belles Lettres, 1972), 1: xi–xii; Robert S. Westman, "Magical Reform and Astronomical Reform: The Yates Thesis Reconsidered," in *Hermeticism and the Scientific Revolution: Papers Read at a Clark Library Seminar, March 9, 1974* (Los Angeles: William Andrews Clark Memorial Library, 1977), pp. 5–91; Ingrid Merkel and Allen G. Debus, eds., *Hermeticism and the Renaissance: Intellectual History and the Occult in Early Modern Europe* (Washington, D.C.: Folger Shakespeare Library, 1988); Brian Vickers, ed., *Occult and Scientific Mentalities in the Renaissance* (Cambridge: Cambridge University Press, 1984); Brian P. Copenhaver, *Hermetica: The Greek Corpus Hermeticum and the Latin Asclepius in English Translation, with Notes and Introduction* (Cambridge: Cambridge University Press, 1991), pp. xl–xli; Copenhaver, "Magic and the Dignity of Man: De-Kanting Pico's *Oration*," in *The Italian Renaissance in the Twentieth Century: Acts of an International Conference, Florence, Villa I Tatti, June 9–11, 1999*, ed. Allen J. Grieco, Michael Rocke, and Fiorella Gioffredi Superbi (Florence: Leo S. Olschki, 2002), pp. 311–20; and Copenhaver, "Natural Magic, Hermetism, and Occultism," pp. 261–6, 289–90.

[8] *Mercurii Trismegisti liber de potestate et sapientia dei: Corpus Hermeticum I–XIV, versione latina di Marsilio Ficino, Pimander*, ed. Sebastiano Gentile (Treviso, 1471; repr. Florence: Studio per Edizioni Scelte, 1989); A.-J. Festugière, *La révélation d'Hermès Trismégiste*, 3 vols. (Paris: Belles Lettres, 1981), vol. 1: *L'Astrologie et les sciences occultes*, pp. 67–88; Garth Fowden, *The Egyptian Hermes: A Historical Approach to the Late Pagan Mind* (Cambridge: Cambridge University Press, 1986), pp. 1–11; and Copenhaver, *Hermetica*, pp. xxxii–xl.

adviser.[9] But once Ficino discovered the *Pimander*, making Hermes as canonical as Plato (427–ca. 348 B.C.E.) or Plotinus (ca. 204–270 C.E.), the old god was there to be exploited by new magicians, who read Agrippa and applied the Hermetic pedigrees less scrupulously than Ficino. When Agrippa listed the first authors of magic, he put Hermes among "the more distinguished masters," setting him alongside the neo-Platonic philosophers whom Ficino rediscovered – Plotinus, Porphyry (ca. 233–305), Iamblichus (ca. 250–330), and Proclus (410–485). But Damigeron, Gog Graecus, and Germa Babylonicus turn up on the same page – barbaric names that Agrippa considered fit company for Hermes.[10]

Unlike Agrippa, Ficino was a careful explorer of the borderland between myth and history. From deep reading in ancient sources, especially the Church Fathers, he derived a scheme of religious and intellectual history, the ancient theology (*prisca theologia*). Prominent in this story was the Hermes of Ficino's *Pimander*; unlike the obscure Gog and Germa, this Hermes was the reputable author of a pious spirituality, as any reader of Ficino's translation could see. But Hermetic genealogies were deceptive; Cicero had counted four distinct deities called Hermes (Mercurius in Latin), in addition to the Egyptian Trismegistus. The god's multiple personalities – some attached to magical texts, some not – easily fused into a single Hermetic persona during the sixteenth century until Isaac Casaubon (1559–1614) proved that the *Hermetica* were not nearly as old as Ficino had thought.[11]

Ficino believed that Hermes was a contemporary of Moses and that he had founded a tradition of human wisdom that ran parallel to the divine revelation of scripture and led to the teachings of Plato and his successors. After Casaubon devalued the Hermetic works in 1614 by redating them to the early Christian era, Ficino's ancient theology lost its reputation, but only slowly. In the 1690s, Isaac Newton (1642–1727) still found it useful for grounding his views about God and space in mythic tradition, though Newton's published works reveal this interest only in faint allusions. Meanwhile, once Ficino had resurrected it, the ancient theology reinforced one of the three main motives for belief in magic by educated Europeans: the *historical* authority

---

[9] Festugière, *La révélation d'Hermès Trismégiste*, 1: 89–308; Fowden, *The Egyptian Hermes*, pp. 1–4; Copenhaver, *Hermetica*, pp. xxxii–vii, xlv–vii; Copenhaver, "Lorenzo de' Medici, Marsilio Ficino, and the Domesticated Hermes," in *Lorenzo il Magnifico e il suo mondo: Atti di convegni*, ed. G. C. Garfagnini (Florence: Istituto Nazionale di Studi sul Rinascimento, 1994), pp. 225–57; and Copenhaver, "Hermes Theologus: The Sienese Mercury and Ficino's Hermetic Demons," in *Humanity and Divinity in Renaissance and Reformation: Essays in Honor of Charles Trinkaus*, ed. John O'Malley, Thomas M. Izbicki, and Gerald Christianson (Leiden: E. J. Brill, 1993), pp. 149–82.

[10] Agrippa, *Opera*, 1: 4.

[11] Cicero, *De natura deorum*, 3.22.56; Yates, *Giordano Bruno*, pp. 398–440; D. P. Walker, *The Ancient Theology: Studies in Christian Platonism from the Fifteenth to the Eighteenth Century* (London: Duckworth, 1972); Frederick Purnell, "Francesco Patrizi and the Critics of Hermes Trismegistus," *Journal of Medieval and Renaissance Studies*, 6 (1976), 155–78; and Anthony Grafton, *Defenders of the Text: The Traditions of Scholarship in an Age of Science, 1450–1800* (Cambridge, Mass.: Harvard University Press, 1991), pp. 145–77.

of a venerated past. Some of the ancient wisdom that Ficino revived, especially its neo-platonized Aristotelianism, provided authority and content for another basis of occultist belief, which was *theoretical*. Not only the neo-Platonists but also Galen (ca. 129–ca. 199), Avicenna (Ibn Sīnā, 980–1037), Thomas Aquinas (ca. 1225–1274), and other thinkers of the first rank – pagan and Christian, ancient and medieval – contributed to the philosophical theory of magic published by Ficino in 1489 and then popularized by Agrippa. Finally, many readers who found this theory philosophically convincing also took it to be confirmed by experience. *Empirical* information supplied a third basis for belief in occultism.[12]

Indeed, empirical details formed the bulk of Agrippa's compendium, illustrating its theory and making it concrete. Agrippa turned again and again to lists of natural objects long regarded as mysterious because their appearances were strange, their mechanisms unknown, or their effects rapid and unusually strong: the magnet, carbuncle, heliotrope, peony, tarantula, basilisk, dragon, electric ray, remora, and hundreds of others. Without a theory to explain them, however, Agrippa's long lists of magical objects would have been meaningless. Encyclopedias, lapidaries, herbals, and bestiaries, as well as works on alchemy, astrology, and medicine, had supplied such lists for centuries, but the theory behind them was weak because its strongest voices, the ancient neo-Platonists, remained faint until the generation before Agrippa, when Ficino, Pico, and other prominent thinkers developed philosophical conceptions of magic using the most authoritative metaphysical, physical, and cosmological ideas of the day. Agrippa was the beneficiary of this theorizing. Personal experience and popular culture also confirmed his beliefs about magic, which nonetheless remained a learned and philosophical project – an occult *philosophy*.[13]

Claiming to derive natural magic from natural philosophy, Agrippa started with an exposition of physics and matter theory – Aristotelian in

[12] J. E. McGuire and P. M. Rattansi, "Newton and the 'Pipes of Pan'," *Notes and Records of the Royal Society*, 21 (1966), 108–43; Copenhaver, "Astrology and Magic"; Copenhaver, "Natural Magic, Hermetism, and Occultism"; Copenhaver, "Did Science Have a Renaissance?"; Copenhaver, *Hermetica*, pp. xlvii–viii; and Copenhaver, "A Tale of Two Fishes: Magical Objects in Natural History from Antiquity through the Scientific Revolution," *Journal of the History of Ideas*, 52 (1991), 373–98.

[13] Agrippa, *Opera*, I: 21–2, 25–6, 35–6, 39, 45, 47, 51, 57–8, 74, 77, 83, 334; Copenhaver, "Astrology and Magic"; Copenhaver, "A Tale of Two Fishes"; Copenhaver, "The Occultist Tradition and Its Critics," pp. 454–65; Copenhaver, "Natural Magic, Hermetism, and Occultism," pp. 275–80; Copenhaver, "Did Science Have a Renaissance?," pp. 396–8; Copenhaver, "Scholastic Philosophy and Renaissance Magic in the *De vita* of Marsilio Ficino," *Renaissance Quarterly*, 37 (1984), 523–54; Copenhaver, "Renaissance Magic and Neoplatonic Philosophy: *Ennead* 4.3–5 in Ficino's *De vita coelitus comparanda*," in *Marsilio Ficino e il ritorno di Platone: Studi e documenti*, ed. G. C. Garfagnini (Florence: Leo S. Olschki, 1986), pp. 351–69; Copenhaver, "Iamblichus, Synesius, and the *Chaldaean Oracles* in Marsilio Ficino' *De vita libri tres*: Hermetic Magic or Neoplatonic Magic?," in *Supplementum Festivum: Studies in Honor of Paul Oskar Kristeller*, ed. James Hankins, John Monfasani, and Frederick Purnell, Jr. (Binghamton, N.Y.: Medieval and Renaissance Texts and Studies, 1987), pp. 441–55; and Copenhaver, "Hermes Trismegistus, Proclus, and the Question of a Philosophy of Magic in the Renaissance," in Merkel and Debus, eds., *Hermeticism and the Renaissance*, pp. 79–110.

its terminology and framework but with neo-Platonic elements as well. His physical primitives are the four elements (fire, air, water, and earth) and their haptic qualities (hot, cold, wet, and dry). These primary qualities of the elements give rise to secondary qualities that account for physical processes that are important to physicians and natural philosophers: softening and hardening, retaining and expelling, attracting and repelling, and so on. Secondary qualities act on parts of bodies to produce tertiary qualities and a myriad of wonders, natural and artificial, from unquenchable fires to perpetual lamps. Emerging from matter and accessible to the senses, all these qualities are called *manifest* and, however wonderful, were not considered magical. Other qualities called *occult*, however, arise not from matter but from specific or substantial form – the immaterial form that accounts for a thing's belonging to its species or kind. Except that they derive from form, the causes of occult qualities are unknown; only the magical phenomena caused by them are perceptible, not the occult qualities themselves. These sources of magical power are hidden both to reason and to sense, which is why they are called occult.[14]

Because they do not depend on matter, occult qualities produce strange effects that are out of place or out of proportion to the size of the objects containing them: stones sing in the earth, tiny fish stop great ships in the water, birds of the air eat iron, and lizards live in fire. But even the elements themselves are involved in magical action. Fire is helpful for ritual magic because it attracts good spirits of light. Earth, implanted celestially with seminal forms, spontaneously generates worms and plants. Air transmits celestial influence and reflects virtual images (*species*) of natural objects, conveying telepathic powers that Agrippa himself claimed to have mastered. And "the wonders of water are countless," as Agrippa noted, even in the Gospel, where an angel stirs a pool of water to cure the incurable.[15]

The forms that give rise to occult qualities are celestial, descended from God's ideas and seeded in lower nature. They reflect the figures of the stars and imprint them as characters, seals, or signatures on natural objects: "Every species has a heavenly figure that matches it," says Agrippa, "from which a wondrous power of action also comes into it." Magical objects are thus marked by signs of their celestial origins that the magus can detect, just as the astronomer can read the stars and planets. In Agrippa's catalog of planetary signatures, for example, one category of objects descends from Saturn – earthy and watery in its elements, melancholic in humor, sympathetic with lead and gold, with sapphire and the magnet, with mandrake, opium, hellebore, and dragon's wort, with "crawling animals that keep to themselves, solitary, nocturnal, gloomy, ... slow-moving, eating filth, consuming their young, ...

---

[14] Agrippa, *Opera*, 1: 5–22.
[15] Agrippa, *Opera*, 1: 5–17; John 5:2–9.

Figure 22.1. Heinrich Cornelius Agrippa von Nettesheim's lunar dragon. In Agrippa, *De occulta philosophia*, in *Opera quaecumque hactenus vel in lucem prodierunt vel inveniri potuerunt omnia*. 2 vols. (Lyon: Beringi fratres, n.d.), 1: 272.

the mole, ass, wolf, hare, mule, cat, camel, bear, pig, monkey, dragon, basilisk and toad."[16]

The scores of such lists in Agrippa's book have a practical point. The magician who knows that the constellation Draco and the planet Saturn rule the dragon-plant, for example, can use this information to attract or repel saturnine influence. Natural objects imprinted with forms by the heavens, signed with celestial seals, and charged with occult power thus become magical objects when the magus discovers and uses them, concentrating them to attract one influence, separating them to avoid another, creating congruities or incongruities to induce the desired form and make matter fit to receive it. Up to a point, the magic works within nature's domain, which extends through the elementary and celestial levels of Agrippa's world. His various devices to produce magical effects – amulets, rings, charms, drugs, unctions, potions, lamps, lights, and fumigations – could, in theory, be wiser, deeper, secret ways to use *natural* objects, avoiding the theologically and morally risky world of demonic minds.[17]

But Agrippa's cosmos is a continuum, where bodies link sympathetically with minds, and nature merges into supernature through the medium of spirit, the power of imagination, and the transmission of forms. One of the many pictures in Agrippa's book, showing a dragon (Figure 22.1), illustrates the perils of magical continuity, which lets demons slip into the magician's practice. Summarizing earlier literature on astrological images, Agrippa notes that its authors

[16] Agrippa, *Opera*, 1: 23–4, 50–1, 56–62.
[17] Ibid., 1: 57–67, 70–85.

... made an image of the Moon's Dragon with Head and Tail, a depiction of that serpent between circles of fire and air. . . . They made it when Jupiter and the Head ruled the middle of the sky, . . . and through this image they wanted to signify a good, lucky demon, depicting its image with serpents. The Egyptians and Phoenicians thought this animal divine above all others . . . [because] its spirit was sharper and its fire fuller. . . . But when the Moon was eclipsed in the Tail or badly situated with Saturn or Mars, they made a similar image of the Tail to cause anxiety and weakness and bring on bad luck, and they called it an evil spirit. A Jew put an image like this on a belt of gold and jewels, which Blanche, daughter of the Duke of Bourbon, gave to her husband Peter, King of Spain, . . . and when he put the belt on, he seemed to have a snake around him. When the magic power implanted in the belt was discovered, he rejected his wife because of it.[18]

Because angels and demons ruled the upper stories of Agrippa's sympathetic cosmos, whereas stones, plants, and animals lay in the basement but still within reach, the magus who tapped the hidden powers of natural objects ran the risk of attracting angelic or demonic attention, benevolent or malevolent.

THE CREDIBILITY OF MAGIC: TEXT, IMAGE, AND EXPERIENCE

Words, images, and experience, especially vicarious experience stored in books, confirmed the magical powers of physical objects – natural objects such as magnets, peonies, and dragons, and artificial objects such as rings, amulets, and automata. The credibility of such objects was rooted in ancient texts, and humanists who recovered and preserved those texts left their magic intact. Faced with Pliny's ancient encyclopedia, for example, with its mass of evidence for magic, most Renaissance editors wanted to strengthen Pliny's authority, not weaken it. Taking up where philology left off, sixteenth-century natural historians from Pierre Belon (1517–1564) and Hans Weiditz (fl. early sixteenth century) to Charles de L'Écluse (1526–1609) and Ulisse Aldrovandi (1522–1605) cited the texts improved by humanist scholarship, thus authenticating the ancient wisdom that legitimized belief in magic. Relying more on old books than on new observations, the best that erudition could do was to expose contradictions in the classics, a sure solvent of belief but a slow one. Moreover, some appeals to personal experience actually reinforced the old tales with current examples. Few followed the advice offered by the French essayist Michel de Montaigne (1533–1592) to verify the facts about marvels before trying to explain them.[19]

---

[18] Agrippa, Ibid., 1: 68–70, 272–3.
[19] Montaigne, *Essais*, 3.11; Charles G. Nauert, "Humanists, Scientists, and Pliny: Changing Approaches to a Classical Author," *American Historical Review*, 84 (1979), 72–85; G. E. R. Lloyd, *Science, Folklore,*

With no strict regime of correspondence between objects described in books and objects seen in nature, the textual manifestation of magical objects came not merely to represent the evidence but actually to constitute the evidence, which was displayed in words and, more and more, in images. The continuities of the magical universe were marked, and its powers were activated by visual signs – by Agrippa's picture of a dragon (Figure 22.1), for example. Since antiquity, such images had been part and parcel of magic; pictures such as those in Agrippa's book worked together with words for mutual confirmation. Through the sixteenth century, the new technology of printing strengthened this partnership by multiplying, stabilizing, and disseminating images on the printed page. New techniques of picturing (such as perspective, shading, woodcuts, and engraving) dazzled the eye with magical sights seldom seen before, picturing them naturalistically and broadcasting them in books, broadsheets, and prints. As magical objects proliferated in word and image, the new learning and the new art made them more believable.[20]

Consider the monster purportedly born in Ravenna in 1496: an armless hermaphrodite with wings, a horn on its head, an eye on its knee, and one eagle's talon in place of a foot. Broadsheets depicting this prodigy (and many others) had been circulating for years in Italy and Germany when one came to the attention of a Florentine apothecary, Luca Landucci, in 1512. The image itself compelled belief. "I saw it painted [*dipinto*]," Landucci exclaimed, "and anyone who wanted could see the painting in Florence" – pictorial proof of nature's horrors and God's impending wrath.[21] Agrippa's world was full of such wonders.[22]

---

and Ideology: Studies in the Life Sciences in Ancient Greece (Cambridge: Cambridge University Press, 1983), pp. 135–49; Lorraine Daston and Katharine Park, *Wonders and the Order of Nature, 1150–1750* (New York: Zone Books, 1998), pp. 24, 27, 63, 287; Copenhaver, "A Tale of Two Fishes"; Copenhaver, "The Occultist Tradition and Its Critics," pp. 457–63; and Brian P. Copenhaver and Charles Schmitt, *Renaissance Philosophy* (A History of Western Philosophy, 3) (Oxford: Oxford University Press, 1992), pp. 24–37, 196–209, 239–60.

[20] Agrippa, *Opera*, I: 272; Hans Dieter Betz, ed., *The Greek Magical Papyri in Translation, Including the Demotic Spells* (Chicago: University of Chicago Press, 1986), pp. 17–23, 102, 125, 134, 143–50, 167–71, 268–99, 318–21; Elizabeth Eisenstein, *The Printing Press as an Agent of Change: Communications and Cultural Transformations in Early Modern Europe* (Cambridge: Cambridge University Press, 1979), pp. 67–70, 254–72, 467–70, 485–8, 555–6; and Brian P. Copenhaver, "A Show of Hands," in *Writing on Hands: Memory and Knowledge in Early Modern Europe*, ed. Claire Richter Sherman (Washington, D.C.: Folger Shakespeare Library, 2000), pp. 46–59.

[21] Daston and Park, *Wonders and the Order of Nature*, pp. 177–90; Ottavia Niccoli, *Prophecy and People in Renaissance Italy*, trans. Lydia Cochrane (Princeton, N.J.: Princeton University Press, 1990), pp. 30–60; and Luca Landucci, *Diario fiorentino dal 1450 al 1516, continuato da un anonimo fino al 1542, pubblicato sui codici della Comunale di Siena e della Marucelliana*, ed. Iodoco del Badia (Florence: Studio Biblos, 1969), p. 314.

[22] Daston and Park, *Wonders and the Order of Nature*, pp. 67–75, 145, 199; Anthony Grafton, "Humanism and Science in Rudolphine Prague," in Grafton, *Defenders of the Text*, pp. 178–203; Copenhaver, "A Tale of Two Fishes"; Keith Thomas, *Religion and the Decline of Magic* (New York: Charles Scribner's Sons, 1971), pp. 212–52; Jean Céard, *La Nature et les prodiges: L'Insolite au XVIe siècle*, 2nd ed. (Geneva: Droz, 1996); Richard Kieckhefer, *Magic in the Middle Ages* (Cambridge: Cambridge University Press, 1989), pp. 16–17, 56–94; William B. Ashworth, "Natural History and the Emblematic

More elegant evidence from the notebooks, drawings, and paintings of Leonardo da Vinci (1452–1519) also shows how picturing made magical objects more credible. Leonardo compiled a bestiary, a file of the allegories and emblems that were the court painter's stock in trade, which included over a hundred species, some of them magical. One of its sources was the work of a fourteenth-century astrologer, the *Acerba* of Cecco d'Ascoli, who describes the dragon, greatest of all serpents and famed among magicians, armed with a poisonous tail and monstrously cruel.[23] Leonardo not only described and drew the magical dragon; he also built one.

According to Giorgio Vasari (1511–1571), the Florentine painter and academician who wrote the first great history of art, Leonardo actually assembled a little living dragon: "On a very peculiar green lizard ... he put wings made out of scales taken from other lizards ... so that they quivered from the movement when it walked; he made eyes, a horn and a beard for it, tamed it and kept it in a box, and it made all his friends run away afraid when he showed it to them." Leonardo's procedure recalls instructions that he left for inventing images of animals. "You cannot make any animal unless each of its own limbs by itself resembles a limb from one of the other animals," he wrote. "Thus, if you wish to make an animal that you have devised seem natural – a dragon, let's say – take the head from a mastiff or hound, the eyes from a cat, the ears from a porcupine, the nose from a greyhound, the brow from a lion, the temples from an old rooster, the neck from a water-turtle."[24]

Dragons of this sort appear in Leonardo's drawings, some of which were made as studies for paintings. The background of the unfinished *Adoration of the Magi* (ca. 1481), for example, shows two men in combat riding horses, long admired as effective statements of equine anatomy. But a preparatory sketch reveals that Leonardo had conceived these two believable animals as fighting a dragon. Other drawings (Figure 22.2) show forms of dragons

---

World View," in Lindberg and Westman, eds., *Reappraisals of the Scientific Revolution*, pp. 303–32; Richard Gordon, "Aelian's Peony: The Location of Magic in Graeco-Roman Tradition," *Comparative Criticism*, 9 (1987), 59–95; William Eamon, *Science and the Secrets of Nature: Books of Secrets in Medieval and Early Modern Culture* (Princeton, N.J.: Princeton University Press, 1994); and David Freedberg, *The Eye of the Lynx: Galileo, His Friends, and the Beginnings of Modern Natural History* (Chicago: University of Chicago Press, 2002), pp. 1–3, 186–94.

[23] Jean Paul Richter, *The Literary Works of Leonardo da Vinci*, 2nd ed., 2 vols. (Oxford: Oxford University Press, 1939; orig. publ. 1883), 1: 382 (670); 2: 262 (1224), 264–5 (1231–2), 266–8 (1234, 1239–40), 270–1 (1248–9); Cecco d'Ascoli, *L'Acerba, secondo la lezione del Codice eugubino dell'anno 1376*, ed. Basilio Censori and Emidio Vittori (Verona: Valdonega, 1971), p. 125; Martin Kemp, *Leonardo da Vinci: The Marvellous Works of Nature and Man* (Cambridge, Mass.: Harvard University Press, 1981), pp. 152–7, 164–7, 281; Martin Kemp and Jane Roberts, *Leonardo da Vinci* (New Haven, Conn.: Yale University Press, 1989), pp. 155–7; Wilma George and Brunsdon Yap, *The Naming of the Beasts: Natural History in the Medieval Bestiary* (London: Duckworth, 1991), pp. 66–8, 89–90, 192–3, 199–203; Lynn Thorndike, *A History of Magic and Experimental Science*, 8 vols. (New York: Columbia University Press, 1923–58), 2: 948–68; Daston and Park, *Wonders and the Order of Nature*, pp. 39–43, 52, 76; Pamela Gravestock, "Did Imaginary Animals Exist," in *The Mark of the Beast: The Medieval Bestiary in Art, Life, and Literature*, ed. Debra Hassig (New York: Garland, 1999), pp. 119–39.

[24] Giorgio Vasari, *Le vite de' piu eccellenti pittori, scultori e architettori nelle redazioni del 1550 e 1568*, ed. Rosanna Bettarini and Paola Barocchi, 6 vols. (Florence: Studio per Edizioni Scelte, 1966–97), 4: 21, 34–5; and Richter, *The Literary Works of Leonardo da Vinci*, 1: 342 (585).

flowing from images of horses and cats or linked to heraldic griffins and schemas from pattern books. In such images, the magical dragon draws its credibility not only from juxtaposition with familiar animals such as horses but also from Leonardo's meticulous control of anatomy, as in the preceding description of his compositional process.[25] Art in Leonardo's manner helped people accustomed to reading dragons in the world – as though they were a text – to picture them as well, and the plausibility of such pictures, which were windows into the world of magic, was indistinguishable from the credibility of other natural objects skillfully drawn or painted (see Niekrasz and Swan, Chapter 31, in this volume).[26]

MAGIC ON TRIAL

Powerful evidence of how seriously magic was taken in Leonardo's day was the vehemence of religious opposition to it. Texts that incriminated certain objects as magical had long been feared as dangers to faith and morals, which is why a book that Leonardo used, Cecco d'Ascoli's *Acerba*, was burned along with its author in 1327. Far away, but not long after this double execution, another court sat in Constantinople's church of Hagia Sophia around 1370 to hear testimony about such books from one Phoudoulos. Having been accused of possessing "unclean" books, Phoudoulos confessed and named a physician, Syropoulos, as his source. Syropoulos led the court to another physician, Gabrielopoulos, whose residence was searched, and whole boxes of books were discovered. One suspicious work was called *Kyranides*; another was a book of spells by Demetrios Chloros, who like Gabrielopoulos was a cleric and physician. When Chloros claimed that his magic books were no different from medical texts, other physicians cried in outrage that Chloros disgraced the art of medicine, insulting their heroes, the ancient physicians Hippocrates and Galen, by calling them magicians.[27]

---

[25] Vasari, *Vite*, 4: 22–5, 31; Pietro C. Marani, *Leonardo da Vinci: The Complete Paintings*, trans. A. L. Jenkens (New York: Abrams, 2003), pp. 101–17; Kemp and Roberts, *Leonardo da Vinci*, pp. 23–65, 54, 66, 96, 145; Arthur Ewart Popham, *The Drawings of Leonardo da Vinci* (New York: Reynal and Hitchcock, 1945), pp. 32–8, 109, 112–13, 116–22, 125, plates 62, 80, 86–8, 104–14, 125; Popham, "The Dragon-Fight," in *Leonardo: Saggi e ricerche*, ed. Achille Marazza (Rome: Libreria dello Stato, 1954), pp. 223–7; and Kemp, *Leonardo da Vinci: The Marvellous Works of Nature and Man*, pp. 54–8.

[26] Michel Foucault, *Les mots et les choses: Une archéologie des sciences humaines* (Paris: Gallimard, 1966), pp. 13–14, 34–59, 128–32; and Copenhaver, "Did Science Have a Renaissance?," pp. 403–7. The notion of *picturing* used here is adapted from its application by Svetlana Alpers to Dutch art in *The Art of Describing: Dutch Art in the Seventeenth Century* (Chicago: University of Chicago Press, 1983). Freedberg, *The Eye of the Lynx*, pp. 5–6, 284–6, stresses both the limitations of picturing and its importance for the Lincean Academy; see also David Freedberg, *The Power of Images: Studies in the History and Theory of Response* (Chicago: University of Chicago Press, 1989), esp. chaps. 9 and 10; and Caroline Jones and Peter Galison, eds., *Picturing Science, Producing Art* (London: Routledge, 1998), especially the essays by Daston, Freedberg, Koerner, Park, Pomian, and Snyder.

[27] Antonio Rigo, "Da Costantinopoli alla biblioteca di Venezia: I libri ermetici di medici, astrologi e maghi dell'ultima Bisanzio," in *Magia, alchimia, scienza dal '400 al '700: L'Influsso di Ermete Trismegisto*, ed. Carlos Gilly and Cis van Heertum, 2 vols. (Venice: Centro Di, 2002), 1: 69–70.

Figure 22.2. Dragons, horses, and cats. Leonardo da Vinci, ca. 1517, drawing on paper. Reproduced by permission of the Royal Library, Windsor Castle, Windsor Leoni volume 12331. The Royal Collection © HM Queen Elizabeth II.

What was so alarming about these books? The *Kyranides* might be just a crude natural history, harmlessly listing plants, stones, and animals under letters of the Greek alphabet, but it was ascribed to Hermes Trismegistus and advertised magical plants and animals that, like Agrippa's dragon, threatened to attract the attention of demons. When the *Kyranides* told its readers how to put things together to make medicine (a hoopoe's heart, hair from a seal, green jasper, and peony root, for example, are items in one remedy of great value), it showed them magical objects with alarming powers.[28]

Another Hermetic book describes the simple peony as "a sacred plant, revealed by God to Hermes Trismegistus as a remedy for mortals, . . . as noted in the holy books of Egypt," recommending it for fever and epilepsy and as a fumigant against demonic possession: "Whoever has some part of its root, if the unutterable names of God are inscribed on it [with magic signs], need have no fear of demons." Galen, who was not interested in smoking out demons, had seen a boy cured of epilepsy by the peony amulet, which shows how faint the line between medicine and magic could be. In any case, making certain magical cures out of natural objects became an offense against the Christian religion.[29]

Yet the attractions of magic were powerful enough to tempt many early modern writers to risk religious persecution. Even after the Catholic Inquisition made oppression more efficient, the literature of magic kept growing – in Venice, for example. The Inquisition arrived there in 1547 but paid little attention to magic until papal opposition intensified in the 1580s. The Index of Forbidden Books was in place by 1559, but its main effect was regional, on printing and selling books in Italy. Books were still smuggled, and manuscripts were still copied. Commerce in magical texts flourished: Clergy were active in the trade and found customers everywhere in Venetian society, people moved by love or hate or mere curiosity.[30]

During the sixteenth century and long afterward, books about magic, some of them illustrated like Agrippa's, poured from the presses. Yet something changed as Agrippa's book grew old and famous – ridiculed by the French satirist François Rabelais (1494–1553), put on stage by the English playwright Christopher Marlowe (1564–1593), attacked by the leading critics of magic, and constantly copied by its foremost advocates. By the time an English translation of the *De occulta philosophia* appeared in 1651, the pioneers of a new science had turned against traditional wisdom and the magical principles that Agrippa derived from it.[31]

---

[28] *Kyranides*, 21, in *Die Kyraniden*, ed. Dimitris Kamaikis (Beiträge zur klassischen Philologie, Heft 76) (Meisenheim: Anton Hain, 1976), 55.96–102; and Festugière, *La révélation d'Hermès Trismégiste*, 1: 214–15.

[29] *Catalogus codicum astrologorum graecorum*, 8.2, ed. C. A. Ruelle (Brussels: Lamertin, 1911), 169–70; Galen, *De simplicium medicamentorum temperamentis* (Kühn, 11: 792, 858–61); Festugière, *La révélation d'Hermès Trismégiste*, 1: 77, 157; G. E. R. Lloyd, *Greek Science after Aristotle* (London: Chatto and Windus, 1973), pp. 136–53; Lloyd, *Magic, Reason, and Experience: Studies in the Origin and Development of Greek Science* (Cambridge: Cambridge University Press, 1979), pp. 42–9.

[30] Federico Barbierato, "La letteratura magica di fronte all'inquisizione veneziana fra '500 e '700," in Gilly and van Heertum, eds., *Magia, alchimia, scienza*, 1: 135–75.

[31] Henry Cornelius Agrippa von Nettesheim, *Three Books of Occult Philosophy Written by Henry Cornelius Agrippa of Nettesheim, Counsellor to Charles the Fifth, Emperor of Germany, and Judge of the Prerogative Court, Translated out of the Latin into the English Tongue by J. F.* (London: Gregory Moule, 1651). In Thorndike's *A History of Magic and Experimental Science*, two volumes cover the seventeenth century, leaving only six for the previous sixteen centuries. For the decline of magic's reputation, see Copenhaver, "The Occultist Tradition and Its Critics"; see also François Rabelais, *Tiers livre*, 25, in

There were many reasons for their disenchantment, but low on the list was ecclesiastical censure. Galileo Galilei (1564–1642) was more at risk for contradicting Aristotle than for casting horoscopes in an age when popes still wanted to know what the stars had in store for them. In part, the erosion of belief in magic reflected a general decline of the physics of qualities and its metaphysical foundations, which, whether Aristotelian or Platonic, gave magic its theoretical grounding in Agrippa's books and other sixteenth-century works.[32] In part, however, it was also new criteria of intelligibility, expressed in new forms of visualization, that caused magical objects and images to lose their credibility and eventually fall out of sight.

VIRTUES DORMITIVE AND VISUAL

Before Galileo, Francis Bacon (1561–1626) and René Descartes (1596–1650) had undermined the foundations of Agrippa's magical cosmos, however, others were shoring them up. One such effort, remarkable for its rich learning, acute philosophizing, and explanatory ambition, appeared in 1548: *De abditis rerum causis* (On Hidden Causes).[33] This very influential book by Jean Fernel (ca. 1497–1558), a French physician, was still finding buyers when Bacon, Galileo, and Descartes were all dead and famous; it had seen nearly thirty editions in the century since its first publication. Born in the late fifteenth century and educated when Paris was the citadel of late scholasticism, Fernel was a great innovator in the theory and practice of medicine, and he interpreted ancient medical texts by using the new philology. The old authorities persuaded him that occult qualities were powerful tools for explaining and treating human illness.

Fernel's book therefore exalts occult forces in medicine, making an expert case for a rationalized occult therapy on principles taken from the best classical sources – Hippocrates, Plato, Aristotle, Galen, and many others. To construct a method for occult medicine, he repudiated medical empiricism and advocated rationalism. Fernel was no patient student of pathological particulars, accumulating observations to wear down the theory of magical medicine. On the contrary, he embraced that theory and sought to extend and refine it, not destroy it. In the process, he not only defended occult qualities but also claimed that they were no less intelligible than their manifest counterparts.[34]

Rabelais, *Oeuvres de François Rabelais*, ed. Abel Lefranc, 6 vols. (Paris: Honoré Champion, 1913–31; Geneva: Librairie Droz, 1955), 5: 188–200; and Christopher Marlowe, *Doctor Faustus*, I.i.111.

[32] Walker, *Spiritual and Demonic Magic*, pp. 205–12; Copenhaver, "The Occultist Tradition and Its Critics"; and Stillman Drake, *Galileo at Work: His Scientific Biography* (Chicago: University of Chicago Press, 1978), pp. 35–6, 55, 169–190, 236, 278–88, 313.

[33] Jean Fernel, *De abditis rerum causis libri duo* (Venice: Andrea Arrivabene, 1550).

[34] For a summary of more recent literature on Fernel, see John M. Forrester and John Henry, ed. and trans., *The Physiologia of Jean Fernel (1567)* (Philadelphia: American Philosophical Society, 2003),

Although Hippocrates was his first inspiration, Fernel knew that medical confidence in elements and qualities had become firm only with Galen, who developed his medical system in a post-Aristotelian framework. The universe in which a wet, cool, watery drug cures a dry, hot, fiery disease is the world in which the four elements constitute and account for everything beneath the moon. Fernel realized that Galen's extension of the Aristotelian and Hippocratic project was incomplete, however, and that Galen himself had to look – timidly – beyond the elements in order to explain common but perplexing medical phenomena. (One such puzzle was the "French disease" that killed Fernel's royal patient, Francis I, which was an ailment of the new age and unknown to Galen.) In effect, Fernel wanted to improve on Galen by deriving a more effective therapy and a more rigorous nosology from the physics of qualities on which magical theory was based. Fernel's key conviction, briefly stated, was that occult forces – arising from form, not from matter – should cure contrary occult diseases, just as the manifest powers of the four material elements cure contrary manifest diseases.[35]

For Fernel, the paradigm of a manifest and material cause was hot, dry, light fire. Our perception that fire has such features arises from sensation; what we really sense, however, are not features of the object but its effects on us. "Because you have sensed that fire burns, you judge it hot," he explains, "in the same way, because you have often observed that a magnet attracts iron, you should conclude from the result which you see that there must have been something antecedent."[36] Fernel maintained that access to qualities of any kind – material or formal, in the case of fire, or manifest or occult in the case of the magnet – is by inference rather than sensation. Beyond inferring that a burning sensation is caused by a feature of the object and then calling this feature "hot," one can say nothing more about the cause of the sensation.

What about opium? Do we sense its occult, dormitive virtue? No: In Fernel's theory, we infer that opium has such a virtue because it makes us feel drowsy. We perceive neither opium's dormitive virtue nor the hot quality of fire. "If I ask for the cause of fire's burning," he explains,

> ... you can say no more than that it comes from intense heat and that this is its nature and property. Having given this confident answer, you will seem to have replied fully and learnedly. Yet when I say that the magnet attracts iron or that peony stops epilepsy by an innate property, according to you I have not expressed the cause clearly enough. Why so? Why make what is

pp. 1–12; for the older view of Fernel as a crusader against magic, see Charles Sherrington, *The Endeavour of Jean Fernel* (Cambridge: Cambridge University Press, 1946).

[35] Hippocrates, *De morbo sacro*, 1–5, 18; [Galen], *De affectuum renibus insidentium dignotione* (Kühn, 19: 643–98); Fernel, *De abditis*, pp. 5–7, 101, 109, 120, 153–6, 204–6, 217–23, 235–6, 249, 280–2, 292–3, 304–5; Julius Röhr, *Der okkulte Kraftbegriff im Altertum* (*Philologus*, Supplementband 17.1) (Leipzig: Dieterich, 1923), pp. 96–133; Lloyd, *Magic, Reason, and Experience*, pp. 15–29; and Lloyd, *Greek Science after Aristotle*, pp. 136–53.

[36] Fernel, *De abditis*, pp. 17–23, 64–5, 82, 149–50, 159, 173–9, 284, 294.

common to both cases special to one, as if it were privileged? Perhaps this is the difference: the property of fire, because it is more familiar, is defined by the special names "heat" and "lightness," while no name has yet been applied to the properties of the magnet, peony and things of that kind. . . . Primary qualities do not explain everything, and . . . we should be no more amazed by the characteristics of occult properties than by those of the elements. . . . [Such] properties arise not from the elements or from matter but from form alone.[37]

For Fernel, then, the difference between manifest and occult was merely nominal. Qualities traditionally called occult differ from those called manifest only because we encounter the latter more often and give them common names such as "heat" instead of ungainly labels such as "dormitive virtue." The distinction, which makes no real difference, arises only from habit, taxonomy, and method, not from physics, ontology, or epistemology. In reality, all qualities are imperceptible whether they are called occult or manifest.

But Fernel did not stop with denying the usual distinction between manifest and occult qualities. He also attempted to replace epistemological puzzles with clinical data by invoking occult *faculties* rather than occult *qualities*. Plain facts of clinical experience were that opium makes people sleepy and that hemlock kills them. By isolating faculties as efficient causes of clinical facts that involved both drugs and patients, Fernel could make sense of opium's effect while evading the epistemological gap between the drug's (objective) dormitive quality and the patient's (subjective) dormitized experience. The drug's faculty is just the efficient cause of narcotic effects observed in the patient. About faculties Fernel could say no more except that they were products of divine form.[38]

As long as the debate stayed fixed on qualities, the epistemological impasse scouted by Fernel blocked further movement within the Aristotelian-Galenic framework. In the next century, Galileo, Descartes, and others would eliminate the obstacle by discarding that framework along with one of its root metaphors: qualities (hot, cold, wet, and dry) that were haptic rather than visual, felt rather than seen. Trying to cover the whole world – from remote stars to miniscule corpuscles – with the same physics and geometry, Descartes pictorialized its tiniest parts, just as Galileo had published pictures of moons circling Jupiter that were hidden to the naked eye. To show that all nature's works are effects of the same material causes – the shape, size, position, and motion of its smallest parts – Descartes depicted invisible micro-objects as they might appear in the macro-world: Special grooved particles (Figure 22.3), for example, solve the puzzle of magnetism by screwing their way

[37] Fernel, *De abditis*, pp. 285–7.
[38] Fernel, *De abditis*, pp. 151–6, 173–9; Lloyd, *Greek Science after Aristotle*, pp. 141–3.

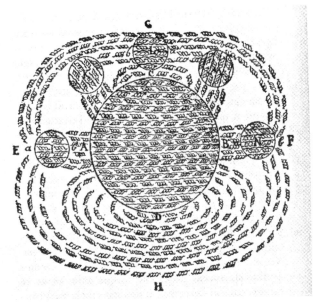

Figure 22.3. Descartes' illustration of magnetic action. In René Descartes, *Principia philosophiae* (Amsterdam: Elzevir, 1644), p. 212. By permission of the Department of special collections, Charles E. Young Research Library, University of California, Los Angeles. [Q155 D43p 1644].

mechanically through the earth.[39] Many such explanatory pictures clarify mechanical arguments made by Descartes that would otherwise be harder to grasp, and the language of vision (ideas are clear and distinct; a natural light illuminates understanding) also pervades Cartesian philosophy. Images made it more plausible to talk about invisible mechanisms working beneath visible phenomena – including effects commonly treated as magical, such as magnetism, sympathies, and antipathies.[40] Whereas for Descartes knowing was like seeing, Fernel had been committed to a knowing-as-touching metaphor. His physical primitives were not shape, size, position, and motion but fire, air, water, and earth – felt but not seen as hot, cold, wet, and dry.[41]

When new theories, with pictures clarifying words and numbers, displaced the old physics, likewise displaced were the old occult qualities that were

---

[39] René Descartes, *Oeuvres*, 11 vols. in 13, ed. Charles Adam and Paul Tannery (Paris: J. Vrin, 1964–74), 8A: 283–315.
[40] J. F. Scott, *The Scientific Work of René Descartes* (London: Taylor and Francis, 1952), pp. 71–81; William R. Shea, *The Magic of Numbers and Motion: The Scientific Career of René Descartes* (Canton, Ohio: Science History Publications, 1991), pp. 129–47, 205–18, 228–49; and Copenhaver, "The Occultist Tradition and Its Critics," pp. 469–73.
[41] Aristotle, *De anima*, 435a11–b25; cf. Aristotle, *De sensu*, 436b13–37a31; Aristotle, *Metaphysics*, 980a23–7; Fernel, *De abditis*, pp. 10–13; and David Lindberg and Nicholas Steneck, "The Science of Vision and the Origins of Modern Science," in *Science, Medicine, and Society in the Renaissance: Essays to Honor Walter Pagel*, ed. Allen Debus, 2 vols. (New York: Science History Publications, 1972), 2: 29–45.

rooted in intuitions that were more haptic than visual. The same use of pictures that served magic so well in the sixteenth century was thus turned against it by mechanical philosophers of the seventeenth century, who made images of what they could not see but had to assert: the microscopic particles and mechanisms that caused macroscopic phenomena. Although magical objects and their sensible effects could be observed, the occult (literally, hidden) causes of those effects had always eluded observation. To account for such things, insofar as experience made them veridical, proponents of the mechanical philosophy visualized invisible mechanisms and published pictures of them. Hence, pictures worked both for and against magic in early modern Europe, at first promoting the reality of magical objects but eventually eroding their theoretical foundations.

New modes of visualization, assisted by new arts of picturing, eventually helped to make magic a mere spectacle, an illusion, ready for the hilarious disclosure of its emptiness by the French dramatist Molière (1622–1673). Molière gave the history of science its best joke in *Le malade imaginaire* (1673), the comedy that exploded the pretense of occult qualities. In the end, laughter was a powerful force in driving magic offstage – laughter that marked a fundamental shift in standards of intelligibility.[42]

One of the themes of *The Imaginary Invalid* is false learning, a pose that Molière found especially funny in physicians, who made their patients pay for empty talk. Diafoirus, the learned healer of the play, describes the ideal physician as one who "won't budge from an opinion, . . . binds himself *blindly* to the opinions of the ancients and has never wished to understand . . . the pretended discoveries of our century on the circulation of the blood and other notions of that ilk."[43] Argan, the hypochondriac of the play's title, reveals his faith in the old occult qualities, as opposed to the newer mechanical notions, while he catalogs his medicines: "a little enema, insinuative, preparative and emollient, . . . a hepatic julep, soporific and somniferous, . . . a nice purgative . . . to flush out and evacuate the bile."[44] The play's finale in macaronic Latin is a song and dance burlesque of the granting of a medical diploma. The *primus doctor* asks: Why does opium bring on sleep? The candidate replies that opium's effect comes from its narcotic power:

> And to your quiz
> My answer is
> A virtue dormitive,
> Whose nature is
> To soften up the senses.

"Good answer – oh good, good, good!" is the verdict: The candidate has responded learnedly; his ordeal is over, and his new colleagues exult.[45]

---

[42] Molière, "Le malade imaginaire," I.i in *Oeuvres complètes* (Paris: Éditions du Seuil, 1962), p. 628.
[43] Molière, "Le malade imaginaire," II.v, pp. 642–3.
[44] Ibid., I.i, p. 628.
[45] Ibid., pp. 659–61.

The traditional philosophy and medicine ridiculed by Molière had treated qualities of objects – their powers, colors, and other features – as entities apart from matter, but the mechanical philosophers saw them as nothing but material structures. In fact, because they could not really see these structures, they posited invisible microstructures, picturing them in visual metaphors based on artificial macroscopic objects – balls, canes, keys, screws, locks, clocks – that blurred the line between nature and art and demystified the origin of qualities. For objects that could not be sensed, they postulated features by analogy between the seen and the unseen. If something could not be visualized in this way, it could be dismissed as occult, which, for those who abandoned the old physics of quality, had come to mean "unintelligible" rather than "hidden."[46]

The corpuscles theorized by Robert Boyle (1627–1691), though endowed with picturable properties of size, shape, and motion and redefined as *primary* qualities, were just as hidden as occult qualities and no more observable. Boyle argued that observable properties emerged only when these least bodies aggregated in structures; however, the resulting *secondary* qualities, such as color or odor, were not the scholastic entities that Molière mocked and Boyle found incomprehensible. Never entirely escaping the world of magic, Boyle improved on occult qualities by replacing them with other indiscernibles, tiny bodies to which he imputed properties such as those of ordinary objects – ground glass, for example. Physicians traditionally treated this substance as a poison and therefore labeled it occult, as possessing a "deleterious faculty, . . . a peculiar and superadded entity." But for Boyle there was "nothing distinct from the glass itself, . . . [whose] sharp points and cutting edges are enabled by these mechanical affections to pierce or wound . . . the stomach and guts."[47]

For Boyle, in other words, the toxic virtue in glass is just its structure, not a separate faculty or quality. To explain the effects of a poison by qualities only posited ingredients and tagged them with the very features whose explanation was sought, such as the dormitive virtue of opium; naming notional entities in this way explained nothing. "What is it to me to know that such a quality resides in such a principle or element whilst I remain altogether ignorant of the cause?" Having framed this question in the *Sceptical Chymist* (1661), Boyle went on to ask, "How little does the chymist teach the philosopher of . . . purgation if he only tells him that the purgative vertue of medicines resides in their salt? . . . 'Tis one thing to know a man's lodging, and another to be

---

[46] Keith Hutchison, "What Happened to Occult Qualities in the Scientific Revolution?," *Isis*, 73 (1982), 233–53; Hutchison, "Dormitive Virtues, Scholastic Qualities, and the New Philosophies," *History of Science*, 29 (1991), 245–78. Compare Copenhaver, "The Occultist Tradition and Its Critics," pp. 457–60, 503, which maintains that occult qualities were *not* unintelligible in the traditional Peripatetic framework. On art and nature, see Daston and Park, *Wonders and the Order of Nature*, pp. 260–5, 280–93, 298–9, 314.

[47] *The Works of the Honourable Robert Boyle*, ed. Thomas Birch, 6 vols. (London: W. Johnston et al., 1772); 3: 4, 11, 13, 18–25, 46–7; Peter Alexander, *Ideas, Qualities, and Corpuscles: Locke and Boyle on the External World* (Cambridge: Cambridge University Press, 1985), pp. 5–9, 18, 39, 61–3, 85; and Copenhaver, "The Occultist Tradition and Its Critics," pp. 488–90.

acquainted with him." Looking for substance and quality while overlooking structure was like trying to explain a clock by telling whether its works are brass or steel while ignoring their configuration.[48]

According to the mechanical philosophers, the structures underlying phenomena that were previously understood as occult are not imperceptible in their nature, unlike the old occult qualities that eluded observation because they arose from immaterial forms. Minute structures are not perceived simply because the human senses are weak. Because we cannot see or feel "the Bulk, Texture, and Figure of the minute parts of Bodies, . . . we are fain to make use of their secondary Qualities, . . . which . . . are nothing but bare Powers. For the Color and Taste of Opium, are, as well as its soporific and anodyne Virtues, mere Powers depending on its primary Qualities." This was the view of Boyle's contemporary John Locke (1632–1704), who helped Boyle shut the door on the old physics of qualities while using pictorial keys and locks to open the way to a sharper, visual understanding of matter.[49]

By reducing causality to structure, Locke and Boyle brought occult phenomena within the scope of the new science. Boyle even proposed a theory to cover action-at-a-distance and its unobservable agents. Rather than attributing an electrical property to amber in order to explain its power of attracting chaff when rubbed, he argued that this familiar but puzzling effect resulted from an effluvium, a structure of imperceptible particles with no properties but size, shape, and motion. Although their smallest parts were ultimately no more perceptible than Agrippa's occult qualities, Boyle's effluvia had two advantages: An imputed structure made them seem concrete and intelligible; and an analogy with visible vapors brought them within the range of everyday experience.[50]

MAGIC OUT OF SIGHT

With his theory of occult faculties, Fernel anticipated the use that mechanical philosophers made of the term "power" to distinguish features of an object from its ability to affect an observer – or a patient.[51] But several

---

[48] Robert Boyle, *The Sceptical Chymist* (London: Dent, 1911), pp. 178–83; and Alexander, *Ideas, Qualities, and Corpuscles*, pp. 37–40, 50–2.

[49] John Locke, *An Essay Concerning Human Understanding*, ed. Peter Nidditch (Oxford: Clarendon Press, 1975), 2.8.8–10, pp. 134–5; 2.8.23, pp. 140–1; 2.23.8–9, pp. 300–1; 4.3.25, pp. 555–6; Alexander, *Ideas, Qualities, and Corpuscles*, pp. 48, 55–9, 61–88, 115–25, 131–4, 139, 150–1, 162–74; Copenhaver, "Scholastic Philosophy and Renaissance Magic," pp. 524–8, 538–46; Copenhaver, "Astrology and Magic," pp. 274–87; and Copenhaver, "The Occultist Tradition and Its Critics," pp. 454–60, 490–3.

[50] Boyle, *Works*, pp. 660, 678–89; Boyle, *Sceptical Chymist*, pp. 104–5; Agrippa, *Opera*, 1: 25, 38, 45, 274, 465–6; and Alexander, *Ideas, Qualities, and Corpuscles*, p. 64. Catherine Wilson, *The Invisible World: Early Modern Philosophy and the Invention of the Microscope* (Princeton, N.J.: Princeton University Press, 1995), pp. 229–32, stresses the weakness of the argument from analogy – its disregard of differences between macro- and micro-objects – as some of the mechanical philosophers realized.

[51] Locke, *Essay*, 2.8.2, pp. 7–10, 15, 17, 22–3, 26; and Alexander, *Ideas, Qualities, and Corpuscles*, pp. 115–22, 131–4, 150–67.

things made the mechanical philosophy and its microscopic particles more credible than Fernel's occult faculties and thereby weakened the theory of magic: One strength of the new science was its claim that unseen structures are intelligible by pictorial analogy with gross phenomena, natural or artificial; another was its confidence that new instruments could reveal features of the world never seen before because they were too small or too far away.

When Galileo, Descartes, and their successors looked for distant moons or hidden microstructures, new tools – telescopes and microscopes – equipped them better than Fernel. Agrippa had found mirrors and lenses merely spellbinding, but Della Porta speculated in his *Magia naturalis* of 1589 (an expanded edition of the 1558 first edition) about optical instruments, whose scientific use awaited the next century. Meanwhile, the influence of ancient atomist texts, whose recovery started in 1417 and opened several lines of attack on Aristotle, grew with no help from optical instruments, though it took a long time for atomism to penetrate natural philosophy. Near the end of this slow process came Galileo's account of particulate matter in *Il Saggiatore* (The Assayer, 1623), which also mentions a magnifying device. Galileo then built the instrument used to produce the first scientific illustration made with a microscope, a 1625 broadsheet showing three magnified bees. Another atomist, Pierre Gassendi (1592–1655), looked through a microscope at crystals and saw their geometry, and Descartes theorized about magnifying lenses and imagined pictures of an invisible microworld.[52]

"By means of the Telescopes," wrote Robert Hooke (1635–1703) in 1665 in his *Micrographia*, "there is nothing so far distant but may be represented to our view, and by the help of Microscopes, there is nothing so small as to escape our inquiry. Hence there is a new visible world discovered, . . . all the secret workings of Nature." Encouraged by their new instruments, and falsely assuming that pictures of the world could go all the way down, the mechanical philosophers thought it possible to shift from haptic to visual primitives, at least in theory. Just as the telescopic sight of moons circling Jupiter extended terrestrial physics to the whole cosmos, microscopic views of minute structures elicited analogies from macro-objects to micro-objects. Depicting real things never seen before, Hooke used the engravings of his lavishly illustrated *Micrographia* to picture cheese mold that looked to him

---

[52] Agrippa, *Opera*, 1: 12–13, 32–3, 153–4; cf. 2: 60–1; Galileo Galilei, *Opere*, ed. Ferdinando Flora (Milan: Ricciardi, 1953), pp. 171–2, 311–2; Descartes, *Oeuvres*, 6: 93, 196–211; Locke, *Essay*, 2.13.19, pp. 175, 2.23.2, p. 295; Daston and Park, *Wonders and the Order of Nature*, pp. 300, 323; Freedberg, *The Eye of the Lynx*, pp. 7, 33, 41, 71, 101–8, 114, 142, 151–4, 160–3, 219, 222–32, 276; Wilson, *The Invisible World*, pp. 57–79, 85–8, 216, 238–43; Norma E. Emerton, *The Scientific Reinterpretation of Form* (Ithaca, N.Y.: Cornell University Press, 1984), pp. 42–3, 129–35, 148–53, 248; Copenhaver and Schmitt, *Renaissance Philosophy*, pp. 198–201; Copenhaver, "The Occultist Tradition and Its Critics," pp. 475–80; and Howard Jones, *The Epicurean Tradition* (London: Routledge, 1992), pp. 142–65.

like "microscopical Mushroms" and a gnat's antennae that seemed like "the horns of an Oxe."⁵³

By the time Hooke published his microscopic investigations, the mechanical philosophy had established itself as the new standard of intelligibility in natural philosophical explanation. But fascination with the microscope was more the effect than the cause of new ways of explaining natural phenomena by picturing them. If the new apparatus of microscopy came too late to explain the transition from Fernel's occult physics of haptic qualities to a mechanical physics of pictured particles, where should we look? I suggest that Europeans in the interim were stimulated by medical debates about occult properties, provoked by atomist materialism, and seduced by the pictorial arts to prepare the way for that momentous change. This conjunction did not rid natural philosophy of wonders – far from it, as witness the revelations of strange new worlds under the microscope. Nor were hidden causes banished, though respectable researchers no longer called them occult. On the contrary, the physics of force and the preformationist embryology of *emboîtement* gave unseen objects a greater role than ever.⁵⁴ But debates about magic and occult qualities had also turned up new criteria for explaining the world of nature, which helped natural philosophy picture a way out of magic.

---

⁵³ Robert Hooke, *Micrographia; or, Some Physiological Descriptions of Minute Bodies Made by Magnifying Glasses with Observations and Inquiries Thereupon* (London: Martyn and Allestry, 1665), sig. aii, pp. 125–6, 185–6. Note the different, though not entirely contrary, point made by Wilson, *The Invisible World*, pp. 57–63, that "the microscope takes away the privilege of surface. What the object looks like on the outside is no guide to what it is. . . . And in the interior of things there is no resemblance . . . even if we must call in the language of every day – of ropes, fibers, globules, forests, looms and children's toys – to describe it."

⁵⁴ Wilson, *The Invisible World*, pp. 113–37; Copenhaver, "The Occultist Tradition and Its Critics," pp. 493–502; and Clara Pinto-Correia, *The Ovary of Eve; Egg and Sperm and Preformationism* (Chicago: University of Chicago Press, 1997).

23

ASTROLOGY

H. Darrel Rutkin

As is well known, astrology finally disappeared from the domain of legitimate natural knowledge during the seventeenth and eighteenth centuries, although the precise contours of this story remain obscure. It is less well known, albeit clearly documented, that astrology was taught from the beginning of the fourteenth century as an important part of the arts and science curriculum at the great medieval and Renaissance universities, including Padua, Bologna, and Paris. There, astrology was studied within three distinct scientific disciplines – mathematics, natural philosophy, and medicine – and served to integrate several highly developed mathematical sciences of antiquity – astronomy, geography, and geometrical optics – with Aristotelian natural philosophy. This astrologizing Aristotelianism provided fundamental patterns of interpretation and analysis in pre-Newtonian natural knowledge. Thus, the history of astrology – and, in particular, the story of its protracted criticism and ultimate rejection as a source of what the learned considered legitimate natural knowledge – is central for understanding the transition from medieval and Renaissance natural philosophy to Enlightenment science. The role of astrology in this transition was neither obvious nor unproblematic. Indeed, astrology's integration of astronomy and natural philosophy under the aegis of mathematics had much in common with the aims of the "new science" of the seventeenth century. Thus it becomes necessary to explain why this promising astrological synthesis was rejected in favor of a rather different mathematical natural philosophy.

Some historians have claimed that Giovanni Pico della Mirandola's extensive attack on astrology, *Disputationes adversus astrologiam divinatricem* (Disputations against Divinatory Astrology), convinced the astrologers to stop practicing upon its publication in 1496. Others have stressed the role of Copernicus's *De revolutionibus orbium coelestium* (On the Revolutions of the Celestial Spheres, 1543) in completing astrology's undoing. Both claims are belied, however, by the fact that two of the most important warriors for the Copernican cause, Galileo Galilei (1564–1642) and Johannes Kepler

(1571–1630), were both practicing astrologers. Indeed, influenced by Pico's *Disputationes*, Kepler set out to reform astrology on a sounder natural philosophical and mathematical foundation, and the English natural philosopher Francis Bacon (1561–1626) proposed his own reform of astrology in one of his last scientific works, the *De augmentis scientiarum* of 1623.

This evidence suggests that the role of astrology in the learned culture of the sixteenth and seventeenth centuries and the story of its ultimate fall from intellectual grace are more complex and important than generally thought. This chapter will sketch that role and that story. In it, I set the historical stage by describing astrology's place on the map of knowledge of the years around 1500 – its relationship to mathematics, natural philosophy, and medicine – and summarize the ideas that structured astrological theory and practice, situating them in their primary institutional location, the university. I then describe astrology's place in the work of major figures of the Scientific Revolution, including Galileo, Bacon, and Robert Boyle (1627–1691). Finally, I ask how, when, and ultimately why astrology came to be removed from its central place in the premodern understanding of nature over the course of the seventeenth and eighteenth centuries.[1]

ASTROLOGY CIRCA 1500: INTELLECTUAL AND INSTITUTIONAL STRUCTURES

In 1474, Galeazzo Maria Sforza, Duke of Milan, wrote a threatening letter to Girolamo Manfredi, professor of astrology, philosophy, and medicine at the University of Bologna. Desist from publishing negative prognostications about me or my realm, he insisted, or I will send two of my men to cut you into pieces.[2] The annual prognostication was in fact a normal part of Manfredi's job description, although it rarely elicited such extreme reactions. In addition to teaching the standard curriculum, the professor of astrology was also required by statute to participate in at least three public astrological disputations per year and to supply the prognostication, together with other astrological services – free and in a timely manner – to the university community. By 1474, annual prognostications regularly appeared in print,

---

[1] For various useful but incomplete discussions of this topic, see, among others, Keith Thomas, *Religion and the Decline of Magic: Studies in Popular Beliefs in Sixteenth and Seventeenth Century England* (New York: Scribners, 1971); Patrick Curry, *Prophecy and Power: Astrology in Early Modern England* (Princeton, N.J.: Princeton University Press, 1989); Mary Ellen Bowden, "The Scientific Revolution in Astrology," Ph.D. dissertation, Yale University, New Haven, Conn., 1975; and Elide Casali, *Le spie del cielo: Oroscopi, lunari e almanacchi nell'Italia moderna* (Turin: Einaudi, 2003). For a fuller discussion (with bibliography) of this and many other issues in this chapter, see H. Darrel Rutkin, "Astrology, Natural Philosophy, and the History of Science, c. 1250–1700: Studies Toward an Interpretation of Giovanni Pico della Mirandola's *Disputationes adversus astrologiam divinatricem*," Ph.D. dissertation, Indiana University, Bloomington, 2002.

[2] Ferdinando Gabotto, "L'Astrologia nel Quattrocento in rapporto colla civiltà: Osservazioni e documenti storici," *Rivista di filosofia scientifica*, 8 (1889), 373–413.

allowing them to reach a broader audience, which undoubtedly sharpened Galeazzo Maria's concern.[3]

Like medicine, one of Manfredi's other subjects, astrology had both a theoretical and a practical dimension. For the definition and outlines of this discipline as it was taught and practiced in the years around 1500, we can look to the Hellenistic astronomer-mathematician Ptolemy (ca. 100-ca. 170) and the German natural philosopher Albertus Magnus (ca. 1200–1280), arguably the two most influential authorities in the field. According to Ptolemy's *Tetrabiblos*, there were two primary kinds of what he called "foreknowledge through the science of the stars." The first – what we would call astronomy – studied the motions of the heavenly bodies mathematically and was an exact science. The second – astrology – investigated the influences of the heavenly bodies on the earth; because of the perturbing interaction with matter, its knowledge was not exact.[4] Although the terminology became increasingly differentiated during the early modern period, the Latin terms *astronomia* and *astrologia* (and their vernacular variants) were often used interchangeably to refer to *both* sciences of the stars. For the sake of clarity, however, I will follow modern usage, except when I use "the science of the stars" to refer to both together.[5]

Building on Ptolemy's distinction, Albertus Magnus described the four types of astrological praxis in his *Speculum astronomiae* (Mirror of the Science of the Stars, ca. 1260s):[6] "Revolutions" were concerned with large-scale changes, primarily in the weather but also in state affairs. This was the major feature of the annual prognostications found in almanacs and elsewhere, and it was what ruffled Galeazzo Maria's feathers.[7] "Nativities," on the other hand, involved the astrological configuration at a person's birth. "Interrogations" entertained questions on matters of concern, including personal, medical, and business affairs. Finally, "elections" determined the most propitious moment to begin an enterprise or perform an activity, such as crowning

---

[3] Albano Sorbelli, "Il 'Tacuinus' dell' università di Bologna e le sue prime edizioni," *Gutenburg Jahrbuch*, 13 (1938), 109–14; Alberto Serra-Zanetti, "I pronostici di Girolamo Manfredi," in *Studi riminesi e bibliografici in onore di Carlo Lucchesi* (Faenza: Fratelli Legov, 1952), pp. 193–213.

[4] Claudius Ptolemaeus, *Apotelesmatike*, 1.1, ed. Wolfgang Hübner (Stuttgart: Teubner, 1998). Claudius Ptolemaus, *Tetrabiblos*, trans. Frank E. Robbins (Cambridge, Mass.: Harvard University Press, 1940), is still useful.

[5] For discussions of terminology, see the index to Lynn Thorndike, *A History of Magic and Experimental Science*, 8 vols. (New York: Columbia University Press, 1923–58); and Brian P. Copenhaver, "Astrology and Magic," in *The Cambridge History of Renaissance Philosophy*, ed. Charles B. Schmitt, Quentin Skinner, and Eckhard Kessler, with Jill Kraye (Cambridge: Cambridge University Press, 1988), pp. 264–300. All translations are mine unless otherwise indicated.

[6] I accept Albertus's authorship of the *Speculum astronomiae*; see Paola Zambelli, *The Speculum astronomiae and Its Enigma: Astrology, Theology, and Science in Albertus Magnus and His Contemporaries* (Dordrecht: Kluwer, 1992).

[7] For annual prognostications primarily as printed in yearly almanacs, see Casali, *Le spie del cielo*, for Italy, and Bernard Capp, *Astrology and the Popular Press: English Almanacs, 1500–1800* (London: Faber, 1979), for England and America.

a ruler, passing the baton of command to a general, or laying the cornerstone of an important building.[8]

Albertus wrote the *Speculum astronomiae* to distinguish between legitimate and illegitimate forms of astrological practice. For Christians such as Albertus, elections in particular raised sensitive issues because they related to the science of images, which occupied the borderlands between what we would call magic, science, and religion. What Albertus considered the legitimate science of images, which included the making of talismans, belonged to the realm of astrological magic. This science was developed by the Arabs, based in part on Greek sources, and the relevant texts were translated into Latin during the twelfth and thirteenth centuries. It worked by means of natural astrological mechanisms and was presumably included among elections because talismans needed to be made at astrologically propitious times.[9] Bad image magic, on the other hand, worked only because it had traffic with demons and thus crossed the line into execrable and theologically suspect practices.[10]

Established in the thirteenth century as a legitimate branch of natural knowledge by Albertus Magnus, among other European scholars, this astrologizing natural philosophy was developed by Pietro d'Abano (1257–ca. 1316), professor of medicine at the University of Padua, into a coherent and systematic educational program with a medical purpose.[11] In his influential textbook, the *Conciliator* – published several times in the late fifteenth and sixteenth centuries – Pietro stated, following Hippocrates and Galen, that

---

[8] See, for example, Mary Quinlan-McGrath, "The Foundation Horoscope(s) for St. Peter's Basilica, Rome, 1506: Choosing a Time, Changing a *Storia*," *Isis*, 92 (2001), 716–41. Keith Thomas's chapters on astrology in his *Religion and the Decline of Magic*, pp. 283–385, are probably still the best introduction to the subject in English. Good general scholarly accounts of astrology are: Tamsyn Barton, *Ancient Astrology* (London: Routledge, 1994); Jim Tester, *A History of Western Astrology* (Woodbridge: Boydell Press, 1987); and J. C. Eade, *The Forgotten Sky: A Guide to Astrology in English Literature* (Oxford: Clarendon Press, 1984). Important contributions on Renaissance astrology include Anthony Grafton, *Cardano's Cosmos: The Worlds and Works of a Renaissance Astrologer* (Cambridge, Mass.: Harvard University Press, 1999), and Steven vanden Broecke, *Limits of Influence: Pico, Louvain, and the Crisis of Renaissance Astrology* (Louvain: E. J. Brill, 2003), both with rich up-to-date bibliographies.

[9] See Charles Burnett, "Talismans: Magic as Science? Necromancy among the Seven Liberal Arts," in Burnett, *Magic and Divination in the Middle Ages: Texts and Techniques in the Islamic and Christian Worlds* (Aldershot: Variorum, 1996), pp. 1–15; David Pingree, "The Diffusion of Arabic Magical Texts in Western Europe," in *La diffusione delle scienze islamiche nel Medio Evo europeo* (Rome: Accademia Nazionale dei Lincei, 1987), pp. 57–102; and Frank Klaassen, "English Manuscripts of Magic, 1300–1500: A Preliminary Survey," in *Conjuring Spirits: Texts and Traditions of Medieval Ritual Magic*, ed. Claire Fanger (University Park: Pennsylvania State University Press, 1998), pp. 3–31.

[10] There is much excellent scholarship on demons and demonology in the early modern period. See, for example, Stuart Clark, *Thinking with Demons: The Idea of Witchcraft in Early Modern Europe* (Oxford: Clarendon Press, 1997), and the volumes in the Routledge series *Witchcraft and Magic in Europe*, all with rich bibliographies. I have avoided configuring astrology within the so-called occult sciences because this category is fundamentally anachronistic and dependent on too many misleading assumptions to be conceptually or historically useful.

[11] See Nancy G. Siraisi, *Arts and Sciences at Padua: The Studium of Padua before 1350* (Toronto: Pontifical Institute of Medieval Studies, 1973). For a similar process with physiognomy, see Jole Agrimi, *Ingeniosa scientia nature: Studi sulla fisiognomica medievale* (Millennio Medievale 36, 8) (Florence: Sismel, 2002).

the complete physician must be well versed in astrology.[12] Thus, at Padua, astrology was taught not only in the mathematics course with mathematical astronomy and in the natural philosophy course with Aristotle but in the medical course as well.[13]

In 1464, the influential mathematician and astronomer Johannes Regiomontanus (1436–1476) praised astrology as the queen of the mathematical disciplines in his inaugural oration for a course in mathematical astronomy at Padua[14] (see Donahue, Chapter 24, this volume). Evoking Padua's distinguished tradition of teaching the sciences of the stars, Regiomontanus located himself in the same tradition as his fellow German Albertus Magnus, a student at Padua in the 1220s,[15] and Pietro d'Abano himself. Regiomontanus was also an important early figure in scientific publishing; while based in Nuremberg, between 1472 and 1475, he composed and printed ephemerides, calendars, and the *Novae theoricae planetarum* (New Theories of the Planets) of Georg Peurbach, his teacher at Vienna, as well as an ancient Latin astrological text, Manilius's hexameter *Astronomica*.[16] After Regiomontanus's death, the German printer Erhard Ratdolt continued this tradition of scientific publication at Venice and Augsburg, but with a more explicitly practical astrological orientation, publishing a variety of fundamental ancient and medieval texts.[17]

Although Padua was in some sense the wellspring of academic astrology, it was taught at other universities as well. As prescribed in the 1405 statutes for the study of arts and medicine at the University of Bologna, still the foundation of instruction in the late fifteenth century, the four-year course in mathematics and astrology began with two years of prerequisites in arithmetic, geometry, and astronomy, including lectures on Euclid, the *Sphere* of Sacrobosco, and the Alphonsine tables.[18] With this background, students

---

[12] Pietro d'Abano, *Conciliator*, diff. 10. I used the facsimile of the Venice, Giunta edition of 1565: Pietro d'Abano, *Conciliator*, ed. E. Riondato and L. Olivieri (Padua: Editrice Antenore, 1985).

[13] See Rutkin, "Astrology, Natural Philosophy, and the History of Science," chaps. 3 and 7. For related issues, see also Robert S. Westman, "The Astronomer's Role in the 16th Century: A Preliminary Study," *History of Science*, 18 (1980), 105–47; and Mario Biagioli, "The Social Status of Italian Mathematicians, 1450–1600," *History of Science*, 27 (1989), 41–95.

[14] Noel M. Swerdlow, "Science and Humanism in the Renaissance: Regiomontanus' Oration on the Dignity and Utility of the Mathematical Sciences," in *World Changes: Thomas Kuhn and the Nature of Science*, ed. P. Horwich (Cambridge, Mass.: MIT Press, 1993), pp. 131–68. See also Paul L. Rose, *The Italian Renaissance of Mathematics: Studies on Humanists and Mathematicians from Petrarch to Galileo* (Geneva: Droz, 1975).

[15] Regiomontanus considered Albertus the author of the *Speculum astronomiae*.

[16] See Ernst Zinner, *Regiomontanus: His Life and Work*, trans. E. Brown (Amsterdam: North Holland, 1990).

[17] See G. R. Redgrave, *Erhard Ratdolt and His Work at Venice* (London: The Bibliographical Society, 1894–5).

[18] See Lynn Thorndike, *University Records and Life in the Middle Ages* (New York: W. W. Norton, 1975), pp. 279–82. For a fuller discussion, see Graziella Federici Vescovini, "I programmi degli insegnamenti del Collegio di medicina, filosofia e astrologia dello statuto dell'università di Bologna del 1405," in *Roma, magistra mundi: Itineraria culturae medievalis, Mélanges offerts au Père L. E. Boyle*, 2 vols. (Louvain: La Neuve, 1998), 1: 193–223.

began their study of astrology proper in the third year with Alcabitius's influential textbook the *Liber isagogicus* and (pseudo-)Ptolemy's *Centiloquium*, a fundamental astrological text for medical practitioners. In the fourth year, the student moved on to more advanced works in both astrology and astronomy – Ptolemy's *Tetrabiblos* and *Almagest* – and the *De urinis non visis* (On Unseen Urines), an astrological treatment of medical questions. This curricular framework informed Regiomontanus's and Ratdolt's publication program.

Astrology was also taught in the natural philosophy course at Bologna, which involved studying the core texts of Aristotelian natural philosophy, including those that provided astrology's natural philosophical foundations: Aristotle's *De caelo* (On the Heavens), *De generatione et corruptione* (On Generation and Corruption), and the *Meteorologia*. Finally, the four-year medical course focused on medical texts. This included some with a strongly astrological character, notably Galen's *De criticis diebus* (On Critical Days), which was studied during each of the first three years. As at Padua under Pietro d'Abano, medical training was the primary end toward which the arts and medicine curriculum at Bologna was oriented.[19]

As taught in Italian universities in the years around 1500, the astrological system focused on three main aspects of the topic: natural philosophy, cosmography, and a geometrical-optical model of planetary influences.[20] According to Aristotle, the movements of the heavens, especially the sun, were fundamental to life on earth. In *De generatione et corruptione*,[21] Aristotle argued that the sun, as it moved around the ecliptic in the course of the year, was the universal efficient cause, ontologically prior to and necessary for generation and corruption; that is, for things to come into being and pass away. Thus, processes of generation, human and otherwise, required the sun (as efficient cause); the male, who provided the formal cause in his seed; and the female, who supplied the material cause in the menses of her womb. In his commentary on this passage, Albertus Magnus expanded the realm of the efficient cause to include the rest of the planets in addition to the sun.

In this understanding of celestial action, natural philosophical structures were fitted onto a fundamentally Ptolemaic cosmographic framework composed of mathematical astronomy calibrated with mathematical geography. This framework allowed the planetary motions to be uniquely determined

---

[19] See Nancy G. Siraisi, *Taddeo Alderotti and His Pupils: Two Generations of Italian Medical Learning* (Princeton, N.J.: Princeton University Press, 1981).

[20] For a much fuller reconstruction, see Rutkin, "Astrology, Natural Philosophy, and the History of Science," chap. 2, where I also discuss important contributions to this system by al-Kindi, Robert Grosseteste, and Roger Bacon. For different interpretations of astrology's relationship with Aristotelian natural philosophy, see Edward Grant, *Planets, Stars, and Orbs: The Medieval Cosmos, 1200–1687* (Cambridge: Cambridge University Press, 1994), pp. 569–617; John D. North, "Celestial Influence: The Major Premiss of Astrology," in *"Astrologi hallucinati": Stars and the End of the World in Luther's Time*, ed. Paola Zambelli (Berlin: Walter de Gruyter, 1986), pp. 45–100; and vanden Broecke, *Limits of Influence*, index s.v. "astrological physics."

[21] *De generatione et corruptione*, 2.9–10.

for any time at any place by means of the horizon. This is important because place was essential for analyzing the role of celestial influences in generation, as Albertus made clear:

> If anyone wished to understand all the natures and properties of particular places, he would know that there is not a point in them that does not have a special property from the virtue of the stars, . . . for the circle of the horizon is varied in relation to each point of the habitation of animals, plants and stones; and the entire orientation of the heavens . . . is varied in relation to the varying circle of the horizon. From which cause their natures, properties, customs, actions and species, which seem to be generated in the same perceptible place, are varied, to such an extent that diverse properties and customs are attributed to twins' seeds, both for brute animals and for men, from this different orientation. And this is reasonable because it has been learned that the heavens pour forth formative virtues into everything that exists. Moreover, it mostly pours them forth by means of rays emitted by the lights of the stars, and therefore it follows that each pattern and angle of rays causes different virtues in things below.[22]

Once the planets' motions were mapped in this way and related to each place on earth, their influences could then be analyzed using the other primary feature of this system, a geometrical-optical model of planetary influences. These irradiating influences were thought to act in straight lines on the model of the planets' other mode of influence, light. The angular relationship of the planets to each other – the planetary aspects – and their collective relationship to each place on earth could then be fully articulated. When the different qualitative natures of each planet – in themselves and as modified by each sign of the zodiac – were taken into account, as well as the variation in effect from their varying angular relationships, the result was an integrated mathematical natural philosophy of richness and sophistication.

ASTROLOGICAL REFORMS

Despite its elegance and power, this explanatory system was not without its critics. The most famous fifteenth-century attack on astrology was that of Giovanni Pico della Mirandola (1463–1494) in his extensive *Disputationes adversus astrologiam divinatricem* (Disputations against Divinatory Astrology, composed in 1493–4 and published posthumously in 1496).[23] This seems to have been at least partially directed against Marsilio Ficino's influential *De*

---

[22] *De natura locorum*, in Albertus Magnus, *Opera omnia* . . . (Cologne: Aschendorff, 1951–), vol. 5, pt. 2, p. 8, ll. 43–62.
[23] Giovanni Pico della Mirandola, *Disputationes adversus astrologiam divinatricem*, ed. and trans. Eugenio Garin, 2 vols. (Florence: Vallecchi, 1946–52). For their heavy editing, see Paola Zambelli, "La critica dell'astrologia e la medicina umanistica," in her *L'Ambigua natura della magia: Filosofi, streghe, riti nel Rinascimento*, 2nd ed. (Venice: Marsilio, 1996), pp. 76–118 (orig. publ. 1965). See also Stephen

vita (On Life, 1489), which included a summa of astrological magic as part of its third book.[24] Pico aimed to remove – or radically redefine in a cabbalistic direction – the astrological superstructure that was traditionally erected on the Aristotelian-Ptolemaic foundation that he accepted as sound. By excising what he saw as outdated Arabo-Latin accretions, Pico hoped to restrict celestial influences to the realm of universal efficient cause, as originally in Aristotle, thus removing the possibility of particular predictions.[25]

In many respects, however, Pico's voice was an isolated one, and the astrological system elaborated by Albertus Magnus and Pietro d'Abano continued to flourish in early modern Europe throughout the sixteenth century and into the seventeenth century at the universities of Bologna, Pisa, Padua, and Louvain (among others).[26] Indeed, Philip Melanchthon (1497–1560) used precisely this astrologizing Aristotelianism to reform the German Lutheran universities, which produced many practicing astrologers, including Tycho Brahe (1546–1601) and Johannes Kepler.[27] Moreover, reacting partly to Pico's

---

A. Farmer, *Syncretism in the West: Pico's 900 Theses (1486)* (Tempe, Ariz.: Medieval and Renaissance Texts and Studies, 1998), pp. 151–76, whose reconstruction should be treated with caution.

[24] It is striking to note that both works use *spiritus* as the physiological quasi-material substance that connects the heavens to the earth. See Marsilio Ficino, *Three Books on Life*, ed. and trans. Carol V. Kaske and John R. Clarke (Binghamton, N.Y.: Medieval and Renaissance Texts and Studies, 1989), with an important introduction and commentary. For the classic discussion of this work, see Daniel P. Walker, *Spiritual and Demonic Magic from Ficino to Campanella* (London: Warburg Institute, 1958).

[25] For an excellent orientation to the complexities of Pico's thought, see Anthony Grafton, "Giovanni Pico della Mirandola: Trials and Triumphs of an Omnivore," in his *Commerce with the Classics: Ancient Books and Their Renaissance Readers* (Ann Arbor: University of Michigan Press, 1997), pp. 93–134, with a useful bibliography.

[26] For Bologna, see Angus Clarke, "Giovanni Antonio Magini (1555–1617) and Late Renaissance Astrology," Ph.D. dissertation, University of London, 1985. For Pisa, see Charles B. Schmitt, "The Faculty of Arts at Pisa at the Time of Galileo," and "Filippo Fantoni, Galileo Galilei's Predecessor as Mathematics Lecturer at Pisa," in his *Studies in Renaissance Philosophy and Science* (London: Variorum Reprints, 1981), pp. 243–72 and 53–62. For Padua, see Antonio Favaro, "I lettori di matematiche nella università di Padova dal principio del secolo XIV alla fine del XVI," *Memorie e documenti per la storia della università di Padova*, 1 (1922), 1–70. For Louvain ca. 1450–1580, see vanden Broecke, *Limits of Influence*, and his "Dee, Mercator, and Louvain Instrument Making: An Undescribed Astrological Disc by Gerard Mercator (1551)," *Annals of Science*, 58 (2001), 219–40. There is excellent scholarship reconstructing astrological culture in particular locations. For Vienna, see Helmut Grössing, *Humanistische Naturwissenschaft: Zur Geschichte der Wiener mathematischen Schulen des 15. und 16. Jahrhunderts* (Baden-Baden: Valentin Koerner, 1983); and Michael H. Shank, "Academic Consulting in Fifteenth-Century Vienna: The Case of Astrology," in *Texts and Contexts in Ancient and Medieval Science*, ed. Edith Sylla and Michael R. McVaugh (Leiden: E. J. Brill, 1997), pp. 245–71. For France, see Jean Patrice Boudet's important introduction to his edition of Simon de Phares, *Le recueil des plus célèbres astrologues de Simon de Phares*, ed. J.-P. Boudet, 2 vols. (Paris: H. Champion, 1997–99); Pierre Brind'Amour, *Nostrodame astrophile: Astrologie dans la vie et l'oeuvre de Nostradamus* ([Ottawa]: Presses de l'Université d'Ottawa, 1993); and Hervé Drevillon, *Lire et écrire l'avenir: Astrologie dans la France du Grand Siècle, 1610–1715* (Seyssel: Champ Vallon, ca. 1996).

[27] See Sachiko Kusukawa, *The Transformation of Natural Philosophy: The Case of Philip Melanchthon* (Cambridge: Cambridge University Press, 1995). Melanchthon was deeply influenced by his esteemed teacher at Tübingen, Johannes Stöffler, who played an important role in the predictions concerning the flood of 1524; see Zambelli, ed., "*Astrologi hallucinati*." Stöffler himself was deeply influenced by Albertus Magnus and his astrologizing Aristotelianism; see Wilhelm Maurer, *Der junge Melanchthon zwischen Humanismus und Reformation*, 2 vols. (Göttingen: Vandenhoeck and Ruprecht, 1967–69), 1: 136.

critique and partly to their own experience, Tycho and Kepler devoted much of their labors to reforming the astronomical and natural philosophical foundations of astrology in order to improve the system.[28]

In August 1563, while studying at the University of Leipzig, a German Lutheran university reformed under Melanchthon's guidance, the sixteen-year-old Tycho had observed an astrologically significant great conjunction of Saturn and Jupiter. Finding it fully a month off from the predictions in the tables, he resolved to establish more firmly the observational basis of the science of the stars, which he ultimately did by founding his famous observatory at Hven. In Tycho's view, a more accurate observational foundation would produce more accurate tables, which would lead in turn to locating the planets more accurately in annual ephemerides. On this basis, more accurate horoscopes could be constructed, which could then be more accurately interpreted. Tycho also proposed a reform of the mundane houses, a central feature of astrological interpretation, moving from a square figure with twelve house divisions to a round figure with eight. This reform, however, found limited acceptance.[29]

Although Kepler is better known for his *Astronomia nova* (New Astronomy) of 1609, which used Tycho's observations of Mars to introduce elliptical planetary orbits, he was also an avid astrologer (see Donahue, Chapter 24, this volume). Indeed, his dedication of this epoch-making text to his patron, the Holy Roman Emperor Rudolf II (1552–1612), associated Mars with Rudolf's natal horoscope. Following Kepler's lead six months later in his *Sidereus Nuncius* (Sidereal Messenger, 1610), Galileo dedicated his equally epoch-making telescopic discovery of Jupiter's four moons to his own patron, Cosimo II de' Medici, by emphasizing the exalted placement of Jupiter in Cosimo's natal horoscope.[30]

Although astrology had long been a key instrument in a courtier's toolkit,[31] these dedications were not merely isolated gestures of courtly compliment. As imperial mathematician (and in other less distinguished roles), Kepler engaged in various sorts of astrological practice, from weather prediction to providing counsel for the emperor.[32] In addition to reforming astronomical

---

[28] For fuller treatments, see Rutkin, "Astrology, Natural Philosophy, and the History of Science," chap. 7; and Bowden, "The Scientific Revolution in Astrology."

[29] Victor Thoren, *The Lord of Uraniborg: A Biography of Tycho Brahe* (Cambridge: Cambridge University Press, 1990). On Tycho's reforms, see John D. North, *Horoscopes and History* (London: Warburg Institute, 1986), pp. 175–7; and vanden Broecke, *Limits of Influence*, pp. 263–9.

[30] See Mario Biagioli, *Galileo, Courtier: The Practice of Science in the Culture of Absolutism* (Chicago: University of Chicago Press, 1993), pp. 127–33; and H. Darrel Rutkin, "Celestial Offerings: Astrological Motifs in Galileo's *Sidereus Nuncius* and Kepler's *Astronomia Nova*," in *Secrets of Nature: Astrology and Alchemy in Early Modern Europe*, ed. William R. Newman and Anthony Grafton (Cambridge, Mass.: MIT Press, 2001), pp. 133–72.

[31] See, for example, Frederick H. Cramer, *Astrology in Roman Law and Politics* (Philadelphia: American Philological Association, 1954), for imperial Rome.

[32] Max Caspar, *Kepler*, trans. C. D. Hellman (New York: Dover, 1993); Gérard Simon, *Kepler astronome astrologue* (Paris: Gallimard, 1979); and Barbara Bauer, "Die Rolle des Hofastrologen und

theory, Kepler also worked to shore up astrology's natural philosophical foundations, inspired in part by Pico's critique. Indeed, Kepler's rejection of the traditional astrological signification of the zodiacal signs and the mundane houses – and thus their panoply of derived doctrine, including planetary rulerships – was every bit as radical a move in astrology as his rejection of uniform circular motion in the realm of mathematical astronomy.[33]

Like Kepler, Galileo was active in astrology during most of his career as a student, teacher, and practitioner of the mathematical disciplines, though he does not seem to have shared Kepler's interest in fundamental astrological reform.[34] Most of our evidence for his astrological practice comes from his period as professor of mathematics at the University of Padua, where he constructed and interpreted horoscopes for both patrons and students.[35] One of his manuscripts contains twenty-seven autograph horoscopes, including ones for Giovanfrancesco Sagredo, one of his principal patrons, as well as for his own now famous daughters, Virginia and Livia, and for himself.[36] Indeed, Galileo had his first brush with the Inquisition at this time for supposedly practicing a *deterministic* astrology out of his house in Padua – a charge the Venetian authorities summarily dismissed.[37]

Francis Bacon proposed a serious reform of astrology in the *De augmentis scientiarum* of 1623, a Latin enlargement and reworking of his *Advancement of Learning* (1605), written during the forced retirement that followed his impeachment as Lord High Chancellor. Bacon began by identifying the many superstitions and lies that needed to be removed from its domain, including the individual planetary rule of each hour of the day and the astrological figure constructed for precise points of time. Reviewing the four principal divisions of astrological practice – revolutions, nativities, elections, and interrogations – he argued that the last three had little if any foundation, whereas he described revolutions as much more sound, though nonetheless in need of attention.[38]

Hofmathematicus als fürstlicher Berater," in *Höfischer Humanismus*, ed. August Buck (Weinheim: VCH, 1989), pp. 93–117.

[33] Judith V. Field, "A Lutheran Astrologer: Johannes Kepler," *Archive for History of Exact Sciences*, 31 (1984), 189–272; and Sheila J. Rabin, "Two Renaissance Views of Astrology: Pico and Kepler," Ph.D. dissertation, City University of New York, 1987.

[34] Galileo's interests in this area seem to have been confined to his discovery of Jupiter's four moons and his brief but persuasive arguments to establish their influence. See his letter to Piero Dini, dated 21 May 1611, in Galileo Galilei, *Le opere di Galileo Galilei*, 20 vols. in 21 (Florence: Giovanni Barbéra, 1890–1909), 11: 105–16.

[35] H. Darrel Rutkin, "Galileo, Astrologer: Astrology and Mathematical Practice in the Late-Sixteenth and Early-Seventeenth Centuries," *Galilaeana*, 2(2005), 107–43. See also Noel M. Swerdlow, "Galileo's Horoscopes," *Journal for the History of Astronomy* 35(2004), 135–41.

[36] Biblioteca Nazionale Centrale, Florence, *Galilaeana* 2(2005), 107–43. MS Galileiana 81.

[37] Antonino Poppi, *Cremonini, Galilei e gli inquisitori del Santo a Padova* (Padua: Centro Studi Antoniani, 1993).

[38] Francis Bacon, *De augmentis scientiarum*, in *The Works of Francis Bacon*, ed. J. Spedding et al., 14 vols. (London: Longmans, 1857–74; repr. Stuttgart: Frommann, 1963), 1: 554–60.

To make the doctrine of revolutions optimal in practice, Bacon prescribed five general rules: most notably, that the operation of the heavens is not strong on all kinds of bodies but only on finer ones, such as humors, air, and *spiritus*; and that every operation of the heavens extends more to the mass of things than to individuals, and only to larger spaces of time, not points of time or precise minutes. Retaining the traditional doctrines of planetary natures and aspects (the astrologically significant angular relationships between planets) – and affirming the principle that the heavenly bodies have in themselves other influences beyond heat and light[39] – Bacon turned to predictions. His considered views, after a lifetime as courtier and politician, are striking: Astrological predictions can profitably be applied to both natural and political domains, including weather, epidemics, and war, but only within a prescribed degree of specificity. Bacon made two important points in this regard, noting that the domain in which astrological predictions may be applied is virtually infinite, but that knowledge of the celestial configuration (the active – meaning efficient – causes) does not suffice.[40] One must also have a deep knowledge of the subjects at issue, the recipients of the active planetary influences. This knowledge seems to be precisely what Bacon's broader natural-historical research program was intended to provide.

In the final section of his discussion of astrology, Bacon responded to a hypothetical challenge by indicating how this research program should be instituted to develop and clarify his earlier claims. Rejecting the collection of future experiences as futile, he suggested historical research to investigate precisely those features already mentioned as legitimate subjects of astrological prediction and election. In particular, astrologers should comb the best histories to locate important events, which should then be compared with the heavenly situations at those times in accordance with his reforming precepts. Where there is a clear correlation, a rule of prediction may be inferred, he concluded, thus applying his inductive procedure to astrological reform.

I have seen no discussion of Bacon's own use of astrology, but given the tenor of his statements, his ever-precarious situation at court, and his passionate political ambitions, it seems likely that he would have pursued this avenue of possible insight – and thus advantage – in the many areas in which he actively participated.[41] Given his breadth and depth as a scholar, it also seems likely that he would have ventured his own historical investigation of these matters so clearly and forcefully advocated. At the very least, it is clear that Bacon, whose writings would have so much influence on the research of the Royal Society of London, provided strong encouragement for a program

---

[39] Ibid., 1: 556: "... quod nobis pro certo constet, Coelestia in se habere alios quosdam influxus praeter Calorem et lumen...."
[40] In this he followed Ptolemy in his *Tetrabiblos*, 1.2–3.
[41] See Lisa Jardine and Alan Stewart, *Hostage to Fortune: The Troubled Life of Francis Bacon* (New York: Hill and Wang, 1999), for a lively account, albeit with no reference to astrology.

of scientific research in a reformed astrology with a broad range of practical utility.

Bacon's vision of a reformed astrology was further developed by the English natural philosopher and alchemist Robert Boyle (1627–1691), one of the founders of the Royal Society of London.[42] In his "Suspicions about some Hidden Qualities of the Air" (1674), Boyle adapted Bacon's position, declaring that the luminaries, planets, and fixed stars likely emit subtle but corporeal emanations beyond light and heat, which reach to our air. Developing this idea in the appendix, "Of Celestial and Aerial Magnets," Boyle suggested experiments to test apparently good magnets by varying the air in which they were located according to different times, temperatures, and aspects of the planets. In this way, the different natures of the air could be discovered along with possible correspondences between the terrestrial and celestial realms.[43]

Another text long attributed to Boyle, a letter printed in his *General History of the Air* (1692) entitled "Of Celestial Influences or Effluviums in the Air," addressed to Samuel Hartlib, provided a fuller treatment of astrology's natural philosophical foundations. After stating that the purpose of accurate astronomical theory is to support accurate astrological practice, the author of this text – now accepted as Boyle's associate Benjamin Worsley (ca. 1620–1673) – discussed celestial influences that descended along with light to the thin and subtle air of earth's atmosphere.[44] Even more like light than the air, he argued, human spirits were all the more impressed and moved by celestial influences, which then affected human bodies. As I will show, these views continued to circulate under Boyle's name in Ephraim Chambers's vastly influential *Cyclopedia* (1728).

THE FATE OF ASTROLOGY

Although astrology received robust proposals for reform from major astronomers and natural philosophers throughout the sixteenth and seventeenth centuries, these ultimately were of no avail, and the field lost its intellectual legitimacy during the eighteenth century. This process is not yet fully

---

[42] See in general Rose-Mary Sargent, *The Diffident Naturalist: Robert Boyle and the Philosophy of Experiment* (Chicago: University of Chicago Press, 1995); and Lawrence Principe, *The Aspiring Adept: Robert Boyle and His Alchemical Quest* (Princeton, N.J.: Princeton University Press, 1998).

[43] [Benjamin Worsley], *Tracts: containing I. Suspicions about Some Hidden Qualities of the Air . . .* (London: W. G., 1674).

[44] The *General History of the Air* was published posthumously and edited by John Locke. Antonio Clericuzio has shown that this essay was in fact written in 1657 by Benjamin Worsley and that Boyle endorsed Worsley's views: See Clericuzio, "New Light on Benjamin Worsley's Natural Philosophy," in *Samuel Hartlib and Universal Reformation*, ed. Mark Greengrass, Michael Leslie, and Timothy Raylor (Cambridge: Cambridge University Press, 1994), pp. 236–46. This letter was republished under Boyle's name in his collected works of 1744: *Works of Honorable Robert Boyle*, ed. T. Birch, 5 vols. (London: A. Millar, 1744), 5: 124–8.

understood. One promising approach is to examine changing disciplinary patterns and their reflection in university curricula. Thus, in this section I will examine continuities and ruptures in this domain, focusing primarily on mathematics and natural philosophy, with a quick glance at medicine, in which astrology seems to have survived longest.

For mathematics, similar patterns may be observed in Italy and England. An important moment in astrology's separation from astronomy was astrology's explicit removal from mathematics instruction by the primary developer of and textbook writer for the mathematics curriculum of the vast Jesuit educational empire, Christoph Clavius (1537–1612), who discussed his decision to exclude astrology in his influential textbook version of Sacrobosco's *De sphaera* (first edition, 1570, with many later editions).[45] It should be noted that although Clavius eliminated astrology from the mathematics course, this does not imply that he thereby wholly rejected it. In addition, his action by no means indicated astrology's overall demise in mathematics education, even in Italy. For example, Giovanni Antonio Magini (1555–1617) taught astrology and wrote influential textbooks for mathematics and medical courses at the University of Bologna until his death in 1617, and Andrea Argoli taught astrology at the University of Rome, *La Sapienza*, from 1622 to 1627, and then at the University of Padua from 1632 to 1656, where he also wrote mathematical textbooks and compiled ephemerides.[46]

These conflicting patterns also appear in England, where the 1619 Savilian statutes of Oxford University expressly forbade the newly established professor of astronomy from teaching the construction of nativities or judicial astrology.[47] There is significant evidence, nevertheless, that astrology was studied and taught at English universities up to and beyond this time.[48] Astrology was also found in non-university mathematics textbooks, including those of Charles II's hydrographer, Joseph Moxon, whose *A Tutor to Astronomy and Geography* (1674) taught the easy construction of horoscopes.[49]

These contradictory impulses were embodied in the activities of John Flamsteed (1646–1719), the first Astronomer Royal and Regius Professor of

---

[45] See James M. Lattis, *Between Copernicus and Galileo: Christoph Clavius and the Collapse of Ptolemaic Cosmology* (Chicago: University of Chicago Press, 1994).

[46] See Clarke, "Giovanni Antonio Magini." For Argoli, see the article, s.v., *Dizionario biografico degli Italiani* (Rome: Treccani, 1960–), 4 (1962), pp. 132–4.

[47] Mordechai Feingold, "The Occult Tradition in the English Universities," in *Occult and Scientific Mentalities in the Renaissance*, ed. Brian Vickers (Cambridge: Cambridge University Press, 1984), pp. 73–94, at p. 78 (with n. 25).

[48] See, for example, Hilary M. Carey, *Courting Disaster: Astrology at the English Court and University in the Later Middle Ages* (Houndsmill: Macmillan, 1992); and Mordechai Feingold, *The Mathematicians' Apprenticeship: Science, Universities, and Society in England, 1560–1640* (Cambridge: Cambridge University Press, 1984).

[49] Joseph Moxon, *A Tutor to Astronomy and Geography* (London: Thomas Roycroft, 1674). I use the reprint: New York: Burt Franklin, 1968, pp. 122–35. In 1682–3, Moxon, who became a Fellow of the Royal Society in 1678, briefly revived the Society of Astrologers, which had become moribund since its heyday in the Interregnum. See Curry, *Prophecy and Power*, p. 77.

Astronomy at the newly erected Royal Greenwich Observatory.[50] Soon after completing an intensive critique of astrology in 1673 – which he conspicuously never published – Flamsteed constructed a horoscope for the foundation of the Greenwich Observatory (1675). Indeed, he and Edmund Halley both provided the most accurate astronomical data available for the tables printed in George Parker's popular astrological almanacs in the 1690s.[51]

We can also trace the removal of astrology from the mathematical disciplines through an important printed genre, ephemerides: multiyear collections of planetary tables. Because sixteenth- and seventeenth-century ephemerides were overwhelmingly astrological – including (inter alia) extensive astrological introductions, horoscopes for the year (useful for making annual prognostications) and tables of planetary aspects – one may ask when this astrological component disappeared?[52] The several Bolognese ephemerides I have examined through 1666 are all astrological, as are Argoli's for Rome and Padua, which provide tables with horoscopes through 1700.

In the first quarter of the eighteenth century, however, there are signs of change. In the introduction to his *Ephemerides motuum coelestium* (Ephemerides of Celestial Motions) for 1715 to 1725, Eustachio Manfredi stated that he was deliberately removing the stain, astrology, from his ephemeris, unlike his predecessors Regiomontanus, Magini, and Kepler. Only tables of planetary aspects remained.[53] Once again, this does not mean that astrology was finished at Bologna, as we can see in Antonio Ghisilieri's *Ephemerides motuum coelestium* for 1721–56, drawn from the tables of de la Hire, Streete, and Flamsteed (1720).[54] Ghisilieri's ephemerides provided horoscopes for eclipses and the four cardinal points of the year (namely, the two solstices and equinoxes), as well as full tables of planetary aspects. Of most interest, however, is Ghisilieri's extensive introductory letter to the reader, wherein he outlined his own reform of astrology, discussing, among others, Descartes and Newton. By 1750, the only remaining astrological vestiges in the Bolognese ephemerides I have examined were the planetary aspects,[55] whereas the first English naval ephemerides (1767) have nothing astrological

---

[50] Flamsteed's complex and ambivalent relation to astrology has been studied in Michael Hunter, "Science and Astrology in Seventeenth-Century England: An Unpublished Polemic by John Flamsteed," in *Astrology, Science, and Society: Historical Essays*, ed. P. Curry (Woodbridge: Boydell Press, 1987), pp. 261–300.

[51] Hunter, "Science and Astrology," pp. 250–1.

[52] I focus here on Italian ephemerides, primarily from Bologna. Other ephemerides that I have examined from France, Germany, England, and elsewhere in Italy corroborate this picture. I used the collections of the Houghton Library of Harvard University, the Burndy Library of the Dibner Institute, and especially Owen Gingerich's private collection. My sincere thanks to Professor Gingerich for granting me access to his important collection of astronomical and astrological publications.

[53] Eustachio Manfredi, *Ephemerides motuum coelestium* (Bologna: Constantinus Pisanus, 1715).

[54] Antonio Ghisilieri, *Ephemerides motuum coelestium* (Bologna: apud successores Benatti, 1720).

[55] Eustachio Zanotti, *Ephemerides motuum caelestium ex anno 1751 in annum 1762: ad meridianum Bononiae ex Halleii tabulis supputatae . . . ad usum Instituti* (Bologna: C. Pisarri, 1750); Eustachio Zanotti, *Emphemerides motuum caelestium ex anno 1775 in annum 1786: ad meridianum Bononiae ex Halleii tabulis supputatae . . . ad usum Instituti* (Bologna: Laelius a Vulpe, 1774).

in them whatsoever.⁵⁶ Further research should clarify the chronological outline indicated here.

Astrology's removal from the natural philosophy curriculum was also complex, as we can see from popular textbooks by the Protestant Johannes Magirus (first edition, 1597) and the Jesuit Coimbra Aristotle commentaries, especially the *De caelo* (first edition, 1592), both of which discussed celestial influences at the appropriate Aristotelian loci. These works were influential throughout the seventeenth century. Magirus's *Physiologia peripatetica* (Aristotelian Natural Philosophy) came out in sixteen editions through 1642, and the Coimbra commentary on *De caelo*, taught within and beyond the Jesuit educational network, saw thirteen editions by 1631.⁵⁷ Major figures such as René Descartes (1596–1650) and Marin Mersenne (1588–1648) learned their natural philosophy in this manner at the Jesuit college of La Flèche.⁵⁸ Moreover, as an undergraduate at Trinity College, Cambridge, in 1661, Newton began his study of Aristotelian natural philosophy by reading Magirus, before turning to Descartes.⁵⁹ Magirus continued to be read at Cambridge into the 1670s, at Harvard into the 1690s, and at Yale into the 1720s.⁶⁰

Descartes and Newton, however, lead us into a new phase in the history of science and philosophy, with the systematic formulation of competing traditions of inquiry that would ultimately complete the uprooting and replacement of the Aristotelian-Ptolemaic worldview. Although Newton was a devoted alchemist, he and Descartes were both antiastrological, or, perhaps more accurately, nonastrological,⁶¹ even though there came to be Cartesian astrologers (for example, Claude Gadrois)⁶² and Newtonian ones, at least after a fashion, as we will see with Richard Mead. This raises an important question: Because mechanisms that stressed the impact of celestial influences on subtle effluvia should have worked well for both Descartes and Newton, why did they not embrace a reformed astrologizing natural philosophy along

---

⁵⁶ *Nautical Almanac and Astronomical Ephemeris for the Year 1767* (London: Commissioners of Longitude, 1766).

⁵⁷ See Charles H. Lohr, *Latin Aristotle Commentaries II: Renaissance Authors* (Florence: Leo S. Olschki, 1988), pp. 235–6 (Magirus) and pp. 98–9 (Conimbricensis Collegii Societatis Jesu commentarii). For their influence, see P. Reif, "Natural Philosophy in Some Early Seventeenth-Century Scholastic Textbooks," Ph.D. dissertation, St. Louis University, St. Louis, Mo., 1962; in general, see Charles B. Schmitt, "The Rise of the Philosophical Textbooks," in Schmitt et al., eds., *The Cambridge History of Renaissance Philosophy*, pp. 792–804.

⁵⁸ Peter R. Dear, *Mersenne and the Learning of the Schools* (Ithaca, N.Y.: Cornell University Press, 1988); and Stephen Gaukroger, *Descartes: An Intellectual Biography* (Oxford: Clarendon Press, 1995).

⁵⁹ J. E. McGuire and Martin Tamny, *Certain Philosophical Questions: Newton's Trinity Notebook* (Cambridge: Cambridge University Press, 1983), pp. 15 ff.

⁶⁰ Samuel E. Morison, *Harvard College in the Seventeenth Century*, 2 vols. (Cambridge, Mass.: Harvard University Press, 1936), 1: 226–7. Astrology was finally removed from Harvard disputations in 1717; ibid., 1: 214.

⁶¹ For Descartes, see Bowden, "Scientific Revolution in Astrology," pp. 197–8. The references for Newton are given in note 68. For the unexpected disjunction between alchemy and astrology, see William R. Newman and Anthony Grafton, "Introduction: The Problematic Status of Astrology and Alchemy in Premodern Europe," in *Secrets of Nature*, pp. 1–37, at pp. 14–27.

⁶² Bowden, "Scientific Revolution in Astrology," pp. 198–202.

the lines laid down by Bacon and Boyle? Perhaps the effluvial mechanisms did not work at the great distances now understood to exist in the solar system.[63] Unfortunately (as far as I know), neither Descartes nor Newton explicitly discussed their views on astrology – a fact significant in its own right.

An important figure in the acceptance of the antiastrological stance of Descartes – and also, ironically, of Newton – was the influential textbook writer Jacques Rohault (1620–1675), who published his Cartesian *Traité de Physique* in 1671, with at least ten editions in Paris alone by 1730. He explicitly rejected astrology in a chapter entitled "Of the Influences of the Stars, and of Judicial Astrology,"[64] where he asked "whether the stars act as the cause of or contribute to effects we see produced on earth." Because he granted causal efficacy to the sun's light and heat (as did Pico), the question applied only to the other stars, whose light moves only our optical fibers and elements in the air. Because the sun's light is infinitely greater than all the stars put together, Rohault argued, also echoing Pico, it should be understood as the sole cause of all these effects. Rohault noted further, thus dissenting from the views of his contemporary Robert Boyle, that if the air is different at different times, we should not look for the cause of this in the heavens but in the air itself or the earth.

After this brief natural-philosophical critique, Rohault addressed judicial astrology, simply asserting that it cannot be proved by any reason. Turning to experience, he argued that astrological rules cannot be based on one-time experiences because it would take many thousands of years for the same celestial configuration to recur to test the rule. This is impracticable, as Bacon had also understood. Furthermore, even if it were possible, the rule would apply only to the country where the experience happened, not universally. Finally, even if the astrologers are sometimes correct in their predictions, this is attributable only to luck. Rohault's conclusion leaves no doubt about his definitive rejection of astrology: "Not to insist any longer upon this Subject, which does not deserve to have any more said of it, and which is not worth being seriously treated by any Philosopher."

Rohault's Latin translator and Newton's disciple, Samuel Clarke (1675–1729), appended only one explanatory footnote to this entire section; namely, to clarify the nature of the moon's influence:

> But as to the real Power of the Moon; since it is evident, that it causes a greater Flux and Reflux in the Air than in the Sea, it must certainly produce some

---

[63] Albert Van Helden, *Measuring the Universe: Cosmic Dimensions from Aristarchus to Halley* (Chicago: University of Chicago Press, 1985).

[64] The references are to John Clarke's English translation of Samuel Clarke's Latin translation with Newtonian commentary of Rohault's *Treatise*; *Rohault's System of Natural Philosophy* . . . , 2 vols. (London: James Knapton, 1723), reprinted in Jacques Rohault, *A System of Natural Philosophy*, 2 vols. (The Sources of Science, 50) (New York: Johnson Reprint Co., 1969), with an informative introduction by Larry Laudan. The chapter (*System*, 2.27), appears in 2: 86–91. My thanks to Domenico Bertoloni Meli for insisting on Rohault's importance.

Alterations in the Temperature of the Heavens, and this may make some Alteration in the Bodies of Animals. But as to any other Effects commonly ascribed to the Moon and Planets, beyond what are owing to these Causes, they are meer Trifles.[65]

Rohault's textbook had an extraordinary influence both in its original French editions and in the Clarke brothers' Latin (1697) and English (1723) annotated translations.[66] Samuel Clarke's three later Latin editions, culminating in 1713, became increasingly Newtonian as he augmented his apparatus of critical explanatory notes, attacking Rohault's Cartesian exposition from the foot of the page. The English translation, which incorporated these notes, served at least into the 1740s as a natural philosophy textbook with an antiastrological bent at Cambridge, as well as Harvard and Yale.[67]

Beyond Rohault's textbook, Newton himself contributed to astrology's undoing, although this was not necessarily his intent. Although few contemporaries actually read his *Philosophiae naturalis principia mathematica* (Mathematical Principles of Natural Philosophy, 1687), Newton's synthesis seems to have undercut the natural-philosophical foundations of astrology in two fundamental and interrelated ways, one natural-philosophical and the other heuristic. First, he did away with the entire hylemorphic fourfold causal framework with Aristotle and his followers' characteristic views of substance, accidents, and change (see Joy, Chapter 3, this volume). Without this causal structure, the natural-philosophical foundations for astrology simply disappeared. Heuristically, Newton radically reoriented the nature of legitimate scientific questions, focusing first of all on what can be observed, quantified, and measured.[68]

Astrology seems to have persisted in medicine longer than in any other branch of learning. Richard Mead (1673–1754), Newton's and Halley's personal physician, a graduate of the University of Padua, wrote *De imperio solis et lunae* (Concerning the Influence of the Sun and Moon upon Human Bodies and the Diseases thereby Produced) in 1704, with a revised edition in 1748 after forty years of medical practice.[69] In this work, Mead grafted a Newtonian iatromechanical explanatory framework onto traditional astrologically informed Galenic praxis, but without the usual emphasis on natal

---

[65] Rohault, *Rohault's System*, 2: 90, n. 1.
[66] Laudan, introduction to Rohault, *Rohault's System*, 1: ix–x.
[67] Ibid., 1: x–xiii.
[68] This can be seen clearly in query 31 of the *Opticks*. See Isaac Newton, *Opticks; or, a Treatise of the Reflections, Refractions, Inflections and Colours of Light* (based on the fourth edition, London, 1730; repr. New York: Dover, 1952), pp. 401–2. Newton assiduously promoted this reoriented research program as president of the Royal Society from 1703 to 1727. See John L. Heilbron, *Physics at the Royal Society during Newton's Presidency* (Los Angeles: William Andrews Clark Memorial Library, 1983); and Richard S. Westfall, *Never at Rest: A Biography of Isaac Newton* (Cambridge: Cambridge University Press, 1980).
[69] It was also published posthumously in his collected works (1775).

horoscopes or medical interrogations.[70] One of his principal authorities in this area was John Goad, who published the rich results of his extensive astrologically informed meteorological observations in his *Astro-Meteorologica* (1686).[71] Later still, Franz Anton Mesmer (1733–1815), who received his medical degree from the University of Vienna in 1766, wrote a dissertation analyzing the influences of the heavens on the health of human beings. Inspired primarily by Mead's treatise, this dissertation was informed by the latest theories of gravitation, magnetism, and electricity and discussed Newton's views in the *Principia*.[72]

THE EIGHTEENTH CENTURY AND BEYOND

In the preface to his popular *Cyclopedia* (1728), Ephraim Chambers declared that he would treat astrology because, in addition to its erstwhile popularity, it still contained something worth preserving:

> The heavenly Bodies have their Influences: The Foundation, therefore, of Astrology is good: but those Influences are not directed by the Rules commonly laid down, nor produce the Effects attributed to 'em: so that the Superstructure is false. Astrology, therefore, ought not to be exploded, but reformed.[73]

He concluded the prefatory statement with a sympathetic nod to Bacon's astrological reform as an *astrologia sana*.

In his article on astrology, which he defined as "the Art of foretelling future Events, from the Aspects, Positions, and Influences of the Heavenly Bodies," Chambers clearly distinguished natural from judicial astrology.[74] In discussing the former, he argued for fundamental features of astrology's natural philosophical foundations. Noting that planetary configurations affect the weather, Chambers cited John Goad. Noting that heavenly bodies affect all physical bodies, in both their rarefaction and condensation and their generation and corruption (the old Aristotelian categories), he drew on Robert

---

[70] Anna Marie Roos, "Luminaries in Medicine: Richard Mead, James Gibbs, and Solar and Lunar Effects on the Human Body in Early Modern England," *Bulletin of the History of Medicine*, 74 (2000), 433–57.

[71] On Goad, see Curry, *Prophecy and Power*, pp. 67–72; and Bowden, "Scientific Revolution in Astrology," pp. 176–87.

[72] See the introduction to Franz Anton Mesmer, *Mesmerism: A Translation of the Original Scientific and Medical Writings of F. A. Mesmer*, trans. G. Bloch (Los Altos, Calif.: W. Kaufmann, 1980); and Robert Darnton, *Mesmerism and the End of Enlightenment in France* (Cambridge, Mass.: Harvard University Press, 1968). Much additional research is needed to complete the story of the fate of astrology in learned medicine during the Enlightenment.

[73] Ephraim Chambers, *Cyclopedia; or, An Universal Dictionary of Arts and Sciences* . . . (London: J. and J. Knapton, etc., 1728), 1: xxviii. My thanks to Richard Yeo for directing me to this material; see in particular his *Encyclopedic Visions: Scientific Dictionaries and Enlightenment Culture* (Cambridge: Cambridge University Press, 2001).

[74] Chambers, *Cyclopedia*, 1: 162–4.

Boyle's *History of the Air*. Finally, Chambers argued that each planet has its own specific quality of light and that the angular relationship of the planets to each other, and to the Earth – the astrological aspects – thus have a particular qualitative effect on sublunary things.

Chambers's treatment of judicial astrology, on the other hand, was wholly dismissive: "Judiciary, or Judicial ASTROLOGY, which is what we commonly call Astrology, is that which pretends to foretell moral Events; i.e. such as have a Dependence on the Will and Agency of Man; as if they were directed by the Stars." Referring to it as superstition, he declared that "the chief Province now remaining to the modern Professors, is the making of Calendars or Almanacks."[75] Chambers's treatment of astrology remained exactly the same at least through the revised edition of 1741, and his discussion, translated into French, was taken over whole in the *Encyclopédie*'s first edition of 1751 by Denis Diderot and Jean d'Alembert.[76] By 1786, however, in Abraham Rees's posthumous edition of Chambers's *Cyclopedia*, the treatment had changed significantly. Although the prefatory discussion remained the same, the description under the term was significantly reduced. Chambers's three original authorities, Boyle, Goad, and Mead, were reduced to Boyle alone.[77]

Chambers's reforming treatment should be sharply contrasted with the abruptly dismissive definition in the *Encyclopaedia Britannica*'s first edition of 1768, which I quote in full:

> ASTROLOGY, a conjectural science, which teaches to judge of the effects and influences of the stars, and to foretel future events by the situation and different aspects of the heavenly bodies. This science has long ago become a just subject of contempt and ridicule.[78]

Taken as a group, these eighteenth-century encyclopedias illustrate, albeit with some delay, the various trends already visible in more specialized seventeenth-century discussions of astrology: its severing from astronomy among the mathematical disciplines and its elimination from the mathematical curriculum; the gradually widening fissure between natural and judicial astrology, together with attempts to incorporate natural astrology into the new sorts of natural philosophy; and a concomitant rejection of judicial astrology's panoply of rules and aims. These are fundamental transformations in the transition from premodern to modern disciplinary configurations.

---

[75] Ibid., 1: 163.
[76] Denis Diderot and Jean d'Alembert, *Encyclopédie, ou Dictionnaire raisonné des arts et des métiers* (Paris: Briasson, etc., 1751), 1: 780–3.
[77] Ephraim Chambers, *Cyclopedia . . . with the Supplement, and Modern Improvements . . . by Abraham Rees* (London: J. F and C. Rivington, 1786).
[78] *Encyclopaedia Britannica; or, a Dictionary of Arts and Sciences . . .* , 3 vols. (Edinburgh: A. Bell and C. Macfarquhar, 1771), 1: 433.

These patterns of rejection and reconfiguration do not, however, appear to extend to lay culture. From 1640 on, even as learned writers increasingly limited astrology's claim to explanatory power, translation of astrological texts into English made in-depth knowledge of astrology accessible to a much broader social spectrum than before.[79] Thus, what remained of judicial astrology in the mid-eighteenth century seems to have found a home primarily within popular culture, especially in almanacs, many of which retained their astrological character.[80] This situation changed again in the nineteenth century, when astrology was integrated into a new configuration of the so-called occult sciences (which included magic, alchemy, witchcraft, and the kabbalah), where we still find it today.[81]

Although I have traced the outlines of the process whereby astrology disappeared from the domain of legitimate natural knowledge as defined by the learned, some of the biggest questions still remain. In particular, why was such a promising astrological synthesis uprooted during the seventeenth and eighteenth centuries? In addition to natural philosophical motivations, part of the answer must also lie in the political domain, where the role played by astrology – still vital in the seventeenth century – lost its importance in the eighteenth.[82] Likewise, astrology increasingly became the brunt of biting satire at this time, as in Jonathan Swift's vehement attack on the real astrologer John Partridge under the fictional guise of another astrologer, Isaac Bickerstaff.[83] This seems to have reflected and reinforced the increasing vulgarization of astrology and its demotion from the world of elite to popular culture.[84]

Although finally severed from its intellectual and institutional roots in Aristotelian-Ptolemaic-Galenic natural knowledge and university culture, astrology continued to exist as a system of thought, even after the middle of

---

[79] See Curry, *Prophecy and Power*, pp. 19 ff.
[80] Capp, *Astrology and the Popular Press*, pp. 238–69. For the somewhat different situation in Italy, see Casali, *Le spie del cielo*, pp. 203–70.
[81] This configuration is reflected in Wayne Shumaker, *The Occult Sciences in the Renaissance: A Study in Intellectual Patterns* (Berkeley: University of California Press, 1972).
[82] For the astrological political think-tank run by Orazio Morandi in Rome in the 1620s, see Brendan Dooley, *Morandi's Last Prophecy and the End of Renaissance Politics* (Princeton, N.J.: Princeton University Press, 2002), which must be treated with caution. For France, see Drevillon, *Lire et écrire l'avenir*. For the great role astrological prophecy played in the English Civil War and Interregnum of the 1640s and 1650s, see Ann Geneva, *Astrology and the Seventeenth Century Mind: William Lilly and the Language of the Stars* (Manchester: Manchester University Press, 1995). For Elias Ashmole and William Lilly as Charles II's court astrologers in the 1660s and 1670s, see C. H. Josten, *Elias Ashmole (1617–92) . . .* , 5 vols. (Oxford: Clarendon Press, 1966). For the eighteenth century, see Curry, *Prophecy and Power*, pp. 95–137.
[83] For this episode, see Curry, *Prophecy and Power*, pp. 89–91. For the general movement "from popular to plebeian," see Curry's section under this title, pp. 109–17.
[84] For a parallel discussion of this process as it concerned prodigies and natural wonders, see Lorraine Daston and Katharine Park, *Wonders and the Order of Nature, 1150–1750* (New York: Zone Books, 1998), chap. 9.

the eighteenth century, as it still does today.[85] Nonetheless, the sixteenth and seventeenth centuries marked a crucial turning point in learned attitudes toward astrology and in its role in learned understandings of the natural world. The exclusion of astrology from these understandings was intimately connected with the undermining and removal of Aristotelian natural philosophy itself and its eventual replacement by Newtonian science.

[85] For a contemporary example of astrological practice in the highest corridors of power, see Joan Quigley, *What Does Joan Say?: My Seven Years as White House Astrologer to Nancy and Ronald Reagan* (Secaucus, N.J.: Carol Publishing, 1990). People still try to justify astrology scientifically, perhaps most famously Michel Gauquelin, who has written such books as: *Birthtimes: A Scientific Investigation of the Secrets of Astrology*, trans. S. Matthews (New York: Hill and Wang, 1983); *Cosmic Influences on Human Behavior*, trans. J. E. Clemow (London: Garnstone, 1974); and *The Scientific Basis of Astrology*, trans. J. Hughes (New York: Stein and Day, 1969).

24

ASTRONOMY

William Donahue

In the late Middle Ages, astronomy, unlike most other natural sciences now recognized, had been studied and practiced for over two millennia. Together with the other ancient sciences of harmonics, optics, and mechanics, it was considered to be a mixed mathematical science, differing from the pure mathematical sciences – arithmetic and geometry – in that astronomy considered number and magnitude in bodies and not in themselves. In the application of this division (which was not always strictly followed), astronomy could only develop and apply mathematical hypotheses: Pronouncements about the true nature of the heavens lay within the province of natural philosophy.[1] Thus astronomers were not recognized as having the authority to decide whether the earth is moving or at rest, or whether comets are celestial or atmospheric. Astronomy's function was only to describe the apparent positions of the heavenly bodies for the purposes of timekeeping, calendar making, and prediction of celestial influences. (This last task was the function of astrology, which was a respected science in the late Middle Ages, dealing with the effects of the celestial motions, just as natural philosophy treated its causes.)

This division of the science was established on philosophical grounds, and was used by philosophers and physical theorists to keep astronomy and the other mathematical sciences in their place. Astronomers, on the other hand, were never entirely content with their marginalization, and, while they improved the predictive power of their science, they strove to show the natural philosophers that the claims of astronomy could not be ignored. In this the astronomers achieved remarkable success, and by the end of the seventeenth century, astronomy had become a near neighbor, or even a branch, of natural philosophy, even as natural philosophy itself came to be mathematized (see Blair, Chapter 17, this volume). At the same time, astrology gradually lost its

---

[1] See, for example, Aristotle, *Physics*, 2.2 (194a7); and James A. Weisheipl, "The Nature, Scope, and Classification of the Sciences," in *Science in the Middle Ages*, ed. David C. Lindberg (Chicago: University of Chicago Press, 1978), esp. pp. 474–80.

academic respectability, largely because of its inability to adapt to the new astronomy and its adherence to an increasingly antiquated cosmology (see Rutkin, Chapter 23, this volume).

Both astronomy and astrology were academic subjects, taught and studied in the universities (though often outside of the regular curriculum). Although the astronomical tradition at the University of Paris, which had flourished in the thirteenth and early fourteenth centuries, had languished by 1500, the teaching of astronomy thrived at Oxford, Cracow, Prague, Vienna, and Bologna. Vienna in particular developed a strong astronomical tradition, beginning in the fourteenth century with Henry of Langenstein (ca. 1325–1397) and continuing with John of Gmunden (ca. 1380–1442), Georg Peurbach (1423–1469), and Johannes Regiomontanus (1436–1476). The latter two figure prominently in this chapter.[2]

The practice of astronomy, which was as a rule ancillary to the practice of astrology, was in contrast often carried on outside the universities. Rulers needed astrological predictions, which required knowledge of planetary positions, and physicians used the stars to plot the likely course of a disease. There was thus employment for astronomers at court, and the medical use of astrology greatly promoted the study and teaching of astronomy. Indeed, many of the most prominent astronomers of the period – one need mention only Nicholas Copernicus (1473–1543), Tycho Brahe (1546–1601), Johannes Kepler (1571–1630), and Galileo Galilei (1564–1642) – worked primarily or entirely outside the universities although all were university-educated (see Moran, Chapter 11, this volume).

Although there were numerous remarkable developments in astronomy in the sixteenth and seventeenth centuries, a broad theme that was common to many of them was the trend toward treating things in the sky as physical objects no different in principle from terrestrial objects. In this trend, there were two central figures: Galileo and Kepler. Galileo's first telescopic observations, published in *Sidereus nuncius* (Sidereal Messenger, 1610), gave a view of the heavens that was never before possible, affording close-up scrutiny of stars and planets. Kepler's theory of Mars's motion, *Astronomia nova* (New Astronomy, 1609), first introduced physical forces into a mathematically precise predictive apparatus. Each of these publications had a profound effect on the development of astronomy in the seventeenth century, but both in turn occurred in the context of well-developed astronomical and physical traditions.

Accordingly, this chapter begins by describing the sixteenth-century context in which questions about the heavens were formulated in the crucial period around 1610. Especially important was the significant role played by humanism in the works of Copernicus and other astronomical reformers, which recovered alternative ancient philosophical traditions that rivaled

[2] Olaf Pedersen, "Astronomy," in Lindberg, ed., *Science in the Middle Ages*, pp. 329–30.

those of the Aristotelian schools and gave the mathematical sciences a more decisive role.

The themes that developed in the sixteenth century played out in different ways. Galileo's observations created an immediate sensation and were soon repeated and elaborated by other observers. Kepler's difficult theories, in contrast, were only gradually and partially accepted as their accuracy became recognized. The cosmological ideas of René Descartes (1596–1650) also did much to further the acceptance of physical arguments in astronomy. At the same time, the discovery of several evidently "new" stars (*novae*) prompted a search for more, which led to the discovery of variable stars and to the development of stellar astronomy as a distinct field of study. This chapter concludes with the publication of Isaac Newton's (1642–1727) *Principia mathematica philosophiae naturalis* (Mathematical Principles of Natural Philosophy, 1687 and 1713), which completely changed the way planetary theory was to develop.

ASTRONOMICAL EDUCATION IN THE EARLY SIXTEENTH CENTURY

In the early sixteenth-century university, astronomy was usually taught in two courses. The introductory course was based on the thirteenth-century *Sphere* of Johannes de Sacrobosco (d. 1244 or 1256), which described the parts of the celestial and terrestrial globes. More advanced instruction usually began with a study of a work called *Theorica planetarum* (Theories of the Planets), often attributed to Gerard of Cremona (1114–1187).[3] The *Theorica* and the tables derived from them were based on planetary models developed by the second-century Alexandrian astronomer Claudius Ptolemaeus (Ptolemy), sometimes with additions or modifications by later astronomers. In Ptolemy's models, a planet (P in Figure 24.1) moves on a circle (the epicycle, PQ in Figure 24.1) whose center, F is carried around a larger circle (the deferent, ABF), whose center, C, is fixed at a point near the earth (located at D). The center of the epicycle's motion on the deferent is not uniform about the deferent's center but moves with uniform angular motion about another point (the equant, E in Figure 24.1), which is twice as far from the earth, D, as the deferent's center, C (that is, $ED = 2CD$), and in the same line (the apse line, AB). Despite the availability of Ptolemy's major work, the *Almagest*, in Latin translations

---

[3] The *Sphere* is translated in Lynn Thorndike, *The Sphere of Sacrobosco and Its Commentators* (Chicago: University of Chicago Press, 1949). Selections are included in Edward Grant, ed., *A Source Book in Medieval Science* (Cambridge, Mass.: Harvard University Press, 1974), pp. 442–51. A complete translation (by Olaf Pedersen) of *Theoricae planetarum* is included in Grant, ed., *Source Book*, pp. 451–65. For more on the *Sphere*, the *Theoricae*, and astronomy in general, see the chapter on "Mediaeval Astronomy" in Olaf Pedersen and Mogens Pihl, *Early Physics and Astronomy: A Historical Introduction* (History of Science Library) (London: MacDonald, 1974), pp. 243–77.

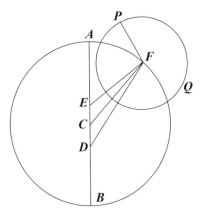

Figure 24.1. Ptolemaic planetary model.

from both Arabic and the original Greek versions, it was seldom studied in university courses because of its difficulty.

The *Theorica* presented models of planetary motions in magisterial style, without explaining how they were arrived at or even whether they should be considered to represent anything real. An important advantage of this text was that it allowed the student to begin using astronomical tables, usually for astrological purposes, almost immediately. Such tables, in their simplest form, would give mean motions, positions of nodes and apses, and other parameters as a function of time, presented at convenient intervals. The practitioner would have to use one set of tables to find the position of the center of the epicycle on the deferent, then another to find the planet's place on the epicycle, and would then find the apparent position from these two by trigonometry. More advanced tables, such as the early fourteenth-century revision of the thirteenth-century Alfonsine Tables by John of Lignières, combined the two operations into a single large table, a great convenience. Early introduction to the use of tables was especially useful for medical students, who would need astrology for diagnostic and prognostic purposes.[4]

RENAISSANCE HUMANISM AND *RENOVATIO*

Peurbach, though often regarded as the first Renaissance astronomer, continued to develop the traditions of medieval astronomy. Nevertheless, in one respect at least, his work represented a departure from that of his predecessors. He was as much a humanist classicist as a mathematician, and endeavored to restore ancient learning from the ravages of scholastic mishandling. His

---

[4] Nancy G. Siraisi, *Medieval and Early Renaissance Medicine: An Introduction to Knowledge and Practice* (Chicago: University of Chicago Press, 1990), pp. 67–8. For the Alfonsine Tables, see John North, *The Norton History of Astronomy and Cosmology* (New York: W. W. Norton, 1994), pp. 217–22.

most widely read work, *Theoricae novae planetarum* (New Theories of the Planets, 1472), aimed at replacing the medieval *Theorica planetarum*. In this, Peurbach was spectacularly successful: The *Theoricae novae* was in use well into the seventeenth century, with over fifty editions, translations, and commentaries published between 1472 and 1653.[5]

Peurbach had hoped to do more than just replace the standard textbook. Encouraged by his friend Cardinal Basilius Bessarion (ca. 1395–1472), he began work on a summary of Ptolemy's *Almagest* that was designed to introduce students to the study of the *Almagest* itself (although Peurbach was still relying on Gerard of Cremona's Latin retranslation from an Arabic version). Bessarion, Greek by birth, was papal legate to the imperial court in Vienna, where, in addition to his diplomatic duties, he worked to encourage the study of Greek classics. Peurbach died leaving this work unfinished, but it was taken up by his student Johannes Müller (1436–1476) of Königsberg (in what is now northern Bavaria), better known by his adopted surname Regiomontanus (Latinized from his birthplace). The *Epitome in Almagestum Ptolemaei* (Epitome on Ptolemy's Almagest), as the work was called, finally appeared in 1496. It was not just a summary of Ptolemy's masterpiece but included later observations, new computations, and some critical comments.[6] The classical tradition was thus being revived as a living body of theory, to be built upon or altered as circumstances required.

The importance of relating astronomy to the *Almagest* rather than to the tradition of the *Theorica* lay chiefly in the different ways of approaching the subject. Whereas the *Theorica* presented completed models without describing how they were constructed, Ptolemy showed how the orbital parameters could be derived from observations. This gave students of astronomy the ability to criticize existing models and revise or replace them.

A further instance of the humanist attempt not just to emulate classical models but to enter into their spirit and to improve them was the revival of the use of homocentric spheres in astronomy. Since the time of Ptolemy, an inconsistency had existed between Aristotle's physical account – homocentric because it restricted heavenly bodies to motion around the center of the universe (with which the center of the earth coincides) – and Ptolemy's use of eccentric circles and epicycles, which implied a very considerable variation in earth–planet distances. Aristotle had mentioned the efforts of Eudoxus to construct a viable homocentric theory, and Calippus's improvements of Eudoxus's models, without providing any details.[7] Some Islamic astronomers and philosophers, most notably al-Bitrūjī (ca. 1100–1185), known to the West by his Latinized name, Alpetragius, attempted a revival of the homocentrics.

---

[5] E. J. Aiton, "Peurbach's *Theoricae novae planetarum*: A Translation with Commentary," *Osiris*, ser. 2, vol. 3 (1987), pp. 5–9.

[6] Joannes Regiomontanus, *Epitome in Almagestum* (Venice: Johannes Hamman, 1496); Aiton, "Peurbach's *Theoricae*," pp. 5–6; North, *The Norton History of Astronomy and Cosmology*, pp. 254–5.

[7] Aristotle, *Metaphysics*, 12.8 (1073b1–1074a18).

The work of al-Bitrūjī was translated into Latin in 1217 and spawned a number of other homocentric theories, among them one by Regiomontanus.[8]

In the early sixteenth century, there were two published attempts at homocentric theories: *De motibus corporum caelestium* (On the Motions of the Heavenly Bodies, 1536), by Giovanni Battista Amico (1512–1538), and *Homocentrica* (1538), by Girolamo Fracastoro (1478–1553). Although entirely inadequate for predicting planetary positions, these theories are notable in that they deliberately attempted to reconstruct the lost theories of Eudoxus. Fracastoro, whose work attained some degree of recognition, added the requirement that adjacent spheres have their axes at right angles – a fine example of the Renaissance tendency to want to go the ancients one better, particularly in adherence to principle – in this case, the Aristotelian principle that planetary motions must be simple and uniform.[9]

A similar motivation is evident in the work of Copernicus, who was a contemporary of Amico and Fracastoro and had received a thorough humanist education at Cracow and Bologna. In his *De Revolutionibus orbium coelestium* (On the Revolutions of the Heavenly Spheres, 1543), Copernicus objected in particular to two things in the *Almagest*: the introduction of nonuniform motions, and Ptolemy's failure to put the separate planetary models together into a systematic whole. On the former, he wrote: "Those who had devised eccentrics . . . had nonetheless admitted many things which appeared to go against the first principles concerning uniformity of motion."[10] That the heavens move with uniform circular motions was the most widely accepted axiom of the astronomers and was derived from the natural philosophical principle of celestial incorruptibility. Yet Ptolemy had supposed centers of uniform motion that were different from the centers of the circles on which the motions took place. Such motions, Copernicus complained, are really nonuniform in that they are in fact faster at one part of the circle and slower at another.

Copernicus's alternative model involved adding another small circle that adjusted the planetary motion enough to obviate the need for nonuniform motions. Similar models had in fact been proposed by Arabic astronomers, and Copernicus hints that he had used the Arabic models,

---

[8] Pedersen and Pihl, *Early Physics and Astronomy*, pp. 266–7, 351, reissued as Olaf Pedersen, *Early Physics and Astronomy: A Historical Introduction*, rev. ed. (Cambridge: Cambridge University Press, 1993), pp. 235–6, 318; and N. M. Swerdlow, "Regiomontanus's Concentric-sphere Models for the Sun and Moon," *Journal for the History of Astronomy*, 30 (1999), 1–23.

[9] J. L. E. Dreyer, *A History of Astronomy from Thales to Kepler* (New York: Dover, 1953), pp. 296–304.

[10] Nicholas Copernicus, *De Revolutionibus* (Nuremberg, 1543), "Preface and Dedication to Pope Paul III," fol. iiiv; 4.2, fol. 99r-v, and 5.2, fol. 140v; Copernicus, *On the Revolutions*, translation and commentary by Edward Rosen (Baltimore: Johns Hopkins University Press, 1992) pp. 4, 176, 240. For a thorough discussion of the issues of uniformity and regularity, see Edward Grant, *Planets, Stars, and Orbs: The Medieval Cosmos, 1200–1687* (Cambridge: Cambridge University Press, 1994), pp. 488–94.

without mentioning specific names.¹¹ Evidently, like Fracastoro, Copernicus saw himself as not just reviving the ancient tradition but building upon it, not in a spirit of iconoclasm but of *renovatio*, of making it new once more.

Concerning the lack of a system in Ptolemy's universe, Copernicus used a simile drawn from Horace's *Ars Poetica*. "What happened to [the mathematicians]," he wrote, "is exactly as if one were to take from various places hands, feet, head, and other members, depicted indeed very well, but not for the composition of a single body, corresponding to each other not at all, so that a monster rather than a human being would be composed from them."¹² An explicitly humanist search through classical sources showed that some of the ancients had explained the heavenly phenomena by ascribing motion to the earth, so he thought that he, too, might be allowed that liberty. What Copernicus found was that "the order and sizes of the heavenly bodies and of all the orbs, as well as the heaven itself, are so connected that in no part of it can anything be displaced without disordering the remaining parts and the whole universe."¹³ Here was an astronomical system that not only accounted for the phenomena of planetary motion but could lay claim to representing the physical truth.

To Copernicus's contemporaries, his achievement was not the theory of the earth's motion, but rather the composition of a modern rival to Ptolemy's *Almagest*, which in some respects surpassed its prototype. In its completeness and (not particularly successful) attempts at using recent observations to improve the accuracy of the theories, *De Revolutionibus* had no rival. It was used by Erasmus Reinhold (1511–1553), professor of mathematics at the University of Wittenberg, to construct a new set of planetary tables, the *Prutenic Tables* (1551), which were widely used and were reprinted in 1585. However, during the entire sixteenth century, hardly any readers of *De Revolutionibus* had anything good to say about its assertion of earth's triple motion (rotation on its axis, revolution about a point near the sun, and directional rotation of its axis).¹⁴ This was partly because of the preface "To the Reader, on the Hypotheses of this Work," which claimed that astronomy "does not think up [hypotheses] in order to persuade anyone of their truth, but only

---

¹¹ Copernicus, *De Revolutionibus*, 34. The manuscript version of this chapter has a passage, later deleted, that ascribes one of these models to "certain people" (*aliqui*). See *Nicolaus Kopernikus Gesamtausgabe, Band I: Opus De Revolutionibus Caelestibus Manu Propria Faksimile-Wiedergabe* (Munich: Oldenbourg, 1944), fol. 75v; Rosen translation, p. 126.
¹² Copernicus, *De Revolutionibus*, fol. iiiv; Horace, *Ars Poetica*, lines 1–13; cf. Robert S. Westman, "Proof, Poetics, and Patronage: Copernicus's Preface to *De revolutionibus*," in *Reappraisals of the Scientific Revolution*, ed. David C. Lindberg and Robert S. Westman (Cambridge: Cambridge University Press, 1990), pp. 179–84.
¹³ Copernicus, *De Revolutionibus*, fol. ivr.
¹⁴ This last motion was an artifact of astronomers' use of an early version of polar coordinates, that required the axis to be cranked around in order to make the north pole face toward the sun at one side of its orbit and away from it at the opposite side. Kepler was the first to point out that this was not a real motion. See Kepler, *Epitomes astronomiae Copernicanae*, I (Linz, 1618), para. 5 sect. 5, pp. 113–4, in *Johannes Kepler Gesammelte Werke* (Munich: C. H. Beck, 1937–), vol. 7 (1953), pp. 85–6.

in order that they may provide a correct basis for calculation." The preface was in fact written by the Lutheran priest and theologian Andreas Osiander (1498–1552), who was in charge of the final stages of publication and made the insertion without Copernicus's knowledge. Because the authorship of the preface was not revealed in print until 1609, early readers were led to think that Copernicus himself regarded his work as merely a hypothesis.[15]

But even without Osiander's preface, the idea of a moving earth was unlikely to convince many readers. It too blatantly contradicted current theories of motion and left too many questions unanswered. How could the sluggish earth carry out the complicated triple motions assigned to it? Why would God create so much empty space between Saturn and the fixed stars?[16] Drastic changes in cosmological ideas, and in the way natural philosophy was done, would have to occur first in order to provide a context in which a moving earth made sense.

CRACKS IN THE STRUCTURE OF LEARNING

One important source of change was the assortment of alternatives to scholastic natural philosophy, and, indeed, the more subtle changes that were taking place in scholastic natural philosophy itself. Of particular interest are authors of philosophical works who believed it to be within the competence of mathematics to make valid judgments about physical reality. Nicholas of Cusa (1401–1464), a Catholic cardinal, was especially remarkable for his view that the mind and the truth it grasps are, at the deepest level, based on magnitude (either numerical or geometrical). Interestingly, he argued, a century before Copernicus, that the earth moves, though the basis for his claim had little to do with astronomy and much to do with theology. Evidently, at least

---

[15] For Reinhold, see C. C. Gillispie, ed., *Dictionary of Scientific Biography* (New York: Scribners, 1970–90), 11: 365–7; and Owen Gingerich, "The Role of Erasmus Reinhold and the Prutenic Tables in the Dissemination of Copernican Theory," in *Colloquia Copernicana*, II, ed. Jerzy. Dobrzycki (Wrocslashelaw;Ossolineum 1973), pp. 43–62, 123–5. For Osiander, see Edward Rosen's notes to Copernicus, *On the Revolutions*, pp. 333–5; Bruce Wrightsman, "Andreas Osiander's Contribution to the Copernican Achievement," in *The Copernican Achievement*, ed. Robert S. Westman (Berkeley: University of California Press, 1975), pp. 213–43; and Nicholas Jardine, *The Birth of History and Philosophy of Science: Kepler's "A Defence of Tycho against Ursus,"* with Essays on Its Provenance and Significance (Cambridge: Cambridge University Press, 1984), pp. 150–4. Kepler's publication of the evidence of Osiander's authorship is on the verso of the title page of *Astronomia nova* (Heidelberg: Vögelin, 1609); see Johannes Kepler, *New Astronomy*, trans. William H. Donahue (Cambridge: Cambridge University Press, 1992), pp. 28–9.

[16] If we are really viewing the stars from a moving platform, their angular distances from each other should change as earth approaches or recedes from them. Because this phenomenon, known as "annual parallax" (see Figure 24.4, p. 590), had not been observed, it would have to be very small, and this would require the stars to be at a very great distance, at least three orders of magnitude farther than the distance generally accepted at that time, even under the assumption of the very low estimate of the sun's distance that was then usual.

the upper echelons of the Church in the fifteenth century could prove more congenial to bold speculation than did the universities.[17]

But despite the freedom afforded by his secure position, Nicholas of Cusa enjoyed only moderate success in finding an audience for his ideas. One influential follower was the Carthusian prior Gregor Reisch (ca. 1467–1525), whose philosophical textbook, *Margarita philosophica* (The Philosophical Pearl, 1503), was widely used and often reprinted throughout the sixteenth century. In most respects, the *Margarita* was anything but revolutionary. But Reisch cited Nicholas of Cusa in support of the inclusion of astronomy and the other mathematical arts under philosophy, classifying them as "speculative philosophy of [material] things."[18] A similar, and perhaps related, tendency may be seen in the teaching of Reinhold at Wittenberg in 1536. Introducing Euclid and Peurbach to his students, he said that "that part of Philosophy that is called 'Physica' takes its origin from Geometry." This thoroughly un-Aristotelian opinion, adumbrating a new, mathematically based view of the entire physical cosmos, could have come directly from Nicholas of Cusa's *Idiota de mente* (The Layman: About Mind), published in 1450.[19]

In the decades following the publication of *De Revolutionibus*, a lively variety of publications appeared that challenged conventional physical theory, including the physics of the heavens. Authors who challenged Aristotle's radical distinction between the pure eternal celestial aether and the four "sublunar" elements of air, earth, fire, and water include the Milanese physician, astrologer, and mathematician Girolamo Cardano (1501–1576), the iconoclastic German physician Theophrastus Bombastus von Hohenheim, known as Paracelsus (1493–1541); Francesco Patrizi (1529–1597), professor of Platonic philosophy at La Sapienza (the University of Rome); Bernardino Telesio (1509–1588), lecturer at the University of Naples; and the Dominicans Tommaso Campanella (1568–1639) and Giordano Bruno (1548–1600) (see Garber, Chapter 2, this volume). Perhaps following Stoic authors, Cardano denied that fire is an elementary substance, thus reducing the number of elements to three. The heavens, in his view, were the source of warmth. By thus removing the sphere of fire that Aristotle had located beneath the lunar spheres, Cardano created a continuity that brought heaven and earth closer together. Accordingly, he believed that comets, which were traditionally

---

[17] A useful biographical sketch and introduction to Nicholas of Cusa's work, with many references, is included in Nicholas de Cusa, *Unity and Reform: Selected Writings*, ed. John Patrick Dolan (Notre Dame, Ind.: University of Notre Dame Press, 1962), pp. 3–53. A succinct statement of Nicholas's view on the mind's operation is presented in Nicholas de Cusa, *Idiota de mente (The Layman: About Mind)*, trans. and intro. Clyde Lee Miller (New York: Abaris Books, 1979), chap. 10, p. 75.

[18] Gregor Reisch, *Margarita philosophica* (Freiburg im Breisgau: J. Schott, 1503), fol. [ ] 4r; cf. bk. 4, pt. 1, chap. 1 for the importance of mathematics and a citation of Nicholas of Cusa.

[19] *Scriptorum publice propositorum a professoribus in Academia Witebergensi ab anno 1540 usque ad 1553, tomus primus* (Wittenberg: G. Rhaw, 1560), quoted in Sachiko Kusukawa, *The Transformation of Natural Philosophy: The Case of Philip Melanchthon* (Cambridge: Cambridge University Press, 1995), p. 180; cf. Nicholas de Cusa, *Idiota de mente*, chap. 10, p. 75.

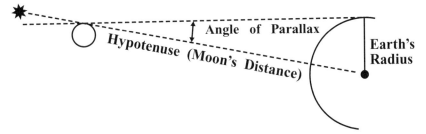

Figure 24.2. Diurnal parallax of the moon.

supposed to be atmospheric, could enter the celestial regions and might even originate there. Cardano proposed measuring their distance by means of parallax observations, which were well known to astronomers because the lunar parallax had to be considered in predicting eclipses. (Parallax is the measure of how much a relatively nearby object appears to move against a distant background because of a displacement of the observer; Figure 24.2 illustrates how diurnal parallax affects the observed position of the moon.) Cardano does not appear to have made any such observations himself, however.[20]

Paracelsus, though an eclectic, drew his worldview primarily from the tradition of alchemy (see Newman, Chapter 21, this volume). Although he had little directly to say about the heavens, his magico-chemical worldview, with three principles (identified with sulfur, salt, and mercury), attracted many followers, including the influential Danish astronomer Tycho Brahe, who was also interested in alchemy. Paracelsus held that the heavens are not radically different from the terrestrial and atmospheric regions and believed the stars to be fiery in nature.[21]

Patrizi, Campanella, and Bruno were all strongly influenced by the supposedly ancient philosophical and mystical tradition stemming from the legendary Hermes Trismegistus, the central characteristic of which was a belief that nature, both as a whole and in each part, is alive and possessed of a soul that governs all of its operations[22] (see Copenhaver, Chapter 22,

---

[20] William H. Donahue, *The Dissolution of the Celestial Spheres, 1595–1650* (New York: Arno Press, 1981), pp. 51–2; and C. Doris Hellman, *The Comet of 1577: Its Place in the History of Astronomy* (Columbia University Studies in the Social Sciences, 510) (New York: Columbia University Press, 1944; New York: AMS Press, 1971), pp. 90–6.

[21] For Paracelsus, see Walter Pagel, "Paracelsus," in Gillispie, ed., *Dictionary of Scientific Biography*, 10: 304–13. On the celestial generation of comets, see Paracelsus, *Opera* (Geneva: J. Anton & S. De Tournes, 1658), 2: 318, cited in *Tychonis Brahei Dani opera omnia*, ed. J. L. E. Dreyer, 15 vols. (Copenhagen: Libraria Gyldendaliana, 1913–29), vol. 4 (1922), pp. 511–12. For Brahe's Paracelsian education, see Victor Thoren, *The Lord of Uraniborg: A Biography of Tycho Brahe* (Cambridge: Cambridge University Press, 1990), pp. 24–5.

[22] Frances A. Yates, *Giordano Bruno and the Hermetic Tradition* (London: Routledge and Kegan Paul, 1964), is a thorough treatment of Renaissance Hermeticism and includes discussions of all of these authors. A briefer account, focused more closely on cosmology and astronomy, is in Donahue, *The Dissolution of the Celestial Spheres*, pp. 41–53. For Bruno, see Paul-Henri Michel, *The Cosmology of Giordano Bruno* (Ithaca, N.Y.: Cornell University Press, 1973); and Giordano Bruno, *The Ash Wednesday Supper*, ed. and trans. Edward A. Gosselin and Lawrence S. Lerner (Hamden, Conn.:

this volume). Despite considerable differences in their development of this philosophy, they were easily identifiable by their contemporaries as followers of Hermetic neo-Platonism, and this identification inevitably colored the way in which their works were read.[23] The idea of a world soul, in particular, was theologically suspect, and all of these authors were charged with various degrees of heresy during the last decade of the sixteenth century. So even though Campanella wrote a book in support of Galileo and Bruno argued for a heliocentric planetary system within an infinite universe, such heterodox writers were generally opposed by proponents of the new approach to nature, including Kepler, Galileo, and Descartes. Moreover, the difference here was not only a matter of political prudence (Bruno had been burned at the stake for heresy): The Hermetics tended to be more interested in numerology than mathematics and scorned the mixed mathematical sciences, astronomy in particular.[24]

The importance of all of these alternative theories was not their explanatory power, or their provision of a possible replacement for scholastic natural philosophy but simply their existence. They represented a sense of the richness of the various traditions that were being rediscovered and the comparative limitations of what was taught in the schools. And even academic accounts of the nature of the heavens were beginning to change, incorporating Stoic ideas and scripturally based alternatives to Aristotle (see Blair, Chapter 17, this volume).[25]

It is clear that whatever the causes, during the sixteenth century there was an increasing tendency on the part of natural philosophers as well as astronomers to consider the heavens as being made of stuff not radically different from earthly matter. At the beginning of the century, scholastic discussions of celestial matter centered around such questions as whether the heavens consist of form and matter or are mere forms (being immaterial). The question of whether the celestial orbs (if they are real) were solid (i.e., hard)

Archon Books, 1977); the introduction to this work is a good brief account of Bruno and his thought. For Campanella, see Tommaso Campanella, *A Defense of Galileo, the Mathematician from Florence*, trans. Richard J. Blackwell (Notre Dame, Ind.: University of Notre Dame Press, 1994).

[23] Patrizi acknowledges his Hermeticism in his dedication to Pope Gregory XIV, *Nova de universis philosophia* (Venice: Robertus Meiettus, 1593). For Campanella's and Bruno's Hermetic ideas, see Yates, *Giordano Bruno and the Hermetic Tradition*, esp. chap. 20.

[24] For Patrizi's view of astronomy, see Patrizi, *Nova de universis philosophia* "Pancosmia," bk. 12, fol. 91, col. 2. For Bruno, see Michel, *The Cosmology of Giordano Bruno*, esp. pp. 190–8; and Yates, *Giordano Bruno and the Hermetic Tradition*, pp. 235–41. On Hermetic mathematics in general, see Johannes Kepler, *Pro suo opere Harmonices mundi apologia* (Frankfurt: G. Tampach, 1622), p. 34, in Kepler, *Gesammelte Werke*, vol. 6: *Harmonices mundi* (1940), p. 432; and W. Pauli, "The Influence of Archetypal Ideas on the Scientific Theories of Kepler," in C. G. Jung and W. Pauli, *The Interpretation of Nature and the Psyche* (London: Routledge and Kegan Paul, 1955), pp. 190–200.

[25] Peter Barker and Bernard R. Goldstein, "Is Seventeenth Century Physics Indebted to the Stoics?," *Centaurus*, 27 (1984), 152–4; Peter Barker, "Stoic Contributions to Early Modern Science," in *Atoms, Pneuma, and Tranquillity: Epicurean and Stoic Themes in European Thought*, ed. Margaret J. Osler (Cambridge: Cambridge University Press, 1991), pp. 135–54; Grant, *Planets, Stars, and Orbs*, p. 267; and Donahue, *The Dissolution of the Celestial Spheres*, pp. 53–9.

or fluid was not even raised in natural philosophy courses before the 1580s. By the early seventeenth century, this question became quite common in school texts. Even for those who followed Aristotle, the heavens were by this time thought to be composed of something that is extended, noninterpenetrable, and (usually) hard and rigid.[26]

THE REFORMATION AND THE STATUS OF ASTRONOMY

Moreover, the schools themselves were undergoing change, especially as a result of the Reformation. Martin Luther (1483–1546) was no admirer of the scholastic philosophies that had served to buttress what he saw as corrupt Catholic doctrines, and the universities in Lutheran lands seemed for a time in danger of falling apart. Seeing the negative consequences that would result from such an event, Luther charged his friend and associate, the humanist scholar Philip Melanchthon (1497–1560), with the mission of developing a reformed system of higher education. Melanchthon's system emphasized ethics. It involved the elimination of much of philosophy and favored sciences that had practical value. Astronomy was retained largely because of its utility in supporting astrology, particularly medical astrology. Melanchthon himself wrote in praise of astronomy, and in his reorganization of the University of Wittenberg he founded a distinctly Lutheran tradition of astronomical instruction and theory.[27]

The Wittenberg school characteristically held Copernicus in high regard while, on the whole, rejecting his systematic claims and his moving earth. Copernicus was seen as having superseded Ptolemy in the consistency and accuracy of his planetary models and so was not to be ignored. This approach suggested that the individual planetary models be revised in accord with the supposition of a central and motionless earth. The *Prutenic Tables* of Wittenberg mathematics professor Reinhold were constructed in this way. Similar views were expressed by Reinhold's successor and Melanchthon's son-in-law Caspar Peucer (1525–1602), Peucer's student Michael Praetorius (1537–1616), and others.[28] This "Wittenberg interpretation" had the effect of encouraging the study of Copernicus's work (which had been reprinted in 1566), thus preparing the way for a serious consideration of the systematic advantages of heliocentrism.

---

[26] Léon Blanchet, *Les antecédents historiques du je pense, donc je suis* (Paris, 1920), p. 69; and Sister Mary Richard Reif, "Natural Philosophy in Some Early Seventeenth Century Scholastic Textbooks," Ph.D. dissertation, St. Louis University, St. Louis, Mo., 1962, pp. 83–97.
[27] Kusukawa, *The Transformation of Natural Philosophy*, pp. 27–74, 171–200.
[28] Robert S. Westman, "The Wittenberg Interpretation of the Copernican Theory," in *The Nature of Scientific Discovery*, ed. Owen Gingerich (Washington, D.C.: Smithsonian, 1995), pp. 393–429; Robert S. Westman, "Three Responses to the Copernican Theory: Johannes Praetorius, Tycho Brahe, and Michael Maestlin," in Westman, ed., *The Copernican Achievement*, pp. 285–345; and Thoren, *The Lord of Uraniborg*, p. 86.

Although the Lutherans were by no means the only cultivators of astronomy, this tradition is worthy of special attention because the most significant astronomers of the late sixteenth and early seventeenth centuries, Brahe and Kepler, were grounded in this tradition. Brahe had begun his studies at the University of Copenhagen, where Melanchthon's influence was strong, and continued them at the University of Leipzig, studying astronomy under Bartholomaeus Scultetus (1540–1614), protégé and successor of Wittenberg alumnus Johannes Homelius (1518–1562). Kepler was a student of Michael Maestlin (1550–1631), and both had been educated at the Lutheran University of Tübingen.[29]

We do not have much evidence of Brahe's early opinions of the heliocentric system, though it is clear that, like the Wittenberg astronomers, he admired the specific Copernican arrangements of circles in the planetary models. However, later publications and letters show that he was troubled by the physical and scriptural obstacles to a moving earth, and by the enormous distances of the fixed stars that Copernicus's system implied. In response to these difficulties, Brahe proposed a compromise system, announced in a chapter added to a work on the comet of 1577 published in 1588 (Figure 24.3).[30] In the so-called Tychonic system, which was intended to satisfy both the astronomers and the philosophers, all planets circle the sun, which in turn orbits around a motionless earth. With this schema as a guide, Brahe hoped to construct better models of planetary motions. Using both his income as a Danish nobleman and supplementary royal grants, he established a major observatory and fitted it with instruments of unprecedented size and accuracy, complementing his superb skill as an observer. From Uraniborg, "The Castle of Urania" (muse of astronomy), as Brahe called his island observatory, he carried out an unprecedented series of observations involving multiple sequential sightings of planets at important points on their paths, as well as a completely new set of coordinates for more than a thousand fixed stars. The planetary theories that he hoped would be built on these observations would not only be far more accurate than their predecessors but could also plausibly lay claim to being true, or at least more nearly true than Ptolemy's or Copernicus's constructions.

Yet Brahe's work also contained within itself the means by which his system was to be undone. He was familiar with Cardano's views on comets, and, as mentioned previously, was partial to Paracelsian ideas.[31] Furthermore, his

---

[29] Thoren, *The Lord of Uraniborg*, pp. 9–12, 42; Westman, "The Wittenberg Interpretation," pp. 393–429; and Westman, "Three Responses," pp. 329–30.

[30] Tycho Brahe, *De mundi aetherei recentioribus phaenomenis* (Hven: Christophorus Vveida, 1588), chap. 8, in *Opera omnia*, vol. 4 (1922), pp. 155–70. For Brahe's arguments against the heliocentric system, see J. L. E. Dreyer, *A History of Astronomy from Thales to Kepler* (New York: Dover, 1953), pp. 360–5, and the sources cited therein. Brahe's development of his system is described in Thoren, *The Lord of Uraniborg*, chap. 8, pp. 236–64.

[31] Hellman, *The Comet of 1577*, p. 92; Brahe, *De cometa anni 1577*, in *Opera omnia*, vol. 4 (1922), pp. 382–3.

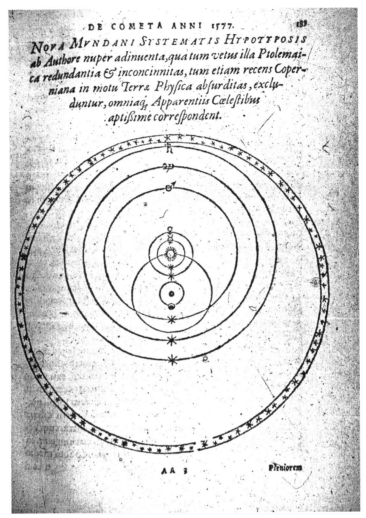

Figure 24.3. Tychonic system. In Tycho Brahe, *De mundi aetherei recentioribus phaenomenis* (Uraniborg: [By the author, 1588]), p. 189. Photograph courtesy of Owen Gingerich.

planetary system required the circles of several planets, and of Mars in particular, to cross the orb of the sun. Brahe was therefore inclined to suspect that the spheres of the astronomers might not be physically real. Accordingly, some of his earliest and most important observations were careful measurements of the parallax of comets, with the aim of determining their altitude above earth. Suspecting that the standard Aristotelian view of comets as vaporous terrestrial exhalations might be wrong, and that they might in fact be celestial bodies, Brahe applied astronomical instruments and methods to them. Having found the parallax to be very small, he concluded that comets must be in the heavens and that the comet of 1577 passed through the region that, according to traditional astronomy, is occupied by the set of spheres carrying the planet Venus. Here was the proof Brahe sought that no such spheres exist: The planets were free to follow their appointed paths in an unobstructed space.[32]

Although this solved Brahe's immediate problem of the intersecting orbs, it raised a new one: How, in the absence of the orbs to carry them, could the planets be made to move in the complex paths required of them by this system? Brahe himself addressed this question only obscurely, combining Stoic and Paracelsian ideas in making the planets and stars self-moving fiery beings.[33] This did little to settle the matter, and the question of the causes of planetary motion became an important question, perhaps the most important question, for astronomers and natural philosophers of the seventeenth century. And, as it turned out, no one was able to give a satisfactory answer in the context of the Tychonic planetary arrangement.

But although his system was ultimately unsuccessful, Brahe's role in the development of astronomy can hardly be overstated. His skill as an observer and instrument designer was celebrated, and he was especially innovative in setting up his own workshop where observers and artisans could work together (see Cooper, Chapter 9, this volume). On the theoretical level, Brahe built upon the Wittenberg tradition of competence in mathematical astronomy while seeking to harmonize astronomy and natural philosophy.

This attempt to forge a synthesis was by no means unprecedented. Numerous astronomers proposed planetary theories that were intended to be physically sound. On a more empirical level, there was a long tradition of comet observations, going back at least to the fourteenth century, which led to the formulation of a non-Aristotelian theory that allowed comets to exist in the planetary regions.[34] Such theories had been ignored by earlier natural philosophers on the grounds that their discipline, being the superior one, was under no obligation to accommodate itself to astronomy.

Yet Brahe was remarkably successful in winning acceptance for his conclusions, even among academic natural philosophers. This was partly because, as already mentioned, the natural philosophers were ready to make the accommodation of co-opting non-Aristotelian ideas and making their heavens a little more earthlike. But Brahe's success also depended on his unusual status as a nobleman and an independent practitioner of his art. He, as well as

---

[32] Brahe, *De mundi aetherei recentioribus phaenomenis*, chap. 6, pp. 89–158, in *Opera omnia*, 4: 82–134; cf. the diagram on p. 203 (chap. 7) showing the comet's position in relation to Venus's orb. See also Hellman, *The Comet of 1577*, pp. 121–37.

[33] Brahe, *Epistolae astronomicae liber primus* (Uraniburg, 1596), in *Opera omnia*, vol. 6 (1919), pp. 166–7; cf. Peter Barker, "Stoic Contributions," p. 146.

[34] Peter Barker and Bernard R. Goldstein, "The Role of Comets in the Copernican Revolution," *Studies in the History and Philosophy of Science*, 19 (1988), 299–319.

his contemporary William IV (1532–1616), Landgrave of Hesse, reinvented the state-supported observatory as an institution, which had been invented by Islamic astronomers in the thirteenth century.[35] These two independent astronomers had the economic and social wherewithal to produce unimpeachable results and to promulgate them to a wide audience. Thus Brahe's careful collation of his own observations of the 1577 comet with those of many other observers came to be cited by later writers, both astronomers and philosophers, as establishing the celestial position of comets and the nonexistence of planetary spheres.[36] A profound change had occurred in astronomy's standing as a natural science: After the second decade of the seventeenth century, astronomy's authority in the realm of physics was never seriously questioned.

In bringing about this change, a crucial role was played by the educational system built by the Jesuits, and especially their leading institution, the Collegio Romano, and its mathematics professor, Christoph Clavius (1538–1612). Clavius had done much to advance the standing of mathematics, including astronomy, in the Jesuit curriculum, and wrote one of the most influential introductory astronomical textbooks of the late sixteenth century, a lengthy commentary on the *Sphere* of Sacrobosco.[37] While adhering to a traditional Aristotelian framework and style of reasoning, Clavius and his Jesuit colleagues and students kept up with new discoveries and theories, such as the relocation of comets to the heavens, the celestial position of new stars (*novae*), and the invention of the telescope. Their influence is indisputable: By 1600 there were over 250 Jesuit colleges in Europe, and Jesuit missionaries were traveling the world, spreading Brahe's, and later Newton's, astronomical theories, and Galileo's telescope, as far as Japan and China. Nor were the Jesuits the only proponents of this moderate reformist position: Several professors at Louvain, a small group in Copenhagen, and other individuals elsewhere were expressing similar views in the first two decades of the seventeenth century.[38]

ASTROLOGY

Astrological theory was quite distinct from astronomy, taking the planetary motions as given and considering their supposed effects (see Rutkin, Chapter 23, this volume). However, because some degree of technical competence was required to cast a horoscope, astrology was usually practiced by those with astronomical training. Kepler likened astrology to a foolish but

---

[35] North, *The Norton History of Astronomy and Cosmology*, pp. 192–202.
[36] Donahue, "The Solid Planetary Spheres," pp. 259–63; Grant, *Planets, Stars, and Orbs*, pp. 345–61.
[37] For Clavius and the Jesuits, see James M. Lattis, *Between Copernicus and Galileo: Christopher Clavius and the Collapse of Ptolemaic Cosmology* (Chicago: University of Chicago Press, 1994).
[38] Donahue, *The Dissolution of the Celestial Spheres*, pp. 53–9, 114–24; North, *The Norton History of Astronomy and Cosmology*, pp. 147–53.

well-off daughter, without whose help mother astronomy would starve. In astrology, as in astronomy, Ptolemy was the chief authority, although the ninth-century Persian astronomer Abū Maʿshar (Albumasar) made substantial contributions to astrological theory and practice. Its theoretical basis was in natural philosophy: In the Aristotelian cosmos, the function of the planetary motions was to stir up the terrestrial elements and make things happen on earth and in the air:[39] There was therefore general agreement, among proponents and opponents of astrology alike, that the stars influenced events on earth. Indeed, some of these influences are obvious, such as those of the sun and moon on days and nights, seasons, and tides. Criticism of astrology focused on whether astral influences could be known and whether astrology as then practiced represented the influences correctly. This was difficult to determine in any theoretical way because astrological principles (such as the properties of the zodiacal signs) were stated dogmatically rather than being deduced from natural causes. Astrology's putative veracity depended upon testimony of many authorities (Plato, Aristotle, Virgil, Ovid, Albumasar, Pliny, the legendary Hermes Trismegistus, Thomas Aquinas, and many others) and upon experience. Thus, oddly, it was one of the most nearly empirical of the sciences; then, as now, one would come to believe in it by practicing it.[40]

Astrology figured prominently in the reformed Lutheran curriculum at the University of Wittenberg that was developed by Melanchthon. In the early stages of this reform, astrology was included because of its practical value in medicine, and in turn it provided the justification for the teaching of astronomy and physics. Later, Melanchthon defended the study of astrology because of its demonstration of the workings of divine providence. His support of astrology thus helped establish a tradition of cultivating astronomy and astrology in Lutheran universities. Astrology was also often studied outside the universities through the reading of texts and the sharing and discussion of horoscopes of notable persons.[41]

Astrology had a number of different branches, some of which were more widely accepted than others. One of its most common uses was in weather

---

[39] Aristotle, *On the Heavens*, 2.3; *On Generation and Corruption*, 2.10. For Kepler's simile, see *De stella nova* (Prague, 1606), chap. 12, in *Gesammelte Werke*, vol. 1 (1937), p. 211. For a more complete account of the roots of European astrology, see J. D. North, "Medieval Concepts of Celestial Influence: A Survey," in *Astrology, Science, and Society*, ed. Patrick Curry (Woodbridge: Boydell Press, 1987), pp. 5–17.

[40] See, for example, Kepler's quotation of Brahe's remark (in Kepler's unpublished *De directionibus* of 1601) that "in astrology, as in theology, one must not seek for reasons, but only believe, insofar as the latter is from authority, and the former from experience." In *Ioannis Kepleri astronomi opera omnia*, ed. C. Frisch, 8 vols. (Frankfurt and Erlangen: Heyder and Zimmer, 1858–71), vol 8.1 (1870): 295. For authorities cited in support of astrology, see Wayne Shumaker, *The Occult Sciences in the Renaissance: A Study in Intellectual Patterns* (Berkeley: University of California Press, 1972), pp. 27–30, 35–6.

[41] Siraisi, *Medieval and Early Renaissance Medicine*, pp. 67–8; Kusukawa, *The Transformation of Natural Philosophy*, pp. 61, 129–59; and Anthony Grafton, *Cardano's Cosmos: The Worlds and Works of a Renaissance Astrologer* (Cambridge, Mass.: Harvard University Press, 1999), pp. 35–7, 42–3, 71–5.

prediction, an application very important to agriculture. The writing of calendars and almanacs containing astrological weather forecasts was one of the functions of mathematicians with official appointments. Kepler, who held a number of posts, including that of imperial mathematician, wrote many such calendars and was proud of the success of his predictions. Another common use was in medical diagnosis. Medical astrology had a long and distinguished history and appears to have motivated a number of early observers. For instance, some of the earliest comet observations were made by physicians (e.g., by Geoffrey of Meaux in 1315, Jacobus Angelus in 1402, Peter of Limoges in 1299, and Paolo Toscanelli from 1433 to 1472), who usually included astrological considerations in their accounts.[42]

The most frequently condemned branch of astrology was so-called judicial astrology, which involved drawing up "nativities," or horoscopes, for individuals. This was suspect not because it had a less solid theoretical foundation than the other branches but because it encroached on the delicate matter of free will and responsibility for sin. Accordingly, those who rejected it often did so for religious reasons: The Swiss religious reformer John Calvin (1509–1564), for example, accepted medical and natural uses of astrology but condemned judicial astrology. Those who defended judicial astrology stressed that the stars incline but do not compel.[43]

The most effective critique of astrology, however, came from the humanist philosopher Giovanni Pico della Mirandola (1463–1494). Pico's two main arguments were, that even among astrologers there was much disagreement about effects and that attempts to explain astrological influences were mostly based on fanciful analogies rather than sound physical reasoning. Although he believed that the motions of the heavens had effects on earth, he also believed that the human soul was free and not subject to astral influences.[44]

Pico's arguments attracted considerable attention throughout the sixteenth century and were carefully considered by Kepler in formulating his revision of astrology at the beginning of the seventeenth century. Defenders of astrology, interestingly, often adduced empirical criteria in their defense, giving examples of astrology's successful predictions and the overwhelming support of those who had studied the art. It was noted that Pico himself

---

[42] Max Caspar, *Kepler*, trans. C. Doris Hellman (New York: Dover, 1993), pp. 58–60, 154–6, 172; Siraisi, *Medieval and Early Renaissance Medicine*, pp. 68, 135; Barker and Goldstein, "The Role of Comets in the Copernican Revolution," pp. 308–10; and Lynn White, Jr., "Medical Astrologers and Medieval Technology," in White, *Medieval Religion and Technology* (Berkeley: University of California Press, 1978), pp. 297–315.

[43] North, *The Norton History of Astronomy and Cosmology*, pp. 265–71; Shumaker, *The Occult Sciences in the Renaissance*, pp. 38, 44–8.

[44] Shumaker, *The Occult Sciences in the Renaissance*, pp. 16–27; North, *The Norton History of Astronomy and Cosmology*, p. 272. Lest the reader form the opinion that Pico was a rationalist skeptic, it should be pointed out that his main interest was in combining Hermetic magic and the Cabala into a sort of Christian gnosis, and that he objected to any claims that would restrict such a magical transformation. For an account of Pico's magical beliefs, see Yates, *Giordano Bruno and the Hermetic Tradition*, pp. 84–116.

was ignorant of both astronomy and astrology. In contrast, opponents of astrology included many, such as Luther, Calvin, and the Florentine radical Girolamo Savonarola (1452–1498), who assailed it from a religious perspective.[45] A very important exception was Luther's friend Melanchthon. He saw astrology as a clear illustration of divine providence, while the technical side of astronomy was simply a computational system subservient to astrology.[46]

Despite the ongoing debate about astrology's validity, the art itself remained stubbornly resistant to change. There was a lively but ultimately fruitless debate around the periphery regarding such issues as how to determine the divisions of the "houses" or positions of the planets with respect to the horizon at a given moment, with different nationalities promoting their respective champions. But teachings about the significance of the planets, the signs of the zodiac, the "aspects" or geocentric angles between the planets, and the houses continued to rely on authoritative texts, such as Ptolemy's *Tetrabiblos* and Albumasar's *Flores astrologici* (The Flowers of Astrology, 1488). Even the cosmological changes that were taking place around the turn of the seventeenth century had little effect on astrological theory, as demonstrated by the English nobleman Christopher Heydon's up-to-date fluid-filled heavens in his otherwise traditional *A Defence of Judiciall Astrologie* (1603).[47] Late seventeenth-century innovations in natural philosophy did, however, produce some interesting new twists on the subject of the terrestrial influences of heavenly bodies (see Rutkin, Chapter 23, this volume).

One of the most ambitious attempts to reform astrology came from Kepler (see Rutkin, Chapter 23, this volume). Kepler had read Pico carefully and agreed with much of what he had to say. But, as he wrote in the title of one of his works, one must be careful not to throw the baby out with the bath water. Kepler proposed a stripped-down astrology based on the new physical astronomy he was developing. He threw out all of the zodiacal signs on the grounds that there was no comprehensible way they could have any effect. What remained were the planetary qualities and their aspects, or angular configurations as seen from earth. These he tested empirically, using weather observations, and on the basis of these and his harmonic principles, he introduced some new aspects. The aspects were effective, he believed, because of the natural ability of souls (of humans, of animals, and even of

---

[45] Shumaker, *The Occult Sciences in the Renaissance*, pp. 42–54.
[46] Kusukawa, *The Transformation of Natural Philosophy*, pp. 134–44.
[47] Heydon was replying to an attack by the Oxford mathematician John Chamber, whose views were conventionally Aristotelian. See Donahue, *The Dissolution of the Celestial Spheres*, pp. 69–72; Christopher Heydon, *A Defence of Judiciall Astrologie* (Cambridge, 1603), chap. 12, p. 302, and chap. 18, p. 370; and John Chamber, *Treatise Against Judicial Astrology* (London: John Harison, 1601), chap. 20, pp. 100–2. For controversies among astrologers, see North, *The Norton History of Astronomy and Cosmology*, pp. 259–61.

the earth as a whole) to perceive the angular relations of rays.[48] Despite Kepler's efforts, his new astrology failed to catch on. Astrologers were too firmly convinced of the real effects of the zodiacal signs to abandon them, and Kepler's physical notions proved as troublesome to the astrologers as they did to the astronomers.

In the sixteenth century, astrology, especially regarding the casting of horoscopes, was therefore not exactly intellectually disreputable but was certainly controversial. To make predictions involving political leaders could be embarrassing or even dangerous. However, making predictions was practically unavoidable for persons with astronomical skills, and a clever (and lucky) forecaster could profit substantially from it.

From the patrons' perspective, on the other hand, astrology was not simply a means of telling fortunes but also a way of understanding the cosmos and mankind's place in it. Such curiosity played a substantial role in supporting astronomical research and providing a market for accurate planetary tables and ephemerides. Without this support, the work of astronomers such as Kepler and Galileo would not have been possible, or at least would have taken a very different turn.

KEPLER'S REVOLUTION

Kepler, in addition to being one of the best-known astrologers of his day, was the first of a new breed of astronomers, a second-generation Copernican. At Tübingen he had studied under Maestlin, who was educated in the Lutheran astronomical tradition established by Melanchthon. However, unlike most in that tradition, Maestlin accepted Copernicus's physical claims, and although his textbook was conventionally geocentric, his students were encouraged to explore the physical and cosmological questions raised by Copernicus. Kepler took full advantage of this opportunity, arguing Copernican positions in public disputations.[49]

Although Kepler had intended to become a Lutheran minister, he accepted a post teaching mathematics at the Stiftsschule in Graz, Austria. While there, he began to ask questions that no one had previously raised, such as how the planets' distances from the sun were established and what, in the absence of real spheres and orbs, determines the planets' motions. After a remarkable flash of insight in the summer of 1595, he published some preliminary answers in his first major work, the *Mysterium cosmographicum* (The Cosmographical

---

[48] Johannes Kepler, *Tertius interveniens* (Frankfurt: Tampach, 1610), in *Gesammelte Werke*, vol. 4, *Kleinere Schriften* (1941), p. 147; Judith V. Field, *Kepler's Geometrical Cosmology* (London: Athlone, 1988), pp. 127–42; and Field, "A Lutheran Astrologer: Johannes Kepler," *Archives for History of Exact Sciences*, 31 (1984), 189–273, which includes a translation of Kepler's *De fundamentis astrologiae certioribus* (Prague, 1602).
[49] Hellman, *The Comet of 1577*, pp. 137–44; and Caspar, *Kepler*, pp. 46–7.

Mystery, 1596). Kepler found that the five regular solids (tetrahedron, cube, octahedron, dodecahedron, and icosahedron), if arranged in the correct order, would fit between the orbits of the six planets with fair accuracy, and he argued that there must be a motive power in the sun that makes planets move faster when they are closer to it. This speed rule, which was an expression of qualitative observations and speculative physical reasoning, was later given a more precise mathematical form in Kepler's *Astronomia nova* (New Astronomy, 1609) and developed into what later came to be called Kepler's Second Law.[50]

The *Mysterium* was a direct attack on the conventional division of the sciences, in that Kepler argued (from scripture) that God created everything according to certain fundamental magnitudes, and that mathematics, being the science of magnitude, is fully qualified to establish physical truths. Although Kepler's views evolved with time, he later acknowledged that all of his later astronomical work had sprung from one or another chapter of that first book.[51]

When the *Mysterium cosmographicum* came to Tycho Brahe's attention, he recognized Kepler's brilliance but saw that he needed to learn the discipline of working with good observations. He invited Kepler to Prague (where Brahe had taken the post of imperial mathematician at the court of Rudolf II) to work with him on planetary theory. Meanwhile, the Counter-Reformation had made life difficult (and eventually impossible) for Lutherans in Graz, so Kepler accepted Brahe's offer, visiting Prague in the spring of 1600 and moving there with his family in October. Once there, he worked on studying Mars, which was a fortunate choice because the large eccentricity of its orbit made it more appreciably elliptical than the orbits of most other planets. The ensuing campaign (as Kepler described it) resulted in Kepler's greatest work, *Astronomia nova*, in which he continued his efforts to construct an astronomy based on physics.

Kepler's physics, however, was something quite different from what was taught in the schools: It involved invisible forces and powers that took on mathematical dimensions, and thus it brought the qualitative physics of the schools together with the quantitative precision of Brahe's observations. Kepler's theoretical method involved three distinct levels: The general physical principles were used to develop geometrical models of planetary motion, which were capable of generating planetary positions that could be checked against the observations. The discrepancies that were noted would then in

---

[50] Caspar, *Kepler*, pp. 46–71; Johannes Kepler, *Mysterium cosmographicum: The Secret of the Universe*, trans. A. M. Duncan (New York: Abaris Books, 1981), chap. 14, pp. 155–9, and chaps. 20–22, pp. 197–221; William H. Donahue, "Kepler's Invention of the Second Planetary Law," *British Journal for the History of Science*, 27 (1994), 89–102; and E. J. Aiton, "Kepler's Second Law of Planetary Motion," *Isis*, 60 (1969), 75–90. The first published statement of this law is in Kepler's *Astronomia nova* (Heidelberg: Vögelin 1609), chap. 40, p. 193.

[51] Johannes Kepler, *Mysterium cosmographicum*, "Dedicatory Epistle" to the second edition, p. 39.

turn help Kepler revise the geometrical models. But any such revision would have to be done in conformity with the physical principles, whose operation would sometimes have to be reformulated to fit the new models. It was an entirely unprecedented way of building a theory; Kepler called it "astronomy without hypotheses," not because no assumptions were made but because each assumption was tested against the observations and against the restraints of physical possibility.[52]

In *Astronomia nova*, Kepler proposed a more exact version of the speed rule that he had originally stated in the *Mysterium cosmographicum* and showed that the orbit of Mars is elliptical. He later extended these principles to the other planets. But he still had no satisfactory expression for the relationship between orbital period and mean distance for multiple planets. He had come up with a provisional proportion in the *Mysterium cosmographicum*, but it failed to fit the observationally determined values. Nearly two decades after the publication of the *Mysterium*, in another flash of insight, Kepler came upon the accurate expression he had been seeking: In modern terms, it stated that the squares of the periods of the planets are proportional to the cubes of their distances from the sun. Kepler was unable to deduce this from physical principles as a general truth but saw it as an artificial proportionality that God had contrived by adjustment of the planets' sizes and densities.[53] Here, as in his elaborately deduced system of celestial harmonies in *Harmonices mundi* (Harmony of the World, 1618), Kepler was thinking along lines that seem strange to modern readers largely because the kinds of questions he was asking, such as what reasons God had for making things the way they are, do not seem meaningful in the context of modern science.

Kepler's astronomy was so radically different from anything that had preceded it, and so idiosyncratic, that he had difficulty convincing others of its truth. Although his introductory textbook, the *Epitome astronomiae Copernicanae* (Epitome of Copernican Astronomy, 1618–21), helped, it was the accuracy of his predictions that ultimately prevailed. Kepler's last great work, the *Tabulae Rudolphinae* (Rudolphine Tables, 1627), had been planned for nearly thirty years and was intended as a fitting tribute to his patron, Emperor Rudolf II, by whose appointment Kepler succeeded Brahe as imperial mathematician in 1601. These tables made the details of Kepler's planetary theories available to other astronomers, and comparisons of predictions, especially of

---

[52] Bruce Stephenson, *Kepler's Physical Astronomy* (New York: Springer-Verlag, 1987; Princeton, N.J.: Princeton University Press, 1994); Kepler, *New Astronomy*, trans. Donahue; and Rhonda Maartens, *Kepler's Philosophy and the New Astronomy* (Princeton, N.J.: Princeton University Press, 2000).

[53] The first statement of this relationship appears in Kepler's *Harmonices mundi*, 5.3, translated in *Great Books of the Western World*, 54 vols. (Chicago: Encyclopaedia Britannica, 1955), vol. 16: *Ptolemy, Copernicus, Kepler*, p. 1020. Compare Field, *Kepler's Geometrical Cosmology*, pp. 142–63; and Bruce Stephenson, *The Music of the Heavens: Kepler's Harmonic Astronomy* (Princeton, N.J.: Princeton University Press, 1994), pp. 140–5.

a transit of Mercury across the sun in 1631 and a transit of Venus in 1639, clearly revealed Kepler's superiority.[54]

GALILEO

Kepler's older contemporary Galileo disagreed with him about almost everything except the physical reality of heliocentric astronomy. Galileo did not care much about the details of planetary motions and to his death continued to think of them as practically circular. His mathematical work dealt with terrestrial motions, though he extended some of his conclusion to the heavens. And aside from his mathematical treatment of motion (see Bertoloni Meli, Chapter 26, this volume), his most important work involved observations: the reinvention of the "optical tube" and transformation of this erstwhile toy into a scientific instrument, the telescope. These two aspects of Galileo's work, mathematical and empirical, were brought into harmony in his radically new approach to the understanding of nature.

Galileo had a thorough training in the natural philosophy of the schools. However, his main interest was in mathematics, especially practical mathematics. His earliest publications involved instruments, and he had a fondness for artisans, whose insights he often valued above those of his fellow academics. Galileo had not particularly cultivated astronomy, though by 1596 he was convinced of the truth of the Copernican system, and his observations of the new star (a galactic supernova) of 1604 show that he agreed with the astronomers' view that measurements of celestial phenomena could prevail over the arguments of the natural philosophers that the heavens must be unchangeable.[55]

Nevertheless, the publication that made Galileo famous showed none of this: It was a simple and straightforward account of his first observations with telescopes he had made himself, after hearing a report of the invention of such an optical device in Holland (see Bennett, Chapter 27, this volume). Galileo turned his new instrument, the design of which he had greatly improved, on the heavens, first observing the moon carefully and noting that its surface appeared to be rough and mountainous. Later, looking at Jupiter, he noticed what he thought were three small stars with which the planet was aligned. Subsequent observations showed that these stars, together with a fourth companion not seen at first, moved with Jupiter and appeared to be orbiting it. Observing the fixed stars, he found the Milky Way to be a vast field of tiny, hitherto unknown stars, and he noted that the telescope did not magnify the bodies of the stars, showing that their angular sizes were much less than

---

[54] Wilbur Applebaum, "Keplerian Astronomy after Kepler: Researches and Problems," *History of Science*, 34 (1996), 462–4.
[55] Stillman Drake, *Galileo at Work* (Chicago: University of Chicago Press, 1978), chaps. 1–6, esp. pp. 106–8.

had been previously thought.⁵⁶ Although he did not draw any conclusions here, this observation countered one common objection to the Copernican arrangement: If the stars were at the huge distances implied by the lack of observable annual parallax, they would have to be almost unimaginably large.

These startling discoveries, published in the *Sidereus nuncius* (1610), illustrated most dramatically the power of observations to affect the plausibility of physical theories. Galileo's own claims were modest: He merely remarked that the discovery of Jupiter's moons removed one common objection to the Copernican system by showing that earth was not the only planet to have a satellite. But the observations spoke for themselves. Of course, there were some who doubted the reality of what appeared in the telescope, but within a year or two they had been discredited or convinced, especially when the observations were confirmed by independent observers in England, France, and other countries.⁵⁷ It was also clear that these new phenomena presented grave difficulties for traditional physics and cosmology. If the moon is rough and mountainous, like earth; if the stars, despite their brightness, are angularly tiny (and might therefore be very distant); if other planets have moons; and if (as Galileo soon announced) Venus goes through phases of illumination, like the moon, then how is the distinction between the changeable earth and the perfect, eternal heavens to be maintained?⁵⁸

In the two decades following publication of his *Sidereus nuncius*, Galileo became a more outspoken proponent of the Copernican system. Although controversy about Jupiter's satellites was still raging, Galileo noticed that Venus was going through phases similar to the moon's, and he argued, in letters that were soon made public, that this phenomenon was inconsistent with the Ptolemaic arrangement of the planets but confirmed heliocentrism. He boldly raised the theological issues, repeating Cardinal Baronius's remark that "Scripture tells us how to go to heaven, not how the heavens go."⁵⁹ In a calculated act of rashness, he picked a lengthy fight with the Jesuits, who had been among the most receptive toward the new discoveries among the traditional philosophers. Shortly after his first telescopic observations, Galileo had found dark spots on the face of the sun, which he believed to be changing blotches on the sun's surface, evidence of alteration in the supposedly

---

⁵⁶ Galileo Galilei, *Sidereus nuncius; or, the Sidereal Messenger*, trans. Albert Van Helden (Chicago: University of Chicago Press, 1989).
⁵⁷ For an account of the reception of the telescopic observations, see Albert Van Helden's conclusion to his translation of Galileo's *Sidereus nuncius*, pp. 90–113. Prominent skeptics who came to accept the observations were Christoph Clavius and Giovanni Antonio Magini (1555–1617), professor of mathematics at Bologna.
⁵⁸ Galileo, *Sidereus nuncius*, pp. 84, 104–6.
⁵⁹ Galileo announced the discovery of Venus's phases in December 1610 in letters to Giuliano de' Medici and Benedetto Castelli: see Van Helden's conclusion to *Sidereus nuncius*, pp. 105–9; see also Galileo Galilei, *Discoveries and Opinions of Galileo*, trans. Stillman Drake (Garden City, N.Y.: Doubleday Anchor Books, 1957), pp. 74–5. Galileo ascribed the epigram to Baronius in a marginal note to his "Letter to the Grand Duchess Christina," written in 1615 but not published until 1636: see Galileo, *Discoveries and Opinions*, p. 186.

unchanging heavens. The Jesuits at the Collegio Romano, although accepting the observations, set about explaining them as occultations by numerous small planets orbiting the sun, thus maintaining the heavens' incorruptibility, in accord with Aristotelian teaching. Such interpretations attempted to bend scholastic physical reasoning to accommodate Galileo's discoveries, dulling their revolutionary impact.

Galileo, in turn, believed that a mere revision of the qualitative physics of the schools would not do and proposed a "reading" of nature, after the humanist manner of reading and emulating classical texts, in which one would seek analogies rather than deductions.[60] A good example of this is his most famous work, the *Dialogo sopra i due massimi sistemi del mondo* (Dialogue on the Two Chief World Systems, 1632), a major part of which is devoted to a consideration of whether things in the heavens are like or unlike the things around us. It is conducted with experiments and examples that hold scholastic opinions up to ridicule.[61]

The effect of this "reading" on astronomy was profound. One would no longer begin, as Ptolemy had, with questions about the shape of the universe and the earth's place in it. Such questions were left open: Galileo emulated Socrates in stressing the importance of admitting ignorance as a first step toward learning.[62] The motion of the heavens was to be investigated not by drawing distinctions, refining definitions, and formulating hypotheses but by careful observation and analogical argument.

DESCARTES' COSMOLOGY

Although the French mathematician and philosopher René Descartes (1596–1650) did no significant work in astronomy, his cosmological theories shaped the way in which generations of astronomers viewed the world. In his *Principia philosophiae* (Principles of Philosophy, 1644) and in the posthumous *Le Monde* (The World, 1664),[63] he outlined a fanciful cosmogony in which a few very simple rules about collisions could lead to the formation of the universe as we see it (see Bertoloni Meli, Chapter 26, this volume). Variously sized chunks of matter are set variously in motion; and they begin to collide with one another and so are deflected into curved paths. This evolves into a

---

[60] For an account of Galileo's polemics against the Jesuits, with references to source documents, see Pietro Redondi, *Galileo Heretic* (Princeton, N.J.: Princeton University Press, 1987), pp. 28–67. English translations of the central published works in this debate appear in Stillman Drake and C. D. O'Malley, trans., *The Controversy on the Comets of 1618* (Philadelphia: University of Pennsylvania Press, 1960).
[61] Galileo Galilei, *Dialogue on the Two Chief World Systems – Ptolemaic and Copernican*, trans. Stillman Drake (Berkeley: University of California Press, 1953), First Day, esp. pp. 60–101.
[62] Galileo, *Discoveries and Opinions*, pp. 256–8.
[63] *Le Monde* was written before the *Principia*, between 1629 and 1633, but Descartes decided not to publish it when he heard of Galileo's condemnation. Nevertheless, Descartes wryly claims that the world he is describing is imaginary.

large number of whirlpools or vortices, which serve to sort out the matter by size, and the more subtle matter is squeezed in toward the center and the very rapid motion it acquires generates light. Thus there is a star at the center of each vortex, with planets, comets, and so on, carried around as flotsam with the whirling fluid.[64]

Descartes' cosmology appealed to a generation of students raised on a modernization of school philosophy that had already accepted Brahe's fluid-filled heavens and Galileo's telescopic observations. It had the advantage of coherence and simplicity, and it kept religious questions mostly separate from scientific matters by creating a separate category for minds, souls, and spirits. In its superficial aspects, the vortex theory resembled the cosmic swirl of Hermetic neo-Platonist philosophers such as Telesio and Patrizi, a comparison to which Descartes strongly objected because he had banished the world soul and other animate agents. Nevertheless, the vortex model inevitably appealed to those who might otherwise have been drawn to the Hermeticists. Further, the vortex model suggested a way of explaining and predicting planetary motions that avoided the somewhat occult appearance of Kepler's invisible forces, powers, and planetary souls.[65]

THE SITUATION CIRCA 1650: THE RECEPTION OF KEPLER, GALILEO, AND DESCARTES

After the publication of Kepler's *Rudolphine Tables*, their superior accuracy made it only a matter of time before they came to be generally accepted. But tables do not themselves give planetary positions; they only provide the means for calculating them. And astronomers found the Rudolphine computations, especially the area rule, which related a planet's position on the orbit to the area of the sector it has traversed (for which there was no direct solution), cumbersome. So even those sympathetic to Kepler's melding of astronomy and physics found the idea of a geometrical approximation of the area rule attractive. Nevertheless, although a number of such shortcuts were

---

[64] E. J. Aiton, *The Vortex Theory of Planetary Motions* (History of Science Library) (London: MacDonald, 1972), pp. 30–64.
[65] The resemblance was pointed out to Descartes by Isaac Beeckman (1588–1637). The exchange is in Beeckman, *Journal de 1604 à 1634; publié avec une introduction et des notes par C. de Waard*, 4 vols. (La Haye: Martinus Nijhoff, 1939–53), 1: 260–1 and 360–1; 4: 49–51. Natural philosophers who mingled Cartesian with Hermetic ideas include Henry More (1614–1687), Daniel Lipstorp (1631–1684), and Athanasius Kircher (1601–1680). Accounts of at least four other contemporary authors who saw Cartesian and Hermetic ideas as congenial are provided by P. M. Rattansi, "The Intellectual Origins of the Royal Society," *Notes and Records of the Royal Society*, 23 (1968), 129–43, at pp. 131–6; and Charles Webster, "Henry Power's Experimental Philosophy," *Ambix*, 14 (1967), 150–78, at p. 153. A full account of mingled Cartesian theories is found in Donahue, *The Dissolution of the Celestial Spheres*, pp. 137–42, 287–92.

developed, the two fundamentally Keplerian ideas of nonuniform motion and noncircular orbits had become firmly established by mid-century.[66]

There was less agreement about the physical causes of the motions, Kepler had opened the debate in his *Astronomia Nova* (1609), developing a theory in which the rotating sun sweeps the planets around by means of quasi-corporeal "fibers," while the planets are moved toward and away from the sun by magnetic attractions and repulsions. Some astronomers, especially early in the century, were as a matter of principle unwilling to accept such physical explanations in astronomy.[67] However, by 1650 there were at least six competent mathematicians and astronomers in Germany, France, the Low Countries, and England who fully endorsed and expounded Kepler's theories in published works or in other ways.[68] Others mingled Keplerian ideas with Galilean or Cartesian physics. One difficulty with Kepler's physical ideas was that he believed, along with Aristotle, that motions would not continue without some moving cause. Those who accepted Galileo's and Descartes' arguments that things in uniform rectilinear motion tend to keep moving by themselves were faced with the challenge of finding a replacement for Kepler's tangential forces and magnetic attractions and repulsions. Suggestions came from the Dutch Cartesian mathematician Christiaan Huygens (1629–1695); Gilles Personne de Roberval (1602–1675), professor of mathematics at the University of Paris and also a Cartesian; the English natural philosophers Robert Hooke (1635–1703) and Thomas Hobbes (1588–1679); the Italian physiologist and mathematician Giovanni Alfonso Borelli (1609–1679); and others. The English mathematician and astronomer Jeremiah Horrox (1619–1641), though a confirmed Keplerian, noticed that Jupiter and Saturn appeared to be interacting, and this suggested that the planets might be attracting each other. Nevertheless, only one of these theorists, Borelli, came up with an alternative physical account that could generate planetary positions before the publication of Newton's *Principia*.[69]

The main alternative to Kepler's theories was their geometrical reworking by the French astronomer Ismael Bouillau (1605–1694). He set out systematically to refute Kepler's physics, replacing it with an elaborately argued theory that, in essence, ascribes to each planet a natural form that governs its motion, that form being the section of a cone, around the axis of which the planet moves uniformly. Bouillau's astronomy was well received, especially in England, and was commended by Newton. However, at least part of its success must be attributed to the substitution of an old-style equant for Kepler's

---

[66] Applebaum, "Keplerian Astronomy," pp. 484–5.
[67] Ibid., p. 459.
[68] Donahue, *The Dissolution of the Celestial Spheres*, pp. 249–50 (citing Noel Duret, Johannes Hainlin, Pierre Herigone, Johannes Phocylides Holwarda, and Jeremiah Horrox), and p. 293 (citing Johannes Hevelius).
[69] Aiton, *The Vortex Theory of Planetary Motions*, pp. 90–8, 126–7; and Alexandre Koyré, *The Astronomical Revolution: Copernicus, Kepler, Borelli* (New York: Dover, 1992).

cumbersome area calculation. Astronomers continued to use whatever system or combination of systems appealed to them, and in principle thought of them as more or less equivalent. However, in 1670, Nicolaus Mercator (ca. 1619–1687), a Danish astronomer living in England, took up the question of the accuracy of the geometrical methods. He showed that all such methods amounted to a solution that Kepler had tried and rejected, which involved uniform angular motion around the empty focus of the ellipse. In Mercator's view, Kepler's ellipses and his area rule must stand or fall together. This proved to be a turning point in the further history of Kepler's planetary theories.[70]

With regard to the physical system into which these planetary orbits fit, Galileo's heresy trial in 1633 changed everything. The question of heliocentrism versus geocentrism (in Catholic countries at least) ceased to be a matter of physics and became a matter of faith. This had the effect of stifling debate and encouraged the adoption of the Tychonic system as a prudent compromise. Influential Jesuits such as Christoph Scheiner (1573–1650), professor of mathematics at the University of Ingolstadt, favored the Tychonic system as early as 1612, and after Galileo's condemnation it was universally espoused by the Jesuits. As late as 1728, the Italian Jesuit Francesco Bianchini (1662–1729) described it as the commonly accepted system. In Lutheran Germany and Anglican England, this arrangement had few adherents – new physical and cosmological ideas had made the idea of a moving earth much less problematic.[71]

The increasing acceptance of the heliocentric (or semiheliocentric) system raised another question: the dimensions of the planetary system and the corresponding distance of the stars. The Copernican arrangement gave relative distances of the planets from the sun but did not relate those distances to familiar terrestrial units. Such a calibration would require the measurement of the diurnal parallax of the sun or a planet. (Diurnal parallax is the parallax that results from our being on the earth's surface rather than at its center.) Early attempts to find a parallax, such as those by Brahe and Kepler, produced mixed but ultimately negative results, which could nevertheless be useful in determining a lower limit for the distance. On the basis of the failure to find any parallax for Mars at opposition (when it is closest to earth), Kepler estimated the solar parallax as no more than one arc minute, making the sun's distance at least 22,000,000 km, about three times the generally accepted

---

[70] Aiton, *The Vortex Theory of Planetary Motions*, p. 91; Ismael Bullialdus [Ismael Boullian], *Astronomia Philolaica* (Paris: S. Piget, 1645), bk. 1, chaps. 12–14, pp. 21–37; North, *The Norton History of Astronomy and Cosmology*, pp. 356–9; Applebaum, "Keplerian Astronomy," pp. 470–1; and Curtis A. Wilson, "From Kepler's Laws, So-Called, to Universal Gravitation: Empirical Factors," *Archive for History of Exact Sciences*, 6 (1970), 106–32.

[71] Christine Schofield, "The Tychonic and Semi-Tychonic World Systems," in *The General History of Astronomy*, vol. 2A, ed. René Taton and Curtis Wilson (Cambridge: Cambridge University Press, 1989), pp. 42–4.

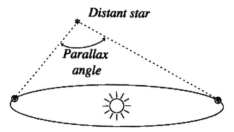

Figure 24.4. Annual parallax.

Ptolemaic distance. In the next half-century, a number of attempts were made (based on Keplerian speculations and unreliable observations) to find the distance more accurately. The resulting distances were all substantially greater than Kepler's lower limit. However, the first reliable measurements were made in 1672, applying the newly developed micrometer to the opposition of Mars in that year. Parallaxes measured by the English Astronomer Royal John Flamsteed (1646–1719) and the French astronomer Jean Richer (1630–1696) yielded distances of 132,000,000 km and 110,000,000 km, respectively. (The modern value is about 150,000,000 km.)[72]

This considerable enlargement of the planetary system made the apparent absence of annual parallax of the fixed stars (see Figure 24.4) all the more surprising and prompted attempts to measure it. Hooke claimed positive results in observations of the star Gamma Draconis in the 1670s using a telescope built into his house. However, his conclusions met with a mixed reception, and similar observations after 1692 by Danish astronomer Ole Römer (1644–1710) produced a negative result. Reliable measurements of stellar parallax, which is less than one arc second even for the closest stars (other than the sun), was beyond the capability of seventeenth-century instruments and had to await the discovery of the aberration of light by English astronomer James Bradley (1693–1762) in 1729 and of the proper motions of the stars, first proposed by English astronomer and natural philosopher Edmund Halley (1656–1743) in 1618.[73]

NOVAE, VARIABLE STARS, AND THE DEVELOPMENT OF STELLAR ASTRONOMY

Although the cosmological changes implicit in the works of Galileo and Descartes did not have an immediate effect on planetary theory, they helped

---

[72] Albert Van Helden, "The Telescope and Cosmic Dimensions," *General History of Astronomy*, 2A: 108–17; and Van Helden, *Measuring the Universe: Cosmic Dimensions from Aristarchus to Halley* (Chicago: University of Chicago Press, 1985), pp. 105–47.

[73] Schofield, "The Tychonic and Semi-Tychonic World Systems," p. 41. Römer, it should be noted, was a proponent of the Tychonic planetary arrangement, and thus saw a small parallax as favoring Tycho over Copernicus. For aberration and proper motions, see North, *History of Astronomy and Cosmology*, pp. 383, 395.

inspire a new interest in the stars. Even those who still held that the stars swirl around the earth once a day no longer felt the need to restrain them to a single spherical surface, and it was becoming common to think of the stars as being like the sun.[74] This raised questions about the distribution and distances of the stars that encouraged more careful attention to them. Furthermore, the telescope had revealed the existence of a great many previously unknown stars, so there was more to look at and much mapping to be done.

But the greatest spur to the study of the stellar region was the possibility of discovering a "new star" – that is, a nova (or supernova) such as the one that had so startlingly appeared in 1572. This nova, which at its peak brightness was visible even during the day, was studied by Brahe and many other astronomers, the prevailing opinion being that it was a real star, or possibly a celestial comet, and not an atmospheric phenomenon. The existence of this nova was very troublesome to those who followed Aristotle in believing the heavens to be eternal and unchanging.[75]

The question naturally arose whether any new stars had ever been previously reported and whether a more careful search would find more of them. A historical search turned up a few candidates, and observers found a new incentive for remapping the heavens in order to be sure of the stars that were already there. Brahe and Dutch cartographer and instrument maker Willem Blaeu (1571–1638) responded to this substantial challenge. The search for new novae was also successful: Less prominent new stars were found by the Frisian astronomer David Fabricius (1564–1617) in 1596 and by Kepler and others in 1600 (and perhaps also in 1602), and a much brighter one (another supernova) was observed at the end of 1603 and through much of 1604.[76]

Over the next three decades, there were a few doubtful or spurious sightings (including one that was in fact the Great Andromeda Galaxy), and then a reliable one in 1638 by the Frisian astronomer Johannes Phocylides Holwarda (1618–1651). In 1640, while his book reporting the discovery was in press, Holwarda was astonished to discover that the star had reappeared. This was reported in a hastily added appendix. Subsequently, careful study of the star by the Danzig astronomer Johannes Hevelius (1611–1687) identified it as probably the same as Fabricius's "nova" of 1596, as well as the star Omicron Ceti in Blaeu's listing. After further observation and reference to earlier accounts, Bouillau found that the fluctuations were periodic, with a period of about 333 days, and he successfully predicted future maxima. He also proposed a physical explanation involving a lucid region of the star that was periodically displayed by the star's rotation.[77]

---

[74] Donahue, *The Dissolution of the Celestial Spheres*, pp. 298–300.
[75] Hellman, *The Comet of 1577*, pp. 111–17.
[76] This paragraph as well as the following account of variable stars is based on Michael A. Hoskin, "Novae and Variables from Tycho to Bullialdus," *Sudhoffs Archiv für Geschichte der Medizin und der Naturwissenschaften*, 61 (1977), 195–204.
[77] Ismael Bullialdus, *Ad Astronomos Monita Duo* (Paris, 1667).

The discovery of the periodicity of Omicron Ceti (or Mira, "the Wonder," as it was called) initiated a widespread interest in the stars that marked the beginning of stellar astronomy as a separate field of study. Thus there were many reports of variable stars in the next few years, and by the end of the century astronomers were beginning to suspect that even the old novae of Brahe and Kepler were really variable stars with periods of many years or even centuries.

NEWTON

The physical theories of Newton had a profound influence on planetary astronomy for centuries to come, establishing the ground rules for the construction of planetary orbits. Nevertheless, his direct contributions to astronomy were few. He made his scientific debut before the Royal Society in 1672 with a reflecting telescope that he had designed and built, but this was merely a by-product of his interest in optical theory. Newton later observed the comet of 1680–1 and the 1682 appearance of what came to be known as Halley's Comet, observations that played an important role in the development of his account of orbital motions in his most celebrated work, *Philosophiae Naturalis Principia Mathematica*, usually called simply *Principia*.[78] And between the first edition of *Principia* and the 1713 second edition, Newton developed a lunar theory, published in 1702.

The *Principia* itself, though not primarily an astronomical work, promised to transform planetary astronomy by providing a physical account of all planetary motions. In this, Newton nearly reached Kepler's goal of creating an "astronomy without hypotheses" to be based on physical forces. Much of Book III of *Principia* is devoted to explaining the various lunar wobbles and finding the orbits of comets. Furthermore, the theory of universal gravitation had implications that led to speculation about, and study of, the distribution of the stars.

When it came to the details, however, Newton's mathematics was not up to the task. It was not until the middle of the eighteenth century that gravitationally based lunar theories began to produce usable numbers. Meanwhile, there was a practical demand for a more accurate lunar theory. If the moon's position could be known accurately for any given time, it would become in effect a universal clock, allowing navigators to determine their longitude by comparing the moon's observed position among the stars with local time.

---

[78] For general information about Newton's life and works, see Richard S. Westfall, *Never at Rest: A Biography of Isaac Newton* (Cambridge: Cambridge University Press, 1980). An account of the telescope appears at pp. 232–40, and Newton's comet observations are at pp. 391–7. Newton's knowledge of contemporary astronomy is described in Wilson, "From Kepler's Laws, So-called, to Universal Gravitation," sec. 8, pp. 89–170. The importance of Newton's observations of Halley's Comet is argued in Nicholas Kollerstrom, "The Path of Halley's Comet, and Newton's Late Apprehension of the Law of Gravity," *Annals of Science*, 56 (1999), 331–56.

The determination of longitude was the most pressing practical scientific problem of the day, for the solution of which the British Parliament in 1714 offered a huge reward. By this time, the commercial importance of accurate astronomical measurements had long been evident, having led to the foundation of the Paris Observatory in the 1660s and the Greenwich Observatory in 1675.

It was partly in response to the longitude problem that Newton wrote *Theory of the Moon's Motion* (1702). This curious little work, first published as an addendum to David Gregory's *Astronomiae elementa* (Elements of Astronomy, 1702), was a purely kinematic and geometric theory, based on Horrox's modification of Kepler's lunar theory, using devices that dated back to ancient Greece. In fact, it was probably the last serious astronomical work to make use of the Ptolemaic epicycle.

Another reason for creating this theory was to establish accurately the various components of the moon's complicated orbit in order to begin the job of explaining it physically. A new version of the lunar theory, heavily disguised and with some corrections, but still recognizable, was included in the 1713 edition of *Principia*. In this revision, Newton attempted to give a physical explanation of as many of the equations as he could, but he finally had to abandon all pretense of a dynamic account – writing, "The computation of this motion is difficult," and continuing with a purely geometrical and kinematic description of the final inequality.[79]

The 1702 lunar theory, retrograde though it was, was remarkably influential. Its basic form (with various adjustments to the parameters) was used in publications by at least a dozen astronomers before 1750, many of whom used the theory to compute lunar tables and ephemerides. The theory fell into disuse after 1753 as a result of the publication of gravitationally based lunar theories by German cartographer and astronomer Tobias Mayer (1723–1762) and Swiss mathematician Leonhard Euler (1707–1783).[80]

The enormous significance of Newton's work for astronomy lay not in his direct contributions but in the way his establishment of universal gravitation and mathematical methods created a new context in which astronomical problems were investigated. In planetary astronomy, the most important question became how to apply gravitational theory to explain ever more subtle anomalies. In stellar astronomy, the great distance of the stars, which had once seemed such an implausible concomitant to the heliocentric planetary

---

[79] Isaac Newton, *The Principia*, trans. I. Bernard Cohen and Anne Whitman (Berkeley: University of California Press, 1999), bk. 3, prop. 35, Scholium, pp. 869–74. The epicycle is described at pp. 871–2, and the quoted phrase is at p. 873. The complete text of the *Theory of the Moon's Motion*, with a complete commentary and assessment of its importance, is in Nicholas Kollerstrom, *Newton's Forgotten Lunar Theory: His Contribution to the Quest for Longitude* (Santa Fe, N.M.: Green Lion Press, 2000). An account of the circumstances of the composition and publication of the lunar theory is in I. Bernard Cohen, *Isaac Newton's Theory of the Moon's Motion (1702), with a Bibliographical and Historical Introduction* (New York: Neale Watson, 1975).

[80] Kollerstrom, *Newton's Forgotten Lunar Theory*, pp. 205–23.

arrangement, was now seen as a necessary consequence of universal gravitation, a means of keeping the universe from collapsing on itself. Thus, the attempt to measure the annual stellar parallax continued, but on a much more refined scale. The stars themselves were now seen as independent bodies, possibly moving through space. (The existence of independent proper motions was soon confirmed, initially by Halley in 1718 and later by French astronomer Jacques Cassini II (1677–1756) in 1738 using more precise measurements.[81]) Newton's accomplishment was the completion of Kepler's quest to turn astronomy into physics, and, in the course of this, the recasting of physics as mathematics.

For observational astronomy, by the end of the seventeenth century it was clear that the current set of problems often required a degree of precision beyond the reach of individual investigators. The Royal Greenwich Observatory, in particular, became the center of a small instrument-making industry that supplied observatories all over Europe.[82] Although there was still room for the amateur observer (as indeed there is today), much of the most important astronomical observation would henceforth have to be done in government-supported institutions.

CONCLUSION

During the sixteenth and seventeenth centuries, astronomy had evolved from a mathematical discipline (with an observational component) into a true physical science. Initially, this transformation was the result of a change in context. Because of developments in philosophy and theology, especially the trend (exemplified by the philosophy of Descartes) toward banishing minds and souls from the physical universe, the heavens came to be seen as not radically distinct from the earth. Planetary astronomy accordingly changed from a purely descriptive geometrical construction into an ongoing project in gravitational dynamics involving physical bodies. Although the geometrical constructs were still widely used, Kepler's challenge of finding the true, physical orbits of the planets and their causes became recognized as the chief goal of planetary theory.

At the same time, stellar astronomy, which had previously consisted of charting the supposedly unchanging positions and magnitudes of the fixed stars, had become a discipline in its own right. Again, it was the Cartesian-Galileian-Newtonian context that helped transform the stars from points of light upon a spherical surface into great luminous bodies in an indefinitely large space. Good luck, in the form of several very prominent "new" stars that

---

[81] North, *History of Astronomy and Cosmology*, chap. 14, esp. pp. 383–4 and 395–6.
[82] Ibid., p. 381.

appeared in the late sixteenth and seventeenth centuries, also drew attention to the stellar region as worthy of more careful study.

The rise of stellar astronomy, as well as the increasing demands of planetary astronomy for better data, proved a powerful incentive for the development of better instruments (see Bennett, Chapter 27, this volume). Instrument making had always played an important role in astronomy, but the invention of the telescope, and the discoveries it immediately revealed, showed dramatically what instrumental innovation could achieve. The result was an enlarged and often publicly supported instrument-making industry that was required to produce devices whose precision and elaborate construction far surpassed that of earlier instruments. And as the use of the new instruments became more technically demanding, observational astronomy became a specialty in its own right.

25

ACOUSTICS AND OPTICS

Paolo Mancosu

During the early modern period, music (of which acoustics is an offspring) and optics belonged to the "mixed mathematical" sciences. "Mixed mathematics" refers here to those physical disciplines that could be treated by extensive use of arithmetic or geometric techniques, such as astronomy, mechanics, optics, and music (see Andersen and Bos, Chapter 28, this volume).

The study of sound in the sixteenth and seventeenth centuries cannot be properly considered to belong to any single discipline but rather is found at the intersection of several fields, including music theory, mechanics, anatomy, and natural philosophy. Thus, no single mixed mathematician of the sixteenth or the seventeenth century can be properly said to have specialized in acoustics. Among the early modern scholars who contributed to the study of sound were mixed mathematician Giovanni Battista Benedetti (1530–1590), musician Vincenzo Galilei (1520–1591), and natural philosopher Robert Boyle (1627–1691), which gives some idea of the variety of disciplinary approaches. It is nonetheless safe to say that the study of music theory provided the common background on the basis of which further studies on sound phenomena would be undertaken. Moreover, in the area of natural philosophy, the classical treatises *De sensu* (On the Senses), *De audibilibus* (On Things Audible), *De anima* (On the Soul), and the *Problemata* (Problems), all attributed to Aristotle at the time, contained material pertaining to acoustic phenomena and were well known to sixteenth- and seventeenth-century scholars.

The practitioners involved in the study of sound were socially disparate as well, ranging from the choirmaster of St. Mark's in Venice, Gioseffo Zarlino (1517–1590); to a schoolteacher in the Netherlands, Isaac Beeckman (1588–1637); to a Minim friar in Paris, Marin Mersenne (1588–1648). Others taught

---

I would like to thank Fabio Bevilacqua, Lorraine Daston, Toni Malet, Katharine Park, Neil Ribe, and an anonymous referee for several suggestions that greatly improved the first draft of this chapter. I would also like to express my gratitude to the Wissenschaftskolleg zu Berlin for the fellowship I was granted in 1997–8, during which time the first draft of the chapter was written.

in the universities, including the English mathematicians and natural philosophers John Wallis (1616–1703) and Isaac Newton (1642–1727). In the second part of the seventeenth century, the new scientific academies also encouraged experimental studies of acoustic phenomena.

In contrast, the subject matter of early modern optics constituted a much more unified discipline. Although it encompassed a wide variety of phenomena, including theories of vision, the rainbow, and the nature of light, its practitioners could count on a shared understanding of its characteristic problems and approaches that harked back to the study of *perspectiva* in the Middle Ages. As in the case of acoustics, however, the practitioners of optics came from very different backgrounds. Johannes Kepler (1571–1630) was imperial mathematician and astronomer in Prague, René Descartes (1596–1650) an independent scholar living in Amsterdam and Paris, and Francesco Grimaldi (1618–1663) a Jesuit mathematician teaching in Bologna. Artisans were also important in the development of acoustics and optics. Mersenne regularly consulted instrument makers for details concerning the workings of musical instruments, and Descartes worked closely with lens grinders (see Bennett, Chapter 27, this volume).

Despite the diversity of these mixed mathematicians, they shared some common background; most would have had an education encompassing *musica*, *perspectiva*, and Aristotelian natural philosophy. They communicated with each other using a variety of means, including letters – Mersenne, for instance, was at the center of a wide epistolary network of over seventy scholars – and through their published works.

The fortunes of optics and music diverged from one another both in the early modern period and in modern histories of science. Optics, together with astronomy and mechanics, has enjoyed pride of place in the works of historians of science. In contrast, historical interest in the study of music as a scientific subject is a much more recent development, beginning only in the 1960s; it is only since the 1980s that substantial new research has began to dramatically increase our knowledge of the relationship between music and the Scientific Revolution. Among the goals of this chapter are to call attention to the richness of the material published in the area of music and the study of nature and to review the substantial historiography of early modern optics.

MUSIC THEORY AND ACOUSTICS IN THE EARLY MODERN PERIOD

Several major trends mark the development of music theory and acoustics in the sixteenth and seventeenth centuries. Music theory saw a shift from a generalized Pythagoreanism, which emphasized abstract numerical considerations as central to the foundations of the science of music, to a new style of analysis characterized by increased experimentation and a physical approach to sound. In the process, the science of acoustics emerged as a discipline independent of the mathematical field of music theory and dedicated to the experimental investigation of sound phenomena. Moreover, these investigations no longer focused on the "sonorous number" of the Pythagorean tradition; rather, numbers became merely instruments to measure physical phenomena and relationships, such as the frequency of vibration of a string. Indeed, Mersenne's rule (*regle*) or law for vibrating strings is one of the major achievements of acoustics in the seventeenth century. One major consequence of the emergence of acoustics as an independent branch of physics was the progressive separation of issues relating to the physics of sound from psychological issues relating to sound perception, such as the pleasurable perception of consonant sounds.

In order to understand these shifts, it is first necessary to understand the classical traditions against which the new experimental science of sound developed, specifically the Pythagorean and Aristoxenian traditions in music theory, which came into conflict in the late Renaissance – a development central to the "scientific" aspects of music theory. Against this background, I will describe the major achievements in seventeenth-century acoustics and its relationship to music theory, with particular attention to the problem of consonances. In addition to specific experimental results, I will survey a variety of contemporary theories elaborated to account for sound propagation and sound perception. Here the most important development consisted in the coupling of physical models of sound propagation and perception with a mechanical philosophy of nature, a combination that encouraged research in the area of the physiology and anatomy of the ear.[1] Lack of space unfortunately precludes a discussion of the interaction of music theory (including theories of sound) and musical practice, a topic that is analogous to the influence of the theories of optics and perspective on painting.[2]

THE SIXTEENTH CENTURY: PYTHAGOREAN AND ARISTOXENIAN TRADITIONS

Before the recovery in the fifteenth and sixteenth centuries of most Greek texts dealing with the science of music (called by the Greeks *musike* or *harmonike*), the traditional source for music theory in the period before 1500

---

[1] For the history of the anatomy and physiology of the ear, the classic reference is A. Politzer, *Geschichte der Ohrenheilkunde*, 2 vols. (Stuttgart: Enke, 1907–13, repr. Hildesheim: Georg Olms, 1967). For a lively account of developments in the area of the anatomy and physiology of the ear, see Chapter 13 of Alistair C. Crombie, *Styles of Scientific Thinking in the European Tradition: The History of Argument and Explanation Especially in the Mathematical and Biomedical Sciences and Arts*, 3 vols. (London: Duckworth, 1994), 2: 1154–66.
[2] For a first orientation, see Claude V. Palisca, *Humanism in Italian Renaissance Musical Thought* (New Haven, Conn.: Yale University Press, 1985); and H. Floris Cohen, *Quantifying Music: The Science of Music at the First Stage of the Scientific Revolution, 1580–1650* (Dordrecht: Reidel, 1984).

was the Roman mathematician and philosopher Boethius's (ca. 480–ca. 525) *De institutione musica* (The Fundamentals of Music, sixth century),[3] whose contents were strongly influenced by Pythagorean and Platonic conceptions of music.[4] According to the Pythagorean worldview, the essence of musical phenomena, indeed of all cosmological phenomena, consisted of numerical ratios. Medieval and early Renaissance music texts described time and time again how Pythagoras discovered the numerical ratios underlying musical intervals by observing that the sound emitted by hammers of different weights striking an anvil was proportional to their weights, together with several other alleged observations, which involved glasses partially filled with water, bells, pipes, and strings stretched by weights (Figure 25.1).

The last device inspired a rich iconographical tradition, epitomized in the Portail Royal of the Chartres Cathedral (twelfth century), which shows Pythagoras plucking the string of an instrument known as the monochord, which consisted of a single string stretched over a resonating chamber.[5] A movable bridge allowed for a lengthening or shortening of the vibrating segment of the string, permitting the study of pitches and their relationships. In particular, the Pythagoreans observed that the octave corresponded to a numerical ratio of 2:1 between the lengths of the vibrating segments of the string required to produce the two sounds. For example, one might first pluck the open string and then move the bridge to the exact middle and pluck one of the segments, corresponding to half of the entire string. Other intervals easily characterized in terms of numerical ratios between string lengths were the fifth, which was given by the ratio 3:2, and the fourth, which corresponded to the ratio 4:3. Figure 25.2 represents a more elaborate monochord obtained by using two movable bridges, indicating how to obtain a fifth (*diapente*), a fourth (*diatessaron*), a major third (5:4) (*dytonus*), and a minor third (6:5) (*semidytonus*).[6]

The Pythagoreans accepted as perfect intervals only those intervals that could be obtained by means of ratios involving the integral numbers 1 to 4.

---

[3] Whereas the main source for music theory up to the early part of the fifteenth century had been Boethius's *De institutione musica* (sixth century), humanist interest in Greek culture revived the theoretical texts of, among others, Cleonides, Ptolemy, Aristoxenus, Aristides Quintilianus, and Plutarch. Claude Palisca has reconstructed in detail the transmission of the ancient Greek sources during the fifteenth and sixteenth centuries; see Palisca, *Humanism in Italian Renaissance Musical Thought*.

[4] On Boethius's musical theory, see J. Caldwell, "The *De institutione arithmetica* and the *De institutione musica*," in *Boethius: His Life, Thought and Influence*, ed. M. T. Gibson (Oxford: Oxford University Press, 1981), pp. 135–54. The standard translation of Boethius's *De institutione musica* is *The Fundamentals of Music*, introduction and notes by Calvin M. Bower, ed. Claude V. Palisca (New Haven, Conn.: Yale University Press, 1989).

[5] For the iconographical tradition in the Middle Ages of Pythagoras as a musician, see Barbara Münxelhaus, *Pythagoras Musicus: zur Rezeption der pythagoreischen Musiktheorie als quadrivialer Wissenschaft im lateinischen Mittelalter* (Bonn: Verlag für Systematische Musikwissenschaft, 1976).

[6] As will become clear, the major third and the minor third were not considered consonant intervals by Pythagorean music theorists before Zarlino.

Figure 25.1. Pythagoras with musical devices. In Franchino Gaffurio, *Theorica musice* (Milan: Filippo Mantegazza for G.P. da Lomazzo, 1492; repr. Bologna: Forni, 1969). Reproduced by permission of the Staatsbibliothek zu Berlin – Preussischer Kulturbesitz/bpk 2004.

Insisting upon whole ratios of numbers from 1 to 4 made the theory of music a very abstract discipline that was often remote from musical practice.[7]

---

[7] In addition to mathematical developments, there were cosmological and ethical aspects of music that were given pride of place in the Pythagorean-Platonic tradition. In Boethius, these took the form of a distinction between *musica mundana*, *musica humana*, and *musica instrumentalis*. *Musica mundana* corresponds to the idea of world harmony and investigates, among other things, the movements of the heavenly bodies. To this category belonged the tradition of the "music of the spheres," which postulated that the planets emit sounds in their motions and that the sounds emitted by different planets harmonize. This idea greatly influenced Kepler, as the title of his work *Harmonices mundi* (1619) reveals. *Musica humana* (harmony of the body) was concerned, among other things, with the harmonious relationship between body and soul and the balance between the different humors in the body; it thus had connections to medicine and suggested to humanists, such as Marsilio Ficino, who was also a physician, the possibility of altering the feelings and balance of elements in the human being. This is the great theme of the marvelous effects of music, which is also of ancient origin but was given new life by the humanists. Finally, *musica instrumentalis* dealt with the practice of music itself. There is a vast body of literature in this area. For an overview, see Eberhard Knobloch,

Figure 25.2. Monochord with two movable bridges. Lodovico da Fogliano, *Musica theorica* (Venice: Ioannem Antonium & fratres de Sabio, 1529), fol. 13r. Reproduced by permission of the Bibliothek der Staatlichen Hochschule für Musik, Trossingen.

"Harmony and Cosmos: Mathematics Serving a Teleological Understanding of the World," *Physis*, nuova ser., 32 (1995), 55–89. On Kepler, see Bruce Stephenson, *The Music of the Heavens: Kepler's Harmonic Astronomy* (Princeton, N.J.: Princeton University Press, 1994); and Daniel P. Walker, *Studies in Musical Science in the Late Renaissance* (Leiden: E. J. Brill, 1978). On magic, Ficino, and related topics, see G. Tomlinson, *Music in Renaissance Magic* (Chicago: University of Chicago Press, 1993); and Daniel P. Walker, *Spiritual and Demonic Magic from Ficino to Campanella* (London: Warburg

The extreme abstraction of the Pythagorean-Platonic tradition had in antiquity already been challenged by a more empirical approach to music. Plato, in the *Republic*,[8] wrote of those "excellent fellows who torment their strings, torturing them, and stretching them on pegs." They put "their ears to their instruments like someone trying to overhear what their neighbors are saying" – an approach Plato condemned because its proponents "put ears before understanding" (*Republic*, 531a). The tradition described by Plato soon found an eminent theorist in Aristoxenus (fourth century B.C.E.), whose work emphasized the discerning ability of the ear against the abstract, and often numerological, explanations of the theorists of the Pythagorean and Platonic camps. In particular, Aristoxenus thought of intervals as segments of a continuum, thus rejecting the Pythagorean account, based as it was on a discrete analysis of pitches and their intervals. This led him to a division of the octave that was identical, for all practical purposes, with that given by equal temperament (i.e., the division of the octave into twelve equal semitones). In such a division, consonant intervals such as the fourth or fifth could only be expressed by irrational numbers, a consequence that no Pythagorean music theorist could have accepted.[9]

Although the Pythagorean tradition in music must once have had a powerful influence on composers, with the passage of time it fell out of step with musical practice. The first challenge to Pythagorean music theory came in the fifteenth century from the increasing use in polyphonic music of the minor and major thirds and sixths as consonants. These consonants cannot be expressed by means of ratios of whole numbers from 1 to 4 and thus according to strict Pythagorean standards cannot be accepted in composition. Second, the development of instrumental music made the issue of tuning, Pythagorean or otherwise, quite urgent. This clash between musical theory and practice forced a reexamination of the theoretical problems of the science of music.

One possible bridge over the gap between theory and practice was a sort of "rubber" Pythagoreanism. In his influential work *Le istitutioni harmoniche* (The Fundamentals of Harmony, 1558), for instance, Zarlino extended the range of allowable numbers for determining the consonants from 4 to 6.[10]

Institute, 1958). For the connection to encyclopedism and universal knowledge in the seventeenth century, see Eberhard Knobloch, "Musurgia Universalis," *History of Science*, 17 (1979), 258–75; and Daniel P. Walker, "Leibniz and Language," in *Music, Spirit, and Language in the Renaissance*, ed. Penelope Gouk (London: Variorum Reprints, 1985). A more recent and important contribution is Penelope Gouk, *Music, Science, and Natural Magic in Seventeenth Century England* (New Haven, Conn.: Yale University Press, 1999).

[8] Plato, *Republic*, translated by G. M. A. Grube, revised by C. D. C. Reeve (Indianapolis: Hackett, 1992), p. 203.

[9] Dividing the octave into twelve equal semitones yields the ratio $\sqrt[12]{2}:1$ for the semitone. Aristoxenus's *Elementa harmonica* is in *The Harmonics of Aristoxenus*, ed. and trans. Henry S. Macran (Oxford: Oxford University Press, 1902).

[10] Gioseffo Zarlino, *Le istitutioni harmoniche* (Venice, 1558). A translation of Book 3, by Guy A. Marco and Claude V. Palisca, is given in *The Art of Counterpoint: Part Three of Le istitutioni harmoniche*

He was thus able to count as consonants the major third (5:4), the minor third (6:5), the major sixth (5:3), and, by means of a dubious argument, also the minor sixth (8:5), for he claimed that 8 was potentially contained in the number 4. He called this system of numbers from 1 to 6 the *senario*. This extension of Pythagorean theory enjoyed, however, only limited success. By the beginning of the seventeenth century, composers were freely using augmented fourths (10:7) and diminished fifths (7:5), which fell outside Zarlino's *senario*. Moreover, the *senario* had already been challenged by several authors who, probably influenced by the Aristoxenian revival, were not convinced by the numerological arguments of the Pythagoreans and their latest exponent, Zarlino. Chief among such objectors were Vincenzo Galilei, father of Galileo Galilei (1564–1642), and Benedetti.[11]

Vincenzo Galilei had been a student of Zarlino in the early 1560s. In the 1570s, he challenged Zarlino's harmonic theories in the context of an extended polemic between the two that concerned, among other things, the problem of intonation and the nature of the consonants. Regarding the latter, Vincenzo Galilei claimed that he had been a committed Pythagorean himself until he "ascertained the truth by means of experiment, the teacher of all things."[12] The experiments in question aimed to test Pythagoras's alleged discovery of the relationship between pitches and weights attached to strings. According to writers in the Pythagorean tradition, if two weights in the ratio of 2 to 1 were attached to two strings that were equal in all other respects, then the pitches generated by the two strings would produce the octave, just as 2:1 was the ratio between string segments on the monochord that was needed to produce the octave – in other words, to go up an octave it is necessary to divide the length by 2 or attach a weight twice as heavy. The same principle applied to the generation of intervals such as the fourths and fifths. However, Vincenzo Galilei was able to show experimentally that the weights had to increase as the square of the length, so that the weights had to be in a proportion of 4 to 1 to produce an octave, 16:9 to produce a fourth, and 9:4 to produce a fifth. On the basis of this observation, he also called into doubt the rationality of the *senario*. After all, there would seem to be nothing sacrosanct about the numbers of the *senario*, which applied only to the length and not the tension of the strings. Vincenzo Galilei had other reasons that made him favor a

---

(New Haven, Conn.: Yale University Press, 1968). A translation of Chapter 4, by Claude V. Palisca, appears in *On the Modes: Part Four of Le istitutioni harmoniche* (New Haven, Conn.: Yale University Press, 1983). On Zarlino, see references in note 2.

[11] Claude V. Palisca brought to the attention of historians of science the important role played by Vincenzo Galilei's and Benedetti's reflections on music for the development of modern science. See Claude V. Palisca, "Scientific Empiricism in Musical Thought," in *Seventeenth Century Science and the Arts*, ed. H. H. Rhys (Princeton, N.J.: Princeton University Press, 1961), pp. 91–137; Palisca, "Was Galileo's Father an Experimental Scientist?," in *Music and Science in the Age of Galileo*, ed. Victor Coelho (Dordrecht: Kluwer, 1992), pp. 143–51; and Walker, *Studies in Musical Science in the Late Renaissance*, chap. 2.

[12] Vincenzo Galilei, *Discorso intorno all'opere di Gioseffo Zarlino* (Florence: G. Marescotti, 1589), pp. 103–4.

more empirical approach to the science of music that reflected an approach to music theory more consonant with the Aristoxenian tradition.[13]

Other sixteenth-century mixed mathematicians, including Benedetti, developed an approach to musical theory that emphasized experimental methods over the Pythagorean tradition.[14] Although I will deal primarily with the former, it should be pointed out that the Pythagorean paradigm continued to strongly influence authors such as Kepler, Descartes, Simon Stevin (1548–1620), Gottfried Wilhelm Leibniz (1646–1716), and Newton.

THE BIRTH OF ACOUSTICS IN THE EARLY SEVENTEENTH CENTURY

Many of the most eminent natural philosophers of the early seventeenth century showed a keen interest in music and acoustics, including Francis Bacon (1561–1626), Descartes, Beeckman, Kepler, Mersenne, Stevin, and Galileo Galilei.[15] Of these, the most influential were Bacon, Galileo, and Mersenne.

Bacon's acoustical investigations formed part of his plan to master the forces of nature by means of natural knowledge. His unfinished natural history, the *Sylva sylvarum* (1626), in particular, sets out an ambitious program for the investigation of acoustic phenomena.[16] Bacon's natural philosophy of sound reflected an eclectic approach. Although Bacon rejected numerical speculations and emphasized the physicality of the phenomenon of sound,

---

[13] In 1961, Palisca, in his "Scientific Empiricism in Musical Thought," drew attention to the existence of two manuscripts that detailed Vincenzo's experiments in this field. Taking his start from Palisca's work, Stillman Drake has proposed in a provocative article "that the origins of modern science are to be sought in sixteenth-century music, just as its mathematical origins have been traced to the ancient Greek astronomers and to Archimedes." See Stillman Drake, "Renaissance Music and Experimental Science," *Journal of the History of Ideas*, 31 (1970), 483–500, at p. 483.

[14] For Benedetti's work in music, see the letters to Cipriano de Rore, probably written in 1563, and published in *Diversarum speculationum mathematicarum & physicorum liber* (Turin: Successors of Nicola Bevilaqua, 1585), pp. 277–83. The letters are reprinted in Josef Reiss, "Jo. Bapt. Benedictus, De intervallis musicis," *Zeitschrift für Musikwissenschaft*, 7 (1924–5), 13–20. For a detailed treatment and further references, see Palisca, *Humanism in Italian Renaissance Musical Thought*, pp. 257–65; and H. Floris Cohen, "Benedetti's Views on Musical Science and Their Background in Venetian Culture," in Instituto Veneto di Scienze Lettere edArti, *Atti del Convegno Internazionale di Studio "Giovan Battista Benedetti e il suo tempo" (Venezia, 1987)* (Venice: Il Istituto, 1987), pp. 303–10.

[15] No complete account is yet available. The best book on the topic is Cohen, *Quantifying Music*. See also the articles contained in Coelho, ed., *Music and Science in the Age of Galileo*; Paolo Gozza, ed., *La musica nella rivoluzione scientifica del Seicento* (Bologna: Il Mulino, 1989); and Gozza, "La musica nella filosofia naturale del Seicento in Italia," *Nuncius*, 1 (1986), 13–47.

[16] The *Sylva Sylvarum* (1626) is published in *The Works of Francis Bacon*, ed. J. Spedding et al., 14 vols. (London: Longmans, 1857–74), 2: 331–680. On Bacon's work on acoustics and music, see Penelope Gouk, "Music in Francis Bacon's Natural Philosophy," in *Francis Bacon: Terminologia*, ed. Marta Fattori (Rome: Edizioni dell'Ateneo, 1984), pp. 139–54; Mordechai Feingold and Penelope Gouk, "An Early Critique of Bacon's *Sylva Sylvarum*," *Annals of Science*, 40 (1983), 139–57; and Penelope Gouk, "Some English Theories of Hearing in the Seventeenth Century: Before and after Descartes," in *The Second Sense: Studies in Hearing and Musical Judgement from Antiquity to the Seventeenth Century*, ed. Charles Burnett, Michael Fend, and Penelope Gouk (London: Warburg Institute, 1991), pp. 95–113. The latter also investigates lesser known contemporaries of Bacon, such as Helkiah Crooke and Thomas Wright.

some of his arguments tied him to the Aristotelian and neo-Platonic traditions.[17] His theory of sound, for example, was of Aristotelian origin.[18] According to Bacon, sound is conveyed by a "species"; that is, an immaterial entity that originates in the sounding body and carries with it all the qualities of the sounding body. Hearing takes place when the species, after traveling through the air, enters into contact with the ear; like Aristotle, Bacon believed that the internal air of the ear was the organ of hearing. Although Bacon's reflections on acoustics were not very original, they had the merit of proposing a wide-ranging program for the study of sound, which included the production of sound in humans and animals, the study of musical instruments, and studies on echoes and the speed of sound. He also suggested the development of instruments to extend the power of the voice and hearing. These proposals set the stage for the acoustical investigations carried out by the Royal Society of London in the second half of the seventeenth century.

In contrast with Bacon's "quality" theory of sound, the account of consonances proposed by Galileo in the important section on music that concludes the first day of *Discorsi e dimostrazioni matematiche intorno a due nuove scienze* (Discourses and Mathematical Demonstrations Concerning Two New Sciences, 1638) develops a mechanistic account of sound transmission and perception. After a summary of criticisms of traditional Pythagorean views, already implicit in the work of his father Vincenzo, Galileo laid out the theory of consonances that became dominant in the seventeenth century:

> Going back to our original purpose, I say that the length of strings is not the direct and immediate reason behind the forms of musical intervals, nor is their tension, nor their thickness, but rather, the ratio of the numbers of vibrations and impacts of air waves that go to strike our eardrum, which likewise vibrates according to the same measure of times. This point established, we may perhaps assign a very congruous reason why it comes about that among sounds differing in pitch, some pairs are received in our sensorium with great delight, others with less, and some strike us with great irritation.[19]

---

[17] Among Aristotelian sources, *De anima* (II, 8, 419b3 to 421a6) and the pseudo-Aristotelian *Problemata* (Books XI and XIX) are particularly relevant here. For a source influenced by the neo-Platonic tradition, see Giambattista della Porta's *Magia naturalis*, first published in four books in 1558 and then expanded into 20 books in 1589. Both editions were first published in Naples. The neo-Platonic influence is especially visible in Bacon's appeal to "spiritus," a pneumatic substance infused through all bodies, which he used to explain how the immaterial species conveying sound affects the perceiver. Bacon's appeal to "spiritus" reveals the influence of thinkers such as Bernardino Telesio, Marsilio Ficino, and Tommaso Campanella. On the notion of "spiritus" in the sixteenth and seventeenth centuries, see Walker, *Spiritual and Demonic Magic*.

[18] On this tradition, see Michael Wittman, *Vox atque sonus*, 2 vols. (Pfaffenweiler: Centaurus, 1987); and C. Burnett, "Sound and Its Perception in the Middle Ages," in *The Second Sense: Studies in Hearing and Musical Judgement from Antiquity to the Seventeenth Century*, ed. Burnett, Fend, and Gouk, pp. 43–69.

[19] Galileo Galilei, *Two New Sciences, including Centers of Gravity & Force of Percussion*, translated with notes by Stillman Drake (Madison: University of Wisconsin Press, 1974), p. 104.

The basis for a theory of consonances is no longer to be found in abstract numerological considerations but in the physics of frequencies. According to Galileo, if the vibrations of two different sounds strike the ear in a certain proportion, or regularity, then they are perceived as pleasant, whereas if they strike the ear without proportion (*sproporzionatamente*) – for instance, if they have incommensurable frequencies – then the two sounds are perceived as dissonant. In this way, Galileo went a long way toward accounting for the classical ratios associated with the consonant sounds. The vibrations of two sounds an octave apart are such that "for every impact that the lower string delivers to the eardrum, the higher gives two, and both go to strike unitedly in alternate vibrations of the high string, so that one-half of the total number of impacts agree in beating together."[20] In the case of the fifth, the pulses will agree every sixth vibration, and in the case of the fourth, every twelfth.

The theory itself did not originate with Galileo, as it was foreshadowed by Benedetti, but these passages by Galileo became the classical exposition of the theory. In particular, Galileo's approach differed from that of Benedetti in that he supported the relationship between pitch and frequency by experiments. Although Galileo's approach showed considerable novelty in moving from numerological arguments to physical and physiological ones in accounting for consonances and dissonances, it remained heavily dependent on the Pythagorean model. Galileo went so far as to claim that the vibrations of two sounds that strike the ear in irrational proportions would produce dissonances. But in any tempered tuning, such as those used in Galileo's time for fretted instruments and on modern pianos, the ratios between consonances (except for the octave) will be irrational – in other words, the ratio of the frequencies of the two sounds cannot be expressed by two whole numbers. It is worth noting that Galileo is unlikely to have performed the experiments he adduced as corroboration for his theory, which he thus seems to have conceived prior to any experimental verification.[21]

The emphasis on a physical account of the consonances is also found in the work of Mersenne, who had close contacts with Galileo; his work, which involved extensive acoustical experiments, can be regarded as the birth of acoustics as a scientific discipline.[22] Mersenne's most comprehensive

---

[20] Ibid.
[21] Through his experiments, Galileo sought to explain sympathetic resonance. One of his experiments involved a water-filled glass that produced a note by rubbing the rim. Galileo claimed that the note would sometimes jump an octave and that when this happened the waves produced by the first sound in the water divided into two, thereby showing "the form of the Octave to be the Double." A second experiment purported to prove the same fact by scraping a copper plate with a chisel. For skepticism about whether the experiments could have been performed, see Walker, *Studies in Musical Science in the Late Renaissance*, chap. 3.
[22] On Mersenne and music in the early modern period, see Cohen, *Quantifying Music*; and Crombie, *Styles of Scientific Thinking*, 2: 783–894. For more technical aspects of Mersenne's work, see Sigalia Dostrovsky, "Early Vibration Theory: Physics and Music in the XVIIth Century," *Archive for History of Exact Sciences*, 14 (1975), 169–218; and Clifford A. Truesdell, "The Rational Mechanics of Flexible or Elastic Bodies, 1638–1788," in Leonhard Euler, *Leonhardi Euleri Opera Omnia*, ed. Ferdinand

contribution to acoustics and music theory, *Harmonie Universelle* (Universal Harmony, 1636–7), treated topics that ranged from combinatorial studies to the physics of musical instruments. His advance over his predecessors is revealed once again by focusing on the problem of consonance.

Mersenne's account of the consonances rejected purely numerical speculations in favor of arguments derived from the physics of sound. He often wrote of sound as a wave, analogous to water waves, which originated by a percussion (*battement*) of the air. According to Mersenne, the pitch of a sound is determined by the frequency of the vibrations, that is, the number of pulses per unit time. Mersenne defended a version of the "coincidence" theory of consonance, whereby the consonances are determined by the relative agreement of the percussions that make up the two sounds. For example, if two strings are a fifth apart (that is, are in a ratio of 3:2), the percussions of the first string will coincide with the percussions of the second string every six times. By contrast, the percussions of two strings an octave apart coincide every second time. However, Mersenne was aware that the hierarchy of consonances obtained in this way makes the fourth (4:3) more consonant than the major third (5:4) because their percussions agree every 12 and 20 times, respectively – despite the fact that in musical practice the major third was perceived as more pleasant than the fourth. This discrepancy led Mersenne to draw up, alongside a list of the consonances according to the physical criteria already mentioned, which he described in terms of "sweetness," a different list of the consonances, sorted according to their auditory agreeableness. He devoted much effort to reconciling this gap between physical and psychological criteria of consonance.

Mersenne's contribution to acoustics included, among other things, studies concerning the determination of relative frequencies (such as those characterizing musical intervals), the absolute frequency of vibrating strings, the speed of sound, and overtones and harmonics. He explored these topics through precise and varied experiments. In the case of the speed of sound, for example, these experiments involved measuring the time elapsed between seeing the flash of a weapon fired at a distance and hearing the shot. A different set of experiments exploited the time elapsed between the emission of a sound and hearing its echo. The inaccuracy of time measurements, which often relied on inaccurate clocks, or even on the human pulse, partly accounts for the inconsistency of the numerical values Mersenne obtained in these measurements.[23]

Mersenne's greatest achievement in the field of acoustics was what is now known as Mersenne's law for vibrating strings. According to Mersenne's law, the frequency of a vibrating string is directly proportional to the square root

---

Rudio, Adolf Krazer, and Paul Stacckel, second ser. (Leipzig: B. G. Teubneri, 1911–), vol. II (1960), pp. 15–141.
[23] See Frederick V. Hunt, *Origins in Acoustics: The Science of Sound from Antiquity to the Age of Newton* (New Haven, Conn.: Yale University Press, 1978), pp. 85–100.

of the tension and indirectly proportional to the length times the square root of the cross-sectional area of the string. Although Mersenne was not the first to notice this relation, he is generally recognized as the one who established it on the basis of several experiments aimed at showing the interdependence of the different components of the relationship. Mersenne, who did not use the word "law" (but rather *regle* or "rule") in this context, presented his result in a series of rules, which show that he understood that it was necessary to make various adjustments in practice and that, in comparing different strings, the density of the string plays a role.[24]

Mersenne engaged in a long correspondence with Descartes and Beeckman, which often touched upon music. Beeckman held an atomistic theory of sound and attempted a derivation of the inverse proportionality between the frequency and the length of a vibrating string.[25] His influence on the young Descartes is well known, and Descartes' *Compendium musicae* (Compendium of Music, 1618) was dedicated to Beeckman.[26] Both men attempted to explain sound perception by analyzing the phenomenon purely in terms of matter and motion, and both paid much more attention to the anatomy of the ear and its role in sound perception than had their predecessors. However, the ambitious project of accounting for sound perception in purely mechanical terms was limited by the lack of exact knowledge of the anatomy and physiology of the nervous system and the brain.

DEVELOPMENTS IN ACOUSTICS IN THE SECOND HALF OF THE SEVENTEENTH CENTURY

In the second half of the seventeenth century, the range of empirical acoustic inquiry widened, as did the quality of the experimental work. For example, the variety of acoustic investigations carried out by members of the Royal Society[27] focused on topics such as the propagation of sound, including speed and echo phenomena, instruments for the improvement of hearing, and the relationship between string vibrations and pendulums. The instruments studied in the period included the otacousticon, a trumpet for improving

---

[24] The rules, amounting to what we call Mersenne's law, are found in *Harmonie universelle*, 3 vols. (Paris: Sebastien Cramoisy, 1636–7), vol. 3, prop. VII, pp. 123–7.

[25] Beeckman's scientific work is found in *Journal tenu par Isaac Beeckman de 1604 à 1634*, ed. C. de Waard, 4 vols. (The Hague: Martinus Nijhoff, 1939–53). On Beeckman, see F. De Buzon, "Science de la nature et theorie musicale chez Isaac Beeckman (1588–1637)," *Revue d'histoire des sciences*, 38 (1985), 97–120; and Cohen, *Quantifying Music*. Cohen has argued that Descartes and Beeckman represent a mechanistic tendency in the seventeenth-century study of music.

[26] On Descartes, see Cohen, *Quantifying Music*, and references therein. See also Paolo Gozza, "Una matematica media gesuita: la musica di Descartes," in *Christoph Clavius e l'attività scientifica dei gesuiti nell'età di Galileo*, ed. Ugo Baldini (Rome: Bulzoni, 1995), pp. 171–88.

[27] Penelope Gouk has studied English contributions to acoustics in the seventeenth century. In addition to her publications cited previously, see Gouk, "Acoustics in the Early Royal Society, 1660–1680," *Notes and Records of the Royal Society of London*, 36 (1982), 155–75; and Gouk, "The Role of Acoustics and Music Theory in the Scientific Work of R. Hooke," *Annals of Science*, 37 (1980), 573–605.

hearing, and the speaking trumpet (tuba stentoro-phonica) for the amplification of sounds, which was invented by English mathematician and instrument maker Samuel Morland (1625–1695) in 1670.[28] Contemporaries often compared the invention of the tuba with Newton's invention of the reflecting telescope. Other investigations concerning the vibrational properties of sound included natural philosopher Robert Hooke's (1635–1703) experiments for the determination of absolute frequencies (by means of a device known as Hooke's wheel) and the discovery of nodes (that is, stationary points) in the vibrating string, which was made independently by Thomas Pigot, William Noble, and Wallis.[29]

By the late seventeenth century, it was universally accepted that sound travels like a wave in the air at a finite speed. But exactly how fast does sound travel? And what role does the air play in sound transmission? Attempts to answer these questions led to sophisticated experiments. Experiments on the speed of sound were carried out by members of the Accademia del Cimento in Florence,[30] by the Académie Royale des Sciences in Paris, and by the Royal Society in London. For instance, the measurement of the speed of sound by natural philosophers affiliated with the Académie Royale des Sciences in 1677 yielded an approximate value of (in modern units of measurement) 356 m/s.[31] In this case, as in others, the tendency toward quantitative experimentation was more important than the actual measurements, which eventually made more reliable and precise results possible.

Naturalists also performed experiments with sonorous devices isolated in an ambient, such as a vessel, from which the air had been removed. Such experiments were carried out by, among others, the Jesuit polymath Athanasius Kircher (1602–1680), German engineer and diplomat Otto von Guericke (1602–1686), and English natural philosopher Robert Boyle (1627–1691). Boyle, with the help of Hooke, improved the air-pumps developed by von Guericke.[32] In *New Experiments Physico-Mechanical Touching the Spring of the Air* (1660), Boyle described how, after having suspended a watch from a

---

[28] Sir Samuel Morland, *Tuba Stentoro-Phonica, An Instrument of Excellent Use, As well at Sea, as at Land; Invented and variously experimented in the Year 1670* (London: Godbid, 1672).

[29] John Wallis, "Dr Wallis' letter to the publisher, concerning a new musical discovery," *Philosophical Transactions of the Royal Society*, 12 (1677–8), 839–42. The French academician Joseph Sauver made important contributions to these topics at the beginning of the eighteenth century. See Sigalia Dostrovski and John T. Cannon, "Entstehung der musikalischen Akustik (1600–1750)," in *Geschichte der Musiktheorie*, ed. Frieder Zaminer (*Hören, Messen und Rechnen in der frühen Neuzeit*, Band 6) (Darmstadt: Wissenschaftliche Buchgesellschaft, 1987), pp. 7–79; and Dostrovsky, "Early Vibration Theory".

[30] Accademia del Cimento, *Saggi di naturali esperienze fatte nell'Accademia del Cimento* (Florence: Cocchini, 1667). English translation by Richard Waller, *Essayes of Natural Experiments made in the Accademia del Cimento* (London: Alsop, 1684).

[31] See Hunt, *Origins in Acoustics*, p. 110.

[32] For further references to the primary and secondary literature, see Steven Shapin and Simon Schaffer, *Leviathan and the Air-Pump: Hobbes, Boyle, and the Experimental Life* (Princeton, N.J.: Princeton University Press, 1985). Kircher's most important work in this connection is the *Musurgia universalis* (Rome: Francisco Corbelletti, 1650).

thread in the cavity of a vessel, he pumped the air from the vessel. When the air was completely removed, the watch could no longer be heard, although it was obvious that it was still ticking. When air was let back into the vessel, the watch could be heard again, "Which seems to prove, that whether or no the Air be the onely, it is at least, the principal medium of Sounds," Boyle concluded.[33] In *Traité de la lumière* (Treatise on Light, 1690, written in 1678), Dutch natural philosopher Christiaan Huygens (1629–1695) interpreted Boyle's experiment as showing that light and sound travel in different media: "One sees here not only that our air, which does not penetrate through glass is the matter by which sound spreads; but also that it is not the same air but another kind of matter in which light spreads; since if the air is removed from the vessel the light does not cease to traverse it as before."[34] Although Huygens believed that light and sound traveled as waves, he thought that the mechanism involved in the two types of propagation differed greatly.[35] Sound was caused by the bouncing of the particles of air, which are agitated very rapidly, so that "the cause of the spreading of sound is the effort these little bodies make in collisions with one another."[36] In contrast, the spreading of light could not be explained by such a propagation of motion on account of its velocity. (Huygens's account of the nature of light is discussed later in this chapter.)

Isaac Newton (1642–1727)[37] also gave, in Book II of his *Principia mathematica philosophiae naturalis* (Mathematical Principles of Natural Philosophy, 1687), an analysis of sound propagation in terms of dynamic transmission of sound waves through a compressible elastic medium (the air).[38] Using this analysis, he attempted to provide new theoretical determinations of the speed of sound to account for the latest experimental results.[39] At the same time, he used musical theory to develop some of his theories of light, postulating, for example, in his 1675 *Hypothesis of Light* an analogy between the

---

[33] Robert Boyle, *New Experiments Physico-Mechanicall, Touching the Spring of the Air and Its Effects* (Oxford: H. Hall, 1660), p. 110.

[34] Christiaan Huygens, *Treatise on Light*, trans. Silvanus P. Thompson (London: MacMillan, 1912). Page numbers refer to the edition in *Great Books of the Western World*, 54 vols. (Chicago: University of Chicago Press, 1952), 34: 545–619, quotation at p. 558.

[35] Chapter 1 of *Treatise on Light* develops an extended comparison of the analogies (and lack of analogies) between light and sound.

[36] Huygens, *Treatise on Light*, p. 558.

[37] For a general introduction to Newton's musical theories, see Penelope Gouk, "The Harmonic Roots of Newtonian Science," in *Let Newton Be! A New Perspective on His Life and Work*, ed. John Fauvel, Raymond Flood, Michael Shortland, and Robin Wilson (Oxford: Oxford University Press, 1988), pp. 101–25.

[38] By the late seventeenth century, most scholars accepted wave theories of sound propagation. In addition to the references in note 27, see also Clifford A. Truesdell, "The Rational Mechanics of Flexible or Elastic Bodies, 1638–1788," in *Leonhardi Euleri Opera Omnia*, 11: 15–141; and Truesdell, "The Theory of Aereal Sound, 1687–1788," in *Leonhardi Euleri Opera Omnia*, vol. 13 (1960), pp. xix–lxxii.

[39] In the first edition of *Principia* (1687), Newton's theoretical determination for the speed of sound, 968 ft/sec, was reasonably close to the available empirical measurements. By the second edition of *Principia* (1713), Newton had changed his theoretical result to account for new experimental results, due to William Derham, according to whom the speed of sound in air was 1142 ft/sec.

division of the musical scale and the color spectrum. Finally, he also held fast to the idea of a universal harmony and went so far as to claim that Pythagoras's theory regarding the relationship between pitches and weights attached to the strings was a cryptic reference to the inverse-square law of universal gravitation, camouflaged to hide the secret from the uninitiated.[40]

By the end of the seventeenth century, the acoustical study of sound phenomena looked remarkably different from its Renaissance predecessor. The arithmetical speculations of the music theorists had yielded to an experimental science that was striving to provide a physical account of the nature, propagation, and perception of sound. This process was paralleled in the coeval development of optics.

OPTICS IN THE EARLY MODERN PERIOD: AN OVERVIEW

The development of optics after 1600 was one of dramatic empirical discoveries and theoretical innovations. Major advances came early in the seventeenth century, when Kepler proposed a new theory of vision based on understanding the eye as an optical instrument. This led to a gradual separation of the problems pertaining to the psychology of vision from those related to the physico-geometrical aspects of vision. The development of the telescope spurred rapid progress in the theory of lenses, leading to a renewal of geometrical optics, which had also been made possible by detailed quantitative experimental work on refraction, culminating in the precise determination of the law of refraction. Although the law could be stated in purely geometrical terms, Descartes' attempts to justify it in physical terms points to another major feature of seventeenth-century optics: the dramatic development of physical optics. Several theories of light were put forward – including those of Descartes, Huygens, and Newton – that gave "mechanical" explanations for its behavior in terms of matter and motion. Spurred by the observations of Hooke on thin-plate colors and those of Grimaldi on diffraction, it was Newton who achieved an impressive systematization of physical optics. Although Newton's theory of light and color was arguably the most important achievement of seventeenth-century optics, it is important to note that by the end of the century optics had also become an extremely sophisticated experimental science.

Unlike the science of music, whose ancient texts were mostly ignored until Renaissance humanist scholars launched a revival of the musical tradition, the sixteenth-century science of optics already enjoyed an established classical textual tradition. Scholars interested in optics had access to the Greek texts, such as those of Euclid (ca. 280 B.C.E.) and parts of Ptolemy's (ca. 100–170) works, and most of the central texts concerning vision from medieval

[40] See, in particular, J. R. McGuire and P. N. Rattansi, "Newton and the 'Pipes of Pan'," *Notes and Records of the Royal Society of London*, 21 (1966), 108–43. See also Gouk, *Music, Science, and Natural Magic*.

Arabic and Latin sources, including works by Ibn al-Haytham (Alhazen, 965–ca. 1040), Roger Bacon (ca. 1219–1292), John Pecham (ca. 1230–1292), and Witelo (ca. 1230–1275). In addition, they could also consult Galen's writings and a long tradition of physiological studies of the eye, as well as new fifteenth-century work related to linear perspective.[41]

Like acoustics, optics addressed a variety of topics. A conservative list would include the nature of the sources of light, the geometrical and physical nature of light propagation, theories of vision, the psychology of perception, and the anatomy and physiology of the eye. In what follows, I will touch upon problems of geometrical optics (radiation through small apertures, refraction, diffraction, and image location), problems of physical optics (physical theories of light and colors, emissionist versus continuum theories of light), and theories of vision (Kepler's double-cone model of vision). Constraints of space have unfortunately precluded a comparable treatment of the applied aspects of optics and developments in the physiology of the eye and perceptual theories in the seventeenth century.[42]

OPTICS IN THE SIXTEENTH CENTURY

Before Kepler's groundbreaking contributions to optics in the years around 1600, only a few scholars made original contributions to received ideas concerning optics, notably Francesco Maurolico (1494–1575), a Sicilian abbot, and Giambattista della Porta (1535–1615), a Neapolitan polymath and natural magician.[43] Maurolico's *Photismi de lumine et umbra* and *Diaphaneon* (*Illuminations of Light and Shadow* and *Transparent Substances*, published posthumously in 1611) fit squarely in the medieval *perspectiva* tradition, offering a deductive approach to optical topics modeled on Euclid's axiomatic

---

[41] The standard secondary account is David C. Lindberg, *Theories of Vision from Al-Kindi to Kepler* (Chicago: University of Chicago Press, 1976). For an updated bibliography concerning the medieval tradition, see Lindberg, *Roger Bacon and the Origins of Perspectiva in the Middle Ages* (Oxford: Oxford University Press, 1996). For the Arabic tradition, see the *Encyclopedia of the History of Arabic Science*, ed. Roshdi Rashed, 3 vols. (London: Routledge, 1996), vol. 2. Alhazen's *De aspectibus* and Witelo's *Perspectiva* were printed in *Opticae thesaurus*, ed. Friedrich Risner (Basel: Episcopios, 1572), the main "classical" source of late sixteenth- and seventeenth-century optics. For the modern reprint, see *Opticae thesaurus*, introd. David C. Lindberg (New York: Johnson Reprint Co., 1972). The literature on linear perspective is immense. A good study on the relationship between the science of optics and art is Martin Kemp, *The Science of Art: Optical Themes in Western Art from Brunelleschi to Seurat* (New Haven, Conn.: Yale University Press, 1990).

[42] The standard reference on the history of the physiology of the eye is Julius Hirschberg, *Geschichte der Augenheilkunde* (Hildesheim: Georg Olms, 1977; orig. publ. 1899 (vol. 1) and 1908 (vol. 2)). For an account of developments in the area of the anatomy and physiology of the eye, see A. C. Crombie, *Styles of Scientific Thinking*, 2.13.1106–54.

[43] Giambattista della Porta, *Magiae naturalis libri XX* (Naples: Horatio Salvianum, 1589). For the best account of the period, see David C. Lindberg, "Optics in XVIth Century Italy," in *Novità celesti e crisi del sapere* (Florence: Giunti Barbera, 1983), pp. 131–48. See also Lindberg, "Laying the Foundations of Geometrical Optics: Maurolico, Kepler, and the Medieval Tradition," in David C. Lindberg and Geoffrey Cantor, *The Discourse of Light from the Middle Ages to the Enlightenment* (Los Angeles: William Andrews Clark Memorial Library, 1985), pp. 3–65.

approach. A novelty in Maurolico's work, however, was the emphasis given to the notion of illumination, as in his analysis of shadows projected on a screen by a body of a certain shape when illuminated by a source of light; in this connection, Maurolico developed an account of the penumbral region between the areas that are fully illuminated and those that are fully shaded. Maurolico also analyzed an apparent paradox known as the problem of radiation through small (finite) apertures (described later in this chapter in connection with Kepler's solution). The *Diaphaneon* dealt with problems of refraction – the phenomenon of deviation of luminous rays when passing through media of a different density; for example, from air to water – and developed some interesting reflections on the nature of lenses. Maurolico believed that the angle of refraction was proportional to the angle of incidence; some of the applications seek to determine the causes of myopia and hypermetropia, treating the crystalline of the eye as a lens.

Della Porta's optical works popularized certain new optical subjects, such as the analysis of radiation through lenses and the camera obscura. A camera obscura is a device, such as a darkened chamber, with a small aperture in one of the walls, through which external illumination passes and projects an image, upside down, on a screen. Although the camera obscura was not a new invention, della Porta was instrumental in drawing attention to the device and in emphasizing the analogy between the workings of the eye and the camera obscura. Although a variety of factors account for the renewal of interest in the camera obscura in the seventeenth century, including its relevance for astronomical observations, della Porta's interest seems to have been motivated by the courtly market for optical wonders.[44]

KEPLER'S CONTRIBUTIONS TO OPTICS

Johannes Kepler (1571–1630) extended the boundaries of prior optical knowledge to effect an optical revolution. Kepler's motivation for studying optics came from astronomy. Previous observations made by the Danish astronomer Tycho Brahe (1546–1601), for whom Kepler worked as an assistant in Prague, using a pinhole camera had established that the lunar diameter appeared smaller during a solar eclipse than at other times, although, according to theory, the distance from the moon to the Earth was the same. The occurrence of a solar eclipse in 1600 spurred Kepler to investigate the phenomenon through the classical perspectivist theory of radiation.[45] His treatment, presented in Chapter 2 of *Ad Vitellionem paralipomena* (Supplements to Witelo, 1604), broke new ground; elaborating an ingenious theory of radiation through

---

[44] William Eamon, *Sciences and the Secrets of Nature: Books of Secrets in Medieval and Early Modern Culture* (Princeton, N.J.: Princeton University Press, 1994), pp. 221–33.
[45] See Stephen Straker, "Kepler, Tycho, and the 'Optical Part of Astronomy': The Genesis of Kepler's Theory of Pinhole Images," *Archive for History of Exact Sciences*, 24 (1981), 267–93.

small finite apertures, he was able to account for the puzzling phenomenon observed by Brahe.

The main problem in the theory of radiation through small apertures was the following. Consider a source of light, and assume that the illumination it produces passes through a small aperture having a shape different from the source of light. Immediately behind the aperture, the illumination conforms to the shape of the aperture, whereas at a more remote distance from the aperture, it conforms to that of the source. Because this phenomenon seemed to contradict the rectilinear propagation of light, it became an important problem in studies of optics. Kepler proposed a three-dimensional model. He considered as the source of light a book placed high in the air. Between this and the floor, Kepler set a table that had a many-cornered aperture. He then fixed threads to the corners of the book and along its edges, which he pulled through the aperture, grazing its edges. Although each such thread described on the floor a figure similar to the shape of the aperture, the totality of all such shaped figures described on the floor produced a figure having the shape of the book. Kepler was then able to show that this model accounted for the observational data without compromising the principle of the rectilinear propagation of light. Figure 25.3 provides a simplified illustration of Kepler's idea.

Kepler also realized that optics was intimately connected with astronomy because the dependence of astronomy on observations required a solution to the problem of vision, which he addressed in Chapter 5 of *Ad Vitellionem paralipomena*.[46] There he criticized the theory of the medieval perspectivists laid out, for example, in the books of Pecham and Witelo. This theory, whose sources stemmed from Arab mathematicians al-Kindī (ca. 801–ca. 866) and Ibn al-Haytham,[47] postulated that light travels in rectilinear paths and that from every point of a shining object luminous rays depart in all directions. Thus each point on the eye's surface receives rays from all points of the visual object. However, for clear vision to emerge out of this chaos of rays, some order must be imposed between the continuum of rays incident on the eye and the clear picture of the visual field we perceive. In other words, the problem is how to ensure that there is a one-to-one correspondence between points in the visual field and stimulated points in the eye. The standard answer in the medieval Arabic and Latin perspectivist tradition was that only rays perpendicular to the eye could affect vision, whereas oblique

---

[46] Among the best sources for Kepler's theory of vision are Lindberg, *Theories of Vision*, chap. 9; and Antoni Malet, "Keplerian Illusions: Geometrical Pictures vs. Optical Images in Kepler's Visual Theory," *Studies in History and Philosophy of Science*, 21 (1990), 1–40. For a "mechanistic" reading of Kepler, see S. Straker, "The Eye Made 'Other': Durer, Kepler, and the Mechanization of Light and Vision," in *Science, Technology, and Culture, in Historical Perspective*, ed. Louis A. Knafla, Martin S. Staum, and T. H. E. Travers (Calgary: University of Calgary Press, 1976), pp. 7–25; and Crombie, *Styles of Scientific Thinking*, 2: 1125–43. See also C. Chevalley's annotated translation of *Ad Vitellionem paralipomena: Les Fondements de l'Optique Moderne: Paralipomenes à Vitellion (1604)* (Paris: J. Vrin, 1990).

[47] On these authors, see Rashed, ed., *Encyclopedia of the History of Arabic Science*, vol. 2.

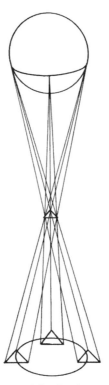

Figure 25.3. Johannes Kepler's model of radiation through small apertures. In David C. Lindberg and Geoffrey Cantor, *The Discourse of Light from the Middle Ages to the Enlightenment* (Los Angeles: William Andrews Clark Memorial Library, University of California, Los Angeles, 1985).

rays, however small their deviation from the perpendicular, could not. This in effect reestablished a one-to-one correspondence between the visual field and its image in the eye. Furthermore, this tradition held that the sensitive organ of sight was the crystalline lens of the eye.

Kepler criticized both the anatomical and the geometrico-physical claims of the perspectivist tradition.[48] Against the idea that the crystalline lens could act as the seat of vision, he appealed to recent anatomical investigations by Félix Platter (1536–1614), professor of medicine at the University of Basel, who had shown that the crystalline lens was not connected to the optic

---

[48] The nature of Kepler's revolution in the history of optics has been the subject of much debate. His work was certainly connected to the perspectivist tradition. His solutions to some of the problems emerging from this tradition, however, are so original that he is held to be the founding father of modern optics and theories of vision. Despite such debate, most scholars agree that he was one of the most important contributors to optics in the seventeenth century. See, for example, David C. Lindberg, "Continuity and Discontinuity in the History of Optics: Kepler and the Medieval Tradition," *History and Technology*, 4 (1987), 431–48.

nerve.[49] Moreover, the perspectivists had erred regarding the shape of the crystalline lens; specifically, Kepler criticized Witelo for having claimed that the posterior surface of the crystalline lens is flat, whereas the anatomical evidence showed it to be rounded. This called for a radical revision of the geometry of visual perception. The perspectivists had also been inconsistent concerning the role of oblique rays in vision, granting that luminous points located at the periphery of the visual field were perceived by means of oblique rays. Kepler found this drastic contrast between oblique and perpendicular rays to be physically unacceptable, arguing that these lay on a continuum of inclinations and that it was unlikely that a ray that deviates slightly from the perpendicular had no role in the process of vision.

Kepler's own solution to the problem of vision consisted in two main moves. (Figure 25.4 shows Descartes' drawing illustrating Kepler's theory.) First, he accepted the starting point of the perspectivists; that is, that from every point in the visual field an infinity of luminous rays departs, without, however, excluding the oblique rays from playing a role in the process of vision. For Kepler, the role of the crystalline lens was to make the rays emitted from the same point source converge again in the same stimulus point in the eye. Second, Kepler identified the retina (as opposed to the crystalline lens) as the principal organ of sight. In conclusion, Kepler explained the process of vision by a double-cone model: Each point in the visual field is the apex of a cone with the surface of the eye as its base, while each point of the retinal image is the apex of a cone whose base is the posterior surface of the crystalline lens. Between the surface of the eye and the posterior surface of the crystalline lens, the rays are refracted according to the different density of the media they traverse (e.g., aqueous humor, crystalline lens). The geometry of radiation thus causes the image on the retina to be reversed and inverted with respect to the original. Thus, in Figure 25.4, rays coming from point V of the visual body VXY hit every point of the surface of the eye, such as B, C, and D. The rays are then refracted and converge to point R of the retina. Similarly, points X and Y converge to points S and T on the retina.

That the visual image was projected onto the retina as an inverted image could be empirically verified in anatomical dissections by scraping the back surface of an eyeball, leaving only a thin layer; it could then be observed that the retina projected an inverted image of the visual figure facing the eye. Christoph Scheiner (1573–1650), a Jesuit mathematician active in Rome, described the experiment in *Oculus hoc est: Fundamentum opticum* (The Eye, 1619), which seems to have played an important role in ensuring the success of Kepler's theory of vision. However, a central problem was still left unresolved. Why do we not see the world upside down? Kepler had no real answer to this problem, but he claimed that the job of optics stops with the *pictura*

---

[49] Félix Platter, *De corporis humani structura et usu libri III* (Basel: König, 1603).

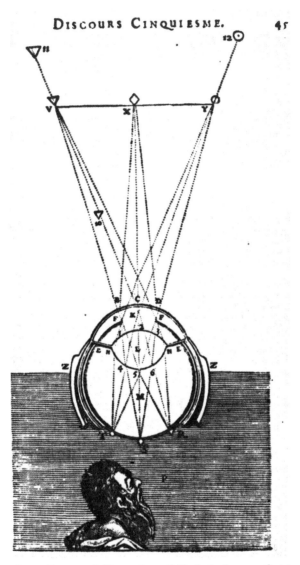

Figure 25.4. René Descartes' illustration of Kepler's theory of vision. In René Descartes, *Discourse de la méthode* (Leiden: I. Maire, 1637). Reproduced by permission of the Library of the Max Planck Institute for the History of Science, Berlin.

in the retina. How the brain interprets this *pictura* was a problem left for physicians.

Kepler's analysis of the eye and his interpretation of the crystalline lens as a focusing lens brought to the fore the need for a theory of lenses, a problem also made pressing by developments in optical instruments (see

Bennett, Chapter 27, this volume). Although eyeglasses came into use in the late thirteenth century, it had taken more than three hundred years to develop the first telescopes and microscopes.[50] Galileo's announcement of his astronomical discoveries, including the four satellites of Jupiter, in *Sidereus nuncius* (The Starry Messenger, 1610) gave particular urgency to the correct analysis of the geometrical optics on which telescopic observations were based[51] (see Donahue, Chapter 24, this volume). After careful reflection, Kepler, in his *Narratio de Jovis satellibus* (Account of the Moons of Jupiter, 1611), sided with Galileo, confirming the validity of Galileo's observations, and in the same year he published *Dioptrice* (Dioptrics), a scientific treatment of lenses. This work can be seen as the outcome of his intensive search for the law of refraction (in Chapter 4 of *Ad Vitellionem*). Although Kepler did not manage to find the quantitative law (Thomas Harriot in England already knew it), he produced a successful geometrical (but nonquantitative) analysis of lenses individually and of the telescope with its multiple lenses. *Dioptrice* marks an important epistemological moment because it points the way to an understanding of lenses not as epistemologically unreliable devices because they are artificial but as tools that improve the power of sight. From this point forward, the geometrical theory of lenses became one of the central chapters of optics.[52]

For both Maurolico and Kepler, refraction was a key phenomenon. Although the phenomenon had been known since antiquity, the search for a precise quantitative analysis of it had not been successful. Given the explanatory role played by refraction in Kepler's theory of vision, this problem became even more pressing.

REFRACTION AND DIFFRACTION

Descartes' greatest contribution to optics is *Dioptrique* (Dioptrics), an essay appended to the *Discours de la méthode* (Discourse on Method, 1637).[53] Most

---

[50] On the history of eyeglasses, see Edward Rosen, "The Invention of Eyeglasses," *Journal of the History of Medicine and Allied Sciences*, 11 (1956), 13–53, 183–218. The complicated history of the discovery of the telescope has been beautifully recounted by Albert van Helden, "The Invention of the Telescope," *The American Philosophical Society*, 67 (1977), 3–67. On the history of the microscope, see S. Bradbury, *The Evolution of the Microscope* (Oxford: Pergamon, 1967).

[51] On Galileo and the telescope, see Vasco Ronchi, *Il cannocchiale di Galileo e la scienza del Seicento* (Turin: Boringhieri, 1958); and Albert van Helden, "Galileo and the Telescope," in *Novità celesti e crisi del sapere*, ed. Paolo Galluzzi (Florence: Giunti Barbera, 1983), pp. 149–58.

[52] On the epistemological significance of optical instruments in the seventeenth century, see Philippe Hadou, *La Mutation du visible: Essai sur la portée épistémologique des instruments d'optique au XVIIe siècle* (Villeneuve d'Ascq: Presses Universitaire du Septentrion, 1999).

[53] The literature on Descartes' optical theories is quite extensive. Fundamental studies are A. I. Sabra, *Theories of Light from Descartes to Newton* (Cambridge: Cambridge University Press, 1981); Pierre Costabel, "La refraction de la lumière et la *Dioptrique* de Descartes," in Pierre Costabel, *Démarches originales de Descartes savant*, (Paris: J. Vrin, 1982), pp. 63–76; and A. Mark Smith, "Descartes's Theory of Light and Refraction: A Discourse on Method," *Transactions of the American Mathematical Society*, 77 (1987), 1–92. For an evaluation of the *Dioptrique* as a whole and for further references, see Neil M.

of the book is devoted to vision and how to improve it by means of optical instruments, and it was in this context that Descartes described telescopes and lenses and how to grind them (Chapters 7–10). The best-known parts of *Dioptrics*, however, lay out a theory of vision (Chapters 3–6) – largely of Keplerian inspiration – and derive the law of refraction (Chapter 2).

The first problems in optics to attract Descartes' attention were finding a mathematical law for the phenomenon of refraction and determining the nature of the anaclastic. The latter problem consists in finding the exact shape of the refracting surface that focuses parallel rays into a single point. Previous writers had posed both problems (e.g., Kepler in *Ad Vitellionem paralipomena*) but had not achieved a precise solution. Descartes managed to solve both problems. By the end of the 1620s, he already had found a general solution to the problem of the anaclastic by means of a class of curves, now called "Cartesian ovals," that includes ellipses and hyperbolas. The solution of the anaclastic problem was probably made possible by Descartes' discovery of the law of refraction. Another striking application of the law was the account of the rainbow that Descartes provided in *Meteors*, one of the other essays accompanying the *Discourse on Method*.

The derivation of the law of refraction in the *Dioptrique* repays close attention.[54] In Book 1, Descartes avoided characterizing the nature of light and investigated the main properties of light by using mechanical analogies such as that of a blind man guiding himself with a cane and, especially, a tennis racquet striking a ball. The latter is the basis for the solution to the problem of refraction presented in Book 2. According to this analogy, as light passes from a less dense to a more dense medium, it behaves like a tennis ball passing, for example, from air to water. There are also contrasting analogies, however. Descartes pointed out that whereas light tends to move toward the normal to the surface, the tennis ball moves away from the normal. Part of the justification for this different behavior, according to Descartes, lies in the fact that, unlike the tennis ball, light travels faster in denser media.

Consider now a ball going from *A* toward *B* (Figure 25.5). Descartes analyzed the ball's direction of motion into a horizontal component, represented by *AF*, and a vertical component, represented by *AC*. When the ball hits the water at *B*, it loses some of its velocity – say, half. According to Descartes, this encounter with the water will affect the vertical component of motion but not the horizontal. From these assumptions, he concluded, using elementary geometrical reasoning, that once it has hit point *B*, the ball will be deflected toward *I*.

Ribe, "Cartesian Optics and the Mastery of Nature," *Isis*, 88 (1997), 42–61. For Descartes' account of the rainbow, see Charles Boyer, *The Rainbow: From Myth to Mathematics* (New York: Sagamore Press, 1959), pp. 200–32.

[54] How Descartes arrived at the law is still a matter of scholarly debate. For a survey of the different accounts, see Antoni Malet, "Gregorie, Descartes, Kepler, and the Law of Refraction," *Archives Internationales d'Histoire des Sciences*, 40 (1990), 278–304.

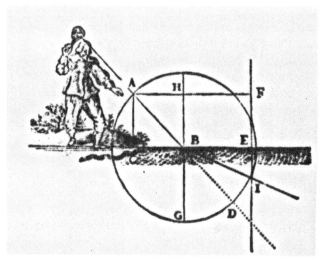

Figure 25.5. Deflection of ball's trajectory passing from air to water. In René Descartes, *Discourse de la méthode* (Leiden: I. Maire, 1637). Reproduced by permission of the Library of the Max Planck Institute for the History of Science, Berlin.

In the case of a ray of light going from air to water (Figure 25.6), Descartes modified the preceding model by assuming that while hitting B the ball is hit again in such a way that its force increases by one-third; it will thus traverse in two instants the same distance that it used to traverse in three instants. (This amounts to claiming that light travels faster in water than in air.) Using the same geometrical reasoning as in the previous case, Descartes then concluded that the ray of light is deflected toward *I*. Thus, the force with which the ball penetrates the water is to the force with which it leaves the air as the line *CB* is to *BE*. This formulation of the law of refraction is equivalent to its now standard formulation, which states that when refraction occurs, the relationship between the sines of the angles of incidence and refraction is a constant (i.e., $\sin i / \sin r = k$, where k is a constant that depends on the medium). Descartes' derivation of the law of refraction was based on two major assumptions: first, that the velocity of the ray of light traversing two different optical media increases or decreases by a constant factor that depends only on the media; and, second, that only the vertical component of velocity is affected in the process, the horizontal component remaining unaffected.

Descartes' derivation of the law of refraction was far from compelling. Already in the seventeenth century, mathematician and jurist Pierre de Fermat (1601–1665) had strongly criticized Descartes' proof for its implausible assumption that light travels faster in denser media. Fermat provided alternative proofs of the law of refraction starting from a least-time principle:

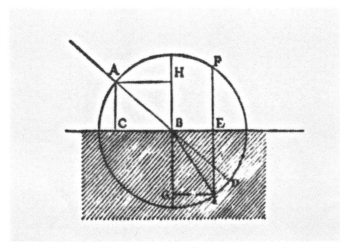

Figure 25.6. Refraction of light ray passing from air to water. In René Descartes, *Discourse de la méthode* (Leiden: I. Maire, 1637). Reproduced by permission of the Library of the Max Planck Institute for the History of Science, Berlin.

The time needed by light to travel through two optical media must be minimized.[55]

With phenomena such as reflection, refraction, and the passage of light through small apertures, an ingenious analysis could succeed without sacrificing the age-old principle that the propagation of light takes place in rectilinear fashion. The first optical phenomena that seemed prima facie to challenge this principle were the diffraction phenomena investigated by Grimaldi and published in his *Physico-mathesis de lumine, coloribus et iride* (A Physico-mathematical Treatment of Light, Colors, and the Rainbow, 1665).[56]

Grimaldi experimented with very small beams of light. For instance, in one experiment (Figure 25.7), he let a ray of sunlight pass through a very small aperture AB in an otherwise dark chamber and observed the behavior of light when a very small obstacle, say a needle FE, was inserted into the cone of light with its vertex in the aperture AB and its base on the screen CD. Grimaldi observed that the shadow projected on the screen by the small obstacle, FE, was much wider than $IL - IL$ being the shadow one would have expected "if everything were to happen by straight lines." He also observed that the strongly illuminated areas CM and ND presented bands of colored light, each

---

[55] On Fermat's objections and his derivation of the sine law, see Kirsti Andersen, "The Mathematical Technique in Fermat's Deduction of the Law of Refraction," *Historia Mathematica*, 10 (1983), 48–62; Michael S. Mahoney, *The Mathematical Career of Pierre de Fermat, 1601–1665* (Princeton, N.J.: Princeton University Press, 1973), pp. 375–90; Sabra, *Theories of Light*, pp. 116–58; and Smith, *Descartes's Theory*, pp. 81–2. The latter provides an interpretation of Descartes' derivation of the sine law against the background of medieval "perspectivist" optics.

[56] No complete account of Grimaldi's book is available in the secondary literature. See, however, Vasco Ronchi, *The Nature of Light* (London: Heinemann, 1970), pp. 124–49.

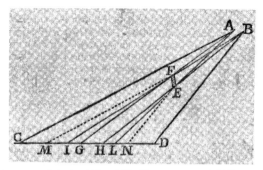

Figure 25.7. Francesco Maria Grimaldi's illustration of diffraction through one small aperture with needle point. In Francesco Maria Grimaldi, *Physico-mathesis de limine, coloribus et iride* (Bologna: Bernia, 1665), p. 2. Reproduced by permission of the Staatsbibliothek zu Berlin – Preussischer Kulturbesitz/bpk 2004.

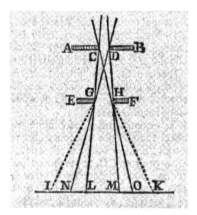

Figure 25.8. Francesco Maria Grimaldi's illustration of diffraction through two small apertures. In Francesco Maria Grimaldi, *Physico-mathesis de limine, coloribus et iride* (Bologna: Bernia, 1665), p. 2. Reproduced by permission of the Staatsbibliothek zu Berlin – Preussischer Kulturbesitz/bpk 2004.

band being white in the center and bordered on the sides by blue and red. In a different experiment (Figure 25.8), Grimaldi put a second surface with a small aperture through the cone of light and studied the second cone of light thus created. Let *CD* be a small hole that allows sunlight to pass through a shutter penetrating in an otherwise dark room. The light entering through *CD* will form a cone of light. At a great distance from *CD*, let a board *EF*, with a small hole *GH*, intersect the cone of light perpendicular to the cone. Let the board be inserted in such a way that the base of the cone, *NO*, is much larger than *GH*. Now the cone of light created by the light going through *GH* is observed to be *IK*, whereas one would, according to the rectilinear propagation of light rays, expect it to be equal to *NO*. Moreover, Grimaldi noticed that the base *IK* on the screen presents white light in the center and is bordered at the extremities by red and blue light.

In both experiments, the study of the illuminated region, the shadow, and the penumbra cast on the screen seemed to reveal a fractioning of light in different components that remained separated. Grimaldi considered this to be a new mode of the propagation of light and called it diffraction.

GEOMETRICAL OPTICS AND IMAGE LOCATION

Historians have shown a renewed interest in one long-neglected part of geometrical optics, the theory of optical images.[57] The idea of optical images that emerged in the third quarter of the seventeenth century was quite different from previous notions, such as Kepler's *pictura*. The most important texts in this development are James Gregory's (1638–1675) *Optica promota* (Advances in Optics, 1663), Isaac Barrow's (1630–1677) *Lectiones opticae* (optical Lectures, 1669), and Newton's *Optical Lectures* (delivered at Cambridge University, 1670–2).[58] These works exhibited a high level of mathematization and downplayed physical explanations of the nature of light. In this context, optical images appeared as mathematical constructs whereby the eye locates an object in physical space by exploiting the divergence of the light rays stimulating the eye. Historian of science Alan E. Shapiro summarized this principle of image location as follows: "An image is located at the place from which the rays entering a single eye diverge; that is, the image is perceived at the place of the real or virtual geometrical image."[59] Shapiro shows how Barrow was able to develop an extremely sophisticated mathematical theory using this principle, thereby solving the problem of determining the location of optical images for any refraction or reflection on plane and spherical surfaces.

The principle as such has empirical content, and much of the discussion to which it gave rise depended on the fact that a physico-geometrical notion of an image (given in terms of convergence of rays) was paired with an empirical claim concerning the identity of the location of the image perceived with the image defined in physico-geometrical terms. The question of the empirical adequacy of the principle of image location came up also in connection with

---

[57] See Alan E. Shapiro, "The *Optical Lectures* and the Foundations of the Theory of Optical Imagery," in *Before Newton: The Life and Times of Isaac Barrow*, ed. Mordechai Feingold (Cambridge: Cambridge University Press, 1990), pp. 105–78; and Antoni Malet, "Isaac Barrow and the Mathematization of Nature: Theological Voluntarism and the Rise of Geometrical Optics," *Journal of the History of Ideas*, 46 (1995), 1–33.

[58] Barrow was Lucasian Professor of Mathematics at Cambridge University. James Gregorie (or Gregory) was professor of mathematics at St. Andrews. On Gregory's work on optics and mathematics, see Antoni Malet, Ph.D. dissertation, Princeton University, 1989.

[59] Shapiro, *The Optical Lectures*, p. 107. The principle of image location is called "Barrow's principle" by Shapiro and the "Gregorie-Barrow principle of image location" by Malet.

the age-old cathetus rule, according to which the image is always located on the perpendicular connecting the light source and the reflecting surface or the refracting medium. Because Barrow's principle of image location and the cathetus rule were often in conflict, Barrow tried to show the invalidity of the cathetus rule by means of experiments.[60]

THE NATURE OF LIGHT AND ITS SPEED

At the beginning of the seventeenth century, there were various competing views concerning the nature of light. Atomist thinkers thought of light as a flux of material particles, whereas Aristotelian philosophers saw light and color as the modification of the diaphanous (i.e., transparent) medium. Descartes described light as a tendency toward motion that could be explained by an instantaneous pressure communicated through the medium instantaneously; that is, with infinite speed. Kepler also maintained that light traveled with infinite speed and defended a spiritual theory of light inspired by neo–Platonic ideas,[61] whereas others held that the speed of light was very high but finite.

The issue of whether the velocity of light was finite or infinite was empirically resolved in 1675, on the occasion of an eclipse of one of Jupiter's satellites, through the experimental observations of the Danish astronomer Ole Römer (1644–1710). Römer predicted that the eclipse would take place approximately ten minutes later than the time predicted by the prevailing theory. Checking Bolognese astronomer Gian Domenico Cassini I's (1625–1712) tables for the periodic eclipses of Jupiter's satellites, Römer had noticed systematic anticipations and retardations of the observed times that correlated with Jupiter's conjunctions and oppositions. He correctly interpreted them in terms of the time needed by light to travel from the satellite to the earth. From these data and from tentative values of the dimensions of planetary orbits, Römer deduced a first numerical approximation of the velocity of light.[62]

By the end of the seventeenth century, theories of light came in two major varieties, emission and continuum. The former had as their central explanatory feature the actual transport of matter. The latter explained light

---

[60] Barrow also proposed a rather difficult problem for his theory, which became widely known as the "Barrovian case." See Shapiro, *The Optical Lectures*, pp. 159–65; and Malet, *Isaac Barrow*, pp. 28–9. The epistemological importance of these works is also confirmed by George Berkeley's criticism of this tradition in his *New Theory of Vision* (1706). On Berkeley's theory of vision and its relationship to English geometrical optics, see Margaret Atherton, *Berkeley's Revolution in Vision* (Ithaca, N.Y.: Cornell University Press, 1990).

[61] On this aspect of Kepler's thought, see David C. Lindberg, "The Genesis of Kepler's Theory of Light: Light Metaphysics from Plotinus to Kepler," *Osiris*, 2 (1986), 5–42.

[62] "Demonstration touchant le mouvement de la lumiere trouvé par M. Römer de l'Academie Royale des Sciences," *Journal des sçavans* (1676), 233–6. On the history of the problems connected to the velocity of light, see *Roemer et la vitesse de la lumiere*, ed. René Taton (Paris: J. Vrin, 1978).

in terms of the "propagation of a state, such as pressure or motion, through an intervening medium."[63] The best-known representatives of the two traditions are Newton (discussed in the next section) and Huygens, respectively.

Huygens presented the results of his investigations in *Traité de la lumière* (Treatise on Light, completed in 1678 but published in 1690). Huygens was convinced that a satisfactory theory of light could be obtained only through the principles of "true philosophy," according to which "light consists in a movement of matter which exists between us and the luminous body."[64] At the same time, however, he held that light cannot be accounted for by any transport of matter to us from the object. Huygens believed that the great speed of light – which he maintained to be finite, following Römer's experiments – and the fact that different rays of light "traverse one another without hindrance" are two of the strongest pieces of evidence against a transport theory. Exploiting an analogy with sound waves, Huygens explained light in terms of spherical waves that travel with finite speed through an ether, an invisible medium that he conceived as being made of small, moving particles of equal size. Each point of a luminous source is the center of a wave. But how do waves move? The answer provided by Huygens is commonly known as Huygens's principle. Consider a wave DCF originating at a luminous point A (Figure 25.9). A transmits its motion to all contiguous particles. Each such particle in turn is the center of a wave. For instance, B will be the center of a wave KCL, which "will touch the wave DCF at C at the same moment that the principal wave emanating from the point A has arrived at DCF."[65] Thus the wave KCL contributes to the wave DCF in the point C, and similarly for all other points such as b, d, etc. This principle explains how the wave moves from the front HBG to DCF and how the wave front DCF is constituted. These waves propagate with a velocity that depends only on the degree of elasticity of the compressed particles that constitute the ether. Using this principle, Huygens proceeded to give a unified account of reflection and refraction, and in particular of the double refraction in calcite (Iceland spar), a phenomenon discovered by Danish mathematician Erasmus Bartholin (1625–1698) in 1669.[66] Although impressive in its achievements, Huygens's theory was quickly forgotten and almost ignored until the beginning of the nineteenth century.[67] It was Newton's theory of

---

[63] For continuum ("wave") theories of light in the seventeenth century, see Alan E. Shapiro, "Kinematic Optics: A Study of the Wave Theory of Light in the Seventeenth Century," *Archive for History of Exact Sciences*, 11 (1973), 134–266, quotation at p. 136.
[64] Huygens, *Treatise on Light*, p. 554.
[65] Ibid., p. 562.
[66] Erasmus Bartholin, *Experimenta crystalli Islandici disdiaclastici, quibus mira et insolita refractio detegitur* (Hafniae: Danielis Paulli, 1669).
[67] On Huygens's theory of light, see Shapiro, *The Optical Lectures*, pp. 159–65; and Sabra, *Theories of Light*, pp. 198–230. Among other things, these works treat the continuum theories of light proposed by Hobbes, Pierre Ango and Ignace Pardies, and Hooke. On Hobbes's theory of vision and light, see Franco Giudice, *Luce e visione: Thomas Hobbes e la scienza dell'ottica* (Florence: Leo S. Olschki,

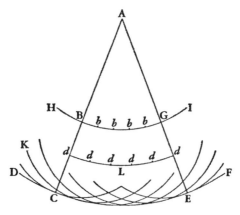

Figure 25.9. Christiaan Huygens's illustration of his principle of secondary wavelets. In Christiaan Huygens, *Traité de la lumière* (Leiden: Pierre vander Aa, 1690). From *Newton, Huygens*, Great Books of the Western World (Chicago: Encyclopaedia Britannica, 1952), p. 526. Reproduced by permission of Great Books of the Western World, © 1952, 1990 Encyclopaedia Britannica, Inc.

light that was given pride of place by late seventeenth- and eighteenth-century thinkers.

NEWTON'S THEORY OF LIGHT AND COLORS

Although Newton's optical researches went back to 1664, his first publication, "New Theory About Light and Colours," appeared in the *Philosophical Transactions of the Royal Society of London* in 1672. In the meantime, he had offered an account of his theories in the *Optical Lectures*. His theories received their most developed form in *Opticks; or, a treatise of the reflections, refractions, inflections and colors of light* (1704 and subsequent editions).[68]

Newton's early work in optics related directly to the most debated optical topics of the seventeenth century. He carefully read Descartes, Boyle, and Hooke, and his interests ranged from the theoretical to the practical.[69] One

---

1999). On Huygens's studies in optics, see also Christiaan Huygens, *Oeuvres*, 22 vols. (The Hague: Martinus Nijhoff, 1888–), vol. 13 (1916).

[68] The literature on Newton is too large to survey. A few standard works and some more recent contributions include: Sabra, *Theories of Light*, pp. 231–342; Alan E. Shapiro, ed., *The Optical Papers of Isaac Newton*, vol. I: *The Optical Lectures, 1670–1672* (Cambridge: Cambridge University Press, 1984), pp. 1–25; Shapiro, *Fits, Passions, and Paroxysms* (Cambridge: Cambridge University Press, 1993); Dennis L. Sepper, *Newton's Optical Writings: A Guided Study* (New Brunswick, N.J.: Rutgers University Press, 1994); and A. Rupert Hall, *All Was Light: An Introduction to Newton's Opticks* (Oxford: Oxford University Press, 1993).

[69] The literature on the theories of colors of Descartes, Hooke, and Boyle is similarly extensive. Hooke's most important work on the subject is his *Micrographia* (1665), and Boyle's is the *Experiments and Considerations touching Colours* (1664). See Hall, *All Was Light*, for a clear introduction to these works in connection with Newton's work on colors.

of the first problems to attract Newton's attention was chromatic aberration in telescopic lenses, the phenomenon whereby the image of a point in a lens is surrounded by an iridescent halo. This was a problem of great practical and theoretical interest, and the attempt to avoid chromatic aberration in telescopes led to experiments with nonspherical lenses. However, Newton soon convinced himself that the phenomenon was unavoidable on account of the unequal refrangibilities of light rays of different colors. This led him in 1668 to develop a reflecting telescope, which made use of a concave spherical mirror instead of lenses.

Newton's practical work was deeply intertwined with his analyses of white light and color, begun in "A New Theory About Light and Colours." Although many before Netwon had observed the colors obtained by letting a ray of sunlight pass through a prism, Newton was puzzled by a very simple fact that he observed when replicating "the celebrated Phaenomena of Colours." After having darkened his room, he let sunlight pass through a small hole in the shutters of a window. Behind the small hole he put a glass prism, so that light could be refracted to the opposite wall. The refraction gave rise to the spectrum of colors, and Newton noticed something puzzling about the spectrum. Relying on his geometrical knowledge of refraction, he expected the refracted light to give rise to a circular image when the prism was placed at minimum deviation,[70] whereas the image refracted on the wall was actually oblong, the width being approximately one-fifth of the length. Determined to find out what could account for such a divergence between the expected shape of the spectrum and what he observed, Netwon carried out a number of experiments that ruled out a number of possible causes of the effect, such as the thickness of the glass or the size of the hole in the shutters. He then devised an experiment – which he considered as the crucial experiment for his theory – using a combination of two prisms. Consider a prism *ABC* through which a ray of sunlight passes (Figure 25.10). Let the spectrum of colors obtained be projected onto a board *DE*, placed behind the prism, which has a small hole *G*. A second board *de*, also with a small hole *g*, is placed at a distance of approximately 12 feet from the first board and behind it a second prism *abc*. By accurately rotating the first prism about its axis, Newton was able to let only rays belonging to a certain section of the spectrum reach the second board, and then he observed on a screen how the rays passing through the hole *g* in the second board were refracted by the second prism. Thus, he noticed that when light from the red end of the spectrum passed through *g*, it would be refracted at point *M* in the screen, whereas light from the blue end of the spectrum going through *g*

---

[70] In 1672, Newton wrote, "I became surprised to see them [colours] in an *oblong* form; which, according to the received laws of Refraction, I expected should have been circular." From Isaac Newton, "A New Theory about Light and Colours," in *The Correspondence of Isaac Newton*, ed. H.W. Turnbull and J.-F. Scott, 7 vols. (Cambridge: Cambridge University Press, 1959), 1: 92.

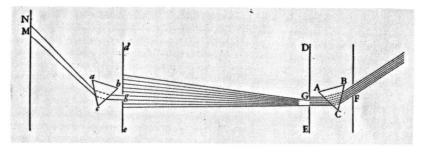

Figure 25.10. Isaac Newton's prism experiment on spectral colors. In Isaac Newton, *Opticks* (London: Samuel Smith and Benjamin Walford, 1704). From *Newton, Huygens*, Great Books of the Western World (Chicago: Encyclopaedia Britannica, 1952), p. 562. Reproduced by permission of Great Books of the Western World, © 1952, 1990 Encyclopaedia Britannica, Inc.

would be refracted at point N.[71] In short, the experiment allowed Newton to single out beams of rays of a certain color by screening them through the first hole and then to observe their refrangibility by means of the second prism. Because the boards and the second prism were fixed, the rays of light always hit the second prism with the same angle of incidence. Consequently, the fact that different colors were refracted into different parts of the screen could only depend on the different refractions they underwent in the second prism.

According to Newton, the experiment made evident the cause for the oblong figure of the spectrum, which sprang from the fact that white light is an aggregate of rays of different degrees of refrangibility: "Light it self is a Heterogeneous mixture of differently refrangible Rays."[72] Moreover, according to Newton, to each color there corresponds a unique degree of refrangibility and vice versa. He was thus led to reject most of the previous theories of colors, which had claimed that colors were somehow modifications of white light caused by refraction or reflection by other bodies, and to assert that colors are innate properties of white light. In particular, by demonstrating that white light is merely the composite of light of all colors, Newton showed that the colors are primary in nature and white light is secondary. This represented a radical revision of traditional views of light and color.

Newton thought this experiment was conclusive evidence for his theory, and this claim led to debates with Hooke, the Jesuit physicist Ignace Gaston Pardies (1636–1673), and Huygens. The topics of contention included Newton's tentative assertion of the corporeal nature of light. The ensuing

---

[71] Newton does not actually mention colors in his 1672 account of the *experimentum crucis* but only speaks in general of the different refrangibility of white light. He specifically mentions colors, however, in his account of the experiment in the *Opticks*.

[72] Newton, "A New Theory about Light and Colours," 1: 95.

discussion was of great methodological importance and led Newton to write his "Hypothesis Explaining the Properties of Light Discoursed of in my Several Papers" (read to the Royal Society in 1675), which took a more tolerant stance toward vibration (or wave) theories of light. Dating from the same period is his "Discourse of Observations" (read to the Royal Society in 1676 and later rewritten as parts 1, 2, and 3 of Book II of the *Opticks*), in which he described the discovery of rings of color – now called "Newton's rings" – obtained by letting light pass through a "lamina" (thin layer) of air contained between two pieces of glass.

The final presentation of Newton's optical theories is found in the *Opticks*, which is divided into three books. The first book deals with problems of geometrical optics, reflection, and refraction of white light and color. The second book treats problems of interference, such as Newton's rings. The experiment consists of a plano-convex lens placed on top of a double convex lens so that the flat side of the plano-convex lens is directed downward. This creates at the point of contact between the two lenses a space of air, a "lamina," of variable thickness. By illuminating the system, one obtains concentric rings of color around the point of contact between the two lenses. Newton's observations showed that a certain periodicity is displayed by the colors that make up white light. In order to account for this periodicity, he proposed a theory of "fits of easy transmission and reflection" that purported to explain the phenomena by a combination of corpuscular and wave conceptualizations of light. Finally, in the third book of the *Opticks*, Newton dealt with diffraction phenomena such as those investigated by Grimaldi, although he used the word "inflexion" instead of "diffraction." He extended to these phenomena the theory that he had already presented in the *Principia* for reflection and refraction, according to which bodies interact at a distance with rays of light and bend them from their rectilinear course. Because his treatment of inflection was inconclusive, Newton concluded the *Opticks* with a set of queries in which he discussed a variety of issues concerning the interaction of light and matter, colors, and vision. Newton's optical theory dominated the scientific scene for most of the eighteenth century, and it is only with Swiss mathematician Leonhard Euler (1707–1783), in the eighteenth century, that optics was substantially extended.[73]

---

[73] On the fortunes of Newton's *Opticks* in England and in continental Europe, see Hall, *All Was Light*. On post-Newtonian optics, see Henry John Steffens, *The Development of Newtonian Optics in England* (New York: Science History Publications, 1977); and Geoffrey Cantor, *Optics after Newton: Theories of Light in Britain and Ireland, 1704–1840* (Manchester: Manchester University Press, 1983). On Euler, see Casper Hakfoort, *Optics in the Age of Euler* (Basel: Birkhäuser, 1995). Despite the influence of the Newtonian legacy in the eighteenth century, some more recent scholarly work has put into doubt the alleged predominance of Newton's theory of light – at least in the German and French contexts. For an overview of such views, see Franco Giudice, "La tradizione del mezzo e la *Nuova teoria della luce* di Leonhard Euler," *Nuncius*, 15 (2000), 3–49.

CONCLUSION

In a famous article, historian of science Thomas Kuhn discussed the development of two traditions in physical science, the mathematical and the experimental.[74] The former included fields – such as astronomy, harmonics, mathematics, optics, and statics – that in antiquity had already been the focus of sustained intellectual activity. Of these, only astronomy, optics, and statics are today still considered parts of physics. The experimental, or Baconian, tradition "created a different sort of empirical science, one that for a time existed, side by side with, rather than supplanting, its predecessor."[75] This distinction is helpful in summarizing early modern developments in acoustics and optics.

Harmonics belonged to the classical mathematical sciences. Although its contents were still studied in the seventeenth century, its significance to the development of acoustics in the sixteenth and seventeenth centuries consisted mainly in providing a set of problems, such as the nature of the consonances and intonation, that came to be treated by physical techniques. The new physical study of sound presented many of the features of a Baconian science. Although the new study of sound had no real "predecessor" to supplant, as Kuhn put it, it grew alongside the classical theory of harmonics. In some cases, it is hard to evaluate the exact nature of the experimental work carried out – this is especially true for the alleged experiences recounted by Galileo in *Two New Sciences* – but there is no doubt that the experiments described by Vincenzo Galilei, Mersenne, and Boyle, among others, were qualitatively different from anything seen before. By the end of the seventeenth century, the accuracy of measurements, the systematic use of instruments such as the air-pump, and the sophistication of experimental work had reached levels undreamed of even at the beginning of the century (see Dear, Chapter 4, this volume). Much of the experimental work in acoustics would not find its theoretical, mathematized treatment until the eighteenth and nineteenth centuries. The study of the vibrating string by Euler, Daniel Bernoulli, and Jean d'Alembert eventually led to the definitive constitution of the theory of classical acoustics in the nineteenth century in the works of Ernst Chladni (1756–1827), Siméon-Denis Poisson (1781–1840), Hermann von Helmholtz (1821–1894), and John William Strutt, third Baron Rayleigh (1842–1919).

In the study of optics (or *perspectiva*, as it was then called), the classical discipline of geometrical optics underwent substantial development during the seventeenth century, including both the solution of problems already raised in the classical and medieval traditions – Kepler's solution to the problem of radiation through small, finite apertures, the law of refraction,

---

[74] Thomas Kuhn, "Mathematical Versus Experimental Traditions in the Development of Physical Science," *The Journal of Interdisciplinary History*, 7 (1976), 1–31.
[75] Kuhn, "Mathematical Versus Experimental Traditions," p. 8.

and the geometry of vision – and the exploration of entirely new areas, such as Kepler's theory of lenses and Barrow's development of the theory of image location. These extensions were in part made possible by the systematic exploitation of new instruments, such as the telescope and the microscope. Moreover, experimentation with optical phenomena unknown to antiquity, including diffraction, "Newton's rings," and double refraction, can be seen as part of a Baconian strand in the science of optics. In providing a first synthesis of the new field of physical optics in the *Opticks*, Newton, as Kuhn recognized, participated significantly in both the classical and the Baconian traditions. In the case of physical optics, as in that of acoustics, the theoretical, mathematized treatment of the phenomena experimentally discovered in the early modern period found its classical systematization only at the beginning of the nineteenth century in the works of Thomas Young (1773–1829) and Augustin Jean Fresnel (1788–1827).

26

MECHANICS

Domenico Bertoloni Meli

This chapter is devoted to mechanics in the sixteenth and seventeenth centuries. Following a distinction traceable at least to Hero of Alexandria (first century) and Pappus of Alexandria (third century), mechanics can be divided into rational and practical (or applied). The former is a mathematical science normally proceeding by demonstration, the latter a manual art with practical aims. Here I privilege rational over practical mechanics, which is discussed elsewhere in this volume (see Bennett, Chapter 27).[1]

A major problem with writing a history of mechanics during this period concerns the changing disciplinary boundaries and meaning of the term "mechanics." Traditionally, mechanics had dealt with the mathematical science of simple machines and the equilibrium of bodies. In the second half of the seventeenth century, however, mechanics became increasingly associated with the science of motion. Therefore, in dealing with an earlier period, it is useful to chart not simply the transformations of mechanics as it was understood before the second half of the seventeenth century but also the relevant transformations in the science of motion that belong more properly to natural philosophy.

Mechanics and natural philosophy differed widely intellectually, institutionally, and socially in the period covered by this chapter. Even rational mechanics retained a practical and engineering component but it was also progressively gaining a higher intellectual status with the editions of major works from antiquity and with a renewed emphasis on its utility; initially its role in the universities was at best marginal, however. By contrast, natural

---

[1] Pappus, *Mathematicae colletiones*, translated by Paul ver Eecke as *La collection mathématique* (Paris: Desclée de Brouwer, 1933), p. 810. See also Isaac Newton, *Principia*, new translation by I. Bernard Cohen and Anne Whitman (Berkeley: University of California Press, 1999), pp. 381–2. On this distinction, see G. A. Ferrari, "La meccanica 'allargata'," in *La scienza ellenistica*, ed. Gabriele Giannantoni and Mario Vegetti ([Naples]: Bibliopolis, 1984), pp. 225–96.

I wish to thank Karin Eckholm and Allen Shotwell for their helpful comments on an earlier draft of this work.

philosophy had been a major academic discipline for centuries and had closer links to theology than to the practical arts. It is therefore necessary to chart the changing contours and domains of mechanics by paying attention to how scholars at the time understood it, lest one write a history of a fictitious discipline by projecting a modern vision of that discipline onto the past.[2]

This chapter starts by examining the impact of the recovery of ancient and medieval learning both in what was understood to belong to mechanics proper and in those portions of natural philosophy dealing with motion. The critical editions and the assimilation of those sources in the sixteenth century led to a transformation of mechanics on many levels.[3] I then move to a brief characterization of some of the leading scholars on motion and mechanics in the sixteenth century that culminates with the work of Galileo Galilei (1564–1642). The interplay among mechanics, the philosophical study of motion, and quantitative experiment is at the center of my reading of Galileo's groping toward a new mathematical science of local motion.

The work of Galileo marks a turning point that organizes the chapter. Although disciplinary contours did not change overnight, a shift occurred between the first and second parts of the seventeenth century. I shall use the debates and controversies triggered by Galileo's major works as a guide to later developments. Moreover, more recent studies have emphasized that in his effort to formulate a new science of motion, Galileo remained enmeshed in portions of the old worldview. By relying on his work, the generation after him could more easily free itself from the past. This is an additional reason that justifies my partition.[4]

Philosophers who studied motion were almost invariably professors at a university or Jesuit college. Practitioners of mechanics, however, had a more varied professional profile. Niccolò Tartaglia (1506–1557) was a teacher of mathematics at the University of Venice who never attained high social standing. The Urbino mathematician Federico Commandino (1509–1575) was a refined humanist scholar, held a medical degree, and moved around papal and princely courts, especially the Farnese. His pupil Guidobaldo dal Monte (1545–1607) was the brother of a cardinal and a marquis himself, with close ties to the della Rovere in Urbino and the Medici in Florence. In his youth, dal Monte had been a military man and in the late 1580s became superintendent

---

[2] For an example of such an anachronistic projection, see Marshall Clagett, *The Science of Mechanics in the Middle Ages* (Madison: University of Wisconsin Press, 1959). An excellent account that is sensitive to these concerns is John E. Murdoch and Edith D. Sylla, "The Science of Motion," in *Science in the Middle Ages*, ed. David C. Lindberg (Chicago: University of Chicago Press, 1978), pp. 206–64.

[3] A useful source is *Mechanics in Sixteenth-Century Italy*, translated and annotated by Stillman Drake and I. E. Drabkin (Madison: University of Wisconsin Press, 1969) on which the following two sections rely. See also the essay review by Charles B. Schmitt, "A Fresh Look at Mechanics in 16th-Century Italy," *Studies in the History and Philosophy of Science*, 1 (1970), 161–75.

[4] For details, see Peter Damerow, Gideon Freudenthal, Peter McLaughlin, and Jürgen Renn, *Exploring the Limits of Preclassical Mechanics* (New York: Springer, 2004).

of Tuscan fortifications. The Venetian Giovanni Battista Benedetti (1530–1590), a student of Tartaglia, became court mathematician first at Parma under the Farnese and then at Turin under the Savoy. The Dutch mathematician Simon Stevin (1548–1620) was a military man and an engineer who was made quartermaster of the army of the Low Countries in 1604. Many of these men had an interest in practical as well as theoretical matters in mechanics.[5]

MECHANICAL TRADITIONS

The main works on mechanics can be associated with a number of texts and traditions beginning shortly after Aristotle's death. I identify four main traditions – those of pseudo-Aristotle, Archimedes, Alexandria (especially Pappus) – and the science of weights.

The first of these traditions is associated with *Quaestiones mechanicae* (Mechanical Problems), traditionally attributed to Aristotle (384–322 B.C.E.) but now considered to be an early product of his school. The work deals mostly with applications of the doctrines of the lever, which depends upon the balance, and of the balance, whose properties are analyzed by imagining that it rotates around its fulcrum so as to describe a circle; hence the use of motion in the study of equilibrium and the strange idea that the properties of the balance depend on the circle. The author claimed that nearly all mechanical problems depend on the lever and in some cases he provided some form of explanation of how some machines, such as the windlass, for example, operate and relate to the circle. One cannot say that there is a systematic and rigorous attempt in this direction, however, but only a number of appropriate remarks. A few passages deal with seafaring, others with the resistance of beams or the force exerted by a moving body on a wedge (later known as the force of percussion). In the age of the printing press, the text went through many editions and translations, often with valuable commentaries, starting in 1497.[6]

Archimedes (287–212 B.C.E.) wrote two major works on mechanics, *De centris gravium* (On the Equilibrium of Planes) and *De insidentibus aquae* (On Floating Bodies). Both survived in several manuscript copies and were first published in 1543 by Tartaglia.[7] Superior editions were later produced by

---

[5] Mario Biagioli, "The Social Status of Italian Mathematicians, 1450–1600," *History of Science*, 27 (1989), 41–95; Paul L. Rose, *The Italian Renaissance of Mathematics* (Geneva: Droz, 1975); and Simon Stevin, *The Principal Works of Simon Stevin*, 5 vol. (Amsterdam: Swets and Zeitlinger, 1955), 1: 1–24.

[6] Paul L. Rose and Stillman Drake, "The pseudo-Aristotelian *Questions of Mechanics* in Renaissance Culture," *Studies in the Renaissance*, 18 (1971), 65–104.

[7] Niccolò Tartaglia, ed., *Opera Archimedis* (Venice: Per Venturinum Ruffinellum, 1543). Tartaglia's edition contains only Book I of *Floating Bodies*. The chief source on the fortune of the Archimedean corpus through to the Renaissance is Marshall Clagett, *Archimedes in the Middle Ages*, 5 vols. (vol. 1, Madison: University of Wisconsin Press, 1964; vols. 2–5, Philadelphia: American Philosophical Society, 1976–84).

Commandino and his student dal Monte.[8] In *On the Equilibrium of Planes* Archimedes produced an axiomatic theory of the balance, thus introducing into mechanics a style based on pure mathematics that constituted a model for many later works. He also determined the centers of gravity of several plane figures.[9] In the sixteenth century, Commandino and Galileo, among others, extended those investigations. *On Floating Bodies* deals with hydrostatics and contains the celebrated statement that a body in a fluid receives an upward thrust equal to the weight of the volume of the displaced fluid, known as Archimedes' principle. The treatise provides equilibrium conditions for bodies with different shapes in a fluid. Archimedean mechanics, whether dealing with the balance or with bodies in a fluid, was based on equilibrium rather than motion.

Hero wrote *Spiritalia* (Pneumatics)[10] and a treatise on mechanics that was known only in part through references in Pappus's *Collectiones mathematicae* (Mathematical Collections), whose eighth book is devoted to mechanics. Following Hero, Pappus argued that all machines could be reduced to the five simple machines (balance or lever, pulley, wheel and axle, wedge, and screw), and the last four in turn could be reduced to the balance. In some instances, Pappus tried to show by means of a geometrical construction how a machine could be reduced to the balance. He sought to determine in this way the equilibrium conditions for a weight on an inclined plane, for example. Although his solution was problematic, it was an attempt to establish mechanics on a solid basis and clear first principles. In 1588, dal Monte supervised a printing of a Latin translation by Commandino from an imperfect manuscript of Pappus's text.[11]

Lastly, during the thirteenth century, several Latin authors contributed treatises in the tradition known as *Scientia de ponderibus* (Science of Weights) dealing with the equilibrium of bodies. The names of many of those authors have not survived, the main exception being Jordanus of Nemore (fl. early thirteenth century), who wrote a *Liber de ponderibus* (Book on Weights). After the Nuremberg *editio princeps* by the German cosmographer Petrus Apianus (Apian, 1495–1552), a new edition was published in Venice in 1565 from the papers of Tartaglia, who had already included some results in his own previous publications. Although works of this tradition generally lack the rigor and elegance found in Archimedean treatises, they contain original notions and valuable results. For example, the treatise *De ratione ponderis* (On the Theory of Weight) did not attempt to rely systematically on the lever

---

[8] Frederico Commandino, *Archimedis de iis quae vehuntur in aqua libri duo* (Bologna: Ex officina A. Benacii, 1565); and Guidobaldo dal Monte, *In duos Archimedis aequiponderantium libros paraphrasis* (Pesaro: Apud Hieronymum Concordiam, 1588).

[9] Archimedes did not define the expression "center of gravity." Later scholars defined it as that point such that a body suspended from it remains in equilibrium.

[10] Marie Boas, "Hero's *Pneumatica*: A Study of Its Transmission and Influence," *Isis*, 40 (1949), 38–48.

[11] Pappus, *Collectiones mathematicae* (Pesaro: Concordia, 1588); and L. Passalacqua, "Le Collezioni di Pappo," *Bollettino di Storia delle Scienze Matematiche*, 14 (1994), 91–156.

but provided a more satisfactory solution to the problem of equilibrium of a weight on an inclined plane than that of Pappus.[12]

STUDIES ON MOTION

Texts dealing with the problem of motion go back to Aristotle and the host of commentators from antiquity to the Middle Ages. My chief concern here is not with all the topics pertinent to the study of motion in Aristotle and his commentators but only with those aspects relevant to the sixteenth-century development of the science of motion. Even within this limitation, my account remains highly selective.

The analysis of motion occupies a central position in Aristotle's study of nature, especially in *Physica* and *De caelo* (On the Heavens). By "motion" Aristotle understood virtually all change occurring in nature, whereas by "local motion" he meant something closer to our understanding of the term. Here and throughout I shall simply refer to "motion," meaning "local motion." Aristotelians drew a basic distinction between natural and violent motions. The former is the downward motion of heavy bodies, endowed with gravity, or the upward motion of light bodies, endowed with levity.[13] The latter is exemplified by the motion of projectiles. With regard to natural motion, Aristotle argued that the speed of a falling body is proportional to its weight and is the inverse of the resistance of the medium.[14] As a consequence, when the resistance tends to zero, as in a void, the velocity becomes infinite, a paradoxical result used by Aristotle to refute the existence of a void. With regard to violent motion, Aristotle argued that after the moving body has left the projector, the body is moved by the surrounding medium. This view derived from his principles that everything that moves is moved by something else and that the mover must be in contact with the moved body. These principles imply that a body set in motion requires an external cause to continue its motion, an opinion that was debated until the seventeenth century.

Starting in late antiquity, several commentators examined Aristotle's views on motion with a critical eye. Themistius (317–387), for example, argued that all bodies would fall in a void with the same speed and that this speed would be finite, not infinite, as Aristotle had claimed. Simplicius (d. after 533) often defended Aristotle's conclusions, such as the denial of the existence of a void,

---

[12] Ernest A. Moody and Marshall Clagett, eds., *The Medieval Science of Weights* (Madison: University of Wisconsin Press, 1960); J. E. Brown, "The Science of Weights," in Lindberg, ed., *Science in the Middle Ages*, pp. 179–205; Jordanus Nemorarius, *Liber de ponderibus*, ed. Petrus Apianus (Nuremberg: Iohannes Petreius, 1533); and Nemorarius, *Opusculum de ponderositate* (Venice: Curtius Troianus, 1565). On Tartaglia's publications, see the section on "Motion and Mechanics" in this chapter.

[13] Levity was understood by Aristotle as an independent quality, not as being caused by extrusion of a specifically lighter body in a specifically heavier medium, on an Archimedean model.

[14] Murdoch and Sylla, "The Science of Motion," p. 224.

but he was critical of Aristotle's proofs. Philoponus (d. ca. 570) was the most thorough ancient critic of Aristotle's physics and especially of his views on motion. He admitted that motion in a void could occur and argued that a medium would increase the time of fall of a body over the time of fall in a void. Philoponus also argued against Aristotle that according to experience the time of fall of two bodies differing greatly in weight is very small. Greek and Latin editions of the works of Themistius, Simplicius, and Philoponus appeared in print in the first half of the sixteenth century and contributed to an erosion of confidence in Aristotle's doctrines coming from a learned and ancient tradition of scholarship. Galileo was familiar with all three of them.[15]

In the Islamic world, one author in particular needs to be singled out here, the Spanish philosopher Avempace, or Ibn Bājja (d. 1138), who was sympathetic to Philoponus. His views became known in the West through the citations and criticisms of Averroes (Ibn Rushd, 1126–1198), whose works were well known in the Latin West.[16]

In medieval Europe, Aristotle became the cornerstone of university education, and the number of his commentators grew considerably. Here I wish to mention the Parisian John Buridan (ca. 1295–ca. 1358) and a group of authors known as *calculatores*. Buridan addressed the problem of projectile motion in a fashion different from Aristotle's. Instead of arguing that the medium had a role in propelling the projectile, Buridan claimed that the projectile moved because of a quality called *impetus* transmitted by the projector. His notion remained in vogue until the seventeenth century. The fourteenth-century study of motion witnessed a remarkable increase in the use of logic and mathematics, especially the theory of proportions. The protagonists of this tradition included Thomas Bradwardine (d. 1349) and Richard Swineshead (fl. 1340–1355) at Oxford and Nicole Oresme (ca. 1325–1382) at Paris. In addition to showing that motion could be treated mathematically, they developed sophisticated geometrical treatments, coined a refined terminology, and attained important results, such as the mean-speed theorem.[17]

However, there are important qualifications. The methods of inquiry developed by the *calculatores* were applied not only to local motion but also to a wide range of topics spanning medicine, theology, and natural philosophy. Secondly, with only one known exception, those methods of inquiry were applied to ideal imaginary entities, not to nature, in the style of a logical exercise.[18] Despite the heavy use of mathematics, the arrangement

---

[15] On the three Greek commentators, see Paolo Galluzzi, *Momento: Studi galileiani* (Rome: Edizioni dell'Ateneo, 1979), pp. 98–106.

[16] The classic work here is Ernest Moody, "Galileo and Avempace: The Dynamics of the Leaning Tower Experiment," *Journal of the History of Ideas*, 12 (1951), 163–93, 375–422.

[17] The theorem states that the space traversed with a uniformly accelerated or decelerated motion is the same as that traversed with a uniform speed equal to the mean degree of speed, namely the degree of speed in the mean instant of time.

[18] The exception was the Spanish Dominican Domingo de Soto (1495–1560).

of the surviving codices suggests that these works were considered part of natural philosophy rather than the mathematical disciplines. Texts from this tradition were published in about 1500.[19]

MOTION AND MECHANICS IN THE SIXTEENTH CENTURY

Despite their brevity, the previous sections show that in the sixteenth century a number of key works on mechanics and motion became available in print. Never before had scholars been able to access such a wealth of intellectual resources on these topics with such ease. This section presents the main works by some of the leading figures in the sixteenth century. Whereas some, such as Tartaglia and Benedetti, sought to build bridges between mechanics and motion, others, such as dal Monte and Stevin, conceived the two domains as separate and saw little hope for a rapprochement.[20]

Tartaglia was a major figure in more ways than one. Besides his editorial work mentioned earlier, he published *La nova scientia* (The New Science, 1537) and *Quesiti et inventioni diverse* (Various Questions and Inventions, 1546). These are composite works dealing largely with the mathematical disciplines, such as gunnery, the science of weights, and the pseudo-Aristotelian *Quaestiones*, but they also include issues dealing with gunpowder and other military matters. Tartaglia sought to determine the trajectory of a projectile shot at different angles and claimed from dubious assumptions that the longest range was shot at 45° above the horizon. Several treatises and manuals on ballistics followed in the sixteenth and seventeenth centuries.[21]

In yet another work, *La travagliata inventione* (The Troubled Invention, 1551), Tartaglia hinted at a proportion applicable to bodies falling in water, suggesting that bodies sink faster in proportion to how much they are specifically heavier than water.[22] The same idea of extending Archimedean hydrostatics to account for motion was developed by Benedetti in a number of

---

[19] Christopher Lewis, *The Merton Tradition and Kinematics in Late Sixteenth and Early Seventeenth Century Italy* (Padua: Editrice Antenore, 1980).

[20] Other relevant figures include the Padua professor of mathematics Giuseppe Moletti (1531–1588), on whom see Walter R. Laird, *The Unfinished Mechanics of Giuseppe Moletti* (Toronto: University of Toronto Press, 2000), and the physician Girolamo Cardano (1551–1576), whose work in mechanics still awaits systematic investigation. Important works on sixteenth-century mechanics include Walter R. Laird, "The Scope of Renaissance Mechanics," *Osiris*, 2nd ser., 2 (1986), 43–68; and Laird, "Patronage of Mechanics and Theories of Impact in Sixteenth-Century Italy," in *Patronage and Institutions: Science, Technology, and Medicine at the European Court, 1500–1750*, ed. Bruce Moran (Rochester, N.Y.: Boydell Press, 1991), pp. 51–66.

[21] A. Rupert Hall, *Ballistics in the Seventeenth Century* (New York: Harper, 1969); and Serafina Cuomo, "Shooting by the Book: Notes on Niccolò Tartaglia's *Nova Scientia*," *History of Science*, 35 (1997), 155–88. Extensive translations from Tartaglia's works can be found in Drake and Drabkin, *Mechanics in Sixteenth-Century Italy*, pp. 63–143.

[22] Clagett, *Archimedes in the Middle Ages*, 3: 3, 574.

works published in the 1550s dealing with falling bodies. One of them bears the none too subtle title *Demonstratio proportionum motuum localium contra Aristotilem et omnes philosophos* (Demonstration of the Proportions of Local Motions, against Aristotle and All Philosophers, 1553).[23] Benedetti expanded his reflection in his magnum opus, *Diversarum speculationum mathematicarum et physicarum liber* (Book of Different Speculations on Mathematical and Physical Matters, 1585), a composite work dealing with mechanics and the pseudo-Aristotelian *Quaestiones*, criticisms of Aristotle's views about motion, and including excerpts from his (Benedetti's) correspondence.[24]

Archimedes was a major source for sixteenth-century scholars of mechanics, but dal Monte had in addition a predilection for Pappus, whose work he knew before 1588 through Commandino's manuscripts. The marquis published the main work of its time on mechanics, *Mechanicorum liber* (Book of Mechanics, 1577), where he examined all the simple machines and, following Pappus, tried to show that they work in accordance with the principle of the balance, as if there were balances in disguise that had to be unmasked. He was concerned not just with results but also with foundations and proofs; being able to go back to the balance meant that he could rely on Archimedes' work on the equilibrium of planes and therefore solve the problem of foundations. The marquis loathed the medieval science of weights and those who worked in that tradition, including Tartaglia, because their proofs lacked the rigor found in the texts from antiquity. In some respects, dal Monte saw an unbridgeable gap between equilibrium and motion. The science of equilibrium could be formulated in mathematical fashion because of its regularity, whereas motion was subject to so many vagaries that mathematics generally remained out of the picture. In dealing with the wedge, however, dal Monte hinted at a proportion that involved motion: He argued that a body hitting a wedge produces a greater effect the greater the height from which it falls. In this way, height and speed of fall were linked to the effect they produce. This problem of determining the force of percussion was part of the classical repertoire from the time of the pseudo-Aristotelian *Quaestiones* and was later discussed by Galileo and other seventeenth-century mathematicians.[25]

Despite his geographical distance from Italian mechanicians, Stevin, too, relied on the editions of Commandino, much as did dal Monte and a host of other contemporary mathematicians. Stevin's main work in mechanics was a collection of treatises with separate title pages that was published in Leiden by Christoffel Plantijn in 1586. They include *De Beghinseln der Weeghconst* (The

---

[23] Carlo Maccagni, *Le speculazioni giovanili "de motu" di Giovanni Battista Benedetti* (Pisa: Domus Galilaeana, 1967).
[24] Extensive translations can be found in Drake and Drabkin, *Mechanics in Sixteenth-Century Italy*, pp. 166–237.
[25] Also in this case extended translations can be found in Drake and Drabkin, *Mechanics in Sixteenth-Century Italy*, pp. 241–328; and Domenico Bertoloni Meli, "Guidobaldo dal Monte and the Archimedean Revival," *Nuncius*, 7, no. 1 (1992), 3–34.

Elements of the Art of Weighing), *De Weeghdaet* (The Practice of Weighing), and *De Behinselen de Waterwichts* (The Elements of Hydrostatics).[26] At the end of the last work, there are two short additions, *Preamble of the Practice of Hydrostatics* and *Appendix to the Art of Weighing*.[27] Stevin also performed experiments by dropping heavy bodies from high places, but his work focused primarily on equilibrium, and his fame rests on his extension of Archimedean hydrostatics and his brilliant solution to the problem of the inclined plane.

GALILEO

Galileo's contributions to the mathematical disciplines and philosophy range from his telescopic findings to his onslaught on the Peripatetic school. Yet, it was the science of motion that he considered to be his most treasured investigation and that represents his most sustained and remarkable intellectual effort. Galileo's main work on motion and mechanics falls into three periods: at Pisa, Padua, and Florence. During his three years as professor of mathematics at the University of Pisa (1589–92), he probably drafted a dialogue, an essay, and a few fragments on motion, all collectively known as *De motu antiquiora* (The Older [Manuscripts] on Motion). All of this material remained unpublished at the time. As professor of mathematics at the University of Padua (1592–1610), Galileo worked intensively on the science of motion and the science of the resistance of materials, and he composed a short tract for his university lectures called *Le mecaniche* (On Mechanics). In this second period also, his works remained unpublished at the time. Lastly, after his return to Florence in 1610, Galileo started publishing on mechanical subjects, first with a treatise on hydrostatics, *Discorso intorno alle cose, che stanno in sù l'acqua, ò che in quella si muouono* (Discourse on Bodies on Water, or that Move in It, 1612), then with his masterpieces, the *Dialogo sopra i due massimi sistemi del mondo, Tolemaico e Copernicano* (Dialogue Concerning the Two Chief World Systems, Ptolemaic and Copernican, 1632) and the *Discorsi e dimostrazioni matematiche intorno a due nuove scienze* (Discourses and Mathematical Demonstrations about Two New Sciences, 1638).[28]

In *De motu antiquiora*, Galileo sought to formulate a science of motion by extending Archimedean hydrostatics, arguing that the speed of a body falling in a medium is proportional to the difference in specific density between the body and the medium. This means that, apart from an initial

---

[26] Stevin, *Works*, vol. 1, *De Beghinselen der Weeghconst*, pp. 35–285; *De Weeghdaet*, pp. 287–373; *De Behinselen des Waterwichts*, pp. 375–483.
[27] Stevin, *Works*, vol. 1, *Anvang der Waterwichtdaet*, pp. 484–501; *Anhang van de Weeghconst*, pp. 503–521.
[28] Several scholars have investigated Galileo's reflections on motion, from Alexandre Koyré and Winifred Wisan to Paolo Galluzzi and Enrico Giusti. For an excellent bibliography, see Damerow et al., *Exploring the Limits*.

period, the speed of falling bodies is constant. Although there are some similarities to Benedetti's work, it is unclear whether Galileo knew it at that time. In addition, following Pappus's and dal Monte's approach, Galileo attempted to use the balance and, more successfully, the inclined plane to account for hydrostatics. Even in this early work, Galileo showed familiarity with a number of authors we have already encountered, from Themistius and Philoponus to Avempace and Avicenna (Ibn Sina, 980–1037). But other more recent and geographically closer sources require attention, too. The University of Pisa professors of philosophy Girolamo Borro (1512–1592) and Francesco Buonamici (1533–1603) engaged in a dispute about motion that lasted for several years and covered the entire period beginning in 1580 from Galileo's education at Pisa to his teaching at the university. In *De motu gravium et levium* (On the Motion of Heavy and Light Bodies, 1576), Borro sided with Averroes and referred to experiments where heavy bodies were dropped from a high window. In his huge *De motu* (On Motion, completed in 1587 but first published in 1591), Buonamici defended Simplicius. Moreover, Buonamici attacked the views of more recent mathematicians, who defended an Archimedean approach. Their dispute was probably not a private affair but rather spilled over into university life, including lectures and the annual series of public disputations known as the *circuli*. A few references point to that dispute as the immediate context for *De motu antiquiora*.[29] At about this time, Galileo embarked on an extended study of philosophy on the basis of lecture notes originating from the chief Jesuit school, the Collegio Romano, probably to strengthen his knowledge of philosophy as well as his dialectical skills.[30]

In *De motu antiquiora* there are several references to experiments, including one of dropping weights from a high tower, probably the leaning tower at Pisa. Overall, however, the experiments performed by Galileo did not conform to his expectations. For example, Galileo investigated the inclined plane and believed he had determined the relationship between its inclination and the (constant!) speed of a body falling along it. By combining this result with his buoyancy theory of fall, Galileo sought to find a plane with the

---

[29] Michele Camerota and Mario Helbing, "Galileo and Pisan Aristotelianism: Galileo's *De motu antiquiora* and the *Quaestiones de motu elementorum* of the Pisan Professors," *Early Science and Medicine*, 5 (2000), 319–65; Mario O. Helbing, *La filosofia de Francesco Buonamici* (Pisa: Nistri-Lischi, 1989), chap. 6; and Charles B. Schmitt, "The Faculty of Arts at Pisa at the Time of Galileo," *Physis*, 14 (1972), 243–72.

[30] William Wallace, *Galileo and His Sources* (Princeton, N.J.: Princeton University Press, 1984). At pp. 91–2, Wallace suggests that the lecture notes were sent by Christophorus Clavius (1537–1612), professor of mathematics at the Collegio Romano, to Galileo in connection with a dispute over the center of gravity, but evidence for his claim is lacking. Lecture notes often circulated at the time, and Galileo may have obtained them from a student at the Collegio. See Corrado Dollo, "Galilei e la fisica del Collegio Romano," *Giornale Critico della Filosofia Italiana*, 71 (1992), 161–201. For a comparison between the Padua philosopher Jacopo Zabarella (1533–1589) and Galileo, see the classic by Charles B. Schmitt, "Experience and Experiment: A comparison of Zabarella's views with Galileo's in *De motu*," *Studies in the Renaissance*, 16 (1969), 80–138.

appropriate inclination whereby a body would fall along it in the same time as another body of different material falls along the vertical. His attempt failed but may have been at the root of his later findings both that falling bodies accelerate and that the acceleration is the same for all bodies.[31] By exploring motion along inclined planes, Galileo came to appreciate that a body needs no force to be set in motion along a plane with zero inclination, by which he meant the horizon. This thought remained one of the characteristic features of his reflections on motion.[32] About 1592 or slightly earlier, Galileo performed some important experiments with his mentor dal Monte. They threw inked balls across an inclined plane and found that their trajectories were symmetrical, resembling a hyperbola or parabola. A similar curve was described by a chain hanging from two nails fixed in a wall. A reference to that experiment can be found in the *Discorsi*, but it is unclear what the two mathematicians would have made of the result in 1592.[33]

Two differences spring to mind when comparing Galileo's early speculations with the works discussed in the previous section: the deep interplay with philosophy and the role of experiments. His experiment of dropping heavy bodies from high places was associated with the philosophical dispute at Pisa, but it also went hand-in-hand with the practice of performing trials that was typical of the mathematical disciplines, from weighing and surveying to music. Those trials would have been familiar to Galileo, who had started his career in mechanics with a short essay on accuracy in weighing and whose father was a musician (see Mancosu, Chapter 25, this volume).[34] Other experiments, however, even if they were problematic and unsuccessful, such as the one with spheres of different materials going down inclined planes with different inclinations, show Galileo seeking regularities in nature by means of contrived experiments, which revealed a sophistication that went beyond contemporary standards.

Between 1592 and 1620, Galileo taught mathematics at the University of Padua, a position, like the previous one at Pisa, he owed to dal Monte's support. At the university, he taught a number of subjects, including fortification and mechanics: His lecture notes show that his course was modeled on dal Monte's *Mechanicorum liber*. Galileo worked on mechanical issues having to do with oars and the size of galleys in collaboration with scholars

---

[31] Galileo Galilei, *On Motion and on Mechanics*, translated with introductions by Israel E. Drabkin and Stillman Drake (Madison: The University of Wisconsin Press, 1969). At p. 69, Galileo states that "the ratios that we have set down are not observed."

[32] At this stage, in all likelihood Galileo believed that a body set in motion would come naturally to rest. See Drake and Drabkin, *Mechanics in Sixteenth-Century Italy*, p. 379; and Galileo, *On Motion and Mechanics*, pp. 66–7.

[33] Damerow et al., *Exploring the Limits*, pp. 158–64.

[34] Galileo's father was a musician. See Stillman Drake, *Galileo at Work* (Chicago: University of Chicago Press, 1978), pp. 15–17; and Claude V. Palisca, *Humanism in Italian Renaissance Musical Theory* (New Haven, Conn.: Yale University Press, 1985), pp. 265–79. For *La Bilancetta*, see Galileo Galilei, *Opere*, ed. A. Favaro, 20 vols. (Florence: Giunti Barbera, 1890–1909), 1: 215–20; Drake, *Galileo at Work*, pp. 6–7; and Jim A. Bennett, "Practical Geometry and Operative Knowledge," *Configurations*, 6 (1998), 195–222.

and technicians at the Venice Arsenal, the city's chief military and industrial facility. It is partly from his work at the Arsenal that his science of the resistance of materials and of scaling originates.[35] At Padua, Galileo resumed his experimental and mathematical investigations on motion and realized a number of important features about falling, oscillating, and projected bodies. He seized on objects such as the inclined plane and the pendulum to investigate motion and realized that constant speeds are unsuitable for describing free fall because bodies accelerate. This finding led Galileo to some results from the *calculatores* tradition, which enabled him to treat acceleration in an elementary fashion. Both the terminology and the visual tools of representations used by Galileo testify to his reliance on this tradition, despite the fact that his itinerary started elsewhere and included a variety of sources.[36]

Galileo argued that a falling body goes through all the infinitely many degrees of speed, a belief that put considerable strain on the limited mathematical resources of his time. He further believed that, apart from small perturbations caused by air resistance, all bodies accelerate in the same way regardless of their weight or specific gravity. Moreover, Galileo realized that the acceleration is uniform and the spaces traversed are proportional to the squares of the times. He believed further that the oscillations of a pendulum are very nearly isochronous, a claim that is quite accurate for small oscillations but whose inaccuracy increases with the amplitude of the oscillations. Galileo also believed, quite erroneously, that the circle arc described by the bob was the curve of fastest descent. Several decades later, he still thought he could produce a proof of this.[37] Galileo further came to appreciate that a body set in motion on a horizontal plane does not stop as long as all accidental perturbations are removed. He finally realized that horizontal projection and vertical fall are independent and that each can be composed as if the other did not exist. This composition gives rise to parabolic trajectories, as Galileo's experiment with dal Monte (ca. 1592) had suggested. Although the fragmentary manuscript record from this period does not allow a detailed reconstruction of Galileo's intellectual itinerary in all circumstances, in some cases specific results have been achieved.[38]

Galileo's findings, remarkable as they were, did not constitute a science based on evident principles and rigorous proofs. In other words, Galileo had found a series of propositions and relations that needed to be given order and structure. When he realized that the balance could not be used

---

[35] Jürgen Renn and Matteo Valleriani, "Galileo and the Challenge of the Arsenal," *Nuncius*, 16, no.2 (2001), 481–504.
[36] Edith D. Sylla, "Galileo and the Oxford *Calculatores*: Analytical Languages and the Mean-Speed Theorem for Accelerated Motion," in *Reinterpreting Galileo*, ed. William A. Wallace (Washington, D.C.: The Catholic University of America Press, 1986), pp. 53–108.
[37] Galileo Galilei, *Two New Sciences*, trans. Stillman Drake (Madison: University of Wisconsin Press, 1974), pp. 212–3.
[38] A detailed analysis of this period can be found in Damerow et al., *Exploring the Limits*, chap. 3.

to found a science of motion, he started to look for a suitable axiom or principle to replace, in an Archimedean fashion, Archimedes' axioms in *On the Equilibrium of Planes*. Galileo was not seeking to establish his science on contrived and elaborate experiments but rather on axioms of the same nature as those of Archimedes, who had postulated at the outset of *On the Equilibrium of Planes* that in a balance equal weights at equal distances are in equilibrium, whereas equal weights at unequal distances are not in equilibrium and incline toward the weight at the greater distance. Letters from this period testify to Galileo's long search for new self-evident principles.[39]

Following his spectacular astronomical discoveries concerning Jupiter's moons and other celestial objects Galileo was called to Florence in 1610 as philosopher and mathematician to the Grand Duke of Tuscany, a highly paid position created especially for him.[40] The first area of mechanics on which Galileo worked after his return to Florence was hydrostatics. As part of a dispute with Aristotelian philosophers, Galileo published two editions in 1612 of the *Discorso intorno alle cose, che stanno in sù l'acqua, ò che in quella si muouono*. The philosophers argued that shape was a decisive factor in buoyancy, whereas Galileo followed Archimedes in identifying specific gravity as the decisive factor. In the controversy, Galileo was aided by his former student Benedetto Castelli (ca. 1577–1643), a Benedictine monk who was then professor of mathematics at the University of Pisa. It was Castelli who went to work on water flow and water management with a pioneering mathematical treatise, *Della misura dell'acque correnti* (On the Measurement of Running Waters, 1628). The problems associated with the motion of waters were Castelli's domain, but Galileo also worked on them. He had already expressed views on river flow while at Padua and continued to do so until the 1630s. There are obvious connections between the science of motion and the cluster of issues linked to river flow and water management, commonly referred to as the science of waters. Galileo considered water in a river to be like a body moving down an inclined plane and tried to apply the corresponding rules, with little success.[41]

After a long period of gestation caused by intellectual as well as political and religious matters, in 1632 Galileo published his scientific and literary masterpiece, the *Dialogo*, wherein three interlocutors discuss over four days the merits of the two chief world systems. Galileo's clumsy attempts at covering his Copernican views allowed the book to make it past the censors but

---

[39] The main letters are to Paolo Sarpi (1552–1623) in 1604 and Luca Valerio (1552–1618) in 1609. They are discussed in Galluzzi, *Momento*, pp. 269–76, 303–7.

[40] The social and intellectual implications of this move are discussed in Mario Biagioli, *Galileo Courtier* (Chicago: University of Chicago Press, 1993).

[41] Richard S. Westfall, "Floods along the Bisenzio: Science and Technology in the Age of Galileo," *Technology and Culture*, 30 (1989), 879–907; and Cesare S. Maffioli, *Out of Galileo: The Science of Waters, 1628–1718* (Rotterdam: Erasmus, 1994).

did not prevent it from being banned by the Inquisition in 1633. Galileo was put under arrest, which was later commuted to house arrest, for vehement suspicion of heresy until the end of his days.[42] The *Dialogo* deals with cosmological matters, but Galileo's defense of Copernicanism relied on the science of motion for the study of the behavior of objects on a moving earth. An important portion of Galileo's strategy was to argue that the motion of the earth would not produce any visible effects on falling or projected bodies on the earth; therefore it would be impossible to determine whether the earth moves in that fashion. His chief argument in favor of Copernicanism was that of tides, which he thought were an effect of the double rotation of the earth on its axis and around the sun. The *Dialogo* contains several passages relevant to the science of motion, but it is primarily in the second day where Galileo discusses relative motion and the effects of the earth's motion. It is in the *Dialogo* that Galileo stated for the first time a number of propositions about falling bodies, such as the odd-number rule or the proportionality between speed and time. Galileo discussed many problems related to the earth's rotation, such as why projectiles, birds, and clouds are not left behind by the earth's rotation, or why bodies on the earth's surface are not projected into the air by its rotation. Moreover, the *Dialogo* contains the first references to the isochronism of the pendulum's oscillations, though not the relation between period and length.[43]

Besides putting Galileo under arrest, the Inquisition prevented him from publishing on any subject. Therefore, the manuscript of his next and final work had to be smuggled out of Italy and published in the Protestant Low Countries by the Elzeviers. When the *Discorsi* appeared in 1638, Galileo was seventy-four years of age and had been working on the problem of motion for about half a century. The book is in the form of a dialogue among the same three personages as the *Dialogo*, but whereas the *Dialogo* was written in a style that imitated the open-ended discussions among the protagonists, the *Discorsi* contained portions structured in a more formal way. The first two days contain many digressions, notably on the nature of the continuum and the physical cause of cohesion, which Galileo believed was caused by infinitely many interstitial vacua. The first new mathematical science is the resistance of materials, whose principles are discussed mainly in day two. Distant roots of these problems can be found in some of the pseudo-Aristotelian *Quaestiones*,[44] but the more immediate context was Galileo's work at the Venice Arsenal. The problem consists in determining the resistance to rupture of a loaded beam of certain dimensions, knowing the resistance to rupture of a similar beam

---

[42] On the relationships between Galileo and the Church, including the 1616 ban of Copernicanism, see Annibale Fantoli, *Galileo: For Copernicanism and for the Church*, trans. George V. Coyne, S.J. (Vatican City: Vatican Observatory Publications, 1994).
[43] Peter Dear, *Mersenne and the Learning of the Schools* (Ithaca, N.Y.: Cornell University Press, 1988), p. 165.
[44] *Quaestiones*, numbers 14 and 16. I wish to thank Antonio Becchi for having pointed this out to me.

with different dimensions. Obviously, if length is unchanged, a thicker beam resists more than a thinner one, whereas if thickness is unchanged, a longer beam resists less than a shorter one. The exact proportions depend on the width, height, and length of the beam and on whether one considers it to be heavy or whether its weight is so much smaller than the loads that it can be neglected. Galileo believed that the foundations of this science lie in the doctrine of the lever; therefore the resistance of materials was seen as part of mechanics.

Days three and four of the *Discorsi* are devoted to the science of motion and are arranged in an unusual fashion: A formal treatise in Latin is interspersed with elucidations and comments in Italian by the three interlocutors. The Latin treatise consists of three parts: on uniform motion, uniformly accelerated motion, and projectile motion. Uniform motion was well understood, but Galileo needed to establish some basic proportions about it as a basis for the following parts. This example highlights a major problem in Galileo's treatment of accelerated motion, namely his lack of suitable mathematical tools, apart from some propositions derived by the *calculatores* tradition. Galileo would often use the mean-speed theorem to move from uniformly accelerated motion to an equivalent uniform one, and then apply the theorems about uniform motion from part one. He remained always doubtful of the theory of indivisibles, a remarkable mathematical achievement by his follower Bonaventura Cavalieri (1598–1647) that would have helped him in some respects.[45] It is in days two and three that Galileo presented some of the results attained at Padua and expanded on them. Although Galileo struggled to find out that in free fall the speed is proportional to time, in the published form he put this statement as a definition, seeking to present it as natural and simple. Galileo thought that he could found his new science on only one postulate, namely that the degrees of speed acquired by a body in falling along planes with different inclinations are equal whenever the heights of those planes are equal.[46] He sought to offer additional underpinning for this statement by claiming that a body falling either along an inclined plane or attached to the string of a pendulum acquires enough impetus to rise back to its original height. Galileo discussed an experiment where a sphere rolls down an inclined plane about twelve *braccia* (1 braccio = 550–655 millimeters) long and raised at one end by one or two *braccia*. Time was measured with a water clock, letting water out of a large container through a tap and then weighing it. The experiment showed that the distance traveled is as the square of the time. Galileo attributed no role to this experiment in the formal establishment of his science. Rather, he formulated the science of motion as a mathematical

---

[45] A useful introduction is Kirsti Andersen, "The Method of Indivisibles: Changing Understanding," *Studia Leibnitiana*, Sonderheft 14 (1986), 14–25.
[46] Galileo, *Two New Sciences*, p. 162.

construction and then used the experiment only at a later stage to show that the science he had formulated corresponded to nature's behavior. In this rather contrived construction, his science would retain a role as a purely mathematical exercise even if bodies fell according to a different rule. In the fourth day, Galileo presented his theory of projectile motion, arguing that trajectories are parabolic. In this case, too, Galileo claimed that his science would remain valid as a purely mathematical exercise even if nature behaved differently.

With regard to the role of experiment in his career, it seems helpful to draw a distinction between private research and public presentation. Private experiments, such as some of those found in *De motu antiquiora* and especially the Padua manuscripts, appear to have had a major heuristic role for Galileo. He seems to have performed them in order to gather quantitative information, especially in the form of proportions among variables, such as time and distance. In some cases, Galileo probably had some ideas as to the outcome and saw them confirmed, but in others the result probably came as a surprise to him, such as the experiment with dal Monte on projectile trajectories on an inclined plane. At times the manuscripts show Galileo calculating and comparing data from experiments to predicted values, his main aim being to determine proportions between variables rather than numerical values for their own sake. Galileo often sought to separate the fundamental features of a phenomenon from what he called accidental perturbations. It was this strategy that often enabled him to provide mathematical formulations of complex phenomena.[47] In print, Galileo's experimental reports vary enormously in style and scope. We have already seen how contrived his report of the inclined-plane experiment in the *Discorsi* was. In the *Discorso* on bodies in water, his reports sometimes have a legalistic tone associated with the nature of the dispute. In that case, Galileo knew from the start the principles of hydrostatics and was seeking a powerful rhetorical presentation. The *Dialogo* includes a large number of informal presentations of experiments, with a dazzling range of rhetorical styles. It is extremely difficult to pinpoint a general pattern, except to remark that Galileo was writing a masterpiece in scientific rhetoric wherein he was referring to experimental trials as if in informal conversation. At times he would claim great accuracy for an experiment, and in other instances he would say that the outcome was so certain that there was no need to perform it, even when we know that Galileo had in fact performed it. For example, Galileo discussed the experiment of dropping a weight from the mast of a moving ship as part of his discussions of motion on a moving earth, arguing that the weight falls at the foot of the mast regardless of whether the ship is in motion. Although in the *Dialogo* he claimed that the outcome could be determined even without an experiment,

---

[47] Damerow et al., *Exploring the Limits*, pp. 208–36; and Noretta Koertge, "Galileo and the Problem of Accidents," *Journal of the History of Ideas*, 38 (1977), 389–408.

we know from a previous letter that in fact he had performed this experiment a few years before.[48] This case highlights some of the problems in reading and interpreting Galileo.

Although Galileo proudly proclaimed at the outset of the third day of the *Discorsi* that he was putting forward a wholly new science about a most ancient topic, scholarly opinions about his actual achievements differ. Some see him as a key figure in careful experimentation, others in the application of mathematics to the study of nature. For some he resolutely broke with the past, whereas others detect long threads from classical and medieval times still entangling his thought and preventing him from offering a full formulation of a new science.[49] Either way, Galileo represents a nodal point in the history of mechanics and science. He had a major role in redefining questions and research topics and in setting the agenda for the following decades.

Several related processes took place in the science of mechanics during the period between the publication of the *Discorsi* in 1638 and the early eighteenth century. Initially, mechanics was situated among the mixed mathematical disciplines and had close ties with engineering, but from the middle of the century, it became progressively more integrated with natural philosophy. During the first part of the century, its practitioners interacted through correspondence networks, such as that centered on the French Minim Marin Mersenne (1588–1648). The second half of the century brought major changes to this landscape. Galileo died in 1642, followed in 1643 by his former student and professor of mathematics at Rome, Benedetto Castelli, and soon after by Galileo's successor at the Tuscan court, Evangelista Torricelli (1607–1647). Mersenne and René Descartes (1596–1650) died within a couple of years of each other. In addition to individual scholars, communication networks disappeared and had to be rebuilt by the new generation. Informal correspondence networks were replaced in the second half of the century by more formal scientific academies, such as the Royal Society in London and the Académie Royale des Sciences in Paris, which became major venues of research and debate on mechanics.[50] Lastly, the audience for works on mechanics changed from mathematicians and engineers to a broader public with philosophical and cosmological interests. The remainder of this chapter addresses in particular how the key works in the discipline were read by other practitioners and the broader intellectual public.

---

[48] Drake, *Galileo at Work*, pp. 84, 294.
[49] Classic interpretations include Drake, *Galileo at Work*; Galluzzi, *Momento*; and Damerow et al., *Exploring the Limits*, chap. 3.
[50] In the large literature on this theme, see Lorraine Daston, "Baconian Facts, Academic Civility, and the Prehistory of Objectivity," *Annals of Scholarship*, 8 (1991), 337–63; and Mario Biagioli, "Etiquette, Interdependence, and Sociability in Seventeenth-Century Science," *Critical Inquiry*, 22 (1996), 193–238.

READING GALILEO: FROM TORRICELLI TO MERSENNE

A scholar of mechanics and the science of motion in about 1640 would have found the field to be in one of its most creative and exciting periods. After having discussed aspects of the science of motion in the *Dialogo*, in 1638 at age seventy-four, Galileo finally produced his masterpiece, the *Discorsi*. The first two days contained, among many digressions, the principles of the new science of the resistance of materials, which dealt especially with the problem of scaling applied to the transition from models of machines to real ones. Days three and four dealt with the science of motion, including falling and projected bodies. The *Dialogo* had received a first Latin translation in 1635 that was often reprinted, and in 1639, Mersenne put forward a free French translation of Galileo's *Discorsi*. In 1639, Castelli had published the second edition of his own work on water flow, *Della misura dell'acque correnti*. Outside Italy, Mersenne had published in 1636 the gigantic and labyrinthine *Harmonie universelle* (Universal Harmony), a work devoted to musical matters. Because sound is produced by motion, he included an extensive discussion of motion that was largely based on Galileo's *Dialogo*. All these works announced the emergence of new relationships between mathematics and the physical world: The science of motion was becoming an integral part of the mathematical disciplines and was tied to mechanics in multiple ways.[51]

The three new mathematical disciplines had technical and engineering roots: The origins of the science of waters were tightly bound to the problem of river flow in central Italy, especially the areas between Bologna and Ferrara, and of the Venetian lagoon; the science of resistance of materials and the problem of scaling were common concerns among engineers, so much so that Galileo introduced them in the *Discorsi* with a discussion inspired by a visit to the Venice Arsenal; and the science of motion had roots in gunnery. Despite such links, the university mathematicians and the philosophers promoting and discussing those disciplines had a more philosophical and learned audience in mind than that of technicians and engineers. On the one hand, their works emphasized not just utility but the importance of the reform of knowledge, especially natural philosophy; on the other hand,

---

[51] Although it is problematic to include the science of motion within mechanics throughout the seventeenth century, it is more problematic to exclude it. During the century, scholars increasingly took equilibrium, or statics, and the science of motion as two sides of the same coin. Alan Gabbey has produced several thoughtful articles on this issue: See Gabbey, "Newton's *Mathematical Principles of Natural Philosophy*: A Treatise on 'Mechanics'?" in *The Investigation of Difficult Things*, ed. Peter M. Harman and Alan E. Shapiro. (Cambridge: Cambridge University Press, 1992), pp. 305–22; Gabbey, "Descartes's Physics and Descartes's Mechanics: Chicken and Egg?" in *Essays on the Philosophy and Science of René Descartes*, ed. Stephen Voss (Oxford: Oxford University Press, 1993), pp. 311–23; and Gabbey "Between *ars* and *philosophia naturalis*: Reflections on the Historiography of Early Modern Mechanics," in *Renaissance and Revolution*, ed. Judith V. Field and Frank A. J. L. James (Cambridge: Cambridge University Press, 1993), pp. 133–45.

during the seventeenth century, the world, or at least significant portions of it, was seen more and more in mechanical terms, and therefore discussions of machines became colored with cosmological implications and natural philosophy. Occasionally Galileo's views were already being discussed and criticized in university textbooks of natural philosophy around the middle of the century.[52]

For several reasons, the science of motion attracted the lion's share of the interest. Galileo thought that he had established the science of the resistance of materials on the principle of the lever. Because it was thought to rely on mathematical and mechanical foundations, which were less problematic than the science of motion,[53] the science of resistance of materials generated fewer controversies and generally did not inspire broader philosophical debates. The mathematical treatment of the resistance of materials attracted the attention of the engineer and mathematician François Blondel (1618–1686), of Galilean mathematicians Alessandro Marchetti (1633–1714) and Vincenzo Viviani (1622–1703), of the Jesuit Honoré Fabri (1607–1688), of the experimental philosopher at the Paris Académie Edme Mariotte (ca. 1620–1684), and of mathematicians such as Gottfried Wilhelm Leibniz (1646–1716) and Jakob Bernoulli (1654–1705).[54] Although the science of waters was exceedingly complex, Castelli's fundamental proposition on water flow was quite straightforward. Descartes and Isaac Newton (1643–1727) implicitly used it with cosmological implications, but the science of waters remained largely a technical matter rooted in Italy.[55]

Let us now consider the science of motion. Galileo had provided different presentations: piecemeal, so to speak, in the *Dialogo*, where the motion of bodies on the earth was tied to Copernicanism, and structured in axiomatic form with definitions and theorems in the *Discorsi*. Even taking this major difference into account, it is striking to notice how differently his works were read by scholars in the late 1630s and 1640s. For some, Galileo had provided a series of propositions to be tested experimentally or examined in their mathematical or mechanical deductions on a one-by-one basis. Part of this process consisted in finding numerical values in which Galileo seemed

---

[52] See, for example, Niccolò Cabeo, *In quatuor libros Meteorologicorum Aristotelis commentaria* (Rome: Typis heredum Francisci Corbelletti, 1646). On this topic, the classic study is Charles B. Schmitt, "Galileo and the Seventeenth-Century Text-book Tradition," in *Novità celesti e crisi del sapere*, ed. Paolo Galluzzi (Florence: Giunti Barbera, 1984), pp. 217–28. For France, see Laurence W. B. Brockliss, *French Higher Education in the Seventeenth and Eighteenth Centuries: A Cultural History* (Oxford: Oxford University Press, 1987).
[53] Aspects of those problematic foundations are discussed later in this chapter.
[54] Edoardo Benvenuto, *An Introduction to the History of Structural Mechanics*, 2 vols. (Berlin: Springer, 1991), vol. 1.
[55] Castelli's proposition was first published in the 1628 *editio princeps* of *Della misura*. It states that, in a river in stationary flow, the areas of the cross sections are inversely as the speeds of the water flowing through them. Descartes, *Principia philosophiae*, pt.3, paras. 51, 98; pt. 4, para. 49. Newton, *Principia*, new translation, pp. 789–90. The main work on the Italian hydraulic tradition is Maffioli, *Out of Galileo*.

to have no particular interest, such as the length of the seconds pendulum or the distance traversed by a falling body in one second. Others were interested in the axiomatic structure of the new science as a whole, in the choice of definitions and axioms, and in the ensuing proofs. Still others became concerned with mathematico-philosophical aspects, such as the nature of the continuum, with special regard to time and speed. Lastly, some scholars objected to the very nature of Galileo's science, arguing that he had neglected physical causes and built an abstract science with no bearings on the real world. Those different readings provide valuable insights on the many perspectives from which mechanics and motion were studied in the mid-seventeenth century.

Galileo and his disciples Torricelli and Viviani were mainly concerned with the formulation of a science in imitation of Archimedes' work on the equilibrium of the balance. Generally, they were already convinced of the truth of individual propositions, but they worried about the overall structure and especially the choice of axioms. Ideally, in their opinion an axiom did not have to be established by experiments but rather had to be chosen as a principle of reason to which the mind naturally agrees. Soon after the publication of the *Discorsi*, Viviani pressed Galileo over the choice of his axiom, and Galileo conceived a way to prove it on the basis of received mechanical principles. The new proof appeared in the second edition of *Discorsi* in 1656, published together with other works by Galileo, excluding the *Dialogo*.[56] Torricelli also moved along similar lines, and in his reformulation and extension of Galileo's science in *De motu* (On Motion, 1644) he introduced a new principle, namely that two joined bodies do not move unless their common center of gravity descends. Although Torricelli instantiated it by mentioning bodies attached to a balance or pulley, it is clear that this principle was not based on experiments and had general validity beyond the specific cases mentioned.[57]

It would be erroneous to generalize these concerns to other quarters. Several readers questioned specific empirical claims and performed experiments that challenged Galileo's statements and results. For example, the Genoa patrician Gianbattista Baliani (1582–1666) and Mersenne in Paris expressed surprise and incredulity at Galileo's claim that in five seconds a body falls only one hundred *braccia*, less than seventy yards. Quite rightly, both believed the real distance to be far greater. We now know that Galileo had extrapolated results from fall along inclined planes to bodies in free fall. Such extrapolations were problematic for reasons unknown at the time, and when Mersenne tested Galileo's claims he found systematic errors for virtually all inclinations. A body falling along the inclined plane covered a distance noticeably shorter

---

[56] The axiom stated that the degrees of speed acquired by the same body over planes of different inclinations are equal whenever the heights of those planes are equal. Galileo, *Two New Sciences*, pp. 206, 214–18, includes the new proof in a footnote.

[57] *De motu* was part of Torricelli's *Opera geometrica* (Florence: Typis Amatoris Masse and Laurentii de Landis, 1644).

than that predicted by Galileo.⁵⁸ Galileo had also claimed that as a body shot upward falls back, it goes through the same degrees of speed as when it was going up. In particular, the speed with which it reaches the ground would be the same as that with which it was shot. Yet experiments based on the force of percussion reported by Mersenne and the Paris mathematician Gilles Personne de Roberval (1602–1675) showed the speed of the body after the fall to be much smaller than that with which it was shot. Experiments performed initially by gunners at Genoa and then at various locations in Europe tested Galileo's claims about parabolic trajectories. Were the trajectories parabolic? Was it true that the longest shot occurred with an inclination of 45°? Could one measure and predict the effect of air resistance? Lastly, probably the most extensive and accurate set of experiments inspired by Galileo was performed from many of the Bologna bell towers as well as civic towers under the direction of Jesuit mathematician Gianbattista Riccioli (1598–1671). Over approximately a decade, Riccioli dropped spheres of lead, clay (empty and full inside), wax, and different types of wood, finding that they followed Galileo's odd-number rule – namely that the distances traversed in successive time intervals are as 1, 3, 5, and so on – though it was not quite true that all bodies fell with the same acceleration. Riccioli also experimented on the length of the seconds pendulum, yet another common theme for Galileo's readers. At least some of these experiments were performed keeping in mind the issue of Copernicanism and the behavior of bodies on a moving earth, which was a major concern especially among Jesuit authors.⁵⁹

Additionally, several readers were concerned that embedded in Galileo's views about motion were both specific and general philosophical propositions. Prominent among the former was the problem of continuity. Galileo had claimed on several occasions that a falling body goes through all the degrees of speed. Because the degrees of speed are infinitely many and the body falls in a finite time, it seems to follow that the body must go through each of them in an instant and that a finite time is composed of infinitely many instants. The composition of the continuum touched on many other themes as well, notably condensation, rarefaction, the cohesion of bodies, and their resistance to rupture – Galileo's main theme in the first day of the *Discorsi*. But even within the science of motion, Galileo's claim was

---

⁵⁸ Marin Mersenne, *Harmonie universelle contenant la théorie et la pratique de la musique* (Paris: Editions du CNRS, 1986), bk. 2, prop. 7, esp. pp. 111–12. The discrepancy was caused by the fact that a sphere rolling down an inclined plane behaves like a rigid body, not a point mass, and follows more complex laws because it rotates. A sphere rolling along an inclined plane covers only five-sevenths of the distance Galileo and Mersenne had in mind.

⁵⁹ Giovanni B. Riccioli, *Almagestum novum* (Bologna: Ex Typographia Haeredis Victorij Benatij, 1651), pt. I, pp. 84–91 and pt. II, pp. 381–97. Classic studies by Alexandre Koyré are: "An experiment in measurement," *Proceedings of the American Philosophical Society*, 97(1953), 222–37; "A Documentary History of the Problem of Fall from Kepler to Newton," *Transactions of the American Philosophical Society*, 45 pt. 4 (1955). More recently, Peter Dear has studied the role of experiments in the mathematical disciplines and the science of motion in *Discipline and Experience: The Mathematical Way in the Scientific Revolution* (Chicago: University of Chicago Press, 1995).

controversial and was challenged by several scholars on several grounds. The Society of Jesus was extremely cautious when it came to matters about the composition of the continuum, potentially impinging on such crucial issues as the dogma of transubstantiation of the Eucharist. They prohibited several propositions, such as "An infinity in number and magnitude can be contained between two unities or points," and "The continuum is composed of a finite number of indivisibles." Two Jesuit scholars, Honoré Fabri and Pierre le Cazre (1589–1664), were among those who became entangled in metaphysical disputes on continuity and defended propositions against both Galileo and surprisingly some of the views of their order. Fabri in particular argued that time was not continuous but consisted of tiny but finite instants. At a later stage, he also claimed that the instants were of variable size. Baliani also argued in a similar vein, possibly inspired by Fabri. In their opinion, a time interval would consist of a finite number of finite instants, and thus speed would change not continuously but discretely at each new instant. Fabri agreed that Galileo's rule was empirically adequate but denied that Galileo had provided a foundation for his science and that experience could serve this purpose. The advantage of his view would be to provide a different and more solid philosophico-metaphysical justification for the science of falling bodies. Additionally, if the instants are very small, the difference between Galileo's odd-number rule and Fabri's can become so small as to be empirically undetectable. Therefore, one would save the experimental side of Galileo's work while providing it with a more solid foundation as to the composition of the continuum. Not surprisingly, the reviver of ancient atomism, Pierre Gassendi (1592–1655), was also involved in these debates.[60]

Other scholars had other fundamental objections concerning the very nature of Galilean science. Descartes was the most prominent among those readers concerned not simply with physical causes but with the architecture of Galileo's science. Whereas Galileo had modeled his sciences on Archimedes and often emphasized his reliance on sensory experiences and mathematics as the new sources of knowledge, Descartes aimed more broadly at a reform of knowledge from its metaphysical foundations.

DESCARTES' MECHANICAL PHILOSOPHY AND MECHANICS

One way to discuss some of the differences between Galileo's and Descartes' perspectives is to focus on their reading of Aristotle and their relation to the

---

[60] C. R. Palmerino, "Two Jesuit Responses to Galileo's Science of Motion: Honoré Fabri and Pierre Le Cazre," in *The New Science and Jesuit Science: Seventeenth Century Perspectives*, ed. Mordechai Feingold (Dordrecht: Kluwer, 2003, *Archimedes*, vol. 6), pp. 187–227, at p. 187. See also Paolo Galluzzi, "Gassendi and *l'Affaire Galilée* on the Laws of Motion," in *Galileo in Context*, ed. Jürgen Renn (Cambridge: Cambridge University Press, 2001), pp. 239–75.

Peripatetic tradition. Both rejected Aristotle and especially the Peripatetic learning of the universities, but whereas Galileo offered a powerful but narrower alternative that focued on motion, matter theory, and cosmology, Descartes offered a broader alternative that included the nature and foundations of our knowledge and aimed at replacing the Aristotelian worldview among scholars as well as in university teaching. Galileo's views on the relationship between mathematics and physical causes varied considerably even within the same work. In the *Discorsi*, for example, he tried to account for the resistance of bodies to rupture in terms of infinitely many interstitial vacua, joining matter theory with the analysis of the continuum. With regard to the science of motion, however, Galileo claimed through his spokesperson Filippo Salviati (1582–1614) that he did not wish to investigate causes and presented a new science in the Archimedean tradition of the science of equilibrium or statics.[61]

Descartes, by contrast, after a crucial collaboration with the Dutch scholar Isaac Beeckman (1588–1637), aimed at creating a new worldview where a mathematical description of nature, and especially motion, was joined to a physical causal explanation. The key to this link was the microworld of corpuscles and subtle fluids responsible for physical phenomena. Through a combination of his study of impacts among particles and of the behavior of fluids, Descartes was able to produce a new physico-mathematics by which he and Beeckman meant a science joining a mathematical description with a physical causal account. The mixed mathematics dealt with the properties of the lever, for example, without exploring the cause of gravity, whereas Descartes made that exploration a major feature of his work.[62] Descartes did not think much of Galileo's *Discorsi* and in a letter to Mersenne he disagreed with the physical account of the resistance of materials and the analysis of the continuum because he thought that Galileo had built a science of motion without foundations by failing to provide a deeper philosophical and causal analysis. Descartes argued that Galileo had solved easy mathematical exercises devoid of physical significance.[63] Descartes' view of gravity as caused by a stream of particles made Galileo's abstraction of the motion of falling bodies in a void meaningless because he was removing the very cause of motion.

Both Galileo and Descartes were familiar with *Quaestiones mechanicae*, a text then believed to be by Aristotle but now attributed to his school. *Quaestiones* relies on the principle of the lever and presents a series of cases and examples dealing with it. Whereas Galileo read it within the tradition of

---

[61] Galileo, *Two New Sciences*, pp. 109, 158–9.
[62] Stephen Gaukroger, *Descartes* (Oxford: Oxford University Press, 1995), chap. 3. Descartes and Beeckman used the term "physico-mathematics" in this sense. The emergence of the same term among Jesuit writers with a somewhat different meaning has been studied in Dear, *Discipline and Experience*. See also Stephen Gaukroger and John Schuster, "The Hydrostatic Paradox and the Origins of Cartesian Dynamics," *Studies in History and Philosophy of Science*, 33 (2002), 535–72.
[63] The relevant portion of the letter is translated in Drake, *Galileo at Work*, pp. 386–93.

mechanics, Descartes found in it and its interpretative tradition references to slings, whirlpools of water, and pebbles rounded on the seashore, three key elements of his worldview that we will encounter again in this section.[64]

Descartes' work, together with some of Galileo's passages on the constitution of matter and Gassendi's Christianized atomism, constitute the pillars of the so-called mechanical philosophy. In *Il Saggiatore* (The Assayer, 1623), Galileo had drawn a distinction between properties of matter such as shape, which is independent of human perceptions, and color, which resides only in the human mind. They were later called primary and secondary qualities, respectively. In a number of works, culminating with the imposing *Syntagma philosophicum* (Philosophical Work, 1653), Gassendi attempted to revive the ancient philosophy whereby elementary constituents of matter or atoms move through empty space. The mechanical philosophy was a heterogeneous collection of views with a common core based on the belief in the fundamental role of the size and shape of particles in motion. Descartes believed neither in the existence of empty space nor in the indivisibility of matter or atoms, unlike Gassendi.

Descartes first outlined his system in *Le monde* (The World), but this work appeared posthumously in 1664. The main text known to his contemporaries was *Principia philosophiae* (Principles of Philosophy), first published in Latin in 1644 and then in 1647 in a French translation by the abbé Claude Picot (d. 1668) partly under Descartes' supervision. This work consists of four parts, on the principles of human knowledge, the principles of material things, the visible world, and the earth. Descartes dedicated it to his friend and correspondent Princess Elizabeth of Pfalz (1596–1662). Descartes claimed that she had a detailed knowledge of all the arts and sciences and therefore was the only person to have understood all the books he had published. In the letter to Picot prefaced to the French edition, Descartes suggested that his book be read first as a novel to figure out what it was about. Only on a second reading should one examine it in greater detail and grasp the order of the whole. The letters to Elizabeth and Picot suggest that Descartes' work was extraordinarily ambitious in the range of matters it treated and in the broad audience it intended to address.[65]

I focus here on Parts 2–4, outlining a world system based on relatively simple laws but with a dazzlingly complex series of interactions among particles. In Part 2, Descartes argued that space could not be distinguished from corporeal substance and that a vacuum and atoms did not exist. Descartes also tried to define the motion of a body with respect to the bodies that touch it, so that if the earth is carried by a vortex, one could say that truly it is at

---

[64] H. Hattab, "From Mechanics to Mechanism: The *Quaestiones mechanicae* and Descartes' Physics," to appear in *Australasian Studies in the History and Philosophy of Science*. I am grateful to the author for having provided me with a draft of her work and for permission to refer to it.

[65] Gaukroger, *Descartes*, chaps. 7 and 9.

rest with respect to its surrounding bodies. Scholars disagree as to whether his definition was genuinely part of his views or was intended to protect his system from the charge of Copernicanism by the Catholic Church.[66]

In the study of motion, Descartes formulated three laws, a significant departure from the term "axiom" used in the mathematical disciplines. He justified them partly on physical and partly on theological grounds. He considered motion and rest as modes of a body and, at least to some extent, equivalent ones, whereby every body perseveres in its state of motion or rest indefinitely. This is his first law of motion. The second specifies that motion in itself is rectilinear and that everything moving circularly tends to escape from the center along the tangent. Descartes often talked of "determination," a notion close to that of direction. Later in the century, the ideas embodied in the first two laws would become known as the law of inertia, despite the fact that the term inertia had been initially employed by Johannes Kepler (1571–1630) with quite a different meaning, namely of a body's innate tendency to come to rest. Descartes tried to provide examples for his first two laws from objects of common experience, the motion of projectiles in the first case and that of a sling in the second.[67]

Traditionally, Descartes' reconceptualization of motion and geometrization of space have been considered as the major event in the history of seventeenth-century mechanics and natural philosophy.[68] According to Kepler and to traditional Aristotelian doctrines, a body in motion would naturally come to rest. In the work of Galileo, horizontal motion often is the limit of a motion along an inclined plane with zero inclination and is truly horizontal in the sense that it coincides with the horizon. Thus, over a short distance it may appear straight, but over a longer one it is circular because of the earth's curvature. Galileo seemed to extend similar views to orbital motions, implying that circular motion is natural. Descartes, by contrast, extended rectilinear uniform motion indefinitely in a Euclidean space. Despite the great significance of his and Gassendi's views on this matter, the transformations occurring in mechanics in mid-century were more broadly based. Mechanics and the science of motion involved a richer set of notions, practices, and problems than was suggested by Descartes' first two laws, such as falling bodies, the motions of strings and pendulums, the resistance of materials, and water flow.[69]

---

[66] Daniel Garber, *Descartes' Metaphysical Physics* (Chicago: University of Chicago Press, 1992), especially chap. 6.

[67] Garber, *Descartes' Metaphysical Physics*, pp. 188–93; Damerow et al., *Exploring the Limits*, pp. 103–23; Descartes, *Principia*, pt. 2, paras. 37–9; and Alan Gabbey, "Force and Inertia in Seventeenth-Century Dynamics," *Studies in History and Philosophy of Science*, 2 (1971), 1–67. See the entry on inertia by Domenico Bertoloni Meli in *Encyclopedia of the Scientific Revolution from Copernicus to Newton*, ed. W. Applebaum (New York: Garland, 2000), pp. 326–8.

[68] The most influential proponent of this view was Alexandre Koyré. See, for example, his *From the Closed World to the Infinite Universe* (Baltimore: Johns Hopkins University Press, 1957).

[69] We lack a comprehensive study of mechanics in the seventeenth century. On the science of waters, see Maffioli, *Out of Galileo*. On the resistance of materials, see Benvenuto, *Structural Mechanics*, vol. 1.

Descartes' third law of nature concerns collision and states that a body hitting a stronger one loses no part of its motion, whereas a body hitting a less strong one loses the same amount that it transmits to the other. Put another way, the third law states the conservation of motion, by which Descartes meant the product of the size of a body and its speed, taken with no regard to direction. Clearly this law required justification and qualification. Descartes claimed that quantity of motion is conserved not in individual bodies but in the universe as a whole because of God's plan and action. In the ensuing explanation, Descartes distinguished between hard and soft bodies, so one gets the impression that the strength of a body is associated with its hardness. Later on, however, he defined a body's strength in terms of its size, surface, and speed, and the nature of the impact. These conditions are quite complex, and when it comes to providing specific rules for actual impacts, Descartes introduced simplifications, such as that the bodies are perfectly hard and are taken in isolation from all surrounding bodies.[70]

These simplifications are quite radical because they describe abstract situations that cannot occur in the Cartesian world, which is a plenum. Even with them, Descartes' seven rules appear problematic, despite his claim that they were self-evident and required no proof. Rules 1–3 deal with bodies moving in opposite directions, and rules 4–6 deal with collisions between a body in motion and one at rest. The last rule examines the various cases of collisions between two bodies moving in the same direction. Descartes' rules do not lack surprising and unconvincing features. For example, in rule 4 he argued that a body at rest could not be set in motion by a smaller one, regardless of its speed. Another example of the problems associated with Descartes' rules emerges from those numbered 2 and 5. According to rule 2, if two bodies, one slightly larger than the other, collide with equal and opposite speeds, the smaller rebounds and both move in the same direction with the same speed. According to rule 5, if a body in motion collides with a smaller body at rest, after the impact both will move in the direction of the impinging body with the same speed, a speed smaller than the impinging body's original speed. In both cases, their common final speed can be determined from the conservation of a Cartesian or scalar quantity of motion, which is larger in rule 2 than in rule 5 for equal bulks. Thus it appears that a body in motion is affected more by colliding with one at rest than with the same body moving with an opposite speed, an unconvincing result or at least one far from self-evident.[71]

Part 3 of Descartes' *Principia* deals with the sensible world and outlines a cosmogony based on the size and shape of particles in motion. He provided an account of the formation of the universe and of the motions of celestial bodies such as planets and comets. The heavens are fluid, and there are three types of

---

[70] Descartes, *Principia*, pt. 2, paras. 40–55.
[71] There is a vast body of literature on the rules of impact. See Damerow et al., *Exploring the Limits*, pp. 91–102; and Garber, *Descartes' Metaphysical Physics*, chap. 8.

matter or elements depending on shape and size. Over time, the particles of matter become rounded like pebbles on a beach, and their minute fragments form different types of matter. The first element forms the sun and fixed stars, and is fine-textured and moves very fast; the second element is coarser and forms the fluid filling the heavens; and finally, the third element is rather gross and forms planets and comets. Light is the pressure resulting from the endeavor of small particles to escape from a rotating vortex. From his second law of motion to his analysis of light, Descartes relied on motion being rectilinear. Curvilinear motion is caused by an external agent and generates a tendency to escape along the tangent. Descartes did not provide a quantitative measure of this outward tendency but laid the conceptual foundations for explaining it that remained in place for several decades. A body tending to escape along the tangent also tends to escape from the center. Thus a rotating stone pulls the sling retaining it and is counterbalanced by the hand holding the sling. Similarly, in the universe, orbiting bodies have a tendency to escape along the tangent, and those with a stronger tendency push the others toward the center. Thus, bodies appearing to have a tendency toward the center are in reality only the losing ones in the competition to move outward. In a universe with no empty space, for every particle moving outward there must be a corresponding one moving the opposite way. Descartes illustrated this with the example of straws floating in a whirlpool of water and being pushed by the rotation toward the center. In dealing with the real world, Descartes did not rely on his impact rules but considered factors such as the structure of matter, the nature of its pores, and the size and speed of the fluid particles flowing through them. In Descartes' eyes, the virtue of this type of explanation is that it accounts for all phenomena with philosophically acceptable notions, avoiding inexplicable attractions and repulsions and all actions not based on direct contact.[72]

Part 4 is devoted to the earth and examines a range of phenomena about its formation and features, extending among others to gravity, tides, chemical phenomena and reactions, the origin of flames, magnetism, and the elasticity of air and other substances. Concerning gravity, for example, Descartes argued that because of the different sizes and speeds of the particles generating it and flowing through terrestrial bodies, the weight of a body is not proportional to its quantity of solid matter or the grosser matter of Descartes' third element.[73]

Descartes' formulation of the first two laws, his insistence on conservation in the third, and his posing the problem of curvilinear motion and impact changed the landscape of the science of motion. By identifying matter with extension, Descartes set the scene in principle for a radical geometrization of the universe. In practice, however, the complexity of the interactions

---

[72] Descartes, *Principia*, pt. 3, paras. 48–63.
[73] Descartes, *Principia*, pt. 4, para. 25. Later in the century, Newton was to address precisely this point.

among streams of particles moving in all directions meant that the actual formulation of a mathematical description of nature could be accomplished only in limited areas.[74]

READING DESCARTES AND GALILEO: HUYGENS AND THE AGE OF ACADEMIES

Although probably only a few devoted followers accepted Descartes' views in their entirety, the intellectual world of the second half of the seventeenth century was dominated by criticism, responses, reformulations, and refinements of his doctrines. No one who read his works went away unchanged. Already during Descartes' lifetime, and even more so in the second half of the century, university textbooks based on his philosophy began to appear, marking a major change in higher education. Henri Regis (1598–1679) and Jacques Rohault (1620–1675), just to mention two of the most prominent authors, wrote influential textbooks that went through many editions stretching, in the latter case, to the eighteenth century.[75]

Let us consider how Descartes' *Principia* was read, especially his laws of nature and impact rules. Among the community of mathematicians, the first two laws of motion fared much better than the third one and its instantiation in the impact rules. Scholars accepted that undisturbed motion is uniform and rectilinear, at least in principle, even if at times they forgot to apply it in practice, as did Giovanni Alfonso Borelli (1608–1679), holder of Galileo's former chair of mathematics at Pisa, in his study of falling bodies on a rotating earth.[76] The conservation of quantity of motion and the rules of impact were more problematic and were subjected to criticism and refutation. Both peaked in the late 1660s.

It may be useful to begin with a brief historiographic consideration. Traditionally, the study of impact has been considered simply as a search for the correct rules.[77] The rules, however, depend on the different types of bodies involved, and therefore the problem is twofold: to determine the rules and to classify bodies appropriately. In order to have meaningful rules, one must know the meaning of terms such as soft, hard, and elastic for impact phenomena. Thus, the study of impact involves the properties of material

---

[74] Optics was an area where Descartes was especially successful in this regard. See Gaukroger, *Descartes*, pp. 256–69.

[75] See, for example, Henricus Regis, *Fundamenta physices* (Amsterdam: Apud Ludovicum Elzevirium, 1646); Jacques Rohault, *Traité de physique* (Paris: Chez la Veuve de Charles Savreux, 1671); and Brockliss, *French Higher Education in the Seventeenth and Eighteenth Centuries*.

[76] On subtle differences between Descartes' laws and later formulations, see Gabbey, "Force and Inertia in Seventeenth-Century Dynamics"; Govanni A. Borelli, *De vi percussionis* (Bologna: Ex typographia Iacobi Montii, 1667), pp. 107–8; and Koyré, "Documentary History," pp. 358–60.

[77] See, for example, Richard S. Westfall, *Force in Newton's Physics* (London: Macdonald, and New York: American Elsevier, 1971), ad indicem.

bodies. The most significant and wide-ranging of those properties was elasticity, a property amenable to mathematical description and linked to a large number of instruments, phenomena, and experiments such as the wind gun, pneumatic fountains, the spring, the Torricellian tube, and the air-pump.

In the late 1650s, the Dutch scholar Christiaan Huygens (1629–1695) had come to mistrust Descartes' rules and proceeded to formulate new ones, but his work was known in part only to a few correspondents, and the treatise he composed remained unpublished at the time. Huygens used pendulums to study impact, an effective technique whereby the speeds before and after the impact can be ascertained by the bobs' heights. His final formulation was axiomatic, in the style of Galileo's science of motion, and was based not on experiments but on the skillful use of general principles. Among the most famous were relativity of motion to disprove some of Descartes' rules, and a generalization of Torricelli's principle whereby the center of gravity of two colliding bodies cannot rise.[78] If one thinks of colliding pendulums, it is quite natural to consider the height the bodies can reach and to introduce an expression based on the square of their speed times their weight, a notion later known, following Leibniz, as *vis viva*, or living force.[79] In 1660 and 1661, Huygens demonstrated his prowess at solving impact problems in Paris and London in front of several members of the Royal Society.

In 1666 and 1667, Borelli published two works, on the motion of the Medicean planets and on the force of percussion, two exquisitely Galilean topics. In his works, he criticized some of Descartes' rules and proposed some of his own. For example, Borelli argued, *pace* Descartes, that a smaller body could set a larger one in motion, and he studied impact for what he called perfectly hard bodies that do not rebound. The key notion in impact for Borelli was *vis motiva*, or speed taken with its direction times the body, which he argued is conserved. Borelli had difficulties in dealing with the bodies' rebound, a common phenomenon in impact that was difficult to explain if the colliding bodies were conceived to be inflexible. He also dealt with elasticity, but did not formulate rules for elastic impacts, possibly because he did not believe they could be given in mathematical form.[80]

In the late 1660s, the Royal Society investigated the problem of motion and addressed the issue of impact at several of its meetings, where experiments and discussions took place. In 1668, the Society invited contributions on the problem, to which the Oxford Savilian professor John Wallis (1616–1703), the architect Christopher Wren (1632–1723), and Huygens provided solutions.

---

[78] Christiaan Huygens, *Oeuvres Complètes*, 22 vols. (The Hague: Martinus Nijhoff, 1888–1950), 16: 21–5 and 95 n. 10. This is a generalization of Torricelli's principle because the two colliding bodies are no longer joined.
[79] *Vis viva* first appeared in print in Leibniz's *Tentamen de motuum coelestium causis*, published in the *Acta eruditorum* for 1689. See Domenico Bertoloni Meli, *Equivalence and Priority* (Oxford: Oxford University Press, 1993), pp. 86–7; the relevant passage from the *Tentamen* is translated at p. 133.
[80] Westfall, *Force*, pp. 213–18.

Wallis's essay was very brief and did not address the main issue. He later expanded his discussion considerably in his *Mechanica, sive de motu tractatus geometricus* (Mechanics, or Geometrical Treatise on Motion, 1670–1), where he classified bodies as hard, soft, and elastic. These are abstractions because real bodies are not perfectly hard or elastic. Clay, wax, and lead are examples of soft substances, and steel and wood are examples of elastic ones. By hard bodies Wallis probably meant the ultimate constituents of matter, but he provided no example, despite the fact that he provided impact rules for them. Wallis excluded soft bodies from his rules, arguing that a portion of their quantity of motion is lost in impact. Thus he did not present a universal conservation rule that was valid for all bodies.

Wren provided a brief and rather cryptic essay quite similar to that of Huygens, who later complained that Wren had gotten his idea in 1661 when they had discussed the matter at the Society. Huygens was irritated that the contributions by Wren and Wallis appeared before his own, and he published a version of his paper first in the *Journal des sçavans* and then in the *Philosophical Transactions of the Royal Society*. He pointed out that Descartes' conservation law was not valid because the Cartesian quantity of motion can increase, remain constant, and decrease. What remains constant in impact for all types of bodies is the quantity of motion in one direction. Huygens distinguished between hard and soft bodies and argued that for the former impact is instantaneous, whereas for the latter it occurs over time. He also claimed that hard non-elastic bodies rebound like elastic ones. Examples of hard bodies are atoms, whose existence Huygens tentatively accepted, and Descartes' subtle matter. Here it would be impossible to understand the terms of the debate without realizing its Cartesian roots.[81]

In the study of curvilinear motion once again, Descartes provided an important conceptual framework with the example of the sling, and, once again, it was Huygens who had the vision and mathematical skill to offer a solution to the problem. As in the case of the impact rules, Huygens reached his important results in the late 1650s, but it took several years before they were published in curtailed form. The initial stimulus to work on this topic came somewhat indirectly from Mersenne. As we have seen, the French Minim's reading of Galileo was often aimed at determining numerical values. The distance fallen by a body in free fall in one second was one of the values sought. Tackling the question directly proved problematic because bodies fall very fast, and thus the problem was best rephrased. Huygens was led to consider the action of gravity in a conical pendulum, whose bob rotates on a plane parallel to the horizon, to be counterbalanced by centrifugal force. Relying on Galileo's science of motion and Descartes' notions, Huygens produced

---

[81] A. Rupert Hall, "Mechanics and the Royal Society, 1668–1670," *British Journal for the History of Science*, 3 (1966), 24–38, is still a useful essay. See also Westfall, *Force*, pp. 231–43.

a remarkable series of theorems, which he then published without proof in his masterpiece, the *Horologium oscillatorium* (The Pendulum Clock, 1673), one of the most original works in the history of mechanics as a whole and a model for Newton's *Principia mathematica philosophiae naturalis* (Mathematical Principles of Natural Philosophy, 1687).[82]

Mersenne had come to doubt on purely empirical grounds Galileo's claim that the oscillations of a pendulum are isochronous regardless of their amplitude. He surmised that this may be the case in a vacuum but was not true in air. While seeking the perfect or isochronous clock, Huygens moved the research from the world of experiments to a more theoretical level, producing a work addressing technical issues of horology and joining the new science of motion with higher mathematics. He proved that the problem was not simply air resistance because pendular oscillations depend on their amplitude, with greater ones being slower. The issue was not one of pure theory but was linked to the problem of finding the longitude at sea, a major concern for the burgeoning colonialism of European states.[83]

Horologium oscillatorium is closer to a Galilean tradition than a Cartesian one. Of course, it was Galileo who had started to use the pendulum as a time-measuring device and who had thought of building a clock regulated by the pendulum's oscillations. Much as Galileo had done with regard to the science of motion in the *Discorsi*, Huygens avoided issues such as the cause of gravity.[84] Rather, he sought to promote the virtues of his clocks, explain their principles of operation and construction, and produce new theories both in mechanics and mathematics. Much as Galileo had done at the end of the first day of the *Discorsi*, when he had tried to show technicians how to draw parabolas by throwing balls along inclined planes or hanging the extremes of a chain from two nails, Huygens showed practical ways for drawing a cycloid.[85] The cycloid was a new curve in the seventeenth century and the double mathematical protagonist of his treatise because in order to have isochronism the curve described by the bob had to be a cycloid, and in

---

[82] Joella G. Yoder, *Unrolling Time* (Cambridge: Cambridge University Press, 1988), provides a detailed account of Huygens's research. Henk J. M. Bos, M. J. S. Rudwick, H. A. M. Snelders, and R. P. W. Visser, *Studies on Christiaan Huygens* (Lisse: Swets and Zeitlinger, 1980), contains many valuable contributions. See also Michael S. Mahoney, "Huygens and the Pendulum: From Device to Mathematical Relation," in *The Growth of Mathematical Knowledge*, ed. Herbert Breger and Emily Grosholz (Dordrecht: Kluwer, 2000), pp. 17–39.

[83] Christiaan Huygens, *The Pendulum Clock*, trans. R. J. Blackwell (Ames: University of Iowa Press, 1986), p. 19; and William J. H. Andrews, ed., *The Quest for Longitude: The Proceedings of the Longitude Symposium, Harvard University, Cambridge, Massachusetts, November 4–6, 1993* (Cambridge, Mass.: Harvard University Press, 1996).

[84] It should be remembered, however, that although he expressly ruled out discussions on the cause of gravity in day three of the *Discorsi*, Galileo embarked on extensive discussions on the cause of cohesion in day one. Thus, it would be inaccurate to take his attitude toward the cause of gravity as representative of his views on physical causes in general.

[85] Galileo, *Two New Sciences*, pp. 142–3. The curve traced by a sphere rolling on an inclined plane is a parabola, whereas that described by a hanging chain resembles the parabola but is more complex. Huygens, *Pendulum Clock*, pp. 21–4.

order to make the bob move along a cycloid, it is necessary to constrain it between cheeks that are also cycloid arcs. Also in Part II of the *Horologium oscillatorium*, on falling bodies and their motion in a cycloid, Huygens often referred to Galileo's proofs in the *Discorsi* and rephrased them.[86] Of course, Huygens expanded on Galileo's results by including a mathematical treatment of motion along a cycloid, something beyond Galileo's capabilities.

Part III of the *Horologium oscillatorium* is a mathematical investigation of curves generated by unrolling a thread on another curve and their mutual relationships. With Part IV, we return to mechanics proper and to a debate involving Mersenne, Descartes, and Roberval dating from the 1640s, namely to find the center of oscillation for a compound pendulum. A simple pendulum has all its mass concentrated in one point, whereas in a real physical pendulum the mass is distributed over a finite area. Whereas the period of a simple pendulum can be determined by its length, in a real pendulum there is no obvious point whereby the length associated with its period can be determined. Finding the center of oscillation means determining that point in a real pendulum. Huygens's success in finding a procedure to determine that point counts as one of the finest achievements of seventeenth-century mechanics.

In Huygens's work in mechanics there is an interesting dichotomy between Galilean and Cartesian approaches. Whereas the *Horologium oscillatorium* clearly looked to Galileo's science of motion as a model for content and structure, other works looked more to Descartes' *Principia philosophiae*. At a debate in 1669 at the Paris Académie Royale des Sciences on the cause of gravity, Huygens proposed a mechanism emphasizing the account of physical causes over mathematical accuracy. He took a bucket of water with a small sphere floating in it that was constrained by two strings stretched between opposite sides of the rim. By setting the water in circular motion and then stopping it, the sphere moved toward the center, thus showing an effect analogous to gravity. Unlike Descartes, Huygens did not believe that the universe was a plenum but accepted empty space and argued that the matter of the vortex rotates in all directions. Heavy bodies do not follow the motion of the particles of the fluid but are pushed toward the center because of their lack of centrifugal force. Huygens also attempted a quantitative estimation of the speed of the particles of the fluid based on his theorems on centrifugal force. He found that the speed of a particle of fluid required to produce gravity was seventeen times the speed of a point on the earth's equator.[87]

With mechanical explanations extended to all types of phenomena, from falling bodies to magnetism, fluids and vortices became common explanatory models, but they were not the only ones. At times elasticity, for example, was explained in terms of subtle fluids, but it was also considered an autonomous property of matter accountable mathematically in mechanical terms and

---

[86] Huygens, *Pendulum Clock*, esp. pp. 40–5.
[87] Westfall, *Force*, chap. 4.

capable of explaining a number of other phenomena. Robert Hooke (1635–1703), curator of experiments at the Royal Society and professor of geometry at Gresham College, was one of the most prominent scholars of elasticity, and the author of *Lectures de potentia resitutiva, or of springs* (1674). Elasticity was tied not only to mathematics and physical explanations but also to more practical concerns such as horology. Huygens and Hooke realized that the oscillations of a spring were isochronous and tried to construct clocks based on that principle.

Huygens was one among many scholars of his age who sought to combine explanations of physical phenomena in Cartesian terms with mathematical descriptions. For several years, Isaac Newton, since 1669 the Cambridge Lucasian Professor of Mathematics, also followed a similar approach. On the one hand, Newton speculated on the specific mechanisms causing gravity, and on the other he calculated that the terrestrial vortex was compressed by the solar vortex by approximately 1/43 of its width.[88]

NEWTON AND A NEW WORLD SYSTEM

The dichotomy between mathematical and physical explanations of gravity mentioned earlier was not unique to Huygens. Hooke also studied certain problems with a similar dual approach whereby mathematical and physical concerns were not always present at the same time. In his study of the motion of celestial bodies, Hooke talked of attractions and provided several inspiring comments. His analysis of the role of force in curvilinear motion differed from that of most Continental scholars. Whereas on the Continent mathematicians favored the idea of an imbalance between opposing centrifugal and center-seeking tendencies, Hooke explored the combination of a rectilinear uniform motion with a center-seeking tendency. It seems plausible that Hooke developed this approach in the mid-1660s while studying the bending of light rays. Curiously, it was the same comet of 1664 that first aroused Newton's interest in astronomy. Hooke saw an imperfect but revealing analogy between the motion of celestial bodies and that of the bob of a conical pendulum. In both cases, a central attraction deflects a body from its rectilinear path, but whereas in the conical pendulum the force increases with distance, in celestial bodies the central force was likely to decrease.[89]

Starting from a celebrated correspondence with Hooke in 1679 on falling bodies on a moving earth, Newton began working at the problem of curvilinear, and especially planetary, motion following Hooke's approach. Shortly

---

[88] Eric J. Aiton, *The Vortex Theory of Planetary Motion* (New York: American Elsevier, and London: Macdonald, 1972); and Derek T. Whiteside, ed., *The Preliminary Manuscripts for Isaac Newton's 'Principia', 1684–1686* (Cambridge: Cambridge University Press, 1989), p. x.
[89] Jim A. Bennett, "Hooke and Wren," *British Journal for the History of Science*, 8 (1975), 32–61; Ofer Gal, *Meanest Foundations and Nobler Superstructure* (Dordrecht: Kluwer, 2002).

thereafter, Hooke began to discuss mathematical problems of planetary orbits with London mathematicians Edmond Halley (1656–1742) and Christopher Wren. Thus, in the first half of the 1680s, several English mathematicians debated problems of celestial motion, such as the elliptical orbits of planets, from a mathematical standpoint, without paying immediate attention to physical causes. Moreover, they were using analogous conceptual tools, without having recourse to a Huygensian centrifugal force. Only Newton succeeded in finding an answer to the problem of the attractive force required to produce Keplerian elliptical orbits.[90]

In 1681, Newton engaged in a correspondence on the huge comet of 1680–1 with John Flamsteed (1646–1719), the Astronomer Royal at Greenwich. Initially, like most of his contemporaries, Newton believed in the existence of two comets, one approaching the sun and another regressing from it. Flamsteed made some clumsy attempt to convince him of the contrary by arguing that the comet had turned in front of the sun and that it was attracted by its magnetic virtue while approaching it and repelled when moving away, but to no avail. Newton pointed out that the comet could not possibly have turned in front of the sun but had to move behind it, and objected that the sun could not be magnetic because magnets are known to lose their power when heated. Despite Newton's rejection of Flamsteed's views, it is easy to see how crucial those views were to become in just a few years, when comets became assimilated with other celestial bodies such as planets and satellites moving under the action of universal gravity.

It is not clear when Newton attained his first result, namely that for elliptical orbits the force is inversely proportional to the square of the distance. Most likely this occurred at the time of his correspondence with Hooke, but thereafter Newton let the matter sleep. By the fall of 1684, following a visit to Cambridge by Halley, Newton produced his first tract on the subject, *De motu corporum in gyrum* (On the Motion of Bodies in a Circle), which was registered at the Royal Society. Newton was able to account also for the two other Keplerian laws of planetary motion besides the first, which states that the orbits are ellipses, where the sun occupies one of the foci. He proved that trajectories described under a central force describe areas proportional to the times, Kepler's second law, and that the squares of the revolution periods of both planets and satellites are as the third power of the major semiaxis of the ellipse – Kepler's third law.

In the following months, Newton went through an extraordinarily creative period during which he accounted for a huge number of phenomena on the basis of his inverse-square law of gravitational attraction and underwent a radical transformation in his views about nature and its creator. The

---

[90] Derek T. Whiteside, "The Prehistory of the *Principia* from 1664 to 1686," *Notes and Records of the Royal Society of London*, 45 (1991), 11–61, provides an excellent account. See also D. Bertoloni Meli, "Inherent and Centrifugal Forces in Newton," forthcoming in *Archive of History of Exact Sciences*.

mathematical and physical results of his research appeared in 1687 as *Principia mathematica*, a large 500-page book published under the auspices of the Royal Society and seen through the press by Halley. What Newton achieved in a couple of years would have more commonly been the result of as many decades and not surprisingly was to prove exceedingly challenging to most of his contemporaries and even his immediate successors.

The work starts with a set of definitions and laws. Especially prominent among Newton's definitions are those of mass, separating it conceptually from weight, and centripetal force. Newton later established with a famous experiment reported in Book 3 that weight and mass are proportional, most likely in response to Descartes, who had denied as much in Part IV of *Principia philosophiae*.[91] Centripetal force was a neologism that became a symbol of Newtonianism. Among the laws of motion, the first, known as the law of inertia, states that a body preserves its state of rest or rectilinear uniform motion unless it is acted upon, and expressed a notion that was generally accepted by 1687. The second law was valid both for attractions and for impulses and stated that the change in quantity of motion is proportional to the motive force impressed and is directed in the same line. The third law, stating that action equals reaction, was the only law Newton attempted to prove experimentally both for collisions, using pendulum bobs, and attractions, using magnetic bodies. It is equivalent to the conservation of quantity of motion in one direction, or as we would say, vectorially.[92]

Books 1 and 2 deal with the motion of bodies in spaces void of resistance and in resisting media, respectively. Book 1 is almost exclusively mathematical, whereas Book 2 provides a mathematical account of motion in resisting media and a refutation of the existence of an aethereal fluid medium filling the spaces and penetrating bodies on the earth. In Book 3, Newton moved to the system of the world and stated the law of universal gravity, according to which all parts of matter attract each other with a force inversely proportional to the square of their distance.

We have seen in the previous section that up to the late 1670s, and probably until the beginning of the 1680s, Newton subscribed to a view of nature that was dominated by subtle fluids and vortices largely inspired by Descartes' *Principia philosophiae*. Probably late in 1684 or in 1685, Newton's views changed dramatically, and he rejected those physical explanations he had followed for decades. It is likely that Newton chose the title of his work to mark his rejection of Cartesianism. His emphasis on mathematical principles highlights a key difference from Descartes. The latter developed philosophical principles, as his title suggests, starting from the principles of human knowledge. Although Descartes stated that the principles of his *physica* were

---

[91] Newton, *Principia*, new translation, pp. 403–4, 806–7. I. Bernard Cohen has provided a detailed and reliable account of the contents of the *Principia* in the guide accompanying the new translation.
[92] Newton, *Principia*, new translation, pp. 416–7.

the same as those of mathematics,[93] in practice most of the mathematics amounted to accounts of the shape and size of particles. In the work of Newton, by contrast, mathematics occupied a leading position in a very different sense. The mathematical principles prominently advertised in the title examined theoretically a series of situations. Experiments and observations then often served to select from those possible mathematical constructions those applying to the real world. At times, Newton tried to argue that this was a more secure method of inquiry in natural philosophy,[94] but elsewhere he implied that his method required due caution. In Book 2, for example, he examined the properties of hypothetical fluids composed of particles repelling each other with forces varying as a power of their distances. One of those repulsion laws led, with significant mathematical simplifications, to Boyle's law for gases, whereby the density is proportional to the compression. Newton showed, assuming that fluids consist of particles repelling each other, that the converse was also true. Yet he added a significant qualification: "Whether elastic fluids consist of particles that repel one another is, however, a question for physics. We have mathematically demonstrated a property of fluids consisting of particles of this sort so as to provide natural philosophers with the means with which to treat the question."[95] A similar reasoning could be easily applied to the universal attractive force at the center of his treatise.

By investigating more and more areas, Newton realized that the phenomena of the heavens, as well as tides and the shape of the earth, came under the compass of his inverse-square law. Celestial motions were especially significant because they had been observed for millennia: The motions of planets, for example, were known to be exceedingly regular, a sign for the astute mathematician that force decreases exactly as the inverse square of the distance. The regularity of the motions of planets and satellites, and the motion of comets in all directions, led Newton to suspect that celestial motions were not due to a fluid vortex, which would have hindered them. The initial suspicion became more and more ingrained with a pincer movement, on the one hand explaining more and more phenomena from the same assumptions and on the other showing the contradiction arising from the hypothesis of vortices.

The structure of *Principia mathematica* reflects Newton's methodological predicament. Book 1 can be seen as a carefully contrived *pars construens*, whereas Book 2 was intended as a lengthy *pars destruens*, clearing the way for his system of the world in Book 3. Book 2 ends with a refutation of Cartesian vortices, arguing that they were incompatible with Kepler's laws of planetary motion, but the entire book was geared toward an attack on vortices

---

[93] Descartes, *Principia philosophiae*, p. 2, para. 64.
[94] George Smith, "The Methodology of the *Principia*," in *The Cambridge Companion to Newton*, ed. I. Bernard Cohen and G. Smith (Cambridge: Cambridge University Press, 2002), pp. 138–73.
[95] Newton, *Principia*, new translation, pp. 588–9, Scholium to sec. 11, and pp. 696–9, quotation at p. 699. Here, by "physics" Newton meant experiment as well. See Smith, "The Methodology of the *Principia*."

even in innocent-looking parts. For example, Newton tried to determine the speed of sound to show, contrary to Robert Boyle (1627–1691), that air is the only medium through which sounds propagate and that no other medium is required.[96] By removing from the heavens the material fluid commonly believed to carry the planets, Newton left open the problem of the cause of gravity. His opinion oscillated somewhat in later years, but at the time of composition of the first edition of *Principia mathematica* it appears that Newton believed God was responsible for gravity through his presence in space. Gravity would thus be caused by an immaterial divine agent immediately present and acting on all the bodies in the universe. Although Newton was not so explicit in *Principia mathematica*, he believed he had said enough for those wishing to understand to realize that the cause of gravity he had envisaged was not material.

These preliminary observations and the contrast between Descartes and Newton highlight both the high methodological profile Newton gave to mathematics and also the range of his investigations. Fundamental as the link between ellipses and an inverse-square attraction was, it proved to be only one of the wealth of results attained by Newton in *Principia mathematica*. In Book 3, Newton put forward his demonstration about universal gravity and was able to account for the motion of planets and satellites, especially the moon, and also for the precession of the equinoxes, tides, and the motion of comets. While writing Book 3, Newton collaborated extensively with Flamsteed, who willingly provided a wealth of astronomical data on the moon, the satellites of Jupiter and Saturn, the shape of Jupiter, and the trajectory of comets.

Although Newton was acutely aware of the importance of styles and methods in mathematics, in *Principia mathematica* the main emphasis was on attaining results. Newton used a heterogeneous set of tools, including the method of first and last ratios – a form of infinitesimal geometry – series expansions, and occasionally the calculus of fluxions. One of the most remarkable features of Newton's work was its use of a wide range of mathematical tools and techniques to produce quantitative predictions and assess orders of magnitude. He did so even for such famously intractable cases as the three-body problem, namely the determination of the motions of three reciprocally attracting bodies.

READING NEWTON AND DESCARTES: LEIBNIZ AND HIS SCHOOL

Unlike Descartes, Newton made sure his *Principia* could not be read as a novel. Descartes could address in print Princess Elizabeth as the ideal reader, whereas probably the first female reader who could truly understand – and

---

[96] Newton, *Principia*, new translation, pp. 776–8. The relevant passage from the first edition is translated in a footnote. See also Boyle, *New Experiments Physico-Mechanical Touching the Spring of the Air*, experiment 27.

indeed translate into French – Newton's work was the Marquise du Châtelet (1706–1749), over a half-century after the book was first published. For a number of reasons, male readers did not fare much better. Even though Newton had kept the calculus of fluxions at the margin of the *Principia*, there was still plenty of cutting-edge mathematics in his work to make it exceedingly challenging for anyone who was not an expert mathematician. A leading philosopher such as John Locke (1632–1704) had to ask Huygens whether he could trust Newton's theorems because he was unable to assess them on his own.[97] Even Huygens and Leibniz, the two leading mathematicians on the Continent, found the work daunting. Because they were not prepared to accept universal gravity on philosophical and, in the case of Leibniz, also theological grounds, they were reluctant to follow page after page of challenging mathematics: Why go through them all if Newton's system was based on the absurd principle of attraction? Wren, too, expressed doubts about Newton's apparent rejection of a physical cause for gravity. Until the turn of the eighteenth century, the problem of a physical cause for gravity was the major concern of the few readers who could follow Newton.[98]

In many respects, reading *Principia mathematica* was colored by contemporary readings of *Principia philosophiae* and developments of Cartesianism, broadly conceived. Physical causes were not the only problem, conservation being another prominent issue. The third law of nature in Descartes' *Principia philosophiae* stated the conservation of quantity of motion in the universe and specifically in impact. Others, too, had relied on different notions of conservation in a range of contexts. Galileo, for example, had claimed that a pendulum displaced from the equilibrium position could rise back to its original height. In the second half of the seventeenth century, several scholars worked with the notion of conservation, but Newton was not among them. Whereas Descartes had seen in conservation a sign of divine order, Newton saw with equal if not greater commitment the lack of conservation in the form of a constant decay and the appearance of periodic phenomena such as comets as a sign of God's intervention and action in the world. These radically different views were prominently debated in 1716–7 by Samuel Clarke (1675–1729), a theologian allied with Newton, and the German polymath Leibniz, councilor and librarian to the Duke of Hanover. Their exchange went through Caroline, Princess of Wales (1683–1737), a woman with deep theological concerns. From our perspective here, it is worth highlighting

---

[97] Niccolò Guicciardini, *Reading the 'Principia': The Debates on Newton's Mathematical Methods for Natural Philosophy from 1687 to 1736* (Cambridge: Cambridge University Press, 1999), explores the range of mathematical methods used by Newton and the way his work was read by mathematicians. On reading Newton's *Principia* largely in an English context, see Rob Iliffe, "Butter for Parsnips: Authorship, Audience, and the Incomprehensibility of the *Principia*," in *Scientific Authorship: Credit and Intellectual Property in Science*, ed. Mario Biagioli and Peter Galison (London: Routledge, 2003), pp. 33–65.

[98] Isaac Newton, *The Correspondence of Isaac Newton*, ed. Herbert W. Turnbull, J. F. Scott, A. R. Hall, and L. Tilling, 7 vols. (Cambridge: Cambridge University Press, 1959–77), 4: 266–7. See also Bertoloni Meli, *Equivalence and Priority*.

a curious feature of Newton's views, namely his belief that motion in the universe would decay were it not for the presence of active replenishing principles such as gravity and fermentation.[99] Newton's emphasis on the decay of motion caused by the lack of elasticity of bodies means that although he disagreed with Descartes on the conservation of quantity of motion – in a Cartesian sense, not including direction – he still considered it a meaningful notion.

Leibniz was probably the most ardent advocate of conservation principles and at the same time paradoxically the most prominent critic of Descartes, because he disagreed with Descartes on what was conserved. Leibniz did not consider quantity of motion as a very significant notion, either in the Cartesian sense without direction or in the Huygensian sense with direction. Rather, he believed that he had identified the conservation of a different notion as a key law of nature. The new conservation law concerned "force," which Leibniz claimed was proportional either to the square of the speed or to the height to which a body can rise. In the case of impact, Leibnizian force was called *vis viva*, or living force, and was proportional to the body's mass and square of the velocity, or mv^2. The problem is that when the colliding bodies are not elastic, *vis viva* is not conserved. In those cases, Leibniz argued that the portion of *vis viva* that appeared to be lost was in fact absorbed by the small components of the colliding bodies. Leibniz provided no direct empirical justification for his claim, but rather he seems to have established the conservation principle in general terms as a law of nature and then found ways to apply it to all cases, including problematic ones. In 1686, Leibniz published in the *Acta eruditorum* a brief essay designed to enrage the Cartesians, *Brevis demonstratio erroris memorabilis Cartesii* (A Brief Demonstration of Descartes' Celebrated Error). His plan succeeded probably beyond his own expectations, and the controversy over the conservation of force became a major feature of mechanics in the first half of the eighteenth century.[100]

With the new century, interest shifted to Newton's mathematics for two main reasons. With the explosion of the priority dispute over the invention of calculus between Newton and Leibniz, Continental mathematicians such as Johann (1667–1748) and Niklaus (1687–1759) Bernoulli started combing through *Principia mathematica* in search of errors showing Newton's inadequate knowledge of calculus. Secondly, in 1700, French mathematician Pierre

---

[99] Samuel Clarke, *A Collection of Papers, which Passed between the Late Learned Mr. Leibnitz and Dr. Clarke in the Years 1715 and 1716* (London: Printed for J. Knapton, 1717); E. Vailati, *Leibniz & Clarke: A Study of Their Correspondence* (Oxford: Oxford University Press, 1997); D. Bertoloni Meli, "Caroline, Leibniz, and Clarke," *Journal of the History of Ideas*, 60 (1999), 469–86; and Isaac Newton, *Opticks* (New York: Dover, 1952), Query 31, pp. 397–401, at p. 398.

[100] An excellent account is Daniel Garber, "Leibniz: Physics and Philosophy," in *The Cambridge Companion to Leibniz*, ed. Nicholas Jolley (Cambridge: Cambridge University Press, 1995), pp. 270–352.

Varignon (1654–1722), a member of the Paris Académie Royale des Sciences, started publishing a series of memoirs on motion under central forces and in resisting media where he translated several propositions by Newton into the language of the differential calculus. Leibniz had published the key rules of calculus in a series of essays in the German journal *Acta eruditorum*, starting with the celebrated *Nova methodus pro maximis et minimis* (New Method for Finding Maxima and Minima, 1684). Although Varignon's work did not contain new results and was largely dependent on Newton, he made some of Newton's results more accessible by systematizing mathematical procedures and notation. Thus his originality cannot be assessed so much with new theorems as with a new style to deal with the science of motion using the differential calculus. In particular, Varignon wrote equations of motion where time appears prominently rather than being swallowed by other symbols.[101] Varignon was careful to treat Newton's work in a purely mathematical fashion, maintaining a noncommittal attitude toward the issue of physical causes. He managed to remain on good terms with both the Newtonian and Leibnizian camps. Despite the talent of Continental mathematicians, the effectiveness of the differential calculus, and their tireless efforts, it is probably fair to say that the results they achieved in competition with *Principia mathematica* were negligible, amounting to the correction of a few inaccuracies and the tightening of some theorems.

Continental mathematicians, however, also worked on other themes relevant to mechanics, such as the study of new curves described by bodies under given conditions and elasticity. Prominent among the new curves were the catenary, described by a chain hanging perpendicular to the horizon from two nails fixed in a wall, and the curve of fastest descent between two points, which remarkably turned out to be the same cycloid that Huygens found to make pendulum oscillations isochronous. Jakob Bernoulli's works on the elastic beam were especially noteworthy.[102]

In 1713, Newton published a second edition of his *Principia*, and the third edition followed in 1726. The second was seen through the press by the Cambridge Plumian Professor of Astronomy Roger Cotes (1682–1716), a very talented mathematician, who transformed large portions of the work and corrected several mistakes. Cotes was the sort of editor whose letters authors open with a shaky hand. He combed the text with unparalleled acumen and patience and never let the matter rest even when Newton made it clear that he was unwilling to embark on major revisions. Newton reshaped large sections of Book 2 especially and performed many new experiments on the motion of bodies in resisting media. From the standpoint of the role of experiments, the second edition – and the third, too – appears like a different book.

---

[101] See Michael Blay, *La naissance de la mécanique analytique* (Paris: Presses Universitaires de France, 1992), for an account of Varignon's achievements and bibliography.
[102] Benvenuto, *Structural Mechanics*, ad indicem.

Similar considerations apply to astronomical data coming from Flamsteed and others, especially on the motion of the moon and comets.

These features of *Principia mathematica* highlight a major difference between Newton's own views of his work and the way Continental mathematicians read it. Newton saw his own work as ultimately relevant for understanding nature. In Book 2, he embarked on elaborate experiments on resisting media, as we have seen, and in Book 3 he needed extensive and up-to-date astronomical data for his system of the world. Despite the bitter and acrimonious dispute with Flamsteed over the use of astronomical data, the coming together of mechanics and astronomy mirrored in the collaboration with astronomers was a novel feature of Newton's work. Cotes and Halley can also be seen as taking part in this enterprise with their interest in tides and comets. No contemporary Continental mathematician showed a comparable interest and proficiency at joining mechanics and astronomy or in engaging in collaborative work with astronomers, as is evident from relevant works by Huygens, Leibniz, Jakob Bernoulli, Johann Bernoulli, Niklaus Bernoulli, and Varignon. When the Padua professor of mathematics Jakob Hermann (1678–1733) published his *Phoronomia* (Law-bearer, 1716), a mathematical work in two books on the motion of bodies in nonresisting and resisting media, he did not include a third book on the system of the world. It was only later in the eighteenth century that mathematicians such as Alexis Clairaut (1713–1765), Jean d'Alembert (1717–1783), and Leonhard Euler (1707–1783) relied on and collaborated with astronomers such as Joseph Delisle (1688–1768), Pierre Charles Le Monnier (1715–1799), James Bradley (1693–1762), and Tobias Mayer (1723–1762) and started producing novel results. Up to that period, around the 1740s, *Principia mathematica* remained the leading work on celestial mechanics.

The transformations of mechanics during the seventeenth century can be characterized in different ways, but one of the most significant was undoubtedly the audience reached. Starting early in the century as a rather narrow field encompassing the science of machines and the science of equilibrium of bodies and liquids, mechanics was studied mainly by mathematicians and engineers. By the turn of the eighteenth century, it had merged with large portions of natural philosophy and, despite its growing mathematical complexity, had become an important subject of study and debate among philosophers and theologians, among scholars interested in cosmology and the system of the world, and in the universities.

27

THE MECHANICAL ARTS

Jim Bennett

In the sixteenth and seventeenth centuries, the term "mechanical" had three main senses – all interconnected and all relevant to the history of science. The traditional meaning referred to activities that were practical or manual. In the sixteenth century, the word acquired a new meaning, a revival of a classical sense, connecting it specifically to machines and their design and management. Finally, in the seventeenth century, "mechanical" came also to refer to a doctrine about the natural world. The phrase "mechanical arts" – *artes mechanicae* in postclassical Latin – had equivalents in a number of European languages. When linked to the first two of these senses, it referred to the skillful practice of a particular practical discipline or handicraft, including the working of machines.

The disciplinary relationships and boundaries observed in the activities and writings of contemporary practitioners of the mechanical arts confirm the relationship of machinery to a wider context of practical work. They also show the importance, increasing over the course of the sixteenth and seventeenth centuries, of bringing mathematics into the characterization of the "mechanical." This is both because the design and management of machines came to be regarded as a mathematical art and because mathematics became engaged in a range of other practical work. It would be difficult and anachronistic, for example, to define a boundary between the mechanical arts and practical mathematics, as carried on by people we might more readily call "mathematical practitioners" than mechanicians or mechanics, as the practical mathematical disciplines, such as architecture, engineering, gunnery, and surveying (often referred to in English as "the mathematicals"), were directly concerned with machines.

When John Dee (1527–1608) tried to capture and characterize the shifting field of mathematics in his famous preface to the English translation of Euclid's *Elements* in 1570, he sought to pin down this slippery terminology so that he could delineate more precisely and carefully his understanding of the role of mathematics in a range of practical arts.[1] For Dee, an "Art Mathematical deriuatiue" was a practical discipline whose methods were grounded in arithmetic and geometry, whereas he tried to restrict the term "Mechanicien" or "Mechanicall workman" to someone whose work was carried out with the requisite skill but with no knowledge of mathematical demonstration. At the same time, he acknowledged that this distinction ran against common usage, which tended to conflate the two: "Full well I know, that he which inuenteth, or maketh these demonstrations, is generally called A speculatiue Mechanicien: which differreth nothyng from a Mechanicall Mathematicien." Already the meanings of mechanical and mathematical – at least at the practical level, the level of mechanical art and mathematical art – were thoroughly entangled.[2]

Dee's acknowledgment that the "speculative mechanician" was typically identified with the "mechanical mathematician" is revealing. A mechanician might be thought of as one with artisanal knowledge and skill, Dee's "mechanical workman." If he is "speculative," he has moved beyond the merely empirical to engage with what we might call theoretical, analytical, generalizing, or inventive aspects of his work or – to use the terminology of the period – his art. Mathematics, on the other hand, has its mechanical aspect; that is, it can be practical in its activities and utilitarian in its outcomes. The areas where mathematics can be practical are numerous, as the developments of the sixteenth century were to prove, including the design, construction, and use of machines, as well as other arts. Thus mathematics could be "mechanical" in two different but related ways: generally, by being practical; and specifically, in dealing with machinery. Thus the speculative mechanician "differed nothing" from the mechanical mathematician because mathematics – in particular, geometry – was the vehicle for the theoretical or speculative aspects of his activity. Mathematics was the way to move beyond the empirical, to secure practice in some systematized and generalized account, to ground an art in a structure of assured knowledge or (in the terminology of the period) a "science."

The mechanical art of machines was one of the first areas of early modern practice to embrace what might be called the "mathematical program," wherein the speculative aspect of a discipline takes a mathematical form. The character and content of the program were shaped within the development of this early instance, namely, the art of mechanics. As it became more clearly formulated and self-aware, over the course of the sixteenth century, mathematical practitioners found opportunities to apply the same program to other

---

[1] Euclid, *The Elements of Geometrie of the Most Auncient Philosopher Euclide of Megara*, trans H. Billingsley (London: John Daye, 1570), preface by John Dee, sigs. aiiijr–aiiijv.
[2] For a valuable discussion of the meaning of terms in the period, see Alan Gabbey, "Between *Ars* and *Philosophia naturalis*: Reflections on the Historiography of Early Modern Mechanics," in *Renaissance and Revolution: Humanists, Scholars, Craftsmen, and Natural Philosophers in Early Modern Europe*, ed. Judith V. Field and Frank A. J. L. James (Cambridge: Cambridge University Press, 1993), pp. 133–45.

practical arts, which would be revised and refounded through mathematics. Advocates of reform promised that they would become more efficient and reliable, and their results more confident and secure. They would become mathematical arts grounded on mathematical science.

The mathematics that would effect this change had a number of characteristics that may seem surprising today and indeed may not appear to be "mathematical" at all. It is important to accept this early modern mathematical discipline on its own terms. Historians of mathematics have tended to look for past work that conforms to modern mathematical standards, and their understanding of "the mathematicals" has suffered from this kind of prejudice.[3] Instruments, for example, played an important and ubiquitous role across the early modern mathematical arts – not simply mathematical instruments in our restricted sense, applied to drawing and calculation, but also those adapted to astronomy, navigation, surveying, warfare, and other practical activities of this sort. It seems as though rendering a new technique accessible through designing an instrument was part of what it meant to do mathematics in sixteenth- and seventeenth-century Europe.

At the beginning of the sixteenth century, mathematics, especially geometry, was understood to be a mode of action in disciplines that were important to specialized practitioners and to the cities and states they served, primarily in Italy. The field grew over two centuries as further mechanical arts took on the character of practical mathematics, and as additional nations, courts, armies, civil administrations, and individual entrepreneurs acquired familiarity with this geometry of action. As a result of this success of the mathematical program, a range of arts made the transition to mathematical practice, sharing the rhetoric of reform through mathematics and adapting common geometrical and instrumental techniques to particular areas of work.

Although this mathematics engaged with material things, it was not understood as having any relevance to natural philosophy, or the explanation of natural processes and phenomena. In terms of the academic hierarchy of learning, natural philosophy was a separate and superior discipline. The nature and structure of the world was understood through the teachings of Aristotle; this "science" – in the sense of a secure, systematized account – was not expected to involve geometry (see the following chapters in this volume: Garber, Chapter 2; Andersen and Bos, Chapter 28).

In some respects, this disjunction from natural philosophy was advantageous to the imaginative development of practical mathematics, which aimed to be reliable, efficient, and useful but not to give insights into the causal or material nature of things. These mathematical arts could be true, albeit according to a different set of criteria: Their truth was not measured

---

[3] Jim Bennett, "Geometry in Context in the Sixteenth Century: The View from the Museum," *Early Science and Medicine*, 8 (2002), 214–30.

by verisimilitude with nature but rather by efficacy in practice, and their principles and practices did not need to be consistent with the doctrines of the natural philosophers. Eventually, however, practical mathematicians built such an extensive network of mathematical practice that their engagement with natural philosophy became inevitable. Through this engagement, the techniques of practical mathematics – mechanical manipulation, mathematical generalization, and the use of instruments – came to be applied to the study of the natural world.

The successes and ambitions of the mechanical arts impressed several contemporary witnesses who were indifferent to the mathematical program, and some, most influentially the English statesman and philosopher Francis Bacon (1561–1626), even suggested that natural philosophy be reformed on their model. In his survey of the state of the sciences in *The Advancement of Learning* (1605), Bacon called for a supplement to traditional natural history in the form of a comprehensive description of what he called "Nature wrought or mechanical." He was confident that this "History Mechanical" would not only be of practical value but also "give a more true and real illumination concerning causes and axioms" sought by natural philosophers.[4] He drove the point home in his blueprint for a renovated natural philosophy, the *Novum organum* (New Organon, 1620), favorably comparing the recent progress of the mechanical arts with the stasis of natural philosophy.[5] For Bacon and a growing number of other seventeenth-century natural philosophers, such as René Descartes (1596–1650), the rising fortunes of the mechanical arts served to undermine the Aristotelian distinction between art (in the sense of things made by humans) and nature. Bacon objected strenuously to this ancient opposition in the context of his plan for a history of the mechanical arts:

> [W]e willingly place the history of arts among the species of natural history, because there has obtained a now inveterate mode of speaking and notion, as if art were something different from nature, so that things artificial ought to be discriminated from things natural, as if wholly and generically different; . . . Whereas, on the contrary, that ought to be sunk deep that things artificial do not differ from natural in form or essence but in efficients only; that in reality man has no power over nature, except that of motion, namely to apply or remove natural bodies; but nature performs all the rest within herself.[6]

In this description, the manipulations of the practitioners of the mechanical arts exploited natural causes and were therefore of direct relevance to the causal explanations of natural philosophy.

---

[4] Francis Bacon, *The Advancement of Learning*, in *The Works of Francis Bacon*, ed. Basil Montagu, 16 vols. (London: William Pickering, 1825–34), 2: 105–6.
[5] Francis Bacon, *Novum organum*, 1.74, in *Works*, 9: 225.
[6] Francis Bacon, "Description of the Intellectual Globe," in *Works*, 15: 153–4. I owe the example of Bacon's attitude toward this question to the volume editors.

THE MECHANICAL ARTS IN 1500

The robust, worldly character of this area of mathematical work, as it took shape in the sixteenth century, was in large part the legacy of fifteenth-century Italian architect-engineers. The work of building was closely linked with the design and management of large machines because many kinds of hoists and cranes were required for moving, lifting, and positioning stone and other materials in substantial building projects. The work of architects went far beyond the limited responsibility for the design of a building to include the planning and management of its construction. Both civil and military architecture and engineering provided opportunities for talented practitioners to sell their services to ambitious princes bent on increasing their territories, revenues, and prestige. Hydraulics (the management of rivers, canals, and aqueducts), the building of bridges, and the draining of land all fell within the technical aspects of governance in war and peace. The architectural, pyrotechnical, and mechanical skills that might deliver victory in warfare or respect in civil administration could also be turned to spectacles in pageants and festivals that reinforced and sustained the power of the successful prince or republic in the fragmented Italian political world (see DeVries, Chapter 14, this volume). In a climate where rulers competed for specialist skills and for the advantage offered by technical innovation, individuals could become known for particular achievements or establish reputations through written compilations describing established or novel techniques.[7] This contrasted markedly with earlier periods, in which master masons or military bombardiers labored in the anonymity characteristic of their crafts.

One of the earliest of these architect-engineers, Filippo Brunelleschi (1377–1446), achieved celebrity through meeting one of the most famous and prominent technical challenges of the Renaissance – throwing a dome over the unfinished crossing of the cathedral in Florence without using central scaffolding. Brunelleschi worked in the design of machines, in fortification, in hydraulics, in theatrical effects, in perspectival painting, and, according to one unconfirmed early authority, in clockwork. This early reference to clockwork as part of the design and management of machines is noteworthy because the designs of clocks and other automata was to feature prominently in the work of later practitioners, such as the Florentine artist and engineer Leonardo da Vinci (1452–1519).[8]

Many fifteenth-century artist- and architect-engineers followed in the tradition of Brunelleschi, and their achievements include not only palaces, bridges, fortifications, aqueducts, navigable waterways, and machines of war

---

[7] Pamela O. Long, *Openness, Secrecy, Authorship: Technical Arts and the Culture of Knowledge from Antiquity to the Renaissance* (Baltimore: Johns Hopkins University Press, 2001), pp. 175–243.
[8] Paolo Galluzzi, *Renaissance Engineers from Brunelleschi to Leonardo da Vinci* (Florence: Giunti, 1996).

but also drawings, paintings, and treatises, in which they recorded their work. These men – examples include Marino di Iacopo, known as Taccola (1382–1458) and Francesco di Giorgio (1439–1501), as well as Leonardo – produced texts of a particular kind, where drawings were not embellishments or ornaments but were indispensable to the content, integrity, and effectiveness of the whole. Buildings and machines, as well as fortresses, clocks, and instruments, could not be recorded without drawing, and the searching observation required for accurate representation became part of a distinctive methodology of understanding and recording.[9] The tradition of treatises on engineering, initially associated with fifteenth-century Siena, continued there into the sixteenth century, spreading also to nearby Florence, which saw a vogue for printed compilations of mechanical illustrations known as "theaters of machines."[10] In this mathematical culture, viewing and drawing became linked as cognitive tools within a profession whose business was the management of mechanical action in the physical world.

A species of mathematics was shaped by the requirements of the architect-engineers. They needed to work on paper, whether to offer a project to a patron, to design a complex machine or an innovative piece of civil engineering, to explain a mechanical solution to a problem in civil or military matters, or to communicate designs to fabricators and constructors. They had to deal in measurement and needed to epitomize their designs in a manner that was consistent and unambiguous. For a machine to work, every part of the specification had to be congruent with the whole, which encouraged the technique of drawing to scale. Drawing in perspective was also a professional asset to the architect-engineer and was achieved by adapting the kind of geometrical projection already used in the geometry of the astrolabe.[11] There were ancient antecedents for these geometrical techniques, in the works of the ancient Greek astronomer and cosmographer Ptolemy, for example, and particularly in the *De architectura* of the ancient Roman architect-engineer Vitruvius, which treats buildings and machines together; this treatise survived without drawings, the absence of which was itself a stimulus not only to master but also to supplement this ancient source with illustrations that reconstructed Vitruvius's intent.[12] While the development of new techniques of recording and systematizing was an asset to individual careers and reputations, it also elevated the status of the discipline to something more distinctive and orderly.

---

[9] Paolo Galluzzi, "Art and Artifice in the Depiction of Renaissance Machines," in *The Power of Images in Early Modern Science*, ed. Wolfgang Lefèvre, Jürgen Renn, and Urs Schoepflin (Basel: Birkhäuser, 2003), pp. 47–68; and Wolfgang Lefèvre, "The Limits of Pictures," in ibid., pp. 69–88.
[10] A. G. Keller, "Renaissance Theaters of Machines," *Technology and Culture*, 19 (1978), 495–508.
[11] Samuel Y. Edgerton, *The Renaissance Rediscovery of Linear Perspective* (New York: Basic Books, 1975), pp. 37–9, 92–104; Martin Kemp, *The Science of Art* (New Haven, Conn.: Yale University Press, 1990), pp. 9–98; and Kim H. Veltman, with Kenneth D. Keele, *Linear Perspective and the Visual Dimensions of Science and Art* (Munich: Kunstverlag, 1986), pp. 19, 42–4.
[12] Galluzzi, "Art and Artifice," p. 47.

From these components – machines, drawing, mathematics, building, representation, and warfare – held together in both the careers and the treatises of the engineers, grew a buoyant and confident domain of activity over the course of the sixteenth century that can loosely be called "practical mathematics." (In Dee's words, "a speculatiue Mechanicien . . . differreth nothyng from a Mechanicall Mathematicien.") The vigor of this discipline was bolstered by its enthusiastic use of printing and encouraged by warfare and commercial and territorial ambition. Instruments came to play an increasingly significant role, with the appearance of many printed books on their design and use and with the establishment of important commercial workshops to produce them. The publishing genre was established by Johann Stöffler in 1513 with his book on the astrolabe,[13] which set a pattern of dealing with construction and application. Among the celebrated workshops of the period were those founded by Lorenzo della Volpaia (1446–1512) in Florence, Georg Hartmann (1489–1564) in Nuremberg, Reiner Gemma Frisius (1508–1555) in Louvain, Christoph Schissler (ca. 1531–1608) in Augsburg, Thomas Gemini (ca. 1510–1562) in London, Philippe Danfrie (ca. 1532–1606) in Paris, and Michael Coignet (1549–1623) in Antwerp. The role of geometry in grounding and improving the arts was established and extended, and its embrace soon included mechanics, architecture, shipbuilding, painting, gunnery, fortification, surveying, navigation, cartography, and horology. If the list seems ambitious, it is substantiated in the careers of any number of practitioners.

CLOCKS AND OTHER CELESTIAL INSTRUMENTS

Mechanical clocks are machines on a smaller scale than those used in architecture, hydraulics, or warfare, but their inclusion within the mechanical arts immediately links this field to astronomy as a branch of practical mathematics. Early clocks were astronomical machines, tracking and displaying the phenomena that depended on processes in the heavens. The first clocks were mechanical astrolabes – the astrolabe being a traditional instrument, circular in shape, that performed a number of functions, including telling time by measuring the altitude of the sun – where the celestial planisphere or *rete* of the astrolabe, which rotates on top of it, contains a circle that is the annual path of the sun (the ecliptic) through the pattern of the fixed stars. Because the astrolabe indicates the sun's position in the sky, moving along the ecliptic as the ecliptic is carried around by the machine once a day, the time is easily registered by such a clock, even though it displays much additional information about the heavens as well.[14] Other celestial machines were even

---

[13] Johann Stöffler, *Elucidatio fabricae ususque astrolabii* (Oppenheim: J. Köbel, 1513).
[14] Anthony John Turner, *Astrolabes: Astrolabe Related Instruments* (The Time Museum, vol. 1, pt. 1) (Rockford, Ill.: The Time Museum, 1985), pp. 1–57.

more ambitious and included the planets as well as the sun and stars – that is, they were self-moving versions of the astronomical instrument known as the equatorium. A famous example was the "astrarium" of Giovanni de' Dondi (1318–1389), built in Padua in the fourteenth century and admired well into the sixteenth.[15] Still other machines might dispense with both the stars and the planets and, mimicking only the progression of the sun, simply present the solar time. These last we are inclined simply to call clocks, but they depended no less on astronomical phenomena than their more ambitious counterparts.

In the seventeenth century, mechanical horology and practical mathematics maintained their close association. According to the Italian mathematician Vincenzo Viviani (1622–1703), it was in 1641 that Galileo Galilei (1564–1642), who had earlier been interested in the pendulum as exemplifying a species of motion, had the idea of using it to regulate the timekeeping of a clock. Viviani made this claim, buttressed by his own drawing of Galileo's design – still the only record of Galileo's invention – in the context of a posthumous priority claim on Galileo's behalf after Christiaan Huygens (1629–1695) had published his design for a pendulum clock in 1658.[16] Viviani's account of Galileo's idea recalls Bacon's insistence that the categories of art, especially the mechanical arts, and nature be merged. According to Viviani, Galileo was "hoping that the very even and natural motions of the pendulum would correct all the defects in the art of clocks."[17] It became accepted that there was something "natural" in the motion of a pendulum, so that it was later proposed as the basis of a "natural" standard of length. In the original suggestion for a pendulum clock, we already see the notion that whereas clockwork exhibits all the defects of art – all the inadequacies and imperfections of the artificial work of man – it can be corrected and regulated by nature. In this way, the pendulum clock challenged the traditional conceptual scheme that opposed nature and art as distinct ontological categories; the assimilation of a patently artificial machine such as the pendulum to nature would previously have been unthinkable.

Huygens, like Galileo, knew that a pendulum swinging in a circular arc was not fully isochronous – a fact that limited its utility as a timekeeping device. To correct this irregularity, Huygens devised a general method of obliging the bob of the pendulum to take a different path: to swing higher by making its suspension string wind itself along the surfaces of two curved "cheeks" extending from the point of suspension for the vertical pendulum.[18]

---

[15] For the astrarium and other elaborate clocks of the Renaissance, see H. C. King and J. R. Millburn, *Geared to the Stars: The Evolution of Planetariums, Orreries and Astronomical Clocks* (Bristol: Adam Hilger, 1978), esp. chap. 3. On the equatorium, see ibid., pp. 19–20.
[16] Silvio A. Bedini, *The Pulse of Time* (Florence: Leo S. Olschki, 1991).
[17] Stillman Drake, *Galileo at Work: His Scientific Biography* (New York: Dover, 1995), p. 419.
[18] Michael S. Mahoney, "The Measurement of Time and Longitude at Sea," in *Studies on Christiaan Huygens*, ed. H. J. M. Bos, M. J. S. Rudwick, H. A. M. Snelders, and R. P. W. Visser (Lisse: Swets

The curve he was aiming at – the truly isochronous arc for all amplitudes of swing, as announced in his *Horologium oscillatorium* (Oscillatory Clock) of 1673 – was the cycloid, and he was able to prove that for this to be achieved, the cheeks themselves had to be cycloids. Huygens then linked this truly isochronous pendulum to the escapement of the clock in such a way that it both controlled the release of the force transmitted by the weight that provided the force to drive the clock through the train of wheels and was sustained in its motion by this force.

One reason why Huygens sought a general solution to the problem of "circular error" (the nonisochronous motion of a circular pendulum) was his ambition to build a longitude clock as a navigational instrument, as motion at sea would certainly have interfered with the regular amplitude of a swinging pendulum. So the mechanical art of clockwork remained within the domain of practical mathematics in two respects: Design developments continued to be informed by geometry, and the outcome was to have been an instrument of navigation. Huygens went on to reinforce the link between horology and practical mathematics through his work on a spring-regulated watch, which seemed a more plausible seagoing timekeeper. Here he clashed with the doyen of English experimental mechanics, Robert Hooke (1635–1703), who also was developing a spring-regulated watch for finding longitude at sea.[19]

The Latin word *horologium* is ambiguous, referring to either a clock or a sundial. That can lead to uncertainties in interpreting early texts, but it also indicates the close connection between the two. Throughout the sixteenth and seventeenth centuries, makers of astronomical and other instruments were also often involved with designing and producing clocks. Examples include the workshop of the della Volpaia family of Florence at the beginning of the period, and, toward its end, the celebrated London clock and watch maker Thomas Tompion (1639–1713), who also worked for Hooke on astronomical instruments.[20] Jost Bürgi (1552–1632), one of the most talented clockmakers of the age, also made astrolabes, globes, armillary spheres, quadrants, sextants, and surveying instruments while in the service of Landgrave Wilhelm IV in Kassel and Emperor Rudolf II in Prague.[21]

The design and manufacture of sundials was one of the major preoccupations of practical mathematicians in the sixteenth and seventeenth centuries. It is also the area where the priorities of the period are most conspicuously ignored by modern historians. Although there is a network of dialing enthusiasts, mainstream historians of science have remained largely unmoved by

---

and Zeitlinger, 1980), pp. 234–70; and J. G. Yoder, *Unrolling Time: Christiaan Huygens and the Mathematization of Nature* (Cambridge: Cambridge University Press, 1988).

[19] M. Wright, "Robert Hooke's Longitude Timekeeper," in *Robert Hooke: New Studies*, ed. Michael Hunter and Simon Schaffer (Woodbridge: Boydell Press, 1989), pp. 63–118.

[20] Robert W. Symonds, *Thomas Tompion, His Life and Work* (London: Spring Books, 1969).

[21] Klaus Maurice, "Jost Bürgi, or On Innovation," in *The Clockwork Universe: German Clocks and Automata, 1550–1650*, ed. Klaus Maurice and Otto Mayr (New York: Neale Watson, 1980), pp. 87–102.

the commitment to this field of so many eminent early mathematicians, including Johannes Regiomontanus (1436–1476), Peter Apian (1497–1552), and Sebastian Münster (1489–1552). A great many books were published on new designs for sundials, and encyclopedic works on mathematics would always include a substantial section on dialing. The common horizontal or garden dial, which shapes and limits modern understandings of the subject, is an impoverished relic of a geometrical discipline that was formerly complex, challenging, and enormously inventive.[22]

Most sixteenth- and seventeenth-century European sundials were portable, serving the function of everyday timekeeping. They generally marked either the hour angle (right ascension) of the sun or its daily progress in altitude. They might work in a single latitude, in a restricted range of latitudes, or be universal (adjustable to any latitude). The plane of the hour lines could be horizontal, vertical, equinoctial (parallel to the equator), or in some other inclined plane. Sundials could tell the time in one or more of a variety of hour systems, including "unequal hours," where the periods of daylight and darkness were each divided into twelve; "common hours," where two sets of equal twelve-hour divisions began at midday and midnight; "Italian hours," where twenty-four hours began at sunset; and "Babylonian hours," where they began at sunrise. They might rely on a magnetic needle to find the meridian or use some means of self-orientation. In addition to these technical parameters, mathematicians might arrange for the pattern or "projection" of hour lines to take a particular shape – such as the "poplar leaf" of Peter Apian[23] – as a technical or aesthetic challenge, or to honor a patron by replicating some heraldic device. In other words, there was great scope for variety and invention, and this was exploited to the fullest.

The field continued to develop into the seventeenth century, when some of the most popular dials were the work of significant mathematicians. Edmund Gunter (1581–1626), professor of astronomy at Gresham College, who brought logarithmic calculation into everyday use through the design of logarithmic rules and devised instruments for navigation and surveying, also invented a new altitude quadrant that became widely used and that included many astronomical functions in addition to telling time.[24] His contemporary William Oughtred (1575–1660), whose "circles of proportion" were an alternative form of logarithmic calculator, published a universal equinoctial ring dial that remained very popular well into the eighteenth century.[25] Sundials were often included alongside other instruments in multifunctional devices

---

[22] Penelope Gouk, *The Ivory Sundials of Nuremberg, 1500–1700* (Cambridge: Whipple Museum, 1988); and H. Higton, *Sundials at Greenwich* (Oxford: Oxford University Press, 2002).
[23] Peter Apian, *Folium populi* (Ingolstadt, 1533).
[24] Edmund Gunter, *De sectore et radio* (London: William Jones, 1623).
[25] W. Oughtred, *The Circles of Proportion and the Horizontal Instrument* (London: Augustine Mathewes for Elias Allen, 1632).

known as compendia. For example, it was attractive to combine a sundial with a nocturnal, which would tell the time at night from the orientation of the pattern of stars around the pole.[26]

Such work should be treated under the mechanical arts for a variety of reasons. For one thing, the practitioners themselves saw dialing as integrally connected with the other arts. As one of the most consumer-oriented products of practical mathematics, dials sustained the instrument workshops that were essential to the practice of the mathematical arts as they were understood in the period because instrumentation, including design, manufacture, and use, was considered a central and characteristic feature of practical mathematics. Furthermore, some of the largest sundials were directly linked to the work of the architect-engineers in that they were parts of buildings. The Florentine astronomer and astrologer Paolo dal Pozzo Toscanelli (1397–1482), an associate of Brunelleschi, set up a meridian line in the north transept of the cathedral in Florence in 1475; this received light from a hole in the recently completed lantern that topped Brunelleschi's dome. The Dominican mathematician Egnatio Danti (1536–1586) began work on a meridian line in the Florentine church of Santa Maria Novella in 1574 and completed one in the Bolognese cathedral of San Petronio in 1576.[27] Such instruments were not, of course, for everyday timetelling but for making more precise and specialized observations to improve the calendar; marking the noon projection of the sun's light at different positions on the line in an annual cycle allowed observers, for example, to determine the intervals between the equinoxes.

MATHEMATICAL AND OPTICAL INSTRUMENTS

Sundials were not the only instruments to carry the practical ambitions of mathematicians and mechanics in early modern Europe and to offer significant evidence of the development of the mechanical arts. Over the course of the sixteenth century, practical mathematics saw the appearance of many new devices, as well as a burgeoning of publications and a significant growth in the number of workshops and in their geographical distribution. In Nuremberg, for example, the legacy of Regiomontanus encouraged an astronomical focus among some skilled craftworkers, and the mathematical workshops of Florence, Rome, Paris, and Antwerp indicate the importance of city commerce for a successful instrument business. In Louvain, we see the influence of able and innovative individuals, notably Gemma Frisius and Gerard Mercator (1512–1594). By the late sixteenth century, the commercial

---

[26] For the nocturnal, see Higton, *Sundials at Greenwich*, pp. 387–406.
[27] John L. Heilbron, *The Sun in the Church: Cathedrals as Solar Observatories* (Cambridge, Mass.: Harvard University Press, 1999), pp. 68–81.

trade in instruments had spread to London, stimulated by the immigration of trained practitioners from the Continent.[28]

In addition to novel types of sundials, new sixteenth-century instruments included surveying tools, rangefinders for warfare, gunner's sights and calculating devices, angle-measuring instruments for seamen, and drawing instruments and calculating tools for all sorts of practitioners. Each of these categories saw the appearance of new instruments, whose designers announced their advantages in practical handbooks on their construction and use. Their makers were specialist craftsmen who formed an identifiable trade: makers of mathematical instruments. Their skills lay mostly in metalwork (though some worked in wood and others in ivory), in engraving, and in aspects of geometry. The instruments they created offered solutions to practical problems in the mathematical and mechanical arts; they were not tools of discovery, and it was none of their business to reveal the truths of nature.

The coherence of this discipline, despite the variety of its practical outcomes, is seen not only in the careers of the practitioners but also in their shared material and geometrical techniques. Instruments were adjusted to different contexts; the astronomer's astrolabe, for example, was simplified and adapted to both navigation and surveying, as was the astronomical cross-staff. The technique of geometrical projection, to which I have already alluded several times, was one of the familiar and transferable techniques of the mathematical arts. In a sundial, the hour lines represent the projection of the motion of the sun through a style (gnomon) or a point onto a specified surface. In the astrolabe, the celestial sphere – the positions of the stars and the apparent annual path of the sun, together with such local features as the observer's horizon and zenith – are projected onto the plane of the equator. Several different astrolabic projections were introduced to western astronomers in the sixteenth century as ways of making the astrolabe "universal"; that is, capable of solving astronomical and astrological problems for any latitude. Similar projective techniques formed the basis of the new mathematical art of drawing in perspective, where the point of projection was the observer's eye and the plane of projection was the picture itself. Projection was also fundamental to cartography, as we shall see. It was the common employment of such intellectual tools and techniques that sustained and unified the mathematical arts into a single domain of practice and allowed its practitioners

---

[28] Anthony John Turner, *Early Scientific Instruments, Europe 1400–1800* (London: Sotheby's, 1987); Jim Bennett, *The Divided Circle: A History of Instruments for Astronomy, Navigation, and Surveying* (Oxford: Phaidon Christie's, 1987); and Gerard L'Estrange Turner, *Elizabethan Instrument Makers: The Origins of the London Trade in Precision Instrument Making* (Oxford: Oxford University Press, 2000).

to range across projects as diverse as mapmaking, dialing, and perspective drawing.[29]

In the seventeenth century, practical mathematics continued to extend its range of instruments, adding, among others, logarithmic slide rules and other forms of calculators, new types of surveyor's levels, and novel navigational instruments such as the backstaff. At the same time, however, a different class of instruments appeared from within another branch of the mechanical arts. Spectacle makers had long ground and polished both convex and concave glass lenses as aids to vision, but in 1608 two rival craftsmen from Middelburg petitioned the States-General of the Netherlands for patent rights to a different use of their lenses: One concave and one convex lens, it appeared, placed at either end of a tube, could make distant things appear closer (see Donahue, Chapter 24, this volume). The military advantages of such a device were obvious, but the States-General was unable to determine which of the claimants should receive the patent, and none was ultimately granted.[30]

The telescope, and shortly afterward the microscope, moved the ambitions of instrument users into a new arena. It appeared that optical instruments, unlike the traditional instruments of mathematics, could indeed enhance natural knowledge, contribute to questions of natural philosophy, and make discoveries in the natural world. The telescopes were evidently objects of art – they had been made by mechanical artists – which raised the traditional question: Could they engage with nature in a reliable and truthful way? Such, however, was the effectiveness of Galileo's promotion of his telescopic discoveries, in his *Sidereus Nuncius* (Sidereal Messenger, 1610), and his insistence on the relevance of his observations of the heavens to the cosmological debate of the time, that the telescope rapidly became a central tool not for the traditional astronomy of measurement (at least at first) but for the natural philosophy of the heavens.[31]

From this point on, natural philosophers concerned with cosmology would be obliged to engage with the mechanical arts, even if this meant only the purchase of telescopes from makers. For others it meant more: either working in direct collaboration with a lens maker, or even grinding and polishing for themselves. Despite the ambitious catholicity of a mathematician and natural philosopher such as Galileo, the epistemological distance between the disciplines of mathematics and natural philosophy remained at issue, while for the tradesmen themselves there was no connection between the

---

[29] Jim Bennett, "Projection and the Ubiquitous Virtue of Geometry in the Renaissance," in *Making Space for Science: Territorial Themes in the Shaping of Knowledge*, ed. C. Smith and J. Agar (London: Macmillan, 1998), pp. 27–38.
[30] Albert van Helden, "The Invention of the Telescope," *Transactions of the American Philosophical Society*, 67 (1977), pt. 4.
[31] Galileo Galilei, *Sidereus Nuncius; or, the Sidereal Messenger*, trans. Albert van Helden (Chicago: University of Chicago Press, 1989).

two separate types of arts. Makers of mathematical instruments had long been established as a specialist trade, and the spectacle makers, who were diversifying into optical instruments, belonged to a separate practice that was carried on in different workshops and regulated by different guilds or companies. Because of the customary separation of mathematics and natural philosophy, the clients who used these instruments also accepted these distinct categories of manufactured objects.

NAVIGATION, SURVEYING, WARFARE, AND CARTOGRAPHY

Beyond the nascent interest of a few natural philosophers, the worldly ambitions of princes, merchants, and landowners meant that practical mathematics maintained its fair share of clients. Through navigation, surveying, and the arts of war, the story of the "speculatiue Mechanicien" or "Mechanicall Mathematicien" was linked to dramas played out on a much larger stage than the pursuit of natural knowledge; rather, the development of the geometrical arts was driven by the powerful engine of European ambition in exploration, territory, commerce, and empire (see Vogel, Chapter 33, this volume).

European navigation in the sixteenth century was beginning to acquire confidence as a mathematical art. Traditional sailing in known waters had relied on techniques of bearing and distance, using the magnetic compass, plane chart, and sounding line in the experience-based practice of "dead reckoning." This meant recording course, estimating speed, allowing for wind and current, and calculating position. The method lacked accuracy even over known routes, and for long and unfamiliar voyages, something more reliable was urgently needed – preferably a method of *finding* a position; dead reckoning could only deduce a position from keeping a record, and if the record was wrong, there was no way to recover one's course.[32]

Astronomers were already familiar with relationships between position on earth and the appearance of the heavens, and latitude adjustments to allow for such changes in appearance were built into many portable instruments, such as astrolabes, sundials, and quadrants. It was not difficult to devise techniques for moving in the reverse direction: for finding the latitude from the appearance of the heavens. Thus a quadrant could be used to find the altitude of the polestar, which would be a direct measure of latitude. Such applications required the development of new forms of instruments and new procedures for using them, however; it was difficult to use a quadrant on a moving ship, and it and other astronomical instruments, such as the

---

[32] E. G. R. Taylor, *The Haven-Finding Art* (London: Institute of Navigation, 1956); and D. W. Waters, *The Art of Navigation in England in Elizabethan and Early Stuart Times* (London: Hollis and Carter, 1958).

astrolabe and the cross-staff, were simplified and adapted to conditions at sea. Observations of the polestar required a correction to allow for its distance from the celestial pole. The altitude of the sun at noon was also latitude-dependent; this demanded a more complicated adjustment, which depended on the annual cycle of the sun's displacement from the equator and required the use of a table of this value through the year.[33]

Routines such as these, and their embodiment in the design of instruments, linked navigation more closely with geometry; this drew the generality of instrument makers more firmly into an engagement with elementary astronomy – a process already initiated by the growing importance of the sundial. Makers and users of these sorts of instruments were not simply passive recipients of designs from astronomers. Innovative contributions from the practitioners of the mechanical arts included, for example, the development of the backstaff, which was used for finding the latitude by measuring the meridian altitude of the sun. Using the cross-staff for solar observation had several disadvantages: The observer had to look in the direction of the sun, and the instrument could not be placed at the center of the required angle because that was at the center of the observer's eye. The backstaff solved these problems by turning away from the sun and using a shadow instead of a direct observation. The general idea for this innovation first appeared in a treatise published by John Davis, an English navigator, in 1595, and the instrument had achieved its standard form by the mid-seventeenth century.[34] Keeping his back to the sun, the user of the backstaff positions a shadow vane to an appropriate value on a 65° arc divided, at most, into single degrees. Sighting the horizon and keeping the shadow cast by this vane on the foresight, he moves his near-sight along a second arc, of 25°, to complete the quadrant, until he has measured the maximum altitude reached by the sun at noon. The 25° arc is drawn to a much larger radius and so can be divided more finely. This magnification of the section of the quadrant where the reading will be taken, without making the whole instrument impossibly large, is a clever idea, and the smaller arc was given the transversal scale of the type used by astronomers such as Tycho Brahe (1546–1601), yielding a precision of one minute of arc. Although this degree of accuracy could not be achieved aboard ship, the ingenious design is a telling illustration of the ambitions of the makers.

Ambition in the mechanical community is shown also in contemporary work on the magnetic compass and the dip circle. Since the early sixteenth century, mariners had been aware that the variation of the magnetic needle (its deviation from true north) was different in different locations and that some account had to be taken of this when using the steering compass. From

---

[33] Waters, *The Art of Navigation in England*, pp. 43–57.
[34] John Davis, *The Seamans Secrets* (London: Thomas Dawson, 1595); and Waters, *The Art of Navigation in England*, pp. 205–6, 302–6.

the growing knowledge of variation, the idea developed that it might be possible to use its global distribution for position finding; for example, the Spanish author on navigation Martin Cortés (d. 1592) held that there was a predictable relationship between latitude, longitude, and variation. Because two of these variables could be measured, magnetic variation might be the basis for a method of finding longitude at sea. Unfortunately, the reports of navigators did not support Cortès's intuition, and his method was not practicable.[35]

Magnetic compasses and variation were generally the business of practical seamen and of the mathematical practitioners who supplied their instruments and instructed them in their use. Global accounts of variation, however, seemed to impinge on questions of natural philosophy, so it is not surprising that it was in this context that the methods of the mathematical arts – involving instrumentation and mechanical work – began to infiltrate into other types of study. In 1581, Robert Norman (fl. 1560–1596), an English compass maker – in his own words, "an unlearned Mechanician" – published a study of magnetic dip, while his associate, the navigator William Borough (1536–1598), added a treatise on variation.[36] Both employed instruments and measurement, and both emphasized experience as the guide to reason and, at times, advocated something rather like experimentation. Norman realized that there were tensions in such work, but he argued with some force that the mechanical artists might be better able to deal with this subject than more theoretical men because "Mechanicians . . . haue the vse of those artes at their fingers endes," and their practical experience differs from "the learned in those sciences beeyng in their studies amongst their bookes."[37] Thus, a decade after Dee had tried, contrary to general usage, to restrict the meaning of "mechanicien" to those ignorant of mathematics, Norman defied him, asserting that there were mathematical mechanics who had benefited from the English textbooks of Robert Recorde, and even from the English translation of Euclid that had carried Dee's preface. Indeed, the mechanical genre had an effect on learned treatments of the subject: When William Gilbert came to write *De Magnete*, published in 1600, he constructed a magnetic natural philosophy that incorporated instruments and experiment.[38]

Navigation continued to concern the practical mathematicians in the seventeenth century. Mercator had devised a projection of the globe that was suited to navigators, where compass bearings or rhumb lines were projected

---

[35] Jim Bennett, "The Mechanics' Philosophy and the Mechanical Philosophy," *History of Science*, 24 (1986), 1–28.
[36] Robert Norman, *The Newe Attractive* (London: Jhon Kyngston for Richard Ballard, 1581); and William Borough, *A Discours of the Variation of the Cumpas, or Magneticall Needle* (London: Jhon Kyngston for Richard Ballard, 1581).
[37] Norman, *The Newe Attractive*, sig. B.iv.
[38] William Gilbert, *De magnete, magneticisque corporibus, et de magno magnete tellure* (London: Petrus Short, 1600); Paolo Rossi, *Philosophy, Technology, and the Arts in the Early Modern Era*, trans. Salvator Attanasio (New York: Harper and Row, 1970); and Bennett, "The Mechanics' Philosophy."

as straight lines so that a course to be steered could readily be laid down on a chart with a straight rule.[39] In contrast, calculations of distance were complex, depending on trigonometric functions unheard of in everyday life at sea. These routines of calculation were rendered instrumental by the work of Edmund Gunter, whom I have already mentioned as a designer of logarithmic rules, where multiplication and division of trigonometric functions were reduced to adding and subtracting lengths using a pair of dividers. Gunter also devised a special form of sector – a common instrument applying similar triangles to proportional calculations and adapted to other types of work – equipped with the functions required for Mercator sailing.[40] The logarithmic rule, though initially offered to the navigator, would prove applicable to all manner of calculations and could be customized to the needs of different professions.

The prehistory of Gunter's sector included many particular versions and protocols of application for, among others, surveyors and gunners. Both of these arts, whose geometrical practices became closely linked in the sixteenth century, had come to the attention of the practical mathematicians. Both were drawn into the program of turning a practical or mechanical art into a mathematical art by founding it anew on mathematical science. In surveying, this was attempted in two fundamental ways: through the introduction of maps drawn to scale and the introduction of angle measurement. Traditional surveying had principally concerned inventorying and valuing content, including land recorded through linear measurement, and did not include drawing plans or maps to scale. In the 1570s and 1580s, English surveyors began regularly to create scaled estate maps.[41] Such maps were geometrical tools and can be regarded as mathematical instruments like so many others. Making them required a number of measurements, but using them yielded many more. This was the typical feature of the geometrical "theoric," such as the geometrical construction of planetary motion, or a table of relationships between the range and elevation of a gun. The inventory yielded only the original measurements to the user, but the map could be interrogated in whatever way the user required.

Like scaled maps, angle measurement, as promoted in manuals of mathematical practitioners, challenged the capacities of the ordinary surveyor. The quadrant, astrolabe, and cross-staff were adapted to the requirements of surveying, as they had been to navigation. Of these various instruments, it was the modified astrolabe that survived, evolving into a "simple theodolite" for measuring horizontal angles, also called azimuths or bearings.[42] The

---

[39] Leo Bagrow, *History of Cartography*, revised by R. A. Skelton (London: C. A. Watts, 1964), pp. 118–19.
[40] E. G. R. Taylor, *The Mathematical Practitioners of Tudor and Stuart England* (Cambridge: Institute of Navigation, 1970), pp. 60–4, 196.
[41] P. D. A. Harvey, *Maps in Tudor England* (Chicago: University of Chicago Press, 1993); and E. G. R. Taylor, *Tudor Geography, 1485–1583* (London: Methuen, 1930), pp. 140–61.
[42] Bennett, *The Divided Circle*, pp. 39–44.

great advantage claimed for the new angular survey involved triangulation. Traditionally, measurement meant tramping across every distance that was required and laying out lengths using poles, ropes, or chains. With triangulation, only one "baseline" needed to be measured; all other visible positions could be located on the map by taking angles from both ends of the baseline. Of course, this involved the measurement and recording of many angles, as well as the subsequent construction of a map through a process known as "protraction," which surveyors found unfamiliar and tedious.[43]

A second instrument that became common among surveyors – especially in lands less settled and cultivated – was the circumferentor. This was a form of magnetic compass with sights, which used the magnetic meridian to give a reference line to a map in the relative absence of boundaries and buildings. The altazimuth theodolite (one with circles or arcs for both horizontal and vertical angles) made an occasional appearance, but only in the hopeful projections of mathematicians and the rhetoric of ambitious surveyors. It was scarcely used and did not become common until the nineteenth century. Its promotion in some textbooks reflects instead the ideology of the mathematical program – to maximize the geometrical content of surveying practice.[44]

The surveyors themselves answered with the design and use of quite a different instrument, the plane table, introduced in the late sixteenth century. Like the navigators with their backstaff, the community of practitioners contributed to the development of this instrument. The plane table comprised a flat board on which was secured a sheet of paper, above which was a detached alidade (a rule with vertical sights). It was necessary to orient the table with a magnetic compass. The surveyor would begin at a significant point – say, the corner of a field – align his alidade with the principal features, including the other corners, and draw lines in these directions on the paper. He would then move to the next station, measuring the distance and marking this off to scale on the relevant line. After orienting his table, he would again draw lines in the directions of the same positions, which were located by the intersections of the pairs of lines on the sheet. In effect, he was drawing his map as the survey proceeded and had no need to measure any angles because everything was done graphically. This seemed like a disaster to the mathematicians and a threat to their program: The graphical method of the plane table allowed surveyors with no knowledge of angular measurements access to triangulation, so they could produce a map without the trouble and mystery of protraction.[45] The title page of Aaron Rathborne's book *The Surveyor* (1616) shows one of these bogus surveyors using a plane table with

---

[43] Jim Bennett, "Geometry and Surveying in Early-Seventeenth-Century England," *Annals of Science*, 48 (1991), 345–54.
[44] Bennett, *The Divided Circle*, pp. 44–50.
[45] Bennett, "Geometry and Surveying."

The Mechanical Arts 691

his foot planted on a book, such was his disdain for learning.[46] Although the mathematicians tried to ignore the plane table or to insult its users – Thomas Digges called it "an Instrument onely for the ignorante and unlearned, that have no knowledge of Noumbers"[47] – it became very popular.

Practical mathematicians also promoted triangulation to gunners as a means of range-finding. A number of specific instruments were designed for triangulation – one by Jost Bürgi,[48] for example – and they were generally promoted for use in warfare, where measurement at a distance could be presented as an obvious advantage. Measuring the baseline and taking sights at either end was again the routine, but in this case one arm of a jointed triangle could be adjusted to the scaled baseline while the other arms were clamped to coincide with the sight lines. The scaled distance was yielded directly by the instrument because, like the plane table, the triangulation instrument employed a graphical method. But here the similarities end, for this does not seem to have been a practicable technique in the urgent, noisy, and messy business of warfare.[49]

There were a number of ways in which warfare seemed open to the advances of the mathematicians in the sixteenth century.[50] Technical innovations in the manufacture of large guns – made in a single casting, equipped with trunnions so that their inclination could be adjusted, and mounted on movable carriages – made their effect more reliable and targeting more realistic. It became worthwhile to consider the relationships between the weight of the shot, the charge of gunpowder, the elevation or inclination of the gun, and the range of fire. Mathematicians seized the opportunity to reform the practice of gunnery by providing such relationships, in the form of tables, for example, along with instruments for measuring the shot (calipers), calculating its weight (sectors and rules), taking the distance to the target (rangefinders), and setting up the gun (sights, levels, and clinometers, in many different designs and combinations).[51]

The whole management of gunnery was rapidly repackaged as a mathematical art, and in the process the practical mathematicians once again encountered questions that traditionally belonged to natural philosophy. Projectile motion had a long history of natural-philosophical discussion and dispute. Now, in practical mathematics, it was subject to a different form of discourse,

---

[46] Aaron Rathborne, *The Surveyor* (London: W. Stansby for W. Barre, 1616), title page.
[47] Thomas Digges, *A Geometrical Practical Treatise Named Pantometria* . . . (London: Abell Jeffres, 1591), p. 55.
[48] Benjamin Bramer, *Bericht zu Jobsten Burgi seligen geometrischen triangular Instruments* (Kassel: Jacob Gentsch, 1648).
[49] Bennett, *The Divided Circle*, pp. 44–6.
[50] Jim Bennett and Stephen Johnston, *The Geometry of War, 1500–1750* (Oxford: Museum of the History of Science, 1996); and J. Büttner, Peter Damerow, Jürgen Renn, and Matthias Schemmel, "The Challenging Images of Artillery," in Lefèvre, Renn, and Schoepflin, eds., *The Power of Images in Early Modern Science*, pp. 3–27.
[51] Bennett and Johnston, *The Geometry of War*, pp. 22–68.

which prioritized practical experience, measurement, instrumentation, geometrical theorics, and predictable outcomes. The founding document for this discourse was *La nova scientia* (The New Science) of Niccolò Tartaglia (1499–1557), first published in Venice in 1537; the stress on novelty in its title announces that the art of gunnery will be refounded on a mathematical science. This program was continued by Guidobaldo dal Monte (1545–1607), a friend and patron of Galileo, who was concerned also with perspective and mathematical instruments – a set of interests he shared with Galileo himself, as I have already indicated. Galileo would go on to write a book on his own "two new sciences," one of which was a geometrical science of motion.[52]

Another military art subject to a new practical geometry was fortification. The walls of the medieval fortress provided high platforms for repelling attempts to scale them, but they presented easy targets for the new weaponry, which could be toppled or punctured by heavy artillery fire. Low, stout walls were needed to withstand bombardment, but these were vulnerable to being overrun by infantry. Thus, each straight tract of wall could be no longer than could be defended by sidelong fire from gun emplacements or bastions at either end, so the overall design of the fort was constrained to be some form of regular polygon with projecting bastions at the corners. These bastions themselves became the objects of geometrical design in order to maximize the protective cover they provided to each other and to the curtain walls between them, and the geometry of the approach to the fortress, or "outworks," was also configured to the advantage of the defenders. The whole art of military architecture became geometrized on the basis of a new mathematical science of fortification. Its practitioners included fifteenth-century Italian geometers such as di Giorgio, Brunelleschi, and Leonardo da Vinci, and later practitioners such as Albrecht Dürer (1471–1528) and Simon Stevin (1548–1620).

Many of the practical mathematicians I have discussed were active in cartography, an art relevant to surveying, astronomy, navigation, and warfare, whose sustaining skills included the engraving of the instrument makers and the projection of the geometers. As both emblems of power and practical instruments of governance, maps consolidated even more firmly the worldly character of Renaissance and early-modern mathematics.[53] Thus, when Cosimo I de' Medici renovated the Guardaroba of the Palazzo Vecchio in Florence in the 1560s to accommodate his store of possessions, the room was designed as an instrument of cosmography. Maps of every region in the world adorned the cupboards on the walls, the ceiling displayed the constellations of the heavens, large geographical and celestial globes served to

---

[52] Niccolò Tartaglia, *La nova scientia* (Venice, 1537); Galileo Galilei, *Discorsi e dimostrazioni matematiche intorno à due nuoue scienze attenenti alla mecanica & i movimenti locali* (Leiden: Elsevier, 1638).
[53] Lloyd Arnold Brown, *The Story of Maps* (New York: Dover, 1979), pp. 150–79.

integrate the maps and charts while the celestial globe also functioned as a spherical astrolabe, and a clock by Lorenzo della Volpaia turned the spheres of the planets. Machines, mechanical ingenuity, and theater were all incorporated into the working of this cosmography, as a hidden mechanism made the globes descend from compartments in the ceiling. One of the mathematicians engaged in the execution of this *sala delle carte geografiche* was Egnatio Danti – maker of sundials and meridian lines, theorist on perspective, and writer on the astrolabe – who was responsible for most of the surviving maps.[54]

Ptolemy's *Geographia* was an influential vehicle of cartographic development in the sixteenth century (see Vogel, Chapter 20, this volume). New discoveries had to be incorporated into its many successive editions, and the choice of projections grew as more were devised to meet the challenge of accommodating an entire sphere on a flat surface. The world maps of the sixteenth century came from such celebrated practitioners as Martin Waldeseemüller (ca. 1480–ca. 1521), Apian, Münster, and Mercator,[55] and because the business of these men was mathematics and not natural philosophy, the coexistence of a variety of projections presented no difficulties; their aim was not verisimilitude but utility, and different users required maps of different shapes and properties. Thus, the Mercator projection, for example, catered to the needs of navigators (see Harris, Chapter 16, this volume). Like other kinds of theorics, the map epitomized information in a systematic or rule-governed way, and the rules were geometrical. But because mathematicians were not supposed to engage with natural philosophy, with its commitment to a consistent, true, and overarching account of the physician world, they had a license for creative diversity. In the words of the physical and practical mathematician Gemma Frisius, writing of the variety of projections used in parallel in both maps and astrolabes, "we can do by geometrical invention what is not permitted in the natural world."[56]

ART AND NATURE

Despite their disjunction from natural philosophy at the beginning of the early modern period, the mechanical arts and the mathematical sciences came eventually to engage with investigating, understanding, and explaining the natural world. The most famous scientific book of the period, Newton's *Philosophiae naturalis principia mathematica* (1687), claimed in its title to

---

[54] Jim Bennett, "Cosimo's Cosmography: The Palazzo Vecchio and the History of Museums," in *Musa Musaei: Studies on Scientific Instruments and Collections in Honour of Mara Miniati*, ed. M. Beretta, P. Galluzzi, and C. Triarico (Florence: Leo S. Olschki, 2003), pp. 191–7.
[55] Bagrow, *History of Cartography*, pp. 77–140.
[56] Gemma Frisius, *De astrolabo catholico liber* (Antwerp, 1556), f. 4v.

address the "mathematical principles of natural philosophy."⁵⁷ The method that characterized the new experimental natural philosophy of the late seventeenth century relied on the kind of manipulative techniques previously employed by mechanicians who had, as the compass-maker Robert Norman had put it in 1581, "the vse of those artes at their fingers endes." The tools of this new natural philosophy were instruments, which had formerly been the trademark of practical mathematics but were now admitted to the study of nature. The explanations offered by this natural philosophy spoke of tiny machines and mechanical action at the micro-level – which explains why it was described as "mechanical" by proponents and practitioners such as Robert Boyle.⁵⁸

This brings us to the third and most recent meaning of the word "mechanical," with which I began. The mechanical philosophy took the machine as the paradigm of action throughout the natural world (see the following chapters in this volume: Joy, Chapter 3; Bertoloni Meli, Chapter 26). Nature could now be investigated and understood – not only epitomized and managed – using mechanical instruments because nature itself was a machine. The difference between art and nature was no longer profound and uncompromising as had been true for earlier writers; it was now a difference of degree rather than of kind. It lay in the scale of the machines involved and also in their perfection, because nature's mechanic was divine and even the tiniest machines of nature were flawless. Hooke made this point at the very beginning of his *Micrographia* (1665) by showing the gross imperfections revealed by the microscope in the point of a needle, the edge of a razor blade, and a printed period. But however closely he examined the parts of even the most sordid of tiny creatures – such as the eye of a fly – he found that they were perfectly formed.⁵⁹

If Hooke's best-known book involved an optical instrument, his early instrumental work concerned that archetypal instrument of experimental natural philosophy, the air-pump he made for Robert Boyle.⁶⁰ Hooke placed his hopes for advancing natural knowledge in instruments, especially the microscope, because, as he wrote in the *Micrographia*, it is through them that "we may perhaps be inabled to discern all the secret workings of Nature, almost in the same manner as we do those that are the productions of Art, and are manag'd by Wheels, and Engines, and Springs, that were devised

---

[57] Issac Newton, *Philosophiae naturalis principia mathematica* (London: Royal Society and Joseph Streater, 1687).

[58] Compare Robert Boyle, *New Experiments Physico-Mechanicall, Touching the Spring of the Air* (Oxford: H. Hall for Thomas Robinson, 1660), and his *Experiments, Notes &c. about the Mechanical Origine or Production of Divers Particular Qualities* (London: E. Flesher for R. Davis, 1676).

[59] Robert Hooke, *Micrographia; or, Some Physiological Descriptions of Minute Bodies with Observations and Enquiries Thereupon* (London: J. Martyn and J. Allestry, 1665).

[60] Steven Shapin and Simon Schaffer, *Leviathan and the Air-Pump: Hobbes, Boyle, and the Experimental Life* (Princeton, N.J.: Princeton University Press, 1985), pp. 22–79.

by humane Wit."[61] In the late seventeenth century, it was in clockwork that the most remarkable things were done by the artful deployment of wheels, engines, and springs; as a result, the clock became the favorite rhetorical resource for assertions of a mechanical nature. If clocks in the gross world of human art could be as ingenious as they were in the hands of the leading makers of the time, what might the divine clockmaker not be able to arrange in the faultless machine of nature?

I have shown that within the mechanical and mathematical arts, sites of engagement with natural philosophy arose in astronomy, navigation, and gunnery. Similar sites can be found elsewhere, such as in pictorial representation (see Niekrasz and Swan, Chapter 31, this volume).[62] Under these circumstances, it was possible for the methods of the practical mathematicians to intrude into the study of nature. The mechanical arts were flourishing and confident at a time when natural philosophy seemed to be plagued by problems and crises. By the time nature, as well as art, became conceived of in a mechanical discourse, machines had been developed of sufficient ingenuity to support a mechanical ontology of nature – the finest of which were clocks and watches. For a time it would seem that the natural world and the products of the mechanical arts differed only in the skills and resources of their respective makers.

[61] Hooke, *Micrographia*, preface, sig. 2av.
[62] See note 11.

28

PURE MATHEMATICS

Kirsti Andersen and Henk J. M. Bos

During the early modern period, "mathematics" was generally understood to mean the study of number and magnitude, or of quantity in general. There were two varieties: "pure" mathematics and, using a term that became common around 1600, "mixed mathematics."[1] The former studied number and magnitude in abstraction, whereas the latter studied them in composite occurrence; that is, linked to (mostly material) objects. By 1700, mixed mathematics was extensive indeed: In the German philosopher Christian Wolff's (1679–1754) paradigmatic *Elementa matheseos universae* (Elements of All Mathematics, 3rd ed., 1733–42), it comprised mechanics, statics, hydrostatics, pressure in air and fluids, optics, perspective, spherical geometry, astronomy, geography, hydrography, chronology, sundials, explosives, and architecture, both military and civil. By 1500, most of these fields were small if they existed at all; the rapid expansion of "mixed mathematics" is a characteristic feature of the early modern period. Compared with the mixed variety, pure mathematics had fewer domains. Wolff summarized it under the headings arithmetic, geometry, plane trigonometry, analysis of finite quantities (i.e., letter algebra and analytic geometry), and analysis of infinite quantities (i.e., differential and integral calculus); the last two were created in the seventeenth century.

In this chapter, we follow this early modern demarcation of pure mathematics; when using the term "mathematics," unless explicitly indicated otherwise, we refer to pure mathematics so defined. The demarcation was in terms of the subject matter; it did not correspond to professional dividing lines. Few if any scholars identified themselves exclusively as pure mathematicians. Yet the principal stimuli for development in early modern pure mathematics were internal to its own traditions, stemming from classical and medieval pure mathematics. Incentives from mixed mathematics

[1] H. M. Mulder, "Pure, Mixed and Applied Mathematics: The Changing Perception of Mathematics through History," *Nieuw Archief voor Wiskunde*, 8, no. 4 (1990), 27–41.

were few and incidental. In the seventeenth century, mechanics prompted essential mathematical innovations, notably in the infinitesimal calculus, but the subjects in question (kinematics of accelerated motion and dynamics) were themselves rather new and nearer to pure mathematics than to mixed mathematics in style; the mechanics classified as part of mixed mathematics hardly exceeded statics and the theory of simple and compound machines.[2]

Pure mathematics followed the lines of geographical diffusion characteristic of early modern scholarly activity in general: around 1500, primarily northern Italy and Rome; around 1550, Italy and France (especially Paris), with certain southern German and Swiss towns emerging as centers; and around 1600, the Netherlands and England were included as well. Throughout the seventeenth century, pure mathematics flourished in these regions, with Paris as probably the most important center; Scandinavia and the Iberian peninsula were in the periphery.

THE SOCIAL CONTEXT

Although pure mathematics to a large extent had its own internal dynamic, the development of the field was of course dependent upon the possibility of exposure to and work in mathematics. In this section, we survey the education and employment as well as the lives of some early modern mathematicians. During the sixteenth and seventeenth centuries, mathematics was not as central to school curricula as it later became, and mathematicians did not come from a well-defined group that earned their living from mathematics.[3] At the elementary level, arithmetic was most often taught at special schools such as the Italian *scuole d'abbaco* (abacus schools) and the institutions run by German *Rechenmeister* (reckoning masters). At some of these schools, the students also learned a few basic geometrical constructions and a bit of algebra. Mathematics could also be part of professional training: For example,

---

[2] For more details on pure mathematics of the sixteenth and seventeenth centuries, see Paul Lawrence Rose, *The Italian Renaissance of Mathematics: Studies of Humanists and Mathematicians from Petrarch to Galileo* (Geneva: Droz, 1975); Derek Thomas Whiteside, "Patterns of Mathematical Thought in the Later Seventeenth Century," *Archive for History of Exact Sciences*, 1 (1960–2), 179–388; Douglas M. Jesseph, "Philosophical Theory and Mathematical Practice in the Seventeenth Century," *Studies in History and Philosophy of Science*, 20 (1989), 215–44; Jesseph, *Squaring the Circle: The War between Hobbes and Wallis* (Chicago: University of Chicago Press, 1999); Paolo Mancosu, *Philosophy of Mathematics and Mathematical Practice in the Seventeenth Century* (New York: Oxford University Press, 1996); and Michael S. Mahoney, "The Mathematical Realm of Nature," in *The Cambridge History of Seventeenth-Century Philosophy*, ed. Daniel Garber and Michael Ayers, 2 vols. (Cambridge: Cambridge University Press, 1998), 1: 702–55.

[3] Ivo Schneider, "Forms of Professional Activity in Mathematics Before the Nineteenth Century," in *Social History of Nineteenth Century Mathematics*, ed. Herbert Mehrtens, Henk Bos, and Ivo Schneider (Boston: Birkhäuser, 1981), pp. 93–104.

apprentices in geodesy, fortification, navigation, and the art of painting were taught those parts of geometry that they needed in their work.

Not every one learned basic mathematics as a child – or for that matter during their further education. The famous English diarist Samuel Pepys (1633–1703) described how at the age of twenty-nine, after grammar school and graduation from Cambridge University, he had to learn how to calculate when he was appointed clerk of the king's ships.[4] The range of mathematical education available to early modern Europeans varied widely from the extreme case of Pepys to the education at Jesuit schools, where students often learned a good deal of mathematics – an example being the French philosopher and mathematician René Descartes (1596–1650), who attended the Jesuit school at La Flèche.[5] At the university level, there was also considerable variety. In general, the curriculum for the introductory studies – taking place at the so-called *artes* faculty, also known as the philosophical faculty – included arithmetic and geometry. In most places, courses in these subjects dealt with properties of numbers and the elementary parts of theoretical geometry. Some universities occasionally offered lectures on more advanced mathematics, but it was not possible to graduate in the subject; degrees could be earned only in the liberal arts more generally and in the graduate faculties of law, theology, and medicine.

Despite the deficient education in mathematics, many scholars and mathematical practitioners were attracted to the field and managed to become well versed in the discipline. To give an impression of the different backgrounds and settings of mathematical activity, we sketch the careers of eight mathematicians, selected with an eye toward heterogeneity in geographical locations and social contexts. We have largely avoided the most famous mathematicians – the Frenchman Pierre de Fermat (1601–1665) being an exception. Mathematicians from the centers as well as the periphery – the latter including Edinburgh, Ulm, Toulouse, and Copenhagen – are represented in our sample. The reader will meet gentleman scholars who worked on their subject in their spare time or who were men of independent means, men who lived from mathematics for a short period of their lives, an engineer, and full-time teachers – typical career patterns for early modern mathematicians. One further category of mathematicians was men attached to a religious order.

The full-time teachers were employed at institutions where mathematics was taught, such as the various schools mentioned earlier, the universities,

---

[4] D. J. Bryden, *Napier's Bones* (London: Harriet Wynter, 1992), p. 14.

[5] For the Jesuit contributions to mathematics and the mathematical sciences, see, for example, Peter Dear, "Jesuit Mathematical Science and the Reconstitution of Experience in the Early 17th Century," *Studies in History and Philosophy of Science*, 18 (1987), 133–75; Joseph MacDonnell, *Jesuit Geometers: A Study of Fifty-Six Prominent Jesuit Geometers during the First Two Centuries of Jesuit History* (Studies in Jesuit Topics, 11) (St. Louis, Mo.: Vatican, Institute of Jesuit Sources/Vatican Observatory Publications, 1989).

or the Collège Royal in Paris, which was founded in the sixteenth century to provide general education without preparation for a special profession such as law or medicine. A few persons earned their living by giving private lectures in mathematics, and a few were mathematicians at courts. The latter mainly worked as astronomers and astrologers, the famous German astronomer Johannes Kepler (1571–1630) being a well-known example. At the end of the seventeenth century, the Paris Académie Royale des Sciences also offered positions for mathematicians, but it was only in the eighteenth century that scientific academies came to play a significant role in mathematics.

Although many mathematicians were geographically isolated, they were knowledgeable about the mainstream mathematics that interested them, learning about it through books and – very importantly – through correspondence with colleagues. Sometimes letters with mathematical content were copied and circulated. The French Minim friar Marin Mersenne (1588–1648) was instrumental in organizing such contacts, as was the English philomath John Collins (1625–1683). Scientific journals only appeared at the end of the seventeenth century.

We start by focusing on one region, northern Italy, with two scholars who both graduated in medicine and published influential books on mathematics but nonetheless had very different careers.[6] During much of Girolamo Cardano's (1501–1576) life, he was a highly esteemed physician in demand at several royal courts, but he also experienced severe poverty. In one such period, he taught mathematics and wrote successful textbooks on the subject. His reputation was established by his book *Ars magna* (The Great Art, 1545), which became a classic in algebra and to which we shall return. Cardano wrote extensively on a wide range of subjects besides algebra,, including astrology, astronomy, medicine, philosophy, and gambling mathematics, a field that later became probability theory.

Federico Commandino (1509–1575) grew up in a noble family in Urbino, was educated in the humanist tradition, and became strongly interested in classical works. After the death of his wife and son, Commandino devoted his time to editing, translating, and commenting upon ancient Greek mathematical and astronomical texts. His accomplishments won him so much fame that he first became tutor and medical adviser to the Duke of Urbino and later joined the duke's brother-in-law in Rome as his personal physician.

Turning to the northwestern part of Europe, we find another gentleman mathematician, namely John Napier (1550–1617). Besides taking care of his domestic affairs at Merchiston Castle near Edinburgh, he wrote on theological matters, invented war machines, and engaged in mathematics, particularly methods to facilitate calculations. He designed the so-called Napier bones,

---

[6] Biographies of the mathematicians mentioned in this section can be found in the *Dictionary of Scientific Biography*, ed. Charles C. Gillispie, 16 vols. (New York: Scribners, 1970–80).

rods with a multiplication table printed on each side in such a way that, by combining the tables, the multiplication of any number by a single digit is reduced to an addition. He pursued the idea of reducing a multiplication to an addition much further in his influential invention of logarithms, with which he worked extensively before publishing his new concept in 1614.

Napier belonged to the group of early modern mathematicians – which includes the English mathematician and natural philosopher Isaac Newton (1642–1727) as its most distinguished member – who received some academic training but did not graduate in one of the major faculties (law, medicine, and theology). Another in this group, Simon Stevin (1548–1620), presumably had acquired considerable skills in practical mathematics when he matriculated at the university in Leiden at the mature age of 35.

Stevin was born in Bruges; otherwise not much is known about him before the last part of his career, when he came into close contact with the stadtholder of the Dutch republic, Prince Maurice of Nassau. Stevin taught the prince pure and mixed mathematics and was inspired by him to write a number of textbooks; he also advised Prince Maurice on military and navigational issues. Stevin was eager to spread knowledge, and he published influential treatises on decimal fractions, arithmetic, algebra, the mathematical theory of perspective, mechanics, the tides, and the Copernican astronomical system.

Whereas Stevin combined practical and pure mathematics with the natural sciences, Johannes Faulhaber (1580–1635) associated them with natural magic.[7] Faulhaber was trained in his family's craft of weaving but decided to become a *Rechenmeister*, opening his own school in Ulm in Germany in 1600. His inclination toward occult matters often brought him into conflict with the authorities, as when he published a book on the cabbalistic art. Not all his work was controversial; thus he found some purely number-theoretical results, for example formulas for the sums of the powers of natural numbers up to the thirteenth power. Faulhaber possibly also had some influence on Descartes' interest and competence in algebra. Faulhaber was one of the last *Rechenmeister* to contribute new results to mathematics, but he was not the last mathematician to succumb – like the Italian cleric Luca Pacioli (ca. 1445–1517) and Kepler – to the fascination of attributing extramathematical properties to arithmetical and geometrical objects. Pacioli regarded the ratio, which later was called the golden section, as divine in part because its three segments reflected the Trinity,[8] while Kepler linked mathematical objects and God's harmonic creation.[9]

---

[7] Ivo Schneider, *Johannes Faulhaber, 1580–1635: Rechenmeister in einer Welt des Umbruchs* (Basel: Birkhäuser, 1993).
[8] For another aspect of mathematics and religion, see Herbert Breger, "Mathematik und Religion in der frühen Neuzeit," *Berichte zur Wissenschaftsgeschichte*, 18 (1995), 151–60.
[9] Charles B. Thomas, "Magic and Mathematics at the Court of Rudolph II," *Elemente der Mathematik*, 50 (1995), 137–48.

A typical example of a mathematician who was geographically isolated but in good contact with the mathematical world was the very creative mathematician Fermat. Born into a wealthy family of merchants, he chose law as his profession. In 1631, Fermat graduated and settled as a lawyer and counselor to the parliament in Toulouse. He also studied mathematics and was acquainted with the classics of Greek mathematics and the works of François Viète (1540–1603), a French mathematician and magistrate. From around 1635 on, Fermat made several mathematical innovations in his spare time. His approach to analytic geometry was remarkable but was overshadowed by Descartes' better-published methods. Fermat did, however, receive recognition for his contributions to the early calculus and probability theory. The French philosopher and scientist Blaise Pascal (1623–1662) had awakened Fermat's interest in probabilistic problems. Fermat tried in return to get Pascal interested in number theory, but in vain. Although he lacked prominent correspondents in this field, Fermat figures prominently in its history,[10] not least because of the theorem known as Fermat's last theorem.[11] He did not prove this theorem, nor did he prove many of his other results, relying instead on an impressive intuition.

Among the few who made a living from pure mathematics was John Wallis (1616–1703). He graduated with a degree in theology from Cambridge University and served for some years as a minister in London. During the English Civil War, he deciphered coded letters, but otherwise he had not distinguished himself in the area of mathematics when in 1649 he was appointed as professor of mathematics at Oxford. However, soon afterward he showed that he was in fact a very able mathematician, producing a considerable and solid mathematical oeuvre that contributed to the early calculus. He kept his chair until his death and was at the same time engaged in several other projects – including the foundation of the Royal Society of London, one of the earliest scientific societies.

Although many early modern mathematicians worked in isolation, they were not spread over all of Europe but tended to concentrate in certain places, such as Paris. The importance of living in surroundings where mathematics was esteemed is illustrated by the career of the Dane Erasmus Bartholin (1625–1698). He studied mathematics and medicine at home and abroad, attending lectures by the Dutch mathematician Frans van Schooten (ca. 1615–1660) in Leiden and contributing to the latter's Latin editions (1649 and 1659) of Descartes' *Géométrie* (Geometry), which also contains commentaries and

---

[10] Catherine Goldstein, *Un théorème de Fermat et ses lecteurs* (Saint-Denis: Presses Universitaires de Vincennes, 1995).
[11] Since antiquity, mathematicians had worked on finding Pythagorean triples (i.e., three integers x, y, and z fulfilling the equation $x^2 + y^2 = z^2$, an example being 3, 4, 5). Fermat examined the equation $x^n + y^n = z^n$ for n larger than 2. He claimed that there exist no triples of nonzero integers that solve this equation. Fermat's statement was proven by the British mathematician Andrew J. Wiles in the second half of the 1990s.

essays by van Schooten and his pupils. Bartholin graduated with a degree in medicine from the university in Padua and then returned to a professorship in mathematics at the university in Copenhagen. He soon moved to the better-paid chair in medicine, but he was free to choose his research topics, and for a while remained active in pure mathematics, though he became ever more involved in astronomy. The reason seems to be that since the days of the successful Danish astronomer Tycho Brahe (1546–1601), astronomy enjoyed much more prestige in Copenhagen than did mathematics.

STIMULI: METHODS AND PROBLEMS

Initially, the most important stimulus for pure mathematics in early modern Europe came from the continued assimilation of the achievements of two earlier mathematical cultures, the classical Greek and the medieval Arabic. The former was primarily geometrical in nature, whereas the latter had developed an effective arsenal of algebraic methods and techniques. The dual assimilation brought about a merging of algebra and geometry, leading to the two highlights of early modern mathematical invention: The techniques of analytic geometry introduced in the 1630s by Fermat and Descartes; and the versions of the differential and integral calculus elaborated by Newton and the German philosopher Gottfried Wilhelm Leibniz (1646–1716), both first published in the 1680s.

The principal dynamic in these developments was the deliberate search for general methods. Method was a strong preoccupation of late Renaissance and early modern intellectuals in general,[12] and mathematics provided three distinct models: combinatorial, axiomatic, and analytical.[13] The combinatorial model was inspired by the example of the *Ars brevis* (Short Art) of the Catalan encyclopedist Ramon Lull (ca. 1232–1316), which featured techniques for enumerating all possible combinations from predefined lists of beings, concepts, qualities, and so on. In the later Hermetic and magical traditions, such techniques were deployed in the quest for fundamental truths and insights. However, this fascination with combinatorics had hardly any influence on early modern pure mathematics. The *Elements* of the Greek mathematician Euclid (active around 300 B.C.E.) was the paradigmatic example of the axiomatic method, much admired and imitated outside mathematics, the *Ethica ordine geometrico demonstrata* (Ethics Exposed in Geometrical Order; 1677) of the Dutch philosopher Benedict (Baruch) de Spinoza

---

[12] Peter Dear, "Method and the Study of Nature," in Garber and Ayers, eds., *The Cambridge History of Seventeenth-Century Philosophy*, 1: 147–77.
[13] H.-J. Engfer, *Philosophie als Analysis: Studien zur Entwicklung philosophischer Analysiskonzeptionen unter dem Einfluss mathematischer Methodenmodelle im 17. und frühen 18. Jahrhundert* (Forschungen und Materialien zur Deutschen Aufklärung, Abteilung 2, Monographien 1) (Stuttgart: Fromman-Holzboog, 1982).

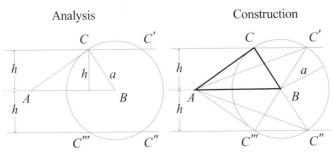

Figure 28.1. Analytical solution of a geometrical construction problem. See the text for explanation.

(1632–1677) being the most illustrious example.[14] Early modern mathematicians also esteemed Euclidean axiomatics, but they applied it less frequently than their Greek paragons had and valued it more for its expository and didactic utility than for its logical rigor. It did not constitute an important factor in the development of early modern mathematics. In contrast, the interest in analytical methods was an essential innovating force, fed by ancient texts (notably Pappus's *Collection*, first printed in 1588) that fragmentarily documented a special method for geometrical problem-solving called "analysis." This method contrasted with the rigorous axiomatic-deductive style of presentation and demonstration that was generally recognized as characteristic of classical Greek mathematics.

Analysis was a method of invention primarily, if not exclusively, applied in solving geometrical problems.[15] Because of the crucial role of the method in the development of early modern mathematics, we will illustrate it in some detail with an example, for which we take a relatively simple geometrical problem, namely to construct on a given base a triangle for which the height and the length of one side are given (see Figure 28.1):

> *Problem*: Let points A and B and two lengths a and h be given; it is required to construct a triangle ABC with height h and such that $BC = a$.
> *Analysis*: Assume that the problem is solved and consider the triangle ABC. The point B and the distance a of C to B are given, and therefore C is on a circle with given center (namely B) and given radius (a); this circle is

---

[14] Hans Werner Arndt, *Methodo scientifica pertractatum: mos geometricus und Kalkülbegriff in der philosophischen Theorienbildung des 17. und 18. Jahrhunderts* (Berlin: Walter de Gruyter, 1971); and H. Schüling, *Die Geschichte der axiomatischen Methode im 16. und beginnenden 17. Jahrhundert* (Hildesheim: Georg Olms, 1969).
[15] Wilbur R. Knorr, *The Ancient Tradition of Geometric Problems* (Boston: Birkhäuser, 1986), p. 358.

therefore given. Moreover, the distance h of C to the line through A and B is given, and therefore C is on a line parallel to AB at that given distance; there are two such lines and they are also given. Hence the point C is on one of the intersections of a given circle and two given lines; the possible positions of C (namely C, C', C'', and C''') are therefore given as well. This concludes the analysis argument.

Note that the term "given" is used synonymously with "known," and in effect synonymously with "geometrically constructible by ruler and compass." The analysis has provided a chain of arguments that transfers the status "given" from the originally given elements of the triangle (A, B, a, h) to the at first unknown point C, which therefore has become known and given as well. Herewith the solution of the problem, the construction of the triangle, is indeed found; it can be read off from the analysis and constitutes the corresponding *synthesis*:

> *Construction*: Construct the circle with center B and radius a; construct the two lines parallel to AB at distance h; these constructions yield the points of intersection of the circle and the two lines. Choose any of these points C, C', C'', and C''' as the vertex and complete the triangle. *Quod erat faciendum* (QEF, done as required).

Two more arguments are needed to complete the solution of the problem: the (usually straightforward) *proof* that the constructed triangle indeed satisfies the requirements, and the so-called *diorismos*, a discussion of the different cases that may arise as to the possibility of solving the problem. (In this case, if $h > a$ there are no solutions, if $h = a$ there are two, and if $h < a$ there are four – note that lengths a and h are assumed to be positive.)

The essential steps in the method of analysis are then as follows. One starts by assuming that the problem is solved so that one can show the as yet unknown solution (the triangle) in a drawing and argue about its parts and properties. Then one explores various chains of linked arguments of the form "if these elements of the figure are given, then those elements are given, too." The analysis is concluded once such a chain is found whose successive arguments link the given elements in the problem to the elements of the required figure. With this chain, the formal "synthesis" (that is, the construction and the proof of its correctness) is achieved by constructing the elements in the order of the chain from the primary givens to the final figure and proving along the same lines that the figure constructed has the required properties.[16]

The preoccupation with method, and in particular with the method of analysis, led to the three major mathematical achievements, letter-algebra,

---

[16] There is an extensive body of literature on the historical and logical interpretation of analysis, centered around J. Hintikka and U. Remes, *The Method of Analysis, Its Geometrical Origin, and Its General Significance* (Dordrecht: Reidel, 1974).

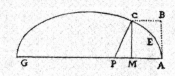

Figure 28.2. Descartes' illustration of an ellipse and its equation. In René Descartes, *La géométrie*, in *Discours de la methode . . . Plus La dioptrique, Les meteores, Et la géométrie* (Leiden: I. Maire, 1637), p. 343; facsimile ed. with English translation, ed. and trans. D. E. Smith and M. L. Latham (New York: Dover, 1954).

analytic geometry, and the calculus; it also gave early modern mathematics its two most distinctive stylistic characteristics: the blending of algebra and geometry and the relaxation of rigor in proof. Early modern mathematicians saw an analogy between the method of geometrical analysis and the techniques of algebra. The use of a symbol (such as x) for the unknown enabled the algebraist to calculate with the as yet unknown solution of an algebraic equation; similarly, in a geometrical analysis, the assumption that the problem was already solved enabled the geometer to argue about the relations between the given elements and the as yet unknown solution of a geometrical problem. This analogy strongly informed the early modern merging of algebra and geometry.[17] Analytic geometry and the calculus combined algebraic and geometrical techniques. The combination permitted a smooth back-and-forth between algebraic and geometrical arguments, between figures and formulas, and between curves and their equations. Formulas became the typographical hallmark of mathematical writing (see Figure 28.2), but they generally referred to geometrical objects. Only during the eighteenth century did these geometrical references gradually subside.

The style of the paradigmatic Greek mathematical treatises – those by Euclid, Apollonius of Perga (late third century B.C.E.), and Archimedes (third century B.C.E.) – required rigorous proofs and constructions, but it did not demand explanations of how the results were found. For instance, Archimedes' proof that the area of a parabola segment is equal to 4/3 of the

---

[17] Indeed the term "analysis" came to denote subjects (e.g., analytic geometry and calculus) with a strongly algebraic nature. After 1700, the meaning of "analysis" shifted further toward subjects that involved infinite processes, particularly limits.

area of its inscribed triangle was eminently cogent but did not explain how this result and its proof were found. In contrast, the new analytical methods were methods of invention; the transformation of the finding process into a construction and/or a proof was often straightforward and in general provided no new insights. Thus, letter-algebra and analytic geometry stimulated no special interest in proof. The new methods of dealing with infinite processes, however, generated more concern about their validity and spurred attempts at rigorously proving either the methods themselves or their particular results. Although these attempts failed, the success of the new methods made a relaxation of rigor acceptable to most mathematicians.

The search for new methods was nourished by a rich supply of challenging mathematical problems. Until the 1650s, these problems originated primarily within pure mathematics itself; indeed, many of them were generalizations of classical Greek problems (e.g., trisection of the angle, duplication of the cube, quadrature of the circle, determination of tangents to curves, and areas and volumes of figures). Later, the new natural philosophy provided an external source of problems. The idea – to use a phrase of the Italian natural philosopher Galileo Galilei (1564–1642) – that the book of nature was written in the language of mathematics implied that relations between variables in physical processes (especially time, velocity, and distance traversed in processes of motion) should be represented mathematically. In Galileo's investigations, these relations were simple proportionalities, as for example in his rule that velocity is proportional to time in free fall. Much more intricate relations, represented by curves and equations, appeared in the theories of later luminaries of rational mechanics, such as those of the Dutch mathematician and natural philosopher Christiaan Huygens (1629–1695), Newton, and Leibniz. The study of dynamic processes in mechanics (e.g., fall, oscillation, and motion under resistance) required that these relations be determined from mechanical laws or assumptions, which, expressed in terms of calculus, were differential equations. The main test cases of the new methods of calculus concerned the solution of differential equations arising in the solution of problems such as the determination of the shape of a hanging chain (the catenary), or of an elastic beam under stress, and the trajectory of a projectile.

Besides the main stimulus – the search for methods, nourished by classical problem types and new problems from natural philosophy – two lesser incentives should be mentioned that led to two different developments close to the core of early modern pure mathematics. In the 1630s, the French engineer Girard Desargues (1591–1661) developed a new approach to conic sections by means of methods now recognized as belonging to projective geometry. As a young, promising mathematician, Pascal also experimented with this approach. Probably because of their divergence from the then fashionable analytical methods, the ideas fell into oblivion, and projective methods were reintroduced in geometry only at the beginning of the nineteenth century. Historians have linked this episode to the theory of perspective, perhaps because Desargues himself indicated that his theory could be applied to perspective, stone cutting, and the construction of sundials. However, contemporary practitioners as well as mathematicians found it impossible to derive applications from Desargues' theory as he had formulated it. An examination of Desargues' work suggests that the main stimulus came from mathematics and once again from one of the classical Greek works, namely Apollonius's theory of conic sections, which Desargues wanted to simplify.[18] Perspective was treated by some sixteenth-century mathematicians, and around 1600 the Italian scientist Guidobaldo dal Monte (1545–1607) created a unified mathematical theory of the discipline. His presentation was soon improved upon by Stevin. Their influence can be traced through a series of seventeenth- and eighteenth-century treatises on perspective, which, however, had no recognizable influence on contemporary pure mathematics.[19]

Another lesser incentive for pure mathematics came from astronomical practice, both theoretical and applied (e.g., in the service of rapidly expanding navigation). Here the constant transfer of data between different coordinate systems (from observations in time- and place-dependent coordinates to tables in fixed celestial coordinates, and vice versa) led to countless multiplications and divisions of large numbers dictated by the rules of spherical trigonometry. Around 1600, a number of scholars, most successful among them Napier, hit upon the idea of logarithms to reduce these multiplications to additions (and the divisions to subtractions); tables of logarithms became available already in the 1610s and 1620s.[20] By the mid-seventeenth century, the logarithmic relationship implicit in these tables was recognized as important for the study of particular types of motion and special types of curves and thus entered the core of early modern pure mathematics.[21]

---

[18] Jan Hogendijk, "Desargues' *Brouillon Project* and the *Conics* of Apollonius," *Centaurus*, 34 (1991), 1–43; and Kirsti Andersen, "Desargues' Method of Perspective," *Centaurus*, 34 (1991), 44–91.
[19] Martin Kemp, "Geometrical Perspective from Brunelleschi to Desargues: A Pictorial Means or an Intellectual End?," *Proceedings of the British Academy*, 70 (1984), 91–132; Kirsti Andersen, "Some Observations Concerning Mathematicians' Treatment of Perspective Constructions in the 17th and 18th Centuries," in *Mathemata, Festschrift für Helmuth Gericke*, ed. Menso Folkerts and Uta Lindgren (Stuttgart: Franz Steiner Verlag, 1985), pp. 409–25; Judith V. Field, "Perspective and the Mathematicians: Alberti to Desargues," in *Mathematics from Manuscript to Print*, ed. Cynthia Hay (Oxford: Clarendon Press, 1988), pp. 236–63; and Kirsti Andersen, "Stevin's Theory of Perspective: The Origin of a Dutch Academic Approach to Perspective," *Tractrix*, 2 (1990), 25–62.
[20] Wolfgang Kaunzner, "Logarithms," in *Companion Encyclopedia of the History and Philosophy of the Mathematical Sciences*, ed. Ivor Grattan-Guinness, 2 vols. (London: Routledge, 1994), 1: 210–28.
[21] External stimuli of a social rather than technological nature initiated mathematical probability theory. Fermat, Pascal, and Huygens mathematized intuitive concepts of probability and expectation by formulating axioms and calculation rules for fair division of the stake in case a game of chance was interrupted.

THE INHERITED ALGEBRA AND AN INHERITED CHALLENGE

An essential part of early sixteenth-century pure mathematics concerned algebra, which had ancient roots. Presumably out of a wish to give their students challenging problems, teachers of mathematics in Mesopotamia around 1800 B.C.E. created problems that led to second-degree equations. About two-and-a-half millennia later, the theory of quadratic equations became an appreciated discipline in the same part of the world, now under Islamic influence, and was first presented in a systematic form around 850 by al-Khwārizmī (ca. 800–ca. 850), who worked as a mathematician in Baghdad. The theory was enlarged to include third-degree equations treated comprehensively by the astronomer, mathematician, and poet 'Omar Khayyam (ca. 1048–ca. 1131) in the second part of the eleventh century. He described how the roots of third-degree equations can be found as line segments (corresponding to coordinates) determined by the points of intersection of two conic sections. He also tried, in vain, to find an algebraic algorithm for determining the roots.

European mathematicians took over the algebra of their Islamic predecessors, and by the sixteenth century the discipline had in Europe regained its earlier two functions, namely as a test of skills, as it had been in Mesopotamia, and as a topic of research, as it had been in medieval Arabic culture. Pupils who would never apply algebra for practical purposes learned the discipline because a reputation for solving algebraic problems improved their career chances in other domains. At the same time, many gifted mathematicians sought an algebraic solution to the third-degree equation – an attractive problem because it had remained unsolved for centuries.

In the early sixteenth century, the professor of mathematics at the University of Bologna, Scipione del Ferro (1465–1526), discovered an algebraic rule for determining the positive root in the equation we write as $x^3 = bx + c$ ($b, c > 0$). Although he had solved a famous problem, he did not announce his finding, but kept it secret. Perhaps he had the idea that he would use it later in a competition for an academic job or for students, or perhaps in a mathematical duel. The latter were quite common in Italy in Ferro's time: Two mathematicians would demonstrate publicly how many problems posed by the opponent they could solve. The questions were posed some time before the meeting, and the participants were supposed to be able to solve their own problems. To do well at such a competition was part of a ranking system that influenced job possibilities.

As far as is known, Ferro never demonstrated his rule in public, but he did not want his secret to be buried with him, so he told it to at least two of his students, his son-in-law, Annibale dalla Nave, and Antonio Maria Fior. The latter tried to gain fame from the rule by inviting one of Italy's well-known mathematicians, Niccolò Tartaglia (1500–1557), to participate in a

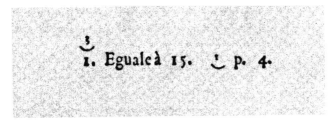

Figure 28.3. A formula as written by Rafael Bombelli, *L' Algebra parte maggiore dell'arimetica* . . . (Bologna: Rossi, 1572), p. 294. Reproduced by permission of The Danish National Library of Science and Medicine.

mathematical dispute. Tartaglia accepted the offer, presumably wondering how Fior dared challenge him. The tables were turned on Tartaglia by Fior's problems concerning third-degree equations. Tartaglia overcame his surprise and managed to deduce a rule for solving the problems posed. Because Fior was not very successful at solving Tartaglia's problems, in the end it was Tartaglia and not Fior who emerged the victor. The dispute was celebrated enough to reach Cardano's ears. Exerting pressure, Cardano managed to persuade Tartaglia to tell him his rule. Cardano included it in his *Great Art* (1545), in which he referred to Ferro and Tartaglia. The result nevertheless became known as Cardano's rule.[22]

The next natural step was to search for a rule that gives the roots of fourth-degree equations. Cardano did just that together with his student Lodovico Ferrari (1522–1565); the latter found a partial solution that Cardano included in his *Great Art* and his Italian successor in algebra, the engineer Rafael Bombelli (1526–1572), generalized in his *L'algebra* (1572).[23]

Bombelli also took up a problem that Cardano had avoided, namely one for which the rule for solving a third-degree equation leads to expressions involving the square root of negative numbers. In particular, Bombelli looked at the equation, which he wrote as shown in Figure 28.3 and we write as $x^3 = 15x + 4$. Cardano's rule gave him the result $x = \sqrt[3]{2 + \sqrt{-121}} + \sqrt[3]{2 - \sqrt{-121}}$. Knowing that the equation in question has the root 4, Bombelli showed how it is possible to interpret the preceding expression as being equal to 4. Bombelli thereby started to work with what later became known as complex numbers. He did not consider them as numbers that could serve as roots

---

[22] For more technical details, see Kirsti Andersen, "Algebraische Lösung der Gleichungen dritten und vierten Grades in der Renaissance," in *Geschichte der Algebra: Eine Einführung*, ed. Erhard Scholz (Mannheim: Wissenschaftsverlag, 1990), pp. 157–81.

[23] Once they were able to solve fourth-degree equations, the mathematicians continued with equations of degree five. Much time was spent on this topic in the seventeenth and eighteenth centuries. In the beginning of the nineteenth century, the Italian and the Norwegian mathematicians Paolo Ruffini (1765–1822) and Niels Henrik Abel (1802–1829), respectively, showed that for equations of degree higher than four it is not in general possible to determine the roots by an algebraic formula.

in equations but only as means to discover the genuine roots. The latter were for Bombelli and his contemporaries the positive roots because they did not conceive of negative numbers and zero as proper mathematical entities. In a geometrical problem, a line segment with a negative length was not considered an acceptable solution.

This situation changed in 1629 when Albert Girard (1595–1632), a French-born mathematician active in Leiden, formulated what later became known as the fundamental theorem of algebra. It states that an equation of degree n has n roots. This theorem is true only if complex and negative numbers as well as zero are accepted as roots. Girard illustrated his theorem with some examples, but gave no proof – a valid proof was first provided by Carl F. Gauss (1777–1855) in 1799.

Apart from his theoretical contribution to the solution of polynomial equations, Bombelli strengthened the field by developing the notation shown in Figure 28.3, which provided an overview of the problem and was in fact quite handy as long as only one unknown was involved. Bombelli's symbols were adopted by Stevin, but soon afterward they were replaced first by Viète's more powerful notation (see Figure 28.4) and then by Descartes' notation (to which we shall return), which was so efficient that it is still used.

THE RECEPTION OF EUCLID'S *ELEMENTS*

Like algebra, Euclid's *Elements* was an important source of inspiration for sixteenth-century pure mathematics. The Latin translations of Arabic texts of the *Elements*, primarily by the early twelfth-century translator and natural philosopher Adelard of Bath and the thirteenth-century ecclesiastic, astronomer, and mathematician Campanus of Novarra, formed the basis of the main family of manuscripts that during the late Middle Ages spread the Euclidean corpus throughout Western Europe. The first printed edition of the *Elements*, in the Campanus version, appeared in 1482. Increased philological interest and expertise established, via a few Greek manuscripts, a link uninterrupted by translation to the fourth-century (C.E.) version of the text by the Greek mathematician and astronomer Theon. The Greek text was printed in 1533; its translation into Latin appeared as early as 1505.[24] The two

---

[24] The editions of Euclid in the sixteenth century provide an impressive illustration of the extraordinary vigor of Renaissance book production. The first printed version of the *Elements*, a Latin text in the Campanus tradition, appeared in Venice in 1482 and was edited by E. Ratdolt, with immediate re-editions in Venice (1482), Ulm (1486), and Vicenza (1491). B. Zamberti published his Latin translation of the Greek text in Venice in 1505; J. Lefèbre d'Étaples based his Latin edition of 1516 published in Paris on both the Campanus and the Zamberti versions; the first edition of the Greek text, by S. Grynaeus, appeared in Basel in 1533; Commandino's improved translation from the Greek into Latin was printed in Pesaro in 1572; and Clavius's edition appeared in 1574 in Rome. Many of these editions were reprinted (or offered for sale with a new frontispiece) several times, and they were not the only editions available. Vernacular editions also soon appeared: Italian (Venice,

text traditions differed considerably; they were combined in the last third of the sixteenth century. Commandino's Latin edition of 1572, based on both traditions, markedly improved the comprehension of the text. The extensive Euclid edition of Christoph Clavius (1537–1612), the German-born professor of mathematics at the Jesuit College in Rome, first published in 1574 and repeatedly reprinted, collected much of the extant philological and mathematical expertise about the text and constituted the basic edition of the *Elements* for more than a century.

A prime example of the improvements brought by these philological endeavors concerned the crucial definition of the equality of ratios in Book 5 of the *Elements*.[25] In the Campanus text, the definition was incomprehensible; only during the second half of the sixteenth century did it come to be gradually understood. The clarification was of great importance because the theory of ratios and proportions was the basis for the use of mathematics in the new natural philosophy; Galileo took a strong interest in the matter.[26]

Forces other than philological care also influenced the early modern transformations of the Euclidean text. Many editors adjusted it for educational purposes or for special audiences. This led to abbreviated editions (characteristically containing only the first six books), in which proofs were often removed or replaced by explanatory examples. Other editors based their adjustments of the text on principled critique: The French philosopher and mathematician Peter Ramus (1515–1572), professor at the Collège Royal in Paris, for example, attacked Euclid's axiomatic deductive presentation as uselessly meticulous. He considered many theorems too obvious to need proof and scorned the (indeed very abstract) theory of irrational ratios in Book 10 of the *Elements* as the "cross of mathematicians," whose content, as Stevin showed somewhat later, could be summarized far more briefly by means of irrational numbers.

1543, by N. Tartaglia), German (Tübingen, 1558, by Scheybl, and Basel, 1562, by W. Holzmann, both partial editions), French (Paris, 1564–6, by P. Forcadel), and English (London, 1570, by H. Billingsley). Indeed, apart from the Bible, no text was edited and translated as often as Euclid's *Elements*. See Euclid, *Les éléments*, ed., trans., and with commentary by B. Vitrac, 4 vols. (Paris: Presses Universitaires de France, 1990–2001), 1: 74–83.

[25] This definition concerned the ratios of (continuous) magnitudes as opposed to ratios of (whole) numbers. In Greek mathematics, numbers referred to discrete objects; they indicated a multitude. Thereby the product of any two numbers a and b was well-defined: The multitude a indicated how many times the multitude b had to be taken to achieve the product $a \times b$. Ratios of numbers were understood and handled in terms of this conception of multiplication, which, however, is not applicable to continuous magnitudes; a line segment between two points P and Q, for instance, does not correspond to a multitude according to which a multiplication can be performed. Classical mathematicians (notably the mathematician and cosmologist Eudoxus of Cnidus [early fourth century B.C.E.], who devised the definitions of Book 5) had therefore elaborated an entirely different theory of ratios for continuous magnitudes. This theory had the advantage that it could deal with irrational ratios, which cannot be expressed by means of whole numbers.

[26] Enrico Giusti, *Euclides reformatus: La theoria delle proporzioni nella scuola Galileiana* (Turin: Bollati Borighieri, 1993).

THE RESPONSE TO ADVANCED GREEK MATHEMATICS: THE APOLLONIAN, ARCHIMEDEAN, AND DIOPHANTINE TRADITIONS

By 1550, the Euclidean corpus had been sufficiently assimilated that mathematicians could turn with profit to the classical sources containing advanced mathematics. The four main sources of this nature were (with the dates of their first publication in print): the *Conica* (Conics, 1537) of Apollonius, the works (1544) of Archimedes, the *Arithmetica* (1575) of Diophantus (3rd century C.E.), and the *Collectiones mathematicae* (Mathematical Collection, 1588) of Pappus (early 4th century C.E.).[27] They gave rise to three research traditions: Apollonian, Archimedean, and Diophantine. The first, based on the *Conics* and on reports about lost classical works in Pappus's *Collection*, focused on conic sections, locus problems, and geometrical construction problems. At first, the mathematicians involved, such as Clavius, Viète, Marino Ghetaldi (ca. 1566–1626), patrician and mathematician of Ragusa (Dubrovnik), and Willebrord Snel (1580–1626), professor of mathematics at Leiden, reconstructed the classical Greek practice of geometrical problem-solving.[28] They surveyed the various types of problems and examined the different constructions proposed in antiquity for problems that could not be constructed by ruler and compass (notably the duplication of the cube, the trisection of the angle, and the quadrature of the circle). Several of these methods involved curves (primarily conics, but also others); they thereby constituted an essential stimulus for the early modern interest in curves. Geometrical problems, together with Diophantus's techniques, also stimulated interest in the methodological aspects of solving problems. Thus, the

---

[27] Because of their advanced nature, these works had a much smaller circulation in print than Euclid's *Elements* (see note 24). Their printing history illustrates the crucial role of Commandino in the Renaissance project of recovering the classical mathematical heritage. The first Latin translation of Books 1–4 of Apollonius's *Conics* appeared in Venice in 1537 in a rather poor edition by G.-B. Memmo; Commandino provided a translation with commentary of much higher philological and mathematical quality that was printed in Bologna in 1566. The Greek text of Books 1–4 became available only in an Oxford edition by Edmond Halley in 1710. Latin versions of Books 5–7, which are preserved only via the Arabic tradition, became available in the seventeenth century. (The final book, Book 8, is lost.) Greek texts and Latin translations of Archimedes' works began circulating in the fifteenth century, and a first Latin edition of some completeness was achieved by Tartaglia in Venice in 1543; the *editio princeps* of the Greek text (with Latin translation) was published by Th. Geschauff in Basel in 1544. A much-improved Latin text of the Archimedean corpus was provided by Commandino (Venice, 1558). Diophantus was little studied before the first printed Latin version of the *Arithmetic*, published by W. Holzmann in Basel in 1575; the Greek text became available in print in C. G. Bachet de Méziriac's edition (Paris, 1621). Copies of the Greek text of Pappus's *Collection* had circulated for quite some time (see A. P. Treweek, "Pappus of Alexandria: The Manuscript Tradition of the Collectio Mathematica," *Scriptorium*, 2 (1957), 195–233) before the Latin version, prepared by Commandino, appeared in 1588 in Pesaro. Although parts of the Greek text became available after about 1650, the first complete edition of the work (by F. Hultsch) only appeared in 1876–8.

[28] Including the actual reconstruction of the lost works mentioned by Pappus, see, for example, Aldo Brigaglia and Pietro Nastasi, "Le riconstruzioni Apolloniane in Viète e in Ghetaldi," *Bollettino di storia delle scienze matematiche*, 6 (1986), 83–134.

Apollonian tradition provided a most fertile ground for the later elaboration (by Viète, Fermat, and Descartes) of new analytical techniques.

Some of the most elegant mathematics inherited from the classical Greek period concerned problems that today are considered to be a part of the differential and integral calculus. The early integration problems were, among others, the so-called problems of quadrature and cubature – that is, determinations of areas and volumes of curved figures. The early differentiation problems concerned determinations of tangents and minimal or maximal line segments fulfilling given properties. Already in antiquity, some of these problems had been treated with great virtuosity, especially by Archimedes.

In working on quadratures, Archimedes used an approach that had been under development for more than a hundred years: He considered two areas, for instance the surface of a sphere and the area of a great circle in the sphere, in order to determine their ratio. His procedure was first to get an idea of what the ratio is, in this case 4:1, and then to prove that he had found the correct value. His proof consisted in showing that a denial of his assumption leads to contradictions. Using this technique of reductio ad absurdum, Archimedes and other ancient Greek mathematicians managed to base their arguments entirely on finite processes. Proofs in the Archimedean style were based on an elegant idea, but in practice a lot of work was involved because many calculations were needed before the required contradictions were obtained. It was also often difficult to find the result that was supposed to be proved (the ratio 4:1 in the preceding example was not intuitively obvious). In proving his theorems, Archimedes did not reveal how he had discovered them and hence gave no guidelines as to how new results could be obtained. He had actually presented the method he had followed in a separate essay – now known as the *Method* – but this was rediscovered only in 1906.

In the second part of the sixteenth century, a number of scholars, including Commandino, dal Monte, Stevin, and Clavius's pupil Luca Valerio (1552–1618), had the training necessary to understand Archimedes and the inclination to read him. Most of his readers admired his precise style, but many also wanted to simplify his method of quadrature and cubature. They started to look for what Galileo's pupil the Italian mathematician Evangelista Torricelli (1608–1647), referring to a legend about Euclid, called "a royal road."[29] The readers of Archimedes sought procedures that combined the steps of finding a result and proving that it is correct, preferably in a way requiring less calculation than the Greek method. In the first decade of the seventeenth century, Valerio made an attempt in which he avoided the reductio ad absurdum technique and yet seemingly kept to finite processes. A bit later, Kepler, fearing that the common procedure applied by Austrian wine dealers was incorrect, wanted to find a technique for determining the volume of wine casks. He took his starting point in Archimedes' method of cubature and

---

[29] E. Torricelli, *Opere*, 1.1 (Faenza: Montanari, 1919), p. 140.

then suggested an alternative based on a more intuitive technique. His solution, published in 1615, was a rather daring maneuver with infinitely small quantities. The works of Valerio and Kepler were the first in a long series, to which we shall return.

Diophantus's work concerned natural and positive rational numbers and was rather isolated in the classical Greek mathematical tradition, in which geometrical subjects dominated. A characteristic example of a Diophantine problem, in modern notation, is: Find pairs of numbers x and y such that $x^3 + y$ is a cubic number and $x + y^2$ is a square number. Diophantus had developed techniques that were strongly algebraic, involving calculations with unknown numbers and abbreviated notations for writing equations. These techniques attracted the interest of algebraists (Bombelli, for instance) and devotees of recreational number problems, such as the French nobleman-scholar Claude-Gaspar Bachet de Méziriac (1581–1638), whose notes on Diophantus inspired much of Fermat's work in number theory. The techniques also inspired Viète to develop his letter algebra.

THE MERGING OF ALGEBRA AND GEOMETRY

During the seventeenth century, the ancient mathematical styles gave way to modern ones.[30] The transition from diligent assimilation of classical mathematical learning to innovative creation that transcended the Greek heritage started in the last decade of the sixteenth century. In 1591, François Viète published his *In artem analyticen isagoge* (Introduction to the Analytical Art), the first of a series of books in which, he claimed, he reconstructed the ancient method of analysis.[31] Viète was convinced that this method had contained a part that was algebraic in nature (but different from the algebra of Arabic origin[32]) and had been kept secret. The idea accorded with a strong contemporary interest in the methods of classical rhetoric, which earlier French algebraists had connected to the techniques of handling

---

[30] On the conceptual innovations beyond classical mathematics, see Jakob Klein, *Greek Mathematical Thought and the Origin of Algebra*, trans. Eva Brann (New York: Dover, 1992; orig. ed. 1934–6); Michael S. Mahoney, "The Beginnings of Algebraical Thought in the Seventeenth Century," in *Descartes' Philosophy, Mathematics, and Physics*, ed. Stephen Gaukroger (Totowa, N.J. and Brighton: Barnes and Noble/Harvester, 1980), pp. 141–56; Henk J. M. Bos and Karin Reich, "Der doppelte Auftakt zur frühneuzeitlichen Algebra: Viète und Descartes," in Scholz, ed., *Geschichte der Algebra* pp. 183–234; Henk J. M. Bos, "Tradition and Modernity in Early Modern Mathematics: Viète, Descartes and Fermat," in *L'Europe mathématique, histoires, mythes, identités: Mathematical Europe, History, Myth, Identity*, ed. Catherine Goldstein, Jeremy Gray, and Jim Ritter, (Paris: Maison des sciences de l'homme–Bibliothèque, 1996), pp. 183–204; and Henk J. M. Bos, *Redefining Geometrical Exactness: Descartes' Transformation of the Early Modern Concept of Construction* (New York: Springer-Verlag, 2001).

[31] Warren van Egmond, "A Catalog of Viète's Printed and Manuscript Works," in Folkerts and Lindgren, eds., *Mathemata*, pp. 359–96.

[32] Giovanna C. Cifoletti, "The Creation of the History of Algebra in the Sixteenth Century," in Goldstein et al., eds., *L'Europe mathématique*, pp. 121–42.

> **THEOREMA II.**
>
> Si A cubus —B—D—G in A quad. + B in D + B in G + D in G in A, æquetur B in D in G: A explicabilis est de qualibet illarum trium B, D, vel G.

Figure 28.4. François Viète's algebraic notation. François Viète, *Opera mathematica* (Leiden: Bonaventura and Abrahamus Elzeviri, 1646; repr. Hildesheim: Olms, 1970), p. 158.

equations.[33] Thus inspired by ancient examples, Viète elaborated a new algebra that was applicable in arithmetic as well as in geometry. This required in particular a new interpretation of multiplication that covered the product of numbers as well as those of line segments or other geometrical magnitudes.[34] Viète solved this difficulty radically by disregarding the particular nature of the objects of his algebra. His algebra dealt with unspecified magnitudes, denoted by letters, whose only further properties were that they obeyed the algebraic operations and that they had dimensions similar to the geometrical dimensions (i.e., length, area, solid) but abstractly extended to higher dimensions beyond geometrical interpretability. Thus, Viète introduced operations into mathematics whose actions were defined only by the rules they obeyed. This was quite uncharacteristic for the age; only in the twentieth century did such a high level of abstraction become a routine phenomenon in mathematics. For early modern mathematicians, Viète's thought was evidently too abstract; they extended the field of algebra by generalizing the number concept rather than by abstracting from the nature of the objects of algebra and concentrating solely on the operations.[35]

Whereas earlier algebra had used symbols only for unknown numbers and their powers, Viète's algebra was a genuine letter-algebra in which unknown as well as known (but indeterminate) magnitudes were represented by letters. He was thereby able to derive essentially new results, such as the relations between the roots and the coefficients of an equation. His formulation in the case of the cubic equation is in Figure 28.4. It is a good example of the power of his notation as well as its difference from modern algebraic notation, which owes

---

[33] Giovanna C. Cifoletti, "La question de l'algèbre: Mathématiques et rhétorique des hommes de droit dans la France du XVIe siècle," *Annales: Histoire, Sciences Sociales*, 50 (1995), 1385–416.
[34] In the traditional conception, the product of two numbers was a number, but the product of two line segments was a rectangle; see note 25.
[35] Helena M. Pycior, *Symbols, Impossible Numbers, and Geometric Entanglements: British Algebra Through the Commentaries on Newton's Universal Arithmetick* (Cambridge: Cambridge University Press, 1997); and Jacqueline Stedall, *A Discourse Concerning Algebra: English Algebra to 1685* (Oxford: Oxford University Press, 2002).

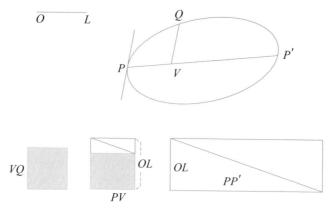

Figure 28.5. Apollonian construction of an ellipse.

much more to Descartes (see Figure 28.2). In present-day notation and choice of letters, and replacing A by x, B by a, D by b, and G by c, Viète's statement can be rendered as: "If $x^3 - (a+b+c)x^2 + (ab+ac+bc)x = abc$, then x can be taken as equal to any of the three values a, b, and c." This implies that in a cubic equation $x^3 + Px^2 + Qx + R = 0$ with roots a, b, and c, the coefficients P, Q, and R can be expressed in terms of the roots as: $P = -(a+b+c)$, $Q = ab + ac + bc$, and $R = -abc$.

In geometry, Viète proved that problems reducible to third- or fourth-degree equations could be geometrically reduced to either the trisection of a given angle or the determination of two mean proportionals between two given lines. Remarkably, Viète did not apply his new algebra to the study of curves. Thus, although he provided all the necessary algebraic apparatus, it fell to later mathematicians, notably Fermat and Descartes, to develop analytic geometry.

In the Apollonian theory of conic sections, it was common to express characteristic properties of a conic section in terms of certain relations between line segments that are defined by a point on the curve with respect to one of its diameters. Figure 28.5 illustrates the procedure in the case of the ellipse, with a diameter (i.e., any line through the center of the ellipse) PP', the tangent at P, and a fixed line segment OL called the *parameter* of the curve with respect to the diameter. Any point Q on the ellipse defines two line segments, its *ordinate*, QV, drawn parallel to the tangent, and its *abscissa*, PV. The ellipse is characterized by the property that:

> For any point Q on the ellipse, the square on the ordinate VQ (the shaded area in the lower left diagram in Figure 28.5) is equal to a part (shaded in the lower middle diagram in Figure 28.5) of a rectangle with sides equal to the corresponding abscissa PV and the parameter OL; this part is determined by the requirement that the remaining part (the nonshaded rectangle in Figure 28.5)

should be similar (in shape) to the rectangle (in the lower right diagram in Figure 28.5), with sides equal to the diameter PP' and the parameter OL.

Fermat and Descartes realized the advantage of expressing such properties in terms of letter-algebra, namely as equations in which the *abscissa* and the *ordinate* of a point on the curve occur as unknowns. In the case of the ellipse, and in Descartes' notation (Fermat used a variant of Viète's notation), with $PV = x$, $VQ = y$, $PP' = q$, $OL = r$, the various areas involved in the property can be expressed as follows: The square on the ordinate VQ is y^2. The rectangle with sides PV and OL is rx. The remaining part of that area (unshaded in the lower middle diagram in Figure 28.5) is a rectangle with one side equal to x and similar in shape to the rectangle in the diagram to the lower right in Figure 28.5, which has sides q and r. Therefore, its sides will have the ratio $\frac{q}{r}$, and so its second side will be $\frac{q}{r}x$ and its area $\frac{q}{r}x^2$. Hence the shaded area in the lower middle diagram in Figure 28.5 will be $rx - \frac{q}{r}x^2$. The equality mentioned in the property is then represented by the equation $y^2 = rx - \frac{q}{r}x^2$ in the two unknowns x and y. Note that the origin of the coordinate system is taken in P and that the directions of the axes are not perpendicular. (The ellipse equation in Descartes' text shown in Figure 28.2 differs from the one here in that the origin G is taken in a point of the ellipse where the tangent is perpendicular to the diameter and that y and x represent the abscissa and the ordinate rather than vice versa.)

Thus Fermat and Descartes inaugurated what may be called the principle of analytic geometry: Curves have equations in two unknowns, and their geometrical properties correspond to the algebraic properties of their equations. Fermat was the first to formulate this principle, doing so around 1636 in a treatise on locus problems that he circulated in a few manuscript copies. Descartes, in his *Geometry*, used the principle but formulated it less explicitly. Because Fermat's work was published in print much later, the principle and the techniques of analytic geometry became generally known among European mathematicians through Descartes' treatise.[36] Descartes' influence was enhanced also by his fortunate choice of notation in algebra; his use of fully symbolic equations and of lowercase letters (a, b, c, etc., for the given quantities and z, y, x, etc., for the unknown ones; see Figures 28.2 and 28.4) became widely accepted in the late seventeenth century. It constitutes one of the prime examples of the advantage of appropriate notation for the development of mathematics.

---

[36] On Descartes' *Géométrie*, one of the three "essays" appended to his *Discours de la méthode*, see Henk J. M. Bos, "On the Representation of Curves in Descartes' *Géométrie*," *Archive for History of Exact Sciences*, 24 (1981), 295–338; Bos, "The Structure of Descartes' *Géométrie*," in *Descartes: Il metodo e i saggi: Atti del Convegno per il 350 Anniversario della Pubblicazione del Discours de la méthode e degli Essais*, ed. Giulia Belgioioso, 3 vols. (Rome: Istituto della Encyclopedia Italiana, 1990), 2: 349–69; Bos, *Redefining*, pp. 285–397; E. R. Grosholz, "Descartes' Unification of Algebra and Geometry," in Gaukroger, ed., *Descartes' Philosophy, Mathematics, and Physics*, pp. 156–69; and Vincent Jullien, *Descartes' La Geometrie de 1637* (Paris: Presses Universitaires de France, 1996).

In the opening sentence of the *Geometry*, Descartes claimed that by his new techniques "all problems in geometry," in particular locus problems and geometrical construction problems, could be solved. His interest in the latter had a strong philosophical component: The analysis and construction of a geometrical problem was for him the prototype of a procedure to achieve certainty in philosophy and the sciences. Descartes therefore elaborated a general set of rules for construction that covered and classified all geometrical problems.[37] This classification of constructions involved a classification of curves: Descartes considered those with algebraic equations to be truly geometrical and grouped them according to the degrees of their equations. He expelled other curves (now called "transcendental") from geometry, excluding explicitly Archimedes' spiral, the quadratrix, and the cycloid.[38]

THE CALCULUS

Although successful in solving standard geometrical problems, analytic geometry was not powerful enough to deal with problems concerning tangents and areas of curves. To solve these problems, several new methods were created in the period 1630–60 by, among others, Torricelli, in Italy, Descartes, Fermat, and Pascal, in France, and in England Wallis and the theologian Isaac Barrow (1630–1677), who for a time held the Lucasian Chair in Mathematics at Cambridge University – the one Newton would later assume. None of the new methods were as general as their authors would have wanted, and all were rather ad hoc.[39] Yet they provided new results, such as rules for calculating the area under the curves $y = x^n$ and under the sine curve – results that correspond to $\int x^n dx = \frac{x^{n+1}}{n+1}$ and $\int \sin x\, dx = -\cos x$.

Similarly, a rule for determining the tangents to some curves was found that implicitly determined a differential quotient, such as for instance $\frac{dx^n}{dx} = nx^{n-1}$.

---

[37] For the rise and decline of this Cartesian mathematical theory, see Henk J. M. Bos, "Arguments on Motivation in the Rise and Decline of a Mathematical Theory: The 'Construction of Equations,' 1637-ca. 1750," *Archive for History of Exact Sciences*, 30 (1984), 331–80.

[38] Descartes based this classification on a conviction that the motions involved in tracing algebraic curves were essentially different from those for tracing the quadratrix, the spiral, and similar curves; see Bos, *Redefining*, pp. 335–354.

[39] Margaret E. Baron, *The Origins of the Infinitesimal Calculus* (Oxford: Pergamon, 1969); Kirsti Møller Pedersen (later Andersen), "Techniques of the Calculus, 1630–1660," in *From Calculus to Set Theory, an Introductory History*, ed. Ivor Grattan-Guinness (London: Duckworth, 1980), pp. 10–48; Kirsti Andersen, "The Method of Indivisibles: Changing Understandings," in *Studia Leibnitiana* (Sonderheft 14, *300 Jahre "Nova Methodus" von G. W. Leibniz [1684–1984]*) (Wiesbaden: Franz Steiner Verlag, 1986), pp. 14–25; Andersen, "Precalculus," in Grattan-Guinness, ed., *Companion Encyclopedia of the History and Philosophy of the Mathematical Sciences*, 1: 292–307; and Jan van Maanen, "Precursors of Differentiation and Integration," in *A History of Analysis*, ed. Hans Niels Jahnke (Providence, R.I.: American Mathematical Society, 2003), pp. 41–72.

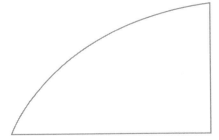

Figure 28.6. An area whose quadrature is wanted.

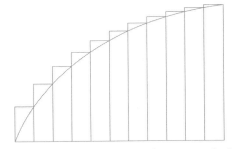

Figure 28.7. The area of Figure 28.6 with circumscribed rectangles.

The various methods had one thing in common, namely that they lacked a rigorous foundation. For area problems, this can be illustrated by looking at the area depicted in Figure 28.6. Intuitively, this area can be understood as being composed by an extremely large number of rectangles (Figure 28.7 in which only a small number of rectangles have been drawn). Each of these rectangles has a side that is exceedingly small. In order that the sum of the rectangles not only approximates the given area but really equals the area, it is not enough that the number of rectangles be extremely large; it must in some sense be infinite, and correspondingly, each of the small sides of the rectangles has to be an infinitely small segment – called an infinitesimal.[40] The methods for determining tangents were also derived from intuitive arguments involving infinitesimals, albeit hidden in the presentations of several authors.

There was no tradition of handling an infinitely large number or an infinitely small quantity. The Greek mathematicians had explicitly excluded infinitesimals from their concepts of quantities and similarly avoided infinite processes. This attitude presumably was taken to avoid logical contradictions such as Zeno's paradox: Swift Achilles would never be able to overtake a slow tortoise with a head start. Nor did the successors to the Greek mathematicians offer a solution as to how to deal with the infinite. So it became a real challenge for seventeenth-century mathematicians to tame the infinite and

---

[40] Later generations of mathematicians would say that the given area is equal to the limit of the sum of the areas of the rectangles when their number increases and their bases decrease.

the infinitely small.[41] They never succeeded, and they were aware of their failure. They were educated in the paradigm of the rigor of Greek mathematics, and they did not seek a revolution. However, they realized that in order to obtain quadrature and cubature procedures that were more convenient than the Greek methods, they had to abandon the admired rigor. They proceeded courageously with techniques that worked but for which they could find no foundation that satisfied the standards set by Greek mathematics.

The methods developed in the period 1630–60 for solving problems of quadrature and cubature, tangents, etc., did not in general survive, but they bore fruit in the form of the creation of the calculus, or rather of two calculi.[42] Building upon the ideas of their predecessors, Newton and Leibniz independently created methods whose generality fulfilled one of the desiderata for a new method. Moreover, both realized that determinations of quadratures and tangents are reciprocal procedures. Barrow had earlier noted in one example that the two procedures were inverses of one another, but Newton and Leibniz were the first to stress this phenomenon and to apply it extensively. The other desideratum – that of finding a sound foundation for the calculus – remained unsolved by Newton and Leibniz.[43]

Newton developed his calculus, which he called the method of fluxions, in the years 1665–6. He added to it and revised it a couple of times but did not publish a separate treatise on the method before 1704. In the meantime, his great contribution to celestial and terrestrial mechanics, the *Principia mathematica philosophiae naturalis* (Mathematical Principles of Natural Philosophy, 1687), had appeared, in which Newton mentioned his method of fluxions but did not in general base his derivations upon it because he preferred a synthetic style resembling the Greek methods. In a number of cases, he included results he had obtained by his new method but left out the proofs.[44]

Newton's method of fluxions got its name from the way he conceived of mathematical quantities, as being generated by a continuous motion. At any moment of this motion, he considered its velocity – which he called "fluxion" – as an intuitively understood concept. He invented a procedure for finding the ratio between the fluxions of two algebraic expressions that for instance gives the ratio between the fluxions of x^n and x as $nx^{n-1}:1$. There was no explicit differentiation involved in Newton's method, nor was there in

---

[41] For various mathematicians' and scientists' use of the infinite, see, for example, Judith V. Field, *The Invention of Infinity: Mathematics and Art in the Renaissance* (Oxford: Oxford University Press, 1997); and Michel Blay, *Reasoning with the Infinite*, trans. M. B. DeBevoise (Chicago: University of Chicago Press, 1998; orig. publ. 1993).

[42] Henk J. M. Bos, "Newton, Leibniz and the Leibnizian Tradition," in Grattan-Guinness, ed., *From Calculus to Set Theory*, pp. 49–93; and Niccolò Guicciardini, "Newton's Method and Leibniz's Calculus," in Jahnke, ed., *A History of Analysis*, pp. 73–103.

[43] The first step toward a solution was taken by the French scientist Augustin Cauchy (1789–1857) around 1820 and the final one by German mathematicians in the 1870s, among them Karl Weierstrass (1815–1897).

[44] Niccolò Guicciardini, "Did Newton Use His Calculus in *Principia*?" *Centaurus*, 40 (1998), 303–43.

principle any infinitesimal involved because his fluxions were finite quantities. However, in his arguments, Newton often resorted to very small intervals of time, which he treated as his contemporaries treated infinitesimals. In the further development of his method, Newton called the flowing quantity a fluent and then set up a system $\ddot{z}, \dot{z}, z, \dot{z}, \ddot{z}$, in which each quantity is preceded by its fluent and followed by its fluxion.

While working on quadratures in 1675, Leibniz got the idea of introducing the signs $\int$ and d. In the beginning, his thoughts on how to deal with these signs were unclear and he changed his mind several times.[45] Leibniz ended up with an interpretation in which he conceived of dx as the difference between two very close x's – that is, as a kind of infinitesimal – and defined $\int$ as the opposite operator in the sense that $\int dy = y$. He called his new discipline a calculus to stress the fact that it involved calculation with symbols. In the process that led Leibniz to the creation of his calculus, his symbols played an important role because they helped him to clarify his thoughts. Later, they were also instrumental in persuading others of the virtues of his new calculus. This is another illustration of how new systems of notation were a powerful part of early modern mathematics.

Leibniz's differential calculus concerns the determination of differences – or differentials, as they were later called – of various expressions, such as for instance $d(xy) = xdy + ydx$. He published this calculus in 1684. Two years later, Leibniz presented his integral calculus to the public. The two new branches of mathematics caught the interest of some Continental mathematicians and were in particular supported by the Swiss mathematicians and brothers Jakob (1654–1705) and Johann (1667–1748) Bernoulli. Before the end of the seventeenth century, Leibniz's calculus had become an established discipline. Whereas the Continental mathematicians adopted Leibniz's method, admiring its efficiency, some British mathematicians worried about its foundations and considered Newton's method more sound – as it contained no explicit infinitesimals. Partly for this reason, and partly because of national pride, Newton's method remained dominant in Britain for more than a hundred years.

The main motivation for developing the calculus had been internal, namely a wish to improve various ancient methods. The reasons for its success despite its shaky foundations were linked to developments in mechanics. The calculus, especially in Leibniz's version, proved very helpful in solving mechanical problems, which could be expressed as equations between differentials. Work on these equations became a major area of mathematical activity in the eighteenth century.

[45] Joseph E. Hofmann, *Leibniz in Paris, 1672–1676*, trans. Adolf Prag (Cambridge: Cambridge University Press, 1974; original German edition Munich, 1949); Henk J. M. Bos, "Differentials, Higher-Order Differentials and the Derivative in the Leibnizian Calculus," *Archive for History of Exact Sciences*, Sonderheft 14 (1974), 1–90; Bos, "Fundamental Concepts of the Leibnizian Calculus," in *Studia Leibnitiana*, pp. 103–18.

CONCLUSION: MODERNITY AND CONTEXT

During the early modern period, pure mathematics, after a period of assimilation of the classical heritage, consciously surpassed ancient achievements and became in that sense modern. Yet "modern" is an elusive term, and some of its senses are an ill fit to early modern pure mathematics. In fact, the new notations and the formulas give seventeenth-century mathematics texts a misleadingly familiar appearance that leads today's mathematicians into neglecting the fundamental differences between modern and early modern mathematics, especially the pervasiveness of the geometrical reference frame in the latter. Thus, analytic geometry treated curves, not an abstract plane of coordinated points, and the calculus treated variable quantities in diagrams involving curves, not functions – indeed, the function concept was virtually absent.[46] Newly discovered curves, although represented by formulas, were accepted only on the basis of explicit geometrical constructions.[47] It was only in the eighteenth century that the formulas themselves became the subject matter of analysis rather than the geometrical objects they had represented earlier. Still less applicable is the term "modern" in the sense of the axiomatic structural style epitomized by Bartel Leendert van der Waerden's (1903–1996) *Modern Algebra* (1930–1) and generalized to wider ranges of mathematics by "Bourbaki"[48] around 1960. Despite Viète's remarkably abstract approach to algebra, the introduction of mathematical concepts solely on the basis of sets and axiomatically (i.e., implicitly) defined operations remained alien to early modern mathematics.

As a historical category, pure mathematics in the early modern period is primarily defined by its subject matter, not by the social cohesion of the group who taught or developed it. Pure mathematics did not constitute a profession, and few if any scholars restricted themselves to it. As a consequence, the field had no single social or institutional base; it resided in many different settings.[49] At schools for practical mathematical trades, pure mathematics was the virtuoso fringe of mixed mathematics, providing fame for the teacher; at universities, it satisfied the logical interest of Aristotelian philosophers and, later, the desire of natural philosophers for a quantitative description of nature; at courts, it was a means of gaining patronage through intellectual brilliance; in genteel society, it was a respectable erudite interest, first because of the humanist prestige of its classical ancestry and later because of its inti-

---

[46] Bos, "Differentials," esp. p. 6.
[47] Henk J. M. Bos, "Tractional Motion and the Legitimation of Transcendental Curves," *Centaurus*, 31 (1988), 9–62.
[48] The name "Bourbaki" was used as a collective pseudonym by a group of French mathematicians.
[49] The great variety of settings of pure and mixed mathematics in Italy is well documented in Mario Biagioli, "The Social Status of Italian Mathematicians, 1450–1600," *History of Science*, 75 (1989), 41–95; see also Ronald Calinger, *A Contextual History of Mathematics to Euler* (London: Prentice-Hall, 1999), pp. 395–655.

mate alliance with the new philosophies; at the Jesuit colleges, it marked the novelty and discipline of their teaching; in the emerging scientific societies and academies, it was a recognized expertise, of interest particularly to the disciplines of astronomy and mechanics. During the early modern period, these institutional and social frameworks underwent significant changes (see Part II of this volume). For pure mathematics, the definitive contextual changes came after 1700: The eighteenth-century academies provided the setting for pure mathematical achievements within the conglomerate of mathematics and rational mechanics; a recognizable profession of pure mathematicians first arose in the context of the nineteenth-century universities.

Part IV

CULTURAL MEANINGS OF NATURAL KNOWLEDGE

29

RELIGION

Rivka Feldhay

Three kinds of narratives have shaped the historiography concerning the relationships between science and the Christian religion.[1] Stories about the "conflict between religion and science," in the words of J. W. Draper, or the "warfare of science and theology," in the words of A. D. White, captured the imagination of Western secular intellectual elites in the nineteenth century.[2] As Draper put it in 1875, "The history of Science is not a mere record of isolated discoveries; it is a narrative of the conflict of two contending powers, the expansive force of the human intellect on one side, and the compression arising from traditional faith and human interests on the other."[3] In stories of this sort, the victory of science over religion lies at the heart of the admirable march of reason that began in Greek antiquity and culminated in the scientism of the nineteenth century. This historiographical tradition rests on a selective, and highly moralized, presentation of a few episodes of real clash between scientific ideas and religious authority, such as the Counter-Reformation Church's condemnation of Galileo Galilei (1564–1642) or nineteenth- and twentieth-century Christian rejections of evolutionary theory, framed by an essentialized understanding of science and theology conceived in terms of the self and its enemies.

Although this story remains surprisingly influential, especially in the popular historiography of science, more recent scholars have developed two alternative and contrasting narratives. A number of theologians, scientists, and

---

[1] For a few general introductions to the whole topic, see David C. Lindberg and Ronald L. Numbers, eds., *God and Nature: Historical Essays on the Encounter between Christianity and Science* (Berkeley: University of California Press, 1986), pp. 1–18; David C. Lindberg and Ronald L. Numbers, *When Science and Christianity Meet* (Chicago: University of Chicago Press, 2003), pp. 1–5; and John H. Brooke, *Science and Religion: Some Historical Perspectives* (Cambridge: Cambridge University Press, 1991), pp. 1–11.

[2] John W. Draper, *History of the Conflict between Religion and Science* (New York: D. Appleton, 1875); and Andrew D. White, *A History of the Warfare of Science with Theology in Christendom* (New York: D. Appleton, 1896; repr. New York: Dover Publications, 1960).

[3] Draper, *History*, p. vi.

some historians have argued that the more typical – and more commendable – relationship between religion and science has involved a separate and peaceful coexistence. Adherents to this view include Protestants and Catholics, conservatives and liberals, the devout and the secular, and they present science and religion as two autonomous enterprises that deal with nature and human life from different perspectives.[4] Science provides specialized technical knowledge about the natural world but is morally neutral, whereas religion supplies a system of norms.

A third way of conceptualizing the relationship of science to religion treats them as addressing related problems within a common framework and thus emphasizes the affinities and interactions between them. Beginning with sociologist Robert Merton's seminal monograph *Science, Technology, and Society in Seventeenth-Century England* (1938), sociologists and historians of early modern English science have explored relationships between the Calvinist attitudes of many English Puritans and the ethos of mainstream natural philosophers.[5] Another, more general tendency within this historiographical tradition emphasizes metaphysics, identifying broad conceptual categories (e.g., the "laws of nature") from which patterns of ideas in both religious and scientific domains were derived.[6] One of the most ambitious practitioners of this strand of historiography was the historian Amos Funkenstein, who made the bold claim that in the seventeenth century, "science, theology and philosophy [were] seen as almost one and the same occupation."[7]

---

[4] From the Catholic point of view, see Pierre Duhem, "Notice sur les titres et travaux scientifiques de Pierre Duhem," in *Mémoires de la Société des Sciences Physiques et Naturelles de Bordeaux*, 7th ser., 1 (Paris: Gauthier-Villars, 1917), pp. 71–169, trans. Y. Murciano and L. Schramm, rev. Pierre Kersberg, in *Science in Context*, 1 (1987), 333–48. From the Protestant point of view, see Ian G. Barbour, *Issues in Science and Religion* (Englewood Cliffs, N.J.: Prentice-Hall, 1966); and more recently Barbour, "Ways of Relating Science and Theology," in *Physics, Philosophy, and Theology: A Common Quest for Understanding*, ed. Robert J. Russell, William R. Stoeger, and George V. Coyne (Vatican City: Vatican Observatory, 1988), pp. 21–48. Examples of historical narratives written from this perspective include Richard S. Westfall, *Science and Religion in Seventeenth-Century England* (Ann Arbor: University of Michigan Press, 1973); J. J. Langford, *Galileo, Science, and the Church* (Ann Arbor: University of Michigan Press, 1971); and John Dillenberger, *Protestant Thought and Natural Science: A Historical Study* (New York: Abingdon Press, 1960).
[5] Robert K. Merton, *Science, Technology, and Society in Seventeenth-Century England* (New York: H. Fertig, 1970), originally published in *Osiris*, 4 (1938), 360–632. In the same tradition, see R. Hooykaas, *Religion and the Rise of Modern Science* (Grand Rapids, Mich.: Eerdmans, 1972); and Colin A. Russell, ed., *Cross-currents: Interactions between Science and Faith* (Grand Rapids, Mich.: Eerdmans, 1985).
[6] Alfred North Whitehead, *Science and the Modern World* (New York: Macmillan, 1925); and John H. Brooke, Margaret J. Osler, and Titse M. van Meer, eds., *Science in Theistic Contexts: Cognitive Dimensions*, *Osiris*, 2nd ser., 16 (2001). Historical narratives written from this perspective include Francis Oakley, *Creation: The Impact of an Idea* (New York: Scribners, 1969); Eugene M. Klaaren, *Religious Origins of Modern Science: Belief in Creation in Seventeenth-Century Thought* (Grand Rapids, Mich.: Eerdmans, 1977); Owen Hannaway, *The Chemists and the Word: The Didactic Origins of Chemistry* (Baltimore: Johns Hopkins University Press, 1975); and Margaret J. Osler, *Divine Will and the Mechanical Philosophy: Gassendi and Descartes on Contingency and Necessity in the Created World* (Cambridge: Cambridge University Press, 1994).
[7] Amos Funkenstein, *Theology and the Scientific Imagination from the Middle Ages to the Seventeenth Century* (Princeton, N.J.: Princeton University Press, 1986), p. 3.

This chapter draws upon all three of these historiographical traditions. It is based on the rich body of historical knowledge accumulated within their various frameworks but is reducible to none of them. Neither simple notions of conflict or separation nor general invocations of "interaction" are powerful enough to capture the subtlety and complexity of the transformations of early modern European culture. In the early modern period, Christianity was shaken to its very core. The novel theology elaborated by Martin Luther (1483–1546) was indicative of the profound changes involved, but other traditions, both religious and intellectual, also combined to reshape the mental world of early modern Europeans and produced dramatic modifications in religious conceptions of salvation as well as in philosophical ideas concerning being and knowledge. At the same time, natural knowledge gained a new significance and status. Both historical processes were shaped by, and also helped to shape, major political and economic developments, such as the emergence of absolutist nation-states, geographical exploration and imperial conquest, and the rise of early modern urban middle-class as well as court societies (see the following chapters in this volume: Vogel, Chapter 33; Moran, Chapter 11). Both occurred in the context of broad intellectual movements – humanism and neo-Platonism in particular – that opened up new perspectives on God, man, and nature. Both entailed new relationships between "high" and "low" culture and between theoretical and practical knowledge. These developments interacted to transform the world of early modern Christianity – in its political and intellectual as well as its theological aspects. One result was the crystallization of a cultural form now recognizable as "science" – in contrast with the older *scientiae* – which had a relationship to religion different from that which had previously obtained between the *scientiae* and theology.[8]

I begin my discussion of these developments by laying out the context that shaped and ultimately transformed older views of the relationship between natural knowledge and theology, describing major cognitive developments as well as the new educational institutions and programs that developed in both Protestant and Catholic areas of Europe. In the second half of this chapter, I use three broad analytical categories to analyze the discourses about God and nature from the Middle Ages to the seventeenth century in terms of the construction of their objects, the constant contestation and redefinition of boundaries among them, and the authority and legitimization of their speakers. (1) Among *objects of discourse*, I discuss the motion of the earth, for example, whose construction by Copernicans as invisible in principle was closely connected to the Copernicans' new methods of investigation and explanation. (2) By the *boundaries between the domains of discursive communities*, I refer to those differences invoked in the speech of mathematicians, natural

---

[8] See Rainer Berndt, Matthias Lutz-Buchmann, Ralf M. W. Stammberger, Alexander Fidora, and Andreas Niederberger, eds., *"Scientia" und "Disciplina": Wissenstheorie und Wissenschaftspraxis im 12. und 13. Jahrhundert* (Berlin: Akademie Verlag, 2002).

philosophers, and theologians that reflect the different criteria used within each community to certify in the hierarchy of truth and value. (3) By *practice of authorization*, I mean the ways in which mathematicians, natural philosophers, and other natural inquirers attempted to legitimize and authorize their claims concerning natural facts as they related to religious authority, which was itself simultaneously attempting to reestablish its lost hegemony in a changing intellectual world.

Early modern Europe witnessed the first attempts to institutionalize boundaries between cultural spheres – specifically the religious, the political, and the scientific – rather than simply between disciplines. Francis Bacon's (1561–1626) insistence on the need to study the book of scripture and the book of nature not only separately but with different methods, Galileo's claim (quoting Cardinal Baronius) that the scriptures "teach us how to go to heaven, not how heaven goes,"[9] and René Descartes' (1596–1650) extreme dualism of mind and matter all appear to signal a drive toward separation. Over the course of the seventeenth century, however, these preliminary attempts had a paradoxical effect, increasing the interdependence of the two domains of religion and natural knowledge: The only way to legitimize natural knowledge without invoking the authority of religion was to sacralize that knowledge, leading naturalists to couch their own claims in a new kind of religious rhetoric. At the same time, however, religious establishments, threatened by the fragmentation of Christianity, attempted to reclaim authority by connecting themselves to the new natural philosophy. Although expressed in different ways, this double dynamic was common both to Catholicism and the reformed religions, and it shaped the collective enterprise of reformed natural knowledge as well as its relations to religious communities.

THEOLOGICAL AND INTELLECTUAL CONTEXTS: SACRED MESSAGE AND BODIES OF KNOWLEDGE

The emergence of this characteristic early modern dialectic between religion and science should be understood against the background of a persistent tension regarding salvation that lay at the heart of medieval Christian culture.[10] Medieval scholars and theologians inherited from the Church Fathers – and in particular from Augustine of Hippo (354–430) – a commitment to faith as a special, protected, and superior form of cognition. This view involved a deep ambivalence toward the natural and secular knowledge of the Greek and

---

[9] Galileo Galilei, *Letter to the Grand Duchess Christina*, in Maurice A. Finocchiaro, *The Galileo Affair: A Documentary History* (Berkeley: University of California Press, 1989), p. 96.
[10] On soteriological tension and the organization of religious cultures, see S. N. Eisenstadt and Ilana Friedrich Silber, eds., *Cultural Traditions and Worlds of Knowledge: Explorations in the Sociology of Knowledge* (London: Jai Press, 1988), particularly Eisenstadt's chapter, "Explorations in the Sociology of Knowledge: The Soteriological Axis in the Construction of Domains of Knowledge," pp. 1–71.

Roman worlds. Augustine and others thought that knowledge of God and the redemptive role of Christ – that is, the knowledge necessary for salvation – could not be completely disengaged from other forms of knowledge, so that grammar and rhetoric, as well as parts of Greek philosophy and science, were relevant for understanding the Bible and receiving the Christian message.[11] Their utility for the life of a Christian community was also recognized. This view coexisted uneasily with the attitude of some earlier Church Fathers who were mostly indifferent to or uninterested in pagan philosophy and science, and sometimes even hostile toward them.[12]

On the one hand Medieval Christian culture remarkably absorbed classical Greek bodies of knowledge translated and developed by Arabic and Hebrew commentators in all areas of natural philosophy and mathematics, including the complete Aristotelian corpus, Ptolemaic mathematical astronomy, geometrical optics, and the medical treatises of antiquity.[13] On the other hand, cultural mechanisms were established in order to contain and control bodies of knowledge and prevent them from undermining or interfering with the sacred message. This was accomplished in two ways: epistemologically, by developing principles to organize the various forms of knowledge in a hierarchy, with Christian theology lodged securely at its apex; and institutionally, by the creation of a class of Church intellectuals – mainly theologians – who were responsible for the selection, legitimization, use, and suppression of other bodies of knowledge in accordance with Christian concerns. Thus, philosophy was baptized as the "handmaid" to theology in the Christian tradition.[14]

In spite of the tensions that pervaded the Christian tradition, writers of the High Middle Ages also bequeathed to their early modern successors a

---

[11] Augustine, *De doctrina Christiana*, ed. Josef Martin (Corpus Christianorum Series Latina, 32) (Turnholt: Brepols, 1962), bk. 2, 39–40, pp. 72–4. For the complex attitude of Augustine and his generation toward profane knowledge, see Peter Brown, *The World of Late Antiquity* (Cambridge, Mass.: Harvard University Press, 1987); Brown, *Augustine of Hippo: A Biography* (London: Faber and Faber, 1967); Norman Kretzmann, "Faith Seeks, Understanding Finds: Augustine's Charter for Christian Philosophy," in *Christian Philosophy*, ed. Thomas P. Flint (University of Notre Dame Studies in the Philosophy of Religion, 6) (Notre Dame, Ind.: University of Notre Dame Press, 1990), pp. 1–36; and John M. Rist, *Ancient Thought Baptized* (Cambridge: Cambridge University Press, 1994).

[12] See Lindberg, "Science and the Early Church," in Lindberg and Numbers, eds., *God and Nature*, pp. 19–48; Lindberg, "The Medieval Church Encounters the Classical Tradition: Saint Augustine, Roger Bacon, and the Handmaiden Metaphor," in Lindberg and Numbers, eds., *When Science and Christianity Meet*, pp. 7–32.

[13] For detailed description and analysis of this process of absorption and development, see John E. Murdoch and Edith D. Sylla, eds., *The Cultural Context of Medieval Learning: Proceedings of the First International Colloquium on Philosophy, Science, and Theology in the Middle Ages – September 1973* (Boston Studies in the Philosophy of Science, vol. 26) (Dordrecht: Reidel, 1975); David C. Lindberg, *The Beginnings of Western Science: The European Scientific Tradition in Philosophical, Religious, and Institutional Context, 600 B.C. to A.D. 1450* (Chicago: University of Chicago Press, 1992); Edward Grant, *The Foundations of Modern Science in the Middle Ages: Their Religious, Institutional, and Intellectual Contexts* (Cambridge: Cambridge University Press, 1996); Grant, *God and Reason in the Middle Ages* (Cambridge: Cambridge University Press, 2001).

[14] Lindberg, *The Beginnings of Western Science*, chap. 10.

framework of thought within which natural knowledge was related to the sacred message in an impressive synthesis. In the work of the Dominican theologian Thomas Aquinas (ca.1225–1274), this synthesis rested on ontological as well as epistemological bridges that bound together the transcendental, human, and physical realms of the universe. For Aquinas, the unity of essence and existence represented the highest degree of being – the fullest metaphysical reality – and this belonged to God alone. Creatures differed according to their degree of participation in God's being, ranging from angels and spirits downward, through the human composite of rational soul and body, to animals, plants, and minerals, and ultimately to the elements.[15] This metaphysical hierarchy of beings underpinned the idea of a hierarchy of knowledge,[16] which situated knowledge of physical reality in a relatively low position and displaced mathematics to the margins (see Smith, Chapter 13, this volume). As a Christian theologian and commentator on Aristotle, Aquinas saw Aristotelian physics – the nonmathematical science of motion and change – as necessary for understanding the lower degrees of being in God's creation, but he believed that the main goal of physical knowledge was to underpin the metaphysics that dealt with higher degrees of being and created a bridge to theology, at the center of which was God, the ultimate object of human intellect and desire.[17] Yet natural knowledge, anchored in sense perception, as Aristotle had argued, was more than just preparation for divine knowledge because the explanatory concepts that organized it – individual natures, matter and form, substance and accidents – applied to every kind of reality, even the reality of Christ's presence in the Eucharist.

This intricately structured synthesis did not long outlive Aquinas' death in 1274. Shaken in the early fourteenth century by John Duns Scotus's (ca. 1266–1308) emphasis on God's radical otherness and by William of Ockham's (ca. 1285–1349) nominalist critique of the ontological bond uniting creation with its creator, it met its ultimate match in the new theological vision of Luther, crystallized in the theses he nailed to the doors of the cathedral in

---

[15] On the distinction between essence and existence and its significance, see Thomas Aquinas, *De ente et essentia*, chap. 4, in *Le "De ente et essentia" de S. Thomas d'Aquin: Texte établi d'après les manuscrits parisiens*, ed. Marie Dominique Roland-Gosselin (Le Saulchoir, Kain, Belgium: Revue des sciences philosophiques et théologiques, 1926), p. 36.10–15: "Est ergo distinctio earum ad invicem secundum gradum potentiae et actus ... Et hoc competur in anima humana etc." See also Aquinas, *Summa theologica*, I q. 4 a. 3: "Et hoc modo illa quae sunt a Deo, assimilantur ei inquantum sunt entia, ut primo et universali principio totius esse" (quoted in Funkenstein, *Theology and the Scientific Imagination*, p. 51). For a sympathetic reading, see Étienne Gilson, *The Christian Philosophy of St. Thomas Aquinas*, trans. L. K. Shook (London: V. Gollancz, 1967); Gilson, "Medieval Science and Its Religious Context," *Osiris*, 2nd ser., 10 (1995), pp. 61–79.

[16] Thomas Aquinas, *Commentary on the De trinitate of Boethius*, q. 5 a. 1, in *Thomas Aquinas: The Division and Methods of the Sciences*, ed. and trans. Armand Maurer (Toronto: Pontifical Institute of Mediaeval Studies, 1986), pp. 9–24. For a modern interpretation of the organization of knowledge in the Middle Ages, see James A. Weisheipl, "Classification of the Sciences in Medieval Thought," *Medieval Studies*, 27 (1965), 54–90.

[17] Aquinas, *Summa theologica*, I q. I a. 1, 2, 5.

Wittenberg in 1517. Following his voluntarist and nominalist predecessors, Luther emphasized God's sovereignty and radicalized the suspicion of knowledge as a bridge to salvation. He also attempted to eradicate all trust in the natural moral qualities of man, creating a radical rift between theology and philosophy.[18] Luther thought that where logic applied, one dealt with knowledge and not with faith; the articles of faith are "not against dialectical truth, but rather outside, under, above, below, around, and beyond it."[19] In this view, the nature of God could not possibly be imagined as an object of theological discourse around which scientific disciplines could be structured, bounded, and legitimized.

Luther's theological challenge to the epistemological and metaphysical foundations of the Thomistic synthesis of faith and knowledge, grace and nature, did not take place in a philosophical vacuum. While the different confessions were struggling over people's souls, the intellectual culture of early modern Europe was being transformed by the humanistic revival of classical antiquity, begun a century or more before Luther's time (see Blair, Chapter 17, this volume). The new editions and translations of Platonic, neo-Platonic, and Hermetic works, as well as of the writings of the Roman naturalists and Greek mathematicians, combined with the explosion of contemporary interest in the vast field of rhetoric, which emphasized praxis, persuasion, and prudence at the expense of logic, to provide new nodal points around which to reorganize the relationship of God, human beings, and nature (see Serjeantson, Chapter 5, this volume).[20] Substantive challenges to Aristotelian natural philosophy as well as to orthodox theology arose from the works of Marsilio Ficino (1433–1499), Theophrastus Bombastus

---

[18] Steven Ozment, *The Age of Reform (1250–1550)* (New Haven: Yale University Press, 1980), pp. 231–44; and Lindberg and Numbers, *God and Nature*, pp. 167–91.

[19] Ozment, *The Age of Reform*, p. 238.

[20] On the impact of Renaissance Platonism, see Eugenio Garin, *Studi sul Platonismo medievale* (Florence: F. Le Monnier, 1958); and James Hankins, *Plato in the Italian Renaissance*, 2 vols. (Leiden: E. J. Brill, 1990). For the "mathematical" Renaissance, see Edward W. Strong, *Procedures and Metaphysics: A Study in the Philosophy of Mathematical-Physical Science in the Sixteenth and Seventeenth Centuries* (Berkeley: University of California Press, 1936); and Paul L. Rose, *The Italian Renaissance of Mathematics: Studies on Humanists and Mathematicians from Petrarch to Galileo* (Geneva: Droz, 1975). For developments in mechanics, see Stillman Drake and I. E. Drabkin, ed. and trans., *Mechanics in Sixteenth-Century Italy: Selections from Tartaglia, Benedetti, Guido Ubaldo and Galileo* (Madison: University of Wisconsin Press, 1969); and Alan Gabbey, "Between ars and philosophia naturalis: Reflections on the Historiography of Early Modern Mechanics," in *Renaissance and Revolution: Humanists, Scholars, Craftsmen, and Natural Philosophers in Early Modern Europe*, ed. Judith V. Field and Frank A. J. L. James (Cambridge: Cambridge University Press, 1993), pp. 133–45. On the Renaissance influence of Pliny, see Roger French and Frank Greenaway, eds., *Science in the Early Roman Empire: Pliny the Elder, His Sources and Influence* (London: Croom Helm, 1986). On rhetoric and its emphasis on praxis, see Garin, *Italian Humanism: Philosophy and Civic Life in the Renaissance*, trans. P. Munz (Oxford: Blackwell, 1965); Nancy S. Struever, *Theory as Practice: Ethical Inquiry in the Renaissance* (Chicago: University of Chicago Press, 1992); and Victoria Kahn, *Rhetoric, Prudence, and Skepticism in the Renaissance* (Ithaca, N.Y.: Cornell University Press, 1985).

von Hohenheim, or Paracelsus (1493–1541),[21] Heinrich Cornelius Agrippa von Nettesheim (1486–1535),[22] John Dee (1527–1608), Tommaso Campanella (1568–1639), and Giordano Bruno (1548–1600) (see the following chapters in this volume: Garber, Chapter 2; Copenhaver, Chapter 22; Newman, Chapter 21). Working for the most part outside the framework of the universities, these writers elaborated ideas that rendered meaningless the boundaries between theory and practice, nature and culture, and philosophy and theology that structured the university curriculum, especially in their diverse and daring conceptions of new objects that infused matter with spirit. Some even recruited the authority of the early reformers, especially Luther, who expressed sympathy for the practices of alchemy.[23] The compliment, however, was rarely returned. In most instances, the reaction of both Catholic and Protestant religious establishments to these new ideas was harshly intolerant. Protestants invoked the literal interpretation of scripture to reject these new approaches to natural knowledge, while Catholics, especially after the Council of Trent (1545–63), sat down to rewrite Inquisitorial manuals in order to eradicate what they deemed a new wave of heresies; the condemnation and execution of Bruno in 1600 was a case in point.

The sense of crisis fostered by these intellectual challenges was intensified by concerns regarding the irrelevance of theological discussion to Christian piety; these were intensified by Luther's break with Rome and ultimately led to deep changes within early modern Catholicism as well.[24] A new balance had to be achieved on the one hand between the need to educate the priesthood and the Catholic public – and more generally to make traditional Catholicism more appealing – and, on the other, the desire to secure the authority of the

---

[21] Walter Pagel, "Paracelsus and the Neoplatonic and Gnostic Tradition," *Ambix*, 8 (1960), 125–66; Allen G. Debus, *The Chemical Philosophy: Paracelsian Science and Medicine in the Sixteenth and Seventeenth Centuries*, 2 vols. (New York: Science History Publications, 1977); Massimo L. Bianchi, "The Visible and the Invisible: From Alchemy to Paracelsus," in *Alchemy and Chemistry in the 16th and 17th Centuries*, ed. Piyo Rattansi and Antonio Clericuzio (Dordrecht: Kluwer, 1994), pp. 17–50.

[22] Agrippa wrote the *De occulta philosophia* in 1509 and 1510; it was first published in 1531 in Antwerp by John Grapheus. Ficino himself elaborated and applied his own cosmology in *De vita coelitus comparanda* (1489); for a modern translation, see *The Book of Life*, trans. Charles Boer (Woodstock, Conn.: Spring Publications, 1996). For a study of Ficino's and Agrippa's recommended practices, see Wayne Shumaker, *The Occult Sciences in the Renaissance* (Berkeley: University of California Press, 1972), chap. 3.

[23] See S. F. Mason, "The Scientific Revolution and the Protestant Reformation – II: Lutheranism in Relation to Iatrochemistry and the German Nature-Philosophy," *Annals of Science*, 9 (1953), p. 155.

[24] *Manifestations of Discontent in Germany on the Eve of the Reformation: A Collection of Documents*, selected, translated, and introduced by Gerald Strauss (Bloomington: Indiana University Press, 1971); Johann Huizinga, *The Waning of the Middle Ages: A Study of the Forms of Life, Thought, and Art in France and Netherlands in the XIVth and XVth Centuries* (London: E. Arnold, 1924); R. R. Post, *The Modern Devotion: Confrontation with Reformation and Humanism* (Leiden: E. J. Brill, 1968); Charles Trinkaus, *"In Our Image and Likeness": Humanity and Divinity in Italian Humanist Thought*, 2 vols. (Chicago: University of Chicago Press, 1970); Henry Outram Evenett, *The Spirit of the Counter-Reformation* (Cambridge: Cambridge University Press, 1968); and Steven Ozment, ed., *The Reformation in Medieval Perspective* (Chicago: Quadrangle Books, 1971); Jürgen Helm and Annette Winkelmann, eds., *Religious Confessions and the Sciences in the Sixteenth Century* (Leiden: Brill, 2001).

Roman Church.[25] In the end, the Council of Trent reached a consensus around Thomist theology, renewing the Roman Church's commitment to theological knowledge together with all the philosophical background that this traditionally presupposed. In addition – and most significantly – the Council of Trent limited its authority to "matters of faith" and did not pretend to extend it to matters of philosophy.[26]

RELIGIOUS IDENTITIES AND EDUCATIONAL REFORMS

Just as sixteenth-century Protestantism and Catholicism responded to similar religious sensibilities, so too they invented similar practical solutions to cope with the need to "modernize" society.[27] All over Europe, the disruption of the sacred order that followed Luther's break with Rome threatened the social and political orders as well. In the Protestant territories, critiques of clerical institutions were joined by more radical calls from below to apply the binding norms of the Bible to social and political life, not only during the peasant wars and the Anabaptist rebellions of the 1520s and 1530s but also in Wittenberg, Zurich, Geneva, Strasbourg, and other cities in the German world; as a result, the reformers focused increasingly on restoring social harmony without disrupting the existing distribution of power and wealth.[28] Confessionalization – the process whereby European societies

---

[25] The historiographical debate on the nature of early modern Catholicism is still open, revolving around issues of Catholic revival versus Catholic repression, and continuing the work of Hubert Jedin on the one hand and Delio Cantimori on the other. Consistent with old paradigms, however, a new view of early modern Protestant and Catholic "reformations" as manifesting the construction of new religious identities through confessionalization and disciplining is gaining much influence. For a succinct analysis of the historiography, see William V. Hudon, "Religion and Society in Early Modern Italy – Old Questions, New Insights," *American Historical Review*, 101 (1996), 783–804; John W. O'Malley, "The Historiography of the Society of Jesus: Where Does It Stand Today?," in *The Jesuits: Cultures, Sciences, and the Arts, 1540–1773*, ed. John W. O'Malley, Steven J. Harris, and T. Frank Kennedy (Toronto: University of Toronto Press, 1999), pp. 3–37.

[26] For the original text of the decree on tradition and interpretation, with an interpretation different from the one suggested here, see Richard J. Blackwell, *Galileo, Bellarmine, and the Bible* (Notre Dame, Ind.: University of Notre Dame Press, 1991), chap. 1, appendix 1, pp. 12–13.

[27] For some prominent studies that paved the way toward, and constituted the view of, Protestant and Catholic reforms as forms of modernization, confessionalization, and disciplining, see S. N. Eisenstadt, ed., *The Protestant Ethic and Modernization: A Comparative View* (New York: Basic Books, 1968); R. W. Green, ed., *Protestantism, Capitalism, and Social Science: The Weber Thesis Controversy* (Lexington, Mass.: Heath, 1973); W. Reinhard, "Gegenreformation als Modernisierung? Prolegomena zur einer Theorie des konfessionellen Zeitalters," *Archiv für Reformationsgeschichte*, 68 (1977), 226–52; Reinhard, "Reformation, Counter-Reformation, and the Early Modern State: A Reassessment," *Catholic Historical Review*, 75 (1989), 383–404; Paolo Prodi and W. Reinhard, eds., *Il Concilio di Trento e il Moderno* (Bologna: Il Mulino, 1996); Paolo Prodi and Carla Penuti, eds., *Disciplina dell'anima, disciplina del corpo e disciplina della società tra medioevo ed età moderna* (Bologna: Il Mulino, 1994); and R. Po-Chia Hsia, *Social Discipline in the Reformation: Central Europe, 1550–1750* (London: Routledge, 1989).

[28] Ozment, *The Age of Reform*, pp. 264–9, 362, 366–7, and 372. See also Ozment, *The Reformation in the Cities: The Appeal of Protestantism to Sixteenth-Century Germany and Switzerland* (New Haven, Conn.: Yale University Press, 1975); and Ozment, *Protestants: The Birth of a Revolution* (New York: Doubleday, 1992).

were politically and culturally structured along contrasting religious lines – required much more than the formulation of dogma. It also involved a long, slow process in which religious identities were crystallized through a wide variety of practices, including discipline by force, censure, persuasion, and ritual; propaganda through catechism and the arts; and education in schools, colleges, and universities.[29]

In many respects, educational reforms in both Protestant and Catholic areas responded to similar concerns and followed similar patterns. In the largely Protestant German world, the key institution that channeled and controlled religious, social, and political energy was the university, and the sixteenth century saw a series of wide-ranging reforms in university education that were modeled on the changes instituted at the University of Wittenberg by Luther's collaborator and successor, Philip Melanchthon (1497–1560). Melanchthon strengthened the control of teaching masters over students and charged the university's rector with enforcing Lutheran orthodoxy. He aimed to secure simultaneously the intellectual and the moral character of graduates by imposing on them the Augsburg Confession (the Lutheran profession of faith), by emphasizing the study of scripture, the works of Augustine, and the history of church councils, and by mandating the practice of annual disputations over theses censured by the rector.[30] Other Protestant universities implemented educational reform on the Wittenberg model and began to produce a new elite of professors, priests, and counselors to princes, who became the principal administrators in Protestant lands.

In Catholic Europe, the most ambitious innovators in the area of education were the Jesuits, the new religious order founded by Ignatius Loyola (1491–1556) in 1540. The Jesuits aimed to reform the Catholic world through what they called "studies and moral formation," and they developed an educational program intended not only to train – or "form" – priests but also to educate the entire Catholic population, not least generations of future rulers.[31] The

---

[29] For all aspects of social disciplining in Italy, see Prodi and Penuti, *Disciplina dell'anima*; for the Central European case, see Hsia, *Social Discipline in the Reformation*, with a comprehensive list for further reading at pp. 188–90. For a revisionist view of the Italian Inquisition with references to the Spanish case, see Hudon, "Religion and Society," notes 28, 38, and 39.

[30] See Gian Paolo Brizzi's remark in "Da 'domus pauperum scholarium' a collegio d'educazione: Università e collegi in Europa (secoli XII–XVIII)," in Prodi and Penuti, *Disciplina dell'anima*, pp. 809–40; and Riccardo Burigana, "La disciplina nelle università tedesche della prima Riforma: Il modello di Wittenberg," in Prodi and Penuti, *Disciplina dell'anima*, pp. 841–62.

[31] For the widening vision of Jesuit education in the sixteenth century, see, for example, the document written by the Rector of the Collegium Germanicum, M. Lauretano, "Utrum convictus iuvenum nobilium in collegio germanico conservandus sit?," in *Monumenta Paedagogica Societatis Iesu*, ed. Ladislaus Lukács (Rome: Monumenta Historica Societatis Iesu, 1974), pp. 995–1004. For the system of values behind Jesuit education, see Steven J. Harris, "Transposing the Merton Thesis: Apostolic Spirituality and the Establishment of the Jesuit Scientific Tradition," *Science in Context*, 3 (1989), 29–65. For the Jesuit educational ideology, see Rivka Feldhay, *Galileo and the Church: Political Inquisition or Critical Dialogue?* (New York: Cambridge University Press, 1995), chap. 6. On "studies and moral formation," see Feldhay, *Galileo and the Church*, p. 147. On the development of an "art of governing" in the educational context, see Gian-Mario Anselmi, "Per un'archeologia della

Jesuits administered a network of hundreds of colleges, centered in Rome, unified by common curricular goals and pedagogical practices (see Vogel, Chapter 33, this volume).[32] They aimed to integrate the humanistic studies into a more traditional program of scholastic learning, and many of their colleges offered a course in the humanities, followed by a course in philosophy. Others developed into full-fledged universities that included a faculty of theology. The Jesuits' success in the field of education brought under their direction not only teaching colleges for novices and lay students but also special colleges for the nobility and seminaries for the priesthood, such as the German College in Rome.[33] These schools were the Catholic answer to the problem of discipline and control that preoccupied Melanchthon, and, like the reformed universities he envisaged, they were to be responsible not only for the intellectual formation of their students but also for their religious observance, social conduct, and moral education.

In neither the Protestant nor the Catholic world, however, was educational reform confined to issues of discipline and administration; it involved profound curricular changes as well. Here, however, the interests and concerns of Protestant and Catholic educators diverged with respect to the relationship between religious instruction and natural knowledge. One of Melanchthon's central concerns was to recast the curriculum to protect the authority of Protestant theologians over the interpretation of scripture – now open to the individual reading of all believers – and to develop a moral theology that would require obedience to civil rulers. Historian Sachiko Kusukawa has argued that Melanchthon used Luther's distinction between divine law – under which Melanchthon included natural philosophy – and scripture to legitimize the new educational program.[34] But natural philosophy had another, broader function as well. God was understood to have left signs to be discerned in the nature that was his creation. Traceable a posteriori through experience and reasoning, these signatures yielded an important but limited kind of knowledge that, when combined with dedication to the message of scripture, supported and sustained faith.

The Protestant quest for divine signatures in nature created new priorities among the objects of natural knowledge. Among the most striking examples was Melanchthon's new reading of one of the basic texts of medieval natural

---

Ratio: Dalla 'pedagogia' al 'governo'," in *La "Ratio Studiorum": Modelli culturali e pratiche educative dei Gesuiti in Italia tra Cinque e Seicento*, ed. Gian Paolo Brizzi (Rome: Bulzoni, 1981), pp. 11–42.

[32] On the educational models of the Jesuit colleges and their pedagogy, see G. Codina Mir, *Aux sources de la pédagogie des Jésuites: Le "modus parisiensis"* (Rome: Institutum Historicum Societatis Iesu, 1968).

[33] G. Angelozzi, "'La virtuosa emulazione.' Il disciplinamento sociale nei 'seminaria nobilium' gesuiti," in *Sapere e/è potere: Discipline, dispute e professioni nell'Università medievale e moderna: Il caso bolognese a confronto*, ed. A. De Benedictis, 3 vols. (Bologna: Istituto per la Storia di Bologna, 1990), 3: 85–108.

[34] Sachiko Kusukawa, *The Transformation of Natural Philosophy: The Case of Philip Melanchthon* (Cambridge: Cambridge University Press, 1995); and Kusukawa, "The Natural Philosophy of Melanchthon and His Followers," in *Sciences et Religions de Copernic à Galilée, 1540–1610* (Rome: École Française de Rome, 1999), pp. 443–53, esp. pp. 443–4.

philosophy, Aristotle's *De anima* (On the Soul).[35] Unlike most earlier commentators on this work, Melanchthon was not interested in its metaphysical aspects. Instead, he complemented the Aristotelian text with a reading of the *De humani corporis fabrica* (On the Fabric of the Human Body, 1543) by the anatomist Andreas Vesalius (1514–1564), which provided the anatomical knowledge he considered necessary to understand the soul's relation to the body (see Cook, Chapter 18, this volume).

A second theme in the new Protestant curricula was a deep interest in studying the heavens, which were considered, even more than other parts of creation, to be imprinted with God's signs.[36] One result of this emphasis on astronomy was what historian Robert Westman has called the "Wittenberg interpretation" of Copernicanism (see Donahue, Chapter 24, this volume).[37] The "Wittenberg interpretation" enabled astronomers to use the Copernican models to determine and predict the angular positions of planets without changing the picture of the universe organized around a central, stationary earth. Members of the circle of brilliant students around Melanchthon – including Joachim Camerarius, Jacob Heerbrand, and Samuel Eisenmenger (teacher of Johannes Kepler) – carried this interpretation from Wittenberg to Leipzig, Tübingen, and Heidelberg and turned Germany into what the French pedagogical reformer Petrus Ramus called "the nursery of mathematics."[38]

The Jesuit colleges that rivaled the Protestant universities stressed useful knowledge; this included rhetoric, also prized by Protestant educators, as well as "mixed mathematics" – mathematics applied to the physical world – a field of knowledge at best marginal in the context of medieval natural philosophy and theology.[39] In this area, the Jesuits were building on new humanist editions and translations of mathematical texts, as well as new genres of writing about practical geometry and the mechanical arts. Scholars such as Niccolò Tartaglia (1505–1557), Federico Commandino (1509–1575), and Francesco Maurolico (1494–1575) had begun the work of recovering the lost works of Greek mathematicians (see Andersen and Bos, Chapter 28, this volume). At the same time, mathematicians from academic backgrounds began to take an increasing interest in the work of practitioners of the mechanical arts in areas such as measuring and surveying. In that context, they applied geometrical theorems and techniques – triangulation, projection, squaring,

---

[35] Kusukawa, *Transformation of Natural Philosophy*, chap. 3.
[36] Ibid., pp. 126–42.
[37] Robert S. Westman, "The Melanchthon Circle, Rheticus, and the Wittenberg Interpretation of the Copernican Theory," in *The Scientific Enterprise in Early Modern Europe: Readings from Isis*, ed. Peter Dear (Chicago: University of Chicago Press, 1997), pp. 7–36.
[38] Westman, "The Melanchthon Circle," p. 15.
[39] Steven J. Harris, "Les chaires de mathématiques," in *Les Jésuites à la Renaissance: Système éducatif et production du savoir*, ed. Luce Giard (Paris: Presses Universitaires de France, 1995), pp. 239–61.

ratio, and proportion – to physical objects in the real world (see Bennett, Chapter 27, this volume).[40]

Jesuit mathematicians also combined mathematics with experimental techniques, arguing that the resulting natural knowledge glorified God through the study of his creation. But whereas the central role accorded the mathematical discipline of astronomy at Wittenberg grew out of the Protestant preoccupation with the specific signs of God's providence manifested in creation, the Jesuits formulated their reasons for introducing mathematics into the philosophy curriculum in epistemological and metaphysical terms. To this end, they elaborated the arguments laid out by Tartaglia in his First Lesson on the works of Euclid, according to which "the naturalist differs from the mathematician in that he considers things clothed, whereas the mathematician considers them as bare of any visible material."[41] The Jesuits interpreted this argument in terms of the Aristotelian-Thomistic framework of thought, arguing that because mathematical entities lay higher on the chain of being than physical entities, mathematical demonstrations enjoyed a higher degree of certainty than physical ones.[42] Inspired by this vision, many Jesuits worked to develop not only astronomy, but also optics and mechanics in physico-mathematical directions.[43] These men included Christoph Clavius (1537–1612), the architect of the general enterprise of "Jesuit science," Christoph Scheiner (1573–1650), who did original work on sunspots, Josephus Blancanus (1566–1624), who wrote a treatise on mechanics, and Paul Guldin (1577–1643), who investigated centers of gravity.[44]

The Jesuits used medieval principles for the organization of knowledge, such as "subordination" and "metabasis" (the rule forbidding the transmission of methods from one discipline to another), to prevent unregulated conflict between human knowledge and divine scripture.[45] The Jesuits organized education around hierarchies and boundaries, such as that which classified

---

[40] Strong, *Procedures and Metaphysics*, pp. 94–6; and J. A. Bennett, "The Challenge of Practical Mathematics," in *Science, Culture, and Popular Belief in Renaissance Europe*, ed. Stephen Pumfrey, Paolo Rossi, and Maurice Slawinski (Manchester: Manchester University Press, 1991), pp. 175–90.
[41] Niccolo Tartaglia, *Euclide Megarense acutissimo philosopho, solo introduttore delle scientie mathematice* (Venice: Giovanni Bariletto, 1569), quoted in Strong, *Procedures and Metaphysics*, pp. 61–2, from the edition of 1560.
[42] See, for example, Clavius's Prolegomena to his *Commentaria in Euclidis Elementorum Libri XV*, 3rd ed. (Cologne: Iohannes Baptista Ciotti, 1591), p. 5.
[43] Hugo Baldini, *Legem impone subactis: Studi su filosofia e scienza dei Gesuiti in Italia, 1540–1632* (Rome: Bulzoni, 1992), pp. 19–73; Feldhay, "The Cultural Field of Jesuit Science," in O'Malley et al., eds., *The Jesuits*, pp. 107–26.
[44] On Clavius, see James M. Lattis, *Between Copernicus and Galileo: Christoph Clavius and the Collapse of Ptolemaic Cosmology* (Chicago: University of Chicago Press, 1994); and Baldini, *Legem impone subactis*, pt. II. For the practice of the mathematical sciences and the Jesuits, see Peter Dear, *Discipline and Experience: The Mathematical Way in the Scientific Revolution* (Chicago: University of Chicago Press, 1995). On Guldin, see Rivka Feldhay, "Mathematical Entities in Scientific Discourse: Paulus Guldin and His Dissertatio De motu terrae," in *Biographies of Scientific Objects*, ed. Lorraine Daston (Chicago: University of Chicago Press, 2000), pp. 42–66.
[45] See Feldhay, *Galileo and the Church*, chap. 11.

astronomy under mathematics rather than natural philosophy, as a strategy for containing tensions in the curriculum. These strategies ultimately prevented them from developing their physico-mathematics in the directions taken by Galileo and Newtonian experimental science later in the seventeenth century. Still, Jesuit institutions trained generations of naturalists in new methods and approaches as they gained hegemony over the education of the religious and political Catholic elites.

FROM COPERNICUS TO GALILEO: SCIENTIFIC OBJECTS, BOUNDARIES, AND AUTHORITY

The publication of Nicholas Copernicus's (1473–1543) *De revolutionibus orbium celestium* (On the Revolutions of the Heavenly Spheres, 1543) marks the intersection of two distinct Renaissance intellectual traditions, those of neo-Platonism and mixed mathematics.[46] Whereas the neo-Platonists – the Florentine philosopher Ficino and his followers – had conceived of matter and soul as being unified by a tenuous but material spirit (*spiritus*), the mathematicians had treated entities abstracted from matter. Copernicus's *De revolutionibus* was first read by contemporaries within the existing tradition of mathematical astronomy, as offering only geometrical models for planetary motions, especially because five of the six books dealt with mathematical models and manifested Copernicus's supreme mastery over the techniques of his métier.[47] However, the text could be read in light of other intellectual traditions, notably ones that encouraged a realist reading, giving birth to a variety of different Copernican discourses.[48] With the moving earth, Copernicus introduced a controversial new kind of scientific object, at once physical and mathematical. The idea of the moving earth could also be seen as echoing Ficinian ideas about a motive soul dwelling in all earthly and heavenly bodies, whereas the justification of the central position of the sun resonated with certain elements in sixteenth-century Hermeticism. This plurality of possible readings of Copernicus undermined any attempt to reduce the meaning of *De revolutionibus* to one canonical understanding and limited the possibilities of its control by university and Roman Church elites.

Copernicus's twofold strategy to establish the authority of mathematicians also foreshadowed controversies to come. On the one hand, he claimed a kind of autonomy to the mathematical sciences and combined this claim

---

[46] For an English translation, see Edward Rosen, *On the Revolutions* (Baltimore: Johns Hopkins University Press, 1992).
[47] See Thomas S. Kuhn, *The Copernican Revolution: Planetary Astronomy in the Development of Western Thought* (Cambridge, Mass.: Harvard University Press, 1957), pp. 185–7.
[48] I am using Robert S. Westman's phrase and argumentation from his chapter, "The Copernicans and the Churches," in *God and Nature*, pp. 76–113.

with the contention – already made by Tartaglia and others before him – that mathematics was necessary for all of the other sciences, including philosophy and theology. On the other hand, he appealed to Pope Paul III as the highest arbiter[49] in scholarly disputes in order to defend mathematicians against the attacks of philosophers and theologians.[50] This complicated strategy demonstrates the difficulty of breaking through the traditional patterns of the organization of learning, where astronomy and mechanics were considered to be subordinate sciences,[51] lacking the high status enjoyed by philosophy and theology in medieval universities and in the Roman Church.

The first anti-Copernican critique, *De coelo et elementis* (On the Heavens and Elements), written by the Dominican Giovanni Maria Tolosani[52] (ca. 1470–1549) sometime between 1544 and 1548 but never published, pointed out these ambiguities. He accused Copernicus of transgressing the principles of physics, logic, the hierarchy of the sciences, and scripture all at once.[53] Tolosani's critique testifies to the recognition, even among contemporaries, of the cultural challenge posed by *De revolutionibus*. By positing the motion of the earth as the basis for a new astronomy, Copernicus had pushed into the center of discussion a motion that he presented as the physical effect of a mathematical property; that is, the earth's sphericity. This motion was invisible to any observer on the earth and was thus inaccessible to direct perception by the senses. Copernicus's model also undermined a primary physical intuition about the gravity of the earth, which seemed to require its immobility. Thus, the motion of the earth foreshadowed the construction of reality through an object whose fundamental properties were mathematical. The primary physical property of such an object – its motion – was hardly accessible to the senses in simple, direct ways. Lastly, the materiality of such an object – heavy matter inherently suffused with motion – was also amenable to spiritual interpretations. The emergence of a new physico-mathematical object in the context of a fairly traditional astronomical text sparked contention among mathematicians, natural philosophers, and theologians with regard to the boundaries and hierarchies among their disciplines. In the work

---

[49] See the letter of dedication in Rosen's translation of *On the Revolutions*. For a broad view of Copernicus's preface to his work, see Robert S. Westman, "Proof, Poetics, and Patronage: Copernicus's preface to *De revolutionibus*," in *Reappraisals of the Scientific Revolution*, ed. David C. Lindberg and Robert S. Westman (Cambridge: Cambridge University Press, 1990), pp. 167–205.

[50] On the role of arbiters in scholarly disputes, see Mario Biagioli, *Galileo, Courtier: The Practice of Science in the Culture of Absolutism* (Chicago: University of Chicago Press, 1993), chap. 3.

[51] James G. Lennox, "Aristotle, Galileo, and Mixed Sciences," in *Reinterpreting Galileo*, ed. William A. Wallace (Washington, D.C.: Catholic University of America Press, 1985), pp. 29–51; and Peter Dear, "Jesuit Mathematical Science and the Reconstitution of Experience in the Early Seventeenth Century," *Studies in History and Philosophy of Science*, 18 (1987), 133–75.

[52] For the text, see Eugenio Garin, "Alle origini della polemica anticopernicana," *Studia Copernicana*, 6 (1973), 31–42.

[53] Westman, "The Copernicans and the Churches," pp. 87–9; see also Salvatore I. Camporeale, "Giovanmaria dei Tolosani O.P., 1530–1546: Umanesimo, riforma e teologia controversista," *Memorie domenicane*, 17 (1986), 184–8.

of mechanical philosophers and different types of experimentalists in the seventeenth century, the status of such new scientific objects and the boundaries between disciplines once again became matters of contention.

Planetary circles – whose shape was first described by Johannes Kepler (1571–1630) as elliptical planetary orbits – had been an archetypal kind of scientific object since the time of Plato. Like their early modern successors – the motion of the earth, corpuscles, forces, or atoms – they were invisible in principle, functioning in astronomical discourse as geometrical representations that generated planetary positions whose fit with observation was acceptable. Finally, as their physical status remained ambiguous, they were amenable either to material or to spiritual interpretations.[54] All these features account for the way they were accommodated in a universe of knowledge dominated by religious consciousness. It was always possible to argue that heavenly bodies were moved by a prime mover as a final cause, or alternatively by angels, without interference with the geometrical models that produced the motions. In addition, it could be claimed that these models, with their eccentric circles and epicycles, were just mathematical hypotheses, capable of "saving the phenomena" without physically explaining them. These were the strategies that subordinated astronomy as a mathematical discipline to natural philosophy while guaranteeing it autonomy within certain limits.[55]

The logic of invisible objects in Copernicus's invocation of the motion of the earth imposed constraints on Kepler's treatment of planetary orbits. Invisibility meant the lack, in principle, of direct experience of the object. Kepler's compensatory strategy was to emphasize the physical reality of the orbits as resulting from the interaction of two opposite forces: one located in the sun itself and attracting the planet to it, the other belonging to the planet and deflecting it from the sun.[56] In the context of traditional astronomical techniques, this constituted a constraint insofar as the center of the curves that represented planetary motions could no longer be chosen arbitrarily, as Copernicus had done, but had to express some kind of physical reality. The positing of forces, however ill-defined, as the physical causes of these motions had deep implications for the understanding of astronomy as a physico-mathematical science, not just a mathematical discipline. Furthermore, in

---

[54] On the status of planetary models in antiquity, see G. E. R. Lloyd, "Saving the Appearances," *Classical Quarterly*, 28 (1978), 202–22. On the debate over planetary models in early modernity, see Nicholas Jardine, "The Forging of Modern Realism: Clavius and Kepler against the Sceptics," *Studies in History and Philosophy of Science*, 10 (1979), 143–73.

[55] On the status and contents of traditional astronomy, see Pierre Duhem, *To Save the Appearances*, trans. E. Doland and C. Maschler (Chicago: University of Chicago Press, 1969); James A. Weisheipl, "Classification of the Sciences in Medieval Thought," *Medieval Studies*, 27 (1965), 54–90; Robert S. Westman, "The Astronomer's Role in the Sixteenth Century: A Preliminary Study," *History of Science*, 18 (1980), 105–47; and Nicholas Jardine, *The Birth of History and Philosophy of Science: Kepler's "A Defence of Tycho against Ursus" with Essays on Its Provenance and Significance* (Cambridge: Cambridge University Press, 1984), chap. 7.

[56] On the discovery of the elliptical shape and its implications, see Kuhn, *The Copernican Revolution*, pp. 209–19.

a paradoxical way, an invisible object was much more dependent on observational evidence to guarantee the claim for its reality than visible objects. Kepler's strict demand for precision in the fit between geometrical model and observational data may be understood against this background; in this way, a new type of object established a new standard of precision (see Donahue, Chapter 24, this volume).

Ultimately, however, Kepler conceived the structure of the universe as a reflection of God's plan for creation, emanating from the geometrical nature of God's intellect, or inscribed as an archetypal model in God's mind. Placing himself within the Protestant tradition of Melanchthon, Kepler thought of the universe as being imprinted by God's signatures, especially that of the Trinity:

> For in the sphere, which is the image of God the Creator and the archetype of the world ... there are three regions, symbols for the three persons of the Holy Trinity – the center, a symbol of the Father; the surface, of the Son; and the intermediate space, of the Holy Spirit. So too, just as many principal parts of the world have been made – the different parts in the different regions of the sphere: the sun in the center, the sphere of the fixed stars on the surface, and lastly the planetary system in the region intermediate between the sun and the fixed stars.[57]

Here Kepler identified God with geometry; the astronomer was able to intuit the model in God's mind. In Kepler's terms, studying the book of nature was like a form of prayer.[58]

Like Kepler, Galileo challenged the traditional boundary and hierarchical relationships between the physical truths of natural philosophy and the hypotheses of mixed mathematics. As he put it in his *Istoria e dimostrazioni intorno alle macchie solari e loro accidenti* (History and Demonstrations Concerning Sunspots and Their Phenomena, 1613), proponents of his "philosophical astronomy,"

> going beyond the demand that they somehow save the appearances, seek to investigate the true constitution of the universe – the most important and most admirable problem that there is. For such a constitution exists; it is unique, true, real, and could not possibly be otherwise; and the greatness and

---

[57] Quoted in Westfall, *Science and Religion*, p. 222.
[58] Charlotte Methuen, *Kepler's Tübingen: Stimulus to a Theological Mathematics* (Aldershot: Ashgate, 1998), p. 206. An authoritative monograph on the relationship of Kepler's theology and natural philosophy is Jürgen Hübner, *Die Theologie Johannes Keplers zwischen Orthodoxie und Naturwissenschaft* (Beiträge zur historischen Theologie, 50) (Tübingen: Mohr, 1975); see also Hübner, "Kepler's Praise of the Creator," *Vistas in Astronomy*, 18 (1975), 369–85; Max Caspar, *Kepler*, trans. C. Doris Hellman (New York: Abelard-Schuman, 1959); and Bruce Stephenson, *The Music of the Heavens: Kepler's Harmonic Astronomy* (Princeton, N.J.: Princeton University Press, 1994).

nobility of this problem entitle it to be placed foremost among all questions capable of theoretical solution.[59]

Galileo's rhetorics sometimes seemed to imply that the only path to knowledge was through the geometrical properties of objects, progressing to ever more complex phenomena, in an endless process of approximation to the infinite mathematical truths known to the divine intellect. Yet his treatment of sunspots clearly indicates that the study of phenomena, as he himself envisaged it, entailed more than the reduction of appearances to their mathematical features: sunspots as scientific objects for the philosophical astronomer – Galileo's favorite description of his role – had to be not only visible but also, in some sense, touchable. To achieve this goal, he constructed an analogy between the manipulation of earthly matter and the mutability of the sun:

> I liken the sunspots to clouds or smokes. Surely if anyone wished to imitate them by means of earthly materials, no better model could be found than to put some drops of incombustible bitumen on a red-hot iron plate. . . . It might even be that if we more accurately observed the bright spots on the sun that I have mentioned, we should find them occurring in the very places where large dark spots had been a short time before.[60]

The justification of his alternative program was formulated as a philosophical broadside against the possibility of knowing "essences" of things, thus destroying the traditional bridges between the senses and the intellect, between God and nature, and between natural knowledge and religion:

> For in our speculating we either seek to penetrate the true and internal essence of natural substances, or content ourselves with a knowledge of some of their properties. The former I hold to be as impossible an undertaking with regard to the closest elemental substances as with more remote celestial things . . . [yet] although it may be vain to seek to determine the true substance of the sunspots, still it does not follow that we cannot know some properties of them, such as their location, motion, shape, size, opacity, mutability, generation, and dissolution. These in turn may become the means by which we shall be able to philosophize better about other and more controversial qualities of natural substances. And finally by elevating us to the ultimate end of our labors, which is the love of the divine Artificer, this will keep us steadfast in the hope that we shall learn every other truth in Him, the source of all light and verity.[61]

Galileo's project to unify natural philosophy and mixed mathematics alienated university philosophers, whose subject matter and methodology he had

---

[59] Galileo Galilei, "History and Demonstrations concerning Sunspots and Their Phenomena," in *Discoveries and Opinions of Galileo*, trans. Stillman Drake (New York: Doubleday, 1957), p. 97.
[60] Ibid., p. 140.
[61] Ibid., pp. 123–4.

sharply criticized and to whose philosophy he had presented an alternative.[62] It also alienated his interlocutors and competitors, the Jesuit mathematicians, who had themselves developed a project of physico-mathematics that did not ignore traditional natural philosophy and metaphysics.[63] It was, however, his ideas about the two books – the books of nature and the Bible – that most boldly defied the authority of theologians. The two books, he contended in his famous *Lettera a Madama Cristina di Lorena Granduchessa di Toscana* (Letter to the Grand Duchess Christina, 1615), differed in their subject matter, their language, their audiences, and their goals.

> [T]he authority of Holy Scripture aims chiefly at persuading men about those articles and propositions which, surpassing all human reason, could not be discovered by scientific research or by any other means than through the mouth of the Holy Spirit himself. . . . [T]he writers of Holy Scriptures not only did not pretend to teach us about the structure and motions of the heavens and the stars, and their shape, size, and distance, but they deliberately refrained from doing so.[64]

Not only was the Bible concerned with problems of faith and salvation, which Galileo separated from problems of understanding nature and its truths, but the difference was as deep as language itself. "[I]t is appropriate for Scripture to say many things that are different (in appearance and in regard to the literal meaning of the words) from absolute truth. . . . [N]ot every scriptural assertion is bound to obligations as severe as every natural phenomenon."[65] Nature, on the other hand, in whose actions God is not "any less excellently revealed," and which "proceed[s] alike from the divine Word," is "inexorable and immutable; she never transgresses the laws imposed upon her, or cares a whit whether her abstruse reasons and methods of operation are understandable to men."[66] Thus philosophy, as Galileo stated in *Il Saggiatore* (The Assayer, 1623), "is written in this grand book, the universe, which stands continually open to our gaze. . . . It is written in the language of mathematics, and its characters are triangles, circles, and other geometric figures." This is the reason the book of nature can only be read by experts: "But the book cannot be understood unless one first learns to comprehend the language and read the letters in which it is composed."[67]

Galileo drew daring conclusions from the differences between the subject matter, language, and aims of the books of nature and scripture, anchoring

---

[62] See Galileo's *Dialogo di Cecco di Ronchitti*, in Stillman Drake, *Galileo Against the Philosophers, in his Dialogue of Cecco di Ronchitti (1605) and Considerations of Alimberto Mauri (1606)* (Los Angeles: Zeitlin and Ver Brugge, 1976).
[63] For Galileo and the Jesuits, see William A. Wallace, *Galileo and His Sources: The Heritage of the Collegio Romano in Galileo's Science* (Princeton, N.J.: Princeton University Press, 1984).
[64] Galileo, *Letter to the Grand Duchess*, in Finocchiaro, *The Galileo Affair*, pp. 93–4.
[65] Ibid., p. 93.
[66] Ibid.
[67] Galileo, *The Assayer*, in Drake, *Discoveries and Opinions*, pp. 237–8.

in these differences his radical claim about the necessity of separating the authority of philosophers, who are acquainted with the mathematical language of nature, from that of theologians, who are trained to interpret the Bible. Moreover, in matters of natural philosophy, he asserted the priority of philosophers over theologians: "I think that in disputes about natural phenomena one must begin not with the authority of scriptural passages but with sensory experience and necessary demonstrations," as he put it in the *Letter to the Grand Duchess Christina*.[68] Galileo petitioned those authorized to interpret the Bible, whom he held to be of supreme authority – "I consider it rank temerity for anyone to contradict them," he wrote – to suspend their judgment about the motion or rest of the Earth, for "it would be proper to ascertain the facts first, so that they could guide us in finding the true meaning of Scripture; this would be found to agree absolutely with demonstrated facts, even though prima facie the words would sound otherwise, since two truths can never contradict each other."[69] Quoting the witty words of a contemporary ecclesiastic, Cardinal Baronius, Galileo summed up his argument by pointing out not only the different contents, language, and audiences of the two books but also their different intentions: "The intention of the Holy Spirit is to teach us how one goes to heaven, not how heaven goes."[70]

Galileo's trial before the Roman Inquisition in 1633 followed the publication the previous year of the *Dialogo sopra i due massimi sistemi del mondo* (Dialogue Concerning the Two Chief World Systems, 1632), in which he presented his arguments for and against the Copernican and Ptolemaic systems in the form of an open conversation, without leaving any doubt, however, about his own sympathies with the Copernican view. Galileo was aware of the risk he was taking in writing and publishing the book. Sixteen years before, in 1616, he had been informed of the Holy Office's decision to prohibit Copernicus's book until it should be corrected, and had been admonished by Cardinal Robert Bellarmine (1542–1621) not to hold or defend Copernicanism because it appeared to contradict scripture. Contrary to Galileo's expectations, he was unable to defend his subversive publication through the support of friends and admirers in the high ranks of the Church, including Pope Urban VIII (1568–1644), who became hostile in reaction to political pressure from various sides. Galileo was forced to abjure, his book was prohibited, and he was sentenced to life imprisonment – a sentence that was commuted to house arrest at Arcetri. The Catholic rejection of Copernicanism remained formally in force until 1835.

More than any other single decision made by the Roman Church during centuries of symbiotic coexistence with natural philosophy, the rejection of Copernicanism and the silencing of Galileo by a judicial act represented the repressive face of the early modern Catholic Church. Thus,

---

[68] Galileo, *Letter to the Grand Duchess*, in Finocchiaro, *The Galileo Affair*, p. 93.
[69] Ibid., p. 104.
[70] Ibid., p. 93.

the trial of Galileo was transformed from a historical event into a powerful cultural symbol, which loomed large in the nineteenth-century treatments of Draper and White with which I began. However, twentieth-century investigations of the trial on the basis of the Inquisitorial documents have pointed out the importance of the specific historical and political circumstances in which the trial took place.[71] Without reaching agreement about the causes of the trial, such studies have been effective in eroding, though not erasing, the belief in an inevitable conflict between science and religion that grew out of the nineteenth-century understanding of the trial of Galileo.

Descartes' decision, after hearing of the trial of Galileo, to suppress the Copernican cosmology of his *Le Monde* (The World), an early treatise published only after his death, became emblematic of the intimidation of Catholic intellectuals. Nevertheless, in his published works, Descartes attempted a much more radical separation of the material world from the divine mind than had either Kepler or Galileo, while still deriving the fundamental principles of the universe from God's nature and making him the sole guarantor of the validity of human knowledge.[72] According to Descartes' *Meditationes* (1641), God's creation of eternal truths depended on his will, which is inseparable from his intellect. There was therefore no room for human intellect to intuit the divine plan because there could have been an infinite number of possible plans if God had so willed. Once the eternal truths were created, however, God was committed to their conservation. As the first cause of all motion in the universe, God's complete transcendence and total sovereignty were secured. In a Cartesian universe, inert matter could not be suffused with spirit. Because of God's immutability, motion in the universe was conserved, thus specifying the fundamental laws of inertia and impact. "It is evident," Descartes wrote, "that there is none other than God who by his omnipotence created matter with motion and who now conserves in the universe by his ordinary concourse as much motion and rest as he placed in creating it."[73]

---

[71] For various contextualizations of the trial of Galileo, see Giorgio de Santillana, *The Crime of Galileo* (Chicago: University of Chicago Press, 1955); Jerome J. Langford, *Galileo, Science, and the Church* (Ann Arbor: University of Michigan Press, 1966); Pietro Redondi, *Galileo Heretic*, trans. Raymond Rosenthal (Princeton, N.J.: Princeton University Press, 1987); Biagioli, *Galileo, Courtier*; Annibale Fantoli, *Galileo: For Copernicanism and for the Church*, trans. George V. Coyne (Vatican City: Vatican Observatory Publications, 1994); and Feldhay, *Galileo and the Church*. The documents were first published by Antonio Favaro in *Le opere di Galileo: Edizione Nazionale sotto gli auspicii di Sua Maestà il Re d'Italia*, 20 vols. (Florence: Giunti Barbèra, 1890–1909; repr. Florence: Giunti Barbèra, 1968), vol. 19. For a more recent collection of documents related to the trial, see Finocchiaro, *The Galileo Affair*.

[72] In the section on Descartes, I have relied upon the following: Margaret J. Osler, *Divine Will and the Mechanical Philosophy* (Cambridge: Cambridge University Press, 1994), chap. 5; Amos Funkenstein, "Descartes, Eternal Truths, and the Divine Omnipotence," *Studies in History and Philosophy of Science*, 6 (1975), 185–99; Gary Hatfield, "Reason, Nature, and God in Descartes," *Science in Context*, 3 (1989), 175–201; and Daniel Garber, *Descartes' Metaphysical Physics* (Chicago: University of Chicago Press, 1992).

[73] Quoted in Osler, *Divine Will*, p. 137, taken from Charles Adam and Paul Tannery, eds., *Oeuvres de Descartes*, 11 vols. (Paris: J. Vrin, 1897–1938), 8: 1.61 and 9: 2.83.

According to Descartes, God, through his will, also guaranteed the truth of clear and distinct ideas, the basis of all human knowledge: "[T]he faculty of knowing which he has given to us, which we call natural light, never perceives any object which is not true in that it perceives it clearly and distinctly; because we would then have to believe that God is a deceiver, if he had given it to us."[74] The principles of physics rooted in Descartes' metaphysics became the locus classicus for the ideal types of objects and explanations of the mechanical philosophy (see Garber, Chapter 2, this volume).

Objects of scientific discourse inaccessible to the senses originated in the mathematical tradition, where they initially functioned as abstractions. Reconceived as hidden mechanisms behind the manifest operations of nature, they transgressed the traditional boundaries between the investigation of nature and the investigation of machines, between mathematics and physics, and often between philosophy and theology. The need for religious justifications, however, was not eliminated from the new emerging discourses. On the contrary, for Copernicus, Kepler, and Descartes this need intensified, although it was satisfied in nontraditional ways: for Copernicus by invoking the Pope as the arbiter among scientific theories, for Kepler by identifying God's intellect with geometry, and for Descartes by appealing to God's nature to guarantee the basic principles of philosophy and the truth of clear and distinct ideas.

In contrast, it was the practice and rhetoric of Galileo around which a narrative of necessary and inevitable conflict between science and religion later crystallized. Galileo sketched the outlines of a unified science of celestial and terrestrial mechanics, a well-bounded field of mathematical physics, and a community of mathematical philosophers with a distinct professional identity and authority. This vision brought him into conflict with university professors, Catholic natural philosophers, and theologians embedded in intricate systems of interests, culminating in his trial by the Roman Inquisition in 1633, which reverberated throughout Catholic Europe and beyond.[75]

AUTHORIZATION AND LEGITIMATION: SCIENCE, RELIGION, AND POLITICS IN THE SEVENTEENTH CENTURY

By the mid-seventeenth century, the metaphor of nature as a clock – a complex mechanism created by the divine artificer – had become the expression

---

[74] Adam and Tannery, eds., *Oeuvres de Descartes*, 8: 1.16 and 9: 2.38.
[75] The impact of the trial, according to the standard view, had been disastrous. For a formulation of that view, see Leonardo Olschki, *The Genius of Italy* (New York: Oxford University Press, 1949). For a more balanced view, see William B. Ashworth, Jr., "Catholicism and Early Modern Science," in Lindberg and Numbers, *God and Nature*, pp. 136–66.

of a new relationship between the senses and the intellect, God and his creation, and natural philosophy and the mechanical arts.[76] The clock metaphor related observable natural phenomena (clock hands) to invisible mechanisms (clock weights and wheels) variously posited as "explanations" or "causes" of the phenomena, or at least as "hypotheses" about those causes. Moreover, the clock metaphor also suggested that the investigation of the universe should be likened to the investigation of a machine. This meant that natural philosophy could no longer be conceived as nonmathematical in principle because mechanics was part of mixed mathematics. It also meant that knowledge about machines – artificial beings without "natures" and therefore not objects of traditional philosophical discussion – was elevated to a status that challenged the traditional hierarchy of disciplines and their boundaries (see Bennett, Chapter 27, this volume). Finally, the clock metaphor entailed a reformulation of ideas about God's intervention in regulating his creation – now conceived as a perfectly ordered machine. If God's perfection could be demonstrated by the perfection of a universe in no need of divine intervention to maintain its regular course, this meant it was necessary to reformulate the limit of authority or principles of faith and the Bible in natural knowledge, at least as far as the interventions of divine providence were concerned.

Francis Bacon (1561–1626) attempted to distinguish the role and aims of knowledge in natural philosophy, as opposed to religion, and to define those goals.[77] The focus of Bacon's critique of traditional forms of learning – humanist, scholastic, and neo-Platonic – was their failure to account for the communication of the human intellect with the world, "the commerce between thoughts and things," as he described it. From this critique stemmed his judgment that all forms of inherited knowledge – "delicate learning," "contentious learning," and "fantastic learning," as he characterized them in the *Advancement of Learning* (1605)[78] – remained on the superficial level of words, unable to penetrate the secrets of things. Thus remedies had to be found to secure the development of knowledge rooted in "doing," not only in thought and speech. The alternative that Bacon envisioned was a program for cooperation among artisans, artists, merchants, philosophers, and other men of learning concerned with improving the lot of humanity and imbued with Christian charity useful for the progress of human society.

---

[76] Steven Shapin, *The Scientific Revolution* (Chicago: University of Chicago Press, 1996), chap. 1.
[77] On Baconian discourse, my discussion draws from Benjamin Farrington, *Francis Bacon: Pioneer of Planned Science* (New York: Praeger, 1963); Paolo Rossi, *Francis Bacon: From Magic to Science*, trans. Sacha Rabinovitch (London: Routledge, 1968). Also see Markku Peltonen, ed., *The Cambridge Companion to Bacon* (New York: Cambridge University Press, 1996).
[78] Francis Bacon, *Advancement of Learning* (1605), in *Francis Bacon: A Selection of His Works*, ed. Sidney Warhaft (Indianapolis: Odyssey Press, 1965), pp. 221–2.

In the first part of his *Great Instauration, De dignitate et augmentis scientiarum* (On the Dignity and Increase of the Sciences, 1623),[79] Bacon articulated the new map of knowledge that grew out of his critique of the role of reason in constructing knowledge about God, society, and nature. The most radical aspect of Bacon's reorganization of learning – which he had started to elaborate many years before in *The Advancement of Learning* – was his attempt to differentiate the role of natural knowledge from knowledge based on scripture in order to legitimize it. According to Bacon, the Bible was the product of revelation, written as a guide to salvation, whereas the book of nature taught the dominion of the universe for the benefit of mankind: "The true and lawful goal of the sciences is none other than this: that human life be endowed with new inventions and power." Most importantly, the purposes of the two books should not be confused: "[W]e do not presume by the contemplation of nature to attain to the mysteries of God."[80]

Despite his distinction between divine and natural knowledge, Bacon still understood the ultimate goal and justification of all "scientific" activity in religious terms. Thus, the goal of natural philosophy was to understand the "true forms of things," or the laws that govern natural phenomena, identified with God's signatures.[81] In the preface to the *Great Instauration*, Bacon prayed to God to allow him to lead people "to the things themselves or to the concordances of things."[82] In his unfinished, utopian vision, *New Atlantis* (posthumously published in 1627), Bacon envisioned the House of Salomon, where all manner of experiments were performed, in terms of a temple and the community of researchers it housed as a form of priesthood.

During the English Revolution of the 1640s, Bacon's project gained popularity among Puritans, millennarians, and other radicals. In the context of the millennarian eschatology that swept England during those years, the promotion of knowledge and the control of nature were seen as a means to bring about God's kingdom on earth. Appealing to Bacon's own interpretation of a biblical prophecy – "Many shall pass to and fro, and science shall be increased" (Daniel 12:4) – Bacon's followers concluded that cautious investigation was a way to glorify God and to restore man's dominion over nature, lost after the fall of Adam and Eve. The Puritans also used Bacon's program as a weapon with which to attack the university system and as a promise of the triumph of the Protestant reformation. Thus Bacon's writings were canonized as fundamental to the Puritan Revolution in the work of

---

[79] Whereas the first book of the *Advancement of Learning* was first published together with the *Novum organum* in 1620, the second book constituted the first part of the *Great Instauration*, first published in 1623.
[80] Bacon, *Advancement of Learning*, p. 204.
[81] Mary Hesse, "Francis Bacon's Philosophy of Science," in *Essential Articles for the Study of Francis Bacon*, ed. Brian Vickers (Hamden, Conn.: Archon Books, 1968), p. 116.
[82] Bacon, *The Great Instauration*, preface, in Warhaft, ed., *Francis Bacon: A Selection of His Works*, p. 309.

clergymen, academics, and schoolmasters, such as John Stoughton, George Hakewill, and William Twisse. Likewise, in his *Pansophiae prodromus* (*Introduction to Universal Knowledge*, 1639), the Moravian reformer John Amos Comenius (1592–1670) outlined a program for an educational reform modeled on Bacon's vision.[83]

By the late 1650s, the destructive potential of the religious radicals' claims to direct communion with God and unmediated access to the secrets of nature became increasingly evident.[84] It was against this background that English experimentalists, notably Robert Boyle (1627–1691) and other members of the Royal Society, developed their project. On the one hand, they attempted to reappropriate Bacon from the radicals by stressing his distinction between the books of nature and scripture and his skepticism concerning direct access to the secrets of nature. On the other hand, they relied upon Bacon to legitimize their own alternative to Aristotelians, Cartesians, and Hobbesians. The experimentalists' philosophical project and their attempt to save the political and religious order of Restoration England should thus be seen as two faces of one enterprise.

The new understanding of experience in terms of particular facts that could only be established under special conditions – and allegedly provided a more secure basis for natural philosophy – distinguished the discourse of the English experimentalists from that of traditional natural philosophers and religious enthusiasts, as well as from that of mechanical philosophers, whether Cartesian or Hobbesian.[85] Against the enthusiasts, empirics, and pansophic reformers, the defenders of the experimentalists and the spokesmen of the Royal Society sought to control experience and to differentiate well-established facts from other, less secure items of knowledge. As opposed to the Cartesians, who appealed to introspection to establish first principles, and the Hobbesians, who rejected experiment as a source of certain knowledge,[86] the experimentalists emphasized the need to secure a basis for philosophy in matters of fact.

The experimentalists also believed that their new "way of life" would provide a solution to problems of social and religious order. After the restoration of the British monarchy, the Civil War and the period of Oliver Cromwell's Republic were understood as a state of civic conflict brought about by the inability to overcome disputations over contested knowledge. Hobbes's

---

[83] Charles Webster, *The Great Instauration: Science, Medicine, and Reform, 1626–1660* (London: Duckworth, 1975), chaps. 1–3.
[84] Charles Webster, *The Intellectual Revolution of the Seventeenth Century* (London: Routledge and Kegan Paul, 1974); and Michael Heyd, *"Be Sober and Reasonable": The Critique of Enthusiasm in the Seventeenth and Early Eighteenth Centuries* (Leiden: E. J. Brill, 1995), chap. 5.
[85] For the cognitive and social meaning of the experimentalists' practices, see Steven Shapin and Simon Schaffer, *Leviathan and the Air-Pump: Hobbes, Boyle, and the Experimental Life* (Princeton, N.J.: Princeton University Press, 1985). On the origins of experimentalism, see Dear, *Discipline and Experience*, and Heyd, *"Be Sober and Reasonable,"* chap. 5.
[86] Shapin and Schaffer, *Leviathan and the Air-Pump*, chap. 1.

patron, William Cavendish, the Earl of Newcastle (1593–1676), succinctly expressed this view: "[C]ontroversy Is a Civill Warr with the pen which pulls out the sorde soone afterwards."[87] The experimental philosophers of the Royal Society sought to contain conflict through universal assent to matters of fact. In order to ensure that disagreement did not spill over from the philosophical to the political and religious spheres, they enforced clear rhetorical boundaries between these discourses. Thus the charter of the Royal Society insisted that its members restrict their inquiries to natural knowledge, "not muddling with Divinity, Metaphysics, Moralls, Politics, Grammar, Rhetoric or Logick."[88] Yet things were different in practice, as Bishop Thomas Sprat's official *History of the Royal Society* (1667) made clear:

> And the Experimenter . . . must judge aright of himself, he must misdoubt the best of his own thoughts; he must be sensible of his own ignorance, if ever he will attempt to purge and renew his Reason. . . . [I]t may well be concluded that the doubtful, the scrupulous, the diligent Observer of Nature, is neerer to make a modest, a sever, a meek, an humble Christian, than the man of Speculative Science, who has better thoughts of himself and his own Knowledge.[89]

Thus English society witnessed the crystallization of certain religious-scientific interests around a group of experimentalists of the Royal Society, among them Robert Boyle, John Wilkins, Edward Stillingfleet, Henry Oldenburg, Peter Pett, John Evelyn, John Wallis, Thomas Sprat, and Joseph Glanvill.[90] Many of these Fellows of the early Royal Society also supported the latitudinarian movement within the Church of England. Within this framework, patient, collective inquiry into nature was held up as a model for resisting religious and philosophical enthusiasm on the one hand and dogmatic submission to authority on the other. As members of the Royal Society and latitudinarians, the group around Boyle promoted the idea that reflecting on God was part of natural philosophy, which revealed the glory of God through the investigation of his creation. In Boyle's words, in his tellingly named treatise *The Excellency of Theology* (1674),

> [N]either the fundamental doctrine of Christianity nor that of the powers and effects of matter and motion, seems to be more than an epicycle . . . of the great and universal system of God's contrivances and makes but a part of the more general theory of things knowable by the light of nature, improved

---

[87] Cited in ibid., p. 290.
[88] Henry George Lyons, *The Royal Society, 1660–1940: A History of Its Administration under Its Charters* (New York: Greenwood Press, 1968), p. 41.
[89] Thomas Sprat, *The History of the Royal Society of London, for the Improving of Natural Knowledge*, ed. J. I. Cope and H. W. Jones (St. Louis, Mo.: Washington University Studies, 1958), p. 367, quoted in Heyd, *"Be Sober and Reasonable,"* p. 156.
[90] Barbara Shapiro, "Latitudinarianism and Science," *Past and Present*, 40 (1968), 16–41; James R. Jacob and Margaret C. Jacob, "The Anglican Origins of Modern Science: The Metaphysical Foundations of the Whig Constitution," *Isis*, 71 (1980), 251–67.

by the information of the scriptures. So that both these doctrines, though very general in respect of the subordinate parts of theology and philosophy seem to be but members of the universal hypothesis, whose objects I conceive to be the nature, counsels, and works of God as far as they are discoverable by us . . . in this life.[91]

CONCLUSION

At the end of the first day of Galileo's *Dialogo*, Salviati, Galileo's spokesman, draws a striking analogy between human and divine knowledge of the mathematical sciences: "Divine intellect indeed knows infinitely more propositions, since it knows all. But with regard to those few which the human intellect does understand, I believe that its knowledge equals the Divine in objective certainty."[92] Here, Galileo proclaimed a new and ambitious ideal of human knowledge. Yet whereas in many senses his project was actually accomplished by Isaac Newton (1642–1727), who succeeded in unifying celestial and terrestrial mechanics in his aptly named treatise the *Principia mathematica philosophiae naturalis* (Mathematical Principles of Natural Philosophy, 1687) (see the following chapters in this volume: Bertoloni Meli, Chapter 26; Donahue, Chapter 24), Newton firmly rejected Galileo's aspirations toward divine knowledge, however partial.

Despite the stark contrast between their positions relative to the development of natural knowledge, their institutional, political, and religious situations, and their intellectual temperaments, the juxtaposition of Galileo and Newton is illuminating. Both attempted to deduce the mathematical regularities in nature from the phenomena of moving bodies and the forces that produced them. Both aspired to certain knowledge of the physical world and hoped to achieve it through the use of mathematics. Both subsumed physics and astronomy under the mathematical sciences. Newton, however, insisted upon the limits of the new mathematical natural philosophy and on the essential role of God in a mechanical universe: Nature was in perpetual need of God's active interventions in order to keep the universe from running down. In contrast with Galileo, Kepler, and Descartes, for whom the mathematical structure of the world was an a priori certainty, Newton believed that the mathematical laws of nature were only deducible from experiments and that no deduction from principles collected by induction was valid without experimental verification. He wished to protect this kind of knowledge from speculative hypotheses as the solid core of science: "For whatever is not deduced from the phenomena is to be called a hypothesis," he

---

[91] Robert Boyle, *The Excellency of Theology*, in *The Works of the Honourable Robert Boyle*, ed. Thomas Birch, 6 vols. (London: J. and F. Rivington et al., 1772; repr. Hildesheim: Georg Olms, 1965–6), 4: 19.
[92] Galileo Galilei, *Dialogue Concerning the Two Chief World Systems*, trans. Stillman Drake (Berkeley: University of California Press, 1962), p. 103.

wrote in the "General Scholium" to the *Principia*, "and hypotheses, whether metaphysical or physical, whether of occult qualities or mechanical, have no place in experimental philosophy."[93]

The limits of scientific discourse, however, were no less dictated by Newton's religious sensibilities. He understood the universe of matter and motion as a product of God's will, created for a purpose, and not as purely mechanical. "God in the beginning formed matter in solid, massy, hard, impenetrable, movable particles, of such sizes and figures, and with such other properties, in such proportion to space, as most conduced to the end for which he formed them."[94] Newton did not banish the creator from his creation, contrary to the clock metaphor of a perfectly self-sustaining machine. In a letter to Cambridge classicist Richard Bentley, Newton argued that: "It is inconceivable, that inanimate, brute Matter should, without the Mediation of something else, which is not material, operate upon, and affect other Matter without mutual Contact. . . . Gravity must be caused by an Agent acting constantly according to certain Laws: but whether this Agent be material or immaterial, I have left to the Consideration of my Readers."[95] Likewise, in a famous query to the *Opticks* (London, 1704), Newton spoke of absolute space as God's "sensorium,"[96] thus construing space not only as a physical concept but also as the expression of God's omnipresence, the "infinite scene of the divine knowledge and control."[97] And in the final query to the same text, Newton envisioned God as a "powerful ever-living Agent, who being in all places is more able by his will to move the bodies within his boundless uniform sensorium, and thereby to form and reform the parts of the universe."[98]

For Newton, the investigation of essences was both intellectually presumptuous, transcending "the comprehension of man," and theologically wrong, suggesting restrictions to the absolute freedom of God's will. For "the wisest of beings required of us to be celebrated not so much for his essence as for his actions, the creating, preserving, and governing of all things according to his good will and pleasure."[99] Adoring God meant investigating real things in nature – for "we know him only by his most wise and excellent contrivance of

---

[93] Isaac Newton, *The Principia: Mathematical Principles of Natural Philosophy*, trans. I. Bernard Cohen and Anne Whitman (Berkeley: University of California Press, 1999), p. 943. The *Opticks* contains the same insight, stating: "For Hypotheses are not to be regarded in experimental philosophy." See Isaac Newton, *Opticks; or, a Treatise of the Reflections, Refractions, Inflections and Colours of Light* (New York: Dover, 1952), bk. 3, pt. 1, p. 404.
[94] Newton, *Opticks*, bk. 3, pt. 1, p. 400.
[95] I. Bernard Cohen, ed., *Isaac Newton's Papers & Letters on Natural Philosophy and Related Documents*, 2nd ed. (Cambridge, Mass.: Harvard University Press, 1978), pp. 303–4.
[96] Newton, *Opticks*, bk. 3, pt. 1, p. 377.
[97] Burtt, *The Metaphysical Foundations of Modern Physical Science* (Garden City, N.Y.: Doubleday, 1954), p. 260.
[98] Newton, *Opticks*, bk. 3, pt. 1, p. 403.
[99] Newton Papers, Yahuda Collection, Hebrew University, Jerusalem, Yahuda MS. 15.7, fol. 154r.

things"[100] while at the same time obeying his commandments in scripture, "for we adore him as his servants."[101] Interpreting biblical prophecies by means of astronomically based chronologies paralleled his reading in the book of nature. For Newton, the two readings were equal in importance and guided by the same principle: "to choose those constructions which without straining reduce things to the greatest simplicity."[102]

In his *Letter to the Grand Duchess Cristina*, Galileo had insisted that scriptural verses touching upon natural phenomena be interpreted in the light of natural philosophy: "One must search for the correct meaning of Scripture with the help of demonstrated truth, rather than taking the literal meaning of the words, which may seem the truth to our weak understanding, and trying somehow to force nature and deny observations and necessary demonstrations."[103] Here he challenged religious authorities to recognize the autonomy of mathematics and natural philosophy. Yet Galileo's own recommendations in the same text testified to the difficulty of disentangling theological and natural philosophical authority. Newton's pronouncements in the *Opticks* on the proper relations between natural knowledge and religion, written a half-century later, testify to analogous difficulties in drawing and maintaining sharp boundaries: ". . . whereas the main business of natural philosophy is to argue from the phenomena without feigning hypotheses, and to deduce causes from effects, till we come to the very first cause, which certainly is not mechanical."[104] Historian Frank Manuel's observation about late seventeenth-century England provides an apt conclusion: "In the England of the Restoration . . . where so many divines doubled as scientists, the coexistence in one head of expert knowledge in both books came to be respected, and the capacity of a man to reveal the glory of God in both spheres was taken for granted."[105] This assessment might be extended to early modern relations between Christian theology and natural knowledge more generally, underscoring the mutual dependence of religious order and scientific authority on one another in this period of unprecedented transformation for both.

[100] Quoted by Richard S. Westfall, "The Rise of Science and the Decline of Orthodox Christianity: A Study of Kepler, Descartes, and Newton," in Lindberg and Numbers, *God and Nature*, p. 229.
[101] Westfall, "Rise of Science," in Lindberg and Numbers, *God and Nature*, p. 229.
[102] Newton Papers, Yahuda Collection, Hebrew University, Jerusalem, Yahuda MS. 1.1, fol. 14r.
[103] Galileo Galilei, *Letter to the Grand Duchess*, in Finocchiaro, *The Galileo Affair*, p. 111.
[104] Newton, *Opticks*, bk. 3, pt. 1, p. 369.
[105] Frank E. Manuel, *The Religion of Isaac Newton* (Oxford: Clarendon Press, 1974), pp. 32–3.

30

LITERATURE

Mary Baine Campbell

Science and imaginative literature have made a dynamic pair of objects for study ever since they were sharply and categorically separated as activities of the mind and kinds of representations: To study them in tandem, at least for literary historians and critics, is to confront the embarrassing question, What is "literature"? – a question harder and harder to answer, and not to be answered here. The relations between science and literature (and early printed book production) have seemed especially interesting since about 1980, as scholars in historical fields have come more and more to poach on each other's lands and goods. During the advent of cultural studies, especially in the work and thought of certain French historians and philosophers interested in science (e.g., Gaston Bachelard, Georges Canguilhem, Michel Foucault, Michel Serres, and Michel de Certeau),[1] the canons of literary history expanded as the study of "discourse" and "representation" relieved it of an older focus restricted to particular authors and genres.

Marjorie Hope Nicolson, perhaps the greatest student of "science and literature" writing in English in the first half of the twentieth century, is known above all for her work on the opening up of "space" (in its modern sense) to the

---

[1] For example, Gaston Bachelard (the precursor), *L'Expérience d'espace dans la physique contemporaine* (Paris: F. Alcan, 1937) and *La psychanalyse du feu* (Paris: Gallimard, 1938); Michel Serres, *Hermes, I–V* (Paris: Editions Minuit, 1968–80); Michel Foucault, *Les mots et les choses* [1966], English trans., *The Order of Things* (New York: Random House, 1970); Michel de Certeau, "History: Science and Fiction," in *Heterologies*, trans. Brian Massumi (Minneapolis: University of Minnesota Press, 1986); de Certeau, "Ethno-Graphy: Speech, or the Space of the Other: Jean de Léry," in *The Writing of History*, trans. Tom Conley (New York: Columbia University Press, 1988); and Georges Canguilhem,"*The Pathological and the Normal* [1943]," trans. Carolyn R. Fawcett and Robert Cohen (New York: Zone Books, 1989). For a wonderful example, see Fernand Hallyn, *Structure poétique du monde* [1987], English trans., *The Poetic Structure of the World: Copernicus and Kepler*, trans. Donald M. Leslie (New York: Zone Books, 1993). More generally, and polemically, see Mary Midgeley, *Science and Poetry* (London: Routledge, 2001). See also Ilse Vickers, *Defoe and the New Sciences* (Cambridge: Cambridge University Press, 1996); and Claire Jowitt and Diane Watt, eds., *The Arts of Seventeenth-Century Science: Representations of the Natural World in European and North American Culture* (Aldershot: Ashgate, 2002).

literary imagination in such books as *Newton Demands the Muse* (1946), *Voyages to the Moon* (1948), and the articles collected in *Science and Imagination* (1956).² Nicolson's primary interest was in canonical English literature and the opportunities provided for it by the materials and potential metaphors of the "new science." Her contemporary, the British historian Frances Yates, brought a similar sense of the relationship of (pan-European) scientific activity and imaginative literature to her account of *Love's Labours Lost* (1598) and its real-life model (according to Yates), London's late sixteenth-century "School of Night," to which such Renaissance luminaries as Walter Raleigh (ca. 1554–1618), mathematician, linguist, and colonialist Thomas Harriot (1560–1621), the poet George Chapman (ca. 1559–1634), and the renegade philosopher Giordano Bruno (1548–1600) belonged or were visitors.³ Both the historian and the critic take the implicit position that contemporary scientific activity and enthusiasm provided topics, metaphors, world pictures, and satirical butts for ambitious writers.

This is true. But the impact of science on literature is not the only (indeed is itself a somewhat problematic) relation to posit in a field of study that less often draws routine or meaningful distinctions between literary and nonliterary texts. Many works important to the history of early science have long been difficult to classify or to restrict to the attentions of a single branch of history: To which disciplinary apparatus would Robert Burton's (1577–1640) *Anatomy of Melancholy* (1621), Girolamo Fracastoro's (1478–1553) epillion *Syphilis* (1530), Johannes Kepler's (1571–1630) *Somnium* (1634), or even Jacques Cartier's (1491–1557) accounts of his New World voyages in the 1540s (sometimes thought to have been written by François Rabelais!) best be assigned?⁴ The preinstitutional state of the sciences (in most cases) and their lack of mathematical grounding (in many cases) leave the works of scientist-authors of the sixteenth, seventeenth, and even early eighteenth centuries particularly open to critical interpretation as multiply significant representations. Poetry and science had not yet diverged and polarized; in the sixteenth century at least, verse could still be used to convey new discoveries in medicine and natural philosophy. There was no sense of contradiction when Sir Phillip Sidney (1554–1586) wrote in his *Defense of Poesy* (1595), "[n]ow therein of all Sciences . . . is our Poet the Monarch." Writers interested in documenting natural phenomena and communicating an analysis of them, as well as writers with more purely rhetorical or sensational aims, were united in several

---

² Marjorie Hope Nicolson, *Newton Demands the Muse* (Princeton, N.J.: Princeton University Press, 1946); Nicolson, *Voyages to the Moon* (New York: Macmillan, 1948); and Nicolson, *Science and Imagination* (Ithaca, N.Y.: Great Seal Press, 1956).
³ See Frances Yates, *A Study of "Love's Labors Lost"* (Cambridge: Cambridge University Press, 1936).
⁴ Robert Burton, *Anatomy of Melancholy* (Oxford: John Lichfield and James Short, for Henry Cripps, 1621); Girolamo Fracastoro, *Syphilis, sive morbus gallicus* (Verona: [S. De Nicolini da Sabbio], 1530); Johannes Kepler, *Somnium, sive Astronomia Lunae* (Sagan and Frankfurt: Ludwig Kepler, 1634). For Cartier, see note 35.

European countries by the desire to forge lexicons and prose styles, increasingly in the vernaculars, suited to increasingly differentiated and specialized purposes. Thus, the history of literary development and the history of science overlap not only with each other but also with the history of the book, which, thanks to the spread of printing technology, experienced its most exciting times in the period covered in this volume.[5]

The concerns of feminist and postcolonial studies have enlarged the context for discussion of texts analyzing the so-called Book of Nature with particular urgency.[6] The consolidation of modern gender roles, nation-states, and colonial empires was a social and political process that facilitated the aims and representations of early modern scientists and writers and that were in turn nourished or justified by them. The privatization of women's social functions and their specularization as passive objects of desire and medical analysis, for instance, could be seen as at least suggestively coeval with the intense feminization of the natural world, taken by the new scientists as a new kind of object, penetrable and knowable by those higher on the scale of being, and the formal exclusion of women from scientific academies (see Outram, Chapter 32, this volume).[7] The mutual benefits provided by the codependence of global colonialism and universalist projects in natural history and ethnography have often been discussed (see Vogel, Chapter 33, this volume). Less often noticed are overlaps between new tendencies toward realistic detail in fiction and the increased powers of naming and identifying opened up by natural history, or between the social mobility facilitated by colonial exploration and trade and the enlarged stable of potential narrative actors (e.g., in picaresque fiction and ethnographic travel accounts). Early modern scientific thought not only had "cultural meanings" and "uses"

---

[5] In relation to issues of book technology and commerce, see Lucien Febvre and Henri-Jean Martin, *L'apparition du livre* (Paris: A. Michel, 1958), English trans., *The Coming of the Book: The Impact of Printing, 1450–1800*, trans. David Gerard (London: Verso, 1984); Elizabeth Eisenstein, *The Printing Press as an Agent of Change: Communications and Cultural Transformations in Early Modern Europe* (Cambridge: Cambridge University Press, 1979); Adrian Johns, *The Nature of the Book: Print and Knowledge in the Making* (Chicago: University of Chicago Press, 1998); Amy Boesky, "Bacon's *New Atlantis* and the Laboratory of Prose," in *The Project of Prose in Early Modern Europe and the New World*, ed. Elizabeth Fowler and Roland Greene (Cambridge: Cambridge University Press, 1997); and Mario Biagioli, "Rights and Rewards: Changing Frameworks of Scientific Authorship in Early Modern Natural Philosophy," in *Scientific Authorship: Credit and Intellectual Property in Science*, ed. Mario Biagioli and Peter Galison (London: Routledge, 2002).

[6] See, for instance, such classics as Talal Asad, *Anthropology and the Colonial Encounter* (London: Ithaca Press, 1973); Evelyn Fox Keller, *Reflections on Gender and Science* (New Haven, Conn.: Yale University Press, 1985); and Mary Louise Pratt, *Imperial Eyes: Travel Writing and Transculturation* (London: Routledge, 1992).

[7] See Carolyn Merchant, *The Death of Nature* (San Francisco: Harper and Row, 1980); Susan Bordo, *The Flight to Objectivity* (Albany: State University of New York Press, 1987); Londa Schiebinger, *The Mind Has No Sex?* (Cambridge, Mass.: Harvard University Press, 1989); Schiebinger, "Feminine Icons: The Face of Early Modern Science," *Critical Inquiry*, 14 (1988), 661–91; Helen Longino, "Subjects, Power, and Knowledge: Description and Prescription in Feminist Philosophies of Science," in *Feminism and Science*, ed. Helen Longino and Evelyn Fox Keller (New York: Oxford University Press, 1996); and Keller, *Reflections*, pt. 1.

but was itself cultural product and producer, or shared emerging conceptual paradigms with other forms of attention.[8]

The historian Margaret Jacobs, for instance, reads the texts and tenets of both mechanistic materialism and pornography for what they can render about the sense of being alive in the seventeenth-century city.[9] Of course, cultural studies will take any text, written or otherwise, as an object of productive exegesis. But it is not merely the potential of early modern scientific texts for multiple interpretations by twenty-first-century readers that makes their relation to literature richer than the influence model would suggest. This chapter will focus on a few important *joint* projects undertaken by sixteenth-century and, differently, by seventeenth-century literary and philosophical writing and writers, closing with some remarks on the development of an explicit professional antagonism between the two kinds of writers and their ways of handling these projects.

LANGUAGE

The most obviously intimate affinity among the texts and practices surveyed in this chapter involves written and, usually, printed language, including issues of style and method (or genre). At the most fundamental level, the texture of informational and philosophical prose was modified for the purposes of scientific reporting: Sixteenth-century allegorism, paradox, intentional obscurity, and such rhetorical "flowers" as metaphor were explicitly eschewed by influential members of the first scientific academies.[10] In the famous words of Thomas Sprat (1635–1713) in his *History of the Royal Society* (1667):

> [The members of the Royal Society] have been most rigorous in putting into execution the only Remedy, that can be found for this extravagance: and that has been, a constant Resolution, to reject all the amplifications, digressions, and swellings of style: to return back to the primitive purity and shortness, when men deliver'd so many things, almost in an equal number of words. They have extracted from all their members, a close, naked, natural way

---

[8] See Pratt, *Imperial Eyes*; and Margaret C. Jacobs, "The Materialist World of Pornography," in *The Invention of Pornography: Obscenity and the Origins of Modernity, 1500–1800*, ed. Lynn Hunt (New York: Zone Books, 1993), pp. 157–202. See also Evelyn Fox Keller's essays "Baconian Science" and "Spirit and Reason at the Birth of Modern Science," in Keller, *Reflections*, pt. 1; and Stephen Greenblatt, "Learning to Curse: Aspects of Linguistic Colonialism in the Sixteenth Century," in *First Images of America*, ed. Fredi Chiappelli (Berkeley: University of California Press, 1976), pp. 561–80.
[9] Jacobs, "The Materialist World of Pornography."
[10] On obscurity in early modern technical and scientific writing, see William Eamon, *Science and the Secrets of Nature: Books of Secrets in Early Modern Culture* (Princeton, N.J.: Princeton University Press, 1994). Neil Kenny has written a close and careful account of the "philosophic" poetry of sixteenth-century France, *The Palace of Secrets: Béroald de Verville and Renaissance Conceptions of Knowledge* (Oxford: Clarendon Press, 1991).

of speaking; positive expressions; clear senses; a native easiness: bringing all things as near the Mathematical plainness as they can.[11]

Although many major sixteenth- and early seventeenth-century works do not manifest this suspicion of the figurative, such as Kepler's allegorical *Somnium, sive Astronomia lunae* (Dream; or, Lunar Astronomy, 1634) and Galileo Galilei's (1564–1642) cryptic and metaphorical private announcement of the phases of Venus (an anagram resolving to *Cynthiae figuras aemulatur mater amorum*, "the mother of loves [Venus] emulates the figures of Cynthia [the Moon]"), by the mid-seventeenth century, new technical lexicons were being forged or expanded in the interests of precision, and the polysemy of natural language (despite the flowers of metaphor in Sprat's recommendations cited earlier) was rejected in principle.[12]

Figures were not the only linguistic resource to occupy the attention of natural philosophy. Narrative dimensions of reporting observations were gradually restricted and formalized: Passive verbs or verbs in the first-person plural, often in the present tense, purified reports sent to scientific journals of such distracting features of ordinary narration as characters (singular agents for the predicates) and historicity.[13]

Distinctions of genre were harder to make in the sixteenth century than by the end of the seventeenth century, or at any rate verse was seen as a more capacious vehicle than it has since become. Fracastoro's sixteenth-century epillion, *Syphilidis, sive morbum gallicum* (Syphilis; or, the French Disease, 1530), communicating not only a New World discovery narrative but an important physician's etiology and suggestions for treatment of the new infectious disease (first named "syphilis" in his poem), is a famous illustration of the use of poetry to transmit serious scientific thinking. Verse continued to be a vehicle for what we might call popularization, or at any rate dissemination, of new information and theories – in the nineteenth century, an Oxford biologist could still find it worth his while to versify Charles Darwin's *The Origin of Species* (1859) – but the most important kind of scientific writing, the announcement of experimental results and new data or theories, was relegated in the seventeenth century to prose, usually carefully differentiated from what was emerging concomitantly as "belles lettres." Writers of drama, fiction, and poetry maintained the depth and polysemy rejected by the experimental philosophers as their particular province, but hoax (especially in geographical writing) and parody (see the section on "Antagonisms" in this chapter) also helped pave the way for a kind of fictional realism in which

---

[11] Thomas Sprat, *The History of the Royal Society* (London: Printed by T. R. for J. Martyn, 1667), p. 113.
[12] For Galileo's anagrams, see Albert van Helden, "The Reception of *Sidereus Nuncius*," in his translation *Sidereus Nuncius; or, The Sidereal Messenger* (Chicago: University of Chicago Press, 1989), pp. 87–113.
[13] See Peter Dear, "*Totius in Verba*: Rhetoric and Authority in the Early Royal Society," *Isis*, 76 (1985), 144–61.

some features we associate with the rhetoric of information transmission were adopted for the purposes of narrative suspense and pleasure. Aphra Behn's (1640–1689) novel *Oroonoko* (1688), set in Surinam, in its highly detailed and sensational empiricism is as far a cry from Thomas Nash's (1567–1601) euphuistic picaresque *The Unfortunate Traveller* (London, 1594; "Why should I go gadding and fizgigging after firking flantado amphibologies?") as William Harvey's (1578–1657) *De motu cordis* (1628, On the Motion of the Heart, 1653) is from the Virgilian *Syphilis*.[14] I do not mean to make literary "realism" sound parasitic on new scientific rhetorical manners but to suggest that fiction writers felt imperatives parallel to those that guided, at the deepest level, the shifts in scientific representation. On both fronts, we find responses to a sense that the world must be newly and, as anthropologist Clifford Geertz would say, *thickly* articulated.

Simultaneous with the effective creation of new diction and a chastened style for natural philosophy was the quest for what English philosophers termed a "real character" – an artificial language, composed according to strict taxonomic principles, of characters themselves information-bearing and biased toward no particular natural vernacular: a language of dense denotative capacity suited to the aims of an international scientific community. A large number of scholars and philosophers in several countries wrote on this topic and designed frameworks, even (in the case of George Dalgarno, ca. 1626–1687) lexicons, for use.[15] The idea of a language that represents in its rules and categories not language itself, but the rest of the world, seems peculiar to most intellectuals now; early modern imaginative writers, whose primary commitment was to the powers of natural language and exploitation of the archaeological wealth of etymology, were similarly uncompelled and many (as we will note) even hostile. Jonathan Swift's (1667–1745) ludicrous virtuosi in *Gulliver's Travels* (1726) carry huge bags of objects, which they hold up in sequence instead of speaking to one another: Natural and scientific languages have dissonant functions and look profoundly disabled from each other's point of view.[16] Nevertheless, in the flowerless prose of a Hemingway or a Camus we see the descendants of the new style put to use in the service of an ambiguity at least as powerful as those maintained by the increasingly figural language of early modern poetry.

---

[14] Aphra Behn, *Oroonoko; or, The Royal Slave: A True History* (London: Will Canning, 1688); Thomas Nash, *The Unfortunate Traveller* (London: Cuthbert Burby, 1594); and William Harvey, *De motu cordis* (Frankfurt: William Fitzer, 1628).

[15] George Dalgarno, *Ars signorum, vulgo character universalis et lingua philosophica* (London: J. Hayes, 1661). A thorough survey of these works is James Knowlson's *Universal Language Schemes in England and France, 1600–1800* (Toronto: University of Toronto Press, 1975). See also Lia Formigari, *Linguistica ed empirismo nel Seicento inglese* (Bari: Laterza, 1970), English trans., *Language and Experience in Seventeenth-Century British Philosophy*, trans. William Dodd (Studies in the History of the Language Sciences, Series 3, Volume 48) (Amsterdam: John Benjamins, 1988).

[16] Jonathan Swift, *Travels into Several Remote Nations . . . of Lemuel Gulliver* (London: Benjamin Motte, 1726).

Cryptography, and especially real characters (or at least rational and invented languages), made significant appearances as fictional phenomena, too, not only as pure parody in Swift but as objects of readerly wonder and pleasure in, for instance, Francis Godwin's (1562–1633) *Man in the Moon* (1638) and Savinien Cyrano de Bergerac's (1619–1655) *Histoire comique de les estats et empires de la lune*, or, as it is traditionally referred to in French, *L'Autre monde* (Voyage to the Moon, ca. 1648).[17] As fictional objects, these "languages" are deliciously complex, postulating the possibility of almost unmediated ("naked") transmission irreconcilable with the gaps and longings of natural language that generate figurative and literary representation in the first place.

TELESCOPE, MICROSCOPE, AND REALISM

The traditional wisdom in literary studies holds that the early modern period in several Western European nations saw the first development of prose fiction in the mode of realism, a term that denotes a style as well as a representational bias in favor of the *representandum*, so to speak (cf. Niekrasz and Swan, Chapter 31, in this volume). Astronomical works employing telescopic observations, along with works produced in the initial enthusiasm for micrographic observation, belong to the very relevant history of the detail (the term is first seen in French and then English in the sixteenth century) – a history that expressed itself in the late sixteenth and seventeenth centuries in the emergence of pornographic poetry and prose as well as in the development of fictional realism and natural history. (All of these forms of inquiry and documentation are obviously also involved in the history of the visual arts in this period.) Detail in European narrative had tended before this time to be allegorical or ceremonial, and scant; travel books usually produced details of a wonderful or sensational turn. For many reasons, the value of "gratuitous" detail returned in the late Renaissance, at a level familiar to a modern reader of novels (or botany). It may be that the erotic dimension of empirical gazing was the first to be exploited. At any rate, Pietro Aretino's (1492–1556) pornographic dialogue between an Ovidian Dipsa figure and a young concubine, the *Ragionamenti* (Dialogues, 1534–6), preceded such major works of scientific empiricism as Galileo's *Sidereus Nuncius* (Sidereal Messenger, 1610), in which the first inventive user of the telescope traces the movements of his newly discovered "Medicean stars" through nightly observations recorded both verbally and graphically.[18] Robert Hooke's (1635–1703) *Micrographia*

---

[17] Francis Godwin, *The Man in the Moon* (London: John Norton, 1638); Savinien Cyrano de Bergerac, *Histoires comiques de les estats et empires de la lune* (Paris: Le Bret, 1657). Cyrano's text has a vexed history: He wrote and circulated it in manuscript around 1648; it was first published posthumously and "sans privilege" in 1650 and then in an authorized but expurgated edition in 1657. The unexpurgated text was not published until modern times.

[18] Galileo Galilei, *Sidereus nuncius* (Venice: T. Baglionum, 1610). Pietro Aretino, *Ragionamento della Nanna, et della Antonia* was published in two parts (Venice: Francesco Marcolini, 1534–6) with false

(1665), the earliest of the really sumptuously illustrated micrographic texts, had undoubted power over literary imaginations – Margaret Cavendish (ca. 1624–1674), Duchess of Newcastle, was explicit and overtly disapproving in her 1666 scientific novel *A New World Called the Blazing-World*, and Swift makes significant allusions to it in Part 2 of *Gulliver's Travels*, wherein Gulliver is grossly offended by the microscopic close-ups his gigantic Brobdinagian hosts inadvertently afford him.[19]

The conservative and even aristocratic positions of those (such as Cavendish and Swift) offended by, in Samuel Johnson's famous phrase, "numb'ring the streaks on the tulip," is suggestive: The most graphic inspector of the tiny and lowly was the Dutch artisan-genius Antonie van Leeuwenhoek (1632–1723), who was not embarrassed to analyze or describe his own feces, and who was the first, and for some time the only, observer to see what he identified as "animuncules" in highly magnified semen.[20] One finds abundant physical detail as well in the picaresque tradition (closely linked, as I have pointed out, to the voyage and travel genres) and in pornography, which, after some sixteenth-century Italian ventures, came into its own in the late seventeenth century; government bans and seizures often included gynecological handbooks as well as fictional narratives such as Charles Cotton's (1630–1687) *Erotopolis, The Present State of Betty-land* (1684) or the anonymous French *L'École des filles* (The School for Girls, 1655).[21]

The change of values and focus involved in the flourishing empiricism of both fiction and natural history in the sixteenth and seventeenth centuries, illustrated as well by English "metaphysical poetry," the drawings of Leonardo

---

publication data on the title pages for reasons of censorship: for the first volume, "Paris 1534," and for the second, "Turin: P. M. L., 1536" – see Paul Larivaille, *Pietro Aretino* (Rome: Salerno Editrice, 1997), p. 199. Some of Galileo's courtly language may strike a modern reader as intermediate between Aretino and the empiricist purism of scientific reporting: At the beginning of *Sidereus Nuncius*, he tells us that "[i]t is most beautiful and pleasing to the eye to look upon the lunar body" (van Helden, *Sidereus Nuncius*, p. 35), and the theme of erotic gazing is maintained, for example, in his defense of earthlight: "But what is so surprising about that? In an equal and grateful exchange the Earth pays back the Moon with light equal to that which she receives from the Moon almost all the time in the deepest darkness of the night" (van Helden, *Sidereus Nuncius*, p. 55).

[19] Robert Hooke, *Micrographia; or, Some Physiological Descriptions of Minute Bodies* (London: J. Martyn and J. Allestry, 1665); Margaret Cavendish, *Description of a New World Called the Blazing-World* (London: A. Maxwell, 1666). For a comparison of these two texts, see Mary Baine Campbell, *Wonder and Science: Imagining Worlds in Early Modern Europe* (Ithaca, N.Y.: Cornell University Press, 1999), chap. 6.

[20] For Leeuwenhoek and his analyses of sperm and feces, see Clifford Dobell, *Antony van Leeuwenhoek and His "Little Animals"* (New York: Staples Press, 1932). Svetlana Alpers, *The Art of Describing: Dutch Art in the Seventeenth Century* (Chicago: University of Chicago Press, 1983) connects Northern European developments in optics and optical instruments with the painting of the period. On the difficulty of compelling virtuosos' interest in the grime of the nonwondrous, see Lorraine Daston and Katharine Park, *Wonders and the Order of Nature, 1150–1750* (New York: Zone Books, 1997), chap. 8.

[21] Charles Cotton, *Erotopolis, The Present State of Betty-land* (London: Thomas Fox, 1684). The stated place of publication for *L'École des filles* (1655) was the fictional "Cythère." Pornographic texts tend to give imaginary places of publication and no publishers. On pornography, see Jacobs, "The Materialist World"; and Paula Findlen, "Humanism, Politics and Pornography in Renaissance Italy," in Hunt, ed., *Invention of Pornography*, pp. 49–108. (The notes to Findlen's essay are a useful bibliographical tool.)

da Vinci (1452–1519) and Dutch and Flemish painting, is the most profound (and overdetermined) of the period. Conservative writers were not wrong to suspect it was connected to demographic shifts of political and economic power away from those whose situations and actions lend themselves so easily to allegory and the almost spontaneous symbolizations of chivalric romance.

PLURALITY OF WORLDS: FROM ASTRONOMY TO SOCIOLOGY

Realism, as a stylistic predisposition to the material world and the infinite differences it offers to be articulated (see Findlen, Chapter 19, this volume), is in some ways homologous with a newly salient plurality at the larger scales of globe and cosmos. The social sciences, like any disciplinary language, institute fine-grained systems of differentiation, which in turn provide discrete objects for interpretation and generalization. These sciences did not of course exist in early modern Europe, but we can see a process of differentiation under way as the earth's supply of inhabited and enculturated islands was multiplied in European discourse and knowledge. Astronomical theory, provoked by discoveries of the uneven surface of the moon and by telescopic evidence of Copernican heliocentrism, shared with the subgenre of the lunar voyage an interest in the philosophical implications of the plurality of worlds. And it also required an Archimedean perspective exploited by poets as well as astronomers and cosmographers: As historian Frank Lestringant puts it, "Without some voyage of the soul, there can be no instantaneous point of view over the cosmos. The kinship between cosmography and sacred poetry was therefore essential and primary."[22] Kepler's *Somnium* (completed in 1621) managed to be both cosmography (not to mention astronomy) and poetry at once; John Milton (1608–1674) in *Paradise Lost* (1667) provided a "voyage of the soul" of the sort that is logically prior to the global geography of Martin Waldseemüller (1470–ca. 1521), Gerardus Mercator (1512–1594), and Abraham Ortelius (1527–1598)[23] (see Vogel, Chapter 20, this volume).

The most interesting and richly signifying texts at the intersection of the astronomical, ethnological, and purely fictional are the moon voyages. The *Somnium* is the most densely informational as well as the most visionary, and it was the first to see publication; it responds in at least one of its hundreds of discursive and exegetical notes to John Donne's (1572–1631) *Ignatius His Conclave* (1611), a satirical colloquy held in a "jesuitical hell" on the moon, and more immediately to Kepler's own Latin translation of Plutarch's seriously

---

[22] Frank Lestringant, *L'Atelier du cosmographe ou L'Image du monde à la Renaissance* (Paris: A. Michel, 1991), English trans., *Mapping the Renaissance World*, trans. David Fausett (Berkeley: University of California Press, 1994), p. 21.
[23] John Milton, *Paradise Lost* (London: Peter Parker, and Robert Boulter and Matthias Walker, 1667).

speculative work *De facie lunae* (On the Face in the Moon, ca. 100).²⁴ While offering a comprehensive and accurate account of the moon's "selenography," as well as of the Earth and Copernican solar system as seen from a lunar perspective, it also plays with the conventions of serious voyage literature such as Harriot's *A Brief and True Report of the New Founde Land of Virginia* (1588), ending as that work had with an account of the inhabitants (of the Moon) and their forms of adaptation to the climate and physiography of their home.²⁵ The social speculations are unobtrusively utopian: Among these variously reptilian and even plantlike indigenes, some have the knack of making boats and, in the "Selenographical Appendix," fortifying their cities against natural disasters, whereas others are not toolmakers; nonetheless, there is no hierarchy among sentient beings, nor is any strife alluded to over the bimonthly motion of all the lunar water to one of the two inhabited hemispheres (that facing the earth and thus more subject to its gravitational pull). Even more interesting, for our purposes here, is the allegorical structure of the narrative (a dream vision, which invites multilayered interpretation generically) and the generic chaos of the Notes (which outnumber pages of main text six to one) – many of them strictly mathematical, some autobiographical, others jocosely exegetical or satirical, or both. The *Somnium* was not just a bijou: It had started life as a dissertation too dangerously Copernican to defend at Tübingen and, having been converted into an apparently autobiographical dream vision, provoked charges of witchcraft against Kepler's mother that took five years to settle in court.²⁶ Then the Notes were added, for reasons both legal and mathematical, and the whole thing was finally published posthumously.²⁷

The two most famous of many subsequent fictional meditations on the possibilities suggested by the thought of inhabited "worlds" are Francis Godwin's parody of Spanish picaresque, *The Man in the Moon* (1638, translated into French in 1648), featuring an upwardly aspiring Spanish midget named Gonsales, and Cyrano's libertine response, *L'Autre monde*, in which Gonsales appears in a cage, where he is joined by the narrator, whom the lunar empress fondly supposes Gonsales can impregnate; this was followed a little later by the unfinished *Estats et empires du soleil* (Estates and Empires of the Sun, 1662).²⁸ All these voyages were published posthumously, in Cyrano's case expurgated. And all involved an imaginary shift of perspective on the

---

²⁴ John Donne, *Conclaue Ignati* (London: W. Hall, 1611); the vernacular version followed immediately: Donne, *Ignatius His Conclave* (London: Richard More, 1611).
²⁵ Thomas Harriot, *A Briefe and True Report of the New Founde Land of Virginia* (London: R. Robinson, 1588; Frankfurt: Theodor de Bry and Sons, 1590).
²⁶ See the introduction to Edward Rosen's translation, *Kepler's "Somnium: The Dream, or Posthumous Work on Lunar Astronomy"* (Madison: University of Wisconsin Press, 1967), and, especially, John Lear's long introduction to the translation of Patricia Frueh Kirkwood, *Kepler's "Dream"* (Berkeley: University of California Press, 1965).
²⁷ On the *Somnium* in relation to both scientific and literary discourses of the seventeenth century, see Timothy Reiss, *The Discourse of Modernism* (Ithaca, N.Y.: Cornell University Press, 1982), chap. 4; Hallyn, *Poetic Structure of the World*, chap. 11; and Campbell, *Wonder and Science*, pp. 133–43.
²⁸ Savinien Cyrano de Bergerac, *Estats et empires du soleil* (Paris: Charles de Sercy, 1662).

part of the voyager from observer to observed (see Nicolson's *Voyages to the Moon*, mentioned at the beginning of this chapter).

It is possible to glean from the moon voyages a complex sense not only of what shifting astronomical paradigms might have meant to early modern literate Europeans but also of what fantasias (and guilts) the increasingly colonized "other worlds" of America and the East Indies provoked. Such imaginative displacement almost invariably focused on the moon itself, long associated with the displacements and projections of erotomania: "With how sad steps, O Moon, thou climb'st the skies," begins one sonnet from Sidney's *Astrophel and Stella* (1591), in which the speaker assumes emotion parallel to his own in the planet as arrestingly visible and yet out of reach to him as "Volva," the shining Earth, is to one hemisphere of Kepler's fascinated moondwellers.[29] But many of the new fantastic voyages, neither in verse nor lunar in inspiration, tend to be about islands in the western and southern oceans. They are largely utopic, or triangulated with that genre: It is the background of utopian voyage writing that makes Henry Neville's (1620–1694) *Isle of Pines* (1668) so funny. The vision of Pines populating his entire newfound island through enthusiastic polygamy is comically desirable (for some at least) but also a meditation on the population dynamics of imported species that Europeans might have done well to consider seriously for a moment. The imagined islands of these fantastic voyages, more clearly possible than inhabited worlds or imaginary planets, are easy to see as, if not foundational for the sciences of sociology and ethnology, at least a playground for inventing and clarifying the objects of such sciences.[30]

GEOGRAPHY, ETHNOGRAPHY, FICTION, AND THE WORLD OF OTHERS

Certain scientific pursuits, or at any rate certain pursuits in search of data later to be organized in the modern scientific disciplines, more clearly than others share space with the aims and materials of poetry, fiction, and "belles lettres." Geography, ethnography, anthropology, and the Renaissance discipline (if it could possibly be called that) of cosmography share with the emerging literary form of the novel an interest in the description of human relations

---

[29] Sir Phillip Sidney, *Astrophel and Stella* (London: Printed [by John Charlewood] for Thomas Newman, 1591).

[30] Henry Neville, *The Isle of Pines* (London: Printed by S. G. for Allen Banks and Charles Harper, 1668). On fantastic voyages, the best bibliographical works are still Geoffroy Atkinson, *The Extraordinary Voyage in French Literature before 1700* (New York: Columbia University Press, 1920); Atkinson, *The Extraordinary Voyage in French Literature from 1700 to 1720* (Paris: É. Champion, 1922); and Philip Babcock Gove, *The Imaginary Voyage in Prose Fiction . . . from 1700 to 1800* (New York: Columbia University Press, 1941). On utopias with an eye toward the development of sociology, start with Frank E. Manuel and Fritzie P. Manuel, *Utopian Thought in the Western World* (Cambridge, Mass.: Belknap Press, 1979).

and cultural formations along with a reliance on the trope of the journey: Travel literature, one of the great literary (and subliterary) forms of the period, occupied an important intermediate ground.[31]

From the earliest ocean voyages enabled by the navigational advances of the fifteenth century, humanists were collecting texts and oral information and publishing it in large, usually vernacular tomes. Some of them, such as Richard Hakluyt's (ca. 1552–1616) *Principall Navigations . . . of the English Nation* (1589) were nationalist in theme, others encyclopedic (the *Decades*, as the English called it, of Peter Martyr of Anghiera [1457–1526], the de Brys' *Grands voyages*, and Samuel Purchas's [ca. 1577–1626] ethnologically inflected expansion of Hakluyt's final collection, *Purchas His Pilgrimes* [1625]).[32] These collections provided access to critical masses of data, informing cartographers and the great atlas-makers Ortelius, Mercator, and the Blaue family, and preceded the appearances of more specialized and synthetic global accounts, such as Francis Willughby (1635–1672) and John Ray's (1627–1705) *Ornithologiae libri tres* (Three Books on Ornithology, 1678), Joseph-François Lafitau's (1681–1746) *Moeurs des sauvages amériquaines* (Customs of the American Savages, 1724, more global in scope than the title indicates), and eventually the *Systema Naturae* (System of Nature, 1735) and *Critica Botanica* (1737) of the great Swedish traveler and naturalist, Carl von Linné, or Carolus Linnaeus (1707–1778).[33]

The individual accounts in the voyage collections inspired hoaxers and writers of fiction (and less often of verse narrative) as well as cosmographers, geographers, ethnologists, and natural historians. Most obviously, they offered examples of interesting and significant action, neither legendary nor necessarily aristocratic, often narrated in the first person (thus providing a subjective dimension not usually found outside the aristocratic and courtly

---

[31] The most comprehensive account of the early modern and eighteenth-century corpus, for England and France especially, is Percy G. Adams, *Travel Literature and the Evolution of the Novel* (Lexington: University Press of Kentucky, 1983). For early modern relations among the enterprises named in this section's subheading see Valerie Wheeler, "Travelers' Tales: Observations on the Travel Book and Ethnography," *Anthropological Quarterly*, 59: 2 (April 1986), 52–63; Mary Baine Campbell, "The Illustrated Travel Book and the Birth of Ethnography," in *The Work of Dissimilitude*, ed. G. Allen and Robert A. White (Newark: University of Delaware Press, 1992), pp. 177–95; and Campbell, *Wonder and Science*, pt. III.

[32] Richard Hackluyt, *Principall Navigations . . . of the English Nation* (London: Printed by George Bishop and Ralph Newbury, for Christopher Barker, 1589); Pietro Martire d'Anghiera, *De rebus oceanicis et orbe novo decades tres* (Basel: Johannes Bebelius, 1533) (one of many editions, ever expanding the earliest, published in Seville in 1511; it was influentially translated and published in England by Richard Eden in 1555, one of the very first "english'd" reports on New World exploration); Théodor de Bry and sons, *Grands voyages . . .* (Frankfurt: Theodor de Bry, 1595–1634); and Samuel Purchas, *Purchas: His Pilgrimes* (London: Printed by William Stansby for Henrie Fetherstone, 1625).

[33] Frances Willughby and John Ray, *Ornithologiae libri tres* (London: John Martyn, 1676) and in English, *The Ornithology of Francis Willughby of Middleton . . . By John Ray* (London: John Martyn, 1678); Joseph Lafitau, *Moeurs des sauvages ameriquaines*, 2 vols. (Paris: Saugrain laîné, for Charles Étienne Hochereau, 1724); Carolus Linnaeus, *Systema naturae* (Leiden: Printed by J. W. de Groot for Theodor Haak, 1735); and Linnaeus, *Critica Botanica* (Leiden: Conrad Wishoff, 1737).

forms of romance and lyric poetry).[34] Picaresque fiction, particularly in Spain (the first major colonial power), shared with many narratives collected orally by such avid interviewers as Peter Martyr and Hackluyt the features of physical mobility in pursuit of class mobility, located in protagonists of little or no status. The romance narrative structure of *aventur* had been opened to the working and vagrant classes of a disintegrating feudalism by way of colonial explorations and settlements. Such New World travel accounts/colonial reports as, for instance, those of Alvar Nuñez Cabeza de Vaca, Hans Staden of Hesse (ca. 1525–ca. 1576), Harriot, Jean de Léry (1534–1611), Raleigh, and Cartier, with their grounding in the conditions of imperial conquest and settlement, offer representation of communities and cultural groups in customary interaction – a familiar prospect from the vantage point of a culture, like our own, possessed of realistic fiction, especially the novel.[35] Several late seventeenth- and early eighteenth-century novels – Gabriel de Foigny's (ca. 1630–1692) *Terre australe connue* (The Southern Land, Known, 1676) and Swift's *Travels into Several Remote Nations . . . of Lemuel Gulliver* – for instance, are precise formal imitations of the voyage genre, as in many ways had been the briefer and more whimsical moon voyages.[36] Such boundary crossing was evident between several domains. Many writers of early fiction also wrote travel and/or linguistic works, as did many remembered largely for their efforts in natural history or experimental science. Hoaxes were increasingly frequent (see, for example, Louis Hennepin and George Psalmanazar), and many obvious (to us) fictions were read *as* hoaxes.[37] André Thevet (1502–1590), royal cosmographer to the Valois kings, in a twist on this

---

[34] See Mary B[aine] Campbell, *The Witness and the Other World: Exotic European Travel Writing, 400–1600* (Ithaca, N.Y.: Cornell University Press, 1988), esp. pt. 2, "The West"; Margaret Hodgson's comprehensive survey, *Early Anthropology in the Sixteenth and Seventeenth Centuries* (Philadelphia: University of Pennsylvania Press, 1964); and, on the structure of Thevet's cosmographical opus, Lestringant, *Mapping the Renaissance World*.

[35] Alvar Nuñez Cabeza de Vaca, *La Relación que dio Alvar Nuñez Cabeza de Vaca de lo acaescido en las Indias* (Zamora: Augustin de Pazy Juan Picardo, 1542); Jean de Léry, *Histoire d'un voyage faict en la terre du Bresil* (La Rochelle: Ant. Chuppin, 1578); Hans Staden of Hesse, *Warhaftige Historie und Beschreibung eyner Landschafft der . . . Menschfresser Leuten* (Marburg: Andres Kolben, 1557); Harriot, *A Briefe and True Report of the New Found Land of Virginia* (London, 1588; as *America, Part I*, Frankfurt: Théodor de Bry, 1590); Walter Raleigh, *Discoverie of The Large Rich and Bewtiful Empire of Guiana . . .* (London: Robert Robinson, 1596); and Jacques Cartier, *Brief récit, et succincte narration, de la navigation faicte es ysles de Canada, Hochelage, etc.* (Paris: Ponce Roffet, 1545).

[36] Gabriel de Foigny, *Terre Australe connue* (Vannes [Genève]: Jacques Venevil, 1676).

[37] Louis Hennepin's "fictitious" account of finding the source of the Mississippi River appeared in his popular and much believed *Nouvelle découverte d'un tres grand pays* (Utrecht: Guillaume Broedelet, 1697). The imposter Psalmanazar was exposed within a few years of publication of his immortal *Historical and Geographical Description of Formosa* (London: Printed for Daniel Brown, G. Strahan and W. Davis, Fran. Coggan and Bernard Lintott, 1704) and *Description de l'île Formosa en Asie* (Amsterdam: Estienne Roger, 1705), but only after the author had taught Oriental Languages at Oxford and dined out on his Formosan stories with most of London's great and powerful! For Hennepin (seventeenth century) and Psalmanazar (1679–1763), see (but do not always trust) Percy G. Adams, *Travelers and Travel Liars, 1660–1800* [1962] (New York: Dover, 1980). For some famous "hoaxes" and fictions now being read as in some ways true, see David Fausett, *Writing the New World: Imaginary Voyages and Utopias of the Great Southern Land* (Syracuse, N.Y.: Syracuse University Press, 1993).

latter dissonance, fumed in print at François Rabelais (ca. 1490–ca. 1553) for scooping him on the *paroles gelées* (frozen words) of the northern Atlantic in the *Quart Livre* (1548)![38] (Rabelais had in fact borrowed from one of Thevet's own half-concealed sources, the cosmographer's "great friend" Cartier.)

But there was a deeper level of mutual involvement than that of genre and its narrative conventions. Voyages and early novels shared an absorption, eventually competitive, in representing "the Other" – behavior and lifeways alien and unknowable enough to be objectified in many ways, from ethnology (including books of fashion plates and "ballets" of ethnic dances in costume) to utopia to fictions of human (European) encounters with idealized, demonized, or primitivized alternate selves. The early entanglement of these differently developed kinds of writing has implications for our sense of the functions of anthropology, sociology (heralded by the utopia), and the novel. Behn's *Oroonoko* is the story of an African prince enslaved and sold to a West Indian administrator; the prince Oroonoko kills his wife and meets a gruesome and terrible end after the failure of a slave rebellion he has led. The novel, usually considered one of the first in the modern genre, is at once an account of the natural history and cultural groupings of colonial Surinam and an examination of the narrator's (putatively Behn's) phantasmatic encounter with the Noble Savage during her residence in Surinam. Behn has been falsely accused of plagiarizing George Warren's *Impartial Description of Surinam* (1667) for much of the natural history and ethnography that make up the novel's dense setting.[39] But the level of detail in the presentation has more self-reflexive uses than does the informational detail of a factual account such as Warren's. Above all, perhaps, it establishes the powerfully knowing gaze of the narrator, thus ensuring the credibility of her observations of the characters' interior experience, the objecthood of all narrated others, and the verisimilitude of the narrative.

This "objecthood of all narrated others," a potential feature of any realistic, observational account of human bodies and action, is related to the development of instruments that register sensational facts. "[A]s *Glasses* have highly promoted our *seeing*, so . . . there may be found many *Mechanical Inventions* to improve our other Senses, of *hearing*, smelling, tasting, *touching*:"[40] Prosthetic "enlargements" of the senses provided or at least promised by the new scientific equipment facilitated and expressed the bodily alienation so profoundly marked in René Descartes' (1596–1650) *Meditations* (1641), or in his acquaintance Margaret Cavendish, Duchess of Newcastle's *Blazing-World*, whose virtuoso Empress spends a lot of satisfying time sharing her body with the spirit of her guest and "scribe," the Duchess of Newcastle, or,

---

[38] François Rabelais, *Quart livre* (Lyon, 1548 [partial]; Paris: Michel Fezandat, 1552).
[39] George Warren, *An Impartial Description of Surinam . . . with a History of Several Strange Beasts, Birds, Fishes, Serpents, Insects and Customs of that Colony . . . from Experience . . .* (London: Nathaniel Brooke, 1667).
[40] Hooke, *Micrographia*, sig. aiir.

together with the Duchess, quitting her body altogether and inhabiting the Duke's body back in England.[41] This proto-novel, a feminist utopia in which Cavendish's alter ego and protagonist *runs* a version of the Royal Society rather than being barred from membership, spells out in fantastical terms the new interest in deep (and hidden, occult) interiors that will be the focus of the "novel of manners" or the "domestic novel." Although "manners" in that phrase derives from the ethnological "manners and customs" (or "moeurs," as in Lafitau's *Moeurs des sauvages amériquaines*), which had been described in terms of rule-bound behavior and such surface phenomena as costume, body language, and tattoo, the novel of manners locates its reader in the hidden psychic spaces understood to be routinely and suspensefully belied by the "manners and customs" of the embodied social person. Even pornography, the narrative of bodily experience, narrates what happens to Cartesian characters, whose intentions and experiences may well be at odds with their bodily activity and dissonant with their sensations, and operates, as do many travel accounts and ethnographies, by objectifying a protagonist's co-actors as (merely) bodies.

ANTAGONISMS

As evidenced by the moon voyages especially, it remains a fact, however differently or scantly postmodern theory and criticism might draw the lines of distinction between categories of text and intellectual practice, that science, new or old, figured as subject matter in early modern drama and narrative forms. Alchemical poems abounded, especially in the sixteenth century, as well as Lucretian versification of metaphysical and physical science – such as Guillaume Du Bartas's *Sepmaine* (Divine Weeks, 1578) and Guy Lefèvre de la Boderie's *Galliade; ou, de la revolution des arts et des sciences* (Galliade; or, the Revolution in the Arts and Sciences, 1578) – and such medical poetry as Fracastoro's *Syphilis*.[42] Sprat's *History of the Royal Society* invites poets and imaginative writers to celebrate and exploit the abundances of information and physical detail being made accessible by the new science and the voyages of mercantile and colonial exploration: "Another benefit of *Experiments* . . . is, that their discoveries will be very serviceable to the *Wits*, and *Writers* of this, and all future Ages" – wit being "founded on such images which are generally known, and are able to bring a strong, and a sensible impression on the *mind*."[43]

But among the most interesting instances of writers' inclusion of science and scientifically derived information as material are the polemical – that is,

---

[41] René Descartes, *Meditationes de prima philosophia* (Paris: Michael Soly, 1641).
[42] Guillaume Du Bartas, *Semaine [Sepmaine]* (Paris: Michel Gadolleau, 1587); Lefèvre de la Boderie, *Galliade; ou, de la revolution des arts et sciences* (Paris: Guillaume Chaudiée, 1578).
[43] Sprat, *History of the Royal Society*, p. 114.

the attacks on the new science and (especially for the dramatists) the new scientists. Book III of *Gulliver's Travels* is the classic instance (and also provides an example of utopia and imaginary voyage, including ethnographic concerns with alternate human social formations). But the playwrights Molière (1622–1673), Thomas Shadwell (ca. 1642–1692), and Susanna Centlivre (ca. 1667–1723) all joined in the dramatic burlesquing of the virtuosi by literary authors, foreshadowed perhaps by the more tragic undermining of Christopher Marlowe's (1564–1593) *Doctor Faustus* (ca. 1588). At the other end of the period, the tragedy is replayed as force in Laurent Bordelon's (1653–1730) parody *L'Histoire des imaginations extravagantes de Monsieur Ouflé* (History of the Extravagant Fantasies of Mister Ouflé, 1710), which offered up a fan of outmoded Renaissance science and magic to the amusement of French readers of fiction. William Shakespeare's (1564–1616) *Love's Labours Lost* (1598) and *The Tempest* (1623) are more ambivalent, but in *The Tempest's* happy ending, the magus Prospero vows to "burn [his] books." Major speculative works by natural philosophers that return the polemical compliment in complaints about slippery, figurative thinking and/or writing in literary or popular forms include Thomas Browne's (1605–1682) *Pseudodoxia epidemica* (also known as Vulgar Errors, 1646) and Francis Bacon's (1561–1626) *Novum organum* (New Organon, 1620), themselves, paradoxically, works of ecstatic prose that have long been studied as examples of literary greatness. Whatever sympathies of project or perspective may have been shared by the scientific and literary enterprises of seventeenth-century culture, the encroaching literalism of natural history and its denotative language was felt as a threat to the essential indeterminacy of literary language, and the emerging sciences felt for a long time to come the urgency of maintaining denotative clarity in their communications of new knowledge.[44]

CONCLUSION

The blurring of generic boundaries promoted by cultural studies, and supported as well by anachronistic attempts to apply those boundaries to a corpus historically prior to the institutionalization of modern genres and disciplines, should not obscure the fact – explicit in the works just discussed – that a dichotomy *was* emerging between the cultural aims and means of the scientific and the imaginative writers, a division that would have meant rather little to Lucretius (ca. 99–55 B.C.E.), or even Geoffrey Chaucer (ca.

---

[44] Christopher Marlowe, *The Tragedie of Dr. Faustus* (London: Printed by V. Simmes for Thomas Bushell, 1604); Laurent Bordelon, *L'Histoire des imaginations extravagantes de Monsieur Ouflé* (Paris: N. Gosselin and C. Leclerc, 1710, and Amsterdam: Estienne Roger, Pierre Humbert, Pierre de Coup and les Frères Chatelain, 1710); William Shakespeare, *Love's Labours Lost* (London: Cuthbert Burby, 1598), *The Tempest* (London: Isaac Jaggard and Edward Blount, 1623); Thomas Browne, *Pseudodoxia epidemica* (London: Edward Dodd, 1646); Francis Bacon, *Novum organum* (London: B. Norton, 1620).

1340–1400). Stylistic changes and the search for the "real character" (eventually realized, in part, in the mathematization of the natural sciences, not to mention the totalizing intentions of such descriptive enterprises as botany and zoology) are fundamentally at odds with the literary investment in dense signification, ambiguity, and polysemy. Language as an object of interest, as a creature developing in time (rather than the set of rules elaborated by grammarians and linguists), became the province of writers cut off in this way from the objectives of thinkers for whom natural language was nothing but a medium – and increasingly not a medium of choice.

Tables, taxonomies, graphs, and mathematical formulas are all more concise, more determinate, and more pragmatic than the sentences and paragraphs (never mind the poetic lines!) of living languages. These attributes are important to a discipline that sees its texts as a means to immediate collective ends. They belong to the economy of "getting things done." The alchemical text that tested the mind and purified the spirit while giving chemical recipes did not often succeed at facilitating quick success in the last of these areas. With the speedy growth and maturity of modern scientific publishing, the literary became a category not of knowledge but of recreation (in many senses), and the Poet became not the Monarch "of all Sciences" but their Other: moral, tenuous, unsystematic, individual, florid, and trivial.

The literary antagonism I have been discussing, to science in general and the "new science" in particular, was based partly, perhaps, on a dawning anxiety among many writers and readers over the loss of cultural power and status in the domain such loss carved out and would in the late eighteenth century honorifically label "literature." But this antagonism can also be seen as an emergence of the literary function of critique. If these two fundamentally different and deeply related modes of observing and representing are mutually excluded, then they can (or will) observe and represent each other: science representing literature through philology, linguistics, and criticism; and "literature," more evaluatively, representing science through parody, hoax, satire, "true histories," fantasy, and pornography – forms that test and mock and transgress the boundaries of the real and the true. The poet and artist William Blake's (1757–1827) scorn for Isaac Newton (1642–1727), who saw, as did so many of the objects of Blake's scorn, with his "corporeal eye," is unfair but illuminating. It was clear enough by the end of the eighteenth century what the now normative respect for category, measurement, and boundary had cost, and how reduced was the scope or social function of nonscientific representation: A mimetic struggle would henceforth maintain the divided world of the "two cultures."[45]

---

[45] For an elaborated and profound account of a related contest (between the "religious and poetic institutions") for "possession of the sufficient means of representation" (p. 53) – less historical but attending in part to the phenomenon as manifested in Europe's Enlightenment – see Allen Grossman, "The Passion of Laocoön: Warfare of the Religious against the Poetic Institution," *Western Humanities Review*, 56 (2002), 30–80.

31

ART

Carmen Niekrasz and Claudia Swan

During the epistemic shift conventionally called the Scientific Revolution, the study of nature came to depend on images. Investigation of the plant world, which was still tied to medical aims but was beginning to take shape as the morphological discipline we now call botany, is a case in point. The implementation of new printing techniques in the late fifteenth century enabled the production of publications that featured images that were precisely reproducible, at least in theory, and therefore understood as trustworthy.[1] Gradually, standard classical texts such as herbals, which had previously circulated as hand-copied manuscripts, were made available in printed form and came to be heavily illustrated (Figure 31.1). The accessibility of standard visual references in relatively affordable printed editions permitted enterprising doctors, pharmacists, and amateurs of the plant world to compare the plants they had at hand and that grew in their native lands with the plants described by classical authorities, among them the Greek naturalists Theophrastus (third century B.C.E.), Dioscorides (first century C.E.), and the Roman encyclopedist Pliny the Elder (d. 79 C.E.). Numerous varieties not contained in the classical texts were "discovered" by learned botanists throughout Europe. Like prints, drawings also served as a basis for comparison of local varieties with the plants the classical authors had described and, in those cases where the plants at hand could not be matched with plants previously described, came to serve as means for recording and cataloguing them.[2]

Beginning in the mid-fifteenth century, European artists increasingly engaged in recording nature. Artistic interest in depicting the natural world

---

[1] See William M. Ivins, Jr., *Prints and Visual Communication* [1953] (Cambridge, Mass.: MIT Press, 1969).
[2] On illustrated botany, see Agnes Arber, *Herbals: Their Origin and Evolution, a Chapter in the History of Botany, 1470–1670* [1912] (Cambridge: Cambridge University Press, 1986); William T. Stearn, *The Art of Botanical Illustration: An Illustrated History* [1950] (New York: Antiquarian Society, 1994); and David Landau and Peter Parshall, *The Renaissance Print, 1470–1550* (New Haven, Conn.: Yale University Press, 1994), esp. "Printed Herbals and Descriptive Botany," pp. 245–59.

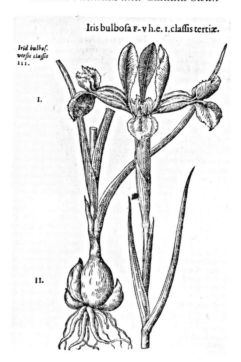

Figure 31.1. *Iris bulbosa*. In Carolus Clusius, *Rariorum plantarum historia* (Leiden: Plantin, 1601), p. 214. Reproduced by kind permission of the National Herbarium Nederland.

coincided with the sorts of close observation and morphological description that structured many scientific endeavors. Beyond illustrated natural history, other areas in which artistic and scientific interests overlapped in the early modern period include the collection and organization of naturalia in cabinets, *musea*, gardens, or *vivaria* and menageries, in which princes, pharmacists, academic doctors, and artists alike took interest. The microscope, the telescope, and the mobile *camera obscura* are optical devices whose features and uses were exploited for artistic and experimental ends. The rise of botanical still-life painting, the involvement of artists in natural history and ethnography both at home and abroad, and widespread interest in the material and metaphysical effects of foreign and exotic products are some of the other relevant areas of intersection. The laboratory and the artist's workshop were both spaces for the intensive exploration of nature, and each borrowed tools, technologies, materials, and even methods of observation from the other. In this chapter, we have chosen to focus upon the variety of standards for naturalism in the early modern period, the features and functions of scientific illustration, the changing role of images in anatomical instruction, and shared practices, as well as potentially instructive ways of excavating the connections between early modern art and science in future scholarship.

NATURALISM

Of all the paradigms for artistic achievement under which early modern visual artists labored, none had as venerable a heritage and was as consistently reiterated as the encouragement to imitate nature. Over and over, the mimetic feats of the late fifth-/early fourth-century B.C.E. Greek painters Zeuxis and Parrhasius (none of whose works are known to survive) were recounted in biographical and theoretical works of the sixteenth and seventeenth centuries. Pliny the Elder had first told of Zeuxis stunning observers by painting grapes so naturalistically that birds swooped down to eat them, and of Parrhasius astonishing Zeuxis in turn by painting a curtain so convincingly that Zeuxis demanded it be pulled aside to reveal the picture beneath.[3] (Grapes and fictive curtains are a topos in early modern painting, which frequently cited these precedents directly.)[4] From the Italian sculptor and author Lorenzo Ghiberti (1378–1455) to the Dutch theorist and painter Samuel van Hoogstraten (1627–1678), countless early modern authors cited Zeuxis and his fellow artist Parrhasius as paragons of mimetic accomplishment, spurring contemporaries to similar feats. "Le Chevalier," an interlocutor in a treatise by the French man of letters Charles Perrault (1628–1703), relates an encounter between an animal and a painting, which he calls "as flattering to modern painting as Zeuxis's grapes is to that of the Ancients, and vastly more entertaining."

> The door of Monsieur Le Brun [Charles le Brun, 1619–1690] was open, and a freshly painted picture had been taken out into the courtyard to dry. In the foreground of this painting was a perfect representation of a large thistle. A woman came past leading an ass, which, when it saw the thistle, plunged into the courtyard; the woman, who was hanging on to its bridle, was dragged off her feet. If it hadn't been for a couple of sturdy lads who gave it some fifteen or twenty blows each with their sticks to force it back, it would have eaten the thistle – and I say *eaten*, because the paint was fresh and would all have come off on its tongue.[5]

Even ironic invocations of the Zeuxian/Parrhasian paradigm worked to instill it. "L'Abbé," "Le Chevalier's" interlocutor, replies by disparaging "such *trompe l'oeil* effects ... commonly found today in works of no repute whatsoever." "Often enough," he explains, "cooks have reached out for accurately represented partridges or capons, intending to put them on the spit, and

---

[3] Pliny, *Natural History*, 35.36, trans. H. Rackham, 10 vols. (Cambridge, Mass.: Harvard University Press, 1986), 9: 308–11.
[4] See Sybille Ebert-Schifferer, *Deceptions and Illusions: Five Centuries of Trompe l'Oeil Painting* (Washington, DC.: National Gallery of Art, 2002).
[5] Charles Perrault, *Parallèle des anciens et des modernes*, 4 vols. (Paris: Jean-Baptiste Coignard, 1688–97), 1: 189.

what happens? Everyone laughs, and the painting stays in the kitchen."[6] "L'Abbé" insists that however convincing and indeed praiseworthy depictions of "stuffs" may be, they only amount to "stuffs" – mere things of the world. A painted thistle provides as little nourishment to the ass as the painted capons. (The reference here may be to paintings of game or *xenia*, the Latin term for gifts of food offered by hosts to their guests, which were hung in the dining and kitchen quarters of early modern homes.)

The drive to render nature in images ran deep, and it found differing expressions among artists, theorists, observers, and others.[7] Leonardo da Vinci (1452–1519) championed and practiced several complementary approaches to the imitation of nature. "Painting," he wrote, "which is the sole imitator of all the manifest works of nature, [is] a subtle invention, which with philosophical and subtle speculation considers all manner of forms: sea, land, trees, animals, grasses, flowers, all of which are enveloped in light and shade."[8] His defense of painting is amply supported by his numerous landscape and meteorological studies, his drawings of the movement of water and of plants, and his anatomical studies. Elsewhere, Leonardo expressed bemusement at the uncanny ability of paintings to deceive their viewers, and he described a mimesis-induced confusion of "dogs barking and trying to bite painted dogs," while "swallows fly and perch on iron bars which have been painted as if they are projecting in front of the windows of buildings."[9] Albrecht Dürer (1471–1528) pronounced of nature that the painter should "diligently . . . orient [himself] to it and not neglect it. . . . In truth, art is lodged in nature; he who can extract it has it."[10] Art was seen to "imitate" nature in a number of senses: It might mimic and by so doing supplant nature, as in Perrault's observations; or render the underlying "truth" of nature, for example, as suggested by Dürer.

Fidelity to nature was not always an end in itself, however; artists were encouraged to temper their observations of nature with ideal conceptions. Those who strove to represent ideal forms relied on their mental and manual powers of analysis and synthesis in selecting elements of existing entities and recomposing them to form a perfect whole. As the English statesman and natural philosopher Francis Bacon (1561–1626) observed, however, the ideal was not necessarily beautiful: "A man cannot tell whether [the Greek painter] Apelles or Albrecht Dürer were the more trifler; whereof the one would make a personage by geometrical proportions; the other, by taking the best parts

---

[6] Ibid., p. 190.
[7] Lorraine Daston and Katharine Park, *Wonders and the Order of Nature, 1150–1750* (New York: Zone Books, 1998), chap. 7: "Wonders of Art, Wonders of Nature," pp. 255–301.
[8] Martin Kemp, ed., *Leonardo on Painting*, trans. Martin Kemp and Margaret Walker (New Haven, Conn.: Yale University Press, 1989), p. 13.
[9] Ibid., p. 34.
[10] *Dürer: Schriftlicher Nachlass*, ed. Hans Rupprich, 3 vols. (Berlin: Deutscher Verein für Kunstwissenschaft, 1956), 3: 295.

out of divers faces, to make one excellent. Such personages ... would please nobody but the painter who made them."[11]

The imitation of nature was not always a straightforward endeavor, given the inevitable gap between the actual forms of the observable world and their two- or (in the case of sculpture) three-dimensional representations. It has been pointed out that many of the early modern images we hold up as being most naturalistic – watercolor depictions of animals by Dürer (Figure 31.2) and pen drawings of anatomical structures by Leonardo, for example – offer highly mediated views of observable reality.[12] The morphological detail provided by a drawing was far more comprehensive than what a viewer perceived when faced with a live specimen, a paradox of which early modern artists were wholly aware. We might think of early modern artists as aiming not to transcribe reality but rather to achieve pictorial "truth to nature," a kind of inspired interpretation of the subject shaped by the artist's selection and synthesis of observations.[13]

Paradoxically, the effect of these calculated manipulations of appearances was an astonishing mimesis. "A perfect painting," wrote Samuel van Hoogstraten, "is like a mirror of Nature, in which things that are not there appear to be there, and which deceives in an acceptable, amusing, and praiseworthy fashion."[14] The Bolognese naturalist and professor Ulisse Aldrovandi (1522–1605) described the Medici court artist Jacopo Ligozzi (1547–1626) as "an excellent painter, who day and night devotes himself exclusively to depicting plants, animals of all sorts ... and birds ..., [some of them] from India, and snakes, from Africa ... and all that is missing from these images is the soul of the thing represented."[15] Among other things, Aldrovandi's praise for

---

[11] Francis Bacon, "Of Beauty," in *The Major Works*, ed. Brian Vickers (Oxford: Oxford University Press, 1996), pp. 425–6.

[12] See James S. Ackerman, "Early Renaissance 'Naturalism' and Scientific Illustration," in Ackerman, *Distance Points: Essays in Theory and Renaissance Art and Architecture* (Cambridge, Mass.: MIT Press, 1991), pp. 185–207, at pp. 187–8; Martin Kemp, "'The Mark of Truth': Looking and Learning in Some Anatomical Illustrations from the Renaissance and Eighteenth Century," in *Medicine and the Five Senses*, ed. W. F. Bynum and Roy Porter (Cambridge: Cambridge University Press, 1993), pp. 85–121; and Kemp, "Temples of the Body and Temples of the Cosmos: Vision and Visualization in the Vesalian and Copernican Revolutions," in *Picturing Knowledge: Historical and Philosophical Problems concerning the Use of Art in Science*, ed. Brian Baigrie (Toronto: University of Toronto Press, 1996), pp. 40–85.

[13] Peter Galison, "Judgment against Objectivity," in *Picturing Science, Producing Art*, ed. Caroline A. Jones and Peter Galison (New York: Routledge, 1998), pp. 327–59, at p. 328.

[14] Samuel van Hoogstraten, *Inleyding tot de Hooge Schoole der Schilderkonst: anders de Zichtbaere Werelt* (Rotterdam: Fransois van Hoogstraeten 1678), p. 25. See Celeste Brusati, *Artifice and Illusion: The Art and Writing of Samuel van Hoogstraten* (Chicago: University of Chicago Press, 1995).

[15] Ernst Kris, "Georg Hoefnagel und der wissenschaftliche Naturalismus," in *Julius Schlosser: Festschrift zu seinem 60sten Geburtstag*, ed. A. Weixlgärtner and L. Planiscig (Vienna: Amalthea, 1927), pp. 243–53, at p. 251. On Ligozzi, see also Lucia Tongiorgi Tomasi and Gretchen Hirschauer, *The Flowering of Florence: Botanical Art for the Medici* (Washington, DC.: National Gallery of Art, 2002); Lucia Tongiorgi Tomasi, *I ritratti di piante di Iacopo Ligozzi* (Pisa: Ospedaletto, 1993); and Tongiorgi Tomasi, "The Study of the Natural Sciences and Botanical and Zoological Illustration in Tuscany under the Medicis from the Sixteenth to the Eighteenth Century," *Archives of Natural History*, 28 (2001), 179–93.

Figure 31.2. *The Young Hare.* Albrecht Dürer, 1502, watercolor and gouache on paper. Reproduced by permission of Graphische Sammlung Albertina, Vienna. Photograph courtesy of Marburg/Art Resource, New York.

Ligozzi provides evidence that artistic naturalism was aligned with scientific interests of the time. Students of Dutch art may be struck by the similarity between Aldrovandi's recommendation and the terms in which the statesman and poet Constantijn Huygens (1596–1687) praised Dutch landscape painting. Huygens wrote in 1629 that "It can even be said, as far as naturalism is concerned, that in the works of these clever [landscape painters] nothing is lacking but the warmth of the sun and the movement caused by the gentle breeze."[16]

Art historians have noted that whereas some artists applied mathematical methods such as perspectival or optical theory for artistic ends, others put art – its mimetic capacity to replicate the forms of nature, for example – to empirical use. Dürer's splendid natural historical watercolors of grasses, birds, or a hare are a case in point. One central line of argument, for example, holds that Dürer's naturalism inspired the tremendous increase in the use of images

[16] Constantijn Huygens, *Mijn Jeugd*, trans. C. L. Heesakkers (Amsterdam: Querido, 1987), p. 79.

by students of nature in the sixteenth century and beyond.[17] Another focuses on the utility for some natural inquirers of training in artistic techniques and perceptual skills. Thus, Galileo Galilei's (1564–1642) first sketches of the pocked lunar surface reveal his understanding of chiaroscuro; indeed, it was the astronomer's extensive training as a draftsman (he was admitted to the Accademia del Disegno in Florence in 1613) that enabled him, unlike his contemporary the mathematician and astronomer Thomas Harriot (1560–1621), to comprehend the strange pattern of shadows at the other end of his telescope as three-dimensional forms.[18]

By the mid-seventeenth century, some naturalists, immersed in the attempt to represent natural appearances pictorially, had become as keenly aware as artists of the instability underlying all efforts to see and represent nature "as she is," even with the aid of new optical technologies that extended vision to the realms of the very small and the very far away. In the preface to his *Micrographia* (1665), the first illustrated account of observations made through the microscope, Robert Hooke (1635–1703) acknowledged of his tiny specimens that:

> Of these kinds of objects there is much more difficulty to discover the true shape, than of those visible to the naked eye, the same Object seeming quite different in the one position of the light, from what it really is. . . . And, therefore I never began any draught before by many examinations in several lights, and in several positions to those lights, I had discovered the true form.[19]

Like Galileo, Hooke was an accomplished draftsman, which sharpened his eye for the potential deceptions of light and shadow and the need to interpret, not just reproduce, visual appearances in order to find "the true shape."

SCIENTIFIC ILLUSTRATION

The same Pliny who recommended the virtuoso mimetic feats of Zeuxis and Parrhasius also condemned artists' efforts to render plants for scientific purposes, arguing that nature is too mutable to be fixed in images and that images are less reliable than text in any case. Pliny referred to the use of illustrations in natural-historical accounts as

---

[17] See Fritz Koreny, *Albrecht Dürer and the Animal and Plant Studies of the Renaissance*, trans. Pamela Marwood and Yehuda Shapiro (Munich: Prestel-Verlag, 1985).

[18] Samuel Y. Edgerton, Jr., "Galileo, Florentine 'Disegno,' and the 'Strange Spottedness' of the Moon," *Art Journal* (Fall 1984), 225–32; see also Horst Bredekamp, "Gazing Hands and Blind Spots: Galileo as Draughtsman," *Science in Context*, 13 (2000), 423–62.

[19] Quoted in Martin Kemp, "Taking It on Trust: Form and Meaning in Naturalistic Representation," *Archives of Natural History*, 17 (1990), 127–88, at pp. 131–2.

a most attractive method, though one which makes clear little else except the difficulty of employing it. . . .[N]ot only is a picture misleading when the colors are so many, particularly as the aim is to copy Nature, but besides this, much imperfection arises from the manifold hazards in the accuracy of copyists. In addition, it is not enough for each plant to be painted at one period only of its life, since it alters its appearance with the fourfold changes of the year.[20]

Classical disparagements notwithstanding, numerous early modern artists and scientists participated actively in the representation of the natural world. The German naturalist Otto Brunfels (1488–1534) was the first of several authors to publish extensive accounts of local flora that included systematically descriptive images.[21] In its organization and text, Brunfels's *Herbarum vivae eicones ad naturae imitationem* (Living Images of Plants in Imitation of Nature, 1530–6) differed little from its classical sources (principally Pliny and Dioscorides), but the images – the very subject of the book's title – heralded an entirely new form of engagement with nature. In 1542, Brunfels's compatriot Leonhart Fuchs (1501–1556) published *De historia stirpium* (On the History of Plants), in which roughly 550 plants were recorded and illustrated. Fuchs outlined his descriptive project in a page-long qualifying subtitle, where he explained that his (verbal) descriptions of the habitats, nature, and medicinal properties of plants were accompanied by the most artful and expressive illustrations, made "*ad naturam.*"[22] Like Brunfels, who referred to the images of plants his book contained as "portrayed with great diligence and artifice," Fuchs advertised the artistic quality of the woodcuts he published (Figure 31.3). The degree of artistry was closely monitored, however. Fuchs specified that "shading and other less crucial things with which painters sometimes strive for artistic glory" have been discouraged in the interest of making "the pictures correspond [more] to the truth."[23]

As the triple portrait Fuchs included in his herbal shows, the production of early modern scientific illustration involved complex negotiations – among

---

[20] Pliny, *Natural History*, 25.4, 9: 140–1. To the first two of Pliny's complaints, woodcuts and engravings offered a remedy. See also Karen Meier Reeds, "Renaissance Humanism and Botany," *Annals of Science*, 33 (1976), 519–42, at p. 530; David Freedberg, "The Failure of Color," in *Sight and Insight: Essays on Art and Culture in Honour of E. H. Gombrich at 85*, ed. J. Onians (London: Phaidon, 1994), pp. 245–62, esp. pp. 245–8; and Freedberg, *The Eye of the Lynx: Galileo, His Friends, and the Beginnings of Modern Natural History* (Chicago: University of Chicago Press, 2002), esp. pt. IV, pp. 347–416.

[21] The inclusion of roots, the description of surface texture, and the effort to show leaves and flowers from a variety of angles distinguish Brunfels's work from all previous publications of the sort. Emphasis on the specific characteristics of individual specimens and the inflections of shadows were criticized, and avoided in later publications, demonstrating that the literal demands of working from life were tempered by the necessity of communicating visual information. See Landau and Parshall, *The Renaissance Print*, pp. 254–5.

[22] See T. A. Sprague and Ernest Nelmes, *The Herbal of Leonhart Fuchs* (London: Linnean Society of London, 1931).

[23] Leonhart Fuchs, *De historia stirpium* . . . (Basle: Isingrin, 1542), fol. 7v; see also Arber, *Herbals*, p. 206.

Figure 31.3. Self-portraits of the artists Heinricus Füllmaurer, Albertus Meyer, and Vitus Rudolph Speckle. In Leonhart Fuchs, *De historia stirpium commentarii insignes* (Basel: Isingrin, 1542). Reproduced by permission of the McCormick Library of Special Collections, Northwestern University Library.

individual producers, artists, blockcutters, authors, editors, publishers, and readers, for example, as well as between word and image on the one hand and classical learning and empirical evidence on the other.[24] By the mid-sixteenth century, so many new plant species had been identified (thanks in large part to the power of pictures to communicate ever finer morphological

---

[24] See Sachiko Kusukawa, "Illustrating Nature," in *Books and Sciences in History*, ed. Marina Frasca-Spada and Nick Jardine (Cambridge: Cambridge University Press, 2000), pp. 90–113; and Lucia Tongiorgi Tomasi, "L'Illustrazione naturalistica: Tecnica e invenzione," in *Natura-Cultura: L'Interpretazione del mondo fisico nei testi e nelle immagini*, ed. Giuseppe Olmi, Lucia Tongiorgi Tomasi, and Attilio Zanca (Florence: Leo S. Olschki, 2000), pp. 133–51.

distinctions) that naturalists began to suffer from "information overload."[25] To some extent, morphological images enabled students of the natural world to organize their experience.[26] Although complaints about the usefulness of static images to capture the variable forms of nature continued to be lodged, nonetheless a characteristically pictorial natural history asserted itself in the sixteenth and seventeenth centuries.[27]

The compulsion to record nature in the early modern period generated not only a substantial corpus of images of plants, humans, and animals but also a remarkably diverse one, incorporating a variety of media, from woodcuts and engravings to watercolor drawings, oil paintings, and even tapestry. These images served a variety of uses, among them description, identification, instruction, substitution for the real item (in collections, for example, where images supplanted unavailable specimens), and elicitation of wonder. Whereas Dürer might aim by rendering nature to capture the external appearance of flora and fauna, Leonardo appears to have used drawings as vehicles for understanding the laws and universal forms embodied within particular phenomena.[28] For Leonhart Fuchs, naturalistic images of plants enabled students of botany to learn to grasp in a glance the essence of a species, distilling its taste, smell, and medicinal powers into a single mnemonic sign.[29] For a collector such as the Holy Roman Emperor Rudolf II (1552–1612), lavish albums of zoological and botanical paintings celebrated the virtuosity and inventiveness of the Divine Creator even as they honored the patron's own erudition, power, and wealth.[30]

ANATOMY LESSONS

Among the most renowned illustrated works of early modern science are the publications of Andreas Vesalius (Andreas van Wesele, 1514–1664) – his *Tabulae sex* (Six Figures) of 1538 and his *De humani corporis fabrica* (On the Fabric of the Human Body) and *Epitome* of 1543. The innovative method of anatomical instruction developed by Vesalius, in which Galenic texts were explored and sometimes corrected through comparison with a human body,

---

[25] Brian Ogilvie, "The Many Books of Nature: Renaissance Naturalists and Information Overload," *Journal of the History of Ideas*, 64 (2003), 29–40, at pp. 32–3.
[26] See Claudia Swan, "From Blowfish to Flower Still Life Painting: Classification and its Images ca. 1600," in *Merchants and Marvels: Commerce, Art, and the Representation of Nature in Early Modern Europe*, ed. Pamela Smith and Paula Findlen (New York: Routledge, 2002), pp. 109–36.
[27] David Topper, "Towards an Epistemology of Scientific Illustration," in *Picturing Knowledge*, pp. 215–49.
[28] Dagmar Eichberger, "*Naturalia* and *Artefacta*: Dürer's Nature Drawings and Early Collecting," in *Dürer and His Culture*, ed. Dagmar Eichberger and Charles Zika (Cambridge: Cambridge University Press, 1998), pp. 13–37, at p. 15.
[29] Sachiko Kusukawa, "Leonhart Fuchs on the Importance of Pictures," *Journal of the History of Ideas*, 58 (1997), 403–27, at pp. 412–16.
[30] See *Le Bestiaire de Rodolphe II: Cod. min. 129 et 130 de la Bibliothèque Nationale d'Autriche*, ed. Manfred Staudinger, H. Haupt, and Thea Vignau-Wilberg, trans. Léa Mavcou (Paris: Citadelles, 1990).

permanently changed the way that medicine was taught starting in the midsixteenth century. Performing dissections in university lecture rooms and temporary structures, Vesalius instituted a form of anatomical instruction in which the formerly triangulated practices of the professor (who presided *ex cathedra*), the *demonstrator* (who performed the dissection according to the order of the text read aloud by the professor), and the *ostensor* (who pointed out the various parts of the body as they were uncovered) were carried out by the demonstrator/anatomist alone.[31] The body itself became the prevailing authority when discrepancies arose between text and physical evidence; Vesalius's close attention to empirical fact is in turn reflected in the prominent role illustrations played in his anatomical publications.

By the middle of the sixteenth century, empirical observation in medicine had become de rigueur. The causes – and the effects – of the shift many European medical curricula underwent are complex. Karen Meier Reeds and Roger French, writing about early modern botany and anatomy, respectively, have described the subtle interplay between the humanist culture devoted to the revival of classical texts in the sixteenth century and the new practices of observation and demonstration.[32] Just as Vesalius insisted on corroborating the dissected body with the written word, privileging the physical specimen over inherited texts when faced with inconsistencies, so, too, did botanical study come to involve direct and sensory study of its objects. Simples (the makings of medicines) were gathered for and by professors of medicine and their students and were cultivated in the gardens newly attached to universities, and their properties were demonstrated in the course of lectures.

Vesalius's professor at the University of Paris, Jacobus Sylvius (Jacques Dubois, 1478–1555), was the most celebrated lecturer on anatomy in the 1530s; his classes were typically attended by as many as 400 or 500 students and in effect spawned the subsequent generation of medical studies. In his *In Hippocratis et Galeni physiologie partem anatomicam isagoge* (Introduction to the Anatomical Part of the Physiology of Hippocrates and Galen), published in 1555 but written earlier, Sylvius recommended the following:

> I would have you look carefully and recognize by eye when you are attending dissections. . . . For my judgment is that it is much better that you should learn the manner of cutting by eye and touch than by reading and listening. For reading alone never taught anyone how to sail a ship, to lead an army, nor to compound a medicine, which is done rather by the use of one's own sight and the training of one's own hands.[33]

---

[31] For the observations of Baldasar Heseler, a German student present at one of Vesalius's dissections, see Ruben Eriksson, *Andreas Vesalius' First Public Anatomy at Bologna 1540: An Eyewitness Report* (Uppsala: Almqvist and Wiksell, 1959).

[32] Reeds, *Botany in Medieval and Renaissance Universities*; and R. K. French, *Dissection and Vivisection in the European Renaissance* (Aldershot: Ashgate, 1999), chaps. 3 and 4.

[33] Jacobus Sylvius, *Opera medica* (Geneva, 1635), p. 127, as cited in M. F. Ashley Montagu, "Vesalius and the Galenists," in *Science, Medicine, and History: Essays on the Evolution of Scientific Thought and*

Sylvius also taught materia medica and cultivated a medicinal garden of both indigenous and foreign varieties for the benefit of his students, who might learn by inspection and observation.[34] Vesalius echoed his teacher's values in his criticism of a senior colleague at Bologna, Matthaeus Curtius (ca. 1474–1544): "The essential thing is to teach the contents and to speak more clearly than eloquently. We cure with things or herbs, not with verbs."[35] Vesalius vehemently rejected textual study in favor of empirical study, where firsthand evidence challenged accepted learning. In his influential illustrated treatises as well as the larger, evolving practice of medical instruction, images proved crucial to this new empirical practice. Early modern scientific images both spurred and attest to investment in autoptic observation of phenomena.

A close reading of a landmark painting by Rembrandt reveals the ways in which Vesalius's example both complicated and bolstered the role played by images in early modern anatomical instruction and, by extension, scientific study generally. *The Anatomy Lesson of Dr. Tulp* may be the most renowned early modern image of medicine in action and as such represents, literally and emblematically, a significant point of intersection between early modern science and art (Figure 31.4).[36] Completed in 1632 for the Surgeons' Guild of Amsterdam, the canvas features a pyramidal grouping of staid professionals, gathered attentively around the body of a cadaver the workings of whose left arm a black-hatted doctor expertly demonstrates. The painting has been understood to represent an actual event that can be located in historical time and place. That time is the winter of 1631–2 and the place the newly founded Atheneum – later the University – of Amsterdam. In combination with the fact that the painting is signed and dated on the scroll hanging on the far wall, the records of the Amsterdam Surgeons' Guild have been adduced to demonstrate that it documents an actual dissection conducted by the praelector of the guild, the prominent Amsterdam physician Nicolaas Tulp (1593–1674). Because it is known that Dr. Tulp demonstrated that winter on the body of the miserable criminal Adrian Adrianszoon, alias het Kint, a multiple offender who was hanged on 31 January 1632, it has been presumed that Rembrandt's painting records the appearances not only of the seven members of the guild in the presence of their praelector, but also of Aris 't Kint. Avidly craning and bending to view the events at hand, the seven surgeons portrayed are respectably engaged in the pursuit of the knowledge on which their trade depended.

Medical Practice Written in Honour of Charles Singer, ed. E. Ashworth Underwood, 2 vols. (London: Oxford University Press, 1953), 1: 374–85, at p. 378.
[34] See Montagu, "Vesalius and the Galenists," 2: 378.
[35] Eriksson, *Andreas Vesalius' First Public Anatomy*, pp. 54–5.
[36] See Norbert Middelkoop, Ben Broos, Jorgen Vadum, and Petria Noble, *Rembrandt under the Scalpel: The Anatomy Lesson of Dr. Nicolaes Tulp Dissected* (The Hague: The Mauritshuis, 1998); W. S. Heckscher, *Rembrandt's Anatomy of Dr. Nicolaes Tulp: An Iconological Study* (New York: New York University Press, 1958); and W. Schupbach, *The Paradox of Rembrandt's Anatomy of Dr. Tulp* (London: Wellcome Institute for the History of Medicine, 1982).

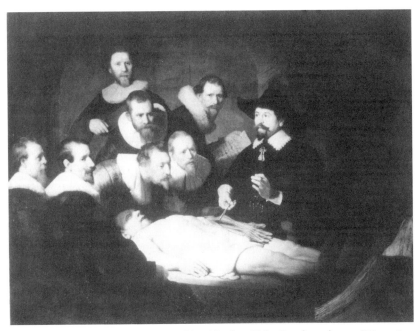

Figure 31.4. *The Anatomy Lesson of Dr. Nicolaes Tulp*. Rembrandt van Rijn, 1632, oil on canvas. Reproduced by permission of The Mauritshuis, The Hague. Photography courtesy of Erich Lessin/Art Resource, New York.

Although Rembrandt's naturalism – his ability to capture the gestures, postures, facial expressions, and unique characteristics of the men gathered here – may corroborate the sense that this painting records an actual event, numerous aspects of the work are entirely artificial. Even if Dr. Tulp is meant to be conducting this particular demonstration in a makeshift location (the official Amsterdam Atheneum anatomical theater was not built until 1639), the setting is quite generic and the standard accoutrements of such demonstrations – linens, candles, lavender for the stench – are conspicuously absent. Moreover, as has frequently been noted, anatomical dissections never begin in the limbs of their subjects but in the abdominal cavity. (Indeed, the records indicate that the abdominal cavity was the primary focus of Dr. Tulp's second public demonstration of the winter of 1631–2.) What does this artifice convey? Paradoxically, the painting might more accurately reflect the teaching and practice of anatomy precisely by virtue of its being so artificially constructed. The multiple and ricocheting gazes of the men at work establish three foci: the pallid expanse of the corpse; the weighty folio-size volume propped up at 't Kint's feet, in the lower right of the composition; and the sheet of paper that the doctor to the left of Dr. Tulp is holding. This sheet of paper contains the sketchy lines of an anatomical drawing, painted to simulate a chalk drawing (occluded by the later addition of a list of names of the doctors present). The faint outlines of at least two arms can still be made out. This group portrait depended originally on and worked to iterate a three-way relationship among what were in the early modern period the fundamental components of medical study – namely, the corpse, the relationship between the dissection being performed and the text at the corpse's feet, and the relationship between the drawing and the actual observed specimen.

The relationship among the book, which we may presume to be an anatomical treatise, a generic but clearly anatomical drawing, and the body under examination is crucial. The published text and the drawing, or on-the-spot record, frame the body. In sum, what Rembrandt has engineered here is not merely a portrait of a number of the members of the guild, in the presence of their praelector, but an enactment of the production of medical knowledge as it was institutionalized starting in the mid-sixteenth century. Rembrandt's *Anatomy Lesson* demonstrates more than just biographical and professional associations among medical practitioners in the Netherlands. It illustrates the structure of anatomical study in the early modern period. The surgeons of the Amsterdam guild who elected to be painted here are portrayed in the act of producing medical knowledge, even as we observe them.

THE ARTIST AS SCIENTIST

Leonardo da Vinci exemplifies perhaps better than anyone else the close relationship between art and science in the early modern era, as well as the importance of trying to understand what these terms meant to contemporaries rather than taking them as self-evident and transhistorical categories. In his notes toward a projected treatise "On Painting," he observed:

> He who despises painting loves neither philosophy nor nature.... Truly this is science [*scientia*], the legitimate daughter of nature, because painting is born of that nature; but to be more correct, we should say the granddaughter of nature, because all visible things have been brought forth by nature and it is among these that painting is born. Therefore we may justly speak of it as the granddaughter of nature and as the kin of god.[37]

Although *scientia* has the general meaning of "knowledge" in this passage, it seems clear that Leonardo meant further to associate the art of painting with one type of knowledge in particular: the disciplined examination of nature. Among the features he emphasized in this latter connection were both mathematical demonstrations and experience, "without which nothing can be achieved with certainty."[38] By mathematical demonstrations, he meant to invoke arithmetic and geometry; from these, he wrote, perspective is born,

---

[37] Kemp, *Leonardo on Painting*, p. 13.
[38] Ibid., p. 14.

which is "devoted to all the functions of the eye and to its delight with various speculations."[39]

In the early modern era, accomplishments of all sorts were measured against a standard of artistic achievement: To be proclaimed the "Michelangelo" of one's field was to be accorded the highest honor. In 1612, the Florentine painter known as Cigoli (Ludovico Cardi, 1559–1613) compared Galileo with Michelangelo Buonarotti (1475–1564) in a letter to the mathematician in which he speaks of the propensity of both men to break rules and set new standards.[40] In many other respects, too, Galileo and his career bear comparison with his artist contemporaries; scholars have called him "close to being an artist in social terms and in practice"[41] and the "Michelangelo of mathematicians."[42] (When he was named court philosopher in 1610, he was granted a form of freedom very like that of the court artist). When early modern Europeans reflected on the enormous advances in human knowledge associated with their own period, they did not invoke theoretical innovations, such as Copernicus's rearrangement of the planets, but rather progress in technology and the arts. Commenting on the stasis of natural philosophy, in his *Novum organum* (New Organon, 1620), Bacon contrasted the plight of philosophy with the flourishing state of the "mechanical arts," which, he noted, "are always thriving and growing."[43]

Throughout the early modern era, artists were celebrated for their inventions: Jan van Eyck (1390–1441), the progenitor of early Netherlandish painting, was hailed repeatedly as the inventor of oil paint. Similarly, Filippo Brunelleschi (1377–1466) is still considered the inventor of pictorial perspective; and artists throughout the early modern era laid claim to their works by signing them with some variations on the term "invented."[44] In these cases as in others, artistic inventions were the result of experimentation. Countless authors have attempted to reconstruct Brunelleschi's every step as he produced the first perspectively structured pictures (now lost) in the square of Florence's Duomo in the second decade of the fifteenth century. These were not self-sufficient paintings but components of an exercise or an experiment involving optics and pictorial representation.

Giorgio Vasari (1511–1574), best known for his monumental *Vite* (Lives) of Italian artists (first edition 1550), and the Dutch painter and author Karel

---

[39] Ibid.
[40] Bredekamp, "Gazing Hands and Blind Spots," p. 425. A long-standing historiographical tendency to equate Michelangelo and Galileo in particular is traced here and at p. 426.
[41] Ibid., p. 426.
[42] Mario Biagioli, *Galileo, Courtier: The Practice of Science in the Culture of Absolutism* (Chicago: University of Chicago Press, 1993), pp. 86–7. See also Martin Warnke, *The Court Artist: On the Ancestry of the Modern Artist*, trans. David McLintock (Cambridge: Cambridge University Press, 1993).
[43] Francis Bacon, *Novum organum*, trans. and ed. Peter Urbach and John Gibson (Chicago: Open Court, 1994), p. 84.
[44] See Evelyn Lincoln, *The Invention of the Italian Renaissance Printmaker* (New Haven, Conn.: Yale University Press, 2000), p. 6.

van Mander (1548–1606), who produced equally heroizing biographies of local artists half a century later in his *Schilder-Boeck* (Book of Painting, first edition 1604), wove yarns about the process by which van Eyck made his signal discovery. Van Mander wrote of his compatriot:

> What was never granted to either the Greeks, Romans or other peoples to discover – however hard they tried – was brought to light by . . . Johannes van Eyck. When he had finally thoroughly explored many oils and other natural materials, he found that linseed and nut-oil dried best of all; by boiling these along with other substances which he added, he made the best varnish in the world. And since such industrious, quick-thinking spirits strive toward perfection through continually researching, he discovered by much experimentation that paint mixed with such oils blended very well, dried very hard and when dry resisted water well; and also that the oil made the colors much brighter and shinier in themselves without having to be varnished. . . . Johannes was highly delighted with this discovery and with good reason: for an entirely new technique and way of working was created, to the great admiration of many.[45]

Van Mander credited van Eyck's invention as a crucial means to a venerable end: "Our art needed only this invention to approximate to, or be more like, nature in her forms."[46]

The attribution to van Eyck of the invention of oil glazes (the technique that enabled him and his followers to produce such striking effects of light and color and, in sum, such mirror-like painted surfaces) held fast in early modern artistic legend. The Flemish-born printmaker and draftsman Johannes Stradanus (Jan van der Straet, 1523–1605) included a visual account of van Eyck's invention in his series of prints illustrating new discoveries (the *Nova reperta*, ca. 1600) (Figure 31.5). The fourteenth plate in a series of twenty, which opens with the discovery of America and includes such other discoveries as the compass, the astrolabe, windmills, book printing, copperplate engraving, the calculation of longitude, and gunpowder, depicts the master painting in his commodious studio, surrounded by various assistants and apprentices (see Park and Daston, Chapter 1, this volume).[47]

In their encomia of van Eyck, Vasari and van Mander described the process of experimentation that led to his discovery. Vasari actually compared him to an alchemist, writing that he "set himself to make trial of various sorts of

---

[45] Karel van Mander, *Het Schilder-Boeck* (Haarlem: Paschier van Wesbusch, 1604), fol. 199r–199v.
[46] Ibid.
[47] The engraving is implausible in numerous respects: Jan van Eyck is not known to have painted St. George slaying the dragon, as he is depicted doing, nor did he ever paint on canvas or at the scale of the work propped up before him. The staging of van Eyck's invention in this setting is telling, nonetheless, on account of these discrepancies with the actual record. As the inscription implies, van Eyck's invention was of use to painters at large ("*Colorem oliui commodum pictoribus, Invenit insignis magister Eyckius*"); in this sense, his studio may represent more the general expectations for how a painter worked in around 1600 than the Flemish master's actual domain.

Figure 31.5. *The Invention of Oil Paint by Jan van Eyck*. Jan Baptist Collaert after Johannes Stradanus (Jan van der Straet), ca. 1580, engraving, plate 14 of *Nova reperta*. Reproduced by permission of the McCormick Library of Special Collections, Northwestern University Library.

colours, and, as one who took delight in alchemy, to prepare many kinds of oil for making varnishes and other things dear to men of inventive brain, such as he was."[48] Was the artist's studio a sort of laboratory? The scientific laboratory had yet to be institutionalized in the sixteenth century, but the arenas of empirical observation and experiment had much to do with the domain of artistic production: Both were, in general, well-lit spaces in which material objects mediated between thought and praxis (see Smith, Chapter 13, this volume). Both were spaces for simulation and staging, often using elaborate full-size or small-scale models; both could be found in urban areas, attached to courts, or later as part of academies. In some cases, the substances and the practices were similar as well: It is difficult not to think of early modern science when we read recipes for paints or varnish, for example, or about devices such as a board featuring a peephole and an elaborate system of threads for constructing perspectival views.[49] The elevation of individuals deemed capable of invention and discovery through long experimentation,

---

[48] Giorgio Vasari, "Antonello da Messina," in *Lives of the Painters, Sculptors and Architects*, trans. Gaston du C. de Vere (New York: Alfred A. Knopf, 1996), pp. 424–9, at p. 425.
[49] Jane Turner, ed., *The Grove Dictionary of Art*, 34 vols. (New York: Grove's Dictionaries, 1996), 29: 850–5.

and the communication of ideals, techniques, and practical advice through what historian Paula Findlen has called the "paper republic" of European naturalists (or the generations of art theorists cited in this chapter) are factors that shaped both disciplines (see Bennett, Chapter 27, this volume).[50]

Stradanus's print (Figure 31.5) shows an environment set up for collaborative work. Making paintings (preparing drawing tools and brushes, grinding and mixing pigments, and constructing and assembling panels, canvases, and frames) involved multiple individuals, as the image shows. The workshop was also a place for education and training; assistants were learning the craft that they themselves aspired to master eventually. Apprentices and assistants such as are shown busy at work in van Eyck's studio were ultimately painted out of the master's work: They would have learned to work in the master's style, so that collaborative products would be seamless. Generally speaking, commissions were obtained and works conceived and supervised by the master. Where signed, they would bear only the name of the master. Authorship came to be negotiated carefully, particularly in the wake of the development of printing techniques.

With regard to the division of labor, the relationship between master and pupil or technician, and the attribution or proclamation of authorship, the operations of early modern artists' studios bear comparison with the structure of contemporary scientific endeavors. Both Rembrandt van Rijn (1606–1669) in Amsterdam and Peter Paul Rubens (1577–1640) in Antwerp accepted numerous apprentices into their care in exchange for which the masters were handsomely paid. A late seventeenth-century account of Rembrandt's studio attributes his financial and critical success at mid-career to the presence of so many ("all but countless") pupils in the master's studio.[51] Rembrandt and Rubens alone signed their works, however. In general, the visual arts revolved around a single-author model in spite of (and because of) the contribution of what the historian of science Steven Shapin, writing about early modern laboratories, has called "invisible technicians." The "collective character of empirical knowledge-making and knowledge-holding" in experimental science, whereby technicians' (or pupils') "skilled manipulations," "records," and "occasional inferential corollaries" were part and parcel of the finished work of the laboratory, maps very efficiently onto the production of art in the

---

[50] Paula Findlen, "The Formation of a Scientific Community: Natural History in Sixteenth-Century Italy," in *Natural Particulars: Nature and the Disciplines in Renaissance Europe*, ed. Anthony Grafton and Nancy Siraisi (Cambridge, Mass.: MIT Press, 1999), pp. 369–400.

[51] A. R. Peltzer, ed. *Joachim von Sandrarts Academie der Bau-, Bild-, und Mahlerey-Künste* (Munich: TK, 1925), p. 203. See Josua Bruyn, "Rembrandt's Workshop: Its Function and Production," in *Rembrandt: The Master and His Workshop*, ed. Christopher Brown, Jan Kelch, and Pieter van Thiel (New Haven, Conn.: Yale University Press, 1991), pp. 68–89. For a more daring interpretation, see Svetlana Alpers, *Rembrandt's Enterprise: The Studio and the Market* (Chicago: University of Chicago Press, 1988).

context of early modern studios.⁵² Moreover, just as Robert Boyle's (1627–1691) and Robert Hooke's laboratories were extensions of domestic space, so, too, Rembrandt and Rubens accommodated the teaching and collaboration with pupils and apprentices in their homes (see Cooper, Chapter 9, this volume).⁵³

Collaborative production took place under the shadow and in the domestic sphere of the master and was geared toward earning prestige and recognition for him alone. Evidence of the prevalence of this model of authorship and production in art is legion; its actual legislation is recorded in a guild regulation issued in 1651 in Utrecht that expressly forbade masters "to keep or employ any persons (whether foreign or native) as disciples or painting for them if they work in another [than the master's] manner (*handelinge*) or sign their own name."⁵⁴ Another, final example of ways in which artistic and scientific labor and practice resembled one another lies in the status of secrets in these respective domains.⁵⁵ Numerous celebrated early modern artists – Michelangelo, Hendrick Goltzius (1558–1617), and others – were adamant that they not be observed at work. Rembrandt is said to have devised a technique for the production of his etchings that allowed him to obtain the extraordinary effects for which they are still hailed and to have taken the technique with him to his grave.⁵⁶

SCIENTIFIC NATURALISM

Despite its centrality to early modern explorations of nature, whether characterized as artistic or scientific, natural history illustration has occupied only a marginal place in the study of art history.⁵⁷ The art historian E. H. Gombrich referred disparagingly to early modern images intended to impart truthful records of their subjects as "illustrated reportage."⁵⁸ Erwin Panofsky wrote

---

[52] Steven Shapin, *A Social History of Truth: Civility and Science in Seventeenth-Century England* (Chicago: University of Chicago Press, 1994), p. 358.
[53] On Boyle and Hooke, see Steven Shapin, "The House of Experiment in Seventeenth-Century England," *Isis*, 79 (1988), 387–420.
[54] As cited in S. Muller Fz., *Schilders-vereenigingen in Utrecht* (Utrecht: Beijers, 1880), p. 76.
[55] Jacopo Ligozzi developed a set of secret techniques for layering colors and varnish in his botanical and zoological paintings for the Medici family; see Tongiorgi Tomasi, "The Study of the Natural Sciences," pp. 182–3. See also William Eamon, *Science and the Secrets of Nature: Books of Secrets in Medieval and Early Modern Culture* (Princeton, N.J.: Princeton University Press, 1994).
[56] Arnold Houbraken, *De groote schouburgh der Nederlantsche konstschilders en schilderessen, waar van 'er veele met hunne beeltenissen ten toneel verschynen, en welker levensbedrag en konstwerken beschreven worden: zynde een vervolg op het SchilderBoek van K. v. Mander*, 3 vols. (Amsterdam: For the author, 1718–21), 1: 271.
[57] See, for a formidable exception, Eugenio Battisti, *L'Antirinascimento* (Milan: Feltrinelli, 1962), esp. chap. 9: "L'Illustrazione scientifica in Italia."
[58] Ernst Gombrich, *Art and Illusion: A Study in the Psychology of Pictorial Representation* (Princeton, N.J.: Princeton University Press, 1960), pp. 78–83.

of Leonardo's drawings of embryos that they "defy the borderline between scientific illustration and 'art'" – thereby excluding the former from the province of the latter.[59] Art historians have conventionally maintained that works of art and scientific images differ from one another constitutionally and irreconcilably: The former are vehicles for aesthetic expression, whereas the latter convey information, not truth or even style, quasi-anonymously. The distinction between the documentary status of scientific representation and the fictive or aesthetic potential of art and the impulse to categorize works as either artistic or scientific as such are legacies of Kantian aesthetics, which was extremely influential in shaping the discipline of art history. Relative values assigned to the two sorts of images are consistent with the hierarchy of the academic ranking of genres of painting, which prizes narrative compositions (*istorie*) over modes of representation concerned "merely" with mirroring the world at hand and the stuff of nature. In addition, the canon of fine arts has long depended on the Kantian notion that interest precludes the aesthetic. That is, the use to which scientific images are put largely disqualifies them from inclusion in studies of fine art.

We have seen that relationships between early modern art and science were rife at the level of praxis. However, the mid-nineteenth-century opposition between art and science, in which polarities such as subjective art versus objective science emerged, and the development of distinct disciplinary models in art and science (as well as in the history of art and the history of science) have occluded significant areas of overlap. More recent developments in the history of science encourage further examination of how intersections between art and science might be traced. Sociologist of science Bruno Latour has subsumed the abundance of pictures, words, diagrams, and signs produced as part of the working life of a laboratory under the elastic heading of "immutable mobiles," offering a way to treat diverse representational practices without the interference of familiar distinctions such as art/science or image/fact.[60] Scholarly attention has also been devoted to the spectacularization of experimentation and demonstration, and to social practices associated with the investigation of the natural world.[61] In both the history of art and the history of science, more recent scholarship has emphasized the importance of excavating the categories of the historical actors themselves in order to avoid anachronistic projections of later models of art, science, and their interactions onto phenomena in early modern Europe. Our conclusion

---

[59] Erwin Panofsky, "Artist, Scientist, Genius: Notes on the Renaissance-Dämmerung," in *The Renaissance: A Symposium* (New York: Metropolitan Museum of Art, 1953), pp. 77–93, at p. 87.
[60] Bruno Latour, "Drawing Things Together," in *Representation in Scientific Practice*, ed. Michael Lynch and Steve Woolgar [1988] (Cambridge, Mass.: MIT Press, 1990), pp. 19–68.
[61] In her analysis of the art collection of the Leiden physicist Franciscus dele Boë Sylvius (1614–1672), Pamela Smith has demonstrated that the relationships between art and science in the early seventeenth century by no means traveled a one-way street. See Smith, "Science and Taste: Painting, Passions, and the New Philosophy in Seventeenth-Century Leiden," *Isis*, 90 (1999), 421–61.

illustrates the difference in interpretation this can make by comparing two models for interpreting early modern relationships between art and science. One is a critical category introduced by a twentieth-century scholar, and the other is a term used by early modern artists and their audiences to describe a particular mode of representation.

The Austrian art historian and psychoanalyst Ernst Kris (1900–1957) coined the term "scientific naturalism" (*wissenschaftlicher Naturalismus*) in reference to works by the Flemish artist Joris Hoefnagel (1542–1601), who was renowned for the extraordinary books he illuminated for Rudolf II in the 1580s and 1590s.[62] Kris showed that Hoefnagel's celebrated attentiveness to the characteristic morphology of naturalia such as plants and insects is as much a feature of his courtly works as of his early chorographic works, which were done in the employ of the Antwerp cartographer Abraham Ortelius (1527–1598). Kris identifies Hoefnagel's interest in nature as "the most acute and characteristic expression of an attitude" more widely held during the period; indeed, he proposes that we speak of "a naturalistic style circa 1600." Citing Dürer as the spiritual patron of this style, Kris defined it as follows: "For the first time it was meaningful to represent a piece of turf or an animal as a picture unto itself, with no aim other than to penetrate as deeply as possible the characteristics of nature."[63] Under the rubric of *wissenschaftlicher Naturalismus* Kris also cited the naturalist painters Hans Hofmann (ca. 1545–ca. 1591) and Daniel Froeschl (1563–1613), who worked at the Hapsburg court in Prague and for the Medici in Florence, and still life and nature painters Roelant Savery (1576–1639; Savery also painted for Rudolf II in Prague), Ambrosius Bosschaert (1573–1621), and Jan Brueghel the Elder (1568–1625).[64] Kris traced the practice of *wissenschaftlicher Naturalismus* south of the Alps as well by adducing the extraordinary watercolors of birds and plants by Ligozzi. Although Kris stated that the style practiced by Hoefnagel and his contemporaries had its roots outside of the domain of art (and, instead, in intellectual and cultural developments), his interest in scientific naturalism lay ultimately in its usefulness in describing a form of artistic expression that is more committed to nature than to the precepts and conventions of Renaissance and Baroque art.

"Scientific naturalism" is certainly useful in helping to organize a broad and frequently overlooked domain of artistic production. Yet *wissenschaftlicher Naturalismus* is a stylistic diagnostic applied a posteriori that fails to take into account the conditions under which these works were produced and the extent to which the naturalism in question served the ends of science.

---

[62] Kris, "Georg Hoefnagel und der wissenschaftliche Naturalismus."
[63] Ibid., p. 252.
[64] See Koreny, *Albrecht Dürer and the Animal and Plant Studies of the Renaissance*; Thomas DaCosta Kaufmann, *The School of Prague: Painting at the Court of Rudolf II* (Chicago: University of Chicago Press, 1988); and Paul Taylor, *Dutch Flower Painting, 1600–1720* (New Haven, Conn.: Yale University Press, 1995).

It serves instead as an additional stylistic category that might usefully supplement existing art-historical taxonomies. A more probing investigation of the area of overlap between artistic and scientific modes of picturing the world is to be found in art historian Svetlana Alper's interpretation of seventeenth-century Dutch art. The distinctively Netherlandish descriptive mode of picturing, she argues, is inherently compatible with early modern empiricism: "Already established pictorial and craft traditions, broadly reinforced by the new experimental science and technology, confirmed pictures as the way to new and certain knowledge of the world."[65] Alpers, however, is explicitly uninterested in images actually employed in early modern science; her purview is, ultimately, fine works of art.[66] A potentially more helpful tool for analyzing artistic and scientific naturalism and the countless early modern images that instantiate it may lie in the conception of works made *ad vivum* or from the life. Numerous early modern authors and purveyors of images promised – in print and in inscription – that their pictures were made "from the life."[67] The phrases *ad vivum, naer het leven, nach dem Leben, au vif*, and *al vivo* were widely applied to portraits, maps, and botanical and other natural-history images. Such works, claimed to have been done from life, did not all look alike: What conjoins them are the claims made for how they were produced. The mimetic potential of images made in this way came to be exploited for artistic as well as documentary ends: From 1530 on, illustrated natural history depended heavily on this qualifying term, and by 1600 the phrase had been integrated into art theory.

The rapid spread of the phrase *ad vivum* is a curious and crucial fact.[68] It was, moreover, associated with special skills. In 1591, Flemish botanist Joseph Goedenhuize (Giuseppe Casabona, 1535–1595) wrote a letter to his patron the Grand Duke of Tuscany, Ferdinand de' Medici (d. 1609), in which he mentioned a "young German artist" (Georg Dyckman, dates unknown), whom he had paid to paint all the plants of Crete *al vivo*. Goedenhuize added that the young artist was "rather talented in this profession."[69] Goedenhuize

---

[65] Svetlana Alpers, *The Art of Describing: Dutch Art in the Seventeenth Century* (Chicago: University of Chicago Press, 1983), p. xxv.
[66] "I shall not take up here the interesting problem of the nature and role of illustration in the works of the Dutch naturalists." Ibid., p. 84.
[67] *The Oxford English Dictionary*, 2nd ed., ed. J. A. Simpson and E. S. C. Weiner, 20 vols. (Oxford: Clarendon Press, 1989), 8: 911, where "the life" is defined as "the living form or model" or "living semblance," and "after, from the life" are defined as "(drawn) from the living model." See "to the life": "with life-like presentation of or resemblance to the original (said of a drawing or painting); with fidelity to nature; with exact reproduction of every point or detail." The earliest examples of the use of the English phrases are 1599 (William Shakespeare) and 1603 (Ben Jonson), respectively. See also Claudia Swan, "*Ad vivum, naer het leven*, from the Life: Considerations on a Mode of Representation," *Word and Image*, 11 (1995), 353–72.
[68] This claim was already made in the text of the early *Gart der Gesundheit* (Mainz, 1484); these woodcuts are vastly more schematic than those published by Brunfels and his followers, all of whom invoke the same phrase.
[69] Lucia Tongiorgi Tomasi, "Daniel Froeschl before Prague: His Artistic Activity in Tuscany at the Medici Court," in *Prag um 1600: Beiträge zur Kunst und Kultur am Hofe Rudolfs II*, 2 vols. (Freren:

deemed working "from the life" a profession, but it was the great Bolognese naturalist and collector Aldrovandi who offered the most sustained commentary on images done *al vivo*.[70] Aldrovandi wrote: "There is nothing on earth that seems to me to give more pleasure and utility to man than painting, and above all paintings of natural things: because it is through these things, painted by an excellent painter, that we acquire knowledge of foreign species, although they are born in distant lands."[71] Aldrovandi also declared that painting is most honorable as an art because it can imitate "the product[s] of nature *al vivo*."[72] "I say," he wrote of images, "that they are of great utility to students, when they are painted *al vivo*, as are also other images, of fish and terrestrial animals and birds."[73] His views on the usefulness of painting and, in particular, painting that does not suffer stylistic or otherwise "artistic" embellishments, are borne out by the truly vast collection of images of the natural world he commissioned, assembled, and published.[74]

Within the context of the rapid development of illustrated natural history in the sixteenth century, the phrase *ad vivum* and its vernacular renderings served to assure viewers or readers of the documentary value of images so described. It guaranteed the images' trustworthiness by vouching for their direct connection to the world observed. Indeed, within a culture that valued the exchange of spectacular information, to make the claim that an image was done "from the life," whether true or not, secured an audience and may have increased its economic value.[75] More broadly speaking, to invoke such terms as *ad vivum* or *naer het leven* was to exercise an internationally valid password in a network of naturalists joined by correspondence and publications. In the early modern period, especially prior to the formation of scientific societies that controlled the flow of information, that community depended upon

Lura Verlag, 1988), 2: 289–98, at pp. 289–91. However, for the correct identification of the artist as Georg Dyckman, see Giuseppe Olmi, "'Molti amici in varij luoghi': Studio della natura e rapporti epistolari nel secolo XVI," *Nuncius*, 6 (1991), 3–31, at p. 25.

[70] On Aldrovandi, see Sandra Tugnoli Pattaro, *Metodo e sistema delle scienze nel pensiero di Ulisse Aldrovandi* (Bologna: Cooperative Libraria Universitaria Editrice Bologna, 1981); Olmi, *L'Inventario del Mondo: Catalogazione della natura e luoghi del sapere nella prima età moderna* (Bologna: Il Mulino, 1992), esp. pp. 21–117 and bibliography; and Findlen, *Possessing Nature*.

[71] Bibliotheca Universitaria, Bologna, MS Aldrovandi, 6 vols., 1: fol. 35r, as quoted in Olmi, *L'Inventario del Mondo*, p. 24.

[72] Biblioteca Universitaria, Bologna, MS Aldrovandi, 6 vols., 2: fol. 129v, as quoted in Olmi, "Osservazione della natura e raffigurazione in Ulisse Aldrovandi (1522–1605)," *Annali dell'Istituto Storico Germanico Italiano in Trento*, 3 (1977), 105–81, at p. 109.

[73] Biblioteca Universitaria, Bologna, MS Aldrovandi, 6 vols., 1: fol. 35r, as quoted in Olmi, "Arte e natura nel Cinquecento bolognese: Ulisse Aldrovandi e la raffigurazione scientifica," in *Le arti a Bologna e in Emilia dal XVI al XVIII secolo*, ed. Andrea Emiliani, 4 vols. (Bologna: Cooperativa Libraria Universitaria Editrice Bologna, 1982), 4: 151–73, at p. 155.

[74] Biblioteca Universitaria, Bologna: Fondo Ulisse Aldrovandi Tavole di Piante, Fiori e Frutti, vol. 01–1, fol. 76, and vol. 04-Unico, fol. 35. The BUB maintains a Web site containing reproductions of all of Aldrovandi's natural-history drawings: http://www.filosofia.unibo.it/aldrovandi.

[75] On the representation of natural wonders and other miraculous or curious phenomena, see especially Peter Parshall, "Imago Contrafacta: Images and Facts in the Northern Renaissance," *Art History*, 16 (1993), 554–79, esp. pp. 564 ff. See also Jean Céard, *La nature et les prodiges: L'Insolite au XVIe siècle, en France* (Geneva: Droz, 1977); and Daston and Park, *Wonders and the Order of Nature*.

what Shapin has called "epistemological decorum."[76] Of the expansion of factual knowledge that was a hallmark of the period, and of how it came to be managed, he has observed: "The work of prying open the inherited box of plausibility and restocking it with new things and phenomena was fundamental to the emergence of new intellectual practices. In the process, new and modified forms for the making and warranting of empirical truth had to be proposed and put in place."[77]

The two-step movement of questioning ancient authority and, at the same time, gathering the evidence of natural particulars, impartially and in accordance with what Lorraine Daston has branded "the factual sensibility," underlies much early modern natural history, as systematized by Bacon and his many followers.[78] The natural historians who practiced their discipline correctly were, Bacon declared, "faithful secretaries [who] do but enter and set down the laws themselves of nature and nothing else," who compile a warehouse of particulars from which true induction might proceed.[79] The catalogues of the plant world described *ad vivum*, the numerous drawings by Dürer and Leonardo, among others, and so many other early modern scientific images amount to warehouses, inscribed by faithful secretaries – or, at least, scribes who declared their epistemological decorum by way of this mode of representing nature's particulars.

---

[76] Shapin, *A Social History of Truth*, esp. chap. 5: "Epistemological Decorum: The Practical Management of Factual Testimony," pp. 193–242.

[77] Ibid., p. 195.

[78] Lorraine Daston, "Baconian Facts, Academic Civility, and the Prehistory of Objectivity," *Annals of Scholarship*, 8 (1991), 337–63. On Bacon's scientific method, see Paolo Rossi, *Francis Bacon: From Magic to Science* [1957], trans. Sacha Rabinovitch (Chicago: University of Chicago Press, 1968), esp. chaps. 2 and 4; and *The Cambridge Companion to Bacon*, ed. Markku Peltonen (Cambridge: Cambridge University Press, 1996).

[79] Francis Bacon, "Preparative towards a Natural and Experimental History," 10 (on "scribae fideles") and 3 (on the repository of knowledge), in *The Works of Francis Bacon, Baron of Verulam, Viscount of St. Alban, and Lord Chancellor of England [1857–74]*, ed. James Spedding, R. L. Ellis, and D. D. Heath, 14 vols. (New York: Garrett Press, 1968), 4: at 262 and 254–5, respectively.

32

GENDER

Dorinda Outram

Historians have often linked two quite separate phenomena: the gendering of early modern natural inquiry as a masculine form of activity in theory and, to a large extent, in practice, and the gendering of nature as female in many early modern texts and images. There is no necessary logical connection between these two phenomena, despite persistent and profound historiographical investments in their linkage, most notably as part of broader critiques of the scientific enterprise by writers with feminist commitments. But there are important and interesting historical connections, which this chapter seeks to explore.

The critical focus on the masculine nature of scientific activity has had the longer history. Antivivisection campaigns in nineteenth-century Britain and America, for example, often (though not always) overlapped with feminist concerns. Antivivisectionists saw biological science in particular as indelibly marked by cruelty toward the animals it used as experimental subjects and by an attitude toward nature that placed more emphasis on advancing scientific knowledge than on respect for the natural world.[1] Others claimed more generally that certain qualities of the scientific enterprise reflected its "masculine" character, that is, were rooted in force and power, as were gender relations in society as a whole. One such writer was Clémence Royer (1830–1902), the first French translator of the works of Charles Darwin (1809–1882), a member of Paul Broca's (1824–1880) Anthropological Society, and a lifelong activist for feminist and other movements of social reform. In her *Le bien et la loi morale* (The Good and the Moral Law) of 1881, she described science as masculine in its practitioners and thereby "masculine" in its practices.[2]

---

[1] Coral Lansbury, *The Old Brown Dog: Women, Workers, and Vivisection in Edwardian England* (Madison: University of Wisconsin Press, 1985); and Roger French, *Anti-Vivisection in Victorian Society* (Princeton, N.J.: Princeton University Press, 1975).
[2] Clémence Royer, *Le bien et la loi morale; Éthique et téléologie* (Paris: Guillaumin, 1881); see also Joy Harvey, *"Almost a Man of Genius": Clémence Royer, Feminism, and Nineteenth-Century Science* (New Brunswick, N.J.: Rutgers University Press, 1997).

In the second half of the twentieth century, these attitudes toward science came to be subsumed into broader antiscientific currents that themselves had little to do with gender issues. The detonation of atomic bombs in 1945 over Hiroshima and Nagasaki, as well as the role played by technology in assisting the Holocaust, were cited as evidence for the claim that science and scientists were motivated by an irresponsible desire for natural knowledge at all costs and were easily manipulated by criminal governments for lethal purposes. By the late 1960s, this antiscience strand of opinion had been reinforced by emphatic criticism of what the French philosopher and theologian Jacques Ellul, influenced by the German-American philosopher Herbert Marcuse, called "the technological society" in his 1964 book of the same name.[3] Such points of view were echoed in feminist accounts, also rooted in the social protests of the 1960s, which described early modern science as essentially masculine and technological.

These ideas shaped even the work of writers who acknowledged no formal debt to the Frankfurt School of social criticism to which Marcuse had belonged. According to the ecofeminist historian Carolyn Merchant, for example, writing in her classic book *The Death of Nature: Women, Ecology, and the Scientific Revolution* (1980), "as Western culture became increasingly mechanized in the 1600s, the female earth and virgin earth spirit were subdued by the machine."[4] Merchant went on to argue, with particular reference to mining and agriculture, that Western science has been driven by attempts to exploit nature in man's interest, and has been understood as an attack on the earth, which has long been seen not only as a female principle but as the mother of humankind. In this way, Merchant linked to great effect the "masculine" enterprise of science, shaped by male practitioners and focused on the domination of nature, with the long-standing literary and artistic practice of personifying nature as female.[5]

By the last two decades of the twentieth century, views of this sort, which linked the early modern period with the origins of a "masculine" science, were common in feminist historiography of science. With varying degrees of subtlety, the English philosopher Francis Bacon (1561–1626) was identified as the archetype of the modern, manipulative, empirical approach to nature, which equated experimental inquiry into nature with the possession of the

---

[3] Jacques Ellul, *The Technological Society*, trans. John Wilkinson (New York: Random House, 1964); and Herbert Marcuse, *One-Dimensional Man: Studies in the Ideology of Advanced Industrial Society* (London: Routledge and Kegan Paul, 1964).

[4] Carolyn Merchant, *The Death of Nature: Women, Ecology, and the Scientific Revolution* (New York: Harper and Row, 1980), p. 4.

[5] On the literary and artistic traditions, see, for example, George Economou, *The Goddess Natura in Medieval Literature* (Cambridge, Mass.: Harvard University Press, 1972); and Mechthild Modersohn, *Natura als Göttin im Mittelalter: Ikonographische Studien zu Darstellungen der personifizierten Natur* (Berlin: Akademie Verlag, 1997). For the importance of such personifications in early modern science, see Londa Schiebinger, *Nature's Body: Gender in the Making of Modern Science* (Boston: Beacon Press, 1993), pp. 56–9; Schiebinger, *The Mind Has No Sex? Women in the Origins of Modern Science* (Cambridge, Mass.: Harvard University Press, 1989), pp. 136–50.

female by the male. Baconian experimental science was often portrayed by subsequent feminist analysis, whether correctly or not, as the point of origin for the allegedly gendered nature of modern science and the latter's role in the construction of modernity.[6]

This historiography was influenced by essentialist ideas of gender: the assumption that if most professional practitioners of science were male, then science must have been "masculine" and would have been conducted very differently if most practitioners had been female. It also takes on board the assumption, still common in the historiography of science, that the early modern period was constitutive of modernity. In this reading, a decisive break occurred in the patterns of scientific inquiry in the seventeenth century that was embodied in the term "Scientific Revolution." Located above all in the physical and cosmological sciences, in the work of Nicholas Copernicus (1473–1543), Johannes Kepler (1571–1630), Robert Boyle (1627–1691), and Isaac Newton (1642–1727), the "Scientific Revolution" was understood as instituting a new order of scientific practice, which was the basis of modern science.[7] Bacon appears as a pivotal figure in both feminist and standard accounts of this shift from premodern to modern science, though for very different reasons.

This historiographical complex has come under increasing attack since the 1980s (see Park and Daston, Chapter 1, this volume). Historians of science and gender such as Londa Schiebinger have rejected essentialist arguments and placed less emphasis on the Scientific Revolution as constitutive of modernity and representative of modern science.[8] Others have questioned the very existence of a "Scientific Revolution," emphasizing the importance of the life sciences and nonexperimental practices such as collecting, as well as the role of women as an audience for new ideas about nature and as practitioners of natural inquiry.[9] It is still not clear whether the seventeenth century is in fact a turning point in the history of early modern categories

---

[6] See, for example, Evelyn Fox Keller, "Baconian Science: The Arts of Mastery and Obedience," in *Reflections on Gender and Science* (New Haven, Conn.: Yale University Press, 1985), pp. 33–42; and Keller, "Spirit and Reason at the Birth of Modern Science," in ibid., pp. 43–65, esp. pp. 53–4. Mary Tiles links Bacon with a view of science as mastery over nature, rather than truth achieved through contemplation in "A Science of Mars or of Venus?," *Philosophy*, 62 (1987), 293–306, esp. pp. 301 and 305–6. See also Kathleen Okruhlik, "Birth of a New Physics or Death of Nature?," in *Women and Reason*, ed. Elizabeth D. Harvey and Kathleen Okruhlik (Ann Arbor: University of Michigan Press, 1992), pp. 63–76.

[7] Classic expositions of this older view include Herbert Butterfield, *The Origins of Modern Science, 1300–1800* (New York: Macmillan, 1951); E. A. Burtt, *The Metaphysical Foundations of Modern Physical Science: A Historical and Critical Essay* (London: Routledge and Kegan Paul, 1964); and multiple works by Alexandre Koyré, including *From the Closed World to the Infinite Universe* (Baltimore: Johns Hopkins University Press, 1955).

[8] Schiebinger, *The Mind Has No Sex?*.

[9] See, for example, Paula Findlen, *Possessing Nature: Museums, Collecting, and Scientific Culture in Early Modern Italy* (Berkeley: University of California Press, 1994); Lorraine Daston and Katharine Park, *Wonders and the Order of Nature, 1150–1750* (New York: Zone Books, 1998), esp. chaps. 4–8; and Erica Harth, *Cartesian Women: Versions and Subversions of Rational Discourse in the Old Regime* (Ithaca, N.Y.: Cornell University Press, 1992).

of gender, of the early modern gendering of nature, or of the gendering of early modern natural inquiry. In this chapter, I have left aside the last matter, which is the subject of Schiebinger's chapter (see Chapter 7) in this volume, and will attempt to provide at least a partial answer to the first two questions, contending that although the sixteenth and seventeenth centuries marked an important and discernable stage in the development of European ideas about the differences between men and women, the same was true, but to a lesser degree, in the development of European ideas about the gendering of nature itself.

Any analysis of this topic is complicated by the fact that early modern understandings of sex differences were fundamentally different from those of twenty-first-century scholars. Like most modern writers, I make a clear distinction between sex differences and gender differences and assume that masculinity and femininity are categories that have been historically constructed and whose meanings have thus varied over time. I use the term "sex difference" to signify the physical differences between men and women and "gender differences" to refer to divisions between the social roles and characteristics of the sexes.[10] This distinction, however, is a product of the 1950s and would have made little sense to early modern writers, for whom the physical and the social orders were more closely linked, both reflecting the structure of nature – including the differences between men and women – which they understood to be divinely ordained. Thus, what we would call gender differences had long been naturalized (that is, ascribed to the natural order); hence attempts to change or criticize them were regarded as at best foolish and at worst morally wrong.

What changed in the early modern period was the understanding of the natural order, which was now seen less as a principle of plenitude and variety, shaped at best by regularities, and more as a uniform mechanical system, governed by unbreakable laws. Sixteenth- and early seventeenth-century Europeans could imagine the differences between men and women – social and corporeal – as flexible, even labile; the social characteristics of masculinity and femininity usually mapped onto the physical ones, but this was not inevitable, and some even believed that women might change into men. By the middle of the eighteenth century, however, it was impossible to conceive of this kind of fluidity because the nature of nature had changed. The physical and social differences between men and women were proposed not only as thoroughly entangled, which had long been the case, but also as unalterable,

---

[10] The sex/gender distinction can be traced back to the American surgeon John Money, who used it to argue for the surgical treatment of intersexuals in the 1950s; see Bernice L. Hausman, *Changing Sex: Transsexualism, Technology, and the Idea of Gender* (Durham, N.C.: Duke University Press, 1995), p. 94. Medical writers in the 1940s had initially called "psychological sex" or the "sex of the mind" what Money and his colleagues referred to as "gender"; see Joanne Meyerowitz, *How Sex Changed: A History of Transsexuality in the United States* (Cambridge, Mass.: Harvard University Press, 2002), pp. 111–12.

discoverable, and predictable as the structures of nature itself. This is a case in which conceptions of the specific "nature" – the defining essence of a thing, here of being a man or woman – interacted with more general ideas of nature as the order of the entire universe, revealing critical ambiguities and insecurities in the conceptual structure of science itself.

SEX AND GENDER DIFFERENCE IN THE EARLY MODERN PERIOD

Ideas of nature and definitions of the differences between men and women entered a period of flux in the early modern period. The connection here is obvious: Because both those definitions relied on nature for their legitimation, changes in understandings of the natural order meant changes in the definitions themselves, or at least in their implications. Furthermore, the ideas concerning the differences between male and female natures inherited by early modern writers from their predecessors were themselves diverse, inconsistent, and unstable. This was partly because ideas of sex and gender difference drew on many different sources of legitimation in this period whose relationships with one another varied. Some remained as pervasive in the early modern period as they had been in the Middle Ages. For example, scriptural commentary on the first two chapters of the book of Genesis, which describes the creation of the first woman out of the body of the first man, had reflected for centuries on the relationship between the sexes; this tradition lost its influence and authoritative status only very gradually, and incompletely, over the course of the seventeenth and eighteenth centuries. No less important was the description of the first woman's temptation in the third chapter of Genesis, which was used to ascribe a separate and inferior moral and intellectual character to women, on the basis of the lustfulness, disobedience, and overweening curiosity of their ancestress, Eve. The same message came from the classical Pauline texts on women's inferiority in the New Testament, such as Colossians 3:18 and Ephesians 5:22.[11]

A second source of legitimation lay in the large body of ancient and medieval natural philosophical and medical texts and commentaries associated with Aristotle and his followers, which continued to be mobilized to assert and account for differences between men and women in the sixteenth, seventeenth, and, to a much lesser extent, the eighteenth centuries. Aristotle had described women as "defective males," marked by anatomical and physiological differences traceable to a lower degree of heat in the heart,[12] and

---

[11] On this tradition, see Ian Maclean, *The Renaissance Notion of Woman: A Study in the Fortunes of Scholasticism and Medical Science in European Intellectual Life* (Cambridge: Cambridge University Press, 1980), chap. 1.

[12] Aristotle, *On the Generation of Animals*, 737a28, in *The Complete Works of Aristotle: The Revised Oxford Translation*, ed. Jonathan Barnes, 2 vols. (Princeton, N.J.: Princeton University Press, 1984),

this theory was reinforced by the Galenic theory that understood the appearance and function of bodies as largely based on the balance of the "primary qualities" of hot, cold, wet, and dry that characterized all matter, including the four elements and the four humors. Generally speaking, females were thought to have colder and wetter complexions than males, which led in turn to a variety of physical characteristics, ranging from softer skin and internal genitals to the need to menstruate as the result of imperfect digestion.[13] In this view, sex difference arose in the first instance not so much from the body's shape or its ability to perform certain reproductive functions but from its elemental constitution. This led to a set of distinctive doctrines about women's physical and mental weakness and vulnerability to physical, emotional, and mental dysfunction, which remained a mainstay of thinking about sex difference throughout the Middle Ages and into the early eighteenth century and seemed to define the sexes in oppositional rather than complementary ways.[14]

These sources of authority slowly lost their commanding position during the early modern period. Even for many who never accepted the religious teachings of Martin Luther (1483–1546) and John Calvin (1509–1564), the Protestant Reformation of the sixteenth century unleashed ideas such as the primacy of individual response to foundational texts. The Reformers' battle cry of *sola scriptura*, or reliance on the biblical text alone, divorced from the authoritative exegesis of the Roman Church, set running many different interpretations of biblical texts and consequent conflicts about the nature of truth in religion. Nor were medical and natural philosophical texts exempt from this new kind of scrutiny. Galen's thinking was increasingly open to challenge, and struggles over the authority of Aristotle also became increasingly frequent, leading in the seventeenth century to the gradual decline of Aristotelian teaching in most Western European centers of learning outside the Iberian peninsula and its replacement by the ideas of Descartes and other contemporary philosophers (see Blair, Chapter 17, this volume).[15] None of these struggles over traditional sources of authority was conclusively or quickly decided before the early eighteenth century. References to the authority of the Bible and of Aristotle and Galen continued to be routine

---

1: 1144. See Maryanne Cline Horowitz, "Aristotle and Women," *Journal of the History of Biology*, 9 (1976), 183–213.

[13] Maclean, *The Renaissance Notion of Woman*, chap. 3. On the complexions and humors in general, see Nancy G. Siraisi, *Medieval and Renaissance Medicine: An Introduction to Theory and Practice* (Chicago: University of Chicago Press, 1990), pp. 101–6.

[14] On this set of ideas, see in general Joan Cadden, *Meanings of Sex Difference in the Middle Ages: Medicine, Science, and Culture* (Cambridge: Cambridge University Press, 1993), chap. 4; Thomas Laqueur, *Making Sex: Body and Gender from the Greeks to Freud* (Cambridge, Mass.: Harvard University Press, 1990), esp. chap. 2; and Maclean, *The Renaissance Notion of Woman*, chap. 3.

[15] On this set of ideas, see in general Joan Cadden, *Meanings of Sex Difference in the Middle Ages: Medicine, Science, and Culture* (Cambridge: Cambridge University Press, 1993), chap. 4; Thomas Laqueur, *Making Sex: Body and Gender from the Greeks to Freud* (Cambridge, Mass.: Harvard University Press, 1990), esp. chap. 2; and Maclean, *The Renaissance Notion of Woman*, chap. 3.

in many quarters at the same time that they were being challenged or even discarded in others. Different areas of Europe, different social classes, different churches, and different teaching institutions accepted these changes at very different rates. This led to prolonged instability and conflict that not surprisingly also manifested themselves in discussions of the differences between the natures of men and women.

An early example of such instability appears in the well-known work *De nobilitate et praecellentia foeminei sexus* (On the Nobility and Preeminence of the Female Sex) by the itinerant German polymath Henricus Cornelius Agrippa von Nettesheim (1486–1535). Published in 1529, early in the Reformation conflict, the book was quickly translated into French, English, Italian, and German.[16] In it, Agrippa radically challenged widely accepted views. His 1526 discourse *De incertitudine et vanitate scientiarum atque artium declamatio* (Declamation on the Uncertainty and Vanity of the Sciences and Arts) had already asserted that the "sciences" (bodies of authoritative knowledge) had no intrinsic certainty and were nothing but human opinions and decisions.[17] Three years later, his oration on women began by attacking the tradition of scriptural commentary on the subject. Reversing the misogynistic interpretation of Genesis, which typically took Eve's creation after and out of Adam as a sign of her inferiority, Agrippa noted:

> How far Woman surpasses man in nobility of race by reason of the order in which she was created the sacred word bears witness most abundantly to us. Woman in fact was fashioned with the angels in Paradise, a place absolutely full of nobility and delight, while man was made outside of Paradise in the countryside among brute beasts and then transported to Paradise for the creation of woman.

Therefore, it is "right that every creature love, honor and respect her; right also that every creature submit to and obey her, for she is the queen of all creatures, and their end, perfection and glory, absolute perfection."[18]

As more recent editors have pointed out, in this treatise Agrippa also challenges Aristotle's arguments (though not Galen's) on the inferiority of women shown in procreation: "Should we pass over the fact that in the procreation of the human race nature has preferred women to men?" he wrote, noting,

---

[16] Henricus Cornelius Agrippa von Nettesheim, *De nobilitate et praecellentia foeminei sexus: Édition critique d'après le texte d'Anvers 1529*, ed. Charles Béné, trans. O. Sauvage (Geneva: Droz, 1990); Agrippa, *Declamation on the Nobility and Pre-Eminence of the Female Sex*, ed. and trans. Albert Rabil, Jr. (Chicago: University of Chicago Press, 1996).

[17] Henricus Cornelius Agrippa von Nettesheim, *De incertitudine et vanitate scientiarum et atrium atque excellentia Verbi Dei declamatio* (Antwerp: Joannes Grapheus, 1530); English trans. in Agrippa, *Of the Vanitie and Uncertaintie of Artes and Sciences* [1575], ed. Catherine M. Dunn (Northridge: California State University Press, 1974).

[18] Agrippa, *Declamation*, trans. Rabil, p. 48.

> This is particularly evident in the fact that only the female seed (according to the stated opinions of Galen and Avicenna) provides matter and nourishment for the fetus, while that of the man intervenes only a little because it affects the fetus rather as accident to substance.... The reason we see so many sons resemble their mothers is that they have been procreated by her blood.... If the mothers are stupid, the sons are stupid; if the mothers are wise, the sons breathe wisdom.... That is why mothers love their children more than fathers do, because they recognize and find in them much more of themselves than their fathers do.[19]

This is a reversal of Aristotle's arguments; the female physical constitution that Aristotle had seen as inescapably inferior was reinterpreted by Agrippa to argue for women's superiority. Against Aristotle's idea that women had cold dispositions, for example, Agrippa pointed to the biblical story that described how King David was warmed by a young woman in his old age.[20] Agrippa's achievement was not only to challenge the authority of Aristotle but also to demonstrate, implicitly, that the so-called authorities often contradicted each other and that the Bible and Aristotle could be made to point in opposite directions on the fundamental question of the physical constitution of woman. Even the Bible itself could be self-contradictory. Thus Agrippa countered biblical statements concerning the inferiority and subordination of women with the argument, drawn from Genesis itself, on the equality of souls: "God has a preference for no one."[21] In such claims, he also demonstrated the growing instability of categories of gender and sex difference in the early modern period.

The appearance of Agrippa's work signals the possibility of challenging traditional ideas concerning the differences between men and women. It became increasingly difficult to make these differences, typically alleged to be immutable parts of divine and natural order, appear to be securely so. It is not surprising that, in the early modern period, stories of individuals who crossed boundaries between the sexes circulated widely and caused sharp debate. Some dismissed such stories, whereas others gave them considerable credence. One of the latter was the French essayist Michel de Montaigne (1522–1592). As Montaigne relates in his essay "De la force de l'imagination" ("On the Power of Imagination"),

> I was travelling through Vitry-le-François when I was able to see a man to whom the bishop of Soissons had given the name of Germain at his confirmation: until the age of twenty-two he had been known by sight to all the townsfolk as a girl called Marie. He was then an old man with a full beard; he remained unmarried. He said that he had been straining to jump

---

[19] Ibid., pp. 56–57.
[20] Ibid., p. 53.
[21] Ibid., p. 96. Compare Genesis 1:27 (Revised Standard Version): "So God created man in his own image, in the image of God he created him; male and female he created them."

when his male organs suddenly appeared. (The girls there still have a song in which they warn each other not to take great strides for fear of becoming boys, 'like Marie Germain'.) It is not surprising that this sort of occurrence happens frequently. For if the imagination does have any power in these matters, in girls it dwells so constantly and so forcefully on sex that it can (in order to avoid the necessity of so frequently recurring to the same thoughts and harsh yearnings) more easily make that male organ into a part of their bodies.[22]

In Montaigne's account, sex difference is so frail that it can easily, in fulfilling itself (the girls fulfill their female nature by longing for the male), reverse itself (the girls acquire male organs). Imagination is strong enough to subvert distinctions of sex, such as those described by Aristotle and Galen, based on complexion and physical conformation. Montaigne's story made sense in an era in which many believed that even in the genital area, which most strongly seemed to show sex difference, women were really shaped like men, the difference being that what men showed on the outside of the body women kept inside. This idea rested on Galen's almost topological view of the male and female organs of generation. "All the parts, then, that men have, women have too," he had written in his second-century work *On the Usefulness of the Parts of the Body*, "the difference between them lying in only one thing . . . namely, that in women the parts are within the body, whereas in men they are outside. . . . Consider first whichever one you please, turn outward the woman's, turn inward and fold double the men's, and you will find then the same in both in every respect."[23] Thus the sixteenth-century French surgeon Ambroise Paré (ca. 1510–1590) was only echoing Galen in his aphorism, "That which man hath apparent without, that women have hidden within."[24] Because of such teachings, it was not difficult for many people in this era to believe that a blow or sudden lesion, or violent exercise, as in Montaigne's anecdote, might occasion a sudden change of sex by causing a woman's internal organs to be forced to the outside of the body. The division between the sexes lay not so much in an absolute difference of conformation, but in defining the female body as being a place of concealment of what was revealed in the male.

---

[22] Michel de Montaigne, *The Essays of Michel de Montaigne*, ed. and trans. M. A. Screech (London: Penguin, 1993), p. 111. A version of this story also appears in Montaigne, *Journal de voyage*, ed. François Rigolot (Paris: Presses Universitaires de France, 1992), pp. 6–7.

[23] Galen, *On the Usefulness of the Parts of the Body*, ed. and trans. Margaret Tallmadge May (Ithaca, N.Y.: Cornell University Press, 1968), p. 628.

[24] Ambroise Paré, *The Anatomy of Man's Body*, in *The Works of that Famous Chirurgeon Ambrose Parey*, trans. Thomas Johnson (London: Richard Cotes and Willi Dugard for John Clarke, 1649), p. 128. See in general Gianna Pomata, "Uomini mestruanti: Somiglianze e differenze fra i sessi in Europa in età moderna," *Quaderni storici*, n.s., 79 (1996), 51–103; and Katharine Eisaman Maus, "A Womb of His Own: Male Renaissance Poets and the Female Body," in *Sexuality and Gender in Early Modern Europe: Institutions, Texts, Images*, ed. James Grantham Turner (Cambridge: Cambridge University Press, 1993), pp. 266–88.

Montaigne's account in fact contains two very different ways of explaining the incident at Vitry-le-François. Montaigne ascribed what happened to Marie Germain first to violent exercise, which ended up revealing what was already there. He did not explicitly refer to the Galenic theory of complexions, so we do not know if he also believed that exercise had heated up the young girl's body to the male threshold, or that exertion had loosened the ligaments that held the uterus inside Marie's body. Second, Montaigne described spontaneous sex changes as being caused by the power of the imagination, particularly imagination driven by strong sexual desire. There was no sense of incompatibility between these two explanations. Montaigne admitted and pondered the idea that sex change can happen, and that female and male natures can be changeable rather than immutable. Yet his story also reveals the strength of the forces that assign even creatures of changed sex to one category or the other.[25]

These complex perceptions about the mutability of corporeal sex coexisted in the early modern period with the commonplace assertion and widespread acceptance of a hierarchical gender order.[26] The early modern discourse on femininity was preoccupied with the problem of how to keep clear the divisions of class and rank around which the early modern social order was constructed while still maintaining women as subordinate to men. Was a woman of high social status still inferior to a man of lower rank? If so, status stability was threatened. If not, the gender order was weakened. These themes were very present to contemporaries. The sixteenth and early seventeenth centuries saw an unusually large number of women in positions of supreme authority, either as rulers in their own right, such as Elizabeth I of England (ruled 1558–1603), or as surrogates and advisors, such as the mid-sixteenth-century regent of France, Catherine de' Medici (1519–1589), who was the force behind her son Charles IX (1550–1574) from his accession to the throne in 1560 to his death in 1574. There was acute awareness among contemporaries that a woman, the heiress of Eve, defined as legally subordinate to man and as an inferior version of him physically, might yet wield sovereign power, or – a matter of daily observation – be as intelligent or more so than her husband.

The matter was made more pointed by the confessional and dynastic complexities of marriage among contemporary rulers. In England, for example, the marriage of Mary Tudor (1516–1558) to Philip II of Spain (1527–1598) had

[25] Similar concerns played upon the contemporary fascination with hermaphrodites. See Katharine Park and Lorraine Daston, "Hermaphrodites in Renaissance France," *Critical Matrix*, 1 (1985), 1–19; Park and Daston, "The Hermaphrodite and the Orders of Nature: Sexual Ambiguity in Early Modern France," *GLQ: A Journal of Lesbian and Gay studies*, 1 (1995), 419–38; and Katharine Park, "The Rediscovery of the Clitoris: French Medicine and the Tribade," in *The Body in Parts: Discourses and Anatomies in Early Modern Europe*, ed. Carla Mazzio and David Hillman (New York: Routledge, 1997), pp. 171–93.
[26] This discussion owes much to Constance Jordan, "Renaissance Women and the Question of Class," in Turner, *Sexuality and Gender*, pp. 90–106; see also Maryanne Cline Horowitz, ed., *Race, Gender, and Rank: Early Modern Ideas of Humanity* (Rochester, N.Y.: University of Rochester Press, 1992).

led to well-founded fears that if the queen were, as was theoretically proper, subordinate to her husband, this would pull England into the orbit of Spain. During the long reign of Elizabeth I, the queen walked a tightrope between her need on the one hand to find necessary marriage alliances for England in a world increasingly polarized along religious lines and on the other to conserve her own independence and supreme authority. After considering many suitors, she remained unmarried.

Discussions of marriage were also much affected by another version of the problem. Writing at almost the same moment as Agrippa, for example, the famous humanist Desiderius Erasmus (ca. 1466–1536) admitted, in his *Institutio matrimonii christiani* (The Institution of Christian Marriage) of 1526, that women are created spiritually equal with men, because of their common creation in the image of God. Yet even he still insisted on their lack of moral or practical autonomy in relation to their husbands.[27] The hierarchical relationship between husband and wife was at odds with the assertion of spiritual equality. This problem was exacerbated by the emphasis of Protestant reformers on the importance of the wife in the spiritual education of the family.[28] Struggles over marital authority were topical enough for William Shakespeare (1564–1616) to devote his *Taming of the Shrew* (1593–4) to this theme. But the set of puzzles caused by the lack of fit between theoretical accounts of gender and the daily observation of the powers and intelligence of actual individuals of either sex was impossible to resolve. It was certainly no answer to argue, as did the Italian poet Torquato Tasso (1544–1595) in his 1572 *Discorso della virtù feminile e donnesca* (Discourse on the Strength of Women and Ladies), that all women of high rank should be considered exceptions to the general nature of woman, provided they exhibited heroic virtue of a masculine kind.[29]

Not just social and political power but also the question of women's intellectual power and authority were affected by these problems of producing stable and mutually consistent categories of sex and gender. This was demonstrated in early modern debates on the moral and intellectual capacities of women, including their capacity to conduct inquiry into the natural world. The very existence of such debates, the so-called *querelle des femmes* in seventeenth-century France, is an indication of the way in which concepts of sex and gender refused to interact in a stable way and could be made to demonstrate the inferiority of women to men, or men to women, or equality

---

[27] Desiderius Erasmus, *Encomium matrimonii*, ed. J. C. Margolin, in *Opera omnia Desiderii Erasmi Roterdami*, 25 vols. to date (Amsterdam: North-Holland, 1969–), ordo I, vol. 5, pp. 333–416; see also Eleanor McLaughlin, "Equality of Souls, Inequality of Sexes: Medieval Theology," in *Religion and Sexism: Images of Women in the Jewish and Christian Traditions*, ed. Rosemary Ruether (New York: Herder and Herder, 1974), pp. 218–47.

[28] Joel Harrington, *Reordering Marriage and Society in Reformation Germany* (Cambridge: Cambridge University Press, 1995).

[29] Torquato Tasso, *Discorso della virtù feminile e donnesca*, ed. Maria Luisa Doglio (Palermo: Sellerio, 1997).

between the two. Gender definitions were constantly rehearsed, only to be challenged anew.[30]

On the matter of women's intellectual abilities, the *querelle* mobilized writers who believed that Eve's sin in the Garden of Eden, which caused the corruption of humankind from the perfect state in which God had created it, had demonstrated the perils of female curiosity – a demonstration only reinforced by the classical myth of Pandora. They also invoked the medical theory of the complexions, to which I referred earlier, arguing for woman's unfitness for intellectual pursuits from the cold and moist character of her body, and of her womb in particular. In the words of the Spanish writer Juan Huarte Navarro (ca. 1529–1588), whose phenomenally influential *Examen de los ingenios* (Examination of Wits, 1582) reflected what was still the contemporary consensus,

> To think that a woman can be hot and dry, or endowed with wit and ability conformable to these two qualities, is a very great error; because if the seed of which she was formed, had been hot and dry in their formation, she should have been born a man and not a woman.... She was by God created cold and moist, which temperature is necessary to make a woman fruitful and apt for childbirth, but an enemy to knowledge.[31]

On the other side, more radical and innovative minds found it perfectly possible to turn the theory of the complexions on its head. The French physician and *savant* Samuel Sorbière (1615–1670), for example, wrote to Princess Elisabeth of Bohemia (1618–1680) in 1660, arguing that women in fact ought to excel in knowledge over men, for "our doctors, who consider the brain the seat of reason and learning, find it as large in women as in men, and also claim that the softness of their constitution ... is much more suitable to the actions of the mind than the dryness and hardness of ours."[32] Other authors went even further and argued that scholars themselves were typically delicate and moist in complexion, and hence more like women. Thus, medical evidence about the natural constitution of women – and about sex difference in general – whether based on the theory of the complexions or on anatomy, was increasingly no longer seen as compelling evidence one way or the other

---

[30] One of the principal spokesmen for the equality of women was François Poullain de la Barre; see his *De l'égalité des deux sexes: Discours physique et moral* (Paris, 1673; repr. Paris: Fayard, 1984) and *De l'éducation des dames pour la conduite de l'esprit dans les sciences et dans les moeurs, De l'égalité des deux sexes, discours physique et moral, où l'on voit l'importance de se défaire des préjugez* (Paris: Jean Du Puis, 1673). See in general Ian Maclean, *Woman Triumphant: Feminism in French Literature, 1610–1652* (Oxford: Oxford University Press, 1977); Mirjam de Baar, ed., *Choosing the Better Part: Anna Maria van Schurman, 1607–1678*, trans. Lynne Richards (Dordrecht: Kluwer, 1996); and Lorraine Daston, "The Naturalised Female Intellect," *Science in Context*, 5 (1992), 209–35.

[31] Juan de Dios Huarte Navarro, *Examen de ingenios: The Examination of Men's Wits*, trans. Camillo Camilli and R. C. Esquire (London: Adam Islip, 1604), p. 274; quoted in Maus, "A Womb of His Own," p. 268.

[32] Samuel Sorbière, *Relations, lettres, et discours de M. de Sorbière sur diverses matières curieuses* (Paris: Robert de Ninuille, 1660), p. 71; quoted in Schiebinger, *The Mind Has No Sex?*, p. 167.

for the nature of women's intelligence. It could not ground gender differences convincingly, durably support their inevitability and ineluctability, or use them to ground arguments for gender hierarchy. In this sense, early modern theories of sex and gender had become fluid and labile in their own right and in relationship to one another in a way that differed distinctly from the situation either in the late Middle Ages or in the Enlightenment.

As literary historian Erica Harth has argued, it was no accident that those who argued for the equality of the mental powers of women in the *querelle des femmes* often rejected the Aristotelian tradition in philosophy for the new philosophy of René Descartes (1596–1650). Descartes' *Discours de la méthode* (Discourse on Method) of 1637 seemed to make a distinction between mental ability and bodily characteristics not only possible but necessary. Descartes did not explicitly participate in the debate over gender and intelligence. Nonetheless, he grounded the successful use of rationality not on the gender characteristics of the person exercising rationality but on the method by which the person thought. Reason was the universal possession of humankind, male or female, and needed only to be pursued by correct methods. Nor did Descartes relate the exercise of rationality to sexual temperament. His work reflects the abandonment of the analogies between microcosm and macrocosm that had organized much earlier thinking and that I will discuss. He made instead a sharper distinction than had been made before between inanimate nature, including the matter of the body, on the one hand, and the human and divine mind on the other. Universal knowledge claims thus need not take into account the different bodily constitutions of men and women.[33]

Descartes' belief in innate ideas, and in the separation of mind and body, seemed to cut the ground out from under those who wanted to base statements about women's intellectual inferiority on their physical conformation. The same effect might have resulted from his emphasis on method, potentially accessible to all, as the basis of correct reasoning, rather than the different physical constitutions of men and women. Some supporters of women in the *querelle des femmes*, such as the radical seventeenth-century French philosopher François Poullain de la Barre (1647–1723), took Descartes' ideas to their logical conclusion and argued that the practice of experimental science should be equally accessible to both genders. Many contemporary women also read Descartes in this way.[34]

---

[33] Harth, *Cartesian Women*. See in general Estelle Cohen, "The Body as a Historical Category: Science and Imagination, 1660–1760," in *The Good Body: Asceticism in Contemporary Culture*, ed. Mary G. Winkler and Letha B. Cole (New Haven, Conn.: Yale University Press, 1994), pp. 67–90.

[34] Harth, *Cartesian Women*, chap. 2; Schiebinger, *The Mind Has No Sex?*, pp. 171–8; Hilda L. Smith, *Reason's Disciples: Seventeenth-Century English Feminists* (Chicago: University of Chicago Press, 1982); Smith, "Intellectual Bases for Feminist Analyses: The Seventeenth and Eighteenth Centuries," in Hervey and Okruhlik, eds., *Women and Reason*, pp. 19–38; and Ruth Perry, "Radical Doubt and the Liberation of Women," *Eighteenth-Century Studies*, 18 (1985), 472–93.

Descartes' legacy, however, was ambiguous. Whereas some feminist assessments of Descartes, such as that by Harth, have interpreted his separation of mind and body as making it possible to endow females as well as males with rationality, other feminist scholars have seen precisely this separation of mind and body enabling ideologies such as that of scientific "objectivity." Objectivity is often defined as a methodological prescription that puts the maximum separation between the observer of the natural world and the object of his observation. This distinction between observer and object was often understood in gendered terms, with the latter gendered as female and the former, inevitably, as male. This acted, paradoxically, to reinscribe older, binary habits of thought about gender and encouraged, once again, ideas of nature as passive: an object to be possessed and controlled, exploited, and explored.[35]

THE PROBLEM OF NATURE

To this point, I have focused on particular natures, specifically the natures attributed by sixteenth- and seventeenth-century writers to men and women. In this section, I consider general nature, which contemporaries understood as the totality of the created world and the object of natural philosophical inquiry, in its relationship to contemporary discourses on sex and gender. If anything, sixteenth- and, especially, seventeenth-century ideas of nature were in a state of confusion and flux even greater than that of contemporary ideas about gender, and they changed in ways that were contradictory, uneven, and uncorrelated with contemporary thinking about gender in any simple or discernible way.

Since the late 1970s, historical scholarship on this topic has been shaped in large part by the work of both the French philosopher Michel Foucault and feminist scholars, of whom Merchant is arguably the most influential. In *The Death of Nature*, Merchant argued that before the late sixteenth century, nature was typically understood through a female personification, as a figure that was intrinsically feminine; she manifested the two sides of women's nature as traditionally understood: its wildness, destructiveness, and unpredictability on the one hand, and its inclination toward maternal nurture on the other. In both incarnations, however, nature was powerful, active, and engaged. This view of nature allegedly found its highest expression in the magical thought of writers such as Tommaso Campanella (1568–1639). Bacon was the first to challenge this acceptance of nature's power and autonomy, according to Merchant, forcing her "to submit to the questions and

---

[35] Merchant, *The Death of Nature*; Susan Bordo, "The Cartesian Masculinisation of Thought," *Signs*, 11 (1986), 439–56; Bordo, *The Flight to Objectivity: Essays in Cartesianism and Culture* (Albany: State University of New York Press, 1987); Geneviève Lloyd, *The Man of Reason: "Male" and "Female" in Western Philosophy* (Minneapolis: University of Minnesota Press, 1984), pp. 38–50.

experimental techniques of the new science."³⁶ For Bacon, nature was not a quasi-divine figure, to be treated with reverence and awe. Rather, she must be tortured and penetrated to reveal her secrets, and constrained and molded by the mechanical arts. This new attitude toward nature was justified and facilitated by her gender. As Merchant put it, "the new image of nature as a female to be controlled and dissected through experiment legitimated the exploitation of natural resources. Although the image of the nurturing earth popular in the Renaissance did not vanish, it was superseded by new controlling imagery. . . . From an active teacher and parent, she has become a mindless, submissive body."³⁷ Merchant identified this change with what she called "the new conceptual framework of the Scientific Revolution– mechanism."³⁸

In *Les mots et les choses* (1966), translated into English as *The Order of Things*, Foucault, too, emphasized a radical shift in views of nature and language in the years around 1650, although he did not interpret this shift with reference to gender. In the sixteenth century, he argued, the world was understood as being structured by analogies, which knit its elements together by a densely woven web of correspondences. One such extended analogy was that between the microcosm (the human body) and the macrocosm (the universe as a whole). The Elizabethan adventurer, poet, and historian Walter Raleigh (ca. 1554–1618) expressed this view with economy and elegance in his *History of the World* (1614):

> . . . whereas God created three sorts of living creatures, to wit, Angelical, Rational and Brutal; giving to angels an intellectual and to beasts a sensual nature, he vouchsafed unto man, both the intellectual of angels, the sensitive of beasts, and the proper rational belonging to man; and therefore . . Man is the bond and chain which tieth together both natures: and because in the little frame of man's body there is a representation of the Universal, and (by allusion) a kind of participation of all the parts thereof, therefore was man called the microcosmos or the little world. . . . His blood, which disperseth itself by the branches of veins through all the body, may be resembled to those waters which are carried by brooks and rivers over all the earth; his breath to the air . . . our generative power, to Nature, which produceth all things . . . the four complexions resemble the four elements, and the seven ages of man the seven planets.³⁹

According to Foucault, such a way of thinking, with its insistence on making analogical links between apparently dissimilar things, on seeing things as analogues of other things, and on hermeneutical methods of knowing,

---

³⁶ Merchant, *The Death of Nature*, p. 164.
³⁷ Ibid., pp. 189–90.
³⁸ Ibid.
³⁹ Sir Walter Raleigh, *History of the World*, in *Selected Writings*, ed. Gerald Hammond (London: Penguin, 1986), p. 154.

was characteristic of Renaissance approaches both to texts and the "book" of nature. In contrast, the "classical" *episteme* of the late seventeenth and eighteenth centuries reconceived both language and natural history in terms of representations and well-ordered series, as in Enlightenment taxonomic systems.

Against Merchant's and Foucault's views of a dramatic change in seventeenth-century views of nature, however, there is much evidence that the shift was in fact far more gradual and uneven, as the ongoing publication of alchemical and magical works would tend to suggest; up to the end of the Enlightenment, these continued to invoke analogies between parts of nature. Nature continued to be personified in ways that suggested her power and autonomy, as well as her passivity, and emphasized both her maternal and her mechanical aspects.[40]

The notion of a very gradual change in the concept of nature is supported by Robert Boyle's lengthy work *Free Inquiry into the Vulgarly Received Notion of Nature* (1682). While attempting to tease out contemporary meanings of nature, Boyle alluded to the ideas of macrocosm and microcosm: "Many things, therefore, that are commonly ascribed to nature, I think, may be better ascribed to the mechanisms of the macrocosm and microcosm, I mean, of the universe and the human body."[41] Yet this mechanistic use of generalized analogies between microcosm and microcosm analogies is very different from the detailed organic parallels drawn by earlier writers such as Raleigh, and it shows the extent to which Boyle tried to incorporate many different and not necessarily compatible meanings of nature. As so often happens, older and newer systems of thinking run in parallel for a long time before one definitively replaces the other; the sharp breaks posited by Foucault are in fact rare. Thus Boyle was able to see nature simultaneously in terms of microcosm and macrocosm and as a law-governed system having nothing to do with older analogical ways of thinking. He described "the frame of the world, already made as a great, and if I may so speak, pregnant automaton, that like a woman with twins in her womb, or a ship fashioned with pumps, ordnance, etc., is such an engine, as comprises or consistes of several lesser engines ... a complex principle whence results the settled order or course of things corporeal."[42]

---

[40] See Schiebinger, *Nature's Body*, pp. 56–9; Katharine Park, "Nature in Person: Renaissance Allegories and Emblems," in *The Moral Authority of Nature*, ed. Lorraine Daston and Fernando Vidal (Chicago: University of Chicago Press, 2004), pp. 50–73. The gradualness of changes in views of the natural order is emphasized in Israel, *Radical Enlightenment*, e.g., pp. 4–7.

[41] Robert Boyle, *Free Enquiry into the Vulgarly Received Notion of Nature*, in *Works*, ed. Thomas Birch, 6 vols. (London, 1772; repr. Hildesheim: Georg Olms, 1966), 5: 158–254, quotation at p. 230. See in general Lorraine Daston, "How Nature Became the Other: Anthropomorphism and Anthropocentrism in Early Modern Natural Philosophy," in *Biology as Society, Society as Biology*, ed. Sabine Maasen, Everett Mendelsohn, and Peter Weingart (Dordrecht: Kluwer, 1995), pp. 37–56.

[42] Boyle, *Free Inquiry*, p. 179.

It is noticeable here that nature is no longer likened solely to the structure of the human body but also to a complex mechanical system. The organic analogy from the human body (the woman pregnant with twins) is quickly replaced by a mechanical one (the armored ship). This system is law-governed rather than analogically ordered. As Boyle noted later, nature is "a rule, or rather a system of rules, according to which those agents, and the bodies they work on, are by the great Author of things, determined to act and suffer."[43] Boyle also took issue with other contemporary attempts to define the nature of nature. He repeatedly cautioned his readers against personifying it, against seeing it as a semi-deity, and against believing that the world has a soul or is alive in the way a single organism is alive, or believing that nature and God are the same. In doing so, he explicitly opposed the Cambridge Platonists Ralph Cudworth (1617–1688) and Henry More (1614–1687), who had credited nature with some kind of soul or "plastic nature."[44]

Yet there are times when Boyle did not take his own advice. He in fact personified nature throughout his *Free Enquiry* as female, describing her as a "nursing mother to living creatures" and as the "common parent of us all."[45] In other words, he presented a picture of nature as still gendered, female, and maternal, although his account lacks the overtones of sexual possession and domination that feminists have often seen as accompanying the gendering of nature as female in the tradition of Bacon; rather, Boyle's nature is an indulgent mother, in contrast with a distant, patriarchal God. Finally, Boyle's account points to the many different concepts of nature during this period, often running parallel to each other, with different sources and implications, simultaneously gendered and ungendered, mechanistic and organic, both linked to the analogy between microcosm and macrocosm and separated from it.

The unself-conscious, self-contradictory swing, so evident in Boyle's essay, between seeing nature as gendered female and seeing it as an impersonal, nonpersonified system of mechanical rules was to continue well into the nineteenth century. This shows the extent to which the history of the definition of gender is not the same as the history of the gendering of nature. For one thing, the two histories do not share a common chronology. As I have argued, views of sex and gender showed strong differences between the Middle Ages and the early modern period and between the early modern period and the Enlightenment. In contrast, the gendering of nature continued to

---

[43] Ibid., p. 219.
[44] See Ralph Cudworth, *The True Intellectual System of the Universe* [1678], in *The Collected Works of Ralph Cudworth*, ed. Bernhard Fabian, 2 vols. (Hildesheim: Georg Olms, 1977), 1: 146–51; and Henry More, *The Immortality of the Soul* (London: James Flesher, 1662), pp. 167–8. On the philosophical and theological context of Boyle's treatise, see Catherine Wilson, "*De ipsa natura*: Sources of Leibniz's Doctrine of Force, Activity, and Natural Law," *Studia Leibnitiana*, 19 (1987), 148–72.
[45] Boyle, *Free Inquiry*, p. 198.

show far stronger continuities, as Boyle's simultaneous endorsement of many different descriptions of nature demonstrates. Nor does either the history of the definition of gender or the history of the gendering of nature share a common chronology with a third separate history, with which they are frequently associated, that of women's participation in scientific activity (see Schiebinger, Chapter 7, this volume).

No more than "gender" was "nature" a stable ontological category in the early modern period.[46] This did not stop Enlightenment writers in the century after Boyle from placing ever more emphasis on nature to legitimate gender differences. This was part of a much broader cultural phenomenon whereby eighteenth-century European intellectuals appealed to nature as the arbiter of everything from aesthetics to the political order, making it the all-purpose resource that religion had been before. The term "nature" also took on more persuasive moral overtones at this time: What was natural was also what was good. Such arguments were easily applied to the differences between men and women as well. Authors such as Jean-Jacques Rousseau (1712–1778), in his *Emile* (1762), alleged that women were overdetermined by their "natural" – that is, their procreative – physical functions. This statement depended on the conflation of normative and descriptive, which allowed nature to carry moral overtones. As Rousseau famously remarked, "The male is male only at certain moments; the female is female her whole life . . . everything constantly recalls her sex to her, and to fulfill its functions, an appropriate physical constitution is necessary to her."[47]

Such arguments recall the attempts of sixteenth- and seventeenth-century writers to ground gender – the social order of masculinity and femininity – in the corporeal realm of sex difference, even though the appeals were no longer to the complexional categories of hot, cold, wet, and dry. But the ties between sex and gender difference posited in the eighteenth century were, if anything, tighter than before, as appeals to the laws of nature replaced earlier appeals to a variety of sources of legitimation – the Bible, Galen, Aristotle, and the Church Fathers – which could be used to counter as well as reinforce one another. Furthermore, Rousseau's words reveal a very different attitude toward the sex mutability that members of Montaigne's generation found invariably fascinating and often plausible. In the Enlightenment, sex difference was more likely to be described in terms of strong and immutable polarities, and writers were increasingly interested in policing the boundaries between the sexes rather than speculating on sex definition as a crossing zone.

---

[46] The ontological instability in the concept of nature itself is discussed in Peter Dear, *Discipline and Experience: The Mathematical Way in the Scientific Revolution* (Chicago: University of Chicago Press, 1995), pp. 18–21, 151–8, 225–6. The problematic character of the concept of nature, and of naturalization projects generally, is discussed in Daston, "The Naturalised Female Intellect"; and Daston and Vidal, eds., *The Moral Authority of Nature*.

[47] Jean-Jacques Rousseau, *Emile, ou de l'éducation* (Paris: Garnier Frères, 1964), bk. V, p. 450.

Thus, discussions of gender difference turned increasingly on descriptions of reproductive functions and organs as opposed to physical complexion, as in the case described by Montaigne; these were assumed, as Rousseau indicated, to be ineluctable, permanent, and a cause of absolute difference between men and women. In the sixteenth and seventeenth centuries, in contrast, gender difference was sometimes seen as mutable and labile. In the Enlightenment, however, it was again presented in a polarized way on the basis of medical discourse and the invocation of nature – but a far more intensely moralized nature than in earlier times.[48]

This had significant implications for understandings of masculinity, as well as for what it meant to be a woman. I have argued in this chapter that definitions of gender were difficult to ground in natural facts during the sixteenth and seventeenth centuries. While the superior place of men in social and legal structures remained largely unchallenged throughout this period, like his claims to superior intelligence, strength, and rationality, definitions of masculinity nonetheless changed. Far less than femininity was masculinity legitimated by nature or by reproductive function alone.[49]

CONCLUSION

The stories of gender and nature that I have told in this chapter are complex. It is difficult to establish clear chronologies for their unfolding and often puzzling to trace the areas of interaction in what are essentially separate histories. Historiographically, the once-clear position of the Scientific Revolution as the turning point for feminist and nonfeminist analyses of early modern science alike has been weakened and modified. It is no longer seen as being exclusively dominated by the mathematized sciences of matter,

---

[48] Karin Hausen, "Die Polarisierung der Geschlechtscharakter," in *Sozialgeschichte der Familie in der Neuzeit Europas*, ed. Werner Conze (Stuttgart: Klett, 1976). See also Cohen, "Body as a Historical Category." Laqueur, in *Making Sex*, puts the process of polarization chronologically later, arguing that "by around 1800, writers of all sorts were determined to base what they insisted were fundamental differences between the male and female sexes, and thus between man and woman, on discoverable biological distinctions, and to express them in a radically different rhetoric" (p. 5). More recently, Michael Stolberg has argued that it is already visible in the years around 1600; see his "A Woman Down to Her Bones: The Anatomy of Sexual Difference in the Sixteenth and Early Seventeenth Centuries," *Isis*, 94 (2003), 274–99, together with replies by Laqueur, "Sex in the Flesh," 300–306, and Schiebinger, "Skelettestreit," 307–13.

[49] The difficulty of finding a secure demonstration of masculinity is discussed in Maus, "A Womb of His Own"; David Kuchta, "The Semiotics of Masculinity in Renaissance England," in Turner, ed., *Sexuality and Gender*, pp. 233–46; and Coppelia Kahn, *Man's Estate: Masculine Identity in Shakespeare* (Berkeley: University of California Press, 1981). For redefinitions of masculinity that were also redefinitions of desire, see Randolph Trumbach, "Erotic Fantasy and Male Libertinism in Enlightenment England," in *The Invention of Pornography*, ed. Lynn Hunt (New York: Zone Books, 1996), pp. 253–82. For definitions of masculinity affecting forms of intellectual association, see Mario Biagioli, "Knowledge, Freedom, and Brotherly Love: Homosociality and the Accademia dei Lincei," *Configurations*, 3 (1995), 139–66.

or by cosmology; nor is its character as a "revolution," a rapid overthrow of preexisting orthodoxies, so secure. At the same time, the older historiographical objective of the recuperation of women's experience in science has been supplemented by a broader historiography of sex and gender difference, that has strong links with the major themes of scientific inquiry. No clear chronological sequence has replaced that once provided by the "Scientific Revolution" or by Foucault's much criticized account of the *âge classique*. Some attempts have been made to link changes in ideas concerning sex and gender difference to the onset of industrialization in the early modern period. But such explanations are unsatisfactory in the light of the uneven progress toward industrialization of different areas of Europe.[50] Similarly, it is difficult to agree with Foucault's thesis that a sudden rapid change in ways of regarding nature and the natural world occurred in Europe around 1650. By the late eighteenth century, the period of flux and ambiguity in gender definition that had marked the sixteenth and seventeenth centuries had been replaced by descriptions of gender difference as ineluctably linked to the reproductive aspects of the physiology of the sexes and of biological character as fixed for life. In the eighteenth century, it also began to be taken for granted that natural facts, or nature itself, could act as the principal or sole legitimization for definitions of gender difference.

We are left with some interesting problems. How and why did these shifts take place? Until we place the history of science and the history of gender in the context of the history of the deep structures of change, we have no means of answering this question. A fruitful approach might be to regard ideas concerning gender as a laboratory space in which to explore the strengths and weaknesses of the claims of natural knowledge. Defining and accounting for gender difference encapsulated difficulties in the establishment of accepted natural facts and the attachment of stable social meanings to them. These problems did not lessen as natural knowledge developed.

We also still have little understanding of the real force of gender ideologies in determining either the participation in scientific inquiry or attitudes toward nature. Neither masculinity nor femininity were unified constructions in the early modern period, and early modern ideas of gender were often very different from current ones. This realization points to a central puzzle posed by current historiography. Gender ideologies were powerful and often constructed in ways that denigrated the intellectual powers of women. But is the minor role played by women in early modern science to be ascribed to these ideologies alone? Chapters in this volume by Schiebinger (Chapter 7) and Alix Cooper (Chapter 9) show the "patchy" nature of women's exclusion.

---

[50] Laqueur makes the industrialization argument in *Making Sex*, pp. 152–4, but he stretches the process between the sixteenth and nineteenth centuries so that clear turning points become difficult to establish. Other authors differ widely on chronology; Daston, "The Naturalised Female Intellect," for example, sees the nineteenth century as the period when anatomy became destiny.

This itself is also a symptom of the instability of early modern notions of gender, and their consequent variability in determining experience and practice. Gender ideologies are important, but they are not hegemonic. They are liable to patchy impacts that can be explained only by reference to other factors, such as class, religion, the nature of public space, and, not least, the internal agendas and problems of natural inquiry itself.

33

EUROPEAN EXPANSION AND SELF-DEFINITION

Klaus A. Vogel

Translated by *Alisha Rankin*

From the beginning of European expansion, natural knowledge was both its precondition and its constantly developing product. Over the course of the fifteenth century, the Portuguese voyages of exploration in the western and southern Atlantic and along the coast of Africa promoted the development of new knowledge regarding marine navigation and orientation, as well as the ability to rule the seas. They led to new experiences as new seas were sailed and new coasts explored, as the equator was crossed, and as the stars of the southern hemisphere were described. Geography emerged as an independent discipline concerned with the systematic description of the inhabited earth. Encounters with previously unknown lands, peoples, animals, plants, and minerals expanded the frontiers of the ancient and medieval knowledge of the world and changed theoretical understandings of nature. As Peter Martyr of Anghiera (1457–1526), chronicler for King Ferdinand of Aragon, put it, "our pregnant ocean here bears new children every hour."[1]

Peter Martyr's early accounts of the "New World" testify to the richness and scope of the European quest for natural knowledge in the first decades of the sixteenth century. In 1493, a few months after the return of Columbus, the Italian scholar began his tenure as royal chronicler by relating reports of the western discoveries. His accumulated works, consisting of letters to his friend Cardinal Ascanio Sforza and other – mostly Roman – personalities, were included in all important European travel collections from 1507 onward. In 1516, he combined the thirty books (in three "decades") of his writings into a single edition, published in Alcalà and dedicated to Charles, the young king of Spain who later became Emperor Charles V.[2]

---

[1] Peter Martyr d'Anghiera, *De orbe novo Petri Martyris ab Angleria Mediolanensis Protonotarii Caesaris senatoris decades* (Compluti: Michael d'Eguia, 1530), fol. 114v.

[2] The first decades were published in Venice (1504), Vicenza (1507), Milan (1508), Seville (1511), Alcalà (1516), and Basel (1521), and the full book was reprinted in Paris (1536), Basel (1537), Antwerp (1537), and Paris (1587); see John Alden, ed., *European Americana: A Chronological Guide to Works Printed in Europe Relating to the Americas*, 6 vols. (New York: Readex Books, 1980), 1: 1493–600.

Natural knowledge plays a central role in Peter Martyr's accounts. His first decade, completed in 1511, described the initial Spanish voyages of exploration up to and including the fourth voyage of Christopher Columbus (1451–1506), who personally recounted his experiences to Peter, as did all subsequent navigators. In this volume, Peter discussed the situation of the islands in the western ocean and described in great detail Hispaniola, Cuba, and "Parias," the coast of the southern mainland of the Caribbean Sea.[3] In his second decade, completed in 1514, he stressed the degree to which the new discoveries superseded ancient knowledge, noting, "Neither with words nor with pen can I express my thoughts concerning these advancements."[4]

According to Peter, the Europeans proved to be in every way superior to the natives of the newfound lands. He named two continuing goals of the voyages of discovery: "to convert the simple natives to our faith" and to further "the investigation of nature in those lands."[5] Thus, in his eyes, the exploration of nature had the same exalted status as religious matters. In the fifth decade, completed in 1523, he praised his friend Pope Hadrian, who had "always labored wisely to study not only the secrets of Mother Nature, but also divine science."[6]

Peter described useful plants, animals, and rivers of the New World and devoted an entire chapter to the "meaningful scholarly question" of how to explain the abundance of water flowing in the numerous rivers on the east coast.[7] He also occasionally pointed to questions for further research, confident that future explorations would provide satisfactory answers. For example, regarding a discussion with two captains about currents in the western ocean, he wrote, "We have listened to what they said and have put their varying viewpoints in writing. We will take up a particular opinion only when we have sufficient reason. For the time being, we must rely on suppositions, until the day comes when nature reveals this secret to us."[8]

The narratives of Peter Martyr show that even the early stages of European overseas explorations – between 1492 and 1526 – were characterized by spontaneous curiosity, practical observation, and learned reflection. The later stages multiplied such initiatives, which had a significant influence on the development of natural knowledge and philosophy in early modern Europe (see Vogel, Chapter 20, this volume). This chapter, however, does not focus on the history of the European travel narrative or the early modern ethnology,[9] nor on changes in European knowledge of the natural world that

---

[3] Peter Martyr, *Acht Dekaden über die Neue Welt*, ed. and trans. Hans Klingelhöfer, 2 vols. (Darmstadt: Wissenschaftliche Buchgesellschaft, 1972–3), 1: 1–130.
[4] Peter Martyr, *De orbe novo*, 1.10, fol. 22r.
[5] Ibid., 2.7, fol. 31v.
[6] Ibid., 5.10, fol. 85r.
[7] Peter Martyr, *Acht Dekaden*, 2.9, ed. and trans. Klingelhofer, 1: 200–8.
[8] Peter Martyr, *De orbe novo*, 3.10, fol. 56r.
[9] See Mary B. Campbell, *The Witness and the Other World: Exotic European Travel Writing, 400–1600* (Ithaca, N.Y.: Cornell University Press, 1988); Stephen Greenblatt, *Marvelous Possessions: The Wonder*

resulted from the voyages of exploration, but on the role of that knowledge in the contact between Europeans and the peoples across the ocean. What was its status and significance in encounters with natives in America and East Asia? What meaning did it have for the self-image of Europeans?

I will explore these questions by focusing on two exemplary yet dissimilar regions: Spanish America and the Far East. Using the example of Mexico, I will trace the ways in which Christian missionaries, especially Franciscans, established natural science as a part of the colonial order beginning in the second quarter of the sixteenth century, and I will describe the ways in which the language, objects, and philosophical system of European natural science were taught and adopted there. In this process, I will argue, native elites were presented with an opportunity for integration and (to a certain extent) for development, which they could not refuse. Although the European colonization of Spanish America was generally complete by the middle of the sixteenth century, it was not until the second half of the sixteenth century that large numbers of Europeans reached East Asia, where, as guests and trading partners in China and Japan, Europeans were subordinated to the native regimes. As I will show, some Europeans – notably Jesuit missionaries – used their knowledge of nature, especially geography, mathematics, and astronomy, to win the attention and recognition of native elites. As a result, the communication of natural knowledge for centuries played a key role in shaping relations between Europe and the Far East – above all, China.

I will conclude by asking how Europeans themselves assessed their natural knowledge in comparison with that of the peoples of the Far West and Far East. How did these comparisons influence Europeans' sense of identity? Natural knowledge had always had an important place in Europeans' self-perception. Peter Martyr and others recognized early on that, as a result of the voyages of exploration, early modern Europeans could put together a more complete picture of nature than had ever been possible for the scholars of antiquity. How, then, did European self-perception change over the course of the sixteenth and seventeenth centuries, first in confrontation with the inhabitants of Spanish America and later in contact with scholars in East Asia? In this connection, two central texts from the last decade of the sixteenth century demonstrate how, after a century of overseas expansion, Europeans used not only Christianity but also the undisputed primacy of European natural knowledge to legitimate their global regency in Africa, Asia, and America.

---

of the New World (Oxford: Clarendon Press, 1991); Anthony Pagden, *European Encounters with the New World: From Renaissance to Romanticism* (New Haven, Conn.: Yale University Press, 1993); Mary B. Campbell, *Wonder and Science: Imagining Science in Early Modern Europe* (Ithaca, N.Y.: Cornell University Press, 1999); and the articles in *Facing Each Other: The World's Perception of Europe and Europe's Perception of the World*, ed. Anthony Pagden, vols. 1–2 (Aldershot: Ashgate, 2000), esp. Joan-Pau Rubiés, "New Worlds and Renaissance Ethnology," 1: 81–121.

NATURAL KNOWLEDGE AND COLONIAL SCIENCE: COLLEGES OF HIGHER EDUCATION AND THE REAL Y PONTIFICIA UNIVERSIDAD DE MÉXICO (1553)

Since antiquity, Europeans had maintained a relatively stable relationship with the wider world. Contacts were mostly indirect, and longer engagements, such as that of Marco Polo at the court of the Mongolian leader Kublai Khan in the second half of the thirteenth century, were few and far between. This changed over the course of the fifteenth century as the Portuguese began to advance strategically around the coast of Africa toward Asia, circumnavigating southern Africa and reaching India in 1498. At the same time, the Spanish and the Portuguese discovered islands and land masses in the western and southern oceans. Even before Europeans came into contact with the large empires of Asia, they had seized sizable parts of Central and South America and set up a functioning colonial system in the erstwhile dominions of the Aztecs and Incas.

Mexico was the first place on the American mainland where European missionaries built schools, colleges, and printing houses and founded a university.[10] The Franciscan monk Pedro de Gante (1490–1572), a native of Ghent in Flanders, arrived in Mexico with a group of missionaries just months after the conquest of the Aztecs. There he opened the first elementary school on the American continent – a school for native children in the region of Texcoco – in 1523. Two years later, in 1525, he founded the Colegio de San José de los Naturales (later Colegio de San Francisco) in Mexico City.[11] Although Pedro de Gante assumed the necessity of educating the Aztecs in Christian belief, he handled the natives with care and respect. At his schools, indigenous children received an education in the European tradition. They were taught fundamentals of the *artes* (the liberal arts with perhaps some basic philosophy), as well as singing, music, and Latin, in order to prepare them for the duties of choral singers and ministrants in church. The Colegio de San Francisco ran workshops for stonemasons, blacksmiths, shoemakers, tailors, and textile workers, and it trained youths in handicrafts that would be necessary to construct and outfit churches and houses. Additional elementary schools for Native American boys and girls were set up in the diocese of Mexico on

---

[10] The University of Santo Domingo, founded by the Dominicans with a 1538 papal privilege on the Caribbean island of Hispaniola and modeled after Alcalá and Salamanca, decreased in importance with the decline of the island in the middle of the sixteenth century. For the foundational privilege, "In Apostolatus culmine" (1538), see *America Pontificia, 1493–1592*, ed. Josef Metzler (Vatican City: Libraria Editrice Vaticana, 1991), pp. 385–8, no. 91. In general, see John Tate Lanning, *Academic Culture in the Spanish Colonies* (Port Washington, N.Y.: Kennikat Press, 1971); Eli de Gortari, *La Ciencia en la historia de México*, 2nd ed. (Mexico City: Editorial Grijalbo, 1980); and Elías Trabulse, *Historia de la ciencia en México: Estudios y textos*, 2 vols. (Mexico City: Fondo de Cultura Economica, 1983), a commented edition of selected sources.

[11] Gortari, *La ciencia en la historia de México*, p. 178.

the initiative of the first bishop, the Franciscan Juan de Zumárraga (1468–1548).[12]

In 1533, the colonizers turned to the creation of an advanced school for the indigenous population, and the Colegio de Santa Cruz de Tlatelolco, founded by the Franciscans, opened its doors in 1536.[13] At first, the Colegio focused on elementary education and religious instruction for the sons of Indian chiefs (*kaziken*) who lived in the vicinity of the capital; the Franciscans instructed them in the elements of theology, chant, and manuscript illumination, and taught them to write their own language using the Latin alphabet. Further education included reading and writing, in both Spanish and Latin; natural philosophy; logic; arithmetic; and music.[14] Because the Franciscan instructors came from Europe, we can assume that lessons in the subjects relating to nature did not differ radically from comparable instruction in European institutions and would have been coordinated with the dissemination of Christian teachings. In his contemporary *Sermonario en Lengua Mexicana* (Sermon in the Mexican Language, 1606), the Franciscan Juan Bautista (1555–ca. 1613) mentioned several of the Colegio's students by name, including a few who later became teachers, and he praised them for their excellent knowledge of Latin.[15] In addition, it is worth noting that Tlatelolco became a center for the research and documentation of indigenous culture, especially the local language and medicine. For this we can thank Bernardino de Sahagún (ca. 1500–1590), a Franciscan who came to Mexico in 1529 and taught at the Colegio for over fifty years.[16]

In the 1530s, at the same time the Colegio was founded in Tlatelolco, the first printing house was set up in Mexico.[17] Before this, printed books had been put together in Spain and transported to America by ship. In June 1539, King Charles I (Emperor Charles V) gave the Sevillian printer Juan Cromberger a mandate to found a local branch of his operation in Mexico. The first work published in the newly founded Mexican print office appeared that same year, a twelve-folio bilingual piece with the title *Breve y mas compendiosa doctrina cristiana en la lengua mexicana y castellana* (Short and Most Complete Christian Doctrine in the Mexican and Castillian Languages). Most of Cromberger's Mexican publications were Christian instructional works in the local language(s), often printed in bi-, tri-, or quadrilingual

[12] Ibid.
[13] Ibid., p. 179.
[14] Fernando Ocaranza, *El Imperial Colegio de Indios de la Santa Cruz de Santiago Tlatelolco* (Mexico City, 1934); and Gortari, *La ciencia en la historia de México*, pp. 171, 179.
[15] Ocaranza, *El Imperial Colegio de Indios de la Santa Cruz de Santiago Tlatelolco*, pp. 27–8.
[16] On Sahagún, see the modern edition of his *Historia general de las cosas de Nueva España*, ed. Ángel María Garibay K., 7th ed. (Mexico City: Editorial Porrúa, 1989), with further literature; and Gortari, *La ciencia en la historia de México*, pp. 169 ff.: "El interés por los conocimientos indígenas."
[17] José Toribio Medina, *Historia de la imprenta en los antiguos dominios españoles de América y Oceanía*, vol. 1 (Santiago de Chile: Fondo Histórico y Bibliográfico, 1958), bk. 1: "El estudio de la primitiva tipografía Mexicana."

editions; he also printed language texts and dictionaries. After the foundation of the university, natural philosophical and medical texts became more numerous. The New World's first publication with a mathematical bent was Juan Díez's *Sumario compendioso* (1556); later works of this sort included Alonso de Veracruz's *Physica speculatio* (1557) and Juan de Cárdenas's *Primera parte de los problemas, y secretos marauillosos de las Indias* (First Part of the Problems and Marvelous Secrets of the Indies, 1591). The first medical work, *Opera medicinalia* (Medical Works, 1570), is considered to be the teachings of the Spanish physician Nicolàs Monardes (ca. 1493–1588), who became known as the first propagator of Nahoan medicine in Europe.[18]

Like the *colegios*, the University of Mexico was first and foremost a humanist operation[19] – its primary purpose being the education of clerics and royal officeholders. Nevertheless, European science played an important role in its curriculum. The first formal proposal for the foundation of a university in Mexico was composed by Bishop Juan de Zumárraga in 1536 but was rejected by the Council of the Indies (Consejo de las Indias), which pointed to the Colegio de Tlatelolco; Viceroy Antonio de Mendoza called the application "premature." It was not until the Mexico City municipal council (Cabildo municipal) filed a new proposal in 1539 calling for the setup of a "general university for the sons of Spaniards and for the natives . . . where they can read the *artes* and theology" that the project found support. The proposal once again used the bishop's name and argued that otherwise the resident Spaniards would have to send their sons on a dangerous voyage through Veracruz and across the open sea to Spain for their studies, that these students would forget the local languages while in Spain, and that many good (indigenous) grammar students and novices would be lost if one could not educate them.[20] At first, only a single professorship of theology was granted. After further applications, the establishment of a Real Universidad de México in the tradition of the University of Salamanca was mandated by royal privilege on 21 September 1551, and given funding. On 25 January 1553, the celebratory opening took place. Four decades later, in 1595, the university was recognized as a *pontificia* by papal privilege.[21]

---

[18] Gortari, *La ciencia en la historia de México*, pp. 187 ff.
[19] On the history of the University of México, see Alberto María Carreño's detailed survey *La Real y Pontificia Universidad de México, 1536–1865* (Mexico: City Universidad Nacional Autónoma de México, 1961). For more basic sources, see John Tate Lanning, ed., *Reales Cedulas de la Real y Pontificia Universidad de México de 1551 a 1816* (Mexico City: Imprenta Universitaria, 1946); Cristobal Bernardo de la Plaza y Jaen, *Cronica de la Real y Pontificia Universidad de México*, ed. Nicholas Rangel (Mexico City: Talleres Gráficos del Museo Nacional de Arqueología, Historia y Etnografía, 1931); and Alberto María Carreño, *Efemérides de la Real y Pontificia Universidad de México segun sus Libros de Claustros* (Mexico City: Universidad Nacional Autónoma de México, 1963), vol. 1.
[20] Sergio Mendez Arceo, "La cedula de ereccion e la Universidad de Mexico," in *Historia Mexicana*, 1, no. 2 (1951), 268–94 at pp. 271–2; and Carreño, *La Real y Pontificia Universidad de México*, pp. 13–19.
[21] *Reales Cedulas de la Real y Pontificia Universidad*, p. xv; and Gortari, *La ciencia en la historia de México*, p. 185.

Eight faculties were created at the university's opening: theology (*Prima de Teología*), Holy Scripture (*Sagrada Escritura*), canon law (*Prima de Cánones*), Roman law (*Prima de Leyes*), administration (*Decreto*), philosophy (*Artes*), and Latin grammar (*Gramática*).[22] The chair of philosophy is especially interesting here: with logic, mathematics, physics, astronomy, and natural philosophy, it encompassed the fundamentals of natural knowledge and included medicine as well.[23] The curricular materials would have largely corresponded with instruction at similar universities in Europe, although there have been no detailed studies on this subject. By the same token, we do not yet know the degree to which the university supported the empirical investigation of nature in the first decades of its existence (see Findlen, Chapter 19, this volume).

Yet we can recognize from the evolution of the faculty chairs at the University of Mexico the increasing significance of medicine and, beginning in the second quarter of the seventeenth century, mathematical subjects and natural philosophy. At first, the chair in *artes* represented one of the least respected elementary subjects, which can be seen in the fact that whereas the chair of theology was conferred for life and those of the remaining subjects for four years, the chair in *artes* was only given for three years.[24] Nevertheless, these basic subjects were popular with students: In the early days of the university, several members of a family would sometimes be listed as *artes* students concurrently.[25] Over the next decades, the university continued to expand; in 1569, two chairs of law were added (*Instituta, Código*), and in 1578, the first chair of medicine (*Prima de Medicina*) was created. Additional chairs of law were set up in 1580. Three chairs of medicine were added in 1599 and 1621, among them anatomy and surgery (*Anatomía y Cirurgía*). In 1626, a chair specifically for Mexican (Mexicano), the language of the Nahoa, was introduced. The university experienced its largest expansion in the following years. In 1637, it added a chair of astrology and mathematics (*Astrología y Matemáticas*), which was needed for the education of physicians. In 1646, it set up four additional chairs in the subjects of natural philosophy (*Prima de Filosofía*), the indigenous language Otomí, the philosophy of Thomas Aquinas, and introductory law (*Vísperas de Leyes*).[26]

By the middle of the seventeenth century, the European study of nature formed its own branch of education at the University of Mexico, encompassing the liberal arts and the medical curriculum as well as natural philosophy and astrology/mathematics. Several important scholars, including Fray Diego Rodríguez (1596–1668) and the Jesuit-educated Mexican Carlos de Sigüenza

---

[22] Gortari, *La ciencia en la historia de México*, pp. 186; and Carreño, *La Real y Pontificia Universidad de México, 1536–1865*, pp. 33 ff.
[23] Carreño, *La Real y Pontificia Universidad de México, 1536–1865*, pp. 34–5.
[24] Ibid., p. 47.
[25] Ibid., p. 49.
[26] Gortari, *La ciencia en la historia de México*, p. 186; and Carreño, *La Real y Pontificia Universidad de México*, pp. 233 ff.

y Góngora (1645–1700), held chairs in the latter. Before taking over the chair in 1672, Sigüenza y Góngora had already made a name for himself as an astronomer, mathematician, geographer, physicist, engineer, artilleryman, historian, poet, and physician.[27] He performed geographical and astronomical observations, participated in scholarly disputes, composed numerous scientific treatises, and had contact with the Europeans Athanasius Kircher, Juan Caramuel y Lobkowitz, Gian Domenico Cassini, Joseph Zaragoza, and John Flamsteed. One of Sigüenza y Góngora's most important and most famous works was his *Libra astronomica y philosophica* (Book of Astronomy and Philosophy, 1690). The volume documented a heated dispute with the Bavarian Jesuit Eusebio Francisco Kino, in which Sigüenza y Góngora emphatically defended his "fatherland Mexico," his mathematical chair, and his university.[28] A nun also belonged to the circle of the New Spanish who concerned themselves with a modern study of nature: Juana Inés de la Cruz (ca. 1648–1695) carried out acoustical studies, criticized Aristotle, and used a drum covered with flour to demonstrate the nature of oscillations; she also wrote poems.[29]

The subject of the Spaniards' reception of indigenous knowledge also offers a wide open field for research. I have already mentioned that indigenous medicine was especially nurtured at the Colegio de la Santa Cruz de Tlatelolco, founded in 1536. Indigenous knowledge of medicinal plants, stimulants, and narcotics was welcomed in European medicine. The chief supporter of the Colegio's research in this area was Bernardino de Sahagún, who gathered a group of native healers from Tepepulco, Tlatelolco, Tenochtitlán, and Xochimilco. The Mexicans gave him information about Nahoan medicine and taught about indigenous medicaments at the Colegio. However, Sahagún's work, *Historia general de las cosas de Nueva España* (General History of the Things of New Spain), did not appear in print and only later made it to Europe.[30]

In addition to the activities at the Colegio, Native Americans who possessed a European education began to describe indigenous nature independently of European instruction. Two native instructors at the Colegio de

---

[27] Carlos de Sigüenza y Góngora, *Obras históricas*, ed. Francisco Pérez Salazar (Mexico City, 1928); Gortari, *La ciencia en la historia de México*, pp. 225–30; Carreño, *La Real y Pontificia Universidad de México*, pp. 320 ff.; María de la Paz Ramos Lara and Juan José Saldaña, "Newton en México en el siglo XVIII," in *The Spread of the Scientific Revolution in the European Periphery, Latin America and East Asia*, ed. Celina A. Lértora Mendoza, Efthymios Nicolaïdis, and Jan Vandersmissen (Turnhout: Brepols, 2000), pp. 91–8, at pp. 91–2.; see also J. M. Espinosa, *La comunidad científica Novohispana ilustrada en la Real y Pontificia Universidad de México*, Tesis de Maestría en Filosofía de la Ciencia, México (Mexico City: Universidad Autónoma Metropolitana Iztapalapa, 1997).

[28] Gortari, *La ciencia en la historia de México*, pp. 226–7.

[29] Juan José de Eguiara y Egurén, *Sor Juana Inès de la Cruz*, ed. Ermilio Abreu Gómez (Mexico City: Antigua Librería Ribrerdo, 1936); and E. Piña, "Comentarios de la historia de la física en México," *Boletín de la Sociedad Mexicana de Física*, 6 (1992), 28, citing Ramos Lara and Saldaña, "Newton en México," p. 92.

[30] See note 16.

Tlatelolco – Martín de la Cruz, who taught medicine, and Juan Badiano, a reader in Latin language – composed a pharmacological-botanical treatise, the *Herbario De la Cruz-Badiano*. Their work included an extensive herbal, with 184 illustrations of native plants, and described the pharmacological treatment of various illnesses. Both a Latin edition of the 1552 manuscript and a contemporary Italian translation exist; it is the only known complete text of this kind written solely by Native Americans, and it documents an important area of indigenous natural knowledge.[31]

Also noteworthy were the famous Nahoan gardens in Azcapotzalco, Texcoco, and in tropical Huaxtepec.[32] These served as a field of investigation for the Spanish research traveler Francisco Hernández (1517–1587), who had been named General Protomédico of the Indies, Islands, and Mainland of the Ocean Sea (*Protomédico general de las Indias, islas, y tierra firma del mar océano*) by the Spanish Council for the Indies. In September 1570, Hernández reached Mexico; he returned seven years later with the work *De historia plantarum Novae Hispaniae* (On the History of the Plants of New Spain) in 16 handwritten volumes, which contained numerous illustrations by the indigenous artists Antón and Baltazar Elías and Pedro Vázquez.[33]

By 1545, Monardes' volume on Nahoan medicine had been published in Spain as *Dos libros el uno que trata de todas las Cosas que traen de Nuestras Indias Occidentales, que sirven al uso de la Medicina*... (Two Books, the First Covering All of the Things Brought from Our West Indies that Are Useful for Medicine). The work was very successful and was reprinted in 1565 and 1569. In 1571, a new, second volume followed, and soon thereafter a second edition of both volumes, an Italian translation, and a Latin translation entitled *De simplicibus medicamentis ex occidentali India delatis* (On Simple Medicines Brought from the West Indies), published in several editions by the eminent University of Leiden botanist Carolus Clusius. An English translation, entitled *Joyfull Newes out of the Newe Founde Worlde*..., was printed by John Frampton in 1577, with further editions in the following years. More than anyone else, Monardes spread the important essentials of Nahoan plant medicine in the European book market.[34]

Significantly more extensive was the *Relaciones geográphicas* project mandated by the Spanish Council for the Indies, which called for a systematic description of colonial culture, the native inhabitants, and the natural attributes of Spanish America[35] (see Harris, Chapter 16, this volume). After a

---

[31] Gortari, *La ciencia en la historia de México*, pp. 171–2, 190–2; and Trabulse, *Historia de la ciencia en México*, p. 43.
[32] Gortari, *La ciencia en la historia de México*, p. 193.
[33] Ibid., pp. 193–4.
[34] Ibid., pp. 172–3.
[35] Howard F. Cline, "The Relaciones Geográficas of the Spanish Indies, 1577–1648," *Handbook of Middle American Indians*, 12 (1972), 183–242. On the maps, see Donald Robertson, "The Pinturas (Maps) of the Relaciones Geográficas, with a Catalog," *Handbook of Middle American Indians*, 12

trial period in 1569, an extensive questionnaire was printed in 1577 and sent through the viceroys of Mexico and Peru to every royal governor. The catalogue contained fifty questions covering almost every aspect of colonial life. It asked for names of plants, both indigenous and imported, especially medicinal plants, as well as mineral sources, tides, soundings, islands, and docking possibilities. A drawing or map was requested for the latter items. Not all governors responded, and reports are not available for every region. Nevertheless, the 208 existing responses and numerous maps offer a highly interesting snapshot – still only partially researched – of the Spanish colonies in the years between 1578 and 1586, and they document the Spaniards' methodical investigation of the new territories.

It is striking how naturally and systematically Europeans exported their religion, culture, and scholarship to Spanish America. From the first decades of the sixteenth century, the founding of educational facilities in America – elementary schools, *colegios*, and universities – was undertaken by monks, nuns, and clerics. The Europeans were completely overpowering on this front. They determined not only the intellectual content of the curriculum but also the media of dissemination: writing, painting, and printing. Natural knowledge may not have played a pivotal role within this curriculum, but it formed an important, well-developed subfield. The foundation of the University of Mexico in 1553 reaffirmed this pattern, as medicine (1578), astrology/astronomy (1637), and natural philosophy (1646) became independent fields of learning. In contrast, indigenous natural knowledge, such as Nahoan plant medicine, was adopted only selectively and separated from its previous contexts. Here, too, the Europeans determined the process of research and reception. In the longer run, however, the investigation of nature united the New World and the Old. As the Spanish colonial system became firmly entrenched in the middle of the seventeenth century, indigenous scholars such as Carlos Sigüenza y Góngora, who was educated in Mexico, participated in learned European discussions as a representative of the Mexican nation.

NATURAL KNOWLEDGE AND THE CHRISTIAN MISSION: THE JESUITS IN JAPAN AND CHINA

In the sixteenth century, three hundred years after Marco Polo's journey to China and Southeast Asia, Europeans reached the Asian coasts for the first time by sea. Defeating the Arab fleets in the Indian Ocean, the Portuguese won domination of the seas and took a large number of strategically and economically important trading cities on the coasts of East Asia, East Arabia,

(1972), 243–78; and Barbara E. Mundy, *The Mapping of New Spain: Indigenous Cartography and the Maps of the Relaciones Geográficas* (Chicago: University of Chicago Press, 1996).

India, Ceylon, and Malaysia. In 1510, on their second attempt, they conquered Indian Goa, which became a missionary center in Asia, and, in 1511, the important trading city of Malakka on the southern coast of Malaysia. Conquering the large empires of East Asia was unthinkable; instead, the Portuguese limited themselves to the development of trade and, where possible, the conversion of the natives to Christian beliefs.

Beginning in the late 1540s, the mission in Asia was driven by the Jesuits. Whereas China initially closed its doors to the outside world – it was not until 1557 that the Portuguese were allowed to set up a trading post in Macao – Japan proved to be more approachable. Francis Xavier (1505–1552) became the first Jesuit to travel as a missionary from Goa to Southeast Asia, reaching Japan in the years 1549–51.[36] He found the Japanese remarkably open, especially with regard to issues of natural knowledge. In a 1552 letter to the founder of the Jesuit Order, Ignatius of Loyola, Xavier wrote that missionaries to Japan should possess an outstanding scientific education because the Japanese were exceptionally interested in the explanation of astronomical, geographical, and meteorological phenomena:

> Two things are especially necessary for the fathers who are sent to Japan in order to educate the inhabitants there in the Christian faith. . . . [T]he first is that they pay no attention to the unjustness of the people . . . and act as an example of every sort of virtue. . . . The second is that they be outstandingly educated. In this way, they will be able to answer easily the frequent questions of the Japanese. It would be very useful if they understood the principles of the dialecticians. In addition, it would help if they have learned astrology and natural sciences (*scientia naturalis*). For [the Japanese] pester us with questions on the movements of the heavenly spheres, the eclipse of the sun, the waning and waxing of the moon, and the origins of water, snow, rain, hail, thunder, lightning, and comets. Our explanations of these things have great influence, so that we win the souls of the people.[37]

Xavier explained the Japanese inferiority to the Europeans in natural knowledge by their limited attention to questions of causality, which he attributed to their ignorance or rejection of the idea of divine creation. Japanese scholarship, he wrote in a letter, contained no teachings on the movement and origin of things because they knew nothing of a creator God:

---

[36] Masao Watanabe, *Science and Cultural Exchange in Modern History: Japan and the West* (Tokyo: Hokusen-Sha, 1997).

[37] "Nonnulla excerpta ex epistola Reverendi P. Magistri Xavieri praesbyteri Societatis Iesu in India Praepositi provincialis, ad Reverendum P. nostrum Magistrum Ignatium de Loiola, Praepositum eiusdem Societatis generalem. Anno 1553," in *Epistolae Indicae de praeclaris, et stupendis rebus, quas divina bonitas in India, et variis insulis per societatem nominis Iesu operari dignata est* . . . (Louvain, 1566), pp. 152–3.

[The Japanese] teach nothing at all of the movements of the world, the sun, the moon, the stars, the heavens, the earth, the sea, and the rest. For they do not believe that all of this has its origin somewhere else. They were most surprised when they heard of the existence of a universal mover of souls, or a Father, by whom they were created.[38]

Thus instruction in European natural knowledge was not simply a strategic tool employed by the European missionaries to win attention. It was self-evident to these missionaries that natural knowledge and theology were interconnected; the deficits of East Asian natural science not only pointed to a weakness in their religious beliefs but also provided a point of entry to prove the superiority of Christian doctrine.

The situation in the Far East was thus significantly different from that in the West. Not until several decades after the violent conquest of Spanish America and the establishment of a colonial culture there did contact between Europeans and East Asians begin to develop. Once it did, Europeans in East Asia focused not on conquest but on trade partnerships and missionary activities. Additionally, the mission into the East was not led by Franciscans – who, following their traditional orientation, concerned themselves especially with salvation and the advancement of medicine – but rather by the militantly missionary Jesuits, who were educated in the emerging mathematical sciences.

Despite strong resistance, the mission in Japan was successful for more than half a century. Under Jesuit leadership, the Christian Church in Japan experienced a golden age. Around 1580, it embraced approximately 150,000 members in 200 congregations and was cared for by sixty-five Jesuit fathers and numerous helpers. The Jesuits founded roughly two hundred elementary schools, which taught basic Christianity, reading and writing in Japanese, composition, arithmetic, music (both vocal and instrumental), and painting. These schools were also open to girls.[39] In 1580, the Jesuits opened a *collegium* to educate priests, instructing Japanese students in theology and European scholarship. The curriculum consisted of theology and philosophy, Japanese language and literature, Latin, Portuguese, history, mathematics, music, art, and copper etching.[40] Under Alessandro Valignano (1539–1606), who came to Japan in 1579 as a *visitator* assigned by the Jesuit general, the propagation of European science in Japan reached its pinnacle.[41] In 1590, a hundred Japanese students studied at the Jesuit *Collegium*.

Mathematics did not play a central role in sixteenth-century Jesuit education in Japan. This changed in the first years of the seventeenth century, after

---

[38] "Xaverii Epistolarum Liber iiii," in *Horatii Tursellini e Societate Iesu de vita Francisci Xaverii qui primus e Societate IESV in Indiam et Iaponiam Evangelium invexit* . . . (Rome: A. Zanetti, 1596), p. 123.
[39] Watanabe, *Science and Cultural Exchange*, p. 188.
[40] Ibid., p. 189.
[41] J. F. Moran, *The Japanese and the Jesuits: Alessandro Valignano in Sixteenth-Century Japan* (London: Routledge, 1993).

the Jesuit father Carlo Spinola observed an eclipse of the moon in 1602 in Nagasaki and, with a longitudinal calculation, laid the foundation for scientific cartography in Japan. In May 1605, missionaries opened an "academy" in the capital, where, among other subjects, they taught geography, navigation, planetary theory, and natural philosophy.[42] It is unclear how this academy was connected to the *Collegium*. In any case, it was too late to develop this groundwork further; following enemy attack and numerous resettlements, the Jesuit *Collegium* had to be given up in 1614. The Japanese mission ended in 1639 with the complete expulsion of all foreigners – with the exception of the Chinese and the Dutch – by the Tokugawa regime.[43] Throughout the rest of the seventeenth century, European natural knowledge reached Japan mostly by way of China, although a limited amount of "Dutch science" (*rangaku*) was allowed.[44]

In China, the Jesuit mission began later than in Japan but lasted longer.[45] At the request of Alessandro Valignano, coordinator of the Jesuit mission in East Asia, Matteo Ricci (1552–1610) came from Goa to Macao in order to learn Chinese. After several failed attempts to enter China, he received permission to spend time in Chao-ch'ing, west of Canton, in 1583. He was later named the leader of the Chinese mission because of his knowledge of the language and his finesse in interactions with the Chinese. Following the experience in Japan, Ricci and his companions appeared publicly as religious individuals in the garb of Buddhist monks. When they later recognized that this caused them to be viewed as outsiders, they abandoned public sermons and took on the clothing and manner of Confucian scholars, who were traditionally well regarded in China.[46]

Although the inhabitants of Chao-ch'ing, fearing conquest by the Portuguese, at first rejected and persecuted the foreigners, Ricci and his companions won the attention of Chinese scholars by their humanity, their commitment to learning the Chinese language, their knowledge of nature, and the scientific objects they had brought with them.[47] In the spring of 1584, Ricci presented the governor of Chao-ch'ing with an enlarged copy of a map of the world with Latin inscriptions that the governor had noticed in the Jesuits' living space. Ricci's map had Chinese inscriptions, and he shifted China to the center of the world. He also gave the governor a sundial. The

---

[42] Henri Bernard, *Le Père Matthieu Ricci et la société chinoise de son temps (1552–1610)*, 2 vols. (Tientsin: Hautes Études, 1937), p. 193.
[43] Watanabe, *Science and Cultural Exchange*, p. 189.
[44] Yabuti Kiyosi, "The Pre-History of Modern Science in Japan: The Importation of Western Science during the Tokugawa Period," in *Scientific Aspects of European Expansion*, ed. William K. Storey (Aldershot: Ashgate, 1996), pp. 258–67.
[45] Bernard, *Le Père Matthieu Ricci et la société chinoise*.
[46] Matteo Ricci, *I Commentari della Cina dall'autografo inedito*, (Macerata: F. Giorgetti, 1911), pp. 103 ff.; and Bonnie B. C. Oh, "Introduction," in *East Meets West: The Jesuits in China, 1582–1773*, ed. Charles E. Ronan, S.J., and Bonnie B. C. Oh (Chicago: Loyola University Press, 1988), p. xx.
[47] Ricci, *I Commentari della Cina*, pp. 138 ff.

governor showed himself to be deeply impressed with both; he had reproductions made of the world map, which he presented to his friends and sent to other provinces.[48]

In his *Commentari della Cina* (Commentaries on the Introduction of the Jesuits and Christianity in China, compiled and printed in 1911), Ricci later remarked that the completion of this world map, with its special remarks and comments on China, was the "best and most useful" action that one could have taken at that time in order to gain the Chinese trust in "the things of holy belief."[49] Traditional Chinese printed maps showed only the fifteen provinces of China, surrounded by a small bit of sea. As Ricci noted, these traditional maps reinforced the Chinese sense of cultural superiority: "With this impression of the immensity of their empire and the diminutiveness of the rest of the world, they were so proud that the whole world seemed to them barbaric and uncultivated in comparison."[50] Less educated Chinese, Ricci reported, laughed at the Jesuits' new map of the world, on which the world seemed so large and China so small in comparison. The more learned Chinese, however, who "saw the beautiful order of the parallels and meridians, with the line of the equator, the tropics, and the five [climatic] zones ... and the whole Earth full of various names," were convinced by the illustration. In addition, Ricci's map of the world had a further important use: It showed the Chinese how far from China Europe really was and that an enormous sea lay in between. Thus "they lost the fear they had at the beginning that our people had come to conquer their empire, which was one of the greatest obstacles to the fathers in converting this nation."[51]

In subsequent years, Ricci, who had studied in Rome with the famous geographer, astronomer, and natural philosopher Christopher Clavius (1537–1612), constructed – in addition to the world map – numerous spheres, terrestrial and celestial globes, and sundials, and gave them to leading Mandarins as gifts.[52] He published several books in the Chinese language, including a catechism, a work on Christian doctrine, a couple of works of moral philosophy, and an art of memory, which found great resonance among the Chinese literati. Among his works on nature were a translation of the first six books of Euclid's *Elements*, based on the Latin version of his mentor Clavius (published in 1607), a partial version of Clavius's *Epitome arithmeticae* (published in 1614), and compendia of practical geometry, geography, and astronomy.[53]

Beginning in 1599, Ricci taught cosmography, mathematics, and Aristotelian physics in Nanking, where a Jesuit residency had been erected. He and his companions won great recognition with their disquisitions on

---

[48] Ibid., p. 141.
[49] Ibid., p. 142.
[50] Ibid.
[51] Ibid.
[52] Ibid., p. 144.
[53] Ibid., pp. 454 ff.

the spherical shape of the Earth, the existence of the antipodes, the order and movement of the stars and planets based on Ptolemy, and the use of globes.[54] "For these learned [Chinese], this was the most meaningful, subtle, and newest thing that they had ever heard," he wrote. "Through it many avowed and avow today that their eyes were opened to very significant things, to which they all previously had been as blind."[55] Ricci taught the technical details only to selected students. In any case, his astronomical knowledge was limited; he acknowledged that he was not in a position to determine the path and the placement of planets, to calculate the dates of eclipses, or to construct ephemerides. Despite this, he remarked, his reputation among the Chinese spread as "the best mathematician in the entire world, due to the small amount that [the Chinese] knew of these things."[56]

In 1598 and 1601, Ricci was called to the emperor's court in Peking.[57] The emperor did not receive him personally but showed great interest in the maps, scientific instruments, and reports Ricci had brought from Europe. In the following years, the center of Jesuit activity shifted to the imperial court, and the first Jesuit residency in Peking was opened with a mass in 1605.[58] This residency became a contact point for numerous high-ranking Chinese, where they attentively studied the Europeans' paintings, maps, books, and scientific objects. When Ricci died in 1610, the emperor transferred ownership of a small palace in the suburbs of Peking to the Jesuits, in the garden of which a burial ground was constructed with a monument to the great scholar. A few years earlier, the Chinese scholar Kuo Tsing-lien, viceroy of Guizhou, wrote of Ricci, "After he stayed so many years in the Middle Empire, he is now no longer a stranger, but rather a Chinese man who belongs to China."[59]

Before his death, Ricci had sent letters to the Jesuit general in Rome asking him to send new scholarly books and astronomers to China. His successors, Sabbatino de Ursis (1575–1620), Johannes Terrentius (1576–1630), Adam Schall von Bell (1592–1666), and Ferdinand Verbiest (1623–1688), were all educated as astronomers. De Ursis exactly predicted the solar eclipse of 15 December 1610, and thereby proved the superiority of European astronomy over the predictions of the Chinese. Terrentius, who exchanged letters with Galileo Galilei (1564–1642), worked with others on a monumental astronomical and calendrical compendium, the *Ch'ung-cheng li-shu* (Ch'ung-cheng Regency Period Treatise on Astronomical and Calendar Science, 1634), which served as the basis for a new calendar. Adam Schall von Bell introduced the telescope to China with the work *Yüan-ching shuo* (On the Telescope, 1630). In 1644, the year the Manchu dynasty took over the empire, Schall

[54] Ibid., pp. 311 ff.
[55] Ibid., p. 455.
[56] Ibid., p. 144.
[57] Ibid., pp. 285 ff., 363 ff.
[58] Ibid., p. 498.
[59] Ibid., p. 623.

von Bell was appointed director of the Chinese Astronomical Bureau. At the same time, the highly traditional "Muslim" section of the observatory, which had existed since the year 1059 and for all intents and purposes had become a center of Chinese science, was dissolved.[60]

This repression of the so-called "Muslim" astronomers ignited a conflict that lasted years and earned Schall von Bell bitter rivals. When he altered the title of the respected compendium *Ch'ung-cheng li-shu* to *Hsi-yang hsin-fa li-shu* (Treatise on Astronomy and Calendrical Science According to the New Western Method) and gave the work to the Manchu emperor Shun-chih, he was heavily criticized as attacking the dignity of Chinese tradition.[61] Later, the leader of the "Muslim" astronomers, Yang Kuang-hsien (1597–1669), who rejected the European image of a spherical Earth in his tract *Pu-te-i* (I Cannot Do Otherwise, 1664) and attempted to prove that the Earth was flat, accused Schall von Bell of high treason, the spreading of a despicable religion, and false astronomical teachings.[62] On his initiative, Schall von Bell, his student Ferdinand Verbiest, and other European missionaries were arrested in the fall of 1664, condemned, and only narrowly escaped execution. A few Jesuits were banned; Schall von Bell died two years later.[63]

After a long trial and open conflict with Yang Kuang-hsien, Verbiest managed to rehabilitate European astronomy and exonerate the Jesuits.[64] In 1669, he took over the position of Director of the Chinese Astronomical Bureau, which the Jesuits held until the second half of the eighteenth century. He became an advisor and tutor to the emperor, and under his watch, European astronomy won over the leading Chinese scholars. In his *Astronomia Europaea* (European Astronomy, 1687), Verbiest described to his European audience the success of the mathematical sciences at the imperial court:

> After Astronomy, marching like a venerable queen between the Mathematical Sciences and rising above all of them, had made her entry among the Chinese and had ever since been received by the Emperor with such an amiable face, all the Mathematical Sciences also gradually entered the imperial Court as her most beautiful companions. They followed Astronomy, adorning themselves with all the extraordinary or beautiful things they carry with them as if they were gold or precious stones, to find more favor in the eyes of such a great majesty. This happened with Geometry, Geodesy, Gnomonics, Perspective, Statics, Hydraulics, Music, and all the Mechanical Sciences, each of them dressed in such precious and skillfully woven clothes that it conveyed the

[60] Compare with "Lebensbild Schalls in der amtlichen Geschichte Chinas unter der Mandschu-Dynastie," in Alfons Fäth, S. J., *Johann Adam Schall von Bell S.J., Missionar in China, Kaiserlicher Astronom und Ratgeber am Hofe von Peking, 1592–1666* (Nettetal: Steyler, 1991), pp. 372–6.
[61] Pingyi Chu, "Trust, Instruments, and Cross-Cultural Scientific Exchanges: Chinese Debate over the Shape of the Earth, 1600–1800," *Science in Context*, 12 (1999), 385–411.
[62] Pingyi Chu, "Trust, Instruments," pp. 397 ff.
[63] Fäth, *Johann Adam Schall von Bell*, pp. 295 ff.
[64] Noël Golvers and Ulrich Libbrecht, *Astronoom van de Keizer: Ferdinand Verbiest en zijn Europese Sterrenkunde* (Leuven: Davidsfonds, 1988); and Pingyi Chu, "Trust, Instruments," p. 398.

impression that they were competing with each other in beauty. However, the aim of their fervent desire to please was not to keep the Emperor's eyes only upon themselves, but to direct them fully towards the Christian Religion, whose beauty they all professed to worship, in the same way as smaller stars worship the sun and the moon.[65]

The public success of European mathematics, geography, and astronomy might have been recognized by Chinese scholars, but it was not attributed solely to the Europeans. While the preface by Xu Guangqi (1565–1633) to Matteo Ricci's 1607 edition of Euclid's *Elements* did laud the innovative work, it simultaneously pointed to analogous queries and methods from Chinese antiquity.[66] In European mathematical, geographical, and astronomical knowledge, the Chinese saw a continuation of and complement to Chinese tradition. As a prime example, Mei Wen-ting (1633–1721), the most important Chinese astronomer of his time, articulated the "doctrine of the Chinese origins of the Western sciences," thereby influencing the subsequent reception of European natural knowledge.[67]

Conversely, not only did European scholars exhibit a growing general interest in China's moral teachings, culture, and politics over the course of the seventeenth century, but Europeans also began to study Chinese natural history and Chinese natural science.[68] On the impetus of Gian Domenico Cassini (1625–1712), director of the astronomical observatory in Paris, and with the personal support of the French King Louis XIV, six French Jesuits, four of them holding the title of Royal Mathematician, were sent to Siam and Peking. There they were to complete astronomical, geographical, and natural observations and study Chinese scholarly works.[69] They reached Peking in 1688, only a few days after the death of Ferdinand Verbiest. Two of them, Joachim Bouvet (1656–1730) and Jean-François Gerbillion (1651–1707), took over his function as imperial adviser. Under the leadership of Bouvet, a further group of French Jesuits arrived in Peking in 1698.

Outside of France as well, scholars initiated contact with China. The German philosopher and mathematician Gottfried Wilhelm Leibniz (1646–1716) attempted to establish a systematic exchange of scholars between China and Europe and wrote to Claudio Filippo Grimaldi, the Jesuit who had taken over leadership of the Astronomical Bureau from Verbiest.[70] Leibniz had long

---

[65] Noël Golvers, *The "Astronomia Europaea" of Ferdinand Verbiest, S.J. (Dillingen, 1687)* (Nettetal: Steyler, 1993), p. 101.
[66] Catherine Jami, "'European Science in China' or 'Western Learning?' Representations of Cross-Cultural Transmission, 1600–1800," *Science in Context*, 12 (1999), 413–34, at pp. 422–3.
[67] Pingyi Chu, "Trust, Instruments," pp. 400–1.
[68] John D. Witek, "Understanding the Chinese: A Comparison of Matteo Ricci and the French Jesuit Mathematicians Sent by Louis XV," in Ronan and Oh, eds., *East Meets West*, pp. 72–3.
[69] Witek, "Understanding the Chinese," p. 73.
[70] Donald F. Lach, "Leibniz and China," *Journal of the History of Ideas*, 6 (1945), 436–55.

shown an interest in China and held a sophisticated view of the relationship between European and Chinese scholarship. Although he saw both as equals in the experimental exploration of nature, he affirmed the superiority of Europeans in the theoretical disciplines:

> In the skills that are required of everyday life, and in the experimental battle with nature, we are – if one makes a uniform comparison – born equal to one another. Each possesses abilities that could be usefully exchanged with the respective other; in the thoroughness of our notional deliberation and in the theoretical disciplines, however, we are superior. For in addition to logic and metaphysics, as well as the cognizance of immaterial things – sciences we rightfully claim as our own – we also without doubt distinguish ourselves in the theoretical ascertainment of forms abstracted through the understanding of materiality, that is to say in mathematics, as one could in fact determine when Chinese astronomy entered into an open competition with ours. To this point they seem not to have known that great illumination of human intelligence, the art of reasoning, and to have satisfied themselves with a sort of mathematics gained by experience, similar to the skills our artisans possess. In the art and science of war as well, they find themselves behind us.[71]

Europeans' self-confidence in this domain remained unbroken at the end of the seventeenth century.[72] In their own eyes, they had won the attention and admiration of the Chinese, first with their language proficiency, their technical virtuosity, and their scientific instruments, and later with their knowledge of geography and astronomy. Since 1644, European Jesuits had been active as the Chinese emperor's official astronomers; in 1669 they won back that post after a hefty competitive dispute. With the arrival of the Royal Mathematicians, the orientation of the Jesuits changed: State-supported research now took its place beside missionary activities. However, the Jesuits' hopes for a broad Christian mission were never fulfilled. Although many of the high Chinese deputies accepted European natural knowledge, and a few converted to Christianity, the Christian religion was generally rejected. The connection between cosmology and Christian doctrine failed to convince the Chinese, who maintained a high reverence for their own traditions. With the doctrine of the Chinese origins of Western science, Chinese scholars qualified the superiority of European natural knowledge. Thus, although European Jesuits were able to remain active as leading astronomers at the imperial court until the second half of the eighteenth century, their influence on Chinese culture remained limited and controlled.

---

[71] Gottfried Wilhelm Leibniz, *Das Neueste von China (1697) Novissima Sinica*, ed. Günther Nesselrath and Hermann Reinbothe (Cologne: Deutsche China-Gesellschaft, 1979), p. 9.
[72] Compare with Carlo M. Cipolla, *Clocks and Culture, 1300–1700* (New York: W.W. Norton, 1977).

NATURAL KNOWLEDGE IN EUROPEAN SELF-DEFINITION AND HEGEMONY

In revisiting the role of natural knowledge in the process of European expansion and its significance for European self-definition, we must first emphasize the multifaceted nature of these issues. The perspective of a learned scholar in Europe differed quite substantially from that of a Franciscan at the University of Mexico or of a Jesuit at the Chinese imperial court. Looking at Spanish America, we see the export of the entire European educational system by the first half of the sixteenth century. In that part of the world, European natural knowledge was an integral part of the religiously inflected colonial system that dominated all areas of life, and it served that system's foundation and development. In contrast, contact with China and Japan developed later, in the second half of the sixteenth century. There, European natural knowledge was introduced in small increments and remained largely separated from its cultural origins. The missionaries used natural knowledge to win over native elites and to explain the distinctiveness of Christian belief. In China, a few missionaries educated as geographers and astronomers took over official posts and stayed at the imperial court for extended periods of time. In sum, European natural knowledge in America was omnipresent and unchallenged in the colonial environment and appeared barely distinguishable from European scholarship in general, whereas it had a more specific focus in East Asia and came into open competition with the knowledge of indigenous scholars.

Over the course of the sixteenth century, Europeans who remained at home or who returned from abroad developed a more detailed conception of the superiority of European natural knowledge over that of the peoples overseas. A prime example of this is the *Historia natural y moral de las Indias* (The Natural and Moral History of the Indies, 1590) by the Jesuit José de Acosta (1539–1600).[73] His work gives perhaps the most reflective depiction of the culture and natural abilities of the Native Americans, comparing them to the cultures of Japan and China. Acosta, who held a leading position in the Jesuit mission, had returned to Europe after a sixteen-year stay in Peru and Mexico to become rector of the University of Salamanca, envoy to the Jesuit general Claudio Acquaviva, and advisor to the Spanish King Phillip II. His work, printed in Seville in 1590 and dedicated to the Infanta Isabella of Austria, had more than just a theological orientation. In the introduction to the *Historia*, Acosta emphasized that previous authors had failed to discuss thoroughly the causes and correlations of phenomena in the New World,[74] and he contrasted

---

[73] José de Acosta, *Historia natural y moral de las Indias, en que se tratan las cosas notables del cielo, y elementos, metales, plantas, y animales dellas: y los ritos, y ceremonias, leyes, y gouierno, y guerras de los Indios* (Seville: Iuan de Leon, 1590).

[74] José de Acosta, *The Natural and Moral History of the Indies* (The Hakluyt Society, 60–61), ed. Clements R. Markham, 2 vols. (London: Hakluyt Society, 1880), 1: xxiv.

the natural knowledge of Spanish America's native inhabitants with that of both the Europeans and the Chinese. (The Jesuits had maintained contact with the latter since the 1580s, and Acosta had personally met a few Chinese in Mexico.)

In two chapters, Acosta first compared the written tradition of the American high cultures with the writing of the Chinese. The use of pictures and symbols by the Aztecs and Incas did not constitute writing, he argued. Nor were the signs of the Chinese equal to European writing because they, unlike the Japanese, could not write any unknown sound combinations.[75] Acosta then discussed the status of knowledge in China.[76] His verdict was clear: Jesuits who were in China had seen no "great schooles or universities of Philosophie, and other naturall sciences." Although some basic schooling existed, students learned only to write the Chinese characters and to read and write stories, legal texts, fables, and moral principles:

> Of divine sciences [theology] they have no knowledge, neither of naturall things, but some small remainders of straied propositions, without art or methode, according to everie mans witte and studie. As for Mathematikes, they have experience of the celestiall motions, and of the stars. And for Phisicke, they have knowledge of herbs, by means wherof they cure many diseases, and use it much.[77]

Overall, the Chinese – who were also great thespians – indicate "much wit and industrie" in their learning. "But all this is of small substance, for in effect all the knowledge of the Chinois tendes onely to read and write, and no farther, for they attaine to no high knowledge." According to Acosta, an Indian in Peru or Mexico who had learned reading and writing from the Spanish knew more than the wisest Mandarin, despite the latter's erudition, for the American could use letters to write all the words and sentences in the world, whereas the Mandarin, with his hundreds of thousands of characters, had difficulty writing even simple names such as Martin or Alonso, especially names of things unfamiliar to him.[78] To illustrate this, Acosta had set up the following experiment: He had known a few Chinese in Mexico and had asked them to write the sentence, "Joseph of Acosta has come from Peru," in their language. As he recalls, one of them had thought about it long and hard and had finally written the sentence, but the rest of the Chinese who read it diverged completely in their pronunciation of the written name.[79]

The methods and conclusions of José de Acosta reflect the prejudices of his time. More interesting is his decision to adopt a global perspective, comparing the abilities and natural knowledge of the Chinese with those

---

[75] Ibid., 2: 397–401.
[76] Ibid., pp. 401–2.
[77] Ibid., p. 401.
[78] Ibid., p. 402.
[79] Ibid., p. 400.

of the inhabitants of Spanish America. European scholarship, which had already been taught in Spanish America for more than two generations, served as the standard for this comparison. Acosta's conclusions were plain: The Europeans were superior to the Chinese not only in theology but also in natural knowledge. Every Indian from Peru or Mexico who learned from the Europeans shared this advantage.

Barely a century after the Europeans had "discovered" America and sailed around Africa and reached Asia, the concrete experience of cultural encounters took the place of uncertain accounts. Educated Europeans – mostly clerics, some of whom had spent a long time overseas – emphasized Europe's advantage in its knowledge of the natural world. Exemplary are the writings of Giovanni Botero (1540–1617), who in 1596 published a description of the entire world, *Le relationi universali* (Universal Accounts), which went through several editions and was translated into other European languages. Europe, he wrote, was not only incomparable in its inventions, printing, artillery, in the use of compasses, and in the art of navigation; it was also a leader in the sciences that had originated in Egypt, Judea, and Greece. Religion also was purer in Europe and the new mission areas. Thus Europe was destined to conquer the ocean and to rule Africa, Asia, and America:

> But what shall we say of the most noble art of printing and of the inestimable invention of firearms, which are Europe's own? . . . Neither Africa nor Asia has anything worthy of comparison with the use of the magnet, which was discovered on the Amalfi coast; and with the excellence of the European peoples in navigation. With the benefit of [these arts], the Spanish, led by an Italian, have discovered a New World: and the Portuguese have sailed around all of Africa, and have discovered passages, and countless lands, which were never known to the Ancients. . . . Finally, the sciences, which were born in Egypt and Judea, from whence they passed to Greece, have now come to rest among us: and the true religion, the belief in Christ, our Lord, is neither pure nor authentic outside of Europe: except in the lands where the European peoples have newly brought it. . . . As if one might say, Europe was destined for this purpose by nature to spread its own riches and to receive the riches of others and to conquer the sea across which it expands its dominion, in order to reign over Africa, Asia and America, which it approaches as if to extend its hand.[80]

The comparative view of the world that developed during the European expansion gave the Europeans confidence both in their global power and in the uniqueness of their natural knowledge. At the end of the sixteenth century, European scholars could recognize that they were superior not only to the ancients but also to the nations of the Far West and the highly regarded

---

[80] Giovanni Botero, *Le relationi universali di Giovanni Botero Benese, divise in quattro parte* (Venice: Nicolò Polo, 1597), pp. 1–3.

cultures of the Far East. For the Europeans, the search for knowledge had become an open process that led from what the English philosopher Francis Bacon (1561–1626) called the *globus materialis* to the *globus intellectualis*. Bacon spelled this out in his *Instauratio magna* (Great Instauration, 1620), where he stressed the significance of the invention of printing, gunpowder, and the compass for the development of literature, warfare, and navigation[81] and emphasized the connection between the opening of the world through overseas travel and exploration and the development of natural knowledge:

> Nor should we ignore the fact that distant voyages and overland travels which have become frequent in our day, have opened up and revealed to us many things in nature which can throw new light on philosophy. And surely it would be disgraceful, in a time when the regions of the material globe, that is, of the earth, the seas and stars, have been opened far and wide for us to see, if the limits of our intellectual world were restricted to the narrow discoveries of the ancients.[82]

The self-assurance with which the Europeans began this process had already been expressed a century earlier by one author, who, as a learned European, knew the important works of antiquity as well as he knew the achievements of his own day. In his *Utopia* of 1516, the first work of European literature in the era of Amerigo Vespucci to reflect the overseas discoveries, the English humanist and politician Thomas More (1478–1535) wrote of Europe's advantages in respect to the "newe lande" discovered by navigators of his time:

> Surely (quod maister Peter) it shal be harde for you to make me beleve, that there is better order in that newe lande, then is here in these countryes, that wee knowe. For good wittes be as wel here as there: and I thinke oure commen wealthes be auncienter than theires; wherin long use and experience hath found out many thinges commodious for mannes lyfe, besides that manye thinges heare amonge us have bene found by chaunce, whiche no wytte coulde ever have devysed.[83]

---

[81] Francis Bacon, *Neues Organon* [1620], ed. Wolfgang Krohn, 2 vols. (Hamburg: Felix Meiner, 1990), 1: 129.
[82] Francis Bacon, *Novum Organum* [1620], 1.84, ed. and trans. Peter Urbach and John Gibson (Chicago: Open Court, 1994), p. 93.
[83] [Thomas More], *More's Utopia* [1516], trans. Raphe Robynson, 2nd ed. [1556] (Cambridge: Cambridge University Press, 1891), p. 64.

INDEX

Abano, Pietro d', 141, 544–5
Abel, Niels Henrik, 709n23
Abū Ma'shar (Albumasar), 578
Academiae Naturae Curiosorum (Schweinfurt), 167, 269, 301, 394
Académie Royale d'Arles, 196
Académie Royale des Sciences (Paris): and acoustics, 609; and Cartesianism, 399; and experimental philosophy, 127, 130–1, 172; and laboratories, 301; and man of science, 185; and mathematics, 699; and mechanical philosophy, 46; and museum, 288–9; and natural history, 467; and natural philosophy, 394, 403–5; and patronage, 269, 270; and popular culture, 221–2; publications of, 170, 171, 325; and selection of members, 405n178; and travel, 356n62; and women, 195–6, 197–8
Académie Royale des Sciences et Belles-Lettres, 198
academies, scientific: and courts, 267–71; and experimental investigations, 130–1; and family model, 231; and laboratories, 301–2; and man of science, 185, 186; and mathematics, 723; and mechanics, 659–64; and museums, 288–9; and natural philosophy, 394; and printing houses, 325–6; rise of in early modern period and issues of proof and persuasion, 169–74; and social conventions, 403–5; and women, 195–8. *See also* Académie Royale des Sciences; Royal Society of London
Accademia del Cimento (Florence), 127, 170, 172–3, 268, 301, 303, 325, 393–4, 609
Accademia dei Lincei (Rome), 268, 269, 393, 452, 467
Accademia dei Secreti (Naples), 170, 301, 325, 393
Accademia Secretorum Naturae (Naples), 269

Accademia Segreta (Naples), 301
accidental causes, and scientific explanation, 92–3
Acontius, Jacopo, 139
Acosta, Cristobal, 453
Acosta, José de, 351n40, 450, 836–8
acoustics: development of in seventeenth century, 604–11; and harmonics, 630; and music theory, 597–8; optics compared to, 597; Pythagorean and Aristoxenian traditions in, 598–604; and study of sound in sixteenth and seventeenth centuries, 596
Acta eruditorum (journal), 167, 200, 331, 394, 671
active principles, and scientific explanation, 93–105
Adelard of Bath, 710
ad vivum, and naturalism in art, 794–5, 796
agriculture, and natural knowledge, 218
Agricola, Georgius, 187, 213, 515
Agrippa von Nettesheim, Henricus Cornelius, 379, 500, 501–2, 514, 519–26, 734, 803–4
Aguilon, François d', 122
Ailly, Pierre d', 478
Alberti, Leon Battista, 295n14, 315, 499
Albert of Saxony, 83, 371n28
Albertus Magnus, 72, 82, 83, 295n13, 543–4, 546, 548n27
Albrecht V, Duke of Bavaria, 264
alchemy: and astrology, 555n61; and classification of knowledge, 4; and courts, 261–2; and experimentation, 516; and gunpowder, 308; and laboratories, 294–5, 299; and medicine, 421, 423, 498; modern distinction between chemistry and, 497–8; and Newton, 94n60; and Paracelsus, 502–10; status of in early sixteenth century, 499–502. *See also* chymistry

841

Aldrovandi, Ulisse: and botanical gardens, 116, 281; and correspondence, 349; and magic, 526; and medicine, 459–60; and museums, 246, 287, 448; and natural history, 437, 444n29, 451, 455, 458, 462, 463; and patronage, 266; and study of fish, 193, 210
Alembert, Jean Le Rond d', 23n7, 672
Aleotti, Giambattista, 315
Alexander of Aphrodisias, 368
Alfonsine Tables, 343, 565
Alfonso V, King of Portugal, 472
Alfonso the Wise, King of Castile, 343
algebra, 705–6, 708–10, 714–18
Alhazen. *See* al-Haytham
alkahest, and alchemy, 513
Almazém de Guiné e India (Lisbon), 471, 483
Alpers, Svetlana, 794
Alpetragius. *See* al-Bitrūjī
Alsted, Johann Heinrich, 405–6
Amico, Giovanni Battista, 567
analysis and synthesis: and mathematics, 703–6, 722; Newton's method of, 94n60, 99–100, 101, 102–3
anatomy: and development of medicine, 415, 430; and executions of criminals, 216; and experience, 111–15; and illustration, 782–6; and optics, 615–18; and theaters, 273–80, 285–6, 415. *See also* ear; eye
Andrade, Antonio de, 359
Andreae, Johann Valentin, 301
Anguillara, Luigi, 445
Anna of Denmark, Queen of England, 254
Annae litterae (Annual Letters), 351, 353
anthropology, and travel literature, 766–70
antipodes, and cosmography, 470, 474, 477, 479
antivivisectionism, 797
Antwerp, and cosmography, 486–7
aphorisms, as genre of early printed book, 167
Apian, Peter (Petrus Apianus), 470, 485–6, 635, 682, 693
Apian, Philip, 486
Apollonius of Perga, 705, 707, 712–13
apothecaries: and global trade, 210–11; and gunpowder, 312; and practice of medicine, 419
Aquinas, Thomas, 8, 27n17, 72, 75n14, 93, 371, 381–2, 523, 732
Archimedes, 634–5, 639, 644, 651, 705–6, 712–14
architecture: of homes and households, 226; and mechanical arts, 677–8; and military science, 315
Aretino, Pietro, 762
Argoli, Andrea, 553
aristocracy: and French salons, 198; and man of science, 189. *See also* class

Aristotelianism: and anti-Aristotelianism in physics, 29–43; and concept of body, 51; critics of and changes in scientific explanation, 71, 72–7, 82–6; definition of, 368n15; historiography of, 70n1; and knowledge of nature, 106–7; and Leibniz, 52; and natural philosophy, 370–95; and physical sciences, 21–8; and rhetoric, 136; and scientific demonstration, 142
Aristotle: and anatomy, 113; and astrology, 546; and classification of sciences, 120; and concept of causal explanation, 73–5, 105; and experience, 106, 108–11; on household and social order, 229; on manual crafts, 293; and mechanics, 634, 636, 637; and medicine, 407; and natural history, 438, 439; and natural philosophy, 366; on sex and gender difference, 801–2, 803–4; on space and the body, 27, 52. *See also* Aristotelianism
Aristoxenus and Aristoxenian tradition, in acoustics, 598–604
armor, and military science, 313–17
Arnaud, Pierre, 366n3
Arnauld, Antoine, 163, 398
art: and artist as scientist, 786–91; and interest in depicting nature, 773–4; and mechanical arts, 693–5; and naturalism, 775–9, 785, 791–6; and natural objects, 529; and optics, 763n20. *See also* illustration
artisans: and development of new epistemology in natural philosophy, 296; and distillation of drugs, 211; and elitism of scientific academies, 222n78; and women in sciences, 199–201. *See also* craft tradition; guilds
Ascoli, Cecco d', 528, 529
Aselli, Gasparo, 429
Ashmole, Elias, 246, 287, 560n82
Ashworth, William, 361–2
astrolabe, 684, 687
astrology: and astronomy, 577–81; continued popularity of, 558–61; and courts, 258–9; history of and transition from Renaissance to Enlightenment science, 541–2; and intellectual and institutional structures in 1500, 542–7; loss of intellectual legitimacy during eighteenth century, 552–8; and medicine, 579; and public performances in piazzas, 215–16; and reforms in sixteenth and seventeenth centuries, 547–52; and universities, 541, 542, 545–6, 553
astronomy: and Aristotelian natural philosophy, 386–7; and astrology, 577–81; celestial physics and heavenly bodies, 28; and changes in scientific explanation, 71; and changes in structure of learning, 569–73; and concept of

Scientific Revolution, 13–14, 16–17; and courts, 255; and Descartes, 586–90, 657–8; and education in early sixteenth century, 564–5; evolution of in sixteenth and seventeenth centuries, 594–5; families involved in, 232, 235; and geography, 491; and Jesuits in China, 833–4; Kepler and revolution in, 581–4, 587–90; and literature, 764–6; and mapping, 344; and mathematics, 41, 119, 120, 121–2, 401–2, 707; and navigation, 686–7; and Newton, 592–4, 664–5, 667–8; and optics, 614; and religion, 573–7, 589–90, 645, 727, 740–8; and Renaissance humanism, 565–9; status of in late Middle Ages, 562–3; and universities, 563; and women, 200–1, 235. *See also* comets; Copernicus; Earth; Galileo; heliocentrism; planetary motion; stars
Athenian Society (London), 335–6
atlases. *See* maps and atlases
atomism: and chymistry, 512; and mechanical philosophy, 47, 53; and natural philosophy in Renaissance, 375–6; and scientific explanation, 87–8
Aubrey, John, 220
Augsburg Confession, 736
Augurelli, Giovanni, 501
August of Braunschweig-Lüneburg, 264
Augustine of Hippo, 78, 294, 730–1
August of Saxony, 264
Austen, Ralph, 218
authority: challenges to traditional sources of, 802–3; and influence of religion on astronomy, 740–8; patterns of in family, 230; proof and argument from, 161; and science, religion, and politics in seventeenth century, 748–53
Auzout, Adrien, 356n62
Avempace. *See* Bajjā, Ibn
Averlino, Antonio Francesco (Filarete), 297, 315
Averroes (Ibn Rushd), 75n14, 83, 368, 637, 641
Avicenna (Ibn Sīnā), 72, 83, 408–9, 499n8, 523

Bachmann, Augustus, 117
backstaff, and navigation, 687
Bacon, Francis: and acoustics, 604–5; and alchemy, 516; and aphorism as genre, 167; and art, 776–7, 787, 796; and astrology, 542, 550–2; and cosmography, 495–6; description of ideal research facility, 272–4, 289, 345, 750; education of, 194; on empiricism in workshop, 212–13, 221; and experience, 106, 110–11, 115, 127, 130; and feminist critiques of science, 798–9; on gender and nature, 810–11; and gunpowder, 307, 308; and history of early modern science, 8; influence of Italian naturalists on, 36; and laboratory, 295, 301; and laws of nature, 45n75; on libraries, 244, 249; and literature, 771; and magic, 532; on man of science, 180, 185, 189; and mechanical arts, 676; and natural history, 116, 458, 463; and natural philosophy, 379, 400; and optics, 612; and physics, 64n144; and religion, 730, 749–51; and scale of practice, 345; and scientific explanation, 83, 84–5; and study of law, 461; and theories of proof and persuasion, 145, 148, 149, 151–2, 158; on travel and exploration, 839
Badiano, Juan, 826
Baer, Nicholas Reymers (Ursus), 258
Baglivi, Giorgio, 431, 432
Bajjā, Ibn (Avempace), 637
Baldwin, William, 370
Baliani, Gianbattista, 651, 653
Barbaro, Ermolao, 441, 499
Barozzi, Francesco, 155
Barrow, Isaac, 121n58, 125–6, 129, 155, 183, 623–4, 718, 720
Bartholin, Erasmus, 701–2
Bartholin, Thomas, 429
Basso, Sebastian, 47, 395
Bauhin, Caspar, 277, 458n83, 459, 464
Bauhin, Jean, 455
Bauhin family, 232, 460n89
Bautista, Juan, 822
Bayern, Ernst von, 506
Bayle, Pierre, 144–5, 331
Becher, Johann Joachim, 217, 261, 302, 304
Bede, Venerable, 474
Beeckman, Isaac, 45, 183, 428, 587n65, 596, 608n25, 654
Beguin, Jean, 304, 508–9, 511
Behn, Aphra, 761, 769
Belgium. *See* Antwerp
belief, and rhetorical account of proof and persuasion, 147–8
Bellarmine, Robert, 377, 746
Belleval, Pierre Richer de, 283
Bellini, Lorenzo, 430
Belon, Pierre, 446, 526
Ben-David, Joseph, 180n4
Benedetti, Alessandro, 275, 441
Benedetti, Giovanni Battista, 596, 603, 604, 606, 634, 638–9
Benedict of Nursia, St., 294
Berengario of Carpi, Jacopo, 236
Berkeley, George, 68, 624n60
Bernoulli, Jakob, 163, 650, 670, 671, 672, 721
Bernoulli, Johann, 672, 721
Bernoulli, Niklaus, 670, 672
Berry, Jean, Duc de, 252
Bérulle, Cardinal Pierre de, 395

Bessarion, Cardinal Basilius, 566
Besson, Jacques, 317
Biancani, Giuseppe, 155, 375n46
Bianchini, Francesco, 589
Bible: and natural philosophy, 371n26, 745–6, 750; and printing houses, 326; sex and gender difference in, 801, 803, 804
Biringuccio, Vannoccio, 210
al-Bitrūjī (Alpetragius), 566–7
Blaeu, Joan, Jr., 492
Blaeu, Joan, Sr., 353, 492
Blaeu, Willem, 591
Blake, William, 772
Blancanus, Josephus, 739
Blondel, François, 650
blood, circulation of, 425–6, 429
Blotius, Hugo, 256
Bodenstein, Adam von, 262, 506
Boderie, Guy Lefèvre de la, 770
Bodin, Jean, 137, 166, 210, 387
Bodley, Thomas, 245
body: and alchemy, 504; Aristotle on space and, 27, 52; and concept of nature, 813; Descartes on extension and, 89n49; and Leibniz's physics, 52n103, 62; mechanist corpuscularianism and Aristotelian concept of, 51; observational account of, 769–70; theories of matter and concepts of, 49–50
Boehme, Jakob, 374
Boerhaave, Hermann, 305, 432
Boethius, 599
Bombelli, Rafael, 709–10
books: in Chinese language, 831; dedication of to courtly patrons, 258; and Galileo, 645; and magic, 529–31; and natural history, 439; printing houses and character of, 323; proof and persuasion in, 164–8. *See also* libraries; literature; printing and printing houses; textbooks
Bordelon, Laurent, 771
Borel, Pierre, 226
Borelli, Giovanni Alfonso, 431, 588, 659, 660
Borough, Stephen, 487
Borough, William, 492, 688
Borro, Girolamo, 641
Bošković, Rudjer, 68
Bosschaert, Ambrosius, 793
botanical gardens: and collection of specimens, 116; colonialism and plants in, 201–2; and courts, 256–7; and curiosity cabinets, 283, 285; emergence and development of, 273–4, 280–3; and natural history, 444, 468
botany: global trade and interest in, 207–8; and illustrations, 116–17, 235n39, 773, 780–2; and medicine, 442, 444–5, 446, 460n89; and natural history, 442–3. *See also* botanical gardens; herbals and herbalists
Botero, Giovanni, 838
Bouchereau, Jacques, 370
Bouillau, Ismael, 588, 591
Boujou, Théophraste, 370n21, 390
Bourne, William, 318
Bouver, Joachim, 834
Boyle, Robert: and acoustics, 609–10; and astrology, 552; and concept of nature, 812–14; and corpuscular theory, 512, 537–8; and discovery of chemical color indicators, 210; and essay as genre, 167; and experience, 129–30, 131, 400–1, 402, 466n114; and foundations of physics, 58; and interest in chymistry, 32; and laboratory, 295, 303; and man of science, 182, 189, 194; and mechanical arts, 694; and mechanical philosophy, 43–4, 47, 49, 64; and medicine, 418; and natural history, 466n114; and natural philosophy, 365, 400–1, 402; and proof or persuasion, 138, 152, 157; and religion, 751, 752–3; and Royal Society, 289; and scale of practice, 345; and scientific explanation, 77, 78–81, 91; and theories of colors, 626n69; and transmutation, 511; and workshops in home, 227
Bracciolini, Poggio, 146, 375
Bradley, James, 590, 672
Bradwardine, Thomas, 637
Brahe, Sophie, 236n41
Brahe, Tycho: and alchemy, 571; and astrology, 548–9; and courts, 257; and design of Uraniborg, 229; and history of astronomy, 574–7, 590n73, 687, 702; and Kepler, 42, 402, 578n40, 582, 613; and natural philosophy, 386; and scientific explanation, 71; and scientific household, 236n41
Brasavola, Antonio Musa, 443n24
Braudel, Fernand, 343
Brazil: and cosmography, 478; natural history of, 453
Breidenbach, Bernhard von, 474
Brendel, Zacharias, 510
Breyne, Jakob, 202
Breyne, Johann Philip, 235
Brissot, Pierre, 412–13
Browne, Thomas, 221, 771
Brueghel, Jan, the Elder, 793
Brunelleschi, Filippo, 677, 692, 787
Brunfels, Otto, 116, 442, 457, 780, 794n68
Bruni, Leonardo, 373, 472
Bruno, Giordano, 33, 35, 36, 375, 377–8, 518, 570, 571–2, 734, 757

Buffon, Georges-Louis Leclerc, Comte de, 465
Buonamici, Francesco, 641
Bureau, Jean and Gaspard, 312n20
Bureau d'Adresse (Paris), 325, 329, 394–5
Bürgi, Jost, 261, 691
Buridan, Jean, 474–5, 637
Burton, Robert, 757
Burtt, E. A., 13
Busbeck, Oliver, 256
Busschof, Herman, 205
Butler, Samuel, 140n10
Butterfield, Herbert, 13

Cabeo, Niccoló, 121–2, 388
Cabeza de Vaca, Alvar Nuñez, 768
cabinets of curiosities. *See* curiosity cabinets
Caboto, Sebastien, 484
Cabral, Pedro Alvarez, 478
calculatores, and mechanics, 637–8
calculators, and mechanical arts, 685
calculus, and mathematics, 671, 705, 706, 718–22
Calvin, John, 579
Calvinism: and natural philosophy, 369, 382; and study of nature, 298n24
Calzolari, Francesco, 210, 460n90
Cambridge University, and natural philosophy, 389–90
Camerarius, Joachim, 738
Camerarius family, 232
Camers, Joannes, 494–5
Campanella, Tommaso, 33–5, 240, 267, 301, 377, 518, 570, 571–2, 605n17, 734, 810
Campanus of Novarra, 710
cannons, and military science, 308–9
Cantimori, Delio, 735n25
Cardano, Girolamo: and astronomy, 570–1; and curiosity cabinets, 286; and development of textbooks, 166, 387; and experience, 115; and magic, 518; and mathematics, 6, 699, 709; and mechanics, 638n20; and medicine, 412; and natural philosophy, 33, 35, 376, 390; and probability, 162; and scientific debates, 339
Cárdenas, Juan de, 823
Caroline, Princess of Wales, 669
Cartesianism: explanations for success of, 397–9; and innovations in natural philosophy, 393; in Italy, 46–7n85; medicine and debate on, 429; Newton's rejection of, 666–7, 669. *See also* Descartes, René
Cartier, Jacques, 757, 768
cartography: and cosmography, 473, 491; and mechanical arts, 689–91. *See also* maps and atlases
Casa de la Contratación (Spain), 358, 471, 483–4

Casaubon, Isaac, 246, 522
Cassini, Gian Domenico, 184, 343, 344, 493, 624, 834
Cassini, Jacques, II, 594
Cassini family, 232
Castagne, Gabriel, 263
Castelli, Benedetto, 644, 648, 649, 650
Castiglione, Baldassare, 253
Catena, Pietro, 156
Catherine the Great, Empress of Russia, 198
Catholic Church: and control of universities, 184; and Galileo, 589–90, 645, 727, 746–8; and libraries, 249; Luther and intellectual challenges to, 734–5; and magic, 529–32; monasteries and alchemy, 498; monastic movements and laboratories, 294; and rejection of Copernicanism, 746. *See also* Inquisition; Jesuits; religion
Cauchy, Augustin, 720n43
causation: and changing forms of scientific explanation, 70–105; and mechanical philosophy, 63–6
Cavalieri, Bonaventura, 646
Cavalieri, Emilio de', 265
Cavendish, Margaret, Duchess of Newcastle, 197, 254, 255, 763, 769–70
Cavendish, William, Earl of Newcastle, 400, 752
Cazre, Pierre le, 653
Cecil, William (Lord Burghly), 260
Celaya, Juan de, 373
celestial instruments, and mechanical arts, 679–83
Cellini, Benvenuto, 296
Centlivre, Susanna, 771
centrifugal force, and mechanics, 663
Ceriziers, René de, 370n21
certainty, and proof in mathematics, 162–4
Cervantes (Saavedra), Miguel, 317
Cesalpino, Andrea, 117, 282, 417n30, 425
Cesi, Federico (Marchese di Monticello), 268, 269
Chalcidius, 99n69
Chamber, John, 580n47
Chambers, Ephraim, 552, 558–9
Champaignac, Jean de, 370n21
Champier, Symphorien, 374
Chapman, George, 757
Charles II, King of England, 185, 560n82
Charles V, Holy Roman Emperor, 461
Charles V, King of France, 252
Charleton, Walter, 58
Chartres Cathedral, and Portail Royal, 599, *600f*
Châtelet, Emilie, Marquise du, 198, 669
Chauvin, Étienne, 115–16

Chavez, Alonso de, 484
chemistry: and alchemy, 29n27; 497–8; and medicine, 422–3, 430–1. *See also* alchemy; chymistry
China, and Jesuits, 827–35
Chladni, Ernst, 630
Choppin, René, 217
chorography, 470
Chouet, Robert, 398
Christina, Queen of Sweden, 194, 254
Christine de Pizan, 195n8
Chronicles of the Indies, 351
chronology, and cosmography, 490–1
chronometer, and cosmography, 492
Church of England, and Royal Society, 752
chymistry: definition of, 509–10; and foundations of physics, 29–33; schools of thought in, 513–17; transmutations and matter theory, 510–13; and universities, 498, 508–9, 510. *See also* alchemy; chemistry
ciarlatani, and performance in town squares, 214, 215, 216
Cicero, 522
Cigoli (Ludovico Cardi), 787
circumferentor, and surveying, 690
Clairaut, Alexis, 672
Clarke, Samuel, 56–7, 556–7, 669
class: and coffeehouses, 333; and concept of man of science, 184–5, 188–91; and cultural performances in piazza, 214; and early modern discourse on femininity, 806; and military science, 317, 318, 319; and participation in natural inquiry, 192–3; and physicians, 416–17; and Royal Society of London, 197. *See also* aristocracy; artisans; elites
classification: and natural history, 467–8; of science in Arabic tradition, 294
Claturbank, Waltho Van, 214
Clauberg, Johann, 61, 399
Clavius, Christoph: and astrology, 553; and astronomy, 577, 585n57; and cosmography, 485; and Euclid's *Elements*, 771; and Galileo, 641n30; and "Jesuit science," 739, 831; and Jesuit universities, 183; libraries and works of, 352; and natural philosophy, 388; on scientific status of mathematics, 120–1, 125, 155
Clement VIII, Pope, 377
clocks: and mechanical arts, 662, 664, 679–83; metaphor of nature as, 748–9
Clusius, Carolus (Charles de l'Écluse), 204–5, 283, 445, 446–7, 453–4, 456, 458n83, 459, 463, 826
coffeehouses, discussion and demonstration of sciences in, 320–2, 332–40

Coigner, Michael, 679
Colbert, Jean-Baptiste, 222, 399, 404, 493
Colden, Jane, 235
Colegio de San Francisco (Mexico City), 821
Colegio de Santa Cruz de Tlatelolco, 822, 825
Collegio Romano, 488, 577, 641
Collegium Experimentale, 394
Collenuccio, Pandolfo, 441, 499
Collins, John, 699
Colombo, Realdo, 112
colonialism: and cosmography, 482–3; and European expansion and self-definition, 818–39; and natural history, 466–7; and travel literature, 768–9; and women in sciences, 201–5. *See also* New World; postcolonialism; Spain
Colonna, Fabio, 461–2
colors, and theory of light, 626–9
Columbus, Christopher, 351, 448–9, 477–8, 819
Comenius, John Amos, 240, 301, 751
comets, and astronomy, 71, 386, 575–6, 591, 592
Commandino, Federico, 633, 635, 639, 699, 711, 712n27, 738
commerce. *See* trade
communication: and fundamental changes in early modern period, 347–55; geography and expansion of scientific, 361. *See also* language; letters; printing and printing houses
Commynes, Philippe de, 309
compass, and mechanical arts, 686, 687–8
confessionalization, and religious identities, 735–6
conic sections, Apollonian theory of, 716–17
consonances, theory of, 605–7
continuity, Galileo's views about motion and, 652–3
Cook, Harold, 193, 355
Copernicanism: and Aristotelian natural philosophy, 375, 385–6; and Church, 746; and Galileo, 223, 645; Wittenberg interpretation of, 738
Copernicus, Nicolas: and astrology, 541; and cosmography, 479–80, 496; and history of astronomy, 14, 567–8, 573, 740–3, 747–8; and man of science, 181; printing of works of, 328; travel and distribution of ideas of, 352. *See also* Copernicanism
Cordemoy, Gérauld de, 61
Cordus, Valerius, 445, 446
corn, and natural history, 449–50
Coronel, Luis, 373
corpuscular philosophy: and Aristotelian concept of body, 51; and chymistry, 512, 517;

and occult theory of medicine, 537–8; rise of mechanical philosophy and, 43–7; and scientific explanation, 87–93; and theories of matter, 49, 85. *See also* mechanical philosophy
correspondence. *See* letters
Cosimo I, Grand Duke of Tuscany. *See* Medici, Cosimo I de'
cosmography: and emergence of geography, 494–6; new discoveries and reorganization of, 469–72, 476–80; and science of description and measurement, 491–3; status of before 1490, 472–6; status of in early sixteenth century, 469, 480–91
cosmology: and Cartesianism, 564; and Copernicanism, 41–3, 375, 385–6, 567–9, 738, 740–2, 747–8; and Descartes, 397, 586–7, 590–1, 747–8, 753; and Galileo, 590–1, 743–7, 753, 755; and homocentric theories, 566–7; and Kepler, 41–3, 580–1, 742–3; and music, 600n7; and telescopes, 685–6. *See also* astronomy; heliocentrism
Cotes, Roger, 67n156, 671, 672
Coto, Francisco, 484
Cotton, Charles, 763
Counter-Reformation: and astronomy, 582; and natural philosophy, 395
courts: and astrology, 549–50; and curiosity cabinets, 263–7; and laboratories, 300; and patronage of science, 253–63; and scientific academies, 267–71; and sociopolitical character of early modern Europe, 251–3. *See also* patronage; state
Cowper, William, 430
craft tradition, and practice of astronomy, 200. *See also* artisans; guilds
Craig, John, 163
Cremonini, Cesare, 381, 424
Croll (Crollius), Oswald, 32, 262, 299
Cromberger, Juan, 822
cross-staff, and navigation, 687
Cruz, Juana Inés de la, 825
Cruz, Martín de la, 826
Cudworth, Ralph, 36, 60n134, 104, 374, 813
culture: and art, 773–96; and European colonialism and self-definition, 818–39; and gender, 797–817; and literature, 756–72; and man of science, 191n24; and religion, 727–55. *See also* etiquette; popular culture
Cunningham, William, 487
curiosity cabinets: and anatomy theaters, 278; and botanical gardens, 283, 285; and courts, 263–7; emergence and development of, 273–4, 286–9; and expansion of travel, 351; and global trade, 209

cycloid, and mechanics, 662–3, 671, 681
Cyrano de Bergerac, Savinien, 762

Dalgarno, George, 761
Daneau, Lambert, 382
Danfrie, Philippe, 679
Danti, Egnatio, 485, 683, 693
Darwin, Charles, 760, 797
Daston, Lorraine, 796
Davis, John, 687
Dawbarn, Frances, 260
Dear, Peter, 209
Dee, Jane, 235
Dee, John, 121, 167, 226, 235, 248, 487, 518, 673–4, 688, 734
DeFoe, Daniel, 206
Delisle, Joseph, 672
demonstration: and laboratory, 302–3; and theories of proof and persuasion, 139–40, 142, 155. *See also* experiment; observation
Derham, William, 610n39
Desargues, Girard, 706, 707
Descartes, René: and acoustics, 608; and astrology, 555–6; and astronomy, 586–90, 747–8; and body in literature, 769; and debate on gender and intelligence, 809–10; education of, 243; and essay as genre, 167; and experience, 123; and foundations of physics, 23, 44, 45, 46, 48, 49, 52–3, 55, 59, 60–1, 63–4, 65–6, 68; and magic, 532, 534–5; and man of science, 182, 194; and mathematics, 698, 715, 716–18; and mechanics, 648, 650, 653–64, 669; and medicine, 427–8; on method and proof, 142; and natural philosophy, 379, 395, 396–9; and networks of correspondence, 349; and optics, 597, 617f, 618–21, 624; principles of organization in works of, 25n11; and religion, 79n25, 397, 730, 747–8; and scientific explanation, 85, 88–9, 91; and theories of proof and persuasion, 142, 151, 152. *See also* Cartesianism
dialectics: and dialogue form, 166; and scientific demonstration, 143–5
dialogue, as genre for transmitting natural philosophy, 166
Diamante di Bisa, 218–19
Dias, Dinis, 477
Diaz, Bartholomeo, 477, 483
Díez, Juan, 823
diffraction, and optics, 618–23
Digby, Kenelm, 46, 51
Digges, Leonard, 318
Digges, Thomas, 260, 691
Dionysius Periegetes, 474

Diophantus, 712–14
Dioscorides, 283, 438, 440, 462n100, 773
discovery, voyages of: and cosmography, 469–72, 476–80; overview of impact on European culture, 818–20; and transformation of science in early modern era, 1–2. *See also* colonialism; New World; travel
distilleries and distillation: and medicine, 421; and role of trade, 210–11
Dodonaeus, Rembertus, 256
domestic spaces, and home or household, 226–9
Dondi, Giovanni de', 680
Donne, John, 764
Dorn, Gerhard, 506
dragons, in Leonardo da Vinci's drawings, 528–9, *530f*
Drake, Stillman, 604n13
Draper, J. W., 727, 747
Du Bartas, Guillaume, 770
Ducci, Lorenzo, 252–3
Duchesne, Joseph (Quercetanus), 262, 263, 508
Dudley, Robert, 260
Du Hamel, Jean-Baptiste, 51
Dunk, Eleazar, 212
Duns Scotus, John, 83, 732
Dunton, John, 335
Duplex, Scipion, 370n21
Dürer, Albrecht, 296, *481f*, 692, 776, 777, 778–9, 782, 793
Du Roure, Jacques, 51
Dutch East India Company, 357, 358–9, 453, 491–2

Eamon, William, 193
ear, physiology and anatomy of, 598
earth, and cosmography, 479–80, *481f*, 493. *See also* planetary motion
East Indies, and natural history, 450. *See also* Dutch East India Company
Ebreo, Leo, 374
ecofeminism, and critiques of science, 798
École Polytechnique, 315
economics: and collections of natural objects, 287–8; courts and patronage system, 260; transformation of in early modern era, 7. *See also* trade
education: and astronomy in early sixteenth century, 564–5; and laboratories, 302n40; and man of science, 182–3; and mathematics, 697–8; and medicine, 187n15, 442, 445–6; and military class system, 319; religious identities and reforms in, 735–40, 750–1. *See also* textbooks; universities

Eisenmenger, Samuel, 738
elasticity, and mechanics, 663–4
elements: and alchemy, 501; Aristotelian doctrine of, 28, 74; and innovations in natural philosophy, 390–1
Elias, Antón and Baltazar, 826
elites, women and learned, 193–9. *See also* class
Elisabeth, Princess of Bohemia, 195
Elizabeth I, Queen of England, 806, 807
Ellul, Jacques, 798
Emerald Tablet, and alchemy, 502, *503f*, 514
Elizabeth of Pfalz, Princess, 655
empiricism, and methodology of natural philosophy, 399–403. *See also* experience; experiment
Encyclopedia Britannica, and astrology, 559
engineering: and courts, 260; and mechanical arts, 677–8; and military science, 312, 314–15, 317–19
England: Civil War and astrology, 560n82; Civil War and influence of religion on science, 750–1, 755; and cosmography, 487–8; ephemerides and astrology, 554; and man of science, 190; and mechanical philosophy, 46; and medical profession, 187n16; Restoration and new understanding of experience, 751–3, 755. *See also* Royal Society of London
Epicureanism, 48, 78–9, 81, 150, 396
Epicurus, 47, 48, 87n44, 373, 376. *See also* Epicureanism
epistemology: and Aristotelian natural philosophy, 108, 110, 111, 112, 122; and induction in Newton's usage, 129–30; and laboratory, 295–300; and scientific explanation, 72. *See also* knowledge
Erasmus, Desiderius, of Rotterdam, 238, 239, 469, 499, 807
Erastianism, 507
Erastus, Thomas, 32, 507
Espagner, Jean d', 366n3
essay, as genre of early printed book, 166–7
essence, distinction between real and nominal, 24
Este, Isabella d', 254
Estienne, Charles, 275–6
Etaples, Jacques Lefèvre d', 487
ethnography, and travel literature, 766–70
etiquette, and scientific academies, 171, 268–9
Euclid, 109, 142, 352, 611, 702, 705, 710–12
Eudoxus of Cnidus, 711n25
Euler, Leonhard, 593, 629, 672
Eustachius a Sancto Paulo, 25n11, 27n17, 166n142
Evelyn, John, 215, 222, 227, 257, 752

experience: and anatomy, 111–15; experiments and "physico-mathematics," 124–6; importance of to conceptions of natural knowledge, 106–8; and mathematics, 119–23; and natural philosophy of Aristotle in early modern era, 108–11; and Newtonianism, 126–30; Restoration England and new understanding of, 751–3; species and taxonomy in natural history and, 115–19; and trade, 212–13. See also demonstration; empiricism; experiment; observation
experiment: and alchemy, 516; and chymistry, 517; and coffeehouses, 320; and concepts of proof and persuasion, 157–62; construal of experience as, 106; and Galileo, 647–8; and laboratory, 302–4; and natural history, 466n114; and Newtonianism, 127; and "physico-mathematics," 124–6; and scientific societies, 130–1. See also demonstration; empiricism; experience; observation
extrinsic causes, and scientific explanation, 77–93
Eyck, Jan van, 787, 788
eye, physiology of, 612, 615–18, 685. See also spectacles

Fabri, Honoré, 650, 653
Fabrici de Aquapendente, Girolamo, 112, 278, 425
Fabricius, David, 591
Fabritius, Paulus, 257
fact: experiments and discourse of, 158–9, 303; and natural philosophy, 160–2; and Pliny's definition of *factum*, 437–8
Falloppia, Gabriele, 460n89
family: forms of in early modern Europe, 234n31; and natural inquiry, 225, 229–33. See also homes and households; marriage
Faulhaber, Johannes, 700
feminism: and critiques of science, 797, 798–9; and studies of science and imaginative literature, 758
Ferdinand II, Archduke of Austria, 264
Fermat, Pierre de, 162, 183, 349, 620–1, 698, 701, 707n21, 714, 716
Fernel, Jean, 423, 424, 532–4
Ferrari, Lodovico, 709
Ficinio, Marsilio: and alchemy, 499–500, 514; and Aristotelian natural philosophy, 733; and concept of "spiritus," 605n17; and cosmology, 734n22; and libraries, 247; and magic, 412, 518, 520, 521, 522–3; and Medici court, 35; Plato and Platonism, 33, 374
field trips, and natural history, 445–7
Finaeus, Orontius, 487
Findlen, Paula, 790

Fioravanti, Leonardo, 210, 211, 217
Fior, Antonio Maria, 708–9
Flamsteed, John, 235, 336–7, 339, 553–4, 590, 665
Flamsteed, Margaret, 235
Fludd, Robert, 32, 37, 328
fluxions, Newton's method of, 720–1
Foigny, Gabriel de, 768
folk medicine, 421
Fontenelle, Bernard le Bovier de, 16, 154, 172, 198, 399, 511
force, and mechanics, 664, 670
Forli, Jacopo da, 8
fortifications, and military science, 313–17
fossils, and natural history, 465
Foucault, Michel, 459n85, 810, 811–12, 816
Foxe, Bishop John, 328
Fracastoro, Girolamo, 33, 34, 35, 567, 757, 760, 770
Framboisier, Nicolas Abraham de la, 263
Frampton, John, 826
France: artisans and French Revolution, 222n78; and Cartesianism, 398; and cosmography, 487; and influence of state on education, 389n108; salons and women in science, 196, 198–9. See also Académie Royale des Sciences
Francesco di Giorgio. See Martini, Francesco di Giorgio
Francis I, King of France, 252, 255
François, Jean, 492
Frankfurt School, of social criticism, 798
Franklin, Benjamin, 405
Frederick III, Prince of Brandenburg, 264
Frederick the Great, King of Prussia, 271
Frederick of Württemberg, Duke, 292
Frederik III, King of Denmark, 264
French, Roger, 783
Fresnel, Augustin Jean, 631
Frey, Jean-Cecile, 36, 388
Friedrich-Wilhelm, Prince of Brandenburg, 264
Fritz, Samuel, 359–60
Froeschl, Daniel, 793
Frytschius, Marcus, 144
Fuchs, Leonhart, 443, 457, 780, 782
Fugger commercial house, 208, 348, 351, 420
Funkenstein, Amos, 728

Gabrieli, Gaspare, 442
Galen: and anatomy, 111, 113, 114; and magic, 523, 531, 533; and medical botany, 413; and natural history, 438–9; and Newton's method of analysis and synthesis, 99n69; and scientific method, 141; and sex difference, 805; as source of authority, 802; and universal antidote, 207; on vision and the eye, 612
Galilei, Vincenzo, 596, 603–4

Galileo Galilei: and acoustics, 605–6; and art, 779, 787; and astrology, 541–2, 549, 550; and Church, 589–90, 645, 727, 746–8; and Copernicanism, 223, 385; and dialogue in early printed books, 166; early career and education of, 6, 194; and experience, 123, 124, 125, 213; and geography, 491; and history of astronomy, 563, 584–6, 587–90, 644, 743–8; and history of early modern science, 8; household of, 232n24; and literature, 760, 762; and magic, 532, 539; and man of science, 181, 183; and mathematical methods, 401–2, 706, 711; and mechanical arts, 680, 685; and mechanical philosophy, 45, 53; and mechanics, 633, 640–64, 669; and model of new natural philosophy, 248; and patronage, 230, 231, 259–60; and Platonism, 375; and religion, 730, 753–5; and scale of practice, 345; and scientific academies, 267; and scientific explanation, 71; and theories of proof and persuasion, 140, 146, 156; and theory of matter, 49
Galle, Jan, 1–2, 5f
Gama, Vasco da, 478
Gans, David, 383–4
Gante, Pedro de, 821
Garth, Samuel, 407–8
gases, Boyle's laws of, 667
Gassendi, Pierre: and anti-Aristotelianism, 36; and libraries, 248; and man of science, 181; and materialism, 59n130; and mechanical philosophy, 46, 47–8, 49, 53, 62, 395, 396, 653, 655, 656; and proof in mathematics, 155; and rejection of metaphysics, 24; and scientific academies, 267; and scientific explanation, 79–80, 81; and syllogism, 152
Gauquelin, Michel, 561n85
Gauss, Carl Friedrich, 710
Gaza, Theodore, 373, 439–40
Geber. See Jābir ibn Ḥayyān
Geertz, Clifford, 761
Gemini, Thomas, 679
Gemma Frisius, Reiner, 412, 486, 492, 679, 683, 693
gender: and concept of man of science, 179; and division of labor in scientific households, 235; of science and nature in early modern period, 797–801, 810–17; and understanding of sex difference in early modern period, 800–10. See also women
genre: and impact of science on imaginative literature, 760–1; and questions of proof and persuasion, 165
Geoffrey, Etienne François, 511
geography: and cosmography, 469, 470, 494–6; and *Relaciones geográphicas* project, 826–7; and travel literature, 766–70. See also maps and atlases
geometry: and algebra, 714–18; and experience, 114; and mechanical arts, 675, 681; and modernity, 722; and navigation, 687; and optics, 623–4, 630–1; proof or persuasion in, 156–7, 704–7. See also mathematics
George of Trebizond, 374
Georgios Gemistos Pletho, 373
Gerard of Cremona, 564, 565
Gerbillion, Jean-François, 834
Germany: educational reforms in Protestant, 736; as intellectual center of chymistry, 32; and mechanical philosophy, 46
Gerritsz, Hessel, 491–2
Gessner, Conrad: and chymistry, 507; and correspondence, 349; and development of discipline-specific books, 353; on distilling of drugs, 211; and natural history, 187, 435, 446, 447, 451, 455–6, 457, 458–9, 460
Ghetaldi, Marino, 712
Gheyn, Jacob de, 313
Ghiberti, Buonaccorso, 315
Ghiberti, Lorenzo, 297, 775
Ghini, Luca, 416, 444, 447
Ghiselin de Busbecq, Ogier, 256
Ghisilieri, Antonio, 554
Gilbert, Adrian, 199
Gilbert, William, 140, 183, 188, 210, 301, 388, 688
Giorgi, Francesco, 378
Giovane, Duchess Juliane, 198
Giovanni di Bartolo, 212
Girard, Albert, 710
Glanvill, Joseph, 60, 189, 208, 302, 752
Glauber, Johann Rudolf, 302
Glisson, Francis, 430
global economy. See trade
Goad, John, 558
Goclenius, Rudolph, 138–9
Godebert, Louis, 243
Godwin, Francis, 762, 765
Goedenhuize, Joseph, 794–5
Goes, Bento de, 359
Goltzius, Hendrick, 791
Gombrich, E. H., 791
Goncalves, Lope, 477
Gonzaga Palace (Mantua), 264
Goorle, David van, 47, 395
Gottfried of Franconia, 218n55
Gournay, Marie le Jars de, 196
Graaf, Reinier de, 187, 213, 430
Graff, Johann Andreas, 203
Grant, Edward, 371
Grassi, Horatio, 71

gravity: and Descartes, 48; and Galileo, 662n84; and mechanics, 635, 643, 658; Newton's law of, 67n156, 101, 102–3, 666, 669
Greece, ancient. *See* Aristotle; Epicurus; Euclid; Hellenism; Plato; Pythagoras and Pythagorean tradition
Greenwich Observatory. *See* Royal Observatory
Gregory, David, 593
Gregory, James, 623
Grew, Nehemiah, 187, 288
Grimaldi, Francesco, 597, 621–3, 629
Grimmelshausen, H. J. C. von, 418
Grotius, Hugo, 151, 244
Grove, Richard, 204
Grynaeus, Simon, 486
guaiacum, and trade in medicinal drugs, 208
Guaineri, Antonio, 219
Guericke, Otto von, 609
Guibert, Nicolas, 508
guilds: and laboratories, 295–6; and women in sciences, 200–1
Guldin, Paul, 739
guns and gunpowder, 307–13, 314, 317, 505, 691–2
Gunter, Edmund, 682, 689

Habermas, Jürgen, 339, 354
Hainhofer, Philipp, 209
Hakewill, George, 751
Hakluyt, Richard, 488, 767, 768
Haller, Albrecht von, 270
Halley, Edmond, 10, 332, 493, 554, 590, 594, 665, 672, 712n27
Halley's Comet, 591
Hanson, Norwood Russell, 108
harmonics, and acoustics, 630
Harrington, James, 320
Harrington, Thomas, 427
Harriot, Thomas, 318, 452, 757, 765, 768, 779
Harris, John, 43n70, 406, 511
Harth, Erica, 809–10
Hartlib, Samuel, 218, 302, 353
Hartmann, Georg, 679
Hartmann, Johann, 261, 300, 510, 511
Harvard University, and astrology, 555n60
Harvey, William, 112–14, 186, 187, 277, 280, 328, 407, 425, 761
Havers, Clopton, 430
al-Haytham (Alhazen), 612, 614
Heerbrand, Jacob, 738
Heinsius, Daniel, 247
Helden, Albert Van, 585n57
heliocentrism: and Aristotelian natural philosophy, 385, 386; and Brahe, 574; and Descartes, 397, 399; and geocentrism, 589–90; and neo-Platonism, 374–5; support of Galileo and Kepler for, 402, 585
Hellenism, and medicine, 411–12
Hellwig, Christoph, 140
Helmholtz, Hermann von, 630
Helmont, Johannes Baptista Van, 32, 47n88, 104, 423, 512–13, 518
Hennepin, Louis, 768n37
Henry III, King of France, 3
Henry IV, King of France, 263
Henry of Langenstein, 563
Henry the Navigator, Prince of Portugal, 350n36, 477
herbals and herbalists: and botanical gardens, 280; herbarium and natural history, 447–8; and illustrations, 773; and natural knowledge in countryside, 218. *See also* botany
Herbert, Edward, Lord of Cherbury, 151
Herbert, Mary Sidney, Countess of Pembroke, 199
Herbst, Johannes, 328
Hermann, Paul, 202
Hermes Trismegistus, 373, 497, 521–2, 571. *See also* Hermeticism
Hermetic Corpus, 295
Hermeticism, and scientific developments of Renaissance, 374
Hernández, Francisco, 452, 462, 826
Hero of Alexandria, 635
Herrera, Antonio de, 353
Herschel, Caroline and William, 232n26
Heseler, Baldasar, 276
Hevelius, Johannes, 183, 200, 201, 227–8, 591
Heydon, Christopher, 580
Highmore, Nathaniel, 430
Hill, Nicholas, 47, 395
Hippocrates, 114, 434, 533
Histoire de l'Académie Royale des Sciences (journal), 331
historiography: of Aristotelianism, 70n1; classrooms and libraries in, 239; and concept of Scientific Revolution, 12–17, 436n4, 799–800; and concept of private versus public space, 224; and debate on nature of early modern Catholicism, 735n25; and feminist critiques of science, 798–9; and gender ideologies, 799–800, 816n50; and literature on coffeehouses, 322n3; and literature on growth of natural history, 436n2; and literature on scientific aspects of music, 600–1n7; and literature on scientific societies, 185–6n13; and Military Revolution, 307n3; of relationship between science and religion, 727–30; and study of impact, 659–60; and study of proof and persuasion, 133

history: change in ideas on in late seventeenth century, 437n7; Ficino's study of myth and, 522–3; and naturalists, 465. *See also* historiography
Hobbes, Thomas: and astronomy, 588; and experience, 126, 400; on geometry, 156; and man of science, 181; and materialism, 59; and mechanical philosophy, 46, 48; and medicine, 427; on proof and persuasion, 142, 152, 157; and rejection of metaphysics, 24; and religion, 751–2; and Royal Society, 329
Hoefnagel, Joris, 793
Hofmann, Hans, 793
Holbein, Hans, 482
Holwarda, Johannes Phocylides, 591
Holy Roman Empire: and cosmography, 485; and scientific societies, 394
Homberg, Guillaume, 131
homes and households: and activities of scientific research in early modern period, 224–5; and domestic spaces, 226–9; and labor in scientific, 233–7; and natural inquiry as family project, 229–33
homocentric theories, in astronomy, 566–7. *See also* heliocentrism
Hoogstraten, Samuel van, 775, 777
Hooke, Robert: and acoustics, 609; and art, 779; and astronomy, 588; and coffeehouses, 336–7, 339, 340; and experience, 128, 400, 402; and journals, 167; and laboratory, 303, 304; and literature, 762–3; on logic and proof, 152, 153; and mechanical arts, 681, 694–5; and mechanics, 664; and medicine, 430; and natural history, 465; and Royal Society, 223, 289, 330; and scientific explanation, 90; on telescope, 539–40; and theories of colors, 626n69
Hooykaas, Reijer, 496, 728
horology, and mechanics, 662–3, 681. *See also* clocks
horoscopes. *See* astrology
Horrox, Jeremiah, 588
Houghton, John, 337, 339
Houlières, Madame de, 196
humanism: and alchemy, 499; and astronomy, 565–9; and courts, 270; and medicine, 411, 413; and music, 600n7; and natural philosophy, 368, 373–4, 380; and new perspectives on God, man, and nature, 729; and questions of proof and persuasion, 135; and Republic of Letters, 349; and revival of Greek and Latin texts, 7–8, 599n3; and travel accounts, 767
Hume, David, 68, 104
humors, and medicine, 410
Huser, Johann, 506

Hutten, Ulrich von, 253
Huygens, Christiaan: and acoustics, 610; on art, 778; and astronomy, 588; and Cartesianism, 397; and cosmography, 493; and empiricism, 131; and foundations of physics, 53; and man of science, 183, 194; and mathematics, 162, 401, 706, 707n21; and mechanical arts, 680–1; and mechanics, 659–64, 669, 672; and optics, 625, 626f; and scientific debates, 330; and scientific explanation, 90, 91; and travel, 10
hydrography, and cosmography, 470
hydrostatics, and mechanics, 635, 640–1, 644
hylemorphism, and natural philosophy, 390, 398
hypothesis, and Newton's methods, 753–4

Iamblichus, 522
Ignatius Loyola, 350
illustration, scientific: and anatomy, 782–6; and botany, 116–17, 235n39, 773, 780–2; and experiments in laboratory, 303; and gendered divisions of labor, 235; and magic, 527–9; and microscope, 539–40; misplaced, mislabeled, or plagiarized, 352n46; and representation of natural world, 779–82. *See also* art
images, theory of optical, 623–4
impact, rules of, 657, 659–61, 670
Imperato, Ferrante, 210, 460n90
Index of Forbidden Books, and Inquisition, 531
India: and Jesuit missionaries, 350; Portuguese and sea route to, 478
inertia: Galileo and law of, 45n75; Newton's definition of, 666
Innocent VIII, Pope, 378
innovation: attitudes toward in mid-seventeenth century, 16; and natural philosophy in Renaissance, 372–9; natural philosophy and resistance to radical, 390–3; and women in sciences, 205
Inquisition: and astrology, 550; and magic, 531; and medicine, 423; and natural knowledge in countryside, 218–19, 220; and trial of Galileo, 746. *See also* Catholic Church
insects, and natural history, 466
Instituto delle Scienze (Bologna), 289
intellectual movements: and Luther's break with Catholicism, 734–5; and theological context of bodies of knowledge, 730–5. *See also* Aristotelianism; Cartesianism; Copernicanism; Epicureanism; humanism; materialism; neo-Platonism; Newtonianism; scholasticism; Stoicism
intrinsic causes, and scientific explanation, 77–93
inventio, and concepts of rhetoric and logic, 149

Index

Ippolito of Montereale, 226–7, 231
Islamic world: and mathematics, 708; and mechanics, 637
Isserles, Moses, 384
Italy: astrology in universities of, 546; and cosmography, 472, 484–5; and libraries, 246; and mathematics, 699; natural history and universities of, 444; Platonism in universities of, 374; Renaissance naturalists and anti-Aristotelianism, 33–6; women and scientific academies in, 198

Jābir ibn Ḥayyān (Geber), 294, 512
Jacobs, Margaret, 759
Jacopo d'Angelo, 472
Jansenism, and Cartesianism, 398
Japan, and Jesuits, 827–35
Jedin, Hubert, 735n25
Jesuits: and Aristotelian natural philosophy, 387–8, 389n108; and astronomy, 577; and authority of Thomist-Aristotelianism, 173; and Cartesianism, 399; colleges and universities of, 241, 367–8, 488–9, 736–7, 738–40; and correspondence, 348, 353; and distinction between natural philosophy and mathematics, 119n49; and geography, 488–9; and journals, 351, 353; and libraries, 352; and mathematics, 121–2, 698, 723, 739, 829–30; and mechanics, 653; missionaries and contributions to scientific knowledge, 202, 350–1, 352, 357, 358, 359–60, 827–35; and scale of scientific practice, 346
Jews and Judaism: and natural philosophy, 383–4; and women in medicine, 236n42
John of Gmunden, 563
John of Lignières, 565
John of Rupescissa, 504
Johnson, Samuel, 763
Jonson, Ben, 794n67
Jordanus of Nemore, 635
Joubert, Laurent, 419n36
Journal des savants (journal), 167, 331
journals: increase in importance of, 167–8; and Jesuits, 351, 353. *See also Acta eruditorum*; *Histoire de l'Académie Royale des Sciences*; *Journal des savants*; *Nouvelles de la République des Lettres*; *Philosophical Transactions*; printing and printing houses

kabbalah, and natural philosophy, 384
Kant, Immanuel, 68
Keckermann, Bartholomaeus, 381
Keill, John, 406n182
Kepler, Johannes: and astrology, 541–2, 548–9, 550, 579, 580–1; and courts, 257–8; and distinction between natural philosophy and mathematics, 119n49; and heliocentrism, 375; and history of astronomy, 563, 574, 581–4, 587–90, 742, 743; and impact of science on literature, 757, 760, 764, 765, 766; and libraries, 248; and man of science, 183; and mathematics, 401–2, 699, 714; and mechanics, 656, 665; and optics, 597, 613–18, 624, 630–1; and patronage, 194; printing of works of, 328; Pythagoreans and foundations of physics, 41–3; and religion, 748; and travel, 10
Khayyām, 'Omar, 708
Khunrath, Heinrich, 290, 299
al-Khwārizmī, 708
al-Kindī, 614
Kino, Eusebius Francisco, 359, 360n77, 825
Kirch, Gottfried, 200, 232
Kircher, Athanasius, 37, 248, 266, 344n14, 349, 353, 360n77, 388, 587n65, 609
knowledge: Aristotelianism and sensory origin of, 107; basis of division in premodern Europe, 6; natural philosophy and distinction between divine and natural, 750; *scientia* and classification of in early modern era, 3–4, 9, 11; Spanish response to indigenous Mexican, 825; theological and intellectual contexts of, 730–5. *See also* culture; natural knowledge
Kolb, Peter, 203
Koopman, Elisabetha, 201
Koyré, Alexandre, 13
Kris, Ernst, 793
Kuhn, Thomas, 154n89, 630, 631
Kuo Tsing-lien, 832
Kusukawa, Sachiko, 737
Kyeser, Conrad, 317

labor, division of: in artists' studios, 790; and laboratories, 292, 294; in scientific household, 233–7
laboratories: artist's studio as, 789–91; description of in early modern period, 290–3; and epistemology, 295–300; evolution of, 300–2; experiment in, 302–4; theory and practice in, 293–5; and universities, 304–5. *See also* workshops
Lactantius, 148–9
Lafitau, Joseph-François, 767
La Forge, Louis de, 61
Laguna, Andrés, 281
Lambert, Madame, 198
Landucca, Luca, 527
language: books in Chinese, 831; and Galileo, 763n18; Latin as scientific, 366n3; and studies of science and imaginative literature, 759–62. *See also* dialogue; narrative; rhetoric

Laqueur, Thomas, 815n48
Lateran Council, 380
latitude, and mechanical arts, 682
Latour, Bruno, 792
Lavanha, João Baptista, 483
Lavoisier, Antoine-Laurent, 510
law and legal studies: and concepts of proof and persuasion, 137–8; and naturalists, 461; and practice of medicine, 419–20
laws, of nature: and changes in scientific explanation, 70–105; and Descartes, 45n75; and mechanical philosophy, 44–5, 656; and Newton's theory of planetary motion, 666; origins of term, 13n19
Le Bossu, René, 51
Leclerc, Georges, 465
Leeuwenhoek, Antonie van, 182, 213, 430, 465, 763
L'Écluse, Charles de. *See* Clusius, Carolus
Le Fèvre, Nicaise, 29–30
Leibniz, Gottfried Wilhelm: and Aristotelianism, 52; on art of invention, 149, 153; and China, 834–5; on connection between science and the state, 270–1; on demonstration of proof, 164, 168; and foundations of physics, 23, 49–51, 55–6, 60–3, 64–5, 67, 68–9; and libraries, 248; and man of science, 185, 194; and mathematics, 702, 720, 721; and mechanics, 650, 670–1, 672; and networks of correspondence, 349; and scientific societies, 394
Leigh, Richard, 335
Lemery, Nicolas, 303–4, 497n1, 511, 515n74
Le Monnier, Pierre Charles, 672
Leonardo da Vinci: and anatomy, 414, 777; and artist as scientist, 786–7; and artistic naturalism, 776, 777, 782; diversity of career of, 5–6; magic and drawings of dragons by, 528–9, *530f*; and mechanical arts, 677; and military science, 315, 317, 318, 692; and natural history, 462; and natural philosophy, 296
Leoniceno, Niccoló, 141, 412, 440–2
Leosel, Johann, 237
Le Roy, Louis, 495
Léry, Jean de, 768
L'Estrange, Sir Roger, 321–2
Lestringant, Frank, 764
letters: and concept of Republic of Letters, 348–51, 354, 360–1; and correspondence among universities, 342; and mathematics, 699; and natural history, 454–6; and printing press, 353–4
lever, doctrine of, 634, 654
Libavius, Andreas, 263, 299, 507, 508

libraries: and classification systems of early modern period, 3; images of in early modern period, 238–40, 244–50; and Jesuits, 352
light: Italian naturalists and concept of, 35; and mechanical philosophy, 658; optics theories of, 611, 614–15, 618–23, 624–9
Ligorio, Pirro, 485
Ligozzi, Jacopo, 777–8, 791n55
Lilly, William, 560n82
Linières, Jean de, 343n8
Linnaeus, Carolus (Carl von Linné), 204, 232, 362, 458n83, 467–8, 767
Lipsius, Justus, 80–1, 104, 376
Lipstorp, Daniel, 587n65
literature: and antagonisms between categories of text and intellectual practice, 770–2; and cosmography, 473–4; and language, 759–62; overview of science and imaginative, 756–9; and plurality of worlds, 764–6; telescope, microscope, and realism in, 762–4; and travel accounts, 766–70; and writers on military topics, 317–19. *See also* books; genre; journals; language; printing and printing houses; textbooks
Locke, John, 24, 53–4, 88, 175, 181, 188, 669
logarithms, 707
logic, and theories of proof and persuasion, 135, 136, 139–46, 151, 152. *See also* methods
Lohr, Charles, 371–2
longitude: and astronomy, 593; and cosmography, 492
López de Velasco, Juan, 208
Lorenzo, Giovanni, 226, 231
Louis XIII, King of France, 264
Lower, Richard, 187
Lucretius, 47
Luizzi, Mondino d', 275
Lull, Ramon, 702
Lusitano, Amato, 218
Luther, Martin, 240, 297–8, 382, 573, 729, 732–5
Lux, David, 355

Machiavelli, Niccolò, 317–18
Maestlin, Michael, 574, 581
Magalotti, Count Lorenzo, 170
Magellan, Ferdinand, 482
magic: and Agrippa von Nettelsheim, 519–26; and astrology, 544; credibility of, 526–9; and medicine, 423, 532–8; in Middle Ages, 518; and natural knowledge in countryside, 220; and new science of early modern period, 538–40; and Pico's beliefs, 579n44; and religion, 529–32. *See also* witchcraft
Magini, Giovanni Antonio, 553, 585n57

Magirus, Johannes, 555
Maharal (Judah Loew ben Bezalel), 384
Maine, Duchess of, 196
maize, and natural history, 450
Malebranche, Nicolas, 61, 62n139, 398
Malpighi, Marcello, 186, 187, 430, 465
Mander, Karel van, 787–8
Manfredi, Girolamo, 542–3, 554
manners, and scientific academies, 171. *See also* etiquette
man of science: and medicine, 186–8; and modern concept of scientist, 179–81; and universities, 182–6, 194
Manuel, Frank, 755
Manutius, Aldus, 411, 440n16
maps and atlases: and expansion of scientific communication in early modern period, 361; of magnetic declination, 344n14; and mechanical arts, 692–3; and scientific practices in medieval period, 342–3; traditional Chinese printed, 831. *See also* cartography; cosmography; geography; navigation
Marchetti, Alessandro, 650
Marcuse, Herbert, 798
Marino di Iacopo (Taccola), 317, 678
market. *See* piazza; trade
Markgraf, Georg, 453
Marlowe, Christopher, 531, 771
Marquette, Jacques, 359
marriage: sex differences and discussions of, 807; and women in sciences, 200–1, 234–6. *See also* family; homes and households
Marriotte, Edme, 131, 650
Marsi, Paolo, 242–3
Martin, Henri-Jean, 323
Martini, Francesco di Giorgio, 315, 317, 678, 692
Marx, Karl, 346
mass, Newton's definition of, 666
materialism: and mechanical philosophy, 59; and medicine, 424–31
mathematics: Apollonian, Archimedan, and Diophantine tradition, 712–14; and astrology, 553, 554; and classification of knowledge, 4; definition of, 696; and Euclid's elements, 710–12; and experience, 119–26; and Galileo, 640, 642–3; and Jesuits, 121–2, 698, 723, 739, 829–30; and mechanical arts, 674–5, 678, 683–6; and mechanics, 637–8; and methodology of natural philosophy, 399–403; methods and problems in, 702–7; and modernity, 722–3; and Newton, 667; probability and certainty in, 162–4; proof and persuasion in traditions of, 154–7; social context for, 697–702; and use of terms, 696–7; and women in sciences, 668–70. *See also* algebra; calculus; geometry; mixed mathematics
matter, theory of: and chymistry, 510–13; and magic, 523–4; and mechanical philosophy, 47–52, 655
Mattioli, Pier Andrea, 207, 218, 256, 450, 460–1, 462n100
Maurice of Nassau, Prince, 700
Mauro, Fra, 472, 473, 476
Maurolico, Francesco, 484, 612–13, 738
Maximillian II, Holy Roman Emperor, 256, 257, 264, 453, 461
Mayer, Tobias, 593, 672
Mayerne, Theodore de, 263
Mayow, John, 187
Mazzoni, Jacopo, 375n46
Mead, Richard, 557–8
measurement, and cosmography, 493
mechanical arts: art and nature in, 693–5; and classification of knowledge, 4; clocks and celestial instruments, 679–83; definitions of, 673; disciplinary relationships and boundaries of, 673–6; mathematical and optical instruments, 683–6; and military science, 319; status of in 1500, 677–9; surveying, warfare, and cartography, 686–93
mechanical philosophy: and foundations of physics, 43–66; natural philosophy and origins of, 395–9; and rational mechanics, 653–9. *See also* Boyle, Robert; Cartesianism; Descartes, René
mechanics: definition of, 632; as distinguished from physics, 24. *See also* mechanical arts; mechanical philosophy; rational mechanics
Medici, Catherine de', Queen of France, 806
Medici, Cosimo I de', Grand Duke of Tuscany, 281, 692
Medici, Ferdinando I de', Cardinal Grand Duke of Tuscany, 265
Medici, Francesco I de', Grand Duke of Tuscany, 265
Medici, Leopoldo de', Grand Duke of Tuscany, 170, 172, 268
Medici family, 246, 255, 264
medicine: and alchemy, 421, 423, 498; and astrology, 579; and botanical gardens, 281–2; changes in early modern period, 432–4; and competition among providers, 211; and courts, 262–3; debate on status of as science, 137–8; global trade and new medicinal drugs, 208; and harmonic balance, 600n7; home and practice of, 226–7; and magic, 423, 532–8; and man of science, 186–8; and materialism,

medicine (*cont.*)
424–31; and military surgeons, 316–17; and natural history, 4, 463n105; New World and expansion of, 416–23; in rural areas, 218–21; Spanish response to indigenous Mexican knowledge of, 825; status of at end of seventeenth century, 407–8; and use of body parts as medications, 216–17; and women, 236n42. *See also* anatomy; apothecaries; physic
Mei Wen-ting, 834
Melanchthon, Philip, 107, 149, 241, 369, 382, 548, 573, 578, 580, 737–8
Melichio, Giorgio, 210
Mendoza, Antonio de, 823
Mentzel, Christian, 231
Mercator, Gerardus, 343, 486, 489–91, 683, 688–9, 693, 764
Mercator, Nicolaus, 589
Merchant, Carolyn, 798, 810–11, 812
mercury, and chymistry, 513–14
Merian, Maria Sybilla, 10, 199, 202–4, 459, 466–7
Mersenne, Marin: and acoustics, 596, 598, 606–8; and astrology, 555; and foundations of physics, 36, 45; and letters, 349; and libraries, 248; and man of science, 181; and mathematics, 699; and mechanics, 648, 649, 651, 662; and natural philosophy, 379, 392; and optics, 597; and "physico-mathematics," 125; and scientific academies, 267; and theories of proof and persuasion, 151
Merton, Robert, 346, 728
Mesmer, Franz Anton, 558
Mesopotamia, and algebra, 708
metallurgy, and military science, 309, 312
metaphysics, and foundations of physics, 22–5, 68
meteorology, and alchemy, 498
methods: and doctrines of scientific proof, 140–2; and mathematics, 702–7, 713. *See also* demonstration; empiricism; experience; observation; scientific explanation
Metzger, Hélène, 511
Meursius, Ioannes, 238–9
Mexico: and natural history, 452; and universities, 821–7
Méziriac, Claude-Gaspar Bachet de, 714
Michelangelo Buonarotti, 787, 791
microscope: and mechanical arts, 684; and medicine, 429–30; and natural history, 465–6; and scientific illustration, 539–40
Mildmay, Grace, 227
Milemete, Walter de, 308
military science: defensive technologies, armor and fortification, 313–17; and engineering, 317–19; offensive technologies and development of guns and gunpowder, 307–13; and state of continuous warfare in early modern period, 306–7. *See also* war and warfare
Milton, John, 73n7, 764
minima, scientific explanation and concepts of, 86
Miscellanea Curiosa (journal), 167
missionaries: and collections of naturalia, 266; and expansion of scientific knowledge, 357, 358, 359. *See also* Jesuits
mixed mathematics: and acoustics, 596; compared to natural philosophy, 4; and debate on scientific knowledge, 120–1; definition of, 696; and natural philosophy, 156, 744–5. *See also* mathematics
mixture, and scientific explanation, 82–6
modernity: and Aristotelianism, 28; and historiography of Scientific Revolution, 13–17; and mathematics, 722–3
Moffet, Thomas, 453
Moletti, Giuseppe, 638n20
Molière (Jean-Baptiste Poquelin), 536, 771
Monardes, Nicolás, 453, 823, 826
monasteries. *See* Catholic Church
monochord, and cosmology, 37, 38f, 599, 600f, 601f
Montaigne, Michel de, 150, 265, 379, 447, 526, 804–6, 815
Montalboddo, Fracanzano da, 484
Monte, Giambattista da, 416
Monte, Guidobaldo dal, 633–4, 635, 639, 643, 692, 707
moon, and lunar theories, 592–3
Morandi, Orazio, 560n82
More, Henry, 36, 54, 59, 104, 374, 397n138, 518, 587n65, 813
More, Thomas, 480, 482, 839
Morin, Jean-Baptiste, 391n117
Moritz of Hesse, 261, 271, 300, 506
Morland, Samuel, 609
motion: Galileo and law of, 45n75; and mechanical philosophy, 52–8; Newton's first law of, 100–1; rational mechanics and studies of, 636–53, 659–68
Moxon, Joseph, 553
Münster, Sebastian, 486, 489, 682, 693
museums: and natural history, 287, 448; and scientific academies, 288–9
music: and acoustics, 596, 597–604; and foundations of physics, 36–43; and mathematics, 120

myth and mythology: alchemical interpretation of ancient, 501n14; Ficino's study of history and, 522–3; Scientific Revolution as, 15–16

al-Nafis, 425
Napier, John, 699–700, 707
narrative: and historiography of relationship between science and religion, 727–30; and impact of science on imaginative literature, 760
Nash, Thomas, 761
Native Americans, and indigenous scientific knowledge, 825–6
natural history: as distinct from natural philosophy, 4; and education of physicians, 187n15; and experience, 115–19; and naturalist, 459–68; and revival of ancient tradition, 437–42; and sharing of information, 454–9; significance of in early modern period, 435–7; travel and expansion of, 361–2; words and things in, 442–8
naturalism, in art, 775–9, 785, 791–6
natural knowledge: in countryside and villages, 217–21; in piazza, 213–17
natural philosophy: and Aristotelianism, 370–95; and astrology, 555; and classification of knowledge, 4; coffeehouses and printing shops, 320–40; and cosmography, 473, 474–6; and distinction between divine and natural knowledge, 750; and education of physicians, 187n15; empirical and mathematical methods in, 399–403; encyclopedic reference works and evolution of, 405–6; and experience, 108–11; and experiment, 157; and forces for change in early modern period, 393–5; and Galileo, 744–5; and magic, 519, 523–4, 540; and mathematics, 706, 723; and "matters of fact," 160–2; and mechanical arts, 675–6; and mechanical philosophy, 395–9; and mechanics, 632–3, 650; and mixed mathematics, 4, 156, 744–5; and physics, 26; Reformation and religious influences on, 379–84; Renaissance Italy and anti-Aristotelianism, 33–6; and resistance to radical innovation, 390–3; and social conventions, 403–5; university context of, 4–5, 366–72; use of term, 365
naturals and nonnaturals, in medicine, 410
nature: changes in concept of, 133–4; clock as metaphor of, 748–9; gendering of as female, 797–801, 810–17; and mechanical arts, 693–5; and natural philosophy, 380n65; Pliny's definition of, 437

Naudé, Gabriel, 249
Navarro, Juan Huarte, 808
Nave, Annibale dalla, 708
navigation: and cosmography, 491–3; and mathematics, 707; and mechanical arts, 682, 684, 685, 686–9; and portolan charts, 342n5. *See also* maps and atlases
neo-Platonism: and acoustics, 605n17; and natural philosophy, 374–5; and new perspectives on God, man, and nature, 729. *See also* Platonism
Netherlands: colonialism and distribution of scientific knowledge, 352; and tulip mania, 209
Neville, Henry, 766
Newton, Isaac: and acoustics, 597, 610–11; and astrology, 555–6, 557; and astronomy, 592–4; and chymistry, 514; and foundations of physics, 32, 54–5, 64, 66–8; and literature, 772; and man of science, 183; and mathematics, 700, 720–1; and mechanical arts, 693–4; and mechanics, 650, 664–70, 671–2; on method and proof, 142; and natural philosophy, 365; and optics, 611, 623, 626–9; publication of works of, 331–2; and religion, 753–5; and Royal Society, 329, 330, 331–2, 402; and scientific explanation, 94–5, 99–105. *See also* Newtonianism
Newtonianism: concepts of space and debate between Leibniz and, 56–7; and experience, 126–30; and physics in eighteenth century, 68–9
New World: and medicine, 416–23; and myth of modernity in early modern period, 16–17; and natural history, 448–54. *See also* Brazil; colonialism; discovery, voyages of; Spain; Surinam; Virginia
Niccoli, Niccolò, 246
Nicholas of Cusa, 374, 569–70
Nicholas of Lyra, 476
Nicholas V, Pope, 439
Nicole, Pierre, 163
Nicolson, Marjorie Hope, 756–7
Nifo, Agostino, 82, 83–4, 85, 142
Noble, William, 609
Noël, Étienne, 58
Norman, Robert, 688
Nouvelles de la République des Lettres (journal), 331
nova (supernova), and stellar astronomy, 591
Nova reperta (New Discoveries), 1–2, 5f, 7, 8–9, 10, 16–17, 788–9
Nuck, Anton, 430
Nuñez, Pedro, 483

objectivity, and discussions of gender, 810
observable effects, and scientific explanation, 103
observation: and astronomy, 585; and experience, 108; and medicine, 432–4; and natural history, 442, 444, 446–7, 466. *See also* demonstration; experience; experiment
Observatoire Royale de Paris, 493, 593
occasionalism, and foundations of physics, 61
oikumene, and cosmography, 470, 473, 474, 476
Oldenburg, Henry, 189, 331, 353–4, 752
Omicron Ceti (star), 591–2
ontology, and transformations in early modern era, 11
optics: and Descartes, 659n74; discipline of acoustics compared to, 597; image location and geometrical, 623–4, 630–1; Kepler's contributions to, 613–18; mechanical arts and instruments for, 683–6; nature and speed of light, 624–6; and Newtonian experience, 127–30; and Newton's theory of light and colors, 626–9; overview of in early modern period, 611–12; painting and Northern European developments in, 763n20; refraction and diffraction in, 618–23; status of in sixteenth century, 612–13. *See also* microscope; spectacles; telescopes
Oresme, Nicole, 637
Orta, Garcia de, 204, 450, 453
Orta, Thomas d', 483
Ortelius, Abraham, 343, 487, 764, 793
Osborne, Francis, 327
Osiander, Andreas, 328
Osser, Josephus, 393n120
Other, representation of in travel literature, 769
Oughtred, William, 683
Ovid, 242–3
Oviedo, Gonzalo Fernández de, 450, 451n51, 461, 462
Oxford University: and astrology, 553; and natural philosophy, 389

Paaw, Pieter, 278, 283
Pacchioni, Antonio, 430
Pacioli, Luca, 700
Pacius, Julius, 75n14
Padua, and botanical garden, 282–3. *See also* University of Padua
Paez, Pedro, 359
Palissy, Bernard, 6, 182, 213, 296, 370, 385, 462
Paludanus, Bernard, 454
Panofsky, Erwin, 791–2
Papin, Denis, 184
Pappus of Alexandria, 99n69, 635, 639, 712

Paracelsus (Theophrastus Bombastus von Hohenheim): and alchemy, 299n28, 502–10; and astronomy, 570, 571; and chymistry, 30–2, 514; education of, 6; and experience, 110; and laboratory, 298–9; and magic, 518; and medicine, 421–2; and natural philosophy, 391; publication of works of, 328; and religion, 734; and scientific explanation, 104; and university classroom, 240
Pardies, Ignace Gaston, 628–9
Paré, Ambroise, 226, 316, 418, 456–7, 805
Paris Observatory. *See* Observatoire Royale de Paris
Parker, George, 554
Parker, Samuel, 164
Parrhasius, 775
Pascal, Blaise, 57–8, 124, 156, 182, 349, 701, 706, 707n21
passive principles, and scientific explanation, 93–105
Patrizi, Francesco, 33, 34, 35, 376–7, 570, 571–2
patronage: and learned elites, 194–5; and man of science, 181, 189; and mathematical and observational work of Galileo and Kepler, 401–2; and military science, 317, 318; and natural history, 453. *See also* courts
Paul de Santa Maria, Bishop (Paul of Burgos), 476
Paul of Venice, 8
Pecham, John, 612
Pecquer, Jean, 429
Pedro, Prince (Portugal), 472
Peiresc, Nicolas-Claude Fabri De, 267
Penny, Thomas, 452
Pepys, Samuel, 698
Pereira, Benito, 155
Périer, Florin, 57, 124
periodicity, of stars, 591–2
Perrault, Charles, 496, 775
Perrault, Claude, 404
Perro, Girolamo, 283, 285
persuasion. *See* proof
Perugino (Pietro di Cristoforo Vannucci), 254
Peschel, Oscar, 472n4, 493n73
Peter Martyr of Anghiera, 478, 767, 768, 818–19
Petrarca, Francesco, 379, 499
Petrarch. *See* Petrarca, Francesco
Petreius, Johannes, 328
Pett, Peter, 752
Petty, William, 164
Peucer, Caspar, 573
Peurbach, Georg, 485, 545, 563, 565–6
Peyer, Johann Conrad, 430
Philip II, King of Spain, 207, 259, 451–2, 806–7

Philip the Good, Duke of Burgundy, 308
Philoponus, 636–7
Philosophical Transactions of the Royal Society of London, 12, 167, 330–1, 332, 354, 404
philosophers' stone: and alchemy, 501–2, 510–11, 513–14; and medicine, 421
philosophy, and theology in Christian tradition, 731. *See also* corpuscular philosophy; mechanical philosophy; natural philosophy
physic: and medicine, 408–16; use of term, 407
physicians, and social class, 416–17. *See also* medicine; surgeons and barber-surgeons
physico-mathematics, and experiments, 124–6, 401–2, 654n62
physics: and Aristotelianism, 21–8; and astronomy, 582–3; and chymistry, 29–33; experiments and "physico-mathematics," 124–6, 401–2, 654n62; and Italian naturalists, 33–6; mathematical order and harmony in, 36–43; modern physics compared to, 21; and mechanical philosophy, 43–66; Newton and foundations of, 32, 54–5, 64, 66–8, 94n60; and Newtonian system in eighteenth century, 68–9
piazza, and natural knowledge, 213–17
Piccolomini, Aeneas Sylvius, 474, 476
Piccolomini, Alessandro, 154–5
Piccolpasso, Cipriano, 296
Pico della Mirandola, Gianfrancesco, 378–9
Pico della Mirandola, Giovanni, 247, 378, 518, 541, 547–8, 579–80
Picot, Claude, 655
Pigot, Thomas, 609
Plancius, Petrus, 491
plane table, and surveying, 690–1
planetary motion: Kepler's theory of, 42, 742; and music, 600n7; Newton's theory of, 664–5, 667–8. *See also* earth
Plantijn, Christoffel, 639
Plato, 33, 293, 602. *See also* Platonism
Platonism: and natural philosophy in Renaissance, 373–7; and scientific explanation, 72n6. *See also* neo-Platonism
Platter, Felix, 233n28, 276–7, 447, 459
Platter, Thomas, Jr., 233n28, 447
Platter, Thomas, Sr., 233n28
Platter family, 233, 237
Plattes, Gabriel, 218
Pliny the Elder, 8, 242, 437–8, 439, 440–2, 462, 478, 526, 773, 775, 779–80
Plot, Robert, 465n111
Plotinus, 374, 522
poetry, and scientific thought, 760, 770
Poisson, Siméon-Denis, 630

politics: courts and patronage system, 260; and influence of religion on science in seventeenth century, 748–53; transformation of in early modern era, 7. *See also* state
Polo, Marco, 342
Pompanius Mela, 474
Pomponazzi, Pietro, 380–1, 416, 424, 518, 520
popular culture: and astrology, 560; coffeehouses and public reason, 339, 340; natural knowledge and intellectuals' contempt for, 221–3. *See also* culture
pornography, and impact of science on imaginative literature, 762, 763, 770, 772
Porphyry, 522
Porta, Giambattista della, 149, 169–70, 220, 269, 393, 518, 539, 605n17, 612, 613
Portugal, and cosmography, 350n36, 472, 477, 483
Possevino, Antonio, 489
postal services, transnational, 341–2
postcolonialism, and study of science and imaginative literature, 758
Poullain de la Barre, François, 192, 808n30, 809
Power, Henry, 188
practice (*praxis*): and growth in scale of science, 344–7; and laboratory, 293–5; and medicine, 408–9
Praetorius, Michael, 573
Presocraticism, and Platonism, 376
primary quality, and scientific explanation, 88–90
primary substances, and Aristotelianism, 74, 76
printing and printing houses: and Aristotle's works, 73n9, 372n31; and coffeehouses, 336, 339; and Cyrano's *Histoire comiques de les estats et empires de la lune*, 762n17; and discussion of science in pamphlets and journals, 321–32; of Euclid's elements, 711n24; and Galileo's works, 762–3n18; and growth in communication in early modern period, 351–5; in Mexico, 822–3; and military arts and sciences, 318n41; and Peter Martyr's accounts of New World, 818n2; and traditions in mathematics, 712n27. *See also* books
private/public space, and science in home or household, 224, 229, 237
probability: mathematical concepts of, 707n21; and theories of proof and persuasion, 139–40, 162–4
Proclus, 374, 522
professionalization model, of man of science, 190
projection, and mechanical arts, 684–5
proof: and Archimedean mathematics, 713; and disciplinary structure in universities, 134–8,

proof (cont.)
  150–4; and experiment, 157–62; importance of in early modern period, 132–4, 174–5; and mathematical traditions, 154–7; and method of analysis in mathematics, 704, 705, 706; in printed books, 164–8; probability and certainty in mathematics, 162–4; and social institutions, 168–74; theories of, 138–49
proper causes, and scientific explanation, 92–3
Protestantism: and educational reforms, 736, 737–8. *See also* Reformation; religion
Psalmanazar, George, 768n37
Ptolemy: and astrology, 543, 546; and astronomy, 564, 566; and cosmography, 469, 470, 473, 474, 488, 693; and experience, 119n49, 120–1, 123n67; and mechanical arts, 678; and optics, 611
public/private space, and science in home or household, 224, 229, 237
Pumphrey, Stephen, 260
Purchas, Samuel, 488, 767
pure mathematics: and astronomy, 707; definition of and use of term, 696–7; and modernity, 722–3. *See also* mathematics
Puritanism, and educational reforms, 750–1
Pythagoras and Pythagorean tradition: in acoustics, 598–604; and foundations of physics, 36–43; and natural philosophy, 376

quadrant, and mechanical arts, 686
quadratic equations, theory of, 708
quadratures, and mathematics, 713
quadrivium, and mathematical sciences, 119–20
Quintilian, 146, 243

Rabelais, François, 206–7, 317, 447, 531, 769
radiation, theory of, 614
Raey, Johannes de, 51
Raleigh, Walter, 452, 757, 768, 811
Ramelli, Agostino, 317
Ramus, Petrus (Pierre de la Ramée), 156, 243, 369, 711, 738
Ramusio, Giambattista, 485
Rand, William, 328–9
Randell, John Herman, Jr., 97n64
Ratdolt, Erhard, 545
Rathbone, Aaron, 690
rational mechanics: and Galileo, 640–53, 659–64; and mechanical philosophy, 653–9; and Newton, 664–70; overview of in early modern period, 632–4; and scientific academies, 659–64; and studies of motion, 636–53, 659–68; texts and traditions of, 634–6. *See also* mechanical philosophy; mechanics

Ratio studiorum (Plan of Studies), 26
Ray, John, 117–19, 344, 458n83, 463, 464, 767
al-Rāzī (Rhazes), 294
realism, and scientific literature, 760–4
Real y Pontificia Universidad de México, 821–7
Rebeschke, Catherina, 201
Rechenmeister, and mathematics, 697, 700
Recorde, Robert, 688
Redi, Francesco, 187, 266, 466
Reede tot Drakenstein, Hendrick van, 204
Reeds, Karen Meier, 783
Rees, Abraham, 559
Reformation: and astronomy, 573–7; and interpretations of Bible, 802; and natural philosophy, 379–84. *See also* Counter-Reformation; Protestantism
refraction, and optics, 618–23
Regiomontanus, Johannes, 324, 545, 563, 566, 682
Regis, Henri, 659
register system, of Royal Society, 330, 331
Regius, Henricus, 397
Reinhold, Erasmus, 568
Reisch, Gregor, 74n10, 77–8, 93–4, 405–6, 570
Relaciones geográphicas project, and geography, 826–7
religion: and astronomy, 740–8; and concept of boundaries between domains of discursive communities, 729–30; and Descartes, 79n25, 397, 730, 747–8; and educational reforms, 735–40; and Epicurus, 48n89; Galileo and Newton compared, 753–5; and God as final cause, 77–81; and influence of politics on science in seventeenth century, 748–53; and missionaries in China, 835; narratives and historiography of, 727–30; and natural philosophy, 379–84; and Newton on nature of God, 99–100; and objects of discourse, 729; in Paracelsus's work, 298–9; and practice of authorization, 730; sex and gender difference in, 801; transformation of in early modern period, 7. *See also* Bible; Calvinism; Catholic Church; Church of England; Inquisition; Islamic world; Jesuits; Jews and Judaism; Protestantism; Reformation; theology
Rembrandt (van Rijn), 784–6, 790–1
Renaudot, Théophraste, 170, 195–6, 302, 325, 395, 430
Republic of Letters, 348–51, 354, 360–1
Reuchlin, Johann, 518
Rhazes. *See* al-Rāzī
Rheticus, Joachim, 328
rhetoric, and issues of proof and persuasion, 135–6, 137, 145–9, 151, 153–4

Ribit, Jean, 263
Ricci, Matteo, 485, 830–2
Riccio, Bartholomeo, 220–1
Riccioli, Giambattista, 124, 492–3, 652
Richelet, Pierre, 196
Richelieu, Cardinal, 395
Richer, Jean, 344, 493, 590
Ringmann, Mathias, 479
Riolan, Jean, the Elder, 263, 426, 508
Riolan, Jean, the Younger, 508
Roberval, Gilles Personne de, 46, 349, 588, 652
Rochefoucauld, Madame, 198
Rodríguez, Fray Diego, 824–5
Rohault, Jacques, 172, 196, 398, 556–7, 659
Rolfinck, Werner, 510, 511
Römer, Ole, 184, 590, 624
Rondelet, Guillaume, 276, 416, 443, 445, 447, 457n79
Rossiter, Margaret, 201
Rothmann, Christoph, 261
Rousseau, Jean-Jacques, 814–15
Royal College of Physicians (London), 186
Royal Observatory (Greenwich), 492, 593, 594
Royal Society of London: and acoustics, 609; and agriculture, 218; and coffeehouses, 336; and experimental investigations, 130–1, 172, 217, 516; and journals, 167; and laboratories, 301; and man of science, 185, 189–90, 190–1; and mechanical philosophy, 46, 660–1; and museum, 288; and natural history, 116, 394, 467; and natural philosophy, 365, 402, 403–5; and Newtonian experience, 126–30; and patronage, 270; publications of, 170–1, 325–6, 329–31; and religion, 751–3; and selection of members, 405n178; Swift's satire of, 118; and travel, 356n62; and women, 197, 255
Royer, Clémence, 797
Rubins, Peter Paul, 790–1
Rudbeck, Olof, 187, 280, 429
Rudolph II, Holy Roman Emperor, 255, 257, 259, 264, 287, 300, 453, 506, 782
Ruffini, Paolo, 709n23
Ruge, Sophus, 472n4, 493n73
rural life: and medicine, 417; and natural knowledge, 217–21
Ruscelli, Girolamo, 301
Rushd, Ibn. *See* Averroes
Ruysch, Frederik, 429

Sabius, Johannes, *481f*
Sacrobosco, Johannes de, 474, 564
Sadeler, Aegidius, 255–6
Sahagún, Bernardino de, 822, 825
St. Pierre, Sieur de, 492

St. Serfe, Thomas, 336
sal nitrum theory, in chymistry, 514–15, 516
Salomon's House (Bacon), 272–3, 289, 345, 750
salons, and women in sciences, 196, 198–9, 235
saltpeter, and early gunpowder, 308, 311, 505
salts, and alchemy, 505, 514
Salviati, Filipo, 654
Sánchez, Francisco, 150–1, 379
Sangallo, Antonio da, 315
Santorio, Santorio, 187, 424–5
Saumaise, Claude de, 246–7
Sauver, Joseph, 609n29
Savery, Roelant, 793
Savile, Henry, 244, 248
Savonarola, Girolamo, 580
Scaliger, Joseph, 244
Scaliger, Julius Caesar, 47n88, 82, 83, 85, 166, 376
Schall von Bell, Adam, 832–3
Schedel, Hartmann, 473–4
Scheiner, Christoph, 115n32, 155, 589, 616, 739
Scherer, Heinrich, 360n77
Scheuchzer, Johann Jakob, 236
Schiebinger, Londa, 799, 800, 816
Schissler, Christoph, 679
Schmitt, Charles, 372–3
scholasticism: and Aristotelian natural philosophy, 106–7, 109, 111, 118, 365, 392; and chymistry, 30; and history of early modern science, 8; and mechanical philosophy, 52, 63–4, 67; and metaphysics as foundation of physics, 22–3, 26; and mixed mathematics, 120; and Newtonian experience, 130; Protestants and medieval legacy of, 382; and separation of philosophy and theology, 380–1; use of term, 368n15. *See also* Aristotelianism; universities
Schooten, Frans van, 701–2
scientia, and classification of knowledge in early modern era, 3–4
scientia de ponderibus (Science of Weights), 635–6
scientific explanation: active and passive principles as model for causation, 93–105; Aristotelianism and intrinsic versus extrinsic causes, 82–6; and causality in Aristotelian tradition, 73–7; corpuscular physicists and intrinsic versus extrinsic causes, 87–93; notable changes in early modern era, 70–3; religion and emergence of laws of nature, 77–81
scientific method. *See* analysis and synthesis; demonstration; experience; experiment; hypothesis; logic; methods; observation; practice; theory

Scientific Revolution, and historiography, 12–17, 436n4, 799–800
scientific societies. *See* academies, scientific
scientist: as artist, 786–91; and man of science in early modern period, 179, 180
Scipione del Ferro, 708
Scribonius, Wilhelm, 369
Scudéry, Madeleine de, 198
Scultetus, Bartholomaeus, 574
secondary qualities, and scientific explanation, 88–90
Sendivogius, Michael, 511, 514
Seneca, 13n19, 80–1
Sennert, Daniel, 32, 84, 85, 395, 507, 512
Sequeira, Rui de, 477
Severinus, Peter (Petrus), 31, 32, 217, 262, 422
Servetus, Michael, 425
Sévigné, Marquise de, 196
sex difference, understanding of in early modern period, 800–10. *See also* gender
Sextus Empiricus, 150, 373
Sforza, Galeazzo Maria, 542–3
Shadwell, Thomas, 771
Shakespeare, William, 253, 771, 794n67, 807
Shapin, Steven, 132n11, 346, 790, 796
Shapiro, Alan E., 623
Shapiro, Steven, 194
Sidney, Sir Phillip, 757, 766
Sigüenza y Góngora, Carlos de, 824–5, 827
Simplicius, 368, 636, 637, 641
Sīnā, Ibn. *See* Avicenna
Sina, Pedro de, 477
slide rules, 685
Sloane, Hans, 203, 335, *338f*, 433
Smith, Samuel, 332
Snel, Willebrod, 712–13
Societas Ereunetica, 394
societies, scientific. *See* Académie Royale des Sciences; academies, scientific; Royal Society of London
society: and context for mathematics, 697–702; and natural philosophy, 403–5; proof and persuasion in institutions of, 168–74; and role of medical man, 187; transformation of in early modern period, 7. *See also* class; family
Solis, Diaz de, 484
Sophie Charlotte, Queen of Prussia, 194–5
Sorbière, Samuel, 173, 808
sound. *See* acoustics
Sozietät der Wissenschaften (Akademie der Wissenschaften, Berlin), 394
space: and Aristotelian philosophy, 27; and mechanical philosophy, 52–8

Spain: and colonialism in New World, 351, 358, 452; and cosmography, 483–4
species, and natural history, 115–19, 465
spectacles, and mechanical arts, 685
speed: of falling bodies, 641–2, 643, 651–2; of light, 624–6; of sound, 610n39
Speroni, Sperone, 166
Spinola, Carlo, 830
Spinoza, Benedict (Baruch) de, 59n130, 64, 157, 158n102, 702–3
Sprat, Bishop Thomas, 153, 171–2, 173, 189, 303, 331, 354, 752, 759–60
Staden of Hesse, Hans, 768
Staehl, Peter, 320
Stahl, Georg, 513, 514
Starkey, George, 32, 511n57, 513
stars, and development of stellar astronomy, 569n16, 590–2, 594–5
state: educational institutions and growth of bureaucracies, 367–8; Leibniz on connection between science and, 270–1; view of as household, 230. *See also* courts; patronage; politics
Stationers' Company (London), 322–3, 329
Steno, Nicolaus, 188, 430, 465
Steuco, Agostino, 378
Stevin, Simon, 241, 634, 639–40, 692, 700, 707, 712
Stillingfleet, Edward, 752
Stöffler, Johann, 548n27, 679
Stoicism: and natural philosophy, 376; and scientific explanation, 78, 79, 80–1
Stoughton, John, 751
Strabo, 474
Stradanus, Johannes, 1, *2f*, *5f*, 8–9, *17f*, 788, 790, 788, 790. *See also Nova reperta*
Strato of Lampsacus, 66n153
Streater, Joseph, 332
Strozzi, Palla, 472
Strozzi, Tito Vespasiano, 499
Strutt, John William, 630
Suárez, Francisco, 74n10, 77n17, 92
sundials, and mechanical arts, 681–3, 684
sunspots, and Galileo, 744
surgeons and barber-surgeons, 419
Surinam, and Merian as naturalist, 203–4, 466–7
surveying: and cosmography, 491–3; and mechanical arts, 684, 685, 689–91
Swammerdam, Jan, 187, 430, 465, 466
Swift, Jonathan, 118, 336, 560, 761, 763, 771
Swineshead, Richard, 637
Sydenham, Thomas, 188
syllogisms, and analysis of propositions, 151–2
Sylvius, François de le Boë, 428, 792n61

Sylvius, Jacobus, 783–4
synthesis. *See* analysis and synthesis
syphilis, and medicine, 420, 760, 770

Taccola. *See* Marino di Iacopo
Talavera, Fernando di, 477
Tannstetter, Georg, 485
Tartaglia, Niccolò, 212, 384–5, 633, 634, 638, 692, 708–9, 712n27, 738
Tasso, Torquato, 807
taxonomy, and concept of experience in natural history, 115–19. *See also* classification; species
telescopes: and astronomy, 539–40, 584; and mechanical arts, 685; and optics, 611
Telesio, Bernardino, 33, 34–5, 377, 570, 605n17
Tellier, Charles Maurice le, 3
Tempier, Étienne, 27, 379n63
Tencin, Madame, 198
Terrentius, Johannes, 832
textbooks: and astrology, 555; and astronomy, 577; development of, 165–6; and natural philosophy, 369–71, 383, 387. *See also* books
theaters. *See* anatomy
Themistius, 368, 636, 637
theology: and alchemy, 504–5; and astrology, 578n40; and intellectual context of bodies of knowledge, 730–5; and Luther, 732–5. *See also* religion
Theon, 711
Theophrastus, 438, 439–40, 773
Theorica planetarum (Theories of the Planets), 564, 565, 566
theory: and laboratory, 293–5; and medicine, 408–9
Thevet, André, 451, 487, 768–9
Thomism: and Jesuits, 173; Luther's theological challenge to, 733. *See also* Aristotelianism
Thorndike, Lynn, 499
Tillyard, Arthur, 320
Titelmans, Frans, 382–3
Toletus, Franciscus, 136
Tolosani, Giovanni Maria, 741
Tompion, Thomas, 681
Torricelli, Evangelista, 57, 648, 651, 660, 713
Toscanelli, Paolo del Pozzo, 472, 683
Tournefort, Joseph Pitton de, 117, 203
Toxites, Michael, 506
trade: and chartered corporations, 356–61; and collections of natural objects, 266, 287–8; expansion of in early modern period, 206–7; in mathematical and optical instruments, 684; role of markets and shops in interest in natural history, 207–13
Tradescant, John, 287

transmutation, and chymistry, 510–13
travel: and expansion of scientific knowledge, 341–4, 350–1, 354–62; and innovations in medicine, 416–23; and innovations in natural philosophy, 387–8; and mobility of practitioners of early modern knowledge, 10–11; and natural history, 446, 448–54, 467. *See also* colonialism; discovery, voyages of; New World
Traversari, Ambroglio, 376
Trew, Abdias, 164
triangulation, and surveying, 690, 691
Tristâo, Nuno, 477
Trithemius, Johannes, 248, 501, 502
trivium, and mathematical sciences, 119–20
Tudor, Mary, 806–7
tulip mania of 1636–7, 209
Tulp, Nicolaas, 784, 785
Twisse, William, 751

universities: and anatomy theaters, 277; and Aristotelian worldview, 107; and astrology, 541, 542, 545–6, 553; and astronomy, 563; and botanical gardens, 281; and chymistry, 498, 508–9, 510; correspondence among, 342; and cosmography, 471; and family, 230–1, 233; and geography, 488–9; and images of classroom, 238–44; and laboratories, 302n40, 304–5; and man of science, 181, 182–6; and mathematics, 119–20, 698–9; and mechanics, 633; in Mexico, 821–7; and natural history, 444–8; and natural philosophy, 4–5, 366–72, 389–90; proof and persuasion in disciplinary structures of, 134–8, 150–4; and women, 193–4, 195. *See also* Cambridge University; education; libraries; Oxford University
University of Basel, 277
University of Bologna, 444, 545–6
University of Coimbra, 387
University of Leiden, 238–9, 241, 247, 278, 279, 283, *284f*, 304–5, 416
University of Marburg, 261
University of Mexico, 823
University of Montpellier, 416
University of Padua, 280, 416, 444, 640
University of Paris, 367, 385, 386, 399, 563, 642–3
University of Pisa, 285
University of Santo Domingo, 821n10
University of Uppsala, 467–8
University of Utrecht, 429
University of Vienna, 485
University of Wittenberg, 385, 573, 578, 736
Uraniborg (Denmark), 229. *See also* Brahe, Tycho

Urban VIII, Pope, 377, 746
Ursus, Sabbatino de, 832

vacuum, and foundations of physics, 57–8
Valerio, Luca, 713, 714
Valignano, Alessandro, 829, 830
Valla, Lorenzo, 135
Valturio, Roberto, 317
van den Zype (Zypaeus), François, 433–4
van der Straet, Jan. *See* Stradanus, Johannes
van der Waerden, Bartel Leendert, 722
Varenius, Bernhardus, 492
Varignon, Pierre, 670–1, 672
Vasari, Giorgio, 528, 787–8, 788–9
Vauban, Sébastien Le Prestre de, 315–16
Vaughan, Thomas, 514
Vázquez, Pedro, 826
Vega, Lope de, 206
Vegio, Maffeo, 499
Veracruz, Alonso de, 823
Verbiest, Ferdinand, 832, 833–4
Vesalius, Andreas: and anatomy, 112, 116–17, 240, 276, 285, 345, 414–16, 782–3; and dedication of books to courtly patrons, 258; on illustrations and experience, 116–17, 240; medicine and man of science, 183–4, 186; printing of works of, 328, 353; and scientific household, 226
Vespucci, Amerigo, 16, 472, 479, 483
Vespuzio, Alberico, 484
Viète, François, 701, 710, 714, 714–16, 722
Vieussens, Raymond, 430
Vigo, Giovanni da, 316
villages, and natural knowledge in countryside, 217–21. *See also* rural life
Vio, Thomas de (Cardinal Cajetan), 381–2
Virginia, Raleigh expedition to, 452
virtual space, and networks of communication, 355–60
Vitruvius, 678
Vives, Juan Luis, 296
Viviani, Vincenzo, 650, 651, 680
Volckamer family, 232
Volder, Burchardus de, 305
Volpaia, Lorenzo della, 679, 681, 693
vortices, and Newton's theory of planetary motion, 667–8

Waldseemüller, Martin, 479, 693, 764
Wallis, John, 155, 597, 609, 660–1, 701, 752
war and warfare: constant state of in early modern period, 306–7, 319; and mechanical arts, 679, 684, 691–2. *See also* military science
Ward, Seth, 152, 183
Warren, George, 769

Watt, Joachim von, 494–5
Webster, John, 302
Weiditz, Hans, 526
Weierstrass, Karl, 720n43
weights, and mechanics, 635–6. *See also scientia de ponderibus*
Westfall, Richard S., 728
West Indies, and natural history, 450
Westman, Robert, 738
Weyer, Johann, 417n30
Wharton, Thomas, 430
White, Andrew Dickson, 727, 747
White, John, 452
White, Thomas, 152
Whitehorne, Peter, 318
Wieland, Melchior, 442
Wiles, Andrew J., 701n11
Wilhelm IV, Landgrave of Hesse-Kassel, 255, 260–1, 264, 577
Wilkins, John, 118, 752
William of Ockham, 732
William of Rubruck, 342
Willis, Thomas, 183, 187, 430, 431
Willughby, Francis, 331–2, 767
Windsor, Ann, 228
Winkelmann, Maria, 200, 232
Winkelmann family, 232
Wirsung, Georg, 430
witchcraft: and charges against Kepler's mother, 765; and magic, 520; and medicine, 417. *See also* magic
Witelo, 612, 616
Wittenberg interpretation, of Copernicanism, 738
Wolff, Christian, 696
Wolff, Jacob, 219
Wolfgang II von Hohenlohe, Count, 290, 506
women: as artisans, 199–201; colonies and colonialism, 201–5; and labor in scientific households, 234–6; and learned elites, 193–9; and mathematics, 668–70; new attitudes toward in early modern period, 192; and science at courts, 254–5; and system of exclusion, 179. *See also* gender; sex difference
Woodward, John, 467
workshops: Bacon on empiricism of, 212–13, 221; and curiosity cabinets, 265; and home or household, 227. *See also* laboratories
Worm, Olaus, 264
Worsley, Benjamin, 552
Wren, Christopher, 330, 660, 661, 665
Wurstisen, Christian, 385–6

Xavier, Francis, 828–9
Xu Guangqi, 834

Yang Kuang-hsien, 833
Yates, Frances, 374, 521, 757
Young, Thomas, 631

Zabarella, Jacopo, 75n14, 82, 85, 94–100, 102, 104, 141–2, 159, 381, 382n76
Zamorano, Rodrigo, 484
Zara, Bishop Antonio, 491

Zarlino, Gioseffo, 596, 602–3
Zeuxis, 775
Zilsel, Edgar, 199
zoology: and Aldrovandi's studies of fish, 193, 210; and difficulty of collecting specimens, 456–7
Zumárraga, Bishop Juan de, 822, 823–7